Springer Water

The book series Springer Water comprises a broad portfolio of multi- and interdisciplinary scientific books, aiming at researchers, students, and everyone interested in water-related science. The series includes peer-reviewed monographs, edited volumes, textbooks, and conference proceedings. Its volumes combine all kinds of water-related research areas, such as: the movement, distribution and quality of freshwater; water resources; the quality and pollution of water and its influence on health; the water industry including drinking water, wastewater, and desalination services and technologies; water history; as well as water management and the governmental, political, developmental, and ethical aspects of water.

Vangala Sunitha · Bandi Muralidhara Reddy ·
Yenugu Sudharshan Reddy · Mannala Prasad ·
Badapalli Pradeep Kumar · Etikala Balaji
Editors

Groundwater Resource Management Planning Strategies

A Geospatial Approach: Volume 1

Springer

Editors
Vangala Sunitha
Department of Geology
Yogi Vemana University
Kadapa, Andhra Pradesh, India

Bandi Muralidhara Reddy
Department of Earth Sciences
Yogi Vemana University
Kadapa, Andhra Pradesh, India

Yenugu Sudharshan Reddy
Department of Civil Engineering
Sai Rajeswari Institute of Technology
Proddatur, Andhra Pradesh, India

Mannala Prasad
Department of Geology
Central Tribal University of
Andhra Pradesh
Vizianagaram, Andhra Pradesh, India

Badapalli Pradeep Kumar
CSIR-National Geophysical Research
Institute
Hyderabad, Telangana, India

Etikala Balaji
Yogi Vemana University
Kadapa, Andhra Pradesh, India

ISSN 2364-6934 ISSN 2364-8198 (electronic)
Springer Water
ISBN 978-3-031-88869-4 ISBN 978-3-031-88870-0 (eBook)
https://doi.org/10.1007/978-3-031-88870-0

This Springer imprint is published by the registered company Springer Nature Switzerland AG
The registered company address is: Gewerbestrasse 11, 6330 Cham, Switzerland

Preface

The book comprehensively explores water quality, encompassing the various types of pollution, including physicochemical parameters, geogenic pollutants like arsenic and fluoride, radionuclides, heavy metals, and microplastics. Through diverse research, reviews, and case studies worldwide, this compilation tackles critical topics such as statistical trends in water quality, economic impacts, groundwater management, and radiation effects on marine ecosystems. Moreover, it also includes studies that explore heavy metal pollution, health risk assessments, and remediation techniques, emphasizing the need for effective solutions while examining the role of geospatial technologies and pollution indices in effectively evaluating and managing groundwater resources. This research also aligns with the United Nations Sustainable Development Goals, specifically Goal 3: Ensure Healthy Lives and Goal 6: Ensure Access to Water and Sanitation, contributing to the global effort to address the water crisis. It is an invaluable resource for students, researchers, policymakers, and planners seeking to ensure safe and sustainable water resources by entailing a broad spectrum of global research.

Kadapa, India — Vangala Sunitha
Kadapa, India — Bandi Muralidhara Reddy
Proddatur, India — Yenugu Sudharshan Reddy
Vizianagaram, India — Mannala Prasad
Hyderabad, India — Badapalli Pradeep Kumar
Kadapa, India — Etikala Balaji

Contents

1 **Contamination of Indian Groundwater and Associated Health Effects** ... 1
Shikha Malik, Sufiyan Mirza, Dhananjay Singh Shyamal, and Absar Ahmad Kazmi

2 **Heavy Metals in Groundwater: Environmental and Health Perspectives** ... 39
Samuel Ukpong Okon, Chiamaka Linda Mgbechidinma, Lynda-Uta Edet Okon, Ita Ewa-Oboho, Unwana Sunday Koffi, and Paul Sunday Udom

3 **Heavy Metals Pollution in the Water, Toxicity and Remedial Measures** ... 71
Esmaeil Allahkarami and Ebrahim Allahkarami

4 **Nature Base Sustainable Solution for Heavy Metals Toxicity and Its Impacts on Humans and Plants Health** ... 109
Ratna Chauhan, Sanjay Sarjerao Kale, Ratan Singh, Dheeraj Rathore, Bernard R. Glick, and Indra Jeet Chaudhary

5 **Heavy Metals in the Environment: Sources, Fate, and Health Implications** ... 135
Mamilla Vijaya Kumar and Hemadri Prasad Raju

6 **Heavy Metal Pollution in Groundwater: Environmental and Health Concerns in Gujarat, India** ... 155
Barathan Balaji Prasath

7 **The Importance of Plants for Environmental Health and Water Quality** ... 173
Mangesh Vedpathak, Balbhim Chavan, and Indra Jeet Chaudhary

8 **Risks of Groundwater Contamination to the Environment and Human Health** 193
Alisha Kakkar and Sudesh Kumar

9 **History of Water Pollutants and Emerging Pollutants** 221
Gagan Matta, Gaurav Pant, Sushil B. Kapoor, Reetesh Sah, Kanchan Deoli Bahukhandi, D. P. Uniyal, Pawan Kumar, Anjali Nayak, and Ujjwal Gulaiya

10 **Approaches to Evaluate the Human Health Risk of Arsenic via Drinking Groundwater** 257
Wiyada Nilkarnjanakul and Srilert Chotpantarat

11 **A Comprehensive Review of Conventional Methods and Some Novel Materials for Removing Arsenic from Contaminated Water** 277
Rupkamal Chetia, Shreemoyee Bordoloi, and Surajit Konwer

12 **Heavy Metal Concentration Along the Musi River—A Case Study** 311
Blessy Ganduri, Udaya Laxmi Gakka, and Vidyasagarachary Dasoju

13 **Groundwater Quality, Pollution by Heavy Metals and Its Implication for Human Health Risk in Onitsha Urban, Southeastern Nigeria** 327
G. N. Anachuna, T. N. Nganje, A. Edet, C. I. Adamu, Lynda-Uta Edet Okon, D. O. Inyang, and E. A. Kudamnya

14 **Source Approximation of Heavy Metals in the Groundwater Regime of Wardha Valley Coalfields, Central India: A Geospatial Approach** 367
Priyadarshan S. Ganvir and Rajeeva Guhey

15 **Heavy Metal Contamination and Health Risk Assessment in and Around the Pallikaranai Marshland, Tamil Nadu, India** 391
N. M. Swetha, S. Maniyarasan, A. Ramachandran, M. L. Anusha, and R. R. Krishnamurthy

16 **Assessing Heavy Metal Contents in Groundwater of South Kamrup Region (Assam), India and Its Impact on Human Health and Environment Using GIS** 405
Jayakrishna Bhuyan, Snehasish Choudhury, and Dharitri Choudhury

17 **Assessment of Various Types of Potable Water Quality and Analyses the Impact of Coal Mining on It Around Raniganj Coalfield, West Bengal in India: Challenges and Strategies** 421
Dibyendu Saha, Md. Nazir, Kushal Roy, and Ayan Saha

18 **Evaluating Groundwater Pollution Risk Vulnerability and Mitigation Techniques in a Changing Climate** 457
K. Nagamani, Mohammad Suhail Meer, and Badapalli Pradeep Kumar

19 **Economic Impacts of Geospatial Groundwater Management in Environmental Health: A Review** 477
Harshit Mishra, Supriya, Sandeep Gautam, Diksha Bohra, and Divyanshi Mishra

20 **Radiation on Marine Ecosystems: Impact of Radionuclides and Their Effect on Life Underwater** 511
Wasswa Shafik

21 **Ascertainment of Microplastic Pollution Load in the Groundwater and Its Antithetical Effects on Human Health: A Systematic Review** 541
Swagata Chakraborty, P. Vignesh, Anas Bin Firoz, and M. Govindaraju

22 **Dwindling Depths: Unraveling the Environmental Health Crisis in Declining Groundwater Quality Referencing SDG 3 and 6** 565
Gagan Matta, Rama Pal, Amit Kumar, Rajesh A. Bobdey, Pawan Kumar, Anjali Nayak, Gaurav Pant, Pooja, and Shivangi Tyagi

23 **Human Health Risk Assessment of Fluoride and Nitrate Using Pollution Index of Groundwater and Water Quality Index in Parts of Semi-arid Regions, Anantapur District, A.P. India** 607
P. Ravi Kumar, S. Srinivasa Gowd, C. Krupavathi, and B. N. Anusha

24 **Evaluation of Water Quality for Drinking, Irrigation, and Health Indices of Palvayi River Basin, Anantapur District, Andhra Pradesh, South India Using Geospatial Techniques** 627
U. Suresh, G. Shoba Rani, B. C. Sundara Raja Reddy, G. Veeraswamy, and Etikala Balaji

25 **A Geo-Chemical Study of Water Pollution Due to the Tanning Industry: A Case Study of Dindigul District, Tamilnadu, India** 651
V. Saravanabavan, C. Vinothini, R. S. Suja Rose, and D. Balaji

26 **Geochemical Studies on the Groundwater of Gundlapochampally Village, Medchal District, Telangana State, India** 685
Musini Venkatesharlu, B. Prasad, Ch. Rajendra Prasad, Padamata Yeswanth Babu, Dosilla Jashuva, and Y. S. Reddy

27 Groundwater Quality in Parts of Hyderabad and Medchal-Malkajgiri Districts, Telangana, India: Hydrochemical and Geospatial Assessment 705
Ramesh Chandra Phani Pothuri, Chinmay Dongare, Rajendra Prasad Kanchi, and Rajesh Kanchi

28 Statistical Trend of Water Quality in Bang Pakong River, Thailand 727
Pawee Klongvessa and Srilert Chotpantarat

29 Application of Water Quality Index, Hierarchical Cluster Analysis, Pollution Index of Groundwater and Geospatial Technologies in Evaluating the Water Quality: A Study from Mangampeta Barytes Mine Area, South India 745
Mannala Prasad, Vangala Sunitha, Etikala Balaji, Bandi Muralidhara Reddy, Yenugu Sudharshan Reddy, and Badapalli Pradeep Kumar

30 Groundwater Quality Assessment of Western Coastal Zone of Kanyakumari District, Tamil Nadu, India, Using Statistical, Water Quality Index and GIS Techniques 775
Annmaria K. George, M. Suresh Gandhi, and Anju Maria Joseph

31 Assessment of Flood Impact on Groundwater Quality in the Coastal Areas of Thiruvallur District, Tamil Nadu, India 795
R. Nishan Raja, P. Nila Rekha, Albin Sunny, Soumyabrata Sarkar, and S. Aravind Kumar

32 Land Use Land Cover and Its Impact on Groundwater Resources in India 809
Y. Shobha Laxmi Kumari, Musini Venkateshwarlu, and Babita Mishra

33 Geospatial Evaluation of SLR and LULC Influence on Groundwater Quality in Coastal Watershed of Thiruvallur District, Tamil Nadu, India 825
P. Nila Rekha, R. Nishan Raja, Albin Sunny, Soumyabrata Sarkar, S. Aravind Kumar, and C. P. Balasubramanian

Contributors

C. I. Adamu Department of Geology, University of Calabar, Calabar, Nigeria

Ebrahim Allahkarami Department of Mining Engineering, Amirkabir University of Technology, Tehran, Iran

Esmaeil Allahkarami Department of Petroleum Engineering, Faculty of Petroleum, Gas and Petrochemical Engineering, Persian Gulf University, Bushehr, Iran

G. N. Anachuna Department of Geology, University of Calabar, Calabar, Nigeria

B. N. Anusha Department of Geology, Yogi Vemana University, Kadapa, AP, India

M. L. Anusha Department of Applied Geology and School of Earth and Atmospheric Sciences, University of Madras, Chennai, India

S. Aravind Kumar ICAR—Central Institute of Brackishwater Aquaculture, Chennai, Tamil Nadu, India

Padamata Yeswanth Babu Department of Civil Engineering, CMR College of Engineering & Technology (A), Kandlakoya (V), Hyderabad, Medchal District, Telangana, India

Kanchan Deoli Bahukhandi Sustainability Cluster, School of Advance Engineering, UPES, Dehradun, India

D. Balaji Department of Geography, School of Science, Tamil Nadu Open University, Chennai, Tamil Nadu, India

Etikala Balaji Department of Geology, Yogi Vemana University, Kadapa, Andhra Pradesh, India

C. P. Balasubramanian ICAR—Central Institute of Brackishwater Aquaculture, Chennai, Tamil Nadu, India

Jayakrishna Bhuyan Department of Geography, B. Borooah College, Guwahati, Kamrup(M), Assam, India

Rajesh A. Bobdey Department of Chemistry, Saibaba Arts and Science College, Parshioni, Nagpur, India

Diksha Bohra Department of Agricultural Economics, College of Agriculture, Maharana Pratap University of Agriculture and Technology, Udaipur, Rajasthan, India

Shreemoyee Bordoloi Department of Chemistry, MDKG College, Dibrugarh, Assam, India

Swagata Chakraborty Centre for Climate Change Research (CCCR), Department of Environmental Biotechnology, School of Environmental Science, Bharathidasan University, Tiruchirapalli, India

Indra Jeet Chaudhary Department of Environmental Science, Savitribai Phule Pune University (University of Pune), Pune, Maharashtra, India

Ratna Chauhan Department of Environmental Science, Savitribai Phule Pune University (University of Pune), Pune, Maharashtra, India

Balbhim Chavan Department of Environmental Science, Dr. Babasaheb Ambedkar Marathwada University, Sambhaji Nagar, Maharashtra, India

Rupkamal Chetia Department of Chemistry, Dibrugarh University, Dibrugarh, India

Srilert Chotpantarat Center of Excellence in Environmental Innovation and Management of Metals (EnvIMM), Sustainable Environment Research Institute, Chulalongkorn University (SERI), Bangkok, Thailand;
Department of Geology, Faculty of Science, Chulalongkorn University, Bangkok, Thailand

Dharitri Choudhury Department of Eco-Restoration, Dimoria College, Khetri, Kamrup (M), Assam, India

Snehasish Choudhury Department of Geography, Dakshin Kamrup College, Mirza, Kamrup (R), Assam, India

Vidyasagarachary Dasoju Department of Geophysics, Osmania University, Hyderabad, India

Chinmay Dongare Department of Geology, Faculty of Science, The Maharaja Sayajirao University of Baroda, Vadodara, India

A. Edet Department of Geosciences, University of Uyo, Uyo, Nigeria

Ita Ewa-Oboho Department of Marine Science, Akwa Ibom State University, Mkpat Enin, Nigeria

Anas Bin Firoz Centre for Climate Change Research (CCCR), Department of Environmental Biotechnology, School of Environmental Science, Bharathidasan University, Tiruchirapalli, India

Udaya Laxmi Gakka Department of Geophysics, Osmania University, Hyderabad, India

M. Suresh Gandhi Department of Geology, University of Madras, Guindy Campus, Chennai, India

Blessy Ganduri Department of Geophysics, Osmania University, Hyderabad, India

Priyadarshan S. Ganvir Department of Geology, M. G. Arts, Science and Late N. P. Commerce College, Armori, Maharashtra, India

Sandeep Gautam Department of Agricultural Economics, College of Agriculture, Acharya Narendra Deva University of Agriculture and Technology, Uttar Pradesh, Kumarganj, Ayodhya, India

Annmaria K. George Department of Geology, University of Madras, Guindy Campus, Chennai, India

Bernard R. Glick Department of Biology, ESC 350, University of Waterloo, Waterloo, ON, Canada

M. Govindaraju Centre for Climate Change Research (CCCR), Department of Environmental Biotechnology, School of Environmental Science, Bharathidasan University, Tiruchirapalli, India

S. Srinivasa Gowd Department of Geology, Yogi Vemana University, Kadapa, AP, India

Rajeeva Guhey Swami Atmanand Government Model College, Jagdalpur, Chhattisgarh, India

Ujjwal Gulaiya Hydrology and Science Communication Research Lab, Department of Zoology and Environmental Science, Gurukul Kangri (Deemed to be University), Haridwar, India

D. O. Inyang Department of Geology, University of Calabar, Calabar, Nigeria

Dosilla Jashuva Department of Civil Engineering, CMR College of Engineering & Technology (A), Kandlakoya (V), Hyderabad, Medchal District, Telangana, India

Anju Maria Joseph Department of Geology, University of Madras, Guindy Campus, Chennai, India

Alisha Kakkar Department of Chemistry, Banasthali Vidyapith, Tonk, Rajasthan, India

Sanjay Sarjerao Kale Department of Environmental Science, Savitribai Phule Pune University (University of Pune), Pune, Maharashtra, India

Rajendra Prasad Kanchi Department of Geology, Osmania University, Hyderabad, India

Rajesh Kanchi Department of Applied Geochemistry, Osmania University, Hyderabad, India

Sushil B. Kapoor Department of Chemistry, Dr. Khatri Mahavidyalaya Tukum, Chandrapur, Maharashtra, India

Absar Ahmad Kazmi Department of Civil Engineering, Indian Institute of Technology, Roorkee, India

Pawee Klongvessa Faculty of Environment, Department of Environmental Technology and Management, Kasetsart University, Bangkok, Thailand

Unwana Sunday Koffi Department of Marine Science, Akwa Ibom State University, Mkpat Enin, Nigeria

Surajit Konwer Department of Chemistry, Dibrugarh University, Dibrugarh, India

R. R. Krishnamurthy Department of Applied Geology and School of Earth and Atmospheric Sciences, University of Madras, Chennai, India

C. Krupavathi Department of Geology, Yogi Vemana University, Kadapa, AP, India

E. A. Kudamnya Department of Geology, University of Calabar, Calabar, Nigeria

Amit Kumar Wadia Institute of Himalayan Geology, Dehradun, India

P. Ravi Kumar Department of Geology, Yogi Vemana University, Kadapa, AP, India

Pawan Kumar Hydrology and Science Communication Research Laboratory, Department of Zoology and Environmental Science, Gurukula Kangri (Deemed to be University), Haridwar, India

S. Aravind Kumar ICAR—Central Institute of Brackishwater Aquaculture, Chennai, Tamil Nadu, India

Sudesh Kumar Department of Chemistry, Banasthali Vidyapith, Tonk, Rajasthan, India

Shikha Malik Himachal Institute of Dental Sciences, Paonta Sahib, India

S. Maniyarasan Department of Applied Geology and School of Earth and Atmospheric Sciences, University of Madras, Chennai, India

Gagan Matta Hydrology and Science Communication Research Laboratory, Department of Zoology and Environmental Science, Gurukula Kangri (Deemed to be University), Haridwar, India

Mohammad Suhail Meer Sathyabama Institute of Science and Technology, Chennai, India

Chiamaka Linda Mgbechidinma School of Life Sciences, Centre for Cell and Development Biology and State Key Laboratory of Agrobiotechnology, The Chinese University of Hong Kong, Shatin, New Territories, Hong Kong, China; Department of Microbiology, University of Ibadan, Ibadan, Oyo, Nigeria

Sufiyan Mirza School of Science and Technology, Glocal University, Saharanpur, India

Babita Mishra Department of Environmental Sciences, Bhartiya Engineering Science & Technology Innovation University (BESTIU), Ananthapur, Andhra Pradesh, India

Divyanshi Mishra Department of Agricultural Economics, College of Agriculture, Acharya Narendra Deva University of Agriculture and Technology, Uttar Pradesh, Kumarganj, Ayodhya, India

Harshit Mishra Department of Agricultural Economics, College of Agriculture, Acharya Narendra Deva University of Agriculture and Technology, Uttar Pradesh, Kumarganj, Ayodhya, India

K. Nagamani Sathyabama Institute of Science and Technology, Chennai, India

Anjali Nayak Hydrology and Science Communication Research Laboratory, Department of Zoology and Environmental Science, Gurukula Kangri (Deemed to be University), Haridwar, India

Md. Nazir Department of Botany, The University of Burdwan, Burdwan, West Bengal, India

T. N. Nganje Department of Geology, University of Calabar, Calabar, Nigeria

P. Nila Rekha ICAR—Central Institute of Brackishwater Aquaculture, Chennai, Tamil Nadu, India

Wiyada Nilkarnjanakul Faculty of Public Health, Thammasat University, Pathum Thani, Thailand

R. Nishan Raja ICAR—Central Institute of Brackishwater Aquaculture, Chennai, Tamil Nadu, India

Lynda-Uta Edet Okon Physical Oceanography Unit, Institute of Oceanography, University of Calabar, Calabar, Nigeria

Samuel Ukpong Okon Suzhou Industrial Technological Research Institute of Zhejiang University, Suzhou, China;
Department of Marine Science, Akwa Ibom State University, Mkpat Enin, Nigeria

Rama Pal ICAR—Indian Institute of Soil and Water Conservation, Dehradun, India

Gaurav Pant Hydrology and Science Communication Research Laboratory, Department of Zoology and Environmental Science, Gurukula Kangri (Deemed to be University), Haridwar, India

Pooja Hydrology and Science Communication Research Laboratory, Department of Zoology and Environmental Science, Gurukula Kangri (Deemed to be University), Haridwar, India

Ramesh Chandra Phani Pothuri Mining and Natural Resources, Cyient Limited, Hyderabad, India

Badapalli Pradeep Kumar CSIR-National Geophysical Research Institute, Hyderabad, Telangana, India

Hemadri Prasad Raju Department of Civil Engineering, Mohan Babu University (Erstwhile Sree Vidyanikethan Engineering College), Tirupati, Andhra Pradesh, India

B. Prasad Department of Civil Engineering, CMR College of Engineering & Technology (A), Kandlakoya (V), Hyderabad, Medchal District, Telangana, India

Ch. Rajendra Prasad Department of Civil Engineering, CMR College of Engineering & Technology (A), Kandlakoya (V), Hyderabad, Medchal District, Telangana, India

Mannala Prasad Department of Geology, Central Tribal University of Andhra Pradesh, Vizianagaram, Andhra Pradesh, India

Barathan Balaji Prasath Coastal and Marine Ecology Division, Gujarat Institute of Desert Ecology, Bhuj, Gujarat, India

R. Nishan Raja ICAR—Central Institute of Brackishwater Aquaculture, Chennai, Tamil Nadu, India

A. Ramachandran Department of Applied Geology and School of Earth and Atmospheric Sciences, University of Madras, Chennai, India

G. Shoba Rani Department of Geology, Sri Venkateswara University, Tirupati, Andhra Pradesh, India

Dheeraj Rathore School of Environment and Sustainable Development, Central University of Gujarat, Gandhinagar, Gujarat, India

Bandi Muralidhara Reddy Department of Earth Sciences, Yogi Vemana University, Kadapa, India

Y. S. Reddy Department of Physics, Chaitanya Bharathi Institute of Technology, Gandipet, Hyderabad, Telangana, India

P. Nila Rekha ICAR—Central Institute of Brackishwater Aquaculture, Chennai, Tamil Nadu, India

Kushal Roy Department of Botany, Ramakrishna Mission Vivekananda Centenary College, Rahara, West Bengal, India

Reetesh Sah History Department, Kumaun University, Nainital, India

Ayan Saha Department of Botany, The University of Burdwan, Burdwan, West Bengal, India

Dibyendu Saha Department of Botany, The University of Burdwan, Burdwan, West Bengal, India

V. Saravanabavan Department of Geography, School of Human & Environmental Sciences, North Eastern Hill University, Shillong, Meghalaya, India

Soumyabrata Sarkar ICAR—Central Institute of Brackishwater Aquaculture, Chennai, Tamil Nadu, India

Wasswa Shafik School of Digital Science, Universiti Brunei Darussalam, Tungku Link, Gadong, Brunei Darussalam;
Dig Connectivity Research Laboratory (DCRLab), Kampala, Uganda

Y. Shobha Laxmi Kumari Department of Environmental Sciences, Bhartiya Engineering Science & Technology Innovation University (BESTIU), Ananthapur, Andhra Pradesh, India;
Annamacharya Institute of Science & Technology, Hyderabad, India

Dhananjay Singh Shyamal School of Science and Technology, Glocal University, Saharanpur, India;
Department of Civil Engineering, Indian Institute of Technology, Roorkee, India

Ratan Singh School of Environment and Sustainable Development, Central University of Gujarat, Gandhinagar, Gujarat, India

Yenugu Sudharshan Reddy Department of Civil Engineering, Sai Rajeswari Institute of Techcnology, Proddatur, Andhra Pradesh, India

R. S. Suja Rose Department of Environmental Remote Sensing and Cartography, School of Earth and Atmospheric Sciences, Madurai Kamaraj University, Madurai, Tamil Nadu, India

B. C. Sundara Raja Reddy Department of Geology, Sri Venkateswara University, Tirupati, Andhra Pradesh, India

Vangala Sunitha Department of Geology, Yogi Vemana University, Kadapa, Andhra Pradesh, India

Albin Sunny ICAR—Central Institute of Brackishwater Aquaculture, Chennai, Tamil Nadu, India

Supriya Department of Agricultural Economics, College of Agriculture, Acharya Narendra Deva University of Agriculture and Technology, Uttar Pradesh, Kumarganj, Ayodhya, India

U. Suresh Department of Geology, Sri Venkateswara University, Tirupati, Andhra Pradesh, India

N. M. Swetha Department of Applied Geology and School of Earth and Atmospheric Sciences, University of Madras, Chennai, India

Shivangi Tyagi Hydrology and Science Communication Research Laboratory, Department of Zoology and Environmental Science, Gurukula Kangri (Deemed to be University), Haridwar, India

Paul Sunday Udom Department of Marine Science, Akwa Ibom State University, Mkpat Enin, Nigeria;
Faculty of Sciences, University of Strathclyde, Glasgow, UK

D. P. Uniyal Uttarakhand State Council for Science and Technology, Dehradun, India

Mangesh Vedpathak School of Technology, Department of Civil Engineering, Sanjay Ghodawat University, Kolhapur, Maharashtra, India

G. Veeraswamy Department of Geology, Sri Venkateswara University, Tirupati, Andhra Pradesh, India

Musini Venkateshwarlu Department of Civil Engineering, CMR College of Engineering & Technology, Hyderabad, Telangana, India

P. Vignesh Centre for Climate Change Research (CCCR), Department of Environmental Biotechnology, School of Environmental Science, Bharathidasan University, Tiruchirapalli, India

Mamilla Vijaya Kumar Department of Civil Engineering, Sri Venkateswara Institute of Science and Technology, Kadapa, Andhra Pradesh, India

C. Vinothini Department of Geography, School of Earth and Atmospheric Sciences, Madurai Kamaraj University, Madurai, Tamil Nadu, India

Chapter 1
Contamination of Indian Groundwater and Associated Health Effects

Shikha Malik, Sufiyan Mirza, Dhananjay Singh Shyamal, and Absar Ahmad Kazmi

Abstract Most of the population depends on groundwater to fulfill various needs in developing countries. In developing nations like India, groundwater contamination from geological and anthropogenic sources has become one of the main issues. In many Indian states, incidences of groundwater contamination have been reported in past years. Therefore, it is necessary to comprehensively know the sources of the contaminant, categories of contaminants, cause of contamination, the affected area and reported health issues in India. So that corrective measures can be taken to deal with the issue. The present study has reviewed the reported groundwater contamination issues in different parts of India. The study also systematically reviewed the sources of contamination, categories of the contaminant and its associated health impacts, and reported health problems and risks in India due to groundwater contamination. Further, the study has summarised the current initiative taken by the government to tackle the situation. Groundwater exceeded the concentration of fluoride, heavy metals, nitrate, and iron in different locations of India due to geogenic and anthropogenic sources. Significant health problems and risks have been identified among people consuming contaminated groundwater in affected areas.

Keywords Groundwater contamination · Sources · Categories · Health risk · Initiatives

---

S. Malik
Himachal Institute of Dental Sciences, Paonta Sahib, India

S. Mirza · D. S. Shyamal (✉)
School of Science and Technology, Glocal University, Saharanpur, India
e-mail: dhananjayshyamal@gmail.com

D. S. Shyamal · A. A. Kazmi
Department of Civil Engineering, Indian Institute of Technology, Roorkee, India

V. Sunitha et al. (eds.), *Groundwater Resource Management Planning Strategies*, Springer Water, https://doi.org/10.1007/978-3-031-88870-0_1

1.1 Introduction

Groundwater is the term used to describe surface water that is held in subsurface rocks and flows downward due to gravity (Katarki, 2021). Groundwater is one of Earth's most valuable natural resources (Shaji et al., 2021). Groundwater is an essential water source for both urban and rural communities worldwide (Balaji et al., 2017; Elangovan et al., 2018). The most important natural resource is groundwater, which serves as the foundation of the ecosystem. Today, it is the principal water source for various operations, including drinking, residential, industrial, agricultural, recreational, and environmental (Selvakumar et al., 2017). The majority of India's suburban areas get their drinking water mostly from groundwater sources (Elangovan et al., 2018). India is the world's largest user of groundwater, supplying about 60% of its agricultural demands and 85% of its drinking water demand (Jain et al., 2021). It is widely used worldwide, with extraction increasing dramatically in recent decades as new and less expensive drilling and pumping techniques become available (Shaji et al., 2021). India's groundwater situation is quite complex due to its large size and a wide range of geographical, geological, and climatic factors. Five main groundwater regions can be used to broadly split the entire country: (a) Coastal regions; (b) alluvial plains of the Indo-Gangetic and Brahmaputra; (c) central alluvial lands; (d) hilly and mountainous regions; (e) Peninsular shield area (Katarki, 2021).

Both natural and human-caused processes impact the physical and chemical elements determining groundwater quality (Selvakumar et al., 2017). There are two main ways that groundwater quality can be degraded: (1) anthropogenic, which refers to man-made activities including mining, sewage and waste landfill discharges from cities, industry, and other sources; and (2) geogenic, which is mostly caused by the interaction of water and rock (Manjula & Kabra, 2021).

Groundwater and surface water quality have declined significantly in many places during the past few decades due to growing industrialization, urbanization, and agricultural activities. This has given rise to grave worries in both developed and underdeveloped countries. Restoring the aquifer for clean water supplies is more difficult when dealing with underground pollution, which is not immediately visible and only becomes obvious after a long time. Surface water contamination, such as in rivers and lakes, is generally easy to detect (Singhal & Gupta, 2010).

An investigation was carried out on the quantity of trace elements in groundwater and the possible health hazards that inhabitants of Southern Sonbhadra, Uttar Pradesh, India, could experience from consuming them. The region's geochemical study revealed that there is significant Fe, Pb, Cd, Cr, As, and Hg contamination in all three clusters. A high quantity of heavy metals suggests that both anthropogenic and natural sources have contaminated the groundwater (Ahamad et al., 2020).

According to a study conducted in Yerraguntla, Kadapa district, Andhra Pradesh, groundwater samples had levels of fluorides, electrical conductivity, magnesium, potassium, calcium, sodium, total hardness, bicarbonates, chlorides, sulfates, nitrates, total dissolved solids, and carbonates that exceeded WHO guidelines. In general, the groundwater quality was unfit for human consumption. The leading causes of the decline in water quality were rock-water interactions, human activity, and chemical weathering (Suvarna et al., 2023).

As, Cd, Ni, Pb, and Sr concentrations in groundwater samples from abandoned mines in the southwest of the Cuddapah Basin, Andhra Pradesh, India, were found to be higher than allowed levels. Non-carcinogenic effects in adults and children and extreme to high ecological risk were identified. In the research region, inactive mining and urban wastes, primary soil erosion, agricultural runoff, and sewage from neighboring towns were the predominant sources of heavy metal pollution (Zn, Sr, As, Pb, and Ni) (Reddy & Sunitha, 2023).

It has been discovered that two herbicides, atrazine and alachlor, and one insecticide, malathion, are present in detectable amounts in India's groundwater. Malathion concentrations in the lower Ganges basin are far greater than allowed. It has also been discovered that significant amounts of fertilizers, including phosphate and nitrate, are present in India's groundwater systems (Dutta Gupta et al., 2018). In certain residential sections of Coimbatore, far from the industrial estate, there have been reports of groundwater contaminated by arsenic and cadmium (Mohankumar, Hariharan & Rao, 2016). An analysis of the Singanallur Sub-basin in Coimbatore, India, revealed significantly elevated salinity concentrations. Sulfate and nitrate are the primary issues for drinking and domestic use, and they are primarily brought on by possible human contributions like improper home sewage discharge and overuse of agricultural fertilizers (Selvakumar et al., 2017).

The data from 15,165 CGWB-tested locations in 32 States for 2015 revealed that the levels of contaminants in groundwater (Fluoride 637-locations, Arsenic 697-locations, Salinity 587-locations, Nitrate 2015-locations, and Iron 1389-locations) were higher than allowed limits. Most States/UTs identified the following contributing to the decline in groundwater quality: improper disposal of municipal and industrial waste, overuse of pesticides and fertilizers, and seawater intrusion. For the broader public, this also presents significant health risks (CAG, 2021). Thermal treatment, stream stripping, chemical oxidation, oxygen sparging, bioremediation, and air stripping are the remediation techniques used for groundwater pollution. Health education, hazardous material monitoring, and recurring environmental audits are preventative measures for groundwater pollution (Ravindiran et al., 2023). Innovative methods for remediating groundwater include rainwater harvesting, groundwater recharge, phytoremediation, bioremediation, ion exchange, nanoscale zero-valent iron technique, ceramic filter module-pilot plant units, Bio-Electrochemical Systems (BES), and membrane-based arsenic treatment plants (ATUs) (Chandnani et al., 2022). This chapter has reviewed the sources and categories of groundwater contaminants, emerging contaminants, health impacts caused by the contaminated groundwater, reported incidences of groundwater contamination in India, and associated health problems and risks.

1.2 Sources of Groundwater Contamination

The chemical composition of groundwater can be altered by natural/geogenic, anthropogenic, and miscellaneous sources. The sources of the groundwater contamination are tabulated in Table 1.1. The details of these sources of groundwater contamination are as follows.

1.2.1 Geogenic Sources

Natural or geogenic sources are the initial sources of contamination. Localized deposits of chlorides, sulfates, nitrates, fluoride, and iron are frequent, and their increased concentration in groundwater may impact the groundwater's quality. Leaching can lead to mineralization, particularly in arid regions. It has been determined that the most common source of contamination is natural leaching (Fuhriman, 1971; Scalf et al., 1973). Since the elemental dominance of groundwater is typically determined by the underlying geological formation and rock contents, these geo-natural factors substantially impact groundwater quality.

1.2.2 Anthropogenic Sources

Anthropogenic pollution, which includes human activity, is the second source of groundwater contamination. Pesticide-intensive farming methods, standalone sewage disposal systems, on-site solid waste disposal, municipal wastewater collection, treatment, and disposal, industrial and other wastewater impoundments, sludge spreading on land, mining waste disposal, petroleum industry brine disposal, animal

Table 1.1 Sources of groundwater contamination

Geogenic	Natural leaching, weathering rocks or minerals, volcanic eruption
Anthropogenic	Municipal (sewer leakage, sewage effluent, sewage sludge, solid wastes, landfills, and poor sanitation)
	Agricultural (fertilizers, soil amendments, pesticides and herbicides, and animal wastes)
	Industrial (toxic and hazardous spills, heavy metals, chemicals, synthetic dyes, paint, pharmaceuticals, organic pollutants, storage tanks, and hydrocarbons)
	Mining wastes (acid mine drainage, heavy metals, processing chemicals, and waste rock)
	Radioactive waste (weapons testing, medical radioisotope disposal, and nuclear power plant waste)
Miscellaneous	Urban runoff, deforestation, irrigation, and sea water intrusion

feedlot waste disposal, deep-well liquid waste disposal, and the disposal of radioactive wastes from various operations are examples of anthropogenic sources. Artificial and Natural contaminants can penetrate groundwater through all of these potential sources of pollution as a result of human activities (Matthess, 1982).

1.2.3 Miscellaneous Sources

Apart from the previously listed sources of contamination, multiple other factors contribute to the decline in water quality. Deforestation, excessive irrigation, accidental oil spills and leaks, road salts, seawater intrusion, and other issues are examples of miscellaneous sources of groundwater contamination.

Groundwater salinity may grow as a result of deforestation, raising the water table (Ruprecht & Schofield, 1991). Groundwater quality is significantly impacted by land use and alterations due to increased economic and population growth (Jiang et al., 2009; Subba Rao et al., 2021). According to Jeong (2001), groundwater contamination is primarily caused by land use and urbanization.

Extensive irrigation raises the water table, which raises the soil and groundwater salinity due to evapotranspiration losses. Numerous pollutants are released into the environment by runoff, and these find their way into groundwater systems through percolation and infiltration.

1.3 Categories of the Contaminants

Groundwater contaminants can be broadly categorized into inorganic, organic, and microbial/biological categories.

1.3.1 Inorganic Contaminants

The following are examples of heavy metals that can be found in excess in groundwater as a result of human and natural activity: lead, cadmium, arsenic, chromium, nickel, cobalt, mercury, radon, and other ions like iron, fluoride, sodium, potassium, etc. Heavy metals fall into four main categories based on their importance to human health: Highly toxic metals (Pb, Cd, and Hg), less toxic metals (As and Sn), nonessential metals (Zr, Li, Al, and Ba), and necessary metals (Fe, Mn, Cr, Co, and Zn). Depending on the kind of industry and the processes employed, the wastewater discharged from them contains a range of suspended and dissolved impurities. Numerous locations release wastewater into the environment without the necessary treatment, contaminating nearby surface and groundwater. For instance, numerous reports of elevated Cr, Ni, Cd, Pb, and Zn concentrations in surface and groundwater originate from north India's industrial areas. Agrochemicals, which include pesticides and fertilizers, irrigate and feed crops. These are used to prevent, eliminate, deter, or lessen weeds and pests (Arias-Estévez et al., 2008). The impacts of fertilizers and pesticides on groundwater are enormous (Issahaku et al., 2017). Crops absorb certain agrochemicals, while other chemicals leach into the ground and contaminate nearby surface waterbodies and groundwater. Approximately 180 g of pesticides are applied to agricultural land every hectare on average in India. There are more than 550 pesticides available in the Indian market (Gupta, 2004).

1.3.2 Organic Contaminants

Particularly in the industrialized nations of Western Europe and North America, the inadvertent discharge of fuel and industrial compounds into groundwater can lead to organic compound contamination, which is extremely concerning and dangerous for drinking water. These contaminants' subsurface migration is influenced by their physical characteristics, such as density, solubility, and vapor pressure, as well as the hydrogeological features of surface and subsurface formations. The most prevalent organic contaminants are petroleum hydrocarbons, however due to their lower aqueous solubility and toxicitiy, they are not causing a serious problem. But because of their toxicity and durability, halogenated hydrocarbons—such as carbon tetrachloride and ethylene dibromide—which have certain hydrogen atoms substituted with bromine, fluorine, and chlorine—are more dangerous. In addition to liver damage, halogenated hydrocarbons can induce brain disorders. In drinking water, they are permitted to be present at less than one part per billion. The majority of organic compounds have a limited ability to dissolve into the aqueous and organic phases of water, making them immiscible. These liquids are referred to as NAPLs, or non-aqueous phase liquids. If the density of petroleum products is lower than that of water, they are classified as light non-aqueous phase liquids (LNAPL), including acetone, gasoline, kerosene, and benzene. If the density of NAPL is greater than that of water, it is classified as a dense nonaqueous phase liquid (DNAPL). Common examples of DNAPLs include coal tar, phenol, chloroform, and chlorinated hydrocarbons. Fawell et al. (2006) and Huang et al. (2014) mentioned that organic chemicals in groundwater, such as benzene, toluene, ethylbenzene, xylenes, and chlorinated volatile organic compounds (Cl-VOCs), are typically released through anthropogenic activities or the breakdown and leachates of naturally occurring organic material.

1.3.3 Microbial Contaminants

The main causes of bacterial contamination of groundwater were poor sanitary practices surrounding wells and urban runoff into wells (Obiri-Danso et al., 2009). Cracks in well slabs, wrong placement of boreholes and wells near garbage disposals and toilets, as well as in places where animals are housed or grazed Groundwater contamination is primarily caused by poorly maintained wells and boreholes (Lutterodt et al., 2018). Groundwater sources in Ghana's Bolgatanga Municipality are likely to be contaminated by landfills and solid waste (SW) management, which is a difficult issue (Akurugu, 2018). In China, Han et al. (2016), described leachate from the landfills seeps into the groundwater systems, sanitary landfills are heavily contaminated. The poisoning of groundwater is caused by this enormous volume of untreated wastewater containing hazardous materials. Additional forms of bacterial contamination of borehole water supplies include open defecation, washing around boreholes, and licking of borehole pipelines by cattle and domestic livestock (Adetunde & Glover, 2010). Contaminated groundwater has high amounts of a variety of microbes, including as Shigalla, *V. cholera*, *E. coli*, *P. aeruginosa*, *S. aureus*, Fecal Coliform, and Total Coliform.

Large amounts of a variety of pathogens are present in different earthquake-affected areas, including *Salmonella* sp., *E. coli*, *V. cholerae*, *A. baumanii*, *A. hydrophila*, *V. parahaemolyticus*, *Klebsiella* sp., *Enterobacter* sp., *P. aeruginosa*, *C. freundii*, *S. dysenteriae*, *Staphylococcus* sp., *S. aureus*, and *E. tarda* (Akbar et al., 2013; Amin & Han, 2009; Rasheed et al., 2009).

1.4 Emerging Contaminants in Groundwater

It has been discovered that Emerging Contaminants (ECs) exist in surface water, groundwater, drinking water, and wastewater treatment plant (WWTP) output (Cabeza et al., 2012; Samaras et al., 2013; Yang et al., 2014). Municipal wastewater, which is one of the main sources of newly emerging contaminants into the environment, includes wastewater from homes, businesses, and water treatment plants (Ternes et al., 2004). The different categories of emerging contaminants have been explained below.

1.4.1 Pharmaceuticals (PhACs)

A group of emerging ecological contaminants known as PhACs are increasingly and widely used in both human and veterinary medicine. These consist of substances which are a risk to the environment, such as steroids, betablockers, analgesics, legal and illegal medications, and antibiotics (Fatta-Kassinos et al., 2011). Their unique manner of action is what makes them persistent in the body. They have been found in drinking water, groundwater, natural waterways, sludge, sediments, and WWTP effluents. They are intended to encourage soil microorganisms to produce genes resistant to antibiotics. Over 160 distinct medicines have been found in aquatic systems at extremely low amounts ranging from low μg/L to ng/L (Kümmerer, 2008).

It has long been recognized that the presence of pharmaceutical compounds in aquatic environments is a cause for concern (Richardson & Bowron, 1985). The main ways that pharmaceuticals enter the environment are through agricultural use, product waste, and human excrement (Poynton & Vulpe, 2009). Numerous pharmaceutical compounds linked to wastewater discharge have been found in surface and groundwater (Barnes et al., 2008; Miller & Meek, 2006; Nikolaou et al., 2007; Pérez & Barceló, 2007; Ternes & Hirsch, 2000; Vulliet & Cren-Olivé, 2011; Watkinson et al., 2009). Veterinary and human antibiotics (such as erythromycin, ciprofloxacin, tetracycline lincomycin, and sulfamethoxazole), non-prescription medications (such as salicylic acid, ibuprofen and paracetamol), and iodinated X-ray contrast media (such as iopamidol and iopromide) are examples of these compounds.

1.4.2 Personal Care Products (PCPs)

Personal care products represent a new category of emerging contaminants that include both human and veterinary medications, both non-prescribed and prescribed, as well as the inert and agile components for individual treatment. To just a few, PCPs include steroids, fragrances, shampoos, cosmetics, and hormone-engineered items. One of the most common PCPs, UV filters, is found in groundwater and other aquatic habitats. These filters are known to have estrogenic activity (Noguera-Oviedo & Aga, 2016). Waste water treatment plants (WWTPs) are a common source of personal care chemical contamination in the aquatic environment which include the following:

Triclosan is a common ingredient in home goods, including toothpaste, soap, and antimicrobial sprays. It also functions as a bactericidal and antifungal agent. Many different cleaning and laundry detergents and personal care products use polycyclic musks—tonalide and galaxolide—as perfumes. N-diethyl-meta-toluamide, or DEET-N, is insect repellents most widely used active component. Since the 1930s, parabens—alkyl esters of p-hydroxybenzoic acid—have been utilized as bacteriostatic and fungistatic ingredients in medications, cosmetics, and food products. UV filters/sunscreen-benzophenones and methoxycinnamates are examples of organic filters (Stuart et al., 2012).

1.4.3 Endocrine Disrupting Chemicals (EDC$_S$)

EDCs are defined as substances that, when consumed by the body, may imitate or obstruct hormones, so affecting the body's regular processes. As stated by the US EPA (Science, 1997), EDCs are defined as external agents that interfere with the body's natural hormones that are responsible for maintaining homeostasis, development, reproduction, and behavior. These hormones can be formed, released, transported, attached, activated, or displaced. Androgenic, thyroidal, and estrogenic are considered by most to be the three main types of EDCs (Snyder et al., 2003). Both naturally occurring and artificially created EDCs are released into the environment by humans, animals, and industry; these releases occur mostly through sewage treatment systems and end up in the soil, surface water, groundwater, and saltwater.

1.4.4 *Life-Style Compounds*

Significant concentrations of cotinine, nicotine, and metabolite have been detected in groundwater contaminated by sewage effluent (Godfrey et al., 2007; Seiler et al., 1999; Teijon et al., 2010). The concentration of artificial sweeteners cyclamate, saccharine, acesulfame and sucralose, was high in groundwater due to infiltration of sewage in ponds (Van Stempvoort et al., 2011). Acesulfame's persistence, motility, and use enable it to be found in various aquatic environments (Buerge et al., 2009).

1.4.5 *Industrial Additives and By-Products*

Industrial chemicals come in a huge range that can be released into the environment. Numerous of these helped to generate well-known compounds, including methyl tertiary-butyl ether, petroleum hydrocarbons, polyaromatic hydrocarbons, nylons, phthalates, adipates, plasticizers/resins, and bisphenols (Garrett et al., 1986; Moran et al., 2005, 2007; Verliefde et al., 2007). Most of these industrial chemicals are not emergent contaminants because they are subject to drinking water limitations or prioritized pollutants. Some breakdown products, though, might be thought of as newly emergent pollutants. Among the industrial emerging Contaminants are:

- When 1,1,1-trichloroethane is utilized, 1,4-dioxane is employed as a stabiliser. It quickly dissolves in groundwater. It also doesn't readily adhere to soils and withstands biodegradation processes that occur naturally (Abe, 1999).
- Benzotriazole compounds present in medications with antifungal, antibacterial, and antihelmintic properties. In an aqueous environment, benzotriazoles are persistent (Giger et al., 2006; Voutsa et al., 2006).
- Breakdown of other micropollutants, like the antibiotic component triclosan, can result in the production of dioxins (ENDS, 2010; Mezcua et al., 2004).

1.4.6 *Water Treatment By-Products*

Two common byproducts of disinfecting water are halo acetic acids and trihalomethanes (Boorman, 1999). N-nitrosodimethylamine (NDMA), which can enter drinking water systems by direct industrial contamination or chlorination reactions, is a contemporary concern. Because of the relatively high amounts of this potent carcinogen created during chlorination, reusing municipal wastewater—whether on purpose or by accident is especially problematic (Mitch et al., 2003).

1.4.7 Fire/Flame Retardants

Polybrominated diphenyl ether flame retardants are commonly found in resins used for commercial applications (Rahman et al., 2001) and they can find their way into the environment through burning or disposal in landfills. Tris-(2-chloroethyl) phosphate, or TCEP, is one phosphate-based retardant that might function by creating a non-flammable barrier and is utilized in both products for consumers and industries (Weil et al., 1996).

1.4.8 Surfactants

Octyl-and nonyl-phenol (OP and NP), priority pollutants, are converted into alkyl phenol ethoxylates (APEs), which are used in the production of surfactants. Alkyl phenols and carboxylic degradation products, as well as the parent ethoxylates and their metabolites, are persistent in the aquatic environment (Montgomery-Brown & Reinhard, 2003; Soares et al., 2008). For more than 50 years, carboxylic acids and perfluorinated sulfonates, such as perfluorooctanoic acid (PFOA) and perfluorooctane sulfonate (PFOS), have been used in food packaging, cosmetics, cookware coatings, surfactants, firefighting foams, and paints. As a result, runoff from major fire sites has introduced these chemicals into the environment. They are highly persistent in the environment and can be found in surface water and WTW effluents (Ahrens et al., 2009; Poynton & Vulpe, 2009). Many industrial coatings and personal care items include siloxanes, raising concerns about their toxicity and environmental mobility.

1.4.9 Hormones and Sterols

Androgens, such testosterone and androstenedione, and oestrogens, like progesterone, oestriol, 17β and 17α oestradiol, and oestrone, are examples of sex hormones. Furthermore, synthetic oestrogens (xenoestrogens) like diethylstilbestrol and 17α-ethinyl oestradiol and synthetic androgens like nandrolone and oxandrolone are utilized as contraceptives. Wastewater and sewage treatment effluent often contain a number of these compounds (Johnson et al., 2000; Standley et al., 2008; Vulliet & Cren-Olivé, 2011).The plant sterols stigmastanol, stigmasterol, and β-sitosterol as well as metabolite (5β-coprostanol) of cholesterol, belong to a similar class of chemicals. Edible plants release plant sterols, also known as phyto-oestrogens, into wastewater, which could be the main environmental source of these substances (Liu et al., 2010).

1.4.10 Ionic Liquids

Low-melting-point salts known as ionic liquids offer a "green" alternative to volatile industrial compounds (Pham et al., 2010; Richardson & Ternes, 2011). These substances contain quaternary ammonium salts and nitrocyclic rings, such as pyridinium, pyrrolidinium, and morpholinium moieties. Ionic liquids have yet to be widely utilized, but the formulations currently on the market have a high water solubility, are likely dangerous, and degrade slowly (Pham et al., 2010).

1.5 Health Effects of Contaminated Groundwater

One of the greatest threats to human health is thought to be the contamination of groundwater sources. Waterborne infections are one of the world's most common causes of death, with 3.4 million deaths documented by the WHO annually. In particular, inadequate sanitary facilities and poor drinking water quality cause 2.2 million deaths yearly in poorer nations (Unies, 2003; well Rainwater, 2006). Fertilizers and pesticides have several detrimental effects on health, including cancer development and problems with reproduction and immunity in living beings (Sanpera et al., 2003). Long-term exposure to certain pesticides can lead to cytotoxic alterations, abnormal liver and kidney function, and a burning sensation in the urinary tract (Azmi et al., 2006). Pesticides are classified as endocrine disruptors by the International Agency for Research on Cancer, with the majority being carcinogens. They affect the human endocrine system, leading to dysfunctional hormones (Ejaz et al., 2004).

1.5.1 Polyaromatic Hydrocarbons

The most common sources of polyaromatic hydrocarbons (PAHs) in water include burning trash or coal, oil spills, natural deposition, landfill leachate, and industrial waste. Polyaromatic hydrocarbons are carcinogenic in nature. PAH contamination could increase due to the aromatic solution found in industrial effluents.

1.5.2 Heavy Metals

There is a remarkable diversity in the biological functions and chemical properties of the heavy metals group's elements. Heavy metals are classified as environmental contaminants because they may harm animals, plants, and humans. Because of mining, smelting, and agricultural operations, dangerously high amounts of heavy metals, including Ni, As, Pb, Cr, Co, and Cd, have been identified in soil. A smaller percentage of heavy metals are found in the atmosphere as vapours or particulate matter, with the majority being found in soil and aquatic environments. Because heavy metals are accumulative in biological systems and persistent in nature, they are detrimental to human health.

1.5.2.1 Zinc

The recommended daily intake of zinc for people is 15 mg. However, pregnant, or breastfeeding women should consume 20–25 mg. Acute zinc toxicity in humans can manifest as nausea, vomiting, renal failure, lethargy, dehydration, lack of muscle coordination, electrolytic imbalance, sleepiness, and stomach pain. An elevated risk of anemia, pancreatic damage, decreased HDL (High Density Lipoprotein) and increased LDL (Low Density Lipoprotein) cholesterol, and maybe worsened Alzheimer's disease symptoms are associated with chronic zinc consumption. Employees who have been exposed to zinc fumes from welding or smelting have had mental fume fever, a transient sickness (Lee et al., 1976).

1.5.2.2 Lead

Children are notably more susceptible to lead poisoning because of their permeable blood–brain barrier and high gastrointestinal absorption, which can have neurotoxic consequences even at very low intake levels. Pb is hazardous because it directly interferes with the function of certain enzymes or displaces necessary metal ions from metalloenzymes. The main ways that inorganic Pb is exposed are by ingestion and adsorption through the respiratory, digestive, and respiratory tracts. Prior to being stored in bones, the kidneys and liver are thought to be possible sites of lead poisoning. Depending on the degree of exposure, lead (Pb) can have various biological effects, such as decreased altered neurobehavioral and psychological functions, renal tubular damage, peripheral neuropathy, hemoglobin production, indirect impact on the heart, and reproductive problems (Brown & Kodama, 1987).

1.5.2.3 Copper

The human body needs copper in order to make enzymes. Excessive use of copper causes a variety of health problems, including central nervous system irritation, hepatic damage, depression, renal damage, capillary damage, and severe mucosal corrosion and irritation. The symptoms of copper poisoning include altered kidney functioning, severe hemolysis, and blue-green diarrheal stools and saliva.

1.5.2.4 Arsenic

Humans can become cancerous when exposed to arsenic or chemicals containing arsenic. Food is the primary source of arsenic exposure for most cultures, with drinking water being the second most common source. Epidermal carcinoma, hyperkeratosis, and changes in pigmentation are the primary outcomes associated with long-term drinking water arsenic exposure. Lung cancer is a result of inhaling dust from the workplace (Järup, 2003). Man is said to be fatally affected by 70–80 mg of trivalent arsenic (III) oxide. Inorganic arsenic may produce localized or systemic acute, subacute, or chronic toxicity effects. Acute toxic outcomes include stomach pains, hyperesthesia in the limbs, abdominal electrocardiograms, and abdominal patellar reflexes (Franzblau & Lilis, 1989).

1.5.2.5 Mercury

Though most people are not greatly affected by methyl mercury, there are some circumstances in which eating a lot of fish that has a high mercury content can raise blood levels linked to a lower risk of adult neurological impairment. The WHO recommends a healthy daily intake of 4.3 μg of mercury. Many fish, such as sharks, swordfish, and tuna, are contaminated freshwater species that are hazardous to people's health. As a result, pregnant women need to avoid consuming excessive quantities of these fish (Järup, 2003).

1.5.2.6 Manganese

An adult's estimated daily requirement for manganese, an important nutrient, is 2.5–5.0 mg (Athar & Vohora, 1995). It is a metal that, in greater quantities, has an acute poisoning effect and is relatively nontoxic to humans. Workers who are exposed to Mn dust fume at work may develop the neurological condition known as manganese (Järup, 2003). Manganese has the following effects: altered erythropoiesis and granulocyte production, lowered systolic blood pressure, and disrupted 17-ketosteroid excretion.

1.5.2.7 Iron

Iron is necessary for the human body to function normally; it serves as a catalyst and is more prevalent than other elements. The WHO established 0.3 mg/L as a standard limit for iron in drinking water. An excess and a deficit of iron can result in a number of major health problems, including diabetes, liver and heart illnesses, and cancer.

1.5.3 Water Chemistry-Controlling Ions

Groundwater chemistry is mostly controlled by naturally occurring ions found in water, such as anions (fluorides, chlorides, carbonates, bicarbonates, nitrates, sulphates, phosphates, and nitrites) and cations (potassium, sodium, calcium, and magnesium). For assessing the water quality these anions and cations are measured typically. The human body needs cations and anions to function properly in several processes. While drinking groundwater has numerous health benefits, excessive consumption might have harmful effects because of calcium and magnesium. To avoid some types of cancer, an increase in bone mass may result from an elevated Ca^{+2} concentration. Nonetheless, extremely high Ca^{+2} and Mg^{+2} concentrations may impair the body's ability to absorb essential minerals (Bacha et al., 2010). Sodium and potassium are important components needed to preserve fluids in the entire human body. In general, surface water has lower Na^+ and K^+ concentrations than groundwater. While the WHO suggested a maximum of 200 mg/L for Na^+ in drinking water, the European Union set a limit of 12 mg/L for K^+. Human hypertension and elevated blood pressure may be caused by excess Na^+ and K^+ levels (Kawasaki et al., 1978).

Fluoride is one of those elements that affects health differently: excess and deficiency. Though an excess or deficit of fluoride might have negative effects, it helps develop a protective layer on teeth and supports human health. It is necessary to reduce dental caries formation if consumed in the range of 0–1.5 mg/L; above that, it produces skeletal and dental fluorosis if consumed in more than 3 mg/L concentration (Raza et al., 2016).

Because of the potential health impacts, examining nitrogenous chemicals, such as nitrites and nitrates, in groundwater is crucial. According to (Majumdar & Gupta, 2000), nitrates can enter the lithosphere and hydrosphere chemically or through the metabolic processes of microorganisms. Many illnesses, including "blue baby syndrome" or methemoglobinemia, chromosomal genotoxicity, severe diabetes, stomach cancer, and hypertension can be brought on by an excessive nitrate intake from drinking water (Valentin et al., 2013; Ward et al., 1996).

1.5.4 Biological Contaminants (Pathogens)

Bacteria, Viruses, Helminths, and Protozoa are examples of pathogenic microorganisms that spread through water. Eighty percent of infections worldwide are caused by unsafe drinking water (Ahmed et al., 2012). Over 80% of human diseases can be caused by poor-quality drinking water. These diseases include typhoid, paratyphoid, dermatitis, hepatitis, dysentery, nausea, and gastrointestinal ailments (Rab et al., 1997). Older adults and children are more susceptible to waterborne diseases. Details of pathogens and diseases caused by them are given in Table 1.2.

1.6 Groundwater Contamination and Reported Health Problems and Risks in India

Groundwater pollution by arsenic and fluoride poses a significant threat to the health and safety of numerous individuals globally, particularly in Southeast Asian nations like Bangladesh, China, and India. Within India, regions such as the middle and lower Gangetic plains and certain areas of Central and South India marked by rugged terrain suffer the most from this contamination. Various sources, both natural and human-made, contribute to the introduction of these contaminants into groundwater. While numerous toxic elements are present, studies predominantly highlight arsenic and fluoride, with a lesser proportion of other harmful substances. Table 1.3 contains the range of different contaminants in Indian groundwater reported by different researchers. The contaminants reported in the Indian groundwater are fluoride, arsenic, nitrate, and iron. These results highlight the urgent need for nationwide comprehensive action to address concerns related to water quality.

In India, consuming contaminated groundwater has been associated with various reported health problems and risks. Table 1.4 gives an overview of health problems and risks experienced by communities due to the consumption of contaminated groundwater in India.

Health problems and risks related to dental and skeletal fluorosis, neurological disorders, renal dysfunction, gastroenteritis, cancer risk, reproductive disorders, laxative effects, congenital heart diseases, arsenic toxicity, chronic lung disease, peripheral neuropathy, skin lesions, keratosis, nervous and musculoskeletal system malformations, arsenicosis, and arsenical neuropathy have been reported in Indian groundwater by various researchers as mentioned in Table 1.4. Figure 1.1 presents reported geogenic and anthropogenic sources of groundwater contamination and associated health problems and risks in India. From health concerns to environmental impacts, the data underscores the urgent need for effective interventions to address water quality issues and safeguard public health.

Table 1.2 Details of the pathogens and diseases caused by them (Ramírez-Castillo et al., 2015)

Pathogen	Associated disease
Bacteria	
Campylobacter spp., *C. jejuni*	Diarrhea, gastroenteritis
Yersinia enterocolitica	Diarrhea, reactive arthritis
Escherichia coli, particularly enterohemorragic *E. coli* (EHEC), and other such as enteropathogenic (EPEC), enterotoxigenic (ETEC), and enteroinvasive (EIEC)	Acute diarrhea, bloody diarrhea and gastroenteritis
Burkholderia pseudomallei	Meliodosis
Legionella pneumophila and related bacteria	Acute respiratory illness, pneumonia (legionellosis)
Non-tuberculous mycobacteria	Pulmonary disease, skin infection
Pseudomonas aeruginosa	Infections on lungs, urinary tract, and kidney Can cause inflammation and sepsis
Salmonella enterica serotype Typhi	Typhoid fever, paratyphoid fever and other serious salmonellosis
Other salmonellae	Gastroenteritis, reactive arthritis
Shigella spp.	Bacillary dysentery or shigellosis
Vibrio cholerae	Gastroenteritis, cholera
Helicobacter pylori	Chronic gastritis, ulcer disease and gastric cancer
Virus	
Adenovirus	Gastroenteritis
Enteroviruses	Gastroenteritis
Hepatitis A virus	Hepatitis
Hepatitis E virus	Infectious hepatitis; miscarriage and death
Rotavirus	Gastroenteritis
Sapoviruses	Acute viral gastroenteritis
Astroviruses	Diarrhea
Norovirus	Gastroenteritis
Protozoa	
Acanthamoeba spp.	Amoebic meningoencephalitis, keratitis, encephalitis
Cryptosporidium parvum	Cryptosporidiosis, diarrhea
Cryptosporidium cayetanensis	Diarrhea
Entamoeba histolytica	Amoeba dysentery
Giardia intestinalis	Diarrhea
Naegleria fowleri	Infection of the brain called primary amebic meningoencephalitis (PAM)
Toxoplasma gondii	Toxoplasmosis, miscarriage, birth defects

(continued)

Table 1.2 (continued)

Pathogen	Associated disease
Helminths	
Dracunculus medinensis	Dracunculiasis (Guinea worm disease, ulcerating skin infection)
Schistosoma spp.	Schistosomiasis, liver and kidney damage, itchy skin, fever, chills, cough and muscle aches

1.7 Initiatives Taken by the Governments to Tackle the Contamination of Groundwater

1.7.1 Indian Initiatives

India is the leading global consumer of groundwater. The escalating urbanization across various regions in India has resulted in deteriorating surface water quality, leading to increased reliance on groundwater for household and agricultural use, raising concerns about its purity (Jain et al., 2021).

Governmental bodies have enacted laws and regulations governing industrial discharge, agricultural runoff, and waste disposal to counteract groundwater contamination. Programs like the National Water Mission emphasize comprehensive water resource management to combat groundwater pollution, promoting water conservation, efficient utilization, and basin-level management strategies to mitigate contamination risks. Technological interventions such as groundwater treatment plants, bioremediation, and phytoremediation are deployed to purify groundwater sources (Chandnani et al., 2022).

Groundwater pollution poses significant challenges to agricultural productivity and public health in India. Acknowledging the severity of the issue, the Indian government has launched various measures to address groundwater contamination and advocate sustainable water management practices.

One crucial measure involves promoting enhanced agricultural techniques and technologies to minimize chemical seepage into groundwater. The government encourages adopting organic farming practices to reduce synthetic fertilizer and pesticide usage, thus mitigating groundwater pollution risks. Initiatives like the Soil Health Card scheme provide personalized recommendations to farmers to optimize fertilizer application, reducing the threat of groundwater contamination. Regulating industrial activities is another priority to prevent the discharge of pollutants into groundwater sources. Stringent environmental regulations and monitoring systems are in place to ensure industrial compliance with water quality standards. Efforts are also underway to encourage the adoption of cleaner production methods and wastewater treatment technologies to curb industrial pollution. Furthermore, in order to deliver piped water to every family, the government is investing in the building of rural drinking water infrastructure through programs like the Jal Jeevan Mission. This

Table 1.3 Contaminants and their range in Indian groundwater

S. No.	State	Contaminant	Concentration range (mg/L)	Source	References
1.	Andhra Pradesh	Fluoride	0.1–9.84	Fluoride extraction from groundwater, Magmatic intrusions, weathering, erosion, ion exchange, alkaline environments, evapotranspiration, rock-water interactions, aquifer materials, agricultural fertilizers, shallow aquifers, and domestic and agricultural waste	(Brindha et al., 2011; Jaganmohan et al., 2010; Rahman & Umamahesh, 2018; Rao, 2009; Reddy et al., 2016, 2022; Subba Rao, 2003; Sujatha, 2003; Yenugu et al., 2020)
2.	Assam	Fluoride	0–15.4	Artificial fertilizers, synthetic farming techniques, and natural geological sources	(Chakrabarty & Sarma, 2011; Kotoky et al., 2008; Lahkar, 2015)
3.	Bihar	Fluoride	0.05–14.4	Erosion, river flooding, calcite precipitation, and fluoride dissolution causing geogenic breakdown	(Aher, 2014; Gouri & Choudhary, 2017; Kumar et al., 2018; Ray et al., 2000; Sinha & Sarthi, 2016; Yasmin et al., 2011)
4.	Chhattisgarh	Fluoride	0.5–11.3	Water evaporation, carbonate precipitation, fluorine-rich minerals, weather-induced silicate mineral alterations, and eroding of fluoride-containing minerals	(Beg, 2009; Behera & Das, 2018; Gitte et al., 2015; Lahkar, 2015; Munna et al., 2016; Patel et al., 2017; Sahu et al., 2017)

(continued)

Table 1.3 (continued)

S. No.	State	Contaminant	Concentration range (mg/L)	Source	References
5.	Gujrat	Fluoride	0–7.62	The erosion process of metamorphic sedimentary rocks, influenced by prolonged and constant irrigation, leads to nutrition depletion and erosion	(Dhiman & Keshari, 2006; Rupal et al., 2012; Trivedi et al., 2012)
6.	Kerala	Fluoride	0.05–6.3	Rock-water interaction, spontaneous mineral breakdown in soils, igneous and sedimentary rocks	(Raj & Shaji, 2017)
7.	Telangana	Fluoride	0.4–4.33	Geogenic sources, such as the breakdown of fluoride-containing minerals and the pulverization of granite bedrock	(Narsimha & Rajitha, 2018; Narsimha & Sudarshan, 2017)
8.	Maharashtra	Fluoride	0.2–4.9	Fluoride-containing minerals are being washed and eroded, releasing fluoride into the environment	(Murkute & Badhan, 2011; Pandith et al., 2017)
9.	Rajasthan	Fluoride	0.6–20.1	Interaction between fluoride-containing minerals in rocks and water, as well as their breakdown and dispersal in deep aquifers	(Sabal & Khan, 2008; Singh, 2017; Suthar et al., 2008)

(continued)

Table 1.3 (continued)

S. No.	State	Contaminant	Concentration range (mg/L)	Source	References
10.	Uttar Pradesh	Fluoride	0.06–6.7	Fluorapatite and biotite minerals' natural origin and erosion process involve breakdown from fluoride-containing minerals, naturally occurring brick and agricultural kilns, and water–rock interactions	(Ali et al., 2017; Ansari & Umar, 2019; Chaurasia et al., 2018; Dey & Raju, 2016; Singh, 2016)
11.	West Bengal	Fluoride	0.1–12.2	Precipitation percolates through fluoride-rich minerals and rocks, with dissolution occurring through erosion and breakdown of these minerals in decreasing conditions	(Mondal & Kumar, 2017; Rahman et al., 2005; Sarkar et al., 2015)
12.	Assam	Arsenic	0.1–569	Kushiara river's natural reserves and industrial and agricultural activities contribute to the pollution of the environment due to the abundance of organic material and sediment	(Borah et al., 2010; Jain et al., 2018; Purkayastha et al., 2015)
13.	Bihar	Arsenic	10–1805	Excessive groundwater extraction, ion exchange processes, dissolution of silicate and carbonate minerals, and interactions between surface and groundwater	(Agrawal et al., 2011; Chakraborti et al., 2016a)

(continued)

Table 1.3 (continued)

S. No.	State	Contaminant	Concentration range (mg/L)	Source	References
14.	Chhattisgarh	Arsenic	148–985	Reductive dissolution associated with dissolved organic matter	(Patel et al., 2017)
15.	Jharkhand	Arsenic	7–115	Phosphate-based fertilizers that sink into the soil and release arsenic	(Alam et al., 2016)
16.	Karnataka	Arsenic	24–235	Arsenic, in addition to alpha brass, enhances its resistance to dezincification due to the presence of sulfide minerals in base metals and their associated mining activities	(Hebbar & Janardhana, 2016; Saba and Umar, 2016)
17.	Punjab	Arsenic	0–688	Over-extraction of groundwater, oxidation, releasing iron, and reducing arsenic-containing minerals with low nitrate concentration, high bicarbonate, and iron levels is easier	(Bhattacharjee et al., 2005; Hundal et al., 2007; Kumar & Singh, 2020)
18.	Tamil Nadu	Arsenic	10–620	Disintegration of rocks and ion release	(Shwetha et al., 2019)
19.	Uttar Pradesh	Arsenic	0–448	Released into the environment when it is oxidized in agricultural and industrial inputs	(Chauhan et al., 2012; Pathak et al., 2013)
20.	West Bengal	Arsenic	0–4622	Inputs used in agriculture and industries	(Bhowmick et al., 2013; Sankar et al., 2014)
21.	Odisha	Nitrarte	2–92	Industrial and mining activities	(Das et al., 2023; Reza & Singh, 2010)

(continued)

Table 1.3 (continued)

S. No.	State	Contaminant	Concentration range (mg/L)	Source	References
22.	Rajasthan	Nitrate	1.2–161.80	Nitrate fertilizer manufacturing, fossil fuel combustion, sewage system releases, and nitrogen-fixing crop modifications	(Rahman et al., 2021)
23.	Tamil Nadu	Nitrate	1–4158	Fertilizers, industrial effluent runoff, and chemical dumping	(Pazhuparambil Jayarajan et al., 2021)
24.	Uttar Pradesh	Nitrate	7.92–298.20	Excessive use of nitrogen-based fertilizers, livestock farming, improper organic waste disposal, industrial activities, sewage discharge, and urban runoff	(Chaudhary, 2011)
25.	Andhra Pradesh	Nitrate	0.14–760.13	Agricultural activities, sewage, fertilizer use domestic, agricultural, and animal waste	(Reddy et al., 2022; Suvarna et al., 2020; Yenugu et al., 2020)
26.	Andhra Pradesh	Iron	0.11–12.8	Dissolution of rocks and ferruginous minerals, iron-related smelting processes, and seepage of domestic sewage effluents	(Raju, 2006)
27.	Uttar Pradesh	Iron	1.03–24.5	Improper disposal of waste and e-waste processing along riverbanks	(Kumar et al., 2017)

reduces dependency on contaminated groundwater for drinking purposes, thereby mitigating health risks associated with groundwater contamination. Furthermore, there's a focus on conserving and rejuvenating water bodies to safeguard groundwater quality. Projects like Namami Gange and the National River Conservation Plan aim to cleanse and revitalize rivers and other water bodies, crucial recharge sources for

Table 1.4 Reported health problems and risks in India due to consumption of contaminated groundwater

S. No.	Location	Contaminants	Health problems and risks	References
1.	Southern Sonbhadra, Uttar Pradesh	Heavy metals	Significant non-cancer health risk	(Ahamad et al., 2020)
2.	Bhojpur, Bihar	Arsenic	Skin lesions linked to arsenic, arsenicosis, and arsenical neuropathy	(Chakraborti et al., 2016a)
3.	India	Various pollutants from landfill leachate	Risks for nervous and musculoskeletal system malformations, skin, hair, and nail conditions	(Kumari et al., 2017)
4.	Rural part of Nalgonda, Telangana	Nitrate, Fluoride	Health risks excluding cancer	(Rao et al., 2021)
5.	Patna, Bihar	Arsenic	Various health effects such as skin lesions and keratosis	(Chakraborti et al., 2016b)
6.	Ballia, Varanasi and Gazipur, Uttar Pradesh	Arsenic	Arsenical toxicity	(Ahamed et al., 2006)
7.	Nadia, West Bengal	Arsenic	Chronic arsenic toxicity, Chronic lung disease and Peripheral neuropathy	(Mazumder et al., 2010)
8.	West Bengal	Arsenic	Non-carcinogenic health problem	(Bhowmick et al., 2018)
9.	Nalgonda, Telangana	Fluoride	Dental and skeletal fluorosis	(Adimalla et al., 2019)
10.	Dindigul, Tamilnadu	Various contaminants from tannery activities	Various health effects as people are consuming contaminated water for drinking purpose	(Mondal et al., 2005)
11.	Ranchi, Jharkhand	Heavy metals	Laxative effects, congenital heart diseases, and kidney problems	(Tirkey et al., 2017)
12.	Brahmaputra Valley	Heavy and toxic metal	Cancer risk	(Saha et al., 2023)
13.	Punjab	Pesticides (e.g., endosulfan, chlorpyrifos)	Cancer risk, reproductive disorders	(Mittal et al., 2021)

(continued)

Table 1.4 (continued)

S. No.	Location	Contaminants	Health problems and risks	References
14.	Rajasthan	Fluoride	Dental fluorosis, skeletal fluorosis and gastroenteritis	(Singh et al., 2022)
15.	Delhi	Leachate heavy metals	Neurological disorders, renal dysfunction	(Guleria & Chakma, 2021)
16.	The semi-arid and cold regions of India	Chemical, biological, and radioactive contaminants	Concerns include bone and dental health risks, methemoglobinemia, and the possibility for cancer development	(Dharmani et al., 2024)
17.	Cuddapah Basin, Andhra Pradesh	Fluoride and nitrate	Risk of fluoride and nitrate exposure	(Reddy et al., 2022)

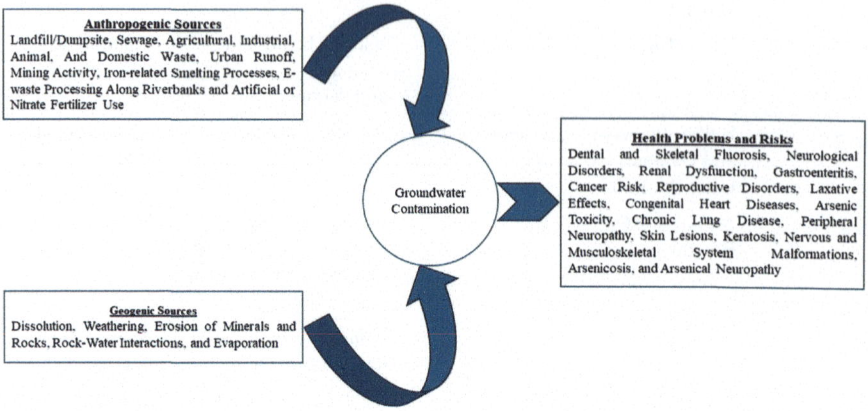

Fig. 1.1 Reported geogenic and anthropogenic sources of groundwater contamination and associated health problems and risks in India

groundwater aquifers. By enhancing surface water health, these endeavors indirectly contribute to improving groundwater quality. Moreover, awareness campaigns and capacity-building programs are launched to educate communities on groundwater conservation and pollution prevention. These initiatives empower local stakeholders to adopt sustainable water management practices, thereby safeguarding groundwater resources (Chaudhuri et al., 2021).

An important endeavor is the National Water Mission, instituted by the Ministry of Jal Shakti, Department of Water Resources, RD & GR. This mission underscores the importance of comprehensive water databases and evaluations to grasp the effects of climate change on water reservoirs. It advocates for citizen and state efforts toward water preservation, particularly in areas facing over-exploitation. Furthermore, the mission seeks to enhance water utilization efficiency by 20% and foster integrated water resources management at the basin level (NWM, 2024).

Integrated Water Resources Management (IWRM) is a pivotal strategy for tackling groundwater contamination. The Central Water Commission (CWC), established by the Government of India, collaborates with state governments to conduct awareness campaigns and enforce IWRM directives. These directives advocate for the synchronized development and supervision of water, land, and associated resources, aiming to optimize economic and social well-being while upholding environmental sustainability (CWC, 2016).

The government has directed attention towards revitalizing conventional water bodies through initiatives such as the Restoration, Renovation, and Repair of Water Bodies under the Pradhan Mantri Krishi Sinchayee Yojana. By prioritizing the enhancement of water bodies and their catchment areas, these initiatives strive to achieve the most efficient utilization of water resources (PMKSY, 2017).

In summary, the Indian government is implementing comprehensive initiatives to combat groundwater contamination and promote sustainable water management practices. Through a multi-dimensional approach involving regulatory measures, infrastructure development, community engagement, and awareness campaigns, the government aims to ensure access to clean and safe water for all citizens while safeguarding groundwater quality.

1.7.2 International Initiatives

The governments of different countries have taken initiatives around the world to tackle the problems related to groundwater. To control water pollution and preserve the integrity of surface and groundwater resources, the United States passed the historic Clean Water Act (CWA). The Clean Water Act (CWA), enacted in 1972, regulates the quantity of contaminants that can be released into water bodies and sets guidelines for wastewater discharge. The CWA's requirements have greatly improved national water quality, decreased industrial pollution, and protected groundwater from contamination (Downing et al., 2003).

The Nitrate Vulnerable Zones (NVZ) program was implemented in the UK to alleviate nitrate pollution in groundwater brought on by agricultural practices. NVZs are regions set aside with more stringent rules governing using nitrogen fertilizers and handling animal waste. The NVZ program, which was put into place in accordance with the EU's Water Framework Directive, has assisted in reducing nitrate pollution in groundwater sources, safeguarding the environment and public health (Osborn & Cook, 1997).

Launched in 1999, China's Grain for Green Project is a massive afforestation and land restoration project meant to counteract desertification, soil erosion, and the loss of water resources. The initiative intends to convert marginal agricultural land into woods and grasslands to lessen soil runoff and groundwater contamination from agricultural practices. The Grain for Green Project has helped to improve groundwater quality and restore ecological balance in troubled parts of China through afforestation and soil conservation methods (Lu et al., 2013).

Passed in 1970, Japan's Water Quality Improvement Act is a comprehensive regulatory framework aimed at preventing water contamination and protecting subterranean water reserves. It establishes standards for water quality, regulates industrial effluents, and promotes pollution control strategies through procedures for licensing and enforcement. This legislation has played a crucial role in maintaining the safety of public health and the integrity of Japan's groundwater through strict rules and monitoring programs (Takeuchi & Tanaka, 2020).

1.7.3 Best Suitable Initiatives for India

Considering the topography of India, the Clean Water Act (CWA) in the United States stands out as being the most significant and successful among them.

The CWA is a strong tool for the overall protection of water resources because it is a complete regulatory framework that targets surface water contamination in addition to groundwater pollution. It creates precise guidelines for the release of wastewater, puts pollution management plans into action, and demands enforcement actions to guarantee adherence. By reducing pollution from industrial, agricultural, and municipal sources, the Clean Water Act (CWA) was established in 1972 and has since greatly improved water quality throughout the United States, including groundwater reservoirs.

Given India's distinct topography, a number of actions can be taken to implement CWA principles there. First and foremost, a national framework for groundwater management needs to be developed, and the legislation now in place needs to be combined. To avoid pollution, this framework ought to establish precise guidelines for agricultural activities, industrial discharge, and municipal wastewater treatment. To guarantee regulatory compliance, it's also essential to bolster enforcement measures and stiffen penalties for infractions (Copeland, 2006).

In addition, India stands to gain by taking proactive steps like creating zones for groundwater conservation and regularly checking water quality. These methods must be customized for India's varied topography, considering elements like aquifer fragility, land use patterns, and local hydrogeological features. Promoting community involvement in groundwater conservation initiatives and sustainable water management techniques depends on highlighting the significance of public participation and awareness.

India may also benefit from the CWA's focus on source management and pollution avoidance by encouraging businesses and farming practices to use cleaner production methods and put in place pollution mitigation measures. Reducing pollution loads and improving groundwater replenishment can also be achieved by investing in infrastructure upgrades, such as wastewater treatment plants and decentralized groundwater recharge systems.

Implementing the Clean Water Act's principles in India necessitates a multifaceted strategy considering the nation's varied topography and socioeconomic variables and addressing institutional, regulatory, and infrastructure issues. India can effectively prevent groundwater contamination and guarantee the supply of clean and safe drinking water for its population by implementing comprehensive groundwater management plans and encouraging sustainable water use practices.

1.8 Conclusions

The chapter has comprehensively reviewed the sources, categories, and health impacts of groundwater contamination. Further, the reported health issues and risks caused by the consumption of groundwater contaminated with heavy metals, fluoride, nitrate, and iron have been reviewed with the source and range of contaminants. Apart from geogenic factors, the main anthropogenic factors contributing to the deterioration in groundwater quality are landfill/dumpsite, sewage, agricultural, industrial, animal, and domestic waste, urban runoff, mining activity, iron-related smelting processes, e-waste processing along riverbanks and artificial or nitrate fertilizer use. There are problems with deteriorating groundwater quality in several Indian states. There is an urgent need for the prevention and remediation of contaminated groundwater in India. However, some initiatives have been taken by the Indian government, such as Integrated Water Resources Management, National Water Mission, groundwater treatment plants, bioremediation, and phytoremediation, Soil Health Card scheme, adoption of organic farming practices, adoption of cleaner production methods and wastewater treatment technologies, National River Conservation Plan, Jal Jeevan Mission, and Namami Gange. Some international initiatives and best suitable initiatives for India have been suggested.

Acknowledgements We would like to thank all the reviewers and editors for their invaluable time and efforts in reviewing this book chapter to enhance its quality.

References

Abe, A. (1999). Distribution of 1, 4-dioxane in relation to possible sources in the water environment. *Science of the Total Environment, 227*(1), 41–47.

Adetunde, L., & Glover, R. (2010). Bacteriological quality of borehole water used by students' of university for development studies, Navrongo campus in upper-east region of Ghana. *Current Research Journal of Biological Sciences, 2*(6), 361–364.

Adimalla, N., Marsetty, S. K., & Xu, P. (2019). Assessing groundwater quality and health risks of fluoride pollution in the Shasler Vagu (SV) watershed of Nalgonda, India. *Human and Ecological Risk Assessment: An International Journal.*

Agrawal, D., Kumar, P., Avtar, R., & Ramanathan, A. (2011). Multivariate statistical approach to deduce hydrogeochemical processes in the groundwater environment of Begusarai district, Bihar. *Water Quality, Exposure and Health, 3*(2), 119–126.

Ahamad, A., Raju, N. J., Madhav, S., & Khan, A. (2020). Trace elements contamination in groundwater and associated human health risk in the industrial region of southern Sonbhadra, Uttar Pradesh, India. *Environmental Geochemistry and Health, 42*, 3373–3391.

Ahamed, S., Sengupta, M. K., Mukherjee, A., Hossain, M. A., Das, B., Nayak, B., Pal, A., Mukherjee, S. C., Pati, S., & Dutta, R. N. (2006). Arsenic groundwater contamination and its health effects in the state of Uttar Pradesh (UP) in upper and middle Ganga plain, India: A severe danger. *Science of the Total Environment, 370*(2–3), 310–322.

Aher, K. (2014). Geochemistry of Groundwater in Basaltic aquifer of Phulambri taluka, District Aurangabad, India. *Research Journal of Chemical Sciences, 1*, 1–124.

Ahmed, K., Ahmed, M., Ahmed, J., & Khan, A. (2012). Risk assessment by bacteriological evaluation of drinking water of Gilgit-Baltistan. *Pakistan Journal of Zoology, 44*(2), 427–432.

Ahrens, L., Felizeter, S., Sturm, R., Xie, Z., & Ebinghaus, R. (2009). Polyfluorinated compounds in waste water treatment plant effluents and surface waters along the River Elbe, Germany. *Marine Pollution Bulletin, 58*(9), 1326–1333.

Akbar, A., Sitara, U., Khan, S. A., Muhammad, N., Khan, M. I., Khan, Y. H., & Kakar, S. U. R. (2013). Drinking water quality and risk of waterborne diseases in the rural mountainous area of Azad Kashmir Pakistan. *International Journal of Biosciences, 3*(12), 245–251.

Akurugu, B. A. (2018). *Hydrogeological and hydrochemical characterisation of crystalline basement rocks in the Upper East Region: The Talensi District.*

Alam, M. O., Shaikh, W. A., Chakraborty, S., Avishek, K., & Bhattacharya, T. (2016). Groundwater arsenic contamination and potential health risk assessment of Gangetic Plains of Jharkhand, India. *Exposure and Health, 8*, 125–142.

Ali, S., Kumari, M., Gupta, S. K., Sinha, A., & Mishra, B. K. (2017). Investigation and mapping of fluoride-endemic areas and associated health risk—A case study of Agra, Uttar Pradesh, India. *Human and Ecological Risk Assessment: An International Journal, 23*(3), 590–604.

Amin, M. T., & Han, M. (2009). Water environmental and sanitation status in disaster relief of Pakistan's 2005 earthquake. *Desalination, 248*(1–3), 436–445.

Ansari, J. A., & Umar, R. (2019). Evaluation of hydrogeochemical characteristics and groundwater quality in the quaternary aquifers of Unnao District, Uttar Pradesh, India. *HydroResearch, 1*, 36–47.

Arias-Estévez, M., López-Periago, E., Martínez-Carballo, E., Simal-Gándara, J., Mejuto, J.-C., & García-Río, L. (2008). The mobility and degradation of pesticides in soils and the pollution of groundwater resources. *Agriculture, Ecosystems & Environment, 123*(4), 247–260.

Athar, M., & Vohora, S. B. (1995). *Heavy metals and environment.* New Age International.

Azmi, M. A., Naqvi, S., Azmi, M. A., & Aslam, M. (2006). Effect of pesticide residues on health and different enzyme levels in the blood of farm workers from Gadap (rural area) Karachi—Pakistan. *Chemosphere, 64*(10), 1739–1744.

Bacha, A. A., Durrani, M. I., & Paracha, P. I. (2010). Physical and bacteriological characteristics of drinking water of Peshawar. *Pakistan Journal of Nutrition, 9*, 1028–1033.

Balaji, E., Nagaraju, A., Sreedhar, Y., Thejaswi, A., & Sharifi, Z. (2017). Hydrochemical characterization of groundwater in around Tirupati Area, Chittoor District, Andhra Pradesh, South India. *Applied Water Science, 7*(3), 1203–1212. https://doi.org/10.1007/s13201-016-0448-6

Barnes, K. K., Kolpin, D. W., Furlong, E. T., Zaugg, S. D., Meyer, M. T., & Barber, L. B. (2008). A national reconnaissance of pharmaceuticals and other organic wastewater contaminants in the United States—I) Groundwater. *Science of the Total Environment, 402*(2–3), 192–200.

Beg, M. (2009). *Geospatial analysis of fluoride contamination in ground water of Tamnar area, Raigarh District, Chhattisgarh State.* ITC.

Behera, B., & Das, M. (2018). Application of multivariate statistical techniques for the characterization of groundwater quality of Bacheli and Kirandul area, Dantewada district, Chattisgarh. *Journal of the Geological Society of India, 91*(1), 76–80.

Bhattacharjee, S., Chakravarty, S., Maity, S., Dureja, V., & Gupta, K. (2005). Metal contents in the groundwater of Sahebgunj district, Jharkhand, India, with special reference to arsenic. *Chemosphere, 58*(9), 1203–1217.

Bhowmick, S., Nath, B., Halder, D., Biswas, A., Majumder, S., Mondal, P., Chakraborty, S., Nriagu, J., Bhattacharya, P., & Iglesias, M. (2013). Arsenic mobilization in the aquifers of three physiographic settings of West Bengal, India: Understanding geogenic and anthropogenic influences. *Journal of Hazardous Materials, 262*, 915–923.

Bhowmick, S., Pramanik, S., Singh, P., Mondal, P., Chatterjee, D., & Nriagu, J. (2018). Arsenic in groundwater of West Bengal, India: A review of human health risks and assessment of possible intervention options. *Science of the Total Environment, 612*, 148–169.

Boorman, G. A. (1999). Drinking water disinfection byproducts: Review and approach to toxicity evaluation. *Environmental Health Perspectives, 107*(suppl 1), 207–217.

Borah, K. K., Bhuyan, B., & Sarma, H. P. (2010). Lead, arsenic, fluoride, and iron contamination of drinking water in the tea garden belt of Darrang district, Assam, India. *Environmental Monitoring and Assessment, 169*, 347–352.

Brindha, K., Rajesh, R., Murugan, R., & Elango, L. (2011). Fluoride contamination in groundwater in parts of Nalgonda District, Andhra Pradesh, India. *Environmental Monitoring and Assessment, 172*, 481–492.

Brown, S. S., & Kodama, Y. (1987). *Toxicology of metals: Clinical and experimental research.*

Buerge, I. J., Buser, H.-R., Kahle, M., Muller, M. D., & Poiger, T. (2009). Ubiquitous occurrence of the artificial sweetener acesulfame in the aquatic environment: An ideal chemical marker of domestic wastewater in groundwater. *Environmental Science & Technology, 43*(12), 4381–4385.

Cabeza, Y., Candela, L., Ronen, D., & Teijon, G. (2012). Monitoring the occurrence of emerging contaminants in treated wastewater and groundwater between 2008 and 2010. The Baix Llobregat (Barcelona, Spain). *Journal of Hazardous Materials, 239*, 32–39.

CAG. (2021). *Report of the comptroller and auditor General of India on ground water management and regulation.* Comptroller and Auditor General of IndiaReport No. 9 of 2021 (Performance Audit).

Chakrabarty, S., & Sarma, H. P. (2011). Fluoride, iron and nitrate contaminated drinking water in Kamrup district, Assam, India. *Archives of Applied Science Research, 3*(4), 186–192.

Chakraborti, D., Rahman, M. M., Ahamed, S., Dutta, R. N., Pati, S., & Mukherjee, S. C. (2016a). Arsenic contamination of groundwater and its induced health effects in Shahpur block, Bhojpur district, Bihar state, India: Risk evaluation. *Environmental Science and Pollution Research, 23*, 9492–9504.

Chakraborti, D., Rahman, M. M., Ahamed, S., Dutta, R. N., Pati, S., & Mukherjee, S. C. (2016b). Arsenic groundwater contamination and its health effects in Patna district (capital of Bihar) in the middle Ganga plain, India. *Chemosphere, 152*, 520–529.

Chandnani, G., Gandhi, P., Kanpariya, D., Parikh, D., & Shah, M. (2022). A comprehensive analysis of contaminated groundwater: Special emphasis on nature-ecosystem and socio-economic impacts. *Groundwater for Sustainable Development*, 100813.

Chaudhary, V. (2011). Nitrate toxicity and possible contaminant sources in groundwater of Western UP, India. *Vegetos, 24*(2), 55–61.

Chaudhuri, S., Parakh, D., Roy, M., & Kaur, H. (2021). Groundwater-sourced irrigation and agro-power subsidies: Boon or bane for small/marginal farmers in India? *Groundwater for Sustainable Development, 15*, 100690.

Chauhan, V. S., Yunus, M., & Sankararamakrishnan, N. (2012). Geochemistry and mobilization of arsenic in Shuklaganj area of Kanpur–Unnao district, Uttar Pradesh, India. *Environmental monitoring and assessment, 184*, 4889-4901.

Chaurasia, A. K., Pandey, H., Tiwari, S., Prakash, R., Pandey, P., & Ram, A. (2018). Groundwater quality assessment using water quality index (WQI) in parts of Varanasi District, Uttar Pradesh, India. *Journal of the Geological Society of India, 92*, 76–82.

Copeland, C. (2006). *Water quality: Implementing the clean water act.*

CWC, Ministry of Water Resources, River Development and Ganga Rejuvenation, Central Water Commission Basin Planning and Management Organization, Government of India. (2016). *Guidelines for integrated water resources development and management*. Central Water Commission.

Das, R., Rao, N. S., Sahoo, H., & Sakram, G. (2023). Nitrate contamination in groundwater and its health implications in a semi-urban region of Titrol block, Jagatsinghpur district, Odisha, India. *Physics and Chemistry of the Earth, Parts A/B/C, 132*, 103424.

Dey, S., & Raju, N. J. (2016). Hydrogeochemical assessment to explore the extent, nature and source of fluoride contamination within the groundwaters of the Panda River Basin, Sonbhadra district, Uttar Pradesh, India. In *Geostatistical and geospatial approaches for the characterization of natural resources in the environment: Challenges, processes and strategies* (pp. 221–229). Springer.

Dharmani, A. B., Verma, M., Rani, S., Narang, A., Singh, M. R., Saya, L., & Hooda, S. (2024). Unravelling groundwater contamination and health-related implications in semi-arid and cold regions of India. *Journal of Contaminant Hydrology*, 104303.

Dhiman, S., & Keshari, A. K. (2006). Hydrogeochemical evaluation of high-fluoride groundwaters: A case study from Mehsana District, Gujarat, India. *Hydrological Sciences Journal, 51*(6), 1149–1162.

Downing, D. M., Winer, C., & Wood, L. D. (2003). Navigating through clean water act jurisdiction: A legal review. *Wetlands, 23*(3), 475–493.

Dutta Gupta, S., Mukherjee, A., Bhattacharya, J., & Bhattacharya, A. (2018). An overview of agricultural pollutants and organic contaminants in groundwater of India. In *Groundwater of South Asia* (pp. 247–255).

Ejaz, S., Akram, W., Lim, C. W., Lee, J. J., & Hussain, I. (2004). Endocrine disrupting pesticides: A leading cause of cancer among rural people in Pakistan. *Experimental Oncology, 26*(2), 98–105.

Elangovan, N. S., Lavanya, V., & Arunthathi, S. (2018). Assessment of groundwater contamination in a suburban area of Chennai, Tamil Nadu, India. *Environment, Development and Sustainability, 20*, 2609–2621.

ENDS. (2010). *River dioxins increased by triclosan*. ENDS Report. Available at: https://www.endsreport.com/article/1566369/river-dioxins-increased-triclosan

Fatta-Kassinos, D., Meric, S., & Nikolaou, A. (2011). Pharmaceutical residues in environmental waters and wastewater: Current state of knowledge and future research. *Analytical and Bioanalytical Chemistry, 399*, 251–275.

Fawell, J., Bailey, K., Chilton, J., Dahi, E., & Magara, Y. (2006). *Fluoride in drinking-water.* IWA Publishing.

Franzblau, A., & Lilis, R. (1989). Acute arsenic intoxication from environmental arsenic exposure. *Archives of Environmental Health: An International Journal, 44*(6), 385–390.

Fuhriman, D. K. (1971). Groundwater pollution in Arizona, California, Nevada, and Utah. In *Copy available from GPO sup doc for $ 2.00*.

Garrett, P., Moreau, M., & Lowry, J. (1986). MTBE as a ground water contaminant. In *Proceedings of the NWWA-API Conference on Petroleum Hydrocarbons and Organic Chemicals in Ground Water-Prevention, Detection and Restoration* (pp. 227–238).

Giger, W., Schaffner, C., & Kohler, H.-P.E. (2006). Benzotriazole and tolyltriazole as aquatic contaminants. 1. Input and occurrence in rivers and lakes. *Environmental Science & Technology, 40*(23), 7186–7192.

Gitte, S. V., Sabat, R., & Kamble, K. (2015). Epidemiological survey of fluorosis in a village of Bastar division of Chhattisgarh state, India. *International Journal of Medicine & Public Health, 5*(3).

Godfrey, E., Woessner, W. W., & Benotti, M. J. (2007). Pharmaceuticals in on-site sewage effluent and ground water, western Montana. *Groundwater, 45*(3), 263–271.

Gouri, K., & Choudhary, S. (2017). Fluoride contamination in groundwater sources of Bhagalpur Municipal Corporation Area, Bhagalpur, Bihar. *IOSR Journal of Environmental Science, Toxicology and Food Technology (IOSRJESTFT), 11*(1), 45–49.

Guleria, A., & Chakma, S. (2021). Probabilistic human health risk assessment of groundwater contamination due to metal leaching: A case study of Indian dumping sites. *Human and Ecological Risk Assessment: An International Journal, 27*(1), 101–133.

Gupta, P. K. (2004). Pesticide exposure—Indian scene. *Toxicology, 198*(1–3), 83–90.

Han, Z., Ma, H., Shi, G., He, L., Wei, L., & Shi, Q. (2016). A review of groundwater contamination near municipal solid waste landfill sites in China. *Science of the Total Environment, 569*, 1255–1264.

Hebbar, A., & Janardhana, M. (2016). Arsenic contamination in groundwater of the areas surrounding Ingaldhal, Chitradurga district, Karnataka state. *International Journal of Geology, Earth & Environmental Sciences, 6*(3), 1–7.

Huang, B., Lei, C., Wei, C., & Zeng, G. (2014). Chlorinated volatile organic compounds (Cl-VOCs) in environment—sources, potential human health impacts, and current remediation technologies. *Environment International, 71*, 118–138.

Hundal, H., Kumar, R., Singh, K., & Singh, D. (2007). Occurrence and geochemistry of arsenic in groundwater of Punjab, Northwest India. *Communications in Soil Science and Plant Analysis, 38*(17–18), 2257–2277.

Issahaku, A., Ampadu, B., & Braimah, M. (2017). The causes, effects, and disease burden attributable to water quality and sanitation conditions in the Kassena Nankana Municipality, Ghana. *Environmental Quality Management, 26*(4), 17–29.

Jaganmohan, P., Narayana Rao, S., & Sambasiva Rao, K. (2010). Studies on the evaluation of angiotensin-I converting enzyme polymorphism under fluorosis mediated renal failures in Nellore District Andhra Pradesh, India P. *Global J Mol Sci, 5*(2), 74–79.

Jain, C., Sharma, S., & Singh, S. (2018). Physico-chemical characteristics and hydrogeological mechanisms in groundwater with special reference to arsenic contamination in Barpeta District, Assam (India). *Environmental Monitoring and Assessment, 190*, 1–16.

Jain, C. K., Sharma, S. K., & Singh, S. (2021). Assessment of groundwater quality and determination of hydrochemical evolution of groundwater in Shillong, Meghalaya (India). *SN Applied Sciences, 3*(1), 33.

Jeong, C. H. (2001). Effect of land use and urbanization on hydrochemistry and contamination of groundwater from Taejon area, Korea. *Journal of Hydrology, 253*(1–4), 194–210.

Jiang, X. W., Wan, L., Wang, X. S., Ge, S., & Liu, J. (2009). Effect of exponential decay in hydraulic conductivity with depth on regional groundwater flow. *Geophysical Research Letters, 36*(24).

Johnson, A., Belfroid, A., & Di Corcia, A. (2000). Estimating steroid oestrogen inputs into activated sludge treatment works and observations on their removal from the effluent. *Science of the Total Environment, 256*(2–3), 163–173.

Järup, L. (2003). *Hazards of heavy metal contamination*. Department of Epidemiology and Public Health.

Katarki, M. V. (2021). Groundwater and the law in India: An overview. In *Groundwater law and management in India: From an elitist to an egalitarian paradigm* (pp. 21–29).

Kawasaki, T., Delea, C. S., Bartter, F. C., & Smith, H. (1978). The effect of high-sodium and low-sodium intakes on blood pressure and other related variables in human subjects with idiopathic hypertension. *The American Journal of Medicine, 64*(2), 193–198.

Kotoky, P., Barooah, P., Baruah, M., Goswami, A., Borah, G., Gogoi, H., Ahmed, F., Gogoi, A., & Paul, A. (2008). Fluoride and endemic fluorosis in the Karbianglong district, Assam. *Fluoride, 40*(1), 42–45.

Kumar, A., & Singh, C. K. (2020). Arsenic enrichment in groundwater and associated health risk in Bari doab region of Indus basin, Punjab, India. *Environmental Pollution, 256*, 113324.

Kumar, S., Venkatesh, A., Singh, R., Udayabhanu, G., & Saha, D. (2018). Geochemical signatures and isotopic systematics constraining dynamics of fluoride contamination in groundwater across Jamui district, Indo-Gangetic alluvial plains, India. *Chemosphere, 205*, 493–505.

Kumar, V., Bharti, P. K., Talwar, M., Tyagi, A. K., & Kumar, P. (2017). Studies on high iron content in water resources of Moradabad district (UP), India. *Water Science, 31*(1), 44–51.

Kumari, P., Gupta, N., & Kaur, A. (2017). A review of groundwater pollution potential threats from municipal solid waste landfill sites: Assessing the impact on human health. *Avicenna Journal of Environmental Health Engineering, 4*(1), 11525–11525.

Kümmerer, K. (2008) *Pharmaceuticals in the environment: Sources, fate, effects and risks*. Springer Science & Business Media.

Lahkar, M. (2015). Fluoride contamination threat in the groundwater of Assam, India: A case study in the high concentration zone of Nagaon. *Journal of Applied and Fundamental Sciences, 1*(1), 85.

Lee, G., Williams, D., Cartwright, G., Prasad, A., & Oberleas, D. (1976). *Trace elements in human health and disease* (Vol. 1). Academic Press.

Liu, Z.-H., Kanjo, Y., & Mizutani, S. (2010). A review of phytoestrogens: Their occurrence and fate in the environment. *Water Research, 44*(2), 567–577.

Lu, Q., Xu, B., Liang, F., Gao, Z., & Ning, J. (2013). Influences of the Grain-for-Green project on grain security in southern China. *Ecological Indicators, 34*, 616–622.

Lutterodt, G., Van de Vossenberg, J., Hoiting, Y., Kamara, A. K., Oduro-Kwarteng, S., & Foppen, J. W. A. (2018). Microbial groundwater quality status of hand-dug wells and boreholes in the Dodowa area of Ghana. *International Journal of Environmental Research and Public Health, 15*(4), 730.

Majumdar, D., & Gupta, N. (2000). Nitrate pollution of groundwater and associated human health disorders. *Indian Journal of Environmental Health, 42*(1), 28–39.

Manjula, S., & Kabra, N. (2021). Groundwater pollution in India and the law. In *Groundwater law and management in India: From an elitist to an egalitarian paradigm* (pp. 77–88).

Matthess, G. (1982). *The properties of groundwater*.

Mazumder, D. N. G., Ghosh, A., Majumdar, K. K., Ghosh, N., Saha, C., & Mazumder, R. N. G. (2010). Arsenic contamination of ground water and its health impact on population of district of Nadia, West Bengal, India. *Indian Journal of Community Medicine: Official Publication of Indian Association of Preventive & Social Medicine, 35*(2), 331.

Mezcua, M., Gómez, M. J., Ferrer, I., Aguera, A., Hernando, M. D., & Fernández-Alba, A. R. (2004). Evidence of 2, 7/2, 8-dibenzodichloro-p-dioxin as a photodegradation product of triclosan in water and wastewater samples. *Analytica Chimica Acta, 524*(1–2), 241–247.

Miller, K. J., & Meek, J. (2006). *Helena Valley ground water: Pharmaceuticals, personal care products, endocrine disruptors (PPCPs), and microbial indicators of fecal contamination*. Citeseer.

Mitch, W. A., Sharp, J. O., Trussell, R. R., Valentine, R. L., Alvarez-Cohen, L., & Sedlak, D. L. (2003). N-nitrosodimethylamine (NDMA) as a drinking water contaminant: A review. *Environmental Engineering Science, 20*(5), 389–404.

Mittal, S., Kumar, R., Sahoo, P. K., & Sahoo, S. K. (2021). Geochemical assessment of groundwater contaminants and associated health risks in the Shivalik region of Punjab, India. *Toxin Reviews, 40*(4), 928–944.

Mohankumar, K., Hariharan, V., & Rao, N. P. (2016). Heavy metal contamination in groundwater around industrial estate vs residential areas in Coimbatore, India. *Journal of Clinical and Diagnostic Research: JCDR, 10*(4), BC05.

Mondal, N., Saxena, V. K., & Singh, V. (2005). Assessment of groundwater pollution due to tannery industries in and around Dindigul, Tamilnadu, India. *Environmental Geology, 48*, 149–157.

Mondal, S., & Kumar, S. (2017). Investigation of fluoride contamination and its mobility in groundwater of Simlapal block of Bankura district, West Bengal, India. *Environmental Earth Sciences, 76*, 1–9.

Montgomery-Brown, J., & Reinhard, M. (2003). Occurrence and behavior of alkylphenol polyethoxylates in the environment. *Environmental Engineering Science, 20*(5), 471–486.

Moran, M. J., Zogorski, J. S., & Squillace, P. J. (2005). MTBE and gasoline hydrocarbons in ground water of the United States. *Groundwater, 43*(4), 615–627.

Moran, M. J., Zogorski, J. S., & Squillace, P. J. (2007). Chlorinated solvents in groundwater of the United States. *Environmental Science & Technology, 41*(1), 74–81.

Munna, K., Guhey, R., & Jhariya, D. (2016). Assessment of groundwater quality, with special reference to fluoride contamination in Bhopalpatnam Block, District Bijapur, Chhattisgarh, India. *IRA-International Journal of Applied Sciences, 5*(2), 74–84.

Murkute, Y., & Badhan, P. (2011). Fluoride contamination in groundwater from Bhadravati tehsil, Chandrapur District, Maharashtra. *Nature, Environment and Pollution Technology, 10*(2), 255–260.

Narsimha, A., & Rajitha, S. (2018). Spatial distribution and seasonal variation in fluoride enrichment in groundwater and its associated human health risk assessment in Telangana State, South India. *Human and Ecological Risk Assessment: An International Journal, 24*(8), 2119–2132.

Narsimha, A., & Sudarshan, V. (2017). Assessment of fluoride contamination in groundwater from Basara, Adilabad district, Telangana state, India. *Applied Water Science, 7*, 2717–2725.

Nikolaou, A., Meric, S., & Fatta, D. (2007). Occurrence patterns of pharmaceuticals in water and wastewater environments. *Analytical and Bioanalytical Chemistry, 387*, 1225–1234.

Noguera-Oviedo, K., & Aga, D. S. (2016). Lessons learned from more than two decades of research on emerging contaminants in the environment. *Journal of Hazardous Materials, 316*, 242–251.

NWM. (2024). *Objective of national water mission*. National Water Mission, Ministry of Jal Shakti, Department of Water Resources, RD & GR, Government of India. Available at: https://nwm.gov.in/objective-national-water-mission. Accessed: June 02, 2024.

Obiri-Danso, K., Adjei, B., Stanley, K., & Jones, K. (2009). Microbiological quality and metal levels in wells and boreholes water in some peri-urban communities in Kumasi, Ghana. *African Journal of Environmental Science and Technology, 3*(3), 059–066.

Office of Science and Technology, US Environmental Protection Agency. (1997). *The Incidence and Severity of Sediment Contamination in Surface Waters of the United States: National sediment quality survey*. Office of Science and Technology, US Environmental Protection Agency.

Osborn, S., & Cook, H. F. (1997). Nitrate vulnerable zones and nitrate sensitive areas: A policy and technical analysis of groundwater source protection in England and Wales. *Journal of Environmental Planning and Management, 40*(2), 217–234.

Pandith, M., Kaplay, R., Potdar, S., Sangnor, H., & Rao, A. (2017). Impact of urbanization coupled with drought situations on groundwater quality in shallow (basalt) and deeper (granite) aquifers with special reference to fluoride in Nanded-Waghala Municipal Corporation, Nanded District, Maharashtra (India). *Environmental Monitoring and Assessment, 189*, 1–14.

Patel, K. S., Sahu, B. L., Dahariya, N. S., Bhatia, A., Patel, R. K., Matini, L., Sracek, O., & Bhattacharya, P. (2017). Groundwater arsenic and fluoride in Rajnandgaon District, Chhattisgarh, northeastern India. *Applied Water Science, 7*, 1817–1826.

Pathak, V. K., Agnihotri, N., Khatoon, N., Khan, A. H., & Rahman, M. (2013). Hydrochemistry of groundwater with special reference to arsenic in Lakhimpur Kheri district, Uttar Pradesh, India. *IOSR Journal of Applied Chemistry, 6*(1), 61–68.

Pazhuparambil Jayarajan, S. K., & Kuriachan, L. (2021). Exposure and health risk assessment of nitrate contamination in groundwater in Coimbatore and Tirupur districts in Tamil Nadu, South India. *Environmental Science and Pollution Research, 28*(8), 10248–10261.

Pham, T. P. T., Cho, C.-W., & Yun, Y.-S. (2010). Environmental fate and toxicity of ionic liquids: A review. *Water Research, 44*(2), 352–372.

PMKSY. (2017). *Guidelines for the scheme on repair, renovation and restoration (RRR) of water bodies under PMKSY (HKKP)*. Government of India Ministry of Water Resources, River Development and Ganga Rejuvenation. Available at: https://pmksy-mowr.nic.in/documents/RRR_PMKSY_Guidelines_2017.pdf. Accessed: 02 June 2024.

Poynton, H. C., & Vulpe, C. D. (2009). Ecotoxicogenomics: Emerging technologies for emerging contaminants 1. *JAWRA Journal of the American Water Resources Association, 45*(1), 83–96.

Purkayastha, S. P., Choudhury, M., Deb, D., & Paul, C. (2015). Arsenic contamination in ground water is a serious threat in the North Karimganj block of Karimganj district, Southern part of Assam, India. *Journal of Chemical and Pharmaceutical Research, 7*(8), 371–378.

Pérez, S., & Barceló, D. (2007). Fate and occurrence of X-ray contrast media in the environment. *Analytical and Bioanalytical Chemistry, 387*, 1235–1246.

Rab, M. A., Bile, M. K., Mubarik, M. M., Asghar, H., Sami, Z., Siddiqi, S., Barzgar, M., Chaudhry, M. A., & Burney, M. (1997). Water-borne hepatitis E virus epidemic in Islamabad, Pakistan: A common source outbreak traced to the malfunction of a modern water treatment plant. *The American Journal of Tropical Medicine and Hygiene, 57*(2), 151–157.

Rahman, A., Mondal, N., & Tiwari, K. (2021). Anthropogenic nitrate in groundwater and its health risks in the view of background concentration in a semi arid area of Rajasthan, India. *Scientific Reports, 11*(1), 9279.

Rahman, F., Langford, K. H., Scrimshaw, M. D., & Lester, J. N. (2001). Polybrominated diphenyl ether (PBDE) flame retardants. *Science of the Total Environment, 275*(1–3), 1–17.

Rahman, M. M., Sengupta, M. K., Ahamed, S., Chowdhury, U. K., Lodh, D., Hossain, A., Das, B., Roy, N., Saha, K. C., & Palit, S. K. (2005). Arsenic contamination of groundwater and its health impact on residents in a village in West Bengal, India. *Bulletin of the World Health Organization, 83*(1), 49–57.

Rahman, S. F., & Umamahesh, M. (2018). A physicochemical investigation of potable water in Adoni, Kurnool District, AP, India. *Journal of Chemistry and Chemical Sciences, 8*, 519–523.

Raj, D., & Shaji, E. (2017). Fluoride contamination in groundwater resources of Alleppey, southern India. *Geoscience Frontiers, 8*(1), 117–124.

Raju, N. J. (2006). Iron contamination in groundwater: A case from Tirumala-Tirupati environs, India. *The Researcher, 1*(1), 28–31.

Ramírez-Castillo, F. Y., Loera-Muro, A., Jacques, M., Garneau, P., Avelar-González, F. J., Harel, J., & Guerrero-Barrera, A. L. (2015). Waterborne pathogens: Detection methods and challenges. *Pathogens, 4*(2), 307–334.

Rao, N. S. (2009). Fluoride in groundwater, varaha river basin, visakhapatnam district, Andhra Pradesh, India. *Environmental Monitoring and Assessment, 152*, 47–60.

Rao, N. S., Dinakar, A., & Kumari, B. K. (2021). Appraisal of vulnerable zones of non-cancer-causing health risks associated with exposure of nitrate and fluoride in groundwater from a rural part of India. *Environmental Research, 202*, 111674.

Rasheed, F., Khan, A., & Kazmi, S. (2009). Bacteriological analysis, antimicrobial susceptibility and detection of 16S rRNA gene of Helicobacter pylori by PCR in drinking water samples of earthquake affected areas and other parts of Pakistan. *Malaysian Journal of Microbiology, 5*(2), 123–127.

Ravindiran, G., Rajamanickam, S., Sivarethinamohan, S., Karupaiya Sathaiah, B., Ravindran, G., Muniasamy, S. K., & Hayder, G. (2023). A review of the status, effects, prevention, and remediation of groundwater contamination for sustainable environment. *Water, 15*(20), 3662.

Ray, D., Ravindar Rao, R., Bhoi, A., Biswas, A., Ganguly, A., & Sanyal, P. (2000). Physico-chemical quality of drinking water in Rohtas district of Bihar. *Environmental Monitoring and Assessment, 61*, 387–398.

Raza, M., Farooqi, A., Niazi, N. K., & Ahmad, A. (2016). Geochemical control on spatial variability of fluoride concentrations in groundwater from rural areas of Gujrat in Punjab, Pakistan. *Environmental Earth Sciences, 75*, 1–16.

Reddy, A., Reddy, D., & Kumar, M. S. (2016). Hydrogeochemical processes of fluoride enrichment in Chimakurthy pluton, Prakasam district, Andhra Pradesh, India. *Environmental Earth Sciences, 75*, 1–17.

Reddy, S., Sunitha, V., & Suvarna, B. (2022). Groundwater quality and its potential health impacts nearby inactive mines using PIG and geospatial technology, Southwestern part of Cuddapah basin, Andhra Pradesh, South India. *Groundwater for Sustainable Development, 17*, 100742.

Reddy, Y. S., & Sunitha, V. (2023). Assessment of heavy metal pollution and its health implications in groundwater for drinking purpose around inactive mines, SW region of Cuddapah Basin, South India. *Total Environment Research Themes, 8*, 100069.

Reza, R., & Singh, G. (2010). Heavy metal contamination and its indexing approach for river water. *International Journal of Environmental Science & Technology, 7*, 785–792.

Richardson, M. L., & Bowron, J. M. (1985). The fate of pharmaceutical chemicals in the aquatic environment. *Journal of Pharmacy and Pharmacology, 37*(1), 1–12.

Richardson, S. D., & Ternes, T. A. (2011). Water analysis: Emerging contaminants and current issues. *Analytical Chemistry, 83*(12), 4614–4648.

Rupal, M., Tanushree, B., & Sukalyan, C. (2012). Quality characterization of groundwater using water quality index in Surat city, Gujarat, India. *International Research Journal of Environment Sciences, 1*(4), 14–23.

Ruprecht, J., & Schofield, N. (1991). Effects of partial deforestation on hydrology and salinity in high salt storage landscapes. I. Extensive block clearing. *Journal of Hydrology, 129*(1–4), 19–38.

Saba, N. U., & Umar, R. (2016). Hydrogeochemical assessment of Moradabad city, an important industrial town of Uttar Pradesh, India. *Sustainable Water Resources Management, 2*, 217–236.

Sabal, D., & Khan, T. (2008). Fluoride contamination status of groundwater in Phulera tehsil of Jaipur district, Rajasthan. *Journal of Environmental Biology, 29*(6), 871–876.

Saha, P., Saikia, K. K., Kumar, M., & Handique, S. (2023). Assessment of health risk and pollution load for heavy and toxic metal contamination from leachate in soil and groundwater in the vicinity of dumping site in Mid-Brahmaputra Valley, India. *Total Environment Research Themes, 8*, 100076.

Sahu, B. L., Banjare, G. R., Ramteke, S., Patel, K. S., & Matini, L. (2017). Fluoride contamination of groundwater and toxicities in dongargaon block, Chhattisgarh, India. *Exposure and Health, 9*, pp. 143–156.

Samaras, V. G., Stasinakis, A. S., Mamais, D., Thomaidis, N. S., & Lekkas, T. D. (2013). Fate of selected pharmaceuticals and synthetic endocrine disrupting compounds during wastewater treatment and sludge anaerobic digestion. *Journal of Hazardous Materials, 244*, 259–267.

Sankar, M., Vega, M., Defoe, P., Kibria, M., Ford, S., Telfeyan, K., Neal, A., Mohajerin, T., Hettiarachchi, G., & Barua, S. (2014). Elevated arsenic and manganese in groundwaters of Murshidabad, West Bengal, India. *Science of the Total Environment, 488* (pp. 570–579).

Sanpera, C., Ruiz, X., Jover, L., Llorente, G., Jabeen, R., Muhammad, A., Boncompagni, E., & Fasola, M. (2003). Persistent organic pollutants in little egret eggs from selected wetlands in Pakistan. *Archives of Environmental Contamination and Toxicology, 44*, 0360–0368.

Sarkar, H., Das, S., & Ray, S. (2015). Scope for development of river bed sub-surface aquifer in fluoride affected blocks of south dinajpur district, West Bengal as sustainable mitigation measures. In *Proceedings of International Seminar on Challenges to Ground Water Management: Vision, 2050* (pp. 423–440).

Scalf, M. R., Keeley, J. W., & LaFevers, C. (1973). *Ground water pollution in the south central states*. National Environmental Research Center, US Environmental Protection Agency.

Seiler, R. L., Zaugg, S. D., Thomas, J. M., & Howcroft, D. L. (1999). Caffeine and pharmaceuticals as indicators of waste water contamination in wells. *Groundwater, 37*(3), 405–410.

Selvakumar, S., Chandrasekar, N., & Kumar, G. (2017). Hydrogeochemical characteristics and groundwater contamination in the rapid urban development areas of Coimbatore, India. *Water Resources and Industry, 17*, 26–33.

Shaji, E., Santosh, M., Sarath, K., Prakash, P., Deepchand, V., & Divya, B. (2021). Arsenic contamination of groundwater: A global synopsis with focus on the Indian Peninsula. *Geoscience Frontiers, 12*(3), 101079.

Shwetha, B., Puttaiah, E., & Nag, B. A. (2019). Prevalance and distribution of arsenic in groundwater in rural parts of Raichur District, Karnataka, India.

Singh, A. (2017). Evaluating fluoride contamination in groundwater of Barmer District, Rajasthan by GIS. *Chemical Science Review and Letters, 6*(21), 31–37.

Singh, R., Upreti, P., Allemailem, K. S., Almatroudi, A., Rahmani, A. H., & Albalawi, G. M. (2022). Geospatial assessment of ground water quality and associated health problems in the Western Region of India. *Water, 14*(3), 296.

Singh, S. (2016). Fluoride contamination in groundwater in some villages of Banda District, Uttar Pradesh, India. *IJIRST—International Journal for Innovative Research in Science & Technology, 3*(06).

Singhal, B., & Gupta, R. (2010). Groundwater contamination. In *Applied hydrogeology of fractured rocks* (2nd ed., pp. 221–236).

Sinha, A., & Sarthi, P. (2016). *Study on iron and fluoride distribution in five districts of Bihar* (pp. 2–3).

Snyder, S. A., Westerhoff, P., Yoon, Y., & Sedlak, D. L. (2003). Pharmaceuticals, personal care products, and endocrine disruptors in water: Implications for the water industry. *Environmental Engineering Science, 20*(5), 449–469.

Soares, A., Guieysse, B., Jefferson, B., Cartmell, E., & Lester, J. (2008). Nonylphenol in the environment: A critical review on occurrence, fate, toxicity and treatment in wastewaters. *Environment International, 34*(7), 1033–1049.

Standley, L. J., Rudel, R. A., Swartz, C. H., Attfield, K. R., Christian, J., Erickson, M., & Brody, J. G. (2008). Wastewater-contaminated groundwater as a source of endogenous hormones and pharmaceuticals to surface water ecosystems. *Environmental Toxicology and Chemistry: An International Journal, 27*(12), 2457–2468.

Stuart, M., Lapworth, D., Crane, E., & Hart, A. (2012). Review of risk from potential emerging contaminants in UK groundwater. *Science of the Total Environment, 416*, 1–21.

Subba Rao, N. (2003). Groundwater quality: Focus on fluoride concentration in rural parts of Guntur district, Andhra Pradesh, India. *Hydrological Sciences Journal, 48*(5), 835–847.

Subba Rao, N., Dinakar, A., Kumari, B. K., Karunanidhi, D., Kamalesh, T. (2021). Seasonal and spatial variation of groundwater quality vulnerable zones of Yellareddygudem watershed, Nalgonda district, Telangana State, India. *Archives of Environmental Contamination and Toxicology, 80*, 11–30.

Sujatha, D. (2003). Fluoride levels in the groundwater of the south-eastern part of Ranga Reddy district, Andhra Pradesh, India. *Environmental Geology, 44*, 587–591.

Suthar, S., Garg, V. K., Jangir, S., Kaur, S., Goswami, N., & Singh, S. (2008). Fluoride contamination in drinking water in rural habitations of Northern Rajasthan, India. *Environmental Monitoring and Assessment, 145*, 1–6.

Suvarna, B., Sunitha, V., Reddy, Y. S., & Reddy, N. R. (2020). Data health risk assessment of nitrate contamination in groundwater of rural region in the Yerraguntla Mandal, South India. *Data in Brief, 30*, 105374.

Suvarna, B., Sunitha, V., Sudharshan Reddy, Y., Muralidhara Reddy, B., Kadam, A., & Ramakrishna Reddy, M. (2023). Groundwater quality assessment using multivariate statistical approach and geospatial modelling around cement industrial corridor, South India. *International Journal of Environmental Science and Technology, 20*(5), 5051–5070.

Takeuchi, H., & Tanaka, H. (2020). Water reuse and recycling in Japan—History, current situation, and future perspectives. *Water Cycle, 1*, 1–12.

Teijon, G., Candela, L., Tamoh, K., Molina-Díaz, A., & Fernández-Alba, A. (2010). Occurrence of emerging contaminants, priority substances (2008/105/CE) and heavy metals in treated wastewater and groundwater at Depurbaix facility (Barcelona, Spain). *Science of the Total Environment, 408*(17), 3584–3595.

Ternes, T. A., Herrmann, N., Bonerz, M., Knacker, T., Siegrist, H., & Joss, A. (2004). A rapid method to measure the solid–water distribution coefficient (Kd) for pharmaceuticals and musk fragrances in sewage sludge. *Water Research, 38*(19), 4075–4084.

Ternes, T. A., & Hirsch, R. (2000). Occurrence and behavior of X-ray contrast media in sewage facilities and the aquatic environment. *Environmental Science & Technology, 34*(13), 2741–2748.

Tirkey, P., Bhattacharya, T., Chakraborty, S., & Baraik, S. (2017). Assessment of groundwater quality and associated health risks: A case study of Ranchi city, Jharkhand, India. *Groundwater for Sustainable Development, 5*, 85–100.

Trivedi, M., Sangai, N., Patel, R., Payak, M., & Vyas, S. (2012). Assessment of groundwater quality with special reference to fluoride and its impact on IQ of schoolchildren in six villages of the Mundra region, Kachchh, Gujarat, India. *Fluoride, 45*(4), 377–383.

Unies, N. (2003). *Water for people, water for life: The United Nations world water development report*. Berghahn Books.

Valentin, L., Nousiainen, A., & Mikkonen, A. (2013). Introduction to organic contaminants in soil: concepts and risks. In *Emerging organic contaminants in sludges: Analysis, fate and biological treatment* (pp. 1–29).

Van Stempvoort, D. R., Roy, J. W., Brown, S. J., & Bickerton, G. (2011). Artificial sweeteners as potential tracers in groundwater in urban environments. *Journal of Hydrology, 401*(1–2), 126–133.

Verliefde, A., Cornelissen, E., Amy, G., Van der Bruggen, B., & Van Dijk, H. (2007). Priority organic micropollutants in water sources in Flanders and the Netherlands and assessment of removal possibilities with nanofiltration. *Environmental Pollution, 146*(1), 281–289.

Voutsa, D., Hartmann, P., Schaffner, C., & Giger, W. (2006). Benzotriazoles, alkylphenols and bisphenol A in municipal wastewaters and in the Glatt River, Switzerland. *Environmental Science and Pollution Research, 13*, 333–341.

Vulliet, E., & Cren-Olivé, C. (2011). Screening of pharmaceuticals and hormones at the regional scale, in surface and groundwaters intended to human consumption. *Environmental Pollution, 159*(10), 2929–2934.

Ward, M. H., Mark, S. D., Cantor, K. P., Weisenburger, D. D., Correa-Villasenor, A., & Zahm, S. H. (1996). Drinking water nitrate and the risk of non-Hodgkin's lymphoma. *Epidemiology, 7*(5), 465–471.

Watkinson, A., Murby, E., Kolpin, D. W., & Costanzo, S. (2009). The occurrence of antibiotics in an urban watershed: From wastewater to drinking water. *Science of the Total Environment, 407*(8), 2711–2723.

Weil, E. D., Zhu, W., Patel, N., & Mukhopadhyay, S. M. (1996). A systems approach to flame retardancy and comments on modes of action. *Polymer Degradation and Stability, 54*(2–3), 125–136.

well Rainwater, S. D. (2006). The global water supply and sanitation assessment 2000 report (WHO and UNICEF 2000), the most recent compilation of global statistics on water supply, changed the way that such data are compiled, from the previous unreliable estimates by provider agencies to consumers' responses in population-based. In *A custom publication of the disease control priorities project* (pp. 229).

Yang, G. C., Yen, C.-H., & Wang, C.-L. (2014). Monitoring and removal of residual phthalate esters and pharmaceuticals in the drinking water of Kaohsiung City, Taiwan. *Journal of Hazardous Materials, 277*, 53–61.

Yasmin, S., Monterio, S., Ligimol, P., & D'Souza, D. (2011). *Fluoride contamination and fluorosis in Gaya Region of Bihar, India.*

Yenugu, S. R., Vangala, S., & Badri, S. (2020) Monitoring of groundwater quality for drinking purposes using the WQI method and its health implications around inactive mines in Vemula-Vempalli region, Kadapa District, South India, *Applied Water Science, 10*, 1–20.

Chapter 2
Heavy Metals in Groundwater: Environmental and Health Perspectives

Samuel Ukpong Okon, Chiamaka Linda Mgbechidinma, Lynda-Uta Edet Okon, Ita Ewa-Oboho, Unwana Sunday Koffi, and Paul Sunday Udom

Abstract Groundwater supplies potable water, particularly in arid and semi-arid regions of the Earth. It is an essential component of the ecological and geological environment, directly influencing human life. While a certain concentration of heavy metals in groundwaters is essential for human metabolism, the excessive consumption of these metals might pose a risk to human health if their concentration is beyond a specific threshold. Exposure to high concentrations of heavy metals can lead to severe neurological disorders and potentially death, following their ability to cause a range of harmful health implications that have attracted global attention. Heavy metal contamination of groundwater frequently evades detection and remains concealed from the public. However, this issue has received significant international attention in recent years, as alarming reports from multiple agencies have highlighted the widespread distribution of heavy metals in groundwater. Considering the increasing significance of examining the potential health risks linked to heavy metals, this chapter reviews groundwater pollution, including sources, associated health risk implications and assessment. Strategic planning and comprehensive management practices are recommended to effectively address the human-health implications of heavy metals in groundwater.

S. U. Okon (✉)
Suzhou Industrial Technological Research Institute of Zhejiang University, Suzhou, China
e-mail: samuelokon@aksu.edu.ng; 21934207@zju.edu.cn

S. U. Okon · I. Ewa-Oboho · U. S. Koffi · P. S. Udom
Department of Marine Science, Akwa Ibom State University, Mkpat Enin, Nigeria

C. L. Mgbechidinma
School of Life Sciences, Centre for Cell and Development Biology and State Key Laboratory of Agrobiotechnology, The Chinese University of Hong Kong, Shatin, New Territories, Hong Kong, China

Department of Microbiology, University of Ibadan, Ibadan, Oyo, Nigeria

L.-U. E. Okon
Institute of Oceanography, University of Calabar, Calabar, Nigeria

P. S. Udom
Faculty of Sciences, University of Strathclyde, Glasgow, UK

V. Sunitha et al. (eds.), *Groundwater Resource Management Planning Strategies*, Springer Water, https://doi.org/10.1007/978-3-031-88870-0_2

Keywords Groundwater · Heavy metals · Water pollution · Health risk assessment

2.1 Introduction

Groundwater, an important and abundant natural resource, has been widely used worldwide for agricultural, industrial, residential, and drinking purposes (Adimalla, 2019; Aghazadeh et al., 2017; Li et al., 2019). It is widely regarded as a reliable source of drinking water for over 2 billion people (Bricker et al., 2017; Qiao et al., 2020) due to its natural filtration process and greater protection against pollution compared to surface water. Although groundwater is generally considered safe for use, numerous studies have demonstrated that it can also be vulnerable to contamination (Chakraborti et al., 2016; Indelicato et al., 2017; Majolagbe et al., 2016). Factors such as rapid industrialization, urbanization, high population density, geological characteristics of the aquifer, environmental inputs, and human activities such as the excessive use of fertilizers and agrochemicals have a negative impact on groundwater quality (Sajil Kumar & Kuriachan, 2022; Wang et al., 2021a, 2021b). Industrial activities serve as major pollution sources, contributing significantly to the widespread contamination of the environment by harmful heavy metals and metalloids. These toxic metals, which are often released directly or indirectly into natural bodies of water, particularly in developing nations, commonly exist in ionic form in soil and groundwater environments. Aluminum (Al), arsenic (As), lead (Pb), cobalt (Co), copper (Cu), cadmium (Cd), and chromium (Cr) are among the most common heavy metals that cause a substantial environmental impact (Luo et al., 2022; Reddy & Sunitha, 2023). However, metalloids of primary interest in polluted groundwater analysis include Manganese (Mn), zinc (Zn), inorganic mercury (Hg), silver (Ag), strontium (Sr), molybdenum (Mo), nickel (Ni), and selenium (Se). Given the persistent problem of groundwater pollution by heavy metals, it is important to re-evaluate and improve methodologies for assessing health risks and developing effective remediation solutions. Figure 2.1 illustrates heavy metal entry pathways and their effects on humans.

Heavy metals have been widely recognized as contaminants because of their toxic properties and ability to bioaccumulate in humans. The toxicity and carcinogenicity of heavy metal ions in groundwater pose a significant risk to human health, even at low concentrations. Depending on the type and quantity of metal ingested, the consumption of hazardous heavy metals can potentially result in life-threatening diseases, including cancer (Ogunfowokan et al., 2013). Even at low concentrations, certain metals can damage multiple organs in the human body and may lead to neurodegenerative disorders (Pinto et al., 2019; Yuan et al., 2016). While low concentrations of certain heavy metals are required for metabolic processes, prolonged or excessive exposure can disrupt the normal functioning of various organs and tissues. Heavy metals in groundwater represent a significant issue, contributing to environmental deterioration and human health concerns. Therefore, continued investigation is necessary to address the persistent risk and increasing complexities associated with heavy

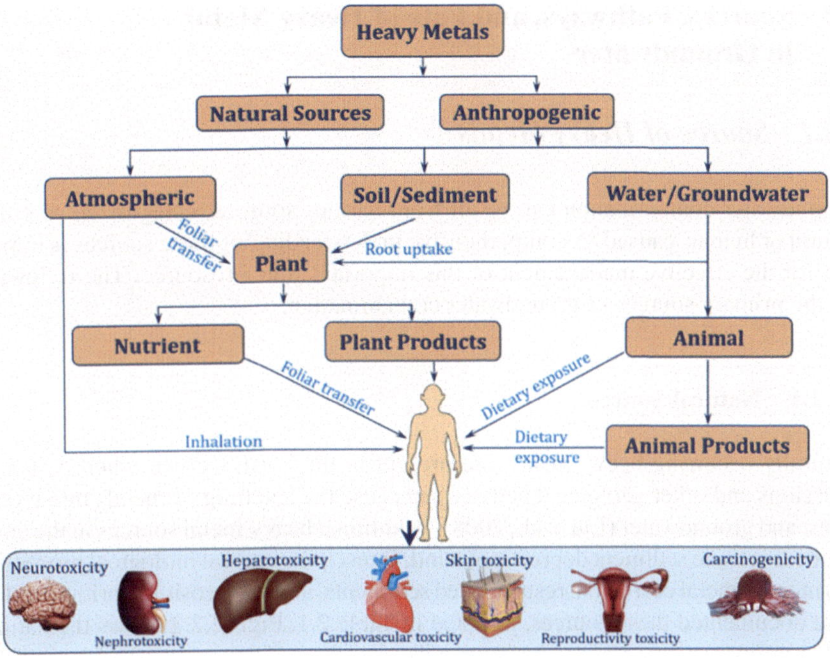

Fig. 2.1 Sources, pathway, and human-health effects of heavy metals in the environment

metals in groundwater. Selenium, an element naturally present in the Earth's crust, is an essential micronutrient for human health, playing an important role in antioxidant defense and thyroid function. However, in large amounts, it can become toxic and pose health risks, including symptoms such as gastrointestinal disturbances and neurological damage. The highly toxic metalloid can accumulate in the form of selenate (SeO_4^{2-}), selenite (SeO_3^{2-}), selenide (Se^{2-}), and elemental selenium (Se^0) in the surface and groundwater of industrialized countries such as the United States, India, China, Japan, and Canada (Guo et al., 2023). The primary route by which selenium enters groundwater is through leaching, in which selenium-rich minerals in soils and rocks release selenium into the water.

After this introduction, the remaining part of this chapter is structured as follows: Sect. 2.2 explores the sources, pathways, and fate of heavy metals in groundwater. Section 2.3 describes the human health implications and toxicity of heavy metal exposure. Section 2.3 presents the monitoring and detection techniques. Section 2.4 introduces indexing approaches and human health risk assessment. Section 2.5 describes risk assessment, human health evaluation, and regulatory framework/policies, while Sect. 2.6 presents the conclusion.

2.2 Sources, Pathways, and Fate of Heavy Metal in Groundwater

2.2.1 Source of Heavy Metals

Groundwater contamination can result from various sources, categorized as either natural or human-caused. A comprehensive understanding of these sources is important for the effective management of this important water resource. The following are the primary sources of groundwater contamination.

2.2.1.1 Natural Sources

Naturally occurring heavy metals are present in the Earth's crust, where volcanic emissions and other geological activity can cause the leaching of metals into rivers, lakes, and groundwater (Liu et al., 2008). Additional heavy metal sources in the environment include sediment debris from wind, forest fires, natural biological processes, marine salt, metal corrosion, resuspended sediments, and soil erosion. Various studies have documented these sources, as listed in Table 2.1. Figure 2.2 shows the natural and anthropogenic sources of heavy metals.

2.2.1.2 Anthropogenic Sources

In recent years, urbanization, industrialization, agricultural and domestic activities have been linked to increased heavy metals in aquatic environments (Gautam et al., 2014; Reddy & Sunitha, 2023; Vareda et al., 2019). Notable anthropogenic sources include metal smelting, burning fossil fuels, industrial waste discharge, inappropriate disposal of domestic waste, emissions from vehicle exhausts, and the use of metal-containing pesticides. Specifically, agricultural activities contribute significantly to heavy metal contamination, primarily due to the use of herbicides, insecticides, fertilizers, and other agrochemicals (Srivastava et al., 2018). These activities are responsible for discharging various quantities of toxic metals, contaminating water sources, and impacting the ecosystem. The increased use of heavy metals has resulted in a rapid increase in metallic compounds within terrestrial and aquatic environments, including groundwater (Fig. 2.2).

2.2.2 Pathways and Fate of Heavy Metals in Groundwater

Whether from human activities or natural processes, heavy metals can find their way into groundwater and interact with other substances in the surrounding environment (Sarwar et al., 2017). Industrial waste that contains toxic heavy metals, including Pb,

Table 2.1 Major sources of some heavy metal ions in water

Heavy metal	Natural sources	Anthropogenic sources	Typical concentrations (mg/L[a] 10^{-3})	WHO[a] permissible limits (mg/l)
Lead (Pb)	Geological weathering, volcanic activity	Industrial effluents, surface runoff, landfills, recycled PVC lead paints	1–15	0.01
Mercury (Hg)	Atmospheric deposition, oceanic sources	Industrial effluents, waste disposal, combustion of coal, and volcanic emissions	0.1–2	0.001
Cadmium (Cd)	Rock weathering, atmospheric deposition	Industrial effluents, agricultural activities, Paints, pigments, electroplated parts, batteries, plastics, synthetic	0.2–10	0.05
Arsenic (As)	Geological processes, volcanic activities	Industrial effluents, pesticides, fossil fuel combustion, poles treated with arsenic-based preservatives, pesticides	1–10	0.01
Chromium (Cr)	Rock weathering, volcanic activities	Industrial effluents, surface runoff, and the leather industry	5–50	0.003
Nickel (Ni)	Geological processes, atmospheric deposition	Industrial effluents, fossil fuel combustion	2–20	0.07
Copper (Cu)	Geological processes	Fertilizers, tanning, and photovoltaic cells		0.02
Zinc (Zn)	Geological processes	Soldering, cosmetics, and pigments	10–40	0.8

[a]WHO (2011)

Cd, Hg, and Cr, can directly flow into rivers, lakes, and seas, resulting in contamination. Runoff from agricultural fields is a major pathway for heavy metals to enter aquatic environments, as fertilizers and pesticides used in farming often include heavy metals. Therefore, companies that generate large quantities of wastewater, residues, and sludges, which could be categorized as hazardous, must implement comprehensive waste treatment processes. The transport and fate of heavy metals in groundwater involves a complex process influenced by a wide range of interconnected physical, chemical, and biological factors (Briffa et al., 2020). Understanding this complex process is important for evaluating the hazards linked to the contamination of heavy metals and formulating efficient remediation strategies. Various pathways

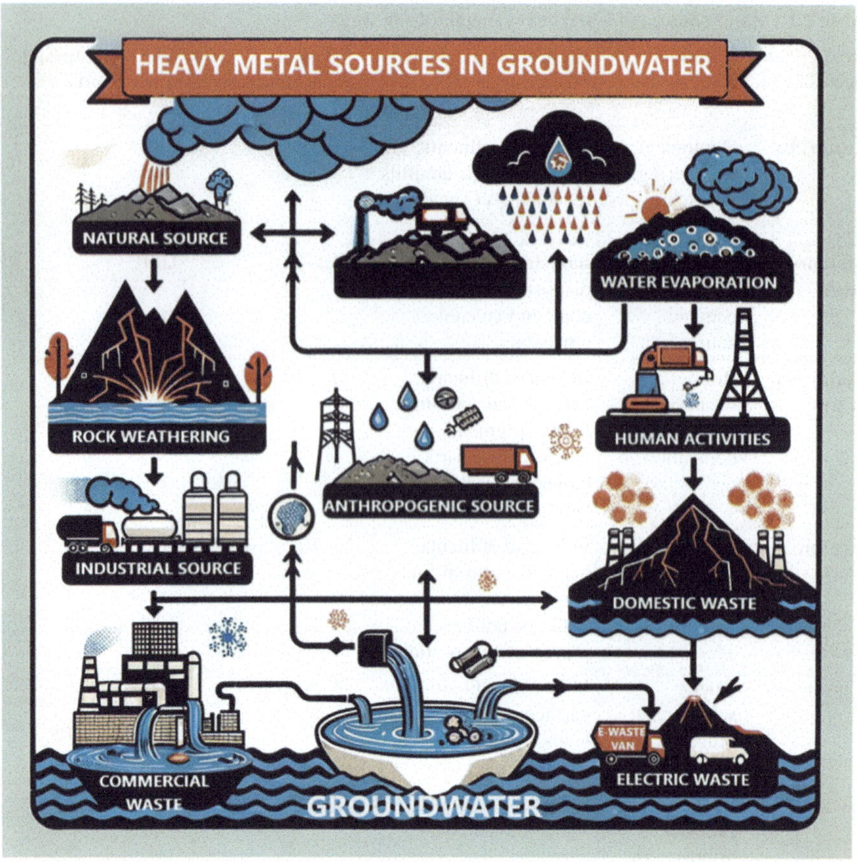

Fig. 2.2 Pictorial illustration of heavy metals sources in groundwater

facilitate the infiltration of heavy metals into groundwater systems, including surface runoff, leaching from contaminated soils, and the percolation of pollutants through the soil. Point sources, such as industrial discharge and landfill leachate, are specific and easily detectable sources of contamination (Boateng et al., 2019; Singh et al., 2022). On the other hand, non-point sources are more widely spread, including agricultural runoff and atmosphere deposition (Kalhor et al., 2019). Once introduced, heavy metals primarily enter groundwater primarily through processes such as dissolution and infiltration. The mobility of heavy metals is significantly influenced by the permeability of the soil and the solubility of the metals. The dynamic of groundwater, including its flow patterns and aquifer characteristics, significantly influences the distribution of toxic heavy metals. Groundwater generally flows through aquifers by following hydraulic gradients. Therefore, the flow rate and direction determine the distance that contaminants can travel. Contaminants such as heavy metals are mainly transported through advection, which involves spreading due to variations in flow velocity within the aquifer. The spread of contaminants is influenced by the features of the aquifer, such as water solubility and chemical reactivity.

Several chemical processes contribute to the dispersion and interaction of these heavy metals in groundwater. The processes involved are adsorption and desorption, redox reactions, complexation and precipitation, bioaccumulation, and biomagnification. The processes of adsorption and desorption, which are governed by the interactions between soil and pollutants, are influenced by factors such as soil pH, organic matter content, and clay mineral composition. Redox reactions entail the exchange of electrons and can alter the chemical conditions of heavy metals, influencing their solubility and mobility in groundwater (Briffa et al., 2020). Microbial activity, particularly in environments with little oxygen, is essential. Heavy metals can form complexes with organic and inorganic compounds, thereby altering their availability and toxicity (Lodygin et al., 2020). Regulated by pH and other ions, these complexes can precipitate out of groundwater. Heavy metals can accumulate from contaminated groundwater (Masindi & Muedi, 2018), with their concentration potentially increasing as they move up the food chain. The impacts of these processes on ecosystems and human health can be significant. The fate and transport of metals in groundwater are controlled by a complex interaction of several biogeochemical and physical factors. Understanding these diverse processes is essential for effectively managing and mitigating the effects of heavy metal contamination in groundwater systems.

2.3 Human Health Implications and Toxicity of Heavy Metal Exposure

Exposure to heavy metals can directly and indirectly affect metabolism and organ function, resulting in various health problems (Al-Ghafari et al., 2019; Okon et al., 2019). The harmful health effects and toxic properties of heavy metals can differ depending on the metal's characteristics, concentration, chemical structure, method, and duration of exposure, among other factors (Cobbina et al., 2015). The toxicity of some heavy metals is described below:

2.3.1 Neurotoxicity

Exposure to heavy metals can cause a variety of neurological and behavioural diseases due to their different neurotoxic properties. While Mn is an essential trace element involved in several physiological and pathological processes, excessive exposure has been associated with neurodegenerative disorders and identified as a potential risk factor for neuronal apoptosis (Liu & Ju, 2023), as well as psychological and neurological illnesses resembling Parkinson's disease (Goldhaber, 2003). Ingesting As can impair cognitive functions in the central nervous system, cause many neurological problems, including changes in neurodevelopment, and contribute to an increased occurrence of neurodegenerative diseases. Exposure to As has been associated with peripheral neuropathy (Helso et al., 2020; Kaur et al., 2024). Several preclinical studies have demonstrated that Cd significantly impairs the functioning of the central- and peripheral nervous system (Marchetti, 2014; Miura et al., 2013). Cu and Zn are essential for the body, but excessive amounts can lead to neurotoxic effects (Grabrucker et al., 2011; Li et al., 2023; Prohaska, 2000). Prolonged exposure to high levels of Se can harm the nervous system (Naderi et al., 2021).

Cd is widely recognized as a nephrotoxicant, bioaccumulating in the body and primarily affecting the kidneys (Hernández-Cruz et al., 2022; Prozialeck & Edwards, 2012; Satarug, 2018; Smereczański & Brzóska, 2023; Wu et al., 2016), by causing a significant accumulation in the urine through damage to the proximal tubular epithelium. Prolonged and excessive exposure might result in renal tubular acidosis, renal failure, and hypercalciuria (Kaewnate et al., 2012). Pb contamination affects various organs, especially the kidneys. Chronic Pb nephropathy is characterized by interstitial fibrosis or renal failure, while acute Pb nephropathy causes proximal tubule dysfunction (Orr & Bridges, 2017). Hg exposure directly affects the kidneys, leading to acute tubular necrosis (Akgül et al., 2016; Gao et al., 2022).

2.3.2 *Carcinogenicity*

Arsenic (As) is a potent human carcinogen that poses a significant cancer risk to individuals exposed to it. Arsenic poisoning increases cancer risk by binding to DNA-interacting proteins, thereby interfering with the DNA repair process (García-Esquinas et al., 2013; Martinez et al., 2011; Park et al., 2015). The carcinogenic process of arsenic includes genotoxicity, which can directly cause DNA damage or generate reactive oxygen species that lead to DNA damage. Pb is a carcinogenic element that hinders the DNA repair process (Hemmaphan & Bordeerat, 2022). Ni functions as a carcinogen by establishing control over many systems that contribute to cancer development, such as gene regulation, management of transcription factors, and creation of free radicals (Guo et al., 2019; Scanlon et al., 2017; Stannard et al., 2017). Some studies suggest that Se supplementation may reduce the risk of certain cancers, while others indicate that excessive quantities may increase the risk of specific types of cancers.

2.3.3 *Hepatotoxicity*

Heavy metals are known for their hepatotoxic effects, which mainly target the liver, which is one of the major organs affected by systemic toxicity. Pb exposure can lead to Hepatomegaly (Islam et al., 2011; Yao et al., 2021), affecting the liver by hindering the activity of different enzymes involved in heme production and triggering oxidative stress. Long-term exposure to lead (Pb) can lead to liver damage, depleting glycogen, causing cellular infiltration, and ultimately resulting in chronic cirrhosis (Hegazy & Fouad, 2014). Exposure to Cd can also have hepatotoxicity implications, manifesting as both acute and chronic liver damage and failure while increasing cancer risk (Hyder et al., 2013). Excessive Cu accumulation can lead to oxidative stress, cytotoxicity, abnormal protein synthesis, and multiple organ damage, as seen in Wilson disease (WD), a metabolic disorder caused by Cu buildup in the liver (Tian et al., 2023; Vo et al., 2024; Zischka & Einer, 2018).

2.3.4 *Immunotoxicity*

Heavy metals can cause immunotoxicity through various mechanisms, disrupting the immune system and increasing vulnerability to infections, autoimmune disorders, and inflammatory diseases. Pb exposure, whether acute or chronic, harms the immune system, leading to various toxic effects and immunological responses, including increased allergies, greater vulnerability to infections, autoimmune diseases, and possible cancer (Dietert et al., 2004). Exposure to Cd might lead to immunosuppressive effects, depending on the specific exposure conditions (Mirkov et al., 2021; Wang et al., 2021a, 2021b). While Hg has no apparent effect on the human immune system, there are concerns about the potential effect of amalgam, a mercury alloy, on the immune system. The immunotoxicity caused by Hg may potentially contribute to the development of autoimmune diseases and allergy reactions (Bjørklund et al., 2020). Several studies have demonstrated that Cr significantly influences the human immune system, leading to various harmful consequences.

2.3.5 *Cardiovascular Toxicity*

Whether acute or chronic, Pb exposure leads to several pathological diseases in humans, including an increased risk of arterial hypertension (Gambelunghe et al., 2016; Tsoi et al., 2021; Vaziri, 2008). Aside from its carcinogenic properties, Cd is also associated with cardiovascular disease, including hypertension and increased risk of cardiovascular events (Hudson et al., 2019; Martins et al., 2021; Tellez-Plaza et al., 2008) and myocardial infarction (Peters et al., 2010). Relevant literature highlights the cardiovascular toxicity properties of Hg (Downer et al., 2017; Genchi et al., 2017), which could impair the normal functioning of the cells lining the blood vessels and contribute to the development of inflammation in the cardiovascular system (Hu et al., 2021). Exposure to Hg has been linked to high blood pressure, heart attack, and negative changes in the structure of the heart (Kim et al., 2016). Exposure to cobalt leads to human health implications known as systolic cardiac depression. Prolonged exposure to high amounts of cobalt can result in a unique type of heart muscle disease called cobalt cardiomyopathy or cobalt-induced cardiomyopathy (Umar et al., 2019).

2.3.6 *Skin Toxicity*

Heavy metals are well-known for their harmful effect on skin health, with various metals having unique mechanisms by which they induce skin toxicity, leading to a range of different clinical symptoms. Prolonged exposure to arsenic (As) can cause various skin conditions, including hyperpigmentation and hypopigmentation, which are the most common skin alterations associated with chronic As exposure (Ötleş and Çağındı, 2010). Cr ingestion leads to a range of acute and chronic severe skin-related effects. Chronic exposure to hexavalent chromium has been linked to a greater vulnerability to skin cancer, typically developing at sites of direct contact, often occurring after forming long-lasting skin ulcers (Shelnutt et al., 2007). Hg toxicity is the leading cause of dermatological issues, causing several skin conditions, such as acrodynia, characterized by pink skin, swelling, and peeling, as well as contact dermatitis and allergic responses from exposure to inorganic mercury. Direct contact with Cd can result in skin irritation and rashes, while prolonged exposure can result in itchy and inflamed wounds.

2.3.7 *Reproductive and Developmental Toxicity*

Heavy metals are widely recognized for their harmful impact on human health, specifically regarding reproductive and developmental processes. Exposure to As can result in reduced sperm production, disrupted menstrual cycles, and increased risk of miscarriages in women. It has been linked to decreased reproductive capacity in both males and females. Specifically, exposure to inorganic As in humans can change testosterone and gonadotropin levels and disrupt steroidogenesis (Kim & Kim, 2015). Exposure to Pb is also associated with a decrease in sperm count and the presence of an abnormal sperm shape in males, while increased Pb levels in women can result in miscarriages, stillbirths, and preterm births. Multiple studies have confirmed that women face the possibility of infertility resulting from being exposed to hazardous heavy metals like Pb, Cd, Hg, etc. These heavy metals are the predominant environmental pollutants that might cause reproductive problems. Table 2.2 shows the global and regional scenario of heavy metal contamination in groundwater and associated risks.

Table 2.2 Heavy metal contamination in groundwater: the global and regional scenario and associated risks

Continent	Country/ State	Study area	Heavy metal analysis/ techniques used	Source of contamination	Remarks on the groundwater quality and associated risk	References
Asia/ Europe	Turkey	Egirdir Lake, Lake District	Cr, Ni, Cu, Zn, As, Se, Mo, Cd, and Pb	Anthropogenic sources	Arsenic was non-carcinogenic; however, chromium and arsenic were carcinogenic in children and adults	(Şener et al., 2017)
Asia	India	Villuppuram District, Tamil Nadu (around Thirukoviloor and near Thenpennai River)	Zn, Cr, Cu, and Fe/ atomic absorption spectroscopy	The release of waste, including industrial and municipal wastewater, the disposal of solid waste in landfills, and other human-caused impacts on the environment	The study region has drinking water that meets WHO standards	(Sirajudeen & Pravinkumar, 2021)
Africa	Nigeria	Calabar Metropolis, Cross River State. The borders are Akpabuyo and Odukpani LGAs	Zn, Fe, and Cr/721D single beam vis spectrophotometer and water testing meters	Weathering and metamorphism are geogenic sources. Insecticides, welding, garbage burning, fertilizers, paint pigments, dye discharges, and open dumpsite leaching are possible anthropogenic sources	Due to localized pollution around some wells, the spatial variability of the groundwater quality index showed that the study's water quality was good (80%), fair (14%), poor (4%), and very poor (2%). Heavy metals were harmful by mean concentration in this order: Zn > Fe > Cr	(Ogarekpe et al., 2023)

(continued)

Table 2.2 (continued)

Continent	Country/State	Study area	Heavy metal analysis/techniques used	Source of contamination	Remarks on the groundwater quality and associated risk	References
North America	Mexico	Nuevo Leon state is located in the northeastern region of Mexico	Al, Si, Cr, Mn, Fe, Zn, As, Ba, Pb, Hg, Se, Ni, Cd, Sr, Li, Cs, Co, Cu, and Ti/Mass spectrometry with induction-coupled plasma	Septic systems, deserted diesel stations, outdated industrial electrical transformers, storage batteries, agricultural equipment, workshop chemicals, electronic and electrical waste, as well as animal and farming activities, are all examples of human-made or human-caused factors	The levels of many heavy metals (Strontium, Lithium, Barium, and Cadmium) are over the maximum allowable limit (4.0 mg/L) set by the USEPA in 2018	(de León-Gómez et al., 2020)
South America	Brazil	It is located north of Campinas County in São Paulo. Atibaia River alluvial plains (east–west flow)	As, Ba, Cd, Cr, Cu, Fe, Mn, Ni and Zn	Natural processes include rock weathering, water–rock interaction, evaporation, atmospheric precipitation, and anthropogenic sources	The water quality index classifies the study area's waters as excellent. However, some wells are contaminated by domestic sewerage systems, making them unfit for drinking	(Rupias et al., 2021)
Australia/Oceania	Australia	North-east Kalgoorlie, Western Australia	Co, Cu, Zn, Cd, Al, and Mn/inductively coupled plasma mass spectrometry (ICP-MS) and inductively coupled plasma atomic emission spectroscopy (ICP-AES)	The tailings wastewater from the mill process undergoes percolation, resulting in the oxidation of the tailings. Simultaneously, the deep groundwater becomes acidic and gets contaminated with heavy metals	Controlling the pH of groundwater can effectively regulate the concentration of heavy metals	(Lei et al., 2010)

2.4 Monitoring and Detection of Heavy Metals in Groundwater

Groundwater is important for drinking, irrigation, and industry because it is better protected from pollution than surface water and has value beyond public health and economic considerations. Groundwater contamination by heavy metals frequently goes undetected. In recent years, alarming studies from various organizations have shown widespread heavy metal contamination in numerous places, causing global concern. As presented below, monitoring and detection techniques for heavy metals in groundwater are important for ensuring the safety and quality of water resources.

2.4.1 Analytical Methods and Techniques for Heavy Metals Detection in Groundwater

Different analytical methods can detect heavy metals in groundwater. These approaches vary in sensitivity, selectivity, accuracy, precision, cost, and complexity. Some processes require sample preconcentration, extraction, or digestion to improve detection limits and reduce matrix interference. Some approaches offer speciation analysis, identifying and quantifying element oxidation states, complexes, and organic and inorganic compounds. Speciation analysis is important because different forms of the same element can have different toxicity, mobility, and bioavailability.

Atomic absorption spectrometry (AAS) is a widespread and sensitive elemental analytical method that can detect metals in samples at extremely low levels, down to picograms. It is sensitive enough to detect most metals and metalloids, making it appropriate for various applications while minimizing interference. Most atom cells used in this technique are flame cells and electrothermal sample cells. When the sample contains sufficient analyte, the flame cell can be used for fast, simple analysis of a few elements. Electrothermal atomic absorption spectroscopy (ETAAS) has better detection limits than flame ASA but requires more experience and is slower to analyze.

Inductively Coupled Plasma (ICP) is an ionization source augmented with different detectors for low-concentration heavy metal quantification. ICP-Mass spectroscopy has been used to detect Mn, Fe, As, and Cr, Hg, As, Pb in various groundwater samples (Sheng et al., 2022; Zhang et al., 2023). Speciation analysis of trace elements like As, Cr, and Hg in groundwater is conducted using ICP-MS, as previous studies have shown that their native structures affect the chemical, biological, and toxicological properties of groundwater, which is becoming increasingly important. Although the Positive Matrix Factorization (PMF) model is known for its quantitative assessment of contamination sources, the Absolute Principal Component Scores-Multiple Linear Regression (APCS-MLR) model, different from the PMF model, indicates that industrial-agricultural activity contributes to the non-cancer and cancer risks linked to heavy metal contamination in groundwater (Sheng et al.,

2022). Heavy metal discharge into the Yangtze River Basin was measured using inductively coupled plasma atomic emission spectrometry (ICP-AES) and compared to WHO and Chinese drinking water standards. 94.74% of the sampled groundwater was hard, having a high conductivity value indicative of a high concentration of ions and/or dissolved solids, mainly Sr > Fe > Ba > Mn > Sc > Cu > Li > Co > Hg (Chen et al., 2023). Exploring the possible transformation of groundwater for sustainable development, inductively coupled plasma optical emission spectroscopy (ICP-OES) and cold vapor atomic absorption spectrometry (CV-AAS) have been recently deployed for targeted screening of groundwater metals and metalloids in Colombia communities striving in farming, fishing, and artisanal/small-scale gold mining (ASGM) (McCourt et al., 2024).

Electrochemical methods measure the current generated by the redox reaction of heavy metals at an electrode surface. Different variants of voltammetry, such as stripping voltammetry and polarography, can be employed for heavy metal detection. Colourimetry methods rely on chemical reactions that change color in the presence of specific heavy metals. The intensity of the color is then correlated with the concentration of the metal. Ion Chromatography is often used to analyze heavy metal ions in water samples. It involves separating ions based on their interaction with an ion-exchange resin, followed by detection using various techniques such as UV–Vis spectrophotometry or conductivity.

Laser-Induced Breakdown Spectroscopy employs a focused laser pulse to vaporize and excite the heavy metal atoms in the sample, followed by measuring the emitted light to identify and quantify the metals. To monitor and detect heavy metals in groundwater, it is necessary to consider the spatial and temporal variations in water quality, including the sources and pathways through which pollutants enter the water. Furthermore, it is essential to consider the hydrogeological and geochemical characteristics of the underground water supply and the possible implications for human health and the ecosystem. Multivariate statistical tools, such as factor analysis, cluster analysis, and principal component analysis, can be used to evaluate the data and identify the patterns, trends, and correlations among the variables. Additionally, risk assessment can be conducted to estimate the extent of exposure and danger posed by heavy metals to consumers through several routes, including ingestion, inhalation, or skin contact.

2.4.2 Exploring GIS in Groundwater Heavy Metal Monitoring

Geographical Information System (GIS), a computerized system, has been widely acknowledged as a valuable tool for assessing groundwater pollution issues and managing water resources. In environmental studies and modeling, it is widely used as a powerful tool because it is integrated with the DRASTIC, APCS-MLR, UNMIX, and PMF models, and other machine learning approaches (Lv et al., 2023; Motlagh et al., 2023). The application of GIS techniques has led to a deeper understanding of the geochemical origin of these contaminants in groundwater sources (Nath et al., 2018). It is widely used to collect spatial data and conduct overlay analysis to represent spatially variable phenomena in a specific area. Various water quality criteria are initially analyzed, and the resulting data is represented through thematic maps and contour maps. Nath et al. (2018) adopted GIS and geochemical techniques to assess heavy metals in groundwater of twenty administrative districts in India with emphasis on the origins, distribution, and level of heavy metal contamination in the area, which is situated in the Brahmaputra River basin. The ecological risk assessment in the central area of a typical heavily industrialized city in southern China was conducted using GIS technology. The levels of pollutants, ranked from highest to lowest, were Hg, Cd, As, Pb, and Zn (Lv et al., 2023). In their study, Long et al. (2021) employed GIS mapping to evaluate the spatial distribution, pollution index, and potential human health implications of heavy metals pollution in the shallow groundwater of the Dongting Lake plain. Therefore, GIS supports the formulation and execution of efficient plans for remediation and mitigation of adverse effects, thus promoting sustainable use of groundwater resources.

2.4.3 *Nanotechnology/Sensors for Groundwater Heavy Metal Detection and Monitoring*

Nanosensors and nanoelectronics are produced from a material with excellent properties such as high mechanical strength, electrical conductivity, and thermal stability. Paper-based electrochemical sensors modified with earth-abundant material have been reported as a novel technique for heavy metal detection in groundwater. Exploring this combined disposable electrode and nanocomposite materials is cost-effective for infield detection. It offers an enhanced sensitive 2.13 μA/ppb (LOD = 0.3 μg/L) and selectivity in the detection of groundwater arsenic, even in the presence of other coexisting ions (Dahake & Bansiwal, 2022). Electroanalytical technologies offer an attractive solution for the real-time monitoring of heavy metal pollution. Electrochemical sensing systems provide high sensitivity and selectivity towards electroactive substances while being user-friendly, compact, rapid, portable, and cost-effective. Cyclic and stripping voltammetry have been established as effective techniques for electrochemical detection of trace amounts of inorganic metals. Xue et al. (2023) demonstrated that a small-sized electrode AuNPs/mesoporous NiO/nickel foam nanocomposite can effectively and accurately detect the presence of lead ions (Pb^{2+}) and copper ions (Cu^{2+}) in groundwater, with high sensitivity for on-site or real-time monitoring of groundwaters. A highly efficient nanosensor for surface-enhanced Raman spectroscopy (SERS) developed using gold/silver bimetallic nanoparticles (Au/Ag-NPs) possessed a face-centered cubic (FCC) structure and core–shell nano-crevice nanoparticles with a 23 nm particle size exhibiting significant plasmonic behaviour, enhancing electromagnetic excitation for achieving SERS detection and plasmonic photocatalysis to detect heavy metals in groundwater (Aarthi et al., 2021).

2.4.4 Artificial Intelligence in Groundwater Heavy Metals Detection and Monitoring

To guarantee the security of the water supply, on-site monitoring and prompt response to high heavy metal concentrations in groundwater, which serves as an essential water source, are necessary. Using fluorescence sensors and smartphones with digital cameras is becoming popular as a novel and growingly favoured method for on-site, real-time environmental monitoring. This method reduces solution brightness by quenching carbon-based quantum dots with the target material. Consequently, different color changes, like shifts from red-blue to blue-yellow, can be detected by radiometric fluorescence optical devices as a response to the presence of the target heavy metal. On-site heavy metal monitoring has deployed fluorescence probes and machine learning (Sajib et al., 2023). An accurate detection technique for Cr(VI) in groundwater was developed by integrating machine learning with high-quantum-yield (QY) N-doped blue-light carbon dots (N-BCDs) (Zhang et al., 2023). Ridge, XGB, SVR, and Linear models achieved 95.2% fitness and 95.71% accuracy at the μg level of Cr(VI) detection level. Deep learning has recently been employed to detect the effectiveness of groundwater heavy metal pollution indices with high accuracy and performance verifiable by artificial neural network-based models for both qualitative and quantitative concerns (Singha et al., 2020). This reveals how deep learning and artificial neural networks can be used to develop supervised learning prediction models to determine the best groundwater heavy metal pollution indexing approach.

2.5 Indexing Approaches of Heavy Metals in Groundwater and the Human Health Risk Assessment

2.5.1 Heavy Metal Pollution Index (HPI)—Indexing Approach

The Heavy Metal Pollution Index (HPI) is a rating system adopted to evaluate the combined effect of specific heavy metals on the overall water quality. The rating is expressed as a numerical number between 0 and 1, which indicates the relative significance/importance of each quality factor. This value is inversely proportional to the accepted standard for each parameter, denoted as Si. The HPI calculation involves (i) establishing the weightage of each parameter (*i*th parameter), (ii) computing the quality rating for each specific metal, and (iii) summing these sub-indices in the overall index, thus:

$$\text{HPI} = \sum_{i=1}^{n}(Q_i W_i) / \sum_{i=1}^{n}(W_i) \tag{2.1}$$

In the formula (Eq. 2.1), Q_i represents the sub-index of the specific parameter, while W_i symbolizes the unit weightage assigned to that parameter. The sub-index (Q_i) and unit weight (W_i) are assigned to specific parameters (ith parameter), while n denotes the total number of parameters being considered. The HPI, a widely embraced approach globally, assigns weights to components based on their inverse relationship with the recommended standard values. Concerning the water quality, it is important to note that an HPI 0–25 is classified as very good, HP1 26–50 is classified as good, an HPI 51–75 is classified as poor, and an HPI above 75 is considered very poor and unsuitable for drinking. This is the established critical pollution index (Sirajudeen & Pravinkumar, 2021).

The sub-index for the ith parameter is determined using Eq. (2.2), whereby the observed, acceptable, and authorized values are correspondingly denoted by Mi, Ii, and Si.

$$Qi = (|Mi - Ii|/Si - Ii) \times 100 \tag{2.2}$$

2.5.2 Contamination Index (CI)—Indexing Approach

The contamination intensity (CI) of water quality, which assesses the overall level of pollution, was determined for each parameter as an individual calculation. The equation used to calculate CI summarizes the water quality characteristics considered unsafe for drinking.

$$CI = \sum_{i=1}^{n}\{(C_{ai}/C_{si}) - 1\} \tag{2.3}$$

The formula for CI is shown in Eq. (2.3), whereby C_{ai} denotes the analytical value of the ith component, and C_{si} represents the permissible value of the ith component.

2.5.3 Contamination Factor (CF)

Calculating the contamination factor (CF) for individual heavy metal pollutants, as initially proposed by Hakanson, quantifies the extent of general pollution in a sampled site. To calculate CF, groundwater samples must be averaged to obtain a mean concentration of the heavy metal pollutants. The mean concentration is compared to a pristine reference level.

$$\mathrm{CF} = C/C^0 \tag{2.4}$$

Following the formula in Eq. (2.4), the CF is calculated by dividing the average concentration of each metal in the sample (C) by the background concentration (C^0), which represents the concentration found in a non-polluted sample.

2.5.4 Degree of Contamination and Modified Håkanson Method

The generalized form for determining the degree of contamination (Cd) is shown in Eq. (2.5). However, the modified level of pollution (mCd) used to determine the total level of pollution at a certain sample location is shown in Eq. (2.6), allowing the inclusion of an unlimited number of metals for analysis in the study. According to Håkanson, the extent of contamination and the modified form categories can be seen in Fig. 2.3.

$$Cd = \sum_{i=1}^{N} CF_i \tag{2.5}$$

$$mCd = \left(\sum_{i=1}^{n} CF_i\right)/n \tag{2.6}$$

where N = number of metals studied, n = number of analyzed elements, i = ith element (or pollutant), and CF = Contamination factor.

2.5.5 Single-Factor Index Analysis and Nemerow Index Methods

Single-factor index analysis provides precise quantitative data of important heavy metal pollution elements and their excessive multiples, as calculated using Eq. (2.7). By analyzing the value of P_{ij}, we can identify the specific pollutants that exceed the acceptable limits and determine the extent of their excess in the study area. This enables us to identify the most significant pollutants and the regions that are most severely affected by pollution. However, various studies indicate that the aforementioned method fails to accurately capture the overall impact resulting from all types

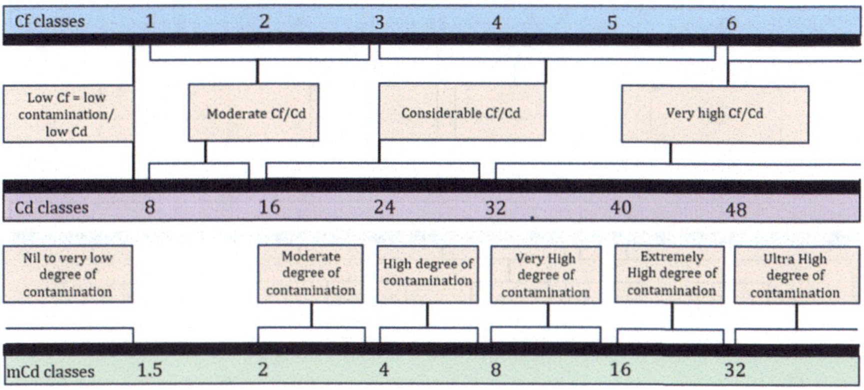

Fig. 2.3 Contamination factors (CF), degree of contamination (Cd), and the modified degree of contamination (mCd) of the Hakanson classification (Kanson, 1980)

of heavy metals. The Nemero index method in Eq. (2.8) does not solely rely on extreme values; instead, it calculates an environmental quality index by considering multiple factors with assigned weights.

$$P_{ij} = C_{ij}/S_i \tag{2.7}$$

$$P_i = \sqrt{\left[\left(P_{ijmax}\right)^2 + \left(P_{ijavg}\right)^2\right]/2} \tag{2.8}$$

where P_{ij} represents the pollution index of the heavy metal j in the ith functional region, C_{ij} denotes the measured contamination value of heavy metal j in the ith functional region, S_{ij} represent background contamination value of heavy metal j, and P_i is the compressive pollution index of the i th functional region, P_{ijmax} refers to the highest value in the single-factor pollution index, while P_{ijavg} denotes the average value in the single-factor pollution index (Fig. 2.4).

2.6 Risk Assessment, Human Health Evaluation and Regulatory Framework/policies

Heavy metals are poisonous elements that can last in groundwater for extended periods. Contaminated groundwater may contain toxic heavy metals, including Pb, As, Hg, Cd, and Cr. Extended exposure to increased concentrations of these metallic elements can significantly affect human well-being. Exposure to lead has been linked to neurological diseases, developmental delays, renal damage, and cognitive impairments, especially in children. Exposure to arsenic through contaminated groundwater can result in the development of skin lesions, cardiovascular ailments, and a higher

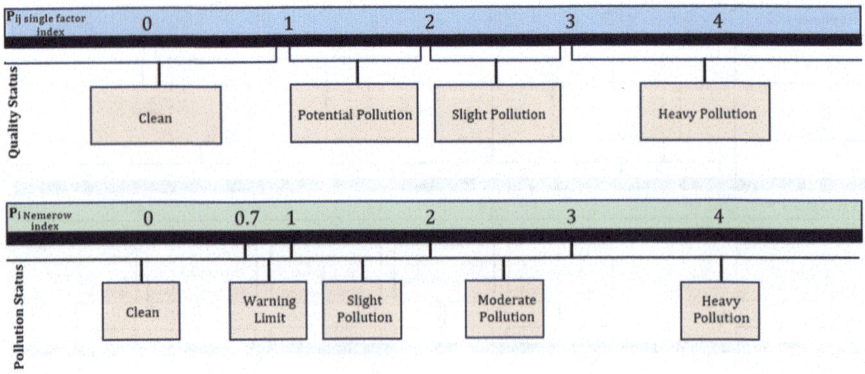

Fig. 2.4 Grading standard of the single-factor index (P_{ij}) and Nemerow index (P_i)

susceptibility to specific kinds of cancer. Mercury exposure can lead to neurological and developmental problems, particularly in newborns and young children. Cadmium is associated with renal impairment and a higher susceptibility to cancer, whereas exposure to chromium can lead to pulmonary carcinoma and skin inflammation. Implementing efficient monitoring and remediation strategies is essential to avoid persistent health effects from exposure to heavy metal contaminants in groundwater.

A major drinking water source in Turkey and the second largest lake, the Eğirdir Lake basin, has been found to have high levels of As and Cr, which pose significant chronic health hazards for both adults and children (Şener et al., 2017). The lake water samples were predominantly observed in the rock–water interaction field using the Gibbs plot, having Ca^{2+}, Mg^{2+}, $HCO_3{}^{2-}$, and $SO_4{}^{2}$ as the major ions with a good to excellent groundwater quality index (WQI). However, the evaluation of the possible negative impact of metals on human health, as determined by hazard quotients (HQ) through consumption and skin contact routes in humans, revealed that As exhibited the highest HQ value (> 1), thus establishing it as the primary pollutant of interest in relation to non-carcinogenic implications. The evaluation of possible hazards to human health arising from non-carcinogenic effects entails analyzing the intake and exposure to heavy metals in the groundwater at a specific research site. Risk assessment refers to the methodologies employed to evaluate the likelihood of harmful health impacts occurring within a specified timeframe. Typically, the evaluation of health hazards related to each contamination involves assessing the degree of risk, which is classified as either carcinogenic or non-carcinogenic (Mohammadi et al., 2019). To evaluate heavy metals contamination and the potential risks to both cancer and non-cancer health resulting from the consumption and absorption of these metals through the skin in the water distribution system, the commonly used indicators include Hazard Quotients (HQ), Hazard Index (HI), and the Incremental Lifetime Cancer Risk (ILCR) (Mohammadi et al., 2019).

2.6.1 Non-carcinogenic Health Risk—Assessing Individual Heavy Metals

The assessment involved the use of Hazard Quotients (HQ) and Hazard Index (HI) to estimate the possible non-cancer health risks linked to skin contact and groundwater consumption (Mohammadi et al., 2019). The targeted population for the human health evaluation used measured and approved values to determine the exposure level to heavy metals through food consumption and cutaneous absorption of water. Equations (2.9) and (2.10) are the U.S. Environmental Protection Agency's equations employed to calculate the Chronic Daily Intake (CDI) of heavy metals via oral consumption and dermal absorption of water.

$$CDI_{oral}\left(mg\ kg^{-1}\ day^{-1}\right) = \frac{(C_{hm} \times IR \times ABS \times EF \times ED)}{(BW \times AT)} \quad (2.9)$$

$$CDI_{dermal}\left(mg\ kg^{-1}\ day^{-1}\right) = \frac{(C_{hm} \times SA \times Kp \times ABS \times ET \times EF \times ED \times CF)}{(BW \times AT)} \quad (2.10)$$

The equation employs the following terminologies: C_{hm} = heavy metal concentration (mg/kg or mg/L), DI = daily average intake of water in the area (L/day), ABS = absorption factor, EF = annual exposure frequency (days/year), ED = exposure duration (years), SA = skin surface contact area (cm^2), Kp = Permeability coefficient (cm/h), ET = exposure time (h/event), CF = conversion factor (L/cm^3), BW = body weight (kg/person), and AT = average time (days).

The health hazard index (HQ) is calculated by comparing the average chronic daily intakes (CDI) of heavy metals to the oral reference dose (RfD) specific to each heavy metal (Mohammadi et al., 2019). The following equations are used to calculate the HQ of heavy metals based on both oral ingestion and dermal water adsorption for the inhabitants of the specified region. Herein, x is either oral or dermal, CDI and RfD are expressed in mg/kg-day.

$$HQ_x = CDI_x/RfD_x \quad (2.11)$$

2.6.2 Non-carcinogenic Health Risk—Assessing for Multiple Heavy Metals

The Hazard Index (HI) encompasses all potential non-carcinogenic health concerns linked to different heavy metals in water (Mohammadi et al., 2019). Moreover, the following equation was used to assess whether individuals residing in the research area were exposed to water consumption and dermal absorption, following the U.S.

Environmental Protection Agency (USEPA) guidelines. Herein, hm is the individual heavy metals.

$$HI_x = \sum_{i=1}^{n} HQ_x + HQ_{hm1} + HQ_{hm2} + \cdots + HQ_{hm\infty} \qquad (2.12)$$

An approach to evaluate the potential health risks related to exposure to heavy metals, which do not cause cancer, is to establish a threshold level of 1.0, where an HI value greater than 1.0 signifies that non-carcinogenic health hazards can be impactful, and when the HI is less than 1, it is unlikely that the exposed individuals will experience noticeable harmful health effects.

2.6.3 Carcinogenic Health Risk

The potential carcinogenic effects linked to a specific dosage of heavy metal exposure can be assessed by calculating the Incremental Lifetime Cancer Risk (ILCR). The ILCR of potential cancer-causing heavy metals can be estimated by multiplying the Chronic Daily Intake (CDI) and Cancer Slope Factor (CSF) to assess the likelihood of causing cancer over a lifetime of exposure (Mohammadi et al., 2019). For every cancer-causing agent, the Incremental Lifetime Cancer Risk (ILCR) was computed using Eq. (2.13). The investigation complied with the permissible range for ILCR established by the United States Environmental Protection Agency (USEPA), which is between 1.0×10^{-6} and 1.0×10^{-4} for both single-element and multi-element substances that cause cancer.

$$\text{ILCR} = \text{CDI} \times \text{CSF} \qquad (2.13)$$

In groundwater management as a shared resource, one of the challenges faced is the issue of free riding, where beneficiaries take advantage of the resource without contributing to its sustainability. To address this problem, frameworks and policies are established to minimize the tragedy by developing governance for the socio-ecological and resilience system, examining the role of resilience in protecting the rights associated with groundwater resources, both public and private. Additionally, the frameworks examine how the persistence and characteristics of heavy metal pollutants influence groundwater control variables, pose challenges to public and private rights, and give rise to controversies. Nevertheless, the resilience of the frameworks or policies can vary and sometimes conflict with each other, posing challenges to the public and private rights associated with groundwater. However, they also emphasize that understanding these dynamics and resilience play a crucial role in fostering trust among stakeholders and encouraging their active participation in water management. The socio-ecological system (SES) framework is employed in planning, policymaking, and determining public and private rights to help strengthen the resilience and assess the dynamics of the groundwater system (Taghilou, 2022). Different countries may adopt the framework illustrated in Fig. 2.5 to enhance groundwater for sustainable development.

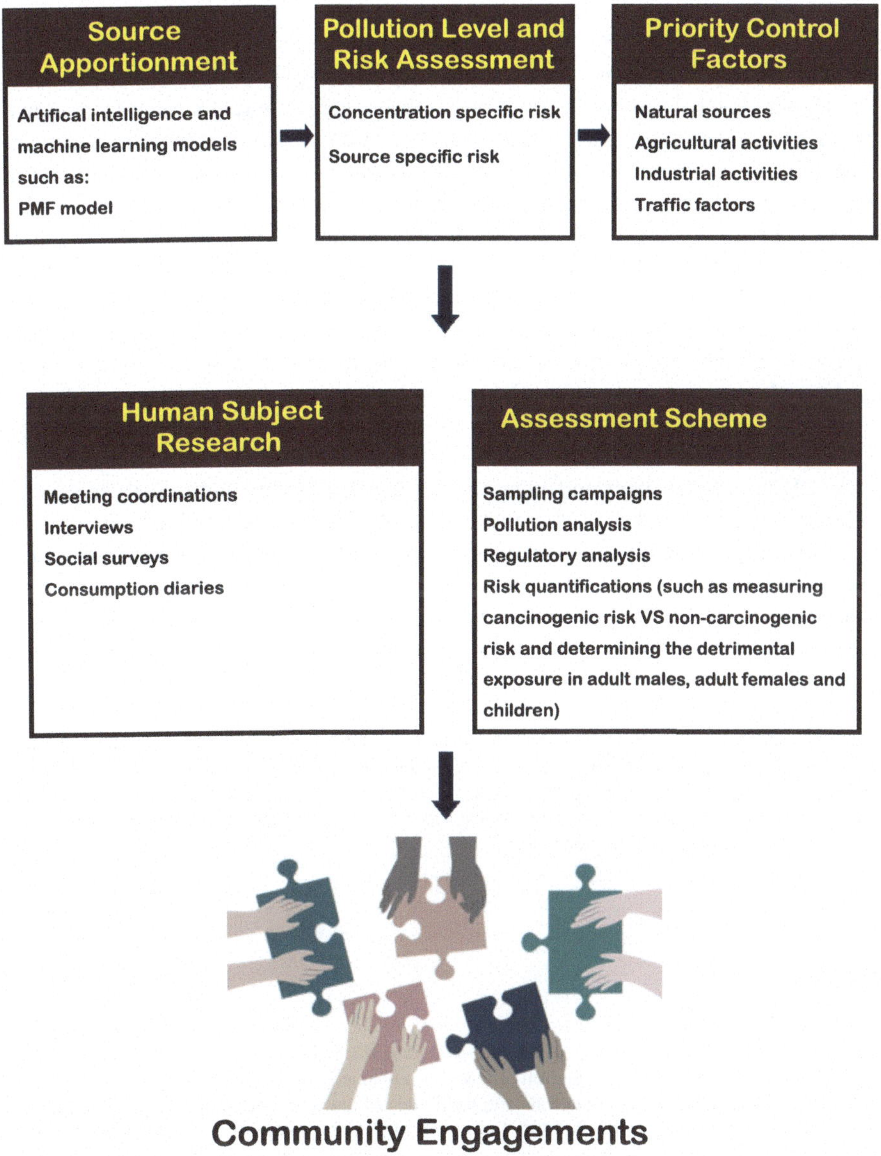

Fig. 2.5 Framework for enhancing groundwater for sustainable development

2.7 Conclusions

Detecting heavy metal contamination in groundwater is important to ensure the safety and quality of water supplies. These contaminants have relatively high densities, atomic weights, or atomic numbers, such as lead, mercury, cadmium, chromium, iron, copper, zinc, and arsenic. Some heavy metals are essential nutrients or relatively harmless, while others are toxic and can cause poisoning. The major natural sources include rocks and soils, while the common anthropogenic sources stream from industrial, urban, and agricultural activities. There are various analytical methods and techniques for the determination of heavy metals in water, such as atomic spectroscopy, electrochemistry, chromatography, and bio-based methods. These methods differ in sensitivity, selectivity, accuracy, precision, cost, and complexity. Some methods require sample preparation, such as preconcentration, extraction, or digestion, to improve the detection limits and reduce matrix interference. Some methods also allow speciation analysis, which involves the identification and quantification of different chemical forms of the same element, such as oxidation states, complexes, or organic and inorganic compounds. Speciation analysis is important because different forms of the same element can have different toxicity, mobility, and bioavailability. Indexing approaches like the Heavy Metal Pollution Index (HPI) and Contamination Index (CI) comprehensively and quantitatively examine water quality for heavy metal contamination. Researchers, policymakers, and the public may understand complex environmental data with these indexes. Using scaled indices to categorize water quality from 'very good' to 'unsuitable for drinking,' these approaches help understand groundwater health concerns. The robust framework of these indices calculates metal contributions to water quality, enabling targeted actions and monitoring pollution management strategies.

Acknowledgements We acknowledge and appreciate the collaborative efforts and shared expertise of our co-authors, as well as all those who contributed to the completion of this chapter. We extend our sincere gratitude to our colleagues who provided feedback during the manuscript preparation process. We are grateful for the insightful feedback and suggestions provided by the three reviewers, which significantly improved the manuscript.

References

Aarthi, A., Bindhu, M. R., Umadevi, M., Parimaladevi, R., Sathe, G. V., Al-Mohaimeed, A. M., et al. (2021). Evaluating the detection efficacy of advanced bimetallic plasmonic nanoparticles for heavy metals, hazardous materials and pesticides of leachate in contaminated groundwater. *Environmental Research, 201*, 111590. https://doi.org/10.1016/j.envres.2021.111590

Adimalla, N. (2019). Groundwater quality for drinking and irrigation purposes and potential health risks assessment: A case study from semi-arid region of South India. *Exposure and Health, 11*(2), 109–123. https://doi.org/10.1007/s12403-018-0288-8

Aghazadeh, N., Chitsazan, M., & Golestan, Y. (2017). Hydrochemistry and quality assessment of groundwater in the Ardabil area, Iran. *Applied Water Science, 7*(7), 3599–3616. https://doi.org/10.1007/s13201-016-0498-9

Akgül, N., Altunkaynak, B. Z., Altunkaynak, M. E., Deniz, Ö. G., Ünal, D., & Akgül, H. M. (2016). Inhalation of mercury vapor can cause the toxic effects on rat kidney. *Renal Failure, 38*(3), 465–473. https://doi.org/10.3109/0886022X.2016.1138832

Al-Ghafari, A., Elmorsy, E., Fikry, E., Alrowaili, M., & Carter, W. G. (2019). The heavy metals lead and cadmium are cytotoxic to human bone osteoblasts via induction of redox stress. *PLoS ONE, 14*(11). https://doi.org/10.1371/journal.pone.0225341

Bjørklund, G., Peana, M., Dadar, M., Chirumbolo, S., Aaseth, J., & Martins, N. (2020). Mercury-induced autoimmunity: Drifting from micro to macro concerns on autoimmune disorders. *Clinical Immunology, 213*, 108352. https://doi.org/10.1016/j.clim.2020.108352

Boateng, T. K., Opoku, F., & Akoto, O. (2019). Heavy metal contamination assessment of groundwater quality: A case study of Oti landfill site, Kumasi. *Applied Water Science, 9*(2). https://doi.org/10.1007/s13201-019-0915-y

Bricker, S. H., Banks, V. J., Galik, G., Tapete, D., & Jones, R. (2017). Accounting for groundwater in future city visions. *Land Use Policy, 69*, 618–630. https://doi.org/10.1016/j.landusepol.2017.09.018

Briffa, J., Sinagra, E., & Blundell, R. (2020). Heavy metal pollution in the environment and their toxicological effects on humans. *Heliyon 6*, (9), E046. https://doi.org/10.1016/j.heliyon.2020.e04691

Chakraborti, D., Rahman, M. M., Ahamed, S., Dutta, R. N., Pati, S., & Mukherjee, S. C. (2016). Arsenic contamination of groundwater and its induced health effects in Shahpur block, Bhojpur district, Bihar state, India: Risk evaluation. *Environmental Science and Pollution Research, 23*(10), 9492–9504. https://doi.org/10.1007/s11356-016-6149-8

Chen, Y., Xiong, F., Zhai, D., Liu, H., Duan, X., Chen, D., et al. (2023). Risk assessment of dissolved trace elements and heavy metals in the upper reaches of the Yangtze River, China. *Water, 15*(7), 1330. https://doi.org/10.3390/w15071330

Cobbina, S. J., Chen, Y., Zhou, Z., Wu, X., Zhao, T., Zhang, Z., et al. (2015). Toxicity assessment due to sub-chronic exposure to individual and mixtures of four toxic heavy metals. *Journal of Hazardous Materials, 294*, 109–120. https://doi.org/10.1016/j.jhazmat.2015.03.057

Dahake, R. V., & Bansiwal, A. (2022). Highly sensitive detection of arsenic in groundwater by paper-based electrochemical sensor modified with earth-abundant material. *Groundwater for Sustainable Development, 19*. https://doi.org/10.1016/j.gsd.2022.100855

de León-Gómez, H., Martin del Campo-Delgado, M. A., Esteller-Alberich, M. V., Velasco-Tapia, F., Alva-Niño, E., & Cruz-López, A. (2020). Assessment of nitrate and heavy metal contamination of groundwater using the heavy metal pollution index: case study of Linares, Mexico. *Environmental Earth Sciences, 79*(18). https://doi.org/10.1007/s12665-020-09164-3

Dietert, R. R., Lee, J. E., Hussain, I., & Piepenbrink, M. (2004). Developmental immunotoxicology of lead. *Toxicology and Applied Pharmacology, 198*(2), 86–94. https://doi.org/10.1016/j.taap.2003.08.020

Downer, M. K., Martínez-González, M. A., Gea, A., Stampfer, M., Warnberg, J., Ruiz-Canela, M., et al. (2017). Mercury exposure and risk of cardiovascular disease: A nested case-control study in the PREDIMED (PREvention with MEDiterranean Diet) study. *BMC Cardiovascular Disorders, 17*(1). https://doi.org/10.1186/s12872-016-0435-8

Gambelunghe, A., Sallsten, G., Borné, Y., Forsgard, N., Hedblad, B., Nilsson, P., et al. (2016). Low-level exposure to lead, blood pressure, and hypertension in a population-based cohort. *Environmental Research, 149*, 157–163. https://doi.org/10.1016/j.envres.2016.05.015

Gao, Z., Wu, N., Du, X., Li, H., Mei, X., & Song, Y. (2022). Toxic nephropathy secondary to chronic mercury poisoning: Clinical characteristics and outcomes. *Kidney International Reports, 7*(6), 1189–1197. https://doi.org/10.1016/j.ekir.2022.03.009

García-Esquinas, E., Pollán, M., Umans, J. G., Francesconi, K. A., Goessler, W., Guallar, E., et al. (2013). Arsenic exposure and cancer mortality in a US-based prospective cohort: The strong heart study. *Cancer Epidemiology Biomarkers and Prevention, 22*(11), 1944–1953. https://doi.org/10.1158/1055-9965.EPI-13-0234-T

Gautam, R. K., Sharma, S. K., Mahiya, S., & Chattopadhyaya, M. C. (2014). *Contamination of heavy metals in aquatic media: Transport, toxicity and technologies for remediation.* www.rsc.org

Genchi, G., Sinicropi, M. S., Carocci, A., Lauria, G., & Catalano, A. (2017). Mercury exposure and heart diseases. *International Journal of Environmental Research and Public Health, 14*(1), 74. https://doi.org/10.3390/ijerph14010074

Goldhaber, S. B. (2003). Trace element risk assessment: Essentiality vs. toxicity. *Regulatory Toxicology and Pharmacology, 38*(2), 232–242. https://doi.org/10.1016/S0273-2300(02)00020-X

Grabrucker, A. M., Rowan, M., & Garner, C. C. (2011). Brain-delivery of zinc-ions as potential treatment for neurological diseases: Mini review. *Drug Delivery Letters, 1*(10), 13–23.

Guo, H., Liu, H., Wu, H., Cui, H., Fang, J., Zuo, Z., et al. (2019). Nickel carcinogenesis mechanism: DNA damage. *International Journal of Molecular Sciences, 20*(19), 4690. https://doi.org/10.3390/ijms20194690

Guo, R., Chen, Q., Wang, F., Fang, M., Li, L., Niu, H., et al. (2023). Selenium-contaminated water: Recent advances in material function and adsorption performance. *Journal of Environmental Chemical Engineering, 11*(5), 110468. https://doi.org/10.1016/j.jece.2023.110468

Hegazy, A. M. S., & Fouad, U. A. (2014). Evaluation of lead hepatotoxicity; histological, histochemical and ultrastructural study. *Forensic Medicine and Anatomy Research, 02*(03), 70–79. https://doi.org/10.4236/fmar.2014.23013

Helso, S. N., Roug, A. S., Mork, M., Maksten, E. F., & Severinsen, M. T. (2020). Severe peripheral neuropathy from treatment with arsenic trioxide in a patient suffering from acute promyelocytic leukemia. *Journal of Hematology, 9*(3), 89–92. https://doi.org/10.14740/jh617

Hemmaphan, S., & Bordeerat, N. K. (2022). Genotoxic effects of lead and their impact on the expression of DNA repair genes. *International Journal of Environmental Research and Public Health, 19*(7), 4307. https://doi.org/10.3390/ijerph19074307

Hernández-Cruz, E. Y., Amador-Martínez, I., Aranda-Rivera, A. K., Cruz-Gregorio, A., & Pedraza Chaverri, J. (2022). Renal damage induced by cadmium and its possible therapy by mitochondrial transplantation. *Chemico-Biological Interactions, 361*, 109961. https://doi.org/10.1016/j.cbi.2022.109961

Hu, X. F., Lowe, M., & Chan, H. M. (2021). Mercury exposure, cardiovascular disease, and mortality: A systematic review and dose-response meta-analysis. *Environmental Research, 193*, 110538. https://doi.org/10.1016/j.envres.2020.110538

Hudson, K. M., Belcher, S. M., & Cowley, M. (2019). Maternal cadmium exposure in the mouse leads to increased heart weight at birth and programs susceptibility to hypertension in adulthood. *Scientific Reports, 9*, 13553. https://doi.org/10.1038/s41598-019-49807-5

Hyder, O., Chung, M., Cosgrove, D., Herman, J. M., Li, Z., Firoozmand, A., et al. (2013). Cadmium exposure and liver disease among US adults. *Journal of Gastrointestinal Surgery, 17*(7), 1265–1273. https://doi.org/10.1007/s11605-013-2210-9

Indelicato, S., Orecchio, S., Avellone, G., Bellomo, S., Ceraulo, L., Di Leonardo, R., et al. (2017). Effect of solid waste landfill organic pollutants on groundwater in three areas of Sicily (Italy) characterized by different vulnerability. *Environmental Science and Pollution Research, 24*(20), 16869–16882. https://doi.org/10.1007/s11356-017-9198-8

Islam, K., Haque, A., Karim, R., Fajol, A., Hossain, E., Salam, K. A. et al. (2011). Dose-response relationship between arsenic exposure and the serum enzymes for liver function tests in the individuals exposed to arsenic: A cross sectional study in Bangladesh. *Environmental Health: A Global Access Science Source, 10*(1). https://doi.org/10.1186/1476-069X-10-64

Kaewnate, Y., Niyomtam, S., Tangvarasittichai, O., Meemark, S., Pingmuangkaew, P., & Tangvarasittichai, S. (2012). Association of elevated urinary cadmium with urinary stone, hypercalciuria, and renal tubular dysfunction in the population of cadmium-contaminated area. *Bulletin of Environmental Contamination and Toxicology, 89*, 1120–1124. https://doi.org/10.1007/s00128-012-0856-8

Kalhor, K., Ghasemizadeh, R., Rajic, L., & Alshawabkeh, A. (2019). Assessment of groundwater quality and remediation in karst aquifers: A review. *Groundwater for Sustainable Development, 8*, 104–121. https://doi.org/10.1016/j.gsd.2018.10.004

Kanson, L. (1980). An ecological risk index for aquatic pollution control—A sedimentological approach. *Water Research, 14*(8), 975–1001. https://doi.org/10.1016/0043-1354(80)90143-8

Kaur, R., Garkal, A., Sarode, L., Bangar, P., Mehta, T., Singh, D. P., et al. (2024). Understanding arsenic toxicity: Implications for environmental exposure and human health. *Journal of Hazardous Materials Letters, 5*, 100090. https://doi.org/10.1016/j.hazl.2023.100090

Kim, K. H., Kabir, E., & Jahang, S. A. (2016). A review on the distribution of Hg in the environment and its human health impacts. *Journal of Hazardous Materials, 306*, 376–385. https://doi.org/10.1016/j.jhazmat.2015.11.031

Kim, Y. J., & Kim, J. M. (2015). Arsenic toxicity in male reproduction and development. *Development & Reproduction, 19*(4), 167–180. https://doi.org/10.12717/dr.2015.19.4.167

Lei, L. Q., Song, C. A., Xie, X. L., Li, Y. H., & Wang, F. (2010). Acid mine drainage and heavy metal contamination in groundwater of metal sulfide mine at arid territory (BS mine, Western Australia). *Transactions of Nonferrous Metals Society of China, 20*(8), 1488–1493. https://doi.org/10.1016/S1003-6326(09)60326-5

Li, P., He, X., Li, Y., & Xiang, G. (2019). Occurrence and health implication of fluoride in groundwater of loess aquifer in the Chinese Loess Plateau: A case study of Tongchuan. *Northwest China. Exposure and Health, 11*(2), 95–107. https://doi.org/10.1007/s12403-018-0278-x

Li, Z., Liu, Y., Wei, R., Yong, V. W., & Xue, M. (2023). The Important Role of Zinc in Neurological Diseases. *Biomolecules, 13*(1), 28. https://doi.org/10.3390/biom13010028

Liu, C., & Ju, R. (2023). Manganese-induced neuronal apoptosis: New insights into the role of endoplasmic reticulum stress in regulating autophagy-related proteins. *Toxicological Sciences, 191*(2), 193–200. https://doi.org/10.1093/toxsci/kfac130

Liu, H., Li, L., Chengqing, Y., & Baoqing, S. (2008). Fraction distribution and risk assessment of heavy metals in sediments of Moshui Lake. *Journal of Environmental Sciences, 20*, 390–397. https://doi.org/10.1016/s1001-0742(08)62069-0

Lodygin, E. D., Alekseev, I. I., Vasilevich, R. S., & Abakumov, E. V. (2020). Complexation of lead and cadmium ions with humic acids from arctic peat soils. *Environmental Research*, 191. https://doi.org/10.1016/j.envres.2020.110058

Long, X., Liu, F., Zhou, X., Pi, J., Yin, W., Li, F., et al. (2021). Estimation of spatial distribution and health risk by arsenic and heavy metals in shallow groundwater around Dongting Lake plain using GIS mapping. *Chemosphere*, 269. https://doi.org/10.1016/j.chemosphere.2020.128698

Luo, D., Liang, Y., Wu, H., Li, S., He, Y., Du, J., et al. (2022). Human-health and environmental risks of heavy metal contamination in soil and groundwater at a riverside site, China. *Processes, 10*(10). https://doi.org/10.3390/pr10101994

Lv, H., Lu, Z., Fu, G., Lv, S., Jiang, J., Xie, Y., et al. (2023). Pollution characteristics and quantitative source apportionment of heavy metals within a zinc smelting site by GIS-based PMF and APCS-MLR models. *Journal of Environmental Sciences*. https://doi.org/10.1016/j.jes.2023.09.033

Majolagbe, A. O., Adeyi, A. A., & Osibanjo, O. (2016). Vulnerability assessment of groundwater pollution in the vicinity of an active dumpsite (Olusosun), Lagos, Nigeria. *Chemistry International, 2*(4), 232–241.

Marchetti, C. (2014). Interaction of metal ions with neurotransmitter receptors and potential role in neurodiseases. *BioMetals, 27*(6), 1097–1113. https://doi.org/10.1007/s10534-014-9791-y

Martinez, V. D., Vucic, E. A., Becker-Santos, D. D., Gil, L., & Lam, W. L. (2011). Arsenic exposure and the induction of human cancers. *Journal of Toxicology, 2011*. https://doi.org/10.1155/2011/431287

Martins, A. C., Lopes, A. C. B. A., Urbano, M. R., Carvalho, M. F. H., Silva, A. M. R., Tinkov, A. A., Aschner, M., Mesas, A. E., Silbergeld, E. K., & Paoliello, M. M. B. (2021). An updated systematic review on the association between Cd exposure, blood pressure and hypertension. *Ecotoxicology and Environmental Safety, 208*, 111636. https://doi.org/10.1016/j.ecoenv.2020.111636

Masindi, V., & Muedi, K. L. (2018). Environmental contamination by heavy metals. In *Heavy metals*. InTech. https://doi.org/10.5772/intechopen.76082

McCourt, K., Sautter, G., Estes, S. L., McMahan, C., Lee, C., Carraway, E. R., et al. (2024). Participatory assessment of pollution and health risk in artisanal and small-scale gold mining communities in Colombia. *Groundwater for Sustainable Development, 25*, 101158. https://doi.org/10.1016/j.gsd.2024.101158

Mirkov, I., Popov Aleksandrov, A., Ninkov, M., Tucovic, D., Kulas, J., Zeljkovic, M., et al. (2021). Immunotoxicology of cadmium: Cells of the immune system as targets and effectors of cadmium toxicity. *Food and Chemical Toxicology, 149*, 112026. https://doi.org/10.1016/j.fct.2021.112026

Miura, S., Takahashi, K., Imagawa, T., Uchida, K., Saito, S., Tominaga, M., et al. (2013). Involvement of TRPA1 activation in acute pain induced by cadmium in mice. *Molecular Pain, 9*(1). https://doi.org/10.1186/1744-8069-9-7

Mohammadi, A. A., Zarei, A., Majidi, S., Ghaderpoury, A., Hashempour, Y., Saghi, M. H., et al. (2019). Carcinogenic and non-carcinogenic health risk assessment of heavy metals in drinking water of Khorramabad, Iran. *Methodsx, 6*, 1642–1651. https://doi.org/10.1016/j.mex.2019.07.017

Motlagh, Z. K., Derakhshani, R., & Sayadi, M. H. (2023). Groundwater vulnerability assessment in central Iran: Integration of GIS-based DRASTIC model and a machine learning approach. *Groundwater for Sustainable Development*, 23. https://doi.org/10.1016/j.gsd.2023.101037

Naderi, M., Puar, P., Zonouzi-Maranda, M., Chivers, D. P., Niyogi, S., & Kwong, R. W. M. (2021). A comprehensive review on the neuropathophysiology of selenium. *Science of the Total Environment, 767*, 144329. https://doi.org/10.1016/j.scitotenv.2020.144329

Nath, B. K., Chaliha, C., Bhuyan, B., Kalita, E., Baruah, D. C., & Bhagabati, A. K. (2018). GIS mapping-based impact assessment of groundwater contamination by arsenic and other heavy metal contaminants in the Brahmaputra River valley: A water quality assessment study. *Journal of Cleaner Production, 201*, 1001–1011. https://doi.org/10.1016/j.jclepro.2018.08.084

Ogarekpe, N. M., Nnaji, C. C., Oyebode, O. J., Ekpenyong, M. G., Ofem, O. I., Tenebe, I. T., & Asitok, A. D. (2023). Groundwater quality index and potential human health risk assessment of heavy metals in water: A case study of Calabar metropolis, Nigeria. *Environmental Nanotechnology, Monitoring and Management, 19*. https://doi.org/10.1016/j.enmm.2023.100780

Ogunfowokan, A. O., Adekunle, J., Oyekunle, O., Olutona, G. O., Ogunfowokan, A. O., Oyekunle, J. A. O., et al. (2013). Speciation study of heavy metals in water and sediments from Asunle River of the Obafemi Awolowo University, Ile-Ife, Nigeria. *International Journal of Environmental Protection, 3*(3), 6–16. https://www.researchgate.net/publication/236199743

Okon, S. U., Atakpa, E. O., & Ewa-Oboho, I. T. (2019). Spatial and seasonal characteristics of heavy metals in Ikpa River, Niger Delta, Nigeria. *International Journal of Fisheries and Aquatic Studies, 7*(2), 193–199.

Orr, S. E., & Bridges, C. C. (2017). Chronic kidney disease and exposure to nephrotoxic metals. *International Journal of Molecular Sciences, 18*(5), 1039. https://doi.org/10.3390/ijms18051039

Ötleş, S., & Çağındı, Ö. (2010). Health importance of arsenic in drinking water and food. *Environmental Geochemistry and Health, 32*, 367–371. https://doi.org/10.1007/s10653-010-9296-8

Park, Y. hee, Kim, D., Dai, J., & Zhang, Z. (2015). Human bronchial epithelial BEAS-2B cells, an appropriate in vitro model to study heavy metals induced carcinogenesis. *Toxicology and Applied Pharmacology, 287*(3), 240–245. https://doi.org/10.1016/j.taap.2015.06.008

Peters, J. L., Perlstein, T. S., Perry, M. J., McNeely, E., & Weuve, J. (2010). Cadmium exposure in association with history of stroke and heart failure. *Environmental Research, 110*(2), 199–206. https://doi.org/10.1016/j.envres.2009.12.004

Pinto, M. M. S. C., Marinho-Reis, P., Almeida, A., Pinto, E., Neves, O., Inácio, M. et al. (2019). Links between cognitive status and trace element levels in hair for an environmentally exposed population: A case study in the surroundings of the Estarreja industrial area. *International Journal of Environmental Research and Public Health, 16*(22). https://doi.org/10.3390/ijerph16224560

Prohaska, J. R. (2000). Long-term functional consequences of malnutrition during brain development: Copper. *Nutrition, 16*(7–8), 502–504.

Prozialeck, W. C., & Edwards, J. R. (2012). Mechanisms of cadmium-induced proximal tubule injury: New insights with implications for biomonitoring and therapeutic interventions. *Journal of Pharmacology and Experimental Therapeutics, 343*(1), 2–12. https://doi.org/10.1124/jpet.110.166769

Qiao, J., Zhu, Y., Jia, X., Shao, M., Niu, X., & Liu, J. (2020). Distributions of arsenic and other heavy metals, and health risk assessments for groundwater in the Guanzhong Plain region of China. *Environmental Research*, 181. https://doi.org/10.1016/j.envres.2019.108957

Reddy, Y. S., & Sunitha, V. (2023). Assessment of Heavy metal pollution and its health implications in groundwater for drinking purpose around inactive mines, SW region of Cuddapah Basin, South India. *Total Environment Research Themes, 8*, 100069. https://doi.org/10.1016/j.totert.2023.100069

Rupias, O. J. B., Pereira, S. Y., & de Abreu, A. E. S. (2021). Hydrogeochemistry and groundwater quality assessment using the water quality index and heavy-metal pollution index in the alluvial plain of Atibaia river- Campinas/SP, Brazil. *Groundwater for Sustainable Development, 15*. https://doi.org/10.1016/j.gsd.2021.100661

Sajib, A. M., Diganta, M. T. M., Rahman, A., Dabrowski, T., Olbert, A. I., & Uddin, M. G. (2023). Developing a novel tool for assessing the groundwater incorporating water quality index and machine learning approach. *Groundwater for Sustainable Development*, 23. https://doi.org/10.1016/j.gsd.2023.101049

Sajil Kumar, P. J., & Kuriachan, L. (2022). Chemometric appraisal of groundwater quality for domestic, irrigation and industrial purposes in Lower Bhavani River basin, Tamil Nadu, India. *International Journal of Environmental Analytical Chemistry, 102*(15), 3437–3460. https://doi.org/10.1080/03067319.2020.1770241

Sarwar, N., Imran, M., Shaheen, M. R., Ishaque, W., Kamran, M. A., Matloob, A., et al. (2017). Phytoremediation strategies for soils contaminated with heavy metals: Modifications and future perspectives. *Chemosphere, 171*, 710–721. https://doi.org/10.1016/j.chemosphere.2016.12.116

Satarug, S. (2018). Dietary cadmium intake and its effects on kidneys. *Toxics, 6*(1), 15. https://doi.org/10.3390/toxics6010015

Scanlon, S. E., Scanlon, C. D., Hegan, D. C., Sulkowski, P. L., & Glazer, P. M. (2017). Nickel induces transcriptional down-regulation of DNA repair pathways in tumorigenic and non-tumorigenic lung cells. *Carcinogenesis, 38*(6), 627–637. https://doi.org/10.1093/carcin/bgx038

Şener, Ş., Şener, E., & Davraz, A. (2017). Assessment of groundwater quality and health risk in drinking water basin using GIS. *Journal of Water and Health, 15*(1), 112–132. https://doi.org/10.2166/wh.2016.148

Shelnutt, S. R., Goad, P., & Belsito, D. V. (2007). Dermatological toxicity of hexavalent chromium. *Critical Reviews in Toxicology, 37*(5), 375–387. https://doi.org/10.1080/10408440701266582

Sheng, D., Meng, X., Wen, X., Wu, J., Yu, H., & Wu, M. (2022). Contamination characteristics, source identification, and source-specific health risks of heavy metal(loid)s in groundwater of an arid oasis region in Northwest China. *Science of the Total Environment*, 841. https://doi.org/10.1016/j.scitotenv.2022.156733

Singh, A., Sharma, A., K. Verma, R., L. Chopade, R., P. Pandit, P., Nagar, V., et al. (2022). Heavy metal contamination of water and their toxic effect on living organisms. In *The toxicity of environmental pollutants*. IntechOpen. https://doi.org/10.5772/intechopen.105075

Singha, S., Pasupuleti, S., Singha, S. S., & Kumar, S. (2020). Effectiveness of groundwater heavy metal pollution indices studies by deep-learning. *Journal of Contaminant Hydrology*, 235. https://doi.org/10.1016/j.jconhyd.2020.103718

Sirajudeen, J., & Pravinkumar, J. (2021). Heavy metal analysis of groundwater of Thirukoviloor, Villuppuram, Tamil Nadu, India. *Materials Today: Proceedings, 36*, 828–831. https://doi.org/10.1016/j.matpr.2020.07.012

Smereczański, N. M., & Brzóska, M. M. (2023). Current levels of environmental exposure to cadmium in industrialized countries as a risk factor for kidney damage in the general population: A comprehensive review of available data. *International Journal of Molecular Sciences, 24*(9), 8413. https://doi.org/10.3390/ijms24098413

Srivastava, V., Gupta, S. K., Singh, P., Sharma, B., & Singh, R. P. (2018). Biochemical, physiological, and yield responses of lady's finger (*Abelmoschus esculentus* L.) grown on varying ratios of municipal solid waste vermicompost. *International Journal of Recycling of Organic Waste in Agriculture, 7*(3), 241–250. https://doi.org/10.1007/s40093-018-0210-1

Stannard, L., Doak, S. H., Doherty, A., & Jenkins, G. J. (2017). Is nickel chloride really a non-genotoxic carcinogen? *Basic and Clinical Pharmacology and Toxicology, 121*, 10–15. https://doi.org/10.1111/bcpt.12689

Taghilou, A. A. (2022). A framework resilient social-ecological system (SES) for groundwater governance and public and private rights. *Groundwater for Sustainable Development, 18*. https://doi.org/10.1016/j.gsd.2022.100788

Tellez-Plaza, M., Navas-Acien, A., Crainiceanu, C. M., & Guallar, E. (2008). Cadmium exposure and hypertension in the 1999–2004 National Health and Nutrition Examination Survey (NHANES). *Environmental Health Perspectives, 116*(1), 51–56. https://doi.org/10.1289/ehp.10764

Tian, Z., Jiang, S., Zhou, J., & Zhang, W. (2023). Copper homeostasis and cuproptosis in mitochondria. *Life Sciences, 334*, 122223. https://doi.org/10.1016/j.lfs.2023.122223

Tsoi, M.F., Lo, C.W.H., Cheung, T.T., & Cheung, B.M.Y. (2021). Blood lead level and risk of hypertension in the United States National Health and Nutrition Examination Survey 1999–2016. *Scientific Reports,* 11, 3010. https://doi.org/10.1038/s41598-021-82435-6

Umar, M., Jahangir, N., Khan, M. F., Saeed, Z., Sultan, F., & Sultan, A. (2019). Cobalt cardiomyopathy in hip arthroplasty. *Arthroplasty Today, 5*(3), 371–375. https://doi.org/10.1016/j.artd.2019.04.010

Vareda, J. P., Valente, A. J. M., & Durães, L. (2019). Assessment of heavy metal pollution from anthropogenic activities and remediation strategies: A review. *Journal of Environmental Management, 246*, 101–118. https://doi.org/10.1016/j.jenvman.2019.05.126

Vaziri, N. D. (2008). Mechanisms of lead-induced hypertension and cardiovascular disease. *American Journal of Physiology. Heart and Circulatory Physiology, 295*, 454–465. https://doi.org/10.1152/ajpheart.00158.2008.-Lead

Vo, T. T. T., Peng, T. Y., Nguyen, T. H., Bui, T. N. H., Wang, C. S., Lee, W. J., et al. (2024). The crosstalk between copper-induced oxidative stress and cuproptosis: A novel potential anticancer paradigm. *Cell Communication and Signaling, 22*, 353. https://doi.org/10.1186/s12964-024-01726-3

Wang, Z., Gao, Z., Wang, S., Liu, J., Li, W., Deng, Q., et al. (2021). Hydrochemistry characters and hydrochemical processes under the impact of anthropogenic activity in Yiyuan City, Northern China. *Environmental Earth Sciences, 80*(2). https://doi.org/10.1007/s12665-020-09361-0

Wang, Z., Sun, Y., Yao, W., Ba, Q., & Wang, H. (2021). Effects of cadmium exposure on the immune system and immunoregulation. *Frontiers in Immunology, 12*, 695484. https://doi.org/10.3389/fimmu.2021.695484

WHO. (2011). *Guidelines for drinking-water quality*, 4th edn. Geneva, Switzerland, 631 pp.

Wu, H., Liao, Q., Chillrud, S. N., Yang, Q., Huang, L., Bi, J., et al. (2016). Environmental exposure to cadmium: Health risk assessment and its associations with hypertension and impaired kidney function. *Scientific Reports, 6*, 29989. https://doi.org/10.1038/srep29989

Xue, B., Yang, Q., Xia, K., Li, Z., Chen, G.Y., Zhang, D., & Zhou, X. (2023). An AuNPs/Mesoporous NiO/Nickel foam nanocomposite as a miniaturized electrode for heavy metal detection in groundwater. *Engineering, 27*, 199–208. https://doi.org/10.1016/j.eng.2022.06.005

Yao, M., Zeng, Q., Luo, P., Sun, B., Liang, B., Wei, S., et al. (2021). Assessing the risk of coal-burning arsenic-induced liver damage: A population-based study on hair arsenic and cumulative arsenic. *Environmental Science and Pollution Research, 28*(36), 50489–50499. https://doi.org/10.1007/s11356-021-14273-y

Yuan, W., Yang, N., & Li, X. (2016). Advances in understanding how heavy metal pollution triggers gastric cancer. *BioMed Research International, 2016*. https://doi.org/10.1155/2016/7825432

Zhang, Y., Zhang, Q., Chen, W., Shi, W., Cui, Y., Chen, L., et al. (2023). Source apportionment and migration characteristics of heavy metal(loid)s in soil and groundwater of contaminated site. *Environmental Pollution, 338*. https://doi.org/10.1016/j.envpol.2023.122584

Zischka, H., & Einer, C. (2018). Mitochondrial copper homeostasis and its derailment in Wilson disease. *The International Journal of Biochemistry & Cell Biology, 102*, 71–75. https://doi.org/10.1016/j.biocel.2018.07.001

Chapter 3
Heavy Metals Pollution in the Water, Toxicity and Remedial Measures

Esmaeil Allahkarami and Ebrahim Allahkarami

Abstract Managing water resources, treating wastewater effectively, and addressing the diverse range of pollutants are critical for achieving sustainable development and ensuring a clean water supply for the growing global population. Heavy metals are widely recognized for their toxicity to living organisms. These metals pose a threat due to their bio-accumulative nature and high solubility in aquatic environments. Human exposure to heavy metals through ingestion, inhalation, and dermal contact poses severe health risks, affecting vital organs and leading to diseases like muscular dystrophy and cancer. It is crucial to manage and treat wastewater effectively to reduce heavy metal concentrations and protect human well-being. Remediation techniques for heavy metals contamination include phytoremediation, bioremediation, chemical precipitation, coagulation and flocculation, ion exchange, electrochemical techniques, membrane, and adsorption. In the pursuit of effective heavy metal removal from wastewater, various methods have emerged, each offering distinct advantages. The choice of technique depends on factors like the type and concentration of heavy metals, site characteristics, and environmental and economic considerations. A combination of these methods may be utilized for effective remediation, focusing on proper waste disposal and compliance with regulatory requirements.

Keywords Heavy metals · Pollution · Health risk assessment · Remediation · Wastewater treatment

---

E. Allahkarami
Department of Petroleum Engineering, Faculty of Petroleum, Gas and Petrochemical Engineering, Persian Gulf University, Bushehr, Iran

Persian Gulf Star Oil Company, Bandar Abbas, Iran

E. Allahkarami (✉)
Department of Mining Engineering, Amirkabir University of Technology, Tehran, Iran
e-mail: ebrahim.allahkarami@aut.ac.ir

V. Sunitha et al. (eds.), *Groundwater Resource Management Planning Strategies*, Springer Water, https://doi.org/10.1007/978-3-031-88870-0_3

3.1 Introduction

Water scarcity is a global issue, attracting increased attention at research and policy levels due to its crucial role in sustainable development (Karri et al., 2021). The human population and resource consumption surge intensifies the challenge of achieving a sustainable lifestyle. Clean water, a vital component for life, faces depletion and pollution, leading to water-related diseases and impacting education in developing countries. Water pollution stems from various sources, including urbanization and industrial processes, endangering human health and aquatic ecosystems (Rezai & Allahkarami, 2021). The growing concern prompts the use of different remediation techniques for water treatment.

Wastewater, a byproduct of water usage, poses environmental and health risks, emphasizing the importance of effective wastewater treatment. Urbanization and industrialization contribute to the exploitation of natural resources, leading to water pollution. Wastewater usually contains different kinds of contaminants that generally classified as radiological, microbiological, chemical, and physical pollutants (Fig. 3.1). Heavy metals, organic compounds, and other pollutants contaminate water, necessitating stringent regulations and advanced treatment technologies (Mojiri and Bashir, 2022). The increasing demand for potable water further exacerbates wastewater production.

Addressing water scarcity involves tackling challenges such as climate change-induced droughts and developing new technologies for wastewater treatment and reuse (Palmate et al., 2017). Automation, wastewater analysis, and toxicity reduction are key considerations for effective water treatment. Additionally, proper management of industrial wastes containing heavy metals is crucial to prevent environmental contamination and associated health risks.

Efforts to improve water quality involve addressing pollutants from natural and anthropogenic sources. The impact of pollutants, such as dyes, heavy metals, and endocrine-disrupting chemicals, poses significant health and environmental risks. Population growth and human activities contribute to global warming and reduce air quality, affecting marine life and the ecosystem.

The term "metal" encompasses a broad category, denoting metals and metalloids with a density exceeding 4 ± 1 g/cm^3. Though lacking a precise definition, it commonly denotes contaminants present in freshwater ecosystems, as stated by Duffus in 2001 (Duffus, 2001). These substances, including cadmium, chromium, copper, mercury, lead, zinc, arsenic, boron, and platinum group metals, exist naturally in the Earth's crust, occurring in soils, rocks, sediments, waters, and even microorganisms at baseline concentrations. Nevertheless, anthropogenic activities contribute to releasing these metals into the environment, resulting in elevated concentrations, as emphasized by Mohammed et al. in 2011 (Mohammed et al., 2011). Table 3.1 provides an overview of the physical properties of certain heavy metals, focusing on those with a density exceeding 5 g/cm^3, which are more commonly encountered in our daily lives.

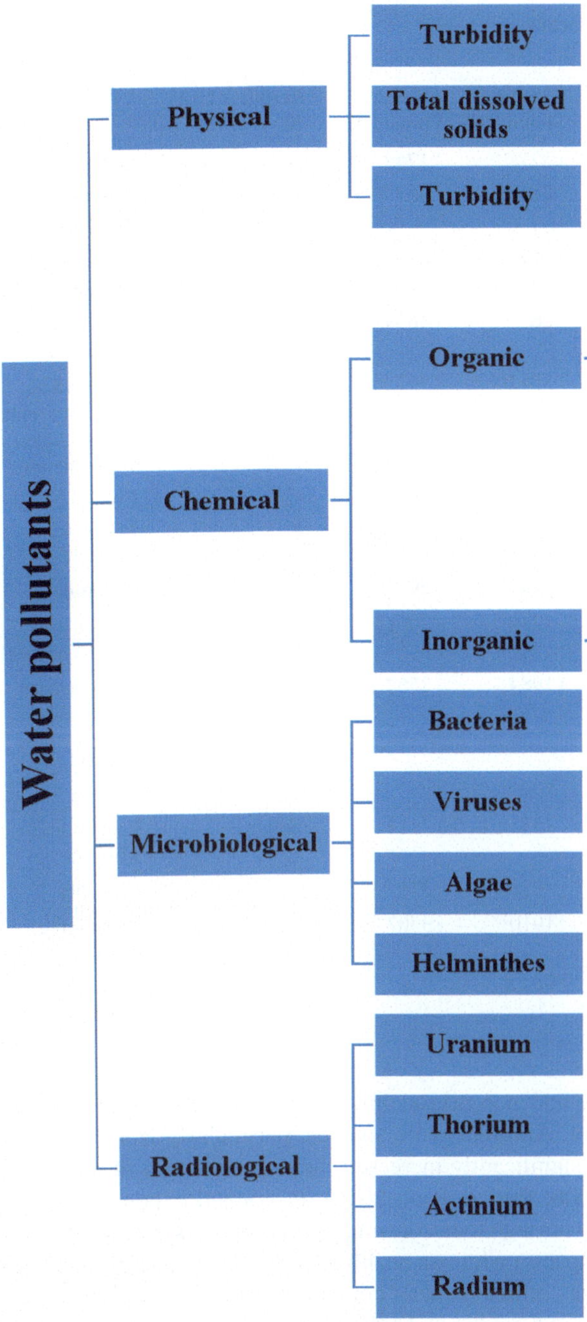

Fig. 3.1 Different kinds of water pollutants

Table 3.1 Properties of heavy metals

Element	Unit				Originally found in the ore
	Density (kg/m^3)	Melting point (°C)	Boiling point (°C)	Resistivity (25 °C) (nΩ m)	
Lead	11,343	327.5	1749	208	Galena
Zinc	7134	419.5	907	59	Sphalerite; Smithsonite
Arsenic (gray)	5.727	817	615	–	Arsenopyrite; Realgar
Molybdenum	10,280	2623	4639	53	Molybdenite; Wulfenite
Cadmium	8650	321.07	767	69	Greenockite; Primarily found in association with zinc ores
Cobalt	8900	1495	2927	62	Cobaltite; Erythrite; Smaltite
Chromium	7140	1857	2672	129	Chromite
Manganese	7213	1246	2061	185	Pyrolusite; Rhodochrosite
Copper	8960	1984	2562	16.8	Chalcopyrite; Chalcocite; Bornite; Covellite; Azurite
Nickel	8908	1455	2913	69.9	Pentlandite; Millerite
Mercury	13,534	– 38.83	356.73	958	Cinnabar
Vanadium	6110	1910	3407	196	Vanadinite; Carnotite
Silver	10,490	961.8	2162	15.9	Argentite; Chlorargyrite

Due to their persistence and non-degradability, heavy metals can have lasting impacts on the environment. Anthropogenic sources, particularly from the global industrial revolution, have significantly increased the dissemination of these toxic substances (Godt et al., 2006; Priti & Paul, 2016). Prolonged exposure to heavy metals, frequently occurring through food consumption, drinking water, or inhalation of air, can exert enduring impacts on human health. Hence, it is imperative to address the origins and consequences of heavy metal contamination, playing a pivotal role in managing both environmental quality and public health (Järup & Åkesson, 2009).

Heavy metals, particularly when present in ionic form in water, are widely recognized for their toxicity to living organisms (Maanan et al., 2015). Even a slightly elevated concentration of a highly toxic metal ion can pose greater risks to water safety than a significantly higher concentration of a less toxic one. To assess these risks and certify the drinking quality of water samples, researchers have considered developing a heavy metal pollution index (HPI). This index is not only applicable to groundwater but also relevant to various environmental matrices, including soil, sediment, and estuarine sediment (Chaturvedi et al., 2018).

The computation of the Heavy Metal Pollution Index (HPI) in water entails evaluating two critical aspects: the quantity of heavy metal ions present and their concentrations relative to the permissible levels established by water quality standards. Various formulations for the HPI have been suggested by different researchers, each exhibiting distinct advantages and limitations. Among the prevalent indexing techniques for heavy metal ions in groundwater, one introduced by Mohan and colleagues in 1996 (Mohan et al., 1996) relies on the highest desirable and maximum permissible concentrations. Another approach involves utilizing the maximum allowable concentration (MAC) (Singh et al., 2017; Venkatramanan et al., 2015). These indexing methods provide a semi-empirical numerical representation to assess and grade the water quality of samples.

Managing water resources, treating wastewater effectively, and addressing the diverse range of pollutants is critical for achieving sustainable development and ensuring a clean water supply for the growing global population. This chapter comprehensively addresses diverse sources of heavy metal pollutants and their environmental ramifications on plants, animals, and humans. It also delves into different methodologies for removing these toxic metals from wastewater, highlighting their advantages and disadvantages.

3.2 Review of Heavy Metals Pollution Sources

Heavy metals found in surface water systems can stem from both natural and anthropogenic origins, as illustrated in Fig. 3.2. Natural sources encompass phenomena like volcanic eruptions, the disintegration of metal-rich rocks, sea-salt sprays, forest fires, and inherent weathering processes. Diverse forms of heavy metals include hydroxides, oxides, sulfides, sulfates, phosphates, silicates, and organic molecules. Notably, volcanic ash generated during explosive volcanic eruptions can introduce contamination to water, leading to cloudiness, acidity, and a decrease in pH levels. This is because the ash contains acidic compounds such as H_2SO_4, HCl, and HF (Nagajyoti et al., 2010; Stewart et al., 2006).

Soils can contain high levels of heavy metals, which can come from both natural sources and human activities. The presence of elevated metal concentrations in soils can develop naturally due to the breakdown of the underlying bedrock (Mohammed et al., 2011). Heavy metals are present in rocks as ores, which can take several chemical forms, including sulfides and oxides. Ore minerals frequently exhibit familial associations, with metals occurring naturally as sulfides commonly coexisting. Lead, cadmium, arsenic, and mercury sulfides can coexist alongside iron sulfides (pyrite,

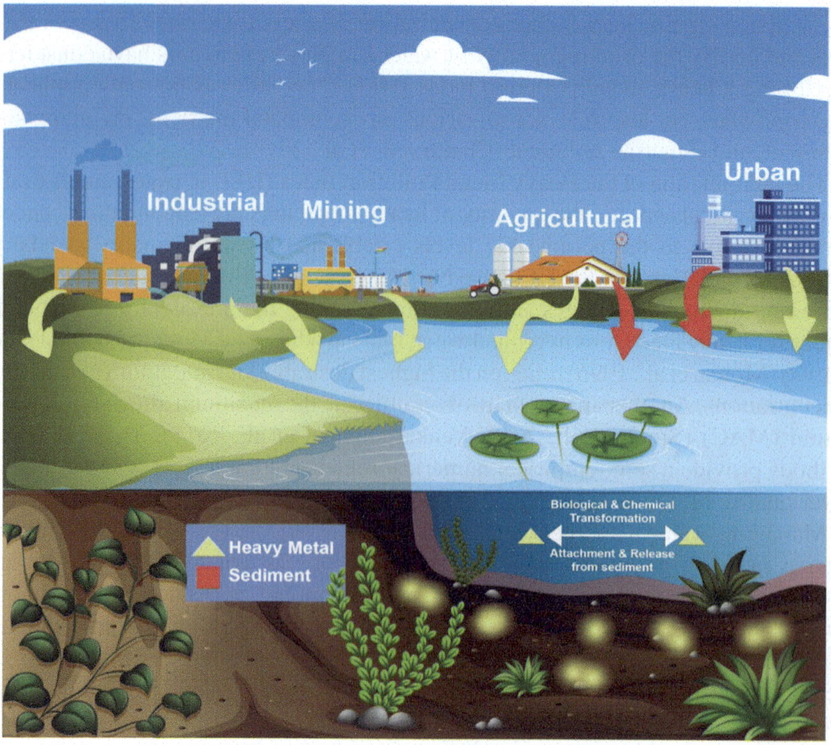

Fig. 3.2 A schematic diagram of heavy metals pollution sources (Mitra et al., 2022)

FeS_2) and copper sulfides (chalcopyrite, $CuFeS_2$). These metals can be extracted as byproducts in several metallurgical processes.

The process of weathering bedrock that initially has low metal concentrations has the potential to lead to an increase in metal content through pedogenic processes. Even in areas where the underlying rock is not recognized for having high metal concentrations, soils may nonetheless exhibit naturally higher amounts of metals. An illustration of this phenomenon may be observed at a secluded location within the Swiss National Park, which is situated a considerable distance away from industrial and traffic lines. At this site, high levels of lead were found in the upper layer of soil, which may have been caused by the long-distance movement of tiny particles in the air, as reported by Nowack et al. in 2001 (Nowack et al., 2001).

Acid drainage occurs when metal sulfides containing high levels of heavy metals or metalloids are exposed to the air. This contact results in the process of oxidation and the generation of water that is either acidic or contains high levels of heavy metals. Acidic drainage can arise from natural phenomena and human actions, notably coal extraction and different metallic ores (Valente & Gomes, 2009). Mining processes can leave behind tailings containing metals, which can be dispersed through wind and water, leading to environmental issues. Additionally, the breakdown of coal mine waste through weathering processes can generate alkaline compounds, heavy metals, and sediments, contributing to water pollution in streams, rivers, and lakes (Dang et al., 2002).

Small particles suspended in the surrounding atmosphere, with diameters ranging from a few nanometers to tens of micrometers, make up atmospheric aerosols, sometimes referred to as particulate matter. Both natural and man-made sources, such as volcanic eruptions, sea salts, forest fires, industrial discharges, incomplete combustion of fossil fuels, traffic emissions, burning biomass, and biological materials like microorganisms, pollen, and plant fragments, can directly emit primary particles and some gaseous pollutants (Wang et al., 2018). The conversion of gas to particles in the environment and gaseous compound condensation on pre-existing particles may both produce secondary particles (Kim & Zhang, 2019). Mass and aerosol concentrations in the lower atmosphere (troposphere) generally range from 1 to 100 μg per cubic centimeter and 10^2 to 10^5 particles per cubic centimeter, respectively (Monks et al., 2009). These aerosols include mostly ammonium, sulfate, nitrate, elemental or black carbon, and organic carbon. About 10–30% of the overall mass load is usually contributed by each of these elements (Ren et al., 2021). The origins, properties, and characteristics of atmospheric aerosols greatly influence air quality, climate, and environmental processes.

Fires play a crucial role in rejuvenating landscapes and maintaining complex habitats, as acknowledged by various studies. However, certain fires can have significant short and long-term impacts (Arkle et al., 2010). In naturally metal-rich soils within forested regions, wildfires have the potential to release metals previously held in check within sediment and plant tissues. Furthermore, in areas marked by historical mining activities, remnants such as concealed mine tailings and waste rock may become uncovered during wildfires. This exposure transforms into a potential source of heavy metal discharge during storm runoff events (Abraham et al., 2017; Murphy et al., 2020).

Regions characterized by elevated metal concentrations in soils and a history of mining operations might be more susceptible to introducing toxic heavy metals post-fire, thereby heightening the ecosystem's vulnerability to the aftermath of such incidents. Metals originating from exposed legacy mining sites can be transported in both particulate and dissolved forms, moving from hillsides to streams during precipitation events subsequent to a wildfire (Rust et al., 2022). This threatens water supplies and the overall health of freshwater ecosystems, emphasizing the importance of understanding and mitigating the environmental impacts of wildfires, especially in areas with a legacy of mining activities.

Industrialization and urbanization on a worldwide scale have led to a rise in the anthropogenic release of heavy metals into the atmosphere (Sonone et al., 2020). Mining and extraction operations emit certain metals into the air, which subsequently contribute to both dry and wet deposition. Industrial operations, such as electroplating, discharge hazardous substances into the environment (Rajoria et al., 2021). In addition, improper handling and disposal of e-waste may lead to the emission of hazardous substances, contributing to various health issues and illnesses (Ahirwar &

Tripathi, 2021). Agricultural pollution has detrimental effects on aquatic environments, resulting in phenomena such as eutrophication and adverse consequences for biodiversity and fisheries (Liu et al., 2021). Mining operations, which involve the extraction of minerals and geological materials, contribute to water contamination through several processes such as mine drainage, cooling, aqueous extraction, and others. The substantial quantities of water generated by these operations significantly enhance the likelihood of chemical pollution in both groundwater and surface water (Gu, 2018).

Another concern is the thermal pollution caused by nuclear and fossil fuel power stations. Discharging hot water into aquatic habitats can result in reduced amounts of dissolved oxygen, increased pH, eutrophication, and negative impacts on aquatic life (Iqbal et al., 2021). Specifically, coal-fired power stations are responsible for the emission of hazardous heavy metals into bodies of water. Furthermore, improper disposal of biomedical waste, which contains significant amounts of heavy metals and toxic substances, can harm water quality, resulting in groundwater pollution (Selvan Christyraj et al., 2021).

Fly ash has high concentrations of arsenic, copper, and selenium, which prompts concerns regarding possible negative effects on soil and water (Zhang, 2014). Aluminum, iron, silicon, and calcium in oxide form comprise about 90% of coal ash. Minor components of fly ash include titanium, potassium, magnesium, and sodium. Cadmium, arsenic, selenium, mercury, and lead are trace metals found in lesser quantities, making up to 1% of the overall ash composition (Ash, 2009). When ash is deposited in landfills and ponds, these trace elements can lead to significant environmental problems, including the generation of leachate, a liquid formed when permeable material percolates with water (Munawer, 2018).

To summarize, both natural and anthropogenic sources may contribute heavy metals to surface water systems (Fig. 3.3). Water bodies are contaminated with heavy metals and other dangerous elements due to various human activities, including mining, agriculture, industrial operations, and improper waste management. This pollution causes significant risks to human health, aquatic ecosystems, and the environment.

3.3 Health Risk Assessment

Water is crucial for life on Earth, yet a significant global population needs clean drinking water. Increased water demand, driven by anthropogenic activities and a growing population, poses a challenge. Reports indicate progress in access to improved drinking water sources, but a considerable number still need access (Xu & Mills, 2018). The demand for water resources is rising due to urbanization, industrialization, and environmental needs, necessitating efficient water purification solutions (Vardhan et al., 2019).

Pollutants enter ecosystems through various pathways, impacting the hydrosphere, lithosphere, and atmosphere. Human activities contribute significantly, with

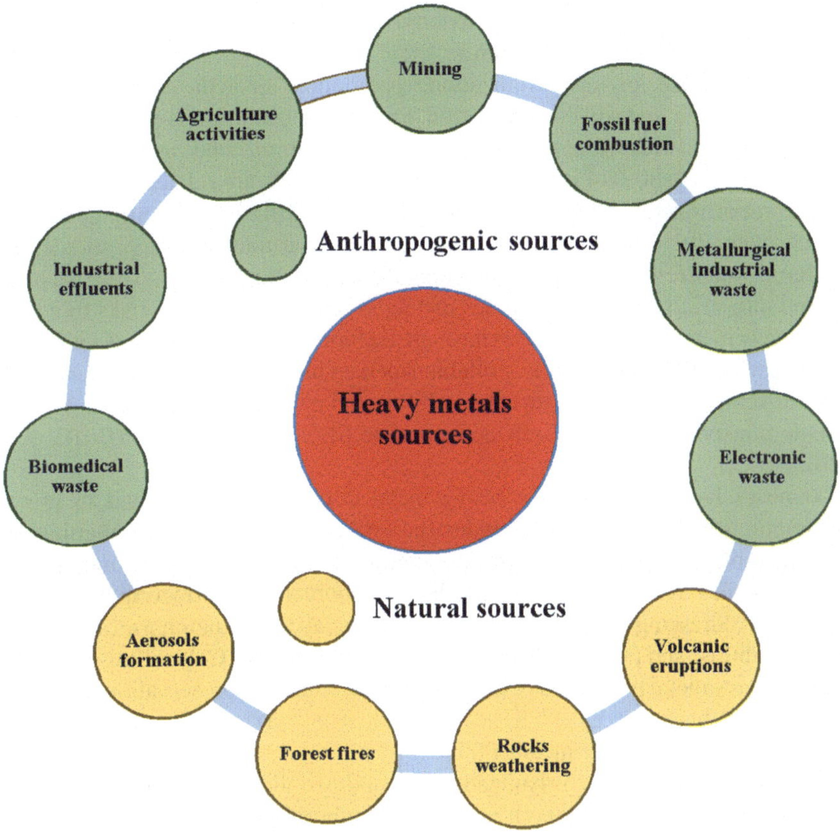

Fig. 3.3 Different sources of heavy metals pollution

unintentional events like oil spills and intentional applications like waste disposal. Heavy metal movement is influenced by factors like temperature, water movements, and air circulation (Walker et al., 2012).

Heavy metals, with a density over 4000 kg/m^3, include toxic elements like copper, cadmium, zinc, chromium, arsenic, and others (Edelstein & Ben-Hur, 2018). While some heavy metals are essential for plant growth, elevated concentrations harm both plants and animals. Soil pollution from industrial wastewater, sewage sludge, and fertilizers affects plant metabolism, leading to reduced biomass growth. Heavy metal pollution not only impacts soil but also influences food quality and human health. In fact, deliberate and inadvertent soil pollution affects ecosystems and compromises soil and water quality (Muchuweti et al., 2006). Certain heavy metals are toxic to plants even at low concentrations, affecting their metabolism and biomass generation. The toxicity of heavy metals depends on factors like oxidation state, soil composition, pH, and structure. High exposure to heavy metals can lead to various health issues, including gastrointestinal and renal toxicity, cardiovascular problems, tumors, and developmental challenges in children.

The rapid pace of industrialization, population growth, and urbanization contribute to water and soil contamination. Freshwater pollution is mainly attributed to untreated industrial waste release, wastewater dumping, and agricultural runoff. Approximately 70–80% of issues in developing countries are related to water pollution, disproportionately affecting women and children (Bhatnagar & Sillanpää, 2010; Vardhan et al., 2019). Pollutants like heavy metals, pharmaceuticals, pesticides, and dyes harm water resources, posing ecological risks. Water pollution from urbanization and industrialization accumulates metals in sediments, posing risks to ecosystems. Heavy metal toxicity cascades through the food chain, with humans vulnerable. Despite sewage treatment efforts, pollutants persist in water bodies. In rivers, pollutants' movement depends on currents, stability, and physical state. In seas, wind and currents transport pollutants, impacting marine life and ecosystems (Briffa et al., 2020).

Countries have set limits for heavy metal concentrations in food to prevent consumption-related health issues. Industrial activities such as mining, pharmaceuticals, electroplating, and others contribute significantly to heavy metal contamination in water sources. Heavy metals can attach to microorganisms, affecting biochemical processes. Addressing heavy metal pollution is crucial due to their harmfulness, non-biodegradability, and potential interference with ecosystems. Controlling and understanding heavy metal pollution is essential for environmental preservation and human well-being. In this context, the US EPA established the maximum contaminated level (MCL) standards for heavy metals (Table 3.2).

Air pollution, driven by urbanization and industrialization, involves pollutants in various forms released by both natural and anthropogenic activities. Particulate matters (PMs) lead to health problems, infrastructure deterioration, acid rain, corrosion, and haze formation. Chimneys and weather conditions affect pollutant travel. The troposphere and stratosphere play crucial roles in pollutant transportation, with global impacts as airborne pollutants react with rain and affect water bodies and land. Despite measures, issues persist with vehicle pollution, pesticide applications, and other sources of atmospheric pollution (Masindi & Muedi, 2018).

The toxicity levels of some heavy metals for humans follow this order: Hg(II) > Cd(II) > Cu(II) > Zn(II) > Ni(II) > Pb(II) > Cr(II) > Al(III) > Co(II) (Mansourri & Madani, 2016). Concentration, emission rate, and length of exposure are some of the factors that affect how harmful metals are to humans. Notably, Hg(II), Cd(II), and Pb(II) have gained greater attention in recent decades (Allahkarami et al., 2023a; Carvalho et al., 2020).

Many of these metal pollutants readily accumulate in plants, entering the food chain and ultimately affecting humans, leading to severe illnesses and diseases (Nedjimi, 2021). Oxidative stress may ensue as a consequence of enzyme and chlorophyll antagonism induced by increased metal concentrations, which disrupted vital ions. Metal toxicity is commonly assessed through the detection of diminished photosynthetic activity and limited plant growth (Shi et al., 2015). Carcinogenic, teratogenic, allergenic, endocrine-disrupting, and mutagenic properties have been

Table 3.2 The maximum contaminated level (MCL) standards for the most important contaminants (Briffa et al., 2020; Rezai & Allahkarami, 2021; U.S. EPA, 2009)

Heavy metals	MCL (mg/l)	Potential health effects
Cadmium	0.005	Cancer; kidney damage; bone fracture; DNA impairment; immune system deficiencies; psychological disorders; central nervous system complications; gastrointestinal disorders
Arsenic	0.05	Increased risk of cancer; gastrointestinal system irritation; skin damage; circulatory system problem; brain damage; lung irritation; heart problems
Molybdenum	0.08	Hypochromic microcytic anaemia; anorexia; decrease in appetite; chest pain; fatigue and weakness; listlessness; headaches; gout; myalgia and arthralgia; copper deficiency; testicular atrophy
Cobalt	No data	Respiratory irritation; cardiomyopathy; congestion; oedema; liver and kidney congestion; ventilator function reduction; conjunctiva; lung haemorrhage; pneumonia and fibrosis; asthma
Chromium	0.1	Liver damage; allergic dermatitis; toxic nephritis; diarrhea; gastrointestinal ulceration; fever; nausea and vomiting; vertigo
Manganese	0.3	Dullness; anxiety; hypotension; akathisia; manganism; dystonia; lethargy; weakness; behavioural changes; tremors
Zinc	No data	Neutropenia; anaemia; nausea and vomiting; stomach cramps; epigastric pain; pancreatic complications; fatigue; impaired immune function; copper deficiency; anaemia; decrease in high-density lipoprotein cholesterol
Copper	1.3	Anxiety; flu-like symptoms; insomnia; diarrhea; agitation; vomiting; irritation of the eyes; restlessness; dizziness; irritation caused in the mouth cavity
Nickel	0.1	Increased possibilities of cancer; asthma; dizziness; allergic reactions; heart disorders; respiratory failure
Lead	0.015	Adults: Miscarriages; iron deficiency; renal impairment; hypertension; abdominal pain; encephalopathic signs; pica; premature and low births; sperm damage; stillbirths; brain injury; peripheral nerve damage; cognitive impairment Infants and children: delays in mental or physical development
Mercury	0.002	Lung irritation; kidney damage; rashes; vomiting and diarrhea; eye irritation
Vanadium	No data	Impairment to the nervous system; tremors; paralyzes and behavioral changes; cardiac disease; liver and kidney hemorrhage; blackening of teeth and tongue; inflammation of the gastrointestinal organs
Silver	0.05	Cardiac abnormalities; respiratory inflammation; skin irritation; unconsciousness; drowsiness; confusion and staggering; allergic dermatitis; dizziness and headaches; stomach discomfort; breathing problems; narcosis; brain damage
Selenium	0.05	Fragile and loss of hair; contorted and loss of nails; nervous system disorders; congenital disabilities; tooth deterioration; bronchitis; rashes; severe pain; headaches and fever; pneumonia; garlic breathe; skin lesions; hepatomegaly; conjunctivitis; reproductive failure; skin swelling; vomiting; abdominal pain and diarrhea; carcinogenic

attributed to certain metals. Furthermore, there have been reports indicating that these substances possess nephrotoxic and hepatotoxic properties, which can have adverse effects on the immune system. Potential complications include bone marrow and central nervous system damage, the development of osteoporosis, behavioral and neurological alterations, and disturbed heart rhythm (Tchounwou et al., 2012).

Heavy metal discharge into aquatic environments may cause a variety of biological, chemical, and physical processes (Guo et al., 2018). These effects fall into two broad categories: the impact of heavy metals on the environment and the influence of the environment on heavy metals (Mandich, 2018). The former category includes changes in population diversity, density, species composition, and community structure, which depend on the concentration of heavy metal species. The latter category emphasizes that conditions in receiving waters can alter the speciation and toxicity of heavy metals influenced by anthropogenic and geochemical factors, industrial effluents, suspended solids, and chelators.

The effects of heavy metals on aquatic plants vary, with population responses influenced by natural parameters such as light and temperature (Singh & Kalamdhad, 2011). There are problems with biological tracking systems utilizing community standards. Additionally, the presence of heavy metals in aquatic environments may modify the size of particles, the organic content of the substrate, and the pH of water. These changes can have an impact on the variety, density, and species composition of plant life (Jaiswal et al., 2018).

The impact of heavy metals on plants is diverse, as these sessile organisms must adapt to varying soil compositions. Essential and non-essential elements in soils can be harmful when present in excess. Metal toxicity effects are further complicated by competition among metals (Jaiswal et al., 2018). Plants acquire nutrients from the soil, categorized as essential nutrients (macronutrients and micronutrients) and non-essential nutrients. High concentrations of non-essential nutrients due to natural availability or industrial activities can be detrimental to plants (Karaca et al., 2010).

As mentioned, heavy metals have widespread environmental occurrences and can lead to severe health issues, prompting regulatory guidelines for their concentrations in soil. Mitigation measures and regulations, such as reducing arsenic in drinking water and banning lead-based paints, aim to address and prevent the adverse effects of these heavy metals on human health and the environment. Furthermore, the toxicity of heavy metals manifests in physiological processes, leading to chlorosis, senescence, reduced biomass production, inhibited growth, and, eventually, plant death. In agricultural soil, heavy metal pollution affects plant growth and metabolic activities and induces oxidative stress. Different heavy metals, such as copper, cadmium, zinc, lead, nickel, chromium, and others, exhibit phytotoxicity at varying concentrations, affecting various physiological functions in plants (Vardhan et al., 2019).

Arsenic (As) is a metalloid and ranks as the top priority on the Agency for Toxic Substances and Disease Registry's (ATSDR) list of hazardous substances (ATSDR, 2007). Arsenic can exist in different valence states, with arsenate (As(V)) dominating in aerobic environments and arsenite (As(III)) dominating under reducing conditions. Chronic exposure to inorganic arsenic can lead to arsenicosis, causing impacts on the nervous, gastrointestinal, and renal systems, as well as skin lesions, diabetes, cardiovascular diseases, and cancer. Lead (Pb), Ranked as the second most hazardous metal by the ATSDR, is a probable human carcinogen. Both natural and anthropogenic sources contribute to lead exposure, with major sources including mining, smelting, lead-based paints, batteries, leaded gasoline, and industrial processes. Chronic exposure to lead, especially in children, can lead to severe neurological damage (Järup, 2003). Mercury (Hg), Ranked third on the ATSDR list, poses severe risks to human health, particularly in early childhood development. Mercury found in elemental, inorganic, and organic forms is released into the environment by coal-fired power plants, waste incineration, oil pipelines, and various industrial activities. Furthermore, methylmercury, formed naturally in the environment, bioaccumulates in fish and shellfish, posing a significant risk through consumption (Pruss & Corralan, 2006). Cadmium (Cd) is a known human carcinogen, and it is ranked seventh on the ATSDR list. Anthropogenic activities, such as Ni/Cd batteries, refineries, electronic products, and industrial waste disposal, significantly increase cadmium levels in the environment. Cadmium toxicity affects the kidneys and skeletal and respiratory systems, and it is persistent and bioaccumulative. Chromium (Cr) occurs naturally and is also found in industrial processes, wood preservatives, textile dying, and leather tanning. Chromium can exist in different forms, including trivalent (Cr(III)), hexavalent (Cr(VI)), and metallic (Cr). Chromium toxicity affects the immune, renal, and respiratory systems, and it is associated with cancer (Zhou et al., 2004).

Oral intake, inhalation, and physical contact with heavy metals may result in serious health problems for humans, including harm to the liver, lungs, the central nervous system, blood composition, kidneys, and others. Cancers, multiple sclerosis, Alzheimer's disease, and muscular dystrophy may all result from prolonged exposure (Asaduzzaman et al., 2017). Oral ingestion is a primary pathway for copper and mercury, particularly through the consumption of fish (Saleh et al., 2018). Inhalation is significant for occupational exposure to metals that exist as vapors, such as mercury. Dermal exposure concerns certain metals, with skin acting as a barrier against absorption. Heavy metals in road dust can also contribute to skin exposure (Tuchman et al., 2015). The food chain is a major pathway for human exposure to soil pollution, posing health risks such as cancer, developmental issues, kidney damage, immunological impact, endocrine disruption, and neurological effects (Sharma et al., 2018; Tepanosyan et al., 2018). Removing toxic heavy metals from water is crucial to mitigate health risks for humans and other organisms. Developing treatment strategies to reduce pollution and remove harmful metal ions is essential for environmental protection (Rezai & Allahkarami, 2021).

3.4 Heavy Metal Remediation Techniques

The excessive presence of heavy metals in water poses a significant concern, necessitating the removal of heavy metals from wastewater before discharge into freshwater sources. Detection methods, including chemical processes and instruments, should be employed in industries to determine heavy metal concentrations, aligning with WHO norms. Strict disposal measures for medical waste are essential to prevent contamination of water bodies. Establishing and wisely utilizing water treatment plants before effluent discharge is crucial. Various treatment methods are employed for removing toxic heavy metals from water and wastewater. These methods include chemical precipitation, coagulation and flocculation, ion exchange, membrane filtration, phytoremediation, bioremediation, electrochemical techniques, and adsorption. Each method aims to effectively eliminate or reduce the concentration of harmful heavy metals, providing diverse options for water treatment based on specific needs and conditions. Table 3.3 presents the advantages and disadvantages of remediation techniques for heavy metals. Conventional ion exchange is commonly employed to separate heavy metal ions from waste streams. However, both membrane and ion exchange approaches become cost-prohibitive when handling substantial volumes of solutions with low concentrations of heavy metal ions. Consequently, these approaches are not economically viable for small-scale industries. Precipitation and electrochemical treatment approaches prove inefficient, particularly when the heavy metal ion is available in trace amounts. The adsorption process offers high removal efficiency, ease of operation, high selectivity, and reduced sludge production. However, it also presents challenges such as adsorbent regeneration issues, pH-sensitive processes, slow kinetics, and scale-up challenges.

3.4.1 Chemical Precipitation

In the process of water and wastewater treatment, dissolved components are transformed into solid particles by chemical precipitation (Benalia et al., 2022). Metal cations, anions (phosphate, cyanide, fluoride), and organic molecules (aromatic amines, phenols, detergents) can be removed by this efficient method. Competing processes like ion exchange, electron precipitation, and reverse osmosis have higher capital and energy costs but are better suited for metal recovery. Chemical precipitation, with low capital costs, is simple but incurs operating costs for chemicals and sludge disposal. Metals are often precipitated as hydroxides, but sulfide and carbonate methods are also used. Phosphate and fluorine removal involve precipitation with iron or aluminum salts and calcium chloride, respectively (Qasem et al., 2021).

Due to its simplicity and cost-effectiveness, chemical precipitation stands out as a highly effective and extensively employed process in various industries (Ku & Jung, 2001). In this process, chemicals react with the ions of heavy metals to generate precipitates that are insoluble. Sedimentation or filtering may then be used to eliminate these precipitates from the water. After treatment, the water is decanted and reused or properly released. According to Fu and Wang (Fu & Wang, 2011), hydroxide and sulfide precipitation are two standard chemical precipitation techniques.

Table 3.3 Advantageous and disadvantageous of different methods for the uptake of heavy metals (Allahkarami & Rezai, 2019; O'Connell et al., 2008)

Methods	Advantageous	Disadvantageous
Chemical precipitation	Versatility; cost-effective; simplicity of application; fast kinetics	Chemical usage; formation of fine particles during the process; impact on water chemistry; pH-sensitive process
Coagulation and flocculation	Effective removal; versatility; reduced sludge volume; enhanced settling; applicability to different water sources	Chemical usage; pH-sensitive process; potential for recontamination; environmental impact (proper disposal or treatment of residual chemicals is essential)
Ion exchange	High selectivity; applicability to low concentrations; regenerability; applicable with limited space	High initial cost; waste generation; sensitivity to pH and ionic strength; complexity of operation; limited capacity
Membrane separation	High removal efficiency; compact design; high selectivity; continuous process; applicability to different water sources	Membrane fouling; high operating costs; limited removal of certain ions; environmental impact of membrane production
Phytoremediation	Cost-effective; environmentally friendly method; Suitable for large-scale applications; minimization of disposal issues; reduced soil erosion	Slow kinetics; soil and climate dependency; potential for plant toxicity; limitations of plants' ability for the uptake of heavy metals
Bioremediation	Environmentally friendly method; cost-effective; applicability to different pollutants; adaptability	Slow kinetics; incomplete removal; climate dependency; continuous monitoring is required; risk of secondary pollution production
Electrochemical techniques	High removal efficiency; fast kinetics; high selectivity; applicability to different water sources; reduced sludge production; continuous process	Electrode fouling; pH-sensitive process; complexity of process design; risk of byproduct formation; scale-up issues; addition of chemicals for enhancing the removal efficiency is required
Adsorption	High removal efficiency; ease of process; high selectivity; reduced sludge production; no chemical additives; versatility; applicability to different water sources	Adsorbent regeneration issues; pH-sensitive process; slow kinetics; scale-up challenges; saturation and breakthrough challenges

Hydroxide precipitation is the most frequently employed method due to its simple nature, cost-effectiveness, and ease of pH regulation (Huisman et al., 2006). Between pH 8.0 and 11.0, metal hydroxides become less soluble. The procedures of flocculation and sedimentation are used to remove metal hydroxides. In commercial hydroxide precipitation, lime is the recommended base because of its low price and simplicity of handling (Baltpurvins et al., 1997).

Favorable results were obtained from an assessment of hydroxide precipitation using calcium hydroxide and sodium hydroxide to remove Cu(II) and Cr(VI) ions from wastewater (Mirbagheri & Hosseini, 2005). The effectiveness of removing heavy metals was significantly improved by using lime as a seed material in the fly ash-lime-carbonation process. Nevertheless, the process of hydroxide precipitation results in the production of significant amounts of sludge with low density, which presents difficulties in terms of both dewatering and disposal. The efficiency of hydroxide precipitation may be limited by the presence of complexing agents in wastewater and the amphoteric characteristics of particular metal hydroxides.

One notable benefit for sulfide precipitation is that metal sulfide precipitates are considerably less soluble in water than hydroxide precipitates. Furthermore, the absence of amphoteric properties in sulfide precipitates enables efficient metal removal across a broader range of pH levels. Furthermore, compared to hydroxide sludges, sulfide sludges demonstrate superior dewatering and thickening properties (Estay et al., 2021). Sulfide precipitation via sulfate-reducing bacteria (SRB) is one of the most recent developments. SRB reacts with divalent soluble metals to produce insoluble metal sulfides by converting sulfates to hydrogen sulfide (Bao et al., 2023). While effective, sulfide precipitation in acidic conditions can release toxic H_2S fumes, requiring neutral or basic mediums.

Chemical precipitation has shown success when combined with other methods. Sulfide precipitation followed by nanofiltration proved effective in reusing and recovering heavy metal ions (González-Muñoz et al., 2006). Similarly, the electro-Fenton process and chemical precipitation combination reduced COD and removed zinc from rayon industry wastewater (Ghosh et al., 2011). When combined with chemical precipitation, ion-exchange treatments demonstrated enhanced nickel removal from wastewater streams (Yan et al., 2020). Conventional chemical precipitation processes have limitations, especially in treating heavy metal wastewater with coordinated agents (Fu & Wang, 2011). Chelating precipitants offer an alternative, with studies examining their effectiveness. Thiol-based compounds (Matlock et al., 2002) and dithiocarbamate-type supramolecular precipitants (Fu et al., 2007; Han et al., 2022) have shown promise in precipitating heavy metals. The xanthate process and new organic chelators have also proven effective in the uptake of heavy metal ions from waste streams (Chang et al., 2002). In conclusion, chemical precipitation remains a versatile and widely used method for heavy metal removal, with ongoing research focused on addressing its limitations and exploring innovative approaches to meet environmental regulations.

3.4.2 Coagulation and Flocculation

In water and advanced wastewater treatment facilities, coagulation and flocculation are vital procedures that improve particle species separation for processes like filtration and sedimentation. Coagulation has been used traditionally to remove turbidity and has shown efficacy against a number of pollutants, including viruses, radionuclides, organic matter, and metals. Enhanced coagulation is particularly valuable for achieving compliance with EPA standards, addressing contaminants such as arsenic, emerging pathogens, and humic substances (Abujazar et al., 2022).

The coagulation process involves coagulant formation, particle destabilization, and interparticle collisions. Flocculation, occurring in a separate tank, optimizes the rate of contact between destabilized particles, promoting their attachment and aggregation into larger, denser flocs. Inorganic coagulants, synthetic organic polymers, and coagulant aids are employed for effective coagulation (Shrestha et al., 2021).

Flocculation can be achieved through mechanical or hydraulic means. Mechanical flocculators use paddle-wheel devices, while hydraulic flocculators utilize horizontal flow with vertical baffles, creating optimal velocity gradients for mixing. Different flow patterns, such as over-and-under and around-the-end configurations, are available. Hydraulically mixed flocculators, lacking mechanical equipment, are popular in developing countries due to their simplicity (El-Gaayda et al., 2021).

Coagulation and flocculation are fundamental processes in wastewater treatment, crucial for both wastewater disposal and the production of safe drinking water. These processes involve using chemicals to induce the aggregation of colloidal materials, forming larger flocs or aggregates that facilitate their removal from the water.

The effectiveness of ferrous sulfate heptahydrate ($FeSO_4{\cdot}7H_2O$) in eliminating total suspended solids (TSS), chemical oxygen demand (COD), and biochemical oxygen demand (BOD) from palm oil mill effluent was examined by Hossain et al. (2019). To explain the coagulation adsorption process, the Freundlich isotherm model was used. This model showed that multiple layers formed with uniform distribution of adsorbed particles. There was a significant increase in removal efficiency with $Fe_SO_4{-}7H_2O$ in a single stage of coagulation treatment; over 80% BOD, about 70% COD, and over 85% TSS were removed. Despite its effectiveness, this method generates a considerable amount of sludge, posing a separation challenge and transferring toxic substances to the solid phase. The technique successfully eliminates suspended particles, dissolved metals, and dyes. While coagulation and flocculation are valuable processes, challenges include the generation of sludge and potential issues with the transfer of contaminants to the solid phase. Continuous research is essential to develop more efficient and environmentally friendly coagulants and flocculants, addressing these challenges.

In summary, coagulation and flocculation play pivotal roles in wastewater treatment by promoting the aggregation of colloidal particles and facilitating their removal. Their effectiveness highlights their importance in addressing various water quality issues (Shrestha et al., 2021).

3.4.3 Ion Exchange

Ion exchange is a traditional procedure that has been used for more than a century and has developed into a vital component of many industrial processes and applications. Through a reversible stoichiometric chemical process, an ion from a solution is exchanged for an ion with an equivalent charge that is affixed to an insoluble and immobile solid substance in this approach. Overall, electro-neutrality is preserved by this procedure (Qasem et al., 2021). Ion-exchange materials often consist of synthetic organic resins, inorganic three-dimensional matrices, and hybrid materials of the newest generation. The development of nuclear technology in the 1940s prompted a need for ion-exchange materials that had excellent sorption capacity, resistance to ionizing radiation, and high thermal stability (Bashir et al., 2019). Organic resins did not possess such features; instead, inorganic exchangers with established ion-exchange properties, such as zeolites, metal titanosilicates, metal phosphates, tungstates, metal sulfides, sodium titanates, layered double hydroxides, and hexacyanoferrates, gained attention. Originally used for water softening, it has evolved into a crucial component of various technical and industrial processes (Bashir et al., 2019).

Utilizing ions as dissolved contaminants, ion exchange technology provides an efficient water-purifying method. Furthermore, it is adaptable to different applications, including nitrate elimination, bed polishing, and the selective discharge of distinct pollutants. The materials used as exchange media for purifying water include montmorillonite, zeolites, synthetic organic resin, humus soil, and polymers in the form of resin. Ion exchangers are commonly classified as cation, anion, or amphoteric species (Saleh et al., 2022). In ion exchange processes, many optimization goals have been investigated. For example, they are minimizing resin quantity to get a greater removal rate and maximizing contact duration to minimize equipment sizing (Torres-Perez et al., 2020). While cationic resins, such as strong and weak acidic resins, are utilized more often, anionic resins tend to be employed for lower concentrations of contaminants (Carolin et al., 2017). Cation exchangers that are frequently employed include weak acidic resins (–COOH) and acidic resins ($-SO_3H$) (Zhao et al., 2016).

Despite its effectiveness, ion exchange has several disadvantages that should be taken into account. Pre-treatment procedures, such as oil or grease removal, are often required (Mahmud et al., 2016). The resins must also be recovered using chemical reagents, which might result in additional contamination. Ion exchange resins may be used to selectively eliminate heavy metals, which can help overcome specific challenges while dealing with various pollutants (Bolisetty et al., 2019). Table 3.4 summarizes various ion exchangers used to remediate heavy metal ions. In summary, ion exchange technology has a rich historical background and serves as a versatile and valuable method in water treatment. Its flexibility in addressing various water quality issues makes it a significant player in modern water purification processes, despite the need for addressing specific challenges associated with its application.

Table 3.4 Summary of different ion exchangers applied for the remediation of heavy metals

Ion exchanger	Heavy metal	Adsorption capacity (mg/g)	References
Lewatit MonoPlus TP220	Cu(II)	10.0	(Wołowicz et al., 2023)
	Zn(II)	8.9	
	Ni(II)	9.2	
Laumontite zeolites	Pb(II)	7.59	(Wangi et al., 2023)
	Zn(II)	7.40	
	Cr(VI)	5.50	
Amberlite IRC747	Cu(II)	83.4	(Lebron et al., 2023)
	Ni(II)	26.3	
	Co(II)	6.5	
Dowex M4195	Cu(II)	103.1	(Lebron et al., 2023)
	Ni(II)	89.9	
	Co(II)	24.4	
Zeolite from coal fly ash	Pb(II)	2.68	(Koshlak, 2023)
	Cd(II)	1.54	
	Ni(II)	1.3	
Chitosan grafted polyvinyl alcohol resin	Pb(II)	6.85	(Patel et al., 2023)
	Cu(II)	243.9	
	Fe(II)	87.72	

3.4.4 Membrane Filtration

In recent decades, membrane technology has emerged as a prominent separation method, finding widespread applications and establishing itself as an indispensable technology in the field (Birniwa et al., 2023). The inherent advantages of membrane technology, such as operating without the need for additional chemicals, low energy consumption, ease of handling, and well-organized process conductions, have contributed to its prominence. Membrane separation demonstrates enhanced efficiency in comparison to traditional techniques, with the efficacy directly tied to the membrane's characteristics (Obotey Ezugbe and Rathilal, 2020). Between the late 1970s and the early 1980s, membrane filtering technologies witnessed a sharp increase in development due to their greater effectiveness in removing pollutants, reduced environmental impact, and sometimes lower energy usage when compared to traditional techniques (Azimi et al., 2017). The simplicity of the separation method is a common thread across various membrane processes, which operate through semi-permeable membranes with different pore sizes, allowing solvent passage while capturing solutes under a higher trans-membrane pressure (Padaki et al., 2015).

Membranes are categorized into three types: hybrid, pressure-driven, and liquid membranes (Singh, 2014). These selective layers with porous or non-porous structures facilitate the contact between two homogeneous phases to uptake different-sized contaminants. Material selection, membrane pore size, and composition significantly influence membrane performance. Various materials, including metallic, ceramic, and composite materials, are employed in membrane fabrication. Polymers, particularly polypropylene, polyvinylidene fluoride, and cellulose acetate, are commonly used due to their cost-effectiveness and porous structure (Koyuncu et al., 2015; Wilderer, 2010). Ceramic materials, such as alumina and zirconia, outperform polymers in certain cases, offering narrow pore size distribution and high stability. The choice of materials aims to produce membranes with chemical resistance, low imperfections, and resistance to fouling (Padaki et al., 2015).

Nanofiltration (NF), an emerging technology, enables the partitioning of bulky molecules by the use of small pores, hence surpassing the limitations of traditional methods. The fabrication and modification of NF membranes, which use techniques like interfacial polymerization and nanoparticles, significantly affect solvents' permeability. Nanofiltration (NF) is becoming more used in environmentally friendly and energy-efficient applications, specifically for water treatment (Al-Rashdi et al., 2013; Zheng et al., 2022). Reverse osmosis (RO), a crucial method of separation, is extensively used in different fields such as food processing, desalination, wastewater treatment, and water reclamation. RO utilizes semi-permeable membranes without distinct pores, relying on hydraulic operating pressure for heavy metal ion removal.

MF and UF, often discussed together, share similarities as pressure-driven processes used for water purification, clarification, and wastewater treatment. They differ in the size of solutes they can remove, with MF addressing suspended particles and UF handling dissolved macromolecules and colloidal materials. Membrane choice significantly influences removal efficiency, with the widely applied hydrophobic and hydrophilic membranes and crystalline polymers (Bardhan et al., 2022; Yaqub & Lee, 2019). ED, developed in the 1950s, is a liquid hybrid membrane process with applications in wastewater treatment and various industrial processes (Benalla et al., 2021). Ion-selective exchange membranes, including cation and anion exchange membranes, facilitate ion transportation through electrical potential or concentration gradients. Electrodialysis reversal (EDR) improves ED by reducing scaling and fouling, although it requires additional plumbing and electrical controls (Juve et al., 2022). Table 3.5 summarizes various membranes used for the remediation of heavy metal ions. In summary, the diverse membrane processes offer efficient water and wastewater treatment solutions, with each type catering to specific applications and separation requirements. The ongoing advancements in membrane technology continue to enhance removal efficiency and broaden their scope of applications in environmental remediation.

Table 3.5 Summary of different membranes applied for the remediation of heavy metals

Membrane	Heavy metal	Rejection efficiency (%)	References
Lamellar graphene oxide-based composite	Cr(VI)	78.0	(Janwery et al., 2023)
	As(III)	80.5	
	Cd(II)	88.0	
	Pb(II)	95.2	
$Ti_3C_2T_x$ MXene@MOF decorated polyvinylidene fluoride	Ni(II)	95.2	(Ghanbari et al., 2023)
	Cd(II)	95.2	
	Mn(II)	97.6	
	Cu(II)	95.2	
	Zn(II)	95.2	
Positively charged nanofiltration	Ni(II)	93.3	(Cheng et al., 2023)
	Mn(II)	92.8	
	Cu(II)	91.0	
	Cd(II)	90.4	
	Pb(II)	89.2	
Sodium alginate composite hydrogel membrane	Cr(III)	96.8	(Cai et al., 2023)
	Cr(VI)	91.4	
Polyphenylsulfone membrane blended with polyaniline	Pb(II)	99.0	(Alhoshan et al., 2023)
	Cd(II)	95.0	
Aminoimidazole ionic liquid (IL) embellishment of nanofiltration	Cu(II)	97.1	(Xing et al., 2023)
	Ni(II)	94.6	
	Cd(II)	95.5	
	Zn(II)	96.4	

3.4.5 Phytoremediation

Phytoremediation, a plant-based technology, has gained significant attention for its ability to address heavy metal contamination (Liu et al., 2020; Shen et al., 2022). This process involves different stages, including transformation, transport, volatilization, accumulation, and translocation (Shah & Daverey, 2020). Remediation potential of more than 400 plant species has been verified (Ali et al., 2013), making it a harmless and viable strategy (Steliga & Kluk, 2020).

Phytoremediation is the process of using plants and the microorganisms in the soil to decrease the levels of pollutants in the environment and mitigate their harmful effects. This method is suitable for eliminating high levels of radionuclides, organic contaminants, and heavy metals. Phytoremediation is regarded as an affordable, effective, eco-friendly, and solar-driven approach. The approach has become popular because of its aesthetic appeal, widespread public acceptability, and suitability for

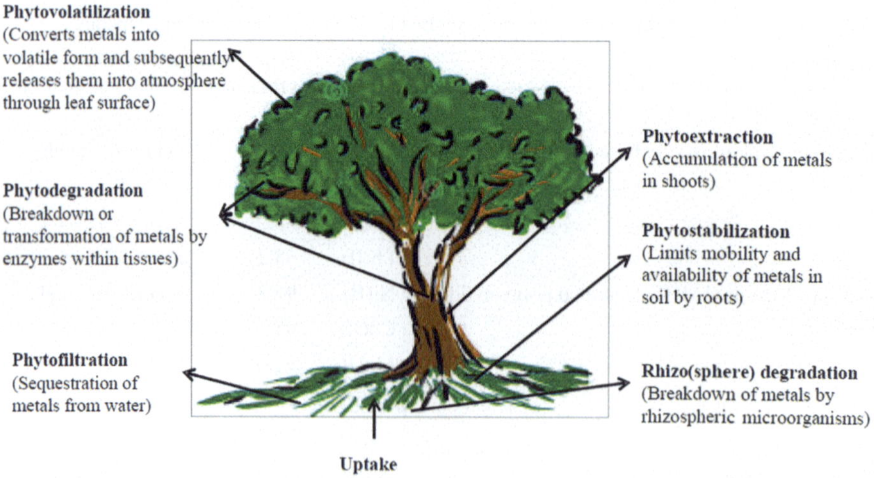

Fig. 3.4 Various strategies of phytoremediation (Dixit et al., 2015)

wide field areas where other decontamination methods are neither feasible nor expensive (Ali et al., 2013). The various techniques/strategies of phytoremediation include (Fig. 3.4):

Phytoextraction: Plants absorb contaminants from soil or water through their roots and accumulate them in aboveground biomass, primarily in shoots (Robinson et al., 2015).

Phytofiltration: Plants are used to remove pollutants from contaminated surface or waste waters, either through root absorption (rhizofiltration), seedling use (blastofiltration), or excised plant shoots (caulofiltration) (Gardea-Torresdey et al., 2004).

Phytostabilization: Certain plants stabilize contaminants in polluted soils, reducing their mobility and bioavailability, thus preventing migration to groundwater or entry into the food chain (Shackira & Puthur, 2019).

Phytovolatilization: Plants take up pollutants from soil, convert them to volatile forms, and release them into the atmosphere. This technique is limited, as it transfers pollutants from soil to the atmosphere without complete removal (Awa & Hadibarata, 2020).

Phytodegradation: Plants degrade organic pollutants through enzymatic activities, particularly dehalogenase and oxygenase, making them effective for organic pollutants (Muthusaravanan et al., 2018).

Rhizodegradation: Soil microorganisms in the rhizosphere (around the roots) break down organic pollutants stimulated by plant root exudates (Verma & Rawat, 2021).

Islam et al. (2019): Halophytic plants are used to remove salts from salt-affected soils, making them suitable for normal plant growth.

Phytoextraction, the main phytoremediation technique for heavy metals, relies on plant species with characteristics such as high growth rate, extensive root systems, and tolerance to the target heavy metals (Sheoran et al., 2016). The choice between hyperaccumulators (less biomass, greater metal accumulation) and non-accumulators (more biomass, comparable overall accumulation) depends on economic considerations and metal recovery objectives. Grasses are more adaptable and grow quickly than shrubs or trees, making them the better choice for phytoextraction. Plants undergoing many harvests during a single growing cycle have a higher capacity for phytoextraction. To reduce the levels of heavy metal contamination, plants may need to go through many cropping cycles when used for phytoextraction, which might contaminate the food chain. Overall, phytoremediation presents an attractive, sustainable, and potentially cost-effective approach for environmental cleanup (Lasat, 2002). However, the disposal of polluted biomass is a challenging endeavor since the current approaches— direct disposal, pyrolysis, composting, leaching, incineration, and compaction —are either costly or may result in secondary contamination (Kovacs & Szemmelveisz, 2017). The disposal of contaminated biomass has not received enough attention despite the increased focus on improving phytoremediation technology, according to a new survey of research studies conducted over the last two decades.

3.4.6 Bioremediation

Bioremediation, a field of green technology that utilizes biological agents such as molds, bacteria, algae, yeasts, and plants to remove and recover heavy metals, has attracted interest. In contrast to physical–chemical approaches, bioremediation provides high removal efficiency, cost-effectiveness, ease of availability, and economic viability. The process transforms hazardous heavy metals into less dangerous ions using the variety of binding sites and high surface area to volume ratio of biological agents (Jeyakumar et al., 2023).

Natural characteristics of biological agents include high photosynthetic efficiency (found in photosynthetic bacteria and plants) and resistance to severe temperatures, high salinity, heavy metal pollution, and nutritional stress (Medfu Tarekegn et al., 2020). The type of biological agent, the properties of the aimed metal, the binding sites involved, the operating parameters (contact time, temperature, solution acidity, concentrations of metal ions and biological agents), and the presence of competing co-ions are some of the factors that affect the efficiency of bioremediation (Cameron et al., 2018). Additionally, biomass waste produced by the bioremediation process may be used as a source of nutrients for the development of plants, which is a unique advantage in improving the method's overall sustainability (Medfu Tarekegn et al., 2020).

The application of bacterial biomass as biosorbents for the uptake of toxic heavy metal ions from aqueous environments has proven effective, even at low concentrations. Both living and dead bacterial biomass have been utilized for metal adsorption. The negatively charged microorganism surfaces, attributed to anionic structures, facilitate the binding of metal cations. Different kinds of bacteria, whether mobile or fixed, may effectively accumulate harmful heavy metals such as uranium, copper, and lead from polluted waste streams (Priyadarshanee & Das, 2021).

Functional groups which are necessary for the adsorption of metals are found in the bacterial biosorbent including sulfydryl, carboxyl, thioether, amino, hydroxyl, imidazole, phosphate, carbonyl, sulfate, amide, and phenol. For instance, carboxyl groups have the ability to bind metal ions. This is shown by the fact that Streptomyces pilosus can bind Cu ions because of the carboxyl groups on its peptidoglycan cell wall. Electrostatic interactions and Chelation are two additional mechanisms by which amine groups facilitate the removal of metal ions (Kumar et al., 2011).

Bacterial biofilm, composed of microbial cells and extracellular polymeric substances (EPS), enhances metal removal. EPS, rich in different functional groups, acts as adsorption sites for heavy metals (Kumari & Das, 2019). Biofilm shows good adsorption capacity and is cheap for wastewater treatment. The EPS layer exhibits diverse mechanisms for metal binding, including complexation, surface precipitation, ion exchange, and native adsorption (Chakraborty & Das, 2014).

The modification of bacterial biosorbents is explored to enhance their metal sorption efficiency. Pre-treatment techniques, such as biological, chemical, and physical procedures, can enhance and activate binding sites on the surface of the biosorbent. The modification of biosorbents aims to overcome limitations associated with untreated biomass, ultimately improving their metal adsorption capabilities (Yang et al., 2015).

However, the utilization of bacterial biomass, biofilm, and EPS as biosorbents provides an efficient, economical, and eco-friendly approach to the uptake of toxic heavy metals from polluted water sources. Modifying biosorbents further enhances their metal adsorption capacity, paving the way for potential applications in field studies and large-scale metal remediation projects.

The biosorption process in bacterial removal of heavy metal ions is influenced by various physico-chemical factors, which are crucial in determining its efficiency. Contact time, temperature, solution acidity, biosorbent dosage, initial concentration of metal ion, and agitation speed are among the key factors impacting biosorption.

pH is vital, affecting the solution's chemistry, metal ion speciation, and the ionization state of available functional groups on the surface of sorbent (Chen et al., 2020b). The optimum pH for effective metal sorption typically ranges from 4 to 8, although specific cases may vary (Abdel-Monem et al., 2010). Temperature directly influences biosorption by affecting liquid viscosity, adsorbate diffusion rates, and solute activity. While higher temperatures generally enhance biosorption due to increased kinetic energy, excessive temperatures can damage binding sites on the surface of sorbent (Aryal et al., 2012).

The dosage of biosorbent is critical. Increased biomass concentration usually improves biosorption up to a limit, after which efficiency may decline due to electrostatic interactions among cells hindering metal ion binding (Aryal et al., 2012). The presence of other ions can reduce the biosorbent's specificity for the target metal by occupying binding sites. Careful management of competing ion concentrations is necessary to avoid hindering metal ion uptake. The nature of biosorbent is crucial, considering size, growth rate, composition, and modification. Small-sized cells with higher surface area are generally more efficient, and the age of the culture and cell wall composition influence biosorption (Priyadarshanee & Das, 2021). Contact time is a key parameter for maximum biosorption. While the rate of metal uptake initially increases, it eventually reaches equilibrium. Different biosorbents exhibit varying contact time requirements, ranging from minutes to hours (Huang et al., 2013).

In summary, effectively controlling and optimizing these parameters, such as maintaining optimal pH and temperature, managing biosorbent dosage, and understanding the biosorbent's nature, are essential for achieving efficient bacterial uptake of heavy metals through biosorption.

3.4.7 Electrochemical Techniques

The electrochemical method is a superior approach for removing heavy metal ions from effluent when compared to other techniques like adsorption and chemical precipitation. It is environmentally friendly, offers flexibility in operation, and has the potential for automation (Feng et al., 2003). The electrochemical technique has specific benefits, such as the production of highly pure recovered products and being able to recycle critical metal components.

The high purity of the recovered products enables direct use in industrial production, saving on post-processing costs. Additionally, electrochemical recovery technology for strategic elements mitigates the complexities and large chemical dosages associated with traditional methods. This results in greater recovery efficacy, assisting in reducing the total expenses related to the electrochemical technique while providing a sustainable and efficient solution (Yang et al., 2021).

Electrochemical engineering, at the intersection of electrochemistry, material science, and electrical engineering, focuses on designing and implementing electrochemical processes with considerations for environmental protection and cost efficiency. It finds applications in various areas, such as inorganic compound synthesis, metal electrocoating, water and wastewater treatment, energy storage, and sensors (Lapicque, 2004).

The electrodeposition process, dating back to 1805, involves using electrical current in order to decrease the concentration of metal ions in solution and facilitate the deposition of a metallic element onto an electrode (Maarof et al., 2017). It is employed for purposes like precious and non-precious metal recovery and metal ion removal from wastewater. While electrodeposition offers advantages like no need for chemical reagents and high selectivity, challenges such as dendrite formation and gas evolution reactions need attention (Chen & Lim, 2005).

Electrocoagulation, an electrochemical water treatment method, utilizes sacrificial metal anodes (usually aluminum or iron) to generate coagulants in situ, facilitating the uptake of heavy metal ions through agglomeration and precipitation. Iron has been reported as more efficient than aluminum for metal uptake. The method has advantages, including lower chemical consumption, but requires careful monitoring of the anode (Bazrafshan et al., 2015).

Electroflotation, involving the production of small gas bubbles through water electrolysis, efficiently removes heavy metals by attaching them to bubbles and lifting them to the water surface. Electroflotation is often combined with electrocoagulation (ECF) for enhanced performance (Kyzas & Matis, 2016).

Electrosorption is exemplified by capacitive deionization (CDI) technology (Chen et al., 2020a). CDI, primarily used for desalination, has been explored for heavy metal removal. Challenges include unwanted reactions at electrodes, high energy consumption, and electrode material limitations (da Costa et al., 2021).

In conclusion, while electrochemical technologies offer environmentally compatible solutions for heavy metal removal, ongoing research aims to address challenges and improve the efficiency of these processes.

3.4.8 Adsorption

Adsorption refers to the process of attracting and retaining molecules of a substance (known as adsorbate) on a surface (known as adsorbent). This process is particularly useful and cost-effective for the uptake of heavy metals (Allahkarami & Rezai, 2019). Clay minerals, activated carbon, metal oxides, polymeric materials, and microbial biomasses are effective adsorbents for heavy metal removal.

The heavy metal removal adsorption mechanism is a complicated process controlled by several parameters involving adsorbent physicochemical features, contaminants, and operating conditions. Temperature, adsorbent amount, solution acidity, adsorption time, and initial metal ion concentration play important roles in the effectiveness of the adsorption process.

Carbonaceous nano adsorbents are widely recognized for their expansive surface areas, making them highly efficient in adsorbing heavy metals (Kumar, 2023). However, the intricate and costly nature of prevalent surface modification methods often curtails their extensive use. These methods, such as nitrogenation, oxidation, and sulfuration, are essential for improving surface area, pore size, mechanical strength, and sorption capability. Researchers continually seek novel, cheap, eco-friendly surface modification techniques to overcome these challenges and facilitate the broader industrial application of carbon-based nanoporous adsorbents.

Chitosan, a natural adsorptive polymer, offers unique properties with its amino ($-NH_2$) and hydroxyl ($-OH$) groups (Allahkarami et al., 2017; Igder et al., 2019). However, it faces challenges such as low mechanical strength and poor stability, limiting its regeneration efficiency. To address these issues, researchers employ structural and chemical modifications, including cross-linking and grafting methods. Despite these advancements, there is an ongoing need for different regeneration approaches as well as adjustments to increase the reusability of chitosan-based adsorbents.

Mineral adsorbents are acknowledged for cost-effectiveness in water purification (Allahkarami et al., 2023c). However, their removal efficiency may decrease after multiple cycles. Researchers explore modification methods, including thermal treatment and acid washing, to enhance their specific surface area and pore size. These modifications aim to overcome limitations and improve the overall adsorption efficiency of mineral adsorbents (Allahkarami et al., 2023a).

Magnetic adsorbents, incorporating iron particles, stand out for their low-cost synthesis and reusability (Allahkarami et al., 2022, 2023b). However, challenges persist, especially regarding the cyclic stability of adsorbents produced through methods like co-precipitation, high-gravity technology, and grafting. The grafting technique is simple and versatile, but it is highly dependent on the quantity of active functional groups and the active hydroxyl available on the iron particles' surface (Khan et al., 2020).

According to Allahkarami and Rezai (2021), biosorption employs biological materials as biosorbents for the elimination of heavy metal ions. Biosorbents are an affordable and environmentally beneficial alternative to readily accessible sorbents. Typical biosorbents include algae, microbial matter, and non-living biomass. Biosorption presents advantages like metal selectivity, high efficiency, minimal sludge formation, and potential for metal recovery, making it a hopeful method for the uptake of heavy metal ions.

The biosorption process is supported by biosorbents, which are obtained from biological materials and have a variety of functional groups, such as hydroxyl, carboxyl, thiol, amino, and phosphate groups (Wang & Chen, 2009). Several mechanisms can facilitate the interaction between pollutants and the surface of the biosorbent. These mechanisms encompass aggregation, reduction, complexation/coordination, microprecipitation, oxidation, and ion exchange. The biosorption process is considerably influenced by the solution acidity, which in turn affects the ionization of functional groups and the surface charge density of the sorbent (Qin et al., 2020).

Metal–organic frameworks (MOFs) present nanomaterials with a high potential for adsorption purposes (Kobielska et al., 2018). However, challenges include micropores' inaccessibility for certain target metals and low stability in water. Future research endeavors should focus on tuning the MOFs' structure, scaling up production, and proposing different functionalization methods to enhance stability and sorption kinetics. These advancements are vital for the effective application of MOFs in various industrial wastewater purposes. Table 3.6 summarizes various adsorbents used for the remediation of heavy metal ions.

Table 3.6 Summary of different adsorbents applied for the remediation of heavy metals

Adsorbent	Heavy metal	Adsorption capacity (mg/g)	References
Schist/alginate composite	Ni(II)	124.79	(Allahkarami et al., 2023a)
	Cu(II)	111.78	
	Cd(II)	119.78	
vermiculite-alginate composite	Ni(II)	47.85	(Allahkarami et al., 2023c)
	Cu(II)	105.26	
Natural dolomite	Cu(II)	0.475	
	Sr(II)	0.378	
	Co(II)	0.220	
	Ag(I)	0.315	
	Cd(II)	0.420	
	Ba(II)	0.295	
	Ni(II)	0.380	
Natural zeolite	Cd(II)	4.73	(Khoshraftar et al., 2023)
	Pb(II)	41.44	
	Zn(II)	3.27	
Orange peel cellulose	Cd(II)	29.0	(Rahman et al., 2023b)
	Cr(VI)	4.9	
	Pb(II)	50.1	
Modified shrimp-based chitosan	As(V)	15.92	(Rahman et al., 2023a)
	Cr(VI)	20.37	
	Ni(II)	7.00	
	Co(II)	6.27	
Fly ash-based composite	AsO_3^{3-}	5.63	(Chen et al., 2023)
	Cd(II)	3.74	
	Pb(II)	15.86	

3.5 Conclusion

The drive for rapid industrialization, aimed at achieving economic prosperity, has led to significant environmental challenges, particularly in the form of heavy metal contamination and wastewater pollution. Heavy metals, particularly when present in ionic form in water, are widely recognized for their toxicity to living organisms. In fact, these metals pose a threat due to their bio-accumulative nature and high solubility in aquatic environments. Heavy metals in surface water have natural sources like volcanic eruptions and anthropogenic origins, including industrial activities, urbanization, and improper waste disposal. Soil contamination arises from natural weathering, acid drainage, and human activities like mining and agriculture. Fires can mobilize metals in metal-rich soils and expose hidden mine tailings. Industrial emissions, electronic waste, agricultural practices, and thermal pollution contribute to water pollution. Coal combustion releases trace metals, posing risks to both air and water quality. This contamination poses significant dangers to aquatic ecosystems, human health, and the overall environment.

Chemical precipitation involves forming insoluble precipitates through reactions, with hydroxide and sulfide precipitation being standard techniques. Chelating precipitants, like thiol-based compounds, provide alternatives and meet environmental regulations. Adsorption relies on the physicochemical properties of adsorbents, with carbon-based nanoporous materials, chitosan, mineral adsorbents, magnetic materials, biosorbents, and metal–organic frameworks explored for their efficiency. Membrane filtration processes, categorized into reverse osmosis, ultrafiltration, microfiltration, nanofiltration, and electrodialysis, revolutionize heavy metal removal with ongoing technological advancements, enhancing their applicability. Ion exchange, dating back a century, utilizes inorganic exchangers like zeolites and metal sulfides for selective and effective heavy metal capture. Coagulation and flocculation processes induce aggregation for efficient removal, with ongoing research focusing on environmentally friendly options. Phytoremediation, an eco-friendly approach, utilizes plants for techniques such as phytoextraction, phytofiltration, phytostabilization, and phytodegradation. Electrochemical treatment stands out as environmentally friendly, with techniques like electrodeposition and electrosorption offering efficient removal. Bioremediation, employing biological agents, provides a cost-effective approach, with bacterial biomass and modification techniques enhancing biosorbents. Integrating these approaches offers a comprehensive sustainable heavy metal removal strategy, contributing to environmental remediation and ecosystem protection. Ongoing research will refine these methods, ensuring their effectiveness in addressing the complexities of heavy metal pollution.

Acknowledgements Our sincere thanks to the editor to be part of this book.

References

Abdel-Monem, M. O., Al-Zubeiry, A. H. S., & Al-Gheethi, A. A. S. (2010). Biosorption of nickel by *Pseudomonas cepacia* 120S and *Bacillus subtilis* 117S. *Water Science and Technology, 61*, 2994–3007.

Abraham, J., Dowling, K., & Florentine, S. (2017). Risk of post-fire metal mobilization into surface water resources: A review. *Science of the Total Environment, 599–600*, 1740–1755.

Abujazar, M. S. S., Karaağaç, S. U., Amr, S. S. A., Alazaiza, M. Y. D., & Bashir, M. J. K. (2022). Recent advancement in the application of hybrid coagulants in coagulation-flocculation of wastewater: A review. *Journal of Cleaner Production, 345*, 131133.

Ahirwar, R., & Tripathi, A. K. (2021). E-waste management: A review of recycling process, environmental and occupational health hazards, and potential solutions. *Environmental Nanotechnology, Monitoring & Management, 15*, 100409.

Al-Rashdi, B. A. M., Johnson, D. J., & Hilal, N. (2013). Removal of heavy metal ions by nanofiltration. *Desalination, 315*, 2–17.

Alhoshan, M., Alam, J., Shukla, A. K., & Hamid, A. A. (2023). Polyphenylsulfone membrane blended with polyaniline for nanofiltration promising for removing heavy metals (Cd^{2+}/Pb^{2+}) from wastewater. *Journal of Materials Research and Technology, 24*, 6034–6047.

Ali, H., Khan, E., & Sajad, M. A. (2013). Phytoremediation of heavy metals—concepts and applications. *Chemosphere, 91*, 869–881.

Allahkarami, E., Allahkarami, E., & Azadmehr, A. (2023a). Enhancing the efficiency of Ni(II), Cd(II), and Cu(II) adsorption from aqueous solution using schist/alginate composite: Batch and continuous studies. *Environmental Science and Pollution Research, 30*, 105504–105521.

Allahkarami, E., Dehghan Monfared, A., Silva, L. F. O., & Dotto, G. L. (2022). Lead ferrite-activated carbon magnetic composite for efficient removal of phenol from aqueous solutions: Synthesis, characterization, and adsorption studies. *Scientific Reports, 12*, 10718.

Allahkarami, E., Dehghan Monfared, A., Silva, L. F. O., & Dotto, G. L. (2023b). Application of Pb–Fe spinel-activated carbon for phenol removal from aqueous solutions: Fixed-bed adsorption studies. *Environmental Science and Pollution Research, 30*, 23870–23886.

Allahkarami, E., Igder, A., Fazlavi, A., & Rezai, B. (2017). Prediction of Co (II) and Ni (II) ions removal from wastewater using artificial neural network and multiple regression models. *Physicochemical Problems of Mineral Processing*, 53.

Allahkarami, E., & Rezai, B. (2019). Removal of cerium from different aqueous solutions using different adsorbents: A review. *Process Safety and Environmental Protection, 124*, 345–362.

Allahkarami, E., & Rezai, B. (2021). A literature review of cerium recovery from different aqueous solutions. *Journal of Environmental Chemical Engineering, 9*, 104956.

Allahkarami, E., Soleimanpour Moghadam, N., Jamrotbe, B., & Azadmehr, A. (2023c). Competitive adsorption of Ni(II) and Cu(II) ions from aqueous solution by vermiculite-alginate composite: batch and fixed-bed column studies. *Journal of Dispersion Science and Technology, 44*, 1402–1412.

Arkle, R. S., Pilliod, D. S., & Strickler, K. (2010). Fire, flow and dynamic equilibrium in stream macroinvertebrate communities. *Freshwater Biology, 55*, 299–314.

Aryal, M., Ziagova, M. G., & Liakopoulou-Kyriakides, M. (2012). Cu (II) biosorption and competitive studies in multi-ions aqueous systems by Arthrobacter sp. Sphe3 and Bacillus sphaericus cells: Equillibrium and thermodynamic studies. *Water, Air, & Soil Pollution, 223*, 5119–5130.

Asaduzzaman, K., Khandaker, M. U., Baharudin, N. A. B., Amin, Y. B. M., Farook, M. S., Bradley, D. A., & Mahmoud, O. (2017). Heavy metals in human teeth dentine: A bio-indicator of metals exposure and environmental pollution. *Chemosphere, 176*, 221–230.

Ash, E. C. (2009). Characteristics, management, and environmental issues. *Technical Results, Program: Coal Combustion Products*, 1019022.

ATSDR, U. S. (2007). *CERCLA priority list of hazardous substances*. http://www.atsdr.cdc.gov/cercla/07list.html

Awa, S. H., & Hadibarata, T. (2020). Removal of heavy metals in contaminated soil by phytoremediation mechanism: A review. *Water, Air, & Soil Pollution, 231*, 47.

Azimi, A., Azari, A., Rezakazemi, M., & Ansarpour, M. (2017). Removal of heavy metals from industrial wastewaters: A review. *ChemBioEng Reviews, 4*, 37–59.

Baltpurvins, K. A., Burns, R. C., Lawrance, G. A., & Stuart, A. D. (1997). Effect of electrolyte composition on zinc hydroxide precipitation by lime. *Water Research, 31*, 973–980.

Bao, Y., Lv, D., Guo, Z., Yun, J., Wang, Y., & Li, D. (2023). Contribution of sulfate-reducing bacteria in the formation of hydrogen sulfide in coal seams of Hancheng mining area, China. *Geomicrobiology Journal*, 1–13.

Bardhan, A., Akhtar, A., & Subbiah, S. (2022). Microfiltration and ultrafiltration membrane technologies. In *Advancement in Polymer-Based Membranes for Water Remediation*. Elsevier.

Bashir, A., Malik, L. A., Ahad, S., Manzoor, T., Bhat, M. A., Dar, G. N., & Pandith, A. H. (2019). Removal of heavy metal ions from aqueous system by ion-exchange and biosorption methods. *Environmental Chemistry Letters, 17*, 729–754.

Bazrafshan, E., Mohammadi, L., Ansari-Moghaddam, A., & Mahvi, A. H. (2015). Heavy metals removal from aqueous environments by electrocoagulation process—A systematic review. *Journal of Environmental Health Science and Engineering, 13*, 1–16.

Benalia, M. C., Youcef, L., Bouaziz, M. G., Achour, S., & Menasra, H. (2022). Removal of heavy metals from industrial wastewater by chemical precipitation: Mechanisms and sludge characterization. *Arabian Journal for Science and Engineering, 47*, 5587–5599.

Benalla, S., Bachiri, B., Touir, J., Tahaikt, M., Taky, M., Ebn Touhami, M., & Elmidaoui, A. (2021). Feasibility of electrodialysis in heavy metals removal from brassware wastewaters. *Desalin. Water Treat, 240*, 106–114.

Bhatnagar, A., & Sillanpää, M. (2010). Utilization of agro-industrial and municipal waste materials as potential adsorbents for water treatment—A review. *Chemical Engineering Journal, 157*, 277–296.

Birniwa, A. H., Habibu, S., Abdullahi, S. S. A., Mohammad, R. E. A., Hussaini, A., Magaji, H., Al-Dhawi, B. N. S., Noor, A., & Jagaba, A. H. (2023). Membrane technologies for heavy metals removal from water and wastewater: A mini review. *Case Studies in Chemical and Environmental Engineering*, 100538.

Bolisetty, S., Peydayesh, M., & Mezzenga, R. (2019). Sustainable technologies for water purification from heavy metals: Review and analysis. *Chemical Society Reviews, 48*, 463–487.

Briffa, J., Sinagra, E., & Blundell, R. (2020). Heavy metal pollution in the environment and their toxicological effects on humans. *Heliyon*, 6.

Cai, R., Chen, Y., Hu, J., Xiong, J., Lu, J., Liu, J., Tan, X., Liu, W., Zhou, Y., & Chen, Y. (2023). A self-supported sodium alginate composite hydrogel membrane and its performance in filtering heavy metal ions. *Carbohydrate Polymers, 300*, 120278.

Cameron, H., Mata, M. T., & Riquelme, C. (2018). The effect of heavy metals on the viability of Tetraselmis marina AC16-MESO and an evaluation of the potential use of this microalga in bioremediation. *PeerJ, 6*, e5295.

Carolin, C. F., Kumar, P. S., Saravanan, A., Joshiba, G. J., & Naushad, M. (2017). Efficient techniques for the removal of toxic heavy metals from aquatic environment: A review. *Journal of Environmental Chemical Engineering, 5*, 2782–2799.

Carvalho, M. E. A., Castro, P. R. C., & Azevedo, R. A. (2020). Hormesis in plants under Cd exposure: From toxic to beneficial element? *Journal of Hazardous Materials, 384*, 121434.

Chakraborty, J., & Das, S. (2014). Characterization and cadmium-resistant gene expression of biofilm-forming marine bacterium Pseudomonas aeruginosa JP-11. *Environmental Science and Pollution Research, 21*, 14188–14201.

Chang, Y.-K., Chang, J.-E., Lin, T.-T., & Hsu, Y.-M. (2002). Integrated copper-containing wastewater treatment using xanthate process. *Journal of Hazardous Materials, 94*, 89–99.

Chaturvedi, A., Bhattacharjee, S., Singh, A. K., & Kumar, V. (2018). A new approach for indexing groundwater heavy metal pollution. *Ecological Indicators, 87*, 323–331.

Chen, J. P., & Lim, L. L. (2005). Recovery of precious metals by an electrochemical deposition method. *Chemosphere, 60*, 1384–1392.

Chen, P., Wu, J., Li, L., Yang, Y., & Cao, J. (2023). Modified fly ash as an effect adsorbent for simultaneous removal of heavy metal cations and anions in wastewater. *Applied Surface Science, 624*, 157165.

Chen, R., Sheehan, T., Ng, J. L., Brucks, M., & Su, X. (2020). Capacitive deionization and electrosorption for heavy metal removal. *Environmental Science: Water Research & Technology, 6*, 258–282.

Chen, X., Zhao, Y., Zhang, C., Zhang, D., Yao, C., Meng, Q., Zhao, R., & Wei, Z. (2020). Speciation, toxicity mechanism and remediation ways of heavy metals during composting: A novel theoretical microbial remediation method is proposed. *Journal of Environmental Management, 272*, 111109.

Cheng, X., Zhang, Y., Shao, S., Lai, C., Wu, D., Xu, J., Luo, X., Xu, D., Liang, H., & Zhu, X. (2023). Highly permeable positively charged nanofiltration membranes with multilayer structures for multiple heavy metal removals. *Desalination, 548*, 116266.

da Costa, J. G. D. R., Costa, J. M., & de Almeida Neto, A. F. (2021). Recent advances and future applications in electro-adsorption technology: An updated review. *Journal of Environmental Chemical Engineering, 9*, 106355.

Dang, Z., Liu, C., & Haigh, M. J. (2002). Mobility of heavy metals associated with the natural weathering of coal mine spoils. *Environmental Pollution, 118*, 419–426.

Dixit, R., Wasiullah, Malaviya, D., Pandiyan, K., Singh, U. B., Sahu, A., Shukla, R., Singh, B. P., Rai, J. P., Sharma, P. K., Lade, H., & Paul, D. (2015). Bioremediation of heavy metals from soil and aquatic environment: An overview of principles and criteria of fundamental processes. *Sustainability* [Online], 7.

Duffus, J. H. (2001). "Heavy metals"—A meaningless term. *Chemistry International—Newsmagazine for IUPAC, 23*, 163–167.

Edelstein, M., & Ben-Hur, M. (2018). Heavy metals and metalloids: Sources, risks and strategies to reduce their accumulation in horticultural crops. *Scientia Horticulturae, 234*, 431–444.

El-Gaayda, J., Titchou, F. E., Oukhrib, R., Yap, P.-S., Liu, T., Hamdani, M., & Akbour, R. A. (2021). Natural flocculants for the treatment of wastewaters containing dyes or heavy metals: A state-of-the-art review. *Journal of Environmental Chemical Engineering, 9*, 106060.

Estay, H., Barros, L., & Troncoso, E. (2021). Metal sulfide precipitation: Recent breakthroughs and future outlooks. *Minerals, 11*, 1385.

Feng, C., Sugiura, N., Shimada, S., & Maekawa, T. (2003). Development of a high performance electrochemical wastewater treatment system. *Journal of Hazardous Materials, 103*, 65–78.

Fu, F., & Wang, Q. (2011). Removal of heavy metal ions from wastewaters: A review. *Journal of Environmental Management, 92*, 407–418.

Fu, F., Zeng, H., Cai, Q., Qiu, R., Yu, J., & Xiong, Y. (2007). Effective removal of coordinated copper from wastewater using a new dithiocarbamate-type supramolecular heavy metal precipitant. *Chemosphere, 69*, 1783–1789.

Gardea-Torresdey, J. L., de la Rosa, G., & Peralta-Videa, J. R. (2004). Use of phytofiltration technologies in the removal of heavy metals: A review. *Pure and Applied Chemistry, 76*, 801–813.

Ghanbari, R., Nazarzadeh Zare, E., Paiva-Santos, A. C., & Rabiee, N. (2023). $Ti_3C_2T_x$ MXene@MOF decorated polyvinylidene fluoride membrane for the remediation of heavy metals ions and desalination. *Chemosphere, 311*, 137191.

Ghosh, P., Samanta, A. N., & Ray, S. (2011). Reduction of COD and removal of Zn^{2+} from rayon industry wastewater by combined electro-Fenton treatment and chemical precipitation. *Desalination, 266*, 213–217.

Godt, J., Scheidig, F., Grosse-Siestrup, C., Esche, V., Brandenburg, P., Reich, A., & Groneberg, D. A. (2006). The toxicity of cadmium and resulting hazards for human health. *Journal of Occupational Medicine and Toxicology, 1*, 1–6.

González-Muñoz, M. J., Rodríguez, M. A., Luque, S., & Álvarez, J. R. (2006). Recovery of heavy metals from metal industry waste waters by chemical precipitation and nanofiltration. *Desalination, 200*, 742–744.

Gu, J.-D. (2018). *Mining, Pollution and Site Remediation*. Elsevier.

Guo, Q., Li, N., Bing, Y., Chen, S., Zhang, Z., Chang, S., Chen, Y., & Xie, S. (2018). Denitrifier communities impacted by heavy metal contamination in freshwater sediment. *Environmental Pollution, 242*, 426–432.

Han, M., He, J., Wei, X., Li, S., Zhang, C., Zhang, H., Sun, W., & Yue, T. (2022). Deep purification of copper from Cu (II)-EDTA acidic wastewater by Fe (III) replacement/diethyldithiocarbamate precipitation. *Chemosphere, 300*, 134546.

Hossain, M. S., Omar, F., Asis, A. J., Bachmann, R. T., Islam Sarker, M. Z., & Ab Kadir, M. O. (2019). Effective treatment of palm oil mill effluent using $FeSO_4.7H_2O$ waste from titanium oxide industry: Coagulation adsorption isotherm and kinetics studies. *Journal of Cleaner Production, 219*, 86–98.

Huang, F., Dang, Z., Guo, C.-L., Lu, G.-N., Gu, R. R., Liu, H.-J., & Zhang, H. (2013). Biosorption of Cd (II) by live and dead cells of Bacillus cereus RC-1 isolated from cadmium-contaminated soil. *Colloids and Surfaces B: Biointerfaces, 107*, 11–18.

Huisman, J. L., Schouten, G., & Schultz, C. (2006). Biologically produced sulphide for purification of process streams, effluent treatment and recovery of metals in the metal and mining industry. *Hydrometallurgy, 83*, 106–113.

Igder, A., Fazlavi, A., Allahkarami, E., & Dehghanipour, A. (2019). Optimization of Ni(II) & Co(II) removal from wastewater and statistical studies on the results of experimental designs. *Geosystem Engineering, 22*, 91–100.

Iqbal, J., Howari, F. M., Mohamed, A.-M. O., & Paleologos, E. K. (2021). Assessment of radiation pollution from nuclear power plants. In *Pollution Assessment for Sustainable Practices in Applied Sciences and Engineering.* Elsevier.

Islam, M. S., Hosen, M. M. L., & Uddin, M. N. (2019). Phytodesalination of saline water using *Ipomoea aquatica, Alternanthera philoxeroides* and *Ludwigia adscendens. International Journal of Environmental Science and Technology, 16*, 965–972.

Jaiswal, A., Verma, A., & Jaiswal, P. (2018). Detrimental effects of heavy metals in soil, plants, and aquatic ecosystems and in humans. *Journal of Environmental Pathology, Toxicology and Oncology,* 37.

Janwery, D., Memon, F. H., Memon, A. A., Iqbal, M., Memon, F. N., Ali, W., Choi, K.-H., & Thebo, K. H. (2023). Lamellar graphene oxide-based composite membranes for efficient separation of heavy metal ions and desalination of water. *ACS Omega, 8*, 7648–7656.

Järup, L. (2003). Hazards of heavy metal contamination. *British Medical Bulletin, 68*, 167–182.

Järup, L., & Åkesson, A. (2009). Current status of cadmium as an environmental health problem. *Toxicology and Applied Pharmacology, 238*, 201–208.

Jeyakumar, P., Debnath, C., Vijayaraghavan, R., & Muthuraj, M. (2023). Trends in bioremediation of heavy metal contaminations. *Environmental Engineering Research,* 28.

Juve, J.-M.A., Christensen, F. M. S., Wang, Y., & Wei, Z. (2022). Electrodialysis for metal removal and recovery: A review. *Chemical Engineering Journal, 435*, 134857.

Karaca, A., Cetin, S. C., Turgay, O. C., & Kizilkaya, R. (2010). Effects of heavy metals on soil enzyme activities. *Soil heavy metals*, 237–262.

Karri, R. R., Ravindran, G., & Dehghani, M. H. (2021). Chapter 1—Wastewater—Sources, Toxicity, and Their Consequences to Human Health. In Karri, R. R., Ravindran, G. & Dehghani, M. H. (Eds.) *Soft computing techniques in solid waste and wastewater management.* Elsevier.

Khan, F. S. A., Mubarak, N. M., Khalid, M., Walvekar, R., Abdullah, E. C., Mazari, S. A., Nizamuddin, S., & Karri, R. R. (2020). Magnetic nanoadsorbents' potential route for heavy metals removal—A review. *Environmental Science and Pollution Research, 27*, 24342–24356.

Khoshraftar, Z., Masoumi, H., & Ghaemi, A. (2023). Experimental, response surface methodology (RSM) and mass transfer modeling of heavy metals elimination using dolomite powder as an economical adsorbent. *Case Studies in Chemical and Environmental Engineering, 7*, 100329.

Kim, H., & Zhang, Q. (2019). Chemistry of new particle growth during springtime in the Seoul metropolitan area, Korea. *Chemosphere, 225*, 713–722.

Kobielska, P. A., Howarth, A. J., Farha, O. K., & Nayak, S. (2018). Metal–organic frameworks for heavy metal removal from water. *Coordination Chemistry Reviews, 358*, 92–107.

Koshlak, H. (2023). Synthesis of zeolites from coal fly ash using alkaline fusion and its applications in removing heavy metals. *Materials* [Online], 16.

Kovacs, H., & Szemmelveisz, K. (2017). Disposal options for polluted plants grown on heavy metal contaminated brownfield lands—A review. *Chemosphere, 166*, 8–20.

Koyuncu, I., Sengur, R., Turken, T., Guclu, S., & Pasaoglu, M. E. (2015). Advances in water treatment by microfiltration, ultrafiltration, and nanofiltration. In *Advances in membrane technologies for water treatment.* Elsevier.

Ku, Y., & Jung, I.-L. (2001). Photocatalytic reduction of Cr (VI) in aqueous solutions by UV irradiation with the presence of titanium dioxide. *Water Research, 35*, 135–142.

Kumar, A., Bisht, B. S., Joshi, V. D., & Dhewa, T. (2011). Review on bioremediation of polluted environment: A management tool. *International Journal of Environmental Sciences, 1*, 1079–1093.

Kumar, S. (2023). Carbon based nanomaterial for removal of heavy metals from wastewater: A review. *International Journal of Environmental Analytical Chemistry, 103*, 7961–7978.

Kumari, S., & Das, S. (2019). Expression of metallothionein encoding gene bmtA in biofilm-forming marine bacterium Pseudomonas aeruginosa N6P6 and understanding its involvement in Pb (II) resistance and bioremediation. *Environmental Science and Pollution Research, 26*, 28763–28774.

Kyzas, G. Z., & Matis, K. A. (2016). Electroflotation process: A review. *Journal of Molecular Liquids, 220*, 657–664.

Lapicque, F. (2004). Electrochemical engineering. *Chemical Engineering Research and Design, 12*, 1571–1574.

Lasat, M. M. (2002). Phytoextraction of toxic metals: A review of biological mechanisms. *Journal of environmental quality, 31*, 109–120.

Lebron, Y. A. R., Moreira, V. R., Santos, L. V. D. S., & Amaral, M. C. S. (2023). Membrane distillation and ion exchange combined process for mining wastewater treatment, water reuse, and byproducts recovery. *Chemical Engineering Journal, 466*, 143181.

Liu, Y., Wang, P., Gojenko, B., Yu, J., Wei, L., Luo, D., & Xiao, T. (2021). A review of water pollution arising from agriculture and mining activities in Central Asia: Facts, causes and effects. *Environmental Pollution, 291*, 118209.

Liu, Z., Chen, B., Wang, L.-A., Urbanovich, O., Nagorskaya, L., Li, X., & Tang, L. (2020). A review on phytoremediation of mercury contaminated soils. *Journal of Hazardous Materials, 400*, 123138.

Maanan, M., Saddik, M., Maanan, M., Chaibi, M., Assobhei, O., & Zourarah, B. (2015). Environmental and ecological risk assessment of heavy metals in sediments of Nador lagoon, Morocco. *Ecological Indicators, 48*, 616–626.

Maarof, H. I., Daud, W. M. A. W., & Aroua, M. K. (2017). Recent trends in removal and recovery of heavy metals from wastewater by electrochemical technologies. *Reviews in Chemical Engineering, 33*, 359–386.

Mahmud, H. N. M. E., Huq, A. K. O., & Binti Yahya, R. (2016). The removal of heavy metal ions from wastewater/aqueous solution using polypyrrole-based adsorbents: A review. *Rsc Advances, 6*, 14778–14791.

Mandich, M. (2018). Ranked effects of heavy metals on marine bivalves in laboratory mesocosms: A meta-analysis. *Marine Pollution Bulletin, 131*, 773–781.

Mansourri, G., & Madani, M. (2016). Examination of the level of heavy metals in wastewater of Bandar Abbas wastewater treatment plant. *Open Journal of Ecology, 6*, 55–61.

Masindi, V., & Muedi, K. L. (2018). Environmental contamination by heavy metals. *Heavy Metals, 10*, 115–132.

Matlock, M. M., Howerton, B. S., van Aelstyn, M. A., Nordstrom, F. L., & Atwood, D. A. (2002). Advanced mercury removal from gold leachate solutions prior to gold and silver extraction: A field study from an active gold mine in Peru. *Environmental Science & Technology, 36*, 1636–1639.

Medfu Tarekegn, M., Zewdu Salilih, F., & Ishetu, A. I. (2020). Microbes used as a tool for bioremediation of heavy metal from the environment. *Cogent Food & Agriculture, 6*, 1783174.

Mirbagheri, S. A., & Hosseini, S. N. (2005). Pilot plant investigation on petrochemical wastewater treatmentfor the removal of copper and chromium with the objective of reuse. *Desalination, 171*, 85–93.

Mitra, S., Chakraborty, A. J., Tareq, A. M., Emran, T. B., Nainu, F., Khusro, A., Idris, A. M., Khandaker, M. U., Osman, H., Alhumaydhi, F. A., & Simal-Gandara, J. (2022). Impact of heavy metals on the environment and human health: Novel therapeutic insights to counter the toxicity. *Journal of King Saud University—Science, 34*, 101865.

Mohammed, A. S., Kapri, A., & Goel, R. (2011). Heavy metal pollution: source, impact, and remedies. *Biomanagement of Metal-Contaminated Soils*, 1–28.

Mohan, S. V., Nithila, P., & Reddy, S. J. (1996). Estimation of heavy metals in drinking water and development of heavy metal pollution index. *Journal of Environmental Science and Health . Part A: Environmental Science and Engineering and Toxicology, 31*, 283–289.

Mojiri, A. & Bashir, M. J. K. (2022). *Wastewater treatment: Current and future techniques*. MDPI.

Monks, P. S., Granier, C., Fuzzi, S., Stohl, A., Williams, M. L., Akimoto, H., Amann, M., Baklanov, A., Baltensperger, U., & Bey, I. (2009). Atmospheric composition change–global and regional air quality. *Atmospheric Environment, 43*, 5268–5350.

Muchuweti, M., Birkett, J. W., Chinyanga, E., Zvauya, R., Scrimshaw, M. D., & Lester, J. N. (2006). Heavy metal content of vegetables irrigated with mixtures of wastewater and sewage sludge in Zimbabwe: Implications for human health. *Agriculture, Ecosystems & Environment, 112*, 41–48.

Munawer, M. E. (2018). Human health and environmental impacts of coal combustion and post-combustion wastes. *Journal of Sustainable Mining, 17*, 87–96.

Murphy, S. F., McCleskey, R. B., Martin, D. A., Holloway, J. M., & Writer, J. H. (2020). Wildfire-driven changes in hydrology mobilize arsenic and metals from legacy mine waste. *Science of the Total Environment, 743*, 140635.

Muthusaravanan, S., Sivarajasekar, N., Vivek, J. S., Paramasivan, T., Naushad, M., Prakashmaran, J., Gayathri, V., & Al-Duaij, O. K. (2018). Phytoremediation of heavy metals: Mechanisms, methods and enhancements. *Environmental Chemistry Letters, 16*, 1339–1359.

Nagajyoti, P. C., Lee, K. D., & Sreekanth, T. V. M. (2010). Heavy metals, occurrence and toxicity for plants: A review. *Environmental Chemistry Letters, 8*, 199–216.

Nedjimi, B. (2021). Phytoremediation: A sustainable environmental technology for heavy metals decontamination. *SN Applied Sciences, 3*, 286.

Nowack, B., Obrecht, J. M., Schluep, M., Schulin, R., Hansmann, W., & Köppel, V. (2001). Elevated lead and zinc contents in remote alpine soils of the Swiss National Park. *Journal of Environmental Quality, 30*, 919–926.

O'Connell, D. W., Birkinshaw, C., & O'Dwyer, T. F. (2008). Heavy metal adsorbents prepared from the modification of cellulose: A review. *Bioresource Technology, 99*, 6709–6724.

Obotey Ezugbe, E., & Rathilal, S. (2020). Membrane technologies in wastewater treatment: A review. *Membranes, 10*, 89.

Padaki, M., Murali, R. S., Abdullah, M. S., Misdan, N., Moslehyani, A., Kassim, M. A., Hilal, N., & Ismail, A. F. (2015). Membrane technology enhancement in oil–water separation. A review. *Desalination, 357*, 197–207.

Palmate, S. S., Pandey, A., Kumar, D., Pandey, R. P., & Mishra, S. K. (2017). Climate change impact on forest cover and vegetation in Betwa Basin, India. *Applied Water Science, 7*, 103–114.

Patel, P. K., Pandey, L. M., & Uppaluri, R. V. S. (2023). Cyclic desorption based efficacy of polyvinyl alcohol-chitosan variant resins for multi heavy-metal removal. *International Journal of Biological Macromolecules, 242*, 124812.

Priti, P., & Paul, B. (2016). Assessment of heavy metal pollution in water resources and their impacts: A review. *Journal of Basic and Applied Engineering Research, 3*, 671–675.

Priyadarshanee, M., & Das, S. (2021). Biosorption and removal of toxic heavy metals by metal tolerating bacteria for bioremediation of metal contamination: A comprehensive review. *Journal of Environmental Chemical Engineering, 9*, 104686.

Pruss, U., & Corralan, C. (2006). *Preventing diseases through healthy environment towards an estimate of the environmental burden of diseases*. World Health Organization.

Qasem, N. A. A., Mohammed, R. H., & Lawal, D. U. (2021). Removal of heavy metal ions from wastewater: A comprehensive and critical review. *Npj Clean Water, 4*, 36.

Qin, H., Hu, T., Zhai, Y., Lu, N., & Aliyeva, J. (2020). The improved methods of heavy metals removal by biosorbents: A review. *Environmental Pollution, 258*, 113777.

Rahman, A., Haque, M. A., Ghosh, S., Shinu, P., Attimarad, M., & Kobayashi, G. (2023a). Modified shrimp-based Chitosan as an emerging adsorbent removing heavy metals (chromium, nickel, arsenic, and cobalt) from polluted water. *Sustainability* [Online], 15.

Rahman, A., Yoshida, K., Islam, M. M., & Kobayashi, G. (2023b). Investigation of efficient adsorption of toxic heavy metals (chromium, lead, cadmium) from aquatic environment using orange peel cellulose as adsorbent. *Sustainability* [Online], 15.

Rajoria, S., Vashishtha, M., & Sangal, V. K. (2021). Review on the treatment of electroplating industry wastewater by electrochemical methods. *Materials Today: Proceedings, 47*, 1472–1479.

Ren, Y., Wei, J., Wu, Z., Ji, Y., Bi, F., Gao, R., Wang, X., Wang, G., & Li, H. (2021). Chemical components and source identification of PM2. 5 in non-heating season in Beijing: The influences of biomass burning and dust. *Atmospheric Research, 251*, 105412.

Rezai, B., & Allahkarami, E. (2021). Chapter 2—Wastewater treatment processes—Techniques, technologies, challenges faced, and alternative solutions. In: Karri, R. R., Ravindran, G., & Dehghani, M. H. (Eds.) *Soft computing techniques in solid waste and wastewater management.* Elsevier.

Robinson, B. H., Anderson, C. W. N., & Dickinson, N. M. (2015). Phytoextraction: Where's the action? *Journal of Geochemical Exploration, 151*, 34–40.

Rust, A. J., Roberts, S., Eskelson, M., Randell, J., & Hogue, T. S. (2022). Forest fire mobilization and uptake of metals by biota temporarily exacerbates impacts of legacy mining. *Science of the Total Environment, 832*, 155034.

Saleh, T. A., Mustaqeem, M., & Khaled, M. (2022). Water treatment technologies in removing heavy metal ions from wastewater: A review. *Environmental Nanotechnology, Monitoring & Management, 17*, 100617.

Saleh, T. A., Tuzen, M., & Sari, A. (2018). Polyamide magnetic palygorskite for the simultaneous removal of Hg (II) and methyl mercury; With factorial design analysis. *Journal of Environmental Management, 211*, 323–333.

Selvan Christyraj, J. R. S., Selvan Christyraj, J. D., Adhimoorthy, P., Rajagopalan, K. & Nimita Jebaranjitham, J. (2021). Impact of biomedical waste management system on infection control in the midst of COVID-19 pandemic. *The Impact of the COVID-19 Pandemic on Green Societies: Environmental Sustainability*, 235–262.

Shackira, A. M. & Puthur, J. T. (2019). Phytostabilization of heavy metals: Understanding of principles and practices. *Plant-Metal Interactions*, 263–282.

Shah, V., & Daverey, A. (2020). Phytoremediation: A multidisciplinary approach to clean up heavy metal contaminated soil. *Environmental Technology & Innovation, 18*, 100774.

Sharma, S., Nagpal, A. K., & Kaur, I. (2018). Heavy metal contamination in soil, food crops and associated health risks for residents of Ropar wetland, Punjab, India and its environs. *Food Chemistry, 255*, 15–22.

Shen, X., Dai, M., Yang, J., Sun, L., Tan, X., Peng, C., Ali, I., & Naz, I. (2022). A critical review on the phytoremediation of heavy metals from environment: Performance and challenges. *Chemosphere, 291*, 132979.

Sheoran, V., Sheoran, A. S., & Poonia, P. (2016). Factors affecting phytoextraction: A review. *Pedosphere, 26*, 148–166.

Shi, G. L., Zhu, S., Bai, S. N., Xia, Y., Lou, L. Q., & Cai, Q. S. (2015). The transportation and accumulation of arsenic, cadmium, and phosphorus in 12 wheat cultivars and their relationships with each other. *Journal of Hazardous Materials, 299*, 94–102.

Shrestha, R., Ban, S., Devkota, S., Sharma, S., Joshi, R., Tiwari, A. P., Kim, H. Y., & Joshi, M. K. (2021). Technological trends in heavy metals removal from industrial wastewater: A review. *Journal of Environmental Chemical Engineering, 9*, 105688.

Singh, J., & Kalamdhad, A. S. (2011). Effects of heavy metals on soil, plants, human health and aquatic life. *Int J Res Chem Environ, 1*, 15–21.

Singh, R. (2014). *Membrane technology and engineering for water purification: Application, systems design and operation.* Butterworth-Heinemann.

Singh, R., Venkatesh, A. S., Syed, T. H., Reddy, A. G. S., Kumar, M., & Kurakalva, R. M. (2017). Assessment of potentially toxic trace elements contamination in groundwater resources of the coal mining area of the Korba Coalfield, Central India. *Environmental Earth Sciences, 76*, 566.

Sonone, S. S., Jadhav, S., Sankhla, M. S., & Kumar, R. (2020). Water contamination by heavy metals and their toxic effect on aquaculture and human health through food Chain. *Letters in Applied NanoBioScience, 10*, 2148–2166.

Steliga, T., & Kluk, D. (2020). Application of *Festuca arundinacea* in phytoremediation of soils contaminated with Pb, Ni, Cd and petroleum hydrocarbons. *Ecotoxicology and Environmental Safety, 194*, 110409.

Stewart, C., Johnston, D. M., Leonard, G. S., Horwell, C. J., Thordarson, T., & Cronin, S. J. (2006). Contamination of water supplies by volcanic ashfall: A literature review and simple impact modelling. *Journal of Volcanology and Geothermal Research, 158*, 296–306.

Tchounwou, P. B., Yedjou, C. G., Patlolla, A. K., & Sutton, D. J. (2012). Heavy metal toxicity and the environment. In *Molecular, clinical and environmental toxicology: Volume 3: Environmental Toxicology*, 133–164.

Tepanosyan, G., Sahakyan, L., Belyaeva, O., Asmaryan, S., & Saghatelyan, A. (2018). Continuous impact of mining activities on soil heavy metals levels and human health. *Science of the Total Environment, 639*, 900–909.

Torres-Perez, J., Huang, Y., Bazargan, A., Khoshand, A., & McKay, G. (2020). Two-stage optimization of Allura direct red dye removal by treated peanut hull waste. *SN Applied Sciences, 2*, 475.

Tuchman, M., Silverberg, J. I., Jacob, S. E., & Silverberg, N. (2015). Nickel contact dermatitis in children. *Clinics in Dermatology, 33*, 320–326.

US EPA. (2009). National primary drinking water regulations: Drinking water regulations for aircraft public water systems. 74 FR 53590. October 19, 2009.

Valente, T. M., & Gomes, C. L. (2009). Occurrence, properties and pollution potential of environmental minerals in acid mine drainage. *Science of the Total Environment, 407*, 1135–1152.

Vardhan, K. H., Kumar, P. S., & Panda, R. C. (2019). A review on heavy metal pollution, toxicity and remedial measures: Current trends and future perspectives. *Journal of Molecular Liquids, 290*, 111197.

Venkatramanan, S., Chung, S. Y., Kim, T. H., Prasanna, M. V., & Hamm, S. Y. (2015). Assessment and Distribution of Metals Contamination in Groundwater: A Case Study of Busan City, Korea. *Water Quality, Exposure and Health, 7*, 219–225.

Verma, P., & Rawat, S. (2021). Rhizoremediation of heavy metal-and xenobiotic-contaminated soil: An eco-friendly approach. *Removal of emerging contaminants through microbial processes* (pp. 95–113).

Walker, C. H., Sibly, R. M., Hopkin, S. P., & Peakall, D. B. (2012). *Principles of ecotoxicology*. CRC Press.

Wang, J., & Chen, C. (2009). Biosorbents for heavy metals removal and their future. *Biotechnology Advances, 27*, 195–226.

Wang, X., He, S., Chen, S., Zhang, Y., Wang, A., Luo, J., Ye, X., Mo, Z., Wu, L., & Xu, P. (2018). Spatiotemporal characteristics and health risk assessment of heavy metals in $PM_{2.5}$ in Zhejiang Province. *International Journal of Environmental Research and Public Health, 15*, 583.

Wangi, G. M., Olupot, P. W., Byaruhanga, J., & Kulabako, R. (2023). Characterization of Natural Zeolite and Determination of its Ion-exchange Potential for Selected Metal Ions in Water. *Environmental Processes, 10*, 53.

Wilderer, P. A. (2010). *Treatise on water science*. Newnes.

Wołowicz, A., Staszak, K., & Hubicki, Z. (2023). Effect of anionic surfactants on the heavy metal ions removal by adsorption onto ion exchangers—Batch and column studies. *Journal of Water Process Engineering, 53*, 103792.

Xing, D. Y., Shen, Q., Tao, S., Zeng, H., Dong, W., & Sun, F. (2023). Aminoimidazole ionic liquid (IL) embellishment of nanofiltration (NF) membrane to simultaneously enhance heavy metals rejection and permeability. *Desalination, 549*, 116345.

Xu, X., & Mills, G. L. (2018). Do constructed wetlands remove metals or increase metal bioavailability? *Journal of Environmental Management, 218*, 245–255.

Yan, P., Mensah, J., Adesina, A., Kennedy, E. & Stockenhuber, M. (2020). Highly-dispersed Ni on BEA catalyst prepared by ion-exchange-deposition-precipitation for improved hydrodeoxygenation activity. *Applied Catalysis B: Environmental, 267*, 118690.

Yang, L., Hu, W., Chang, Z., Liu, T., Fang, D., Shao, P., Shi, H., & Luo, X. (2021). Electrochemical recovery and high value-added reutilization of heavy metal ions from wastewater: Recent advances and future trends. *Environment International, 152*, 106512.

Yang, T., Chen, M.-L., & Wang, J.-H. (2015). Genetic and chemical modification of cells for selective separation and analysis of heavy metals of biological or environmental significance. *TrAC Trends in Analytical Chemistry, 66*, 90–102.

Yaqub, M., & Lee, S. H. (2019). Heavy metals removal from aqueous solution through micellar enhanced ultrafiltration: A review. *Environmental Engineering Research, 24*, 363–375.

Zhang, X. (2014). Management of coal combustion wastes. *IEA Clean Coal Centre*, 2–68.

Zhao, M., Xu, Y., Zhang, C., Rong, H., & Zeng, G. (2016). New trends in removing heavy metals from wastewater. *Applied Microbiology and Biotechnology, 100*, 6509–6518.

Zheng, J., Zhang, X., Li, G., Fei, G., Jin, P., Liu, Y., Wouters, C., Meir, G., Li, Y., & van der Bruggen, B. (2022). Selective removal of heavy metals from saline water by nanofiltration. *Desalination, 525*, 115380.

Zhou, D. M., Hao, X. Z., Xue, Y., Cang, L., Wang, Y. J., & Chen, H. M. (2004). Advances in remediation technologies of contaminated soils. *Ecology and Environmental Science, 13*, 234–242.

Chapter 4
Nature Base Sustainable Solution for Heavy Metals Toxicity and Its Impacts on Humans and Plants Health

Ratna Chauhan, Sanjay Sarjerao Kale, Ratan Singh, Dheeraj Rathore, Bernard R. Glick, and Indra Jeet Chaudhary

Abstract Urbanization and industrial development are the major contributors to global environmental pollution, especially heavy metals. Metals are typically a part of soil; some are beneficial to plants, while some are toxic to both humans and plants. Once higher concentrations of metals enter into the environment, they become toxic. Then only plants can be removed by accumulation and detoxification processes. Nowadays, heavy metal contamination is a global problem, particularly in developing nations like India. Heavy metal toxicity reported worldwide has caused harmful effects on life, including plants and animals. Therefore, the chapter focuses on sources of heavy metals and their influences on plant and human health. This chapter also focuses on metal remediation technologies and the metal absorption capabilities of urban tree species. The technique identification of the higher metal accumulation capability of trees will be helpful for metal remediation from urban areas' environments (air, water, and soil). Because trees can store elements long-term, we can recreate historical changes in heavy metals (HMs) concentrations in ecosystems.

R. Chauhan · S. S. Kale · I. J. Chaudhary (✉)
Department of Environmental Science, Savitribai Phule Pune University (University of Pune), Pune, Maharashtra, India
e-mail: indrajeet.cug@gmail.com

R. Chauhan
e-mail: ratnasingh.bhu@gmail.com

S. S. Kale
e-mail: sanjakale@gmail.com

R. Singh · D. Rathore
School of Environment and Sustainable Development, Central University of Gujarat, Gandhinagar, Gujarat, India
e-mail: ratansingh459@gmail.com

D. Rathore
e-mail: dheeraj.rathore@cug.ac.in

B. R. Glick
Department of Biology, ESC 350, University of Waterloo, Waterloo, ON, Canada
e-mail: glick.bernard@gmail.com

V. Sunitha et al. (eds.), *Groundwater Resource Management Planning Strategies*, Springer Water, https://doi.org/10.1007/978-3-031-88870-0_4

Therefore, this chapter will provide information on minimizing HMs from air, soil, and water, as well as human health impacts and environmental sustainability.

Keywords Heavy metal pollution · Accumulation · Health risks · Remediation technology

4.1 Introduction

Urban areas have significant environmental issues due to unplanned and rapid development, particularly in developing nations (Cheshmehzangi et al., 2021). The biggest problem facing the majority of urban areas is air pollution worldwide. Urban transportation and industrial processes contribute to the discharge of particulate matter containing heavy metals and gaseous pollutants into the atmosphere of cities (Chaudhary & Rathore, 2018, 2019; Jeong et al., 2020). HMs are often absorbed by carbonates, minerals, and organic matter in soil. Therefore, industrial and automobile emissions of heavy metals have a negative impact on the environment (Etikala et al., 2022; Harrison et al., 2021; Peng et al., 2022). People are exposed to heavy metals through air pollution and other components of the polluted environment, including soil and water (Reddy et al., 2022; Suvarna et al., 2023). Since they can affect health, HM concentrations in the environment should be checked (Saroop & Tamchos, 2021).

Heavy metals, which can be toxic or deadly, pollute the air, water, and soil and are highly detrimental to plants and animals. Examples of HMs are Pb, Cd, Hg, As, etc. (which is a metalloid). These metals are considered the "main threats" to human plant and animal health (Reddy & Sunitha, 2023; Suvarna et al., 2023). Plant roots are the key interaction part of the food chain because they are absorbed by the primary producer and consumed by the consumer. HMs, on the other hand, are taken up directly by the leaves in aquatic systems (Nguyen et al., 2021). The nature, characteristics, and influence of heavy metal soil pollution on plant growth are briefly discussed in this chapter.

Plants are constantly utilized as bio-monitors of HMs. Polluted plants exhibit symbols of heavy metal buildup in many regions of their bodies (Kumar et al., 2022). Different plant tissues of various plants can absorb heavy metals more effectively than others. Consequently, choosing plants in pollution remediation programs in polluted areas is crucial (Yan et al., 2020). Almost all human activities, in addition to those that occur naturally, have the potential to produce heavy metals as by-products. Events that contribute to the pollution of ecosystems include the movement of these contaminants into uncontaminated areas by leachates and sewage sludge (Rajendran et al., 2022).

Some metals [K (I), Na (I), Mg (II), and Ca (II)] play vital roles as nutrients in the plant body. Similar micronutrients (Zn, Cu, iron, Mn, etc.) can also be harmful if they exist in excess (Gupta & Gupta, 2020). On the other hand, these ions typically have significant and advantageous effects on human metabolism. In enzymatic pathways, HMs can function as inhibitors. An estimated 50% of all enzymes need a metal cofactor to function and be active (Jomova et al., 2022). Few definitions exist for "heavy metals." According to Khalef et al. (2022), metals and metalloids have comparatively great densities (< 5 g/cm^3) and the capability to bioaccumulate along the food chain. They are highly poisonous to living things. Some researchers propose using "potentially toxic elements" instead of the contentious "heavy metals." This category includes naturally occurring and artificially produced poisonous metals (Cr, Cd, As, Ni). These metals cause problems (intestinal, lung, circulatory, propagative, renal, and neurological) in the human body (Negahdari et al., 2021). Certain heavy metals decrease cancer's sensitivity to treatment and accelerate the growth of the disease through many pathogenetic pathways (Bhat et al., 2024). These metals produce oxidative stress, which increases the oxidative damage in a cell and damages proteins, lipids, and DNA molecules. It also promotes the development of cancer (Kasprzak, 2023). Due to industrialization, Toxic metals in soil, water, and air rise. This results in a direct and indirect hazard to human health. Despite that org, organizations like EEA, WHO, and EMA give their best efforts for their management. Nearby, several procedures are being used to remove environmental contaminants, but most are very expensive and perform much below optimally. Therefore, planting and designing urban trees around and in urban areas are the best tools for sustainable city development. The HMs-contaminated land improved by potential trees provides research progressive knowledge and improvement of urban air quality. This chapter will explain heavy metal management techniques, their health impact, and environmental sustainability.

4.2 Sources of HMs

HMs may come from natural and human activities. HMs have been the subject of extensive study efforts due to their persistence and status as one of the most extensively distributed industrial contaminants. Many components in the atmosphere significantly contribute to particle emissions. Trace metals come from natural sources such as biogenic particles, salt spray, and continental dust (Pd, Cd, Zn, etc.) (Schwanck et al., 2022; Vithanage et al., 2022). Traffic air was shown to be the chief cause of atmospheric metals. Mining operations may be sources of airborne metal and metalloid contamination through direct wind erosion and mining operations (Punia, 2021).

HMs are generally well-defined as elements with an atomic numeral of more than 20 and metallic properties. The best usually observed HM contaminants are Cr, Cd, Cu, Pb, Zn, and Hg (Xiang et al., 2021). Soil naturally contains metals (Zn, Cu, Mn, Ni, and Co), which are a few of these metals that are essential micronutrients for plant progression. While Cd, Hg, and Pb have unclear biological functions (Alengebawy et al., 2021). HMs do not degrade through biodegradation and are harmful to biotic organisms. Since toxic metals like Pb, Co, and Cd cannot be biodegraded, they can instead accumulate in organisms. Even at low concentrations, they cause various diseases and problems (Rahman & Singh, 2020). Higher organisms risk many health complications due to heavy metals, which can persist in soil for thousands of years. They are also known to adversely influence soil microbiota, ground cover, and plant growth (Chinta & Araki, 2023). HMs must either be physically removed or chemically changed because they cannot be chemically broken down (Oyewo et al., 2020). Various sources of HMs were shown in Table 4.1.

4.3 Health Risk Assessment

Metals are dangerous, tenacious in the environment, and non-biodegradable. They harm humans, animals, and plants, causing them to become pollutants (Bharti & Sharma, 2022). In an ecosystem, heavy metals create a cyclical chain of contamination that includes humans, food, water, soil, industry, and the atmosphere (Bharti & Sharma, 2022). There are several ways that heavy metals can enter a person's body, including through contaminated food, drink, skin, and inhalation. Most of these water-soluble metals that are absorbed through the colon are then transported to different organs over the circulatory pathway. However, even at least concentrations of endothelial, epithelial, and other cells in the respiratory tract are impacted by heavy metals (Albano et al., 2022). Plant growth is also restricted by heavy metal pollution. Elevated metal concentrations in the soil harm plants and cause yield loss and physiological, metabolic, genetic, and seed germination processes (Bharti & Sharma, 2022). Plant toxicity varies according to plant characteristics, pH, and soil composition (Fig. 4.1).

4.3.1 Effect of Metals on Plants

Metal pollution is a problem caused by the gradual development of automobiles and industries because of the detrimental effects of heavy metal deposition in soils. The industry has issues with food safety and crop growth due to phytotoxicity (Bali et al., 2020). The root of the plant absorbs the soluble metals in the soil. Some metals are essential to plant growth, and excessive concentrations of these metals may be toxic to plants (Bali et al., 2020). Plants can acquire non-essential metals due to their ability to store them (Nunes da Silva et al., 2022). Since metals cannot be broken down, they negatively affect plants directly and indirectly when their concentrations are higher than ideal. For instance, high metal concentrations can harm cell structure by inhibiting cytoplasmic enzymes and causing oxidative stress (Pisoschi et al., 2021). Plant growth is adversely affected indirectly by the spare of essential nutrients (Gondek et al., 2020) by the cation exchange in plants. Heavy metals' indirect adverse effects negatively impact plant growth on soil microorganism development and function. This may be helpful for soil bacteria due to enhanced metal concentration, which could sluggishly depress the breaking of OM and lower soil fertility. Enzymatic activity is severely vulnerable to HM contacts. Therefore, soil

Table 4.1 Sources of heavy metals

S. No.	Heavy metal sources		References
1	Natural sources	Earth's crust (99% of metals include Si, O, Al, Fe, Ca, K, Na, Ti, Mg, and P) The other elements are referred to as "trace elements" and have concentrations of no more than 1000 mg/kg^{-1} (0–1%)	Abdullahi et al. (2021)
2	Anthropogenic sources	Petrochemical spills, unintentional leaks, agricultural practices, wastewater irrigation, sewage sludge, coal combustion, industrial discharge, activity of mine, dumping wastes, metal wastes disposal, paints, leaded petrol, and atmospheric deposition	Xiong et al. (2019)
3	Industrial sources	Mining, metal treating, refineries, the oil and petrochemical industry, coal burning, nuclear power plants, plastics, microelectronics, textiles, and pulp and paper industries	Alsafran et al. (2023)
4	Agricultural sources	**Fertilizers**: Plants require nutrients (N, P, K, S, Ca, and Mg). Certain soils lack important heavy metals (Co, Cu, Fe, Mn, Mo, Ni, and Zn) **Pesticides**: Approximately 10% of the fungicides and insecticides based on composites that contain Cu, Hg, Mn, Pb, or Zn are allowed for use in the United Kingdom **Biosolids and Manures**: Biosolids, including composts, manures, and municipal sewage sludge, pollute the soil with heavy metals (As, Cd, Cr, Cu, Pb, Hg, Ni, Se, Mo, Zn, etc.)	Alengebawy et al. (2021), Jayakumar et al. (2021)
5	Wastewater	Practice applying industrial, municipal, and related effluents to the land. Irrigation of land with wastewater containing metals might eventually cause heavy metal pollution in the soil	Sarwar et al. (2020)
6	Metal mining	The extensive mining and smelting of metal ores contaminate the land, endangering ecological and human health	Jayakumar et al. (2021)
7	Industrial wastes	Chemicals, textiles, tanning leather, metallurgy, and metal processing produce waste products that contain heavy metals	He et al. (2015)
8	Air-borne sources	Metals can enter the atmosphere through stack or duct emissions of gas vapor streams in the atmosphere. Metals from airborne sources end up as particles in the gas stream	Sankhla et al. (2016)

(continued)

Table 4.1 (continued)

S. No.	Heavy metal sources		References
9	Flooding and Sediment Deposition	Rivers carry suspended sediments containing heavy metals from different sources. When these rivers flood, a large portion of the suspended sediment in the water overflows onto land and deposits itself on alluvial soils	Ciszewski and Grygar (2016)
10	Contamination of urban soils	Urban areas located near multiple sources of pollution, including factories, power plants, heating plants, and roadways, frequently have higher soil concentrations of heavy metals Pb, Cd, and Zn are the most common heavy metals found in urban soils; however, other elements, including Cu, As, Ba, CD, Pb, Sb, Ti, and Hg, are also frequently found there	Liu et al. (2023)

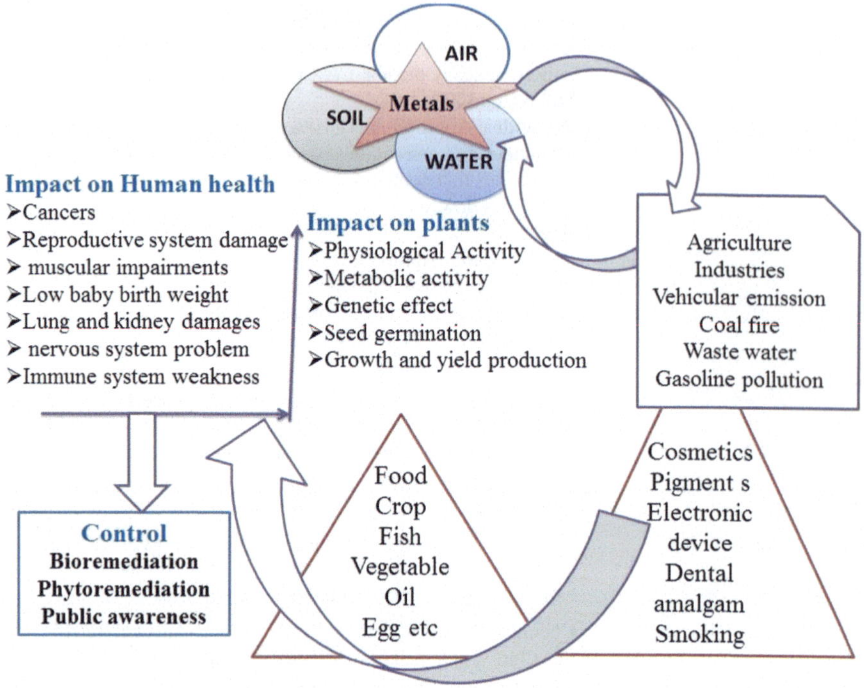

Fig. 4.1 Heavy metal exposure impacts human and plant health and its control technologies

microbial action and plant metabolism are vital. These detrimental effects on plant growth ultimately cause the plant to die (Alengebawy et al., 2021).

Different heavy metals have different effects on plants' growth and development. It has been discovered that certain metals, including Pb, Cd, Hg, and As, have damaging effects on plants, even at deficient concentrations in the different growth stages (Rashid et al., 2023). Many crops (growth and productivity) can also be obstructed by increased concentrations: (1) Zn and Cd cause oxidative damage and reduce the metabolic activity of plants; (2) Cu produces reactive oxygen species and oxidative stress; (3) Hg can cause physiological problems and apparent damage; (4) Cr inhibits the processes of photophosphorylation, electron transport, CO_2 fixation, and enzyme activity during photosynthesis; (5) Pb promotes aberrant plant morphology; (6) Ni throws off the plant's nutritional balance and causes problems with the activities of cell membranes; (7) Fe produces free radicals, which break down membranes, DNA, and proteins and permanently alters cell structures; causes wilting and necrosis of the leaves, which is followed by staining of the roots and a delay in the formation of new shoots (Teschke, 2022; Vaishnavi et al., 2023). For example, Cd toxicity inhibits wheat plant growth when Cd concentrations in the soil are as low (Li et al., 2020). Most minor photosynthetic activities, plant nutrition, and decreased specific enzyme activity may be responsible for most of the decline in plant growth indices in contaminated soils (Li et al., 2020).

The same heavy metal ions at higher concentrations and some others (Cd, Hg) are detrimental to the metabolic routes. Nevertheless, all living things are habitually sensitive to the richness of HMs. Globally, the impacts of metal contamination on plants have been studied (Suvarna et al., 2023). The potential harm that heavy metals can cause to the environment has made the toxic waste of farming soil by these compounds a significant concern of the environment. Such hazardous compounds are frequently present and have detrimental effects on plants grown (both acute and chronic). Metal ions are necessary for both organic and inorganic components of life, and life cannot exist without them. Numerous metals, including Fe, Co, Cu, Se, and Zn, are necessary to life. Plants contain little amounts of these substances, which support growth and metabolism. HMs enter through roots and are subsequently transported to various plant sections. Various HMs and their impact are given in Table 4.2.

4.3.2 Effect of Metals on Soil-Microbe Interaction

Other than pathogens associated with plants, soil microorganisms help them grow and develop correctly (Glick, 2020). Microbes linked with plants transform nutrients unavailable for direct plant use. In addition to aiding in the uptake of nutrients, soil microorganisms can trigger defense responses and suppress diseases that inhibit plant growth. The favorable interactions between soil nutrients, microbes, and plants make further research and the development of microbial inoculant technology in the

Table 4.2 Effects of certain HMs on plants

S. No.	Effect of HMs on plant	References
1	**Impacts of Nickel (Ni)**: Nickel is a small and essential constituent for plant growth. Nickel causes chlorosis and necrosis by interfering with the metabolism of Fe in plants. It also causes biomass loss, chloroplast damage, and reduced root branching	Hassan et al. (2019), Parwez et al. (2023)
2	**Impacts of Arsenic (As):** Arsenic harms plants in various ways, including reducing arsenate to arsenite within plant cells and interfering with aerobic phosphorylation. In addition to causing adverse effects on plant growth, crop yields, seed germination, and fruit yield, it increases leaf senescence and causes plant mortality	Nabi et al. (2021), Bharti and Sharma (2022)
3	**Impacts of Mercury (Hg)**: Mercury stress affects the photosynthetic machinery and reduces the uptake of photosynthetic elements (Fe and Mn), resulting in decreased chlorophyll contents, water absorption, and transpiration rate. Plants have many reactive sites for DNA binding, meaning DNA could bond with Hg and damage plant chromosomes	Khanam et al. (2020), Mao et al. (2021)
4	**Impacts of Lead (Pb)**: Minute amounts of lead penetrating plant cells can negatively impact physiological processes. It changes the membrane permeability, lowers the water potential, and decreases the activity of plant enzymes. Most plants' roots, leaves, stems, and seeds absorb lead. Pb reduces the dry mass of roots and shoots and their length, germination index, and germination percentage	Gholinejad et al. (2020)

agricultural field for better crop quality. The principal mediators of important biological activities in soil, including the breakdown of organic matter and mineralization, are heterotrophic microbial populations in the soil (Raza et al., 2023).

According to Bharti and Sharma (2022), toxic metals denaturize proteins and harm microbial cell peels, thereby intruding on plant cell growth, shape, and metabolism. There are a lot of publications on how heavy metals affect soil microbial processes. Still, there needs to be more data on how coexisting plant species might affect soil microbial dynamics when there is heavy metal contamination. The authors of one study (Bharti & Sharma, 2022) noted that various metal interactions, which could be antagonistic, additive, or competitive, exert a moderating influence on metal toxicity. Furthermore, pH affects the metal species' solubility, mobility, and bioavailability. Potentially highly harmful metals may not have an effect on the soil microbiota depending on the soil properties (such as pH). Because it influences soil microbiota, higher metals in soil are not necessarily a sign of pollution or pressure. It was discovered in several investigations that plant productivity, not plant diversity as was previously thought, was the driver of alterations in microbial richness and composition adversity. Accordingly, variations in plant communities frequently lead to variations in the kind and amount of OMs added to the soil (Gunina & Kuzyakov, 2022).

Soil microorganisms with the proper catabolic capacities are eventually selected to mineralize the wide range of organic products that enter the soil. Microbial biomass, a significant constituent of microorganisms, is frequently thought of as a sensitive indication of changes in the size and composition of microbial communities. Long-term contact with metal pollution has been linked to a decline in soil microbial growth (Marzaioli et al., 2022).

4.3.3 *Effect of Metals on Human*

The environment naturally encompasses HMs. The least amounts of metals are helpful to humans. They are essential metals because they support biological activities such as the synthesis of hemoglobin, the transport of oxygen and electrons, cell metabolism, enzyme regulation, hormone and antioxidant production, and human cell proliferation (Jomova et al., 2022). The higher levels of heavy metals are harmful to humans. HMs adversely affect humans and are primarily ingested and inhaled through various channels (Reddy & Sunitha, 2023; Teschke, 2022). The biological structure and function of the soil deteriorate as a consequence of heavy metal contamination, which also negatively impacts agricultural output (Mao et al., 2021). Secondary ecological and environmental problems might arise from the careless discharging of contaminated soil into the atmosphere and water. Furthermore, the food chain can introduce heavy metals into the bodies of humans or other animals (Table 4.3). Human organs may be severely and fatally affected by excessive HM deposition, which can cause cancer, deformity, and mutation (Song et al., 2022).

4.4 Heavy Metal Remediation Technology

Various technologies are available to remediate HM contamination from soil and water. These technologies have definite outcomes: (i) degradation or destruction of pollutants; (ii) extraction of pollutants (iii) stabilization of pollutants (iv) unraveling uncontaminated constituents; and (v) limit exposure (Naidu et al., 2021; Ossai et al., 2020). Remediation tools for heavy metal-contaminated soils have advanced quickly, leading to the maturity of specific methods and more advanced studies on remediation mechanisms and influencing elements. Furthermore, several novel and efficient combination remediation solutions and an increasing number of successful real-world examples have gradually surfaced. For this reason, it is crucial to thoroughly summarise the developments made in soil and water remediation technology for sites contaminated with heavy metals. Thus, from the perspectives of physical, chemical, biological, and combined remediation are the mainstream use in this chapter. A thorough discussion is held on the influencing elements, benefits, drawbacks, remedial process, and practical application.

Table 4.3 Effects of certain heavy metals on human health

S. No	HMs human health risk	References
1	**Cadmium (Cd)** turns healthy epithelial cells into cancerous cells and becomes a carcinogen. Cadmium affects the manufacture of DNA, RNA, and proteins, causes anomalies related to the circulatory system, modifies hormone levels, and increases male infertility	Xiang et al. (2021)
2	**Nickel (Ni)**: Caused chromosomal damage, mutagenesis, carcinogenesis, hyperplasia, bronchitis, lumen, alveolar congestion, and inhibition of NK cell activity	Rizvi et al. (2020)
3	**Arsenic (As)**: Arsenic causes Diabetes, renal and hepatic failure, and neurological issues. Women are more likely than men to have skin problems caused by arsenic (dermatitis)	Bharti and Sharma (2022)
4	**Mercury (Hg)** is the most hazardous non-essential metal. It causes depression, tremors, skin epidemics, memory loss, cracking of hands and feet, and redness in children	Al Mazed et al. (2022)
5	**Lead (Pb)**: Pb exposure develops hypertension and cardiovascular disease. Pb produces two types of anemia in humans: hemolytic anemia, which occurs when a person is exposed to high levels of lead, and Frank anemia, which results from prolonged blood lead elevation	Sani and Amanabo (2021)

Additionally, the prospects are presented. The references in this chapter will help rationalize the selection and advancement of remediation technologies for sites contaminated with heavy metals. The basic technology used for metal removals is described below (Fig. 4.2).

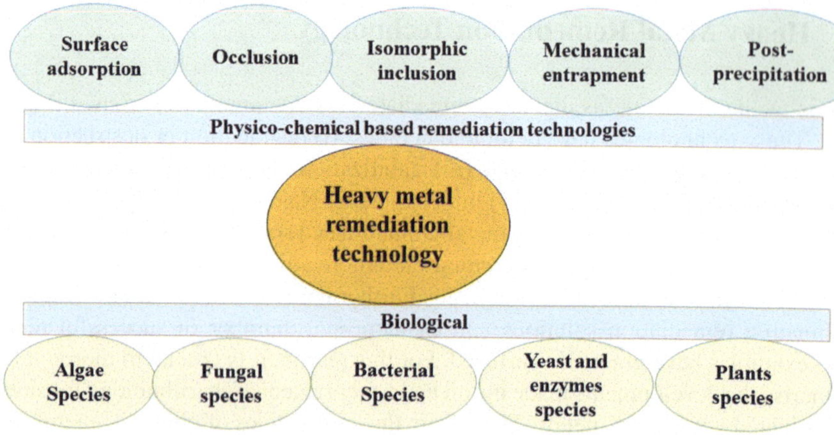

Fig. 4.2 Heavy metal remediation technologies

4.4.1 Physico-Chemical-Based Remediation Technologies

The elimination of HMs from wastewater can be achieved by known and frequently used chemical and physical procedures. This section will examine the approaches based on the physics and chemistry of remediation processes. Chemical precipitation (CP) adds chemicals to the soluble matter and creates an excessively saturated environment to transform it into an insoluble form. Redox combined with precipitation is a genetically comparable process and is the most used method for treating wastewater containing HMs (Singh et al., 2022). CP is a complex phenomenon that happens when hyper-saturated circumstances are generated. 3 phases of precipitation remain (1) nucleation, (2) crystal formation, and (3) flocculation. The challenge in CP operations is for the wastewater producers to first remove organic and inorganic contaminants from the sewage. Then to dispose of the residual ingredients after the wastewater treatment process (Yadav et al., 2019). When a solid segment precipitates contaminants from a solution ordinarily soluble under the precipitation conditions. Because it may attach to nuclei and be removed with a solid basis in a single step, according to Singh et al. (2022), CP/adsorption is a co-removal technology used to remove contaminants from wastewater. Co-precipitation/adsorption and its main types are given in Table 4.4.

Earlier to the biodegradation of potentially dangerous organic composites, CP was active as a pre-treatment technique to eliminate organic and inorganic pollutants. Five main techniques mainly were used for this goal. These processes include, for example, the treatment of xanthate, hydroxides, carbonates, and sulfides of metals. Additionally, these processes were sometimes merged. The most commonly used CP procedure for the removal of dyes and HMs is hydrogen precipitation (Balladares et al., 2018). It is possible to extract HMs and dyes from a solution using flocculation (FoC) or coagulation (CoG). The process of slowly combining destabilized particles to promote accumulation is known as flocculation. According to Iwuozor (2019), the CoG technique uses ferrous sulfate and ferric chloride, which efficiently eliminates organic and inorganic contaminants, especially dyes. Nevertheless, this method fails regarding acidic dyes (Sun et al., 2020). The goal is to reduce the surface charge of dispersed materials in water and cause them to aggregate into larger particles, destabilizing the system. Its processes consist of three stages: flocculation, aggregation, and fast mixing. The coagulants are first thoroughly combined. The coagulants then neutralize the surface charge of the excellently discrete particles. Next, using flocculants and moderate mixing, lesser particles are brought nearer calm to produce bigger particles. Sedimentation is then frequently employed to get rid of the larger particles. Many different kinds of coagulants and flocculants are utilized to treat wastewater. Supplements containing coagulants are commonly used to expedite the remedial process. Coagulants can made from metal, polymers, and naturally occurring substances. In current years, coagulants such as aluminum sulfate, calcium chloride, polymer poly aluminum chloride, poly aluminum chloride sludge, and other naturally occurring compounds have all been employed to eliminate HMs from wastewater (Ang and Mohammad, 2020; Nath et al., 2021).

Table 4.4 Physicochemical-based remediation technologies

No	Co-precipitation / Adsorption	Processes
1	Surface adsorption	Instead of being incorporated into the underlying crystal structure, toxins are adsorbed onto the residues outside the surface. This technique involves two components: (1) refute overlay, which is reserved maximum or least flexibly, and (2) essential ion adsorbed on the surface, which would be recollected strongly. The surface properties of the emerging solid phase attract or repel subsidiary components in the surrounding aqueous media
2	Occlusion	In this instance, the impurities are adsorbed during crystal development rather than absorbed onto the crystal lattice, which produces crystal faults. The majority of the obstruction is dictated by adsorption procedures that take place during crystal formation
3	Isomorphic inclusion	An alternative term for this is diverse crystal development. The impurity integrates keenly with the precipitate's matrix construction, replacing a lattice ion with similar dimensions and chemical characteristics. Dirt is always absorbed into the mineral matrix, creating a heterogeneous crystal
4	Mechanical entrapment	Microscopic flaws or hollows arise through the fast growth and accumulation of the crystals physically enclosing a minor portion of the parental liquid. The primary liquid solution occupies the pores till the precipitate completely enfolds them
5	Post-precipitation	When the residue and parent fluid come into contact, a second substance eventually precipitates with the precipitating mixture. Surface adsorption and this type of precipitate contamination are closely related

Membranes have gained popularity as filtering devices that may remove colored dyes and HMs from wastewater. Ultrafiltration (UF) is employed when transmembrane operating pressures (TMP) are low. Additives may be assured to the metals and dyes to enhance their size since the UF membrane apertures are expected to be larger than the dyes and metals (Singh et al., 2022). Another method to enhance UF is to add surfactants to the mixture. This method is called micellar-enhanced ultrafiltration (MEUF). Micelles are formed when surfactant molecules group; the micelles then charge the molecules. This enables the membrane to retain the contaminants and surfactants readily. The optimal wastewater for MEUF is one with a low concentration of HMs. The performance of MEUF is affected by several factors, including the addition of solutes, the kind of surfactant, the process, and the type of membrane used (Rahmati et al., 2017). Excellent polymer bonding, extraction, sustainability, reutilization of polymers, low energy requirements, and moderate operating costs are all demonstrated by Polymer-Enhanced UF (PEUF) (Yang et al., 2020). Finding suitable water-soluble polymer (WSP) continues to be the most challenging part of developing machinery. Nanofiltration (NF) extracts dissolved materials with more than 200 Da and concentrates ingredients with molecular weights more significant than 1000 Da. Because of this, the NF's operative range is situated in the middle of the UF and RO methods (Qasem et al., 2021). The use of the membrane with microspores to filter out micrometer-sized contaminants and other toxins from wastewater is known as microfiltration. Like other low-pressure membrane processes, the MF technique uses membrane holes ranging in size from 0.1 to 10 μm (Urošević and Trivunac, 2020). However, the application of MF in eliminating contaminants has not gotten much attention due to its poor removal capabilities. RO is a pressure-driven separation technique that uses a semi-permeable membrane (pore size 0.5–1.5 nm) to let mostly tiny particles through. The RO process turns the conventional osmosis process on its head by applying pressure (20–70 bar) more significantly than the osmotic pressure of the feed solution. Samaei et al. (2020) report that the inhibited solutes typically have a molecular size (0.00025–0.003 μm), and its process can remove 95–99% of organic and inorganic salts. The rejection rate of the RO process is high and compact. Membrane fouling and deterioration are the primary issues with RO systems (Samaei et al., 2020).

One effective method for eliminating HMs from wastewater is ion exchange. These techniques are ion exchange (ER, EC, EF, and EO) (Liu et al., 2019; Shahedi et al., 2020). Despite all the methods discussed, this method has drawn the most responsiveness. The chemical reaction is reversible, replacing ions in the wastewater for those on the resin to remove impurities. After submerging, the resin cradle must be replenished through an acid or essential medium to remove the impurity ions. The regenerated brine still needs to be appropriately treated or retrieved even though its volume and waste composition are lower than the original wastewater's. Wastewater with HMs is often passed through an ion exchange resin until all are soaked (Caluori, 2020).

Conversely, dispersed metals are a less desirable treatment option associated with ineffectiveness (Singh et al., 2022). The range that ion exchangers can handle is minimal, and methods for cleaning up have not yet been used extensively or actively. This technology has multiple benefits (refer to the supplementary material) and can be applied to removing HMs (Singh et al., 2022).

4.4.2 *Biological Treatment Techniques*

Utilizing naturally occurring biological processes and biological treatment techniques enables specific plants and microorganisms to eliminate metals. Since the middle of the 1970s, plant-based remediation techniques have been suggested for slurries of scoured substantial soils polluted with metal (Liu et al., 2018). Several studies (Schück & Greger, 2020; Thakare et al., 2021) expressed doubts over plants' capacity to remove metals. Biological activities employ many mechanical tools (oxidation/reduction adsorption and methylation) to terminate HMs from subsurface soil or groundwater (Wang et al., 2020). Tegene (2020) enumerates several methods of in-situ and ex-situ heavy metal bioremediation (Bioaccumulation, bioleaching, phytoremediation, bioventing, bio stimulation etc.).

One inexpensive, environmentally friendly way to reduce sludge generation is through bioremediation. This process is less chemically intensive, uses less energy, and can result in complete color mineralization. It is also more affordable. The quest for a biological wastewater treatment method that is economical, ecologically sustainable, and efficient led to the development of the biological adsorption technique. The most evident benefit of using a biological process to remove metals and dyes is that significant effluents may be treated with low concentrations of biological adsorbent, and no hazardous secondary compounds are produced. The quick operation time is an additional advantage. HM ions are passively taken in by living or dead biological materials. The dissolved molecules are adsorbed in the liquid phase (solvent, usually water) and the solid phase (biological adsorbent). The principal mechanism is the attraction of the adsorbent to the adsorbate, which can then be removed using various techniques, including mineral or organic acids, heat treatment, or electrochemical treatment (Patel, 2021).

Algae have high HM absorption ability, substrate selectivity, and sufficient mechanical properties for biological pathway adsorption. Brown algae were found to be the most productive and promising. Their core biological makeup is the explanation for their better performance. The properties of their cell wall components, in particular, are crucial for HM absorption. The biological process of metallic cation adsorption to a portion of the algal cell wall is a surface process. According to Geng et al. (2022), amino, carboxyl, sulfonate, and sulfhydryl are the main chemicals involved in metal adsorption. Algal cell walls contain structural polymers such as proteins, peptidoglycans, and polysaccharides (sulfated polysaccharides, alginic acid). Ion exchange is one of the main living adsorption tools used for HM by algae. However, other binding activities are also involved in the procedure of HM absorption. Such a type of process is known as complexation and micro-precipitation. A novel depurative technique called biological-based adsorption employing algae uses naturally occurring biological species to clean dye-contaminated water, other heavy metals, and industrial effluent.

Fungal cell walls are mainly made up of polysaccharides, ~ 80% of their dry weight, and sometimes even more than that amount. Due to numerous cell wall mechanisms with potent metal binding abilities, fungi can be active bio-sorbents (Priya et al., 2022). In addition to a small amount of glycoprotein, the cell walls of fungi include bulky volumes of chitosan, chitin, and glucan. These polymers are rich in metal-binding ligands (carboxyl, hydroxyl, amine, and phosphate functional groups). Fungal mycelium, which is made up of hyphae that resemble threads and is the vegetative portion of the fungus, has also been reported to exhibit biological adsorption of Zn^{2+} ions (Wang et al., 2023). Among the many benefits of using fungi in biological adsorption processes is their high biomass yield and short multiplication cycle, which make them easy to cultivate on a large scale. They are easy to grow with inexpensive growing material and straightforward fermentation techniques. Fungal biomass has a significant cost advantage over other biosorbents since it is easily accessible as an industrial waste product (Yaashikaa et al., 2021). Notably, the primary fungi used as biological adsorbents are safe to handle and are non-pathogenic. Numerous factors impact biological adsorption by fungi; each one functions differently and desires to account. The concentration of the initial solute, the type of biomass concentration, and physicochemical characteristics like pH, temperature, and ionic strength are all relevant factors. There are many different fungi, such as living and dead, free and immobilized, raw and pre-treated, lab cultures, industrial waste biomass, etc. Numerous studies have shown that fungi hold significant potential as biological adsorbents, mainly when cleaning industrial wastewater containing toxic heavy metals.

Most biological biomass on Earth comprises bacteria, which are common and diverse microorganisms (Fan et al., 2017). In commercial fermentation processing, bacteria are usually produced as a waste by-product and can be intentionally created in massive amounts. Numerous bacteria (*Pseudomonas*, *Bacillus*, *Streptomyces*, etc.) have had their HM absorption ability investigated (Iravani & Varma, 2020). Bacterial cell walls function as efficient HM chelating mediators. Additionally, polysaccharide gunk layers for HM binding are present in bacteria. Typically, two steps are tangled in the process of HMs attaching to the bacterial cell wall. While subsequent HM species accumulate at increasing concentrations in the second stage, the initial stage includes ions' interaction with reactive species on apparent cells. Generally speaking, HM deposition happens at the glutamic acid carboxyl groups in peptidoglycan. Furthermore, metabolism-independent biological adsorption is a significant pathway for the uptake of HMs in many bacteria. Although the adsorption process is metabolism-independent, metabolism-dependent pathways may promote the buildup of HMs on the surface of bacteria.

Biological pathway clean-up has several benefits for impurities (organic and inorganic). However, all of these have practically universal drawbacks. Not all materials can undergo total degradation in a short period. The by-products of biodegradation may sometimes be more hazardous. Many times, biological procedures are highly particular. Vital site considerations include the metabolically competent, ideal environmental circumstances, and ideal concentrations of contaminants and nutrients. Also, it is challenging to induce findings from pew and pilot-scale to large-scale research.

Further study is required to determine the suitability of biologically based remediation solutions for places with combinations of impurities that are not equally distributed in the environment. It is possible for contaminants to exist as gases, liquids, or solids. Compared to further treatment procedures like excavation and soil removal or incineration, it takes a lot longer. The definition of acceptable performance requirements in bioremediation protocols remains to be determined by regulators. Evaluating the efficacy of biologically based remediation is challenging since there is no consensus on what constitutes "clean" (Table 4.5).

Table 4.5 Biological methods of heavy metal removal

Bio-species	Heavy metal removal	References
Algal species		
Brown alga *S. filipendula*	Pb(II)	Verma et al. (2016)
Coelastrum sp.	Cd	Zheng et al. (2016)
Cystoseiracrinitophylla	Cu(III)	Christoforidis et al. (2015)
Scenedesmus quadricauda	Cd, Pb	Kipigroch et al. (2016)
Spirulina sp.	Hg	Kumar et al. (2020)
Thalassiosira sp./*Microcystis aeruginosa*	Cd(II)	Li et al. (2017)
Fungal species		
Ganoderma lucidum	AR	Yang et al. (2015)
Aspergillus niger/*Penicillium* sp.	RB5, CR, MG	Bilal and Asgher (2015)
Aspergillus niger/*Penicillium* sp.	Cr, Mg	Alam et al. (2018)
Bacterial species		
Sporosarcina kp-4 and kp-22	Cu, Zn, Ni, Cd	Qiao et al. (2021)
Sporosarcina pasteurii, Stenotrophomonas rhizophila, Variovoraxboronicumulans, V. boronicumulans	Zn, Pb, Cd	Jalilvand et al. (2020)
Paenibacillus sp. WP-1 and *Bacillus gaemokensis* strain EB28	Cu, Zn, Pb, Cr, Ni, and Cd	Yang et al. (2019)
Yeast and mushroom species		
Yeast strain W303α	Cu, Zn, Cd, Pb, Hg	Sun et al. (2020)
Saccharomyces cerevisiae	Cr	Ksheminska et al. (2006)
Rhodotorula mucilaginosa	Cr(VI)	Chatterjee et al. (2012)
Phanerochaete chrysosporium	Pb(II), Cd(II), Cu(II), Zn(II), Ni(II)	Huang et al. (2019)
Pleurotus ostreatus	Cd(II) and Cr	Li et al. (2017)

4.5 Metals Accumulation Capability of Urban Trees

The spread of contaminants in natural habitats harms the soil, water, and life, among other ecosystem components (Reddy & Sunitha, 2023; Singh et al., 2019). The chemical makeup, concentration, and stability of the pollutant are some of the aspects that affect the severity of the pollution (Lei et al., 2015). Human activity and HMs have been widely dispersed, resulting in an excess buildup that exceeds the legal limits for these compounds, creating major environmental problems (Akhtar et al., 2021). Despite having similar physical features, HMs are not easily defined. The toxicity of the compounds is frequently used to determine heavy metals. They can also be further categorized according to their weight, which is why they are called "heavy." Certain HMs (Pb, Cd, Hg, and As) continue to be dangerous for humans even at deficient amounts; the uptake of very high doses of all heavy metals is toxic to plants and causes heavy metal poisoning (Balali-Mood et al., 2021). Heavy metals have various characteristics that are exclusive to their category.

Along with their aesthetic value, plants used for treatment in cities with heavy traffic also save soil, act as windbreakers, provide for people and animals, are used as medicine, and lower air pollution (Sabancı, 2017). Plants are constantly utilized as biomonitors for the buildup of heavy metals. Polluted plants exhibit signs of heavy metal buildup in many regions of their structure (Kumar et al., 2022). Consequently, the best plants in contamination remediation programs in dirty areas are crucial (Yan et al., 2020). This work aims to ascertain the levels of HMs in plant leaves from urban areas.

Different amounts of HMs uptake by plants are known bioavailability (Table 4.6). The ensuing gathering of metal in tree tissues can be influenced by the cause of the metal pollution and the site's climatic circumstances (Günthardt-Goerg et al., 2019; Liu et al., 2023). The mobility of HMs in soils is predisposed to various influences (pH, OM biological activity, etc.) (Cui et al., 2021). A connection between metal accumulation and soil variables affects the overall translocation of HMs (Mleczek et al., 2009). Wang et al. (2023) found that spruce fine roots incline to take extra HMs than poplar roots. Ramzan et al. (2022) reported comparable results, identifying notable Cd and Zn absorption in several plant species. *Salix caprea* leaves were found to contain as much as Cd (116 mg) and Zn (4680 mg kg^{-1}). In contrast, metal concentrations in the soil did not relate to leaf concentration (Soares, 2022). According to several studies, Age impacts the quantity of heavy metals that trees absorb. *S. viminalis* accumulates heavy metals (Cd, Co, Cu, Cr, Ni, Pb, and Zn) in soils with varying quantities per study related to tree age (Mleczek et al., 2009). The best trees for rehabilitation were two to three years old since they demonstrated higher stress tolerance and greater adaptability than plants just one-year-old. According to Canals et al. (2017), this was explained by dilution caused by increased biomass making and metal gathering, which did not indicate a decrease in metal acclimatization time. These variations in growth rates over a whole year were connected to these willow and poplar modifications (Phillips et al., 2015). Furthermore, because of a comparatively high food intake growth pace, the early vegetative growth stage usually exhibited higher metal concentrations (Souri et al., 2019).

4.6 Conclusion

Metal pollution in the urban environment (air, water, and soil) is a burning problem worldwide. Heavy metals cause serious health effects on plants and humans. Industrialization and urbanization are the front sources of metal pollution. Once pollution enters the environment, only plants can remove it by bioaccumulation in plant tissues.

Table 4.6 Metal accumulation capability of urban trees

S. No.	Plant species and metal removal efficiency	References
1	Trends of metals (Zn, Cu, Pb, and Cd) accumulation in chronologically *Pinus eldarica* (7.74) > *Wisteria sinensis* (8.82) > and *Nigral morus* (27.15)	Alahabadi et al. (2017)
2	*F. nitida* and *E. globulus* trees absorb Pb and Cd from the air	El-Khatib et al. (2020)
3	*Cerasus yedoensis*, *Salix babylonica*, and *Nerium indicum* are the three plant species with the most significant accumulation of Cu, Zn, Cr, and Pb	Jia et al. (2021)
4	It was discovered that *Alstonia scholaris*, *Ficus religiosa*, *Psidium guajava*, and *Mangifera indica* were effective phytoextractors of copper. However, *Mangifera indica* and *Ficus religiosa* have high metal accumulation efficiency	Roy et al. (2020)
5	According to a study, the leaf's summer and winter HM absorption patterns are Zn > Pb > Cu > Cd. *Nicotiana glauca* (0.77), *Tamarix nilotica* (1.31), *Eucalyptus rostrate* (0.62), and Pb (0.95) had the highest bioconcentration factor values	El-Amier and Alghanem (2018)
6	The accumulation of Pb, Cr, Mn, Cu, and Zn was measured using four species of trees. The *R. typhina* leaves (29.682) and bark (12.407) had the highest metal accumulation index	Liu et al. (2022)
7	*S. babylonica* had the most significant mean concentrations of Cd, Ni, and Zn. At the same time, *A. hippocastanum* had the highest mean values of Pb and Mn. In traffic, *S. babylonica* had the highest Cd, Ni, and Zn concentrations	Turkyilmaz et al. (2020)
8	Accumulation trends of Cd chronologically *E. camaldulensis* > *M. alba* > *C. erectus* > *A. indica* > *P. deltoids* > *C. fistula*	Alam et al. (2022)

Therefore, plants' high accumulation ability and tolerance ability allow them to be used as metal scavengers. Various physical and chemical technologies were utilized for metal removal. The biological technologies for metal removal have several advantages: they are outmoded approaches, low cost, eco-friendly, and safe procedures. Therefore, the chapter concluded that higher sources of heavy metals were seen in industrial and traffic areas, and health issues were also found in those areas due to heavy metals. This chapter also gives the details of various remediation technologies. The method for identifying trees with higher metal accumulation capacities will help remove metals from urban areas' air, water, and soil. We can replicate past variations in the concentrations of heavy metals (HMs) in ecosystems because trees can store elements for an extended period. As a result, this chapter will provide information on how to reduce the amount of heavy metal pollution in the air, soil, and water, as well as how to protect human health and the environment.

Acknowledgements The authors wish to thank the Department of Environmental Science, Savitribai Phule Pune University, Pune, Maharashtra for the platform that enabled this interesting work.

References

Abdullahi, A., Lawal, M. A., & Salisu, A. M. (2021). Heavy metals in contaminated soil: Source, accumulation, health risk, and remediation process. *Bayero Journal of Pure and Applied Sciences, 14*(1), 1–12.

Akhtar, N., Syakir Ishak, M. I., Bhawani, S. A., & Umar, K. (2021). Various natural and anthropogenic factors responsible for water quality degradation: A review. *Water, 13*(19), 2660.

Al Mazed, M., Rahman, M. A., & Ahmed, S. I. (2022). A review on effects of heavy metals on aquatic animals and public health significance. *Veterinary Sciences: Research and Reviews, 8*(2), 96–104.

Alahabadi, A., Ehrampoush, M. H., Miri, M., Aval, H. E., Yousefzadeh, S., Ghaffari, H. R., & Hosseini-Bandegharaei, A. (2017). A comparative study on the capability of different tree species in accumulating heavy metals from soil and ambient air. *Chemosphere, 172*, 459–467.

Alam, M. Z., Khan, M. J. H., Kabbashi, N. A., & Sayem, S. M. A. (2018). Development of an effective biosorbent by fungal immobilization technique for removal of dyes. *Waste Biomass Valorization, 9*(4), 681–690.

Alam, M. R., Rahman, M. M., Tam, N. F. Y., Yu, R. M. K., & MacFarlane, G. R. (2022). The accumulation and distribution of arsenic species and selected metals in the saltmarsh halophyte spiny rush (*Juncus acutus*). *Marine Pollution Bulletin, 175*, 113373.

Albano, G. D., Montalbano, A. M., Gagliardo, R., Anzalone, G., & Profita, M. (2022). Impact of air pollution in airway diseases: Role of the epithelial cells (cell models and biomarkers). *International Journal of Molecular Sciences, 23*(5), 2799.

Alengebawy, A., Abdelkhalek, S. T., Qureshi, S. R., & Wang, M. Q. (2021). Heavy metals and pesticides toxicity in agricultural soil and plants: ecological risks and human health implications. *Toxics, 9*(3), 42. https://doi.org/10.3390/toxics9030042. PMID: 33668829; PMCID: PMC7996329.

Alsafran, M., Saleem, M. H., Al Jabri, H., Rizwan, M., & Usman, K. (2023). Principles and applicability of integrated remediation strategies for heavy metal removal/recovery from contaminated environments. *Journal of Plant Growth Regulation, 42*(6), 3419–3440.

Ang, W. L., & Mohammad, A. W. (2020). State of the art and sustainability of natural coagulants in water and wastewater treatment. *Journal of Cleaner Production, 262*, 121267.

Balali-Mood, M., Naseri, K., Tahergorabi, Z., Khazdair, M. R., & Sadeghi, M. (2021). Toxic mechanisms of five heavy metals: Mercury, lead, chromium, cadmium, and arsenic. *Frontiers in Pharmacology, 12*, 227.

Bali, A. S., Sidhu, G. P. S., & Kumar, V. (2020). Root exudates ameliorate cadmium tolerance in plants: A review. *Environmental Chemistry Letters, 18*, 1243–1275.

Balladares, E., Jerez, O., Parada, F., Baltierra, L., Hernández, C., Araneda, E., & Parra, V. (2018). Neutralization and co-precipitation of heavy metals by lime addition to effluent from acid plant in a copper smelter. *Minerals Engineering, 122*, 122–129.

Bharti, R., & Sharma, R. (2022). Effect of heavy metals: An overview. *Materials Today: Proceedings, 51*, 880–885.

Bhat, A. A., Moglad, E., Bansal, P., Kaur, H., Deorari, M., Thapa, R., Almalki, W. H., Kazmi, I., Alzarea, S. I., Kukreti, N., & Ali, H. (2024). Pollutants to pathogens: The role of heavy metals in modulating TGF-β signaling and lung cancer risk. *Pathology-Research and Practice, 256*, 155260.

Bilal, M., & Asgher, M. (2015). Dye decolorization and detoxification potential of Caalginate beads immobilized manganese peroxidase. *BMC Biotechnology, 15*, 111.

Caluori, M. (2020). Ion exchange equilibrium: Selectivity coefficient and ion exchange capacity, heavy metals removal, and mathematical modelling. Doctoral dissertation.

Canals, R. M., Emeterio, L. S., Durán, M., & Múgica, L. (2017). Plant-herbivory feedbacks and selective allocation of a toxic metal are behind the stability of degraded covers dominated by *Brachypodium pinnatum* in acidic soils. *Plant and Soil, 415*, 373–386.

Chatterjee, S., Chatterjee, N. C., & Dutta, S. (2012). Bioreduction of chromium (VI) to chromium (III) by a novel yeast strain *Rhodotorula mucilaginosa* (MTCC 9315). *African Journal of Biotechnology, 11*, 14920–14929.

Chaudhary, I. J., & Rathore, D. (2018). Suspended particulate matter deposition and its impact on urban trees. *Atmospheric Pollution Research, 9*(6), 1072–1082.

Chaudhary, I. J., & Rathore, D. (2019). Dust pollution: Its removal and effect on foliage physiology of urban trees. *Sustainable Cities and Society, 51*, 101696.

Cheshmehzangi, A., Butters, C., Xie, L., & Dawodu, A. (2021). Green infrastructures for urban sustainability: Issues, implications, and solutions for underdeveloped areas. *Urban Forestry and Urban Greening, 59*, 127028.

Chinta, Y. D., & Araki, H. (2023). Responses of bulk and rhizosphere soil microbiomes to different cover crop inputs and their connection and contribution to soil fertility and plant growth. *Pedobiologia, 101*, 150907.

Christoforidis, A. K., Orfanidis, S., Papageorgiou, S. K., Lazaridou, A. N., Favvas, E. P., & Mitropoulos, A. C. (2015). Study of Cu(II) removal by Cystoseiracrinitophylla biomass in batch and continuous flow biosorption. *Chemical Engineering Journal, 277*, 334–340.

Ciszewski, D., & Grygar, T. M. (2016). A review of flood-related storage and remobilization of heavy metal pollutants in river systems. *Water, Air, and Soil Pollution, 227*, 239. https://doi.org/10.1007/s11270-016-2934-8

Cui, X., Mao, P., Sun, S., Huang, R., Fan, Y., Li, Y., Li, Y., Zhuang, P., & Li, Z. (2021). Phytoremediation of cadmium contaminated soils by *Amaranthus hypochondriacus* L.: The effects of soil properties highlighting cation exchange capacity. *Chemosphere, 283*, 131067.

El-Amier, Y. A., & Alghanem, S. M. (2018). Tree leaves as bioindicator of heavy metal pollution from soil and ambient air in urban environmental. *Plant Archives, 18*(2), 2559–2566.

El-Khatib, A. A., Barakat, N. A., Youssef, N. A., & Samir, N. A. (2020). Bioaccumulation of heavy metals air pollutants by urban trees. *International Journal of Phytoremediation, 22*(2), 210–222.

Etikala, B., Madhav, S., & Somagouni, S. G. (2022). Urban water systems: An overview. *Current Directions in Water Scarcity Research, 6*, 1–19.

Fan, H.-L., Zhou, S.-F., Jiao, W.-Z., Qi, G.-S., & Liu, Y.-Z. (2017). Removal of heavy metal ions by magnetic chitosan nanoparticles prepared continuously via high-gravity reactive precipitation method. *Carbohydrate Polymers, 174*, 1192–1200.

Geng, Y., Cui, D., Yang, L., Xiong, Z., Pavlostathis, S. G., Shao, P., Zhang, Y., Luo, X., & Luo, S. (2022). Resourceful treatment of harsh high-nitrogen rare earth element tailings (REEs) wastewater by carbonate activated *Chlorococcum* sp. microalgae. *Journal of Hazardous Materials, 423*, 127000.

Gholinejad, B., Khashij, S., Ghorbani, F., Bandak, I., & Farajollahi, A. (2020). Effects of lead ions on germination, initial growth, and physiological characteristics of *Lolium perenne* L. species and its bioaccumulation potential. *Environmental Science and Pollution Research, 27*, 11155–11163.

Glick, B. R. (2020). *Beneficial plant-bacterial interactions* (2nd ed.). Heidelberg: Springer.

Gondek, M., Weindorf, D. C., Thiel, C., & Kleinheinz, G. (2020). Soluble salts in compost and their effects on soil and plants: A review. *Compost Science and Utilization, 28*(2), 59–75.

Gunina, A., & Kuzyakov, Y. (2022). From energy to (soil organic) matter. *Global Change Biology, 28*(7), 2169–2182.

Günthardt-Goerg, M. S., Vollenweider, P., Hermle, S., & Schulin, R. (2019). Growth and metal accumulation of young forest trees and understorey plants on contaminated topsoil: Influence of subsoil and time. *Plant and Soil, 437*, 375–395.

Gupta, P. K., & Gupta, P. K. (2020). Metals and micronutrients. In *Problem solving questions in toxicology: A study guide for the board and other examinations* (pp. 165–179)

Harrison, R. M., Allan, J., Carruthers, D., Heal, M. R., Lewis, A. C., Marner, B., Murrells, T., & Williams, A. (2021). Non-exhaust vehicle emissions of particulate matter and VOC from road traffic: A review. *Atmospheric Environment, 262*, 118592.

Hassan, M. U., Chattha, M. U., Khan, I., Chattha, M. B., Aamer, M., Nawaz, M., Ali, A., Khan, M. A. U., & Khan, T. A. (2019). Nickel toxicity in plants: Reasons, toxic effects, tolerance mechanisms, and remediation possibilities—A review. *Environmental Science and Pollution Research, 26*, 12673–12688.

He, Z., Meng, M., Yan, L., Zhu, W., Sun, F., Yan, Y., Liu, Y., & Liu, S. (2015). Fabrication of new cellulose acetate blend imprinted membrane assisted with ionic liquid ([BMIM] Cl) for selective adsorption of salicylic acid from industrial wastewater. *Separation and Purification Technology, 145*, 63–74.

Huang, Z., He, K., Song, Z., Zeng, G., Chen, A., Yuan, L., Li, H., & Chen, G. (2019). Alleviation of heavy metal and silver nanoparticle toxicity and enhancement of their removal by hydrogen sulfide in *Phanerochaete chrysosporium. Chemosphere, 224*, 554–561.

Iravani, S., & Varma, R. S. (2020). Bacteria in heavy metal remediation and nanoparticle biosynthesis. *ACS Sustainable Chemistry and Engineering, 8*(14), 5395–5409.

Iwuozor, K. O. (2019). Prospects and challenges of using coagulation–flocculation method in the treatment of effluents. *Advanced Journal of Chemistry-Section A, 2*, 105–127.

Jalilvand, N., Akhgar, A., Alikhani, H. A., Rahmani, H. A., & Rejali, F. (2020). Removal of heavy metals zinc, lead, and cadmium by biomineralization of urease-producing bacteria isolated from Iranian mine calcareous soils. *Journal of Soil Science and Plant Nutrition, 20*(1), 206–219.

Jayakumar, M., Surendran, U., Raja, P., Kumar, A., & Senapathi, V. (2021). A review of heavy metals accumulation pathways, sources and management in soils. *Arabian Journal of Geosciences, 14*, 1–19.

Jeong, H., Choi, J. Y., Lee, J., Lim, J., & Ra, K. (2020). Heavy metal pollution by road-deposited sediments and its contribution to total suspended solids in rainfall runoff from intensive industrial areas. *Environmental Pollution, 265*, 115028.

Jia, M., Zhou, D., Lu, S., & Yu, J. (2021). Assessment of foliar dust particle retention and toxic metal accumulation ability of fifteen roadside tree species: Relationship and mechanism. *Atmospheric Pollution Research, 12*(1), 36–45.

Jomova, K., Makova, M., Alomar, S. Y., Alwasel, S. H., Nepovimova, E., Kuca, K., Rhodes, C. J., & Valko, M. (2022). Essential metals in health and disease. *Chemico-Biological Interactions, 367*, 110173.

Kasprzak, K. S. (2023). Oxidative DNA damage in metal-induced carcinogenesis. *Toxicology of Metals, 1*, 299–320.

Khalef, R. N., Hassan, A. I., & Saleh, H. M. (2022). Heavy metal's environmental impact. In *Environmental impact and remediation of heavy metals*. IntechOpen.

Khanam, R., Kumar, A., Nayak, A. K., Shahid, M., Tripathi, R., Vijayakumar, S., & Pathak, H. (2020). Metal (loid) s (As, Hg, Se, Pb and Cd) in paddy soil: Bioavailability and potential risk to human health. *Science of the Total Environment, 699*, 134330.

Kipigroch, K., Janosz-Rajczyk, M., & Skowron-Grabowska, B. (2016). The use of algae in the removal of Cd and Cu in the process of wastewater recovery. *Desalination and Water Treatment, 57*(3), 1508–1514.

Ksheminska, H., Honchar, T., Gayda, G., & Gonchar, M. (2006). Extra-cellular chromate reducing activity of the yeast cultures. *Open Life Sciences, 1*, 137–149.

Kumar, P., Goud, E. L., Devi, P., Dey, S. R., & Dwivedi, P. (2022). Heavy metals: Transport in plants and their physiological and toxicological effects. In *Plant metal and metalloid transporters* (pp. 23–54). Springer Nature Singapore.

Kumar, M., Singh, A. K., & Sikandar, M. (2020). Biosorption of Hg (II) from aqueous solution using algal biomass: Kinetics and isotherm studies. *Heliyon, 6*(1), e03321. https://doi.org/10.1016/j.heliyon.2020.e03321

Lei, Y., Chen, C. S., Tu, Y. J., Huang, Y. H., & Zhang, H. (2015). Heterogeneous degradation of organic pollutants by persulfate activated by CuO-Fe_3O_4: Mechanism, stability, and effects of pH and bicarbonate ions. *Environmental Science and Technology, 49*(11), 6838–6845.

Li, L., Zhang, Y., Ippolito, J. A., Xing, W., Qiu, K., & Wang, Y. (2020). Cadmium foliar application affects wheat Cd, Cu, Pb and Zn accumulation. *Environmental Pollution, 262*, 114329.

Li, X., Wang, Y., Pan, Y., Yu, H., Zhang, X., Shen, Y., Jiao, S., Wu, K., La, G., Yuan, Y., & Zhang, S. (2017). Mechanisms of Cd and Cr removal and tolerance by macrofungus *Pleurotus ostreatus* HAU-2. *Journal of Hazardous Materials, 330*, 1–8.

Liu, C., Wu, T., Hsu, P.-C., Xie, J., Zhao, J., Liu, K., Sun, J., Xu, J., Tang, J., Ye, Z., Lin, D., & Cui, Y. (2019). Direct/alternating current electrochemical method for removing and recovering heavy metal from water using graphene oxide electrode. *ACS Nano, 13*, 6431–6437.

Liu, H., Tang, J., Chen, T., Zhu, P., Sun, D., & Wang, W. (2023). Assessment of heavy metals contamination and human health risk assessment of the commonly consumed medicinal herbs in China. *Environmental Science and Pollution Research, 30*(3), 7345–7357.

Liu, L., Li, W., Song, W., & Guo, M. (2018). Remediation techniques for heavy metal-contaminated soils: Principles and applicability. *Science of the Total Environment, 633*, 206–219.

Liu, Y., Zhao, X., Liu, R., Zhou, J., & Jiang, Z. (2022). Biomonitoring and phytoremediation potential of the leaves, bark, and branch bark of street trees for heavy metal pollution in urban areas. *Environmental Monitoring and Assessment, 194*(5), 344.

Mao, Q., Tang, L., Ji, W., Rennenberg, H., Hu, B., & Ma, M. (2021). Elevated CO_2 and soil mercury stress affect photosynthetic characteristics and mercury accumulation of rice. *Ecotoxicology and Environmental Safety, 208*, 111605.

Marzaioli, R., Trifuoggi, M., & Rutigliano, F. A. (2022). Soil microbial biomass, activities and diversity in Southern Italy areas chronically exposed to trace element input from industrial and agricultural activities. *Applied Soil Ecology, 174*, 104392.

Mleczek, M., Rissmann, I., Rutkowski, P., Kaczmarek, Z., & Golinski, P. (2009). Accumulation of selected heavy metals by different genotypes of *Salix*. *Environmental and Experimental Botany, 66*(2), 289–296.

Nabi, A., Naeem, M., Aftab, T., Khan, M. M. A., & Ahmad, P. (2021). A comprehensive review of adaptations in plants under arsenic toxicity: Physiological, metabolic and molecular interventions. *Environmental Pollution, 290*, 118029.

Naidu, R., Biswas, B., Willett, I. R., Cribb, J., Singh, B. K., Nathanail, C. P., Coulon, F., Semple, K. T., Jones, K. C., Barclay, A., & Aitken, R. J. (2021). Chemical pollution: A growing peril and potential catastrophic risk to humanity. *Environment International, 156*, 106616.

Nath, A., Mishra, A., & Pande, P. P. (2021). A review of natural polymeric coagulants in wastewater treatment. *Materials Today: Proceedings, 46*, 6113–6117.

Negahdari, S., Sabaghan, M., Pirhadi, M., Alikord, M., Sadighara, P., Darvishi, M., & Nazer, M. (2021). Potential harmful effects of heavy metals as a toxic and carcinogenic agent in marine food—An overview. *Egyptian Journal of Veterinary Sciences, 52*(3), 379–385.

Nguyen, T. Q., Sesin, V., Kisiala, A., & Emery, R. N. (2021). Phytohormonal roles in plant responses to heavy metal stress: Implications for using macrophytes in phytoremediation of aquatic ecosystems. *Environmental Toxicology and Chemistry, 40*(1), 7–22.

Nunes da Silva, M., Machado, J., Osorio, J., Duarte, R., & Santos, C. S. (2022). Non-essential elements and their role in sustainable agriculture. *Agronomy, 12*(4), 888.

Ossai, I. C., Ahmed, A., Hassan, A., & Hamid, F. S. (2020). Remediation of soil and water contaminated with petroleum hydrocarbon: A review. *Environmental Technology and Innovation, 17*, 100526.

Oyewo, O. A., Elemike, E. E., Onwudiwe, D. C., & Onyango, M. S. (2020). Metal oxide-cellulose nanocomposites for the removal of toxic metals and dyes from wastewater. *International Journal of Biological Macromolecules, 164*, 2477–2496.

Parwez, R., Aftab, T., Khan, M. M. A., & Naeem, M. (2023). Exogenous abscisic acid fine-tunes heavy metal accumulation and plant's antioxidant defence mechanism to optimize crop performance and secondary metabolite production in *Trigonella foenum-graecum* L. under nickel stress. *Plant Science, 332*, 111703.

Patel, H. (2021). Review on solvent desorption study from exhausted adsorbent. *Journal of Saudi Chemical Society, 25*, 101302.

Peng, J. Y., Zhang, S., Han, Y., Bate, B., Ke, H., & Chen, Y. (2022). Soil heavy metal pollution of industrial legacies in China and health risk assessment. *Science of the Total Environment, 816*, 151632.

Phillips, C. J., Marden, M., & Lambie, S. M. (2015). Observations of "coarse" root development in young trees of nine exotic species from a New Zealand plot trial. *New Zealand Journal of Forestry Science, 45*, 13. https://doi.org/10.1186/s40490-015-0043-x

Pisoschi, A. M., Pop, A., Iordache, F., Stanca, L., Predoi, G., & Serban, A. I. (2021). Oxidative stress mitigation by antioxidants—An overview on their chemistry and influences on health status. *European Journal of Medicinal Chemistry, 209*, 112891.

Priya, A. K., Gnanasekaran, L., Dutta, K., Rajendran, S., Balakrishnan, D., & Soto-Moscoso, M. (2022). Biosorption of heavy metals by microorganisms: Evaluation of different underlying mechanisms. *Chemosphere, 307*, 135957.

Punia, A. (2021). Role of temperature, wind, and precipitation in heavy metal contamination at copper mines: A review. *Environmental Science and Pollution Research, 28*(4), 4056–4072.

Qasem, N. A., Mohammed, R. H., & Lawal, D. U. (2021). Removal of heavy metal ions from wastewater: A comprehensive and critical review. *NPJ Clean Water, 4*, 1–15.

Qiao, S., Zeng, G., Wang, X., Dai, C., Sheng, M., Chen, Q., Xu, F., & Xu, H. (2021). Multiple heavy metals immobilization based on microbially induced carbonate precipitation by ureolytic bacteria and the precipitation patterns exploration. *Chemosphere, 274*, 129661.

Rahman, Z., & Singh, V. P. (2020). Bioremediation of toxic heavy metals (THMs) contaminated sites: Concepts, applications and challenges. *Environmental Science and Pollution Research, 27*(22), 27563–27581.

Rahmati, N. O., PourafshariChenar, M., & Azizi Namaghi, H. (2017). Removal of free active chlorine from synthetic wastewater by MEUF process using polyethersulfone/titania nanocomposite membrane. *Separation and Purification Technology, 181*, 213–222.

Rajendran, S., Priya, T. A. K., Khoo, K. S., Hoang, T. K., Ng, H. S., Munawaroh, H. S. H., Karaman, C., Orooji, Y., & Show, P. L. (2022). A critical review on various remediation approaches for heavy metal contaminants removal from contaminated soils. *Chemosphere, 287*, 132369.

Ramzan, M., Ayub, F., Shah, A. A., Naz, G., Shah, A. N., Malik, A., Sardar, R., Telesiński, A., Kalaji, H. M., Dessoky, E. S., & Elgawad, H. A. (2022). Synergistic effect of zinc oxide nanoparticles and *Moringa oleifera* leaf extract alleviates cadmium toxicity in *Linum usitatissimum*: Antioxidants and physiochemical studies. *Frontiers in Plant Science, 13*, 900347.

Rashid, A., Schutte, B. J., Ulery, A., Deyholos, M. K., Sanogo, S., Lehnhoff, E. A., & Beck, L. (2023). Heavy metal contamination in agricultural soil: Environmental pollutants affecting crop health. *Agronomy, 13*(6), 1521.

Raza, T., Qadir, M. F., Khan, K. S., Eash, N. S., Yousuf, M., Chatterjee, S., & Oetting, J. N. (2023). Unrevealing the potential of microbes in decomposition of organic matter and release of carbon in the ecosystem. *Journal of Environmental Management, 344*, 118529.

Reddy, S., Sunitha, V., & Suvarna, B. (2022). Groundwater quality and its potential health impacts nearby inactive mines using PIG and geospatial technology, Southwestern part of Cuddapah basin, Andhra Pradesh, South India. *Groundwater for Sustainable Development, 17*, 100742.

Reddy, Y. S., & Sunitha, V. (2023). Assessment of Heavy metal pollution and its health implications in groundwater for drinking purpose around inactive mines, SW region of Cuddapah Basin, South India. *Total Environment Research Themes, 8*, 100069.

Rizvi, A., Parveen, S., Khan, S., & Naseem, I. (2020). Nickel toxicology with reference to male molecular reproductive physiology. *Reproductive Biology, 20*(1), 3–8.

Roy, A., Bhattacharya, T., & Kumari, M. (2020). Air pollution tolerance, metal accumulation, and dust capturing capacity of common tropical trees in commercial and industrial sites. *Science of the Total Environment, 722*, 137622.

Sabancı, P. (2017). An evaluation of user opinions on landscaping in Saklı Kent Case. Master's thesis, Eastern Mediterranean University (EMU)-Doğu Akdeniz Üniversitesi (DAÜ)

Samaei, S. M., Gato-Trinidad, S., & Altaee, A. (2020). Performance evaluation of reverse osmosis process in the post-treatment of mining wastewaters: Case study of Costerfield mining operations, Victoria Australia. *Journal of Water Process Engineering, 34*, 101116.

Sani, A. H., & Amanabo, M. (2021). Lead: A concise review of its toxicity, mechanism and health effect. *GSC Biological and Pharmaceutical Sciences, 15*(1), 055–062.

Sankhla, M. S., Kumari, M., Nandan, M., Kumar, R., & Agrawal, P. (2016). Heavy metals contamination in water and their hazardous effect on human health—A review. *International Journal of Current Microbiology and Applied Sciences, 5*(10), 759–766.

Saroop, S., & Tamchos, S. (2021). Monitoring and impact assessment approaches for heavy metals. In *Heavy metals in the environment* (pp. 57–86). Elsevier.

Sarwar, T., Shahid, M., Natasha, Khalid, S., Shah, A. H., Ahmad, N., Naeem, M. A., Ul Haq, Z., Murtaza, B., & Bakhat, H. F. (2020). Quantification and risk assessment of heavy metal buildup in soil-plant system after irrigation with untreated city wastewater in Vehari, Pakistan. *Environmental Geochemistry and Health, 42*, 4281–4297.

Schück, M., & Greger, M. (2020). Screening the capacity of 34 wetland plant species to remove heavy metals from water. *International Journal of Environmental Research and Public Health, 17*(13), 4623.

Schwanck, F., Simoes, J. C., Handley, M., Mayewski, P. A., & Bernardo, R. (2022). Anthropogenic trace elements (Bi, Cd, Cr, Pb) concentrations in a West Antarctic ice core. *Anais Da Academia Brasileira De Ciências, 94*, e20210351.

Shahedi, A., Darban, A. K., Taghipour, F., & Jamshidi-Zanjani, A. (2020). A review of industrial wastewater treatment via electrocoagulation processes. *Current Opinion in Electrochemistry, 22*, 154–169.

Singh, A., Pal, D. B., Mohammad, A., Alhazmi, A., Haque, S., Yoon, T., Srivastava, N., & Gupta, V. K. (2022). Biological remediation technologies for dyes and heavy metals in wastewater treatment: New insight. *Bioresource Technology, 343*, 126154.

Singh, S., Chaudhary, I. J., & Kumar, P. (2019). Utilization of low-cost agricultural waste for removal of toxic metals from environment: A review. *International Journal of Scientific Research in Biological Sciences, 6*, 4.

Soares, T. D. M. (2022). Influence of filter cake associated with mineral phosphate sources on soil phosphorus dynamics in sugarcane. Doctoral dissertation, Universidade de São Paulo.

Song, P., Xu, D., Yue, J., Ma, Y., Dong, S., & Feng, J. (2022). Recent advances in soil remediation technology for heavy metal contaminated sites: A critical review. *Science of the Total Environment, 838*, 156417.

Souri, M. K., Hatamian, M., & Tesfamariam, T. (2019). Plant growth stage influences heavy metal accumulation in leafy vegetables of garden cress and sweet basil. *Chemical and Biology Technologies in Agriculture, 6*, 25. https://doi.org/10.1186/s40538-019-0170-3

Sun, Y., Zhou, S., Pan, S.-Y., Zhu, S., Yu, Y., & Zheng, H. (2020). Performance evaluation and optimization of flocculation process for removing heavy metal. *Chemical Engineering Journal, 385*, 123911.

Suvarna, B., Sunitha, V., Sudharshan Reddy, Y., Muralidhara Reddy, B., Kadam, A. K., & Ramakrishna Reddy, M. (2023). Groundwater quality assessment using multivariate statistical approach and geospatial modelling around cement industrial corridor, South India. *International Journal of Environmental Science and Technology, 20*(5), 5051–5070.

Tegene, B. G. (2020). Mode of action, mechanism, and role of microbes in bioremediation service for environmental pollution management. *Journal of Biotechnology & Bioinformatics Research, 2*(3), 1–18.

Teschke, R. (2022). Aluminum, arsenic, beryllium, cadmium, chromium, cobalt, copper, iron, lead, mercury, molybdenum, nickel, platinum, thallium, titanium, vanadium, and zinc: Molecular aspects in experimental liver injury. *International Journal of Molecular Sciences, 23*(20), 12213.

Thakare, M., Sarma, H., Datar, S., Roy, A., Pawar, P., Gupta, K., Pandit, S., & Prasad, R. (2021). Understanding the holistic approach to plant-microbe remediation technologies for removing heavy metals and radionuclides from soil. *Current Research in Biotechnology, 3*, 84–98.

Turkyilmaz, A., Cetin, M., Sevik, H., Isinkaralar, K., & Saleh, E. A. A. (2020). Variation of heavy metal accumulation in certain landscaping plants due to traffic density. *Environment, Development and Sustainability, 22*, 2385–2398.

Urošević, T., & Trivunac, K. (2020). Achievements in low-pressure membrane processes microfiltration (MF) and ultrafiltration (UF) for wastewater and water treatment. In *Current Trends and Future Developments on (Bio-) Membranes* (pp. 67–107). Elsevier.

Vaishnavi, P. A., Prakash, N. U., Jayanthi, M., Suganthi, M., & Thenmozhi, M. (2023). Impact of metal oxide nanoparticles against biotic stress in plants. In *Nanometal Oxides in Horticulture and Agronomy* (pp. 101–140). Academic Press.

Verma, A., Kumar, S., & Kumar, S. (2016). Biosorption of lead ions from the aqueous solution by *Sargassum filipendula*: Equilibrium and kinetic studies. *Journal of Environmental Chemical Engineering, 4*, 4587–4599.

Vithanage, M., Bandara, P. C., Novo, L. A., Kumar, A., Ambade, B., Naveendrakumar, G., Ranagalage, M., & Magana-Arachchi, D. N. (2022). Deposition of trace metals associated with atmospheric particulate matter: Environmental fate and health risk assessment. *Chemosphere, 303*, 135051.

Wang, M., Li, Y., Zhao, D., Zhuang, L., Yang, G., & Gong, Y. (2020). Immobilization of mercury by iron sulfide nanoparticles alters mercury speciation and microbial methylation in contaminated groundwater. *Chemical Engineering Journal, 381*, 122664.

Wang, Y., Tong, L. L., Yuan, L., Liu, M. Z., Du, Y. H., Yang, L. H., & Guo, D. S. (2023). Integration of physiological, transcriptomic and metabolomic reveals molecular mechanism of *Paraisaria dubia* response to Zn_{2+} stress. *Journal of Fungi, 9*(7), 693.

Xiang, M., Li, Y., Yang, J., Lei, K., Li, Y., Li, F., Zheng, D., Fang, X., & Cao, Y. (2021). Heavy metal contamination risk assessment and correlation analysis of heavy metal contents in soil and crops. *Environmental Pollution, 278*, 116911.

Xiong, X., Liu, X., Iris, K. M., Wang, L., Zhou, J., Sun, X., Rinklebe, J., Shaheen, S. M., Ok, Y. S., Lin, Z., & Tsang, D. C. (2019). Potentially toxic elements in solid waste streams: Fate and management approaches. *Environmental Pollution, 253*, 680–707.

Yaashikaa, P. R., Kumar, P. S., Saravanan, A., & Vo, D. V. N. (2021). Advances in biosorbents for removal of environmental pollutants: A review on pre-treatment, removal mechanism and future outlook. *Journal of Hazardous Materials, 420*, 126596.

Yadav, M., Gupta, R., & Sharma, R. K. (2019). Chapter 14—Green and sustainable pathways for wastewater purification. In: Ahuja, S. (Ed.), *Advances in Water Purification Techniques* (pp. 355–383). Elsevier.

Yan, A., Wang, Y., Tan, S. N., Mohd Yusof, M. L., Ghosh, S., & Chen, Z. (2020). Phytoremediation: A promising approach for revegetation of heavy metal-polluted land. *Frontiers in Plant Science, 11*, 359.

Yang, C., Xu, W., Nan, Y., Wang, Y., Hu, Y., Gao, C., & Chen, X. (2020). Fabrication and characterization of a high-performance polyimide ultrafiltration membrane for dye removal. *Journal of Colloid and Interface Science, 562*, 589–597.

Yang, J., Yang, X., Lin, Y., Ng, T. B., Lin, J., & Ye, X. (2015). Laccase-catalyzed decolorization of malachite green: Performance optimization and degradation mechanism. *PLoS ONE, 10*, e0127714.

Yang, K., Zhu, L., Zhao, Y., Wei, Z., Chen, X., Yao, C., Meng, Q., & Zhao, R. (2019). A novel method for removing heavy metals from composting system: The combination of functional bacteria and adsorbent materials. *Bioresource Technology, 293*, 122095.

Zheng, H., Guo, W., Li, S., Wu, Q., Yin, R., Feng, X., Du, J., Ren, N., & Chang, J. S. (2016). Biosorption of cadmium by a lipid extraction residue of lipid-rich microalgae. *RSC Advances, 6*(24), 20051–20057.

Chapter 5
Heavy Metals in the Environment: Sources, Fate, and Health Implications

Mamilla Vijaya Kumar and Hemadri Prasad Raju

Abstract The abundance of heavy metals in the environment can be attributed to their natural occurrence as well as their numerous uses in agriculture, domestic and industrial sectors. Industrial untreated wastewater and other untreated water sources often contain high concentrations of heavy metals. When these untreated water and wastewaters are utilized for irrigation, heavy metals may build up in agricultural soil and poison crop plants. Heavy metal pollution of soils presents major risks to the ecosystem and can be harmful to consumers through skin contact and ingestion. The level, mode of exposure, chemical species, and age, gender, and nutritional status of the individual are among the many factors that influence the toxicity of heavy metals, because of their high level of toxicity, As, Cd, Cr, Pb, and Hg are regarded as priority metals with relevance to public health. Since certain metallic elements are known to harm organs even at lower exposure levels, they are classified as systemic intoxicants. The current book chapter discusses heavy metal pollution sources, the toxicological effects of heavy metals on human health, and the assessment of health risks to humans.

Keywords Heavy metals · Sewage and industrial effluents · Human health · Toxic effect · Risk assessment

M. Vijaya Kumar (✉)
Department of Civil Engineering, Sri Venkateswara Institute of Science and Technology, Kadapa, Andhra Pradesh, India
e-mail: civilvijay1@gmail.com

H. Prasad Raju (✉)
Department of Civil Engineering, Mohan Babu University (Erstwhile Sree Vidyanikethan Engineering College), Tirupati, Andhra Pradesh, India
e-mail: prasadrajuh@gmail.com

V. Sunitha et al. (eds.), *Groundwater Resource Management Planning Strategies*, Springer Water, https://doi.org/10.1007/978-3-031-88870-0_5

5.1 Introduction

The industrial untreated wastewaters and other untreated water sources such as storm water runoff through agricultural fields and near chemical factories often contain high concentrations of heavy metals. These waters may ultimately enter streams, rivers, ponds, lakes and artificial reservoirs. When the waters from these sources utilized for irrigation purpose, heavy metals may build up in agricultural soil and poison crop plants. Heavy metal pollution of soils presents major risks to the ecosystem and can be harmful to consumers through skin contact and ingestion. The level, mode of exposure, chemical species, and age, gender, and nutritional status of the individual are among the many factors that influence the toxicity of heavy metals, because of their high level of toxicity. As, Cd, Cr, Pb, and Hg are regarded as priority metals with relevance to public health. Since certain metallic elements are known to harm organs even at lower exposure levels, they are classified as systemic intoxicants.

Heavy metals are metallic substances with densities greater than those of water (Fergusson, 1990). Extreme industrialization as a strategy for development originated from the desire and greed of people in society being heightened by the recent decades' fast population expansion. Because of this, the world's natural resources have been continuously exploited. Manufacturing industries and home chores are the main drivers of wastewater generation, portable water is one of the mostly utilizes in these areas. Numerous deaths and disabilities are brought on by long-term exposure to heavy metals. A complex global issue is the existence of heavy metals affects people's health from a various source. Farmers everywhere, especially in India, water their crops with industrial effluents since there isn't a substitute supply. Micro and macronutrient concentrations, including heavy metals, are frequently detected in contain substantial quantities in the effluents of wastewater treatment plants that treat industrial and municipal wastewater. Conversely, certain microelements—such as iron, zinc, manganese, copper, and boron—are essential for stimulating plant cell growth. These heavy metals may become hazardous to plants if they build up to a high concentration (Parveen et al., 2014).

High levels of heavy metals are typically produced by significant chemical dumping industries, roads, and impure chemical fertilizer producing facilities. These metals can accumulate in the atmosphere, in plants, and in soil. As a result of burning lead petrol during transportation, lead is a heavy metal. Tires, engine parts, fuel tanks, and lubricating fluids coated with zinc are typical places to find additional toxic metals like nickel, zinc, and cadmium. Effluents from industries provide a new hazard to farming due to their release heavy metals into the soil, where they aggregate to substantial amounts in agricultural soil and water. Plants may be affected by the substantial amounts of toxic metals in these discharges and trash from industries as well as additional sources (Naz et al., 2016).

Living things are highly vulnerable to any chemical or biological alteration brought on by soil contamination. On the ground surface, contamination originates directly from wastewater or solid waste that contains heavy metals and pollutes the soil. The soil in the area deteriorates as a result of contamination from surface contaminants that seep through. Nonetheless, ground surface pollution is the primary source of soil contamination. The sources of this type of surface pollution include spills and dumping activities, leaks, and agricultural practices that seep into the soil (Patil et al., 2014).

Because of the detrimental consequences that heavy metal pollution has on crops and human health, heavy metal contamination in agricultural fields is currently a substantial problem. It is concerning that industrial regions have been established close to rural and agricultural communities, releasing a significant amount of pollutants into the surrounding ecosystem. Their prevalence in agricultural environments is caused by irrigation of fields with heavily polluted river water or industrial and municipal wastewater, and the application of different chemical fertilizers and insecticides. Heavy metals including Nikel, zinc, Chromium, Ferrous, Cobalt, Copper, Cadmium, Arsenic, Mercury and Lead may be present in agricultural soils affected by these operations (Malik & Maurya, 2014; Morais et al., 2012; Kumar et al., 2018). Because these metals are neither biodegradable nor thermally degradable, they remain in the soil for a longer amount of time and build up within the plant bodies (Kayastha, 2014; Shan et al., 2013; Vodyanitskii, 2013; Wu et al., 2010).

Based on their roles in the organisms, metals are classified as either essential or non-essential. They serve significant roles in a variety of oxidation–reduction reactions and function as co-factors in numerous enzymes. Although essential heavy metals are vital to an organism's physiology and biochemistry, insufficient concentrations of these metals can lead to a number of deficiency disorders (WHO, 1996). Mercury, cadmium, and lead are heavy metals that are not required by any species, not even in trace amounts. In contrast, certain heavy metals, including as Mn, Fe, Cu and Zn are vital micronutrients (Laws, 1945). Because of their toxicity and propensity to build up in food systems, certain heavy metals become problematic. Since fish, molluscs, and other aquatic species are at the base of the food chain, they have the potential to collect metals, which can then be absorbed by humans and cause acute or chronic illnesses (Forstner & Wittman, 1981; Jorgensen & Pedersen, 1994; Laws, 1945). Recent studies have revealed that a wide range of health issues can be brought on by even low concentrations of Al, As, Cd, Pb, Cd and Hg (Forstner & Wittman, 1981; Hassaan et al., 2016). It has been found that metal ions interact with DNA and nuclear proteins, causing damage and morphological changes to the latter that may lead to changes in the cell cycle, death, or cancer (Beyersmann & Hartwig, 2008; Chang et al., 1996; Wang & Shi, 2001).

5.2 Sources of Heavy Metal Pollution

Numerous compounds, such as silicates, carbonates, phosphates, hydroxides, oxides, organic matrix, and sulphides, are known to contain heavy metals. They come from both artificial and natural sources (Etikala et al., 2022, 2024; Langmuir et al., 1997; Reddy & Sunitha, 2023; Sparks, 2003; Selim et al., 2001). However, it is believed that man-made sources are the primary source and pathway through which heavy metals are distributed to other ecosystem areas. The three main media that heavy metals travel via are air, soil, and water.

Heavy metals enter the agricultural fields naturally through a variety of processes, including rock weathering, volcanic eruptions, wind-blown dust particles, sea spray, and aerosols (Nagajyoti et al., 2010). Anthropogenic sources include burning fossil fuels, field irrigation with wastewater from buildings and industries, organic manure, motor vehicle emissions, and various amounts of zinc, nickel, cadmium, and other elements in organic and inorganic fertilizers, pesticides, and fungicides (Kelepertzis, 2014; Malik et al., 2017; Nicholson et al., 2003; Toth et al., 2016; Zhang, 2006). Fertilizer and pesticide applications on a regular basis enrich agricultural soils with heavy metals. Several studies show correlation between the presence of heavy metals in food to the watering of farmed vegetables with chicken dung, chemical fertilizers, pesticides, and wastewater (Modaihsh et al., 2004). The following provides an explanation of the main heavy metal contamination sources.

5.2.1 Sources of Arsenic Pollution

Arsenic is a ubiquitous element that has been identified in almost all environmental matrices at low levels. Both natural occurrences like volcanic eruptions and soil erosion as well as human activity contribute to environmental arsenic poisoning. Many compounds containing arsenic are produced by industrial processes and are used for producing agricultural commodities such as dyes, insecticides, herbicides, fungicides, algicides, sheep dips, and wood preservatives (Tchounwou et al., 1999).

5.2.2 Sources of Cadmium Pollution

Cadmium is heavy metal that is extremely dangerous to both environment and human health. Cadmium is found on Earth with an average quantity of 0.1 mg/kg in the crust. Cadmium compounds are found in sedimentary rocks at the greatest concentrations found in the environment; marine phosphates have a cadmium level of approximately 15 mg/kg. In the industrial sector, cadmium is widely utilized to make batteries, alloys, and pigments (Wilson, 1988). Phosphate fertilizers and waste incinerators are major sources of cadmium in the environment (Jarup, 2003).

5.2.3 Sources of Chromium Pollution

The earth crust contains Chromium (Cr), an element that occurs naturally. Its valence states, or oxidation levels, range from Cr (II) to Cr (VI). Natural elemental chromium [Cr (0)] does not exist. Chromium is released into the air, water, and soil via a variety of man-made and natural sources, with industrial facilities being the primary source. The manufacturing of chromate, tannery facilities, stainless steel welding, ferrochrome, and chrome pigments are among the industries that contribute most to the emission of chromium. Airborne and wastewater emissions of Cr, primarily from the metallurgical industries, refractory, and chemical industries. Human activity releases hexavalent forms of chromium into the environment (Guertin et al., 2005).

5.2.4 *Sources of Lead Pollution*

Traces of lead, a bluish-grey metal found naturally in the earth's crust, are present. Although lead is found in the environment naturally, human activities such as mining, manufacturing, and burning fossil fuels release significant amounts of the metal into the atmosphere. Lead has several applications in the domestic, agricultural, and commercial domains. It can be used in many different applications, such as paint, glass, pigments, chemicals, lead-acid batteries, ammunition, metal products and radiation shielding systems. The industrial usage of lead has decreased dramatically in recent years, mainly from caulking, pipe solder, paints, and ceramic items (Gabby, 2003, 2006). Additionally, lead is employed in lead mining, drinking water pipes, and cultural practices (Mahuta, 2020).

5.2.5 *Sources of Mercury Pollution*

There are three distinct types of mercury found in nature: elemental, inorganic, and organic. Each has a unique toxicity profile (Clarkson et al., 2003). The most common organic form of mercury in the environment is methylmercury. It is produced when soil and water-dwelling bacteria methylate inorganic forms of mercury (Dopp et al., 2004). Mercury is formed naturally by volcanoes and forest fires, but it is primarily released into the atmosphere by burning of fossil fuels in power plants (Driscoll et al., 2013).

5.3 Toxicological Effects of Heavy Metals

While some heavy metals, such as Fe, Zn, Cu, and so on, are necessary for human growth in small amounts, others, like Cd, Pb, As, Hg, and Cd mix with various environmental elements, like soil, water, and air, and become deadly when inhaled or consumed by humans (Filipoiu et al., 2022). So, the heavy metals can become medium to heavy toxic to humans and other living organisms when exposed to them through different ways such as food chains and food webs. Classification of heavy metals according to their function in the human body and their consequences.

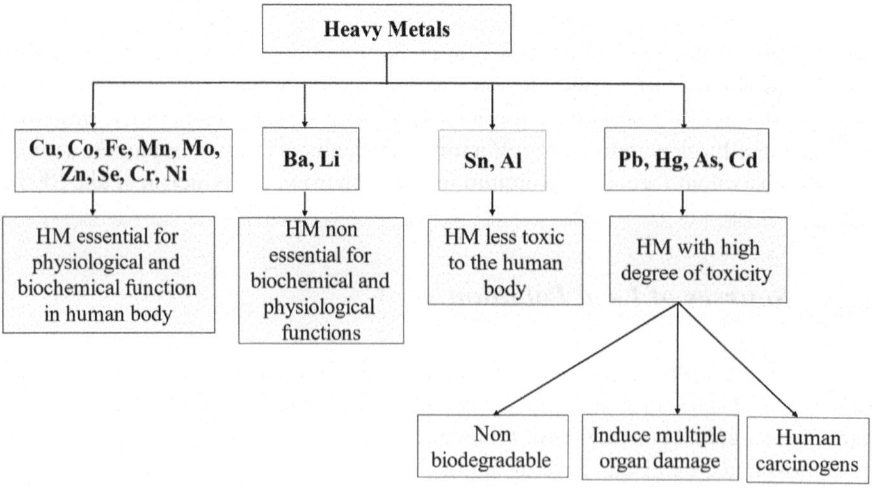

Fig. 5.1 Heavy metals' classification is according to their impact on the human body

Heavy metals are classified according to their function in the human body and their consequences are illustrated in Fig. 5.1.

One of the main concerns for food security and safety has been heavy metal pollution. The potential entry points for heavy metals into food have been mainly soil, agricultural irrigation water, and food processing materials (Huang et al., 2018; Islam et al., 2018). Plants that contain heavy metals acquire these metals, which when ingested by humans through contaminated food are biomagnified and have been associated with a higher risk of neurological, kidney, and cardiovascular illnesses (Yang et al., 2019). The metals accumulate in the edible parts that people eat through food crops when grown in soil that contains heavy metals (Amin et al., 2013).

Toxins containing heavy metals can have several harmful consequences on health. These approximately 20 distinct heavy metal toxins affect human health, and each toxin will cause a person exposed to it to experience unique behavioural, physiological, and cognitive alterations. A heavy metal toxin's impact on a system, organ, tissue, or cell is contingent upon both the toxin and the level of exposure of the individual to it. Although this is only an approximation, the toxicity of heavy metals can be ranked as follows: Hg > Cd > Cu > Zn > Ni > Pb > Cr > Al > Co. However, species' susceptibilities to specific metals vary. Additionally, toxicity varies based on environmental factors that regulate the chemical speciation of metals (Gray, 1999).

It has been discovered that organelles and components of cells, such as the cell membrane, enzymes, lysosomes, mitochondria, and nuclei, are affected by heavy metals. According to recent studies, metal ions interact with DNA and nuclear proteins to create DNA damage that can alter the cell cycle, trigger apoptosis, or result in cancer (Tchounwou et al., 2012). Damage particular to specific sites has been found as a result of associations between metallic substances and proteins in nuclear cells and DNA. Two types of damage can result from heavy metals: "direct" damage, where the metal alters the conformation of biomolecules, and "indirect" damage, where the metal produces reactive oxygen and nitrogen species (Valko et al., 2005).

Heavy metals such as zinc, copper, and nickel, in trace amounts, can be beneficial as micronutrients for human growth (Liang et al., 2019). Even at low concentrations, some toxic non-essential heavy metals can be damaging to human health, including lead (Pb), mercury (Hg), cadmium (Cd), and arsenic (As). Small children and pregnant women are particularly susceptible to the harmful effects of heavy metals which results in iron shortage (Kim et al., 2013; Liang et al., 2019; Reuben et al., 2017).

5.3.1 Effects of Lead (Pb)

Even at low concentrations, lead is toxic to human health and is not a necessary element for our bodily systems. It gets into our bodies through tainted food, drink, water, and air. Lead is a known human carcinogen, according to the EPA. It has a long biological half-life (20–30 years), and it builds up in teeth and bone before entering the bloodstream and attaching itself to erythrocytes. It causes haemoglobin levels in the body to drop causing anaemia by blocking the function of enzymes necessary for haemoglobin production (Bradl, 2005; Landis et al., 2003). Young children and pregnant women are especially vulnerable to lead poisoning which causes an iron shortage (Flora et al., 2006).

5.3.2 Effects of Chromium (Cr)

Diabetes and cardiovascular disorders are brought on by a chromium shortage, which is a necessary heavy element for the metabolism of fats and carbohydrates. While Cr (VI) form poses a greater risk to human health, the Cr (III) form is a necessary micronutrient. The majority of allergic reactions, including skin rashes, nose irritation, and bleeding, afflict those who work in the leather and textile industries (Karadede et al., 2004). According to a study, the body develops genetic abnormalities when chromium concentrations are too high (Kherici et al., 2009). Lung cancer has also been linked to large doses of Cr compounds in humans, according to reports. (Jordao et al., 2002).

Higher Cr (VI) concentrations in the atmospheric air can irritate the nose's lining and result in nasal ulcers. When animals consume Cr (VI) compounds, the most common health issues they experience include anaemia, sperm destruction, stomach and small intestine inflammation ulcers, and harm to the male reproductive system. Compounds containing Cr (III) are far less hazardous and don't seem to be the source of these issues. Some people are very sensitive to Cr (VI) or Cr (III); allergic responses characterized by significant skin swelling and redness have been reported. Stomach tumours are more common in animals and people exposed to Cr (VI) in drinking water. Humans have accidentally or purposely consumed very high levels of Cr (VI), which have led to severe neurological, haematological, hepatic, respiratory, cardiovascular, and gastrointestinal effects. Some patients have even died as a result of these effects, but others have survived thanks to medical intervention (ATSDR, 2008). The mechanism via which chromium causes cancer is not fully known, despite the seemingly substantial evidence of its carcinogenicity in humans and terrestrial mammals (Chen et al., 2009).

5.3.3 Effects of Cadmium (Cd)

According to the ATSDR list, cadmium is the seventh most hazardous heavy metal. It is a by-product of zinc manufacturing that can be found in the environment or the workplace for both humans and animals (Jaishankar et al., 2014). Eating contaminated foods including liver, mushrooms, and seaweeds exposes a person to cadmium (Jarup et al., 2000). It travels to the liver through the bloodstream, where it binds to the protein metallothionein to sequester and transport the substance to the kidney. It accumulates in the kidney and blocks blood filtration mechanisms. It deforms and harms the kidneys, liver, and bones (Abbas et al., 2008). Cadmium is immunosorbent, mutagenic, carcinogenic, teratogenic, and detrimental to developing embryos. Anaemia and hyperglycemia can also result from it (Rehman and Anjum, 2010).

5.3.4 Effects of Copper (Cu)

The body absorbs copper via food, drink, and dietary supplements. The World Health Organization's tolerance limits for copper in drinking water are 2.00 mg/L (WHO, 2004). The Environmental Protection Agency advised a tolerance limit of 1.3 mg/L for copper in the body. Prolonged contact with copper may cause nausea, vomiting, diarrhea, and rashes on the lips, eyes, and nose. Significant levels of copper have been observed to be rapidly accumulated in the liver in cases of juvenile cirrhosis in India. High copper concentrations have been found in certain fruits, nuts, and visceral organs (Pandit & Bhave, 2002). Many pediatric hepatic disorders, including Indian childhood cirrhosis, Wilson's disease, and idiopathic copper toxicosis (ICT), have been linked to copper buildup in the liver (Muller et al., 1998).

5.3.5 Effects of Mercury (Hg)

For both humans and animals, mercury is a very neurotoxic toxin that particularly damages the central nervous system. It enters the body through food consumption, where it can be found in both organic and inorganic forms. Methylmercury (MeHg), the organic form, poses a greater risk to human health. Foods with varying amounts of mercury indicate the degree of environmental pollution in the area (Dudka & Miller, 1999). Even at low exposure levels, there are toxic effects on many physiological systems, such as the digestive and neurological systems as well as organs like the kidney and lungs (Jyothi, 2020). Moreover, it has been demonstrated that mercury compounds raise MDA levels in the testes, livers, kidneys, and lungs of rats given $HgCl_2$ (Lash et al., 2007).

The relationship between exposure to mercury and the onset of cancer is hotly contested. Despite some studies supporting the metal's potential for genotoxicity, there is no evidence linking mercury exposure to genotoxic damage (Valko et al., 2004). The molecular mechanism of mercury-induced genotoxicity has been found through studies using oxidative stress. Consequently, it has been shown that mercury contributes to the generation of reactive oxygen species, which harm cells' DNA and may initiate the development of cancerous processes (Ogura et al., 1996; Valko et al., 2006).

5.3.6 *Effects of Arsenic (As)*

Exposure to arsenic can lead to illnesses of the skin and organs, as well as immune system dysfunction (Duker et al., 2005). Proteins and enzymes that it binds to have their sulfhydryl groups denaturated inside the cell (Gebel, 2004). It causes more reactive oxygen species to develop, which harms cells and their constituent parts (Ahmad et al., 2000). Arsenic can be found in human and veterinary medications, herbicides, and agricultural pesticides (Tchounwou et al., 1999). Leukaemia cells undergo programmed cell death, or apoptosis, as a result of its therapeutic activity (Yedjou & Tchounwou, 2007).

Millions of people are thought to be chronically exposed to arsenic worldwide, especially in countries like India, Uruguay, Taiwan, Bangladesh, Mexico, Chile, and where there is considerable groundwater contamination with arsenic. Exposure to arsenic can occur parenterally, through skin contact, through inhalation, or even through oral ingestion (Tchounwou et al., 1999). Arsenic in water has an average value of less than 10 μg/L, however, it can be found in higher concentrations around mining sites or natural mineral deposits. Natural soil arsenic levels typically range from 1 to 40 mg/kg; however, pesticide use and waste disposal can lead to much higher levels (Morton & Dunnette, 1994).

The digestive, neurological, hepatobiliary, pulmonary, renal, and cardiovascular systems are all impacted by arsenic exposure (Tchounwou et al., 2003). Almost every organ system is impacted by arsenic exposure, including the neurological, hepatobiliary, renal, gastrointestinal, respiratory, dermatological, and cardiovascular systems (Tchounwou et al., 2003). Research has also shown that the standardized death rates for cancers of the bladder, kidney, skin, and liver are significantly higher in many areas where arsenic poisoning is widespread. The strength of the detrimental

Table 5.1 Permissible weekly intake of heavy metals

Heavy metal	Permissible weekly intake (mg/kg body weight)
Mercury (Hg)	0.0016
Lead (Pb)	0.025
Cadmium (Cd)	0.007
Arsenic (As)	0.015

health consequences of arsenic is depending on its chemical form and is both dose- and time-dependent (Tchounwou et al., 2002; Yedjou et al., 2006).

The World Health Organization and the European Food Safety Authority have determined the tolerable weekly intake (TWI) values listed in Table 5.1. These values indicate the amount of each heavy metal that an individual can consume per kilogram of body weight per week without experiencing significant adverse health effects. There is currently no TWI for copper (Cu) or chromium (Cr) specified by WHO/EFSA.

However, it is important to note that these values are subject to updates and may vary based on regional guidelines and specific risk assessments. Individuals should always refer to the latest recommendations and consult with healthcare professionals for personalized advice regarding heavy metal exposure and safety.

5.4 Health Risk Assessment

A description of the possible harm to human health is thought to be included in an assessment of the health hazards linked to exposure to heavy metals (USEPA, 2012). Throughout this procedure, possible exposure paths are found, hazards are quantified, and a numerical value indicating the potential risk is calculated using information from statistics, science, and engineering. Carcinogenic and non-carcinogenic components are classified according to health risk assessment. The method to be used for calculating possible risks is determined by the classification (Lushenko, 2010). Assessments of carcinogenic and non-carcinogenic components in everyday ambient metal exposures were conducted. The two main ways that one is exposed to metals are by bathing, which absorbs the metals through the skin, and by consuming water. Based on USEPA criteria, calculations were performed (USEPA, 1996).

The hazard quotient (HQ) can be estimated to assess non-carcinogenic risks. The quotient of the reference dose (RfD) and the environmental exposure are used to compute it. For each element and exposure pathway, HQ values were found. The HQ of each factor was then assumed to produce the hazard index (HI), which is defined as the overall risk via the health exposure pathway. Ultimately, the HI via oral and cutaneous routes was added up to determine the overall HI. Under unification, HI values are regarded as secure. (USEPA, 1989). The estimated daily intake (EDI) is derived by using the following equation and is used to assess the risk level associated

Table 5.2 Values and units for estimating exposure assessment in Water (Liang et al., 2011; Liu et al., 2007)

Factors of risk exposure	Values	Units
Ingestion rate	2.2	L/day
Exposure frequency	360	days/year
Exposure duration	30	Years
Average time	0.6	h/day
Average body weight	70	kg
Carcinogenic potency slope	Pb = 0.009, Ni = 1.7, Cd = 0.6	$\mu g\ g^{-1}\ day^{-1}$

with pollutant exposure.

$$HQ = \frac{EDI}{RfD} \tag{5.1}$$

The hazard index was determined using the total of all the hazard quotients for all the substances that a person is exposed to (USEPA, 2011).

$$HI = HQ_A + HQ_B + \cdots + HQ_n \tag{5.2}$$

The target hazard quotients for intakes A, B and n are HQ_A, HQ_B and HQ_n respectively. The hazard index is represented by HI. Target cancer risk (TR) was used to assess carcinogenic risk. The USEPA Criteria III Risk-Based Concentration offered the TR estimation method (USEPA, 2011).

$$TR = \frac{(MC \times IR \times 10^{-3} \times CPS \times EF \times ED)}{(BW \times AT)} \tag{5.3}$$

TR stands for target cancer risk; MC stands for metal concentration in the sample (μg /g); IR stands for ingestion rate (g/day); and CPS stands for carcinogenic potency slope; EF stands for exposure frequency; ED stands for exposure duration; BW stands for average body weight; AT stands for average time. Table 5.2 displays the parameters for exposure to water along with their respective values.

5.5 Low-Cost Remedies for the Removal of Heavy Metals

Modern methods for recovering heavy metals from contaminated settings and/or removing them from them include bioremediation. Using microbes and plants, or their products, the technique makes use of natural biological mechanisms to remove harmful chemicals and return contaminated habitats to its pre-polluted state (Akcil et al., 2015; Dixit et al., 2015; Mani et al., 2014). Compared to traditional chemical and physical procedures, which are frequently more expensive and ineffectual, especially for low metal concentrations, it is an environmentally benign and economical method of recovering heavy metals. These traditional techniques also produce a large volume of hazardous sludge.

5.6 The International Agency for Research on Cancer and Regulations to Control Heavy Metals in India

The IARC has been by the international committee for control of heavy metals. However, it focuses primarily on evaluating the carcinogenicity of substances rather than setting regulations for their control. However, its assessments and classifications regarding the carcinogenic potential of heavy metals can inform regulatory agencies and policymakers in developing control measures and guidelines. IARC publishes monographs that summarize its evaluations of carcinogenic risks associated with specific agents. These monographs categorize substances into several groups based on their carcinogenicity.

IARC's assessments and classifications have a significant impact on public health policies, regulatory decisions, and scientific research worldwide. Governments, health organizations, and industries use IARC's findings to develop strategies for cancer prevention, risk reduction, and exposure management. In addition to its role in carcinogen evaluation, IARC conducts research studies on cancer epidemiology, genetics, environmental exposures, and prevention strategies. It collaborates with various institutions, universities, and research centers globally to advance cancer research and control efforts. IARC's evaluations and decisions are based on scientific evidence and follow transparent processes involving expert committees and peer-reviewed publications. The agency maintains independence from commercial and political influences to ensure the integrity of its assessments. Overall, IARC plays a crucial role in identifying and raising awareness about potential cancer risks, guiding public health policies, and promoting scientific research to reduce the global burden of cancer.

Regarding heavy metals usage and control, IARC has evaluated the carcinogenicity of several metals and classified some of them based on the strength of evidence linking them to cancer. For example: IARC has classified arsenic and arsenic compounds as Group 1 carcinogens, indicating that they are carcinogenic to humans. Bladder, lung, and skin cancers are among the cancers linked to arsenic exposure. According to the IARC, Cd and its compounds are Group 1 carcinogens that are mostly associated with prostate and lung cancer in exposed populations. Based on significant evidence of carcinogenicity in animals and insufficient data in people, the IARC has classified lead and its compounds as Group 2A (probably carcinogenic to humans). The IARC has assessed several mercury forms. Group 3 mercury elements cannot be classed as carcinogenic to humans, and compounds containing methylmercury cannot be categorized either because there is not enough data.

Regarding India, the nation has laws and policies in place to limit exposure to and pollution from heavy metals. The Ministry of Environment, Forests, and Climate Change (MoEFCC) and different state pollution control boards are principally responsible for enforcing these policies. Some key initiatives and regulations related to heavy metals in India include: (1) *Environmental Protection Act (EPA), 1986*: The EPA provides the framework for environmental protection and pollution control in India, including regulations related to air and water pollution, hazardous waste management, and industrial emissions, (2) *Water (Prevention and Control of Pollution) Act, 1974*: This legislation regulates water pollution and sets discharge standards for pollutants into water bodies, including heavy metals, (3) *Air (Prevention and Control of Pollution) Act, 1981*: The Air Act addresses air pollution control measures, including emissions from industries and vehicles that may contain heavy metals, (4) *Hazardous Waste Management Rules, 2016*: These rules govern the management, handling, and disposal of hazardous waste, including waste containing heavy metals, and (5) *National Ambient Air Quality Standards (NAAQS)*: India has established NAAQS that sets limits for various pollutants, including heavy metals, in ambient air to protect public health and the environment. Additionally, India has specific regulations and guidelines for industries, mining activities, waste management, and contaminated site remediation to control heavy metal pollution and mitigate health risks associated with exposure. These policies are periodically updated and enforced by regulatory authorities at the national and state levels.

5.7 Conclusion

This chapter elucidated that the primary sources of environmental contamination are both man-made and natural. It was also mentioned that several factors facilitate the heavy metals discharge from various environmental components. The primary cause of pollution due to heavy metals in the environment is human activity. Leaching and weathering various mineral-rich environmental components can cause natural pollution of the environment, thereby enriching it with heavy metals.

Regulations have been put in place to limit the quantity of heavy metals being released into the environment by human activities, thereby preventing their buildup in the environment and reducing human consumption and inhalation. Many technologies are available to regulate the heavy metals released into the environment which include bioremediation and chemical processes. If pollution abatement methods prove ineffective, alternative institutions may employ alternative technology to mitigate the environmental effects of heavy metals.

Legislation, regulations, and the identification of regions with elevated heavy metal concentrations must be effective. The harmful effects that heavy metals impose will cause serious issues down the road if the exposure is not controlled. Engineering solutions can reduce the amount of heavy metal exposure that occurs in the workplace.

There are several ways that heavy metals can enter the human body, such as through air, drink, food, or sometimes even skin contact. Heavy metals are absorbed, retained, and built up within the human body. Hazardous effects on a range of body tissues and organs result from the bioaccumulation of hazardous metals. Prostate, liver, blood, and skin cancers are linked to the harmful effects of arsenic, according to the IARC. Mercury is the primary cause of carcinogenic effects on the skin, brain, lung, and colorectal tissues. Central nervous system, lung, and colon cancers are caused by lead metal. Human cancers of the stomach, breast, lung, and kidney are brought on by cadmium toxicity. The primary cause of Alzheimer's disease is cadmium, which is building up in soil and getting into food chains as a result of phosphate fertilizers being used extensively. According to the WHO, mercury is a dangerous toxic metal that has a severe negative impact on aquatic life. Human ingestion of foods contaminated with mercury can result in several serious diseases, including Minamata, as well as various physiological impacts.

Acknowledgements We thank the reviewers and editors of this book series for their expert feedback and support during the publication of this chapter.

References

Abbas, A. M., Ismail, N., & Easa, A. M. (2008). Assessment of arsenic and heavy metal contents in cockles using multivariate statistical techniques. *Journal of Hazardous Materials, 150*, 783–789.

Agency for Toxic Substances and Disease Registry (ATSDR). (2008). *Toxicological profile for chromium.* U.S. Department of Health and Human Services. Public Health Service.

Ahmad, S., Kitchin, K. T., & Cullen, W. R. (2000). Arsenic species that cause release of iron from ferritin and generation of activated oxygen. *Archives of Biochemistry and Biophysics, 382*, 195–202.

Akcil, A., Erust, C., Ozdemiroglu, S., Fonti, V., & Beolchini, F. (2015). A review of approaches and techniques used in aquatic contaminated sediments: Metal removal and stabilization by chemical and biotechnological processes. *Journal of Cleaner Production, 86*, 24–36.

Amin, N., Hussain, A., Alamzeb, S., & Begum, S. (2013). Accumulation of heavy metals in edible parts of vegetables irrigated with waste water and their daily intake to adults and children, District Mardan, Pakistan. *Food Chemistry, 136*, 1515–1523. https://doi.org/10.1016/j.foodchem.2012.09.058

Beyersmann, D., & Hartwig, A. (2008). Carcinogenic metal compounds: Recent insight into molecular and cellular mechanisms. *Archives of Toxicology, 82*(8), 493–512. (PubMed: 18496671).

Bradl, H. (2005). *Heavy metals in the environment: Origin* (p. 6). Elsevier, Academic Press.

Chang, L. W., Magos, L., & Suzuki, T. (Eds.) *Toxicology of metals.* CRC Press.

Chen, T. L., Wise, S. S., Kraus, S., Shaffiey, F., Levine, K., Thompson, D. W., Romano, T., O'Hara, T., & Wise, J. P. (2009). Particulate hexavalent chromium is cytotoxic and genotoxic to the North Atlantic right whale (*Eubalaena glacialis*) lung and skin fibroblasts. *Environmental and Molecular Mutagenesis, 50*, 387–393.

Clarkson, T. W., Magos, L., & Myers, G. J. (2003). The toxicology of mercury-current exposures and clinical manifestations. *New England Journal of Medicine, 349*, 1731–1737. (PubMed: 14585942).

Dixit, R., Malaviya, D., Pandiyan, K., Singh, U. B., Sahu, A., Shukla, R., Singh, B. P., Rai, J. P., Sharma, P. K., & Lade, H. (2015). Bioremediation of heavy metals from soil and aquatic environment: An overview of principles and criteria of fundamental processes. *Sustainability, 7*, 2189–2212.

Dopp, E., Hartmann, L. M., Florea, A. M., Rettenmier, A. W., & Hirner, A. V. (2004). Environmental distribution, analysis, and toxicity of organometal compounds. *Critical Reviews in Toxicology, 34*, 301–333. (Pubmed: 15239389).

Driscoll, C. T., Mason, R. P., Chan, H. M., Jacob, D. J., & Pirrone, N. (2013). Mercury as global pollutant: Sources, pathways and effects. *Environmental Science and Technology, 47*, 4967–4983.

Dudka, S., & Miller, W. P. (1999). Accumulation of potentially toxic elements in plants and their transfer to human food chain. *Journal of Environmental Science and Health, Part B. Pesticides, Food Contaminants, and Agricultural Wastes, 34*, 681–708.

Duker, A. A., Carranza, E. J. M., & Hale, M. (2005). Arsenic geochemistry and health. *Environment International, 31*, 631–641.

Etikala, B., Madhav, S., & Somagouni, S. G. (2022). Urban water systems: An overview. In *Current directions in water scarcity research* (pp. 1–19). https://doi.org/10.1016/b978-0-323-91838-1.00016-6

Etikala, B., Vangala, S., & Madhav, S. (2024). *Groundwater geochemistry using modified integrated water quality index (IWQI) and health indices with special emphasis on nitrates and heavy metals in southern parts of Tirupati.* South India: Environmental Geochemistry and Health. https://doi.org/10.1007/s10653-024-02229-3

Fergusson, J. E. (Ed.). (1990). *The heavy elements: Chemistry*. Pergamon Press.

Filipoiu, D. C., Bungau, S. G., Endres, L., Negru, P. A., Bungau, A. F., Pasca, B., Radu, A. F., Tarce, A. G., Bogdan, M. A., Behl, T., Nechifor, A. C., Hassan, S. S., & Tit, D. M. (2022). Characterization of the toxicological impact of heavy metals on human health in conjunction with modern analytical methods. *Toxics, 10*(12), 716. https://doi.org/10.3390/toxics10120716

Flora, S. J. S., Flora, G., & Saxena, G. (2006). *Environmental occurrence, health effects and management of lead poisoning in lead chemistry, analytical aspects, environmental impacts and health effects* (pp. 158–228). Elsevier.

Forstner, U., & Wittman, G. (1981). *Metal pollution in the aquatic environment*. Springer. https://doi.org/10.1007/978-3-642-69385-4.

Gabby, P. N. (2003). "Lead." Environmental defense "Alternatives to Lead-Acid Starter Batteries." Pollution Prevention Fact Sheet.

Gabby, P. N. (2006). Lead: In mineral commodity summaries. U.S. Geological Survey.

Gebel, T. (2004). Confounding variables in the environmental toxicology of arsenic. *Toxicology, 144*, 155–162.

Gray, N. F. (1999). *Water technology: An introduction for scientists and engineers*. Eliane Wigzell.

Guertin, J., Jacobs, J. A., & Avakian, C. P. (2005). Overview of chromium (VI) in the environment: Background and history. In *Chromium (VI) Handbook* (pp. 1–22). CRC Press. https://doi.org/10.1201/9780203487969

Hassaan, M., El Nemr, A., & Madkour, F. F. (2016). Application of ozonation and UV assisted ozonation on decolorization of direct yellow 50 in sea water. *The Pharmaceutical and Chemical Journal, 3*(2), 131–138.

Huang, Y., Chen, Q., Deng, M., Japenga, J., Li, T., Yang, X., & He, Z. (2018). Heavy metal pollution and health risk assessment of agricultural soils in a typical peri-urban area in southeast China. *Journal of Environmental Management, 207*, 159–168. https://doi.org/10.1016/j.jenvman.2017.10.072

Islam, M. A., Romic, D., Akber, M. A., & Romic, M. (2018). Heavy metals accumulation in soil irrigated with polluted water and assessment of human health risk from vegetable consumption in Bangladesh. *Environmental Geochemistry and Health, 40*, 59–85. https://doi.org/10.1007/s10653-017-9907-8

Jaishankar, M., Tseten, T., Anbalagan, N., Mathew, B. B., & Beeregowda, K. N. (2014). Toxicity, mechanism and health effects of some heavy metals. *Interdisciplinary Toxicology, 7*(2), 60–72.

Jarup, L. (2003). Hazards of heavy metal contamination. *British Medical Bulletin, 68*(1), 167–182.

Jarup, L., Hellstrom, L., Alfven, T., Carlsson, M. D., Grubb, A., Persson, B., Pettersson, C., Spang, G., Schutz, A., & Elinder, C. G. (2000). Low level exposure to cadmium and early kidney damage: The OSCAR study. *Occupational and Environmental Medicine, 57*, 668–672.

Jordao, C., Pereira, M., & Pereira, J. (2002). Metal contamination of river waters and sediments from effluents of kaolin processing in Brazil. *Water, Air, and Soil Pollution, 140*, 119–138.

Jorgensen, L. A., & Pedersen, B. (1994). Heavy metals in fish used for time trend analysis and as environmental indicators. *Marine Pollution Bulletin, 28*, 235–243. https://doi.org/10.1016/0025-326X(94)90099-X

Jyothi, N. R. (2020). Heavy metals—Their environmental impacts and mitigation. *Intech Open*. https://doi.org/10.5772/intechopen.95370

Karadede, H. L., Oymak, S. A., & Aoenl, A. E. (2004). Heavy metals in mullet, Liza abu and catfish, *Silurus triostegus*, from the Atat Ark Dam Lake (Euphrates), Turkey. *Environment International, 30*, 183–188.

Kayastha, S. P. (2014). Heavy metal pollution of agricultural soils and vegetables of Bhaktapur District, Nepal. *Scientific World, 12*, 48–55. https://doi.org/10.3126/sw.v12i12.13597

Kelepertzis, E. (2014). Accumulation of heavy metals in agricultural soils of Mediterranean: Insights from Argolida basin, Peloponnese, Greece. *Geoderma, 221*, 82–90. https://doi.org/10.1016/j.geoderma.2014.01.007

Kherici, B. H., Kherici, N., Derradji, E., Rousset, C., & Caruba, R. (2009). Behaviour of chromium VI in a multilayer aquifer in the industrial zone of Annaba, Algeria. *Environmental Geology, 57*, 1619–1624.

Kim, Y., Ha, E. H., Park, H., et al. (2013). Prenatal lead and cadmium co-exposure and infant neurodevelopment at 6 months of age: The mothers and Children's Environmental Health (MOCEH) study. *Neurotoxicology, 35*, 15–22.

Kumar, V., Chauhan, R. K., Srivastava, S., Singh, J., & Kumar, P. (2018). Contamination, enrichment and translocation of heavy metals in certain leafy vegetables grown in composite effluent irrigated soil. *Archives of Agriculture and Environ-Mental Science, 3*, 252–260.

Landis, W. G., Sofield, R., & Yu, M. H. (2003). *Introduction to environmental toxicology: Impacts of chemicals upon ecological systems*. CRC Press.

Langmuir, D., Hall, P., & Drever, J. (1997). *Environmental geochemistry*. Prentice Hall.

Lash, L. H., Putt, D. A., Hueni, S. E., Payton, S. G., & Zwicki, J. (2007). Interactive toxicity of inorganic mercury and trichloroethylene in rat and human proximal tubules (Effects of apoptosis, necrosis, and glutathione status). *Toxicology and Applied Pharmacology, 221*(3), 349–362. (PubMed: 17481684).

Laws E. A. (1945). *Aquatic pollution, an introductory text*. Wiley.

Liang, F., Yang, S. G., & Sun, C. (2011). Primary health risk analysis of metals in surface water of Taihu Lake, China. *Bulletin of Environmental Contamination and Toxicology, 87*(4), 404.

Liang, G., Gong, W., Li, B., Zuo, J., Pan, L., & Liu, X. (2019). Analysis of heavy metals in foodstuffs and an assessment of the health risks to the general public via consumption in Beijing, China. *International Journal of Environmental Research and Public Health, 16*, 909.

Liu, C. W., Liang, C. P., Lin, K. H., Jang, C. S., Wang, S. W., Huang, Y. K., & Hsueh, Y. M. (2007). Bioaccumulation of arsenic compounds in aqua cultural clams (*Meretrix lusoria*) and assessment of potential carcinogenic risks to human health by ingestion. *Chemosphere, 69*, 128–134.

Lushenko, M. A. (2010). *A risk assessment for ingestion of toxic chemicals in fish from Imperial Beach*. San Diego State University.

Mahuta, M. A. (2020). An overlook of sources and strategies of minimizing lead poisoning in Nigeria. *Journal of Humanities and Social Sciences, 20*(6), 330–340.

Malik, D. S., & Maurya, P. K. (2014). Heavy metal concentration in water, sediment, and tissues of fish species (*Heteropneustis fossilis* and *Puntius ticto*) from Kali River, India. *Toxicological and Environmental Chemistry, 96*, 1195–1206.

Malik, Z., Ahmad, M., Abassi, G. H., Dawood, M., Hussain, A., & Jamil, M. (2017). Agrochemicals and soil microbes: Inter-action for soil health. *Xenobiotics in the soil environment: Monitoring* (pp. 139–152). Springer International Publishing.

Mani, D., & Kumar, C. (2014). Biotechnological advances in bioremediation of heavy metals contaminated ecosystems: An overview with special reference to Phytoremediation. *International Journal of Environmental Science and Technology, 11*, 843–872.

Modaihsh, A. S., Al Swailem, M. S., & Mahjoub, M. D. (2004). Heavy metal content of commercial inorganic fertilizers used in the Kingdom of Saudi Arabia. *Agricultural and Marine Sciences., 9*, 21–25.

Morais, S., Costa, F. G., & Pereira, M. L. (2012). Heavy metals and human health. In Oosthuizen, J. (Ed.) *Environmental health—Emerging issues and practice* (pp. 227–246).

Morton, W. E., & Dunnette, D. A. (1994). Health effects of environmental arsenic. In J. O. Nriagu (Ed.), *Arsenic in the environment Part II: Human health and ecosystem effects* (pp. 17–34). Wiley.

Muller, T., Muller, W., & Feichtinger, H. (1998). Idiopathic copper toxicosis. *The American Journal of Clinical Nutrition, 67*, 1082–1086.

Nagajyoti, P. C., Lee, K. D., & Sreekanth, T. V. M. (2010). Heavy metals, occurrence and toxicity for plants: A review. *Environmental Chemistry Letters, 8*, 199–216.

Naz, S., Anjum, M. A., & Akhtar, S. (2016). Monitoring of growth, yield, biomass and heavy metals accumulation in spinach grown under different irrigation sources. *International Journal of Agriculture and Biology, 18*, 689–697.

Nicholson, F. A., Smith, S. R., Alloway, B. J., Carlton-Smith, C., & Chambers, B. J. (2003). An inventory of heavy metals inputs to agricultural soils in England and Wales. *Science of the Total Environment, 311*, 205–219. https://doi.org/10.1016/S0048-9697(03)00139-6

Ogura, H., Takeuchi, T., & Morimoto, K. A. (1996). A comparison of the 8-hydroxyl-deoxyguanosine, chromosome aberrations and micronucleus techniques for the assessment of the genotoxicity of mercury compounds in human blood lymphocytes. *Mutation Research, 340*, 175–182. https://doi.org/10.1016/s0165-1110(96)90047-0. (PubMed: 8692180).

Pandit, A. N., & Bhave, S. (2002). Copper metabolic defects and liver disease: Environmental aspects. *Journal of Gastroenterology and Hepatology, 17*, 403–407.

Parveen, T., Hussain, A., & Rao, S. (2014). Growth and accumulation of heavy metals in turnip (*Brassica rapa*) irrigated with different concentrations of treated municipal waste water. *Hydrology Research, 46*(1), 60–71.

Patil, S. S., Gandhe, D. H., & Ghorade, B. I. (2014). Physicochemical properties of soil samples of Ahmednagar Industrial area. *Global Research Analysis (GRA), 3*(4), 81–82.

Reddy, Y. S., & Sunitha, V. (2023). Assessment of Heavy metal pollution and its health implications in groundwater for drinking purpose around inactive mines, SW region of Cuddapah Basin, South India. *Total Environment Research Themes, 8*, 100069. https://doi.org/10.1016/j.totert.2023.100069

Rehman, A., & Sohail Anjum, M. (2010). Cadmium uptake by yeast, *Candida tropicalis*, isolated from industrial effluents and its potential use in wastewater clean-up operations. *Water, Air and Soil Pollution, 205*, 149–159. https://doi.org/10.1007/s11270-009-0062-4

Reuben, A., Caspi, A., Belsky, D. W., et al. (2017). Association of childhood blood lead levels with cognitive function and socioeconomic status at age 38 years and with IQ change and socioeconomic mobility between childhood and adulthood. *JAMA, 317*, 1244–1251.

Selim, H. M. E., Sparks, D. L., & Meeting SSSOA. (2001). Physical and chemical processes of water and solute transport/retention in soils. Proceedings of a symposium sponsored by divisions S-1 and S-2 of the Soil Science Society of America in Baltimore, MD, 18 to 22 October 1998, Soil Science Society of America.

Shan, Y., Tysklind, M., Hao, F., Ouyang, W., Chen, S., & Lin, C. (2013). Identification of sources of heavy metals in agricultural soils using multivariate analysis and GIS. *Journal of Soils and Sediments, 13*, 720–729.

Sparks, D. L. (2003). Environmental soil chemistry. *Academic*. https://doi.org/10.1016/B978-012 656446-4/50001-3

Tchounwou, P. B., Patlolla, A. K., & Centeno, J. A. (2003). Carcinogenic and systemic health effects associated with arsenic exposure—A critical review. *Toxicologic Pathology, 31*(6), 575–588. (PubMed: 14585726).

Tchounwou, P. B., Wilson, B. A., Abdelgnani, A. A., Ishaque, A. B., & Patlolla, A. K. (2002). Differential cytotoxicity and gene expression in human liver carcinoma (HepG2) cells exposed to arsenic trioxide and monosodium acid methanearsonate (MSMA). *International Journal of Molecular Science, 3*, 1117–1132.

Tchounwou, P. B., Wilson, B., & Ishaque, A. (1999). Important considerations in the development of public health advisories for arsenic and arsenic-containing compounds in drinking water. *Reviews on Environmental Health, 14*(4), 211–229. https://doi.org/10.1515/reveh.1999.14.4.211.

Tchounwou, P. B., Yedjou, C. G., Patlolla, A. K., & Sutton, D. J. (2012). Heavy metal toxicity and the environment. *EXS, 101*, 133–164.

Toth, G., Hermann, T., Silva, M. R., & Montanarella, L. (2016). Heavy metals in agricultural soils of the European Union with implications for food safety. *Environmental Pollution, 88*, 299–309.

United States Environmental Protection Agency (USEPA). (1989). *Risk assessment guidance for superfund (Volume I)—Human health evaluation manual Part A interim final*. EPA/540/1-89/002. Office of Emergency and Remedial Response.

USEPA. (1996). *Quantitative uncertainty analysis of superfund residential risk pathway models for soil and ground water: White Paper*. Office of Health and Environmental Assessment.

USEPA. (2011). *USEPA regional screening level (RSL) summary table*.

USEPA. (2012). *Waste and cleanup risk assessment*. http://www2.epa.gov/risk/waste-and-cleanup-risk-assessment

Valko, M., Izakovic, M., Mazur, M., Rhodes, C. J., & Tesler, J. (2004). Role of oxygen radicals in DNA damage and cancer incidence. *Molecular and Cellular Biochemistry, 266*, 79–110. (PubMed: 15646029).

Valko, M., Morris, H., & Cronin, M. (2005). Metals, toxicity and oxidative stress. *Current Medicinal Chemistry, 12*, 1161–1208.

Valko, M., Rhodes, C. J., Monocol, J., & Izakovic-Mazur, M. (2006). Free radicals, metals and antioxidants in oxidative stress-induced cancer. *Chemico Biological Interaction, 160*, 1–40. https://doi.org/10.1016/j.cbi.2005.12.009

Vodyanitskii, Y. N. (2013). Contamination of soils with heavy metals and metalloids and its ecological hazard (analytic review). *Eurasian Soil Science, 46*, 793–801.

Wang, S., & Shi, X. (2001). Molecular mechanisms of metal toxicity and carcinogenesis. *Molecular and Cellular Biochemistry, 222*, 3–9. (PubMed: 11678608).

WHO/FAO/IAEA. (1996). *Trace elements in human nutrition and health*. World Health Organization.

WHO. (2004). *Guidelines for drinking water quality*. World Health Organization.

Wilson, D. N. (1988). *Association Cadmium. Cadmium—Market trends and influences*. Cadmium 87 Proceedings of the 6th International Cadmium Conference (pp. 9–16).

Wu, G., Kang, H., Zhang, X., Shao, H., Chu, L., & Ruan, C. (2010). A critical review on the bioremoval of hazardous heavy metals from contaminated soils: Issues, progress, eco environmental concerns and opportunities. *Journal of Hazardous Materials, 174*, 1–8.

Yang, L., Liu, G. M., Di, L., Wu, X., You, W., & Huang, B. (2019). Occurrence, speciation, and risks of heavy metals in soils of greenhouse vegetable production from the vicinity of industrial areas in the Yangtze River Delta, China. *Environmental Science and Pollution Research, 26*, 8696–8708. https://doi.org/10.1007/s11356-019-04313-z

Yedjou, G. C., Moore, P., & Tchounwou, P. B. (2006). Dose and time dependent response of human leukemia (HL-60) cells to arsenic trioxide. *International Journal of Environmental Research and Public Health, 3*(2), 136–140.

Yedjou, G. C., & Tchounwou, P. B. (2007). In vitro cytotoxic and genotoxic effects of arsenic trioxide on human leukemia cells using the MTT and alkaline single cell gel electrophoresis (comet) assays. *Molecular and Cellular Biochemistry, 301*, 123–130.

Zhang, C. (2006). Using multivariate analyses and GIS to identify pollutants and their spatial patterns in urban soils in Galway, Ireland. *Environmental Pollution, 142*, 501–511. https://doi.org/10.1016/j.envpol.2005.10.028

Chapter 6
Heavy Metal Pollution in Groundwater: Environmental and Health Concerns in Gujarat, India

Barathan Balaji Prasath

Abstract Groundwater is an essential resource found within the Earth's crust in underground reservoirs called aquifers and is necessary for the provision of clean water worldwide. However, there has been a growing concern over the sustainable use of groundwater due to increasing population and industrialization. Over-extraction and pollution of groundwater resources have led to depletion and deterioration causing threats to human life and ecosystem viability. Some human activities such as industrial practices, agricultural practices, wrong waste disposal, and hazardous materials among others have introduced harmful pollutants and chemicals into groundwater leading to long term damage on its quality as well as availability. In Gujarat, heavy metals in ground water pose a high risk to people's health. Neurological disorders, kidney damage even cancer may be caused by exposure to heavy metals like lead, arsenic, cadmium or mercury. For instance, there must be comprehensive health risk assessments carried out so that concentrations of heavy metals can be determined in subsurface water thereby introducing suitable remediation actions towards ensuring safe drinking water sources. To comprehend potential dangers posed by presence of certain heavy metals in drinking water and consequently introduce relevant mitigation measures it calls for a deep evaluation on their health risks associated with their presence. This paper discusses migration pathways of heavy metals in Gujarat's groundwaters taking into account environmental factors that affect their movement. There are several methods used for removing heavy metals including adsorption, membrane based methods, chemical precipitation method involving precipitation coagulation flocculation process, electrochemical method or photo catalytic processes. By addressing these issues, we can develop innovative ideas and research approaches for future investigations.

Keywords Heavy metal · Human health · Contamination · Mitigation

---

B. B. Prasath (✉)
Coastal and Marine Ecology Division, Gujarat Institute of Desert Ecology, Bhuj, Gujarat, India
e-mail: b.balajiprasath@gmail.com

V. Sunitha et al. (eds.), *Groundwater Resource Management Planning Strategies*, Springer Water, https://doi.org/10.1007/978-3-031-88870-0_6

6.1 Introduction

Water is the most critical product of existence on Earth since all living things depend on it. Water has two main sources: natural fresh surface water and groundwater accessed by wells and boreholes (Akhtar et al., 2021; Musie & Gonfa, 2023). The polar nature of water is one of its unique chemical properties that make it combine with different substances and pollutants in its surroundings due to hydrogen bonding. Toxic environmental pollution remains a major concern for local water users because large volumes of such pollutants get into the aquatic environment from various sources including industrial activities, population growth as well as mismanagement of natural resources such as agricultural and domestic waste run off (Etikala et al., 2022; Thushari & Senevirathna, 2020). Heavy metals are one of the contaminants with the highest risk to aquatic organisms due to their toxic nature, persistence and bioaccumulation in food chains (Ali et al., 2019). Some of these metals cause severe toxicity in people like renal damage; cardiovascular disorder; gastrointestinal system dysfunction; central nervous system damage which affects brain and spinal cord among others (Balali-Mood et al., 2021). Even though it may be challenging to completely avoid exposure to hazardous metals as they may be available in food, drinks or air, there are ways through which metal toxicities can be reduced (Briffa et al., 2020). Additionally, both essential and non-essential heavy metals make up biological systems. Although small amounts of essential metals such as iron, zinc, copper, manganese, chromium molybdenum selenium are necessary for human health but higher concentrations can have adverse effects on humans' health if non-essential heavy metals like lead mercury cadmium tin were present in them. It is important to understand how these metals interact with each other so their effect on biological systems and environmental health may be understood completely (Ding et al., 2022). Through a healthy lifestyle that ensures metabolization or excretion of heavy metals from your diet, such risks may be mitigated. The pollution of groundwater is seen as one of the most critical environmental issues. In particular heavy metals are alarming because they can even be toxic at low concentrations (Li et al., 2021). This is driven by the complexity of metal's behavior in ground water which is controlled by different factors including its source and biogeochemical reactions under primary conditions. Therefore, explaining these complex processes is important for professionals in this field to manage and prevent potential environmental hazards effectively. Human activities alter both the quantity and quality of groundwater thus making it unsuitable for drinking or ruining nature preservation areas (Kumar et al., 2022, 2023). It will require significant scientific efforts to assess impacts on groundwater or soil layers beneath the Earth's surface, as well as predict the scope and degree of future changes. Groundwater serves a variety of uses in commercial, farming, and domestic arenas with several human activities like agriculture and household tasks releasing many pollutants into aquatic environments (Carrard et al., 2019). Groundwater has become a major worldwide concern in recent decades due to increased urbanization; intensified agriculture as well as industrialization which has resulted in more dependence on groundwater resources while pollution has left surface waters dry. Usable water available worldwide is quite large but only a small fraction is pure, whereas about 30% of the world's population rely on groundwater (Swain et al., 2022). Over eighty percent of India's populations derive their water from underground sources primarily aquifers. There is also groundwater without which over forty percent food crops globally cannot survive and nearly one-third of industrial water depend on it (Dangar et al., 2021). Nevertheless, climate change has made groundwater less vulnerable to rainfall patterns; altering hydrological rhythms coupled with diminished groundwater recharging in many places. Moreover, there has been an increase in seawater encroachment and low precipitation leading to high conductivity and salinity decrease as well as water depth. High levels of salinity in water can have a negative effect on the growth of crops due to its effect on the soil's permeability and osmotic potential. Additionally, anthropogenic sources including septic tank, chemicals, quarry, landfill sites continue to pollute groundwater with mismanaged industrial hazardous waste being a major factor (Kamran et al., 2019). Heavy metals which include arsenic (As) and cadmium (Cd) among others are some of the groundwater contaminants while sediments together with minerals are both natural ones. In different areas of India, regions have been discovered having toxic metal levels including NO_3^-, Fe, and HCO_3^- that can cause adverse health effects (Sharma et al., 2019). Gujarat's groundwater is contaminated by human induced pollution such as Arsenic and Cadmium that become pollutants in it. Long-term exposure to even small amounts of these heavy metal pollutants can be harmful for human beings. Ankleshwar is reported to have elevated levels of cadmium as well other heavy metal pollutants like molybdenum, zinc, lead, nickel cobalt and iron. Additionally, a global overview of arsenic contamination in aquifers across regions like Gujarat highlights the health risks arsenic pollutants pose in groundwater (Kayastha et al., 2022). Groundwater pollutant with arsenic is an issue that affects the entire planet that can cause neurological, reproductive, immunological and skin cancer. In addition, sediment load and effusive eruptions are the main sources of cadmium, found in groundwater, which is considered to be highly hazardous (Kumar et al., 2021). Given the high risk associated with groundwater contamination, it is essential to develop effective strategies to mitigate the risks and preserve the quality of this vital resource for future generations.

6.2 Gaps in Groundwater Policy and Management

In many parts of India, groundwater extraction is a major water crisis resulting in declining water tables as well as aggravating water scarcity. In some regions of India such as Punjab, Rajasthan, Haryana, Delhi, Madhya Pradesh among others in U.P and Tamil Nadu they are extracting groundwater faster than it can be recharged hence creating a widening gap between the demand and supply of this vital natural resource. This state remains because there is no aquifer management system which has all-encompassing scope. This is causing severe social, economic, and ecological consequences as critical and overexploited units rise (Ingrao et al., 2023). On the other hand unregulated extraction continues to pose a substantial problem despite several discussions concerning it. Currently the only legal regime governing this resource is the Indian Easement Act of 1882 (Wahi, 2022). As per the current legal regime only landowners have exclusive rights to extract groundwater while those who do not own land have no access to this important resource. Tragically there is no law that requires any kind of oversight or regulation on how landowner withdraws and uses water. The central government has recently released a model law for state governments to enhance their regulatory mechanisms. However some states still rely on an outdated version of the model bill to improve its efficacy, model bill should adopt a more participatory approach. A major challenge is that adequate data and knowledge about aquifer systems are lacking. A demand has been emerging persistently in relation to mapping of aquifers so as to allow for better management planning purposes. National or state-level management plans should integrate this approach with regards to groundwater recharge/extraction. The lack of regulations and public finance discourages plans for groundwater management (Shanthakumar et al., 2021). A national-level master plan for artificial recharge of aquifers requires a substantial investment of Rs. 1.33 lakh crore. Encouraging people's participation in planning and execution is crucial for success.

6.3 The Groundwater Reserves Have Been Contaminated with Heavy Metals

Groundwater can contain heavy metals from two primary sources—geogenic and anthropogenic. Geogenic are the parent rocks that weather or leach heavy metals (Alengebawy et al., 2021). Despite the abundance of heavy metals in groundwater from geogenic sources typically falling within acceptable limits set by environmental protection agencies, these have become a major concern in many areas with respect to underlying rock types. On Northern Develi Closed Basin, Kayseri, Arslan et al. (2017) reported their concentrations of Arsenic, Boron and Iron excessive than permissible portable water values for Turkey. This is because volcanic products, altered significantly with water contained metal sources of lead, zinc iron and copper (Parisi et al., 2011). The presence of ions in surface and pore water is due to the weathering and dissolution of silicate, carbonate and sulfide minerals. Weathering or dissolution process which minerals undergo is influenced by climate among others chemical composition. Silicates are more soluble than carbonates e.g. calcite which dissolves most easily. Carbonates display greater reactivity relative to silicates. These include high sulphide content zones where the capacity of carbonate minerals for neutralizing acidity as well as pH levels controls mineral erosion (Al et al., 2000). Carbonates act as buffers against sulfide oxidation-induced acidification that maintains alkalinity in groundwater thereby preventing further sulfide dissolution. However clay has lesser capability than calcite to be neutralized by so that it leads aquifers into acidic conditions (Guo et al., 2020). A aquifers have been discovered to contain naturally-occurring carcinogenic substances like uranium and arsenic, presenting potential health hazards to humans (Kumar et al., 2023). Uranium worldwide is a geogenic source of groundwater contamination leading to health problems. High arsenic concentrations are found in the Gangetic Plain and Bengal, India, due to unweathered alluvial sediments from the Himalayas. Calcium arsenate ($Ca_3(AsO_4)^2$), rocks with coal content and Fe-and Mn-oxi/hydroxides release As during natural weathering (Uddin et al., 2011). In a reducing environment, sediments and organic matter mobilize As. It is important to understand that heavy metal contamination can be a major environmental concern occurring naturally or as a result of human activities (Briffa et al., 2020). Nonetheless, these metals exist naturally in the earth's crust, but have been found in various aquatic environments. Nevertheless, mining constitutes an important source of heavy metal pollution (Khatri & Tyagi, 2015). Also released into water during mining are direct and indirect wastes while heavy metals are also released into water by open fires as well as production of huge amounts of metal. The industrial revolution saw metals being mined from nature and refined in industries making it worse for water resources. Human activities such as smelting or treating ores of metals, mining; burning fossil fuels; discharging industrial or domestic waste; discarding automobile exhausts; using pesticides containing heavy metal salts etc., results to aquifer heavy metals contamination (Dewangan & Bhatia, 2023). Nevertheless, we can address this form of pollution. Appropriate measures such as e-waste management and disposal methods like sanitary landfills as well as runoff controls on agricultural practices may reduce possible incidence of heavy metal pollution (Ferronato & Torretta, 2019). Incorrect disposal of human excrement coupled with certain farming techniques may lead to deadly toxic elements released into underground waters which then pose severe health risks. Meanwhile BIS 10500:2012 has proposed the allowable values for some heavy metals concerning drinking water in India. Total arsenic (As): 0.01 mg/L (10 μg/L), Cadmium (Cd): 0.003 mg/L (3 μg/L), Total Chromium (Cr): 0.05 mg/L (50 μg/L), Copper (Cu): 0.05 mg/L (50 μg/L), Iron (Fe): 0.30 mg/L (300 μg/L), Lead (Pb): No relaxation (Mawari et al., 2022). There is a need to frequently monitor and evaluate the quality of drinking water sources, especially in places that are prone to pollution by heavy metals to ensure that the levels of such toxic substances do not exceed permissible limits. For purposes of safeguarding water resources and guaranteeing public health, it is imperative to control human activities that lead to contamination by heavy metals as shown in Fig. 6.1. Waste from households needs better treatment for metal removal while sludge from waste-water treatment should be properly treated. Urban storm water runoff should be reduced through effective city planning, traffic control and road construction (Müller et al., 2020). Industrial effluent should be treated before discharging into aquatic environment; also monitoring and regulating discharge of metal into the environment is necessary. In conclusion, there is an urgent need for an integrated approach towards effective mitigation of heavy metal pollution. An efficient way to protect our water resources and the environment is by preventing any heavy metal contamination at all stages.

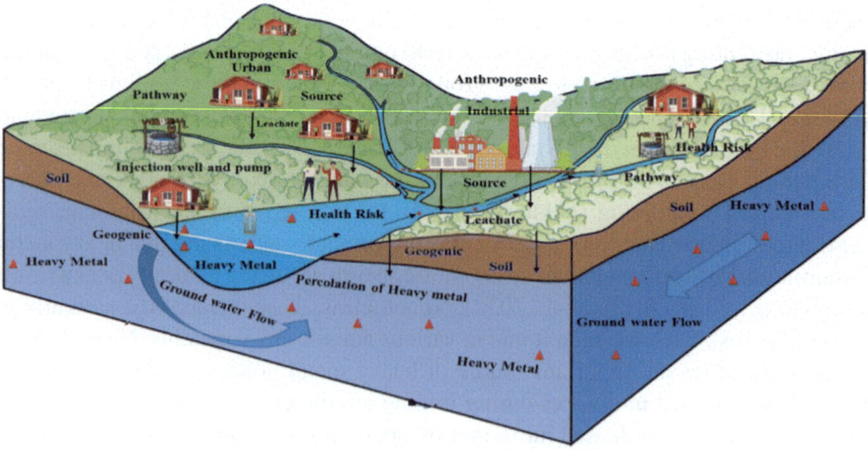

Fig. 6.1 The process of heavy metal transportation to groundwater by human begins

6.4 Conflicting Reports on Gujarat's Groundwater

A western Indian state with a population of about 60 million, Gujarat is the ninth most populous state in India. More than 43% of people residing in this state live in urban areas and that amounts to around 25.9 million people. The State has 167 Urban Local Bodies (ULBs) exclusively responsible for water supply and sanitation services across the entire population of the urban dwellers. Population growth has resulted in increased demand for water and therefore more sewage generation leading to more pressure on it (Kapshe et al., 2013). Future plans are being made to provide sewage treatment plants for all cities and municipal corporations though this are not enough at present since they underestimate its demands from several cities planned out by now. Consequently, untreated water is discharged into the nearest water bodies which harm agricultural sector greatly (Falkenberg & Saxena, 2018; Obaideen et al., 2022). Gujarat generates approximately 5013 MLD (Million liters per day) of wastewater, of which only 68% or 3378 MLD is treated.

The remaining 32% or 1635 MLD of untreated wastewater is either discharged into neighboring water bodies or else used for irrigation. Gujarat boasts a number of industries including textiles, chemicals, pharmaceuticals, electronics and motor vehicles. Arsenic occurs naturally in soil and groundwater as a result of the breakdown of rocks and minerals. Leaching may be caused by arsenic leachate through runoff hence posing serious health problems to millions who might suffer from geogenic arsenic poisoning via groundwater (Wu et al., 2021). Predictions on arsenic pollution were generated at a resolution of 1km using logistic regression models that combined groundwater data with secondary soil, climate and topographic variables. A pseudo-contour map for portable water arsenic pollution could be made by combining models based on different concentrations of arsenic. This is because there are some areas where there is no updating about the issue from private surveys in relation to the Narmada water supply department's report on expert findings regarding the 2020 survey (Fig. 6.2). If Saurashtra, North Gujarat and Central Gujarat districts' groundwater samples consistently show high levels of arsenic (permissible limit being 0.01 mg/L according to WHO standards), appropriate measures should be taken towards this end (Wu et al., 2021). The state government and policy makers need to take an assertive stand to address the problem of groundwater depletion in particular areas. Such steps include cessation of extraction of underground aquifer groundwater through hand pumps or stand posts. Moreover, covering all households with door-to-door audio-visual campaign would sensitize people against this form of water usage that has adverse effects on human life. People living in these regions must also stop using it for drinking water as well as other domestic purposes as advised by villagers' norms. Additionally expanding surface coverage and quality in these regions is critical. This can be done through distributive model using 3 tier Panchayati Raj system. Thus, it is possible to avoid aggravating the problem of groundwater depletion and at the same time ensure that sustainable water supply is available for residents. Moreover, WSMO's wide-ranging role in rural Gujarat's water management could also be used to help improve accessibility and quality of surface water for these regions. However, more research is needed on the impact of sewage irrigation on crops. Studies from Anand and Rajkot had indicated that there were excessive toxic metals in vegetables irrigated with sewage, which may be harmful to human health. Spinach, radish, tomato, chili, cabbage, beans, lady's finger, brinjal and cauliflower had higher concentrations of heavy metals. Heavy metal contamination in tested vegetable samples might have occurred due to sewage irrigation; pesticides/herbicides use; and agricultural chemical fertilizers (Mohanty & Das, 2023). The Indian Government has stated that no habitation in Gujarat has contaminated groundwater. However, reports from the Central Ground Water Board suggest that several districts in Gujarat have groundwater with high levels of fluorides, arsenic, iron, nitrates, and chlorides. As of July 23, 2021, 47,873 rural village stated water pollution issues, according to information provided by the Department of Drinking Water & Sanitation and States/UTs (as on 23.07.2021). Although 19 states and Union territories with contaminated groundwater are mentioned in the annexure. Managing water quality is primarily the responsibility of the states, and the central government is implementing various schemes to improve groundwater quality, including the Jal Jeevan Mission, National Water Quality Sub Mission, Atal Mission for Rejuvenation and Urban Transformation (AMRUT), and National Aquifer Mapping Programme.

6.5 Impact of Heavy Metals in Water on Human Health

It is an examination of the impact of heavy metal on human health particularly as water pollutants. However, some metals are essential for body functions in small quantities but excesses can be harmful (Das et al., 2023). They can accumulate in animals and humans resulting to toxic levels with severe consequences. Heavy metals can lead to

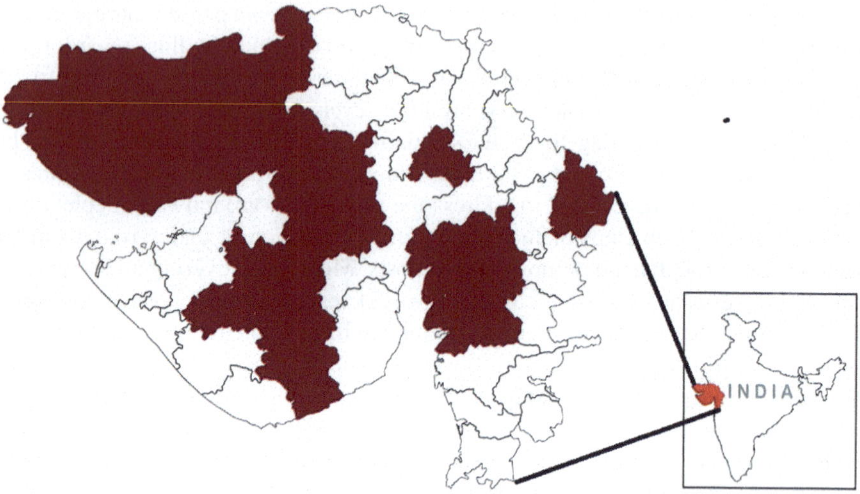

Fig. 6.2 The groundwater in several districts of Gujarat is contaminated with fluorides, arsenic, and iron, particularly in areas with heavy metal presence

fatal diseases and also interfere with hormone metabolism within the body. It may cause degenerative neurological and physical processes which resemble Parkinson's or Alzheimer's diseases. There are 23 out of 35 problem metals that are heavy such as antimony, arsenic, lead, mercury. Although these elements are needed in small amounts for proper health, large volumes may be dangerous or even poisonous (Malyan et al., 2019). A comprehensive understanding of heavy metals and their possible toxicity is necessary. These toxic substances can affect important organs like liver, lungs, kidneys, brain and blood thus reduce energy levels. Hence there is need for steps to prevent overexposure. It should also be noted that sometimes the toxicity level of some metals may exceed natural occurring environmental background levels by a minor degree hence there is need to survey the contents of heavy metals in our environment and diet (Nag et al., 2022). Arsenic is a carcinogenic heavy metal that exists in various forms like oxides sulfides and salts containing other elements such as sodium calcium copper iron etc. (Okoro et al., 2023). It is commonly found in geological and mineral systems in the form of arsenate (H_2AsO_4) and arsenite (H_3AsO_3) complexes, which can be extremely harmful to both the natural and human wellbeing. Arsenic mainly affects the sulfhydryl group of cells, leading to health issues such as hypertension, significant impacts on the cardiovascular structure, hepatic damage, and reduction in spermatogenesis and gonadotropin and testosterone release in humans. Lead is a hazardous component that can cause harm even in small quantities, primarily found in water and food.

The occurrence of lead-bearing neurotoxicants may bring about directly or indirectly toxicities due to the activation of oxidative stress by lipid peroxidation. This can promote accumulation of reactive oxygen species (ROS) that lower the amount of antioxidants present in the body. Lead absorption at high levels in humans' bodies might cause lifelong damage or a person's death; for example, it results in injury or death to brain, central nervous system and kidneys (Collin et al., 2022).

Nickel is common in human tissues, but its concentration increases significantly upon exposure to high levels. However, nickel uptake from air and drinking water is generally less important than dietary intake, which is also the main pathway for this element to get into organisms. Ingestion of nickel leads to general toxicity mostly affecting lungs and upper respiratory tract when it enters through mouth while inhaling it causes kidney problems. Genitalia such as the cardiovascular, immune and blood may also be influenced by nickel release into the system. Exposure to highly polluted water with nickel has various effects on health (Genchi et al., 2020). Chromium serves as essential nutrient at trace amounts for human health but becomes toxic at higher levels. It occurs naturally in minerals like chromite within environment and is also released through mining or industrial activities. Hexavalent chromium presents itself as poison metal which has been associated with skin irritations, lung diseases and cancer issues among others health hazards. Inhalation of Cr (VI) causes lung cancer while ingestion through contaminated water can result in stomach cancer. Chromium VI has some harmful impacts on liver too besides kidney harms. The concentration of trace elements in groundwater is increasing due to human and natural pollution (Balali-Mood et al., 2021). The study area is a flat agricultural terrain with significant industrial development in the past decades. These elements which traces such as arsenic are harmful to human beings when they are in high concentrations in groundwater. The most hazardous health risk is the contamination through arsenic contaminated water used for drinking, food processing and agricultural irrigation. Deposits of sulfide minerals and sedimentary rocks from volcanic rocks can contain high levels of arsenic that poses health risks after ingestion. Compounds containing cadmium can cause acute poisoning resulting in liver and kidney damage instantly, it has been known that these compounds also cause cancer. While copper levels in surface and groundwater are typically low, human activities like mining, farming, manufacturing processes, and wastewater releases can introduce copper into the environment thus contaminating drinking water. Soluble nickel exposure can lead to cancer and other health problems. Iron-rich groundwaters are geogenic contaminants responsible for cardiovascular diseases, hyperkeratosis, pigmentation changes, diabetes mellitus, and neurological disorders. In Gujarat (Fig. 6.3), groundwater is often found contaminated with toxic elements such as arsenic, uranium or cadmium that may pose severe health hazards to humans upon consumption or exposure. Manganese is vital for good health—but too much exposure can be harmful; excess intake gives an unpleasant taste to drink sand stains bathroom fixtures. High levels of aluminum contents in portable water causes chromosome abnormalities and other non-carcinogenic effects on humans' wellbeing. In particular lead is a dangerous element because it's capable of causing permanent damages to organs and systems over time. Twenty percent of human lead exposure comes from drinking water (Eid et al., 2024). High intake rates of zinc can cause acute and chronic adverse health effects. Zinc deficiency is common among humans especially those living in underdeveloped regions who rely on cereal proteins for sustenance. Consuming Chromium

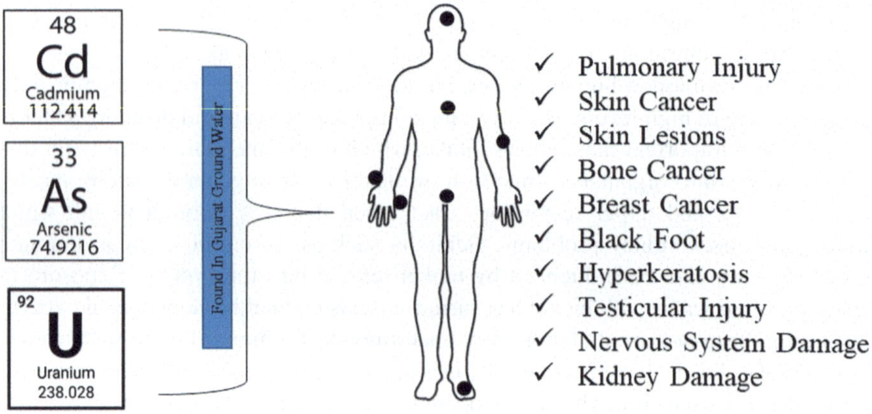

Fig. 6.3 The harmful effects of heavy metals on human health

(VI) contaminated groundwater which has higher levels of chromium than other types of rock can cause intestinal, gastric and liver problems.

6.6 Remediation Methods

A number of methods were devised over the past few decades for removing heavy metals from polluted water. These include ion exchange, phytoremediation, oxidation, coagulation-flocculation, electrokinetics, and membrane filtration (Fig. 6.4). Each method has its merits and demerits. As an example, oxidation involves the use of such oxidants as Cl_2, ClO_2, O_3, H_2O_2, NH_2Cl, MnO_4^-, and FeO_4^{2-} to remove heavy metals from water (Ahmed et al., 2022). UV radiation and other elements are used in photochemical and photocatalytic techniques to produce hydroxyl radicals that can be used to extract pollutants that are dissolved normally (Pavel et al., 2023). Polluted water can also be cleaned using coagulation-flocculation techniques which removes toxic metal from it. In this process a coagulant is added into the polluted water which forms a floc that can possibly remove heavy metals. Coagulants with positive charges like aluminum and iron salts are commonly used thus eliminating the colloids' negative charge because they agglomerate into larger particles called floc which eventually settle in water due to gravity (El-taweel et al., 2023). Phytoremediation is a technique that uses plants to remove toxic heavy metals from the contaminated environment. Phytoextraction, phytomining, phytostabilisation, hemofiltration, and phytovolatilization are the different steps involved in this process. Various plants, such as *Pteris Vittata* and *Vetiver grass*, have been used to remediate toxic metals through phytoremediation. However, this process has the disadvantage of contaminating crop plants that are grown for food may absorb heavy metals from the soil and water, which can make the elimination of these metals from soil and water a time-consuming process (Kafle et al., 2022). The method of remediating metal contaminated groundwater using biotechnology has been found to be most effective. Naturally occurring micro-organisms are deployed to decompose or eliminate environmental contaminants in this technique as defined. In both soil and water, bioremediation methods for heavy metals have been successful in situ and ex situ. Biostimulation, Venting, Sparging, Bioaugmentation and Phytoremediation are the practical ways in which heavy metal bioremediation can be done on site. Such organisms include *Pseudomonas, Sphingomonas, Rhodococcus, Alcaligenes*, and *Mycobacterium* are crucial in bioremediation processes. Ex-situ technologies like land farming, composting, biopiles as well as bio-reactors have also shown good results when it comes to removing heavy metals from sites that have been contaminated with these elements. Although bioremediation has several advantages such as being cost-effective and eco-friendly plus it is also sustainable and easy to implement; it tends to be slow while the byproducts resulting from biodegradation may be more poisonous than the initial compounds. For that reason improvement of those activities is currently under development aimed at more efficient removal of heavy metals from polluted environments (Abo-Alkasem et al., 2023; Sharma, 2020). Membrane filtration seems to work great at decontaminating a variety of heavy metals hence could be said that membrane filtration is an efficient and simple operation technology with a great potential for decontamination of heavy metals as was put forward by Xiang et al (2022). Some techniques used so far include microfiltration (MF), ultrafiltration (UF), reverse osmosis (RO) and electrodialysis (Xiang et al., 2022). UF processes involve a pore size range between 10 and 100 Å which is a low pressure driven membrane process for removing dissolved and colloidal materials. Micellar-enhanced ultrafiltration (MEUF) and polymer-enhanced ultrafiltration (PEUF) are two efficient ways of removing metal ions from solutions. Electrodialysis (ED) is an effective process that uses electric field to separate ions across charged membranes, transferring them from one solution to another. Ion exchange has also been used successfully in eliminating cationic and anionic impurities present in water for several reasons such as high treatment capacity, removal efficiency and kinetics; also the artificial solid resin works better than natural one. This occurs on a cationic column where the metal ions replace hydrogen ions on the resin during passage of heavy metal solution through it (Qasem et al., 2021).

A few of the factors that influence the absorption of heavy metal ions by ion-exchange resins include pH, temperature, initial metal concentration, and contact time (Qasem et al., 2021). Zeolites and montmorillonites are naturally occurring minerals capable of satisfactorily exchanging cations with heavy metal ions in various experimental conditions. Amongst these minerals, clinoptilolite which is a zeolite species has widely been studied and well known for its great affinity to heavy metals. However, there is need for improvement if compared to synthetic resins when using zeolites and montmorillonites as ion exchange resins for heavy metal decontamination. Moreover, their application is restricted to laboratory experiments only (De Magalhães et al., 2022). Electrokinetics remediation (ER) technique was applied to remove heavy metals from polluted fine-grain soils. This method involves three phenomena namely electroosmosis, electromigration and electrophoresis which

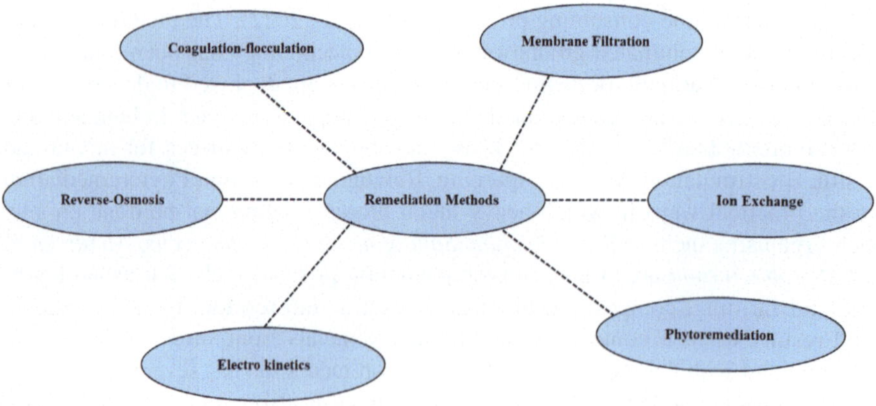

Fig. 6.4 The process of removing heavy metal ions from wastewater through water treatment technologies

define it. It's a controllable method used in situ or ex-situally too. The method is appropriate for the treatment of low permeability soils and removal of organic matters as well as inorganic soil impurities. The electrokinetics approach can eliminate pollutants like pesticides, polychlorinated biphenyls (PCBs) as well as heavy metals among others. Nevertheless, this tool could be less effective in removing non-polar contaminants while the remediation process takes longer than usual.

6.7 Application of GIS in Groundwater Studies

The application of remote sensing and Geographic Information Systems (GIS) have revolutionized the field of hydrogeology by providing the terrain with synoptic view and repetitive data coverage (Nag & Kundu, 2018; Suvarna et al., 2023). This technology is very much cost effective and has proven to be highly effective compared to traditional methods of data capture. Therefore, an accurate identification of controlling features of groundwater can be obtained by analyzing satellite images taken from a distance which may lead to proper classification of terrain according to ground water potentiality and prosperity. By using GIS, integration of various information like geomorphology, drainage, land use and vegetation can help in extraction of groundwater prospectivity at any given area (Berhanu & Hatiye, 2020; Das & Pardeshi, 2018). Groundwater is the major source for water in northern part, particularly arid Gujarat which is the seventh largest state in India. The Aravalli region largely depends on ground water due to its rocky terrains. However, clayey sub-soil has contributed to water logging in some parts of Mahesana District e.g., Kheralu, Vadnagar, Visnagar, Unjha Talukas etc. Visual Modflow software tool will be used for developing hydro-geological model so that it can identify areas where there are imminent risks such as waterlogged area determines rechargeable quantum etc. through lower aquifer recharge. Remote sensing and GIS have provided elaborate understanding on hydrogeology that is vital for groundwater management which aims at sustainable development. Future generations will benefit from adequate groundwater resources if suitable models are developed along with implementing appropriate recharge measures (Pandey et al., 2013) conducted a study on the use of geospatial technology for groundwater mapping, while Akbar et al. (2011) assessed the technology's potential in determining groundwater quality. Another study by Wahyuni et al. (2008) focused on using geospatial technology to identify groundwater storage potential for effective management purposes.

6.8 Future Aspects

Groundwater is vital for many functions such as domestic use, industry and agriculture that seek to ensure ecological sustainability through provision of fresh water. Groundwater resources on the other hand continuously depleting and deteriorating due to over exploitation and pollution posing grave environmental and health complications. Still industrial wastewaters containing heavy metal ions are released into streams, which in turn contaminate groundwater resources despite strict regulatory frameworks. Consequently, there is need to have a real-time monitoring system to keep track of groundwater sources especially those that are contaminated with heavy metals for sustainable management of these resources while fostering socio-economic growth. Furthermore, industrial wastes must be treated before being released into ground and other surface waters using effective strategies. In case of large scale treatment of contaminated groundwater membrane filtration as well as adsorption should be given priority among advanced treatment technologies because they produce purified water and also allow recovery of valuable heavy metals from polluted water. Treatment process should thus include long term planning considering future industrial expansions, irrigation requirement and drinking water supply so as to optimize the groundwater resource.

6.9 Conclusion

Groundwater needs to be managed, monitored and reported on so as to ensure its sustainable use. Unfortunately, groundwater pollution can occur naturally or through human activities. Arsenic, a highly toxic substance causing cancer and mutagensis, is one of the most dangerous pollutants in groundwater. In many cases bioremediation offers an alternative for cleaning up polluted groundwater. This technique destroys harmful substances rendering them harmless through the activity of microorganisms. Besides this phytoremediation is also a very productive method of water cleanse through plants' absorption. This process has been proven extremely successful in restoring contaminated areas because it depends on several root structure, biomass, pollutant toxicity adaptability, and pollutant removal rates. Well this approach is less invasive compared to others; environmentally friendly, and less expensive. Despite its potential, however, education on groundwater is limited due to the multidisciplinary nature of groundwater science. To address groundwater issues, researchers must undertake extensive studies of soil topology, groundwater quality, emerging pollutants, and health risk assessments. Further research is needed to identify the causes and solutions to groundwater contamination. Although the future of groundwater research is uncertain, the scope of research is expanding. This comprehensive overview of groundwater quality factors and associated health risks will broaden the scope of multidisciplinary research and help researchers develop more effective solutions for groundwater contamination.

Acknowledgements The author expresses gratitude to Gujarat Institute of Desert Ecology, Bhuj-Kachchh, for supporting their research, and acknowledges the anonymous reviewers' constructive feedback to enhance the manuscript.

Data Availability Statement Not applicable.

Ethics Approval Not applicable.

Conflict of Interest The author declares no conflict of interest.

References

Abo-Alkasem, M. I., Hassan, N. H., & Abo Elsoud, M. M. (2023). Microbial bioremediation as a tool for the removal of heavy metals. *Bulletin of the National Research Centre, 47*(1), 31. https://doi.org/10.1186/s42269-023-01006-z

Ahmed, S. F., Kumar, P. S., Rozbu, M. R., Chowdhury, A. T., Nuzhat, S., Rafa, N., Mahlia, T. M. I., Ong, H. C., & Mofijur, M. (2022). Heavy metal toxicity, sources, and remediation techniques for contaminated water and soil. *Environmental Technology and Innovation, 25*, 102114. https://doi.org/10.1016/j.eti.2021.102114

Akbar, T. A., Lin, H., & DeGroote, J. (2011). Development and evaluation of GIS-based ArcPRZM-3 system for spatial modeling of groundwater vulnerability to pesticide contamination. *Computers and Geosciences, 37*(7), 822–830. https://doi.org/10.1016/j.cageo.2011.01.011

Akhtar, N., Syakir Ishak, M. I., Bhawani, S. A., & Umar, K. (2021). Various natural and anthropogenic factors responsible for water quality degradation: A review. *Water, 13*(19), 2660. https://doi.org/10.3390/w13192660

Al, T. A., Martin, C. J., & Blowes, D. W. (2000). Carbonate-mineral/water interactions in sulfide-rich mine tailings. *Geochimica et Cosmochimica Acta, 64*(23), 3933–3948. https://doi.org/10.1016/S0016-7037(00)00483-X

Alengebawy, A., Abdelkhalek, S. T., Qureshi, S. R., & Wang, M.-Q. (2021). Heavy metals and pesticides toxicity in agricultural soil and plants: ecological risks and human health implications. *Toxics, 9*(3), 42. https://doi.org/10.3390/toxics9030042

Ali, H., Khan, E., & Ilahi, I. (2019). Environmental chemistry and ecotoxicology of hazardous heavy metals: Environmental persistence, toxicity, and bioaccumulation. *Journal of Chemistry, 2019*, 6730305. https://doi.org/10.1155/2019/6730305

Arslan, Ş, Yücel, Ç., Çallı, S. S., & Çelik, M. (2017). Assessment of heavy metal pollution in the groundwater of the Northern Develi Closed Basin, Kayseri, Turkey. *Bulletin of Environmental Contamination and Toxicology, 99*(2), 244–252. https://doi.org/10.1007/s00128-017-2119-1

Balali-Mood, M., Naseri, K., Tahergorabi, Z., Khazdair, M. R., & Sadeghi, M. (2021). Toxic mechanisms of five heavy metals: Mercury, lead, chromium, cadmium, and arsenic. *Frontiers in Pharmacology*. https://doi.org/10.3389/fphar.2021.643972

Berhanu, K. G., & Hatiye, S. D. (2020). Identification of groundwater potential zones using proxy data: Case study of Megech Watershed, Ethiopia. *Journal of Hydrology: Regional Studies, 28*, 100676. https://doi.org/10.1016/j.ejrh.2020.100676

Briffa, J., Sinagra, E., & Blundell, R. (2020). Heavy metal pollution in the environment and their toxicological effects on humans. *Heliyon, 6*(9), e04691. https://doi.org/10.1016/j.heliyon.2020.e04691

Carrard, N., Foster, T., & Willetts, J. (2019). Groundwater as a source of drinking water in Southeast Asia and the Pacific: A multi-country review of current reliance and resource concerns. *Water, 11*(8), 1605. https://doi.org/10.3390/w11081605

Collin, M. S., Venkatraman, S. K., Vijayakumar, N., Kanimozhi, V., Arbaaz, S. M., Stacey, R. G. S., Anusha, J., Choudhary, R., Lvov, V., Tovar, G. I., Senatov, F., Koppala, S., & Swamiappan, S. (2022). Bioaccumulation of lead (Pb) and its effects on human: A review. *Journal of Hazardous Materials Advances, 7*, 100094. https://doi.org/10.1016/j.hazadv.2022.100094

Dangar, S., Asoka, A., & Mishra, V. (2021). Causes and implications of groundwater depletion in India: A review. *Journal of Hydrology, 596*, 126103. https://doi.org/10.1016/j.jhydrol.2021.126103

Das, S., & Pardeshi, S. D. (2018). Integration of different influencing factors in GIS to delineate groundwater potential areas using IF and FR techniques: A study of Pravara basin, Maharashtra, India. *Applied Water Science, 8*(7), 197. https://doi.org/10.1007/s13201-018-0848-x

Das, S., Sultana, K. W., Ndhlala, A. R., Mondal, M., & Chandra, I. (2023). Heavy metal pollution in the environment and its impact on health: Exploring green technology for remediation. *Environmental Health Insights, 17*, 11786302231201260. https://doi.org/10.1177/11786302231201259

De Magalhães, L. F., da Silva, G. R., & Peres, A. E. C. (2022). Zeolite application in wastewater treatment. *Adsorption Science and Technology, 2022*, 4544104. https://doi.org/10.1155/2022/4544104

Dewangan, S., & Bhatia, A. K. (2023). Heavy metal contamination in air, groundwater, freshwater and soil. In *Heavy Metals in the Environment: Management Strategies for Global Pollution* (pp. 79–101). American Chemical Society

Ding, C., Chen, J., Zhu, F., Chai, L., Lin, Z., Zhang, K., & Shi, Y. (2022). Biological toxicity of heavy metal(loid)s in natural environments: From microbes to humans. *Frontiers in Environmental Science*. https://doi.org/10.3389/fenvs.2022.920957

Eid, M. H., Eissa, M., Mohamed, E. A., Ramadan, H. S., Tamás, M., Kovács, A., & Szűcs, P. (2024). New approach into human health risk assessment associated with heavy metals in surface water and groundwater using Monte Carlo Method. *Scientific Reports, 14*(1), 1008. https://doi.org/10.1038/s41598-023-50000-y

El-taweel, R. M., Mohamed, N., Alrefaey, K. A., Husien, S., Abdel-Aziz, A. B., Salim, A. I., Mostafa, N. G., Said, L. A., Fahim, I. S., & Radwan, A. G. (2023). A review of coagulation explaining its definition, mechanism, coagulant types, and optimization models; RSM, and ANN. *Current Research in Green and Sustainable Chemistry, 6*, 100358. https://doi.org/10.1016/j.crgsc.2023.100358

Etikala, B., Madhav, S., & Somagouni, S. G. (2022). Urban water systems: An overview. In *Current directions in water scarcity research* (pp. 1–19). Elsevier. https://doi.org/10.1016/b978-0-323-91838-1.00016-6

Falkenberg, T., & Saxena, D. (2018). Impact of wastewater-irrigated urban agriculture on diarrhea incidence in Ahmedabad, India. *Indian Journal of Community Medicine, 43*(2), 102–106. https://doi.org/10.4103/ijcm.IJCM_192_17

Ferronato, N., & Torretta, V. (2019). Waste mismanagement in developing countries: A review of global issues. *International Journal of Environmental Research and Public Health, 16*(6), 1060. https://doi.org/10.3390/ijerph16061060

Genchi, G., Carocci, A., Lauria, G., Sinicropi, M. S., & Catalano, A. (2020). Nickel: Human health and environmental toxicology. *International Journal of Environmental Research and Public Health, 17*(3), 679. https://doi.org/10.3390/ijerph17030679

Guo, J.-J., Huang, X.-P., Xiang, L., Wang, Y.-Z., Li, Y.-W., Li, H., Cai, Q.-Y., Mo, C.-H., & Wong, M.-H. (2020). Source, migration and toxicology of microplastics in soil. *Environment International, 137*, 105263. https://doi.org/10.1016/j.envint.2019.105263

Ingrao, C., Strippoli, R., Lagioia, G., & Huisingh, D. (2023). Water scarcity in agriculture: An overview of causes, impacts and approaches for reducing the risks. *Heliyon, 9*(8), e18507. https://doi.org/10.1016/j.heliyon.2023.e18507

Kafle, A., Timilsina, A., Gautam, A., Adhikari, K., Bhattarai, A., & Aryal, N. (2022). Phytoremediation: Mechanisms, plant selection and enhancement by natural and synthetic agents. *Environmental Advances, 8*, 100203. https://doi.org/10.1016/j.envadv.2022.100203

Kamran, M., Parveen, A., Ahmar, S., Malik, Z., Hussain, S., Chattha, M. S., Saleem, M. H., Adil, M., Heidari, P., & Chen, J.-T. (2019). An Overview of Hazardous Impacts of Soil Salinity in Crops, Tolerance Mechanisms, and Amelioration through Selenium Supplementation. *International Journal of Molecular Sciences, 21*(1), 148. https://doi.org/10.3390/ijms21010148

Kapshe, M., Kuriakose, P. N., Srivastava, G., & Surjan, A. (2013). Analysing the co-benefits: Case of municipal sewage management at Surat, India. *Journal of Cleaner Production, 58*, 51–60. https://doi.org/10.1016/j.jclepro.2013.07.035

Kayastha, V., Patel, J., Kathrani, N., Varjani, S., Bilal, M., Show, P. L., Kim, S.-H., Bontempi, E., Bhatia, S. K., & Bui, X.-T. (2022). New Insights in factors affecting ground water quality with focus on health risk assessment and remediation techniques. *Environmental Research, 212*(Pt A), 113171. https://doi.org/10.1016/j.envres.2022.113171

Khatri, N., & Tyagi, S. (2015). Influences of natural and anthropogenic factors on surface and groundwater quality in rural and urban areas. *Frontiers in Life Science, 8*(1), 23–39. https://doi.org/10.1080/21553769.2014.933716

Kumar, A., Ali, M., Kumar, R., Kumar, M., Sagar, P., Pandey, R. K., Akhouri, V., Kumar, V., Anand, G., Niraj, P. K., Rani, R., Kumar, S., Kumar, D., Bishwapriya, A., & Ghosh, A. K. (2021). Arsenic exposure in Indo Gangetic plains of Bihar causing increased cancer risk. *Scientific Reports, 11*(1), 2376. https://doi.org/10.1038/s41598-021-81579-9

Kumar, P., Srivastava, S., Banerjee, A., & Banerjee, S. (2022). Prevalence and predictors of water-borne diseases among elderly people in India: Evidence from Longitudinal Ageing Study in India, 2017–18. *BMC Public Health, 22*(1), 993. https://doi.org/10.1186/s12889-022-13376-6

Kumar, V., Maity, A., Kumar, A., Saha, S., Kay, P., Singh, B., & Mukherjee, T. (2023). Critical review on uranium and arsenic content and their chemical mobilization in groundwater: A case study of the Malwa region Punjab, India. *Science of the Total Environment, 885*, 163885. https://doi.org/10.1016/j.scitotenv.2023.163885

Li, P., Karunanidhi, D., Subramani, T., & Srinivasamoorthy, K. (2021). Sources and consequences of groundwater contamination. *Archives of Environmental Contamination and Toxicology, 80*(1), 1–10. https://doi.org/10.1007/s00244-020-00805-z

Malyan, S. K., Singh, R., Rawat, M., Kumar, M., Pugazhendhi, A., Kumar, A., Kumar, V., & Kumar, S. S. (2019). An overview of carcinogenic pollutants in groundwater of India. *Biocatalysis and Agricultural Biotechnology, 21*, 101288. https://doi.org/10.1016/j.bcab.2019.101288

Mawari, G., Kumar, N., Sarkar, S., Frank, A. L., Daga, M. K., Singh, M. M., Joshi, T. K., & Singh, I. (2022). Human health risk assessment due to heavy metals in ground and surface water and association of diseases with drinking water sources: A study from Maharashtra, India. *Environmental Health Insights, 16*, 11786302221146020. https://doi.org/10.1177/11786302221146020

Mohanty, B., & Das, A. (2023). Heavy metals in agricultural cultivated products irrigated with wastewater in India: A review. *AQUA—Water Infrastructure, Ecosystems and Society, 72*(6), 851–867. https://doi.org/10.2166/aqua.2023.122

Müller, A., Österlund, H., Marsalek, J., & Viklander, M. (2020). The pollution conveyed by urban runoff: A review of sources. *Science of the Total Environment, 709*, 136125. https://doi.org/10.1016/j.scitotenv.2019.136125

Musie, W., & Gonfa, G. (2023). Fresh water resource, scarcity, water salinity challenges and possible remedies: A review. *Heliyon, 9*(8), e18685. https://doi.org/10.1016/j.heliyon.2023.e18685

Nag, R., O'Rourke, S. M., & Cummins, E. (2022). Risk factors and assessment strategies for the evaluation of human or environmental risk from metal(loid)s—A focus on Ireland. *Science of the Total Environment, 802*, 149839. https://doi.org/10.1016/j.scitotenv.2021.149839

Nag, S. K., & Kundu, A. (2018). Application of remote sensing, GIS and MCA techniques for delineating groundwater prospect zones in Kashipur block, Purulia district, West Bengal. *Applied Water Science, 8*(1), 38. https://doi.org/10.1007/s13201-018-0679-9

Obaideen, K., Shehata, N., Sayed, E. T., Abdelkareem, M. A., Mahmoud, M. S., & Olabi, A. G. (2022). The role of wastewater treatment in achieving sustainable development goals (SDGs) and sustainability guideline. *Energy Nexus, 7*, 100112. https://doi.org/10.1016/j.nexus.2022.100112

Okoro, H. K., Orosun, M. M., Oriade, F. A., Momoh-Salami, T. M., Ogunkunle, C. O., Adeniyi, A. G., Zvinowanda, C., & Ngila, J. C. (2023). Potentially toxic elements in pharmaceutical industrial effluents: A review on risk assessment, treatment, and management for human health. *Sustainability, 15*(8), 6974. https://doi.org/10.3390/su15086974

Pandey, V. P., Shrestha, S., & Kazama, F. (2013). A GIS-based methodology to delineate potential areas for groundwater development: A case study from Kathmandu Valley, Nepal. *Applied Water Science, 3*(2), 453–465. https://doi.org/10.1007/s13201-013-0094-1

Parisi, S., Paternoster, M., Perri, F., & Mongelli, G. (2011). Source and mobility of minor and trace elements in a volcanic aquifer system: Mt. Vulture (southern Italy). *Journal of Geochemical Exploration, 110*(3), 233–244. https://doi.org/10.1016/j.gexplo.2011.06.010

Pavel, M., Anastasescu, C., State, R.-N., Vasile, A., Papa, F., & Balint, I. (2023). Photocatalytic Degradation of organic and inorganic pollutants to harmless end products: Assessment of practical application potential for water and air cleaning. *Catalysts, 13*(2), 380. https://doi.org/10.3390/catal13020380

Qasem, N. A. A., Mohammed, R. H., & Lawal, D. U. (2021). Removal of heavy metal ions from wastewater: a comprehensive and critical review. *NPJ Clean Water, 4*(1), 36. https://doi.org/10.1038/s41545-021-00127-0

Shanthakumar, S., Banerjee, M., & Batas, A. (2021). Ground water management in Gujarat: An empirical study. *IOP Conference Series: Earth and Environmental Science, 822*(1), 012022. https://doi.org/10.1088/1755-1315/822/1/012022

Sharma, I. (2020). Bioremediation techniques for polluted environment: Concept, advantages, limitations, and prospects. In M. A. Murillo-Tovar, H. Saldarriaga-Noreña, & A. Saeid (Eds.), *Trace metals in the environment*. Rijeka: IntechOpen.

Sharma, S., Nagpal, A. K., & Kaur, I. (2019). Appraisal of heavy metal contents in groundwater and associated health hazards posed to human population of Ropar wetland, Punjab, India and its environs. *Chemosphere, 227*, 179–190. https://doi.org/10.1016/j.chemosphere.2019.04.009

Suvarna, B., Sunitha, V., Sudharshan Reddy, Y., Muralidhara Reddy, B., Kadam, A. K., & Ramakrishna Reddy, M. (2023). Groundwater quality assessment using multivariate statistical approach and geospatial modelling around cement industrial corridor, South India. *International Journal of Environmental Science and Technology, 20*(5), 5051–5070. https://doi.org/10.1007/s13762-022-04210-y

Swain, S., Taloor, A. K., Dhal, L., Sahoo, S., & Al-Ansari, N. (2022). Impact of climate change on groundwater hydrology: A comprehensive review and current status of the Indian hydrogeology. *Applied Water Science, 12*(6), 120. https://doi.org/10.1007/s13201-022-01652-0

Thushari, G. G. N., & Senevirathna, J. D. M. (2020). Plastic pollution in the marine environment. *Heliyon, 6*(8), e04709. https://doi.org/10.1016/j.heliyon.2020.e04709

Uddin, A., Shamsudduha, M., Saunders, J. A., Lee, M.-K., Ahmed, K. M., & Chowdhury, M. T. (2011). Mineralogical profiling of alluvial sediments from arsenic-affected Ganges-Brahmaputra floodplain in central Bangladesh. *Applied Geochemistry, 26*(4), 470–483. https://doi.org/10.1016/j.apgeochem.2011.01.006

Wahi N (2022) The Evolution of the Right to Water in India. Water 14(3). https://doi.org/10.3390/w14030398

Wahyuni, S., Oishi, S., & Sunada, K. (2008). The estimation of the groundwater storage and its distribution in Uzbekistan. *Doboku Gakkai Ronbunshuu B, 52*, 31–36.

Wu, R., Podgorski, J., Berg, M., & Polya, D. A. (2021). Geostatistical model of the spatial distribution of arsenic in groundwaters in Gujarat State, India. *Environmental Geochemistry and Health, 43*(7), 2649–2664. https://doi.org/10.1007/s10653-020-00655-7

Xiang, H., Min, X., Tang, C.-J., Sillanpää, M., & Zhao, F. (2022). Recent advances in membrane filtration for heavy metal removal from wastewater: A mini review. *Journal of Water Process Engineering, 49*, 103023. https://doi.org/10.1016/j.jwpe.2022.103023

Chapter 7
The Importance of Plants for Environmental Health and Water Quality

Mangesh Vedpathak, Balbhim Chavan, and Indra Jeet Chaudhary

Abstract The plant is one of the essential life forms on Earth for environmental sustainability and groundwater purification. Biotic and abiotic components of environment are relatively connected with each others. Plants of all sizes and shapes were present in both aquatic and terrestrial ecosystems. Some examples include trees, mosses, ferns, vines, grasses, shrubs, herbs, flowering plants, and green algae. The plants contribute significantly to the environment and ecological equilibrium. They are essential elements to our environment and maintain ecological balance. Plants and animals have habitats in terrestrial and aquatic ecosystems, such as the ocean, reservoirs, forests, grasslands, mountains, and deserts. Additionally, plants supply living things with the three fundamental necessities of food, clothing, and shelter. In addition to providing us with food and clean air daily, plants play a crucial role in maintaining a healthy ecology. As a result, trees are the primary food source for all living things. In addition, they provide us with oxygen, medicines, fuel, wood, recreation, industrial goods, preservatives, pesticides, and other supplies. Plants make our environment better by eliminating air pollution and groundwater recharged. This chapter focuses on the role of plants in a sustainable environment and water quality.

Keywords Air pollution · Climate · Environment · Forest · Habitats · Plants · Soil

---

M. Vedpathak
School of Technology, Department of Civil Engineering, Sanjay Ghodawat University, Kolhapur, Maharashtra, India
e-mail: mvedpathak512@gmail.com

B. Chavan
Department of Environmental Science, Dr. Babasaheb Ambedkar Marathwada University, Sambhaji Nagar, Maharashtra, India
e-mail: drblchavan@yahoo.co.in

I. J. Chaudhary (✉)
Department of Environmental Science, Savitribai Phule Pune University (University of Pune), Pune, Maharashtra, India
e-mail: indrajeet.cug@gmail.com

V. Sunitha et al. (eds.), *Groundwater Resource Management Planning Strategies*, Springer Water, https://doi.org/10.1007/978-3-031-88870-0_7

7.1 Introduction

Nature alerts us to its imbalance in several ways, including the rise in global temperature, climate change, shifting seasonal patterns, unpredictable rainfall, excessive rain, and drought (Upadhyay, 2020). Trees significantly influence every one of these natural cycles. It can be seen that all of India's ancient cultures have talked about how excellent tree wealth is. All living things, including humans, require energy to drive their metabolic processes of growth, development, and reproduction. The process by which green plants make their food is called photosynthesis, which organisms use to stay alive. During photosynthesis, photoautotrophs transform carbon dioxide and water into glucose and oxygen. Oxygen and glucose are the two main by-products of photosynthesis, both essential to living things' survival (Madkour, 2019). As a result, photosynthesis is required for all organisms to exist on Earth. Medicinal plants have been used as a source of medicine since prehistoric times of human society. Some plant species removed Heavy metals from polluted soil and water (Awa & Hadibarata, 2020). Because the flora provides the fauna with food, the fauna cannot produce their food. Plants are also used to balance gaseous cycles and reduce air pollution. Plants also play a crucial role in keeping the soil fertile, reducing erosion and landslides, and controlling floods (Saco et al., 2021). To provide food for the growing population, grain crops were grown, which resulted in the development of cultivated varieties with high yields.

Despite air pollution control and groundwater recharge, plants serve many resources, such as food, fodder, vegetables, spices, and fruits, which are extracted from plants and utilized in a variety of ways, including the construction of shelters, animal shelters, house parts, and wild genetic resources for crop plant improvement. These plants are also economically advantageous to farmers as a source of high-quality forage and assistance in crop management, such as pesticides, compost, and green manure (Adegbeye et al., 2020). Without plants, we cannot exist. The United Nations General Assembly established the U.N. Sustainable Development Goals (UN SDG) to improve people's living conditions. Planting trees and protecting them is an excellent way to help achieve these objectives, particularly in urban areas where most people live. Trees are great for well-being and social prosperity since they decrease air contamination, lessen pressure, energize more active work, support significant areas of strength, and give different assets. The main aim of this chapter is to provide knowledge of the importance of plant information on pollution removal from the environment and groundwater recharge and quality control.

7.2 Role of Plants for Environmental Sustainability

Plants provide nutrients to all living organisms directly or indirectly. They balance the ecosystem by releasing oxygen, absorbing carbon dioxide, and improving air quality. Plants also help remove organic and inorganic pollutants from the environment. They

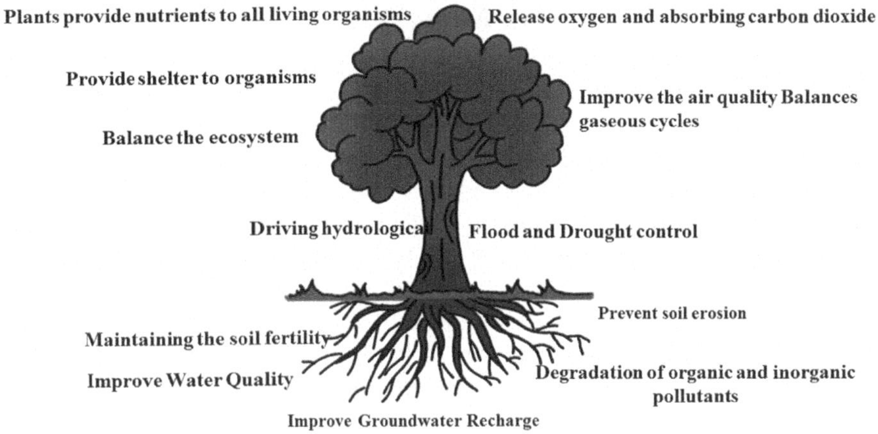

Fig. 7.1 Role of plants for environmental sustainability

provide food and shelter to organisms. Plants can prevent soil erosion, drought, and flood (Fig. 7.1).

7.2.1 Plants for Pharmaceutical Purposes

In the past, most food and medicine came from plants and their products; numerous modern pharmaceuticals have been separated and are now utilized to benefit living things. In today's developing nations, ensuring the safety, quality, and usefulness of herbal products and medicinal plants has become a significant concern. The Vedas, the Quran, and the Bible mention the extensive use of herbal remedies and medical preparations (Ansari, 2021; Tiwari et al., 2019). For a long time, medicinal plants have been used to treat health conditions, preserve food, flavor food, and prevent diseases, including epidemics. Over many centuries, information about their therapeutic properties has been passed down. Plants have been an essential source of natural products for human health for a long time, especially in the last ten years, when more intensive research has been done on natural therapies. Herbal medicine is a very safe and effective method for treating cancer when compared to modern Western medicine. The biological properties of plant species that are utilized worldwide for various purposes, including the treatment of infectious diseases, are typically responsible for the active compounds produced during secondary metabolism (Jain et al., 2019).

7.2.2 Habitats for Flora and Fauna

Plants and animals naturally reside in habitats. There are specific types of habitats in both terrestrial and aquatic habitats, which are habitats with regions of land and water. These areas include the flourished or barren land areas oruncovered or covered by water, such as ponds, rivers, and oceans. Terrestrial habitats include forests, grasslands, and deserts on the earth's surface. Finding out the structure and variety of the plant cover is a crucial step in gathering data regarding the suitability of the habitat, the degree of disturbance, the recovery process, and management efforts. Furthermore, according to Haq et al. (2022), plant diversity reflects the health, productivity, and disturbance of the forest ecosystem at the moment, which are crucial for drawing wildlife species to live, feed, rest, and reproduce. The dry-temperate forest ecosystem's most notable feature is the composition and diversity of its plant cover, which is considered a crucial home for various wildlife and fauna species. The site's quality, aspect, slope, microclimate, terrain, and human influence can all affect the flora of dry-temperate forests, which can differ from place to place. Variations in species composition and structure create heterogeneous habitats rich in food resources and exhibit spatial variability in layers. These habitats support a higher diversity of wildlife species that rely on the structure and composition of plant cover for survival, either directly or indirectly (Russo et al., 2023). Wildlife fauna species have a strong relationship with plant diversity and richness, as they are habitat specialists. This may be because diverse habitats provide a range of food sources, secure breeding grounds, appropriate resting places, and protection from predators and inclement weather (Rajpar & Zakaria, 2014; Rajpar et al., 2018).

7.2.3 Balances with Gaseous Cycles

The water, carbon, nitrogen, phosphorus, and sulfur cycles are important biogeochemical cycles for living things. Energy moves through various mediums in ecosystems through global cycle processes, including organisms, the atmosphere, rocks, bodies of water, and soil (Thies & Grossman, 2023). Due to biogeochemical cycles, essential elements are also available to plants and other species. Decomposers break down organic materials into simpler compounds with calcium, nitrogen, phosphorus, water, and carbon dioxide (Findlay, 2021). The things that plants require to grow are these.

The urban environment exacerbates many environmental problems, including the urban heat island effect, summer heat waves, stormwater runoff and flood risk, and chemical and particle pollution of the air, soil, and water (Varma et al., 2021). It is possible to manage urban trees and the urban forest to influence the cycles of urban heat, water, carbon, and pollutants. However, more and more empirical data is needed to determine the extent of the positive and negative effects of urban trees and how the built environment and climate zone affect those effects. This section presents new research that expands our understanding of urban forests' biological and environmental benefits (Lepczyk et al., 2017). The fourteen papers presented offer an international viewpoint on the function of trees in towns and cities across five continents. Studies show that local microclimates in urban green spaces, both with and without trees, have a cooling effect (Grilo et al., 2020). Some studies only address the mesoscale cooling advantage of transpiration from urban trees or the street and pedestrian scale cooling benefit of canopy shade. Additional research focuses on how tree species intercept rainfall through their canopy, how water is absorbed from biofilter systems, and how stormwater runoff nitrogen intake enhances water quality. The research presented here considers both the advantages and disadvantages of trees for air quality, taking into account their ability to release and remove air pollutants such as ozone and potentially dangerous volatile organic compounds and allergic particles. A transdisciplinary framework for future urban forest research is proposed to understand better and express the role of urban trees in highly controlled, highly disrupted urban biogeochemical cycles critical to human health and well-being.

7.2.4 Reduction in Air Pollution

Air pollution, brought on by industrial processes, power plants, home heating, automobile traffic, and other sources, is the primary issue affecting the environment in every Indian city (Chaudhary and Rathore, 2018a, 2018b). Growing green plants in and around industrial and urban areas is a suitable biological method for reducing air pollutants (Chaudhary and Rathore, 2019). Kapoor and Chittora (2016) in the experiment, concluded that *Mangifera indica* L., *Pongamia pinnata* (L.), *Pierre, Dalbergia sissoo* Roxb, and *Holoptelea integrifolia* L. are helpful for bio-monitoring, the development of green belts around industrial areas and roadside as well as to reduce air pollution from various sources including vehicles and industries. Through the stomata located on the leaf's surface, trees purify the air, remove dust and particulates, and absorb ozone, carbon monoxide, sulfur dioxide, and other pollutants. Plants such as *Mangifera indica, Albizia lebbeck, Magnolia champaca, Psidium guajava, Bougainvillea spectabilis, Thespesia populnea,* and *Terminalia catappa* are categorized as tolerant based on their high air pollution tolerance indices. With low air pollution tolerance indices, the species *Bauhinia varigeta, Tecoma stanus, Tabernaemontana divaricate,* and *Muntingia calabura* are grouped as being more prone to pollution. To increase greenery and reduce air pollution, species with a high air pollution tolerance index are essential and suggested for plantation (Rupa & Venkatachalam, 2018; Chaudhary & Rathore, 2019).

Plants and human life are severely harmed by urban pollution, which includes dust particles and gaseous pollutants (Manisalidis et al., 2020; Chaudhary & Rathore, 2019, 2021a, 2021b, 2021c). Growing quantities of particulate matter have emerged as the primary contaminant in many densely populated cities. Urban trees enhance local air quality by removing air pollutants. In Guangzhou, urban trees removed about 312.03 mg of air pollutants in 2000. The bulk of the cleanup was completed in the winter due to the elevated pollutant concentrations. Moreover, a thick, continuous tree canopy may increase the efficiency of air pollutant removal. The results of previous studies on the removal rates of air pollution include the removal capacity of urban trees in American cities (Nowak et al., 2013). The pollutant removal calculations are regarded as macro-scale estimations since average pollutant deposition velocity, tree cover, and monthly average air pollutant concentration were used in the computations (Table 7.1).

One of the best ways to reduce air pollution in urban areas is to plant trees and shrubs. When choosing plants, consider color, robustness, space, watering requirements, and potential to help improve air quality (Babu et al., 2013). Understanding

Table 7.1 Dust removal efficiency of tree species

S. No.	Plant species	Dust deposition	References
1	*Ficus benghalensis*	1.1–1.3 mg/cm^2	Rai (2016)
2	*Ficus benghalensis*	0.0003–0.001 mg/cm^2	Chaudhary and Rathore (2018a)
3	*Psidium guajava*	1–1.2 mg/cm^2	Rai (2016)
4	*Bougainvillea spectabilis*	1–1.5 mg/cm^2	
5	*Mangifera indica*	1–1.1 mg/cm^2	
6	*Lantana camara*	1–1.1 mg/cm^2	
7	*Hibiscus rosasinesis*	1–1.1 mg/cm^2	
8	*Lagerstroemia speciosa*	0.9–1.1 mg/cm^2	
9	*Artocarpus heterophyllus*	0.9–1.1 mg/cm^2	
10	*Ficus religiosa*	0.0002–0.001 mg/cm^2	Chaudhary and Rathore (2018a)
11	*Ficus religiosa*	0.9–1.1 mg/cm^2	Rai (2016)
12	*Delonix regia*	0.9–1 mg/cm^2	
13	*Cassia auriculata*	0.8–0.91 mg/cm^2	
14	*Bauhinia variegate*	0.7–0.8 mg/cm^2	
15	*Dalbergia sissoo*	0.0006–0.001 mg/cm^2	Chaudhary and Rathore (2018a)
16	*Albizia lebbeck*	0.0004–0.0008 mg/cm^2	
17	*Azadirachta indica*	0.0004–0.0006 mg/cm^2	
18	*Ficus virens*	0.00012–0.00024 mg/cm^2	
19	*Cassia fistula*	0.000005–0.00011 mg/cm^2	
20	*Agrostis stolonifera*	1.81 ± 0.06 g m^{-2} year^{-1}	Speak et al. (2012)
21	*Festuca rubra*	3.21 ± 0.1 g m^{-2} year^{-1}	
22	*Plantago lanceolata*	0.49 ± 0.02 g m^{-2} year^{-1}	
23	*Santalum album*	0.42 ± 0.01 g m^{-2} year^{-1}	
24	*Syzygium cumini*	(4 mg cm^{-2})	Chaturvedi et al. (2013)
25	*Tectona grandis*	(7 mg cm^{-2})	
26	*Anthoceopalus cadamba*	(7 mg cm^{-2})	

Table 7.2 Air pollution tolerance index of tree species

S. No.	Plants species	APTI	References
1	*Ficus benghalensis*	43.35	Chaudhary and Rathore (2018a)
2	*Ficus virens*	38.14	
3	*Dalbergia sissoo*	16.59	Navjot et al. (2016)
4	*Dalbergia sissoo*	26.61	Chaudhary and Rathore (2018a)
5	*Albizia lebbeck*	19.23	
6	*Cassia fistula*	26.96	Chaudhary and Rathore (2018a)
7	*Ziziphus zizyphus*	56.58	Marimuthu et al. (2014)
8	*Azadirachta indica*	65.24	
9	*Azadirachta indica*	14.66	Navjot et al. (2016)
10	*Azadirachta indica*	25.92	Chaudhary and Rathore (2018a)
11	*Ficus religiosa*	47.45	
12	*Juniperus chinensis*	7.72	Bui et al. (2021)
13	*Pinus densiflora*	9.52	
14	*Pinus strobus*	8.12	
15	*Acer triflorum*	8.38	

plant tolerance levels is crucial. Given that different species and types of pollutants have different effects on plants, biomonitoring of plants can be a helpful method for figuring out how susceptible a given species of plant is to air pollution. This may be accomplished by computing the Air Pollution Tolerance Index (APTI). In light of the selection of suitable plant, species for urban planting that have a greater particulate matter removal efficiency and APTI value (Table 7.2).

7.3 Role of Plant for Water Quality

The water's quality greatly depends on the plants. By holding the soil in place, controlling stream flows, and removing sediments from the water, a diverse plant cover contributes to the health of watersheds, lakes, and streams (Baulch et al., 2019). Water is naturally cleansed of pollution by trees. The canopies, trunks, roots, related soil, and other natural landscape features that filter polluted particulate matter out of the flow reduce the amount of pollution washed into a drainage region. For a long time, forests were considered important sources of clean water for drinking. A significant amount of the water required for domestic, agricultural, and industrial purposes is supplied by forested catchments. Through transpiration, trees and forests recycle atmospheric moisture to increase rainfall. The following are examples of the significance of forests and trees to water supply and rainfall: reduction of runoff, improvement of the water cycle, enhancement of water table replenishment, pollution filtering, flood control, and stormwater regulation (Ekhuemelo and Udo, 2016).

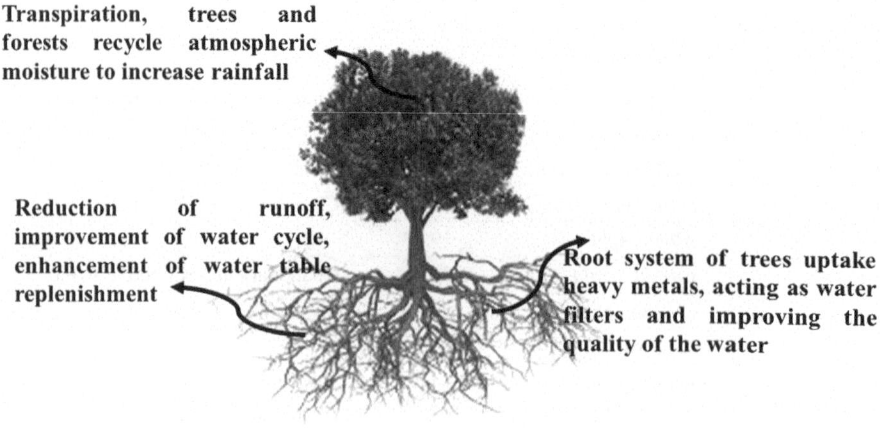

Fig. 7.2 Role of plants for water quality

Through their extensive root system, trees conduct the uptake of heavy metals, acting as water filters and improving the water quality (Fig. 7.2).

7.3.1 Degradation of Organic Pollutants

According to Srivastava et al. (2014), a significant contributor to contaminated environmental media is the widespread field application of organic compounds such as poly-aromatic hydrocarbons (PAHs), chlorinated organic compounds, polybrominated biphenyl ethers (PBEs), and polychlorinated biphenyls (PCBs). The aquatic systems are the most susceptible by these pollutants. Microbes are well-known bioremediation that can break down almost all kinds of organic compounds due to their catabolic activity (Srivastava et al., 2017). One of the main strategies used by bacteria to metabolically break down resistant organic compounds and obtain organic carbon and electron acceptors, which are abundant in the rhizospheric zone of both terrestrial and aquatic macrophytes, is co-metabolism (Shahid et al., 2020). While the microbial population primarily depends on the macrophytic species, the biodegradation rate in natural waters follows second-order kinetics. It is related to the quantity of xenobiotics and the number of bacteria (Shahid et al., 2020). Furthermore, the rhizospheric microorganisms receive organic carbon from the plants, which aids in the breakdown of complex, resistant chemical compounds like pyrenes and PAHs (Urana et al., 2020). The classic example of this concerted mutual benefit was described by Glick (2012), wherein plants receive the growth hormone indole acetic acid (IAA) due to the breakdown of PAHs by rhizospheric microbial activity. Other studies have previously documented similar findings on several aquatic plants and sediments (Kiddet al., 2021; Li et al., 2015). Furthermore, past studies by Ramesh et al.

(2023) indicate that bacteria thrive in the rich aquatic environments of algae, effectively utilizing the carbon generated from the algae for growth and multiplication. Freshwater odor and taste issues are caused by increased microorganisms (Lee et al., 2017). Aquatic plant-associated biofilm can break down dissolved organic matter (DOM), which includes PCBs (polychlorinated biphenyls) and atrazine, as well as algal-derived organics, which are primarily composed of amines, aliphatic aldehydes, and phenolics (Srivastava et al., 2017). Furthermore, the rhizoplane of aquatic plants is abundant in methanotrophs, a class of α and γ proteobacteria that use methane as a carbon source and energy source (Srivastava et al., 2017). Methanotrophs, such as Methylosinustrichosporium OB3b and Methylococcuscapsulatus, generate particulate methane monooxygenase (MMO). Through a series of enzymatic reactions that produce formaldehyde and, ultimately, the terminal compound CO_2, pMMO degrades a wide range of toxic organic compounds, particularly chlorinated ethenes (Hwangbo et al., 2023).

7.3.2 Removal of Inorganic Contaminants

Okereafor et al. (2020) found that the aquatic biota is not negatively impacted by low concentrations of metal ions that naturally occur in aquatic systems due to delayed leaching from rocks and soil. In many world regions, industrial, agricultural, and municipal waste are rivers' principal sources of excess metal ions. Numerous biochemical and biophysical factors affect the mobility of metal ions in water, such as the pH and Eh (redox potential) of the water, the presence of hydrated oxides of iron, metal carbonates, and the interaction between microbes and plants as a biofilm on the rhizosphere of macrophytes (De Giudici et al., 2017). Most metals in water form cations that stick to the negatively charged EPS of the biofilm matrix and keep metal ions from getting into the water and the plants. Most aquatic macrophytes absorb metal ions from water and have iron plaque surrounding their roots and submerged portions (Jiang et al., 2022; Serafini et al., 2023). According to Gagan et al. (2023), iron plaque is a coating of iron (hydr)/oxide precipitate that forms around plant parts as a result of iron being oxidized by molecules of O_2 or by bacteria that oxidize iron, such as Leptospirillum ferroxidans and *Ferroplasma* sp. It has been noted that plant root porosity, which raises the oxygen content at the rhizoplane, influences radial oxygen loss (Li et al., 2011). More iron plaque may build as a result of iron-oxidizing bacteria. The roles of root porosity, ROL, plaque development, and the toxic response to As (arsenic) metalloid have also been shown by Li et al. (2011); however, the latter was found to be significantly reduced with increasing plaque formation. Sulfate reduction, which occurs when bacteria linked to aquatic macrophytes as biofilm convert sulfate into sulfides and lower pH, which is necessary for microbial cells to biosorb the metal ions from the water column, is another significant metal removal process in aquatic systems, following the oxidation of iron (Chen et al., 2016). Furthermore, due to sulfate reduction, metal ions react with hydrogen sulfide in water to form metal sulfide. This precipitates in acidogenic conditions and descends to the sediments, sequestered by metal ions from the water column (Rosendo et al., 2022). In addition to macrophytes, algae can work with microbes to remove contaminants from water. For example, Zhao et al. (2023) found that a microalga called *Chlorella sorokiniana*, which was associated with the bacterium *Ralstonia basilensis* was able to enhance the adsorption of toxic metals like Cu(II), Cd(II), Ni(II), and Zn(II). This was especially true for Cu(II) adsorption, as more Cu binders were present than for the other metals. According to Sarkar et al. (2021), mycorrhizae associate endophytic symbionts with most aquatic plants and facilitate P uptake and translocation of different plant nutrients. Plants are shielded from harmful contaminants like heavy metals by mycorrhizal relationships (Srivastava et al., 2010). In Vetiver grass, a common wetland species found throughout the Indian subcontinent, South East Asia, and Australia, Srivastava et al. (2010) showed the importance of mycorrhizal association in protecting against As (III) by obstructing the membrane transport system of phosphorus.

7.3.3 Role of Plants for Driving Hydrological cycle

Transpiration is the process of releasing the water absorbed by the plants through their root system after utilizing the nutrients for building their tissues in a specified time (Bhatla & Lal, 2023). The numerous growing plants usually play a significant role in driving the hydrologic cycle in nature. The water absorbed by plants drawn into the plant's rootlets from the moisture held in soil is owing to osmotic pressure and moves up through the plants via stems and leaves. Moreover, the water is released as water vapor through the stomatal openings. The amount of such transpiration depends on the density and size of the plant or vegetation existing in that place. The water used for irrigating the crops get transpired into the air. This transpiration is dominant during the growing season of crops in agricultural lands. Most of this happens when photosynthesis is active in plants (Balasubramanian & Nagaraju, 2015).

The most delicate and valuable ecological services, such as habitat support, coastal protection, and sediment trapping, are offered by coastal wetlands (Zhao et al., 2016). These services are globally valued at billions of dollars annually (Costanza et al., 2014). Regretfully, the coastal community has suffered in recent decades due to social development and climate change, as seen by changes in the landscape, rising sea levels, deforestation, etc. (Nguyen et al., 2018; Thorne et al., 2018). This degradation will likely continue into this century (Gabler et al., 2017). Therefore, governments prioritize coastal conservation for sustainable development on a global scale (Neumann et al., 2017).

7.3.4 Trees Improve Groundwater Recharge

Springs are groundwater-dependent ecosystems (GDEs) that exist at the interface of the groundwater table and the surface, where groundwater yields are expressed (Fig. 7.3). By lowering evapotranspiration (E.T.) rates, rainfall-snow intercept by dense tree canopies, and the formation of hydrophobic soil as a result of severe wildfires, forest interventions can enhance watershed performance (Schenk et al., 2020). These effects may directly impact the yields of groundwater and surface water. Previous hydrologic studies (Schenk et al., 2020) for the 4FRI anticipated that noticeable gains in surface water yield required a 20–40% reduction in forest basal area. Nevertheless, the long-term advantages of forest thinning are debatable because any increases in water yield could be quickly obscured by vegetation regrowth and climate change (Schenk et al., 2020; Wyatt et al., 2015). More research has looked at the benefits of vegetation management on groundwater recharge to date, either globally or within the context of the 4FRI project (Wyatt, 2013; Wyatt et al., 2015). Aquifer and hydrologic responses might be minor, but if increased water yield above an environmental threshold, there could be significant, long-term benefits to GDEs, such as springs (Cartwright & Johnson, 2018).

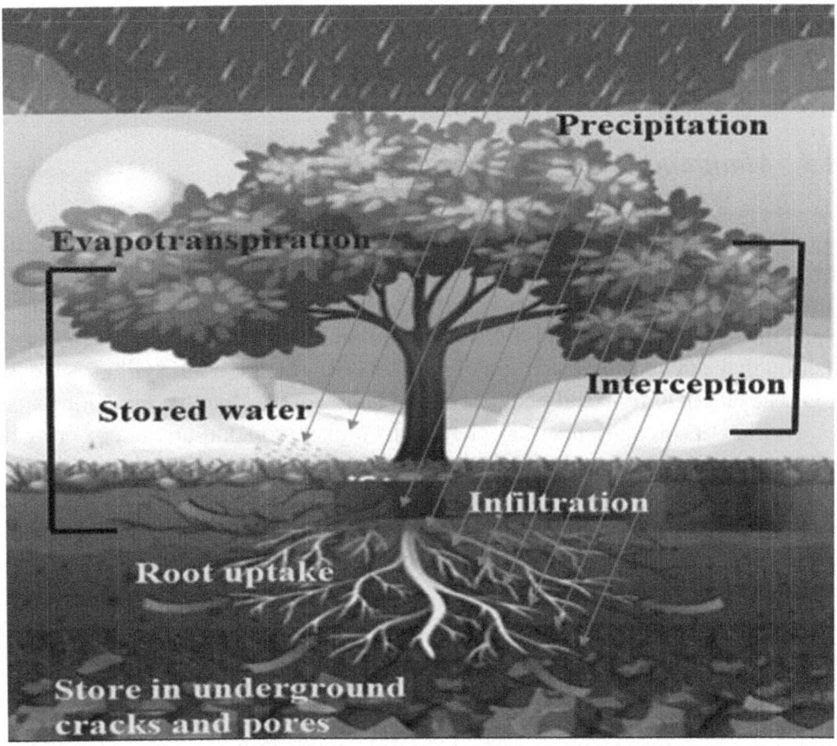

Fig. 7.3 Trees improve groundwater recharge

Trees contribute significantly to groundwater recharge, which is essential for sustaining river flow, especially during dry seasons. This recharge is vital in regions like India, facing worsening groundwater scarcity. Trees aid in preventing surface flooding by facilitating water infiltration into the soil, thus maintaining groundwater levels. Moreover, increased groundwater levels bolster dry season flows, mitigating the impact of prolonged dry spells. Approximately one-third of the world's population lives in poverty due to water shortages. Despite their many advantages, planting trees in arid areas is frequently discouraged due to worries that they will make less water available. However, there is pertinent research from the tropics, and more is needed to know about the effects of intermediate tree cover. An ideal tree cover hypothesis that maximizes groundwater recharge at an intermediate tree density was created and put to the test. The benefits of adding trees to water percolation below this ideal tree density outweigh their additional water demand, resulting in higher groundwater recharge; above the optimum, the opposite happens. Our findings show that groundwater recharge was maximized at intermediate tree densities. They are based on groundwater budgets calibrated using transpiration and drainage data in a farmed West African woodland. Thus, contrary to popular belief, planting trees and using different tree management techniques can enhance groundwater supplies. A moderate amount of tree cover can improve groundwater recharge. We assess what is needed for these results to be valid and propose that they are probably prevalent in tropical regions that experience seasonal dryness. This means there is a chance for broad tree establishment and more significant advantages for hundreds of millions.

7.3.5 Maintaining the Soil Fertility

The soil must supply sufficient amounts of the multiple nutrients for a soil to be fertile, which may limit plant growth (Gianquinto et al., 2013; Singh et al., 2019a, 2019b). In natural ecosystems, plants play a role in the formation of fertile soils by fixing carbon (C) and nitrogen (N), chemically liberating soil minerals from deep to surface soils, and through the uptake and retention of nutrients (Rawat et al., 2018; Shafreen et al., 2021). The output of food and fiber per unit of land will rise due to population expansion. The loss of soil nutrients is accelerated by higher crop output, underscoring the need for comprehensive and efficient nutrient management to restore soil fertility. The capacity of the soil to absorb nutrients lost due to crop removal determines the amount of nutrients found in native soil (Guilherme et al., 2015). Mineral dissolution and surface exchange reactions replenish P, K, Ca, Mg, and micronutrients, whereas mineralization of soil organic components supplies limited amounts of N, S, and micronutrients (Dhaliwal et al., 2019). The assessment of nutrient availability to plants, ion transport to plant roots, and, ultimately, nutrient management decisions are influenced by nutrient mobility in the soil. To implement effective nutrient management, soil testing is necessary to quantify crop nutrient requirements and the soil's potential to supply nutrients (Hopkins et al., 2020). Following their assessment, the nutrient management strategy includes the optimal nutrient rate, suitable nutrient source(s), most efficient nutrient placement method(s), and timing of nutrient treatment. Management choices will change based on the particular nutrient. To maximize agricultural output and minimize environmental effects of fertilizer use, soil nutrients must be managed effectively (Chaudhary & Singh, 2018; Chaudhary et al., 2020; Singh et al., 2019a, 2019b). Resultant plants play a significant role in soil fertility, water storage capability, and hydro and geological cycle maintenance.

7.3.6 Reduce Soil Erosion

Soil erosion is the process by which soil moves from one location to another, either naturally or artificially. Soil that has been transferred might be fertile or not. All plants take in micro- and macronutrients from the soil as they grow on soil media. The below-the-ground parts of plants make up the root system. Some of the examples of this are Rhizomes, tubers, and roots. Each plant's root binds the soil particles tightly together. Therefore, plantation prevents soil erosion. Soil erosion will be minimal in this location if plantation cover is at its highest—the loss of fertile soil due to soil erosion led to lower agricultural yields. Turf captures and slows down stormwater and reduces soil erosion with its dense system of roots and shoots. According to studies, it holds up to as much as 20 times more soil than traditionally farmed cropland during extremely intense rainstorms (three inches per hour) (Okunlola et al., 2019).

7.3.7 Flood Control

Urban impervious surfaces convert precipitation to stormwater runoff, which causes water quality and quantity problems. Cultivating trees and other woody plants serves additional consideration as a stormwater control measure. Trees interact with the urban hydrologic cycle by intercepting incoming precipitation, removing water from the soil via transpiration, and enhancing infiltration (Berland et al., 2017).—The extensive root systems of higher plants and the networks formed by many grasses (e.g. *Cynodon dactylon*) on the ground surface reduce flooding and wind-related damage by holding soils in place and absorbing significant volumes of water through their roots and canopies.

7.3.8 Drought Control

In general, droughts are extreme hydrologic occurrences that result in acute water shortages and last long enough to negatively affect people, plants, animals, and the environment over a wide area (Gautam & Bana, 2014). Trees' moisture released into the air (during transpiration) stabilizes rainfall, decreasing drought episodes.

Drought is caused by changing weather patterns manifested through the excessive build-up of heat on the earth's surface, which results in more excellent evaporation rates. Drought, land degradation, and desertification affect biodiversity. On-site effects include habitat deterioration and species extinction, which result in a general decline in economic and ecological output. Wetlands and water bodies will become depleted and polluted due to soil erosion. Communities are compelled to relocate or engage in various coping mechanisms when biological and economic output declines, contributing to biodiversity deterioration (Reid et al., 2019). In addition to protecting the soil, increasing its capacity to store water, reducing runoff, and regulating drainage (both surface- and subterranean), trees and plants help conserve and enhance the soil's productive potential and the fertility of nearby agricultural land (Hervás-Gámez & Delgado-Ramos, 2019).

7.4 Conclusion

Trees are good for the environment in many ways. The importance of plants is not limited to their role in the environment. Plants must be used more frequently to solve environmental and health issues. Trees are significant to the economy and improve air and water quality. Plants generate oxygen by decarbonization in the atmosphere and play an essential role in removing pollutants and recharging groundwater. Plants also play an important role in sound absorbance, windbreaks, urban cooling, clean water, medicinal value, sociological benefits, and the support of wildlife. Hence, plants improve our environment by maintaining the biogeochemical cycles in the hydrosphere, lithosphere, and atmosphere. Therefore, more studies will be needed to understand the role of plants in groundwater recharge, water quality, and nutrient management.

Acknowledgements The authors wish to thank the Department of Environmental Science, Savitribai Phule Pune University, Pune, Maharashtra and School of Technology, Department of Civil Engineering, Sanjay Ghodawat University, Kolhapur, Maharashtra, India and Department of Environmental Science, Dr. Babasaheb Ambedkar Marathwada University, Chhaatrapti Sambhajinagar, Aurangabad, Maharashtra, India for the platform that enabled this interesting work.

References

Adegbeye, M. J., Salem, A. Z. M., Reddy, P. R. K., Elghandour, M. M. M., & Oyebamiji, K. J. (2020). Waste recycling for the eco-friendly input use efficiency in agriculture and livestock feeding. In *Resources use efficiency in agriculture* (pp. 1–45).

Ansari, S. (2021). Overview of traditional systems of medicine in different continents. In *Preparation of phytopharmaceuticals for the management of disorders* (pp. 431–473). Academic Press.

Awa, S. H., & Hadibarata, T. (2020). Removal of heavy metals in contaminated soil by phytoremediation mechanism: A review. *Water, Air, & Soil Pollution, 231*(2), 47.

Babu, B. G., NazaneenParveen, S., Naveen Kumar, K., & Sridhar, R. M. (2013). Evaluation of air pollution tolerance indices of plant species growing in the vicinity of cement industry and Yogi Vemana University Campus. *Indian Journal of Advances in Chemical Science, 2*(1), 16–20.

Balasubramanian, A., & Nagaraju, D. (2015). *The hydrologic cycle* (1st ed.). UO.

Baulch, H. M., Elliott, J. A., Cordeiro, M. R., Flaten, D. N., Lobb, D. A., & Wilson, H. F. (2019). Soil and water management: Opportunities to mitigate nutrient losses to surface waters in the Northern Great Plains. *Environmental Reviews, 27*(4), 447–477.

Berland, A., Shiflett, S. A., Shuster, W. D., Garmestani, A. S., Goddard, H. C., Herrmann, D. L., & Hopton, M. E. (2017). The role of trees in urban stormwater management. *Landscape and Urban Planning, 162*, 167–177.

Bhatla, S. C., & Lal, M. A. (2023). Secondary metabolites. In *Plant physiology, development and metabolism* (pp. 765–808). Springer Nature Singapore.

Bui, H. T., Odsuren, U., Kwon, K. J., Kim, S. Y., Yang, J. C., Jeong, N. R., & Park, B. J. (2021). Assessment of air pollution tolerance and particulate matter accumulation of 11 woody plant species. *Atmosphere, 12*(8), 1067.

Cartwright, J., & Johnson, H. M. (2018). Springs as hydrologic refugia in a changing climate? A remote-sensing approach. *Ecosphere, 9*(3), e02155.

Chaturvedi, R. K., Prasad, S., Rana, S., Obaidullah, S. M., Pandey, V., & Singh, H. (2013). Effect of dust load on the leaf attributes of the tree species growing along the roadside. *Environmental Monitoring and Assessment, 185*, 383–391.

Chaudhary, I. J., Neeraj, A., Siddiqui, M. A., & Singh, V. (2020). Nutrient management technologies and the role of organic matrix-based slow-release biofertilizers for agricultural sustainability: A review. *Agricultural Reviews, 41*(1), 1–13.

Chaudhary, I. J., & Rathore, D. (2018a). Suspended particulate matter deposition and its impact on urban trees. *Atmospheric Pollution Research, 9*, 1072–1082.

Chaudhary, I. J., & Rathore, D. (2018b). Phytomonitoring of dust load and its effect on foliar micro morphological characteristics of urban trees. *Journal of Plant Science, 2*(3), 170–179.

Chaudhary, I. J., & Rathore, D. (2019). Dust pollution: Its removal and effect on foliage physiology of urban trees. *Sustainable Cities and Society, 51*, 101696. https://doi.org/10.1016/j.scs.2019.101696

Chaudhary, I. J., & Rathore, D. (2021a). Assessment of dose-response relationship between ozone dose and groundnut (*Arachis hypogaea* L.) cultivars using Open Top Chamber (OTC) and Ethylenediurea (EDU). *Environmental Technology and Innovation, 22*, 101494. https://doi.org/10.1016/j.eti.2021.101494

Chaudhary, I. J., & Rathore, D. (2021b). Assessment of ozone toxicity on cotton (*Gossypium hirsutum* L.) cultivars: Its defensive system and intraspecific sensitivity. *Plant Physiology and Biochemistry, 166*, 912–927. https://doi.org/10.1016/j.plaphy.2021.06.054

Chaudhary, I. J., & Rathore, D. (2021c). Micro-morphological and anatomical response of groundnut (*Arachis hypogaea* L.) cultivars to ground-level ozone. *Journal of Applied Biology and Biotechnology, 9*(04), 137–150. https://doi.org/10.7324/JABB.2021.9419

Chaudhary, I. J., & Singh, R. P. (2018). Studies on growth, mobilization of nutrients and yield of wheat (*Triticum aestivum* L. PBW-343) applied with organic matrix based slow release bio fertilizers. *International Journal of Current Microbiology and Applied Sciences, 3*(7), 3221–3238.

Chen, Y., Wen, Y., Zhou, Q., Huang, J., Vymazal, J., & Kuschk, P. (2016). Sulfate removal and sulfur transformation in constructed wetlands: The roles of filling material and plant biomass. *Water Research, 102*, 572–581.

Costanza, R., De Groot, R., Sutton, P., Van der Ploeg, S., Anderson, S. J., Kubiszewski, I., Farber, S., & Turner, R. K. (2014). Changes in the global value of ecosystem services. *Global Environmental Change, 26*, 152–158.

De Giudici, G., Pusceddu, C., Medas, D., Meneghini, C., Gianoncelli, A., Rimondi, V., Podda, F., Cidu, R., Lattanzi, P., Wanty, R. B., & Kimball, B. A. (2017). The role of natural biogeochemical barriers in limiting metal loading to a stream affected by mine drainage. *Applied Geochemistry, 76*, 124–135.

Dhaliwal, S. S., Naresh, R. K., Mandal, A., Singh, R., & Dhaliwal, M. K. (2019). Dynamics and transformations of micronutrients in agricultural soils as influenced by organic matter build-up: A review. *Environmental and Sustainability Indicators, 1*, 100007.

Ekhuemelo, D. O., & Udo, A. M. (2016). Investigation of variations in the fibre characteristics of *Moringa oleifera* (Lam) stem for pulp and paper production. *International Journal of Science and Technology, 5*(1), 19–25.

Findlay, S. E. (2021). Organic matter decomposition. In *Fundamentals of ecosystem science* (pp. 81–102). Academic Press.

Gabler, C. A., Osland, M. J., Grace, J. B., Stagg, C. L., Day, R. H., Hartley, S. B., Enwright, N. M., From, A. S., McCoy, M. L., & McLeod, J. L. (2017). Macroclimatic change is expected to transform coastal wetland ecosystems this century. *Nature Climate Change, 7*(2), 142–147.

Gagan, M., Gaurav, P., Dhingra, G. K., Avinash, K., Anjali, N., & Pawan, K. (2023). Plant-microbe interaction in freshwater ecosystem for improving water quality. In *Current status of fresh water microbiology* (pp. 117–133). Springer Nature Singapore.

Gautam, R. C., & Bana, R. S. (2014). Drought in India: Its impact and mitigation strategies—A review. *Indian Journal of Agronomy, 59*(2), 179–190.

Gianquinto, G., Muñoz, P., Pardossi, A., Ramazzotti, S., & Savvas, D. (2013). 10. Soil fertility and plant nutrition. In *Good agricultural practices for greenhouse vegetable crops* (p. 205).

Grilo, C., Koroleva, E., Andrášik, R., Bíl, M., & González-Suárez, M. (2020). Roadkill risk and population vulnerability in European birds and mammals. *Frontiers in Ecology and the Environment, 18*(6), 323–328.

Guilherme, M. R., Aouada, F. A., Fajardo, A. R., Martins, A. F., Paulino, A. T., Davi, M. F., Rubira, A. F., & Muniz, E. C. (2015). Superabsorbent hydrogels based on polysaccharides for application in agriculture as soil conditioner and nutrient carrier: A review. *European Polymer Journal, 72*, 365–385.

Haq, S. M., Calixto, E. S., Rashid, I., Srivastava, G., & Khuroo, A. A. (2022). Tree diversity, distribution and regeneration in major forest types along an extensive elevational gradient in Indian Himalaya: Implications for sustainable forest management. *Forest Ecology and Management, 506*, 119968.

Hervás-Gámez, C., & Delgado-Ramos, F. (2019). Drought management planning policy: From Europe to Spain. *Sustainability, 11*(7), 1862.

Hopkins, B. G., Stark, J. C., & Kelling, K. A. (2020). Nutrient management. In J. C. Stark, M. Thornton, & P. Nolte (Eds.), *Potato production systems* (pp. 155–202). Springer.

Hwangbo, M., Shao, Y., Hatzinger, P. B., & Chu, K. H. (2023). Acidophilic methanotrophs: Occurrence, diversity, and possible bioremediation applications. *Environmental Microbiology Reports, 15*(4), 265–281.

Jain, C., Khatana, S., & Vijayvergia, R. (2019). Bioactivity of secondary metabolites of various plants: A review. *International Journal of Pharmaceutical Sciences and Research, 10*(2), 494–504.

Jiang, C., Zhang, T., & Yang, Z. (2022). Risk assessment of potential toxic metal pollution in water-sediment-submerged macrophyte systems: A case study of urban shallow lakes in Central China. *Aquatic Ecology, 56*(4), 1001–1017.

Kapoor, C. S., & Chittora, A. K. (2016). Efficient control of air pollution through plants a cost effective alternatives. *Journal of Climatology and Weather Forecasting, 4*(184), 2.

Kidd, P. S., Álvarez, A., Álvarez-López, V., Cerdeira-Pérez, A., Rodríguez-Garrido, B., Prieto-Fernández, Á., & Chalot, M. (2021). Beneficial traits of root endophytes and rhizobacteria associated with plants growing in phytomanaged soils with mixed trace metal-polycyclic aromatic hydrocarbon contamination. *Chemosphere, 277*, 130272.

Lee, J., Rai, P. K., Jeon, Y. J., Kim, K. H., & Kwon, E. E. (2017). The role of algae and cyanobacteria in the production and release of odorants in water. *Environmental Pollution, 227*, 252–262.

Lepczyk, C. A., Aronson, M. F., Evans, K. L., Goddard, M. A., Lerman, S. B., & MacIvor, J. S. (2017). Biodiversity in the city: Fundamental questions for understanding the ecology of urban green spaces for biodiversity conservation. *BioScience, 67*(9), 799–807.

Li, W., Xu, L., Wu, J., Ma, L., Liu, M., Jiao, J., Li, H., & Hu, F. (2015). Effects of indole-3-acetic acid (IAA), a plant hormone, on the ryegrass yield and the removal of fluoranthene from soil. *International Journal of Phytoremediation, 17*(5), 422–428.

Li, Y., Zhou, W., Hu, B., Min, M., Chen, P., & Ruan, R. R. (2011). Integration of algae cultivation as biodiesel production feedstock with municipal wastewater treatment: Strains screening and significance evaluation of environmental factors. *Bioresource Technology, 102*(23), 10861–10867.

Madkour, L. H. (2019). Function of reactive oxygen species (ROS) inside the living organisms and sources of oxidants. *Pharmaceutical Sciences and Analytical Research Journal, 2*, 180023.

Manisalidis, I., Stavropoulou, E., Stavropoulos, A., & Bezirtzoglou, E. (2020). Environmental and health impacts of air pollution: A review. *Frontiers in Public Health, 8*, 14. https://doi.org/10.3389/fpubh.2020.00014

Marimuthu, K., Kalimuthu, R., Ponraj, K., Lavanya, K., Magesh, P., & Jasbin, S. G. (2014). Air pollution tolerance index assessment of Yercaud road side plants. *International Journal of Pharmaceutical Sciences Review and Research, 26*(2), 177–181.

Navjot, K., Simpy, S., & Gupta, N. C. (2016). Study of pollution tolerance index for plant species exposed to vehicular traffic on urban streets. *International Journal of Current Sciences, 19*(4), 181–192.

Neumann, B., Ott, K., & Kenchington, R. (2017). Strong sustainability in coastal areas: A conceptual interpretation of SDG 14. *Sustainability Science, 12*, 1019–1035.

Nguyen, A. T., Vu, A. D., Dang, G. T., Hoang, A. H., & Hens, L. (2018). How do local communities adapt to climate changes along heavily damaged coasts? A Stakeholder Delphi study in Ky Anh (Central Vietnam). *Environment, Development and Sustainability, 20*, 749–767.

Nowak, D. J., Hirabayashi, S., Bodine, A., & Hoehn, R. (2013). Modeled PM2.5 removal by trees in ten U.S. cities and associated health effects. *Environmental Pollution, 78*, 395–402.

Okereafor, U., Makhatha, M., Mekuto, L., Uche-Okereafor, N., Sebola, T., & Mavumengwana, V. (2020). Toxic metal implications on agricultural soils, plants, animals, aquatic life and human health. *International Journal of Environmental Research and Public Health, 17*(7), 2204.

Okunlola, J. O., Oke, D. O., Adekunle, V. A. J., & Owolabi, K. E. (2019). Effects of climate change and coping strategies among crop farmers in South West, Nigeria. *Agroforestry Systems, 93*, 1399–1408.

Rai, P. K. (2016). Impacts of particulate matter on plants: Implications for environmental biomonitoring. *Ecotoxicology and Environmental Safety, 129*, 120–136. https://doi.org/10.1016/j.ecoenv.2016.03.012

Rajpar, M. N., & Zakaria, M. (2014). Mangrove fauna of Asia. In *Mangrove ecosystems of Asia: Status, challenges and management strategies* (pp. 153–197).

Rajpar, M. N., Zakaria, M., Ozdemir, I., Sheryar, S., & Rab, A. (2018). Ascertain the productivity of heterogenous wetland and adjacent habitats through avian foraging guilds.

Ramesh, R., Nelson, D., & Pillepich, A. (2023). The circumgalactic medium of Milky Way-like galaxies in the TNG50 simulation—II. Cold, dense gas clouds and high-velocity cloud analogs. *Monthly Notices of the Royal Astronomical Society, 522*(1), 1535–1555.

Rawat, J., Sanwal, P., & Saxena, J. (2018). Towards the mechanisms of nutrient solubilization and fixation in soil system. In *Role of rhizospheric microbes in soil: Volume 2: Nutrient management and crop improvement* (pp. 229–257).

Reid, A. J., Carlson, A. K., Creed, I. F., Eliason, E. J., Gell, P. A., Johnson, P. T., Kidd, K. A., MacCormack, T. J., Olden, J. D., Ormerod, S. J., & Smol, J. P. (2019). Emerging threats and persistent conservation challenges for freshwater biodiversity. *Biological Reviews, 94*(3), 849–873.

Rosendo, J. C. M., da Paz, G. M., & Rosendo, A. (2022). Constructed wetlands applied on domestic wastewater for decetralized systems: Concepts, processes, modalities, combinations and enhancements; A review. *Reviews in Environmental Science and Bio/technology, 21*(2), 371–397.

Rupa, P., & Venkatachalam, T. (2018). Studies on air pollution tolerance index of native plant species to enhance greenery in industrial area. *Indian Journal of Ecology, 45*(1), 1–5.

Russo, N. J., Davies, A. B., Blakey, R. V., Ordway, E. M., & Smith, T. B. (2023). Feedback loops between 3D vegetation structure and ecological functions of animals. *Ecology Letters, 26*(9), 1597–1613.

Saco, P. M., McDonough, K. R., Rodriguez, J. F., Rivera-Zayas, J., & Sandi, S. G. (2021). The role of soils in the regulation of hazards and extreme events. *Philosophical Transactions of the Royal Society B, 376*(1834), 20200178.

Sarkar, S., Dey, A., Kumar, V., Batiha, G. E. S., & El-Esawi, M. A. (2021). Fungal endophyte: An interactive endosymbiont with the capability of modulating host physiology in myriad ways. *Frontiers in Plant Science, 12*, 701800.

Schenk, E. R., O'donnell, F., Springer, A. E., & Stevens, L. E. (2020). The impacts of tree stand thinning on groundwater recharge in arid land forests. *Ecological Engineering, 145*, 105701.

Serafini, R. J. M., Arreghini, S., Troiani, H. E., & de Iorio, A. R. F. (2023). Copper, zinc, and chromium accumulation in aquatic macrophytes from a highly polluted river of Argentina. *Environmental Science and Pollution Research, 30*(11), 31242–31255.

Shafreen, M., Vishwakarma, K., Shrivastava, N., & Kumar, N. (2021). Physiology and distribution of nitrogen in soils. *Soil nitrogen ecology* (pp. 3–31). Springer.

Shahid, M. J., Al-Surhanee, A. A., Kouadri, F., Ali, S., Nawaz, N., Afzal, M., Rizwan, M., Ali, B., & Soliman, M. H. (2020). Role of microorganisms in the remediation of wastewater in floating treatment wetlands: A review. *Sustainability, 12*(14), 5559.

Singh, S., Chaudhary, I. J., & Kumar, P. (2019). Utilization of low-cost agricultural waste for removal of toxic metals from environment a review. *International Journal of Scientific Research in Biological Sciences, 6*(4), 56–61.

Singh, V., Shah, K. N., Chaudhary, I. J., & Rana, D. K. (2019). Impact assessment of different organic manures on growth, morphology and yield of onion (*Allium cepa* L.) cultivar. *Asian Journal of Agricultural Research, 13*, 20–27.

Speak, A. F., Rothwell, J. J., Lindley, S. J., & Smith, C. L. (2012). Urban particulate pollution reduction by four species of green roof vegetation in a U.K. city. *Atmospheric Environment, 61*, 283–293.

Srivastava, J. K., Chandra, H., Kalra, S. J., Mishra, P., Khan, H., & Yadav, P. (2017). Plant–microbe interaction in aquatic system and their role in the management of water quality: A review. *Applied Water Science, 7*, 1079–1090.

Srivastava, M., Simakov, O., Chapman, J., Fahey, B., Gauthier, M. E., Mitros, T., Richards, G. S., Conaco, C., Dacre, M., Hellsten, U., & Larroux, C. (2010). The *Amphimedon queenslandica* genome and the evolution of animal complexity. *Nature, 466*(7307), 720–726.

Srivastava, N., Hinton, G., Krizhevsky, A., Sutskever, I., & Salakhutdinov, R. (2014). Dropout: A simple way to prevent neural networks from overfitting. *The Journal of Machine Learning Research, 15*(1), 1929–1958.

Thies, J. E., & Grossman, J. M. (2023). The soil habitat and soil ecology. In *Biological approaches to regenerative soil systems* (pp. 69–84).

Thorne, J. H., Choe, H., Stine, P. A., Chambers, J. C., Holguin, A., Kerr, A. C., & Schwartz, M. W. (2018). Climate change vulnerability assessment of forests in the Southwest USA. *Climatic Change, 148*, 387–402.

Tiwari, A. K., Chaudhary, I. J., & Pandey, A. K. (2019). Indian traditional trees and their scientific relevance. *Journal of Medicinal Plants, 7*(3), 29–32.

Upadhyay, R. K. (2020). Markers for global climate change and its impact on social, biological and ecological systems: A review. *American Journal of Climate Change, 9*(03), 159.

Urana, R., Dahiya, A., Singh, N., & Sharma, P. (2020). A review on rhizoremediation: plant-microbe interaction enhances the degradation of polyaromatic hydrocarbons. In *Microbial services in restoration ecology* (pp. 283–295).

Varma, K., Srivastava, V., Singhal, A., & Jha, P. K. (2021). Urban and environmental hazards. In *Recent technologies for disaster management and risk reduction: Sustainable community resilience & responses* (pp. 319–362).

Wyatt, C. J. W. (2013). *Estimating aquifer response following forest restoration and climate change along the mogollon rim, northern arizona (Order No. 1537821).* Available from Publicly Available Content Database. (1367180989). Retrieved from https://www.proquest.com/dissertations-theses/estimating-aquifer-response-following-forest/docview/1367180989/se-2

Wyatt, M. C., Panić, O., Kennedy, G. M., & Matra, L. (2015). Five steps in the evolution from protoplanetary to debris disk. *Astrophysics and Space Science, 357*, 1–20.

Zhao, D., Cheah, W. Y., Lai, S. H., Ng, E. P., Khoo, K. S., Show, P. L., & Ling, T. C. (2023). Symbiosis of microalgae and bacteria consortium for heavy metal remediation in wastewater. *Journal of Environmental Chemical Engineering, 11*(3), 109943.

Zhao, Q., Bai, J., Huang, L., Gu, B., Lu, Q., & Gao, Z. (2016). A review of methodologies and success indicators for coastal wetland restoration. *Ecological Indicators, 60*, 442–452.

Chapter 8
Risks of Groundwater Contamination to the Environment and Human Health

Alisha Kakkar and Sudesh Kumar

Abstract Underground water is a vital element in the progress of any region, serving as a primary source of drinkable, farming, and factory water. Historically, there was a belief that rock deposits and topsoil, functioning as filtration agents, shielded subsurface water from contamination. Groundwater pollution is a widespread issue that profoundly affects the physical health of individuals and the environment. Typically, research on subsurface water contamination encompasses the empirical comprehension of botanical, physical, and chemical methods that govern the course and development of pollutants in the subsurface water. Elevated chemical levels in clean water can present a health risk. Public health research has demonstrated that inadequate potable water quality, acting as the primary transmission pathway, has been accountable for numerous water-related illnesses. Various microbes and artificial chemical agents possess the capacity to cause underground water contamination. Potable water that harbors enzymes and bacteria can result in diseases like jaundice or gastrointestinal infection. Preventive actions are generally more accessible and cost-effective than remedial actions for subsurface water pollution. Selecting suitable treatment methods relies on precise site conditions and frequently considers remedy objectives related to environmental safety and human well-being. Hence, this book chapter focuses on an in-depth analysis of the health and environmental impacts of toxic metals like lead, cadmium, and arsenic. Various techniques to detect groundwater contamination and their remedial methods are also well-documented. It also presents future research directions in the field of groundwater contamination.

Keywords Water · Lead · Cadmium · Contamination · Treatment

A. Kakkar · S. Kumar (✉)
Department of Chemistry, Banasthali Vidyapith, Tonk, Rajasthan, India
e-mail: sudeshneyol@gmail.com

V. Sunitha et al. (eds.), *Groundwater Resource Management Planning Strategies*, Springer Water, https://doi.org/10.1007/978-3-031-88870-0_8

8.1 Introduction

Subterranean water constitutes a vital reservoir of clean water for the worldwide populace and serves household, farming, and commercial needs. Potable water is arguably the most precious among Earth's natural assets. Roughly 30% of the worldwide inhabitants rely on underground water for potable water supply. Subterranean water is essential, especially in dry and semi-dry areas, where surface water and rainfall are scarce. Ensuring a secure and sustainable provision of uncontaminated aquifers for consumption is a fundamental catalyst for a country's progress. Nevertheless, urban expansion, farming techniques, commercial operations, and weather fluctuations collectively present substantial risks to the quality of subterranean water. Groundwater pollution can also be conveyed through interfaces between aboveground water bodies and groundwater, resulting in the decline of water purity on the surface. The overall volume of accessible water on Earth amounts to 1386 million km^3, out of which 10.63 million km^3 is non-polluted (Mukherjee & Singh, 2018). Although precipitation serves as the primary mechanism for recharging aquifers, the effects of weather shifts have diminished aquifer levels' vulnerability and modified patterns of precipitation and hydrology, leading to a decline in aquifer replenishment in numerous areas. Escalated marine water intrusion and inadequate precipitation adversely affect water quality, heightening conductivity and salt levels while diminishing water table depths, respectively. Due to the escalated salt content, underground water becomes saltier. Watering crops with saline-rich water can hamper crop development by altering the osmotic pressure and penetrability of the soil. Although soil effectively filters numerous particulate matter and soluble substances, contaminants from human-made sources persist in infiltrating groundwater, leading to its pollution. Besides sewage facilities, chemical agents, extraction, dump sites, animal manure (Sasakova et al., 2018), and roadway drainage pollute the aquifers (Fig. 8.1). The main human-induced factors contributing to groundwater contamination are agrochemicals and improper management of factory waste, which leads to infiltration into aquifers (Adeyemi & Ojekunle, 2021). Human-induced sources have also altered the pattern of metallic elements, among which lead and arsenic have emerged as significant pollutants in groundwater. Besides pollutants from human-made origins, organic contaminants (particles and rock fragments) pollute groundwater, diminishing its suitability for household uses. Elevated metal levels impact the groundwater's purity, and the uranium, nitrate, surroundings. Furthermore, elevated concentrations of various groundwater pollutants, like iron and bicarbonate have been documented in various regions of India. Prolonged contact with both non-cancerous [nitrates (10 mg/L), fluorides (1.5 mg/L), polycyclic aromatic hydrocarbons (0.1 μg/L)] and cancerous contaminants [benzene (5 ppb), arsenic (10 μg/L), vinyl chloride (2 ppb)] even in minute quantities, can lead to adverse health effects (such as high blood pressure, respiratory issues, carcinogenic vulnerability, diabetes mellitus, skin issues, etc.). Groundwater pollution by arsenic is a worldwide issue. Arsenic contributes to various neural, fertility, immune system disorders, heart conditions, hormonal imbalances, respiratory cancer, and skin malignancies. Health hazards can be evaluated by examining the amount/proportion of metal pollutants. Hence, it is crucial to undertake groundwater restoration because of the elevated level of danger linked with the pollution of groundwater origins. Groundwater restoration is a process that treats polluted groundwater by extracting pollutants and transforming them into benign substances for treatment. Various methods for groundwater restoration have been examined previously, encompassing membrane filtration, sedimentation, precipitation, ion exchange, sorption, biodegradation, and natural approaches (Zhang et al., 2020). Two fundamental classes of groundwater restoration techniques exist Off-site and On-site methods. Off-site methods involve extracting the contaminated groundwater, processing the water above ground using physio-chemical or natural methods, and reintroducing the processed water into the groundwater. The inherent remediation of a location, or the renewal of the aquifer reservoirs and soil to a functional condition, can be achieved through diverse means. On-site degradation is a cheap and versatile approach for eliminating pollutants as it operates directly near the origin of pollution and employs microbial coating or aquatic microorganism populations (Patel et al., 2021). Renewable economic progress necessitates equilibrium between the pace of replenishment of raw materials and human requirements. Persistent pollution of aquifers could diminish the accessibility of clean water, disrupting the equilibrium between water provision and need, potentially resulting in economic upheaval and even conflicts. Water scarcity triggered by pollution might emerge as a catalyst, sparking disputes among residents in the years ahead and potentially impeding a country's economic and societal progress. Groundwater pollution isn't solely an ecological concern but also a societal problem, necessitating cooperation between scientists specializing in nature and those focused on societal dynamics. Previous studies on groundwater contamination concentrated on their sources, environmental and human health impact, and various techniques for removing groundwater pollution (Al-Sudani, 2019; Economou-Eliopoulos & Megremi, 2021; Li et al., 2021; Pradhan et al., 2023). Therefore, this chapter also offers the methods to detect groundwater contamination. It improves the capability to promptly detect contaminants, minimize their effects, and safeguard groundwater. It also emphasizes the utilization of nanotechnology as valuable in mitigating groundwater pollution and future recommendations for protecting groundwater reserves for both present and subsequent generations.

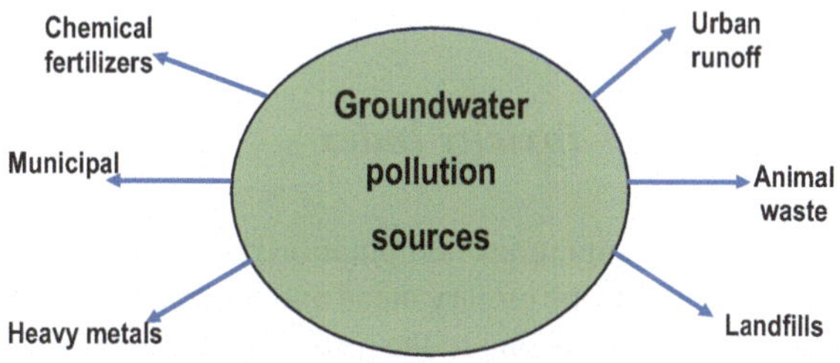

Fig. 8.1 Sources of groundwater pollution

8.2 Health Hazard Evaluation

It is essential to perform a threat analysis of contaminants to gauge their potential impact on health. Evaluating health hazards involves assessing whether the identified pollutants threaten human well-being. The responsibilities of both evaluators and decision-makers in the evaluation procedure are vital for understanding threat analysis (Karunanidhi et al., 2020). This evaluation process includes utilizing ecological data to gauge the extent and characteristics of detrimental impacts on human well-being. Generally, two methods are employed to evaluate the threat posed by a specific contaminant to a given populace: primary (natural) and secondary (ecological surveillance). Nevertheless, analyzing contaminants chemically at their origin can be an exceptional approach to assessing health hazards. Several research efforts have been undertaken in this domain, employing diverse methods and computations for Hazard Quotients, Health Hazard Ratios, and Carcinogenic Risk. The standard water quality attributed are listed below (Table 8.1).

8.2.1 Health Evaluation Metrics

Metal pollution metrics assist in pinpointing the origin, intensity, and extent of metal pollution. Several metrics are utilized in extensive data collection to condense and streamline the impact of toxic metals on humans into a singular value. These metrics are instrumental in identifying the existence of heavy metals and the total extent of pollutants in aquifers (Eldaw et al., 2020).

8.2.2 Risk Indicator

Ingestion of aquifers containing heavy metals presents a non-cancer risk. According to United States Environmental Protection Agency guidelines, nitrate, fluoride, iron, and manganese are considered non-cancerous. Thus, they are taken into account when calculating the risk indicator.

8.2.3 Metal Toxicity Index

The metal toxicity index computation relies on the average of toxic metal amounts. It is determined as the unitary mean and represents the proportion of the total sub-index product and mass value to the total mass value of each variable.

Table 8.1 Standard water quality metrics

Variables	Units	BIS	WHO
Total dissolved gases	mg/L	500	MAC-2000 HDL-500
Total hardness ($CaCO_3$)	mg/L	200	MAC-600 HDL-300
Potassium (K^+)	mg/L	Nil	MAC-Nil HDL-Nil
Sodium (Na^+)	mg/L	Nil	MAC-Nil HDL-Nil
Magnesium (Mg^{2+})	mg/L	0.05	MAC-150 HDL-30
Calcium (Ca^{2+})	mg/L	75	MAC-200 HDL-75
Nitrate (NO_3^-)	mg/L	45	MAC-No relaxation HDL-50
Nitrite (NO_2)	mg/L	Nil	MAC-Nil HDL-Nil
Chloride (Cl^-)	mg/L	200	MAC-1000 HDL-250
Fluoride (F^-)	mg/L	1	MAC-1.5 HDL-1
Sulphate (SO_4^{2-})	mg/L	200	MAC-400 HDL-200
Iron (Fe^{2+})	mg/L	0.1	MAC-1.0 HDL-0.3
pH	Na	6.5–8.5	MAC-No relaxation HDL-6.5–9.5
Manganese (Mn^{2+})	mg/L	0.05	MAC-0.1 HDL-0.1
Chromium (Cr^{3+})	mg/L	Nil	MAC-Nil HDL-Nil

HDL highest desirable limit, *MAC* Maximum acceptable concentration, *BIS* Bureau of Indian Standards, *WHO* World Health Organization

8.2.4 Trace Element Contamination Index

It reflects the comprehensive water quality concerning heavy metals.

8.2.5 Pollution Severity Index

The pollution severity index value delineates the combined impact of various water quality attributes deemed detrimental to water distribution (Reddy et al., 2022; Reddy and Sunitha, 2023; Suvarna et al., 2022; Veeraswamy et al., 2018; Wagh et al., 2018).

8.3 Indicators of Susceptibility to Groundwater Contamination

The affected members encompass the local inhabitants and the secondary assets and endeavors dependent on water reservoirs. Amidst the metrics of susceptibility to threat are aspects related to shifts in availability and positioning of the populace and their possessions about the tainted source and substitute supplies. Hence, metrics that possess a distinctive sensitivity for particular demographic segments (such as age categories) and specific properties and productive pursuits in the direction of resource pollution are considered. Furthermore, metrics that mirror the potential for encountering the risk of pollution concurrently with other hazards impacting water-based assets, precipitating a more profound disaster, like concurrent aridity, are encompassed. It is also emphasized that groundwater's prior pollution status may be crucial to gauge its recuperative capability and resilience. Socially directed elements for safeguarding against threats encompass a range of actions for hazard alleviation, including statutes, contingency schemes, safety measures, data frameworks, advancement of unity in the event of a calamity, and the accountability of every person regarding pollution. To assess the susceptibility to a potential pollution incident, socially directed facets of resilience are considered, like identifying substitute supply origins for resources and community requirements or acquiring insurance to protect impacted resources and pursuits. GOD and DRASTIC are the approaches for assessing the susceptibility of groundwater. DRASTIC assesses groundwater vulnerability based on depth to water, recharge, aquifer type, soil type, topography, impact of vadose zone, and aquifer conductivity. GOD evaluates groundwater vulnerability by considering groundwater occurrence, overall lithology, and depth. Aquifer Vulnerability Index (AVI) is a method used to assess the vulnerability of aquifers to contamination based on factors such as geological and hydrogeological characteristics, land use, and contaminant sources. It integrates these factors into an index to rank the vulnerability of different areas. SINTACS method of groundwater vulnerability assessment is another qualitative method that combines various parameters related to hydrogeology, soil, topography, land use, and contaminant sources to evaluate groundwater vulnerability. It utilizes a scoring system to categorize areas based on their susceptibility to contamination. Both these AVI and SINTACS methods represent diverse approaches to assessing groundwater vulnerability, each tailored to specific geological, hydrogeological, and environmental conditions and integrating GIS technology to enhance spatial analysis and decision-making.

Nevertheless, this isn't applicable when assessing susceptibility in the human surroundings, signifying that two operational scales are necessary. The assessment of contact and susceptibility in the human habitat compels us to consider the populace and resources, notwithstanding their position not directly above the groundwater, along with additional variables that might be nearby or provincially existing.

8.4 Impacts of Groundwater Pollution

Pollution of aquifers can affect human well-being, ecological integrity, and social and economic progress. For instance, numerous research investigations have demonstrated that elevated fluorine compounds, nitrogen oxides, heavy metals, and organic contaminants pose health hazards to humans, like thyroid swelling, tumors, congenital malformations, and high blood pressure (Wu et al., 2020). Water–rock interaction with carbonate rocks and basic flows, like dolerites, can contribute to higher fluoride concentrations in groundwater. This occurs through carbonate leaching, where fluoride-containing minerals may be released into the water. Additionally, using fertilizers can introduce fluoride into groundwater through runoff or leaching processes. Moreover, the alkalinity of groundwater can also influence fluoride levels, as alkaline conditions can enhance the solubility of fluoride-bearing minerals, leading to higher concentrations in the water. These natural and anthropogenic factors can contribute to elevated fluoride levels in groundwater, which can have implications for human health and the environment (Yenugu et al., 2020). This is particularly crucial for newborns and kids, who are more vulnerable to the impacts of these pollutants than mature individuals. For instance, blue baby syndrome is triggered by elevated nitrate levels in the potable water for infant nourishment. The health of individuals can also be impacted by the pollution of aquifers, influencing the crop cultivation processes. Figure 8.2 shows the environmental impacts of groundwater pollution.

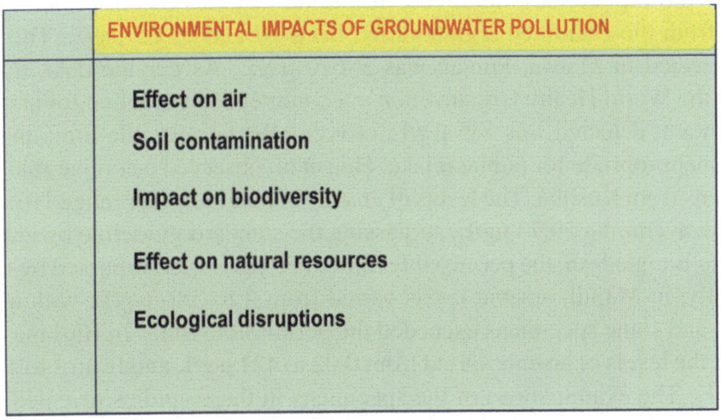

Fig. 8.2 Impacts of groundwater contamination on the environment

Utilizing groundwater tainted with metals for irrigation and polluted water containing enduring pollutants can lead to the buildup of harmful substances in grains and edibles, posing health hazards to people (Jenifer & Jha, 2018). Additionally, groundwater pollution can adversely impact the condition of soils and natural vegetation. Polluted groundwater can result in soil.

Pollution. For instance, in numerous farming regions within dry areas, elevated underground water salinity is a principal element impacting soil contamination. The dissolved salts and additional pollutants, including hazardous heavy metals, have the potential to accumulate in the rooting environment, thereby impacting plant development.

8.4.1 Underground Water Pollution by Arsenic

Groundwater pollution from arsenic remains a significant global issue. Communities susceptible to elevated levels of arsenic in drinking water are vulnerable to adverse health consequences due to the existence of pollutants. Therefore, it is crucial to establish standards that specify an amount deemed safe for regular potable water intake. Utilizing noxious research, the Global Health Authority has set the highest permissible amount of arsenic equal to the threshold of 10 μg/L. Nonetheless, investigations in groundwater reservoirs across different regions have documented arsenic amounts surpassing this benchmark, giving rise to significant health issues (Ali et al., 2018). Researchers documented elevated arsenic concentrations in diverse nations, such as Bangladesh, Argentina, Pakistan, and China. These results support observations signifying substantial metalloid amounts in underground water within these and additional nations, implying that severe conditions related to metal exposure persist in these areas for a long time (Aullón Alcaine et al., 2020). Within Argentina, the levels of arsenic in underground water specimens collected from the examined region varied from 5.9 to 535.1 μg/L, along with a mean concentration of 114 μg/L. Within Darrang, Assam, the levels of arsenic in underground water specimens collected from the examined region varied from 10.1 to 93.05 μg/L. The levels of arsenic detected in Malwa, Punjab was 5–620 μg/L. As per the data, most wells surpassed the World Health Organization's recommended guideline for arsenic. The highest amount detected was 535 μg/L, crossing the permissible limit and making the water inappropriate for public intake. Huq et al. observed a comparable outcome in specimens from Kushtia. The levels of arsenic in potable water ranged from 6.04 to 590.7 μg/L, averaging 59.73 μg/L, surpassing the standard guideline by greater than 50 times. In Bangladesh, the permissible limit of 50 μg/L was surpassed by ten times. Additionally, in Majuli, arsenic levels varied from 5 to 386 μg/L, with a mean of 137 μg/L, and some specimens exceeded the permissible limit. In Burkina Faso and Myanmar, the levels of arsenic varied from 0.02 to 421 μg/L and from 1 to 134 μg/L, respectively. The examinations of the specimens in these studies revealed the existence of supplementary pollutants alongside arsenic, including iron, fluoride, and copper, which frequently surpassed the threshold set by the World Health Organization. Earlier research conducted by Podgorski et al. (2017) revealed elevated arsenic amounts in the drinking water ingested by a significant portion of the global population. Arsenic-polluted underground water was documented to jeopardize 150 million individuals globally. Specifically, tube wells serve as the primary reservoir of polluted water, particularly in rural areas. While arsenic activation occurs biologically due to redox reactions from deposits to aquifers, excavating wells and employing chlorine for cleaning amplifies the discharge of arsenic and various other metals into the water. This disrupts the groundwater and initiates a geochemical imbalance. Erickson et al. discovered disruptions in the aquifers of Florida following the implementation of community wells and household wells (Erickson et al., 2019). Additionally, arsenic levels might escalate over time due to pumping. This phenomenon triggers redox process alterations, consequently influencing arsenic mobility and blending diverse waters towards varied paths leading to the well. While drilling wells might disrupt the geochemical balance of groundwater and elevate arsenic concentrations, alternative pollution origins exist as well. The metalloid's aggregation and activation can happen through biological mechanisms like volcanic eruptions, geothermal processes, erosion of igneous rocks, hydrothermal events, and bacterial processes. However, as previously noted, various human-made actions encompassing mining activities and using arsenic-containing agricultural chemicals also play a role in introducing the metalloid into the subsoil and detritus. Arsenic frequently exists with other pollutants like cadmium, lead, zinc, and iron, mainly from gold extraction activities. Therefore, mining operations worldwide have led to the dissemination of arsenic into aquifers for more than years. Elevated levels of arsenic in aquifers occur primarily due to interventions between minerals and water. In aquifers, the activation and aggregation of arsenic depend on the environmental and chemical aspects of interactions between water and soil, including factors like acidity, reduction–oxidation environments, and the composition of minerals within topsoil and particulates. Hence, examining and appraising various origins of exposure, ecological utilization, and aquifer environments in every global region could facilitate the identification of tailored, efficient medical remedies through technology.

8.4.1.1 Effects of Arsenic Contact on Well-Being

Arsenic-contaminated subterranean water is the primary origin of human exposure to this harmful substance. However, it is crucial to acknowledge other significant origins, like skin contact (Rahman et al., 2018), along with various exposure pathways through the consumption of vegetables. Inside the human organism, arsenite attaches to SH groups in proteins within numerous organs, including the kidney, lungs, and digestive tract lining (Fig. 8.3). Moreover, arsenic can be detrimental even at minimal levels. Additionally, arsenite hampers pyruvate decarboxylation by bonding with the active state of lipoic acid, thus disrupting energy processing in cells. The harmful effects of arsenate occur due to the deactivation of numerous catalysts, particularly those associated with the production and restoration of DNA, as well as the generation of cellular energy. In a recent study, researchers explored the cancer-inducing potential of arsenic in kidney tubular cells and observed negligible impacts on cellular growth rate after a 72-h operation. Prolonged exposure to arsenic has been linked to detrimental consequences for human well-being. Indeed, the existence of pollutants was documented to exacerbate health hazards in human beings. Table 8.2 presents the harm caused to different organs due to arsenic exposure. Arsenic enters the body mainly by breathing and oral intake, and it can also be absorbed through skin contact or epithelial linings. The way this element moves within the body and its toxicokinetics rely on the kind and length of exposure and its physical and chemical properties. Primary buccal contact (consumption of polluted water) and secondary contact (ingestion of crops cultivated with polluted water) stand as the principal methods of exposure out of every pathway accountable for various ailments caused by metalloids globally (Chakraborti et al., 2017). Concerning prolonged contact, non-carcinogenic skin abnormalities are identified as the most significant trait utilized in recognizing people displaying particular signs of arsenic intoxication, known as arsenicosis. In light of this, researchers regard these skin irregularities as a clinical indicator of vulnerability to the pollutant. Contact with arsenic correlates directly with water consumption and the amount of arsenic in water. Intriguingly, the mean of primary and secondary water intake by people from diverse nations displays significant variability. Chakraborti et al. discovered that detrimental skin conditions arise at diverse levels of arsenic in potable water (Patel et al., 2021). Communities exposed to 50 μg/L, the standard set by many progressing nations, displayed intense skin abnormalities. Nonetheless, researchers noted skin abnormalities at even reduced exposure thresholds within the spectrum of 0.005–0.01 mg/L of arsenic in potable water. People enduring prolonged exposure to arsenic form non-malignant conditions as well as malignant skin abnormalities, including rodent ulcers and epidermoid cancer, alongside bowenoid papulosis. The systemic harm provoked by arsenic is linked to the possibility of developing epidermal cancer. And internal malignancies (pulmonary, hepatic, bladder, and renal), as well as breathing and neural ailments. Smith et al. proposed that pulmonary carcinoma stands as the prevalent factor in arsenic-related fatality even after extended periods of minimizing contact with the pollutant. The results concerning the carcinogenic impacts triggered by contact with arsenic align with the perceptions made by Roh et al. (2018) and various other public health investigations documenting the occurrence and death arising from the formation of tumors directly linked to contact with arsenic. Chakraborti et al. reported that inorganic arsenic compounds in females enduring prolonged exposure permeate the fetal membrane, potentially impacting the fertility and growth procedures of the unborn offspring (Sudharshan Reddy and Sunitha, 2023). Various complicating elements might additionally play a role in fertility and growth abnormalities. For example, the duration of contact, the amount at the origin, and the lowest amount of arsenic necessary to cause a congenital disability, repeated deliveries, and undernutrition in communities enduring prolonged exposure to arsenic. Miscarriages, newborns born without life signs, preterm deliveries, low birth weight, and deaths within the first month after birth are more common in people persistently exposed to arsenic. While arsenic is not acknowledged as primary mutagenic, this can prompt

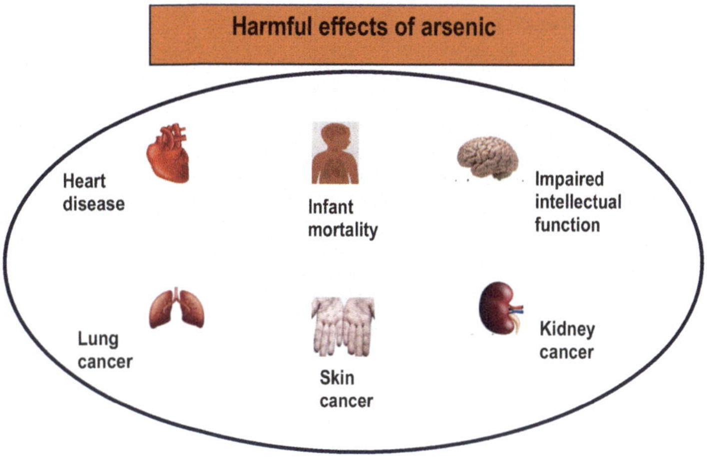

Fig. 8.3 Harmful effects of arsenic on the human body

oxidative damage, swelling, and epigenomic imbalance. Table 8.3 displays disorders triggered by arsenic contact and linked epigenetic consequences. Indeed, when processed within the human system and during the methylation procedure, arsenic produces free radicals, including peroxyl, superoxide, and hydroxyl radicals, and cellular enzyme functions are altered during the transfer of electrons within the electron transport chain. Hence, considering the possible apprehension arising from arsenic exposure to people's well-being, researchers have taken significant steps to enhance the comprehension of this ecological pollutant.

8.4.2 Underground Water Pollution by Lead

Groundwater pollution from lead may cause significant impacts on the intellectual growth of youngsters (Lanphear et al., 2019), while prolonged exposure to minimal lead amounts can result in adverse health consequences for grown-ups, like heart disease and malignancy (Lanphear et al., 2018). The Environmental Protection Agency threshold for lead amounts in potable water stands at 15 μg/L, while the Canadian Health Department recently reduced its highest allowable lead level to 5 μg/L in 2019. Aquifers generally maintain low lead levels, but increased (greater than 7.5 μg/L) amounts are frequently detected in underground water in the Eastern United States (Jurgens et al., 2019). Exposure to lead from potable water is commonly associated with the seepage of lead from pipe fittings with lead content in the supply network (links from water source to residence) and within the building's pipe fitting. The decrease of lead in Clean Water Legislation was established in 2014 to decrease the permissible lead levels in substances, apparatus, and elements of water networks formerly considered "non-lead." Nevertheless, these rules do not apply to individual

Table 8.2 Consequences caused by contact with arsenic

Affected body parts	Impacts	References
Cutaneous system	Skin abnormalities, arsenic-related conditions (indicators: abnormal skin growth, pigmentation changes, thickening of the skin, white or dark spots), Bowen's disorder, cancerous growth of squamous cells	Edmunds et al. (2015)
Pulmonary system	Pulmonary carcinoma, cough, vocal cord carcinoma, modification of the respiratory passages, damaged pulmonary capability, difficulty breathing	Roh et al. (2018)
Vascular system	Peripheral artery disorder, high blood pressure	Suvarna et al. (2022)
Renal system	Renal tumors, bladder tumors	Antoni et al. (2017)
Neurological system	Nerve damage (indicators: peripheral tingling, head pain, limb discomfort, heightened sensitivity, indications of reduced sensation in limbs)	Chakraborti et al. (2017)
Hearing system	Lightheadedness, giddiness	Patel et al. (2021)
Islet cell damage	Hyperglycemia	Reddy et al. (2022)
Reproductive	Elevated occurrence of prostate malignancy, unexpected miscarriages	Patel et al. (2021)

water provisions, and residences depending on these sources for potable water remain susceptible to lead pollution. Pipe erosion causes lead to seep into the potable water, and therefore, when needed, water service providers also administer a corrosion prevention agent, like phosphoric acid (Ng et al., 2012). Acidified aquifer, frequently found in low-depth wells, is linked to elevated lead concentrations in potable water (Pieper et al., 2016). Previous studies examining lead hazards have primarily concentrated on publicly supplied water (often with lower corrosiveness). Therefore, evaluations of risk derived from lead levels might undervalue the potential danger for individuals relying on private wells. Even if states enforce building norms for wells, there are no guidelines for the materials used inside the well. Additionally, water well installers are responsible for advocating optimal methods, such as employing pipe fittings devoid of lead. Residents residing in vintage residences should be urged to examine and purify their potable water. However, even individuals living in recently constructed residences might find it advantageous to conduct tests if they inhabit regions where the geological states are conducive to lead seepage. Geological models have been employed to chart the dissolvability of lead in unprocessed aquifers, which can assist in identifying areas where individuals relying on private wells might be more vulnerable to lead contact. Such models can be integrated into public health programs to pinpoint communities at higher threat.

Table 8.3 Disorders resulting from contact with arsenic

Epigenetic impacts	Chemical cell signaling	References
Renal cell carcinoma	Regulation of proteins associated with cancer-suppressing genes: BAP1, TP53	Antoni et al. (2017)
Urothelial carcinoma	Regulation of proteins associated with cancer-inhibiting genes: PABPC1, KRAS	Suvarna et al. (2022)
Malignant conversion in airway epithelial cells	Suppression of miRNA-31 production, heightened expression of adenine–thymine-rich DNA-binding protein 2	Chen et al. (2018)
Malignant and precancer skin abnormalities	Epigenetic modification: Areas of cancer-inhibiting DAPK and p16 genes	Bjørklund et al. (2017)
Harmful effects on uroepithelial cells	Reduced histone four lysine 16 acetyltransferase, altered miRNA expression	Cardoso et al. (2018)
Stimulation of autophagy in hepatic tissue	Reduced levels of miR-21, elevation of miR-21 concentrations in a dose-dependent fashion	Podgorski et al. (2017)
High blood sugar	Epigenetic modification: Suppression of glucose-triggered insulin release	Khan et al. (2017)
Neural disorders: Alzheimer's syndrome	Form clots and penetrate the substantia nigra, impact on dopamine-related brain activity	Costa (2019)
Arterial plaque buildup	Increased methylation and suppressed cholesterol efflux transporter A1 expression lead to reduced plasma HDL-C levels. Enhance oxidation processes	Song et al. (2019)

8.4.3 Impact of Cadmium Exposure on Human Health

Cadmium hazard to human well-being continues to be a worldwide ecological and health apprehension. Studies have explored the detrimental impacts of cadmium on humans (Xiao et al., 2008). Cadmium has serious health consequences on humans, with the skeletal system and renal organs being the most crucial organs affected (Friberg et al., 2019). Cadmium generally builds up in the human system and eventually leads to diverse detrimental health consequences, notably resulting in bone impairment and kidney cell damage (Nordberg et al., 1992). Approximately fifty percent of the overall cadmium penetrating the human organism accumulates in the renal system, where it attaches to metallothionein. Higher renal retention is noticed in women than men due to elevated absorptive capacity. The decay period of cadmium in the kidney is nearly a decade. The gathering of cadmium in the kidneys and liver results from this organ's capacity to generate metallothionein, which exhibits a strong binding for cadmium. A research performed in Belgium on cadmium-affected individuals revealed a 10% incidence of renal tubular protein leakage at a concentration of 2 nmol of cadmium per millimole of creatinine (Buchet, 1990). Likewise, in another investigation on cadmium-affected individuals, a 50% occurrence of renal tubular protein leakage was detected in the predominantly exposed category. Prolonged exposure to cadmium results in gradual impairment of kidney ductal. In the advanced stages of kidney impairment due to cadmium contact, calcium and phosphate depletion, glucose in urine (glycosuria), and disrupted calcium metabolic activity, leading to complications such as osteomalacia. If there is a kidney tubular disorder, the elimination of cadmium, along with other substances, might escalate. However, cadmium is harmful to the kidneys and can lead to impairment in the kidney tubules, ultimately decreasing renal filtration capacity. Contact with cadmium also diminishes bone strength due to phosphorus and calcium loss. Research in Sweden involving 1021 individuals found a notable adverse relationship between cadmium levels in urine and bone strength. Both cadmium and calcium are assimilated in the intestinal tract. In this context, cadmium directly hampers calcium uptake by vying for binding regions in the intestinal lining projections with calcium ions. Implicitly, it impacts calcium uptake by interfering with vitamin D transformation. Within cells, cadmium disrupts the metabolic activity of calcium, iron, magnesium, copper, and zinc, resulting in mineral loss, bone thinning, and bone-related ailments. Cadmium replaces calcium within the bones, resulting in their fragility and, finally, leading to bone cracks, particularly in youngsters and females (Elinder et al., 1977). Cadmium can also cause disturbances in glycolysis. Reports indicate that cadmium contact is linked to an array of medical conditions, like heart failure, malignancies, loss of smell, stroke, bone weakening, lung condition, the presence of excess protein in the urine, and clouding of the eye lens. Cadmium has been designated as a cancer-causing agent by the Global Cancer Research Organization and triggers the onset of breast neoplasm in women exposed to it. Elevated cadmium contact impacts androgen and protein levels for sex hormones, ultimately influencing hormones regulating reproduction. Contact with cadmium also hampers the operation of the hormonal system. Youngsters are more vulnerable to cadmium harmfulness than grown-ups, and even minimal levels of cadmium can result in detrimental health consequences.

8.5 Techniques Employed to Detect Groundwater Contamination

Assessing various contaminants in groundwater is commonly accomplished using gas chromatography (GC) with mass spectrometry, GC, or liquid chromatography (LC). By coupling chromatography with mass spectrometry, analysts can achieve high sensitivity, selectivity, and accuracy in detecting and quantifying pollutants in polluted water. This allows for the comprehensive analysis of complex mixtures of pollutants and the identification of trace levels of contaminants that may pose risks to the environment and public health. Untargeted screening refers to analyzing a sample without prior knowledge of its components. In tandem mass spectrometry (MS/MS), untargeted approaches involve scanning for all detectable ions in a sample rather than specifically targeting known compounds. While tandem mass spectrometry is highly sensitive and selective, it may not be ideal for untargeted screening due to data complexity. Despite these challenges, advances in analytical techniques, data processing algorithms, and spectral databases are continuously improving the applicability of tandem mass spectrometry for untargeted screening. However, researchers often combine tandem mass spectrometry with other techniques, such as high-resolution or chromatographic separation, to enhance the identification and quantification of unknown compounds in complex matrices. Coupling high-resolution mass spectrometry with LC or GC offers several advantages for analyzing emerging contaminants in complex matrices. Four primary evaluation techniques are employed to evaluate the hazards of pesticide contamination in aquifers (Zhao and Pie, 2012).

8.5.1 Indexing Approach

Indexing methods are the most commonly utilized for evaluating groundwater contamination. These methods rely on essential variables linked to contaminant infiltration into aquifers, even in the saturated and unsaturated zones, to assess contamination risk (Patel et al., 2021).

8.5.2 Mechanism-Oriented Approaches

Contaminants like agrochemicals undergo a sequence of physio-chemical and natural methods within saturated and unsaturated zones. Simulation methodologies utilize mathematical models to depict associated movements and transformations. A quantitative measure can be computed within simulation frameworks to delineate the contamination threat.

8.5.3 Data-Driven Techniques

Using recorded contamination data, quantitative models can be employed to identify correlations between research data and factors associated with contamination. Subsequently, the assessment metrics can be validated, and every metric, determined as significant through mathematical analysis, is allocated a pertinence value.

8.5.4 *Fuzzy Evaluation Approach*

The fuzzy evaluation approach initially necessitates ensuring the contamination categorization. Furthermore, the assessment metrics must be chosen, and then a benchmark value for each metric must be at various pollution levels. Subsequently, the extent to which every specimen pertains to specific amounts of pollution can be established by combining the data into a formula based on comparative membership value. During this stage, the weighting vector for the parameters is also essential. Ultimately, by performing matrix multiplication on the matrix representing the optimum relative rank and the sequence matrix, a unique scalar array can be attained, and the contamination hazard can be evaluated based on this array.

8.6 Groundwater Cleanup Methods

Multiple techniques exist for remediating aquifer pollutants to reduce their amounts to a permissible limit. Groundwater cleanup is typically achieved using physiochemical or natural methods, of which natural methods are mainly used. Crognale et al. (2017) accomplished the remediation of arsenic by deploying microorganisms alongside biological–chemical mechanisms linked to the arsenic mobility pathway. Maity et al. explored arsenic elimination with microbial variants. They concluded that microbial function supports the activation of arsenic in the As^{+3} chemical state, which poses more significant harm to living organisms (Maity et al., 2021). Mukherjee and Singh (2018) identified a pair of fungal variants to assess their effectiveness in arsenic removal and found that 50–76% of the arsenic could be eradicated. Maity et al. employed fulvic acid for arsenic removal from polluted water. Maity et al. (2021) examined processes like sorption, membrane sieving, and oxidation to eliminate arsenic. They determined that environment-friendly natural techniques, like microorganisms and phytoremediation methods, could assist in cleansing water and land. Table 8.4 and Table 8.5 represent various arsenic and cadmium removal methods, respectively. Research has highlighted that biological treatment, or biodegradation, is currently the most environmentally friendly choice. This method is less disruptive, eco-friendly, and cheaper than other options. Biodegradation methods are primarily categorized as off-site and on-site methods. The on-site method is the most common and often employed approach for eliminating pollutants originating from aquifers. This method proves economical and flexible as it occurs precisely where the pollution is present, utilizing biological membranes or aquatic microorganisms to eliminate the pollutants. The pollutants can be broken down by the Indigenous microbial population, which is biologically altered. Microbial metabolic processes lead to the degradation of pollutants in the presence of electron donors, electron receptors, and vital nutrients. Groundwater biodegradation takes place through both biological and tailored routes.

Table 8.4 Various methods for arsenic removal

Methods	Benefits	Disadvantages	Rate of removal	References
Agglomeration-Precipitation	Financially feasible Capable of handling substantial water volume Necessitates the utilization of salts like iron(III) salts and aluminum	Clarification process is necessary following the agglomeration-precipitation process There is generally less efficacy in eliminating arsenic Pre-treatment oxidation is needed to eliminate As^{+3}	More than 70%	Matsui et al. (2017)
Sorption	Quite familiar and readily accessible commercially Doesn't generate substantial amounts of secondary products	Efficient impact of sulfuric acid salts on remediation rate Rejuvenation agents come at a high cost	94%	Xu et al. (2019)
Ionic exchange	Exceptional selectivity for eliminating arsenic Great capability for the capturing	Efficient disruption caused by carbonate and sulfate ions on elimination effectiveness Restricted resin lifespan	Less than 85%	Velazquez-Peña et al. (2019)
Membrane filtration	Chemical supplements are unnecessary Implementation is straightforward Outstanding efficiency	High initial capital outlay Thermal constraints	96.4%	Kukučka et al. (2016)

Table 8.5 Various methods for cadmium removal

Methods	Benefits	Disadvantages	Rate of removal (%)	References
Adhesion	Optimal stability Reliability Economically efficient	Effectiveness relies on the starting metal levels, acidity, dosage of flocculant, temperature	94–96	Shenashen et al. (2013)
Sorption	Outstanding removal efficacy	Varies with acidity and metal ion level	69–93	Chandrashekharaiah et al. (2021)
Agglomeration-Precipitation	Cost-effective and effective unit in elimination	Effectiveness relies on the solution acidity and the flocculant levels	100	Oyewo et al. (2019)

8.6.1 *Biological Remediation*

This method represents a passive approach that relies on biological processes to disperse pollutants from aquifers. Moreover, this technique is also known as self-remediation. This method involves processes through which contaminants transform degradation in low-oxygen and oxygen-rich conditions, movement, dissipation, cooperative metabolic, adsorption, spreading, and scattering (Cecconet et al., 2020). The advantage of biological bioremediation over tailored or man-made on-site methods lies in its utilization of native microbes. It is, therefore, more affordable than the tailored method since there is no need for modification, and the microorganisms naturally adjust to their ecosystem until nutrient and air concentrations reach their minimum limits.

8.6.2 *Tailored Remediation*

Although biological remediation is economical, the tailored method has demonstrated superior outcomes in contaminant removal within the same time frame. This approach involves augmentation, microbial stimulation, microbial suction and venting, microorganism-assisted reactive barrier, and plant-based remediation.

8.6.3 Augmentation and Microbial Stimulation

Augmentation is a technique wherein microbes, either fixed or liberated, serve as a culture, breaking down pollutants and allowing them to persist in unfavorable conditions. Instances arise when native microbes cannot decompose substantial amounts of contaminants present in aquifers; additional substances are introduced to break down compounds like polyaromatic hydrocarbons, perchloroethylene, pharmaceuticals, trichloroethylene, and crude oil (Kong et al., 2018). These specific microbes can move through to enhance their operational capability because engineered microbes or non-native variants are enclosed within the aquifer flow, requiring remediation; the success of this procedure relies heavily on the response of indigenous microbe communities to their presence, primarily for nourishment. Microbial stimulation represents an alternate approach to enhance the pace of remediation. In this technique, the remediation process is expedited by modifying the physio-chemical attributes of the aquifer. To improve the accessibility of electron donors, diverse nourishing resources like biofuel, fertilizer, liquid manure, and organic substances are introduced into the polluted aquifer. A comparative analysis revealed that enhancing microbial activity surpasses introducing additional microorganisms in eliminating petroleum hydrocarbons. However, when utilized together, they yield the most favorable results within a minimal timeframe.

8.6.4 Microbial Suction and Venting

Venting is a technique that integrates remediation in an oxygen environment and subsurface aeration to extract low and medium boiling point hydrocarbon fractions from the unsaturated region of the aquifer level (Tzovolou et al., 2015). This strategy intends to reduce the vaporization of polyaromatic hydrocarbons. In contrast, subsurface aeration eradicates organic vapors, and aerobic degradation transpires by providing an air environment to maintain bacterial function in the unsaturated region. Microbial suction is a method of on-site remediation employed to eliminate volatile and partially vaporizable organic compound pollution from saturated and unsaturated areas. This process consists of two phases: the initial phase concerns the retrieval of freely flowing substances, and the subsequent phase entails the purification of the polluted aquifer after retrieving the freely flowing substances. Due to the heightened water amount in the soil, this procedure lowers the pace of oxygen transfer, reducing bacterial function. Microbial suction involves venting, subsurface vapor removal, and vacuum-aided extraction to remove freely flowing substances from the aquifer. A slurp tube uses the suction pump to draw out the freely flowing substances. This approach segregates petroleum hydrocarbons from water and atmosphere by pulling them from the upper part of the aquifer and capillary zone (Iqbal et al., 2020). The microbial suction technique constitutes a multifaceted removal method that can transition into a regular venting setup once the freely flowing substances are retrieved.

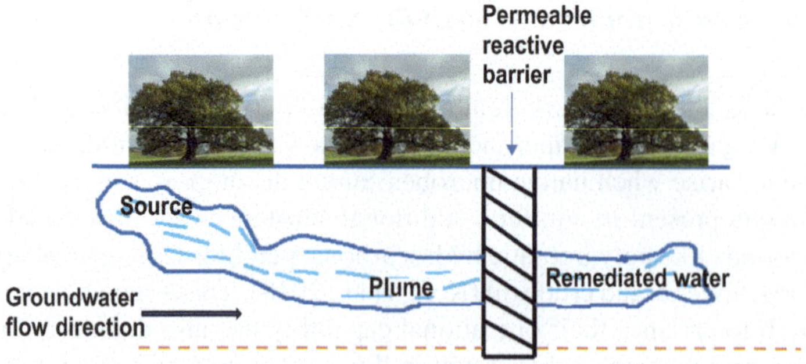

Fig. 8.4 Permeable reactive barrier

8.6.5 Microorganism-Assisted Reactive Barrier

This on-site method eliminates contaminants from aquifers, predominantly toxic metals, and chlorinated compounds. An in-situ chemical barrier is positioned perpendicular to the flow direction of the contaminated aquifer in this process (Fig. 8.4). While the porous reactive barrier process is fundamentally a physical technique, it demonstrates a significant influence from microbes within the chemical barriers. Integrating bacterial function into this process can achieve remediation, wherein microbes act as aquifer purifiers. Permeable reactive barriers can operate in both dynamic and static modes. The dynamic mode employs water control structures, injecting boreholes, or other active systems that guide the aquifer's flow toward on-site management units. A static approach harnesses gradient to channel hydraulic movement to the inactive treatment unit.

8.6.6 Plant-Based Remediation

This is an approach to clean polluted aquifers by harnessing botanic-contaminant interplay to mitigate the harmful impact of pollutants. This method encompasses various mechanisms, such as aggregation, sieving, decomposition, vaporization, and solidification (Rostami & Azhdarpoor, 2019). Aggregation can eliminate hazardous metals, while solidification and degradation can eliminate organic pollutants. Plants with substantial biosorption capability can intake, process, amass, relocate, and endure elevated amounts of toxic metals. Approximately 400 plant species have been recognized as metal-accumulating plants. The root configuration of the vegetation, aerial plant mass, the contaminant's hazardousness, the vegetation's capacity to adapt and thrive, and the anticipated extent of pollutant remediation all influence the remediation method carried out through plant-based remediation.

8.6.7 Nanoscale Technology in Groundwater Pollution

The resolutions to most groundwater purity issues are challenging to identify and address; furthermore, they are considerably costly, laborious, and not consistently effective. Within ecological preservation, a crucial objective of contemporary investigations is to discover innovative and state-of-the-art materials. Lately, a substantial amount of emphasis has been directed toward the suitability of nano-engineered materials as adsorptive materials or accelerators to eliminate poisonous and detrimental substances from the atmosphere and sewage (Yang et al., 2006). Nanoscale engineering involves describing, planning, manufacturing, and using configurations, gadgets, and systems by manipulating structure and dimensions at the nanoscale. Responsive iron nanomaterials were discovered to be valuable in mitigating groundwater pollution and, therefore, could also demonstrate effectiveness in eradicating agricultural chemicals from the surroundings. In the realm of nanoscience, the generation of nanostructures and commodities incorporating them is swiftly evolving in sectors that offer numerous possibilities for modern advancement. In reducing groundwater pollution, the utilization of nanoscience is only starting to be investigated, potentially sparking profoundly transformative shifts in the environment (Masciangioli & Zhang, 2003). The principal element that delineates the potential of nanostructures as an exceedingly adaptable remedial tool encompasses their extraordinarily minute particle dimensions, contrasting with a standard microbial cell with a diameter of 1000 nm. Therefore, nanomaterials can be conveyed efficiently by the aquifer movement. Despite their diminutive stature, particles at the nanoscale possess the capability to tackle specific obstacles in pollution remediation economically. Furthermore, the nanomaterials can stay in solution for extended durations to initiate an on-site remediation area. Consequently, nanomaterials can be affixed onto a sturdy framework like activated charcoal for enhanced water, sewage, or gas phase processing. In the recent several years, the restoration domain has progressed incessantly, concurrently evolving and embracing innovative techniques to enhance the restoration procedure. The pump and treat are regarded as one of the most entrenched systems within those innovative techniques. The fundamental mechanism associated with these systems is the expulsion of the polluted aquifer from the subsurface downstream of the pollution location, followed by its treatment before reintroducing it to the subsurface (Fig. 8.5). Nevertheless, there are some limitations linked with these systems; for instance, the method could be more laborious and costly to function and sustain. In the initial 1990s, with the exploration of the minimizing abilities of metals like elemental iron for addressing a broad spectrum of pollutants in harmful effluent, the prevailing utilization of elemental iron has been developed in the shape of reactive barriers devised to obstruct discharges in the groundwater and thus mitigating them. In 1994, reactive barriers were initially set up at the terrain level to provide an alternative to the more traditional pump and treat techniques. Due to their limited depth-setting capacities, this methodology was confined solely to the superficial streams up to 50 feet below the surface. Additionally, the streams had to be meticulously distinguished and delineated to avert pollutants from bypassing

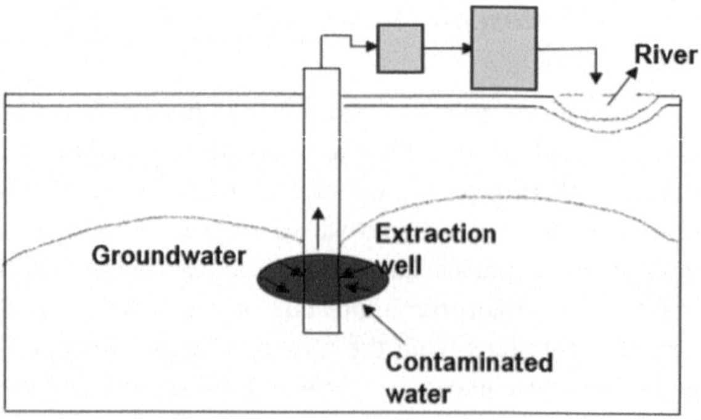

Fig. 8.5 Pump and treat system

the reactive barriers. Issues arise from a decline in iron-induced reactivity by the metallic hydroxide and carbonate deposition onto the iron's surface—the limited responsivity of elemental iron against chlorinated substances generated hazardous secondary products. Just as reactive barriers were crafted to present a superior alternative to the pump and treat method, novel techniques are currently accessible to contend with reactive barriers. Metal nanoparticles and their composites offer a possibility for a more efficient and cost-effective alternative to numerous mitigation techniques. The minute particle dimensions of the metal nanoparticles enable a significantly elevated degree of restorative adaptability.

8.7 Future Perspective

Groundwater pollution is currently a worldwide challenge, and addressing these issues mandates extensive cooperation among academics in institutions, governmental bodies, businesses, and policymakers at every echelon of governance. Resolving the problems related to groundwater pollution demands global cooperation. This necessity is especially pronounced in emerging nations, where ecological means and entry to cutting-edge techniques are not easily accessible. Particular attention should be dedicated to the following facets of education and research:

- The challenges of groundwater pollution across various nations necessitate applying diverse approaches, methodologies, and regulations. While groundwater pollution is a universal concern, its characteristics and determining factors vary significantly among countries, climatic zones, and geomorphological characteristics. There may be more suitable courses of action than employing remedial strategies that have proven effective in different nations. To illustrate, nitrate contamination results from agrochemicals in certain farmlands (Zhang et al., 2018). At the same time, in other regions, it could stem from commercial and household wastewater pollution or munitions utilized in mineral exploration activities. Employing varied strategies might be essential to alleviate distinct forms of nitrate contamination. Even when agrochemical use is the prevalent source of nitrate contamination in equatorial and moderate areas, the mitigation methods could diverge, as climate elements and soil attributes will significantly impact the methods and scope of pollutant movement.

- Given the swift technological advancement, numerous innovative methodologies have emerged for investigating groundwater pollution, encompassing earth science and geospatial technology methods. Geospatial data systems and satellite imagery have expedited the progress of aquifer science (Ahmed et al., 2020). Moving forward, extensive data examination, artificial intelligence, aerial drone investigations, and molecular and isotopic tracing methods will become more extensively accessible for utilization in aquifer studies. Groundwater researchers must embrace and employ these emerging techniques to examine groundwater pollution.
- Governments, especially in emerging nations, must prioritize investments and actively support learning and research initiatives in subsurface hydrology. In many areas, people are compelled to rely on polluted groundwater, lacking viable alternatives for consumption, possibly leading to diverse health consequences. Funding is essential to ascertain the scope of this pollution and develop solutions to mitigate the effects on human well-being or to identify other origins of potable water.

8.8 Conclusion

Groundwater sources are crucial in sustaining human well-being, fostering social progress, and conserving the natural environment. Currently, the deterioration in the quality of groundwater is a significant global concern. The purity of aquifers faces escalating risks due to urban expansion, farming techniques, commercial operations, and weather fluctuations. Groundwater pollution can also be conveyed through interfaces between above-ground water bodies and groundwater, resulting in the decline of water purity on the surface. The main human-induced factors contributing to groundwater contamination are agrochemicals and improper management of factory waste, which leads to infiltration into aquifers. The ramifications of groundwater contamination are far-reaching. Prolonged contact with both non-cancerous and cancerous contaminants, even in minute quantities, can lead to adverse health effects (such as high blood pressure, respiratory issues, carcinogenic vulnerability, diabetes mellitus, skin issues, etc.). Selecting a suitable treatment approach relies on factors particular to the region and frequently considers purification objectives associated with potential hazards to safeguard human well-being and surroundings. The trajectory of aquifer research remains ambiguous despite the expanding scale of research. This study offers a comprehensive overview of the health repercussions resulting from groundwater pollution and the available remediation techniques. Due to its cost-effectiveness, the permeable reactive barrier is one of the most favored techniques. In the future, scientists must embrace and utilize emerging techniques like cognitive computing, extensive aerial drone inspections, and data analytics to investigate groundwater pollution. Lastly, governments should allocate more significant resources towards groundwater initiatives and promote research, training, and development in this vital area.

Acknowledgements The authors are proud privilege and appreciation for providing laboratory and internet facilities to the Department of Chemistry, Vigyan Mandir, Banasthali Vidyapith, Rajasthan.

References

Adeyemi, A. A., & Ojekunle, Z. O. (2021). Concentrations and health risk assessment of industrial heavy metals pollution in groundwater in Ogun state, Nigeria. *Scientific African, 11*, e00666. https://doi.org/10.1016/j.sciaf.2020.e00666

Ahmed, A., Nawaz, R., Woulds, C., & Drake, F. (2020). Influence of hydro-climatic factors on future coastal land susceptibility to erosion in Bangladesh: A geospatial modelling approach. *Journal of Geovisualization and Spatial Analysis*. https://doi.org/10.1007/s41651-020-00050-x

Ali, W., Rasool, A., Junaid, M., & Zhang, H. (2018). A comprehensive review on current status, mechanism, and possible sources of arsenic contamination in groundwater: A global perspective with prominence of Pakistan scenario. *Environmental Geochemistry and Health, 41*(2), 737–760. https://doi.org/10.1007/s10653-018-0169-x

Al-Sudani, H. I. Z. (2019). A review on groundwater pollution. *International Journal of Recent Engineering Science (IJRES), 6*(5), 14–22.

Antoni, S., Ferlay, J., Soerjomataram, I., Znaor, A., Jemal, A., & Bray, F. (2017). Bladder cancer incidence and mortality: A global overview and recent trends. *European Urology, 71*(1), 96–108. https://doi.org/10.1016/j.eururo.2016.06.010

Aullón Alcaine, A., Schulz, C., Bundschuh, J., Jacks, G., Thunvik, R., Gustafsson, J.-P., Mörth, C.-M., Sracek, O., Ahmad, A., & Bhattacharya, P. (2020). Hydrogeochemical controls on the mobility of arsenic, fluoride and other geogenic co-contaminants in the shallow aquifers of northeastern La Pampa Province in Argentina. *Science of the Total Environment, 715*, 136671. https://doi.org/10.1016/j.scitotenv.2020.136671

Bjørklund, G., Aaseth, J., Chirumbolo, S., Urbina, M. A., & Uddin, R. (2017). Effects of arsenic toxicity beyond epigenetic modifications. *Environmental Geochemistry and Health, 40*(3), 955–965. https://doi.org/10.1007/s10653-017-9967-9

Buchet, J. (1990). Renal effects of cadmium body burden of the general population. *The Lancet, 336*(8717), 699–702. https://doi.org/10.1016/0140-6736(90)92201-r

Cardoso, A. P. F., Al-Eryani, L., & States, J. C. (2018). Arsenic-induced carcinogenesis: The impact of miRNA dysregulation. *Toxicological Sciences*. https://doi.org/10.1093/toxsci/kfy128

Cecconet, D., Sabba, F., Devecseri, M., Callegari, A., & Capodaglio, A. G. (2020). In situ groundwater remediation with bioelectrochemical systems: A critical review and future perspectives. *Environment International, 137*, 105550. https://doi.org/10.1016/j.envint.2020.105550

Chakraborti, D., Das, B., Rahman, M. M., Nayak, B., Pal, A., Sengupta, M. K., Ahamed, S., Hossain, Md. A., Chowdhury, U. K., Biswas, B. K., Saha, K. C., & Dutta, R. N. (2017). Arsenic in groundwater of the Kolkata Municipal Corporation (KMC), India: Critical review and modes of mitigation. *Chemosphere, 180*, 437–447. https://doi.org/10.1016/j.chemosphere.2017.04.051

Chandrashekharaiah, P. S., Sanyal, D., Dasgupta, S., & Banik, A. (2021). Cadmium biosorption and biomass production by two freshwater microalgae *Scenedesmus acutus* and *Chlorella pyrenoidosa*: An integrated approach. *Chemosphere, 269*, 128755. https://doi.org/10.1016/j.chemosphere.2020.128755

Chen, X., Zhu, G. Y., & Jin, T. Y. (2008). Progress of the study on toxic effects of cadmium on kidney and bone.

Chen, Q. Y., Li, J., Sun, H., Wu, F., Zhu, Y., Kluz, T., Jordan, A., DesMarais, T., Zhang, X., Murphy, A., & Costa, M. (2018). Role of miR-31 and SATB2 in arsenic-induced malignant BEAS-2B cell transformation. *Molecular Carcinogenesis, 57*(8), 968–977. https://doi.org/10.1002/mc.22817

Costa, M. (2019). Review of arsenic toxicity, speciation and polyadenylation of canonical histones. *Toxicology and Applied Pharmacology, 375*, 1–4. https://doi.org/10.1016/j.taap.2019.05.006

Crognale, S., Amalfitano, S., Casentini, B., Fazi, S., Petruccioli, M., & Rossetti, S. (2017). Arsenic-related microorganisms in groundwater: A review on distribution, metabolic activities and potential use in arsenic removal processes. *Reviews in Environmental Science and Bio/technology, 16*(4), 647–665. https://doi.org/10.1007/s11157-017-9448-8

Economou-Eliopoulos, M., & Megremi, I. (2021). Contamination of the soil–groundwater–crop system: Environmental risk and opportunities. *Minerals, 11*(7), 775. https://doi.org/10.3390/min11070775

Edmunds, W. M., Ahmed, K. M., & Whitehead, P. G. (2015). A review of arsenic and its impacts in groundwater of the Ganges–Brahmaputra–Meghna delta, Bangladesh. *Environmental Science: Processes and Impacts, 17*(6), 1032–1046. https://doi.org/10.1039/c4em00673a

Eldaw, E., Huang, T., Elubid, B., Khalifa Mahamed, A., & Mahama, Y. (2020). A novel approach for indexing heavy metals pollution to assess groundwater quality for drinking purposes. *International Journal of Environmental Research and Public Health, 17*(4), 1245. https://doi.org/10.3390/ijerph17041245

Elinder, C.-G., Piscator, M., & Linnman, L. (1977). Cadmium and zinc relationships in kidney cortex, liver, and pancreas. *Environmental Research, 13*(3), 432–440. https://doi.org/10.1016/0013-9351(77)90023-8

Erickson, M. L., Malenda, H. F., Berquist, E. C., & Ayotte, J. D. (2019). Arsenic concentrations after drinking water well installation: Time-varying effects on arsenic mobilization. *Science of the Total Environment, 678*, 681–691. https://doi.org/10.1016/j.scitotenv.2019.04.362

Friberg, L. T., Elinder, G. G., Kjellstrom, T., & Nordberg, G. F. (Eds.). (2019). Cadmium and health: A toxicological and epidemiological appraisal: Volume 2: Effects and response (Vol. 1). CRC press.

Iqbal, I., Tian, G., Wang, Z., Masood, Z., Liu, Y., Zhao, W., & Li, Y. (2020). Symmetry between theoretical and physical investigation of water contamination using amplitude variation with offset analysis of ground-penetrating radar data. *Symmetry, 12*(6), 991. https://doi.org/10.3390/sym12060991

Jenifer, M. A., & Jha, M. K. (2018). Comprehensive risk assessment of groundwater contamination in a weathered hard-rock aquifer system of India. *Journal of Cleaner Production, 201*, 853–868. https://doi.org/10.1016/j.jclepro.2018.08.005

Jurgens, B. C., Parkhurst, D. L., & Belitz, K. (2019). Assessing the lead solubility potential of untreated groundwater of the United States. *Environmental Science and Technology, 53*(6), 3095–3103. https://doi.org/10.1021/acs.est.8b04475

Karunanidhi, D., Aravinthasamy, P., Deepali, M., Subramani, T., Bellows, B. C., & Li, P. (2020). Groundwater quality evolution based on geochemical modeling and aptness testing for ingestion using entropy water quality and total hazard indexes in an urban-industrial area (Tiruppur) of Southern India. *Environmental Science and Pollution Research, 28*(15), 18523–18538. https://doi.org/10.1007/s11356-020-10724-0

Khan, F., Momtaz, S., Niaz, K., Hassan, F. I., & Abdollahi, M. (2017). Epigenetic mechanisms underlying the toxic effects associated with arsenic exposure and the development of diabetes. *Food and Chemical Toxicology, 107*, 406–417. https://doi.org/10.1016/j.fct.2017.07.021

Kong, F., Sun, G., & Liu, Z. (2018). Degradation of polycyclic aromatic hydrocarbons in soil mesocosms by microbial/plant bioaugmentation: Performance and mechanism. *Chemosphere, 198*, 83–91. https://doi.org/10.1016/j.chemosphere.2018.01.097

Kukučka, M., Kukučka, N., & Habuda-Stanić, M. (2016). Water reclamation during drinking water treatments using polyamide nanofiltration membranes on a pilot scale. *Environmental Science and Pollution Research, 23*(18), 17919–17927. https://doi.org/10.1007/s11356-016-6919-3

Lanphear, B. P., Hornung, R., Khoury, J., Yolton, K., Baghurst, P., Bellinger, D. C., Canfield, R. L., Dietrich, K. N., Bornschein, R., Greene, T., Rothenberg, S. J., Needleman, H. L., Schnaas, L., Wasserman, G., Graziano, J., & Roberts, R. (2019). Erratum: "Low-Level Environmental Lead Exposure and Children's Intellectual Function: An International Pooled Analysis." *Environmental Health Perspectives, 127*(9). https://doi.org/10.1289/ehp5685

Lanphear, B. P., Rauch, S., Auinger, P., Allen, R. W., & Hornung, R. W. (2018). Low-level lead exposure and mortality in US adults: A population-based cohort study. *The Lancet Public Health, 3*(4), e177–e184.

Li, P., Karunanidhi, D., Subramani, T., & Srinivasamoorthy, K. (2021). Sources and consequences of groundwater contamination. *Archives of Environmental Contamination and Toxicology, 80*(1), 1–10. https://doi.org/10.1007/s00244-020-00805-z

Maity, J. P., Chen, C.-Y., Bhattacharya, P., Sharma, R. K., Ahmad, A., Patnaik, S., & Bundschuh, J. (2021). Advanced application of nano-technological and biological processes as well as mitigation options for arsenic removal. *Journal of Hazardous Materials, 405*, 123885. https://doi.org/10.1016/j.jhazmat.2020.123885

Masciangioli, T., & Zhang, W. X. (2003). Peer reviewed: Environmental technologies at the nanoscale. *Environmental Science and Technology, 37*(5), 102A-108A.

Matsui, Y., Shirasaki, N., Yamaguchi, T., Kondo, K., Machida, K., Fukuura, T., & Matsushita, T. (2017). Characteristics and components of poly-aluminum chloride coagulants that enhance arsenate removal by coagulation: Detailed analysis of aluminum species. *Water Research, 118*, 177–186. https://doi.org/10.1016/j.watres.2017.04.037

Mukherjee, I., & Singh, U. K. (2018). Groundwater fluoride contamination, probable release, and containment mechanisms: A review on Indian context. *Environmental Geochemistry and Health, 40*(6), 2259–2301. https://doi.org/10.1007/s10653-018-0096-x

Ng, D.-Q., Strathmann, T. J., & Lin, Y.-P. (2012). Role of orthophosphate as a corrosion inhibitor in chloraminated solutions containing tetravalent lead corrosion product PbO_2. *Environmental Science and Technology, 46*(20), 11062–11069. https://doi.org/10.1021/es302220t

Nordberg, M., Jin, T., & Nordberg, G. F. (1992). Cadmium, metallothionein and renal tubular toxicity.

Oyewo, O. A., Mutesse, B., Leswifi, T. Y., & Onyango, M. S. (2019). Highly efficient removal of nickel and cadmium from water using sawdust-derived cellulose nanocrystals. *Journal of Environmental Chemical Engineering, 7*(4), 103251. https://doi.org/10.1016/j.jece.2019.103251

Patel, G. B., Shah, K. R., Shindhal, T., Rakholiya, P., & Varjani, S. (2021). Process parameter studies by central composite design of response surface methodology for lipase activity of newly obtained Actinomycete. *Environmental Technology and Innovation, 23*, 101724. https://doi.org/10.1016/j.eti.2021.101724

Pieper, K. J., Krometis, L.-A.H., Benham, B. L., & Gallagher, D. L. (2016). Simultaneous influence of geology and system design on drinking water quality in private systems. *Journal of Environmental Health, 79*(2), E1–E9.

Podgorski, J. E., Eqani, S. A. M. A. S., Khanam, T., Ullah, R., Shen, H., & Berg, M. (2017). Extensive arsenic contamination in high-pH unconfined aquifers in the Indus Valley. *Science Advances, 3*(8), 1700935. https://doi.org/10.1126/sciadv.1700935

Pradhan, B., Chand, S., Chand, S., Rout, P. R., & Naik, S. K. (2023). Emerging groundwater contaminants: A comprehensive review on their health hazards and remediation technologies. *Groundwater for Sustainable Development, 20*, 100868. https://doi.org/10.1016/j.gsd.2022.100868

Rahman, M. A., Rahman, A., Khan, M. Z. K., & Renzaho, A. M. N. (2018). Human health risks and socio-economic perspectives of arsenic exposure in Bangladesh: A scoping review.

Ecotoxicology and Environmental Safety, 150, 335–343. https://doi.org/10.1016/j.ecoenv.2017.12.032

Reddy, S., Sunitha, V., & Suvarna, B. (2022). Groundwater quality and its potential health impacts nearby inactive mines using PIG and geospatial technology, Southwestern part of Cuddapah basin, Andhra Pradesh, South India. *Groundwater for Sustainable Development, 17*, 100742. https://doi.org/10.1016/j.gsd.2022.100742

Roh, T., Steinmaus, C., Marshall, G., Ferreccio, C., Liaw, J., & Smith, A. H. (2018). Age at exposure to arsenic in water and mortality 30–40 years after exposure cessation. *American Journal of Epidemiology, 187*(11), 2297–2305. https://doi.org/10.1093/aje/kwy159

Rostami, S., & Azhdarpoor, A. (2019). The application of plant growth regulators to improve phytoremediation of contaminated soils: A review. *Chemosphere, 220*, 818–827. https://doi.org/10.1016/j.chemosphere.2018.12.203

Sasakova, N., Gregova, G., Takacova, D., Mojzisova, J., Papajova, I., Venglovsky, J., Szaboova, T., & Kovacova, S. (2018). Pollution of surface and ground water by sources related to agricultural activities. *Frontiers in Sustainable Food Systems*. https://doi.org/10.3389/fsufs.2018.00042

Shenashen, M. A., El-Safty, S. A., & Elshehy, E. A. (2013). Architecture of optical sensor for recognition of multiple toxic metal ions from water. *Journal of Hazardous Materials, 260*, 833–843. https://doi.org/10.1016/j.jhazmat.2013.06.025

Song, Y., Zhou, T., Zong, Y., Gu, B., Tan, X., & Yang, L. (2019). Arsenic inhibited cholesterol efflux of THP-1 macrophages via ROS-mediated ABCA1 hypermethylation. *Toxicology, 424*, 152225. https://doi.org/10.1016/j.tox.2019.05.012

Sudharshan Reddy, Y., & Sunitha, V. (2023). Assessment of heavy metal pollution and its health implications in groundwater for drinking purpose around inactive mines, SW region of Cuddapah Basin, South India. *Total Environment Research Themes, 8*, 100069. https://doi.org/10.1016/j.totert.2023.100069

Suvarna, B., Sunitha, V., Sudharshan Reddy, Y., Muralidhara Reddy, B., Kadam, A. K., & Ramakrishna Reddy, M. (2022). Groundwater quality assessment using multivariate statistical approach and geospatial modelling around cement industrial corridor, South India. *International Journal of Environmental Science and Technology, 20*(5), 5051–5070. https://doi.org/10.1007/s13762-022-04210-y

Tzovolou, D. N., Theodoropoulou, M. A., Blanchet, D., Haeseler, F., Jeczalik, M., Kasela, T., Slack, W., & Tsakiroglou, C. D. (2015). In situ bioventing of the vadose zone of multi-scale heterogeneous soils. *Environmental Earth Sciences, 74*(6), 4907–4925. https://doi.org/10.1007/s12665-015-4501-9

Veeraswamy, G., Nagaraju, A., Balaji, E., Sreedhar, Y., & Rajasekhar, M. (2018). Water quality assessment in terms of water quality index in Gudur Area, Nellore District, Andhra Pradesh. *International Journal of Technical Research and Science*. https://doi.org/10.30780/ijtrs.v3.i1.2018.020

Velazquez-Peña, G. C., Solache-Ríos, M., Olguin, M. T., & Fall, C. (2019). As(V) sorption by different natural zeolite frameworks modified with Fe, Zr and FeZr. *Microporous and Mesoporous Materials, 273*, 133–141. https://doi.org/10.1016/j.micromeso.2018.07.003

Wagh, V. M., Panaskar, D. B., Mukate, S. V., Gaikwad, S. K., Muley, A. A., & Varade, A. M. (2018). Health risk assessment of heavy metal contamination in groundwater of Kadava River Basin, Nashik, India. *Modeling Earth Systems and Environment, 4*(3), 969–980. https://doi.org/10.1007/s40808-018-0496-z

Wu, J., Zhang, Y., & Zhou, H. (2020). Groundwater chemistry and groundwater quality index incorporating health risk weighting in Dingbian County, Ordos Basin of Northwest China. *Geochemistry, 80*(4), 125607. https://doi.org/10.1016/j.chemer.2020.125607

Xu, F., Chen, H., Dai, Y., Wu, S., & Tang, X. (2019). Arsenic adsorption and removal by a new starch stabilized ferromanganese binary oxide in water. *Journal of Environmental Management, 245*, 160–167. https://doi.org/10.1016/j.jenvman.2019.05.071

Yang, K., Zhu, L., & Xing, B. (2006). Adsorption of polycyclic aromatic hydrocarbons by carbon nanomaterials. *Environmental Science and Technology, 40*(6), 1855–1861. https://doi.org/10.1021/es052208w

Yenugu, S. R., Vangala, S., & Badri, S. (2020). Monitoring of groundwater quality for drinking purposes using the WQI method and its health implications around inactive mines in Vemula-Vempalli region, Kadapa District, South India. *Applied Water Science*. https://doi.org/10.1007/s13201-020-01284-2

Zhang, Y., Wu, J., & Xu, B. (2018). Human health risk assessment of groundwater nitrogen pollution in Jinghui canal irrigation area of the loess region, northwest China. *Environmental Earth Sciences* 77(7). https://doi.org/10.1007/s12665-018-7456-9

Zhang, X., Tian, J., Han, H., Sun, W., Hu, Y., Li Wang, T. Y., Yang, Y., Cao, X., & Tang, H. (2020). Arsenic removal from arsenic-containing copper and cobalt slag using alkaline leaching technology and $MgNH_4AsO_4$ precipitation. *Separation and Purification Technology, 238*, 116422. https://doi.org/10.1016/j.seppur.2019.116422

Zhao, Y. Y., & Pei, Y. S. (2012). Risk evaluation of groundwater pollution by pesticides in China: A short review. *Procedia Environmental Sciences, 13*, 1739–1747. https://doi.org/10.1016/j.proenv.2012.01.167

Chapter 9
History of Water Pollutants and Emerging Pollutants

Gagan Matta, Gaurav Pant, Sushil B. Kapoor, Reetesh Sah, Kanchan Deoli Bahukhandi, D. P. Uniyal, Pawan Kumar, Anjali Nayak, and Ujjwal Gulaiya

Abstract Life on earth depends on water because it is a necessity for life. Several aspects of our life rely on water a lack thereof—and it could be said that the use of water is the foundation upon which our entire civilization has been built. The contamination of water is a big issue that affects the entire world. All across the world, it impacts water used for drinking and different water bodies. As a result, it damages both the environment as a whole and the health and wellbeing of human beings. This chapter looked at how different water resources are used and affected during the different civilizations and how water polltants has affected the quality of water and public health over time. The problem that mankind currently faces is how to obtain drinkable water for people and livestock, in addition to how to handle the wastages we generate. This problem was first brought about by agriculture and the expansion of human habitation.

Keywords Water pollutants · Emerging pollutants · Water conservation · SDG 6

G. Matta (✉) · G. Pant · P. Kumar · A. Nayak · U. Gulaiya
Hydrology and Science Communication Research Lab, Department of Zoology and Environmental Science, Gurukul Kangri (Deemed to be University), Haridwar, India
e-mail: drgaganmatta@gkv.ac.in

D. P. Uniyal
Uttarakhand State Council for Science and Technology, Dehradun, India

R. Sah
History Department, Kumaun University, Nainital, India

S. B. Kapoor
Department of Chemistry, Dr. Khatri Mahavidyalaya Tukum, Chandrapur, Maharashtra, India

K. D. Bahukhandi
Sustainability Cluster, School of Advance Engineering, UPES, Dehradun, India

V. Sunitha et al. (eds.), *Groundwater Resource Management Planning Strategies*, Springer Water, https://doi.org/10.1007/978-3-031-88870-0_9

9.1 Introduction

Almost 97% of the world's freshwater resources are the salty water in the oceans. Freshwater makes up the rest 3% of the total, of which approximately 69% is stored in ice caps, 30% is stored in karst and porous aquifers, and the remaining 7% is stored in surface water reservoirs including lakes, rivers, waterfalls, and marshland. Freshwater's fundamental component of life on this planet is becoming depleted. According to research, 33% of people worldwide and 75% of Europeans utilize groundwater as their main resource of drinkable water. However, anthropogenic activities have damaged the quality and quantity across several nations. The water crisis affects 88 emerging economies with half of the global population (AbuZeid, 2012; Adu, 2018). Any body of water exposed to the environment and prone to the land's runoff is considered surface water.

Consequently, it is highly probable to include microbes that can create illness and, in some situations, more severe, even deadly, diseases. In certain places, a sizable amount of surface drinking water comes through bank filtering, which contains a variety of microbes and pollutants and has to be purified. However, since sediments and soils protect groundwater is thought to be less susceptible than surface water. Even yet, the equipment needed for its abstraction, like as drilling and pumping, is rarely easily accessible or durable, particularly in underdeveloped nations (Aëtius, 1998).

The number of water resources is predicted to decline in the coming decades as the demography expands and groundwater consumption increases. The most critical step in providing a place with safe drinking water is choosing the greatest readily available water supply. Hence, the more source waters are preserved, the simpler and more affordable it is to transform them into clean drinkable water (Agnel et al., 2017). Freshwater is not uniformly available; it changes across time and space. Surface water makes up most of the earth's permanent freshwater, and its distribution is generally unequal. Although most of the groundwater is nonrenewable, fossil water, it is more equally dispersed. Various variables, including the density of population, the land use patterns, macroeconomic variables, and the natural groundwater supply influence a region's water usage.

The yearly rainfall is approximately 110,000 km^3, the primary freshwater source that is recurrent on a global level. A substantial percentage (70,000 km^3 per year) of rain water that falls on land evaporates and dissipates when it passes through vegetation (Hippocrates, 2022). The amount of water that infiltrates a surface depends on the nature of soil, how much moisture is there, how much rain falls, and how long it rains. Overland flow happens when precipitation rates are higher than infiltration rates. Humans may consume around 26% of this cycle's volume (18,000 km^3 per year). A 41,000 km^3 total yearly runoff is thought to occur globally, comprising soil penetration and groundwater recharge. The yearly runoff is limited to 12,500 km3 / annum due to geographic barriers and periodic problems like flooding (Al-Gheethi et al., 2018). Accordingly, it is calculated that around 54% of the usable runoff and 23% of the overall sustainable source (rainfall over land) are now available for humans use in a certain way. Despite its relatively small amount globally, groundwater is a vital water source that can meet human demands since it is a portion of the finite freshwater reserve. Renewable sources are crucial from the standpoint of long-term sustainable development.

Given that the withdrawal rate may be larger than the replenishment rate, it is also possible to consider it a nonrenewable source. The hydrological process is impacted by climate change because it causes the atmosphere to heat. The quantity, frequency, type, and volume of rainfall, the flow of water in drainage, along with the quality of both freshwater and oceanic ecosystems, all alter as a result, based on the region. In addition to altering groundwater temperature and depth, climate change also brings an additional, lesser threat to groundwater and water supplies that requires an understanding to develop adaptive strategies. One of the world's biggest health and ecological issues, especially for underprivileged citizens, is the lack of access to potable water. Viral illnesses such as cholera, cholera, hepatitis A, influenza, typhoid, and polio are all at a higher risk when there is a paucity of accessibility to safe drinking water resources (Aphomis, 1931). Additionally, it makes malnutrition, particularly childhood growth, worse. Figure 9.1 shows that it is ranked as a major worldwide risk factor for mortality. The Global Burden of Disease research shows that contaminated water caused 1.2 million early deaths in 2017. It was equivalent to the number of people killed in road fatalities worldwide and also three times the number of deaths in 2017 (Tiwari et al., 1988).

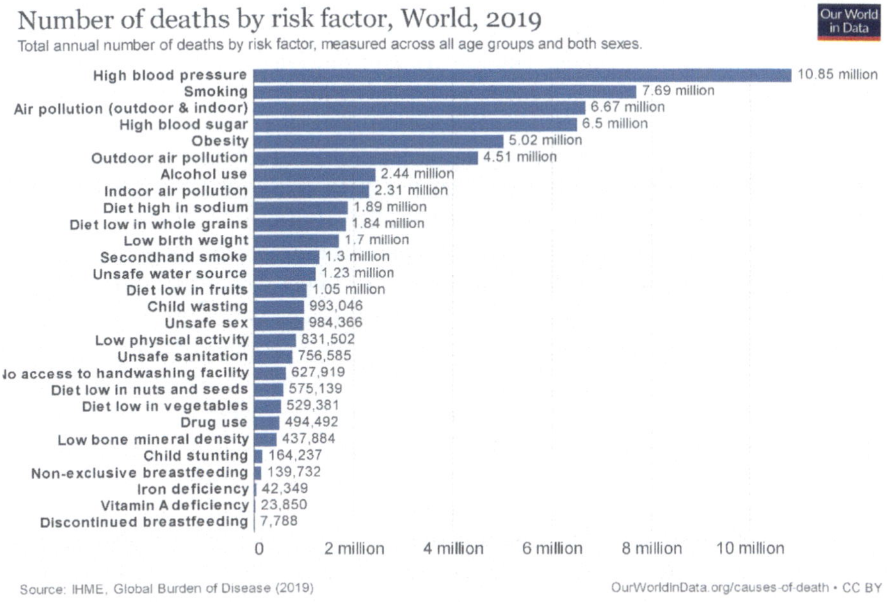

Fig. 9.1 This figure is covered by the Creative Commons Attribution 4.0 International License as mentioned on *Source* https://ourworldindata.org/water-access

According to a 2016 preliminary investigation of the water quality scenario in rivers throughout Africa, Asia and Latin America, extreme infective contamination impacts approximately one third of river systems, extreme organic pollution approximately one-seventh of rivers, and mild to extreme salinity pollution approximately one-tenth. This research from the United Nations Environment Programme (UNEP) demonstrates how water contamination appears in different parts of the world. You may understand what they're talking about by investigating a few nations in the regions mentioned above. Due to the trash produced by industrial facilities and the inappropriate and illegal disposal of that garbage, China is among the world's largest manufacturers and can thus be expected to have problems with both water and air pollution problems. (UNEP). Although several African nations, such Ghana and Libya, not only have polluted water but also a shortage of funds and capacity to address the issue. Because of this, a great number of individuals—particularly small children and older people with weaker immunity, and numerous and sometimes fatal illnesses in these underdeveloped nations, frequently die as a result of it. Nearly half of the nation's rivers and a third of its lakes, as reported by the NRDC, are too contaminated to be used for fishing, swimming, or even drinking. The Environmental Protection Agency (EPA) started a programme called How's My Waterway to focus on allowing Americans to check the cleanliness and pollution levels of their local water sources using EPA databases (Baya, 2008).

About 63 million Americans, or about one-fifth of the population, lack access to clean fresh water, according to a 2017 report by USA Today. The freshwater may contain pollutants like arsenic and compounds from different sources. However, several people who are susceptible to this contaminated water live where they cannot afford the supplies and equipment required to filter it out. Due to this, almost all people are prone to digestive diseases, cancer, and issues with infant growth (Bensoltane, 2019).

An overview on current trends in the accessibility of cleaned drinking water and forecasts for the future are offered in the 2018 issue of the United Nations (UN) World Water Development Report (WWDR) (Bolisetty, 2020). Many people presently lack consistent access to sufficient amounts of water of adequate standard, and in the coming years, the problem will continue to get worse (Broom, 2019). In present era of 7.7 billion population, clean water shortage is a significant problem. When the world population reaches 9.4 to 10.2 billion people in 2050, a 22–34% rise, the burden on the water supply will rise. An: An uneven population increase in various locations irrelevant to community resources will worsen the burden. It is anticipated that most of this population expansion would occur in developing nations, first in Africa and subsequently in Asia, where a lack of clean water is currently a significant problem. Population increase, economic boom, and shifting consumer habits contribute to rising water consumption. Over the past century, the need for water has grown by 600% worldwide (Bruun, 1991). By 2050, the current annual global water requirement for all purposes, which is around 4600 km^3, will have increased by 20–30%, reaching 5500–6000 km^3 (Broom, 1991). The astonishing figure of 5.4 billion individuals without access to public sewerage in 2030—1.1 billion greater than in 2000—is predicted. Compared to 2000, when 71 percent of the world's population lacked access, this number will be marginally lower at 67 percent. Clearly, that's preferable to a percentage increase, but it barely counts as positive news (Burek et al., 2016).

9.2 Water Resources

More than 5 billion people are anticipated to experience insufficient water availability by the year 2050, up from the current 3.6 billion. According to UN-Water, 74% of all natural catastrophes between 2001 and 2018 have a water component. The most recent UN climate change conference, COP27, recommended that states incorporate water into mitigation actions further. This is the first instance freshwater has been included in a COP result statement in acknowledgment of its crucial relevance.

In comparison to the average of the 30-year hydrological base period, significant portions of the world had drier than usual weather in 2021. The South and South-East Amazon, the Rio de la Plata region of South America, the Colorado, Missouri, and Mississippi river basins, as well as other basins in North America, were among these locations. The Rio de la Plata region has been experiencing a severe drought since 2019. In 2021, the flow from African rivers like the Niger, Volta, Nile, and Congo was below average. In a similar manner, rivers in regions of West Siberia, Central Asia, and Russia saw reduced flow in 2021. In several river systems of Northern America, the Northern Amazon, Southern Africa (Zambezi and Orange), China (Amur), and Northern India, river flow was beyond average. In around one-third of the regions examined, the 30-year average was present. Several countries, including China, northern India, Western Europe, and nations affected by tropical storms, including, the Philippines, Mozambique and Indonesia, recorded severe flood occurrences with many fatalities. A regional drought has been caused by multiple years of below-average precipitation in Kenya, Ethiopia and Somalia (Campbell, 2015).

Water on the top of the land and in the subsurface collectively constitute surface water reserve. On the West Coast of the United States, in the centre of South America and Patagonia, in North Africa and Madagascar, Central Asia and the Middle East, Pakistan and Northern India, and in contrast to the average estimated from 2002 to 2020, surface stored water was categorised as below standard in 2021. The Amazon region, central Africa, northern Southern America, and northern China all had above-average temperatures. The analysis identified a number of areas with a long-term downward trend in surface water storage. Some examples are Patagonia, the Ganga and Indus origins, the Rio So Francisco basin in Brazil, and the southwestern part of the United States. Contrarily, the Great Lake area has a strong anomaly, like the Niger basin, East African Rift, and North Amazon basin. The adverse trends are generally more pronounced than the decent ones. Overexploitation of groundwater for agriculture exacerbates some of the hotspots (Fig. 9.2). Alaska, Patagonia, and the Himalayas are among the places where the melting of snow has a considerable influence (Campbell, 2015).

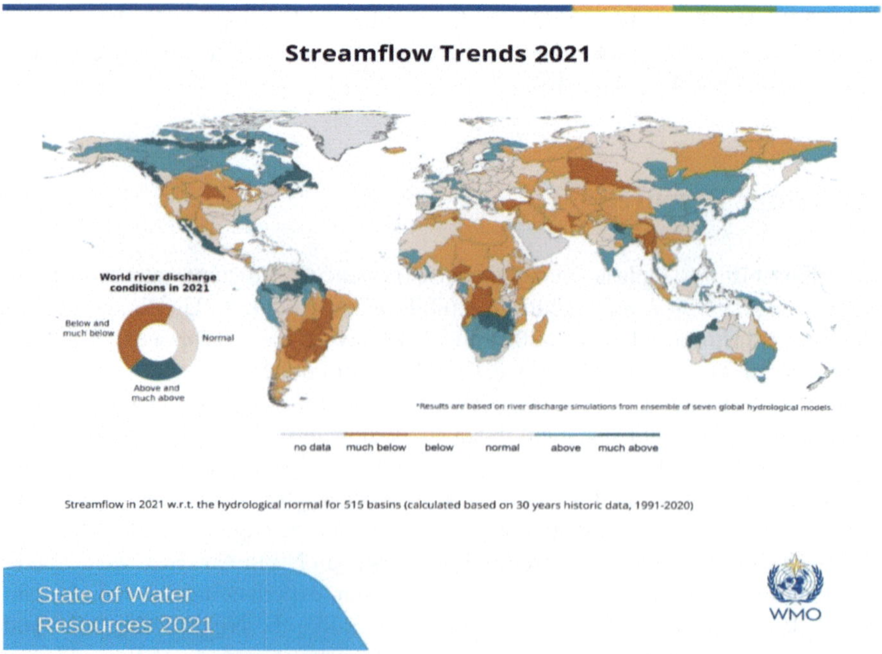

Fig. 9.2 World river discharge conditions in 2021. 2015 *Source* WMO (Campbell,)

The largest natural freshwater reserve in the planet is the cryosphere, which includes glaciers, Snowpacked Mountains, ice caps, and, when present, permafrost (Fig. 9.3). Due to their role as origin of streams and water supply for an approximate 1.9 billion people, mountains are sometimes referred to as natural "water towers." Food security, people's health, environmental balance, and conservation are all significantly impacted by changes to cryosphere water supplies and the growth of the economy and society. River floods and flash floods driven by glacial lake breakouts are further risks such alterations bring (Cambell, 2015). When temperatures rise, the annual glacial runoff normally climbs initially before declining after reaching a tipping point, known as "peak water." Forecasts of alterations in glacial runoff and the duration of peak water heavily influence decisions about long-term adaptability. Future WMO State of Global Water Resources assessments will encourage frequent evaluation of alterations in the cryosphere as well as the instability of water supplies, at the basins and regional scale (https://www.thestar.com/news/world/2014/05/12/china_wakes_up_to_its_water_crisis.html).

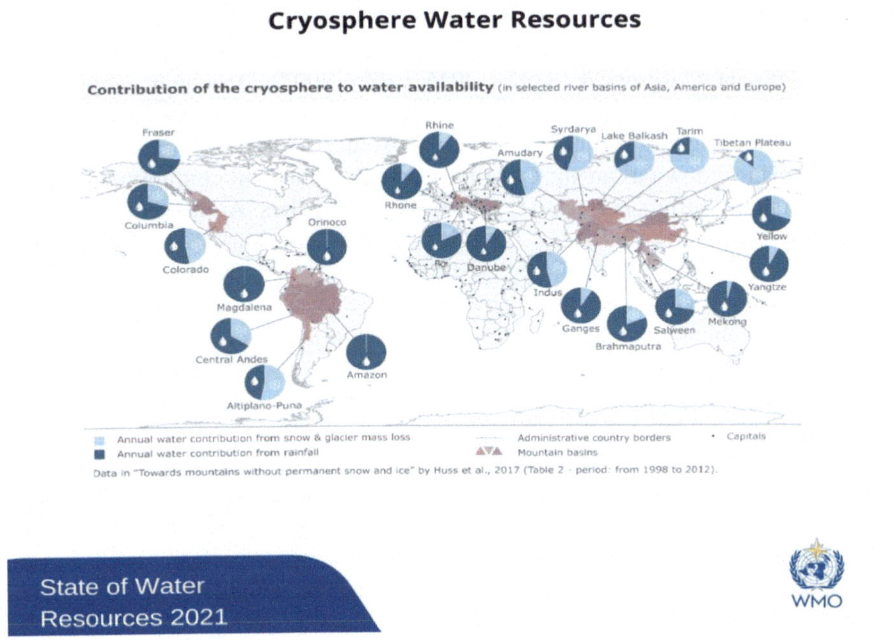

Fig. 9.3 Contribution of the cryosphere to water resources in the basins of major rivers with headwaters in high mountains. 2015 *Source* WMO (Campbell,)

9.3 Water Pollution

One of the biggest problems facing the globe today is water pollution. Tragically, customers continue to make the issue worse. The only realistic option to reduce the majority of water pollution that the world is presently experiencing is for restaurants to say bye to disposable plastics, companies to get rid of plastic wrapping, and for businesses to stop supporting manufacturers who use packaging that are improperly produced, since very few customers are willing to start buying less products from super markets, restaurants, and others. Whereas most people are completely aware that the oceans of entire Planet are contaminated, only a small percentage of people comprehend the complexities of the pollution issue society is now dealing with. Because of the paucity of information, the water quality of the world's streams, reservoirs, and groundwater remains uncertain, putting the health of nearly 3 billion inhabitants at danger. As per latest study conducted by the United Nations Environment Programme (UNEP) and partner organizations, a fifth of the world's largest river systems are undergoing major variability in the availability of water, and 2.3 billion individuals live in countries that are considered to be "water-stressed," which include 721 million in regions where the water issue is "extremely crucial." (Clark et al., 1997).

According to data on water pollution, high-income nations including the United States, Germany, Canada, and Japan recycle about 70% of the effluent they produce as of 2017. However, low-income nations only manage to treat 8% of the municipal and industrial effluent (Cosgrove & Loucks, 2015; Cullet, 2007). Data on water pollution indicate that large levels of nutrients (including nitrogen and phosphorus), typically in the form of agricultural runoff, are the root source of this problem. An intense growth of plants known as an algal bloom is driven by nutrient contamination. As a result, marine life suffers harm and sometimes even perishes. Additionally, it causes the development of "dead zones." (DeGood, 2020; Denchak, 2023; Duke, 2016). Radioactive waste that has been negligently managed or inadvertently spilled can endanger aquatic environment, human health, and create economic loss. Additionally, data on water contamination indicates that radioactive substances can persist inside the water for a long time. The muscular tissues of organisms that dwell in the Pacific Ocean's greatest depths have just been confirmed to have radioactive carbon by nuclear testing conducted in the 1950s (Evans et al., 2019).

Nitrate was the main contamination element in 387 districts across India in 2019 (Fig. 9.4). India was ranked 120th out of 122 nations in terms of water quality, according to Niti Aayog, and 70% of the nation's freshwater sources have been proven to be polluted. Several studies have also indicated the water pollution in India and other parts of the world regarding heavy metal toxicity (Fadlallah, 2001; Fayiga et al., 2018; Frontinus, 1925; Galen, 1951; Gjyli et al., 2018) and other physicochemical factors (GBD, 2019; UNEP, 2021; Hasan er al., 2019; Haseena et al., 2017; Helmer et al., 1997; Hodge, 1981; Howard & Gelo, 2002; Huss et al., 2017; UNICEF & GOI, 2012; Hippocrates, 1988; Jayaswal et al., 2018). Contamination from landfill, sewer systems, leaking subsurface gas tanks, and excessive fertilizer and pesticide usage were the main causes of pollution. Additionally, much unprocessed residential sewage was being dumped into nearby rivers and lakes. The lack of water treatment facilities at the appropriate level was one of the causes of this problem. (https://theoceancleanup.com/great-pacific-garbage-patch/) (Juuti et al., 2006).

The results are horrific despite only 31% of US waterbodies have been examined. Due to nutrient contamination, the EPA has designated 800,000 miles of rivers and tributaries as damaged (Kidd, 1983). In between Hawaii, California, and Japan, there is a trash patch in the Pacific Ocean known as The Great Pacific Garbage Patch. According to statistics on water pollution, it includes 1.8 trillion plastic pieces and is one of the world's largest trash dumps. The GPGP is thought to contain plastic waste that is three times as large as France and weighs more than 80,000 tonnes (Fig. 9.5). In addition to plastic, there is also trash from the 2011 Japanese disaster (20%) and discarded fishing equipment (46%) inside it (Kumar et al., 2021).

According to government figures, China's lakes and rivers are much more than 70% contaminated, and about 50% of them could possess water which is hazardous to human contact or ingestion (Gjyli et al., 2020). Hazardous industrial and municipal wastewater releases into surface waterbodies are Africa's main causes of pollution. Guidelines for the wastewater release into surface waterbodies are present in a few African nations, however it is uncertain if these regulations are followed. Most of the

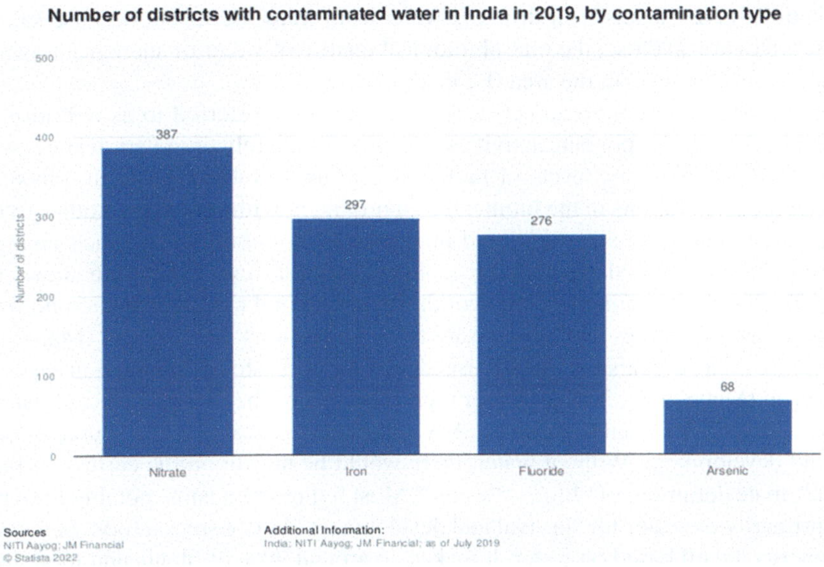

Fig. 9.4 Contamination level in India. *Source* Juuti et al., (2006)

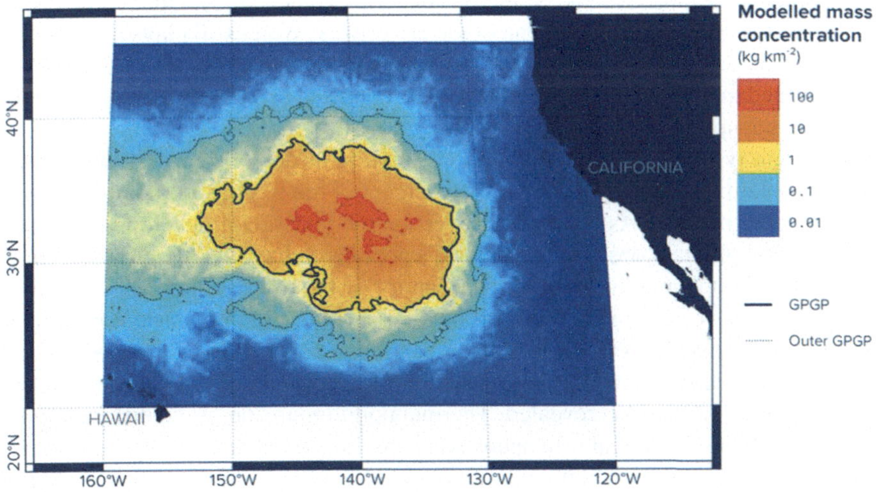

Fig. 9.5 Modelled concentration of plastic garbage in The Great Pacific Garbage Patch. *Source* Kumari and Singh (2016)

population in Africa gets their drinking water from water wells, but because sanitation services are so close, the microbiological quality of water of such groundwater resources is often poor in the area (Le et al., 2017).

An adverse alteration to an ecosystem component is referred to as pollution. It originates from anthropogenic activities, whether completely or in part. It starts with alterations in the energy, levels of radiation, and undesirable biological, physical, and chemical alterations in the biome, the environment within which all other living things exist. Via food, air, water, and numerous farm commodities, such changes may directly or indirectly impact the ecological equilibrium. Figure 9.6 shows the major freshwater consumers of the world. Environmental contaminants come from a wide range of sources and have a variety of perceptions and impacts (Mahmood et al., 2016). The organism's life is harmed, and the ecosystem's production capacity is reduced (Matta et al., 2022a). Water represents one of the most significant natural sources in every financial framework. Any community relies on it as a source of both life and development. Without water, there would be no life on the earth's surface. The ultimate definition of "life" is "water." Most living organisms, notably humans, rely primarily on water for survival and development. Agribusiness, food, beverage, and energy are all based on water. It makes up around 90% of plants and more than 75% of the human body. It is a basic requirement for survival.

Contrarily, water has played a significant part in the development and wellbeing of early societies for long periods. Numerous historic evidence reveal that civilizations developed and grew in significant masses nearby rivers. Water still plays a significant role in society today. Yet, fish, among the planet's more significant food sources, can only be found in water resources. In fact, water connects diverse regions of the globe by marine transportation (Matta, 2020).

Numerous types of pollution harm water supplies. Its depletion will occur quickly if water contamination rises further. Forty percent of the world's population lives in 88 developing nations. As a result of its economic and social development, water scarcity is a significant obstacle. The world's population has steadily grown, as have the variety of human activities and the quantity of fresh water they use for varied needs (Matta et al., 2022a). Solids and oxygen-sucking organic waste are dumped into streams, primarily from unregulated urban sewage and industry drainage, where the expansion of trace contaminants (toxic compounds and metals including mercury, zinc, lead, and cadmium) is revealed. Thermal pollution is one of the different types of water contamination, and it happens when cooling water from industries, power plants, and other sources is dumped into river systems, raising the temperature and disrupting the balance of the water ecosystem (Matta et al., 2022a). There are several factors that affect surface water contamination and its usefulness, along with the following (Matta et al., 2022a):

1. How quickly bacteria can analyze trash and pollutants.
2. Type of trash and contaminants that are discharged into the sea.
3. The stream's water current's flow rate.
4. Water's dissolved oxygen (DO) concentration.

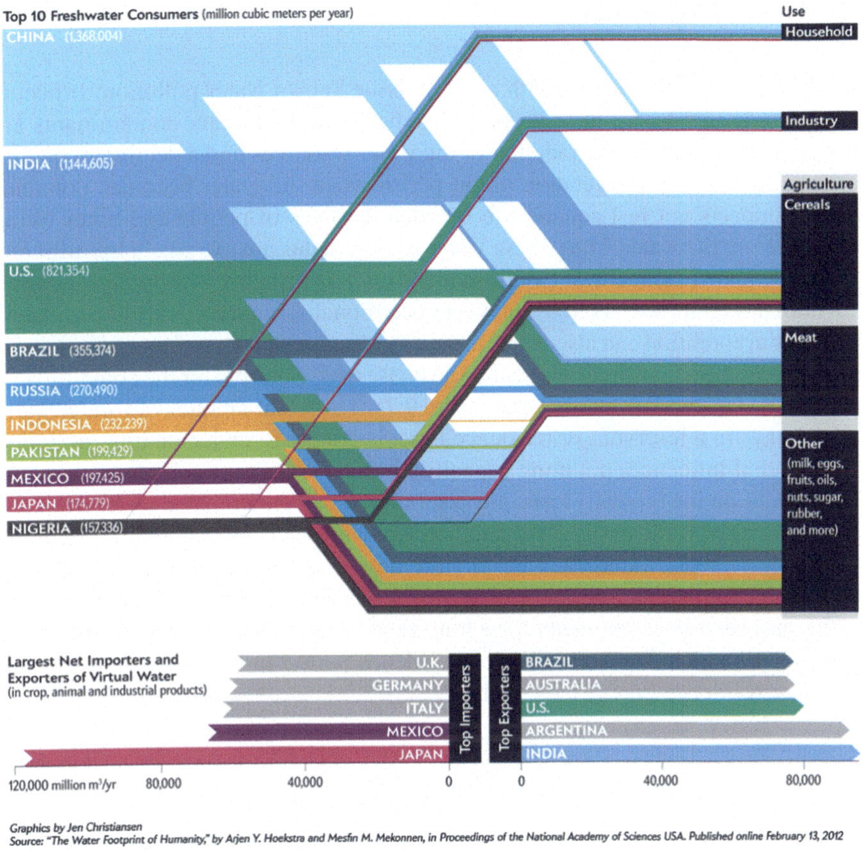

Fig. 9.6 Major consumers of Freshwater in the world. *Source* Matta (2020)

As a matter of fact, the below represent the major crucial characteristics of water that make it usable (Matta, 2020):

1. It is chemically devoid of all kinds, including insecticides, chemical fertilizers, and other things.
2. Acidity and neutralization are completely absent.
3. It must be tasteless, odourless, and colourless.
4. Microbes including germs, fungi, and many others shouldn't be present.

Water parameters vary due to contamination since it makes water less capable of performing its natural function. Because of this, it is no longer suitable for the human, agricultural production, or industrial uses that were assigned to it. It used to be believed that the ideal locations to dispose of manufacturing and consumption trash and remnants were streams, seas, and oceans, among other waterways.

9.4 Water Pollutants

The globe is currently dealing with a severe issue known water pollution. It occurs when garbage that hasn't been properly handled and other water contaminants are released, resulting in the degradation of the water resources and a decline in quality of water. Animals and vegetation might perish when the water becomes contaminated. It impacts not just aquatic life but also the lives of people and other living things who use the water. Thus, these water contaminants are extremely harmful and require adequate management (Matta et al., 2021). The majority of pollution comes from humans, land-based activity, however occasionally, natural events like volcanic activity or evaporation can also result in water pollution. As the water containing the contaminants moves via hydrological cycle phases, it is possible for the contaminants to disperse via several water reserves (Fig. 9.7). The usual amount of time a water particle stays in a reservoir, or residence time, is crucial to understanding pollution issues since it influences polluting capacity. Because of river water's low residence time, contamination typically lasts only a short period. Contaminated water from one source, like a river, can easily go to another, like the oceans, which might pose more issues. Since groundwater often has a slow flow and a prolonged residence time, groundwater contamination can be extremely hazardous. In addition, a contaminant may be absorbed into sediments for a long time or taken up for a long time by the environment, which can lead pollution residence period to be substantially longer than water residence time (Matta et al., 2022b).

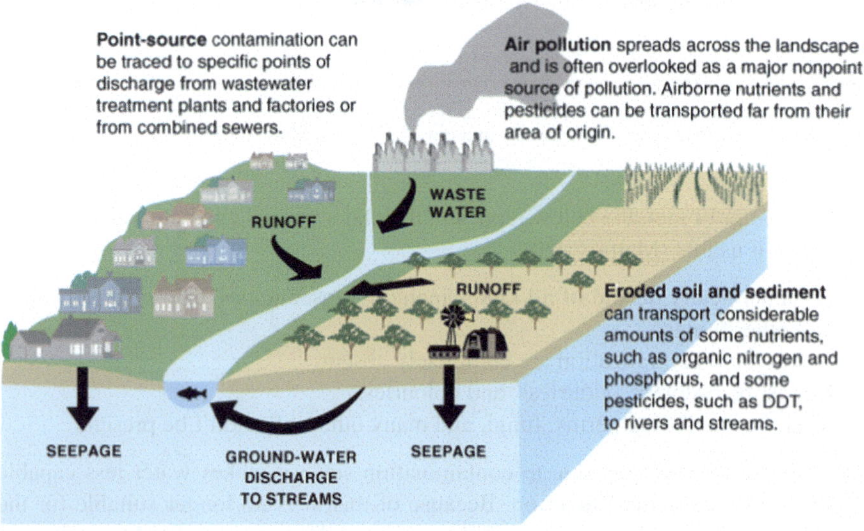

Fig. 9.7 Sources of some water contaminants and how those pollutants are transferred into various water reservoirs during the hydrological cycle (such as rivers and aquifers). 2020 *Source* Matta et al. (a)

Several anthropogenic activities can pollute water, as can preexisting natural characteristics like mineral-rich geological formation. In addition to many other things, effluent processing facilities on both a micro and massive scale, landfills, livestock operations, and commercial and agricultural activities all can contaminate water supplies. Water disperses material left behind by these possible contamination sources when it flows over the land or seeps into the ground. The variety of chemicals present determines the implications and form of treatment for a pollutant.

One distinct source can be identified as the cause of point source contamination. Livestock factories farms, for instance, produce a large quantity and dense population of animals, including cows, pigs, and chickens. Integrated sewage systems, which use a unified system of underneath pipelines to collect both sewerage and storm water discharge from roads for treating wastewater, can also be significant point sources of contaminants. Storm water runoff during periods of heavy rain may surpass a sewer's capacity, resulting in a backup. As a result, untreated wastewater is discharged right into rivers and streams (Fig. 9.8). Pipes from industries, landfills, storage facilities, and chemical contamination are more events (Matta et al., 2022c).

Many widely scattered sources contribute to pollution from nonpoint sources. Pollutant contributions as a whole are detrimental, yet individual elements may not exceed hazardous levels. Urban areas, rural areas, and abandoned mining operations are examples of non-point pollutant sources. As it rains, contaminants that originate all across the watershed are carried via the air as well as the ground (includes lands and shorter streams that discharge into a certain water body). Synthetic pesticides, insecticides, fertilizer from yards and farm land, oil, coolant, animal feces, road salt from cities, and acid and hazardous materials from defunct mines could all be considered among such contaminants. Runoff then carries this contamination into both surface and groundwater basins. Due to its lower quantity, numerous sources, and considerably larger amount of water than point-source contamination, which

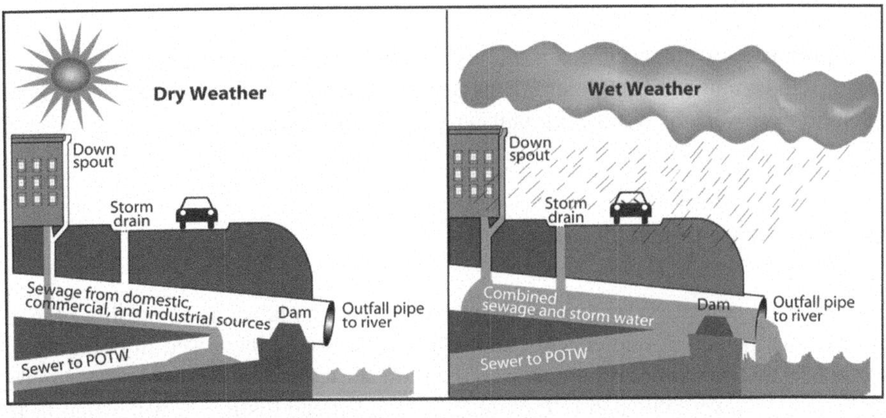

Fig. 9.8 When it rains heavily, an outflow of unprocessed effluent from a integrated sewer system could be a significant point source of water pollution. 2020 *Source* Matta et al. (b)

Table 9.1 Major categories of water pollutants

Category	Examples	Sources
Biological agent	Bacteria, virus, protozoa and parasitic worm	Human and animal waste, domestic sewage
Organic chemicals	Oil, gasoline, plastic, cleaning solvent, and detergents	Industrial effluent, household cleansers, surface runoff from farms and yards
Plant nutrients	Water soluble compounds containing nitrate, phosphate and ammonia	Sewage, manure and runoff of agriculture
Sediments	Soil and silt	Land erosion due to agriculture and strip mining
Inorganic chemicals	Compound of toxic metal, leads, arsenic and fluoride	Industrial effluents household cleansers, chemicals used for killing insects, fungi and weeds in agricultural land
Radioactive substances	Radioactive isotopes of iodine, uranium and thorium	Nuclear power station plant, mining of uranium containing minerals, nuclear weapon production
Heat (Thermal pollution)	Excessive heat	Water used for cooling in industries

is the primary factor causing water pollution in the United States, nonpoint source pollution is typically far more challenging and costly to prevent (Matta et al., 2023).

9.5 Types of Water Pollutants

Land pollution originates from a variety of sources. Microbes, pesticides, heavy metals, sands, and radioactive elements are a few types of contaminants. The type and sources of various water pollutants are shown in Table 9.1.

9.6 History of Water Pollutants

From approximately 200,000 years ago, Homo sapiens have lived on this planet, much of that period as hunter-gatherers who progressively increased in population. Around 50 thousand years ago, early civilization started colonizing every land mass and individuals were constantly moving. The widespread dislike of water that taste foul, smelled bad, and looked unpleasant must have evolved relatively earlier over the cultural and biological development of mankind. Pathogens spread by polluted water often caused problems, but this is rare. According to theories, hunter-gatherers didn't have many water-related health problems. Ancient civilizations obtained their drinking water from water sources including streams and lakes. Nevertheless, the same bodies of water also contained human waste. As a result, rivers became so contaminated that all these ancient societies had to construct dams in order to receive pure water for consumption (Matta et al., 2018a). The main source of contaminated water was fecal matter that found its way into streams and rivers.

People during medieval times didn't know what was beneficial or harmful for them. The majority of individuals and agriculture waste regularly poured mercury into the water, and it was let to start flowing into rivers and ponds. Most city dwellers simply threw their trash and junk on the streets, which undoubtedly smelled awful. The second half of this time saw numerous significant advances in civilization, frequently at the detriment of the environment. Many even ingested toxins at that time since they believed it was beneficial for their health. The impact on the environment was not given much consideration throughout the Industrial Revolution of the 1800s, which benefited industry. The streams have been exposed to a variety of chemicals.

The environmental effects of new discoveries were not a consideration at this time (Matta et al., 2020c). Following the Second World War,, water pollution issues were greatly exacerbated by the growth of factories and industries in addition to pollutants including fecal matter, bio fertilizer, waste from dyeing and butchering leather, and composts. As garbage from industry was easily thrown into rivers without regard for the environment, it had an impact on vegetation, animals, and human ecology (Matta 2018a). The Cuyahoga River saw a spate of fires in 1969, and it was then discovered that combustible industrial debris and oil slicks were the causes of the flames. The government then started the research necessary for adopting the Clean Water Act of 1972 (Matta et al., 2018b).

When humanity transitioned to an agrarian lifestyle about 10,000 years ago, permanent townships were developed. As a result of the widespread adoption of this new way of life, the population started to grow more quickly than before. Construction of towns, cities, and ultimately states—all of them heavily reliant on water—was made possible by passive agrarian lifestyle. As a result, people and water now have a completely new relationship. For passive farmers, pathogens spread by polluted water pose a very real health concern. In this environment, ensuring that everyone has access to clean water has become necessary for effective urbanization and state development. During 8000–7000 B.C., Jericho, which was situated close to springs and other water sources, is the first documented permanent township categorized as urban. Wells dating back as 3000 B.C. were discovered in Egypt, and stone rainwater drains were discovered in Mesopotamia. Researchers have found numbers of historic wellbores, water mains, and bathrooms from beginning of the Bronze Age town of Mohenjo-Daro, which is now in modern-day Pakistan. The Mycenaean civilization of Crete in the 2nd millennium B.C. provides the earliest example of the deliberate development of Europe's water system, restrooms, toilet, and drainage channels.

9.6.1 Ancient Greece and Rome

Between 500 B.C. and 500 A.D., the Mediterranean region experienced the earliest European settlement. Between 10 and 20% of people lived in cities with in years leading up to the birth of Jesus. The fastest rates of urban development were seen in the Eastern Mediterranean, Egypt, North Africa, the Apennine Peninsula, as well as the southern Iberian Peninsula. Just modest rains were also observed in the most of these areas. The abundance of literary and historical documentation from this era makes it easier to investigate the relationship between water and social flourishing.

Alcmaeon of Croton, who lived approximately 470 B.C., became the very first Greek doctor to suggest that bad water quality might affect a person's health (Matta et al., 2020d). The multiple causes, traits, and health effects of water are extensively covered in the 400 B.C. Hippocratic text Airs, Waters, Places (Matta et al., 2019). Precise findings on water's impact on public health can be found in a number of Hippocratic texts, the majority of which were written around 400 B.C. (Matta et al., 2018c).

Beginning first-century philosopher B.C. Vitruvius informed against locating cities near swamps. Pliny the Elder wrote extensively about the varied perspectives on the best kind of water in his texts starting in the first century AD (Matta et al., 2020e). Galen, among the foremost renowned doctors of history (second century A.D.), emphasises the advantages of water (McNeill, 1979).

The characteristics of tasting, aroma, appearance, and heat were used to assess the water's quality. It was also considered (Medema et al., 2003) how a water source would affect living beings. Whether flavorful or flavourless, chilly, odourless, and colourless water were deemed the best in history, while relatively stable, swampy water was to be ignored. These notions, as articulated by Palladius (Miller & Hutchins, 2017) or Paulus Aeginata (Reddy et al., 2019), were held until the end of antiquity. The risks of water flowing from mountainous regions where there was mines were likewise well known to the old Greeks and Romans (Matta et al., 2019; Matta et al., 2018c; Matta et al., 2020e; McNeill, 1979; Medema et al., 2003). Historical authors have discussed the impacts of various types of water on individuals and communities, although this is difficult to determine whether these statements impacted people's health. It's nearly difficult to assess the health of historical civilizations and the contribution of water to that too due to the inadequate sources. Yet, it is very appropriate to say that metropolitan areas faced significant public health issues despite the extraordinary procedures required for getting clean drinkable water. Early Greek or Roman civilization lacked the resources and desire to properly address public health issues (Nayak et al., 2023).

If the freshwater wasn't up to their standards, the Greeks and Romans employed several techniques to enhance its cleanliness. We know that ancient peoples employed settling tanks, separators, filtration, and boiling as well as through written texts and historical investigations. The health consequences of drinking water of poor quality might have been somewhat reduced by boiling, which ancient experts commonly advised. Although while boiling water was probably hygienically acceptable, it was not commercially or environmentally practicable in widespread use because firewood and other flammable materials would eventually run out throughout the Mediterranean.

Throughout history, there was definitely a significant risk to the public's health due to the inefficient management of waste, particularly effluent. For example, there must have been significant sanitary issues. A public washroom might be flushed with the water supplied to the bath. But the Romans didn't have modern toilet paper. They most likely frequently rinsed their bottoms with sponges, moss, or other wet materials in the conduit facing the seat. In public restrooms, amenities were available to everyone; nevertheless, they lacked privacy, were congested, and lacked a suitable basic hygiene facility. The private restrooms were frequently found close to the kitchens yet probably lacked running water. All this created an excellent opportunity for the spreading of intestinal pathogens.

One of the major reasons of mortality must have been water-borne diseases. Several types of diarrhoea and dysentery must have decimated the inhabitants. Even though historical scientific authors documented certain intestinal disorders, it is hard to make a retrospective diagnosis and determine the cause. Given that various passages in the Hippocratic texts describe the prevalence of digestive disorders throughout the summertime and early autumn when freshwater sources were few in the Mediterranean region (Matta et al., 2019; Nayak et al., 2023; Nutton, 2005), that must've been a time when drinkable water was commonly polluted. Children must have died frequently, particularly those who had just been born (https://www.un.org/en/global-issues/oceans-and-the-law-of-the-sea).

- Since ancient Rome, concerns about public health have included references to lead water pollution. Lead pipe usage was questioned by Roman authors, who suggested ceramic pipe usage instead (Medema et al., 2003; Miller, 2017). While ceramic pipes were utilised, it appears that in practise lead pipes were frequently employed to transport water, as indicated by both literary sources (Miller, 2017; Palladius, 1975) and archaeological findings (Paulus Aeginata, 1914; Johnson, 1941). A. Trevor Hodge (Johnson, 1941) noted two arguments to support the notion that water exposure was extremely modest. First, in most instances a calcium carbonate layer separates the lead and the water due to the quality of water. Second, due to the steady flow, there was not enough chance for lead to contaminate the water in the pipeline.
- It's possible that in ancient times, water's secondary impacts on public health were more significant than its direct benefits. Adequate water availability was essential for agriculture. Floods and droughts caused a lack of food or famine. In ancient times, the movement of people, food, and infections was facilitated mainly by water. Around the Mediterranean during the years 200 B.C. to 200 A.D., there was a brisk coastal trade. As a result, viral infections spread more or less widely throughout the Mediterranean region (Postel et al., 1996). Malaria and schistosomiasis were 2 major parasitic illnesses with a close relationship to water and water management in the past.

9.6.2 *Medival Period*

Europe's water distribution and sewage treatment facilities underwent significant alterations following the end of the Roman Empire.. In medieval times, towns, palaces, and monuments always utilized their own ground water, ponds, or reservoirs. Towns typically constructed a few simple toilet facilities for the residents, which were typically insufficient given the population density. Poor sanitary conditions exacerbated the impact of epidemics in mediaeval European towns. When contemporary institutions began to emerge in the thirteenth century, science and technology were institutionalised, and the agrarian sector began its industrialization drive in the eighteenth century. As a result, global population expansion accelerated rapidly. Intellectual citizens were attracted by the notion of advancement as the West became more industrialised and urbanised. The West adopted the washroom as a basic cultural requirement by 1900, much as the Roman Empire had done with irrigation canals. Without considering where human excrement travelled via sewage lines, the washroom was hailed as a triumph for the healthcare system.

Local authorities became more burdened than ever by healthcare and ecological issues due to the beginning of industrial development and urbanization, and modern technologies were frequently considered solutions. Current water distribution and sanitary facilities were first developed in the United Kingdom in the nineteenth century, but the improvements quickly extended to Germany, other regions of Europe, the United States, and other countries. One of several nineteenth century's greatest accomplishments was the improvement of sanitation in cities across Europe. The importance of water in spreading various serious illnesses, including cholera, diarrhea, typhoid, and dysentery, became apparent throughout the period. The identification of the bacteria that cause these disorders provided the conclusive evidence. In the early 1900s, cholera was a reason for the global hygienic revolution.

9.6.3 *Infrastructure in the Modern Towns*

The early twentieth century saw significant demographic expansion, with the world population roughly quadrupling and the urban demography rising by a factor of 13 during this time (Fig. 9.9). Over 50% of the demography in nearly each nation by the year 2000 A.D. resided in urban areas. Energy demand grew by a multiple of ten throughout the century, but industrial productivity expanded by 40. Facilities associated with water and hygiene played a significant part in the world's fast socioeconomic transition.

As bleaching and many other water filtration procedures were discovered and implemented globally in the industrialised nations during the beginning of the twentieth century, the health issues related to water pollution appeared to have been remedied. Water-related microbial issues were mostly thought to be an issue in the underdeveloped countries. But, with in the post-modern West society of the late twentieth

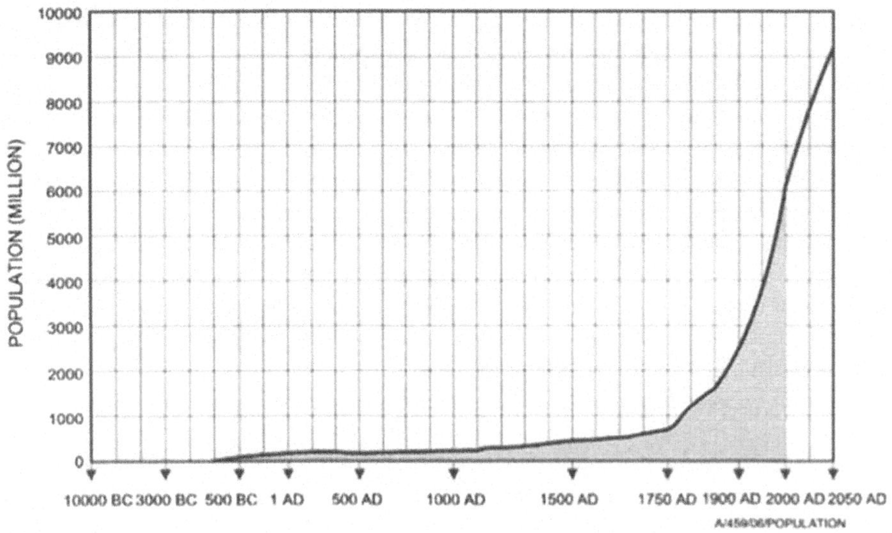

Fig. 9.9 Expected growth in human population from 10 000 B.C. to 2050. *Source* Withington et al. (1927)

century, bacterial risks spread by water returned. In the 1960s, concerns regarding chemicals and radiological risks and their effects on human health increased. The identified biological or chemical health risks that are spread by water have multiplied through the latter part of the twentieth century.

In the modern era, ailments including diarrhea, typhoid, and different diarrhoeal disorders that are brought on by a scarcity of clean water or proper hygiene affect about 10,000 persons per day. However, death rate due to such contaminated water illnesses is quite frequently thought to be inevitable because the majority of those who die suddenly are young babies and seniors,, whose deaths are regarded "natural," or individuals that are largely marginalised in their communities (such as refugees, the poor, or residing beyond the regions that are significant for the world economic growth).

9.6.4 *Future Scenarios*

The expansion of urban areas has always been consistent, sometimes increasing, throughout history. Nowadays, most of them centres are found in emerging world economy, which means that the poor citizens are those who suffer the most as a result. The most serious limitations include inadequate living circumstances, a democratic deficit, poor sanitation, lack of education, corruption, and a lack of appropriate sanitation and water services. These limitations particularly affect women and young kids.

Potable water is currently in low supply everywhere. You must also be ready to invest substantially when making particular choices about water supply and sewage. The current increased operating level of services was not attained quickly or without significant contributions and resources. When evaluating potential course of action and considering necessary tactics, consider that. As much as a society's capacity to assume accountability for creating the living conditions for its inhabitants and appropriate regulations, the degree of water and sanitation services in that civilization is not always dependent on place and time. In certain circumstances, the condition was simpler in the past than it is now. Adopting the washroom as a broad cultural requirement is one example of a decision regarding sanitation and water infrastructure that, due to path dependency, have constrained future possibilities. However, there have been instances where a technological decision was considered hazardous from the start but was made regardless. For example, lead pipes were still employed in home links even though they were deemed unhealthy in ancient times (Rashid, 2017).

The Chisso chemical facility discharged sewage directly carrying unprocessed mercurial to Minamata Bay and Shiranui Sea between 1932 to 1968, causing a polluted sea calamity in Japan. The illness brought on by mercury exposure in this place is known as Minamata disease. Although the initial contamination was found in 1956, it wasn't until 1968 that the government formally judged that the Chisso plant was to be accountable for Minamata. Its effects persisted for more than 3 decades.

Convulsions, feeble limbs, and bafflement are symptoms of contamination. birth problems in foetuses. 10,000 individuals were impacted, and approximately 2000 individuals have died. Both the afflicted dogs and cats become insane and die. Water from the deceased fish has covered the ocean waters floor. Figure 9.1 shows where mercury contamination was in Japan in 1932. Chisso was obligated to clean up damaged maritime regions and had already paid sufferers $86 million as recompense by 2004. One of the four most severe diseases brought on by ecological damage in Japan is still the Minamata disease. Even now, when the sufferers are 40–50 years old, they are confined to their homes, cut off from society, and responsible for taking care of their families.

According to the New York Times, the Deepwater Horizon oil disaster occurred in 2010 when the BP oil rig burst off the Louisiana coast in the United States. The accident happened as the Horizons Deepwater Drilling Rig dug at the Macondo Prospective oilfield services at a depth of 1500 m. 11 people were killed and 17 others were hurt when a really highly pressurized well's exhaust gas burst. Almost 5,000,000 oil barrels flowed into the wide Gulf of Mexico as the rigs burns and sank into the water, devastating ecosystems and harming local aquaculture, ecotourism, and fishing industry. In American historical record, this ecological collapse is the worst. The oil spill impacted more than 400 marine species. Based on the US Department of Meteorology and Hydrology (NOAA), measurements of crude oil levels in Gulf fish 5 years after the disaster show that they are still larger than average, which causes fish to develop congenital cardiac abnormalities and die early. The long-term effects of the spilled oil on the ecology "exceed our expectations." we believe," according to NOAA. Among the 32 dolphins that were seen, many are thinner, have anaemia, lung and liver diseases, and hormone production significantly lower stress and metabolic activity (Rashid, 2017).

9.7 The Causes of the Continuing Growth in Water Contamination

Moreover, it is quite vulnerable to toxicity. This is primarily due to water's ability to dissolve various chemicals as it serves as a general solvent. Although people utilize this excellent quality for daily tasks such as preparing food, housekeeping, and seeking treatment, this is also the same quality that leads to water pollution.

9.7.1 Industrial Waste

Water contamination is primarily caused by industries and industrial facilities all over the globe. Although many construction plants are controlled, others still lack adequate waste management and generate wastage of dangerous chemicals and contaminants. Industrial waste is occasionally thrown into surrounding natural bodies of water. Commercial trash can quite rapidly contaminate aquatic ecosystems if it's not handled correctly or treated in any way (Reddy & Sunitha, 2023; Reddy et al., 2024; Etikala et al., 2024).

River, streams, and other waterbodies that flow straight to the ocean can get contaminated by industrial effluents from mining operations, manufacturing facilities, and agricultural areas. In addition to rendering water hazardous for direct utilization, the dangerous substances in the wastes generated by these operations can affect the temperature of aquatic environments, endangering a variety of aquatic organisms (Sampat, 2000).

9.7.2 Dumping in Oceans

Dumping trash into ocean seas is essentially the same thing as "marine dumping." Although it may sound absurd, many nations worldwide keep collecting household trash and discard it into the oceans (State of Global Water Resources, 2021). For the majority of these things, the full decomposition procedure can last from 2 to 200 years (https://pubs.usgs.gov/circ/circ1225/html/sources.html).

9.7.3 Sewage and Wastewater

Even after treatment, dangerous substances, germs, and pathogens can still be detected in sludge and wastewater. Each residence releases effluent and sludge into the ocean along with clean water. As a result of the microbes and germs present in that effluent, both humans and animals might get illnesses (The Facts about Nutrient Pollution, 2012).

9.7.4 Oil Leaks and Spills

When characterizing two substances that do not combines correctly or in any way, the idiom "like water and oil" has long been used. Water and oil don't really mix, and oils doesn't really disintegrate in water, as the proverb goes. Despite being frequently unintentional, oil spillage and leakages are a major contributor to water contamination. Oil exploratory drill in the water or oil-transporting ships frequently result in leaks and accidents. From 1958, 5.5 million tons of crude oil have been spilled into coastal areas waters with mangroves, potentially contaminating up to 1.94 million hectares of mangrove swamps and destroying at least 126,000 ha of mangroves flora (Tiseo, 2022).

9.7.5 Agriculture

Farmers frequently employ pesticides and insecticides to shield their crops against pests and bacteria. These compounds can be harmful to human beings, vegetation, animal life, and the environment if they leak into groundwater. Furthermore, the toxins combine with rainwater whenever it showers, which then runs through streams and rivers that drain in to the ocean, adding to the water contamination (Tufenkji et al., 2002).

9.7.6 Global Warming

In relation to water contamination, increasing temperatures brought forward by climate warming are indeed a serious issue. seawater organisms may die as a result of rising water temperatures brought about by climate warming. Massive die-offs exacerbate the problem by considerably polluting the water source. Floods and poor water quality significantly threaten urban regions because of climate change and overpopulation (UNESCO World Water Assessment Programme, 2017).

9.7.7 *Radioactive Waste*

The ecosystem can be severely harmed by nuclear waste from nuclear energy production plants, so it must be appropriately discarded. This is due to the very poisonous nature of uranium, the material used to produce atomic energy. Inconveniently, accidents still happen at all these sites, releasing poisonous waste into the ecosystem. The gas and coal sectors aren't much better in many aspects. One of the main drivers behind the production of alternate, cleaner fuels like wind and solar power is this (Sunitha & Reddy, 2022).

9.8 Steps to Be Followed for the Conservation of Water in Current and Prospects

Improved water management practices that reduce or enhance the beneficial use of water is necessary. A water conservation measure is an action, behavioral change, device, technology, or improved design or process implemented to reduce water loss, waste, or use. Water efficiency is a tool of water conservation. That results in more efficient water use and thus reduces water demand (Robertson, 1932).

9.8.1 *Regulations and Policy Action*

During the late nineteenth century, federal legislation governing the management of freshwater sources in the United States and the associated scientific knowledge has evolved concurrently, resulting in regulatory reform that has only sometimes been influenced by research. Although the movement of the legislation known as the Clean Water Act in the 1970s signified a previously unheard-of legislative effort to safeguard American waters, the lack of engagement between legislation and scientific research hasn't just stayed. However, it has also been made worse by several Supreme Court decisions that left a hazy standard for compliance to be pursued by federal authorities, state courts, and farmworkers and developers who mainly rely on there own land for their living standards. When the Judiciary examined the extent of government regulatory jurisdiction in the Clean Water Act in the dispute of Rapanos v. United States in 2006, the discrepancy regarding the structure guiding water resource management may have reached a pinnacle (Wada et al., 2016).

The US Army Corps of Engineers (Corps) was officially given jurisdiction over impediments to the country's water ways at the end of the nineteenth century according to Rivers and Harbors Act of 1899. 2 Throughout the twentieth century, a number of laws were passed, the first of which established the conventional concept of "navigable waters" as the foundation for the national government's duty to oversee the prevention of water contamination throughout the nation. When the Legislature approved the Rivers and Harbors Act of 1899, the phrase "navigable waters," which up to that point had been defined to grant governmental authority over rivers and streams via Federal government jurisdiction under Commerce Clause, came into contact with the first governmental law regulating pollution problems. Under Section 9, establishing the conditions for authority to build dam, embankments, bridges, or passageways in a navigational channel, this law gave the Corps jurisdiction over the obstacle to navigational waters. The Federal Water Pollution Control Act (FWPCA), approved by Congress in 1948, was the first major declaration of the national government's engagement in quality water initiatives after almost 50 years without much to speak for it. The government regulatory jurisdiction was significantly broadened by Water Quality Act of 1965. The 1965 Act requires states to submit action plan to comply with ambient water quality criteria specified for interstate water ways. The next year, the Clean Water Restoration Act was passed, adding to the FWPCA's national government agencies. The Cuyahoga River fire of 1969 and other high-profile occurrences, as well as the flaws and inefficiency of earlier water contamination prevention laws, prompted Congress to approve several modifications in 1972 that drastically restructured and expanded the FWPCA. The FWPCA later picked up the name "Clean Water Act" following the 1972 Amendments. The 1972 Amendments stated that federal objective was to stop dumping pollutants into navigational waterways by 1985 (Nayak et al., 2023), with prime focus being to "regain and preserve the chemical, physical, and ecological balance of the Country's waters." Fig. 9.10 illustrates the legislative structure from the late 1900s to the current circumstances. The Legislation of the Sea Agreement, which was adopted by the UN in 1982, was a game-changer in the way that our planet's enormous, communal water sources came under the purview of universal law. The treaty has settled some significant issues related to ocean use and sovereignty (Wang & Yang, 2016).

The fundamental rules of water law that apply today in India also come from agricultural acts. For example, the earlier Northern India Canal and Drainage Act, 1873 tried to control the region's drainage, transportation, and irrigation. This legislation

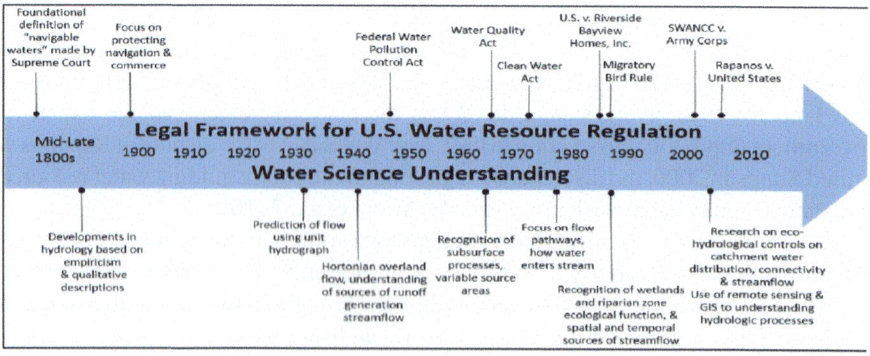

Fig. 9.10 History of parallel legal framework and water scientific understanding (Wada et al., 2016)

established the government's ability to "use and manage for common use the water of all streams and rivers running in natural courses, and of all lakes," which was one of its long-term effects. 15 The 1873 statute refrained from declaring that surface waterways were under governmental control. Yet, this act is significant because it affirmed the government's authority to regulate water consumption for the good of the wider population.

This grew stronger over time. Because of this, the Madhya Pradesh Irrigation Act, 1931 stepped forward and declared explicit governmental jurisdiction over water: "All right in the water of almost any stream, natural stream or existing drainage path, natural lakes and other natural documentation of water must belong in the Government." The Control of Waters Act of 1949 was also reaffirmed in MP, stating once more that "all right in the waters of every natural source of supplies will remain in the Administration." 17 The far more current Bihar Irrigation Act, 1997 nevertheless states that the state owns all rights to surface water. Water legislation is typically mostly based on state laws. It is the result of the legislative framework, that, as of the Government of India Act of 1935, has theoretically granted the provinces the authority to enact laws in this domain. Therefore, the provinces' sole authority to control water resources, irrigation and waterways, drainage and dikes, reservoirs, hydroelectric power, and fisheries remains (https://www.worldbank.org/en/news/press-release/2020/03/19/wastewater-a-resource-that-can-pay-dividends-for-people-the-environment-and-economies-says-world-bank).

The Water Act was enacted by Parliament in 1974 in response to water contamination. This law protects and regulates water contamination while upholding and restoring the water's integrity. Water councils are given authority to establish guidelines and rules for pollution prevention and management (Water Act, 1974).

9.8.2 Conservation Technologies

The term "conservation process" can also refer to protection from damage or wastage. In a nutshell, it refers to utilising all available technology to make the maximum use possible of the nation's water resources. In general, water management tries to balance demand with supply. The solutions for conserving water could be demand-, supply-, management-, or any combination of these. The techniques may change depending on whether the water is being used for household, agricultural, or industrial purposes (Robertson, 1932).

(1) Harvesting rainwater is the practise of extracting rainwater from building rooftops and conserving it underneath to be used after that. This recharge stops groundwater evaporation and elevates the falling water table, or it can increase water supplies. Harvesting rainwater and artificial recharge are topics that are becoming extremely crucial. Groundwater depths must be stabilised, seawater infiltration must be stopped, seawater must not move ashore, and monsoon rainfall surface water runoff must be minimised.

(2) Improved Irrigation Approaches—Water management is crucial in the agriculture industry because it is required to develop plants and crops. Things have gotten severe due to an eroding water table and an increase in salinity brought by an excessive consumption of chemical fertilisers and pesticides. To address the issue, numerous techniques for water harvesting and replenishment have been and are utilised worldwide. Locals have adopted basic methods that are appropriate for their environment to reduce water consumption in areas with little rainfall and scarcity of water.
(3) Despite being freely accessible, seawater is hardly used for agricultural irrigation since it limits the development of plants and their productivity. Crop variants that can endure salt have also recently been produced.
(4) Mulching, or the use of either inorganic or organic matter like manure or plant litter, minimises evaporative loss, enhances soil health, and decelerates surface runoff.
(5) Significant volumes of water are present in fog and mist, which are directly consumable by specialised plant species. Fog and mist can accumulate on man-made surfaces like polyethylene sheets or trap with net surfaces. Plants can be irrigated with the resultant liquid.
(6) In flat and hill locations, contour agriculture is used for paddy fields. Farmers know the effectiveness of contour-based methods for soil and water conservation.
(7) Tippy Tap is a basic gadget that allows for a full hand wash while slowly dispensing a small quantity of water. When using a piped water supply, around 300–500 ml of water are used each time the tap is opened for a hand wash. Tippy Tap allows you to wash your hands effectively with only 60–80 ml of water.
(8) The planting of drought-resistant plants—those which require less water—should be done as a step against conserving water and the proliferation of native plants varieties.
(9) Water runoffs and water logging can be prevented by installing soak pits close to water sources such as pumping stations. This serves as a hygienic treatment and aids in ground water recharging.
(10) Further actions performed to protect our water supplies include planting trees in water catchment areas and riverbeds and organising clean-up campaigns near water sources.
(11) Desalination techniques including the distillation process, electro-dialysis, and reverse osmosis are readily accessible to offset the increased extraction that depletes the supply of clean water in coastline locations. Area dependent selection and application of these techniques.
(12) Another way to significantly enhance the advantages is transporting water from additional basins by building reservoirs properly and connecting different methods.

9.9 Future Perspectives

No single road leads to an ideal future, let's say in 2050. Many routes lead to numerous "ideal" possibilities. Nonetheless, seeing any of these futures avoiding the water shortages is difficult. This will be especially evident in developing nations like China and India, accounting for approximately one-third of the earth's inhabitants by 2050. The presence of thriving habitats will undoubtedly be a characteristic of any "pleasant" future. Water is necessary for a living environment. The necessity for water to support ecosystem life has frequently been seen as conflicting with other water needs. We must also stop disposing of human and industrial effluent in the environment, which is frequently done without thinking about how it will affect habitats, in parallel to allocating water for the conservation of the natural ecosystem. In other words, we need to figure out how to prevent starving environments of water that is crucial to survival and contaminating them with trash (Water Pollution Statistics, 2022).

9.9.1 Water Conservation Efforts for the Future

- It is necessary to implement effective management approaches, such as the employment of new technology like water stress plant types, drip and sprinkle irrigation facilities, and mulching methods in farming, as well as practises that preserve river catchment regions and maintain quality of the water.
- Persistent attempts by the government and many other professional organisations to conserve and manage water.
- Artificial recharging of water table across both rural and urban regions.
- All drainage basin initiatives must be evaluated to comprehend potential advantages and disadvantages and create essential adjustments.
- Encourage farmers to use conservation practices and to reduce wastage of water in their farms.
- R&D on climate change should be encouraged.

9.9.2 Using Science Communication to Reduce Water Contamination

Communicating precise information is made easier through communication, especially amid a pandemic such as COVID-19. The crucial lesson society has realised over the past 2 decades is the need to spread precise data, combat false data, start sharing scientific and WASH information for a good knowledge of precautionary measures and prevention strategies, and boost science education to confront challenges such as pandemics and epidemics. Language, degrees of faith in scientific and educational processes, and communication patterns vary significantly between communities and different ages. Goals, methods, target populations, and government-public–private collaborations, comprising experts and scholars, are all included in WASH communications. Common approaches for building societal trust in science's accurate dissemination of water, sanitation, and hygiene (WASH) knowledge are essential for society. Individual verbal communication was previously the greatest popular form of science communication, followed by television, radio and tv, and newspaper articles among the most popular media.

The primary objective of WASH communication is to positively influence human behaviour in a positive wapositively so that people become more aware of the importance of practising good hygiene, using proper sanitation, conserving water, and acting appropriately in the situation of pandemic outbreaks. The following are the specific goals of WASH communication:

- To establish a friendly atmosphere for community engagement, raise society awareness of safe drinking water, sanitary conditions, and cleanliness.
- In the interest of promoting the construction and occupancy of toilets as well as other sanitary facilities and the elimination of open excretion, it is important to make sure that people in village areas know the correlation among sanitation and clean water. Additionally, increasing education is necessary for ongoing evaluations and the immediate need to determine whether WASH services are functioning properly.
- To improve cooperation between residents and Partner organisations operating for such meetings in order to reduce barriers to effective communication and engage all groups, including those that are disabled or uneducated, at all levels.

A combination of traditional and modern methods of communication must be used to spread awareness of WASH practises among the general public in order to fulfil the goals of WASH communication. These methods fall into three classes: communication stratagems, advocacy, and behavioural modifications.

- Mass media: By decreasing the language and geographic barriers and using a good impression, tools can be utilised to onvey with wider populations clearly. Through a wide range of goods,, such as radio and television advertisements, print articles, and live television programmes, mass media can aid in promoting healthy hygiene behaviours. It can also train members of society who are not communication specialists to serve as communicators and deliver crucial information about Sanitation facilities and practises, particularly throughout pandemics and epidemics. To ensure that content published via the media has been verified and that conversion into any local languages does not modify the context or the public health recommendations, there must be a thorough verification via internal agencies and government organisations (Weis, 2015).
- Public media: Some areas of society and certain people are not known to mainstream media. Examples of public media include wall artworks, indications, flags, and more conventional ways to communicate like folklore. Local languages and

pictures could be used in public media to facilitate communication. Public media could also be captured and distributed on mass media systems.

- Public appeals: Empowering and inspirational messages from public figures, motivational speakers, and spiritual leaders could be created to inform the huge numbers of their supporters on good hygiene, sanitary conditions, and clean water habits. These messages could then be streamed on different mass media platforms, regardless of language or area.
- Mobile campaign: Using the various messages and audiovisual apps accessible on smartphone devices is the simplest means of connecting with the highest number of individuals in the twenty-first century. To promote WASH awareness for sanitation and cleaning and to connect the people of the different nations, voice mails, brief videos, visual messages, and texts may all be easily obtained. Moreover, it could exchange authorised data during an outbreak or pandemic by enlisting network suppliers, telecommunication ministries, and professional communicators.
- Digital media portals: From Facebook to Twitter, YouTube to WhatsApp, advertising on various types of digital media has emerged as the most important method of communication, along with engaging with a skilled and qualified communicator to advance WASH education and understanding. It was essential during the COVID-19 epidemic, when many were quarantined and locked up and millions were affected. Genuine information about safe water, cleanliness, and sanitation was disseminated and false information was refuted through group discussion on social media platforms.

To strengthen the WASH program and continuously improve the readiness policies, urgent support for effective communication for administration and planning, including the administration, press, people from community, governmental agencies, private organisations, and NGOs is required (Weis, 2015). By integrating these groups together, government projects can be implemented effectively with the help of communicators and the public, leading to fact-based advocacy for measures like the following:

- Members of the societies and decision-makers are urged to collaborate and continually promote WASH practises and awareness among society members and in the numerous government goals for planning and execution.
- As part of government-public collaborations, many social groups promise to hold discussions to report on and help implement WASH practises involving specific government entities, such as municipalities, and to go through forthcoming challenges of multiple platforms.
- To improve behaviour towards government agency activities for WASH practises, conventional and new ways of communication are encouraged and aided.
- To support their efforts and inspire others to become interested in such practice, newly qualified communicators are inspired to bring up health hygiene and sanitary conditions problems at the local level while minimising lingual and geographic hurdles, aiding the accessibility of publications and other insightful resources, and aiding acknowledgements after analysing the results.

- The need to advocate for ongoing planning, investigations, and advancement for Sanitary facilities is urgent, given the epidemics and pandemics experienced over the past 20 years. This will help the existing facilities prioritise providing access to clean water, sanitary conditions, and hygienic practices. This involves teaching residents and others to change their attitudes amid epidemics and pandemics, growth in Sanitary facilities, and continuously hiring young communicators from the community.

Every age group must have a strong understanding of science and WASH to promote positive behaviour in the community and a solid environment for future planning. With the use of diverse communication channels and instruments, ongoing WASH communication practises help change the behaviour of the majority in the population. The control measure can be more effectively implemented by grouping the relevant partners and communities, defining the needs and degrees of understanding, and informally addressing the communities to encourage positive behavioural changes for WASH practices and education (WHO, 2016; The United Nations World Water Development Report, 2018).

The present state of national water resource legislation is in flux, which presents both severe regulatory and legal hurdles and a possibility. An excellent opportunity to more closely align research with the legal regime exists given the current state of scientific knowledge and the accessibility of technologies for identifying influential regions on the landscape. Together, the current situation gives administrators, researchers, and legal experts a chance to put together a more logical course of action to safeguard the United States' water supplies.

References

AbuZeid, K. M. (2012). Mediterranean water outlook: Perspective on policies and water management in Arab countries. In: R. Choukr-Allah, R. Ragab, R. Rodriguez-Clemente (Eds.), *Integrated water resources management in the Mediterranean region: Dialogue towards new strategy*. Springer Science & Business Media.

Adu, J. T., & Kumarasamy, M. V. (2018). Assessing non-point source pollution models: A review. *Polish Journal of Environmental Studies, 27*(5).

Aëtius. (1998). On the opinions of the philosophers V.30.1. In J. Longrigg (Eds.), *Greek Medicine. From the Heroic to the Hellenistic Age. A Source Book* (p. 31). Duckworth.

Al-Gheethi, A. A., Efaq, A. N., Bala, J. D., Norli, I., Abdel-Monem, M. O., Ab. Kadir, M. O. (2018). Removal of pathogenic bacteria from sewage-treated effluent and biosolids for agricultural purposes. *Applied Water Science, 8*, 1–25. https://doi.org/10.1007/s13201-018-0698-6

Aphorisms. (1931). *In Hippocrates Volume IV*, with an English translation by W. H. S. Jones. The Loeb Classical Library.

Baya, B. (2008). *Environmental pollution and development in Medea of Biskra* (Doctoral dissertation, Ministry of Higher Education and Scientific Research, University of Matouri Qatnya, Faculty of Humanities and Social Sciences, Department of Sociology and Demography).

Bensoltane, M. A., Zeghadnia, L., Bordji, N., Gagan, M., & Bouranen, S. (2021). Drinking water quality assessment using principal component analysis: Case study of the town of Souk Ahras,

Algeria. *Egyptian Journal of Chemistry, 64*(6), 3069–3075. https://doi.org/10.21608/EJCHEM.2021.53654.3112

Bolisetty, S., Coray, N. M., Palika, A., Prenosil, G. A., & Mezzenga, R. (2020). Amyloid hybrid membranes for removal of clinical and nuclear radioactive wastewater. *Environmental Science: Water Research & Technology., 6*(12), 3249–3254.

Broom, D. (2019). *Environmentalists have removed nearly 40 tonnes of trash from the Pacific.* https://www.weforum.org/agenda/2019/07/environmentalists-have-removed-nearly-40-tonnes-of-trash-from-the-pacific/

Bruun, C. (1991). *The water supply of Ancient Rome. A study of Roman Imperial Administration.* Societas Scientiarum Fennica. Commentationes Humanarum Litterarum 93.

Burek, P., et al. (2016). *Water futures and solution: Fast track initiative (final report). IIASA Working Paper* (International Institute for Applied Systems Analysis (IIASA), Laxenburg, Austria.

Campbell, C. (2015). *US Federal water pollution control: How history has contributed to the mismatch between the legal framework and the current state of the science.*

China Wakes Up to Its Water Crisis. (2014). https://www.thestar.com/news/world/2014/05/12/china_wakes_up_to_its_water_crisis.html

Clark, R. B., Frid, C., & Attrill, M. (1997). *Marine pollution.* Clarendon Press.

Cosgrove, W. J., & Loucks, D. P. (2015). Water management: Current and future challenges and research directions. *Water Resources Research, 51*(6), 4823–4839. https://doi.org/10.1002/2014WR016869

Cullet, P. (2007). *Water law in India overview of existing framework and proposed reforms.* International Environmental Law Research Centre. http://www.ielrc.org/content/w0701.pdf

DeGood, K. (2020). *A call to action on combating nonpoint source and storm water pollution.* https://www.americanprogress.org/article/call-action-combating-nonpoint-source-stormwater-pollution/

Denchak, M. (2023). *Water pollution: Everything you need to know.* https://www.nrdc.org/stories/water-pollution-everything-you-need-know

Duke, N. C. (2016). Oil spill impacts on mangroves: Recommendations for operational planning and action based on a global review. *Marine Pollution Bulletin, 109*(2), 700–715. https://doi.org/10.1016/j.marpolbul.2016.06.082

Etikala, B., Vangala, S., & Madhav, S. (2024). Groundwater geochemistry using modified integrated water quality index (IWQI) and health indices with special emphasis on nitrates and heavy metals in southern parts of Tirupati, South India. *Environmental Geochemistry and Health, 46*(11). https://doi.org/10.1007/s10653-024-02229-3

Evans, A. E., Mateo-Sagasta, J., Qadir, M., Boelee, E., & Ippolito, A. (2019). Agricultural water pollution: Key knowledge gaps and research needs. *Current Opinion in Environmental Sustainability, 1*(36), 20–27. https://doi.org/10.1016/j.cosust.2018.10.003

Fadlallah, S. (2001). Environmental pollution and its impact on agricultural economic development. *Assiut Journal of Environmental Studies, 20*, 82–84.

Fayiga, A. O., Ipinmoroti, M. O., & Chirenje, T. (2018). Environmental pollution in Africa. *Environment, Development and Sustainability., 20*(1), 41–73.

Frontinus, S. J. (1925). *The stratagems, and the aqueducts of Rome* (Vol. 174). Heinemann, W., & Fuhrman, R. E. (1984). History of water pollution control. *Journal (Water Pollution Control Federation), 56*, 306–331.

Gagan, M., Uniyal, D. P., Chadha, K., Sunil, K., Gaurav, P., Avinash, K., Anjali, N., Pawan, K. (2022a). Impact of pandemic COVID19 on air and water quality in India: A systematic review. *International Journal of Engineering and Advanced Technology, 11*, 149–167. https://doi.org/10.35940/ijeat.E3590.0611522

Galen. (1951). De Sanitate Tuenda. A translation of Galen's Hygiene (De Sanitate Tuenda) by Robert Montraville Green. With an introduction by Henry E. Sigerist. Charles C. Thomas.

Gjyli, L., Vlachogianni, T., Kolitari, J., Matta, G., Metalla, O., & Gjyli, S. (2020). Marine litter on the Albanian coastline: Baseline information for improved management. *Ocean & Coastal Management, 1*(187), 105108. https://doi.org/10.1016/j.ocecoaman.2020.105108

Gjyli, L., Mucaj, O., Matta, G., Krasniqi, M. (2018). Inorganic toxicity: Barium. In G. Matta, & L. Gjyli (Eds.), Inorganic toxicity: Environment and human health. Lambert Publication House.

Global Burden of Disease Collaborative Network. Global Burden of Disease Study 2019. (GBD 2019). Results. Institute for Health Metrics and Evaluation (IHME), Seattle, United States.

Global Challenges: Freshwater. Available online: https://www.theworldcounts.com/challenges/freshwater/people-without-sewage-systems

Hasan, M. K., Shahriar, A., & Jim, K. U. (2019). Water pollution in Bangladesh and its impact on public health. *Heliyon, 5*(8), e02145.

Haseena, M., Malik, M. F., Javed, A., Arshad, S., Asif, N., Zulfiqar, S., & Hanif, J. (2017). Water pollution and human health. *Environmental Risk Assessment and Remediation, 1*(3).

Helmer, R., Hespanhol, I., & Supply, W. (1997). *Water pollution control: A guide to the use of water quality management principles.* E & FN Spon.

Hippocrates. (1988). *Diseases 3. Internal Affections. Regimen in Acute Diseases.* Translated by Paul Potter. Loeb Classical Library 473. Harvard University Press.

Hippocrates. (2022). *Ancient medicine. Airs, waters, places. Epidemics 1 and 3. The Oath. Precepts. Nutriment.* Edited and translated by Paul Potter. Loeb Classical Library 147. Harvard University Press.

Hodge, A. T. (1981). Vitruvius, lead pipes and lead poisoning. *American Journal of Archaeology, 85*, 486–491.

How Much Plastic Floats In The Great Pacific Garbage Patch? https://theoceancleanup.com/great-pacific-garbage-patch/

Howard, K. W. F., & Gelo, K. K. (2002). Intensive groundwater use in urban areas: The case of megacities. In M. Ramon Llamas, E. Custodio (Eds.), Intensive use of groundwater: Challenges and opportunities. A.A. Balkema Publishers, CRC Press.

Huss, M., Bookhagen, B., Huggel, C., Jacobsen, D., Bradley, R. S., Clague, J. J., Vuille, M., Buytaert, W., Cayan, D. R., Greenwood, G., & Mark, B. G. (2017). Toward mountains without permanent snow and ice. *Earth's Future, 5*(5), 418–435. https://doi.org/10.1002/2016EF000514

India's 2012 to 2017 National WASH Advocacy and Communications Strategy; UNICEF & Ministry of Drinking Water and Sanitation, Government of India (2012)

Jayaswal, K., Sahu, V., & Gurjar, B. R. (2018). Water pollution, human health and remediation. *Water Remediation,* 11–27.

Johnson, R. P. (1941). *Pliny, natural history.* With an English Translation in Ten Volumes (Vol. III, Libri VIII–XI).

Juuti, P. S., Katko, T. S., & Vuorinen, H. S. (2006). *Environmental history of water—Global view of community water supply and sanitation.* IWA Publishing.

Kidd, D. E. (1983). The history and definition of water pollution: A teaching unit. *Bulletin of Science, Technology & Society, 3*, 121–126.

Kumar, A., Matta, G., & Bhatnagar, S. (2021). A coherent approach of water quality indices and multivariate statistical models to estimate the water quality and pollution source apportionment of River Ganga System in Himalayan region, Uttarakhand, India. *Environmental Science and Pollution Research, 28*, 42837–42852. https://doi.org/10.1007/s11356-021-13711-1

Kumari, M., & Singh, J. (2016). Water conservation: Strategies and solutions. *International Journal of Advanced Research and Review, 1*(4), 75–79.

Le, V. V., Cao, D. N., & Nguyen, X. P. (2017). The overview of water pollution in the world. *International Journal of Scientific & Technology Research, 6*(8), 221–224.

Mahmood, I., Imadi, S. R., Shazadi, K., Gul, A., & Hakeem, K.R. (2016). Effects of pesticides on environment. Plant, soil and microbes: Volume 1: Implications in crop science (pp. 253–269). https://doi.org/10.1007/978-3-319-27455-3_13

Matta, G., Kumar, A., Kumar, P., Nayak, A., Kumar, P., Kumar, A., & Tiwari, A. K. (2021). Evaluation of ground water quality by use of water quality index in the vicinity of the Rajaji National Park Haridwar, Uttarakhand, India. In *Geostatistics and geospatial technologies for groundwater resources in India* (pp. 343–356). https://doi.org/10.1007/978-3-030-62397-5_17

Matta, G., Kumar, A., Nayak, A., Kumar, P., Kumar, A., Naik, P. K., & Singh, S. K. (2023). Assessing heavy metal index referencing health risk in Ganga River System. *International Journal of River Basin Management, 21*(4), 759–769. https://doi.org/10.1080/15715124.2022.2098756

Matta, G., Kumar, A., Nayak, A., Kumar, P., Kumar, A., & Tiwari, A. K. (2020). Determination of water quality of Ganga River System in Himalayan region, referencing indexing techniques. *Arabian Journal of Geosciences, 13*, 1–1. https://doi.org/10.1007/s12517-020-05999-z

Matta, G., Kumar, A., Nayak, A., Kumar, P., & Pant, G. (2022). Pollution complexity quantification using NPI and HPI of River Ganga system in Himalayan Region. *Proceedings of the Indian National Science Academy, 88*(4), 651–663. https://doi.org/10.1007/s43538-022-00111-3

Matta, G., Kumar, A., Nayak, A., & Kumar, P. (2022). Appraisal of spatial–temporal variation and pollution source estimation of Ganga River system through pollution indices and environmetrics in Upper Ganga basin. *Applied Water Science, 12*(3), 33. https://doi.org/10.1007/s13201-021-01552-9

Matta, G., Kumar, A., Tiwari, A. K., Naik, P. K., & Berndtsson, R. (2020d). HPI appraisal of concentrations of heavy metals in dynamic and static flow of Ganga River System. *Environment, Development and Sustainability, 22*, 33–46.https://doi.org/10.1007/s10668-018-01182-3

Matta, G., Kumar, M. A. (2018b). Inorganic toxicity: Environment & human health. In L. Gjyli (Eds.), *LAP LAMBERT*. Academic Publishing. Matta, G., Kumar, A., Kumar, A., et al. (2018). Assessment of heavy metals toxicity and ecological impact on surface water quality using HPI in Ganga River. *INAE Lett, 3*, 123–129. https://doi.org/10.1007/s41403-018-0041-4

Matta, G., Kumar, P., Uniyal, D. P., & Joshi, D. U. (2022). Communicating water, sanitation, and hygiene under sustainable development goals 3, 4, and 6 as the panacea for epidemics and pandemics referencing the succession of COVID-19 surges. *ACS Es&t Water, 2*(5), 667–689. https://doi.org/10.1021/acsestwater.1c00366

Matta, G., Nayak, A., Kumar, A., Kumar, P., Kumar, A., Tiwari, A. K., & Naik, P. K. (2020). Evaluation of heavy metals contamination with calculating the pollution index for Ganga River system. *Taiwan Water Conservancy, 68*(3), 10–6937. https://doi.org/10.6937/TWC.202009/PP_68(3).0005

Matta, G., Nayak, A., Kumar, A., & Kumar, P. (2020). Water quality assessment using NSFWQI, OIP and multivariate techniques of Ganga River system, Uttarakhand, India. *Applied Water Science, 10*(9), 206. https://doi.org/10.1007/s13201-020-01288-y

Matta, G. (2020). Applicability of WQI and scientific communication for conservation of River Ganga System in India. In *Frontiers in Water-Energy-Nexus—Nature-based solutions, advanced technologies and best practices for environmental sustainability: Proceedings of the 2nd WaterEnergyNEXUS conference*, November 2018a, Salerno, Italy 2020 (pp. 265–267). Springer International Publishing. https://doi.org/10.1007/978-3-030-13068-8

Matta, G. (2020). Science communication as a preventative tool in the COVID19 pandemic. *Humanities and Social Sciences Communications, 7*(1), 1–4. https://doi.org/10.1057/s41599-020-00645-1

Matta, G., Naik, P. K., Machell, J., Kumar, A., Gjyli, L., Tiwari, A. K., & Kumar, A. Comparative study on seasonal variation in hydro-chemical parameters of Ganga river water using comprehensive pollution index (CPI) at Rishikesh, (Uttarakhand) India. *Desalination and Water Treatment, 118*, 87–95. https://doi.org/10.5004/dwt.2018.22487

Matta, G., Lu, X. X., & Kumar, A. (2019). Hydrological assessment of river henwal using water quality indices with reference to planktonic composition. In *Advances in water pollution monitoring and control*. Springer Publication. https://doi.org/10.1007/978-981-32-9956-6_18

Matta, G., Rajput, A., Rajput, A., Kumar, P., Kumar, A., Nayak, A., Kumar A., Dhingra, G. K., Chauhan, A., Chadha, S. K., & Wats, M. Assessment of water quality of Choyyia Nadi (River) Catchment area in Bijnor District, Uttar Pradesh, India. *Eco. Env. & Cons., 26*(November Suppl. Issue), S402–S412.

McNeill, W. H. (1979). *Plagues and peoples*. Penguin Books.

Medema, G. J., Shaw, S., Waite, M., Snozzi, M., Morreau, A., & Grabow, W. (2003). Catchment characterization and source water quality. In: *Assessing microbial safety of drinking water; improving approaches and methods.* Chapter 4, World Health Organization.

Miller, J. D., & Hutchins, M. (2017). The impacts of urbanisation and climate change on urban flooding and urban water quality: A review of the evidence concerning the United Kingdom. *Journal of Hydrology: Regional Studies, 1*(12), 345–362. https://doi.org/10.1016/j.ejrh.2017.06.006

Nayak, A., Matta, G., & Uniyal, D. P. (2023). Hydrochemical characterization of groundwater quality using chemometric analysis and water quality indices in the foothills of Himalayas. *Environment, Development and Sustainability, 25*(12), 14229–14260. https://doi.org/10.1007/s10668

Nutton, V. (2005). *Ancient medicine.* Routledge.

Oceans and the Law of the Sea. https://www.un.org/en/global-issues/oceans-and-the-law-of-the-sea

Palladius. (1975). Opus Agriculturae. In *Palladii Rutilii Tauri Aemilian Vri Inlustris.* Opus Agriculturae, De Veterinaria Medicina, De Insitione. Bibliotheca Scriptorum Graecorum et Romanorum Teubneriana. B.G. Teubner Verlagsgesellschaft.

Paulus Aeginata. (1914). Paulos von Aegina des besten Arztes Sieben Bücher. Uebersetzt und mit Erläuterungen versehen von I. Berendes. E.J. Brill.

Postel, S. L., Daily, G. C., & Ehrlich, P. R. (1996). Human appropriation of renewable fresh water. *Science, 271*(5250), 785–788.

Rashid, S. (2017). The implications of the economic problems of environmental pollution in Iraq and ways to address them. *Al-Mustansiriya Journal for Arab and International Studies. Letters, 14*, 105–6.

Reddy, B. M., Sunitha, V., Prasad, M., Reddy, Y. S., & Reddy, M. R. (2019). Evaluation of groundwater suitability for domestic and agricultural utility in semi-arid region of Anantapur, Andhra Pradesh State, South India. *Groundwater for Sustainable Development, 9*, 100262. https://doi.org/10.1016/j.gsd.2019.100262

Reddy, Y. S., & Sunitha, V. (2023). Assessment of heavy metal pollution and its health implications in groundwater for drinking purpose around inactive mines, SW region of Cuddapah Basin, South India. *Total Environment Research Themes, 8*, 100069. https://doi.org/10.1016/j.totert.2023.100069

Reddy, Y. S., Sunitha, V., Kumar, P. (2024). Land use & land cover dynamics and its relationship with nitrate pollution in groundwater around inactive mine areas using geospatial techniques, SW part of Cuddapah Basin, Southern India. *Chemosphere, 365*, 143322. https://doi.org/10.1016/j.chemosphere.2024.143322

Robertson, D. S. (1932). The Loeb Vitruvius-Vitruvius on Architecture. Edited... and translated into English by Frank Granger, D. Lit., ARIBA, Professor in University College, Nottingham. In two volumes. I (Books IV). Pp. xxxvi+ 317; 8 plates.(Loeb Classical Library.) London Heinemann, 1931. Cloth, 10s. net; leather, 12s. 6d. net. The Classical Review, 46(1), 29–31.

Sampat, P. (2000). Deep trouble-the hidden threat of groundwater pollution. *World Watch Paper, 154* (Worldwatch Institute, Washington, DC).

Sources of nutrients and pesticides. U.S. Geological Circular 1225—The Quality of Our Nation's Waters—Nutrients and Pesticides. https://pubs.usgs.gov/circ/circ1225/html/sources.html

State of Global Water Resources. (2022). *World Meteorological Organization.* https://library.wmo.int/doc_num.php?explnum_id=11435

Sunitha, V., Muralidhara Reddy, B. (2022). Geochemical characterization, deciphering groundwater quality using pollution index of groundwater (PIG), water quality index (WQI) and GIS in hard rock aquifer, South India. *Applied Water Sciences, 12*, 41. https://doi.org/10.1007/s13201-021-01527-w

The Facts About Nutrient Pollution. (2012). https://www.epa.gov/sites/default/files/2015-03/documents/facts_about_nutrient_pollution_what_is_hypoxia.pdf

Tiseo, I. (2022). *Number of districts with contaminated water in India 2019, by contamination type.* https://www.statista.com/statistics/1111747/india-number-of-districts-with-contaminated-water-by-contamination-type/

Tiwari, A. K., Kumar, A., Singh, A. K., Singh, T. N., Suozzi, E., Matta, G., & Russo, S. L. (Eds.) (2022). *Water scarcity, contamination and management.* Elsevier. Badran, A. (1988). Environmental pollution its sources and its types. *Journal of Science and Technology, 4*(7).

Tufenkji, N., Ryan, J. N., Elimelech, M. (2002). The promise of bank filtration. A simple technology may inexpensively clean up poor-quality raw surface water. *Environmental Science and Technology, 36*, 422A–428A.

U.S. Environmental Protection Agency, Washington, D.C. (2004). Report to Congress: Impacts and Control of CSOs and SSOs. Document No. EPA 833-R-04-001.

UNEP. (2021). Globally, 3 Billion People at Health Risk Due to Scarce Data on Water Quality. https://www.unep.org/news-and-stories/story/globally-3-billion-people-health-risk-due-scarce-data-water-quality

UNESCO World Water Assessment Programme. (2017). *The United Nations world water development report 2017: Wastewater: The untapped resource; facts and figures.* https://unesdoc.unesco.org/ark:/48223/pf0000247553

Wada, Y., et al. (2016). Modelling global water use for the 21st century: The water futures and solutions (WFaS) initiative and its approaches. *Geoscientific Model Development, 9*, 175–222.

Wang, Q., & Yang, Z. (2016). Industrial water pollution, water environment treatment, and health risks in China. *Environmental Pollution, 1*(218), 358–365.

Wastewater A Resource that Can Pay Dividends for People, the Environment, and Economies, Says World Bank, World Bank Group. https://www.worldbank.org/en/news/press-release/2020/03/19/wastewater-a-resource-that-can-pay-dividends-for-people-the-environment-and-economies-says-world-bank

Water (Prevention and Control of Pollution) Act, 1974. Available at http://www.ielrc.org/content/e7402.pdf

Water Pollution Statistics. (2022). Available online: https://www.sourgum.com/trash-talks-blog/water-pollution-statistics/

Weis, J. S. (2015). *Marine pollution: What everyone needs to know.* Oxford University Press.

WHO. (2019). Fact sheet—Sanitation. Updated March 2022. Online https://www.who.int/news-room/fact-sheets/detail/sanitation

Withington, E. T., Potter, P., & Smith, W. D. (Eds.). (1927). *Hippocrates* (Vol. 3). Harvard University Press.

World Water Assessment Programme (Nations Unies). (2018). *The United Nations World Water Development Report 2018* (United Nations Educational, Scientific and Cultural Organization, New York, United States) www.unwater.org/publications/world-water-development-report-2018/.

Chapter 10
Approaches to Evaluate the Human Health Risk of Arsenic via Drinking Groundwater

Wiyada Nilkarnjanakul and Srilert Chotpantarat

Abstract Groundwater pollution with arsenic (As) is occurrence in many countries in Asia such as China, India, Bangladesh, Thailand, Vietnam, and Cambodia. Drinking water is a major pathway that can cause the potential health effects depending on the As species and its concentration, solubility form, magnitude of exposure, and individual factors. Estimation of risk is mainly considered by health risk assessment. Deterministic risk and probabilistic can be used to estimate risk value following the USEPA guide under the considering situation, but Monte Carlo modelling can simulate many cases of output as randomly sampling from probabilistic distribution to make a decision for minimizing risk. Furthermore, the uncertainty of information analysis can be reduced while sensitivity analysis can generate a high effect on the input variable. Both tools can give some different advantages depending on the selection of user and limitation of data resources. Meanwhile, the investigated bioindicator is better able to know the actual level of As in the receptor. The evidence of exposure to other As sources such as the behavior of groundwater consumption, occupation and food consumption should be concerned to reduce the overestimation. Considering exposure scenarios among sources, groundwater contamination, and receptors is a challenge for the relevant organizations. In addition, more attention should be paid to the monitoring and protection plan for vulnerable groups.

Keywords Health risk assessment · Exposure assessment · Biomarker · Arsenic in groundwater · Groundwater contamination

---

W. Nilkarnjanakul
Faculty of Public Health, Thammasat University, Pathum Thani, Thailand

S. Chotpantarat (✉)
Center of Excellence in Environmental Innovation and Management of Metals (EnvIMM), Sustainable Environment Research Institute, Chulalongkorn University (SERI), Bangkok, Thailand
e-mail: Srilert@chula.ac.th

Department of Geology, Faculty of Science, Chulalongkorn University, Bangkok, Thailand

V. Sunitha et al. (eds.), *Groundwater Resource Management Planning Strategies*, Springer Water, https://doi.org/10.1007/978-3-031-88870-0_10

10.1 Introduction

Arsenic (As) in groundwater can be generated from rock and mineral erosion such as arsenopyrite and realgar (Reddy and Sunitha 2023). Meanwhile, the community area continuously expanded and increased anthropogenic activity as the industrial area, agricultural and mining activities because of the development of economic growth. As accumulation in soil was observed in agricultural area where using chemical fertilizer and pesticide for long-term (Zhou et al. 2018). Besides agricultural application, improper waste management such as municipal solid waste and e-waste might cause As to the environment, particularly soil and groundwater (Chotpantarat et al., 2020; Houessionon et al., 2021; Ravindra & Mor, 2019; Thanh et al., 2022; Thanh et al., 2024; Wisitthammasri et al., 2020). Under different natural conditions such as environmental factors, characteristic of topography, geology, and hydrology are influencing As species, mobilization, toxicity, and concentration in groundwater (Bhowmick et al., 2018; Chotpantarat & Thamrongsrisakul, 2021; Samdang et al., 2023; Thanh et al., 2023). High As concentration is common occurrence according to groundwater flow and also the elevation of the area (Tiankao & Chotpantarat, 2018). The important case study of As contamination in soil and groundwater is namely in Ronphibun district, Nakhon Si Thammarat province, Thailand, where evidence of As poisoning from improper waste management old tin mines (Zhang et al., 2001). As levels of 128 wells in both shallow and deep groundwater wells were varied from 1 to 6400 μg/L throughout the study area. Health effects have been presented as evidence of arsenical skin disorders, including skin lesions and skin cancer, in a variety of age groups since 1987.

Many people mainly use it for drinking purposes. Some of them have been pretreatment, whereas others are untreated. Various reports have been conducted to assess health risk assessment through drinking groundwater while the contribution of dermal contact is a minor route (Mazumder et al., 2009; Nilkarnjanakul et al., 2022). In case of skin exposure, people can come in contact with many activities in daily life such as bathing, cooking, washing, and farming but this pathway can be negligible (Sheng et al., 2021). As can be stored in tissues, particularly hair and nails, while approximately 70% of the total As level can be eliminated in the kidney by methylation (ATSDR, 2007). Hair and nails can be used as the representative of the person who exposed As for several months (Slotnick & Nriagu, 2006). Therefore, these biomarkers were used to estimate actual As exposure and explain the health signs and symptoms (Rahman et al., 2005; Bhowmick et al., 2018). While the biomarker presents as the exposure concentration, it cannot confirm that all of the existing concentrations are caused by only one As source. Therefore, considering the source of the receptor is an effectiveness that should focus on the other exposure factors like behavior, occupation, food and beverage consumption (Rehman et al., 2020). In this chapter, exposure of As through drinking groundwater focusses on the estimation of health risk assessment: (i) deterministic and (ii) probabilistic risk and exposure assessment using biomarkers including the other sources of As that lead to influencing the influence of the level of As in the human body. Linking among

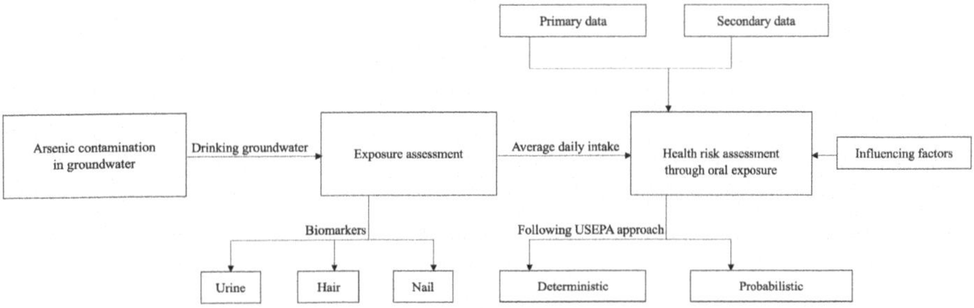

Fig. 10.1 Conceptual framework of As exposure pathway through groundwater drinking

this pathway can for risk management to decide for risk management. Flowchart represents the conceptual of the health risk assessment via groundwater drinking, as presented in Fig. 10.1.

10.1.1 Occurrence and Distribution of Arsenic in Groundwater

Arsenic contamination in groundwater is a prevalent issue in many countries of South Asia, including Thailand, Vietnam, Cambodia, as well as in Southeast Asian countries such as Bangladesh, Pakistan, and India (Boonsrang et al., 2018; Charuseiam et al., 2022; Zhang et al., 2001; Kim et al., 2011; Waqas et al., 2017; Ravindra & Mor, 2019; Saha & Rahman, 2020). Interestingly, most people in these areas usually used groundwater for drinking purposes. As contamination was generated from geological origins, especially granite rock and the sulfide mineral such as pyrite (Smedley & Kinniburgh, 2002; Bhowmick et al., 2018). Additionally, sources of As can be attributed to improper waste management such as landfills, mining activity, and industrial activity (Charuseiam et al., 2022; Wongsasuluk et al., 2021a). A recent study identified that As in groundwater was polluted in the area located near municipal solid waste disposal areas and informal E-waste recycling plants (Ravindra & Mor, 2019). Although agricultural areas, which are mostly in rural areas, have been used fertilizer and pesticide (Shahid et al., 2018). Although the source and concentration are different, the deposition and release to the environment could be affected by the food chain.

Dominant As species in the aquifer are inorganic arsenic (IAs) consisting of arsenate (As^{V}) and arsenite (As^{III}) species. Fate and transport of As in groundwater depend mainly on hydrochemical parameters such as cation and anion parameters that were revealed by many studies (Boonkaewwan et al., 2021; Nilkarnjanakul et al., 2022; Boonkaewwan & Chotpantarat, 2024). The main factor in enhancing arsenic mobilization is the redox potential (Sohrabi et al., 2020). Oxidation–reduction potential (ORP) plays a potential role to control the As mobilization and its species in groundwater. The As^{III} species was rich in the reducing environment as a negative level of ORP. This species has high toxicity and high mobility compared to As^{V}. pH above 8.5 can increase the dissolution of As, particularly As^{V} from Fe oxides. The strongly reducing condition can lead to As^{III} release from Fe and Mn oxides (Smedley & Kinniburgh, 2012; Phan et al., 2013). The dissolution of As from metal oxides can be increased by the reduction condition. However, reduction and oxidation can be exchanged due to the chemical reactions in the aquifers. Moreover, several factors have evidence that also control the mechanism of As mobilization such as pH, Fe, Mn and Al oxides, HCO_3 and degradable organic matter (Saha & Rahman, 2020). Considering the depth of groundwater, the shallow aquifer is more likely to improve the release of As by the surface environment and rainwater (Kumar et al., 2010; Ravindra & Mor, 2019). In contrast, the deep groundwater aquifer found a low variation of the influencing factors to its concentration. Meanwhile, seasonal variation also controlled the groundwater chemistry (Chotpantarat et al., 2020; Reddy & Sunitha, 2023). However, evidence of As-bearing from the mineral dissolution should be observed in the area.

10.1.2 Health Risk Assessment

As is highly toxic to humans, resulting in to both acute and chronic effects. According to the 2007 report by ATSDR, arsenic is classified as both a non-carcinogenic and carcinogenic substance. Contaminated groundwater with As is a serious public health concern. Consuming groundwater from wells with As above than the guideline value of 10 μg/L set by the World Health Organization (WHO) poses potential health risks. Waqas et al. (2017) demonstrated that the resident consuming groundwater experienced harmful health effects. Arsenic poisoning can cause both acute toxic effect such as abdominal pain, vomiting and diarrhea, as well as chronic effects on various organs, including urinary, gastrointestinal, nervous, cardiovascular, reproductive systems (ATSDR, 2007; Ratnaike, 2003). In contrast, even low concentrations of arsenic in routinely consumed water can adversely affect health. Therefore, conducting a health risk assessment is crucial to estimate the extent of risk under given conditions. Generally, two main techniques are widely used: (i) deterministic risk assessment and (ii) probabilistic risk assessment.

10.1.2.1 Estimation of the Health Risk Assessment

Exposure assessments such as deterministic and probabilistic approaches are calculated following the USEPA method (USEPA, 2004) as follows the Eqs. (10.1)–(10.3):

Exposure assessment trough oral route:

$$ADD_{oral} = C_{As} \times IR \times EF \times ED/BW \times AT \qquad (10.1)$$

where:

ADD_{oral} = Average daily dose through oral exposure (mg/kg-d)
C_{As} = Arsenic concentration in groundwater (mg/L)
IR = Intake rate (L/d)
EF = Exposure frequency (d/y)
ED = Exposure duration (y)
BW = Body weight (kg)
AT = Average time (d)

Risk characterization:

$$HQ_{As} = ADD_{oral}/RfD_{oral} \tag{10.2}$$

$$CR_{As} = ADD_{oral} \times SF_{oral} \tag{10.3}$$

where:

HQ_{As} = Hazard quotient of As
CR_{As} = Cancer risk of As
RfD_{oral} = Reference dose of oral exposure for non-carcinogenic risk (0.0003 mg/kg-d); (Iris, 2013)
SF_{oral} = Slope factor of oral exposure for carcinogenic risk (1.5 per mg/kg-d); (Iris, 2013)

When: HQ > 1; It can be interpreted that people are at risk for non-carcinogenic risk (USEPA, 2004).

CR > 10^{-6}; It can be interpreted that people are at risk for carcinogenic risk (USEPA, 2004).

After estimating risk following Eqs. (10.2) and (10.3), the output is a point value. Deterministic risk is preferable for simple use and appropriate for screening risk levels. The result is easy to communicate and understand. The probabilistic technique generates the distribution of exposure by using the mathematical model. The simulation can provide multiple scenarios such as lower and upper population exposure which flexible to make a decision. The information of the two approaches is presented in Table 10.1.

The principle of this modeling is the use of the Monte Carlo simulation technique. Random numbers are drawn from the probability distribution with independent values (ranging from 0 to 1) in each sample of all iterations, as presented in Fig. 10.2. The description methodology for probabilistic risk assessment can be explained according to USEPA (1997).

This technique is more flexible and less complicated for the users. In addition, the different outputs are provided which can prioritize to make the decision for minimizing risk. In this section, an example of Monte Carlo simulation is presented in Fig. 10.3.

Table 10.1 Comparation between deterministic and probabilistic risk assessment

List	Deterministic risk assessment	Probabilistic risk assessment
Equation	Following USEPA guide	Following USEPA guide
Input value	Point value	Min and max Average and standard deviation Fixed value
Calculation	Manual calculation Excel spreadsheet	Computing modelling such as @Risk Software Crystal ball Software
Result	Single value	Graphical data Statistical output Sensitivity and uncertainty output

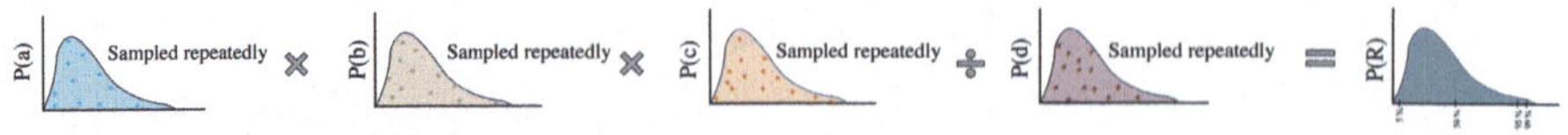

Fig. 10.2 Monte Carlo simulation approach

Step I: Input the value of each parameter such as min–max, average, standard deviation, or constant value (select depending on your characteristic of data).
Step II: Generate the probability distribution of the input parameter except for the constant value.
Step III: Set the iteration and run to simulate data following the equation of exposure assessment.
Step IV: Explore the output scenario such as graphical and statistical results (select the output formatting).
Step V: Analyze the input variables to observe the influence that contributes to the changing of output (commonly used tornado graphs to present).

Monte Carlo can help decrease the uncertainties of information sources and the effect of contribution to variance (Das et al., 2022). Sensitivity analysis can confirm that As concentration and intake rate are the most impact on risk value (Wang et al., 2020). Many studies supported that oral exposure is a primary pathway of As exposure, while dermal exposure is a minor route (Wang et al., 2020; Sohrabi et al., 2021). Studies with deterministic and probabilistic risk have been more observed as presented in Table 10.2.

10.1.2.2 Exposure Assessment of as by Using Biomarker Sample

As mentioned, the estimation of health risk assessment can generate the risk value, but this result is calculated based on the assumption of exposure assessment following the USEPA equation. Taking into account biomarker samples can determine the actual value of As in the human body. The relationship between the investigation of

Fig. 10.3 Monte Carlo simulation process for health risk assessment from @Risk 8.0 software—Palisade

Table 10.2 The study of health risk assessment of As exposure from groundwater

Countries	Estimation of health risk assessment	Exposure assessment using biomarker	Remark	References
Deterministic risk assessment				
Vietnam	• As concentration in treated (using a sand filtration system) and untreated groundwater • As in human hair • Health risk assessment via drinking water (using assumed value)	Male and female hair	• As in female hair was correlated with As level in groundwater • Treated groundwater by sand filtration was not effective in reducing As level • Long-term As exposure (>10 μg/L), especially for untreated groups has a potential health risk both non-carcinogenic and carcinogenic risks	Bang et al. (2009)
Pakistan	• As concentration in groundwater (tube well, electric pumps, and hand pumps) • Health risk assessment through oral and dermal exposures (using the reference value from USEPA and previous studied)	None	• As level depends on depth High As contaminated groundwater resulting in residents in the area are at high risk both non-carcinogenic and carcinogenic risks Oral exposure is the main route • This study is recommended for investigating food and feed crops as the other As sources	Waqas et al. (2017)
Pakistan	• As concentration in groundwater • Health risk assessment via drinking water (using the reference value of USEPA and previous studied)	None	• As concentration is the value of main factor in increasing the health risk • High As contaminated groundwater (>10 μg/L) resulting in residents in the area being at high risk of both non-carcinogenic and carcinogenic risks	Shahid et al. (2018)

(continued)

Table 10.2 (continued)

Countries	Estimation of health risk assessment	Exposure assessment using biomarker	Remark	References
Pakistan	• Total As, As^{III}, As^{V}, MMA, and DMA in drinking water • Health risk assessment via oral exposure (using reference value from USEPA)	• Total As, As^{III}, As^{V}, MMA, and DMA in hair, nail, urine plasma and red blood cells	• The questionnaire survey was used to collect information on age, body weight, drinking water sources, containers, pretreatment, and habits • Health information of the relevant As effects such as cancer, skin diseases, kidney diseases, gastroenteritis, diabetes, developmental abnormalities, cardiovascular diseases, and liver failures were observed from the residents The As level in the biomarker is related to the As level in groundwater • Total As is highest in all biomarkers compared to the other As species • Hair was recommended for use as an indicator of accumulation of As exposure	Rehman et al. (2020)
Thailand	• As concentration in groundwater • Health risk assessment via oral and dermal exposures (using reference value from the other study)	None	This study considered the additional documents from the relevant organization as follows: • Sources of drinking water and their As level (deep and deep wells, rainwater, bottled water, and water provided by Provincial Waterworks Authority) • As level in vegetable and fruit to explain the health risk assessment	Zhang et al. (2001)

(continued)

Table 10.2 (continued)

Countries	Estimation of health risk assessment	Exposure assessment using biomarker	Remark	References
Thailand	• As concentration and other heavy metal in groundwater • Health risk assessment via oral exposures (using actual value in the study area)	Hair, nail and urine	• A high water intake rate was presented with an average of 4.21 L/d because the occupation participants were farmers who had high exposure to sunlight. Furthermore, Thailand has a tropical climate that results in high water consumption The As level in groundwater has a positive correlation with the As level in urine, while the other biomarkers do not have a relationship because it is not the main route of elimination	Wongsasuluk et al. (2021b)
Thailand	• As concentration in groundwater • Health risk assessment via oral exposures (using actual value in the study area)	Urine	• The person who was exposed to high As in groundwater (>10 μg/L) was found to have a high presence of As in urine • The intake rate is 2.4 L/d • Behavior of groundwater consumption socio-demographic and health information were interviewed Oral exposure is the main route	Nilkarnjanakul et al. (2023)
India	• As concentration and other heavy metals in groundwater • Health risk assessment via drinking water (the reference value of USEPA)	None	• High As level presence shallow well compared to the deep well Drinking groundwater in the area can pose a potential health risk for As	Ravindra and Mor (2019)

(continued)

Table 10.2 (continued)

Countries	Estimation of health risk assessment	Exposure assessment using biomarker	Remark	References
China	• As concentration and other heavy metals in groundwater in shallow and deep groundwater • Health risk assessment via drinking water (the reference value of USEPA)	None	• As enrichment in shallow groundwater compared to deep water	Qiao et al. (2020)
Probabilistic risk assessment				
Cambodia	• As concentration in groundwater • Health risk assessment via oral and dermal exposures (using the reference value of USEPA and the other studies)	None	• The most influential parameters are exposure duration (ED) and concentration factor Oral exposure is the main route	Sheng et al. (2021)
Bangladesh	• As concentration in groundwater • Health risk assessment via oral and dermal exposures between children and adult groups (using the reference value of USEPA and the other studies)	None	• The most influential parameters are exposure duration and As level in groundwater, respectively • The adult group poses a high risk compared to the children Oral exposure is the main route	Saha and Rahman (2020)

(continued)

Table 10.2 (continued)

Countries	Estimation of health risk assessment	Exposure assessment using biomarker	Remark	References
China	• As concentration in groundwater • Health risk assessment via drinking water in different age groups (the reference value of USEPA and previous studies)	None	• The most influential parameters are As level in groundwater, ingestion rate (IR), and body weight (BW), respectively • Children group pose a high risk compared to teenagers and adults	Wang et al. (2020)
India	• As concentration in groundwater and raw rice grain • Health risk assessment via oral exposures among differently aged populations (using the actual value in the study area)	Hair, nail, and urine	• The most influential parameters are As level in groundwater and rice grain, ED and BW • As concentration is related among groundwater, rice grain, and biological samples • Different aged groups affect the As level in biomarker due to the time of exposure and exposure frequency • Nail is appropriate biomarker to determine As toxicity	Das et al. (2022)

risk prediction and the existing level of As in biomarkers helps understand the link between the source-pathway-receptor. On the basis of As toxicity, inorganic forms such as As^{V}, As^{III} are high toxic compared to organic forms such as monomethyl arsenic (MMA) and dimethyl arsenic (DMA). Meanwhile, pentavalence is less toxic compared to trivalence. The rate of absorption and elimination is important when As enters the body. The primary target organ is the gastrointestinal tract, which can be absorbed as a compound especially as trivalent species up to 95% of the total concentration (ATSDR, 2007). Then, the other remaining concentration is distributed through the blood system, and metabolism occurs. The metabolism of As depends on the methylation with AS3MT. It is an important process for detoxification from inorganic As species, to organic As species which usually occurs in the liver organ. Moreover, the reaction process also changes the species from the pentavalent compound to the trivalent compound As.

Subsequently, a high proportion of As is typically excreted in urine, which is the primary route of As elimination. This is followed by excretion through feces, hair, nails, skin desquamation, and sweat. The half-life of As is approximately 10 h. In case of low As concentration, it can be excreted via urine for a few days. Contrastingly, As can accumulate in various tissues when the capacity for methylation is exceeded. Mostly of As can bind to the keratin and sulfhydryl groups in tissues such as hair, nails, skin, teeth, and bone. Therefore, several of these biomarkers are used to measure As concentration and their speciation which are available in the human body, as presented in Table 10.2. Phan et al. (2011) suggested that hair is a representative exposure 2–5 months before a haircut. While nail samples such as fingernails and toenails reflect the recent intake of As for 6–12 months and 12–18 months before nail clipping. In case of blood and urine, the presence of As demonstrates around 6 h and 2–6 days from the past exposure, respectively (Pullella & Kotsopoulos, 2020). Additionally, understanding the pathway of exposure to As is necessary to select the appropriate biomarkers. The ingestion route can be absorbed into the gastrointestinal tract and deposited in blood and other organs, while dermal contact directly affected to their skin and most accumulated in hair, nails, and epidermis.

Hair Sample

As level in groundwater and hair samples present a strong correlation (Kim et al., 2011; Phan et al., 2013). It can be confirmed by the report of Cleland et al. (2009) who found a high As concentration in the hair of exposure group compared to the control group. Hair is used to observe As deposition for long period exposure. Bang et al. (2009) demonstrated the As concentration in the hair of the participants who were exposed contaminated drinking groundwater around 8 years which ranging between 0.12 and 1.09 mg/kg while the control group was below 0.09 mg/kg. The variation depends on concentration, water consumption rate, and exposure time. Additionally, gender is an important factor that can be explained by personal characteristics. The male has low As levels in their hair because they have a high frequency of haircuts compared with females. Hence, male hair was an inadequate biomarker to evaluate As exposure. In response to the report of Hanh et al. (2011) who confirmed that drinking groundwater contaminated with As can result in elevated As level in hair. Drinking water that has As above 10 μg/L in groundwater can be found As in hair ranging from 0.12 to 37.7 μg/g. However, the As level could decrease with groundwater consumption.

Urine Sample

Nilkarnjanakul et al. (2023) observed total As in urine of the two participant groups, including high As exposure via groundwater (>10 μg/L) and low As exposure (<10 μg/L). Most of the participants in the high exposure group had a high As in the urine higher than 50 μg/L which is the normal value of the National Health and Nutrition Examination Survey. However, some participants of low exposure group had quite high As concentrations in the urine, which might be because of receiving As from the other source. Meanwhile, species are determined in many studies to identify the concentration, toxic, and probably source. Loffredo et al. (2003) have been evaluating the concentration of As among the different species in urine. The result presents that the DMA compound has the highest concentration compared to other species, which is similar to the report of ATSDR (2007). In addition to species, the different ethnic populations (China, Chile, and Mexico) are related to different As concentration. This might be because of the arsenic methylation gene. It can be concluded that the variation of As concentration and its species in the human body is likely to come from several factors. Therefore, the information from interviews or questionnaire surveys has a benefit in understanding the relationship between source and exposure.

Using Various Biological Indicators

Rehman et al. (2020) have reported the relationship between As concentration in groundwater and biomarker such as hair, nail, urine, and blood. Similar to the study of Wongsasuluk et al. (2021b), who illustrated that concentration and groundwater consumption are the main factors in improving As in urine, but hair and fingernails are not correlation. It can be explained by the biologic mechanism.

Hair and nail samples represent the accumulation of long-term exposure. On the other hand, urine is suitable to identify As exposure as a recent contact. Meanwhile, hair might interfere with its biology and growth rate compared to nail sampling. Therefore, fingernails are also recommended for exposure to observation for As since it has low affecting from external contamination and are easier to draw.

10.1.3 Factors Affecting as in Human Health

Variable inputs are mainly used the reference value from USEPA for health risk assessment. On the other hand, people usually do many activities in daily life, the primary data as an information should be a survey in each study area. The different characteristics of the population and the various age groups should be considered to understand the exposure and those receptors. It can use the idiom 'One size *doesn't* fit all" to explain the influencing factors as follows:

10.1.3.1 Exposure Factors

Concentration and Intake Rate

Compared between regions, the intake rate is importance for the assessment of exposure. Low intake rates might be low risk compared to the person with a high rate. The average intake rate for the adult population is 2 L/d according to USEPA. Contrastingly, the countries where located in the tropical zone generally intake more drinking water than the reference value USEPA (Nilkarnjanakul et al., 2023). Ohno et al. (2007) revealed that the average intake rate of adult people in Bangladesh is 3 L/d with a maximum of 5.7 L/d. Intake rates between different genders were not correlated. The high consumption rate was related to temperature, particularly hot season. According to the report of Biswas et al. (2014) who shows the relationship between a high As level in urinary and high consumption rate. Seasonal variation can be directly affected to the intake rate, particularly dry season. Furthermore, this study suggested that agricultural countries located in tropical areas should use an actual water intake rate to protect the result of underestimation from health risk assessment.

Occupation

In addition to drinking groundwater, high exposure through dermal contact might have occurred in the contaminated area or the person who directly works on As-related activities. In addition, arsenic exposed to working might indirectly affect that can be explained by occupation. Remarkably, outdoor activities such as farmers who have a long time to the sunlight can lead to high average daily water consumption (Wongsasuluk et al., 2021b). Furthermore, other water supplies should be observed because they could expose the drinking water varied to their workplace. For example, male residents in Vietnam usually spent more time working outside the village resulting in low exposure consumption from the residence (Bang et al., 2009). Other water sources include rainwater, bottled water, water from vending machines, and tap water. Furthermore, individuals working in gold mines face a high risk of As exposure through dermal contact with contaminated soil (Wongsasuluk et al., 2021a).

Food and Beverage Consumption

The other major As source might be rice consumption (Mondal et al., 2008; Bhowmick et al., 2018; Das et al., 2022). As in rice might be adsorb by soil which cause from the long term of fertilizer using. In some areas, arsenic contamination in irrigation and soil have been found to be a source of elevated As in the plant. Therefore, rice consumption information should be interviewed as another As source when As concentration from biological indicators. Considering the exposure of As concentration from groundwater consumption and biomarker should be concerned about the arsenic species. IAs commonly found as dominant species in groundwater, rice products, and seaweed. Whereas other food consumptions, particularly fish and seafood can cause arsenobetaine and arsenocholine to be enrichment as the organoarsenic species (Bhowmick et al., 2018). Considering As species in the biological indicator is recommended to reduce the overestimation. The actual concentration of As in each species is helpful in avoiding misinterpretation and understand the As toxicity (Loffredo et al., 2003). In some cases of urine samples, participants should avoid these foods before the planned collection around 48 h (ATSDR, 2007). Meanwhile, check again when taking a urine sample to prevent the increasing As level from other sources. In addition to food and beverage consumption, herbal medicines should be considered. Since traditional medicines commonly made from plants such as yahom, turmeric, and *Andrographis paniculata*, some of them might be contaminated during the cultivation process according to the study of Vattanasit et al. (2020).

Household Treatment Method

Although residents have a common treatment system with sand filtration, the As concentration was still extremely high As concentration (Bang et al., 2009; Nilkarnjanakul et al., 2023). The effective treatment method should be minimized As lower than the drinking water guideline of 10 μg/L.

10.1.3.2 Individual Factors

Epidemiological evidence of As exposure is investigated in many reports (Wongsasuluk et al., 2018; Demissie et al., 2024). In addition to exposure factors, the toxicity depends on variations in As metabolism in the human body should be considered because individual factors are also complex mechanisms. Arsenic methylation is genetically linked (Vahter & Concha, 2001). In addition, signs and symptoms with associated As poisoning are recommended for interview during sample collection. In addition, the association among As in drinking water source, dietary, As speciation in the human biomarker and health information (Shokoohi et al., 2022).

10.1.4 Conclusion and Recommendations

Health risk assessment is an important tool commonly used for predicting the magnitude of exposure. Probabilistic assessment can provide the possible risk level among the different percentile of population exposure which depends on the sample size and input data. While deterministic estimation is represented as a single estimate of exposure that is simple to use and understand. Both approaches are recommended for health risk assessment depending on the objective and the limitation of informational resources. The result from risk calculation varies by the assumed value during the study time. Primary data of exposure assessment is recommended to use for a good representative due to the different population groups and contexts of the area, especially intake rate and exposure duration. Meanwhile the actual exposure to As can be known by observing the biological samples such as urine, hair, and nails. However, the toxicities are varied according to the individual factors, such as the pathway of exposure, metabolism, As speciation, and concentration. Additionally, other As sources such as rice products, seafood, seaweed, and other water sources, can enhance As levels and their different species in biomarkers. Therefore, it can be concluded that the scientific evidence of a link between exposure assessment and their relevant influencing factors is necessary. Future challenges and solutions for health risk management of As-contaminated groundwater via drinking exposure can include:

How to confirm that the treatment process before drinking can reduce or eliminate arsenic?
How to minimize and communicate health risks for different population characteristics which suffer from As toxicity particularly, susceptible groups such as children, the elderly, and those highly exposed?
How to monitor signs and symptoms and share any health information of those population in the area among the relevant organizations?

Acknowledgements The authors gratefully acknowledge the Thailand Science Research and Innovation Fund Chulalongkorn University and Department of Geology, Faculty of Science, Chulalongkorn University for their financial support and laboratory services.

References

Agency for Toxic Substances and Disease Registry (ATSDR). (2007). Toxicological profile for arsenic. Available from: https://www.atsdr.cdc.gov/toxprofiles/tp2.pdf

Bang, S., Viet, P. H., & Kim, K. W. (2009). Contamination of groundwater and risk assessment for arsenic exposure in Ha Nam province, Vietnam. *Environment International, 35*(3), 466–472.

Bhowmick, S., Pramanik, S., Singh, P., Mondal, P., Chatterjee, D., & Nriagu, J. (2018). Arsenic in groundwater of West Bengal, India: A review of human health risks and assessment of possible intervention options. *Science of the Total Environment, 612*, 148–169.

Biswas, A., Deb, D., Ghose, A., Santra, S. C., & Mazumder, D. N. G. (2014). Seasonal perspective of dietary arsenic consumption and urine arsenic in an endemic population. *Environmental Monitoring and Assessment, 186*, 4543–4551.

Boonkaewwan, S., Sonthiphand, P., & Chotpantarat, S. (2021). Mechanisms of arsenic contamination associated with hydrochemical characteristics in coastal alluvial aquifers using multivariate statistical technique and hydrogeochemical modeling: A case study in Rayong province, Eastern Thailand. *Environmental Geochemistry and Health, 43*, 537–566.

Boonkaewwan, S., & Chotpantarat, S. (2024). Impact of ionic strength on goethite colloids co-transported with arsenite (As^{3+}) through a saturated sand column under anoxic condition: Experiment and mathematical modeling. *Environmental Research, 260*, 119660.

Boonsrang, A., Chotpantarat, S., & Sutthirat, C. (2018). Factors controlling the release of metals and a metalloid from the tailings of a gold mine in Thailand. *Geochemistry: Exploration, Environment, Analysis, 18*, 109–119.

Charuseiam, Y., Chotpantarat, S., & Sutthirat, C. (2022). Acid mine drainage potential of waste rocks in a gold mine (Thailand): Application of a weathering cell test and multivariate statistical analysis. *Environmental Geochemistry and Health, 44*, 1049–1079.

Chotpantarat, S., Parkchai, T., & Wisitthammasri, W. (2020). Multivariate statistical analysis of hydrochemical data and stable isotopes of groundwater contaminated with nitrate at Huay Sai Royal Development Study Center and adjacent areas in Phetchaburi Province, Thailand. *Water, 12*(4), 112.

Chotpantarat, S., & Thamrongsrisakul, J. (2021). Natural and anthropogenic factors influencing hydrochemical characteristics and heavy metals in groundwater surrounding a gold mine, Thailand. *Journal of Asian Earth Sciences, 211*, 104692.

Cleland, B., Tsuchiya, A., Kalman, D. A., Dills, R., Burbacher, T. M., White, J. W., Faustman, E. M., & Mariën, K. (2009). Arsenic exposure within the Korean community (United States) based on dietary behavior and arsenic levels in hair, urine, air, and water. *Environmental Health Perspectives, 117*(4), 632–638.

Das, A., Joardar, M., Chowdhury, N. R., Mridha, D., De, A., Majumder, S., Das, J., Majumdar, K. K., & Roychowdhury, T. (2023). Significance of the prime factors regulating arsenic toxicity and associated health risk: A hypothesis-based investigation in a critically exposed population of West Bengal, India. *Environ Geochem Health, 45*(6), 3423–3446.

Demissie, S., Mekonen, S., Awoke, T., Teshome, B., Mengistie, B. (2024). Examining carcinogenic and noncarcinogenic health risks related to arsenic exposure in Ethiopia: A longitudinal study. *Toxicology Reports*

Hanh, H. T., Kim, K. W., Bang, S., & Hoa, N. M. (2011). Community exposure to arsenic in the Mekong river delta, Southern Vietnam. *Environmental Monitoring Assess, 13*(7), 2025–2032.

Houessionon, M. K., Ouendo, E. M. D., Bouland, C., Takyi, S. A., Kedote, N. M., Fayomi, B., Fobil, J. N., & Basu, N. (2021). Environmental heavy metal contamination from Electronic Waste (e-waste) recycling activities worldwide: A systematic review from 2005 to 2017. *IJERPH, 18*(7), 3517.

Integrated Risk Information System (IRIS). (2013). *IRIS Assessment*. IRIS of USEPA. Available from: https://cfpub.epa.gov/ncea/iris2/atoz.cfm

Kim, K. W., Chanpiwat, P., Hanh, H. T., Phan, K., & Sthiannopkao, S. (2011). Arsenic geochemistry of groundwater in Southeast Asia. *Frontiers in Medicine, 5*, 420–433.

Kumar, P., Kumar, M., Ramanathan, A. L., & Tsujimura, M. (2010). Tracing the factors responsible for arsenic enrichment in groundwater of the middle Gangetic Plain, India: A source identification perspective. *Environmental Geochemistry and Health, 32*, 129–146.

Loffredo, C. A., Aposhian, H. V., Cebrian, M. E., Yamauchi, H., & Silbergeld, E. K. (2003). Variability in human metabolism of arsenic. *Environmental Research, 92*(2), 85–91.

Mazumder, D. N., Majumdar, K. K., Santra, S. C., Kol, H., & Vicheth, C. (2009). Occurrence of arsenicosis in a rural village of Cambodia. *Journal of Environmental Science and Health. Part A, Toxic/hazardous Substances & Environmental Engineering, 44*(5), 480–487.

Mondal, D., & Polya, D. A. (2008). Rice is a major exposure route for arsenic in Chakdaha block, Nadia district, West Bengal, India: A probabilistic risk assessment. *Applied Geochemistry, 23*(11), 2987–2998.

Nilkarnjanakul, W., Watchalayann, P., & Chotpantarat, S. (2022). Spatial distribution and health risk assessment of As and Pb contamination in the groundwater of Rayong Province, Thailand. *Environmental Research, 204*, 11183.

Nilkarnjanakul, W., Watchalayann, P., & Chotpantarat, S. (2023). Urinary arsenic and health risk of the residents association in contaminated-groundwater area of the urbanized coastal aquifer, Thailand. *Chemosphere, 313*, 137313.

Ohno, K., Yanase, T., Matsuo, Y., Kimura, T., Rahman, M. H., Magara, Y., & Matsui, Y. (2007). Arsenic intake via water and food by a population living in an arsenic-affected area of Bangladesh. *Science of the Total Environment, 381*(1–3), 68–76.

Phan, K., Sthiannopkao, S., & Kim, K. W. (2011). Surveillance on chronic arsenic exposure in the Mekong River basin of Cambodia using different biomarkers. *International Journal of Hygiene and Environmental Health, 215*(1), 51–58.

Phan, K., Phan, S., Huoy, L., Suy, B., Wong, M. H., Hashim, J. H., Yasin, M. S. M., Aljunid, S. M., Sthiannopkao, S., & Kim, K. W. (2013). Assessing mixed trace elements in groundwater and their health risk of residents living in the Mekong River basin of Cambodia. *Environmental Pollution, 182*, 111–119.

Pullella, K., & Kotsopoulos, J. (2020). Arsenic exposure and breast cancer risk: A re-evaluation of the literature. *Nutrients, 12*(11), 3305.

Qiao, J., Zhu, Y., Jia, X., Niu, X., & Liu, J. (2020). Distributions of arsenic and other heavy metals, and health risk assessments for groundwater in the Guanzhong Plain region of China. *Environmental Research, 181*, 108957.

Rahman, M. M., Sengupta, M. K., Ahamed, S., Lodh, D., Das, B., Hossain, M. A., Nayak, B., Mukherjee, S. C., Pati, S., Saha, K. C., Palit, S. K., Kaies, I., Barau, A. K., & Asad, K. A. (2005). Murshidabad—one of the nine groundwater arsenic affected districts of West Bengal India. Part I: Magnitude of contamination and population at risk. *Clinical Toxicology, 43*, 823–834.

Ratnaike, R. N. (2003). Acute and chronic arsenic toxicity. *Postgraduate Medical Journal, 79*(933), 391–396.

Ravindra, K., & Mor, S. (2019). Distribution and health risk assessment of arsenic and selected heavy metals in Groundwater of Chandigarh, India. *Environmental Pollution, 250*, 820–830.

Reddy, Y. S., & Sunitha, V. (2023). Assessment of heavy metal pollution and its health implications in groundwater for drinking purpose around inactive mines, SW region of Cuddapah Basin, South India. *Total Environment Research Themes, 8*, 100069.

Rehman, U. U., Khan, S., & Muhammad, S. (2020). Ingestion of arsenic-contaminated drinking water leads to health risk and traces in human biomarkers (hair, nails, blood, and urine), Pakistan. *Exposure and Health, 12*, 243–254.

Saha, N., & Rahman, M. S. (2020). Groundwater hydrogeochemistry and probabilistic health risk assessment through exposure to arsenic-contaminated groundwater of Meghna floodplain, central-east Bangladesh. *Ecotoxicology and Environmental Safety, 206*, 111349.

Sumdang, N., Chotpantarat, S., Cho, K. H., & Thanh, N. N. (2023). The risk assessment of arsenic contamination in the urbanized coastal aquifer of Rayong groundwater basin, Thailand using the machine learning approach. *Ecotoxicology and Environmental Safety, 253*, 114665.

Shahid, M., Khalid, M., Dumat, C., Khalid, S., Niazi, N. K., Imran, M., Bibi, I., Ahmad, I., Hammad, H. M., & Tabassum, R. A. (2018). Arsenic level and risk assessment of groundwater in Vehari, Punjab Province, Pakistan. *Exposure and Health, 10*, 229–239

Sheng, D., Wen, X., Wu, J., Wu, M., Yu, H., & Zhang, C. (2021). Comprehensive probabilistic health risk assessment for exposure to arsenic and cadmium in groundwater. *Environmental Management, 67*, 779–792.

Shokoohi, R., Khazaei, M., Karami, M., Seid-Mohammadi, A., Khazaei, S., & Torkshavand, Z. (2022). Application of fingernail samples as a biomarker for human exposure to arsenic-contaminated drinking waters. *Science and Reports, 12*(1), 4733.

Slotnick, M. J., & Nriagu, J. O. (2006). Validity of human nails as a biomarker of arsenic and selenium exposure: A review. *Environmental Research, 102*(1), 125–139.

Smedley, P. L., & Kinniburgh, D. G. (2002). A review of the source, behaviour and distribution of arsenic in natural waters. *Applied Geochemistry, 17*(5), 517–568.

Smedley, P. L., & Kinniburgh, D. G. (2012). Arsenic in groundwater and the environment. *Essentials of medical geology* (pp. 279–310). Springer.

Sohrabi, N., Kalantari, N., Amiri, V., Saha, N., Berndtsson, R., Bhattacharya, P., & Ahmad, A. (2021). A probabilistic-deterministic analysis of human health risk related to the exposure to potentially toxic elements in groundwater of Urmia coastal aquifer (NW of Iran) with a special focus on arsenic speciation and temporal variation. *Stochastic Environmental Research and Risk Assessment, 35*, 1509–1528.

Tiankao, W., & Chotpantarat, S. (2018). Risk assessment of arsenic from contaminated soils to shallow groundwater in Ong Phra Sub-District, Suphan Buri Province, Thailand. *Journal of Hydrology Regional Studies, 19*, 80–96.

Thanh, N. N., Chotpantarat, S., Trung, N. H., & Ngu, N. H. (2022). Mapping groundwater potential zones in Kanchanaburi Province, Thailand by integrating of analytic hierarchy process, frequency ratio, and random forest. *Ecological Indicators, 145*, 109591.

Thanh, N. N., Chotpantarat, S., Ha, N. T., & Trung, N. H. (2023). Determination of conditioning factors for mapping nickel contamination susceptibility in groundwater in Kanchanaburi Province, Thailand, using random forest and maximum entropy. *Environmental Geochemistry and Health, 45*, 4583–4602.

Thanh, N. N., Chotpantarat, S., Ngu, N. H., Thunyawatcharakul, P., & Kaewdum, N. (2024). Integrating machine learning models with cross-validation and bootstrapping for evaluating groundwater quality in Kanchanaburi province, Thailand. *Environmental Research, 252*, 118952.

United States Environmental Protection Agency (USEPA). (2004). *Risk assessment guidance for superfund Volume I: Human health evaluation manual (Part E, Supplemental Guidance for Dermal Risk Assessment)*. EPA/540/R/99/005, OSWER 9285.7-02EP PB99-963312 (Washington, DC, USA).

United States Environmental Protection Agency (USEPA). (1997). *Guiding principles for Monte Carlo Analysis*. EPA/630/R-97/001 (Washington, DC, USA).

Vahter, M., & Concha, G. (2001). Role of metabolism in arsenic toxicity. *Pharmacology and Toxicology, 89*, 1–5.

Vattanasit, U., Nakkong, K., & Kampeerakup, R. (2020). Cancer and non-cancer health risk assessment of arsenic and cadmium exposure from traditional medicine consumption. *Phranakhon Rajabhat Research Journal (Science and Technology), 15*, 166–183. Available from: https://li01.tci-thaijo.org/index.php/PRRJ_Scitech/article/view/190874/168869

Wang, Y., Zhu, G., Engel, B., & Wu, Y. (2020). Probabilistic human health risk assessment of arsenic under uncertainty in drinking water sources in Jiangsu Province, China. *Environmental Geochemistry and Health, 42*, 2023–2037.

Waqas, H., Shan, A., Khan, Y. G., Nawaz, R., Rizwan, M., Rehman, S. U., Shakoor, M. B., Ahmed, W., & Jabeen, M. (2017). Human health risk assessment of arsenic in groundwater aquifers of Lahore, Pakistan. *Human and Ecological Risk Assessment, 23*(4), 836–850.

Wisitthammasri, W., Chotpantarat, S., & Thitimakorn, T. (2020). Multivariate statistical analysis of the hydrochemical characteristics of a volcano sedimentary aquifer in Saraburi Province, Thailand. *Journal of Hydrology: Regional Studies, 32*, 100745.

Wongsasuluk, P., Chotpantarat, S., Siriwong, W., & Robson, M. (2018). Using hair and fingernails in binary logistic regression for bio-monitoring of heavy metals/metalloid in groundwater in intensively agricultural areas, Thailand. *Environmental Research, 162*, 106–118.

Wongsasuluk, P., Tun, A. Z., Chotpantarat, S., & Siriwong, W. (2021a). Related health risk assessment of exposure to arsenic and some heavy metals in gold mines in Banmauk Township, Myanmar. *Scientific Reports, 11*(1), 22843.

Wongsasuluk, P., Chotpantarat, S., Siriwong, W., & Robson, M. (2021b). Human biomarkers associated with low concentrations of arsenic (As) and lead (Pb) in groundwater in agricultural areas of Thailand. *Science and Reports, 11*(1), 13896.

Zhang, J., Chen, X., Parkpian, P., Tabucanon, M. S., & Mongkolsuk, S. (2001). GIS application on arsenic contamination and its risk assessment in Ronphibun, Nakhorn Si Thammarat, Thailand. *Geographic Information Sciences, 7*, 69–78.

Zhou, Y., Niu, L., Liu, K., Yin, S., & Liu, W. (2018). Arsenic in agricultural soils across China: Distribution pattern, accumulation trend, influencing factors, and risk assessment. *Science of the Total Environment, 616*, 156–163.

Chapter 11
A Comprehensive Review of Conventional Methods and Some Novel Materials for Removing Arsenic from Contaminated Water

Rupkamal Chetia, Shreemoyee Bordoloi, and Surajit Konwer

Abstract Arsenic is considered as one of the most carcinogenic water pollutants found in the atmosphere, soils, rocks, natural waters and organisms. Presence of arsenic even in low concentration in drinking water is considered as carcinogen because consume of arsenic for a long term can cause skin, lung, bladder, liver and kidney cancer. This review article focuses on the background history of arsenic occurrences in the environment and examines both established and advanced removal techniques for its removal, including oxidation, coagulation/precipitation, adsorption, membrane processes, ion exchange and other techniques. In addition, a brief account of the application of various cutting-edge materials for arsenic removal, such as conducting polymers, and metal organic frameworks/quantum dots. It also discusses several promising nano-adsorbents such as iron oxide/hydroxide nanoparticles, titanium dioxide, zinc oxide nanoparticles, carbon nanotubes etc. highlighting their pros and cons of each in terms of arsenic removal. The discussion also covers emerging functional materials like graphene, metal organic frameworks, zeolites and other such impending materials. The article also highlights the need for arsenic removal technology that is simple, user-friendly, cost-effective, low-maintenance, highly efficient in the field, and accessible to low-income communities.

Keywords Arsenic · Adsorption · Metal organic framework · Quantum dots · Water purification · Zero valent iron

R. Chetia (✉) · S. Konwer
Department of Chemistry, Dibrugarh University, Dibrugarh, India
e-mail: rupkamalchetia101@gmail.com

S. Konwer
e-mail: surajitkonwer@dibru.ac.in

S. Bordoloi
Department of Chemistry, MDKG College, Dibrugarh, Assam, India

V. Sunitha et al. (eds.), *Groundwater Resource Management Planning Strategies*, Springer Water, https://doi.org/10.1007/978-3-031-88870-0_11

11.1 Introduction

Arsenic (As) is a heavy metal from group V on the periodic table, with an atomic number of 33. It's found in the air, soil, rocks, water, and living things. Even small amounts of arsenic, if taken for a long time, can cause serious health problems. (Tchounwou, 2012). The use of arsenic-contaminated drinking water is the prime source of arsenic exposure in humans and animals. The presence of inorganic arsenic in drinking water has been recognized as a carcinogen for humans even at relatively low concentrations. Drinking water with arsenic over a long period can lead to serious health issues. These include cancers of the skin, lungs, bladder, liver, and kidneys, along with skin changes like dark spots and thickening. It can also cause anemia, eye irritation, swollen legs, liver scarring, chronic lung disease, tissue death in toes (gangrene), nerve problems, muscle weakness, loss of appetite, and nausea (EFSA, 2009; Mazumder, 2000; Argos, 2010). Arsenic can occur in the environment in several oxidation states (− 3, 0, + 3, and + 5) in both inorganic and organic forms. Trivalent arsenic can exist as arsenous oxide (As_2O_3), arsenious acid ($HAsO_2$), arsenite ($H_2AsO_3^-$, $HAsO_3^{2-}$, AsO_3^{3-}) ions, arsenic trichloride ($AsCl_3$), arsenic sulfide (AsS_3) and arsine (AsH_3). Pentavalent arsenic commonly exists as arsenic pentoxide (As_2O_5), ortho arsenic acid (H_3AsO_4), metaarsenic acid ($HAsO_3$), and arsenate ($H_2AsO_4^-$, $HAsO_4^{2-}$, AsO_4^{3-}) ions (Fig. 11.1). It predominantly occurs as inorganic arsenite, As(III), and arsenate, As (V) in drinking water. The inorganic form of arsenic is found to be more toxic than its organic forms such as mono-methyl arsenate (MMA) and dimethyl arsenate (DMA). The inorganic + 3 form is about 60 times more toxic than + 5 state as the + 3 state is more genotoxic, cytotoxic, soluble, and mobile (Khosravi-Darani, 2022; Arcibar-Orozco, 2014; Meez, 2021; Yao, 2014).

Arsenic pollution originates from both natural phenomena and anthropogenic activities. The human body can be affected by arsenic through food, water, soil, and air. Arsenic gets into the environment from both natural and human-made sources (Fig. 11.2). Natural events like volcanic eruptions and rock weathering can release

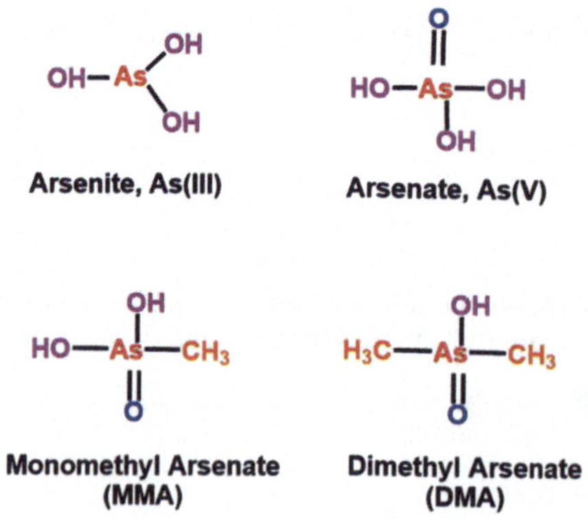

Fig. 11.1 Structure of some arsenic species found in water

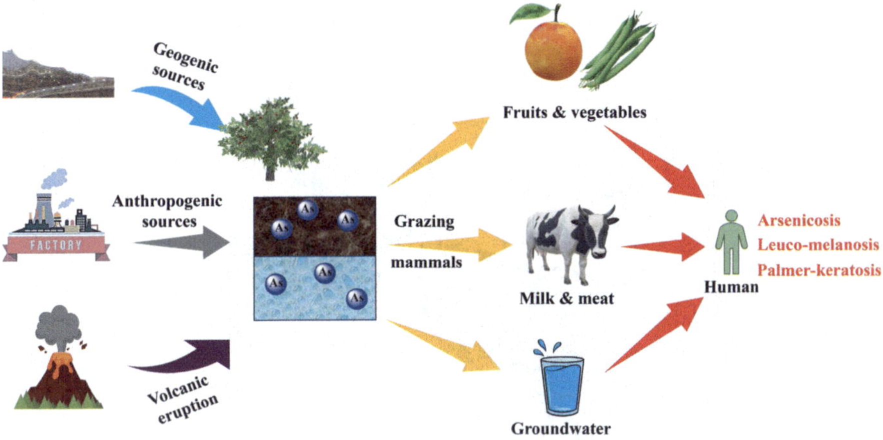

Fig. 11.2 Schematic diagram demonstrating the transfer of arsenic from a different source to the human body causing various diseases

arsenic, while human activities like using arsenic-based herbicides and pesticides, industrial processes, and mining also contribute to its spread (Fierro, 2009). A recent study found that arsenic is the only known substance that can cause cancer when inhaled or ingested (Khosravi-Darani, 2022). Arsenic poisoning has long-term consequences. Various methods such as oxygen therapy, chelation therapy, naturopathy, etc. can be used to diagnose arsenic-poisoned people (Tchounwou, 2004; Abdul, 2015; Mohanty, 2017). However, detoxification of arsenic poisoning is quite complicated to achieve. As a result, prevention is preferable to cure. Arsenic poisoning may therefore be avoided by keeping it from entering the human body by drinking water. Because arsenic is toxic, the World Health Organization (WHO) has set a limit of 0.01 mg/L for arsenic in drinking water (Singh, 2007).

Numerous arsenic-contaminated sites have been reported worldwide, affecting millions of people in countries like India, Bangladesh, the United States, Mexico, Chile, China, Japan, Vietnam, and others, creating health risks for these communities (Fig. 11.3) (Argos, 2010; Meez, 2021; Abdul, 2015). The rising concern regarding the heightened risk of arsenic carcinogenicity in developing countries has impelled increased research efforts in the field of arsenic contamination and its subsequent treatment for removal. The increasing prevalence of arsenic contamination in water sources on a global scale has impelled a concerted effort among scientists and researchers to develop more sophisticated and contextually appropriate solutions to address this burning issue. The ongoing research initiatives are characterized by a commitment to advancing beyond conventional methods and embracing innovative strategies to effectively mitigate arsenic contamination (Alka, 2021).

This comprehensive review aims to explain the effectiveness and limitations of conventional techniques while exploring the potential of novel materials for arsenic removal from contaminated water. By examining the principles, mechanisms, and

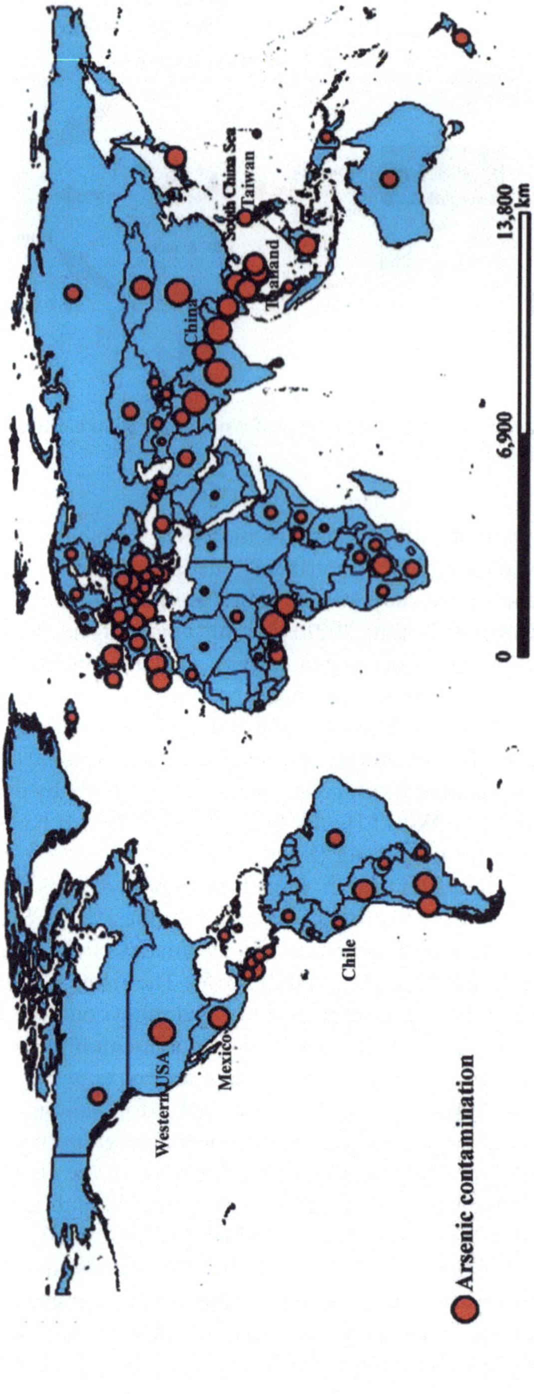

Fig. 11.3 Areas globally affected by arsenic contamination

performance of these approaches, this review seeks to provide insights into optimizing arsenic remediation processes and advancing toward sustainable solutions for ensuring safe drinking water access globally.

11.2 Traditional Methods for Arsenic Remediation

Traditional methods for arsenic remediation include a range of techniques aimed at mitigating arsenic contamination in water sources. These methods operate within the framework of complex chemical interactions influenced by factors such as the oxidation state of arsenic species, pH of the solution, and redox potential (Bhat, 2024). The scientific community has grappled with significant challenges in addressing arsenic contamination, owing to the intricate chemistry involved. Extensive research efforts have been directed towards enhancing the efficacy of arsenic removal strategies, often through field studies to validate their performance in real-world settings.

However, despite advancements, many conventional methods exhibit limited success when applied to real groundwater conditions. To render these methods suitable for rural areas, particularly in low-income regions, several key objectives must be considered.

- First, it's essential to improve arsenic removal methods, especially when other common ions are present in water.
- When choosing arsenic removal methods, it's important to consider factors like cost, technology compatibility, and how well the target community will accept the approach.
- Furthermore, addressing maintenance challenges and ensuring proper handling of generated waste to prevent secondary pollution are essential considerations in the development and implementation of arsenic remediation technologies.

Various techniques are employed for arsenic remediation, including oxidation, coagulation/precipitation, flocculation, membrane technology, ion exchange, and adsorption (Bhat, 2024) (Fig. 11.4). Conventional arsenic removal systems often integrate a combination of these techniques, either sequentially or concurrently, to optimize efficacy. The overarching objective is not only to improve the efficiency of arsenic removal strategies but also to establish sustainable and accessible solutions for communities grappling with arsenic-related water quality concerns.

11.2.1 Oxidation

The primary goal of the oxidation method is to change the highly mobile As(III) into the less mobile As(V), which can then be removed through precipitation. The conversion from As(III) to As(V) is often desirable due to the reduced mobility and increased sorption of As(V) in geological formations. Pre-oxidation of As(III)

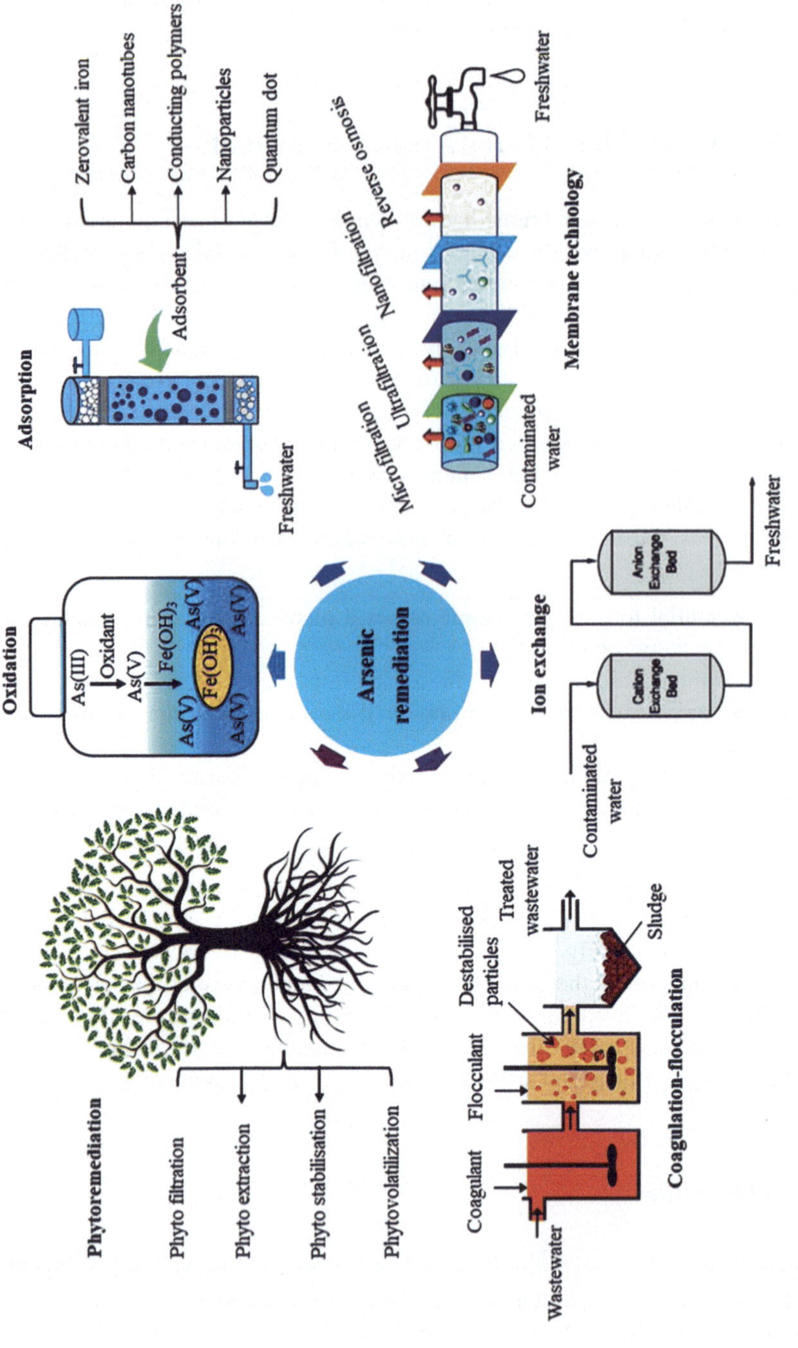

Fig. 11.4 Schematic diagram for various arsenic removal techniques from contaminated water

to As(V) is required in the fact that As(V) is generally more amenable to removal through various treatment methods compared to As(III). Arsenite doesn't stick well to mineral surfaces, but arsenate attaches easily to them. This makes oxidation and precipitation a very effective method for removing arsenic from water. (Nicomel et al., 2016; Ghurye et al., 2004). Chemical oxidants such as chlorine, ozone, chlorine dioxide, chloramine, H_2O_2 (Leupin, 2005) or sand-coated manganese oxide or by other advanced oxidation processes (Yang et al., 2005) like Fenton's reagent, hypochlorite, and permanganate. Chlorine, ozone, permanganate could be preferred over chloramines and H_2O_2 (Bissen, 2003; Lee et al., 2003). According to Kim and Nriagu (2000), the use of air and pure oxygen can result in the oxidation of approximately 54–57% of As(III) to As(V) in contaminated groundwater. On the other hand, complete oxidation of As(III) can be attained by employing ozone as the oxidizing agent (Kim, 2000). Manganese oxides are also recognized for its effective catalytic properties. It facilitates the oxidation of arsenite (As(III)) to arsenate (As(V)) and subsequent scavenging of arsenic from aqueous solutions. In a recent study, Zhang et al. (2018) discussed a metal oxide complex made of Ti(IV) and Mn(IV) with special properties that can both oxidize arsenite and remove it by adsorption at neutral pH. This works effectively with arsenic concentrations ranging from 2 to 60 μg/L (Zhang et al., 2019). Birnessite, a naturally occurring layered manganese oxide mineral, is particularly known for its rapid catalytic activity in promoting the conversion of As(III) to As(V)and subsequent removal. Synthesizing these naturally occurring manganese oxides under laboratory conditions could provide a novel approach to arsenic mitigation is noteworthy. However, this synthesis process under controlled conditions, may involve resource-intensive processes and elevated costs. The development of Cu-doped ZnO through precipitation methods for arsenic oxidation under sunlight introduces a potentially advantageous strategy to mitigate the limitations associated with weak oxidation processes of chlorine and other oxidizing agents (Vaiano et al., 2018). Use of sunlight as an energy source aligns with the principles of green chemistry and may contribute to a more sustainable and efficient arsenic remediation process. Bordoloi et al. (2013) found that oxidation coagulation is effective when the pH is optimized and when different doses of oxidizing agents like potassium permanganate and coagulating agents like ferric chloride are used (Bordoloi et al., 2013). Arsenic removal to levels below 1 μg/L can be achieved by sequentially adding oxidizing agents (such as permanganate) and coagulants (like ferric iron or ferric chloride) to either the influent or effluent of a rapid sand filter. This approach allows effective arsenic reduction (Ahmad et al., 2017).

11.2.1.1 Photochemical Oxidation

The oxidation rate of arsenite [As(III)] in water can be increased using UV light in the presence of oxygen. When UV or sunlight interacts with specific elements, especially Fe(III), it creates hydroxyl radicals, which speed up the oxidation of arsenic. Adding Fe(III) to arsenic-contaminated water and then exposing it to UV or sunlight improves arsenic removal (Emett, 2001). The researchers found that ferric hydroxide and chloride compounds absorb light and produce highly reactive hydroxyl and dichloro radicals. These radicals convert As(III) to As(V). An As(III) solution with 0.06–5 mg/L Fe(II) and Fe(III), when exposed to UV-A light, removed over 90% of 500 μg/L of total arsenic within 2–3 h (Hug et al., 2001). The utilization of solar light for arsenic removal from natural water has been explored in studies by Lara et al. (2006). The combined addition of iron and citrate to natural water is implicated in the removal of arsenic under solar light exposure. The specific roles of iron and citrate are likely associated with their ability to participate in photochemical reactions that contribute to arsenic oxidation. Arsenic removal by photochemical oxidation presents a promising approach for water treatment, especially in areas where conventional methods may be limited. However, considerations such as cost-effectiveness, scalability, and adaptability to specific environmental conditions must be carefully evaluated for practical implementation (Yoon & Lee, 2005).

11.2.1.2 Biological Oxidation

This new treatment method uses natural biological processes, letting specific plants and microorganisms help clean up metals from soil and groundwater. Katsoyiannis and Zouboulis (2004) found that microorganisms Gallionella ferruginea and Leptothrix ochracea could biologically oxidize iron (Katsoyiannis et al., 2004). They investigated how quickly bacteria could oxidize As(III) and then remove As(V) by absorbing it onto naturally formed manganese oxides. Iron-oxidizing bacteria remove arsenic more effectively than manganese-oxidizing bacteria. The oxidation of iron, manganese, and arsenic by these bacteria is faster than traditional chemical oxidation methods (Saalfield & Bostick, 2009; Gupta et al., 2009).

11.2.1.3 Oxidation of Co-existing Iron and Subsequent Removal of As

Naturally occurring iron in groundwater can play a pivotal role in oxidizing arsenic and its subsequent removal from contaminated groundwater. The efficient adsorption and sequestration of arsenic onto precipitates of iron hydroxide [$Fe(OH)_3$] represent a robust mechanism for arsenic removal. In the context of tube wells with higher iron content in groundwater, the combination of aeration and sedimentation proves to be an effective strategy for arsenic removal (Dahi & Liang, 1998).

11.3 Coagulation/Precipitation

Coagulation/precipitation is the oldest and a widely practiced arsenic remediation technique. Coagulants alter the surface charge of solids, allowing particles to cluster together into a flocculated precipitate. This is the most effective method for removing arsenic. In this process, As(V) adsorbs onto the coagulated flocs, which can then be filtered out (Song et al., 2006). The insoluble coagulating products containing arsenic are simply separated by using gravity filtration. Pre oxidation is required as an additional step if arsenic exists in the form of arsenite As (III) before coagulation (Bordoloi et al., 2013). Coagulants are quickly mixed into contaminated water to form tiny hydroxide-based flocs, which encourage particles to clump together. This process allows negatively charged ions like arsenite to bind to the flocs, making it easier to separate them from the water. Both alum and ferric salts are effective at removing As(V), but to get the best results, As(III) needs to be oxidized first (Andrianisa et al., 2008; Baskan & Pala, 2010). Enhanced coagulation is a more effective version of the standard coagulation process. It involves changes like adding more coagulant, lowering the pH, or doing both to improve its performance (Lacasa et al., 2011).

11.3.1 Lime Softening

Hardness in water can be reduced with lime softening, which involves adding lime to alter the carbonate balance. This process converts bicarbonate into carbonate, causing calcium to form calcium carbonate, which can then be removed. Lime softening not only reduces water hardness caused by calcium and magnesium, but also removes As(III) and As(V) within a pH range of 9–9.5 (Fields et al., 2000; Wang et al., 2002). If there's not enough bicarbonate in the water, soda ash (sodium carbonate) is added to help reduce hardness to the desired level. If the softened water's pH is too high (above 9.5), it needs to be neutralized to make it safe to drink. The usual way to adjust the pH in softening plants is to add carbon dioxide, known as recarbonation. The best pH for removing As(V) with lime softening is around 10.5, while for As(III), it's about 11 (Fields et al., 2000).

11.3.2 Microfiltration with Coagulation

Microfiltration is a widely used coagulation method to remove particles, turbidity, and microorganisms from water. It functions similarly to a standard gravity filter. Here are some advantages of microfiltration compared to conventional filtration (Muilenberg, 1997):

- Smaller flocs can be filtered out, reducing the amount of coagulants needed; and
- It allows for increased plant capacity.

Coagulation-assisted microfiltration can lower arsenic levels to under 2 μg/L in water with a pH between 6 and 7, even when the incoming Fe(III) concentration is around 2.5 mg/L. To avoid membrane fouling, it's crucial to ensure all iron and manganese are completely oxidized before reaching the filtration system (Ghurye et al., 2004; Goosen et al., 2005).

11.4 Adsorption

Adsorption is when molecules or particles stick to the surface of a solid or liquid. It's commonly used to clean water by trapping contaminants on the surface of a solid adsorbent material. Adsorption can be either physical or chemical, depending on the type of interactions involved. Because it's straightforward, efficient, and inexpensive, adsorption is widely used to remove arsenic from water. Its effectiveness depends on factors like the initial concentration of arsenic, its oxidation state, temperature, pH, competing ions, and contact time. Here are some common adsorbents used to remove arsenic from contaminated water (Das et al., 2013).

11.4.1 Activated Alumina

Activated alumina is a popular material for removing arsenic from contaminated water. This porous substance is made by heating aluminum hydroxide to high temperatures, turning it into amorphous and gamma alumina oxide. Its effectiveness in removing arsenic comes from its ability to exchange anions with the surface hydroxides on the alumina particles as contaminated water flows through layers of the material (Das et al., 2013).

In practice, contaminated water is continuously circulated through activated alumina packed beds, allowing the contaminant anions, including arsenic species, to adhere to the surface of the alumina particles. Over time, as the adsorption sites on the surface of activated alumina become saturated with arsenic, the bed requires regeneration. Regeneration typically involves a series of steps, beginning with rinsing the bed with a regenerant, such as a strong base, to displace the adsorbed arsenic ions. Subsequently, the bed is flushed with water to remove any remaining regenerant, followed by neutralization with a strong acid, such as sulfuric acid, to restore the bed to its functional state (Camacho et al., 2015).

One notable advantage of activated alumina is its versatility in operating within a wide pH range, typically between 5.5 and 6.0, making it suitable for diverse water sources with varying pH levels. Field studies conducted on activated alumina for arsenic removal have demonstrated promising results, with the technology capable of reducing arsenic concentrations to as low as 3 μg/L. However, these studies have also highlighted the importance of pH adjustments, post-treatment processes, and overall efficiency at the field level in optimizing arsenic removal using activated alumina (Das et al., 2013; Camacho et al., 2015).

Studies by Shi et al. (2015), Karmacharya et al. (2016), Lata and Samadder (2016), and Lee et al. (2017) have provided insights into the use of activated alumina for removing arsenic from contaminated water (Shi et al., 2015; Karmacharya et al., 2016; Lata & Samadder, 2016; Lee et al., 2017). These studies emphasize the need for comprehensive understanding and optimization of operational parameters, including pH adjustment, to maximize the efficiency and effectiveness of activated alumina in real-world applications. Additionally, considerations such as post-treatment methods and overall system performance are crucial for ensuring sustainable and reliable arsenic removal using activated alumina technology.

11.4.2 Iron Based Adsorbents

Iron-based adsorbents have emerged as prominent tools for arsenic remediation, offering efficient and practical solutions for addressing contamination in water sources (Hao et al., 2018). Different types of iron-based materials, such as granular ferric hydroxide, iron-coated sand, and modified iron or iron oxide-based adsorbents, are commonly used to remove arsenic from water. These materials have unique properties and mechanisms that make them effective at adsorbing arsenic. (Dhanasekaran & Sahu, 2021).

Granular ferric hydroxide, introduced by Driehaus et al. in 1998 (Driehaus et al., 1998), has garnered attention for its superior adsorption capabilities, meeting key criteria such as high removal efficiency, operational simplicity, safety, and minimal residual mass. This material operates effectively across a range of water conditions, making it versatile for diverse applications. Additionally, hydrous ferric oxide and goethite are recognized as important iron-based adsorbents due to their proven efficacy in arsenic sorption, as highlighted by Smedley and Kinniburgh in 2002 (Smedley & Kinniburgh, 2002).

Furthermore, advancements in material science have led to the development of highly stable amorphous mesoporous iron oxides, which demonstrate promising efficiency for arsenic removal, as demonstrated by Muruganandham et al. (2010). Additionally, composite materials such as Fe_3O_4-graphene composites exhibit high adsorption capacities for arsenic (V), as reported by Mishra and Ramaprabhu in 2012 (Mishra & Ramaprabhu, 2012). The unique properties of these composites, including high surface area and favorable chemical interactions, contribute to their effectiveness in arsenic removal applications.

Iron oxide nanoparticles are notable for their high surface area-to-volume ratio, which gives them a greater ability to adsorb arsenic from water. This makes them especially effective for arsenic removal. Recent studies have explored innovative approaches, such as ferric-impregnated volcanic ash, as low-cost adsorbents for arsenic removal from aqueous media, demonstrating the potential for utilizing abundant and readily available materials for sustainable remediation efforts, as highlighted by Chen et al. (2011).

In summary, iron-based adsorbents offer a diverse and effective arsenal for arsenic remediation, leveraging various materials and innovative approaches to address contamination challenges in water sources. Continued research and development in this field hold promise for advancing sustainable and accessible solutions to mitigate arsenic-related health risks in communities worldwide.

11.4.3 Zero-Valent Iron

Zero-valent iron (Fe^0) based adsorbents are gaining attention for their effectiveness in removing arsenic from water. They are cost-effective and have a minimal environmental impact. Fe0 is known for efficiently removing both As(III) and As(V) through surface precipitation or adsorption. Its effectiveness is due to several factors: it's non-toxic, widely available, affordable, easy to produce, and requires little maintenance (Lien & Wikin, 2005).

Studies show that Fe^0 can effectively remove arsenic in various settings, including low pH environments and water with high sulfide content. Although Fe^0's reducing power decreases at neutral pH, the hydroxide compounds that form on its surface provide excellent sites to adsorb both As(V) and As(III) at neutral and basic pH levels. This flexibility makes Fe^0-based adsorbents suitable for treating water with a wide range of pH levels (Bretzler et al., 2020; Su & Puls, 2001).

Research conducted by Leupin and Hug (2005), Katsoyiannis et al. (2008), Klas and Kirk (2013), Hussam and Munir (2007), and Neumann et al. (2013) has contributed to understanding the effectiveness of Fe^0in arsenic removal and its applicability under different environmental conditions (Leupin & Hug, 2005; Katsoyiannis et al., 2008; Klas & Kirk, 2013; Hussam & Munir, 2007; Neumann et al., 2013). Additionally, the advent of nanoscale zero-valent iron (nZVI) has further expanded the scope of Fe^0-based arsenic remediation.

Nanoscale zero-valent iron offers higher intrinsic reactivity compared to conventional Fe^0, leading to enhanced arsenic removal efficiency. Studies by Zhu et al. (2009), Baikousi et al. (2015), and Singh et al. (2021) have demonstrated the efficacy of nZVI in arsenic remediation applications (Zhu et al., 2009; Baikousi et al., 2015; Singh et al., 2021). However, challenges such as aggregation and cementation tendencies associated with nZVI have prompted ongoing research aimed at sustaining material efficiency and optimizing its performance in practical applications.

In summary, Fe^0 and nanoscale zero-valent iron represent promising avenues for arsenic remediation, offering effective, affordable, and environmentally benign solutions for addressing arsenic contamination in water sources. Continued research and development efforts are crucial for advancing these technologies and ensuring their practical applicability in diverse environmental settings.

11.5 Membrane Processes

Membrane processes offer a versatile and effective approach for arsenic remediation by leveraging selective barriers to separate contaminants from water. These processes exploit potential differences across membranes to control the movement of constituents, with driving forces categorized based on pressure, concentration, electrical potential, and temperature. Among the methods for arsenic removal, pressure-driven and electrical potential-driven membrane processes are frequently used (Shahrin, 2019; Kobya et al., 2020; Brandhuber & Amy, 1998).

Pressure-driven membrane processes used for arsenic removal are generally divided into four types based on pore size: microfiltration, ultrafiltration, nanofiltration, and reverse osmosis. These processes differ in pore size, which affects their ability to remove various contaminants and their effectiveness in removing arsenic. Membrane processes that operate at higher pressure and have smaller pores are better at removing a wider range of substances from water compared to those with lower pressure and larger pores (Brandhuber & Amy, 1998).

Microfiltration membranes feature relatively larger pore sizes, enabling the removal of larger particles and microorganisms from water. Ultrafiltration membranes possess smaller pores, allowing for the removal of colloidal particles, macromolecules, and some dissolved species. Nanofiltration membranes offer even finer filtration capabilities, selectively removing ions and small organic molecules. Reverse osmosis membranes have the smallest pore sizes, enabling the removal of virtually all dissolved solids, including arsenic ions (Nasi et al., 2022).

High-pressure membrane processes have a major advantage: they can remove a wide variety of substances, including arsenic, from water. This makes them ideal for arsenic cleanup in different environments. Additionally, pressure-driven membrane processes can be customized for specific arsenic removal needs by choosing membranes with the right pore sizes and adjusting the operating conditions accordingly (Frenkel, 2015).

In summary, pressure-driven membrane processes represent a versatile and efficient approach for arsenic remediation, offering advantages such as broad contaminant removal capabilities and adaptability to diverse water sources. Continued research and development efforts in membrane technology hold promise for advancing arsenic remediation strategies and ensuring access to safe drinking water for communities worldwide.

11.5.1 Microfiltration

Microfiltration is a pressure-driven membrane process used in water treatment, but its larger pore sizes mean it isn't very effective at removing dissolved or colloidal arsenic. However, if arsenic is in a particulate form, microfiltration can help remove it, though only to a limited degree. To improve its efficiency, especially when most of the arsenic is dissolved, microfiltration is often used alongside coagulation processes (Cheryan, 1998).

The principle behind this combined approach involves the use of coagulants to promote the aggregation of arsenic particles, thereby increasing their size and facilitating their removal through microfiltration. By forming larger flocs, coagulants aid in the efficient capture of arsenic particles by the microfiltration membrane, resulting in enhanced removal rates. This synergistic combination effectively addresses the challenge posed by low levels of particulate arsenic in source waters (Gao et al., 2019).

A study conducted by Ghosh et al. in 2011 demonstrated the effectiveness of combining electrocoagulation with microfiltration, employing a ceramic membrane, for arsenic removal (Ghosh et al., 2011). In their research, they successfully removed arsenic from feed water containing a concentration of 200 μg/L in the presence of fluoride. The electrocoagulation process facilitated the formation of arsenic-containing flocs, which were subsequently removed through microfiltration using a ceramic membrane. This innovative approach highlights the potential of integrating different treatment methods to achieve efficient arsenic remediation.

Despite its limitations in directly removing dissolved or colloidal arsenic, microfiltration plays a crucial role in arsenic removal strategies when combined with appropriate pre-treatment processes, such as coagulation. By leveraging the advantages of both microfiltration and coagulation, this combined approach offers a practical and effective solution for addressing arsenic contamination in water sources, particularly in scenarios where arsenic is primarily present in particulate form.

In summary, while microfiltration alone may not be sufficient for arsenic removal from water, its integration with coagulation processes enhances removal efficiency, especially in cases where arsenic exists in particulate form. The successful application of combined electrocoagulation and microfiltration strategies underscores the importance of innovative approaches in mitigating arsenic contamination and ensuring access to safe drinking water. Continued research and development efforts in this field hold promise for further advancing arsenic remediation technologies and addressing water quality challenges globally.

11.5.2 Ultrafiltration

Ultrafiltration, a membrane process with smaller pores than microfiltration, offers a unique approach for removing arsenic, especially from surface waters that contain high levels of colloidal and particulate arsenic. Although ultrafiltration might not be ideal for treating groundwater with low arsenic concentrations, its effectiveness in cleaning surface waters from arsenic has been shown in several studies (Koli & Singh, 2023).

One approach to enhancing arsenic removal using ultrafiltration involves the utilization of micellar-enhanced ultrafiltration, as investigated by Iqbal et al. in 2007 (Iqbal et al., 2007). This method leverages the formation of micelles, or aggregates of surfactant molecules, to facilitate the removal of arsenic species from water through ultrafiltration membranes. By enhancing the solubility and mobility of arsenic particles, micellar-enhanced ultrafiltration improves their retention and removal by the ultrafiltration membrane, thereby increasing arsenic removal efficiency.

Additionally, a 2001 study by Brandhuber and Amy explored how the surface charge of ultrafiltration membranes affects their ability to remove arsenic (Brandhuber & Amy, 2001). They found that the electrostatic interaction between arsenic ions and the negatively charged membrane surface was a key factor in how arsenic is removed. This finding suggests that membrane surface properties can be tailored to optimize arsenic removal efficiency through electrostatic interactions, highlighting the importance of membrane design considerations in ultrafiltration processes for arsenic remediation.

While ultrafiltration alone may not be sufficient for arsenic removal from groundwater, its combination with pre-treatment processes or modification of membrane properties can enhance removal efficiency. For instance, incorporating coagulation or adsorption processes upstream of ultrafiltration can aid in aggregating arsenic particles or facilitating their interaction with the membrane surface, thereby improving removal rates. Ultrafiltration presents a promising method for addressing arsenic contamination, especially in surface waters characterized by elevated levels of colloidal and particulate arsenic. Through innovative techniques such as micellar-enhanced ultrafiltration and consideration of membrane surface properties, ultrafiltration processes can be optimized to achieve effective arsenic removal. Continued research and development in this field hold potential for advancing ultrafiltration-based arsenic remediation technologies and addressing water quality challenges in various environmental contexts.

11.5.3 Nanofiltration

Nanofiltration, a high-pressure membrane technique, is a promising way to remove arsenic from contaminated water, especially when the water contains low levels of suspended solids. It uses membranes with lightly charged surfaces, which are essential for separating arsenic from water (Worou et al., 2021).

Studies by Seidel et al. (2001) and Velizarov et al. (2004) have explained how nanofiltration separates arsenic, highlighting the role of electrostatic repulsion between negatively charged arsenic species and the membrane's charge, a process known as Donnan exclusion. Furthermore, nanofiltration membranes have small pores, allowing them to efficiently filter out dissolved arsenic compounds from natural water sources (Seidel et al., 2001; Velizarov et al., 2004).

It's worth mentioning that nanofiltration mainly filters out divalent ions like calcium and magnesium, while monovalent salts like sodium and chloride typically pass through. Despite this, nanofiltration can still effectively remove dissolved As(V) and As(III) due to its size-exclusion properties (Worou et al., 2021).

The application of nanofiltration for surface water treatment often requires extensive pretreatment to remove particles and potentially address dissolved constituents to prevent fouling of the membrane. Operating parameters, membrane properties, and arsenic speciation also influence the effectiveness of nanofiltration for arsenic remediation. In summary, nanofiltration represents a versatile and effective method for arsenic removal, offering advantages such as selective removal of arsenic species and the capability to remove significant portions of dissolved arsenic compounds. However, successful implementation of nanofiltration for surface water treatment necessitates careful consideration of pretreatment requirements and operating parameters to ensure optimal performance and prevent membrane fouling (Amini et al., 2008.). Continued research and development in this field will contribute to further advancing nanofiltration-based arsenic remediation technologies and addressing water quality challenges globally.

11.5.4 Reverse Osmosis

Reverse osmosis (RO) stands as one of the earliest and most established membrane technologies, traditionally utilized for desalination purposes in treating brackish and seawater. Operating on the principle of applying pressure to one side of a selective membrane, RO effectively separates contaminants from water, including arsenic. Numerous bench- and pilot-scale studies have underscored the efficacy of RO as an arsenic remediation technology, highlighting its potential for widespread application (Lee et al., 2011).

One notable advantage of RO is its preferential removal of arsenate [As(V)] over arsenite [As(III)], as documented in various research studies. This selectivity makes RO particularly suitable for treating water contaminated with arsenic species, with arsenate being more effectively removed than arsenite. It's important to understand that RO experiences high osmotic pressure because of the concentration of salts on the feed side of the membrane. Despite this, RO is highly effective at removing arsenic, particularly when its operating conditions are carefully managed (Víctor-Ortega & Ratnaweera, 2017; Siddique et al., 2020).

A 2011 study by Akin et al. explored the factors affecting arsenic removal with reverse osmosis (RO). They found that the level of arsenic in the feed water had little impact on removal rates, but the feed water's pH and the operating pressure had a significant effect on how well arsenic was removed (Akin et al., 2011). Notably, when arsenite [As(III)] predominates in groundwater, arsenic removal efficiency may decline to less than 50%. This decrease in efficiency is attributed to the presence of uncharged As (III) species in solution, particularly under acidic conditions (pH below 10). Consequently, RO may be less effective in treating natural water contaminated with arsenic, especially under reducing conditions where arsenite is the predominant species. In summary, reverse osmosis emerges as a robust and effective technology for arsenic remediation, offering selective removal of arsenic species and demonstrating versatility in treating various water sources. Despite challenges such as osmotic pressure and variations in arsenic speciation, careful control of operating parameters can optimize arsenic removal efficiency using RO. Continued research and technological advancements will further enhance the applicability and effectiveness of RO in addressing arsenic contamination in water sources globally (Gholami et al., 2006).

11.5.5 Electrodialysis

Electrodialysis is an advanced method for removing arsenic. It uses direct electric current to move ions through special membranes that are selectively permeable. This process depends on ion exchange, with membranes designed to let either cations or anions pass through. The membranes are stacked in a sequence between opposite electrodes, alternating between cation and anion exchange membranes (Strathmann, 2010). The operational principle of electrodialysis involves the controlled movement of ions through the membrane stack, resulting in the creation of alternating compartments containing water with varying ion concentrations. This controlled ion mobility generates compartments with low and high ion concentrations, facilitating the separation and removal of contaminants, including arsenic.

One key advantage of electrodialysis is its ability to remove a high percentage of total dissolved solids from water. With a typical recovery rate of 70–80%, it efficiently concentrates dissolved solids, making it effective at cleaning water sources by removing contaminants. While specific studies on the application of electrodialysis for arsenic removal may be limited, its effectiveness in removing total dissolved solids suggests its potential utility for arsenic remediation. By capitalizing on the ion exchange properties of membranes and the controlled ion movement induced by electric current, electrodialysis presents itself as a promising technology for treating arsenic-contaminated water. In summary, electrodialysis represents an advanced and promising approach for arsenic remediation, utilizing ion exchange mechanisms and selective membrane permeability to isolate and eliminate contaminants from water. Continued research and development efforts are crucial to optimize electrodialysis technology for effectively addressing arsenic contamination, thereby ensuring access to safe drinking water for communities worldwide (Juve et al., 2022).

11.6 Ion Exchange

Ion exchange is a key method for removing arsenic, using a physical and chemical process where ions in the feed water are swapped with ions in a solid medium. This medium usually contains synthetic resin beads designed to selectively adsorb certain contaminants, including arsenic. The process starts with feed water flowing through a bed of these resin beads, which can happen in either a down flow or up flow manner. As the water flows through, the resin captures and holds the contaminant ions. When the resin beads become saturated with contaminants, the bed needs to be regenerated to continue the process (Oehmen et al., 2011; Sharma, 2017).

To regenerate the ion exchange column, it's rinsed with a solution containing a high concentration of the original ions that were swapped with those in the resin beads. Sodium hypochlorite ($NaOCl$) can also be used for regeneration. $NaOCl$ not only oxidizes arsenite to improve adsorption but also kills harmful bacteria in the contaminated water. One advantage of ion exchange for arsenic removal is its ability to produce fresh water by adjusting pH and coagulant dose. This method can selectively transport arsenic ions based on their valence state, using iron or aluminum-based coagulants to ensure effectiveness (Cobzaru & Inglezakis, 2015).

Additionally, ion exchange membranes can be chosen to match the specific valence state of arsenic ions, ensuring that arsenic is efficiently removed while limiting the removal of other substances from the water. Moreover, solid suspension, impurities, and filtration processes within the ion exchange system contribute to the elimination of organic compounds that could impair resin efficacy. In summary, ion exchange offers a robust and versatile approach for arsenic remediation, leveraging the selective adsorption properties of resin beads to remove contaminants from water. Through careful regeneration techniques and the use of appropriate coagulants, ion exchange can effectively produce freshwater while mitigating arsenic contamination. Continued advancements in ion exchange technology hold promise for further enhancing its efficacy in addressing arsenic contamination and ensuring access to safe drinking water for communities worldwide (Oehmen et al., 2011; Sharma & Bhattacharya, 2017; Cobzaru & Inglezakis, 2015).

11.7 Other Technologies

Besides the methods discussed above several new methods have been studied recently. Some of the interesting methods are briefly described below.

11.7.1 Electrocoagulation

Electrocoagulation emerges as a highly effective method for arsenic remediation in water, offering the potential to achieve removal efficiencies of up to 99.9%, as demonstrated in studies such as Demirbas et al. (2019). Despite its efficacy, electrocoagulation is associated with certain drawbacks, including the need for design optimization and substantial power consumption, as highlighted by Mohora et al. (2018). However, ongoing research efforts seek to address these challenges and optimize the efficiency of electrocoagulation systems.

One of the distinguishing features of electrocoagulation is its utilization of metals such as iron, aluminum, or titanium for arsenic removal, each offering unique advantages and considerations, as discussed by Kumar et al. (2004). Recent studies have further demonstrated the versatility and applicability of electrocoagulation in arsenic remediation, showcasing its effectiveness in various water treatment systems (Demirbas et al., 2019; Mohora et al., 2018; Kumar et al., 2004). Integration of electrocoagulation with innovative technologies, such as air–fuel cells, presents promising avenues for enhancing arsenic removal efficiency while minimizing electricity consumption, as observed in studies by Kobya et al. (2020). Moreover, the adoption of green energy sources like solar energy can further reduce the energy requirements of electrocoagulation processes, making them more sustainable and cost-effective.

Overall, electrocoagulation is recognized as a successful approach for reducing arsenic levels in water to below the permissible limits set by the World Health Organization (WHO), without imposing long-term adverse health impacts on humans. Despite challenges related to design optimization and energy consumption, ongoing research and innovation hold promise for further advancing the efficiency and applicability of electrocoagulation in arsenic remediation, ensuring access to safe drinking water for communities worldwide.

11.7.2 Different Nano Adsorbents for Arsenic Removal

11.7.2.1 Iron Oxide/Hydroxide Nanoparticles for Arsenic Removal

Iron oxide and hydroxide nanoparticles are promising nano-adsorbents for removing arsenic from water. They are highly efficient at adsorption and have magnetic properties that make them easy to separate using high gradient magnetic separation (HGMS). Several types of iron oxide and hydroxide nanoparticles have been studied for their ability to adsorb arsenic, including hematite (α-Fe_2O_3), maghemite (γ-Fe_2O_3), magnetite (Fe_3O_4), goethite (α-FeOOH), akagneite (β-FeOOH), lepidocrocite (γ-FeOOH), feroxyhyte (δ-FeOOH), ferrihydrite ($Fe_5(OH)_8 \cdot 4H_2O$), and bernalite ($Fe(OH)_3$), have been studied for their arsenic adsorption capabilities (Li et al., 2022).

Studies by Nassar (2012) and Ghanizadeh et al. (2010) have demonstrated the effectiveness of these iron oxide/hydroxide nanoparticles in adsorbing arsenic from water (Nassar, 2012). The magnetic properties of these nanoparticles enable their easy separation from water after adsorption, making them highly suitable for practical applications.

Furthermore, research by Mamindy Pajany et al. (2011) has shown that the adsorption rate of arsenate on various iron oxide/hydroxide nanoparticles follows the order of goethite < hematite < magnetite < zerovalent iron (Mamindy-Pajany et al., 2011). This indicates that different types of iron oxide/hydroxide nanoparticles exhibit varying degrees of arsenic adsorption efficiency, with zerovalent iron showing the highest adsorption capacity among the studied nanoparticles.

Efforts have also focused on improving the dispersibility of iron oxide nanoparticles and preventing iron from leaching into solutions. Feng et al. (2012) found that adding ascorbic acid to aqueous suspensions of Fe_3O_4 nanoparticles not only improved their dispersibility but also effectively stopped iron from leaching into the solution (Feng et al., 2012). This innovation contributes to the stability and efficiency of iron oxide nanoparticle-based adsorbents for arsenic removal applications.

In summary, iron oxide/hydroxide nanoparticles offer a promising solution for arsenic remediation, combining efficient adsorption properties with magnetic characteristics for easy separation from water. Ongoing research aimed at optimizing the adsorption efficiency and stability of these nanoparticles will further advance their practical applicability in addressing arsenic contamination in water sources.

11.7.2.2 Titanium Dioxide Nanoparticles

Titanium dioxide (TiO_2) nanoparticles have garnered significant attention for their potential as nano-adsorbents for arsenic removal, owing to a multitude of favorable properties. These nanoparticles exhibit remarkable chemical stability, low cost, and non-toxicity, making them highly desirable for water treatment applications. Moreover, TiO_2 nanoparticles possess strong oxidizing power, allowing them to effectively oxidize arsenite (As(III)) to arsenate (As(V)), thereby eliminating the need for pre-oxidation steps (Guan et al., 2012).

Research studies, such as those conducted by Hung et al. (2006), have highlighted the advantageous characteristics of TiO_2 nanoparticles, including their resistance to corrosion and high affinity for arsenic (Hung, 2007). Additionally, TiO_2 nanoparticles exhibit inherent photo-oxidizing capabilities, enabling the conversion of As(III) to As(V) under light irradiation. This property has been extensively studied and validated in works by Bang et al. (2005) and Pena et al. (2006), emphasizing the potential of TiO_2 nanoparticles to facilitate arsenic removal processes efficiently (Bang et al., 2005; Pena et al., 2006).

A study by Han et al. in 2010 explored the surface characteristics of nanoporous titanium adsorbents using a surface complexation model (SCM) (Han et al., 2010). Through this analysis, the researchers identified the types of surface species that contribute to the apparent arsenic removal at specified pH levels. This understanding of surface interactions provides valuable insights for optimizing the design and performance of TiO_2 nanoparticle-based adsorbents for arsenic remediation.

In summary, TiO_2 nanoparticles are a promising option for removing arsenic from water thanks to their desirable properties, such as chemical stability, low cost, non-toxicity, and strong oxidizing power. They are also effective at converting As(III) to As(V) when exposed to light, which is a key advantage in arsenic removal. Ongoing research to better understand and optimize the surface characteristics of TiO_2 nanoparticles will help improve their effectiveness in tackling arsenic contamination in water.

11.7.2.3 Zinc Oxide Nanoparticles

Zinc oxide nanoparticles (ZnO) have emerged as promising nano-adsorbents for arsenic removal in water treatment applications, owing to their exceptional properties and versatile functionality. These nanoparticles have garnered significant attention due to their high surface area, reactivity, and stability, making them well-suited for arsenic remediation purposes (Naseem & Durrani, 2021).

One of the notable advantages of ZnO nanoparticles is their ability to operate effectively across a wide pH range of 5.8 to 6.8. This versatility in pH tolerance makes

ZnO nanoparticles suitable for application in diverse water treatment settings, where pH levels may vary. The high surface area of ZnO nanoparticles provides ample active sites for arsenic adsorption, facilitating efficient removal from aqueous solutions. Additionally, the reactivity of ZnO nanoparticles enables them to interact with arsenic species present in water, leading to effective adsorption and immobilization (Zhang et al., 2016).

Furthermore, the stability of ZnO nanoparticles ensures their longevity and effectiveness in arsenic removal processes, even under varying environmental conditions. This stability contributes to the reliability and durability of ZnO-based nano-adsorbents for long-term arsenic remediation applications.

In summary, Zinc oxide nanoparticles offer a promising solution for arsenic removal in water treatment due to their high surface area, reactivity, stability, and wide pH tolerance. Their ability to effectively adsorb arsenic across different pH conditions makes them suitable for diverse water treatment scenarios. Continued research and development efforts aimed at optimizing the synthesis and performance of ZnO nanoparticles will further enhance their efficacy in addressing arsenic contamination in water sources.

11.7.3 Carbon Nano Tubes

Carbon nanotubes (CNTs) present a promising option for arsenic removal through adsorption, primarily due to their unique structural and chemical properties. While limited studies have explored their use in this application, CNTs offer several advantages that make them attractive nano-adsorbents (Arora & Attri, 2020).

One of the main benefits of CNTs is their highly porous and hollow structure, offering a large surface area with plenty of sites for adsorption, as mentioned by Ma et al. (2013). This structure allows CNTs to effectively absorb arsenic from water. CNTs also have a high specific surface area, low mass density, and high purity, which boost their adsorption capacity. Their strong interaction with contaminants, like arsenic, makes them excellent adsorbents.

Additionally, CNTs can be customized to improve their adsorption capabilities, allowing them to target specific contaminants. For example, adding metal oxides to CNTs, merges the adsorption features of metal oxides with the unique traits of CNTs, like their large surface area and surface functional groups (Ma et al., 2013). This modification strategy can improve the arsenic removal efficiency of CNTs by optimizing their adsorption capacity and selectivity. However, it is important to note that practical applications of iron oxide/CNT composites may face challenges due to health and environmental risks associated with these materials. Despite this limitation, ongoing research efforts continue to explore novel approaches and modifications to CNTs to overcome these hurdles and enhance their suitability for arsenic remediation.

In summary, carbon nanotubes offer significant potential as nano-adsorbents for arsenic removal, leveraging their unique structural and chemical properties. With further research and development, CNT-based adsorption technologies have the potential to provide effective solutions for addressing arsenic contamination in water sources, contributing to improved water quality and public health.

11.7.4 Graphene

Graphene, a two-dimensional carbon allotrope, has attracted significant interest as a potential nano-adsorbent for removing arsenic from water due to its remarkable properties and versatile applications. Although research on graphene for arsenic remediation is still in its early stages, it shows great potential for effectively tackling arsenic contamination (Nupearachchi et al., 2017; Yusuf et al., 2015).

A key benefit of graphene is its large surface area, offering plenty of active sites for arsenic adsorption. Because graphene is only one atom thick, it has a high specific surface area, which allows it to effectively adsorb arsenic from water. This high surface area-to-volume ratio boosts its adsorption capacity, making it an excellent choice for arsenic removal. Additionally, graphene is mechanically strong and chemically stable, which means it can withstand harsh conditions and remain effective for a long time. This durability ensures that graphene-based adsorbents maintain their adsorption efficiency over extended periods, even in challenging environments (Rana et al., 2024).

In addition to its superior adsorption properties, graphene can be functionalized or modified to enhance its arsenic removal capabilities. Functionalization of graphene with specific chemical groups or metal nanoparticles can tailor its surface chemistry to optimize arsenic adsorption efficiency and selectivity. This versatility allows for the development of tailored graphene-based adsorbents tailored to the unique requirements of arsenic removal applications.

Furthermore, graphene-based composites, such as graphene oxide (GO) and reduced graphene oxide (rGO), offer enhanced adsorption performance due to their unique physicochemical properties. These composites combine the advantages of graphene with additional functionalities, such as oxygen-containing functional groups and increased surface roughness, further improving arsenic adsorption capacity and kinetics (Das et al., 2020; Yoon et al., 2016).

Despite its impressive properties, there are challenges in producing and using graphene-based nano-adsorbents for large-scale arsenic removal. Concerns about cost, scalability, and environmental impact need to be resolved to make graphene-based technologies viable for widespread water treatment. In summary, graphene has great potential as a nano-adsorbent for removing arsenic, with its large surface area, strength, chemical stability, and versatility (Yoon et al., 2016). Continued research and development are required to overcome technical hurdles and refine graphene-based solutions for effective arsenic remediation. These efforts could help ensure clean and safe drinking water for communities around the world.

11.7.5 Lanthanum Compounds

Lanthanum compounds have gained attention as effective nano-adsorbents for removing arsenic from water, offering a cost-efficient solution to arsenic contamination. Lanthanum, one of the least expensive rare-earth elements, is a practical choice for water treatment. Various lanthanum compounds, such as lanthanum hydroxide, lanthanum carbonate, and basic lanthanum carbonate, have been studied for their ability to remove arsenic(V) from water. The mechanism involves adsorption through the exchange of carbonate (CO_3^{2-}) and/or hydroxide (OH^-) groups with arsenic (V) ions in water (Merodio-Morales et al., 2023; Tan et al., 2020).

Studies have shown that lanthanum compounds exhibit optimal arsenic removal efficiency within a neutral-to-basic pH range. This pH-dependent adsorption process ensures effective arsenic removal under environmentally relevant conditions, making lanthanum compounds suitable for practical applications in water treatment. Lanthanum-based nano-adsorbents offer several advantages for arsenic remediation, including high adsorption capacity, cost-effectiveness, and accessibility. Additionally, lanthanum compounds are relatively stable and environmentally friendly, minimizing potential risks associated with their use in water treatment processes (Dudek, 2020; Huo et al., 2021).

Furthermore, the abundance and low cost of lanthanum make it an attractive option for large-scale arsenic removal projects, particularly in regions facing arsenic contamination challenges with limited financial resources. The affordability of lanthanum compounds facilitates their widespread adoption and deployment in arsenic-affected communities, contributing to the provision of safe and clean drinking water. In summary, lanthanum compounds represent a promising nano-adsorbent for arsenic removal, offering an effective, economical, and environmentally friendly solution to arsenic contamination in water sources. Ongoing research and development are crucial to improving lanthanum-based adsorbents and boosting their effectiveness for real-world water treatment. This work will help ensure safe drinking water for communities worldwide.

11.7.6 Zeolites

Zeolites have emerged as promising nano-adsorbents for arsenic removal from aqueous solutions, leveraging their unique properties such as catalytic, sieve, and exchange capabilities. These naturally occurring minerals are abundant and offer a low-cost alternative for various applications, including water treatment. The porous structure of zeolites enables effective adsorption of contaminants, including arsenic ions, from aqueous solutions. This adsorption process occurs through interactions between the surface of the zeolite and arsenic species, facilitated by the exchange properties of zeolites. The cation exchange capacity of zeolites allows for the removal of arsenic ions by exchanging them with other cations present in the zeolite structure (Buzukashvili et al., 2024).

Although various naturally occurring zeolites have been identified, only a few are utilized for arsenic removal due to their specific properties and characteristics. To enhance their arsenic removal efficiency, zeolites can be modified through chemical treatments or functionalization. Modified zeolites offer improved adsorption properties and selectivity for arsenic ions, making them a better alternative for arsenic remediation applications (Zhou et al., 2021; Hu et al., 2020).

However, it is important to note that large-scale application of modified zeolites for arsenic removal may involve higher costs compared to natural zeolites. The cost associated with the modification process, as well as the availability of raw materials, can impact the feasibility of using modified zeolites for arsenic remediation on a commercial scale. In summary, zeolites represent a promising nano-adsorbent for arsenic removal, offering advantages such as natural abundance, low cost, and effective adsorption properties. While naturally occurring zeolites can be utilized for arsenic removal, modified zeolites hold potential for enhanced performance. Continued research and development efforts are needed to optimize zeolite-based adsorbents for practical applications in arsenic-contaminated water treatment, ensuring access to safe and clean drinking water for communities worldwide.

11.7.7 Manganese Greensand

Manganese greensand, a special type of zeolite-based glauconite mineral, is becoming popular for removing arsenic from water. This material has a unique coating of hydrous manganese dioxide and other high manganese oxides, boosting its adsorption capabilities and making it especially effective at arsenic removal (Habib et al., 2020).

The composition of manganese greensand, with its coating of active hydrous manganese dioxide, creates an environment conducive to arsenic adsorption. The presence of manganese dioxide on the surface of the greensand provides numerous active sites for the adsorption of ions of arsenic from aqueous solutions. This adsorption mechanism relies on the affinity between arsenic species and the active sites on the manganese dioxide coating. Studies have shown that manganese greensand exhibits superior arsenic removal efficiency compared to other adsorbents under certain conditions. For instance, with an iron-to-arsenic ratio of 20:1, manganese greensand has proven to be more effective at removing arsenic than iron-oxide-coated sand or resin-based adsorbents. This showcases the unique ability of manganese greensand to remove arsenic, especially in settings where iron is present (Habib et al., 2020).

The effectiveness of manganese greensand as a nano-adsorbent for arsenic removal is attributed to its specialized composition and surface properties. The active hydrous manganese dioxide coating provides a high surface area and numerous active sites for arsenic adsorption, ensuring efficient removal from aqueous solutions (Hanson et al., 2000). Overall, manganese greensand represents a promising nano-adsorbent for arsenic removal, leveraging its unique composition and surface characteristics to effectively mitigate arsenic contamination in water sources. More research and development are needed to fine-tune the use of manganese greensand for practical arsenic removal, helping to ensure that communities around the world have access to safe and clean drinking water.

11.7.8 Metal Organic Framework (MOF)/ Quantum Dots (QDs)

Metal–organic frameworks (MOFs) and quantum dots (QDs) have emerged as promising nano-adsorbents for arsenic removal, leveraging their unique properties and synergistic effects (Gatou et al., 2023; Wang et al., 2022). MOFs are gaining interest among researchers because of their versatile structures, customizable porosities, and large surface areas. Several stable MOFs with high micropore volumes have been identified, providing plenty of active sites for arsenic adsorption. One example is MIL-53(Fe), a MOF known for its reduced recombination of photogenerated charge carriers and smaller Fe–O cluster size, which enhances its photoactivity. MIL-53(Fe) has shown effectiveness in breaking down contaminants like sulfamethazine when aided by Fe(III) (Chen et al., 2022). However, to fully harness the advantages of the porous structures of MOFs for arsenic removal, some modifications may be necessary.

QDs have become popular as coupling mediators because they can store and transfer electrons effectively. Nitrogen-doped carbon quantum dots (N-CQDs) have special physical and chemical properties, with plenty of functional groups on their surfaces, making them ideal for photocatalytic reactions. By encouraging electron transfer between molecules or creating surface sites that separate electrons, N-CQDs embedded in MOFs can boost photocatalytic performance and lower the rate of charge carrier recombination. (Sengottuvelu et al., 2022).

A recent study by H. Chen et al. showed that embedding N-CQDs into MIL-53(Fe) using a one-step hydrothermal reaction led to high efficiency in removing As(III). This was achieved through photocatalytic oxidation and simultaneous adsorption. (Chen et al., 2022). The synergistic effect of N-CQDs and MIL-53(Fe) led to a six-fold increase in the removal ability of As(III) compared to pure MIL-53(Fe). The improved performance was due to better charge transport and more efficient use of electrons and holes caused by the in-situ embedded N-CQDs. Additionally, MIL-53(Fe)'s properties in As(III) photocatalytic oxidation and the simultaneous adsorption of generated arsenate [As(V)] were also noted. This combined approach illustrates the potential of MOF/QD composites for effective arsenic removal from water, emphasizing the importance of synergy between different nano-adsorbents for enhanced results.

In summary, the combination of MOFs and QDs represents a promising approach for arsenic removal, offering synergistic effects that enhance adsorption and photocatalytic oxidation processes. Continued research and development efforts in this field are crucial to optimize MOF/QD composites and unlock their full potential for practical applications in arsenic-contaminated water treatment, ultimately ensuring access to safe and clean drinking water for communities worldwide.

11.8 Recommendations to Reduce the Risk of Arsenic Contamination

Reducing the risk of arsenic contamination is critical for public health, especially in areas where arsenic occurs naturally in groundwater or soil. Here are some key steps to lower the risk:

- *Test Drinking Water Regularly*: Frequent testing for arsenic contamination in water sources is essential for early detection and risk mitigation.
- *Seek Alternative Water Sources*: In regions with high arsenic levels in groundwater, consider other sources like rainwater harvesting or importing water from less-contaminated areas.
- *Test Soil for Arsenic*: This is particularly important in farming regions, where contaminated soil can affect food crops.
- *Educate the Public*: Raise awareness about the risks of arsenic, promote safe water practices, and guide people on identifying potential contamination.
- *Encourage a Diverse Diet*: Eating a variety of foods helps reduce the risk of arsenic exposure from a single source, such as rice, which can accumulate arsenic.
- *Strengthen Public Health Surveillance*: Implement programs to monitor health effects linked to arsenic exposure and take action when needed.

11.8.1 Recommended Approach for Affordable and Environmentally Friendly Arsenic Removal

- *Support Research and Innovation*: Encourage the development of new, cost-effective technologies for detecting and removing arsenic from water and soil.
- *Implement Arsenic Removal Techniques*: While using surface water can be the best solution to avoid arsenic contamination, switching from groundwater to surface water might be too costly for developing countries. A simpler, more economical alternative is required. A notable example is 'Arsiron Nilogon', developed at Tezpur University in Assam, which can remove both arsenic and iron from water (Bordoloi et al., 2013). This method uses three simple steps:

Step 1: Add a specified amount of sodium bicarbonate (baking soda) to the arsenic-contaminated water to adjust the pH.
Step 2: Introduce a specified quantity of potassium permanganate as an oxidant.
Step 3: Add a specified quantity of $FeCl_3$ as a coagulant to help precipitate arsenic and iron.

This method is low-cost, efficient, easy to use, eco-friendly, and doesn't require electricity, making it ideal for rural households.

Combining these recommendations creates a comprehensive strategy to reduce arsenic contamination and protect public health. Each step should be customized to the specific local context with guidance from environmental science, public health, and policy experts.

11.9 Summary and Future Recommendation

In summary, cleaning arsenic-contaminated water is a major global challenge that requires effective and sustainable solutions. Traditional methods like oxidation, coagulation/precipitation, and ion exchange are commonly used to remove arsenic and have been studied extensively. Yet, the complex chemistry of arsenic and the varying environmental conditions make it difficult to consistently achieve effective and reliable remediation. Recently, researchers have been exploring advanced nano-adsorbents that show promise for improving arsenic removal. Various materials—like metal oxide/hydroxide nanoparticles, carbon-based nanomaterials, zeolites, and MOF/QD composites—have been studied for their ability to adsorb arsenic. These nano-adsorbents have unique characteristics, including high surface area, adjustable pore structures, and effective electron transfer, which make them good options for arsenic remediation.

Metal oxide/hydroxide nanoparticles like iron-based adsorbents and titanium dioxide nanoparticles are highly effective at absorbing arsenic because of their active surfaces and redox characteristics. Likewise, carbon-based nanomaterials like carbon nanotubes and graphene provide large surface areas and strong bonds with arsenic, leading to improved adsorption efficiency.

Zeolites, with their porous structures and ion exchange properties, provide effective arsenic removal capabilities, especially when modified or functionalized to optimize adsorption performance. Additionally, combining quantum dots with metal–organic frameworks (MOFs) has produced promising results in boosting photocatalytic oxidation and adsorption for arsenic removal.

The combined effects of different nano-adsorbents, like embedding N-CQDs into MIL-53(Fe) MOFs, have shown outstanding efficiency in arsenic removal. This happens because of improved charge transportation and better use of electrons and holes. This approach not only boosts adsorption capacity but also supports the photocatalytic oxidation of arsenic, providing a comprehensive solution for treating arsenic-contaminated water.

Despite the significant progress made in nano-adsorbent research for arsenic remediation, several challenges remain. These include scalability, cost-effectiveness, and environmental impact assessment of nano-materials at the industrial scale. Moreover, the long-term stability and regeneration capabilities of nano-adsorbents need further investigation to ensure their practical applicability in real-world scenarios. In summary, advanced nano-adsorbents show great potential for tackling the challenges of arsenic-contaminated water. By using the unique properties and combined effects of nano-materials, we can create innovative solutions to provide safe, clean drinking water to communities worldwide. To turn these technologies into sustainable arsenic-remediation practices, ongoing research and collaboration are crucial to overcoming current limitations.

Acknowledgements The authors acknowledge Dibrugarh University and MDKG College for their outstanding help and resources, which considerably aided in completing of this book chapter.

References

Abdul, K. S. M., Jayasinghe, S. S., Chandana, E. P., Jayasumana, C., & De Silva, P. M. C. (2015). Arsenic and human health effects: A review. *Environmental Toxicology and Pharmacology, 40*(3), 828–846.

Ahmad, A., Van Dijk, T., Gude, J., van de Wetering, S., Groenendijk, M., & Bhattacharya, P. (2017). Chapter A10. *Best practice guide* (p. 219).

Akin, I., Arslan, G., Tor, A., Cengeloglu, Y., & Ersoz, M. (2011). Removal of arsenate [As(V)] and arsenite [As(III)] from water by SWHR and BW-30 reverse osmosis. *Desalination, 281*, 88–92.

Alka, S., Shahir, S., Ibrahim, N., Ndejiko, M. J., Vo, D. V. N., & Abd Manan, F. (2021). Arsenic removal technologies and future trends: A mini review. *Journal of Cleaner Production, 278*, 123805.

Amini, M., Abbaspour, K. C., Berg, M., Winkel, L., Hug, S. J., Hoehn, E., Yang, H., & Johnson, C. A. (2008). Statistical modeling of global geogenic arsenic contamination in groundwater. *Environmental Science & Technology, 42*(10), 3669–3675.

Andrianisa, H. A., Ito, A., Sasaki, A., Aizawa, J., & Umita, T. (2008). Biotransformation of arsenic species by activated sludge and removal of bio-oxidised arsenate from wastewater by coagulation with ferric chloride. *Water Research, 42*(19), 4809–4817.

Arcibar-Orozco, J. A., Josue, D. B., Rios-Hurtado, J. C., & Rangel-Mendez, J. R. (2014). Influence of iron content, surface area and charge distribution in the arsenic removal by activated carbons. *Chemical Engineering Journal, 249*, 201–209.

Argos, M., Kalra, T., Rathouz, P. J., Chen, Y., Pierce, B., Parvez, F., Islam, T., Ahmed, A., Rakibuz-Zaman, M., Hasan, R., & Sarwar, G. (2010). Arsenic exposure from drinking water, and all-cause and chronic-disease mortalities in Bangladesh (HEALS): A prospective cohort study. *The Lancet, 376*(9737), 252–258.

Arora, B., & Attri, P. (2020). Carbon nanotubes (CNTs): A potential nanomaterial for water purification. *Journal of Composites Science, 4*(3), 135.

Baikousi, M., Georgiou, Y., Daikopoulos, C., Bourlinos, A. B., Filip, J., Zbořil, R., Deligiannakis, Y., & Karakassides, M. A. (2015). Synthesis and characterization of robust zero valent iron/mesoporous carbon composites and their applications in arsenic removal. *Carbon, 93*, 636–647.

Bang, S., Patel, M., Lippincott, L., & Meng, X. (2005). Removal of arsenic from groundwater by granular titanium dioxide adsorbent. *Chemosphere, 60*(3), 389–397.

Baskan, M. B., & Pala, A. (2010). A statistical experiment design approach for arsenic removal by coagulation process using aluminum sulfate. *Desalination, 254*(1–3), 42–48.

Bhat, A., Ravi, K., Tian, F., & Singh, B. (2024). Arsenic contamination needs serious attention: An opinion and global scenario. *Pollutants, 4*(2), 196–211.

Bissen, M., & Frimmel, F. H. (2003). Arsenic—A review. Part I: Occurrence, toxicity, speciation, mobility. *Acta hydrochimica et hydrobiologica, 31*(1), 9–18.

Bordoloi, S., Nath, S. K., Gogoi, S., & Dutta, R. K. (2013). Arsenic and iron removal from groundwater by oxidation–coagulation at optimized pH: Laboratory and field studies. *Journal of Hazardous Materials, 260*, 618–626.

Brandhuber, P., & Amy, G. (1998). Alternative methods for membrane filtration of arsenic from drinking water. *Desalination, 117*(1–3), 1–10.

Brandhuber, P., & Amy, G. (2001). Arsenic removal by a charged ultrafiltration membrane—Influences of membrane operating conditions and water quality on arsenic rejection. *Desalination, 140*(1), 1–14.

Bretzler, A., Nikiema, J., Lalanne, F., Hoffmann, L., Biswakarma, J., Siebenaller, L., Demange, D., Schirmer, M., & Hug, S. J. (2020). Arsenic removal with zero-valent iron filters in Burkina Faso: Field and laboratory insights. *Science of the Total Environment, 737*, 139466.

Buzukashvili, S., Sommerville, R., Rowson, N. A., & Waters, K. E. (2024). An overview of zeolites synthesised from coal fly ash and their potential for extracting heavy metals from industrial wastewater. *Canadian Metallurgical Quarterly, 63*(1), 130–152.

Camacho, L. M., Ponnusamy, S., Campos, I., Davis, T. A., & Deng, S. (2015). Evaluation of novel modified activated alumina as adsorbent for arsenic removal. In *Handbook of arsenic toxicology* (pp. 121–136). Academic Press.

Chen, H., Yuan, X., Jiang, L., Wang, H., & Zeng, G. (2022). Highly efficient As (III) removal through simultaneous oxidation and adsorption by N-CQDs modified MIL-53 (Fe). *Separation and Purification Technology, 286*, 120409.

Chen, R., Zhang, Z., Yang, Y., Lei, Z., Chen, N., Guo, X., Zhao, C., & Sugiura, N. (2011). Use of ferric-impregnated volcanic ash for arsenate (V) adsorption from contaminated water with various mineralization degrees. *Journal of Colloid and Interface Science, 353*(2), 542–548.

Cheryan, M. (1998). *Ultrafiltration and microfiltration handbook*. CRC Press.

Cobzaru, C., & Inglezakis, V. (2015). Ion exchange. In *Progress in filtration and separation* (pp. 425–498). Academic Press.

Dahi, E., & Liang, Q. (1998, February). Arsenic removal in hand pump connected iron removal plants in Noakhali, Bangladesh. In *International Conference on Arsenic Pollution of Ground Water in Bangladesh: Causes, Effect and Remedies, Dhaka* (pp. 8–12).

Das, B., Devi, R. R., Umlong, I. M., Borah, K., Banerjee, S., & Talukdar, A. K. (2013). Arsenic (III) adsorption on iron acetate coated activated alumina: Thermodynamic, kinetics and equilibrium approach. *Journal of Environmental Health Science and Engineering, 11*, 1–10.

Das, T. K., Sakthivel, T. S., Jeyaranjan, A., Seal, S., & Bezbaruah, A. N. (2020). Ultra-high arsenic adsorption by graphene oxide iron nanohybrid: Removal mechanisms and potential applications. *Chemosphere, 253*, 126702.

Demirbas, E., Kobya, M., Oncel, M. S., Şık, E., & Goren, A. Y. (2019). Arsenite removal from groundwater in a batch electrocoagulation process: Optimization through response surface methodology. *Separation Science and Technology, 54*(5), 775–785.

Dhanasekaran, P., & Sahu, O. (2021). Arsenate and fluoride removal from groundwater by sawdust impregnated ferric hydroxide and activated alumina (SFAA). *Groundwater for Sustainable Development, 12*, 100490.

Driehaus, W., Jekel, M., & Hildebrandt, U. (1998). Granular ferric hydroxide—A new adsorbent for the removal of arsenic from natural water. *Journal of Water Supply: Research and Technology—Aqua, 47*(1), 30–35.

Dudek, S., & Kołodyńska, D. (2020). Enhanced arsenic (V) removal on an iron-based sorbent modified by lanthanum (III). *Materials, 13*(11), 2553.

EFSA Panel on Contaminants in the Food Chain (CONTAM). (2009). Scientific opinion on arsenic in food. *EFSA Journal, 7*(10), 1351.
Emett, M. T., & Khoe, G. H. (2001). Photochemical oxidation of arsenic by oxygen and iron in acidic solutions. *Water Research, 35*(3), 649–656.
Feng, L., Cao, M., Ma, X., Zhu, Y., & Hu, C. (2012). Superparamagnetic high-surface-area Fe_3O_4 nanoparticles as adsorbents for arsenic removal. *Journal of Hazardous Materials, 217*, 439–446.
Fields, K. A., Chen, A. H., & Wang, L. (2000). *Arsenic removal from drinking water by coagulation/filtration and lime softening plants.*
Frenkel, V. S. (2015). *Advances in membrane technologies for water treatment. Planning and design of membrane systems for water treatment.* https://www.sciencedirect.com/topics/engineering/high-pressure-membrane
Fierro, V., Muñiz, G., Gonzalez-Sánchez, G., Ballinas, M. L., & Celzard, A. (2009). Arsenic removal by iron-doped activated carbons prepared by ferric chloride forced hydrolysis. *Journal of Hazardous Materials, 168*(1), 430–437.
Gao, B., Zhang, D., & Li, Y. (2019). Preparation of PSSS-grafted polysulfone microfiltration membrane and its rejection and removal properties towards heavy metal ions. *Polymers for Advanced Technologies, 30*(4), 1096–1105.
Gatou, M. A., Vagena, I. A., Lagopati, N., Pippa, N., Gazouli, M., & Pavlatou, E. A. (2023). Functional MOF-based materials for environmental and biomedical applications: A critical review. *Nanomaterials, 13*(15), 2224.
Ghanizadeh, G. H., Ehrampoush, M., & Ghaneian, M. (2010). Application of iron impregnated activated carbon for removal of arsenic from water. *Journal of Environmental Health Science & Engineering, 7*(2), 145–156.
Gholami, M.M., Mokhtari, M.A., Aameri, A. & Fard, M.A. (2006). Application of reverse osmosis technology for arsenic removal from drinking water. *Desalination, 200*(1–3), 725–727.
Ghosh, D., Medhi, C. R., & Purkait, M. K. (2011). Treatment of fluoride, iron and arsenic-containing drinking water by electrocoagulation followed by microfiltration. In: *Proceedings of 12th International Conference on Environmental Science and Technology*, Rhodes, Greece.
Ghurye, G., Clifford, D., & Tripp, A. (2004). Iron coagulation and direct microfiltration to remove arsenic from groundwater. *Journal-American Water Works Association, 96*(4), 143–152.
Goosen, M. F. A., Sablani, S. S., Al-Hinai, H., Al-Obeidani, S., Al-Belushi, R., & Jackson, A. (2005). Fouling of reverse osmosis and ultrafiltration membranes: A critical review. *Separation Science and Technology, 39*(10), 2261–2297.
Guan, X., Du, J., Meng, X., Sun, Y., Sun, B., & Hu, Q. (2012). Application of titanium dioxide in arsenic removal from water: A review. *Journal of Hazardous Materials, 215*, 1–16.
Gupta, B. S., Chatterjee, S., Rott, U., Kauffman, H., Bandopadhyay, A., DeGroot, W., Nag, N. K., Carbonell-Barrachina, A. A., & Mukherjee, S. (2009). A simple chemical free arsenic removal method for community water supply–A case study from West Bengal, India. *Environmental Pollution, 157*(12), 3351–3353.
Habib, M. E., Ali, M. A., & Fattah, K. P. (2020). Modified sand for the removal of manganese and arsenic from groundwater. *Journal of Environmental Engineering and Science, 15*(4), 189–196.
Han, D. S., Abdel-Wahab, A., & Batchelor, B. (2010). Surface complexation modeling of arsenic (III) and arsenic (V) adsorption onto nanoporous titania adsorbents (NTAs). *Journal of Colloid and Interface Science, 348*(2), 591–599.
Hanson, A., Bates, J., Heil, D., Bajpai, S., & Chaudhuri, M. (2000). Discussion and closure: Removal of arsenic from ground water by manganese dioxide-coated sand. *Journal of Environmental Engineering, 126*(12), 1160–1161.
Hao, L., Liu, M., Wang, N., & Li, G. (2018). A critical review on arsenic removal from water using iron-based adsorbents. *RSC Advances, 8*(69), 39545–39560.
Hu, G., Mian, H. R., Dyck, R., Mohseni, M., Jasim, S., Hewage, K., & Sadiq, R. (2020). Drinking Water treatments for arsenic and manganese removal and health risk assessment in White Rock, Canada. *Exposure and Health, 12*, 793–807.

Hug, S. J., Canonica, L., Wegelin, M., Gechter, D., & Von Gunten, U. (2001). Solar oxidation and removal of arsenic at circumneutral pH in iron containing waters. *Environmental Science & Technology, 35*(10), 2114–2121.

Hung, W. C., Fu, S. H., Tseng, J. J., Chu, H., & Ko, T. H. (2007). Study on photocatalytic degradation of gaseous dichloromethane using pure and iron ion-doped TiO_2 prepared by the sol–gel method. *Chemosphere, 66*(11), 2142–2151.

Huo, J. B., Yu, G., & Wang, J. (2021). Magnetic zeolitic imidazolate frameworks composite as an efficient adsorbent for arsenic removal from aqueous solution. *Journal of Hazardous Materials, 412*, 125298.

Hussam, A., & Munir, A. K. (2007). A simple and effective arsenic filter based on composite iron matrix: Development and deployment studies for groundwater of Bangladesh. *Journal of Environmental Science and Health, Part A, 42*(12), 1869–1878.

Iqbal, J., Kim, H. J., Yang, J. S., Baek, K., & Yang, J. W. (2007). Removal of arsenic from ground water by micellar-enhanced ultrafiltration (MEUF). *Chemosphere, 66*, 970–976.

Juve, J. M. A., Christensen, F. M. S., Wang, Y., & Wei, Z. (2022). Electrodialysis for metal removal and recovery: A review. *Chemical Engineering Journal, 435*, 134857.

Karmacharya, M. S., Gupta, V. K., Tyagi, I., Agarwal, S., & Jha, V. K. (2016). Removal of As (III) and As (V) using rubber tire derived activated carbon modified with alumina composite. *Journal of Molecular Liquids, 216*, 836–844.

Katsoyiannis, I. A., Ruettimann, T., & Hug, S. J. (2008). PH dependence of Fenton reagent generation and As (III) oxidation and removal by corrosion of zero valent iron in aerated water. *Environmental Science & Technology, 42*(19), 7424–7430.

Katsoyiannis, I. A., Zouboulis, A. I., & Jekel, M. (2004). Kinetics of bacterial As(III) oxidation and subsequent As(V) removal by sorption onto biogenic manganese oxides during groundwater treatment. *Industrial & Engineering Chemistry Research, 43*(2), 486–493.

Khosravi-Darani, K., Rehman, Y., Katsoyiannis, I. A., Kokkinos, E., & Zouboulis, A. I. (2022). Arsenic exposure via contaminated water and food sources. *Water, 14*(12), 1884.

Kim, M. J., & Nriagu, J. (2000). Oxidation of arsenite in groundwater using ozone and oxygen. *Science of the Total Environment, 247*(1), 71–79.

Klas, S., & Kirk, D. W. (2013). Understanding the positive effects of low pH and limited aeration on selenate removal from water by elemental iron. *Separation and Purification Technology, 116*, 222–229.

Kobya, M., Soltani, R. D. C., Omwene, P. I., & Khataee, A. (2020). A review on decontamination of arsenic-contained water by electrocoagulation: Reactor configurations and operating cost along with removal mechanisms. *Environmental Technology & Innovation, 17*, 100519.

Koli, M. M., & Singh, S. P. (2023). Surface-modified ultrafiltration and nanofiltration membranes for the selective removal of heavy metals and inorganic groundwater contaminants: A review. *Environmental Science: Water Research & Technology, 9*(11), 2803–2829.

Kumar, P. R., Chaudhari, S., Khilar, K. C., & Mahajan, S. P. (2004). Removal of arsenic from water by electrocoagulation. *Chemosphere, 55*(9), 1245–1252.

Lacasa, E., Canizares, P., Sáez, C., Fernández, F. J., & Rodrigo, M. A. (2011). Removal of arsenic by iron and aluminium electrochemically assisted coagulation. *Separation and Purification Technology, 79*(1), 15–19.

Lara, F., Cornejo, L., Yanez, J., Freer, J., & Mansilla, H. D. (2006). Solar-light assisted removal of arsenic from natural waters: Effect of iron and citrate concentrations. *Journal of Chemical Technology & Biotechnology: International Research in Process, Environmental & Clean Technology, 81*(7), 1282–1287.

Lata, S., & Samadder, S. R. (2016). Removal of arsenic from water using nano adsorbents and challenges: A review. *Journal of Environmental Management, 166*, 387–406.

Lee, C. G., Alvarez, P. J., Nam, A., Park, S. J., Do, T., Choi, U. S., & Lee, S. H. (2017). Arsenic (V) removal using an amine-doped acrylic ion exchange fiber: Kinetic, equilibrium, and regeneration studies. *Journal of Hazardous Materials, 325*, 223–229.

Lee, K. P., Arnot, T. C., & Mattia, D. (2011). A review of reverse osmosis membrane materials for desalination—Development to date and future potential. *Journal of Membrane Science, 370*(1–2), 1–22.

Lee, Y., Um, I. H., & Yoon, J. (2003). Arsenic (III) oxidation by iron (VI)(ferrate) and subsequent removal of arsenic (V) by iron (III) coagulation. *Environmental Science & Technology, 37*(24), 5750–5756.

Leupin, O. X., & Hug, S. J. (2005). Oxidation and removal of arsenic (III) from aerated groundwater by filtration through sand and zero-valent iron. *Water Research, 39*(9), 1729–1740.

Li, Y., Li, S., Hu, B., Zhao, X., & Guo, P. (2022). FeOOH and nZVI combined with superconducting high gradient magnetic separation for the remediation of high-arsenic metallurgical wastewater. *Separation and Purification Technology, 285*, 120372.

Lien, H. L., & Wilkin, R. T. (2005). High-level arsenite removal from groundwater by zero-valent iron. *Chemosphere, 59*(3), 377–386.

Ma, J., Zhu, Z., Chen, B., Yang, M., Zhou, H., Li, C., Yu, F., & Chen, J. (2013). One-pot, large-scale synthesis of magnetic activated carbon nanotubes and their applications for arsenic removal. *Journal of Materials Chemistry A, 1*(15), 4662–4666.

Mamindy-Pajany, Y., Hurel, C., Marmier, N., & Roméo, M. (2011). Arsenic (V) adsorption from aqueous solution onto goethite, hematite, magnetite and zero-valent iron: Effects of pH, concentration and reversibility. *Desalination, 281*, 93–99.

Mazumder, D. G. (2000). *Diagnosis and treatment of chronic arsenic poisoning*. United Nations Synthesis Report on Arsenic in Drinking Water.

Meez, E., Tolkou, A. K., Giannakoudakis, D. A., Katsoyiannis, I. A., & Kyzas, G. Z. (2021). Activated carbons for arsenic removal from natural waters and wastewaters: A review. *Water, 13*(21), 2982.

Merodio-Morales, E. E., Mendoza-Castillo, D. I., Jauregui-Rincón, J., & Bonilla-Petriciolet, A. (2023). Lanthanum-functionalized lignocellulosic wastes for the arsenate and fluoride depollution of water: Antagonistic adsorption and interfacial interactions. *Industrial Crops and Products, 200*, 116787.

Mishra, A. K., & Ramaprabhu, S. (2012). Ultrahigh arsenic sorption using iron oxide-graphene nanocomposite supercapacitor assembly. *Journal of Applied Physics, 112*(10).

Mohanty, D. (2017). Conventional as well as emerging arsenic removal technologies—A critical review. *Water, Air, & Soil Pollution, 228*(10), 381.

Mohora, E., Rončević, S., Agbaba, J., Zrnić, K., Tubić, A., & Dalmacija, B. (2018). Arsenic removal from groundwater by horizontal-flow continuous electrocoagulation (EC) as a standalone process. *Journal of Environmental Chemical Engineering, 6*(1), 512–519.

Muilenberg, T. (1997, February). Microfiltration basics: Theory and practice. In *Proceedings Membrane Technology Conference*.

Muruganandham, M., Amutha, R., Ahmmad, B., Repo, E., & Sillanpää, M. (2010). Self-assembled fabrication of superparamagnetic highly stable mesoporous amorphous iron oxides. *The Journal of Physical Chemistry C, 114*(51), 22493–22501.

Naseem, T., & Durrani, T. (2021). The role of some important metal oxide nanoparticles for wastewater and antibacterial applications: A review. *Environmental Chemistry and Ecotoxicology, 3*, 59–75.

Nasir, A. M., Adam, M. R., Kamal, S. N. E. A. M., Jaafar, J., Othman, M. H. D., Ismail, A. F., Aziz, F., Yusof, N., Bilad, M. R., Mohamud, R., & Rahman, M. A. (2022). A review of the potential of conventional and advanced membrane technology in the removal of pathogens from wastewater. *Separation and Purification Technology, 286*, 120454.

Nassar, N. N. (2012). Iron oxide nanoadsorbents for removal of various pollutants from wastewater: An overview. In *Application of adsorbents for water pollution control* (pp. 81–118).

Neumann, A., Kaegi, R., Voegelin, A., Hussam, A., Munir, A. K., & Hug, S. J. (2013). Arsenic removal with composite iron matrix filters in Bangladesh: A field and laboratory study. *Environmental Science & Technology, 47*(9), 4544–4554.

Nicomel, N. R., Leus, K., Folens, K., Van Der Voort, P., & Du Laing, G. (2016). Technologies for arsenic removal from water: Current status and future perspectives. *International Journal of Environmental Research and Public Health, 13*(1), 62.

Nupearachchi, C. N., Mahatantila, K., & Vithanage, M. (2017). Application of graphene for decontamination of water; Implications for sorptive removal. *Groundwater for Sustainable Development, 5*, 206–215.

Oehmen, A., Valerio, R., Llanos, J., Fradinho, J., Serra, S., Reis, M. A., Crespo, J. G., & Velizarov, S. (2011). Arsenic removal from drinking water through a hybrid ion exchange membrane–coagulation process. *Separation and Purification Technology, 83*, 137–143.

Pena, M., Meng, X., Korfiatis, G. P., & Jing, C. (2006). Adsorption mechanism of arsenic on nanocrystalline titanium dioxide. *Environmental Science & Technology, 40*(4), 1257–1262.

Rana, K., Kaur, H., Singh, N., Sithole, T., & Siwal, S. S. (2024). Graphene-based materials: Unravelling its impact in wastewater treatment for sustainable environments. *Next Materials, 3*, 100107.

Saalfield, S. L., & Bostick, B. C. (2009). Changes in iron, sulfur, and arsenic speciation associated with bacterial sulfate reduction in ferrihydrite-rich systems. *Environmental Science & Technology, 43*(23), 8787–8793.

Seidel, A., Waypa, J. J., & Elimelech, M. (2001). Role of charge (Donnan) exclusion in removal of arsenic from water by a negatively charged porous nanofiltration membrane. *Environmental Engineering Science, 18*(2), 105–113.

Sengottuvelu, D., Shaik, A. K., Mishra, S., Ahmad, H., Abbaszadeh, M., Hammer, N. I., & Kundu, S. (2022). Multicolor nitrogen-doped carbon quantum dots for environment-dependent emission tuning. *ACS Omega, 7*(31), 27742–27754.

Shahrin, S., Lau, W. J., Goh, P. S., Ismail, A. F., & Jaafar, J. (2019). Adsorptive mixed matrix membrane incorporating graphene oxide-manganese ferrite (GMF) hybrid nanomaterial for efficient As (V) ions removal. *Composites Part B: Engineering, 175*, 107150.

Sharma, S., & Bhattacharya, A. J. A. W. S. (2017). Drinking water contamination and treatment techniques. *Applied Water Science, 7*(3), 1043–1067.

Shi, Q., Yan, L., Chan, T., & Jing, C. (2015). Arsenic adsorption on lanthanum-impregnated activated alumina: Spectroscopic and DFT study. *ACS Applied Materials & Interfaces, 7*(48), 26735–26741.

Siddique, T. A., Dutta, N. K., & Roy Choudhury, N. (2020). Nanofiltration for arsenic removal: Challenges, recent developments, and perspectives. *Nanomaterials, 10*(7), 1323.

Singh, N., Kumar, D., & Sahu, A. P. (2007). Arsenic in the environment: Effects on human health and possible prevention. *Journal of Environmental Biology, 28*(2), 359.

Singh, P., Pal, P., Mondal, P., Saravanan, G., Nagababu, P., Majumdar, S., Labhsetwar, N., & Bhowmick, S. (2021). Kinetics and mechanism of arsenic removal using sulfide-modified nanoscale zerovalent iron. *Chemical Engineering Journal, 412*, 128667.

Smedley, P. L., & Kinniburgh, D. G. (2002). A review of the source, behaviour and distribution of arsenic in natural waters. *Applied Geochemistry, 17*(5), 517–568.

Song, S., Lopez-Valdivieso, A., Hernandez-Campos, D. J., Peng, C., Monroy-Fernandez, M. G., & Razo-Soto, I. (2006). Arsenic removal from high-arsenic water by enhanced coagulation with ferric ions and coarse calcite. *Water Research, 40*(2), 364–372.

Strathmann, H. (2010). Electrodialysis, a mature technology with a multitude of new applications. *Desalination, 264*(3), 268–288.

Su, C., & Puls, R. W. (2001). Arsenate and arsenite removal by zerovalent iron: Kinetics, redox transformation, and implications for in situ groundwater remediation. *Environmental Science & Technology, 35*(7), 1487–1492.

Tan, P., Zheng, Y., & Hu, Y. (2020). Efficient removal of arsenate from water by lanthanum immobilized electrospun chitosan nanofiber. *Colloids and Surfaces A: Physicochemical and Engineering Aspects, 589*, 124417.

Tchounwou, P. B., Centeno, J. A., & Patlolla, A. K. (2004). Arsenic toxicity, mutagenesis, and carcinogenesis—A health risk assessment and management approach. *Molecular and Cellular Biochemistry, 255*, 47–55.

Tchounwou, P. B., Yedjou, C. G., Patlolla, A. K.,& Sutton, D. J. (2012). Heavy metal toxicity and the environment. In *Molecular, clinical and environmental toxicology: Volume 3: Environmental toxicology* (pp. 133–164).

Vaiano, V., Iervolino, G., & Rizzo, L. (2018). Cu-doped ZnO as efficient photocatalyst for the oxidation of arsenite to arsenate under visible light. *Applied Catalysis B: Environmental, 238*, 471–479.

Velizarov, S., Crespo, J. G., & Reis, M. A. (2004). Removal of inorganic anions from drinking water supplies by membrane bio/processes. *Reviews in Environmental Science and Bio/technology, 3*, 361–380.

Víctor-Ortega, M. D., & Ratnaweera, H. C. (2017). Double filtration as an effective system for removal of arsenate and arsenite from drinking water through reverse osmosis. *Process Safety and Environmental Protection, 111*, 399–408.

Wang, L., Chen, A. S., Sorg, T. J., & Fields, K. A. (2002). Field evaluation of As removal by IX and AA. *Journal-American Water Works Association, 94*(4), 161–173.

Wang, X., Kong, L., Zhou, S., Ma, C., Lin, W., Sun, X., Kirsanov, D., Legin, A., Wan, H., & Wang, P. (2022). Development of QDs-based nanosensors for heavy metal detection: A review on transducer principles and in-situ detection. *Talanta, 239*, 122903.

Worou, C. N., Chen, Z. L., & Bacharou, T. (2021). Arsenic removal from water by nanofiltration membrane: Potentials and limitations. *Water Practice & Technology, 16*(2), 291–319.

Yang, J. K., Barnett, M. O., Zhuang, J., Fendorf, S. E., & Jardine, P. M. (2005). Adsorption, oxidation, and bioaccessibility of As (III) in soils. *Environmental Science & Technology, 39*(18), 7102–7110.

Yao, S., Liu, Z., & Shi, Z. (2014). Arsenic removal from aqueous solutions by adsorption onto iron oxide/activated carbon magnetic composite. *Journal of Environmental Health Science and Engineering, 12*, 1–8.

Yoon, S. H., & Lee, J. H. (2005). Oxidation mechanism of As (III) in the UV/TiO_2 system: Evidence for a direct hole oxidation mechanism. *Environmental Science & Technology, 39*(24), 9695–9701.

Yoon, Y., Park, W. K., Hwang, T. M., Yoon, D. H., Yang, W. S., & Kang, J. W. (2016). Comparative evaluation of magnetite–graphene oxide and magnetite-reduced graphene oxide composite for As (III) and As (V) removal. *Journal of Hazardous Materials, 304*, 196–204.

Yusuf, M., Elfghi, F. M., Zaidi, S. A., Abdullah, E. C., & Khan, M. A. (2015). Applications of graphene and its derivatives as an adsorbent for heavy metal and dye removal: A systematic and comprehensive overview. *RSC Advances, 5*(62), 50392–50420.

Zhang, F., Chen, X., Wu, F., & Ji, Y. (2016). High adsorption capability and selectivity of ZnO nanoparticles for dye removal. *Colloids and Surfaces A: Physicochemical and Engineering Aspects, 509*, 474–483.

Zhang, W., Liu, C., Zheng, T., Ma, J., Zhang, G., Ren, G., Wang, L. & Liu, Y. (2018). Efficient oxidation and sorption of arsenite using a novel titanium (IV)-manganese (IV) binary oxide sorbent. *Journal of Hazardous Materials, 353*, 410–420.

Zhang, W., Liu, C., Wang, L., Zheng, T., Ren, G., Li, J., Ma, J., Zhang, G., Song, H., Zhang, Z., & Li, Z. (2019). A novel nanostructured Fe-Ti-Mn composite oxide for highly efficient arsenic removal: Preparation and performance evaluation. *Colloids and Surfaces A: Physicochemical and Engineering Aspects, 561*, 364–372.

Zhou, C., Han, C., Min, X., & Yang, T. (2021). Enhancing arsenic removal from acidic wastewater using zeolite-supported sulfide nanoscale zero-valent iron: The role of sulfur and copper. *Journal of Chemical Technology & Biotechnology, 96*(7), 2042–2052.

Zhu, H., Jia, Y., Wu, X., & Wang, H. (2009). Removal of arsenic from water by supported nano zero-valent iron on activated carbon. *Journal of Hazardous Materials, 172*(2–3), 1591–1596.

Chapter 12
Heavy Metal Concentration Along the Musi River—A Case Study

Blessy Ganduri, Udaya Laxmi Gakka, and Vidyasagarachary Dasoju

Abstract The perineal River Musi is flowing from centre of the state capital city and bifurcating in to twin cities Hyderabad and Secunderabad. Every day the River Musi is carrying loads of domestic and industrial waste (effluents) along with water. The groundwater quality is assessed by determining the content of heavy metals like Al, As, Ba, Co, Cr, Cu, Zn, Pb, Ni and Mn. The life sustainability of the ecological system is dependent on the concentration of these heavy metals, which is life threatening when they are above their standard permissible limits. The discharge from industrial effluents and domestic waste along the Musi River are the main source of heavy metals in groundwater. Added to this the slow movement of surface water in the river accumulates and precipitates the heavy metals.

The adverse impacts on local water bodies and shallow aquifers on untreated and or partially treated waste water used for irrigation have been significantly observed. This effect is observed in the vicinity of Musi River making the ground water unsuitable for domestic or agricultural purposes. Hence, there is an urgent need to find the solution to this problem with a suitable plan and an achievable strategy. Bio-remediation in the surface water bodies could check the depletion and contamination of aquifers for sustainable use. This can be achieved with an experimental strategy of excessive plantation along the river banks.

12.1 Introduction

Hyderabad is the fifth largest metropolitan city in India, and it has diverse cultures entangled and cultivated on the banks of river Musi with a population of 13 million. It has grown to the present state with a rapid pace from Qutub Shahi (Golkonda dynasty) rulers to modern era witnessing monuments, hills, forests and surface water bodies along with industries (Hill, 2008; Srisailam Gogula & Sunder Kumar Koli, 2016).

B. Ganduri · G. Udaya Laxmi (✉) · D. Vidyasagarachary
Department of Geophysics, Osmania University, Hyderabad, India
e-mail: udayalaxmi.g@gmail.com

V. Sunitha et al. (eds.), *Groundwater Resource Management Planning Strategies*, Springer Water, https://doi.org/10.1007/978-3-031-88870-0_12

The Rive Musi is a tributary to river Krishna flowing from the heart of the city and bifurcating into Hyderabad and Secunderabad as twin cities. The river Musi originates in Ananthagiri Hills of Ranga Reddy district 90 kms to the west of the city of Hyderabad and confluences to river Krishna at Wadapally, Nalgonda district on eastern part of the state and it travels to a stretch of around 170 Kms. The water is used for drinking and around one lakh hectares land is cultivated at the downstream of Musi River during the monsoon periods.

The water in the river Musi flows from laterites of Deccan traps to over the plains of granites and attained purity and medicinal properties. This water brings good name and fame to the Hyderabad city over the centuries. Today, that innocuous water became very dangerous and toxic with colour and odour. Rapid urbanization and industrialization around the city of Hyderabad polluted the surface water bodies and damaged to beyond measures (Adimalla 2019). The river receives large amounts of untreated sewage from industrial, domestic disposal dumping caused the water to contaminate to the extreme toxic levels. Along the banks of the Musi River, most of the farmers cultivate a variety of crops, on a vast area of land, which is attributed to the perennial availability of surface and ground water. Due to the economic situation of the farmers their cultivation is carried out even in polluted water or secondary treated water from sewage treatment plant (STP) abutting the River Musi.

The groundwater in the local aquifers surrounding the Musi River, which are irrigated by waste water and ground water coexist and hence the quality of groundwater in the local aquifers deteriorates consistently, especially the salinity of water rises with change in pH (Fig. 12.1). The industrial, domestic and municipal waste water is discharged in to the river Musi and flows downstream. The study is to determine the degree of pollution and its acceptability levels for the agriculture practices and other uses. A broader perspective approach is made to emphasize the lateral and vertical extent of ground water contamination in the area specified. River Musi is a source for irrigation and agrarian produce in the downstream and vegetables grown with acrid stench (Blessy et al., 2019).

12.2 Location

The present area of investigation forms part of toposheet no. 56 K/11 and 56 K/15 and is bounded by $78^0 35'30''$ E to $78^0 57'35''$ E longitudes and $17^0 18'30''$ N to $17^0 28'30''$ N latitudes (Fig. 12.1) and covers an area of 570 square kilometres. The topography is rugged with undulations consisting of structural and residual hills. The highest elevation is 505mts and the lowest is 326mts, with a general slope to the eastern side.

The agricultural practices in these areas dependent on ground water from where the River Musi is flowing. The present catchment area consists of 10 major, 15 medium and around 50 small scale surface water bodies. In turn these surface water bodies are filled and or attached to the river. The investigation is mainly concentrated on the downstream area to affirm the status of ground water.

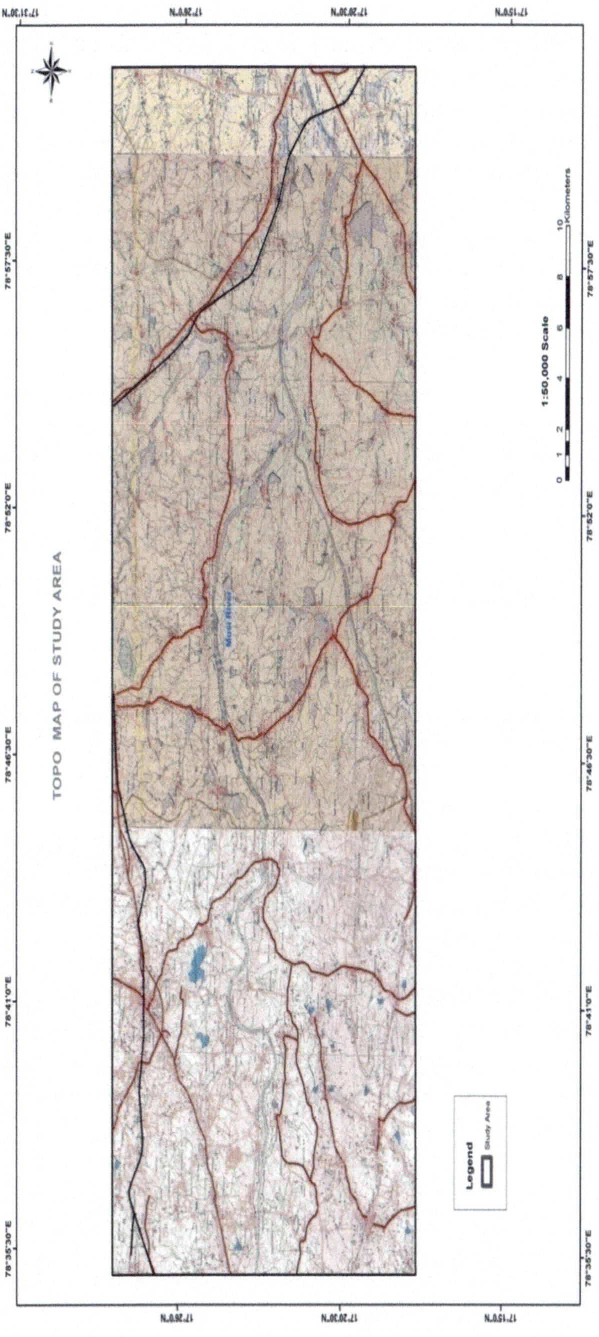

Fig. 12.1 Location map of the study area (SOI Toposheet no. 56 K/11 and 56 K/15)

12.2.1 Geology and Drainage

The present area is part of peninsular gneissic complex of South India of lower Proterozoic era. Peninsular gneisses and granites of Archaean age are the oldest rocks exposed in the area. The dominant type of rocks intruded by closepet granites and dolerite dykes. Dharwar comprising of amphibolite gneisses, schist and quantizes occur as narrow bands within the granites. Feldspar and quartz veins occur as intrusive bodies in the granites (Fig. 12.2). There are two types of granites exposed in the vicinity namely pink and grey. Grey ones are medium to fine grained with less porosity and permeability, where in the pink granites are medium to coarse grained fractured, fissured and sufficiently permeable to accommodate the ground water. The weathering in the area is intense and pronounced, leading to develop secondary porosity and increasing the percolation capacity.

The drainage in the area is dendritic to sub dendritic because of the rugged topography and long continued drainage pattern is not developed. The drainage density is highland third to fourth stream orders are found. No profound lineaments are noticed in the area, and majority are only inferred (Fig. 12.3).

12.2.2 Sample Collection and Analysis

Ground water samples collected along the River Musi from the east of Hyderabad city Peerzadiguda to Valigonda village (Yadadri Bhuvanagiri District) to a length of 50 km, covers both north and south banks. Sampling is being carried out at about 500 m to one km away from the river. The water sampling may help in demarcating the lateral extent of pollution from river Musi. The water is being collected for analyses from the village water supply bore wells (Fig. 12.4). The depths of the bore wells varying from 200 to 300 ft and a few are beyond 500ft. Although, the bore well water is not used for drinking, but used for other activities like gardening, washing and cleaning and also to animal to drink and wash.

The ground water samples collected were analyzed for heavy metals along using the instrument Inductively Coupled Plasma Mass Spectrometry (ICP–MS). Instrument reproducibility is checked with repetition of analyses along with prescribed standard methods. The heavy metals analyzed are Al, As, Co, Cr, Cu, Pb, Zn, Ni and Mn. The analytical data is presented in Table 12.1. The statistical analyses and their permissible limits are shown in Table 12.2.

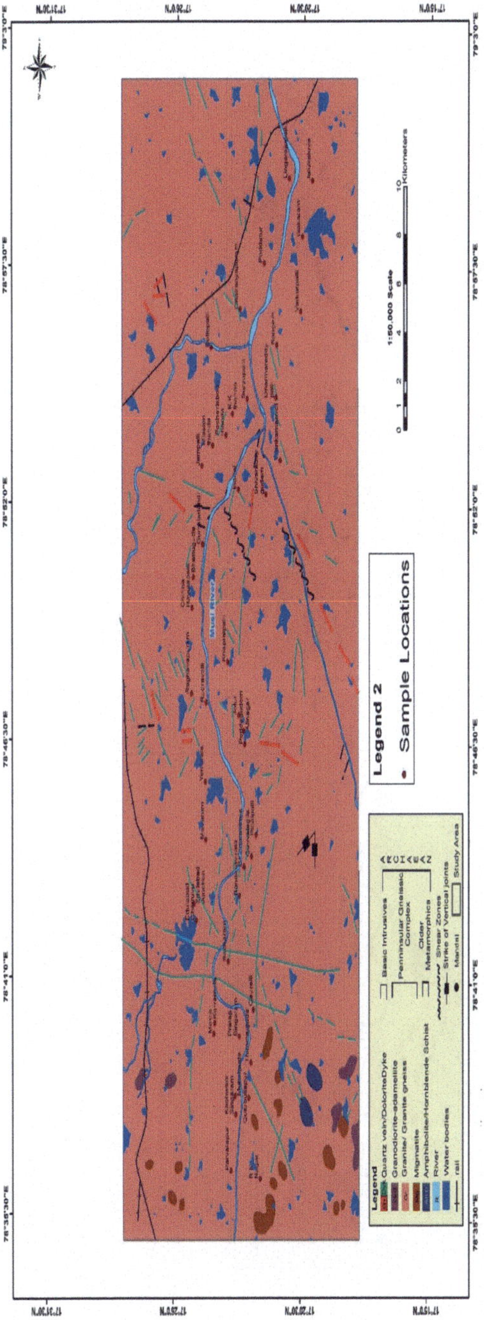

Fig. 12.2 Map showing the Geology and of the study area. GSI, 2002 *Source*

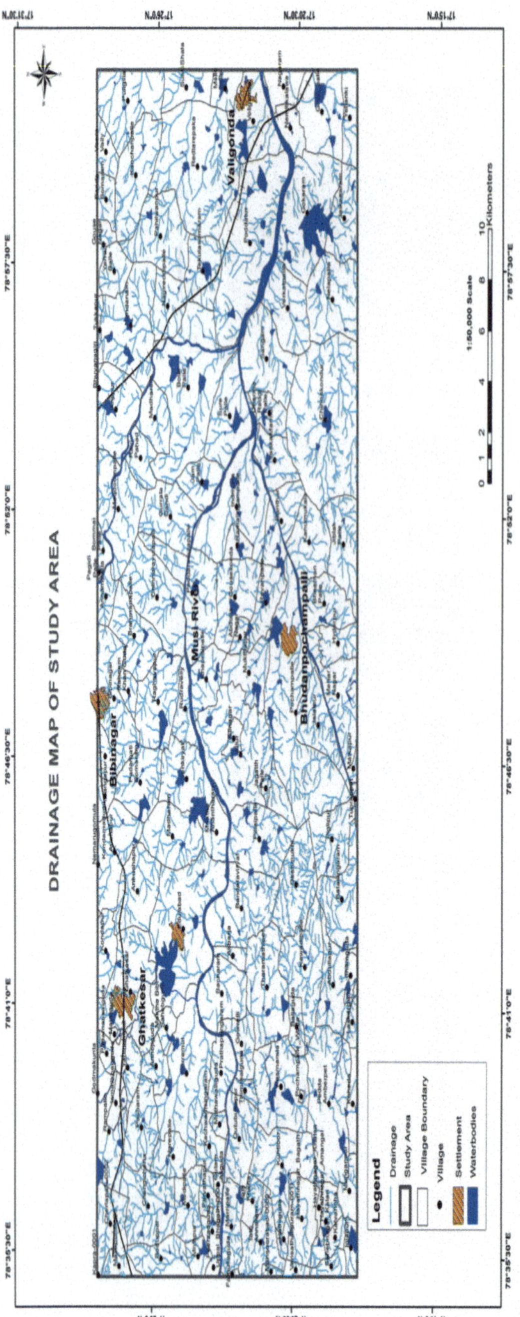

Fig. 12.3 Drainage map of Musi River area

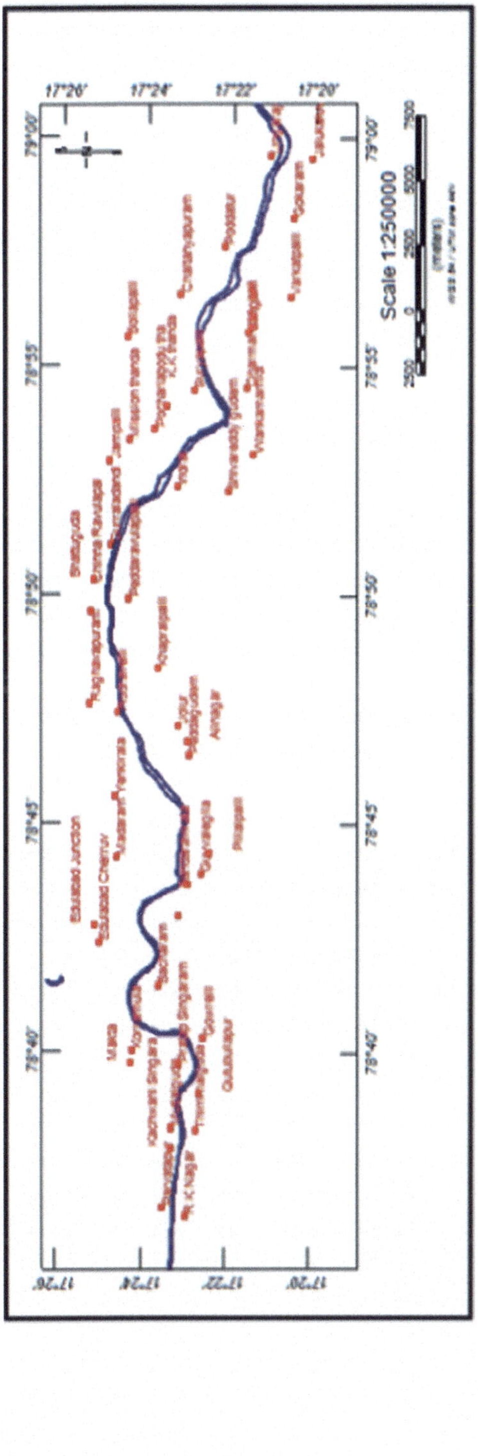

Fig. 12.4 Location map of water sample to the downstream of Musi River

Table 12.1 Elemental concentrations in ground water of River Musi

Sample. No.	Location Names	pH	Al ppm	As ppm	Co ppm	Cr ppm	Cu ppm	Mn ppm	Ni ppm	Zn ppm
1	Parvatapur	10.44	22.16	ND	ND	ND	5.01	29.21	ND	46.99
2	KachwaniSingaram	11.02	39.40	ND	ND	ND	ND	ND	8.05	ND
3	Mutialguda	11.02	29.47	ND	ND	5.84	ND	ND	ND	59.43
4	Pratap Singaram	10.36	9.85	ND	13.62	4.97	ND	ND	13.37	ND
5	Makta	11.25	25.40	ND	6.12	ND	ND	0.64	ND	32.14
6	Korremula	10.06	22.73	ND	9.68	ND	4.27	238.68	14.65	58.95
7	EdulabadCherruvu	11.13	4.91	ND	ND	5.56	ND	ND	ND	32.47
8	Edulabad Junction	10.90	14.11	ND	ND	ND	ND	ND	0.24	29.10
9	Madaram	10.14	13.43	ND	6.26	ND	ND	ND	ND	27.25
10	Yenkirala	10.10	24.20	ND	ND	ND	1.34	ND	ND	ND
11	Rudravelli	10.41	12.97	ND	ND	ND	ND	12.81	ND	ND
12	Raghavapuram	10.61	52.97	ND	2.05	ND	ND	1.09	ND	13.98
13	ChinnaRavulapalli	10.24	25.61	ND	ND	5.94	ND	0.61	ND	ND
14	Bhattuguda	10.24	22.18	6.66	14.61	2.75	ND	0.87	11.49	1217.14
15	Gurraladandi	10.27	17.25	5.14	8.86	ND	ND	ND	ND	168.07
16	Jampalli	11.02	31.50	10.51	9.15	ND	ND	ND	15.88	23.61
17	Mission thanda	11.10	19.12	ND	6.89	ND	ND	ND	ND	ND
18	Pocharlaboduthanda	10.83	27.68	6.75	ND	ND	0.84	ND	8.74	53.29
19	K.K thanda	10.39	30.28	ND	ND	ND	ND	ND	ND	53.67
20	Suryapalli	10.29	13.65	ND	ND	ND	6.86	ND	ND	ND
21	Bollapalli	8.88	9.90	8.46	ND	ND	ND	ND	ND	42.21
22	Chaitanyapuram	10.32	12.07	4.53	ND	ND	ND	ND	6.69	23.81
23	Poddatur	10.59	8.88	3.17	ND	3.21	ND	ND	ND	ND
24	Lingarajupalli	10.34	12.61	ND	4.53	ND	ND	ND	10.45	2436
25	R K Nagar	10.21	28.48	9.78	4.53	ND	ND	42.41	8.05	1296.21
26	Thimmaiguda	10.89	16.74	ND	ND	ND	ND	ND	ND	151.49
27	Qutubullapur	11.00	85.90	ND	ND	ND	ND	ND	8.77	36.79
28	Gaurelli	10.51	21.00	ND	ND	ND	ND	511.41	7.32	39.89
29	Bacharam	10.50	17.17	9.53	9.57	ND	1.34	4.71	3.68	83.23
30	Bandaravirala	10.41	32.54	5.72	ND	2.19	ND	266.76	14.86	1704.02
31	Chinnaravirala	10.85	20.77	9.36	ND	ND	ND	ND	ND	166.72
32	Guvvalegila	10.78	20.67	ND	ND	ND	ND	ND	ND	78.15
33	Pillaipalli	10.35	13.52	ND	ND	4.05	ND	11.54	10.93	75.48
34	Peddagudem	10.19	6.66	ND	ND	ND	ND	2.99	8.62	103.70
35	Alinagar	11.15	12.59	ND	ND	ND	ND	ND	16.29	ND
36	Julur	10.21	16.85	9.90	11.42	ND	7.74	ND	16.45	61.58

(continued)

Table 12.1 (continued)

Sample. No.	Location Names	pH	Al ppm	As ppm	Co ppm	Cr ppm	Cu ppm	Mn ppm	Ni ppm	Zn ppm
37	Khapraipalli	10.45	9.86	ND	8.99	ND	ND	1.11	ND	21.49
38	Peddaravulapalli	10.70	49.49	ND	11.78	ND	ND	ND	ND	40.36
39	Indrial	10.59	11.22	ND	2.89	ND	17.32	2.98	ND	78.66
40	Shivareddygudem	10.31	24.15	ND	11.13	ND	ND	138.35	ND	274.94
41	Wankamamidi	10.75	11.83	ND	ND	ND	ND	ND	ND	40.81
42	Dharmareddypalli	10.32	35.93	ND	9.11	ND	ND	ND	10.11	35.73
43	Sangem	10.18	22.92	ND	1.35	4.82	ND	ND	15.91	42.44
44	Varkatpalli	10.12	47.97	3.21	ND	ND	ND	1.90	ND	43.45
45	Gokaram	10.16	23.56	ND	2.79	ND	ND	94.53	ND	287.04
46	Jalukaluva	10.49	24.18	ND	ND	ND	ND	ND	ND	31.50

Table. 12.2 Statistical analysis of heavy metals

Parameter	Pre-monsoon				Post-monsoon				WHO limits in ppm
	Min ppm	Max ppm	Average ppm	S D	Min ppm	Max ppm	Average ppm	S D	
Cu	0.84	17.32	5.14	5.23	< 1.0	16.96	1.0	2.9	1.0
Pb	15.59	21.74	18.6	4.35	< 1.0	19.96	0.7	3.6	0.05
Zn	13.98	2436	244	530	< 1.0	1703	143	347	5.0
Ni	0.24	16.45	10.39	4.47	< 1.0	16	5.0	6.1	0.05
Co	1.35	14.61	7.94	3.90	< 1.0	13.67	3.0	4.2	NA
Cr	2.19	5.94	4.37	1.39	< 1.0	5.36	0.8	1.8	0.1
Mn	0.61	511	75.70	137	< 1.0	509	29	91.2	0.05
Al	53.00	4.91	21.57	11.00	5.0	48.6	20	10.6	0.2
Ba	369	47.22	153	82.39	47.0	369	153	82.4	2.0
As	10.51	3.17	7.13	2.63	< 1.0	9.84	2.0	3.1	0.05

12.3 Sources of Heavy Metals in Groundwater

The cause of pollution in groundwater can be attributed to hazardous elements and minerals that can arise from natural and anthropogenic sources. Natural sources are seepage of metallic elements through leaching of rocks and minerals in to water and through volcanic activity, forest fires etc. Pollution of surface or ground water is also possible by the partitioning of elements (in clay minerals), between sedimentary rocks and their precursor sediments and water (Verma and Dwivedi 2013), along with rapid industrialization consumerist life style. Pollution of ground water occurs both at the levels of industrial products and the end use of products and run off.

The heavy metals refer to any metal and metalloid element that has a relatively high density ranging from 3.5 to 5.0 g/cm and is toxic or poisonous even at low concentrations, includes arsenic (As), chromium (Cr), barium (Ba), zinc (Zn), nickel (Ni), copper (Cu) Cobalt (CO), manganese (Mn) and lead (Pb). The earth's crust consists of heavy metals generally that are non-biodegradable in nature. These toxic heavy metals enter the human body through air, water and food. However, a few heavy metals in limited minute quantities play an essential role in the metabolism of humans and animals, while higher concentration may cause toxicity and severe health hazards. The toxic heavy metals enter the environment through various human activity, like mining, industrial wastes and household discharges into adjoining water bodies (Samba Shiva Rao et al., 2017).

Greater Hyderabad Metropolitan agglomerate peripheries engulfed with hundreds of large scale, small scale and cottage industries all along the up land and catchment areas of river Musi. Industries like mining, leather, lead acid batteries, e-waste, ceramics, bangle industry, electroplating, paints, incineration, and fuel combustion and among the different pharmaceutical industries responsible for the Musi River water pollution. Most of the industries located along the river Musi are not utilising the sewage treatment plant (STP) to discharge their effluents for many years. All these inhuman activities led to cultivate detrimental organisms in the urban ecosystems with metal contaminants, and get accumulate in soil and sludge.

12.3.1 Cu, Pb and Zn

The elements Cu, Pb and Zn have geochemically close affinity, but in this area Pb is not detected in water, may be due to non-dissolvable and slow geochemical mobility. The distribution of Cu and Zn in the study area pose serious concern to discuss, because of their abnormal concentrations in the water samples. Lead is found to be only at two locations i.e., Yenkirala and Mission Thanda villages with 15.59 and 21.74 ppm respectively to the north of the river Musi (Table. 12.1). Lead is the least mobile element in water. The industrial effluents discharged in to river Musi might have segregated and precipitated in the upstream portion on the north of river. The major contribution of lead in groundwater and soil is through the corrosion of lead pipelines and excessive use of lead in the paint industry. Lead can neither be dissolved nor broken down in order to reduce the toxicity, but may be converted to other forms.

The environment is being polluted by the increasing use of copper. Rivers are depositing sludge on their banks that is contaminated with copper, due to the disposal of copper containing wastewater and will also contaminate the soil. The copper present in the soil is adsorbed by the organic material present in it and thereby further increasing its toxicity. Since copper is adsorbed by organic matter it is not transported further and does not contaminate groundwater. However, in surface water copper is transported to large distances or forms a suspended sludge or exists as free ions. The outcome of this result is that the elemental copper in water is influenced by pH and dissolved oxygen (Mebratu and Zarabruk 2011). The lowest copper content is found at PocharlaboduThanda is 0.84 ppm and the highest is at Indriyal village is 17.32 ppm. In some parts copper is not detected in ground water samples may be due to elevated levels of dissolved oxygen in water.

Zinc is being more soluble element and is accumulated more at Bandaravirala (1704 ppm), Lingarajupalli (2436 ppm) villages and is much higher than its permissible limits. Naturally Zinc must be in low concentrations in surface and ground waters because of its restricted mobility. The burning of coal, mining and metallurgical processing of zinc ores and its industrial application are the major source of zinc to pollute the environment.

12.3.2 Arsenic (As)

Arsenic is the most toxic and dangerous element found in the groundwater. The lowest and highest concentrations are 3.17 to 10.51 ppm respectively (Table. 12.2). High As concentration is found mainly in the southern part of the river at Julur, R.K. Nagar and Jampally villages in the northern area represented with very high values of As.

Arsenic present in the contaminants is extremely difficult to convert to volatile products or to make it water-soluble (Murcott, 2012). Since, arsenic is easily transportable it may not appear in large concentrations at a specific site (Garelick et al., 2008). However, owing to this property it has a negative impact on the environment as it can pollute larger areas (Tuli et al., 2010). Human activities, mostly through mining and smelting, immobilize arsenic compounds and are found in more places than where they would exist naturally (Shankar et al., 2014).

The major sources of arsenic production are from the copper, lead and zinc industries and in agricultural processes. These contaminants cause health hazards as they cannot be destroyed once they enter the environment.

12.3.3 Chromium (Cr)

A small part of the chromium that contaminates water will eventually dissolve. In water, chromium occurs in two oxidation states, + 3 and + 6. In saline water conditions a pH of more than8would suggest that Cr^{+} cation may be present. Oxidized chromium, Cr^{+6}, under the form of anionic compounds is found in its unstable form mainly in the upper layers of the water (shallow aquifers). In this context chromium is not detected in all the samples, and is confined to the villages nearer to the river. In all the instances Cr is found to be above 4 ppm. Generally, in waters, chromium in suspension form constitutes about 34–65% of the total chromium. Concentrations of chromium above 100 ppm are seldom encountered, except in industrial areas (Dojlido and Best 1993).

12.3.4 Cobalt (Co)

Cobalt present in the environment cannot be destroyed. It would however react with other particles or be adsorbed in soil or sediments present in water. Cobalt is mobilized only under acidic conditions; however, most cobalt will culminate in soils and sediments (Malinidevi and Inbanila 2016). The occurrence of cobalt in soils determined to a large extent by its bedrock. Cobalt concentration in the ground water varies from 1.35 to 14.67 ppm (Table. 12.1 and 12.2), and is generally high in the saline soils with warm and dry climate. Higher Cobalt is due to the effluent influence carried from different sources. In, general Cobalt occurs in the surface waters in small amounts, most often less than 0.1 ppm (Turekian and Scott 1994), and whose permissible limit is less than 0.1 ppm.

12.3.5 Nickel (Ni)

The accumulation of Ni in ground water is shown in Table 12.1. The minimum and maximum content of Ni is varying from 0.24 to 16.45 ppm respectively. The high concentration of Ni is noticed in Mission Thanda and Julur villages with more than 16 ppm, Sangam and Jampally are also found with high concentrations of Ni.

The major part of all the nickel compounds released in the environment will be adsorbed by sediments or soil particles and become immobile. In an acidic environment nickel would be more mobile and it will often dissolve in groundwater (Barałkiewicz and Siepak 1999), Nickel concentration is generally small in natural waters because it occurs in a colloidal form and is subject to rapid adsorption by loamy minerals and hydroxides of iron and manganese at the bottom sediments (Dojlido 1995). Nickel is easily accumulated in the biological environment, particularly in the phytoplankton or other aquatic plants.

12.3.6 Aluminium (Al)

Aluminium can be selectively leached from rock and soil to enter any source of water apart from its artificial sources. The ability of aluminium to quickly form insoluble compounds makes it practically safe for plants and animals (Bulanova et al., 2001; Momot and Synzynys 2005). Due to the tendency of aluminiumto form a cation under the action of acid rains it reacts with biological material and alters them or their biological functions.

It is observed that there seems to be variation to pre and post monsoon period. The high Al concentration is found at the stream confluence or meandering points.

Rudravelli, Peddaravualapally and Gokaram are the villages where high Al is concentrated. In the entire study area, the Al distribution is quite high and above W.H.O. permissible limits of 0.2 ppm.

12.3.7 Manganese (Mn)

Manganese a common element on earth is one among three toxic trace elements. It is present in a mineral form in rocks and soil and also in some underground polluted sources (Talaat et al., 2010). manganese does not occur separately in either surface or groundwater and is mostly associated with iron. Manganese is present mostly as manganese ion (Mn^{++}) in water sources and its salts are mostly soluble in acid rather than in alkaline water (Loranger and Zayed 1994).

The distribution of Manganese along the Musi River is shown in Table 12.1. The Mn is concentrated in Gaurelli (511.41 ppm); Bandavaripally (266.76 ppm) and Korremla (238.68 ppm) and is found with abnormal concentrations. The lowest Manganese is concentrated as observed at Kapraipally, Peddagudem and Varkatpally villages. In some of the villages, Mn could not be detected in water samples, these villages are away from river Musi and thus are not influenced.

Manganese often occurs together with iron in groundwater but it usually occurs in much lower concentrations than iron. It is present in soil as a result of mineral weathering and atmospheric depositions, originating from both natural and anthropogenic sources. In the present study the Mn content is very low and Fe could be detected in the water samples.

12.3.8 Barium (Ba)

The occurrence of barium is associated as natural compounds or due to anthropogenic environmental impact. Barium is a useful element and its mineral compounds have been used as a drilling fluid in oil and gas exploration, as a contrast agent in X-ray diagnostic tests and in various products like plastics, rubbers, paint, glass, carpets, ceramics, sealants, furniture, fertilizers and pesticides.

Naturally occurring barium is found in different types of rocks and can be found in soil and groundwater by leaching and eroding from sedimentary rocks. The solubility of barium compounds increases as the pH level decreases (US EPA, 1989). High concentrations are found in drinking water and are likely to be associated with groundwater in alkaline igneous and volcanic rocks and manganese-rich sedimentary rocks associated with a low pH and hence concentrations are, expected to be relatively stable. Organic barium compounds are ionic and are hydrolyzed in water (Cotton & Wilkinson, 1962). The concentration of barium ions in natural aquatic systems is limited by the presence of naturally occurring anions and possibly also by the adsorption of these ions onto metal oxides and hydroxides (Hem, 1959). Highest barium concentration of 369 ppm at R.K. Nagar on the south of river and Jampally, Poddutur, Thimaiguda and Gokarm are also found with above average along the river Musi and the lowest concentration of barium is 47.22 ppm found at Kachiwanisingaram on to the north of the river. The Ba concentration in the present study area is extremely higher than its permissible limits of 2.0 ppm.

12.4 Conclusions

The polluted water flowing through the river Musi seeps in to subsurface and contaminates mainly the shallow aquifers that sustain the water table. There is regular practice of utilizing the Musi water for the cultivation of paddy, vegetables, etc. The untreated waste water used for agricultural practices posing adverse impacts on water bodies and shallow aquifers, which makes the local ground water unsuitable for domestic, drinking and irrigation purposes. The problem is compounded during the monsoon times, where the percolation intensity is higher. The toxic element as soluble or particulate form present in the polluted river water seeps underground and reaches the water table, resulting in changing its texture and composition of both soil and water.

Wastewater has been in use for cultivation in peri-urban Hyderabad for a little over four decades now. The ground water pollution in this region is a major concern which needs a strategic plan and enforcement of strict vigilance. It is customary to handle the discharge of effluents through sewage treatment plants. There is a need to increase the number of sewage treatment plants (STP) to monitor the effluent treatment more effectively. A strategic plan may also be enforced to discourage the discharge of effluent in to the river or water bodies.

Among the emerging technologies bioremediation has been used with other physical and chemical treatment methods for complete management of diverse group of environmental pollutants. It is found to be suitable as a sustainable approach for the environmental pollution and management, hence, there is an urgent need for more research in this area. Efforts need to be made to generate a synergistic interaction between the environmental impact on the outcome and behaviour of environmental contaminants and assortment and performance of the most suitable bioremediation technique and other relevant techniques that can sustain the effective and successful operation and monitoring of a bioremediation process.

Bio-remedial measures may most effectively check the ground water contamination for future sustainable use. All along the banks of River Musi in the downstream region, it is necessary to cultivate pollution absorbent plants and evaporation of effluents and periodic de-silting the river bed must be in practice to control pollution to provide sustainable and portable water for domestic and agricultural use. Remediation of polluted sites using microbial process has proven effective and reliable due its eco-friendly features. There is an urgent need to implement these practices for a sustainable and economic development of the Telangana State.

Acknowledgements The Present Book chapter is an outcome of Ph.D Thesis submitted by Ms Blessy Ganduri. The authors wish to express their gratitude to DST-Inspire Fellowship for financial support. The authors also grateful to the Head Department of Geophysics, Osmania University, Hyderabad, India for the logistic support rendered during the progress of work.

References

Adimalla, N. (2019). Groundwater quality for drinking and irrigation purposes and potential health risks assessment: A case study from semiarid region of south India. *Expo Health, 11*, 109–123.

Barałkiewicz, D., & Siepak, J. (1999). Chromium, nickel and cobalt in environmental samples and existing legal norms. *Polish Journal of Environmental Studies, 8*(4), 201–208.

Blessy, G., Udayalaxmi, G., & Vidyasagarachary, D. (2019) Deleterious concentrations of heavy metals in ground water along river Musi, Telangana State, India. *International Journal of Geology and Earth Science, 5*(2), 146–158.

Bulanova, N. V., Synzynys, B. I., & Koz'min, G. V. (2001). Aluminum induces chromosome aberrations in cells of wheat root meristem. *Russian Journal of Genetics, 37*, 1455–1458.

Cotton, F. A., & Wilkinson, G. (1962). *Advanced inorganic chemistry* (pp. 441–448). NY, Interscience Publishers.

Dojlido J. R. (1995). Chemiawod powierzchniowych, Wyd. EkonomiaiSrodowisko, Bialystok.

Dojlido, J. R., & Best, G. A. (1993). *Chemistry of water and water pollution*. Ellis Horwood Limited.

EPA (U.S. Environmental Protection Agency). (1989). Risk assessments methodology. In *Environmental impact statement, NESHAPS for radionuclides, background information documents* (Vols. 1, 2. EPA-520/1-89/005 and EPA-520/1-89/006-1). U.S. Environmental Protection Agency, Office of Radiation Programs.

Garelick, H., Jones, H., Dybowska, A., & Valsanijones, E. (2008). Arsenic pollution sources. Springer, pp. 17–60.

Mebratu, M., & Zarabruk, S. (2011). Concentration of heavy metals in drinking water from urban areas of the Tigrey Region, northern Ethiopia mamona. *Ethiopia Journal of Science, 3*, 101–121

Geological Survey of India. (2002). Geology map with 1:250000 scale

Hem, J. D. (1959). *Study and interpretation of the chemical characteristics of natural water*. Scientific Publishers.

Hill, K. (2008). *Understanding Environmental pollution* (2nd ed.). University Press, New York.

Loranger, S., & Zayed, J. (1994). Manganese and lead concentrations in ambient air and emission rates from unleaded and leaded gasoline between 1981 and 1992 in Canada: A comparative study. *Atmospheric Environment, 28*, 1645–1651.

Malinidevi K, T. Inbanila (2016): Comparative study on heavy metal concentration in ground water and soil samples in and around Sipcot Industrial complex, Cuddalore district. *International Research Journal of Engineering and Technology (IRJET), 03*(04), 2620–2623.

Momot, O., & Synzynys, B. (2005). Toxic Aluminium and Heavy Metals in Groundwater of Middle Russia: Health Risk Assessment. *International Journal of Environmental Research Public Health., 2*(2), 214–218.

Tuli, R., Chakrabarty, D., Trivedi, P. K., & Tripathi, R. D. (2010). Recent advances in arsenic accumulation and metabolism in rice. *Molecular Breeding, 26*(2), 307–323.

Verma, R., Dwivedi, P. (2013). Heavy metal water pollution-A case study. *Recent Research in Science and Technology, 5*(5), 98–99.

S. Murcott, Arsenic Contamination in the World. (2012). *An International Sourcebook*. IWA Publishing.

Samba Shiva Rao, A., Manjunath, A. K., Ramesh, K., & HimaBindu, Y. (2017). Hydrogeo chemistry of Musi River basin, *International Journal for Research in Applied Science & Engineering Technology (IJRASET), 5(XII)*, 2760–2768.

Shankar, S., Shankar, U., & Shikha. (2014). Arsenic contamination of ground water: A review of source, Prebalence, Health risks and strategies for mitigation. *The scientific world Journal, 2014*(3045241), 18.

Srisailam Gogula and Sunder Kumar Koli. (2016). Effect of Musi River Pollution on Human Anthropogenic Activities. *Res. J. Chem. Sci., 6*(12), 18–24.

Talaat, H. A., Montaser, Y., Ghaly, M. Y., Kamel, E. M., Enas M. Ahmed, E.M., and Ahmed A. M. (2010) Simultaneous removal of iron and manganese from ground water by combined photo-electrochemical method, *Journal of American Science, 6*(12), 1–7.

Turekian, K. K., & Scott, M. (1994). Concentrations of Cr, Ag, Mo, Ni, Co and Mn in suspended material in streams. *Environmental Science and Technology, 1*, 940.

Chapter 13
Groundwater Quality, Pollution by Heavy Metals and Its Implication for Human Health Risk in Onitsha Urban, Southeastern Nigeria

G. N. Anachuna, T. N. Nganje, A. Edet, C. I. Adamu, Lynda-Uta Edet Okon, D. O. Inyang, and E. A. Kudamnya

Abstract This study examines the fluctuation in the quality of groundwater, the degree of pollution caused by Potentially Toxic Metals (PTMs), and possible risks to human health in Onitsha urban, southeastern Nigeria. The results indicated that compared with World Health Organization (WHO) recommended standards, the groundwater is fresh, slightly acidic, and unsuitable for drinking as regards to K^+, NO_3–, Cd, Mn, Pb, and Se. During the dry season, the water is classified as outstanding to good quality while in the wet season; the shallow and a few deep groundwater have poor to extremely poor quality. Based on the degree of contamination/pollution, in the wet season, the groundwater is of low pollution and no risk. However, the water becomes extremely polluted and is considered harmful to the health of children and adults if consumed in the dry season. The calculated indices, including contamination index (C_d), heavy metal pollution index (HMPI), and Water Quality Index (WQI), showed a strong and positive correlation, indicating that similar geological factors and input sources influence them. An assessment of health risks using the hazard index (HI) showed that children had a higher exposure to non-carcinogenic health risks during the wet season. Approximately 45% of the HI values exceeded the danger limit of 1, with nitrate (NO_3^-) pollution making a substantial contribution. The primary sources of nitrate pollution are natural and manmade activities. These include soil leachates resulting from decomposed organic matter, agricultural runoff caused by using fertilizers, and inadequate septic and waste disposal systems. An elevated concentration of NO_3^- exceeding the allowable level, constitutes a major health threat and of concern. Methemoglobinemia is

G. N. Anachuna · T. N. Nganje (✉) · C. I. Adamu · D. O. Inyang · E. A. Kudamnya
Department of Geology, University of Calabar, Calabar, Nigeria
e-mail: nganjetherese001@gmail.com; therese_nganje@yahoo.com

A. Edet
Department of Geosciences, University of Uyo, Uyo, Nigeria

L.-U. E. Okon
Physical Oceanography Unit, Institute of Oceanography, University of Calabar, Calabar, Nigeria

V. Sunitha et al. (eds.), *Groundwater Resource Management Planning Strategies*, Springer Water, https://doi.org/10.1007/978-3-031-88870-0_13

a widely recognized health problem associated with nitrates and is known to affect children. It is a deadly disease characterized by cyanosis and caused by bacteria in the gastrointestinal tract. Hence, treating, monitoring, managing, and enhancing the region's groundwater resources is imperative while safeguarding the well-being of individuals residing within the research area.

Keywords Groundwater · Physicochemical parameters · Heavy metals · Pollution · WQI · HMPI · HI · Nigeria

13.1 Introduction

Groundwater is a major natural resource that provides drinking water to over half of the global population. Groundwater contamination emerged as a significant problem in the late 20th century, mainly due to industrialization and the movement of people from rural to urban areas. The movement has led to a rise in the demand for clean water for consumption. The excessive reliance on groundwater resources, combined with the escalating exploitation, has resulted in the discharge of pollutant loads from both point and non-point sources into these water resources, thus leading to water pollution. (Etikala et al., 2022), states that contaminants emitted from different point and non-point sources possess the capacity to harm the physical, chemical, and biological components of urban water. Aside from point and non-point sources of groundwater contamination, the inherent geochemical composition of groundwater, resulting from chemical reactions with the minerals present in rocks and geological materials, has led to low-quality water due to the release of high concentrations of metals. In addition, Cd, Cr, and Ni have the potential to cause cancer and respiratory issues (Iqbal & Shah, 2013; Suvarapu & Baek, 2017), when their concentrations surpass the maximum permissible values in some countries, including Nigeria. Sudharshan Reddy and Sunitha (2023) have documented that the presence of heavy metals within the human body can lead to a range of health issues, including neurological impairment, developmental setbacks, renal damage, and the development of cancer. Heavy metals are significant contaminants that pose health risks, particularly impacting the chemistry and quality of groundwater (Table. 13.1).

Several authors have documented significant contributions to the quality of groundwater and human health in some areas of the world over the past few years, including N (Nganje et al., 2015; Rezaei et al., 2019; Shuaibu et al., 2024; Singh et al., 2017; Sudharshan Reddy & Sunitha, 2023; Taiwo et al., 2023). These studies have shown that both natural and human activities can introduce inorganic pollutants into groundwater by interacting with weathered minerals, which can affect the chemistry and overall quality of the water. Controlling and limiting the pollution caused by heavy metals is imperative to safeguard public health and the environment (Sudharshan Reddy & Sunitha, 2023). Due to the growing population, there is a higher demand for clean and suitable water. However, the current public water supply cannot match this demand, resulting in significant strain on surface water

Table. 13.1 List of some heavy metals and related health problems

Metal	Health effect	References
Cd	It is a toxic metal and carcinogenic. It has a broad spectrum of adverse health consequences in humans and animals, particularly in the liver and kidneys	Mohammadpour et al. (2023), Hussain and Rao (2014)
Cu	Essential for human health and nutrition. Excess can cause overweight/obesity, heart disease, high blood pressure, high cholesterol levels, and renal disease	ASTDR (2019)
Fe	Essential for human nutrition. Higher Fe intake can cause cardiovascular diseases	Narayanan (2007)
Mn	Essential for human nutrition. Excessive intake of manganese may cause damage to the central nervous system and systematic, immunological, neurological, reproductive, developmental, genotoxic, and carcinogenic effects on humans	Rőllin (2011), Williams (2012)
Pb	It is non-essential, toxic, and a B2 carcinogen. Excess intake of Pb in the human system may cause a breakdown in the biochemical functioning of the liver, kidney, lung, cardiovascular, and nervous toxicity in children. Inhibition of the synthesis of hemoglobin and may cause acute and permanent brain damage	Gall et al. (2015), ASTDR 2019, Boskabady et al. (2018), Sudharshan Reddy and Sunitha (2023)
Se,	Essential for life. Excess over long periods may lead to selenosis, a disease known for hair loss, brittle nails, and numb extremities	ASTDR 2003
Zn	Essential for life. Excess may cause neurological disorders, Osteoporosis, and reproductive effects. system dysfunction, and impairment of growth and reproduction	ASTDR, 2019, Sudharshan Reddy & Sunitha, 2023

sources and leading to their depletion. Etikala et al. (2022) have identified the depletion of water resources as a worldwide issue. This depletion is caused by water loss through seepage, leaks, and pipeline failures, which are attributed to the aging and poor quality of infrastructure.

Onitsha Urban, a prototypical emerging economy, is the most populous metropolis in southeastern Nigeria and is currently grappling with severe surface and groundwater issues that have significant health consequences. Water shortages are common even in the most developed areas of the city. Occasionally, taps experience water scarcity due to inadequate water provision and are threatened by environmental issues and pollution. During severe water scarcity, water from uncontrolled water sources such as wells, boreholes, streams, and rivers are sold by dishonest individuals utilising water tanks. This practice contributes to a higher prevalence of diseases like diarrhea. Orjil et al. (2015) reported a total of 1011–1143 instances of dysentery, 5164–7240 cases of diarrhea, and 5010–6459 cases of typhoid, which are waterborne diseases, in the Onitsha urban area between 2011 and 2014. Diarrhea was the most common disease, and children were the most afflicted. Several studies (Akudinobi & Chinenye, 2013; Ezeabasili et al., 2015; Igbomor & Amadi, 2019; Isikhueme & Omorogieva, 2015) have documented cases of water quality, contamination and pollution in the urban area of Onitsha. These cases have been attributed to water use from unregulated and unprotected sources, known to have poor quality. None of the research conducted in Onitsha considered the seasonal fluctuations in the chemistry and pollution levels of the groundwater. Given the current trajectory of economic growth, urbanization, industrialization, and the need for clean drinking water, it is crucial to understand the quality of groundwater resources and the characteristics of heavy metals in terms of pollution. This study encompasses the following objectives: (1) Analysing the composition and quality of groundwater. (2) Assessing the origins of heavy metals and their appropriateness using contamination indices. (3) Estimating the health risks associated with consuming polluted water. The hydrogeochemical data acquired from this study will contribute to the worldwide database, improve the management and safety of drinking water, evaluate the potential future alterations in groundwater properties, formulate policies, establish monitoring programs, and propose remedial measures.

13.2 Materials and Methods

13.2.1 Study Area Description

Onitsha is located within the latitudes of 06°04′ 5″ and 06°10′ 10.7″ N of the equator and the longitudes of 006°44′ 47.8″ and 006°50′ 31.8″ E of the Greenwich Meridian (Fig. 13.1). The overall area of the span is 100 square kilometers. According to Wikipedia, as of 2022, Onitsha, situated in Anambra State, has a population of 1,499,000. The population of Anambra State as a whole is predicted to be 11,400,000. Onitsha is located in the Nkisi and Idemili watershed, which ultimately drains into the adjacent River Niger.

Onitsha is located in Nigeria's humid tropical rainforest region, which spans from dense forest to savannah. This area has an average annual temperature range of 22–27 °C and an average yearly rainfall between 1500 and 2500 mm with a dry and wet season. The dry season lasts from November to March, and the wet season lasts from April to October. The dry season has a peak temperature of 34 degrees Celsius while towards the end of the rains, it is about 18–21 °C (Offodile, 2002).Onitsha is located in the Anambra basin and is hydrogeologically characterized by coastal plain sands (Benin Formation) near the River Niger, Ogwashi Asaba Formation, which consist of coarse grain, sandstone, clay, and carbonaceous shale (lignite) underlain by Ameki Group consisting of Nanka sand and Nsugbe Formations (Nwajide, 2014). The water table in Onitsha has a high elevation, measuring approximately 20–30 m (Offodile, 2002). Isikhueme and Omorogieva (2015) identified four aquiferous units in Onitsha and surrounding areas. These units are categorized as shallow, middle, upper, and deep, with depths ranging from 15–24 m, 30–32 m, less than 90 m and greater than 90 m, respectively. The aquifers exhibit characteristics such as coarse sand, clay, shale/sandy, and medium-grained sand zones.

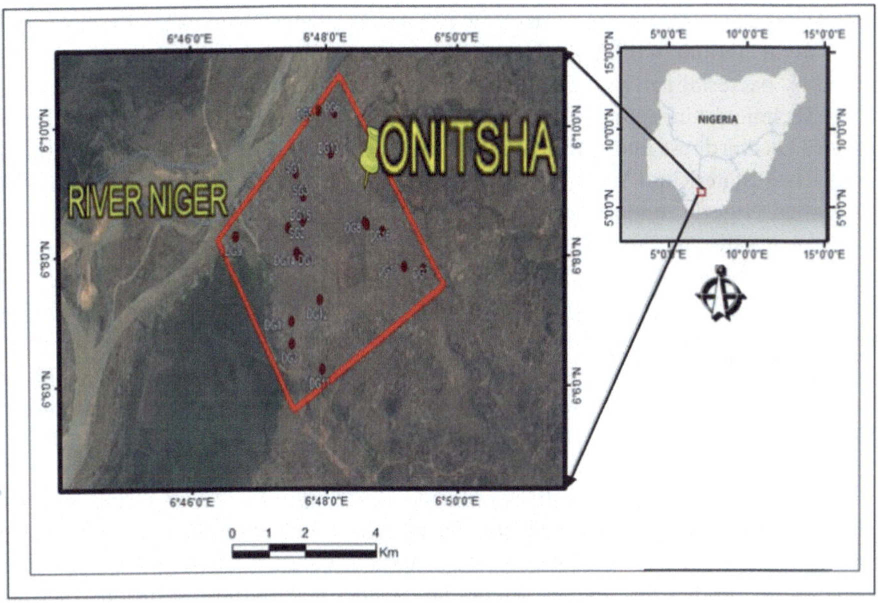

Fig. 13.1 Location of the study area showing sampling sites

Onitsha is primarily characterized by its residential areas, commercial establishments such as markets, filling stations, mechanic workshops, and industrial sectors, including textile, pharmaceutical, brewery, plastic, and paint production, schools and dumpsites. The discharge of waste from industrial facilities, such as corrosion of galvanized plumbing and water main pipes, disposal and spills, leaching from batteries, paints, and fertilizers, as well as commercial and domestic activities, including the disposal of batteries, paint, and gasoline containing lead, are potential pathways for the introduction of inorganic and organic pollutants into groundwater (Ezeabasilli et al., 2015). The unplanned and unregulated expansion has resulted in issues within the study area, making it difficult to distinguish between residential and commercial zones (Izueke & Eme, 2013).

13.2.2 Sampling and Analytical Procedure

Forty (40) samples were taken in 2021 for the two seasons (dry and wet) using prewashed, well-labeled 250 ml polyethylene bottles, as depicted in Fig. 13.1. Two samples were obtained from each location. The physical parameters like electrical conductivity (EC), pH, and Total Dissolved Solids (TDS)) were measured in the field using a Hanna portable meter. The samples for the major anion parameters were collected and stored at a low temperature cooler filled with ice, before being transported to the laboratory. The samples for cation parameter analysis were filtered using Whatman filter papers measuring 125 mm diameter. They were then acidified with a small amount of diluted HNO_3^- to prevent any changes in metal concentration caused by bacterial activity. The samples were sent to the Institute of Bio-sciences at the Technical University Freiberg in Germany for analyses of both anions and cations. The hardness and carbonate levels were measured in facilities hosted by the Cross River State Rural Water Supply and Sanitation Agency (CRS-RUTWATSSA) in Calabar, Nigeria, under typical operating conditions. The hardness of the water sample was estimated using the titration method, while the carbonate content was measured using the Wagtech photometer (7100). The procedure involved filling a test tube with the water sample up to the 10 ml mark and adding an alkaphot tablet. The tablet was then crushed and mixed until all its particles had dissolved. After being left unattended for a minute, it was subsequently remixed. Phot 2 was chosen using the Wagtech photometer, and a measurement was recorded. The outcome was presented as milligrams per litre of calcium carbonate (alkalinity), which was transformed into total alkalinity as bicarbonate (HCO_3^-) by multiplying the outcome by 1.22.

Analyses of the anions such as chloride (Cl^-), nitrate (NO_3^-), and sulphate (SO_4^{2-}) were done using Ion Chromatography (IC) with a Dionex DX-120 instrument. The IC system had an AS4A-AC analytical column, an AMMS-11 (4 mm) regenerating suppressor, and a conductivity detector. Standard solutions of these anions were created by dissolving 1000 mg/l of the respective salts in deionized water. The samples were injected into the system through a 12.5μL sample loop. The elution process used a solution containing 1.8mlM Na_2CO_3 and 1.7 nM $NaHCO_3$. Working standards were established to calibrate the system by diluting the samples further, and the calibration curve of the various anions was created, which followed a linear pattern. The method's sensitivity under the employed working conditions was assessed by examining the limit of detection (LOD). Inductively Coupled Plasma—Mass Spectrophotometer (ICP-MS), specifically the Thermo Scientific Element XSERIES 2 was employed in the analysis of major cations (Na, K, Ca, Mg) and potentially toxic metals (Ag, Cd, Cu, Fe, Mn, Pb, Se, and Zn). Internal standards (10 μg/l of Re (Rhenium) and Rh (Rhadium) were incorporated into the analysis; before the examination, the water samples were diluted to a maximum of 1%. The ICP-MS was configured to measure each sample's average values and percent relative standard deviation (%RSC) of three replicate studies.

Analytical precision was achieved by performing analyses on known standards, duplicates, and blank samples after every ten samples. The precision was quantified as the percent relative standard (%RSC). In addition, calibration check standards were included in the analyses to guarantee quality control and analytical precision of less than 10% for cations, with data reproducibility of within 5% for the anions.

13.3 Data Analysis

Microsoft 10 Excel spreadsheet was used to compute a summary of the statistical parameters (mean, standard deviation, minimum, and maximum values). The concentrations of key elements in various samples were analysed for spatial and temporal changes using Principal Component Analysis (PCA) and Pearson's correlation using OriginPro® 2021 software (Origin Lab., MA, USA). Principal Component Analysis (PCA) is an advantageous technique for streamlining information and uncovering its innate structures. Pearson's correlation detects linear relationships among the measured parameters for interpretation (Edet, 2019).

13.3.1 Potable Water Quality

Numerical measures in terms of the water quality index (WQI) have been extensively utilised to evaluate water quality. These indices were applied in the studies of Ponsadailakshmi et al. (2018) and Nganje et al. (2020). The computation of the water quality index (WQI) was employed as outlined in the study by (Shuaibu et al., 2024). In this computation, based on its significance to human health (Kumar et al., 2024), each parameter was given a particular weight (wi), which varied from 5 to 1. The parameters were assigned a maximum weight of 5 and a minimum weight of 1. Relative weight (Wi) was determined as follows:

$$Wi = \frac{wi}{\sum_{i=1}^{n} wi} \qquad (13.1)$$

As outlined in Shuaibu et al. (2024), Wi represents the relative weight, wi represents the weight assigned to individual parameters, and n represents the sum of the parameters. The parameters analysed in this study included pH, hardness, TDS, nitrate, sulphate, iron, and bicarbonate. These parameters were assigned weights of 5, 5, 5, 3, 3, and 2, respectively. The elements Fe and HCO_3^- were given a value of 2. The pH, hardness, total dissolved solids (TDs), nitrate, sulphate (SO_4^{2-}), iron (Fe), and bicarbonate (HCO_3^-) were assigned relative weights of 0.18, 0.18, 0.18, 0.18, 0.11, 0.11, and 0.07, respectively. These parameters were utilised in the construction of the WQI. The rating (qi), also known as the quality rating for each parameter, was determined as follows:

$$qi = Ci/Si \times 100 \qquad (13.2)$$

The concentration of different parameters, denoted by Ci, is calculated by dividing each quality parameter by its respective upper permissible limit set by WHO (2008, 2011). The limits are as follows: 8.5 for pH, 1000 for TDS, 45 for NO_3^-, 250 for SO_4^{2-}, 0.3 for Fe, and 600 for HCO_3^-. The quality rating, qi, and Si (sub-index), of the *i*th parameter were computed as:

$$Si = Wi \times qi \tag{13.3}$$

$$WQIforeachlocation = \sum Si \tag{13.4}$$

ΣSi is the total sub-index of the *i*th parameter (Shil & Singh, 2019).

Equation (13.3) states that Si equals the product of Wi and qi.

WQI) for each location was calculated by summing up the individual water quality scores (Si) for that location.

The water is classified into different categories based on its Water Quality Index (WQI) values as stated by Shuaibu et al., 2024 as excellent water, < 50, good water, 50–100, poor water, 100–200, very poor water, 200–300, and unsuitable > 300.

13.3.2 Contamination Index (C_d)

The value of C_d was determined using the calculations provided in Eqs. 13.5 and 13.6 (Edet & Offiong, 2002).

$$\text{Cd} = \sum_{i=1}^{n} Cfi \tag{13.5}$$

$$Cfi = \frac{CAi}{CNi} \tag{13.6}$$

The contamination factor (Cfi), as stated by Edet and Offiong (2002), represents the level of contamination for the ith component, CAi denotes the analytical value of the *i*th component, while CNi represents the upper allowable concentration of the *i*th component. N represents the normative value. Classification of C_d levels, as described in Nayak et al., 2023, is based on three categories: C_d < 1 (low contamination), C_d 1–3(medium contamination), and C_d > 3 (high contamination).

13.3.3 Heavy Metal Pollution Index (HMPI)

The HMPI quantifies the overall water quality Majhi and Biswal (2016). The evaluation of the HMPI was conducted using the weighted arithmetic quality mean approach established in two parts, as outlined in (Zakir et al., 2020). Firstly, Qi, the rating scale, is established for each selected parameter by assigning weightage, as shown in Eq. 13.2, and secondly, the pollutant. According to Edet and Offiong (2002), the rating system is a subjective measure ranging from zero to one. The choice of the rating depends on the importance of the parameter's quality. The unit weightage (Wi) value is inversely proportional to the recommended standard of the associated heavy metal (Zakir et al., 2020), as stated in Eq. 13.1. This study awarded the elements Cd and Pb a weight of 5. The PTMs (Ag, Fe, Mn, Se, and Zn) were given a weight of 3, while Cu was assigned 2. The given constants of proportionality are as follows: 0.11 for Ag, Fe, Mn, Se, and Zn, 0.19 for Cd and Pb, and 0.07 for Cu. The calculation of HMPI is determined using the following method:

$$HMPI = \frac{\sum_{i=1}^{n}(Qi \rightleftharpoons xWi)}{\sum_{i=1}^{n} Wi} \tag{13.7}$$

Qi denotes the sub-index of the ith parameter, Wi represents the unit weightage for the ith parameter, and n is the number of examined parameters. According to Majhi and Biswal (2016), a concentration of HMPI of more than 75 is deemed critical and not appropriate for consumption.

13.3.4 Environmental Water Quality Index (EWQI)

EWQI is an additional technique to assess groundwater quality, as Zakir et al. (2020) described. It involved the integration of both WQI and Heavy Metal Toxicity Index (HMTI) (Zakir et al., 2020). To assess the EWQI, TMTI was calculated first as follows:

$$TMTI = \sum_{i=1}^{n} T_f \tag{13.8}$$

$$T_f = C_S \times T_{si} \tag{13.9}$$

Tf is the toxicity factor, n denotes the number of parameters, Cs indicates the concentration of each potentially toxic metal (PTM), and Tsi corresponds to the total point score for each PTM determined from the list provided by Zakir et al., 2020. All heavy metals were investigated in this study except for Fe, which had no total point score, likely due to its aesthetic quality. The Tsi of the following elements are as follows: Ag (604), Cd (1318), Cu (805), Mn (797), Pb (1531), Se (775), Zn (913).

The equation EWQI is defined as the quotient of WQI divided by the sum of the product of Cs and TSi and the number of elements n being considered as presented in Eq. 13.10:

$$EWQI = WQI / \sum ni = 1Cs \times TSi \tag{13.10}$$

The EWQI categorizes water into different classes. A value of less than one (<1) indicates acceptable water quality, whereas a value greater than one (>1) indicates hazardous water quality.

13.4 Assessment for Irrigation Quality for Agricultural Applications

Appendix 1 summarizes data evaluation techniques for irrigation. The irrigation quality was evaluated using Kelly's Index (KI), Permeability Index (PI), Magnesium Hazard (MH), Sodium Adsorption Ratio (SAR), and Sodium percentage (Na%). Sulphate and chloride are vital nutrients for plant growth and were employed by Edet (2019) for irrigation assessment. The summary of the data evaluation methods for irrigation is presented in appendix 1.

13.5 Assessment of Risk to Human Health

This study utilised risk assessment methods, as described by Singh et al., 2017, to evaluate the exposure and non-carcinogenic risk, specifically through water consumption. This investigation included the elements Mn, Zn, Pb, Fe, Cd, Cu, and NO_3^-. To accomplish this, the average daily intake was first calculated as below:

$$ADI = \frac{C \times IngR \times EF \times ED}{BW \times AT} \tag{13.11}$$

ADI stands for Average Daily Intake, which measures the amount of metals humans are exposed to daily. C is the concentration of the contaminant in drinking water (Nganje et al., 2015) and is measured in milligrams/litre (mg/l). The ingestion rate per unit time (IngR) was 0.78 L per day for children and 2.5 L per day for adults. The exposure duration (ED) is six years for children and 70 years for adults. The EF, known as exposure frequency, is calculated as the number of days per year (365). Children's body weight (BW) was 15 kg (Kg), and adults' body weight was 70 kg. The average time (AT) is calculated as the product of the exposure duration (ED) and the conversion factor from years to days (365).

On the other hand, the hazard quotient (HQ) quantifies the health risk posed by pollutants and assesses the magnitude of the negative impact on non-cancer health risks. This evaluation was conducted based on the methodology as outlined in the study of (Agbor et al., 2024), specifically defined in Eq. 13.12.

$$\mathrm{HQ} = \mathrm{ADI} \tag{13.12}$$

RfD

ADI stands for Average Daily Intake, and RfD refers to the Oral Reference Dose (RfD) in milligrams per litre per day (mg/l/day). The Reference Dose (RfD) values are as follows: 20 for manganese (Mn), 300 for zinc (Zn) and iron (Fe), 1.4 for lead (Pb), 0.5 for cadmium (Cd), 40 for copper (Cu), and 1.6 for nitrate (NO_3^-).

The hazard index (HI) examines the possible risks to human health. HI values were calculated by adding the HQ values for each parameter. The calculation was performed independently for adults and children using the equation as shown below:

$$HI = \sum_{i=1}^{n} HQi \tag{13.13}$$

i represents the HQ for each parameter. Values less than 1 indicate no risk, and values greater than 1 depict a high health risk, suggesting the probability of serious health effects.

13.6 Results and Discussions

13.6.1 Overall Physicochemical Parameters

The groundwater physicochemical parameters, including maximum, minimum, and standard deviation, were analysed for seasonal change. The results are summarized in Table 13.2, and more detailed results are presented in Appendix 2. The analysis considered the guidelines of SON (2007) and WHO (2011). The mean values of the parameters do not exhibit a specific pattern of variation, which may be attributed to the peculiarities of the aquifers.

The average pH values (< 6.5) and total dissolved solids (TDs) (< 1000 mg/l) suggest that the groundwater is moderate to mildly acidic and fresh. This finding is consistent with previous studies (Igbomor & Amadi, 2019; Okoyeh et al., 2021). The presence of acidic groundwater in the study area can be attributed to both natural processes and microbiological activity, as well as the oxidation of pyrite from the shale bedrock that is intercalated with the medium and coarse sand. Another possible factor contributing to low pH levels could be the decomposition of plant matter. The average hardness values, K^+, $SO4^-$, NO_3^-, Cd, Mn, Pb, and Se in shallow groundwater (SG) and deep groundwater (DG) surpassed the recommended threshold values set by WHO (2008, 2011) and SON (2007). This suggests that the groundwater needs to meet the criteria for being safe to drink based on these parameters. The chemical parameters decrease in concentration in the following order: $SO_4^{2-} > Cl^- > NO_3^- > HCO_3^- > Na+ > Ca^{2+} > K^+ > Mg^{2+} >$ Zn > Mn > Cu > Fe > Ag > Pb > Cd > Se(SG). For DG, the order is $SO_4^{2-} > NO_3^- > Cl^- > HCO_3^- > Na^+ > Ca^{2+} > K^+ > Mg^{2+} >$ Zn > Mn > Pb > Fe > Ag > Cd > Cu > Se.

Table 13.2 Statistical summary of physicochemical parameters in groundwater

		Shallow ground water (SG)				Deep groundwater (DG)				Drinking water		Irrigation
N		3				17				Quality standard		Standard
		SG				DG						
Parameters	Units	Mean	Std	Min	Max	Mean	Std	Max	Min	WHO (2008, 2011)	SON (2007)	FAO
pH		5.86	1.09	4.62	6.68	5.42	0.78	7.31	4.1	6.5–8.5[b]	6.5–8.5	6.5–8.5
EC	µs/cm	550.92	467.82	200.85	1082.25	178.98	175.87	628.5	0	1500[b]	1000	3000
TDS	mg/l	436.03	54.47	373.19	467.82	77.82	99.27	295.74	6.44	1000[b]	500	2000
Hardness	mg/l	360.07	95.7	304.18	470.57	232.18	168.03	531	7.12	500[a]	150	
Na	mg/l	66.86	10.49	59.6	78.89	17.37	14.28	50.77	1.01	200[b]	200	920
K	mg/l	30.33	3.84	26.96	34.51	6.91	5.79	24.33	0.23	12[b]	–	2
Ca	mg/l	51.67	36.06	22.45	91.97	7.88	7.04	25.34	1.45	100[b]	–	400
Mg	mg/l	12.97	0.49	12.6	13.52	6.58	5.23	16.96	0.33	20[b]	0.2	60.8
Cl	mg/l	219.97	48.32	164.21	249.6	53.82	48.34	170.61	2.56	250[b]	250	1063.5
HCO_3	mg/l	153.62	86.1	56.67	221.19	23.25	25.57	73.99	0	600[b]	–	610
SO_4	mg/l	220.15	115.1	123.92	347.66	146	156.02	619.72	0	200[b]	100	960
NO_3	mg/l	165.85	54.43	107.01	214.39	59.14	71.06	214.54	0.98	50[a]	50	
Ag	µg/l	54.52	28.84	24.51	82.03	55.95	62.25	228	0	100[a]	–	
CD	µg/l	25.34	11.19	13.09	35.02	30.65	23.82	77	0.07	3[a]	3	0.01
CU	µg/l	94.1	80.83	28.84	184.51	24.55	24.53	85.46	0	2000[a]	1000	0.2
Fe	µg/l	78.93	58.7	36.54	145.93	116.69	115.55	451	34.93	300[a]	300	5
Mn	µg/l	1029.77	1406.66	0.1	2632.5	282.75	331.82	1149.8	3.63	200[b]	200	0.2
Pb	µg/l	25.79	42.21	0.71	74.52	137.85	415.25	1716.05	0	10[a]	10	5

(continued)

Table 13.2 (continued)

		Shallow ground water (SG)				Deep groundwater (DG)				Drinking water		Irrigation
N		3				17				Quality standard		Standard
		SG				DG						
Parameters	Units	Mean	Std	Min	Max	Mean	Std	Max	Min	WHO (2008, 2011)	SON (2007)	FAO
Se	μg/l	17.53	3.72	14.61	21.72	6.44	5.81	20	−0.03	10[a]	-	
Zn	μg/l	1194.08	814.02	357.93	1984	355.79	354.42	1165.99	7.01	3000[a]	3000	2

FAO (Ayers &Westcot 1985) values for Ag, Cd, Cu, Fe, Mn, Pb, Se and Zn are in mg/l; [a] WHO (2008); [b] WHO (2011); SON Standard Organisation of Nigeria

Nevertheless, when considering the average concentration, shallow groundwater exhibits higher contamination levels than deep groundwater. This is because shallow groundwater is more susceptible to contamination from unregulated sources than deep groundwater. The chemical parameters with the highest and lowest mean concentration were SO_4^{2-} and Se, respectively, for SG and DG. Furthermore, 45% of the maximum values for SG and 50% for DG surpassed the maximum allowable limits. Fluctuation in the physicochemical properties of water, like EC, acidity and pH, significantly influences the concentration of ions and metals in groundwater (Masood et al., 2024).

13.6.2 Spatial–Temporal Variation

Based on the information provided in Table 13.3 and Fig. 13.2a, b and c, it is evident that there is no discernible pattern of variation with season and location. pH is a crucial factor in evaluating the acidity or alkalinity of water. The average levels in both seasons did not meet the criteria of potable water consumption by humans, as stated by WHO (2011) and SON (2007), which are classified as slightly acidic. The groundwater at greater depths exhibited higher acidity, with pH values as low as 4.1 in the wet season. This may be associated with elevated concentrations of dissolved acids in the relatively reduced deep groundwater. These acids likely originate from decomposed vegetation and pyrite oxidation in the shale rock, which reacts with water and releases $H_2SO_4^{2-}$.into the groundwater. A significant proportion, ranging from 70.59% to 82.5%, of pH levels during the rainy and dry seasons fell below the required range of 6.5 to 8.5. Only a few areas managed to maintain pH levels within the prescribed ranges.

The mean values obtained for electrical conductivity (Ec) in drinking water were within WHO and SON limits of 1500 and 1000 μs/cm, respectively. However, there was one exception during the wet season at DG 1, when the maximum concentration value of Ec was measured to be 2310 μs/cm, exceeding both standards. The rise in the electrical conductivity of water indicates an increase in the concentration of ions (Meride & Ayenew, 2016). The mean hardness values were above the limitations set by the Standards Organization of Nigeria (SON) for SG in both seasons and for DG in the dry season. These findings are reported in Table 13.4. Hard water can negatively affect human health by increasing and contributing to reproductive health problems and constipation. The mean concentration of chemical data decreases in the following order: $SO_4^{2-} > NO_3^- > Na^+ > HCO_3^- > Cl^- > Mg^{2+} > K^+ > Ca^{2+} >$ Pb > Zn > Fe > Mn > Ag > Cu > Cd > Se (dry season). In the wet season, the order is $SO_4^{2-} > Cl^- > HCO_3^- > NO_3^- > Ca^{2+} > Mg^{2+} > Na^+ > K^+ >$ Fe > Cu > Pb > Mn > Zn > Cd > Ag > Se. SG had the highest average concentration of chemical parameters observed during the rainy season. This might be attributed to the impact of human activities, leading to contamination of the surface water into SG. The high vulnerability of SG to surface contamination is due to its proximity to the exposed bedrock. In addition, water infiltrates the pores of weathered exposed bedrock during the wet season, causing

Table 13.3 Statistical summary of physicochemical parameters in shallow groundwater with season

		SG										
N		3								Drinking water quality standard		Irrigation standard
U Parameters	Units	Dry season				Wet season				SON (2007)	WHO (2008, 2011)	FAO
		Mean	STD	MIN	Max	Mean	STD	MIN	Max			
pH		6.22	0.58	5.54	6.56	6.62	0.54	6.02	7.05	6.5–8.5	6.5–8.5 [b]	6.5—8.5
EC	µs/cm	678.33	257.56	398	904.5	643.33	612.39	58.5	1280	1000	1500[b]	3000
TDS	mg/l	17.06	12.40	9.36	31.37	321.61	526.91	11.42	930	500	1000[b]	2000
Hardness	mg/l	452	50.68	393.5	482.5	678.71	190.65	458.64	793.5	150	500[a]	–
Na	mg/l	200.97	230.75	13.38	458.64	44.22	55.84	11.02	108.7	200	200[b]	920
K	mg/l	9.46	1.95	7.21	10.67	19.18	22.09	4.33	44.57	–	12[b]	2
Ca	mg/l	7.29	1.97	5.435	9.35	63.27	92.60	2.08	169.8	–	100[b]	400
Mg	mg/l	12.18	7.36	4.035	18.37	54.73	54.66	12.35	116.43	0.2	20[b]	60.8
Cl	mg/l	129.63	192.58	18.18	352	299.33	246.26	30	513	250	250[b]	1063.5
HCO_3	mg/l	188.78	165.06	13.34	341	41.16	34.23	18.48	80.53	–	600[b]	610
SO_4	mg/l	120.46	171.43	10.98	318.02	346.27	122.50	205.09	424.67	100	200[b]	960
NO_3	mg/l	288.21	307.19	5	614.78	4.27	4.03	0.02	8.02	50	50[a]	
Ag	µg/l	59.66	90.37	6.06	164	93.52	82.02	0.05	153.5	–	100[a]	
Cd	µg/l	63	45.64	26	114	123.00	128.17	6.01	260	3	3[a]	0.01
Cu	µg/l	82	34.60	55	121	315.03	501.06	3.08	893	1000	2000[a]	0.2
Fe	µg/l	227.33	159.93	51	363	3263.07	3378.83	0.20	6747	300	300[a]	5
Mn	µg/l	321	447.80	55	838	229.80	395.95	0.04	687	200	200[b]	0.2

(continued)

Table 13.3 (continued)

		SG										
N		3								Drinking water quality standard		Irrigation standard
U Parameters	Units	Dry season				Wet season				SON (2007)	WHO (2008, 2011)	FAO
		Mean	STD	MIN	Max	Mean	STD	MIN	Max			
Pb	μg/l	1090.67	1889.09	0	3272	273.81	286.19	11.43	579	10	10[a]	5
Se	μg/l	60	78.62	0	149	25.72	15.77	7.5	35	–	10[a]	
Zn	μg/l	244	371.53	27	673	202.2	2699.01	356	5136	3000	3000[a]	2

FAO (Ayers &Westcot 1985) values for Ag, Cd, Cu, Fe, Mn, Pb, Se and Zn are in mg/l; [a] WHO (2008); [b] WHO (2011); SON Standard Organisation of Nigeria

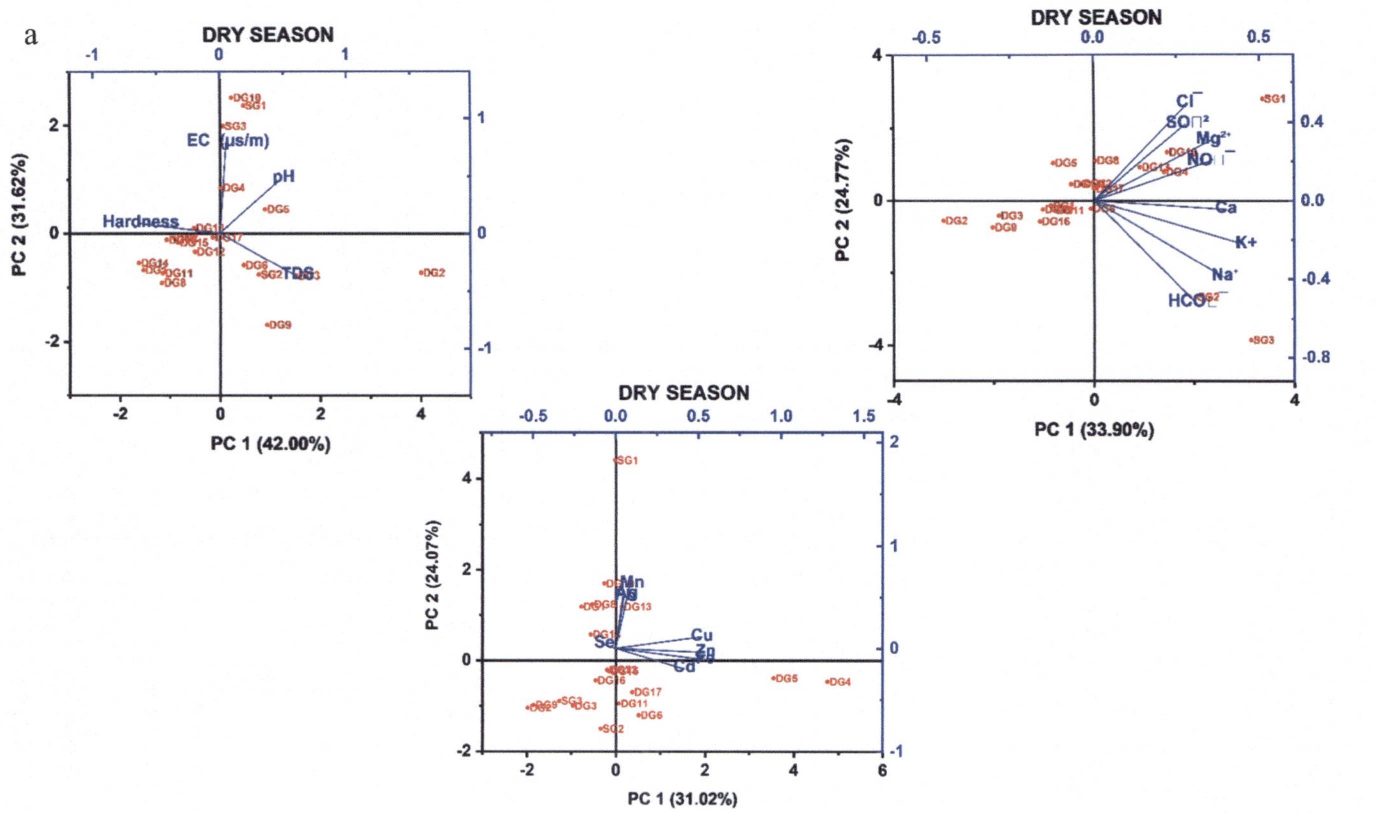

Fig. 13.2 **a** Biplot showing spataial variation of physicochemical parameters for dry season in groundwater **b** biplot showing spatial variation of physicochemical parameters for wet season in groundwater **c** biplot showing Temporal variation of physicochemical parameters in groundwater

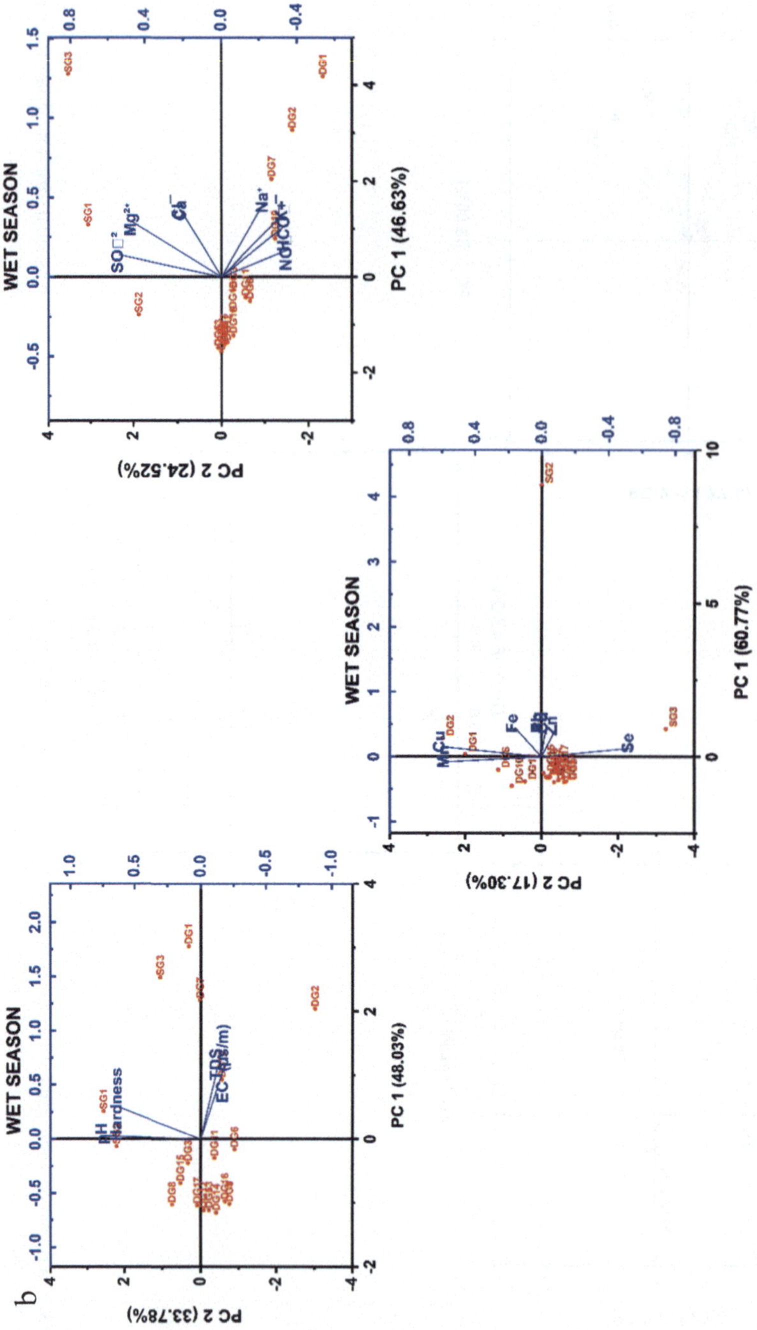

Fig. 13.2 (continued)

Fig. 13.2 (continued)

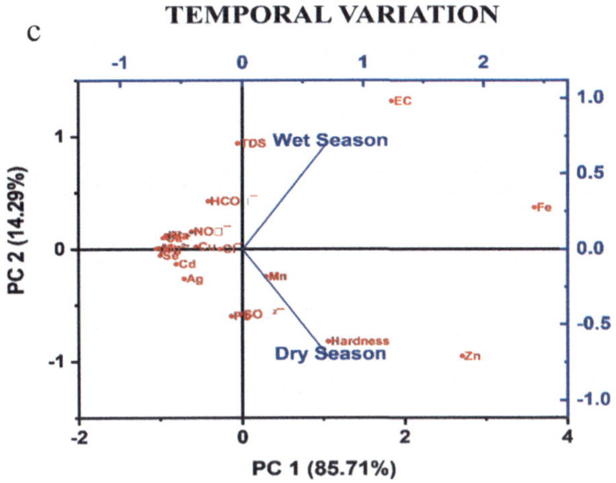

the release of ions and metals into the water. The fluctuation in the concentration of chemical parameters in SG and DG across the seasons can be attributed to diverse seasonal inputs, bedrock geochemistry, and human activities. Furthermore, as stated by Mohammadpour et al. (2023), the accessibility of bedrock, soil composition, regional human activities, and the solubility of heavy metals contribute to a noticeable increase in heavy metal concentrations in water (Bahita et al., 2021; Kazemi et al., 2022).

The studies uncovered two distinct spatial patterns that vary with seasons and groundwater quality. During the dry season, groundwater samples' pH, TDS, EC, and hardness can be used to understand their interactions and determine the differences in concentration between shallow and deep groundwater samples. At locations SG1, SG2, and DG4 during the dry season, the Total Dissolved Solids (TDS) exhibited a robust association with Electrical Conductivity (EC), indicating a strong positive correlation between them. This suggests that these sites with elevated TDS levels also tend to have high EC values. Sunitha et al. (2022) identified several factors that contribute to the higher electrical conductivity (EC) in the Cuddapah region of South India, including a high mineral content, increased salinity, temperature, and the presence of certain ions. These factors are particularly prevalent in hard rock terrain. According to the study, an increase in EC concentration is directly related to an increase in TDS concentration. This suggests that increased changes in electrical conductivity may be attributed to both natural geochemical processes and human activities in the area. The ions NO_3^-, Mg^{2+}, Cl^-, SO_4^{2-}, K^+, Na^+, Ca^{2+}, and HCO_3^- are linked to DG4, DG10, SG2, and SG3 during the dry season. On the other hand, the PTMs (Ag, Cd, Cu, Fe, Mn, Pb, Se, and Zn) are associated with DG1, DG3, DG6, and DG13 during the same period. Furthermore, during the rainy season, pH, Hardness, TDs, Ec, and SO_4^{2-} levels are linked to SG1 and DG1. On the other hand, SO_4^{2-}, NO_3^-, HCO_3^-, Na^+, Cu, Mn, Fe, Cd, Pb, and Zn exhibit a shared correlation and are connected to SG2, DG7, DG2, and DG1 correspondingly. This association

Table 13.4 Statistical summary of physicochemical parameters in deep groundwater with season

N		DG								Drinking water quality standard		Irrigation standard
		17										
		Dry season				Wet season						
Parameters	Units	Mean	STD	MIN	Max	Mean	STD	MIN	MAX	WHO (2008, 2011)	SON (2007)	FAO
pH		5.97	0.48	5.16	6.58	5.10	1.52	0.73	7.2	6.5–8.5 [b]	6.5–8.5	6.5–8.5
EC	µs/cm	182.25	238.60	6.73	904.5	522.92	744.87	8.5	2310	1500[b]	1000	3000
TDS	mg/l	11.60	5.01	4.91	24.56	237.43	311.79	1	920	1000[b]	500	2000
Hardness	mg/l	500.67	148.96	147.62	791	71.78	71.24	5.65	222.11	500[a]	150	
Na	mg/l	10.55	3.03	4.59	16.76	41.63	52.09	1.01	152.29	200[b]	200	920
K	mg/l	5.80	1.94	0.77	8.49	39.58	75.37	0.23	247.82	12[b]	-	2
Ca	mg/l	3.65	2.72	0.348	9.28	22.00	22.01	1.863	74.15	100[b]	-	400
Mg	mg/l	12.16	5.62	3.524	18.7	4.04	5.04	0.241	17.49	20[b]		60.8
Cl	mg/l	136.66	111.08	22.67	352	67.86	112.08	1.31	379.33	250[b]	250	1063.5
HCO_3	mg/l	10.99	3.64	6.32	18.23	158.44	278.54	0.04	1077	600[b]	-	610
SO_4	mg/l	306.53	171.09	105.44	619.72	8.96	16.01	0.11	62.15	200[b]	100	960
NO_3	mg/l	4.96	1.70	1.94	9.5	94.30	122.06	0.42	367.13	50[a]	50	
Ag	µg/l	116.39	55.99	18.01	193	0.025	0.04	0.001	0.13	100[a]	-	
Cd	µg/l	70.53	25.15	26	118.5	0.16	0.21	0.036	0.87	3[a]	3	0.01
Cu	µg/l	82.39	117.31	0	426.78	22.39	31.32	3.32	123.88	2000[a]	1000	0.2
Fe	µg/l	809.47	1321.39	58	4889.5	151.97	193.53	23.85	670.2	300[a]	300	5
Mn	µg/l	263.63	310.29	0	838	126.92	135.32	7.26	408.9	200[b]	200	0.2
Pb	µg/l	92.53	151.01	0	484	2.72	2.74	0.51	11.45	10[a]	10	5

(continued)

Table 13.4 (continued)

N		DG								Drinking water quality standard		Irrigation standard
		17										
		Dry season				Wet season						
Parameters	Units	Mean	STD	MIN	Max	Mean	STD	MIN	MAX	WHO (2008, 2011)	SON (2007)	FAO
Se	μg/l	16.86	6.36	10	31	0.55	0.85	-0.01	2.89	10[a]	1o	
Zn	μg/l	897.28	743.30	78	2467.5	50.83	59.86	7.01	186.19	3000[a]	3000	2

FAO (Ayers &Westcot 1985) values for Ag, Cd, Cu, Fe, Mn, Pb, Se and Zn are in mg/l; [a] WHO (2008); [b] WHO (2011); SON Standard Organisation of Nigeria

signifies areas with a significant concentration influenced by comparable geological, geochemical, and anthropogenic causes. The discharge of effluent from the neighboring River Niger also constitutes a source of pollution in the studied area. The presence of iron (Fe) and manganese (Mn) during the dry season and manganese (Mn) and copper (Cu) during the wet season indicates a positive correlation between these elements and a common origin. Nevertheless, outliers in the biplots, far from the major clusters, could suggest distinctive geological or environmental circumstances that impact the elemental composition, such as industrial influence or natural abnormalities. The regional changes in parameters can be attributed to local differences in geology and aquifer characteristics, as influenced by the mechanism of flow, hydrogeochemical processes, and the residence time (McCall et al., 1996; Zakir et al., 2020).

13.6.3 Correlation

Compared to the wet season, as shown in Table 13.5, the correlation analyses did not reveal any significant correlation between the chemical parameters in the aqueous phase during the dry season. This may be due to the low mineral content in the groundwater in the area, which indicates that the minerals have limited solubility in the water during the study period. The moderate positive association between HCO_3^- and Na^+, K^+, and Ca^{2+} in both seasons indicates that they likely have a common origin, which could be linked to calcite, feldspar, clay, and silicate breakdown of weathered bedrocks. Singh et al., 2013 observed that silicate weathering significantly contributes to Na^+, K^+, Ca^{2+}, and HCO_3^- presence in water. The significant positive correlation observed between Na^+, Cl^-, and NO_3^-, as well as between Cl^- and SO_4^{2-} during the wet season, indicates the presence of both natural (geogenic) and human (anthropogenic) sources. These sources include the dissolution of halite (NaCl) and gypsum, atmospheric deposition, inputs from sewage treatment plants, industrial discharges, runoff, and leaching of fertilizers from agricultural land. Additional potential sources of NO_3^- include leachates from soil resulting from the decomposition of organic materials and industrial and municipal trash. In addition, nitrate is very soluble and can infiltrate the subsurface environment through leaching and surface runoff (Sun et al., 2023). The dry season exhibits a moderate positive correlation, whereas the wet season shows a high positive correlation for heavy metals. This suggests that similar input sources and fluvial-geochemical behavior in both seasons, such as interactions between rocks and water and associated products in the study area (Nganje et al., 2020). While heavy metals can occur naturally, they can also originate from human activities such as the corrosion of galvanized plumbing and water main pipes, discarded batteries, paint, gasoline containing lead, and certain pesticides that contain lead arsenate (Hussain & Rao, 2014). Ezeabasilli et al. (2015) also recognized several anthropogenic sources as input sources.

Table 13.5 Correlation matrix of spatial and temporal variation of physicochemical parameters

Correlation matrix—dry season								
	Na	K	Ca	Mg	Cl	HCO_3	SO_4	NO_3
Na	1	**0.56164**	0.43239	0.30443	–0.04546	**0.52943**	–0.09947	0.02956
K		1	0.20904	0.25133	0.06905	**0.5753**	0.35738	0.26334
Ca			1	0.35078	0.40935	0.31856	–0.06544	0.25147
Mg				1	0.48045	–0.16648	0.35487	0.16211
Cl					1	–0.26767	0.37142	0.34885
HCO_3						1	–0.06494	0.25173
SO_4							1	0.47348
NO_3								1
	Ag	Cd	Cu	Fe	Mn	Pb	Se	Zn
Ag	1	0.10336	–0.02297	–0.12371	**0.58787**	0.27396	–0.27396	0.0776
Cd		1	0.35111	0.41528	0.15791	–0.2157	–0.09915	0.39155
Cu			1	**0.55432**	0.2937	0.15175	0.02223	**0.54694**
Fe				1	0.0393	–0.02584	–0.02932	**0.63341**
Mn					1	0.41349	0.0553	–0.04256
Pb						1	0.07888	
Se							1	–0.14571
Zn								1
					pH	EC	TDS	Hardness
pH					1	0.28866	0.20844	–0.29806
EC						1	–0.19756	–0.01649
TDS							1	–0.48859
Hardness								1

(continued)

Table 13.5 (continued)

Correlation matrix—dry season

	Na	K	Ca	Mg	Cl	HCO_3	SO_4	NO_3
Na	1	0.77727	**0.69923**	0.41184	0.57841	0.61734	0.00667	0.60742
K		1	0.2906	0.16353	**0.59796**	**0.80051**	–0.06939	0.26873
Ca			1	**0.85994**	**0.51582**	0.23658	0.18448	0.22601
Mg				1	**0.64388**	0.06289	0.48708	–0.12702
Cl					1	0.39885	**0.54837**	0.07599
HCO_3						1	0.13131	0.04094
SO_4							1	0.18684
NO_3								1
	Ag	Cd	Cu	Fe	Mn	Pb	Se	Zn
Ag	1	**0.99836**	0.21388	**0.96514**	–0.19599	**0.99811**	0.14726	**0.79417**
Cd		1	0.21178	0.96226	–0.20487	**0.99899**	0.20291	**0.82205**
Cu			1	0.40681	0.22768	0.24973	–0.12519	0.34557
Fe				1	–0.08663	**0.97109**	0.09089	0.8032
Mn					1	0.18525	–0.1657	–0.04475
Pb						1	0.18487	**0.82494**
Se							1	**0.59419**
Zn								1
					pH	EC	TDS	Hardness
pH					1	–0.12557	–0.0021	0.37931
EC						1	**0.83756**	0.22033
TDS							1	0.17045
Hardness								1

Bold values are significant at $p < 0.01$

13.6.4 Water Quality Index (WQI) Evaluation

Figure 13.3 shows that, based on WQI, the water is of good to exceptional quality during the dry season. Also, 66.67% (SG) are classified as poor in the wet season, and 33.3% are in the very poor class. For DG, 47.06% of the sites are in the excellent class, 29.42% in the good class, and 25.53 in the low class. Variations in the water quality for drinking purposes might be connected to the nature and depth of the groundwater

Table. 13.6 Correlation between different indices

Dry season	WQI	C_d	HPI	EWQ I
WQ I	1	0.01257	0.0517	0.05623
C_d		1	**0.90978**	−0.21279
HPI			1	−0.23512
EWQI				1
Wet season				
WQI	1	0.464079	0.420662	0.161331
C_d		1	**0.602787**	−0.42354
HPI			1	−0.44261
EWQI				1

Bold values are significant at $p < 0.01$

and the characteristics of the aquifers. SG1, SG2, SG3, DG1, and DG3 are in poor to extremely poor categories and unsuitable for drinking. These are locations within the dump sites and market areas which are characterized by a lot of anthropogenic activities such as disposal and discharge of effluents which easily leach/wash off from domestic areas and waste disposal sites into the groundwater in the wet season, thereby contributing to people experiencing poor to very poor water quality. Besides, these sites are within a heavily inhabited part of the study area where much waste is generated. The poor-quality water observed by DG 7 and DG during the wet season could result from site-specific variation involving geogenic processes, the nature and depth of water sources, and the geological properties of the aquifer (Shah et al., 2023). However, mixed sources of contamination (Singh et al., 2013) may be responsible for the low quality of water at these sites. A plot of C_d (Fig. 13.3) and HMPI for both seasons has shown that the C_d and HMPI are in the class of high contamination and above the critical level in the dry season, depicting low-quality water and being unfit for drinking.

13.6.5 Environmental Water Quality Index (EWQI)

EWQI values ranged from 0.01(SG2 and DG16) to 0.29(DG 1) dry season) and from 0.00 to 3.83 (wet season). The lowest values were recorded at locations SG2, DG15, DG16, and DG17, with the maximum value obtained from location DG 10. EWQI > 1 represents low/poor quality water. The EWQI data indicated that in the dry season, the quality of the groundwater is good, whereas, in the rainy season, the SG is in a good class, with 35.30% and 64.71% of DG in hazardous and good quality classes, respectively. The EWQI values presented an almost comparable trend with the WQI with a few exceptions (Fig. 13.3). Generally, in comparison, the WQI data indicated DG to be more potable for Ingestion and irrigation than SG for both seasons.

Table 13.7 Classification of groundwater for irrigation purposes in the study area

Percentage (%) of groundwater samples								
Indices	Class	Ranges	SG		DG		Total	
			Dry	Wet	Dry	Wet	Dry	Wet
SAR	Excellent	< 10	100	100	100	100	100	100
	Good	10 –18	–	–	–	–	–	–
	Doubtful	18–26	–	–	–	–	–	–
	Unsuitable	> 26	–	–	–	–	–	–
% Na	Excellent	< 20	66.67	–	–	–	28.57	5.41
	Good	20–40	33.33	33.33	100	26.67	71.43	56.76
	Permissible	40–60	–	66.67	–	60	–	32.43
	Doubtful	60–80	–	–	–	13.33	–	5.41
	Unsuitable	> 80	–	–	–	–	–	–
P.I	Not good	< 75	100	33.33	75	6.67	85.71	16.22
	Good	> 75%	–	66.67	25	93.33	14.29	83.78
MH	Suitable	< 50	–	100	–	100	4.76	100
	Unsuitable	> 50	100	–	100	–	95.24	–
KR	Fit	< 1	100	33.33	100	40	100	59.46
	Unfit	> 1	–	66.67	–	60	–	40.54
SO_4	Excellent	< 192	–	100	8.33	100	9.52	100
	Good	192–336	66.67	–	66.7	–	52.38	–
	Permissible	336–575	–	–	16.7	–	28.57	–
	Precaution usable	575–960	33.33	–	8.33	–	9.52	–
	Unsuitable	> 960	–	–	–	–	–	–
Cl	Excellent	< 142	–	33.33	83.3	100	58.33	86.49
	Good	142–298	33.33	66.67	–	–	9.52	10.81
	Permissible	298–426	66.67	–	16.7	–	19.05	–
	Precaution usable	426–710	–	–	–	–	4.76	–
	Unsuitable	> 710 -		–	–	–	–	2.7

Correlation (Table. 13.4) was done to show the relationship between the calculated indices. The considerable and moderate positive connection between C_d and HMPI in both seasons suggested a common source of the heavy metals. Generally, the positive correlation observed between WQI, C_d, and HMPI indicates similar geochemical controls on the variation in the chemistry (Yang et al., 2015) and other parameters employed in the evaluation of WQI. A similar distribution pattern was also detected between C_d, HMPI, WQI, and EWQI (Fig. 13.3).

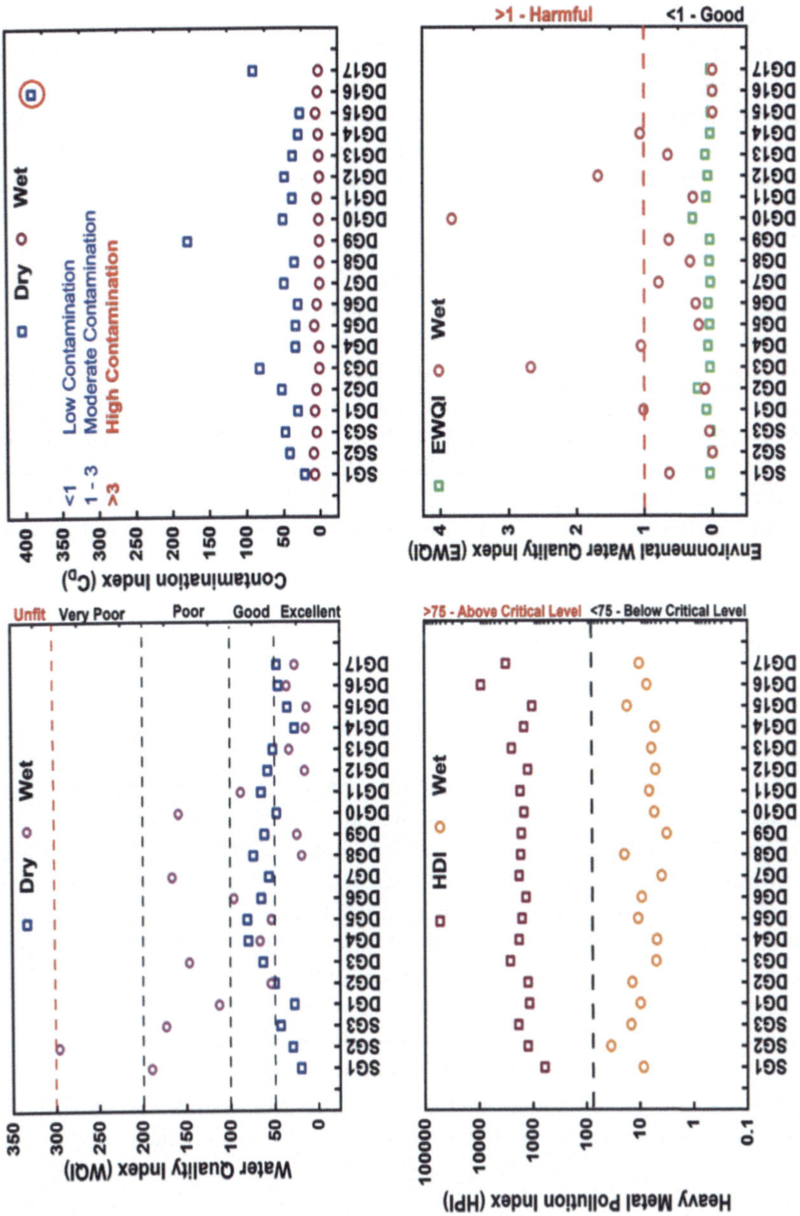

Fig. 13.3 Spatio-temporal variation of WQI, C_d, HMPI and EWQI

13.6.6 Suitability for Irrigation

With some small fluctuations in the wet season in terms of C_d and HMPI, the groundwaters are not appropriate for irrigation during the dry season. In comparison with the FAO values (Ayers & Westcot, 1985) for irrigation purposes, the mean values of pH, K, Cd, and Mn, where above the FAO values of 6.5–8.5, 2.0, 0.01, 0.2, 0.2 indicating poor quality water and not suitable for long term irrigation for both SG and DG in both seasons whereas the other major anions and cations are suitable for irrigation uses (Table 13.2). 80% to 90% of sites had pH, K, and Cd surpassing the FAO limits during the dry season period, while in the wet season, 15% to 75% of sites had pH, K, Cd, and Mn beyond these limits. Also, the maximum HCO_3^-, Fe, and Zn concentrations in the wet season were above the FAO guidelines. SG recorded the highest values of these parameters, and DG recorded the lowest values. Values over FAO standards show that the groundwaters are unsuitable for long-term irrigation. Sites with values within the FAO guidelines are acceptable for irrigation. Similarly, as for water quality for drinking, the wet season samples are of more excellent quality for long-term irrigation than the dry season samples. Regarding MH for both seasons and PI in the dry season, 6.67% of the sites have values for PI outside the acceptable limits and are in the inappropriate class. The SAR (%), Na (%), KI, Cl, and SO_4^{2-} (Table 13.5) for both seasons with slight differences are appropriate for irrigation.

13.6.7 Heavy Metal Pollution

The computed values for C_d indicated that the groundwater was extremely polluted and of poor quality in the dry season based on the (Nayak et al., 2023) classification scheme. 100% of the sites (Fig. 13.3) in the dry season was evaluated as severely contaminated and unsuitable for drinking and irrigation purposes. In the wet season, 55%, 30%, and 15% of the locations were rated as high, medium, and low polluted water. In terms of HMPI, 100% of the sites exceeded the critical limits of 75 in the dry season (Fig. 13.3) as recommended by Majhi & Biswal, 2016 were considered highly polluted and unsafe for drinking and irrigation quality, while all sites in the wet season had HMPI < 75 and were rated as low pollution and suitable for irrigation usage. Generally, the dry season samples were more extensively polluted than the rainy season samples, which may be attributed to the high evaporation rate, which may increase the concentration of the metals. Anthropogenic activities and geogenic to a minor extent, constitute major sources of metal pollution and poor quality of water in the study area (Egbueri, 2020; Ezeabasili et al., 2015; Onwuka et al., 2022).

13.7 Health Implication and Risk Evaluation by PTMs

In this investigation, indices of health significance that surpassed the WHO guidelines are NO_3^-, Cd, Pb, and Se. Frequent Ingestion of contaminated groundwater is one of the common methods of dietary intake of these pollutants by people. Nitrate (NO_3^-) pollution in groundwater is a worldwide environmental problem (Sun et al., 2023). The maximum allowable levels of NO_3^- in drinking water as established by WHO (2011) and SON (2007) is 50 mg/l. Intake of groundwater polluted by nitrate can constitute serious harm to human health. The blue baby disease, also known as Methemoglobinemia, is a health problem linked to nitrate, with children being more prone to this sickness (Sunitha et al., 2022). Another probable nitrate-linked health consequence is cancer (Sunitha et al, 2022).

Cd and Pb are harmful heavy metals designated as B1 and B2 carcinogens (ASTDR, 2018) and are not needed by humans, even in very low quantities. The recommended values by WHO for Cd and Pb in drinking water are 0.003 mg/l and 0.01 mg/l, respectively. Drinking water contaminated by Cd and Pb might result in significant health concerns. According to Sudharshan Reddy & Sunitha, 2023, Pb is a dangerous metal and, through respiration, can enter the human system, causing neurological disorders with children exhibiting behavioral abnormalities (Sudharshan Reddy & Sunitha, 2023). The allowable limit of Se in drinking water, as advised by WHO, is 0.01 mg/l. Se consumption over lengthy periods may lead to selenosis, a disease noted for hair loss, brittle nails, and numbness of extremities (ASTDR, 2018). The hazardous level of metals absorbed by humans and the possibly damaging effects become apparent only after many years of exposure (Agbor et al., 2024).

Mean, STD, and range of ADI, hazard quotient (HQ) with accompanying upper tolerated intake level (UTIL), and oral reference dose are shown in Appendix 3. The mean values of ADI were below the UTIL for adults and children in both seasons. Only the mean nitrate intake was over the UTIL of 3.7 for children in the wet season signaled possible health effects. Also, the ADI values of $\mathbf{NO_3^-}$ and Cd in both seasons and the mean ADI of Pb for the dry season were above the oral reference dose. Values of $\mathbf{NO_3^-}$**,** Cd, and Pb over the UTIL and oral reference dose suggested significant potential health risks through dietary consumption exposure. Values of HQ in the wet season were above > 1 for $\mathbf{NO_3^-}$ implying adverse health effects with children more exposed than adults, while that of other metals were < 1. The water quality is highly affected by nitrate, and the source of the nitrate is most certainly from discharge from waste disposal facilities, sewage, and applications of fertilizers, etc. The estimated HI (Fig. 13.4) showed that all the SG and 35.29% of DG samples had values > 1 for children in the wet season and are not below the permissible range for non-carcinogenic health risk, with $\mathbf{NO_3^-}$ being the most significant contributor to the potential health risk in the study area.

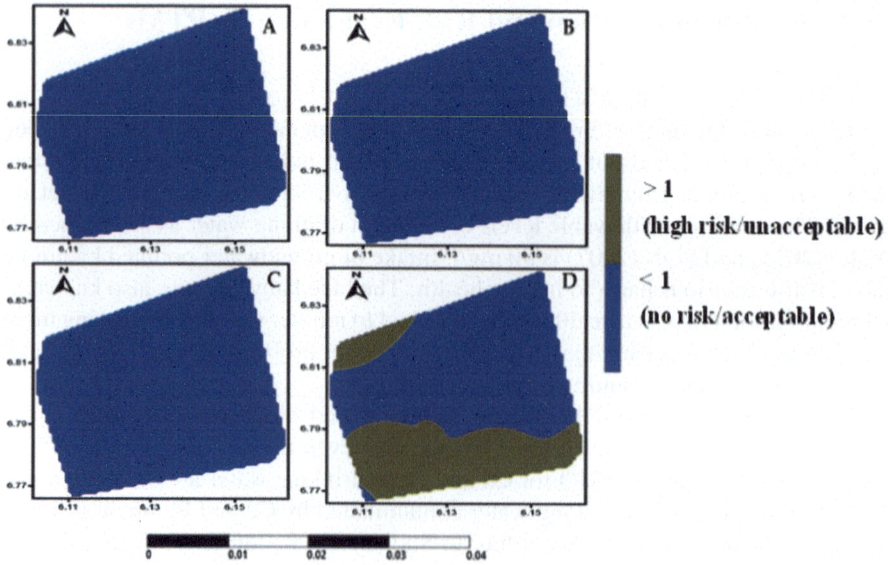

Fig. 13.4 Map showing the distribution of HI (A and B, adult for the dry and wet season, C and D, children for dry and wet season

13.8 Conclusions

The results indicated no definite distribution pattern of the physicochemical characteristics within the season. The pH, hardness, Fe, NO_3^-, Cd, Mn, and Pb were above the WHO and SON guidelines. The computed C_d and HMPI over the acceptable standard revealed that heavy metals severely contaminate the groundwater and exceed the critical level in the dry season. The contamination source is natural and manmade, generated from point and non-point sources, including land disposal sites, sewage disposal, vehicle batteries, petroleum products, agricultural runoff, urban runoff, and atmospheric deposition. The C_d, HMPI, EWQI, and irrigation indices revealed better quality in wet-season samples than in dry-season samples. Hazard index values were all within acceptable class in the dry season for both adults and children with 43.75% in not acceptable class in the rainy season for children due to a rise in NO_3^- concentration.

It is advised that pollution control methods be emphasized and groundwater quality in moderate to high contamination areas be improved. Monitoring and proper treatments of the groundwater, such as boiling, periodic servicing of the pumps, and cleaning of the storage tanks, are encouraged to improve the quality of the water to avert unfavorable health implications for humans living within the study area.

Acknowledgements This study is part of the PhD dissertation of the first author carried out under the supervision of Nganje, T.N and Adamu, C.I with conceptualization by Edet, A. Chemical analyses of the water samples by Edet, A, was made possible by fellowship provided to him by the Alexander—von Humbolt foundation, Bonn (Germany).

Appendix 1

Data Analysis of Water Quality Assessment for Irrigation Purposes

Parameter/Reference	Formula	Range	Class
Sodium Adsorption (SAR) Sadashivaiah et al. (2008)	$SAR = \frac{Na^+}{\{(Ca^{2+}+Mg^{2+})/2\}^{1/2}}$	< 10 10–18 18–26 > 26	Excellent Good Doubtful Unsuitable
Sodium Percentage (Na%) Todd (1980)	$Na\% = \frac{[(Na+ + K+) \times 100]}{(Ca2+ + Mg2+ + Na+ + K+)}$	< 20 20–40 40–60 60–80 > 80	Excellent Good Permissible Doubtful Unsuitable
Permeability Index (P.I) Doneen (1964)	$P.I = \frac{(Na^+ + \sqrt{HCO_3^-}}{Ca^{2+}+Mg^{2+}+Na^+} x100$	< 75% > 75%	Not good Good
Magnesium Hazard (MH) Kumar et al. (2013)	$MH = Mg^{2+} / (Ca^{2+} + Mg^{2+}) \times 100$	< 50 < 50	Suitable Unsuitable
Kelly's Ratio (KR) Kelly (1951)	$K.I = \frac{Na}{Ca+Mg}$	< 1 > 1	Fit Unfit
Sulphate (SO_4^{2-}) Edet (2019)	Concentration Values	< 192 192–336 336–575 575–960 > 960	Excellent Good Permissible Precaution usable Unsuitable
Chloride Cl^- Edet (2019)	Concentration Values	< 142 142–298 298–426 426–710 > 710	Excellent Good Permissible Precaution usable Unsuitable

Appendix 2

Spatial variation of parameters in the dry season

	PH	(MS/M)	(MG/L)	(mg/l)	(MG/L)	(MG/L)	(MG/L)^2	(MG/L)^2	(MG/L)	(MG/L)	(MG/L)^2	(MG/L)	(MG/L)	(MG/L)	(MG/L)	(MG/L)	(MG/L)	(MG/L)	(MG/L)	(MG/L)
	Sample Location	Dry	Dry	Dry	Dry	Dry	Dry	Dry	Dry	Dry	Dry	Dry	Dry	Dry	Dry	Dry	Dry	Dry	Dry	Dry
SG1	6.56	904.50	10.45	393.50	13.38	7.21	7.08	18.37	352.00	13.34	619.72	614.78	193.00	26.00	121.00	268.00	838.00	3272.00	31.00	673.00
SG2	5.54	398.00	31.37	480.00	13.90	10.67	5.44	4.04	18.18	341.00	318.02	244.84	6.06	114.00	55.00	51.00	55.00	0.00	0.00	27.00
SG3	6.55	732.50	9.36	482.50	58.63	10.50	9.35	14.13	18.72	212.00	32.37	5.00	8.92	49.00	70.00	363.00	70.00	0.00	149.00	32.00
DG1	5.16	134.00	7.18	531.00	11.92	5.17	1.45	16.96	77.00	7.48	10.98	4.50	153.00	52.00	0.00	451.00	670.00	0.00	20.00	248.00
DG2	7.20	40.00	36.00	5.65	2.22	0.34	1.86	0.24	3.34	10.00	0.12	5.50	0.02	0.08	4.26	38.13	7.26	1.59	–0.01	8.15
DG3	6.41	31.15	24.56	305.82	4.59	3.47	2.63	3.52	22.67	11.61	128.80	4.74	18.01	38.04	10.63	368.07	3.63	242.80	13.50	559.08
DG4	6.58	349.68	13.25	574.13	8.83	7.91	9.28	18.08	217.00	18.23	307.93	5.82	96.75	118.50	253.00	4889.50	342.25	152.50	18.25	2467.50
DG5	6.33	283.77	9.47	147.62	5.98	0.77	5.89	14.47	206.27	6.32	147.96	1.94	31.11	94.44	426.78	1809.54	280.61	284.71	11.59	1999.58
DG6	6.28	6.73	17.71	438.00	11.89	6.91	4.65	16.09	70.00	10.77	105.44	3.50	58.00	66.00	29.00	2428.00	0.00	0.00	10.00	691.00
DG7	6.01	52.70	4.92	504.00	11.08	6.43	3.10	5.17	270.00	15.22	271.18	5.15	164.00	82.00	71.00	110.00	0.00	0.00	12.50	512.00
DG8	5.16	134.00	12.30	531.00	11.92	5.17	1.45	16.96	77.00	7.48	617.71	4.50	153.00	52.00	49.00	451.00	670.00	0.00	20.00	248.00
DG9	5.20	10.00	18.00	21.45	3.47	1.68	6.92	1.01	4.04	10.00	6.81	0.98	0.00	0.09	25.27	56.71	30.89	0.51	0.02	24.44
DG10	6.56	904.50	7.18	393.50	13.38	7.21	7.05	18.37	352.00	13.34	315.62	5.00	164.00	26.00	121.00	268.00	738.00	0.00	31.00	673.00
DG11	5.61	22.30	10.90	606.00	6.96	6.61	3.40	6.81	42.00	13.22	257.60	3.98	36.00	76.00	17.00	698.00	0.00	484.00	27.00	1110.00
DG12	6.01	52.70	10.45	504.00	11.08	6.43	3.10	5.17	270.00	15.22	271.18	5.15	164.00	82.00	71.00	110.00	0.00	0.00	12.50	512.00
DG13	5.44	447.00	13.12	440.00	16.76	4.43	7.91	18.70	252.00	9.48	246.08	6.25	144.00	72.00	138.00	149.00	683.00	0.00	14.50	405.00
DG14	5.53	128.90	13.25	791.00	11.89	5.20	1.40	16.09	80.00	7.12	314.80	6.48	164.00	60.00	25.50	164.00	331.00	0.00	24.00	546.00
DG15	6.21	15.85	9.47	636.00	11.77	5.43	1.47	7.28	30.00	12.40	303.69	3.54	138.00	61.00	12.00	122.00	0.00	224.00	15.00	1305.00
DG16	5.93	93.50	7.35	588.00	10.39	7.43	0.35	8.05	52.00	10.48	115.08	4.35	124.00	112.00	12.00	58.00	136.00	0.00	12.00	78.00
DG17	6.33	77.00	12.22	520.00	9.90	8.49	1.61	10.65	32.00	6.47	573.22	9.50	109.00	66.00	0.00	66.00	0.00	0.00	11.00	2105.00

Spatial variation of parameters in the wet season

	PH	(MS/M)	(MG/ L)	(MG/ L)	(MG/ L)	K+	CA	(MG/ L)⁻	MG^{2+}	CL^-	HCO**2**	SO4^2	(MG/ L)	(MG/ L)	(MG/ L)	(MG/L)	(MG/ L)	(MG/ L)	(MG/ L)	(MG/L)
Sample Location	Wet	Wet	Wet	Wet	Wet	Wet	Wet	Wet	Wet	Wet	Wet	Wet	Wet	Wet	Wet	Wet	Wet	Wet	Wet	Wet
SG1	7.05	591.50	822.40	784.00	152.29	204.84	60.16	116.43	513.00	1077.00	424.67	106.33	153.50	260.00	893.00	6747.00	687.00	579.00	35.00	5136.00
SG2	6.02	58.50	578.11	793.50	111.46	247.82	2.08	12.35	355.00	452.35	409.03	107.09	127.00	103.00	49.00	3042.00	2.37	231.00	7.50	574.00
SG3	6.80	1280.00	930.00	458.64	108.70	44.57	169.80	35.41	379.35	315.00	205.10	68.09	0.05	6.01	3.08	670.20	0.04	11.43	34.62	356.00
DG1	7.00	2310.00	23.42	222.11	12.96	8.63	17.92	17.49	286.45	24.47	20.04	8.02	0.08	0.45	67.06	0.20	310.90	6.41	0.40	115.78
DG2	6.20	2088.17	11.42	105.78	11.02	4.33	21.21	13.63	30.00	18.48	25.01	54.77	0.13	0.87	123.88	538.92	200.53	11.45	0.34	186.19
DG3	6.10	670.00	1.00	118.16	25.64	22.31	37.95	4.03	53.46	80.53	0.12	0.42	0.01	0.04	4.95	293.00	104.60	0.66	0.17	18.23
DG4	4.20	60.00	52.00	19.60	8.63	1.85	6.09	1.07	5.19	0.04	0.82	34.88	0.01	0.06	9.64	52.96	29.98	1.63	0.06	12.76
DG5	5.20	10.00	18.00	21.45	3.47	1.68	6.92	1.01	4.04	10.00	6.81	347.32	0.00	0.09	25.27	56.71	30.89	0.51	0.02	24.44
DG6	4.10	400.00	291.00	57.15	43.50	11.51	17.48	3.28	45.51	0.16	1.94	214.54	0.01	0.15	15.11	102.00	408.90	3.75	1.60	180.10
DG7	6.00	1260.00	920.00	214.87	144.40	61.82	74.15	7.19	147.19	100.00	62.15	19.07	0.05	0.17	17.01	23.85	75.42	1.42	1.56	42.86
DG8	7.20	40.00	36.00	5.65	2.22	0.34	1.86	0.24	3.34	10.00	0.12	5.50	0.02	0.08	4.26	38.13	7.26	1.59	–0.01	8.15
DG9	4.20	60.00	52.00	19.60	8.63	1.85	6.09	1.07	5.19	0.23	0.82	34.88	0.01	0.06	9.64	52.96	29.98	1.63	0.06	12.76
DG10	4.80	810.00	582.00	134.56	84.78	44.23	42.77	6.74	89.21	40.00	11.45	367.13	0.03	0.16	6.33	35.06	370.30	1.40	2.89	14.37
DG11	4.90	320.00	236.00	88.35	38.02	18.66	29.00	3.87	35.40	15.00	3.44	189.70	0.04	0.12	11.30	70.55	243.80	2.16	1.38	29.20
DG12	5.50	10.00	17.00	15.42	2.63	0.91	5.07	0.67	1.31	135.00	1.48	1.96	0.00	0.04	25.22	33.72	13.11	1.53	0.00	35.65
DG13	5.20	10.00	18.00	21.45	3.47	1.68	6.92	1.01	4.04	10.00	6.81	0.98	0.00	0.09	25.27	56.71	30.89	0.51	0.02	24.44
DG14	4.90	8.50	11.00	7.12	1.01	0.23	2.31	0.33	2.56	15.00	0.11	74.18	0.00	0.07	3.32	34.93	22.45	1.73	0.03	7.01
DG15	6.50	150.00	112.00	79.17	20.66	10.83	27.66	2.44	11.36	230.00	6.32	20.59	0.02	0.13	25.44	209.60	42.49	3.83	0.20	25.24
DG16	4.50	80.00	64.00	11.52	10.33	2.95	3.61	0.61	10.98	135.00	1.83	36.30	0.00	0.07	5.17	45.53	75.62	2.19	0.03	9.38
DG17	5.80	90.00	6.44	27.95	8.46	1.40	9.70	0.90	5.19	0.25	0.95	36.51	0.00	0.07	4.60	173.40	64.43	1.63	0.02	91.21

Appendix 3

Summary of statistical description of daily intake (DI) and hazard quotient (HQ) for adult and children in both seasons

Paramaters		ADI (adult)		ADI (Children)		HQ (Adult)		HQ (Children)		MG DAY-Person	(MG KG-DAY-)
		Dry	Wet	Dry	Wet	Dry	Wet	Dry	Wet		
NO_3^-	Mean	0.187668	2.887612	0.273245	4.204364	0.117293	1.804758	0.170778	2.627727	3.7[a]	1.6
	Std	0.061426	4.360691	0.089436	6.349166	0.038391	2.725432	0.055897	3.968229		
	Max	0.339286	15.09697	0.494	21.98119	0.212054	9.435607	0.30875	13.73824		
	Min	0.126429	0.014286	0.18408	0.0208	0.079018	0.008929	0.11505	0.013		
CD	Mean	0.002691	4.81E–06	0.003918	0.000007	0.005382	9.62E–06	0.007837	1.4E–05	0.064[c]	0.001
	Std	0.000905	5.74E–06	0.001318	8.35E–06	0.00181	1.15E–05	0.002636	1.67E–05		
	Max	0.004143	2.93E–05	0.006032	4.26E–05	0.008286	5.85E–05	0.012064	8.52E–05		
	Min	0.000929	3.57E–07	0.001352	5.2E–07	0.001857	7.14E–07	0.002704	1.04E–06		
CU	Mean	0.002867	0.000605	0.004174	0.000881	7.17E–05	1.51E–05	0.000104	2.2E–05	10[b]	0.04[d]
	Std	0.003371	0.000749	0.004909	0.00109	8.43E-05	1.87E–05	0.000123	2.73E–05		
	Max	0.012964	0.003825	0.018876	0.005569	0.000324	9.56E–05	0.000472	0.000139		
	Min	0	8.83E–05	0	0.000129	0	2.21E–06	0	3.22E–06		
FE	Mean	0.018298	0.013163	0.026642	0.019166	0.000061	4.39E–05	8.88E–05	6.39E–05	45[b]	0.7[d]
	Std	0.030259	0.025658	0.044057	0.037358	0.000101	8.55E–05	0.000147	0.000125		
	Max	0.108643	0.135571	0.158184	0.197392	0.000362	0.000452	0.000527	0.000658		
	Min	0.002071	0.00011	0.003016	0.00016	6.9E–06	3.67E–07	1.01E–05	5.35E–07		
MN	Mean	0.018212	0.006378	0.026517	0.009286	0.000911	0.000319	0.001326	0.000464	11[b]	0.014[d]
	Std	0.029183	0.013695	0.042491	0.01994	0.001459	0.000685	0.002125	0.000997		
	Max	0.116857	0.071179	0.170144	0.103636	0.005843	0.003559	0.008507	0.005182		
	Min	0	7.29E–06	0	1.06E–05	0	3.64E–07	0	5.3E–07		
PB	Mean	0.010076	8.58E–05	0.01467	0.000125	0.007197	6.13E–05	0.010479	8.93E–05	0.24[c]	0.0035[e]
	Std	0.029583	9.73E–05	0.043072	0.000142	0.02113	6.95E–05	0.030766	0.000101		
	Max	0.1225	0.000565	0.17836	0.000823	0.0875	0.000404	0.1274	0.000588		

(continued)

(continued)

Paramaters		ADI (adult)		ADI (Children)		HQ (Adult)		HQ (Children)		MG DAY-Person	(MG KG-DAY-)
		Dry	Wet	Dry	Wet	Dry	Wet	Dry	Wet		
	Min	0	1.36E–06	0	1.98E–06	0	9.69E–07	0	1.41E–06		
ZN	Mean	0.037664	0.002071	0.054839	0.003016	0.000126	6.9E–06	0.001326	1.01E–05	40[b]	0.3[d]
	Std	0.032489	0.004847	0.047304	0.007057	0.000108	1.62E–05	0.002125	2.35E–05		
	Max	0.11425	0.027464	0.166348	0.039988	0.000381	9.15E–05	0.008507	0.000133		
	Min	0.002786	0.000231	0.004056	0.000336	9.29E–06	7.69E–07	0	1.12E–06		

[a] EFSA 1985
[b] FDA
[c] Garcia-Rico et al. (2007)
[d] USEPA (2010)
[e] IRIS 1987

References

Agbor, E., Besong, E., Ebai, P., et al. (2024). Baseline assessment of the health risk of potentially toxic heavy metals in commonly consumed vegetables in parts of Mamfe, Southwest Region. *Cameroon J Trace Elem Miner, 8*, 100115. https://doi.org/10.1016/J.JTEMIN.2023.100115

Akudinobi, B., & Chinenye, O. (2013). Qualitative evaluation of urban water sources in Onitsha area of Anambra state, Nigeria, 35–44.

ASTDR (Agency for toxic substances & Disease Registry) (2018) *Toxicological Profile for hazardous substances*. US Department of Health and Human Service.

Ayers, R. S., & Westcot, D. W. (1985). Water quality for agriculture, 29.

Bahita, T. A., Swain, S., Pandey, P., & Pandey, A. (2021). Assessment of heavy metal contamination in livestock drinking water of Upper Ganga Canal (Roorkee City, India). *Arabian Journal of Geosciences, 14*, 2861. https://doi.org/10.1007/s12517-021-08874-7

Boskabady, M., Marefati, N., Farkhondeh, T., Shakei, F., Farshbaf, A., & Boskabady, M. H. (2018). The effect of environmental lead exposure on human health and the contribution of inflammatory mechanism, a review. *Environmental International, 120*, 404–420.

Edet, A. (2019). Seasonal and spatio-temporal patterns, evolution and quality of groundwater in Cross River State, Nigeria: Implications for groundwater management. *Sustain Water Resource Management, 5*, 667–687. https://doi.org/10.1007/s40899-018-0236-6

Edet, A. E., & Offiong, O. E. (2002). Evaluation of water quality pollution indices for heavy metal contamination monitoring. A study case from Akpabuyo-Odukpani area, Lower Cross River Basin (southeastern Nigeria). *GeoJournal, 57*, 295–304. https://doi.org/10.1023/B:GEJO.000 0007250.92458.de

Egbueri, J. C. (2020). Heavy metals pollution source identification and probabilistic health risk assessment of shallow groundwater in Onitsha, Nigeria. *Analytical Letters, 53*, 1620–1638. https://doi.org/10.1080/00032719.2020.1712606

Etikala, B., Madhav, S., & Somagouni, S. G. (2022). *Urban Water Systems: An Overview, 6*, 1–19. https://doi.org/10.1016/B978-0-323-91838-1.00016-6

Ezeabasili, A. C. C., Anike, O. L., & Okoro, B. U. (2015). Urban water pollution by heavy metals and health implication in Onitsha, Nigeria. *African Journal Environ Science Technology, 9*, 325–331.

Gall, J. E., Boyd, R. S., & Rajakaruna, N. (2015). Transfer of heavy metals through terrestrial food webs: a review. *Environmental Monitoring and Assessment, 187*, 201.

Garcia-Rico, L., Leyva-Perez, J., & Jara-Marini, M. E. (2007) Content and daily intake of copper, zinc, lead, cadmium, and mercury from dietary supplements in Mexico. *Food Chemistry and Toxicology, 45*, 1599–1605.

Hussain, M., & Rao, T. P. (2014). *Toxic trace element contamination in ground water of Bollaram and Patancheru*. Andhra Pradesh.

Igbomor, B. A., & Amadi, A. N. (2019). Investigation of groundwater quality in parts of Onitsha, Southeastern Nigeria. *Pacific Journal Science Technology, 20*, 266–275.

Iqbal, J., & Shah, M. H. (2013). Health Risk Assessment of Metals in Surface Water from Freshwater Source Lakes, Pakistan. *Human and Ecological Risk Assessment an International Journal, 19*, 1530–1543. https://doi.org/10.1080/10807039.2012.716681

Isikhueme, M., & Omorogieva, O. (2015). Hydrogeology and water quality assessment of the middle Aquiferous Horizon of Onitsha and environs in Anambra Basin, Eastern Nigeria. *British Journal Application Science Technology, 9*, 475–483. https://doi.org/10.9734/BJAST/2015/16279

Izueke, E., Eme, O. (2013). Urban Planning Problems in Nigeria : A case of Onitsha metopolis of Anambra State. *Singaporean Journal Bussiness, Economic Management Studies 1*,41–59. https://doi.org/10.12816/0003828

Kazemi, A., Esmaeilbeigi, M., Sahebi, Z., & Ansari, A. (2022). Health risk assessment of total chromium in the qanat as historical drinking water supplying system. *Science of the Total Environment, 807*, 150795. https://doi.org/10.1016/J.SCITOTENV.2021.150795

Kumar, X. R. A., Giridharan, L., Shyamala, J., Velmurugan, P. M., & Jayaprakash, M. (2013). Urbanization impact of groundwater quality in Cuddalore District, East Coast of India. *Journal of Environmental Chemistry and Ecotoxicology, 5*(3), 63–73.

Kumar, P. R., Gowd, S. S., & Krupavathi, C. (2024). Groundwater quality evaluation using water quality index and geospatial techniques in parts of Anantapur District, Andhra Pradesh, South India. *HydroResearch, 7*, 86–98. https://doi.org/10.1016/J.HYDRES.2024.01.001

Majhi, A., & Biswal, S. K. (2016). Application of HPI (heavy metal pollution index) and correlation coefficient for the assessment of groundwater quality near ash ponds of thermal power plants. *International Journal of Science Engineering and Advance Technology (IJSEAT), 4*(8), 395–405.

Masood, N., Zafar, T., Hudson-Edwards, K. A., & Farooqi, A. (2024). Spatial distribution and health risk assessment of toxic metal(oid)s in soils of coal mining areas of the Salt Range, Punjab, Pakistan. *Physics and Chemistry of the Earth, Parts a/b/c, 134*, 103566. https://doi.org/10.1016/J.PCE.2024.103566

McCall, M. G. J., Appleton, J. D., & Fuge, R. (1996). *Environmental geochemistry and health: With special reference to developing countries*. Geological Society.

Meride, Y., & Ayenew, B. (2016). Drinking water quality assessment and its effects on residents health in Wondo genet campus. *Ethiopia. Environment Systems Research, 5*, 1. https://doi.org/10.1186/s40068-016-0053-6

Mohammadpour, A., Rajabi, S., Bell, M., Baghapour, M. A., & Aliyeva, A. M. (2023). Seasonal variations of potentially toxic elements (PTEs) in drinking water and health risk assessment via Monte Carlo simulation and Sobol sensitivity analysis in southern Iran's largest city. *Applied Water Science, 13*. https://doi.org/10.1007/s13201-023-02041-x

Narayanan, P. (2007). *Environmental pollution: Principles, analysis and control*. LBs Publishers and Distribution.

Nayak, A., Matta, G., & Uniyal, D. P. (2023). Hydrochemical characterization of groundwater quality using chemometric analysis and water quality indices in the foothills of Himalayas. *Environment, Development and Sustainability, 25*, 14229–14260. https://doi.org/10.1007/s10668-022-02661-4

Nganje, T. N., Agbor, E. E., Adamu, C. I., Ukpong, A. S., Katte, B. F., Edet, A. E., & Hursthouse, A. S. (2020). Public health challenges as a result of contaminated water sources in Kumba. *Cameroon. Environmental Geochemistry and Health, 42*(4), 1167–1195.

Nganje, T. N., Hursthouse, A. S., Edet, A., et al. (2015). Assessment of the health risk, aesthetic and agricultural quality of rainwater, surface water and groundwater in the shale Bedrock Areas, Southeastern Nigeria. *Water Quality Exposure Health, 7*, 153–178.

Standard Organization of Nigeria (SON). (2007). *Nigeria standard for drinking water quality*. Abuja: Nigerian Industrial Standard.

Nwajide, C. S. (2014). *Geology of Nigeria's Sedimentary Basins*. CSS publishers Nigeria.

Offodile, M. E. (2002). *Hydrogeology: groundwater studies and development in Nigeria* (2nd ed.). Mecon Geology and Engineering Services.

Okoyeh L, Ahaneku C, Utom A, et al (2021) Spatial Distribution of heavy metal contamination in groundwater of Onitsha Urban, Southeast Nigeria. *International Journal of Advanced Academic Research*, 11–22. https://doi.org/10.46654/ij.24889849.e7415

Onwuka, O. S., Kenneth, E. C., & Chikezie, O. K. (2022). Groundwater source evaluation and quality checks, for drinking and irrigation uses in Eha-Amufu and environs, Eastern Nigeria. *Environmental Forensics, 23*, 153–169. https://doi.org/10.1080/15275922.2021.1884625

Orjil, M. U., Okoli, I., & Ezenwaji, E. E. (2015). Four-year retrospective survey of water, sanitation and hygiene associated diseases in Onitsha, Nigeria (2011–2014). *World Rural Observations, 2*(7), 7–11.

Ponsadailakshmi, S., Sankari, S. G., Prasanna, S. M., & Madhurambal, G. (2018). Evaluation of water quality suitability for drinking using drinking water quality index in Nagapattinam district, Tamil Nadu in Southern India. *Groundwater for Sustainable Development, 6*, 43–49. https://doi.org/10.1016/J.GSD.2017.10.005

Rezaei, H., Zarei, A., Kamarehie, B., et al. (2019). Levels, distributions and health risk assessment of lead, cadmium and arsenic found in drinking groundwater of Dehgolan's Villages. *Iran. Toxicology and Environmental Health Science, 11*, 54–62. https://doi.org/10.1007/s13530-019-0388-2

Rőllin, H. B. (2011). *Environmental pollution and health effects* (pp. 617–629). Encyclopedia of Environmental Health: Elsevier

Sadashivaiah, C., Ramakrishnaish, C. R., & Ranganna, G. (2008). Hydrochemical analysis and evolution of groundwater quality in Tumkur Taluk, Karnataka state, India. *Journal of Environmental Research Publishing and Health, 5*(3), 158–164.

Shah, R. A., Shafi, A., Andrabi, S. M. A., et al. (2023). Spatial distribution and health risk assessment of heavy metals in the groundwater of selected watersheds of Srinagar and Pulwama districts of Kashmir Himalaya. *Water, Air, and Soil Pollution, 234*, 665. https://doi.org/10.1007/s11270-023-06615-y

Shil, S., & Singh, U. K. (2019). Health risk assessment and spatial variations of dissolved heavy metals and metalloids in a tropical river basin system. *Ecological Indicators, 106*, 105455. https://doi.org/10.1016/J.ECOLIND.2019.105455

Shuaibu, A., Kalin, R. M., Phoenix, V., Banda, L. C., & Lawal, I. C (2024). Hydrogeochemistry and water quality index for groundwater sustainability in the komadugu-yobe basin, sahel region. *Water, 16*

Singh, A. K., Raj, B., Tiwari, A. K., & Mahato, M. K. (2013). Evaluation of hydrogeochemical processes and groundwater quality in the Jhansi district of Bundelkhand region, India. *Environment and Earth Science, 70*, 1225–1247. https://doi.org/10.1007/s12665-012-2209-7

Singh, R., Venkatesh, A. S., Syed, T. H., et al. (2017). Assessment of potentially toxic trace elements contamination in groundwater resources of the coal mining area of the Korba Coalfield. *Central India. Environmental Earth Sciences, 76*, 566. https://doi.org/10.1007/s12665-017-6899-8

Sudharshan Reddy, Y., & Sunitha, V. (2023). Assessment of Heavy metal pollution and its health implications in groundwater for drinking purpose around inactive mines, SW region of Cuddapah Basin. *South India. Total Environment Research Themes, 8*, 100069. https://doi.org/10.1016/J.TOTERT.2023.100069

Sun, Q., Yang, K., Liu, T., et al. (2023). Health risk assessment of nitrate pollution of drinking groundwater in rural areas of Suihua, China. *Journal of Water and Health, 21*, 1193–1208. https://doi.org/10.2166/wh.2023.069

Sunitha, V., Reddy, Y. S., Suvarna, B., & Reddy, B. M. (2022). Human health risk assessment (HHRA) of fluoride and nitrate using pollution index of groundwater (PIG) in and around hard rock terrain of Cuddapah, A.P. *South India. Environmental Chemistry and Ecotoxicology, 4*, 113–123. https://doi.org/10.1016/J.ENCECO.2021.12.002

Suvarapu, L. N., & Baek, S.-O. (2017). Determination of heavy metals in the ambient atmosphere. *Toxicology and Industrial Health, 33*, 79–96. https://doi.org/10.1177/0748233716654827

Taiwo, A. M., Somade, O. C., Ojekunle, O. Z., et al. (2023). Human health risk assessment of metals and metalloids in groundwater resources around the sanitation facilities in major markets from Abeokuta Metropolis. *Southwestern Nigeria. Journal Trace of Elements and Minerals, 6*, 100105. https://doi.org/10.1016/J.JTEMIN.2023.100105

USEPA. (2010). *Integrated Risk Information system (IRIS). United States Environmental Protection Agency*. Available online at http://www.Epa.Gog/irrris/index.infml. Accessed March, 2016.

WHO. (2011). *Guidelines for drinking-water quality* (4th ed.).

Williams, M. (2012). *Health effects—Toxicological profile for Manganese*. Available online at http://www.ncbi.nlm.nih.gov

World Health Organization (WHO). (2008). *Guidelines for drinking water quality* (3rd ed.).

Yang, X., Duan, J., Wang, L., et al. (2015). Heavy metal pollution and health risk assessment in the Wei River in China. *Environmental Monitoring and Assessment, 187*, 111. https://doi.org/10.1007/s10661-014-4202-y

Zakir, H. M., Sharmin, S., Akter, A., & Rahman, M. S. (2020). Assessment of health risk of heavy metals and water quality indices for irrigation and drinking suitability of waters: A case study of Jamalpur Sadar area. *Bangladesh. Environmental Advances, 2*, 100005. https://doi.org/10.1016/J.ENVADV.2020.100005

Chapter 14
Source Approximation of Heavy Metals in the Groundwater Regime of Wardha Valley Coalfields, Central India: A Geospatial Approach

Priyadarshan S. Ganvir and Rajeeva Guhey

Abstract The groundwater in and around active and abandoned coalmines has been always obstinately under the menace of contamination. The central India's well-established coalfields lineated along the Wardha River are also under similar menace. It is judiciously comprehensible that a sincere approach for the monitoring and management of the groundwater is anticipated. The geospatial techniques have confirmed its potentiality in various sectors including groundwater studies. The present work is to comprehend the heavy metals' contamination within groundwater regime of the study area and its source approximation by means of geospatial techniques. In 2019, 45 groundwater samples for pre and post-monsoon each were considered from the study area to outline the groundwater regime. After pilot investigation, the groundwater was analysed for Aluminium, Cadmium, Chromium, Copper, Iron, Nickel, lead and Zinc. The groundwater samples have evidently manifested the contamination of Cadmium, Iron, Nickel and Lead at certain locations. Towards attainment of an accurate spatial depiction of contamination, an Ordinary Kriging method has been employed for obtaining interpolation and prediction standard error maps for identified contaminants. The base map was so systematized that the metal's enrichment and paucity in the groundwater is easily comparable with the lithology and relief of the study area. The consequent maps indicated the contamination hotspots in the proximity of the active coalfields and lithology constituting rock strata belonging to Barakar formation. These spatial depictions of the contaminations were then confirmed by evidences obtained from the petrology of the selected rock samples from the identified area. The geospatial approach not only elucidated the dispersal of heavy metals in the groundwater but also aided in identifying their both spatial and temporal sources. The heavy metals' toxicity in terms of bioaccumulation is an established fact. The source approximation and elucidated dispersal pattern of

P. S. Ganvir (✉)
Department of Geology, M. G. Arts, Science and Late N. P. Commerce College, Armori, Maharashtra, India
e-mail: priyadarshan.ganvir@gmail.com

R. Guhey
Swami Atmanand Government Model College, Jagdalpur, Chhattisgarh, India

V. Sunitha et al. (eds.), *Groundwater Resource Management Planning Strategies*, Springer Water, https://doi.org/10.1007/978-3-031-88870-0_14

heavy metals in the groundwater, provides greater and better likelihoods to hinder their bioaccumulation.

Keywords Heavy metals · Groundwater contamination · Wardha valley Coalfields · Gondwana supergroup · Geographical information system · Ordinary kriging · Interpolation maps · India

14.1 Introduction

The unceasing demand of natural energy resources accelerates the exploration and excavation of the fossil fuels (Abas et al., 2015). The coal is one of them and is preferred due to its terrestrial abundance and ease access. Since early times, the exploitative convenience for any natural resource leads to the mismanaged usage and undesirable outcomes (Armaroli & Balzani, 2011). The coal resources are no exceptions. In the contemporary scenario, the coalfields are the energy providers and economic builders with no doubt but have also manifested multifaceted environmental problems (Bach, 1981; Balat, 2007). Among which, the groundwater pollution is the prominent one. In the developing countries the groundwater has been underestimated qualitatively and quantitatively as well, to keep the wheel of development in motion (Villarroya & Aldwell, 1998). The absence of any dedicated governance on the groundwater regimes puts them into the state of vulnerability (Conti & Gupta, 2016). Looking at the hazardous impacts of groundwater pollution on the peripheral ecosystems, a continuous monitoring and evaluation is a word of must to conserve the groundwater resources (Mufid al-hadithi, 2012; Nagaraju et al., 2018; Papadkar et al., 2023; Reddy & Sunitha, 2023; Reddy et al., 2022; Singh, 1997). The recent advancement in the geospatial technology has proved to be a saviour in pollution studies including groundwater pollutions (Buys et al., 1992; Maqsoom et al., 2021; Suvarna et al., 2023).

The present study circumscribes the groundwater pollution in and around the Wardha valley coalfields by means of collaboration between the traditional and geospatial tools. The selected study area has already been under scanner for the multifaceted pollution issues including the groundwater (Ganvir & Guhey, 2021; George et al., 2013; Mhaiske & Jain, 2019; Salve et al., 2007; Satapathy et al., 2009; Trivedi et al., 2010; Zadan & Javed, 2022). Though multiple studies have proved the contaminations of trace/heavy metals in the water regimes of the study area, a systematic source approximation of contaminants and revelation of its governing mechanism were indeed needed. The central motivation behind the present study is to evaluate the groundwater quality for selected metals (identified through literature review), either to justify or put caution to its usage for domestic and agricultural purpose. Also, the identification of contaminants' sources (if any) and the understanding of the process responsible have also been kept as a key objective, to provide a forehand for the mitigation policy formulators. The study also strives to assess the association between the coal mining setups and groundwater quality. The present study has a potential to chalk out the plausible threats to the health of the associated biota in specific and ecosystem as a whole on one hand and on the other hand also establish a sound platform for the budding researchers to formulate a mitigation and remediation policies for the groundwater contamination issues in and around the coalfield.

The study systematically advocates and justifies the application of the geospatial approach in groundwater contamination studies for prolific upshots. In the due course, the precisely prepared base map of the study area was used for all sort of interpolation executed by the Ordinary Kriging method. The collective effect of analysed values of metals along with spatial interpolations gave a magnificent and reliable depiction. Eventually, the study manifests the effective monitoring and evaluation on using outputs generated by the groundwater analysis by traditional methods as inputs to the geospatial tools.

14.2 Study Area

The Wardha valley coalfields are the central India's foremost coal generating fields and politically are separated into Chandrapur and Yavatmal; the two districts of Maharashtra (Fig. 14.1a). The study area is lengthened along NW-SW with approximately 894.35 km^2 area. It is observable within the spatial coordinates 19°47′ to 20°10′ N and 78°57′ to 79°23′ E. The toposheets numbered 55L/16, 55 P/4, 55 P/8, 56 I/13, 56 M/1 and 56 M/5 has been used to demarcate the area of interest. Spatially, the study area is planner when considered in totality with 272 m and 104 m as highest and lowest elevations respectively, where the central region is low lying due to the south-east flowing Wardha river (Fig. 14.1b). The majority of the elevations in central regions are of anthropogenic origin and are overburden dumps; whereas, the elevations along the margins are hilly areas. The lowest point belongs to the mining pits. The sub-dendritic drainage pattern in the study area somehow suggests sedimentary terrain. The majority of the study area is rural; while along the margin, few semi-urban and urban settlements are observable (Fig. 14.1c). In terms of land use, most of it is used as agricultural fields along with few insignificant but noticeable patches of waste lands. The coal mining dumps can be observed over a significant expanse of land representing it as a foremost industry in the study area. The vegetal cover is quite rare, except along the eastern side. The climate of the study area can be categorized as subtropical with dry and hot summers and mild winters. The monsoon is the prime source of water for agricultural and other domestic usage. Hence, the pre-monsoons are dry with usual water scarcity and post-monsoons are humid with plenty of water for use.

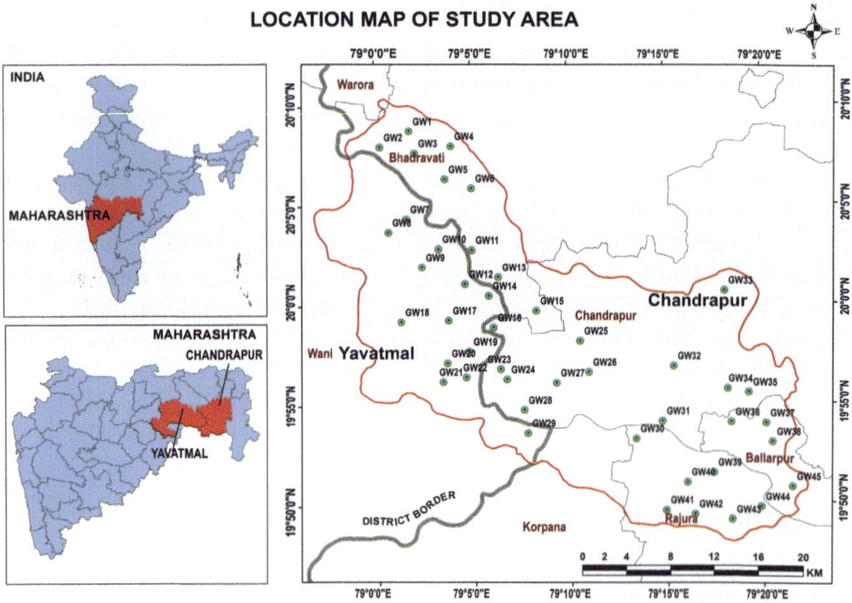

a.

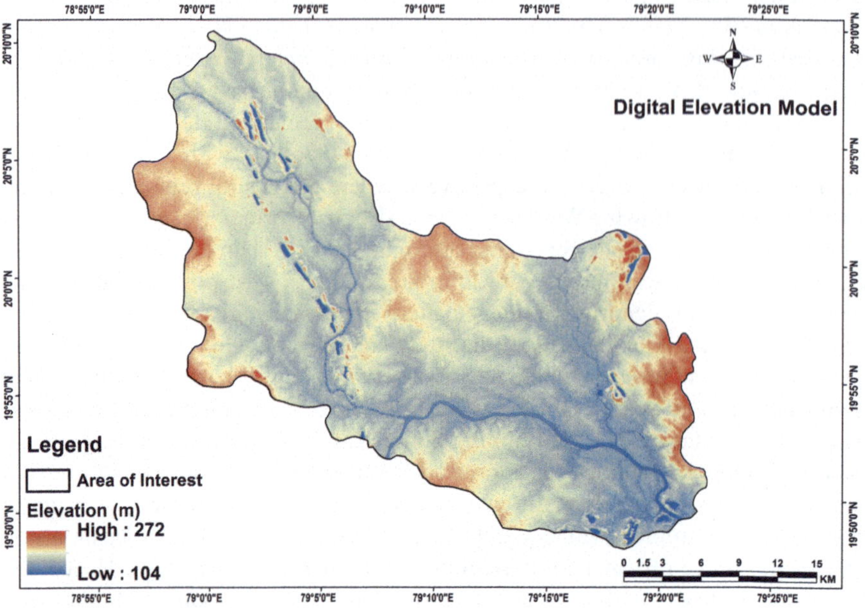

b.

Fig. 14.1 Study area **a** political Location map **b** digital Elevation map and **c** LULC map

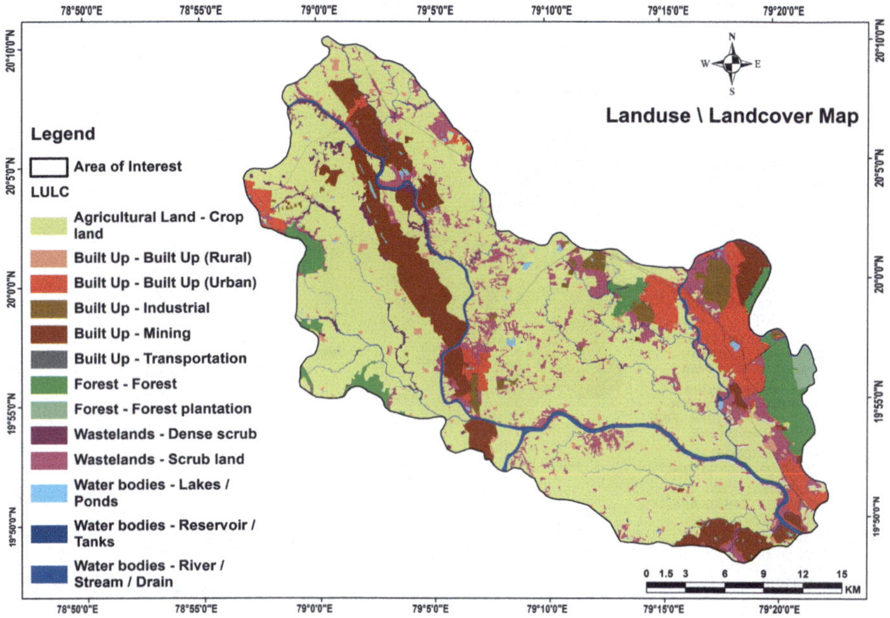

c.

Fig. 14.1 (continued)

14.2.1 General Geology

The study area is along fault oriented due NW–SE in the Pranhita–Godavari sub-basin of Gondwana Basin (Hughes, 1877). The Archean basement is overlaid by the Gondwana supergroup rocks of them the Talchir formation of glacial origin constituting striated pavement, pebbly mudstone and glacial varves is the lowest one (Fedden, 1875; Ghosh, 1962). Overlain to the Talchir rocks, the Barakar formation encompassing rocks like white to grey sandstone, siltstone and varieties of shale is the only geological setup hosting the prominent coal seam of study area (Singh et al., 1982). Being, the top rock cover of the Gondwana supergroup, the Kamthi formation manifests rocks like ferruginous & variegated sandstone and multi-coloured shale (Raja Rao, 1982). Some insignificant outcrops of Lameta limestone aged Late Cretaceous and exposures of Pakhal supergroup also exist in the study area (Ramesh et. al., 2018). Apart from above geological formations, a prolific cover of the alluvium of fluvial origin acts as a blanket over the entire study area (Fig. 14.2).

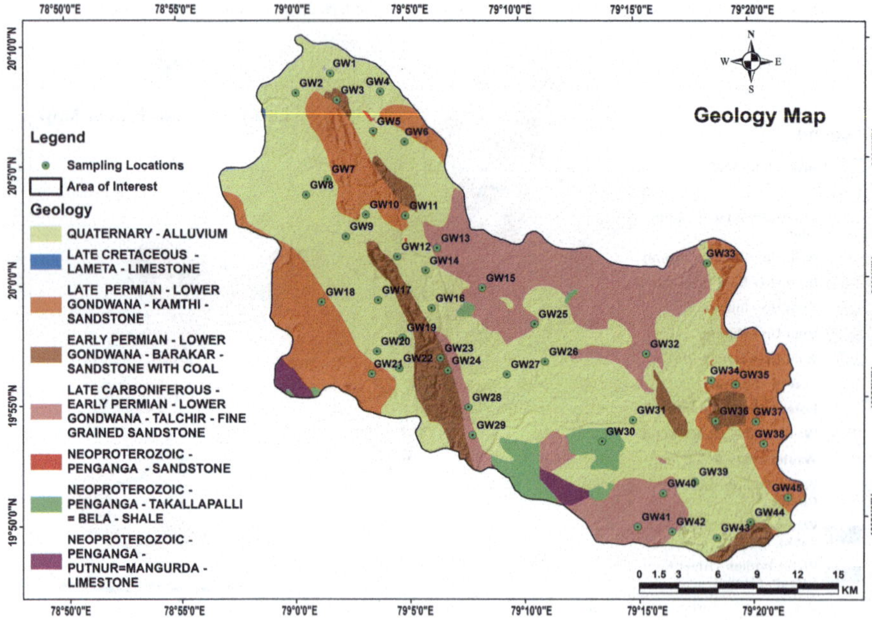

Fig. 14.2 Map of the study area demonstrating geology and sample locations of groundwater (Ganvir & Guhey, 2021)

14.3 Methodology

The primary observations done during the field visits and the add-ons derived from the GIS tools (Arcmap software and ArcGIS software) provided a confidence in pinpointing the groundwater sampling locations. A cluster sampling method was engaged to collect 45 samples of groundwater during pre and post-monsoon seasons of year 2019 by keeping justifiable scattering over suburban, cultivated lands, active and inactive coalfields, etc. (Fig. 14.2). The reliability of the groundwater samples were assured by the guidelines given by APHA (1998). To achieve a proper representation of the sample in terms of quality, the stagnancy factor was eradicated from the bore well and the dug well during the sampling. On primary observation, the turbid samples are intentionally skipped and if not possible a filter paper of 0.45 μm was used. A narrow mouth pre-washed high density polyethylene bottles were used for sampling and were methodically rinsed with the sample water before sampling. The groundwater samples were acidified by the concentrated HNO_3 instantly after assessing the pH, electric conductivity and Total Dissolved Salt during spot analysis. The acidification stabilizes the metal ions and prevents the precipitation. Thereafter, acidified samples were preserved and transported in a cold bath assuring around 4 °C as per the prescribed guidelines. Also, special care has been taken regarding the controlling factors during transportation and preservation to minimize any sort of impurity infiltration or instability, until it is made available to the analysis. Thereafter, the groundwater samples were analysed for the pre-selected metal ions viz., Aluminium (Al), Cadmium (Cd), Chromium (Cr), Copper (Cu), Iron (Fe), Nickel (Ni), lead (Pb) and Zinc (Zn) by employing AAS through standard techniques on recommended wavelengths. To assure the reliability of the analytical results, only ultra-pure base solutions were used to prepare respective standard solutions. The opening results of metals were then interpolated by Ordinary Kriging method using ArcGIS software on the base map prepared by highlighting the lithology and relief for better correlation. The Ordinary Kriging method is one of the most popular and frequently used method in the groundwater studies (ESRI, 1995; Olea, 2001; Yimit et al., 2011). The selected interpolation method has a great potential in portrayal of the known and predicated values on the base map. The predicted standard error maps are spontaneously created along with the interpolation maps proving it a worthwhile method. The disclosed areas with anomalous values of metals obtained by interpolation were marked and rock samples were collected from there to identify the temporal source of contamination. A detailed microscopic study was carried out of the potential rock samples sorted after megascopic observation to identify the probable source of the metals' contamination.

14.4 Heavy Metals' Evaluation in the Groundwater

The evaluated metals' order of average percentage contributions in the analysed groundwater samples for the pre-monsoon was Zn > Fe > Cu > Al > Cr > Ni = Pb > Cd (Table 14.1). The order of average percentage contributions was somewhat same with a minor change in post-monsoon, as the positions are interchanged between Fe and Zn (Table 14.2). Among all observed metals, the concentration of Cd, Fe, Ni and Pb was exceeding the permissible/acceptable value prescribed by BIS (2012) at certain locations and hence, the respective groundwater locations were classified as contaminated. The evaluation details of the contaminant metals are as follows.

14.4.1 Cadmium (Cd)

As per the BIS (2012), the acceptable limit for Cd is 3 μg/l. In the pre-monsoon groundwater samples, nearly 48% were in contaminated category; whereas, the samples of post-monsoon indicated of 6% drop in contaminated samples, when compared to the preceding season. The groundwater sample G.W.11 stood as most contaminated with 52 μg/l and 41 μg/l of Cd concentration in pre and post-monsoon respectively (Fig. 14.3a).

Table 14.1 The detected values of analysed metals from the groundwater samples of pre-monsoon 2019. (N.D.—Not Determined)

Sample	Locations		Al	Cd	Cr	Cu	Fe	Ni	Pb	Zn
G.W.1	20°08′46″N	79°01′48″E	22	N.D	30	62	180	10	N.D	110
G.W.2	20°07′58″N	79°00′16″E	31	N.D	30	190	143	N.D	N.D	34
G.W.3	20°07′39″N	79°02′04″E	66	27	45	255	398	38	16	220
G.W.4	20°07′59″N	79°03′58″E	51	N.D	40	60	209	N.D	N.D	95
G.W.5	20°06′20″N	79°03′38″E	44	8	22	41	180	N.D	N.D	47
G.W.6	20°05′52″N	79°05′00″E	36	N.D	33	114	188	14	8	112
G.W.7	20°04′21″N	79°01′37″E	48	38	66	140	426	54	60	180
G.W.8	20°03′42″N	79°00′40″E	36	N.D	43	208	225	N.D	N.D	40
G.W.9	20°01′57″N	79°02′23″E	180	8	44	620	360	34	14	186
G.W.10	20°02′51″N	79°03′16″E	44	24	42	228	384	40	38	400
G.W.11	20°02′47″N	79°04′59″E	55	52	48	205	618	56	84	194
G.W.12	20°01′05″N	79°04′37″E	144	26	38	455	450	43	22	1620
G.W.13	20°01′25″N	79°06′20″E	48	N.D	N.D	100	120	12	N.D	300
G.W.14	20°00′29″N	79°05′50″E	120	12	49	710	340	36	16	320
G.W.15	19°59′43″N	79°08′17″E	33	N.D	14	165	145	11	2	159
G.W.16	19°58′54″N	79°06′04″E	108	3	38	332	280	22	16	893
G.W.17	19°59′16″N	79°03′45″E	150	14	40	330	424	38	21	515
G.W.18	19°59′13″N	79°01′17″E	89	N.D	N.D	55	160	8	N.D	400
G.W.19	19°57′42″N	79°04′48″E	160	9	30	210	373	33	24	123
G.W.20	19°57′08″N	79°03′40″E	49	N.D	18	178	98	N.D	2	465
G.W.21	19°56′12″N	79°03′26″E	20	N.D	10	N.D	66	N.D	2	328
G.W.22	19°56′25″N	79°04′38″E	168	27	41	540	624	54	22	1050
G.W.23	19°56′49″N	79° 06′25″E	177	22	32	440	540	48	28	830
G.W.24	19°56′18″N	79°06′44″E	160	7	41	530	330	27	20	665
G.W.25	19°58′11″N	79°10′33″E	24	N.D	39	240	76	N.D	N.D	415
G.W.26	19°56′37″N	79°10′59″E	41	N.D	25	145	240	N.D	N.D	330
G.W.27	19°56′06″N	79°09′19″E	40	N.D	22	200	50	19	3	128
G.W.28	19°54′45″N	79°07′37″E	149	26	28	334	460	27	18	220
G.W.29	19°53′35″N	79°07′47″E	60	30	42	240	328	34	41	243
G.W.30	19°53′15″N	79°13′25″E	29	N.D	34	65	220	N.D	N.D	98
G.W.31	19°54′07″N	79°14′47″E	45	N.D	40	45	110	N.D	5	30
G.W.32	19°56′53″N	79°15′24″E	40	N.D	54	81	168	N.D	N.D	74
G.W.33	20°00′38″N	79°18′07″E	98	45	40	200	552	72	52	230
G.W.34	19°55′44″N	79°18′14″E	165	12	45	110	460	45	40	515
G.W.35	19°55′33″N	79°19′18″E	75	N.D	25	47	188	N.D	7	40

(continued)

Table 14.1 (continued)

Sample	Locations		Al	Cd	Cr	Cu	Fe	Ni	Pb	Zn
G.W.36	19°54′02″N	79°18′24″E	110	32	42	75	320	32	25	220
G.W.37	19°53′59″N	79°20′10″E	66	N.D	20	66	300	10	N.D	33
G.W.38	19°53′03″N	79°20′29″E	85	N.D	35	60	170	N.D	9	20
G.W.39	19°51′31″N	79°17′28″E	49	N.D	20	70	70	N.D	5	20
G.W.40	19°51′03″N	79°16′04″E	75	N.D	18	63	130	N.D	10	39
G.W.41	19°49′40″N	79°14′56″E	120	N.D	14	40	115	9	N.D	68
G.W.42	19°49′27″N	79°16′27″E	102	25	48	320	280	60	70	395
G.W.43	19°49′10″N	79°18′24″E	85	40	52	105	440	55	39	400
G.W.44	19°49′48″N	79°19′53″E	120	18	40	130	382	38	45	120
G.W.45	19°50′48″N	79°21′29″E	140	20	42	220	430	50	30	270
Measurement values are in μg/l		Min	20	N.D	N.D	N.D	50	N.D	N.D	20
		Max	180	52	66	710	624	72	84	1620
		Average	83.48	11.66	33.75	200.53	283.33	22.86	17.64	293.2

14.4.2 *Iron (Fe)*

As per the BIS (2012), the acceptable limit for Fe is 300 μg/l. Approximately 44% of the analysed samples were surpassing the prescribed limit with G.W.22 as the highest one having the value of 624 μg/l during pre-monsoon season; whereas, it reduces by 2% in the succeeding season with the G.W.22 at the apex (Fig. 14.3b).

14.4.3 *Nickel (Ni)*

As per the BIS (2012), the acceptable limit for Fe is 20 μg/l. Around half of the analysed samples manifested higher concentration than the prescribed one with G.W.33 having apex value of 72 μg/l during pre-monsoon; whereas, in succeeding season, the contaminated samples reduced by 09%, with G.W.33 again as the sample with highest value (Fig. 14.3c).

14.4.4 *Lead (Pb)*

As per the BIS (2012), the acceptable limit for Fe is 10 μg/l. In the season of pre-monsoon, around half of the groundwater samples demonstrated values surpassing the acceptable limit with G.W.11 having highest concentration. On contrary, in nearly 29% of the groundwater samples, the Pb concentration was below the detectable limit. In succeeding season, approximately 47% of the analysed samples demonstrated

Table 14.2 The detected values of analysed metals from the groundwater samples of post-monsoon 2019. (N.D.—Not Determined)

Sample	Locations		Al	Cd	Cr	Cu	Fe	Ni	Pb	Zn
G.W.1	20°08′46″N	79°01′48″E	20	N.D	24	48	154	N.D	N.D	71
G.W.2	20°07′58″N	79°00′16″E	30	N.D	22	140	132	N.D	N.D	40
G.W.3	20°07′39″N	79°02′04″E	62	14	35	201	318	22	11	200
G.W.4	20°07′59″N	79°03′58″E	50	N.D	N.D	42	180	N.D	N.D	56
G.W.5	20°06′20″N	79°03′38″E	24	N.D	20	30	175	N.D	N.D	33
G.W.6	20°05′52″N	79°05′00″E	28	N.D	24	120	150	10	N.D	102
G.W.7	20°04′21″N	79°01′37″E	42	24	41	122	388	37	22	166
G.W.8	20°03′42″N	79°00′40″E	20	N.D	32	180	200	N.D	N.D	42
G.W.9	20°01′57″N	79°02′23″E	98	3	40	480	298	25	12	144
G.W.10	20°02′51″N	79°03′16″E	38	20	34	214	315	33	24	314
G.W.11	20°02′47″N	79°04′59″E	50	41	35	188	555	49	63	164
G.W.12	20°01′05″N	79°04′37″E	123	12	22	444	360	31	16	870
G.W.13	20°01′25″N	79°06′20″E	56	N.D	N.D	221	47	N.D	N.D	302
G.W.14	20°00′29″N	79°05′50″E	126	6	25	661	322	20	14	300
G.W.15	19°59′43″N	79°08′17″E	20	N.D	N.D	118	130	N.D	2	168
G.W.16	19°58′54″N	79°06′04″E	44	N.D	19	229	305	19	11	210
G.W.17	19°59′16″N	79°03′45″E	134	8	39	279	332	28	20	415
G.W.18	19°59′13″N	79°01′17″E	89	2	12	16	98	16	N.D	325
G.W.19	19°57′42″N	79°04′48″E	10	N.D	24	321	324	N.D	17	120
G.W.20	19°57′08″N	79°03′40″E	81	N.D	26	N.D	168	N.D	2	60
G.W.21	19°56′12″N	79°03′26″E	38	N.D	N.D	N.D	48	N.D	2	400
G.W.22	19°56′25″N	79°04′38″E	102	12	32	364	498	32	10	680
G.W.23	19°56′49″N	79° 06′25″E	75	10	41	167	367	40	23	490
G.W.24	19°56′18″N	79°06′44″E	112	10	42	460	247	20	20	441
G.W.25	19°58′11″N	79°10′33″E	25	N.D	22	300	55	N.D	N.D	365
G.W.26	19°56′37″N	79°10′59″E	33	N.D	17	105	148	N.D	N.D	290
G.W.27	19°56′06″N	79°09′19″E	62	N.D	12	241	24	12	3	55
G.W.28	19°54′45″N	79°07′37″E	70	21	32	122	400	33	17	40
G.W.29	19°53′35″N	79°07′47″E	62	24	40	232	304	30	30	190
G.W.30	19°53′15″N	79°13′25″E	22	N.D	21	49	166	N.D	N.D	38
G.W.31	19°54′07″N	79°14′47″E	30	N.D	32	30	105	7	3	21
G.W.32	19°56′53″N	79°15′24″E	32	N.D	50	66	150	N.D	N.D	55
G.W.33	20°00′38″N	79°18′07″E	90	33	30	195	449	70	48	201
G.W.34	19°55′44″N	79°18′14″E	110	7	33	120	330	33	16	325
G.W.35	19°55′33″N	79°19′18″E	55	N.D	21	45	155	N.D	6	55
G.W.36	19°54′02″N	79°18′24″E	124	20	50	72	304	40	20	225

(continued)

Table 14.2 (continued)

Sample	Locations		Al	Cd	Cr	Cu	Fe	Ni	Pb	Zn
G.W.1	20°08′46″N	79°01′48″E	20	N.D	24	48	154	N.D	N.D	71
G.W.37	19°53′59″N	79°20′10″E	40	N.D	15	68	280	5	5	24
G.W.38	19°53′03″N	79°20′29″E	90	N.D	30	71	144	8	N.D	31
G.W.39	19°51′31″N	79°17′28″E	30	N.D	15	58	50	N.D	N.D	12
G.W.40	19°51′03″N	79°16′04″E	60	N.D	22	50	47	N.D	8	41
G.W.41	19°49′40″N	79°14′56″E	105	N.D	18	40	125	10	N.D	60
G.W.42	19°49′27″N	79°16′27″E	143	10	35	224	365	57	74	262
G.W.43	19°49′10″N	79°18′24″E	72	24	50	110	300	40	14	387
G.W.44	19°49′48″N	79°19′53″E	80	10	22	108	322	42	33	106
G.W.45	19°50′48″N	79°21′29″E	130	12	45	163	368	36	21	241
Measurement values are in μg/l		Min	10	N.D	N.D	N.D	24	N.D	N.D	12
		Max	143	41	50	661	555	70	74	870
		Average	65.26	7.17	26.688	166.97	237.82	17.88	12.6	203.04

contamination and around 33% samples stood blank. The sample G.W.42 replaced the sample G.W.11 (Fig. 14.3d).

As an obvious effect of dilution during the post-monsoon, the contamination intensities were found to be dropping in comparison to pre-monsoon. Though, the percentage of contaminated samples appears to be large in number, in real only around 18% and 09% are of acute intensity in pre and post-monsoon respectively. Technically, majority samples were surpassing the acceptable limits of respective metals by only insignificant margins.

The results significantly elucidates that the groundwater is contaminated by the dissolution of the contaminants leached out at somewhere by a certain mechanism. Looking at the contaminants i.e. cadmium, iron, nickel and lead, their sources could be conjoint or different. In the coalfields, the occurrences of the metals including the above contaminants in the coal and associated strata are evident (Finkelman, 1981; Pires et al., 1997; Saikia et al., 2009). Also, the mechanism driving the leaching, dissolution and migration of these metals under the influence of acidic drainage from the source point to the groundwater regimes has been detailed in multiple works (Chandra, 1992; Ganvir & Guhey, 2022, 2023; Karanth, 1997; Schurch et al., 2004; Zipper et al., 2014). Hence, it would not be exaggeration, if the present results claims the tendency towards the sources of contaminants in the coal and associated strata and are leached, dissolved and migrated under the influence of acid mine drainage.

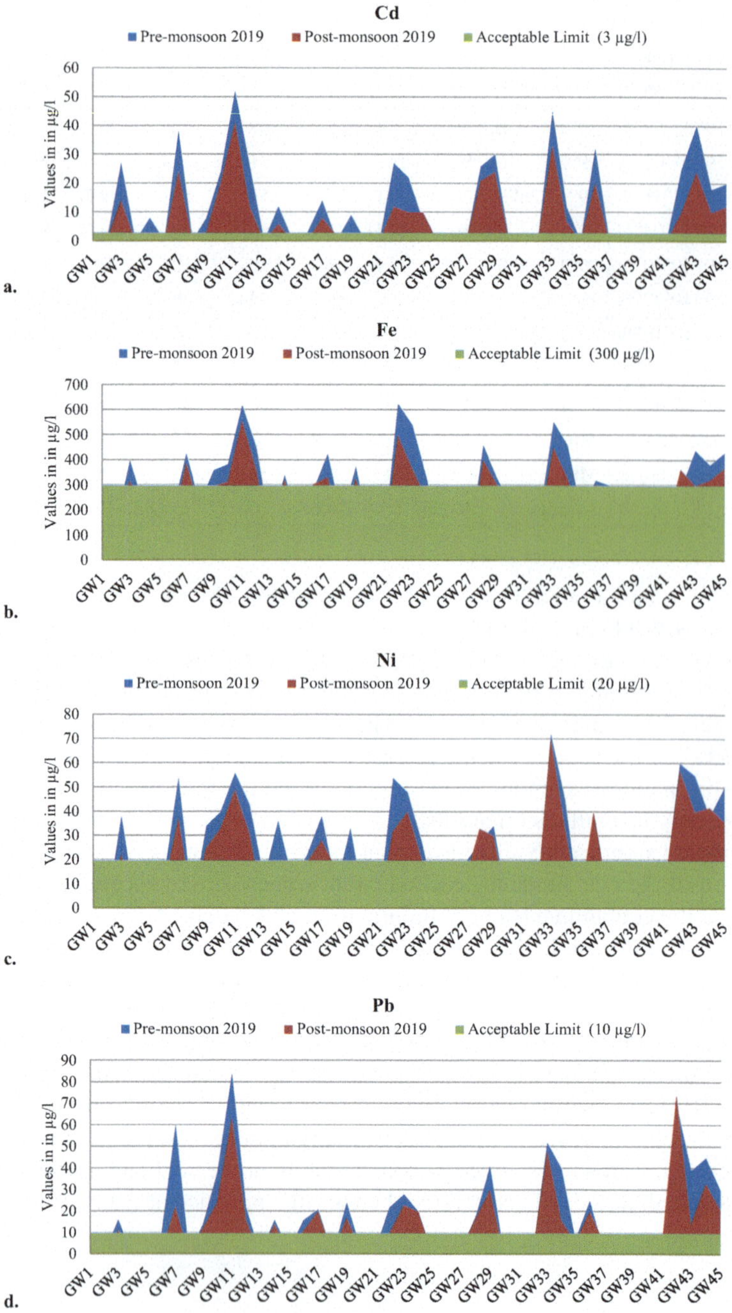

Fig. 14.3 Area Plots **a** Cd **b** Fe **c** Ni and **d** Pb

14.5 Spatial Depiction of the Contamination

The variations of the values of the variables under observation in the natural environment are somewhat intricate to be depicted precisely (Buys et al., 1992). Also, the comprehensive evaluation of the variables under observation is not advisable and feasible. At this point, the geo-statistical approach aids to produce a variogram for establishing spatial correlation. The variogram integrated in an interpolation technique namely ordinary kriging produce impartial linear approximations on the sites where samples have not been taken. The ordinary kriging technique justifies its engagement by its collective generalized linear regression to reduce and assess the variance. Along with this, it spontaneously generates a probabilistic simulation of ambiguity in estimates by anticipated values (Deutsch & Journel, 1998). The simultaneously generated prediction standard error map validates the interpolation outputs.

To sort the contamination hotspots in the study area, a spatial dispersal of the contaminated metals' viz., Cd, Fe, Ni and Pb was obvious. Henceforth, an ordinary kriging method was employed by using the results obtained for metals by groundwater analysis. The statistical accountability was upheld by replacing the non-detected values of metals by the respective background values. The base map used for the interpolation was a shaded relief, superimposed by lithological units. The shaded relief reflects the coal mining pits and the overlying lithology aids in the correlation of the metals' interpolation outcomes with temporal source. The designated colour clearly indicates the enrichment or paucity of the respective metal concentration in the groundwater. Also, in prediction standard error (PSE) maps, the predicted errors are inversely proportional to the number of samples taken. A gradual shade change and instant change in shade reflects steady and abrupt change in values respectively.

14.5.1 Pre-monsoon 2019

1. Cadmium (Cd)—The Cd concentration in the groundwater appears to be amplified in the vicinity of the lithology encompassing Barakar and the Kamthi rocks, where the coal mining pits are also visible (Fig. 14.4a). The error predicted for the Cd from the PSE maps ranges from 03.69–4.13 μg/l to 5.36–06.14 μg/l (Fig. 14.4e).
2. Iron (Fe)—The Fe dispersion in the groundwater is roughly observable in all over the study area with the enrichments around the Barakar and the Kamthi formations (Fig. 14.4b). These enrichment zones are in close periphery of the active coal fields. The PSE map revealed 38.46–41.07 μg/l to 50.85–55.52 μg/l as the range of error (Fig. 14.4f).
3. Nickel (Ni)—The enrichment of the Ni ions was restricted around the coal mines sites and positively associated with the lithology of Barakar and the Kamthi

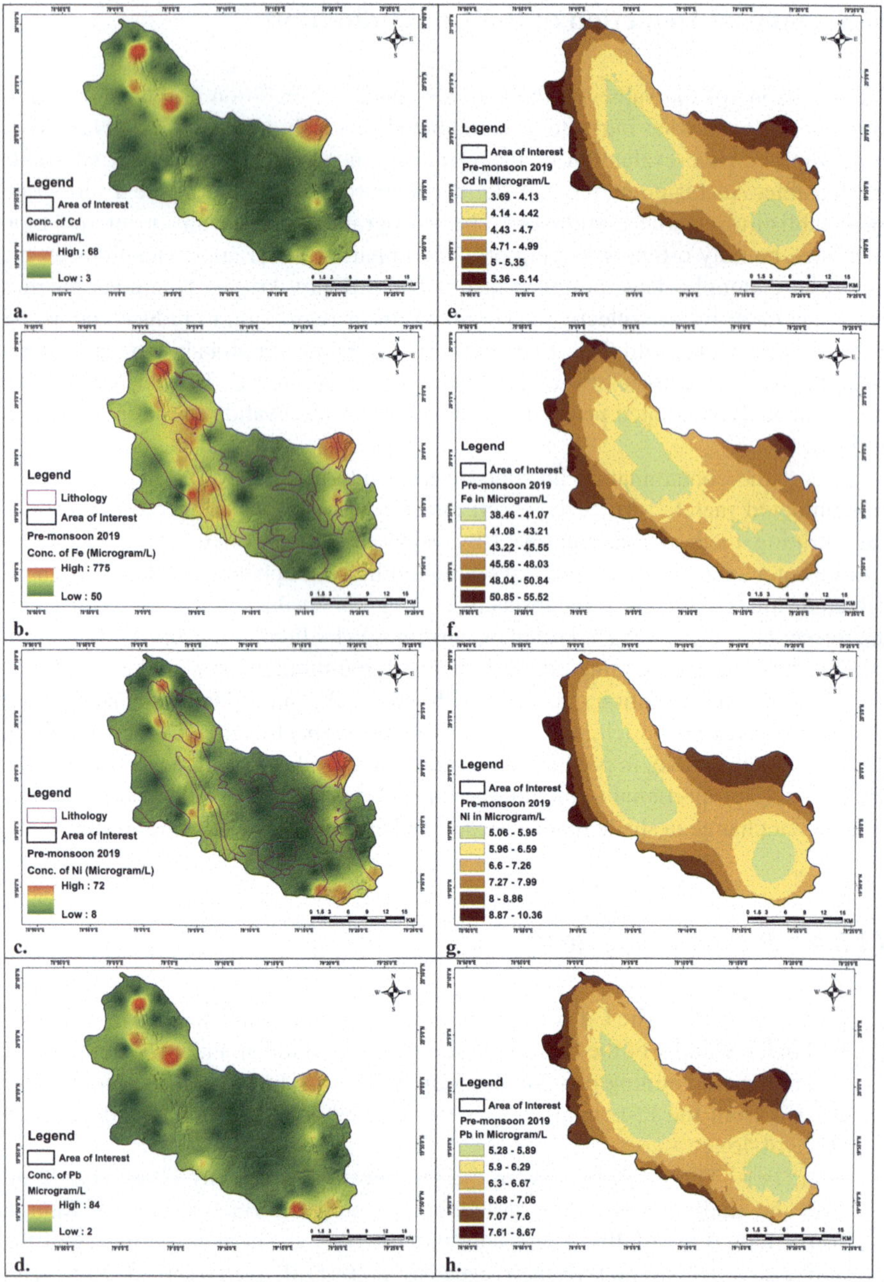

Fig. 14.4 Interpolation maps for the pre-monsoon 2019 using ordinary kriging method for the analysed metals in the groundwater; **a** Cd, **b** Fe, **c** Ni and **d** Pb and the respective prediction standard error maps; **e** Cd, **f** Fe, **g** Ni and **h** Pb

formations (Fig. 14.4c). For the interpolation of Ni, the prediction error ranges from 05.06–05.95 μg/l to 08.87–10.36 μg/l in the PSE maps (Fig. 14.4g).
4. Lead (Pb)—The Pb concentration in the groundwater is intensified around the coal mining regions, where the Barakar and the Kamthi rocks are principal lithological setup (Fig. 14.4d). In PSE maps, the prediction error ranges from 05.28–05.89 μg/l to 07.61–08.67 μg/l (Fig. 14.4h).

14.5.2 Post-monsoon 2019

1. Cadmium (Cd)—In the similar line of preceding season, the hotspots of Cd concentration in the groundwater are certain on and around the coal mining sites and lithology of the Barakar and the Kamthi formations (Fig. 14.5a). The prediction error for Cd ranges from 04.45–06.78 μg/l to 09.05–09.58 μg/l (Fig. 14.5e).
2. Iron (Fe)—The dispersal of the Fe ions in the groundwater is coarsely noticeable all over the study area with the hotspots on the formations of Barakar and the Kamthi; specifically, around the coal mining region (Fig. 14.5b). The prediction standard error map is disclosing the range of error from 35.26–40.32 μg/l to 56.06–63.92 μg/l (Fig. 14.5f).
3. Nickel (Ni)—The concentration of the Ni ions seems to be enriched in the groundwater around the coal mining regions and is in immediacy of lithology constituting the Barakar and the Kamthi rocks (Fig. 14.5c). The prediction error ranges from 05.37–06.03 μg/l to 08.26–09.24 μg/l (Fig. 14.5g).
4. Lead (Pb)—The amplification of the Pb ions in the groundwater regime is adjacent to the coal mining areas, where the Barakar and the Kamthi formations are as major lithology (Fig. 14.5d). The prediction standard error map indicates a steady non-fluctuant value ranging from 03.78–03.94 μg/l to 04.82–05.41 μg/l (Fig. 14.5h).

In both the seasons, the dispersal of Cd, Ni and Pb in the groundwater were quite restricted to the hotspots; whereas, the dispersal of Fe was comparatively extensive. In terms of aquifers, the enriched zones for the groundwater are up to 450 m below the ground level; where, the Kamthi sandstone turns out to be the prime potential aquifer (GSDA, 2013). On obtaining the entire spatial depiction of the contaminants over the study area, logical conjectures for the metals' dispersal could be deduced. The Kamthi sandstone is ferruginous in nature, which surely explains the widespread dispersal of Fe in the groundwater (Raja Rao, 1982). Also, being a common element in many rock forming minerals, the wide-ranging dispersion of Fe in the groundwater could logically be approved; but, the hotspots of Fe around the Barakar formation cannot be overlooked. The excavations done for the sake of the mining has definitely would have exposed the formations concealing the contaminants in one form or another. This somehow backs the speculation that the sources of the Cd, Ni and Pb could be constrained to the Barakar strata; especially in and around the coalfields.

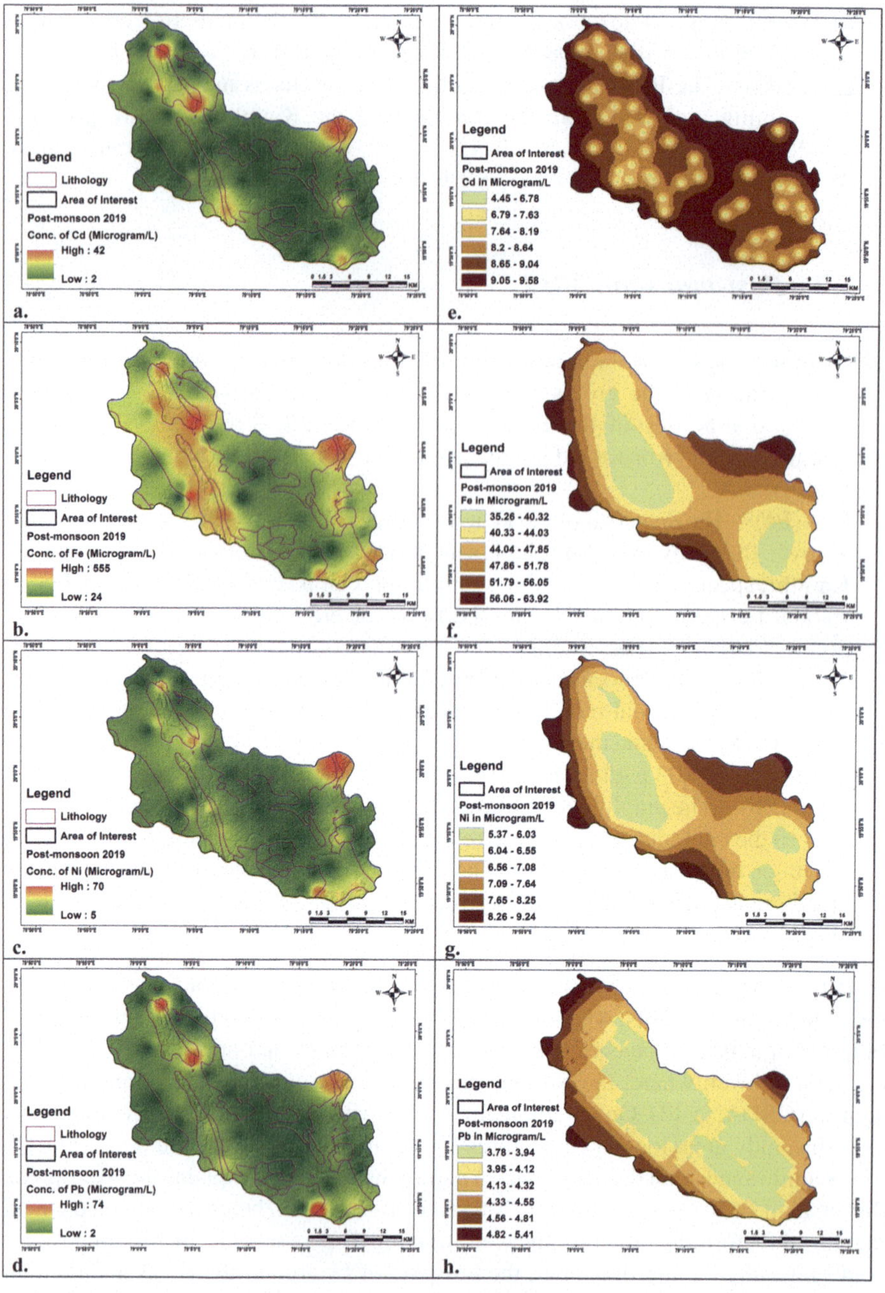

Fig. 14.5 Interpolation maps for the post-monsoon 2019 using ordinary kriging method for the analysed metals in the groundwater; **a** Cd, **b** Fe, **c** Ni and **d** Pb and the respective prediction standard error maps; **e** Cd, **f** Fe, **g** Ni and **h** Pb

The correlation with the lithology, eventually posed the interpolation maps as the passageway through which the pursuit of contaminants' sources will be advanced.

14.6 Source Approximation

The entire evaluation of groundwater and its statistical treatment has been aimed to understand the source of the contaminants. The exact source can be sorted out only by articulating the spatial and temporal data. Following is the portrayal of the source approximation in terms of spatial and temporal domain.

14.6.1 Spatial

The interpolation maps have clearly indicated the positive association between the contaminants and the lithology of the Barakar and Kamthi formation; specifically in the vicinity of the coal mines. The insignificant dispersal of the contaminants in the rest of the study area clearly indicates the localization of the contaminant sources in and around the coal mining regions exposing the rocks of Barakar and Kamthi formation. The interpolation maps prepared out of the factor loading deduced from the principal component analysis of the evaluated metals for the year 2019 in the Wardha valley coalfields has also confirmed the similar spatial sources (Ganvir & Guhey, 2023).

14.6.2 Temporal

The identification of the geological stratum or strata responsible for the contamination is a tedious task and has to be done distinctly; but, the hydro-geochemical elucidations manifest much of the information for sorting the accountable geological setup (Singh et al., 2020). In Wardha valley coalfields, the SO_4^{2-} has been found to be dominant anion in the hydro-geochemical investigation (Ganvir & Guhey, 2022). In the Barakar formation, the incidences of sulphide minerals can be held responsible for the enhanced values of the SO_4^{2-} in groundwater (Karanth, 1997). The pyrite mineral is profuse in the coals of Ghugus which is a major coalfield in the Wardha valley (Singh et al., 1983). In Wardha valley coalfields, the coal has been evident with the disseminations of the mineral pyrite as a framboidal clusters and cavity fillings (Sarate, 2004 and 2010). The pyrite mineral is an evident root for several trace metals' discharge (Saikia et al., 2009). The usual sulphide minerals accompanying with carbonaceous material and coal are pyrite, sphalerite, galena, chalcopyrite and pyrrhotite (Finkelman, 1981). In sulphide minerals, several trace elements are being hosted, comprising Cd, Fe, Ni and Pb (Pires et al., 1997). The high sulphur content in the carbonaceous matter and the dominant sulphide mineralization in the strata of the Barakar formations, advocates the existence of the contaminant sources incorporated within the sulphide phases (Ganvir & Guhey, 2021).

The Cd is the one of the most hazardous contaminant in the groundwater and its source is probably localized in the sulphide mineralization of the Barakar rocks. The Cd containing sphalerite mineral phase is evident with sulphide mineralization (Saikia et al., 2009). The existence of pyrite nodules associated sphalerite within the coal fissures are noticeable (Cobb et al., 1980). Being a usual constituent in common rock forming minerals, the Fe content has been uniform all over the study area. Still, the hotspots are noticeable around the coal mines. The oxidative weathering of the pyrite mineral, which usually occurs as a minor mineral in the coals and associated sedimentary rocks contributes in Fe liberation (Chandra, 1992). This liberated Fe ions, produces a foraging effect of iron oxyhydroxide's on the associated trace metals, consequently pedals the metals' enrichments in the groundwater (Schurch et al., 2004). A substantial incorporation of Ni in the pyrite is evident in the Wardha valley coalfields, justifying the hotspots (Mukherjee et al., 1988). In inorganic minerals like pyrite existing within the coal, the existence of the Pb has been reported in multiple works (Crossley, 1947; Gluskoter et al., 1977; Wagner et al., 1980). In the coal of the Wardha Valley Coalfields, especially around Ghugus, a significant concentration of Pb has been observed with positively associated to sulphide minerals (Singh et al., 1983).

14.7 Implications of Investigation

14.7.1 Controls Over Contaminations

The overall investigation points towards the carbonaceous materials, coal and accompanying rock strata as the plausible point of origin of the heavy metals i.e. contaminants in the groundwater regime of the study area. In customary circumstances, these heavy metals stay concealed and stable in the host rocks; but, on exposure to the extraordinary circumstances may leach out to the environment (Scokart et al., 1983). In case of the current study, the higher percentage of the Fe ions clearly reflects the oxidation of the pyrite or similar sulphide minerals in presence of water producing the acid mine drainage (AMD). The protruding percentage of the SO_4^{2-} ions in the groundwater of the study area has already been proved by similar works and justifies the oxidative weathering of the sulphide minerals (Ganvir & Guhey, 2021 and 2022). In coalfields, the AMD is common and possess potential to dissolve the acid dissolving metals which otherwise remain stable (Singh, 1987). The confluence of the coal mining activities and the exposure of the sulphide minerals bearing rock strata of the Barakar formation in the study area provides a favourable environment for metals' leaching through the AMD. Thereafter, the AMD transports these metals by certain extent to the local water bodies including the groundwater regime.

Being precise, the enhanced weathering of the otherwise concealed rock strata could be prime regulator of the groundwater contaminations (Ganvir and Guhey, 2023; Saikrishna et al., 2023).

14.7.2 *Bioaccumulation and Its Mitigation*

The metal ions dissolved in the groundwater regime of the study area, finds halt in the soil through irrigation done with contaminated water and soil system becomes first ecological unit, to be affected by contaminants (Gupta et al., 2010). The thriving flora on it takes up the contaminant metals and accumulates in their tissues (Jambhulkar & Juwarkar, 2009). These contaminant metals then get an access to the fauna and magnify into the whole food chain (Khan et al., 2015). When the contaminant metals increase in successive levels of the food chain, it is deliberated as bio-magnification (Neely, 1980). The hazardous impacts of these contaminant metals on biotic world are well-known and have to be controlled at the inception (Papadkar et al., 2024).

The removal of the contaminant metals is the only and effective approach to mitigate the bioaccumulation. This has been done by many typical practices like of chemical precipitation, osmosis, solvent extraction, electrochemical technique, etc. (Barakat, 2011). On the other hand, the bioremediation pose an optimistic vision for contaminant metals' removal by using agricultural derivatives, flora, microbes, etc. of which, the adsorption by the agricultural wastes adheres the sustainability. (Akpor et al., 2014; Odumbe et al., 2023; Saxena et al., 2017). Apart from this, phyto-remediation and wetlands has also proved to be efficient in contaminant metals' eradication (Prathap et al., 2023; Zipper et al., 2014).

14.8 Impact on Human Settlements

In human physiology, some trace metals act as vitals in certain circumstances and at definite quantity, but some pose a threat even in a trace concentration. The contaminants observed in the present investigation except Fe are of considerable threat. The toxic effects of Cd on humans include respiratory disorders, kidney damage, carcinogenic effects and even damage to the DNA functioning (Binner et al., 1983; Mouron et al., 2001; Waisberg, et al., 2003). The Ni possess toxic effects on human health such as contact dermatitis, headaches, gastrointestinal manifestations, respiratory manifestations, lung fibrosis, cardiovascular diseases, lung cancer, nasal cancer and epigenetic effects (Genchi et al., 2020). The Pb blocks multiple vital tasks of the human body system including the central nervous, hematopoietic, hepatic and renal system, but the effects on children by leading to the poor development of the grey matter of the brain is the gravest one (Flora, 2009; Udedi, 2003).

Apart from direct health effects, the occurrence and the consumption of contaminated water by the human settlements hold long term impacts on their socio-economic growth (Chandnani et al., 2022; Lima et al., 2015; Parameswari et al., 2012). The area with pollution issues fails to lure the investments from the commercial sectors, ultimately lowering the overall property value and business potential (Nicholls & Cromptonm, 2018). Though, to certain extent the mining sector provides employment to the locals, the cumulative economic turnover does not stand with the trend. The lower pace of economic growth eventually lowers the business and job opportunities instigating a socio-economic lethargy. Along with the above issues the social problems also protrude in the contaminated areas (Chowdhury & Singh, 2019; Elouga et al., 2024).

14.9 Conclusion

The investigation has clearly elucidated the groundwater contaminations of the Cadmium, Iron, Nickel and Lead in the Wardha valley coalfields. The geospatial tools has aided not only in the preliminary investigations but also efficiently dealt with the core process of investigation. In the entire endeavour, the geospatial techniques have ascertained its competence in the source approximation of the contaminant metals. Only on the disclosure of the contaminant metals' spatial dispersal, the temporal sources were investigated. In Wardha valley coalfields, the coal mining region and the strata of the Barakar formations were identified as the plausible spatial and temporal sources of contaminant metals respectively. The identification of contaminants' sources, both spatial and temporal and the regulatory process elucidation regarding contamination has provided a lead over the mitigation polices. The bioaccumulation and bio-magnification are the major threats to the biotic world and these have to be blocked at very early stage. This can only be done by proper understanding of the contaminants' sources and process of migration to ecological units. Hence, looking at the multifaceted upshots, the geospatial approach is strongly recommended in the groundwater contamination studies.

Acknowledgements Authors are thankful to the Principal, Government N. P. G. College of Science Raipur, for providing all facilities during the work. We would also like to extent thanks to Rajiv Gandhi National Groundwater Training & Research Institute and CGWB Raipur, for their valuable support and guidance. We also forward our sincere thanks to Editors of the Book for a magnificent opportunity.

References

Abas, N., Kalair, A., & Khan, N. (2015). Review of fossil fuels and future energy technologies. *Futures, 69*, 31–49.

Akpor, O. B., Ohiobor, O. G., & Olaolu, T. D. (2014). Heavy metal pollutants in wastewater effluents: Sources, effects and remediation. *Advances in Bioscience and Bioengineering, 2*(4), 37–43.

APHA, (1998). Standard methods for the examination of water and waste water (20th ed.). *Washington DC: Water Pollution Control Federation*, pp.10–161

Armaroli, N., & Balzani, V. (2011). The legacy of fossil fuels. *Chemistry–An Asian Journal, 6*(3), 768–784.

Bach, W. (1981). Fossil fuel resources and their impacts on environment and climate. *International Journal of Hydrogen Energy, 6*(2), 185–201.

Balat, M. (2007). Influence of coal as an energy source on environmental pollution. *Energy Sources, Part a: Recovery, Utilization, and Environmental Effects, 29*(7), 581–589.

Barakat, M. A. (2011). New trends in removing heavy metals from industrial wastewater. *Arabian Journal of Chemistry, 4*(4), 361–377.

Binner, W. F., Bridges, W. J., & Rose, J. (1983). *Trace elements in health.* p. 1.

BIS, (2012). *Indian specifications for drinking Water.* IS: 10500, Bureau of Indian Standards pp.1–3.

Buys, J., Botha, J. F., & Messerschmidt, H. J. (1992). Triangular irregular meshes and their application in the graphical representation of geohydrological data. *Water Research Commission*, Report No 271/2/92.

Chandnani, G., Gandhi, P., Kanpariya, D., Parikh, D., & Shah, M. (2022). A comprehensive analysis of contaminated groundwater: Special emphasis on nature-ecosystem and socio-economic impacts. *Groundwater for Sustainable Development, 19*, 100813.

Chandra, D. (1992). Mineral resources of India, 5: Jharia coalfields. *Journal Geological Society of India, Bangalore, 5*, 1–149.

Chowdhury, S., & Singh, P. (2019). Information campaign on water quality and marriage market: The case of arsenic exposure in rural Bangladesh.

Cobb, J. C., Steele, J. D., Treworgy, C. G., & Ashby, J. F. (1980). The abundance of zinc and cadmium in sphalerite-bearing coals in Illinois. *Illinois minerals no.* 074.

Conti, K. I., & Gupta, J. (2016). Global governance principles for the sustainable development of groundwater resources. *International Environmental Agreements: Politics, Law and Economics, 16*(6), 849–871.

Crossley, H. E. (1947). The inorganic constituents of coal (Discussion). *Journal of the Institute of Fuel, 20*, 94–95.

Deutsch, C. V., & Journel, A. G. (1998). *GSLIB: Geostatistical software library and user's guide* (2nd ed.). Oxford University Press.

Elouga, M., Wilson, P. F. & Koultchoumi, B. (2024). Effects of climatic hazards on the geogenic pollution of groundwater in the Mayo Tsanaga sub-basin in Cameroon: Gender inequalities associated with dental fluorosis. In *The Climate-Health-Sustainability Nexus: Understanding the Interconnected Impact on Populations and the Environment* (pp. 421–442). Springer Nature Switzerland.

ESRI (1995). *Cell-based modelling with grid. On-line computer help system.* Environmental Systems Research Incorporated.

Fedden, F. (1875). On the evidences of ground-ice in tropical India during the Talchir period. *Geological Survey of India Records, 8*(1), 16–18.

Finkelman, R.B., 1981. *Modes of occurrence of trace elements in coal.* U.S. Geological Survey Open-file Report, Series 81–99, p. 312.

Flora, S. J. S. (2009). Structural, chemical and biological aspects of antioxidants for strategies against metal and metalloid exposure. *Oxidative Medicine and Cellular Longevity, 2*, 191–206.

Ganvir, P. S., & Guhey, R. (2021). Geochemical studies of some heavy metals' toxicity in groundwater with their plausible sources around Gondwana supergroup, Wardha valley Coalfields, Maharashtra. *Journal Geological Society of India, 97*, 1415–1421.

Ganvir, P. S., & Guhey, R. (2022). Hydro-geochemical elucidation and its implications in the Wardha valley coalfields of central India. *In IOP Conference Series: Earth and Environmental Science, 1032*, 012015.

Ganvir, P. S., & Guhey, R. (2023). An Implication of Enhanced rock weathering on the groundwater quality: A case study from Wardha Valley Coalfields, Central India. *Weathering and Erosion Processes in the Natural Environment*, 215–242. https://doi.org/10.1002/9781394157365.ch9

George, K. V., Patil, D. D., & Alappat, B. J. (2013). PM10 in the ambient air of Chandrapur coal mine and its comparison with other environments. *Environmental Monitoring and Assessment, 185*, 1117–1128.

Genchi, G., Carocci, A., Lauria, G., Sinicropi, M. S., & Catalano, A. (2020). Nickel: Human health and environmental toxicology. *International Journal of Environmental Research and Public Health, 17*(3), 679.

Ghosh, P. K. (1962). Recurrence of glaciation in the Talchir series, North Karanpura coalfield, Bihar. *Geological Survey of India Rec.*, v. 87(4), pp. 751–756

Gluskoter, H. J., Ruch, R. R., Miller, W. G., Cahill, R. A., Dreher, G. B., & Kuhn, J. K. (1977). Trace elements in coal: Occurrence and distribution. *Illinois Geological Survey Circular, 499*, 154.

GSDA. (2013). Groundwater information Chandrapur Maharashtra, pp. 5.

Gupta, S., Satpati, S., & Nayek, S. (2010). Effect of wastewater irrigation on vegetables in relation to bioaccumulation of heavy metals and biochemical changes. *Environmental Monitoring and Assessment, 165*, 169–177.

Hughes, W. T. H. (1877). The Wradha Valley Coalfield. *Memoir of the Geological Survey of India, 13*, 1–154.

Jambhulkar, H. P., & Juwarkar, A. A. (2009). Assessment of bioaccumulation of heavy metals by different plant species grown on fly ash dump. *Ecotoxicology and Environmental Safety, 72*(4), 1122–1128.

Khan, A., Khan, S., & Khan, M. A. (2015). The uptake and bioaccumulation of heavy metals by food plants, their effects on plants nutrients, and associated health risk: A review. *Environmental Science and Pollution Research, 22*, 13772–13799.

Karanth, K. R. (1997). *Groundwater assessment, development and management*. Tata McGrraw Hill, p. 720.

Lima, M. L., Romanelli, A., & Massone, H. E. (2015). Assessing groundwater pollution hazard changes under different socio-economic and environmental scenarios in an agricultural watershed. *Science of the Total Environment, 530*, 333–346.

Maqsoom, A., Aslam, B., Alwetaishi, M., Awais, M., Hassan, U., Maqsoom, S., Alaloul, W. S., Musarat, M. A., Zerouali, B., & Hussein, E. E. (2021). A GIS-based groundwater contamination assessment using modified DRASTIC geospatial technique. *Water, 13*(20), 2868.

Mhaiske, V.M., & Jain, A. (2019). Effect of tree plantations on soil organic carbon in Wardha coalfields of Maharashtra, India. *International Journal of Farm Sciences, 9*(2), 65–71.

Mouron, S. A., Golijow, C. D., & Dulout, F. N. (2001). DNA damage by cadmium and arsenic salts assessed by the single cell gel electrophoresis assay. *Mutation Research, 498*, 47–55.

Mufid al-hadithi. (2012). Application of water quality index to assess suitability of groundwater quality for drinking purposes in Ratmao–Pathri Rao watershed, Haridwar District, India. *American Journal of Scientific and Industrial Research, 3*(6), 395–402.

Mukherjee, K. N., Datta, N. R., Chandra, A. N., Pandalai, H. S., & Singh, M. P. (1988). A statistical approach to the study of the distribution of trace elements and their organic/inorganic affinity in lower Gondwana Coals of India. *International Journal of Coal Geology, 10*, 99–108.

Nagaraju, A., Balaji, E., Sun, L., & Thejaswi, A. (2018). Processes controlling groundwater chemistry from Mulakalacheruvu Area, Chittoor District, Andhra Pradesh, South India: A statistical approach based on hydrochemistry. *Journal of the Geological Society of India, 91*(4), 425–430. https://doi.org/10.1007/s12594-018-0875-0

Neely, W. B. (1980). *Chemicals in the environment: Distribution, Transport, Fate, Analysis* (p. 245). Marcel Dekker.

Nicholls, S., & Crompton, J. (2018). A comprehensive review of the evidence of the impact of surface water quality on property values. *Sustainability, 10*(2), 500.

Odumbe, E., Murunga, S., & Ndiiri, J. (2023). Heavy metals in wastewater effluent: Causes, effects, and removal technologies. *IntechOpen*. https://doi.org/10.5772/intechopen.1001452

Olea, R. (2001). *Geostatistical glossary and multilingual dictionary*. Oxford University Press.

Papadkar, J. N., Ganvir, P. S., Nimbarte, G. R., Patre, P. P., Barsagade, A., & Chandekar, R. (2023). A review over the causal relationship between hydro-geochemistry and bioaccumulation in special reference to coalfields. *European Chemical Bulletin, 12*(4), 15288–15297.

Papadkar, J. N., Ganvir, P. S., Nimbarte, G. R., Patre, P. P., Pusala, S. V., Sakhare, S. K., & Ghagargunde, P. G. (2024). Implications of coalfield dynamics on associated ecosystems: A tangential review. *Uttar Pradesh Journal of Zoology, 45*(20), 34–43.

Parameswari, K., Mudgal, B. V., & Nelliyat, P. (2012). Evaluation of groundwater contamination and its impact: An interdisciplinary approach. *Environment, Development and Sustainability, 14*, 725–744.

Pires, M., Fielder, H., & Teixeira, E. C. (1997). Geochemical distribution of trace elements in coal: Modelling and environmental aspects. *Fuel, 76*, 1425–1437.

Prathap, A., Shaikh, W. A., Baudhh, K., & Chakraborty, S. (2023). Phyto-management potential of naturally thriving plants on the metal contaminated overburden dump of coal mines: A study from Jharkhand, India. *Bioremediation Journal, 27*(3), 290–300.

Raja Rao, C. S. (1982). Coal resources of Tamil Nadu, Andhra Pradesh, Orissa and Maharashtra. *Bulletin of the Geological Survey of India, coalfields of Maharashtra, Wardha Valley Coalfield; Series A., 45*, 85–87.

Ramesh, S., Sharan, P. S., & David, K. (2018). Tectono-stratigraphy of the Pakhal Basin, Warangal (Rural) District of Telangana state with special reference to the Bollapally and the Pattipally Area. *International Journal of Innovative Research in Engineering & Multidisciplinary Physical Sciences, 6*(5), 32–37.

Reddy, Y. S., & Sunitha, V. (2023). Assessment of Heavy metal pollution and its health implications in groundwater for drinking purpose around inactive mines, SW region of Cuddapah Basin, South India. *Total Environment Research Themes, 8*, 100069.

Reddy, S., Sunitha, V., & Suvarna, B. (2022). Groundwater quality and its potential health impacts nearby inactive mines using PIG and geospatial technology, Southwestern part of Cuddapah basin, Andhra Pradesh, South India. *Groundwater for Sustainable Development, 17*, 100742.

Salve, P. R., Satapathy, D. R., Katpatal, Y. B., & Wate, S. R. (2007). Assessing spatial occurrence of ground level ozone around coal mining areas of Chandrapur District, Maharashtra, India. *Environmental Monitoring and Assessment, 133*, 87–98.

Saikia, B. K., Goswamee, R. L., Baruah, B. P., & Baruah, R. K. (2009). Occurrence of Some Hazardous Metals in Indian Coals. *Coke and Chemistry, 52*, 54–59.

Saikrishna, K., Purushotham, D., Sunitha, V., Reddy, Y. S., Brahmaiah, T., Reddy, B. M., & Nallusamy, B. (2023). Deciphering groundwater quality, mechanisms controlling groundwater chemistry in and around Suryapet, Telangana, South India. *Total Environment Research Themes, 6*, 100035.

Sarate, O. S. (2004). Biopetrological investigations on the coals of Kosar, Dongargaon and Mahadoli areas, Wardha Valley Coalfield, Maharashtra. *Journal Geological Society of India, 63*, 406–414.

Sarate, O. S. (2010). Study of petrological characteristics of coal deposits from Marki-Jhari-Jamni area, Wardha Valley Coalfield, Maharashtra. *Journal Geological Society of India, 76*, 361–368.

Satapathy, D. R., Salve, P. R., & Katpatal, Y. B. (2009). Spatial distribution of metals in ground/surface waters in the Chandrapur district (Central India) and their plausible sources. *Environmental Geology, 56*, 1323–1352.

Saxena, A., Bhardwaj, M., Allen, T., Kumar, S., & Sahney, R. (2017). Adsorption of heavy metals from wastewater using agricultural–industrial wastes as biosorbents. *Water Science, 31*(2), 189–197.

Schurch, M., Edmunds, W. M., & Buckley, D. (2004). Three-dimensional flow and trace metal mobility in shallow Chalk groundwater, Dorset, United Kingdom. *Journal of Hydrology, 292*(1–4), 229–248.

Scokart, P., Meeus-Verdinne, K., & DeBorger, R. (1983). Mobility of heavy metals in polluted soils near Zn smelters. *Water, Air, and Soil Pollution, 20*, 451–463.

Singh, G. (1987). Mine water quality deterioration due to due to acid mine drainage. *International Journal of Mine Water, 6*, 49–61.

Singh, M. P., Singh, R. M., Chandra, D., & Somsuddin, A. K. (1982). Petrology, sedimentology and depositional environment of Ghugus coals, Wardha Valley, district Chandrapur and Yeotmal, Maharashtra, India. *Quaternary Journal of Geology, Mining and Metallurgical Society of India, 54*, 1–17.

Singh, N. P., Mukherjee, T. K., & Shrivastava, B. B. P. (1997). Monitoring the impact of coal mining and thermal power industry on landuse pattern in and around Singrauli coalfield using remote sensing data and GIS. *Journal of the Indian Society of Remote Sensing, 25*, 61–72.

Singh, R. M., Singh, M. P., & Chandra, D. (1983). Occurrence, distribution and probable source of trace elements in Ghugus coals, Wardha valley, districts Chandrapur and Yeotmal, Maharashtra, India. *International Journal of Coal Geology, 2*, 371–381.

Singh, S., Tignath, S., Deolia, D. K., Jha, M., & Dixit, R. (2020). Geochemical assessment of groundwater quality for its suitability for drinking and irrigation purpose in Jabalpur, Madhya Pradesh, India. *International Journal Research in Engineering, Science and Management, 3*(08), 130–134.

Suvarna, B., Sunitha, V., Sudharshan Reddy, Y., Muralidhara Reddy, B., Kadam, A. K., & Ramakrishna Reddy, M. (2023). Groundwater quality assessment using multivariate statistical approach and geospatial modelling around cement industrial corridor, South India. *International Journal of Environmental Science and Technology, 20*(5), 5051–5070.

Trivedi, R., Mondal, A., Chakraborty, M. K., & Tewary, B.K. (2010). A statistical analysis of ambient air quality around opencast coal projects in Wardha Valley coalfields, western coal fields limited, India. *Indian Journal of Environmental Protection, 30*(12), 969–977.

Villarroya, F., & Aldwell, C. R. (1998). Sustainable development and groundwater resources exploitation. *Environmental Geology, 34*(2–3), 111–115.

Udedi, S. S. (2003). From guinea worm scourge to metal toxicity in Ebonyi State. *Nigeria. Chemistry in Nigeria, 2*(2), 13–15.

Wagner, P., Williams, J. M., Wewerka, E. M., Bertino, J. P., Wanger, L. W., & Wanek, P. L. (1980). Trace element contamination of drainage from coal cleaning wastes. *American Institute of Chemical Engineer, 21*, 79–86.

Waisberg, M., Joseph, P., Hale, B., & Beyersmann, D. (2003). Molecular and cellular mechanisms of cadmium carcinogenesis. *Toxicology, 192*, 95–117.

Yimit, H., Eziz, M., Mamat, M., & Tohti, G. (2011). Variations in groundwater levels and salinity in the Ili River Irrigation Area, Xinjiang, northwest China: A geostatistical approach. *International Journal of Sustainable Development & World Ecology, 18*(1), 55–64.

Zadan, K., & Javed, A. (2022). Correlation between land surface temperature (LST) and normalized difference vegetation index (NDVI) in Wardha Valley Coalfield, Maharashtra, Central India. *Nova Geodesia, 2*(3), 53–53.

Zipper, C., Skousen, J., & Jage, C. (2014). *Passive treatment of acid mine drainage* (pp. 1–13). Virginia Polytechnic Institute and State University.

Chapter 15
Heavy Metal Contamination and Health Risk Assessment in and Around the Pallikaranai Marshland, Tamil Nadu, India

N. M. Swetha, S. Maniyarasan, A. Ramachandran, M. L. Anusha, and R. R. Krishnamurthy

Abstract The study aimed to evaluate water quality and conduct a health risk assessment based on heavy metal contamination levels in and around the Pallikaranai marshland Chennai. A total of 10 water samples that were collected on september 2022 during Southwest Monsoon, the samples are subjected to analysis using Atomic Absorption Spectroscopy (AAS). Various physico-chemical parameters, including such as Depth, pH, electrical conductivity (Ec), total dissolved solids (TDS), oxidation–reduction potential (ORP), dissolved oxygen (DO), salinity, and temperature, were measured. The observed high pH value indicated an alkaline nature of the wetland water. Additionally, trace elements like iron (Fe), manganese (Mn), copper (Cu), and chromium (Cr) were assessed, with Fe and Cr displaying higher concentrations. The present study investigating the concentrations of these toxic elements in water bodies is a key contributor to water pollution. The findings show that the Perungudi dumps, unchecked sewage flow, and industrial waste are causing a lot of aquatic weeds to grow in the marshes. These weeds are harming the plants and animals living there. These pollutants release toxins that can ultimately enter the human food chain, posing health risks to the local population and broader ecosystem. Despite the essential role of water as a natural resource, it is concerning that some individuals continue to misuse and mismanage this invaluable creation of nature.

Keywords Water quality · Trace elements · Health risk assessment · Ecosystem · Pallikarani marshland · India

N. M. Swetha (✉) · S. Maniyarasan · A. Ramachandran · M. L. Anusha · R. R. Krishnamurthy
Department of Applied Geology and School of Earth and Atmospheric Sciences, University of Madras, Chennai, India
e-mail: swethanmsv@gmail.com

R. R. Krishnamurthy
e-mail: teaching@rrkrishnamurthy.org

V. Sunitha et al. (eds.), *Groundwater Resource Management Planning Strategies*, Springer Water, https://doi.org/10.1007/978-3-031-88870-0_15

15.1 Introduction

Wetlands are characterized as areas that serve as a transition between terrestrial and aquatic environments, where the water table is typically close to the surface or the land is covered by standing water not exceeding 6 m. These regions have been pivotal in supporting life ecosystems that have sustained human communities for countless generations. Wetlands play a vital role in human civilization by fulfilling various essential needs, including providing drinking water, purifying water, storing water, replenishing groundwater, controlling erosion, and stabilizing shorelines (Agarwal, 2008). Wetlands are recognized as the most productive ecosystem globally, holding significant socio-economic, ecological, and bio-aesthetic value. They offer a wide array of tangible and intangible benefits to humanity in a sustainable manner. Functioning akin to natural reservoirs or sponges, wetlands store and release water gradually, effectively lowering flood heights and supporting groundwater recharge. This gradual water release contributes to base flow in surface water systems during periods of low precipitation, showcasing their role in water management. The notable water storage capacity also plays a crucial role in minimizing the risk of flooding. Consequently, wetlands are often likened to the kidneys of the landscape due to their contributions to hydrological and chemical cycles, as well as biological supermarkets due to their support of extensive food webs and rich biodiversity (Klee 1994). Moreover, wetlands play a distinctive role in the global ecosystem by serving as habitats for some of the most endangered animal and plant species. Their significance to humans extends beyond their water management functions, as they contribute to air purification, replenish groundwater resources, and provide a haven for diverse plant and animal life, thereby helping to mitigate climate change and safeguarding coastlines.Query

Heavy metals, characterized by a relatively high density and toxicity even at parts per billion levels, comprise a group of metals and metalloids. This group includes, but is not limited to, elements such as Pb (lead), As (arsenic), Hg (mercury), Cd (cadmium), Zn (zinc), Ag (silver), Cu (copper), Fe (iron), Cr (chromium), Ni (nickel), and Pt (platinum) (Manavi Yadav et al., 2019). These elements are defined as heavy metals based on having an atomic number greater than 20 and an atomic density exceeding 5 g/cm^3, along with displaying metallic properties. While heavy metals are naturally present in the environment and essential for life, they can become detrimental when they accumulate in living organisms. Some of the most common heavy metals that pollute the environment include mercury, cadmium, arsenic, chromium, nickel, copper, and lead (Ali et al., 2019).

15.2 Materials and Methods

15.2.1 Study Area

The Pallikaranai Marsh, one of the last remaining natural wetlands in Chennai, is locally known in Tamil as "Kazhuveli," meaning a flood plain or a water-soaked area. This freshwater marsh is situated within the city of Chennai. It is one of the three wetlands in Tamil Nadu acknowledged by the National Wetland Conservation and Management Program (NWCMP). The Pallikaranai Marsh is geographically positioned at 12.949371 N latitude and 80.218184 E longitude (Fig. 15.1). It is flanked on the east by the Buckingham Canal and the Old Mahabalipuram Road, which is notably the site of the Information Technology (IT) corridor. The marsh is bordered by a mix of residential and institutional areas to the south and west. To the north of the marsh, one can find the mass rapid transit system along with other significant public infrastructure. Spanning an area sof 80 square kilometers, the Pallikaranai Marsh drains into the Bay of Bengal through two outlets, namely Okkiyam Madavu and the Kovalam stream, managing the drainage of a 250 km area in south Chennai that encompasses around 65 wetlands.

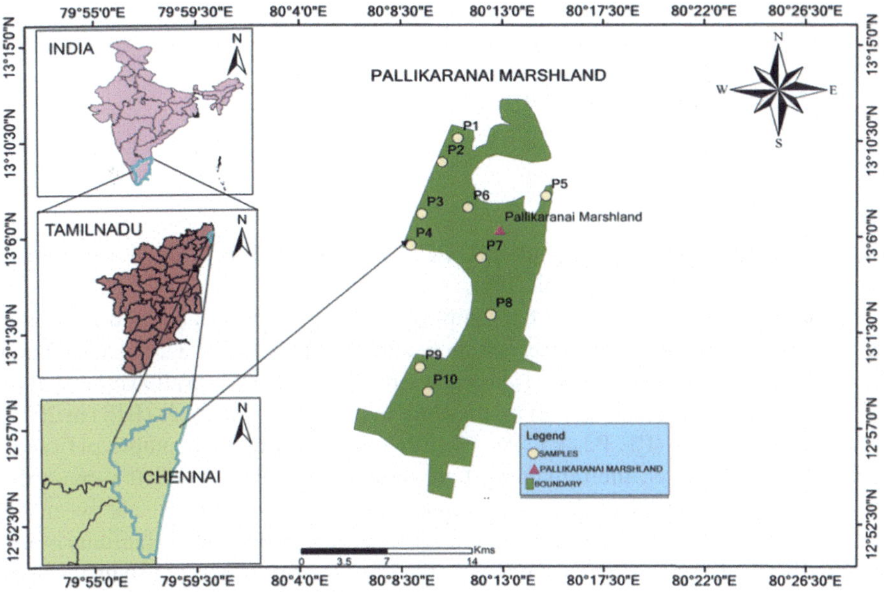

Fig. 15.1 Study area with sampling locations

15.2.2 Collection of Water Samples and Methodology

Surface water samples were methodically collected from the Pallikaranai Marshland at several sites, utilizing bottles for the collection process. These sampling sites were strategically spread throughout the study area and accurately located with the aid of GPS. In total, ten samples were acquired, with each sample comprising 500 ml of water, sourced from various accessible points within the marshland. It is important to note that this area includes key waste management infrastructures such as major garbage dumps and waste treatment plants in Chennai.

The procedure for collecting water samples involved gathering them from the Pallikaranai Marshland. To ensure purity, polyethylene bottles used for the collection were first rinsed with distilled water and then acidified using nitric acid (HNO3). Each of these bottles, intended for collecting 500 ml of water, was obtained from various points across the marshland. A total of ten surface water samples were meticulously collected, aimed at analyzing a range of physical and chemical properties. To minimize contamination, each bottle was pre-rinsed with the water sample itself before being filled. After collection, the samples were securely packaged to prevent exposure to sunlight and were then transported to the laboratory for analysis. Once in the lab, the water samples were stored in a cool environment, and for those designated for trace element analysis, 2 ml of HNO_3 was added to each.

15.3 Results and Discussion

15.3.1 In Situ Patrameters

The in–situ parameters here include pH, EC, TDS, DO, Salinity, and Temperature. Their values are given in the Table 15.1.

In the research zone, a total of ten samples were meticulously gathered for evaluative purposes. The pH levels observed show a minimum value of 8.2 at sites P2 and P9, while the highest value of 8.8 is noted at site P10, with an overall average pH of 8.5. According to the standards set by the Bureau of Indian Standards (BIS) in 2012, six of these samples (P1, P2, P5, P6, P7, P9) fall within the acceptable pH range of 6.5 to 8.5. However, the remaining four samples (P3, P4, P8, P10) surpass the permissible threshold. The pH values for most samples in the area range between 8.5 and 8.8, averaging at 8.5, suggesting the water's alkaline nature. This alkalinity could potentially be attributed to anthropogenic influences and the effects of urban development in the area. The current study reveals that the minimum electrical conductivity (EC) observed among the samples is 2150 µs/cm at site P4, while the highest EC is recorded at 5860 µs/cm in location P9. On average, the electrical conductivity across the ten samples stands at 3849.7 µs/cm (Table 15.2). As per the standards of the Bureau of Indian Standards (BIS) set in 2012, all sampled values exceed the allowable limit of 1500 µs/cm. This elevated conductivity indicates a concerning scenario,

Table 15.1 Measured physio chemical parameters in Pallikaranai wetland

Sample No	pH	EC μs/cm	TDS mg/l	DO mg/l	Salinity ‰	ORP Mv	Temperature °C
P1	8.4	5593.0	2799.0	3.5	30.1	257.0	31.1
P2	8.2	5720.0	3210.0	3.11	33.1	275.8	31.8
P3	8.6	2186.0	1096.0	4.04	11.1	252.4	31.4
P4	8.7	2150.0	1076.0	3.39	10.9	252.2	31.2
P5	8.3	4061.0	2033.0	4.08	21.4	264.4	32.1
P6	8.4	4116.0	2060.0	4.21	21.7	257.1	31.5
P7	8.3	2927.0	1465.0	3.62	15.1	266.2	32.1
P8	8.7	3258.0	1629.0	4.58	16.9	254.0	31.7
P9	8.2	5860.0	3479.0	4.45	34.4	272.6	32.5
P10	8.8	2626.0	1312.0	4.79	13.5	244.2	32.1
Minimum	8.2	2150.0	1096.0	3.11	10.9	244.2	31.1
Maximum	8.8	5860.0	3479.0	4.79	34.4	275.8	32.5
Average	8.5	3849.7	2015.9	3.977	20.8	259.6	31.75

likely resulting from the high concentration of electrolytes originating from leachate formations, a consequence of improper solid waste disposal at dumping sites. Additionally, there appears to be a correlation between the river's flow patterns and the EC values, mirroring a similar trend and relationship with Total Dissolved Solids (TDS). The recent investigation indicates that the lowest Total Dissolved Solids (TDS) measurement is 1096 mg/l at site P3, with the highest being 3479 mg/l at site P9. The average TDS value across all samples is 2015 mg/l (Table 15.2). In accordance with the Bureau of Indian Standards (BIS) guidelines established in 2012, five of these samples (located at P3, P4, P7, P8, and P10) fall within the acceptable TDS limit. However, the remaining samples exceed this threshold. Factors contributing to high TDS levels in urban streams include wastewater seepage, runoff from lawn fertilizers, and accidental spills. Additionally, the influx of seawater from the Bay of Bengal through the Okkiyam channel, influenced by tidal variations, is a contributing factor.

Table 15.2 Comparison of physicochemicl properties values with permissible limit

S. No	Parameters	pH	EC (μs/cm)	TDS (mg/l)	DO (mg/l)	Salinity (‰)	ORP (mv)
1	Minimum	8.2	2150	1096.0	3.11	10.9	244.2
2	Maximum	8.8	5860.0	3479.0	4.79	34.4	275.8
3	Average	8.5	3849.7	2015.9	3.977	20.8	259.6
4	Indian standard (BIS 2012)	6.5–8.5	1500	500–2000	4–6	–	–

Table 15.3 Classification of water based on TDS

Fresh water	0–1000 ppm
Brackish water	1000–10000 ppm
Saline water	10,000–100,000 ppm
Braine water	Above–100,000 ppm

Based on the TDS classification outlined in Table 15.3 by Carroll (1962), all samples from the study area fall under the 'Brackish water' category.

The recent investigation indicates that the lowest Total Dissolved Solids (TDS) measurement is 1096 mg/l at site P3, with the highest being 3479 mg/l at site P9. The average TDS value across all samples is 2015 mg/l. In accordance with the Bureau of Indian Standards (BIS) guidelines established in 2012, five of these samples (located at P3, P4, P7, P8, and P10) fall within the acceptable TDS limit. However, the remaining samples exceed this threshold. Factors contributing to high TDS levels in urban streams include wastewater seepage, runoff from lawn fertilizers, and accidental spills, as noted by Mikalsen et al. (2005). Additionally, the influx of seawater from the Bay of Bengal through the Okkiyam channel, influenced by tidal variations, is a contributing factor.

The current research indicates that Dissolved Oxygen (DO) levels range from a minimum of 3.11 mg/l at location P2 to a maximum of 4.79 mg/l at P10, with an average of 3.977 mg/l. As per BIS 2012 standards, all samples fall within the permissible DO range of 4–6 mg/l. The decline in DO suggests a potential increase in waste and organic pollution in the Marshland, likely attributed to waste leaching from the nearby Perungudi dump yard.

In terms of Salinity, the study reveals a minimum value of 10.9‰ at P4 and a maximum value of 34.4‰ at P9, with an average of 20.8‰. This variation may be attributed to the high evaporation of water within the marshland.

Regarding Oxidation–Reduction Potential (ORP), the recorded values range from a minimum of 244.2 mV at P10 to a maximum of 275.8 mV at P2, with an average of 259.6 mV. This signifies the presence of a high concentration of redox-active species and potent oxidizing (chlorine) and reducing agents (sulphite ion).

15.3.2 *Heavy Metals*

This research mainly focuses on four trace elements, namely Iron, Manganese, Chromium, and Copper. The concentrations of these trace elements in the surface water samples of the study area are provided in Table 15.4.

The study examines four trace elements: Iron (Fe), Manganese (Mn), Chromium (Cr), and Copper (Cu) in surface water samples from various locations (P1 to P10). The concentrations are outlined in Table 15.4.

For Iron (Fe), the minimum recorded value is 0.014 ppm at location P8, while the maximum is 3.92 ppm at location P6. The average Fe concentration is 1.1809 ppm.

Table 15.4 Concentration of heavy metals in water samples

Sample No	Fe (ppm)	Mn (ppm)	Cu (ppm)	Cr (ppm)
P1	0.117	0.092	0.093	0.096
P2	0.129	0.024	0.042	0.172
P3	5.287	0.697	0.025	6.703
P4	0.052	0.007	0.018	2.455
P5	0.039	0.433	0.017	0.088
P6	3.922	0.01	0.023	0.26
P7	1.547	0.011	0.016	0.023
P8	0.014	0.569	0.012	0.046
P9	0.658	0.367	0.027	0.018
P10	0.044	0.054	0.063	0.091

According to BIS 2012 standards, the permissible limit is 0.3 ppm. Among the 10 samples, nine (P1, P2, P3, P4, P5, P7, P8, P9, P10) fall within the limit, while one sample (P6) exceeds the permissible limit of 0.3 ppm.Moving on to Manganese (Mn), the minimum value is 0.007 ppm at P4, and the maximum is 0.697 ppm at P3, with an average of 0.2264 ppm. The permissible limit set by BIS 2012 is 0.05 ppm. Among the 10 samples, four (P2, P4, P6, P7) comply with the limit, while six (P1, P3, P5, P8, P9, P10) surpass the permissible limit.

Regarding Copper (Cu), the minimum recorded value is 0.012 ppm at P8, and the maximum is 0.093 ppm at P1, with an average of 0.0336 ppm. The permissible limit, as per BIS 2012, is 0.07 ppm. Eight samples (P2, P3, P4, P5, P7, P8, P9, P10) conform to the limit, while one sample (P1) exceeds it due to the influence of industrial effluents, air and sewage, and medical waste.

For Chromium (Cr), the minimum value is 0.018 ppm at P9, and the maximum is 6.703 ppm at P3, with an average of 0.9952 ppm. The permissible limit, according to BIS 2012, is 0.05 ppm. Four samples (P2, P7, P8, and P9) adhere to the limit, while six samples (P1, P3, P4, P5, P6, P10) surpass the permissible limit. Elevated Cr values are often associated with basic rock terrains and are influenced by both natural and anthropogenic sources, with industrial waste being a significant contributor. The minimum, maxiumum, average and BIS limits were outlined in Table 15.5.

Table 15.5 Comparison of values with the permissible limit

S. No	Trace metals	Fe	Mn	Cu	Cr
1	Minimum	0.014	0.007	0.012	0.01
2	Maximum	3.922	0.697	0.093	6.703
3	Average	1.1809	0.2264	0.036	0.9952
4	BIS (2012)	0.3	0.05	0.07	0.05

15.3.3 Statistical Analysis

A correlation matrix serves as a tabular representation of correlation coefficients between various variables, offering insights into relationships within a dataset. It efficiently summarizes extensive data, facilitating the identification and visualization of patterns. Comprising rows and columns representing variables, each cell contains a correlation coefficient. This matrix is a valuable tool often used in tandem with other statistical analyses, such as multiple linear regression. In the study area, positive correlations (>0.5) were observed, specifically between EC–TDS, Salinity–ORP, and TDS–Salinity, where the correlation coefficients indicate strong positive associations. Additionally, positive correlations were identified between Fe–Cr and Mn–Cr. Notably, many elements, including TDS, EC, Zn, and Mn, exhibited positive correlations. Mn and Pb, Mn and Cr, as well as Cu and Mn, demonstrated particularly high positive correlations, suggesting shared origins, possibly from mining and industrial waste discharges (Table 15.6).While most heavy metals showed positive correlations with certain elements, the lack of significant correlations among most heavy metals suggests independent or diverse natural anthropogenic sources in the study area, as outlined by Dash et al. (2016).

Factor analysis is a fundamental statistical method employed to examine relationships among multiple variables. The primary objective is to condense extensive information obtained from original variables into smaller factors with minimal loss of information (Stanko Ilić Popov et al., 2014). The results of the sorted rotated factor analysis for hydrogeochemical data are presented in Table 15.7. Through the rotation of the axis defined by factor analysis, a new set of factors is generated, each predominantly encompassing a subset of the original variables with minimal overlap. This process organizes the original variables into distinct groups. Liu et al. (2005) categorized factor loadings as 'strong,' 'moderate,' and 'weak,' corresponding to absolute loading values of 0.75, 0.75–0.50, and 0.50–0.40, respectively.

Table 15.6 Correlation matrix for parameters and trace metals

Parameters	PH	EC	TDS	DO	SALINITY	ORP	Fe	Mn	Cu	**Cr**
PH	1									
EC	−0.70	1								
TDS	−0.75	0.72	1							
DO	0.31	−0.03	−0.08	1						
Salinity	−0.79	0.92	0.93	−0.10	1					
ORP	−0.94	0.56	0.81	−0.35	0.74	1				
Fe	0.12	−0.28	−0.27	0.06	−0.30	−0.17	1			
Mn	0.13	−0.14	−0.10	0.40	−0.15	−0.09	0.30	1		
Cu	0.05	0.47	0.18	−0.06	0.36	−0.19	−0.25	−0.34	1	
Cr	0.41	−0.57	−0.38	−0.16	−0.51	−0.34	0.68	0.50	−0.18	1

Table 15.7 Rotated Component matrix of parameters and trace metals

Rotated component matrix				
Component	1	2	3	4
PH	−0.92	0.10	0.16	0.21
Salinity	0.92	−0.20	0.31	0.02
TDS	0.91	−0.14	0.12	0.02
ORP	0.91	−0.09	−0.31	−0.24
EC	0.80	−0.24	0.45	0.09
Cr	−0.34	0.90	−0.07	−0.07
Fe	−0.11	0.85	−0.08	0.03
Cu	0.06	−0.13	0.97	−0.09
DO	−0.16	−0.14	0.03	0.93
Mn	0.04	0.54	−0.26	0.67
Total	4.14	1.98	1.44	1.43
% of Variance	41.42	19.80	14.42	14.26
Cumulative %	41.42	61.22	75.65	89.90

In total, three factors account for 89.90% of the cumulative variance in the original dataset, offering a comprehensive understanding of the data structure. Factor 1 accounted for 41.42% of the total variance with positive loadings on Salinity, TDS, ORP, EC, and low factor loadings for the remaining parameters. Factor 1 indicates the geogenic activity and it represents the dissolution from the rock-water interaction. Factor 2 records 19.80% of the total variance, with positive loadings of Chromium, Iron simply implies that. the presence of basic rocks and are mainly due to anthropogenic activities. Chromium is known to enter water supply systems through industrial waste from where chromite is used to inhibit metal corrosion. Factor 3 records 75.65% with positive loading of copper. This factor may be attributed to the mixing of sterile copper industrial effluents through the air and sewage and the surface runoff of the fertilizer in the southern part and also the leaching of industrial wastes. Factor 4 records 14.26% of the total variance, with positive loadings of Dissolved oxygen and Manganese simply implying that the origin of these metals may be anthropogenic and industrial runoff that reaches the groundwater system. From the above discussion it could be concluded that factor 1 is attributed to processes like dissolution, weathering, and evaporation and factor 2 are to anthropogenic sources like industrial and agricultural activities and Factor 3 is due to the copper industrial effluents through sewage. Factor 4 is attributed to anthropogenic activities.

Cluster analysis was applied to interpret the data collected in the study area, offering a convenient method for identifying homogeneous groups of objects. This analysis enables the grouping of objects, such as cases or observations, with shared characteristics into specific clusters. Utilizing linkage between groups based on correlation, relationships among the physical–chemical parameters and total trace elements in the water samples were derived through cluster analyses.In the study area, the cluster analysis identified two distinct clusters, namely Cluster 1 and Cluster 2. Cluster 1 encompassed stations P2, P3, P4, P5, P6, P7, P9, and P10, while Cluster 2 was associated with P1 and P8 (Fig. 15.2).

15.3.4 Human Health Risk Assessment

15.3.4.1 Chronic Risk

It is imperative to assess the potential health risks associated with prolonged arsenic ingestion, given its prevalence as the most abundant element in the drinking water of the study area. This study exclusively focused on the ingestion pathway to evaluate both non-carcinogenic (CDI and HQ) and carcinogenic health risks, as detailed in Table 15.8. The outcomes revealed a consistent rise in HQ values across all groups, with arsenic HQ values surpassing 1 for both children and adults at all sampling sites. This indicates a substantial and enduring health risk associated with the consumption of drinking water in these locations. To evaluate the non-carcinogenic effects of arsenic, the Hazard Index (HI) was employed, revealing that arsenic exceeded the USEPA standard, primarily driven by its elevated HQ value. The highest HQ was observed in the drinking water of P1, reaching 16 for adults and 15 for children. The findings from the HI underscore a significant non-carcinogenic health risk for residents in these areas, affecting both children and adults.

For adults, approximately 55% of the samples indicated a moderate level of chronic risk, while the remaining 45% of samples were associated with a low chronic risk level. Conversely, nearly 80% of the samples indicated a medium chronic risk, with the remaining 20% suggesting a lower level of chronic risk. Across all scenarios, for both adults and children, the sampling point with the highest potential chronic risk was consistently identified as P5, whereas the sampling point with the least potential chronic risk was consistently determined to be P8.

15.4 Conclusion

The current investigation furnishes comprehensive scientific insights into Pallikaranai Lake, situated in one of the rapidly urbanizing regions of the Indian subcontinent. The degree of urbanization along East Coast Road and Old Mahabalipuram has escalated from 1998 to the present. Trace metals in natural waters, particularly in rivers and lacustrine areas, originate from diverse sources, encompassing both natural and external inputs. Once introduced into the aquatic medium, these metals undergo transformations, transportation, and often precipitation due to dynamic environmental conditions arising from various geochemical processes. Numerous studies across aquatic regions have been conducted to comprehend the partitioning, transformation, and absorption of elements into various organisms.

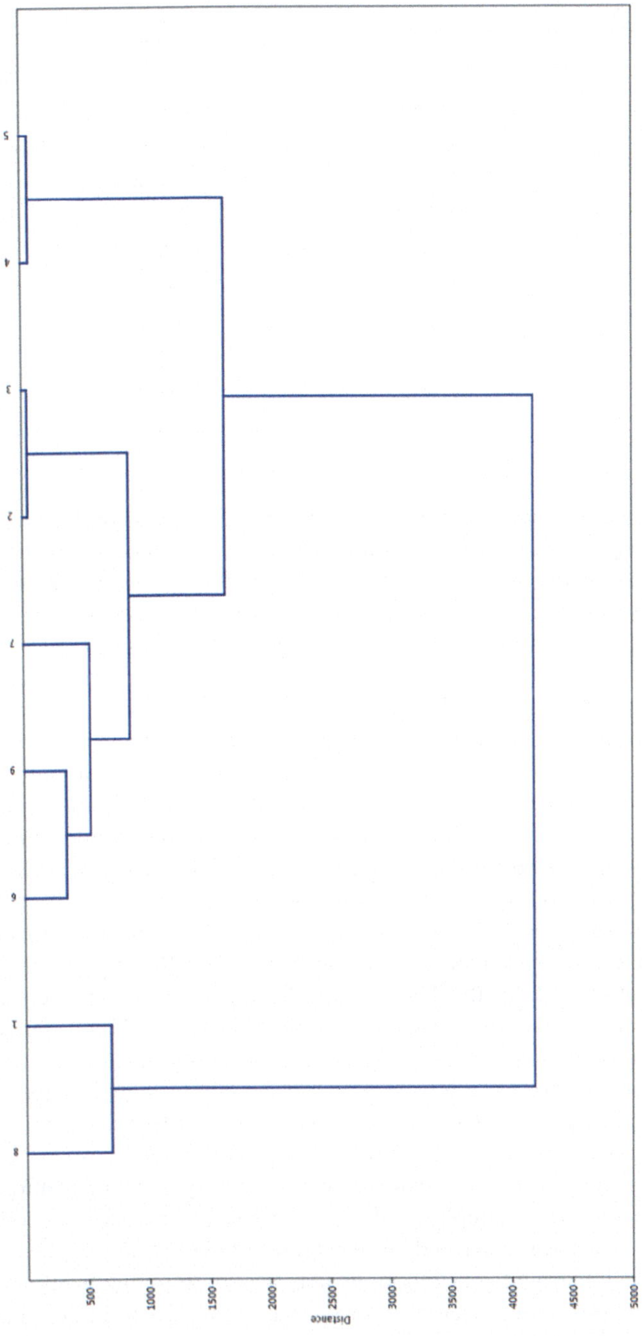

Fig. 15.2 Cluster analysis of parameters and trace elements

Table 15.8 Non-carcinogenic risk according to chronic daily intake (*CDI*) and hazard quotient (HQ) in adults and children

Site	As (Mean Conc.)	Adults		Children	
		CDI	HQ	CDI	HQ
P1	0.068	0.00480	16	0.004533	15
P2	0.054	0.00381	13	0.0036	12
P3	0.058	0.00409	14	0.003867	13
P4	0.057	0.00402	13	0.0038	13
P5	0.03	0.00212	7	0.002	7
P6	0.057	0.00402	13	0.0038	13
P7	0.019	0.00134	4	0.001267	4
P8	0.005	0.00035	1	0.000333	1
P9	0.033	0.00233	8	0.0022	7
P10	0.056	0.00395	13	0.003733	12

The physicochemical analysis unveiled that pH, TDS, EC, ORP, DO, Salinity, and Temperature surpassed permissible limits. The water exhibited an alkaline nature owing to an increase in pH, with EC concentrations escalating towards the southwest regions. Elevated salinity and TDS concentrations were recorded in the P2 and P9 regions, attributed to the inflow of seawater through the Okkiyam channel.

Studies were conducted to decipher geochemical processes, trace metal partitioning, and their connection with dissolved/solid phases. The composition and concentration of trace elements in groundwaters exhibited significant variation due to the diverse geochemical properties of earth materials, coupled with numerous anthropogenic sources and processes. Exceeding permissible limits, the water contained elevated levels of trace metals, including Fe, Mn, Cr, and Cu, stemming from anthropogenic activities, industrial settings, landfills, and rock formations, with a major contribution from waste dumpings in the Perungudi dump yard.

Correlation studies indicated that positively correlated elements originated from similar inputs, likely a consequence of waste disposal in the dump yard. Factor analysis interpretation revealed that Factor 1 signified geogenic activity, representing rock-water interaction dissolution. Factor 2 implicated anthropogenic sources such as industrial and agricultural activities, while Factor 3 was linked to copper industrial effluents via sewage. Factor 4 was attributed to broader anthropogenic activities. The observed water quality posed threats to aquatic and domestic life, necessitating urgent management interventions to mitigate surface water quality concerns.

In essence, this study strives to monitor changes in heavy metal concentrations in Pallikaranai wetlands' water samples. If this escalating trend persists, the local food web ecosystems in these wetlands could face heightened risks of induced heavy metal contamination. Such elevated concentrations may also pose public health concerns for the surrounding population. Hence, rigorous management actions are imperative to safeguard the ecological stability of the wetland.

Acknowledgements The authors would like to express their sincere gratitude to Deparment of Applied Geology, University of Madras for providing the facilities and resources essential for conducting this research. We extend our heartfelt thanks to Chief Conservation of Forest, whose guidance, support, and valuable insights greatly contributed to the successful completion of this work. Finally, we acknowledge the support of our colleagues, friends, and family, who provided encouragement and assistance throughout the preparation of this book chapter. Their contributions, both direct and indirect, are deeply appreciated.

References

Agarwal, S. K., & Gupta, A. (2008). Aquaporins: The renal water channels. *Indian Journal of Nephrology, 18*(3), 95–100.

Ali, H., Khan, E., & Ilahi, I. (2019). Environmental chemistry and ecotoxicology of hazardous heavy metals: environmental persistence, toxicity, and bioaccumulation. *Journal of chemistry* (2019)

Aravind Kumar, J., Saravanakumar, K., Gokulakrishnan, M., & Indira, B. (2014). Assessment of physio-chemical parameters of water at environmentally degraded pallikaranai Marsh Area, Chennai, India. *International Journal of Scientific and Engineering Research, 5*(7), 5–8.

Azeez, P. A., Bhupathy, S., Ranjini, J., Dhanya, R., Nikhil Raj P. P., Azeez, P. A., & Nikhil Raj, P. P. (2007). Management plan for the eco-restoration of pallikaranai reserve forest study team. *Natural History, 62.*

Balakumar, S. & Das, S. (2015). Investigating household preferences for restoring pallikaranai marsh, Working paper 126/2015. Banuraman, S. and Madavan, V. 2011. Study of Ground Water in Perungudi Area of Chennai. Correlation with PhysicoChemical Parameters. *Civil and Environmental Research, 1*(1), 23–32.

Bassi, N., Kumar, M. D., Sharma, A., & Pardha-Saradhi, P. (2014). Status of wetlands in India: A review of extent, ecosystem benefits, threats and management strategies. *Journal of Hydrology: Regional Studies, 2*, 1–19.

Carroll, D. (1962). Rainwater as a chemical agent of geologic processes: A review.

Dash, C. J. et al. (2016). Simulation of nitrate leaching under maize–wheat cropping system in a semiarid irrigated area of the Indo-Gangetic Plain, India. *Journal of Irrigation and Drainage Engineering, 142*(2), 04015053.

Nitin Bassi (2014). Status of wetlands in India: A review of extent, ecosystem benefits, threats and management strategies. 2, 1–19.

Central Ground Water Board. (2008). Technical Report Series of District ground water Brochure Chennai District, Tamil Nadu, 1–6.

Dr. Chandramohan. (2007). Role of public governance in the conservation of urban wetlands, pp. 14–22.

Dash, S., & Kalamdhad, A. S. (2021). Hydrochemical dynamics of water quality for irrigation use and introducing a new water quality index incorporating multivariate statistics. *Environmental Earth Sciences, 80*, 1–21.

Johnpaul, A., Ragunathan, M. B., & Selvanayagam, M. (2010). Population dynamics of freshwater molluscs in the lentic ecosystems in and around Chennai. *Recent Research in Science and Technology, 2*(4), 80–86.

Juniet, J. M. (2016). Current status of pallikaranai wetland: A review. *International Journal of Development Research, 06*(08), 9002–9007.

Karpagavalli, M. S., Malini, P., & Ramachandran, A. (2012). Analysis of heavy metals in dying wetland Pallikaranai, Tamil Nadu India. *Journal of Environmental Biology, 33*(4), 757–761.

Klee, G. A. (1994).Wetlands, by William J. Mitsch and James G. Gosselink. *Journal of Coastal Research, 10*(3)

Lavanya, V., & Ravichandran, S. (2013). Microbial contamination of drinking water at the source and household storage level in the peri-urban area of southern Chennai and its implication on health India. *Journal of Public Health (Germany), 21*(5), 481–488.

Liu, W. H. et al. (2005). Impacts of sewage irrigation on heavy metal distribution and contamination in Beijing, China. *Environment International, 31*(6), 805–812.

Malik, D. S., & Joshi, N. (2013). Habitat selection pattern of Migratory Avifauna in relation to nutrients in Asan wetland at Doon Valley (Garhwal Himalaya), India. *International Journal of Recent Scientific Research, 4*(10), 1470–1475.

Malleshappa, H., & Jayanthi, M. (2013). *State of environment report of chennai metropolitan area.* ENVIS Centre Published by Department of Environment.

Mathiazhaghan. (2013). Geophysical technique for sensing of solid waste dump site-induced groundwater contamination leachate dump site-induced groundwater contamination. *Pollution Research, 32*(3), 509–514.

Millennium Ecosystem Assessment. (2005). *Ecosystems and Human well-being: Wetlands and water synthesis.* World Resources Institute

Mitsch, W. J., Gosselink, J. G., Zhang, L., & Anderson, C. J. (2009). *Wetland ecosystems.* John Wiley & Sons.

Mitsch, W. J., & Gosselink, J. G. (1986). *Wetlands.* Van Nostrand Reinhold.

Mundanthra, B. (1993). Wetlands of India: Their unique potential for Quality life improvement Environmental Problems and Prospects in India, Oxford and IBH Publishing Co. Pvt. Ltd. New Delhi, pp. 21–22

Murugesan. (2015). Drinking water supply and demand management in Chennai City—A literature survey. *International Journal of Innovative Science, Engineering & Technology, 2*(3), 717–719.

Nagajyoti, P. C., Lee, K. D., & Sreekanth, T. V. M. (2010). Heavy metals, occurrence and toxicity for plants: A review. *Environmental Chemistry Letters, 8*(3), 199–216.

Nies, D. H. (1999). Microbial heavy metal resistance. *Applied Microbiology and Biotechnology, 51*, 730–750.

Nikhil Raj, P. P., Ranjini, J., Dhanya, R., Subramanian, J., Azeez, P. A., & Bhupathy, S. (2010). Consolidated checklist 9006 International Journal of Development Research, 06, Issue, 08, 9002-9007, August, 2016 of birds in the Pallikaranai wetlands, Chennai, India. *Journal of Threatened Taxa, 2*(July), 1114–1118.

Popov, S. I., et al. (2014). Applying of factor analyses for determination of trace elements distribution in water from river Vardar and its tributaries, Macedonia/Greece. *The Scientific World Journal.*

Sethuram. (2014). Case study: Water management issues in Chennai, India. *Universal Journal of Environmental Research and Technology, 4*(6), 300–302.

Vencatesan, J., & Daniel, R. (2014). Comprehensive management plan for pallikaranai marsh. Conservation Authority for Pallikaranai Marshland - TNFD and Care Earth Trust, Govt. of Tamil Nadu, pp. 70–82.

Yadav, M., Gupta, R., Arora, G., Yadav, P., Srivastava, A., Sharma, R.K.: Current status of heavy metal contaminants and their removal/recovery techniques. InContaminants in our water: identification and remediation methods 2020 (pp. 41–64). American Chemical Society.

Yadav, M., Gupta, R., & Sharma, R. K. (2019). Green and sustainable pathways for wastewater purification. In *Advances in water purification techniques* (pp. 355–383). Elsevier.

Chapter 16
Assessing Heavy Metal Contents in Groundwater of South Kamrup Region (Assam), India and Its Impact on Human Health and Environment Using GIS

Jayakrishna Bhuyan, Snehasish Choudhury, and Dharitri Choudhury

Abstract Groundwater is indeed vital for ecosystems and socio-economic stability. It supports the growth of plants and provides habitats for various organisms, contributing to ecological balance. Additionally, it serves as a reliable source of drinking water for communities and fuels agricultural activities, thereby enhancing socio-economic stability within nations. Presently, the rapid extension of anthropogenic activities including urbanization, industrialization and agricultural modernization has been continuously deteriorating the physicochemical qualities of underground water resulting adverse effect on human health and environment respectively. The industrial wastes are the principal factors for the degradation of groundwater quality as this act like a source of heavy metal pollutant. This research focused on the assessment of Heavy metal contents in groundwater and its impact on human health and surrounding environment of industrial areas in South Kamrup Region of Assam, India. To achieve this goal, samples of groundwater are collected from various industrial sites of Palasbari,Bijoynagar, Mirza, Loharghat, Rani and Dharapur area of south Kamrup in pre monsoon season, January, 2024 and are analyzed for physiochemical properties such as pH, TDS (Total dissolve Solids) and major Heavy metal contents (As, Cd, Pb, Mn,)in laboratory by applying FL/SOP/ICPMS-01 method based on APHA 23rd Ed.2017 at FARE LABS private Limited, M.G road, Gurugram. Result shows that heavy metals concentration in all groundwater samples is fall under the current drinking water standard given by WHO, 2022. Findings show that the maximum HPI (Heavy Metal Pollution Index) value is estimated from the ground water sample of Mirza is greater than 90 (>90). Thus, it is perceived that the ground

J. Bhuyan
Department of Geography, B. Borooah College, Guwahati, Kamrup(M), Assam, India

S. Choudhury (✉)
Department of Geography, Dakshin Kamrup College, Mirza, Kamrup (R), Assam, India
e-mail: snehasishchoudhury74@gmail.com

D. Choudhury
Department of Eco-Restoration, Dimoria College, Khetri, Kamrup (M), Assam, India

V. Sunitha et al. (eds.), *Groundwater Resource Management Planning Strategies*, Springer Water, https://doi.org/10.1007/978-3-031-88870-0_16

water of Mirza is moderately contaminated while remaining sampling sites fall under low heavy metal contamination category (HPI, < 90). In addition to this, maximum TDS value and pH value were detected from Palasbari is 230 ppm and Mirza that is 8.1 respectively. Consequently, in order to show the heavy metal contents, TDS and pH values, groundwater maps are prepared using ArcGIS software.

Keywords Groundwater quality · Heavy metal · Industrial waste · India · South Kamrup · Human health · GIS (Geographical information System)

16.1 Introduction

Heavy metals are defined as a naturally occurred elements having high relative density and specific gravity with high atomic mass more than 5 g/cm^3 and high atomic number more than 20 (Arora et al., 2021). In environment, heavy metals such as arsenic (As), chromium (Cr), cadmium (Cd), lead (Pb), mercury (Hg), manganese (Mn), nickel (Ni), copper (Cu) are some common but hazardous heavy metals present in contaminated ground water and waste water as environmental pollutants which can cause severe adverse effects on human health and natural ecosystem (Sonone et al., 2021). Most common sources of heavy metals are products of volcanic eruption (natural source), mining industry, petrochemicals, refineries, batteries, application of chemical fertilizers and pesticides in croplands, untreated sewage sludge due to lack of waste water management (UNEP, 2020). Mining industry including urban and inactive mines are consider as a primary source of heavy metal pollution in groundwater (Reddy et al., 2023). Some heavy metals are considered as essential nutrients for human being and other organisms e.g. iron (Fe), cobalt (Co), zinc (Zn) or harmless e.g. silver (Ag), ruthenium (Ru) and indium (In) but can be very toxic if taken in excess amount or in certain forms. Other heavy metals such as mercury, arsenic, cadmium and lead are categorized as highly poisonous.

Heavy metals are also categories as trace elements as these are present in environment in trace form (Arora et al., 2021). Industrial and human activities are primarily responsible for heavy metals contamination in groundwater (Lorestani, et al., 2020). Rapid anthropogenic interferences including urbanization and industrialization accelerated the amount of heavy metal concentration in underground water results in the deterioration of groundwater water quality. As a consequence, several impacts on environment and natural terrestrial and aquatic ecosystem occurs (Ehya, et al., 2016) In addition to this, heavy metal contaminated ground water causes several physiological imbalances in human being including neurological, cardiovascular, immunological, cognitive, reproductive, dermatological and other health problems (WHO, 2022). According to UNEP, exposure to mercury (Hg) contaminated water cause harmful effects on brain cells and cardiovascular systems followed by methyl mercury cause cancer in humans (WHO, 2022), lead (Pb) can cause adverse impact on gastrointestinal, neurological, renal, hematological and cardiovascular system in human being (WHO, 2022). Cadmium contaminated water causes bone marrow weakness and also impact on respiratory, prostate and kidney function of humans (UNEP, 2020). High concentration of lead (Pb) is linked with zinc coated pipes or corrosion of water distribution pipes and geological contaminants (Dilebo et al., 2023). A recent study on sosian river revealed that excessive amount of lead concentration in surface water is connected with various industrial outputs, urban activities including car washes, garages, electronic and battery products etc. (Munene et al., 2023). Similarly, presence of Zinc (Zn) in drinking water is correlated with use of zinc related fertilizers in agricultural farms including wheat and flower farms (Munene et al., 2023). A recent analysis shows that excessive amount of heavy metals in groundwater can stem from a variety of human induced sources such as poultry farming and stone quarrying etc. (Kana et al., 2022). Intensified heavy metal contamination level in groundwater also indicate the deterioration of groundwater quality caused by leachate seepage from landfill site (Boateng et al., 2019; Etikala et al., 2022, 2024). Besides these, new investigation shows that fluoride contamination in drinking water is indeed a widespread issue, especially in developing countries. It's a serious concern for public health, as excessive fluoride intake can lead to various health problems, including dental and skeletal fluorosis (Sunitha et al., 2008). Water with fluoride concentrations exceeding 1.5–2.0 mg/l can indeed lead to dental mottling, which is an early indicator of dental fluorosis. Das et al. (2020) conducted important research on fluoride levels in the water of the Bharalu River and the groundwater beneath Guwahati. Assessing the transfer of fluoride to groundwater aquifers near the river is crucial for understanding potential impacts on water quality and public health in the area (Das et al., 2020). The range of fluoride concentrations observed in both the river water (0.02 to 3.73 mg/l) and underground water (0.04–4.7 mg/l) indicates variability and potential sources of contamination (Das et al., 2020). A recent study found that in Agra city, the regions of Dayalbagh have excessive levels of fluoride contamination, surpassing the permissible limits (Bhaskar et al., 2016).

Many researchers (Borah et al., 2009; Das et al., 2009; Singh, 2004) have reported heavy metal contamination in drinking water of many parts of the world. The North eastern part of India also faces problems associated with ground water pollution. Presence of arsenic in groundwater of 21 out of 24 districts of Assam indicates ground water contamination in the state (Chetia et al., 2008). Borah et al. (2009) estimated high level of iron and cadmium concentration in the drinking water of tea belts of Darrang district, Assam. In Hajo area of Kamrup district, groundwater is contaminated with excessive arsenic concentration (Sarma et al., 2021). In Hajo area it is observed that blood pressure is slightly high among the patients of arsenic related diseases (Sarma et al., 2021).

The selected study sites South Kamrup is situated in kamrup district, Assam. Due to several anthropogenic activities includes industrialization and urbanization the ground water quality of kamrup district and its nearby areas rapidly deteriorated causing several health problems among its people (Chakrabarty et al., 2010). According to central groundwater board, NE region,2013, though groundwater of Kamrup district is fresh and potable for both domestic and agricultural purposes slightly excess amount of iron and fluoride is present in some parts of the district.

The present study focused on the GIS based analysis of spatial variation of heavy metal concentration in South Kamrup area of Assam, India.

16.2 Objectives

The primary goals of this study are to:

1. Assess the extent of heavy metal (Arsenic, Manganese, Cadmium, Lead) contamination in the groundwater of specified research locations in the South Kamrup region of Assam, India utilizing GIS technology.
2. To investigate the impact of heavy metal contamination on human health.

16.3 Materials and Methods

16.3.1 Study Area

The study area (Lat. 25.9790°N to 26.1417°N and Long. 91.2362° E to 91.6202°E) of present research is located within the South Kamrup area of Kamrup district of Assam, India (Fig. 16.1), situated on the southern bank of river Brahmaputra. The state Assam is situated on the north eastern region of India, it is covered by Arunachal Pradesh, Bhutan from north, Nagaland, Manipur from east, Mizoram, Tripura and Meghalaya from south. The kamrup metro and kamrup district is enclosed by morigaon from east, darrang, nalbari from north, barpeta, goalpara from west and the southern part is linked with the Meghalaya border. The area is surrounded by rani hills toward the southern part, where the area is mostly bounded by paddy fields. National Highway 31 passes through south kamrup providing good connectivity with other parts of the state. The climate of the area is classified as subtropical humid climate with heavy rainfall, hot summer and high humidity with average annual temperature ranging from 12 to 38°c during the year. The region receives an average annual rainfall of 1752 mm during the entire year. The area is fusion of both traditional and modern society though traditional way of livelihood is followed more. Most of the population of the area is still dependent on agriculture (Fig. 16.1). But over the years, the area has evolved with many new industries starting from pharmaceutical company, food packaging industries, plywood industries, fertilizer industries, chemical industries, plastic bottle industries and so on. With respect to these heavy industries there are multiple small industries in the areas contributing to such pollution and disturbances of the groundwater level. Industrial wastes such as solid wastes, waste products of mining, wooden wastes, metallic wastes, pharmaceutical products, plastic are some major sources of pollutants. This contaminated water will affect a larger area of the kamrup region, and may lead to severe ill diseases. The continuation of such activities will overall increase the risk of polluting the mass.

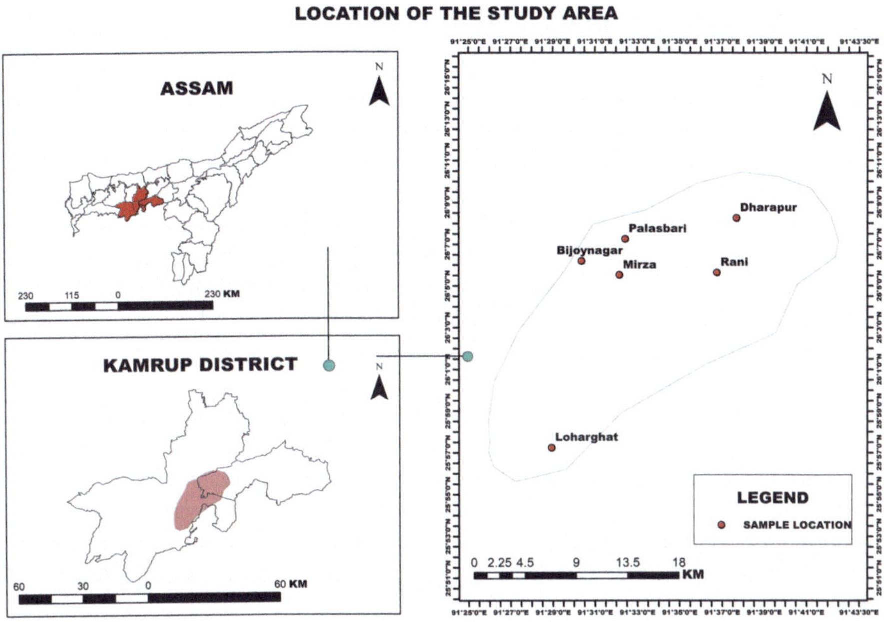

Fig. 16.1 Location map of selected study area (South Kamrup, Assam)

16.3.2 Geology and Geomorphology

The geological settings of south kamrup area are represented by shillong group of rocks of pre-Cambrian age including sedimentary formations such as sandstone, shale, conglomerate belong to Shillong plateau. Geomorphologically the area is diverse due to its location within the Brahmaputra River basin. Its geomorphology is greatly molded by various fluvio geomorphic processes including meandering channel, floodplain, alluvial plain, piedmont, denudational hills composed of granite and gneissic rocks and terraces formed by River Brahmaputra and its tributaries kulsi and kapili River.

16.3.3 Hydrogeology

The entire region is consisting of two broad hydrogeological unit.

Early geological formations such as Pre Cambrian consolidated rocks and quaternary deposits comprises of unconsolidated sediments. Pre Cambrian consolidated earth materials are restricted to hilly areas and inselbergs where groundwater occurs in shallow erosion areas. In the alluvial plain ground water occurs in regionally wide-ranging aquifers down to the depth of 305 m.

16.3.4 Methodology and Data Analysis

For this research, we have located 6 sample sites based on their population concentration, industries, daily transport influx. In the chosen study area, a total of six groundwater samples were gathered using 1-L water bottles from the water sources. each from water pump of selected study sites namely Palasbari, Mirza, Bijoynagar, rani, Loharghat and Dharapur area (Assam, India) in the month of January, 2024. The test was done on the basis of arsenic (As), cadmium (Cd), lead (Pb), manganese (Mn). During the water sample collection, water bottles were pre cleaned with distilled water and after that the six collected water samples for certain heavy metals were analyzed by applying FL/SOP/ICPMS-01 method based on APHA 23rd Ed. 2017 at FARE LABS private Limited, M.G road, Gurugram. Thereafter, we collected the GPS location of the sample area. The location of our study area has recently gained more human impact regarding industries, tourism, transport, settlement which has led to deteriorate the groundwater quality. Based on the water analysis, spatial concentration of heavy metal maps was prepared using ARC GIS software. The level of Heavy metal pollution in groundwater of selected study sites are estimated using Heavy Metal Pollution Index (HPI), statistical analysis are done by using MS Excel software.

16.3.5 Heavy Metal Pollution Index (HPI)

The Heavy Metal Pollution Index (HPI) utilizes measurable limits provided by the World Health Organization (WHO) and other organization such as Bureau of Indian Standards (BIS) for standard permitted values (Si) and ideal values (Ii) to assess water quality in terms of heavy metal concentrations. The HPI equation, typically set to a critical pollution index level of 100 for drinking water, is calculated using the formula:

HPI (Heavy Metal Pollution Index).

$$= \frac{\sum_{i=1}^{n} wiQi}{\sum_{i=1}^{n} wi}$$

where Wi denotes the Unit weight (ith parameter). Qi denotes the Sub index (ith parameter).

Unit weight (Wi) = k /Si, where k denotes the Proportionality Constant.

Si denotes the Standard permissible (ith parameter) in μg/L K = 1/Σ 1/Si.

16.4 Results and Discussion

Present study revealed the heavy metal concentration in groundwater of selected study sites of South Kamrup, Assam, India. The current research shows that almost all the results of groundwater samples are fall under the current drinking water standard given by WHO. Table. 16.1 shows the spatial variation of heavy metal concentration in the ground water of selected study sites represents that the concentration of arsenic (As) in groundwater of Palasbari industrial region is estimated 7.16 ppb that falls under the maximum prescribed limit of arsenic (0.01 mg/l or 10 ppb) in drinking water (WHO, 2022) followed by 6.9 ppb in Loharghat, 6.6 ppb in Dharapur, 6.5 ppb in Mirza and 5.6 ppb in both Rani and Bijoynagar industrial area (Fig. 16.2). Among all the sampling sites, high amount of arsenic in groundwater are detected from Palasbari and Loharghat area that eventually fall under the standard limit of drinking water mention by WHO (2022). According to WHO (2022), arsenic is predominantly found in earth crust in an oxidation state of −3, 0, + 3 and + 5 often as sulfides or metal arsenide or arsenate (+5). In natural water sources it is mostly present in a concentration range of 1–2 ppb (WHO, 2022). Table 16.2 shows the permissible limit of heavy metals according to BIS, 2012. Recent study shows that 94–220 million people on earth are facing health problems due to high level of arsenic concentration in groundwater (UNEP, 2020). Exposure to high quantity of arsenic in drinking water can cause several dermatological problems including skin irritation and skin Cancer (WHO, 2022). From present study it is found that most of the workers living in the industrial area of Palasbari and Loharghat are facing health problems including skin rashes, skin irritation and other dermatological problems due to the contaminated ground water. The occurrence of increasing amount of arsenic in groundwater water of south kamrup region is accelerated by several anthropogenic activities including establishment of factories, industries and application of chemical fertilizers in croplands in recent years. From Table. 16.1, it can be observed that the estimated results of Cadmium (Cd) concentration in groundwater of selected study sites are fall under the acceptable limit of cadmium in drinking water prescribed by WHO as 0.003 ppm except Mirza where concentration of cadmium is little higher in groundwater than the standard limit of WHO (Fig. 16.3). In addition to these parameters, spatial variation of concentration of Lead (Pb) and Manganese (Mn) in groundwater of selected study sites doesn't exceeds the maximum acceptable limit prescribed by WHO in 2022 (Fig. 16.4), (Fig. 16.5). Table 16.1 shows that maximum amount of Total Dissolved Solid (TDS) is present in groundwater sample of Palasbari and the estimated value is 230 ppm. According to WHO guideline for drinking water quality, 2022 the acceptable value of TDS in groundwater is 600 ppm (WHO, 2022).

The Heavy Metal Pollution Index (HPI) serves as a comprehensive tool to assess groundwater pollution by integrating multiple characteristics into a single value. Groundwater quality is categorized into three zones based on HPI values: low (<90), medium (90–180), and high (>180) levels of pollution (Prasad & Kumari, 2008; Reza & Singh, 2010). From Table 16.3 it is observed that all groundwater samples are fall within the acceptable limit of heavy metal pollution. For all selected sampling

Table 16.1 Concentration of heavy metals in groundwater of selected study sites of South Kamrup region, Assam (India) 2024

Sample location	Sources	PH	TDS (ppm)	Arsenic (As), ppb	Cadmium (Cd), ppm	Lead (Pb), ppm	Manganese (Mn), ppm
Palasbari	Tube well	7.9	230	7.16	0.002	0.002	0.005
Bijoynagar	Tube well	6.8	140	5.6	0.001	0.003	0.008
Mirza	Tube well	8.1	135	6.5	0.0037	0.001	0.006
Loharghat	Tube well	6.8	133	6.9	0.001	0.002	0.005
Rani	Tube well	7.0	210	5.6	0.002	0.004	0.003
Dharapur	Tube well	6.7	110	6.6	0.001	0.002	0.005

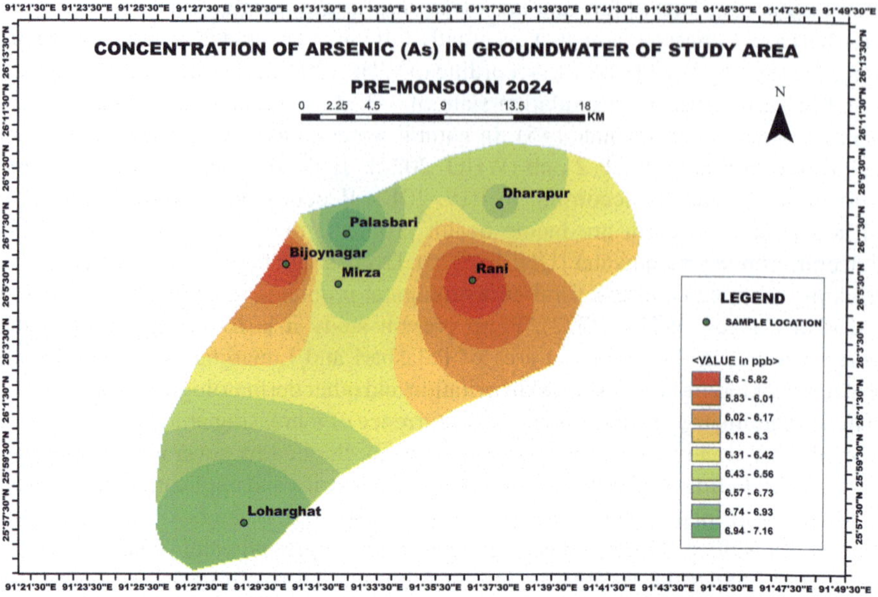

Fig. 16.2 Spatial concentration of Arsenic in groundwater of south Kamrup (Assam), India 2024

Table 16.2 Standard Permissible limit of selected heavy metals concentration in groundwater (Bureau of Indian Standards, 2012)

Heavy metals	Standard Permissible limit (Si) in ppb	Ideal Value (Ii) in ppb
Arsenic	10	0 (No relaxation)
Cadmium	3	0 (No relaxation)
Lead	10	0 (No relaxation)
Manganese	300	100

Source BIS (2012)

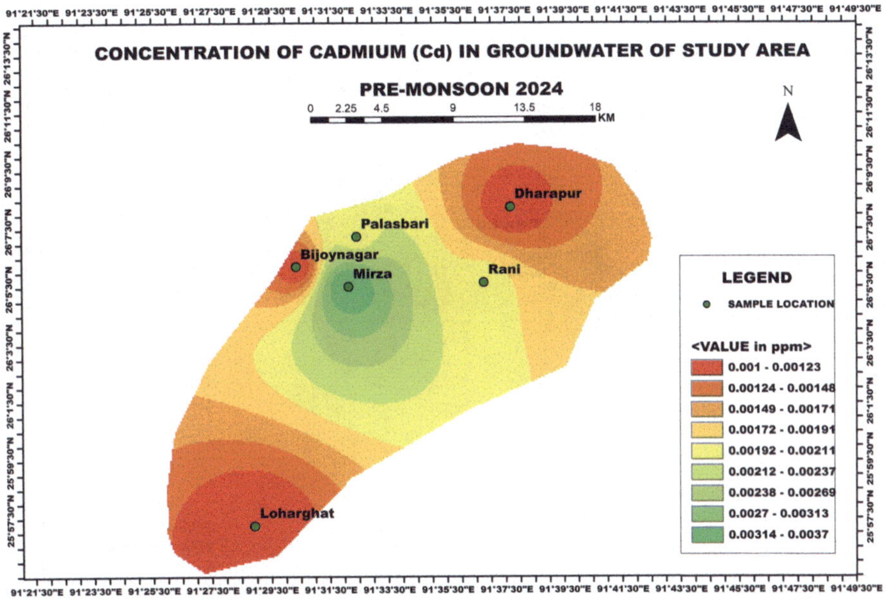

Fig. 16.3 Spatial concentration of Cadmium in groundwater of south Kamrup (Assam), India 2024

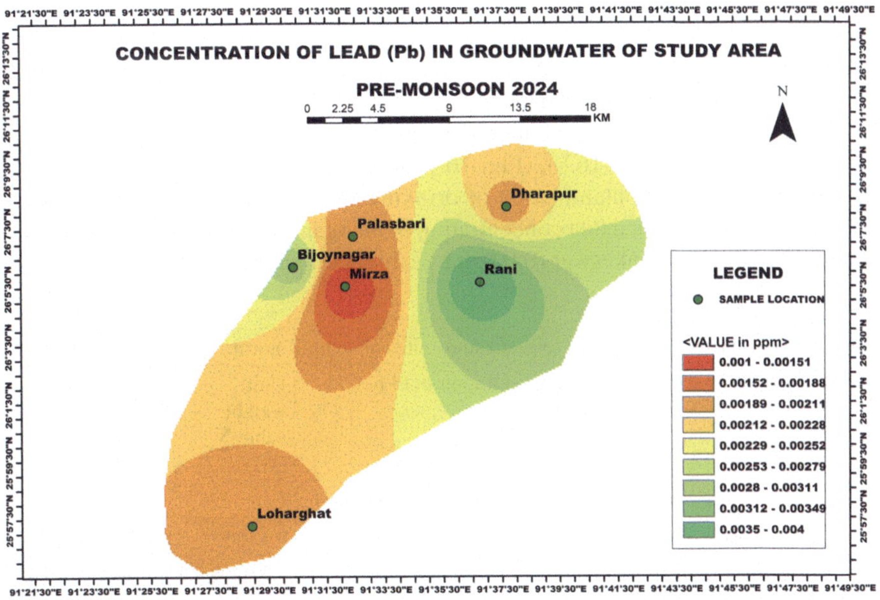

Fig. 16.4 Spatial concentration of Lead in groundwater of South Kamrup (Assam), India 2024

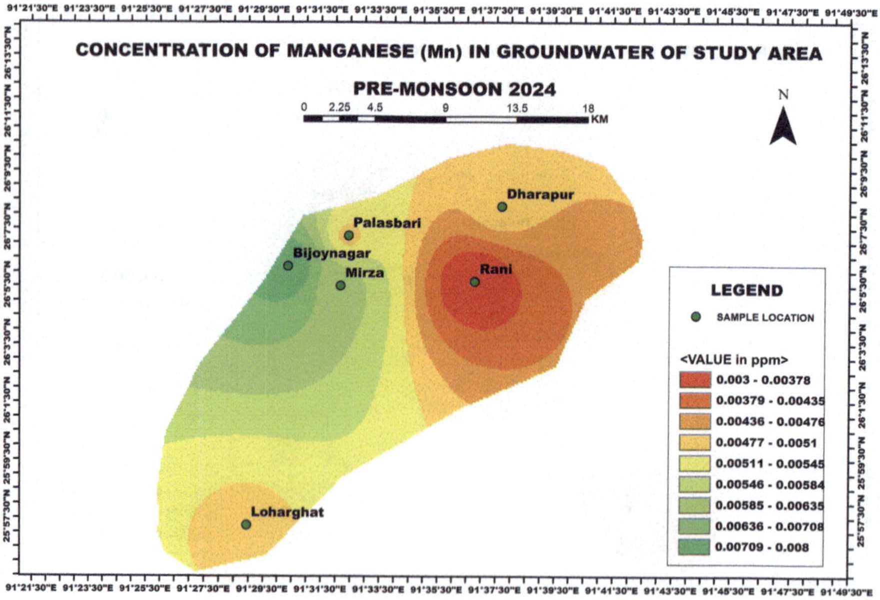

Fig. 16.5 Spatial concentration of Manganese (Mn) in groundwater of south Kamrup area (Assam) India, 2024

sites, the estimated HPI values are less than 90 except Mirza where the estimated HPI value is 90.87 (Table 16.3). Hence, it can be said that the underground water of Mirza is moderately polluted in compare to the other sampling sites.

Previous research (Fig. 16.8), (Fig. 16.9) was conducted by Chakrabarty et al., 2010 on Heavy metal contamination of drinking water in Kamrup district Assam, India (Table 16.4) revealed that concentration of manganese in groundwater of Chaygaon (0.14 ppm) and Lankeswar (0.04 ppm) is higher than the prescribed acceptable

Table 16.3 Estimated HPI value of selected study sites of south Kamrup, Assam, India (2024)

S. No	Sample Sites	Heavy metal concentration in PPb Monitor value (mi)				Unit weight value ∑ (Wi)	∑WiQi	HPI
		Arsenic	Cadmium	Lead	Manganese			
1	Palasbari	7.16	2	2	5	0.53666	31.5405	58.7712
2	Bijoynagar	5.6	1	3	8	0.53666	19.8644	37.0144
3	Mirza	6.5	3.7	1	6	0.53666	48.7677	90.871
4	Loharghat	6.9	1	2	5	0.53666	20.1694	37.5828
5	Rani	5.6	2	4	3	0.53666	31.9838	59.5973
6	Dharapur	6.6	1	2	5	0.53666	19.8694	37.0238

limit (0.02 ppm) of WHO. High concentration of cadmium in drinking water is found in both Chaygaon (0.012 ppm) and Lankeswar (0.021 ppm) region of South Kamrup region that exceeds the maximum prescribed limit by WHO (Chakrabarty et al.,2010) (Fig. 16.8). The same study shows (Fig. 16.8) that the lead concentration in groundwater water of most of the study site exceeds the WHO's prescribed acceptable limit. Figures 16.6, 16.7, 16.8 and 16.9 shows the spatial and temporal fluctuations in concentration of heavy metals in groundwater of south kamrup area in both year 2010 and 2024.

Table 16.4 Concentration of heavy metals in groundwater of selected study sites of South Kamrup region, Assam, India (2010)

Sample stations	Sources	Heavy metal concentration in drinking water			
		Arsenic (ppb)	Cadmium (ppm)	Lead (ppm)	Manganese (ppm)
Palasbari	Tube well	7.054	0.002	0.07	0.02
Boko	Tube well	BDL	0.002	0.07	0.02
Chaygaon	Tube well	BDL	0.012	BDL	0.14
Azara	Tube well	6.941	0.002	0.07	0.02
Lankeswar	Tube well	BDL	0.021	0.02	0.04

Source Chakrabarty et al. (2010)

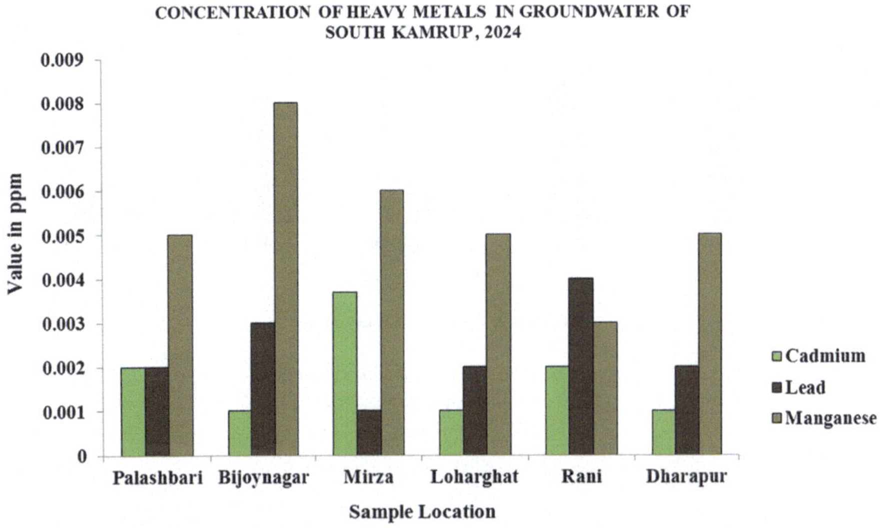

Figure. 16.6 Concentration of heavy metals (Cd, Pb and Mn) in groundwater of selected study sites of South Kamrup, Assam (India), 2024

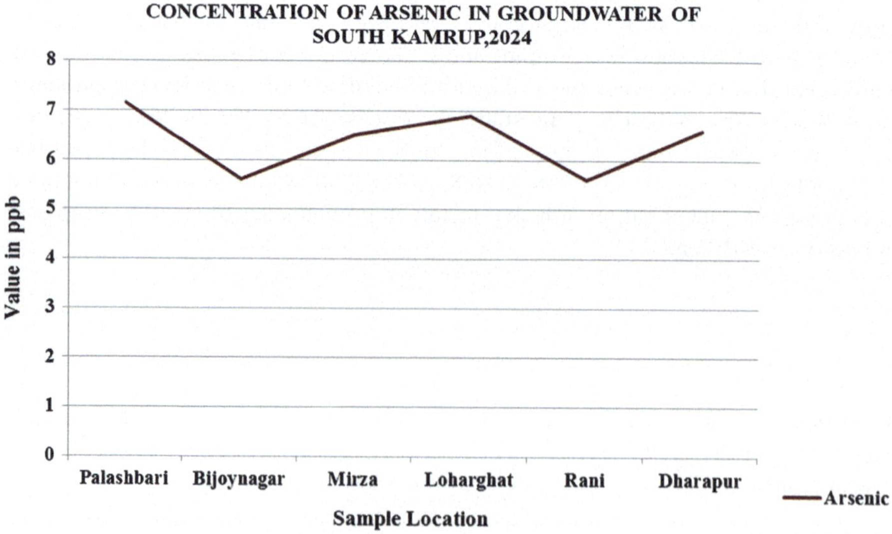

Fig. 16.7 Concentration of arsenic in groundwater of selected study sites of South Kamrup region (Assam), India, 2024

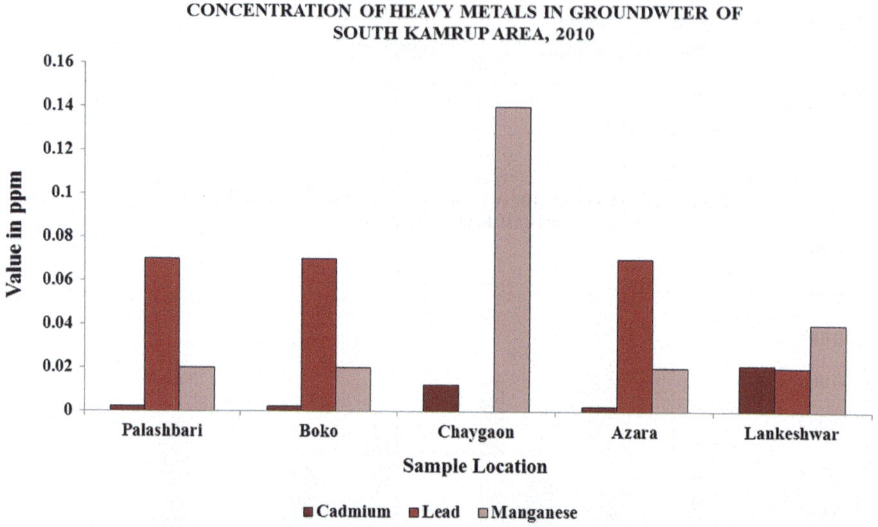

Fig. 16.8 Concentration of heavy metals (Cadmium, lead and manganese) in groundwater of south Kamrup (Assam), India, 2010, (*source* Chakrabarty et.al. 2010)

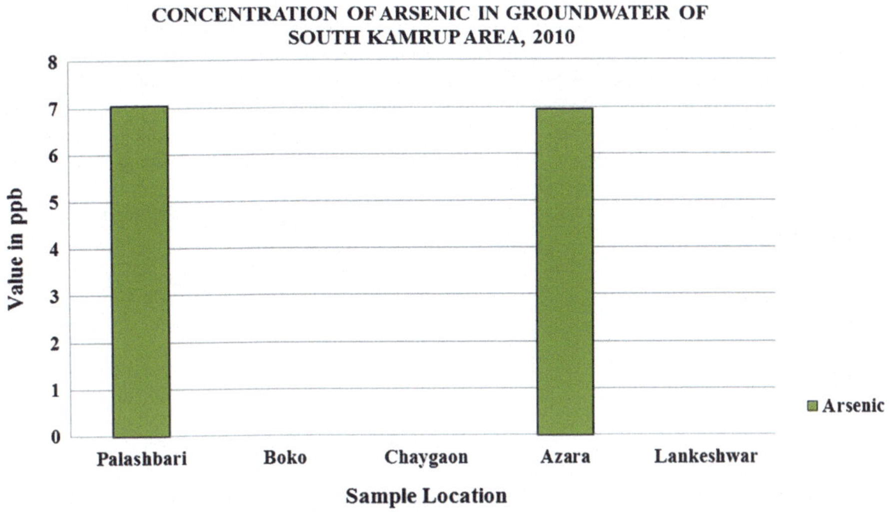

Fig. 16.9 Concentration of arsenic(As) in groundwater of South Kamrup area (Assam), India, 2010 (*Source* Chakrabarty et al. 2010)

16.5 Conclusion

Based upon this study, it is inferred that the ground water quality of South Kamrup region of Assam (India) is moderately contaminated. The heavy metal concentration of groundwater of selected study sites shows excess amount of cadmium and arsenic. Maximum HPI value is estimated in Mirza that is > 90 (90.87) which indicates the gradual pollution of groundwater in Mirza area (South Kamrup, Assam). Furthermore, this study also shows a GIS based analysis of spatial variation of heavy metal concentration in groundwater of South Kamrup area, Assam (India). The primary contributor of heavy metal concentration in groundwater of south kamrup area includes waste products of plastic industries, pharmaceutical industries, brick kiln industries, fertilizer industries etc. Excess concentration of heavy metal in groundwater is also linked with various health problems such as skin rashes, skin irritation, vomiting, diarrhea, nausea and other health problems (WHO, 2022) among the labor class people working in brick kiln industries, medicine industries and plastic industry of Palasbari and Loharghat area (South Kamrup, Assam). Elevated arsenic and cadmium concentration in drinking water lead to several health concerns e.g. skin rashes, renal disease, nephropathy etc. among the labor class people and industrial localities. Spatial variation of heavy metal concentration in underground water is mostly connected with industrial development and urban sprawling in certain areas. In Palasbari industries are situated in a cluster form resulting high degree of groundwater contamination. Therefore, it becomes necessity to test the water quality before using for drinking or other domestic purpose.

Acknowledgements The authors express their heartfelt gratitude to FARE LABS, M.G. Road, Gurugram, Haryana, India for providing laboratory facilities and access to standard instruments essential for this research work. We are deeply grateful to Dr. Niranjan Bhattacharjee, Associate Professor and Head of the Department of Geography, Pandu College, Guwahati, Assam, India, and Mrs Rebecca Kramsapi, Assistant Professor, Department of Geography, Pandu College, Guwahati, Assam, India, for their invaluable guidance and constructive feedback during the course of this study.

We also extend our sincere thanks to Mr Bhabesh Kalita for his generous assistance during the field visits.

References

Arora, K. N., & Chauhan, R. (2021). *2021*. Society for Environmental Sustainability.

Bhaskar, S., Madhav, H., Singh, P., & Jaiswar, G. (2016). Assessment of heavy metals, qualities of groundwater, and its correlation with natural organic matter. *Innovare Journal of Medical Science, 4*(5), 23–29. ISSN: 2321-4406

Boateng, K. T., Opoku, F., & Akoto, O. (2019). Heavy metal contamination assessment of groundwater quality: a case study of Oti landfill site, Kumasi. *Applied Water Science, 9*, 33. https://doi.org/10.1007/s13201-019-0915-y

Borah, K. K., Bhuyan, B., & Sarma, H. P. (2009). Heavy metal contamination of groundwater in the tea garden belt of Darrang District, Assam. *India. E-Journal of Chemistry, 6*(S1), 501–507.

Chakrabarty, S., & Sarma, P. S. (2010). Heavy metal contamination of drinking water in Kamrup district, Assam, India. *Environ Monit Assess, 179*, 479–486. https://doi.org/10.1007/s10661-010-1750-7

Chetia, M., Singh, S. K., Bora, K., Kalita, H., Saikia, L. B., Goswami, D. C., et al. (2008). Groundwater arsenic contamination in three blocks of Golaghat district of Assam. *Journal of Indian Water Works Association, 40*(2), 150–154.

Das, B., Rahman, M. M., Nayak, B., Pal, A., Chowdhury, U. K., Mukherjee, S. C., et al. (2009). Groundwater arsenic contamination, its health effects and approach for mitigation in West Bengal, India and Bangladesh. *Water Quality Exposure and Health, 1*, 5–21.

Das, M., & Bhattacharyya, R. K. (2020). A regression-based analysis to assess the impact of fluoride reach river water on the groundwater aquifer adjacent to the river: A case study in Bharalu River Basin of Guwahati, India. *Pollution, 6*(3), 637–650, Summer 2020; https://doi.org/10.22059/poll.2020.299434.764; Print ISSN: 2383-451X Online ISSN: 2383-4501(https://jpoll.ut.ac.ir)

Deshmukh, K. K., & Aher, S. P. (2016). Assessment of the impact of municipal solid waste on groundwater quality near the Sangamner City using GIS approach. *Water Retour Manag, 30*, 2425–2443.

Dilebo, B. W., Anchiso, D. M., Kidane, T. T., & Ayalew, E. M. (2022). Assessment of selected heavy metals concentration level of drinking water in Gazer Town and Selected Kebele, South Ari District, Southern Ethiopia. Hindaw international Journal of Analytical Chemistry, 2023, Article ID 1524850, 12 pages https://doi.org/10.1155/2023/1524850

Ehya, F., & Marbouti, Z. (2016). Hydrochemistry and contamination of groundwater resources in the Behbahan plain. *SW Iran. Environ Earth Sci, 75*, 45.

Etikala, B., Madhav, S., & Somagouni, S. G. (2022). Urban water systems: An overview. In *Current directions in water scarcity research* (pp. 1–19). Elsevier. https://doi.org/10.1016/b978-0-323-91838-1.00016-6

Etikala, B., Vangala, S., & Madhav, S. (2024). Groundwater geochemistry using modified integrated water quality index (IWQI) and health indices with special emphasis on nitrates and heavy metals in southern parts of Tirupati, South India. *Environmental Geochemistry and Health, 46*(11). https://doi.org/10.1007/s10653-024-02229-3

Groundwater Information Booklet of Kamrup & Kamrup Metro District, Assam, Central Ground Water Board North Eastern Region, Ministry of Water Resources, Guwahati, September 2013.

Kana, A. A. (2022). Heavy metal assessment of groundwater quality in part of Karu. *Central Nigeria, Water Practice and Technology, 17*(9), 1802. https://doi.org/10.2166/wpt.2022.102

Lorestani, B., Merrikhpour, H., & Cheraghi, M. (2020). Assessment of heavy metals concentration in groundwater and their associated health risks near an industrial area. *Environmental Health Engineering and Management, 7*(2):67–77, https://doi.org/10.34172/EHEM.2020.09

Munene, N. E., Hashim, O. N., & Ambusso, N. W. (2023). Human health risk assessment of heavy metal concentration in surface water of Sosian river, Eldoret town, Uasin-Gishu County, Kenya.https://doi.org/10.1016/j.mex.2023.102298

Prasad, B., & Sangita, K. (2008). Heavy metal pollution index of ground water of an abandoned open cast mine filled with fly ash: A case study. *Mine Water and the Environment, 27*(4), 265–267.

Reddy, Y. S., & Sunitha, V. (2023). Assessment of Heavy metal pollution and its health implications in groundwater for drinking purpose around inactive mines. *SW Region of Cuddapah Basin, South India, Total Environment Research Themes, 8*, 100069. https://doi.org/10.1016/j.totert.2023.100069

Reza, R., & Singh, G. (2010). Assessment of heavy metal contamination and its indexing approach for river water. *International Journal of Environmental Science and Technology, 7*(4), 785–792.

Sarma, T., & Saikia, S (2021). Status of groundwater arsenic contamination and human health in Hajo Circle, Assam, India. *Natural Volatiles and Essential Oils Journal, 8*(4), 11297–11305.

Singh, A. K. (2004). Arsenic contamination in groundwater of North Eastern India. Proceedings of national seminar on hydrology held at National Institute of Hydrology, Roorkee.

Sonone, S. S., Jadhav, S., Sankhla, S. M., & Kumar, R. (2021). Water contamination by heavy metals and their toxic effect on aquaculture and human health through food chain. https://doi.org/10.33263/LIANBS102.21482166

Sunitha, V., & Reddy, Y. S. (2019). Hydrogeochemical evaluation of groundwater in and around Lakkireddipalli and Ramapuram, YSR District, Andhra Pradesh India. *Hydro Res, 2*, 85–96.

Sunitha, V., Abdullah Khan, J., Muralidhara Reddy, B., Prasad, M., & Ramakrishna Reddy, M. (2014). Assessment of groundwater quality in parts of Kadapa and Anantapur Districts, Andhra Pradesh, India. *Indian Journal of Advances in Chemical Science, 3*, 96–101.

Sunitha, V., & Reddy, M. R. K. (2008). Fluoride in groundwater and fluorosis http://www.earthscienceindia.info/Sunitha.htm. 31-10-2008.

United Nations Environment Programme. (2020). https://www.unep.org

WHO. (1984). *The guidelines for drinking water quality recommendations*. World Health Organization.

WHO. (2011). Guidelines for drinking water quality, 4th edn WHO press p. 564

WHO. (2022). Guidelines for drinking-water quality, World Health Organization, ISBN 978-92-4-004506-4 (electronic version) ISBN 978-92-4-004507-1 (print version), https://www.who.int

WHO/UNEP GEMS. (1989). Global fresh water quality. Oxford: Blackwell (published on behalf of the World Health Organization/United Nations Environment Programme).

Chapter 17
Assessment of Various Types of Potable Water Quality and Analyses the Impact of Coal Mining on It Around Raniganj Coalfield, West Bengal in India: Challenges and Strategies

Dibyendu Saha, Md. Nazir, Kushal Roy, and Ayan Saha

Abstract In Raniganj coalfield (RCF), West Bengal, India, potable waters are affected by anthropogenic, and mining activities. The present study aims to evaluate the quality of potable water, to study the effects of anthropogenic and mining activities on the water quality of RCF, and also suggest necessary strategies for purification, and sustainability of wastewater of pit lake (PL). Season-wise, nine different samples of surface and subsurface potable water were collected and analyzed from 2019 to 2022. Outcomes show the pH of a small percentage of potable water is nearly neutral and mostly alkaline. Principal Components Analysis (PCA) is applied to understand and identify the interrelationships between the different indicators and the most contributive sources of water quality. It exhibits that the size of the Eigenvalues are > 1 and cumulative values are 91.3, 99.2, 99.7, and 99.9%. Computed water quality index (WQI) ranges from 12.276 to 103.469 indicating that the colliery-related waters are unfit for use, and some are even very poor. The topographic wetness index (TWI) also shows how much topography affects a certain area's hydrological parameters. Acid Mine Drainage (AMD), Potentially Toxic Elements' (PTE) pollution, chemical pollution, Erosion, and sedimentation are the prime types of mining activities that impact potable water quality. To reduce water scarcity, especially in the summer season sustainable water management strategies have been suggested by the authors which are required to be implemented in this area.

D. Saha (✉) · Md. Nazir · A. Saha
Department of Botany, The University of Burdwan, Burdwan, West Bengal, India
e-mail: dsaha@bot.buruniv.ac.in

A. Saha
e-mail: dsaha@bot.buruniv.ac.in

K. Roy
Department of Botany, Ramakrishna Mission Vivekananda Centenary College, Rahara, West Bengal, India

V. Sunitha et al. (eds.), *Groundwater Resource Management Planning Strategies*, Springer Water, https://doi.org/10.1007/978-3-031-88870-0_17

Keywords Anthropogenic · India · Potable water · Raniganj coalfield · Topographic wetness index · Water management strategy · Water quality index

Abbreviations

AMD	Acid mine drainage
ARD	Acid rock drainage
NH_3-N	Ammoniacal nitrogen
Arsenic	As
AMC	Asansol municipal corporation
BCCL	Bharat coking coal limited
Cd	Cadmium
Cl^-	Chloride
Cr (VI)	Hexavalent chromium
ISI	Indian statistical institute
Fe	Iron
Pb	Lead
HGMS	Magnetic separation in high gradient
Mn	Manganese
NTU	Nephelometric turbidity unit
Ni	Nickel
H_2SO_4	Sulfuric acid
UST	Ultrasound technique
UN	United nations
WQI	Water quality index
WWC	World water council
Cl	Chlorine
CIL	Coal India limited
Co	Cobalt
Cu	Copper
CN^-	Cyanide
EMT	Electrodialysis membrane treatment
F^-	Fluoride
GAC	Granulated activated carbon
NO_3^-	Nitrates
PL	Pit lake
PTEs	Potentially toxic elements
PCA	Principal components analysis
RCF	Raniganj coalfield
RO	Reverse osmosis
SO_4^{2-}	Sulfate
S^{2-}	Sulfide
UVR	Ultraviolet radiation

Wn	Unit weight
TWI	Wetness index
WHO	World health organization
Zn	Zinc

17.1 Introduction

The imperative for clean water has intensified due to the tremendous strain that prevalent advancements of populace, industrialization, and urbanization have placed on water resources and an enormous rise in contamination (Etikala et al., 2022; Sunitha & Reddy, 2022). Experts have ascribed contamination of water to its provenance that this fluid can dissolve more compounds than any other fluid on Earth. Water resources including rivers, reservoirs, lakes, and the oceans are brimming with contaminants, trash, plastics, and other hazardous contaminants owing to this anomaly, resulting in water contamination. The World Water Council (WWC) predicts that by 2030, approximately 3.9 billion people will reside in regions lacking sufficient water supplies (Xu et al., 2018). Consumer appetite for wastewater treatment facilities suitable for dwellings, businesses, or agricultural consumption has risen in light of the prevailing and anticipated constraint of water. Around the world, the sprawling sector of coal mining releases a lot of wastewater. For mankind to sustain the explosive rise of civilization among people while minimizing contaminants and health challenges, the administration of wastewater is critical. The leaching of toxic compounds as well as the ensuing pollution they inflict have escalated into an environmental concern that impacts millions of the welfare of individuals. The ingestion of contaminants from coal mining wastewater unveils some particular peril to emerging economies and water resources. Every day, India's largest cities create an astonishing 38 billion liters of wastewater, despite the fact that the country only has 11 billion liters of sanitation capacity. In contrast, just 60% of commercial wastewater is collected, primarily from large firms. India's green technologies sector continues to grow promptly and technological advances in curbing pollution provide obvious financial advantages. The United Nations (UN) (2015: 6.3) implemented an objective of enhancing water excellence by lowering contamination, eradicating discharges, as well as limiting the dispersion of potentially harmful substances, having a proportion of wastewater that is untreated, and radically boosting regenerating and sound recuperation across borders by 2030 (Gautam & Chattopadhyay, 2016). The processing of wastewater becomes critically important in such a scenario. Multiple sectors, notably the fabric, mineral extraction, paper and pulp, dye alongside colorant intermediary substances, drug-based, tanning facility, paint, and bleaching sectors expel an array of organic contaminants into water supplies. Of these, the apparel sector contributes to half of the wastewater-containing dyes currently apparent in the ecosystem. The industries that contribute to 40% of the total effluents are the coloration, paper and pulp, tannery and paint, and dye manufacturers, in that order (Velusamy et al., 2021).

Mining as an industry also severely depletes water supplies. The effluents from these businesses seriously harm the ecosystem. West Bengal, India's Paschim Bardhaman District has the Asansol neighborhood referred to as Raniganj which shares a part of RCF - the oldest and one of the leading Indian coal mines. It is under the jurisdiction of the Asansol Municipal Corporation (AMC). With the discovery of coal, the region became industrialized, clearing most of the forest. During the British East India era, in 1774, coal mining activities started in the RCF area in a spur of the moment. (Nazir et al., 2024; Chattopadhyay, 2001; Hunter, 1886). Subsequently, Bharat Cooking Coal Limited (BCCL), an affiliate of Coal India Limited (CIL) set up in the early 1970s, nationalized the coal mining market. The primary contributors to the damage of the locale's palatable waters are the unearthing and shipment of coal, burning of coal, drainage out of mine waste, cramped dumps, and a dearth of contaminated mine water during extraction operations (Mahato et al., 2016; Tiwari et al., 2016, 2019). Meanwhile, improper and unsound mining actions, inadequate environmental administration, and an overwhelming absence of comprehension of hazardous waste in the environment are all contributors to the deteriorating condition of the water resources in this mining region. Nevertheless, since coal and associated rocks embrace an array of minerals, surface and subsurface water fouling is an inevitable consequence of mineral extraction. Monitoring multiple parameters pollution levels following the established recommended limit along with implementing a feasible ecologically sound and sustainable management plan are two ways of alleviating water pollution (Egbueri, 2019; Egbueri & Unigwe, 2020; Egbueri et al., 2020; Etikala et al., 2024; Mgbenu & Egbueri, 2019). The WQI is a widely acknowledged tool for evaluating the water's quality, data that pertains to several human necessities. The designation and volatility of water quality were mostly assessed using the WQI (Chen et al., 2019; Nong et al., 2019; Singh et al., 2013). Multiple WQI have been deployed extensively; among them, Brown et al. (1972) developed a widely used WQI (Katyal, 2011; Mondal & Palit, 2019; Zotou et al., 2020). With a combination of a variety of multivariate statistical approaches, the current study aims to evaluate the hydrogeochemistry of various potable water quality levels in conjunction with seasonal fluctuations from 2019 to 2022. Additionally, it reviews different purification strategies that can be implemented in the RCF area.

17.2 Methodology

17.2.1 Field of Study

The RCF, which has been prevalent in West Bengal, India's Asansol subdivision (Fig. 17.1), encompasses 23°37′44″ N latitude and 87°06′54″ E longitude, covers 1530 km^2, and includes Paschim Bardhaman, Purulia, Birbhum, and Bankura districts. The topography is gently undulating, with elevations ranging from 60 to 80 m, and Panchet and Biharinath hills are the highest (Masto et al., 2015). Eastern

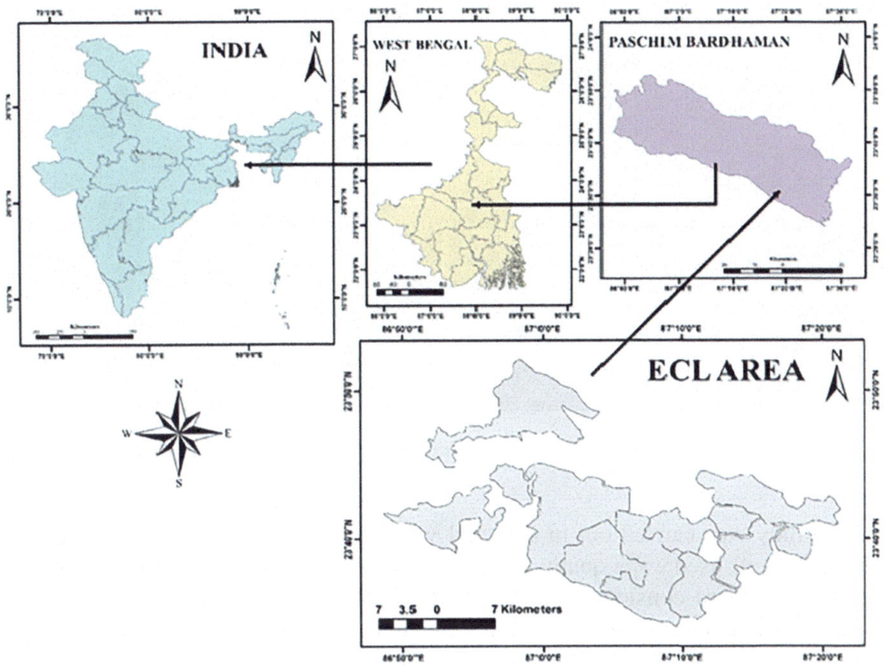

Fig. 17.1 Geographical location of RCF

Coalfields Limited (ECL) has 107 operating mines, 23 billion tonnes of coal reserve, with 6 billion tonnes of proven extractable reserve (Peterson, 1997). The RCF area experiences maximum rainfall during the monsoon season (80–90%), with yearly precipitation ranging from 1150 to 1450 mm. The drainage system is controlled by the Damodar, Ajay, and Barakar rivers. The geological structure of the RCF area, where water table depths typically range from 3 to 15 m below ground level, has a significant impact on the occurrence and storage of groundwater (Banerji & Mitra, 2019; Kumar & Singh, 2020; Manna & Maiti, 2016).

17.2.2 *Materials and Methods*

The present study is purely based on different surface and subsurface water samples collected from an abandoned PL, pressure filter plant of a colliery, colliery, pump house, boring, well, and tap of nine study sites such as Ratibati, Nimcha, Kalipahari, Sanctoria, Raniganj township, Satgram, Chappai Sewra village, Kalla ECL Central Hospital in the RCF (Fig. 17.2).

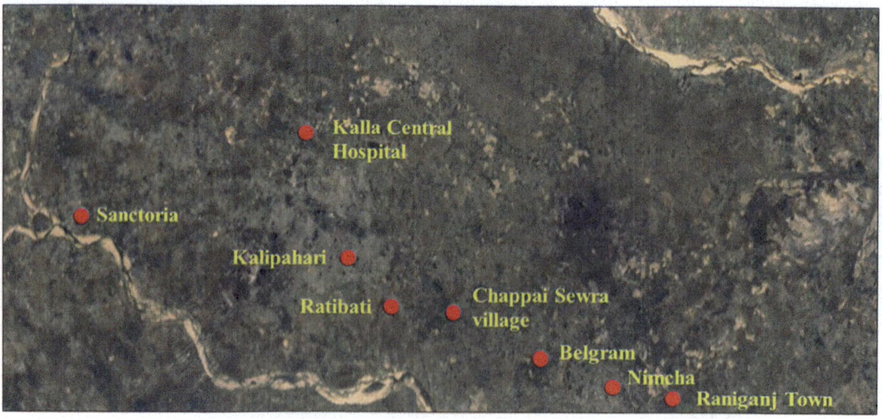

Fig. 17.2 Location of collection sites of different types of water samples from RCF

The study was carried out in seasonal variation from 2019 to 2022. The most commonly well-used water quality parameters as per IS 10500:1991 Drinking Water-Specification were considered to analyze the water quality of the different types of potable waters from the RCF area and then cited the mean values in tabular forms. Multivariate statistical techniques including analysis of PCA were applied for statistical exploration of water quality of potable waters in RCF.

The WQI model is a popular tool for evaluating surface water quality. It uses amalgamation algorithms to simplify immense quantities of data on the health of water to a singular index or number (Uddin et al., 2021). In the present study, pH, turbidity, dissolved solids, oil and grease, the concentration of chlorine (Cl) (mg/l), the concentration of amoniacal nitrogen ($NH_3{}^{-}N$) (mg/l), the concentration of nitrate nitrogen ($NO_3{}^{-}N$) (mg/l), the concentration of arsenic (As) (mg/l), the concentration of chromium (Cr) (mg/l), the concentration of copper (Cu) (mg/l), the concentration of cyanide (CN^{-}) (mg/l), the concentration of fluoride (F^{-}) (mg/l), concentration of sulfate ($SO_4{}^{2-}$) (mg/l), concentration of phenolic compounds (mg/l), concentration of iron (Fe) (mg/l), hardness and alkalinity were used as the parameters to assess the water quality.

pH

pH is a crucial physicochemical parameter that influences water quality and metal concentrations in aquatic habitats. It influences chemical reactions like acid–base, solubility, oxidation–reduction, and complexations. Water bodies near mining operations are susceptible to metal contamination from dumpsite leachate and trash discharge. Metals can bio-accumulate throughout the food chain, making pH a key factor in defining water's chemical and biological characteristics (Ndur et al., 2020; Vinodhini & Narayanan, 2008).

Turbidity

Turbidity is a water quality indicator that detects sediment discharge, organic matter, and pollutants, influencing water purity, aesthetics, and light penetration. It assesses clarity by comparing water clarity to suspended particulates, which might seem foggy, muddy, or hazy (Bozorg-Haddad et al., 202, Sibanda et al., 2014).

Dissolved Solids

Dissolved solids, or the total quantity of inorganic and organic components in water, might indicate contamination from industrial effluents, agricultural runoff, or natural sources, affecting water taste, odor, and appropriateness for various applications (Akhtar et al., 2021).

Oil and Grease

Oil and grease in water are indicators of pollution from industrial activities, urban runoff, or oil spills. They can form surface slicks, coat aquatic organisms, and disrupt ecosystem functioning. Monitoring oil and grease levels helps assess the impact of human activities on water quality (Kanchana et al., 2021).

Cl Concentration

Cl, a disinfectant used in water treatment, may be toxic to aquatic life at high levels, indicating over-chlorination or pollution from chlorinated chemicals (Zammitd et al., 2020).

NH_3–N

NH_3^-N, a consequence of agricultural runoff, wastewater discharges, or organic matter decomposition, can result in elevated amounts of ammonia and ammonium ions in water, which indicates poor water quality.

NO_3–N

NO_3^-N, which occurs naturally in water or is generated from agricultural fertilizers, sewage, or animal waste, may induce eutrophication, algal blooms, and oxygen depletion in aquatic habitats (Bano et al., 2022).

As, Cr, Cu, CN^-, F^-, SO_4^{2-}, Phenolic Compounds, Fe

These parameters refer to the numerous chemical pollutants that may be present in water as a result of industrial discharges, mining operations, or natural sources. Each of these compounds has different health and environmental effects, and their presence in water can endanger human health and aquatic life (Bano et al., 2022; Das, 2023).

Hardness and Alkalinity

Hardness relates to the concentration of calcium (Ca) and magnesium (Mg) ions in water, whereas alkalinity evaluates its ability to buffer pH variations. These factors influence water appropriateness for drinking, irrigation, and industrial activities, as well as the health of aquatic ecosystems (Chhabra & Chhabra, 2021).

The overall, monitoring these qualities provides helpful information about water quality, assists in the identification of pollution sources, and leads management activities to preserve and preserve water.

This study used Microsoft Excel (MS-Excel) to measure it using the Weighted Arithmetic Index approach. The WQI was estimated using the following formula:

1st Step

Calculate the unit Weight (Wn) factor for each parameter by using the formula

$$\mathrm{Wn} = \frac{K}{\mathrm{Sn}}$$

where,

$$1/S1 + 1/S2 + 1/S3 + \cdots + 1/\mathrm{Sn}$$

Sn = Standard desirable value of the *n*th parameters.

On summation of all selected parameters unit weight factors, Wn = 1(unity).

2nd Step

Calculate the Sub-Index (Qn) value by using the formula

$$\mathrm{Qn} = \frac{(\mathrm{Vn} - \mathrm{V0})}{(\mathrm{Sn} - \mathrm{V0})} * 100$$

where,

Vn = mean connection of the *n*th parameters.

Sn = Standard desirable value of the *n*th parameters.

V0 = Actual value of the parameters of the pure water (generally V0 = O2 for most parameters except for pH

$$\mathrm{QpH} = \frac{[(\mathrm{VpH} - 7)]}{[(8.5 - 7)]} * 100$$

3rd Step

Combining 1st Step, and 2nd Step, WQI is calculated as follows

$$\mathrm{WQI} = \frac{\sum \mathrm{WnQn}}{\sum \mathrm{Wn}}$$

Table 17.1 The water may be classified into five types based on computed WQI

WQI level	Water quality status	Feasible utilization
0–25	Excellent	Manufacturing, agriculture, and consumption
26–50	Good	Manufacturing, agriculture, and consumption
51–75	Poor	Manufacturing, and agriculture
76–100	Very poor	Only agriculture
> 100	Not suitable for consumption	The prior treatment is necessary before use

IDW techniques of interpolation method of spatial analyst tools in arc gis 10.8 software have been used to make a map based on WQI of different considered places of the study. Table 17.1 shows the categories of WQI.

TWI

The TWI was applied to measure how much topography affects a certain area's hydrological parameters. TWI analyses the topographically-based spatial distribution of moisture conditions. It determines the propensity for water to accumulate as well as the proclivity for water to flow lower due to gravity. Where TWI is expressed as:

$$\mathrm{TWI} = \mathrm{In}\left(\frac{A}{\tan \beta}\right)$$

where, A is the specific catchment area, and $\tan \beta$ is the local slope degree of the specific grid.

A specific keyword-based search approach was employed to prepare the review on water purification strategies. Wastewater purification, Phytoremediation, Wastewater, and other phrases were used as the keywords in this regard. Utilization of agricultural waste in a sustainable manner, valorization of agricultural waste, bioenergy, and so on. Among those chosen were Research Gate, Science Direct, PubMed, NCBI, and Google Scholar. Results from the last three decades, or from 2011 to 2024, were permitted for the unique search technique.

17.3 Results and Discussion

The geochemical analyses and monsoon-wise mean value of limnology of different types of potable water from the study site are represented in Table 17.2. The pH on average of different potable waters ranges from 6.77 to 7.77 which is within the maximum permissible limit as per ISI. In the RCF mine area, water is mildly acidic to alkaline in nature (Mendhe et al., 2017). The freshwater samples' turbidity ranged from 0.001 to 0.09 NTU, within the prescribed permissible limit, turbidity relates to a major problem in water treatment (Asrafuzzaman et al., 2011). Dissolved Solids, Oil, and Grease ranged from 133–761 mg/l, and 0–0.03 mg/l where the maximum limit was 500 mg/l, and 0.01 mg/l respectively. Sufficient information is currently available on the impact of total residual Cl to determine the highest level of residual Cl that would protect aquatic life, human health, and the environment. (Suzuki et al., 2006), in the samples, the total residual Cl varied from 0–0.2 mg/l within the permissible limit. The concentration of NH_3^-N, and NO_3^-N ranged from 0–3.3 mg/l, and 0–0.03 mg/l, and the described limit are 0.1 mg/l and 3.3 mg/l. In boring water, Raniganj town NH_3^-N is beyond the permissible limit. Coal mining has immense effects on local area water resources, both surface and underneath water (Reddy & Sunitha, 2023). Surface waters are being polluted through mine drainage whereas underneath water is being polluted due to pilferage of the water table. Very low-pH water also pollutes with PTEs. PTE contamination presents the trace occurrences of elements coming from the reservoir rock on its hazard to mine sewerage water in an acidic environment. As the mine water discharges the reservoir rock, metals are percolated into the water and associated with the surface water bodies (Ray & Dey, 2020). As, Cr (VI), Cu, and CN^- present in these water samples are very minimal or absent. Cl^- is regarded as a major water quality parameter. In surface and under surface water the Cl^-, NO_3^-, and Phenolic compounds content occurs mainly from natural sources, wastewater, agricultural, domestic activities, and industrial outflow. Phenolic compounds are toxic and have both serious and enduring effects on humans (Anku et al., 2017; Solanki, 2012). The average Cl^-, NO_3^- and phenolic compound concentrations of the water samples are between 20–188 mg/l, 0–0.01 mg/l, and 0.001–0.01 mg/l which are about within the desirable limit of ISI 250 mg/l, 45 mg/l, and 0.005 mg/l respectively. It is noticeable that water from all sources in the RCF area has F^- provincially above the WHO limit of 1.0 ppm (Gupta & Mondal, 2013). In this study it was observed that the concentration of F^- for all the considered water samples is very minimal with the standard permissible limit except in Ratibati abandoned PL i.e., 185.2 mg/l.

The variation of different water quality parameters at different considered water samples during the study period is cited in Table 17.2. Because of precipitation from the surface during the monsoon season, the mean values of F^- in this particular abandoned PL rose. The range of the ideal limit was likewise matched by the concentrations of Fe and SO_4^{2-} in the water samples, which were 0.001–0.21 mg/l and 0–0.001 mg/l, respectively (Palit et al., 2014). Total alkalinity and total hardness are basic water quality varying in different freshwater sources (Boyd et al., 2016). Here the average Total Hardness and Total Alkalinities ranged from 31–355 mg/l, and 32-365 mg/l when permissible limits for both are 300 mg/l.

WQI of the selected potable waters from RCF in different seasons of the study period are prepared from important and various regularly used parameters (Tables 17.3 and 17.4) and ranged from 12.276–103.469. Prior to any usage, treatment is necessary for the abandoned PL at Ratibati, New Ghusik Colliery, Kalipahari, and Sanctoria 5 No pump house where WQIs were 103.469, 103.177, and 103.223, respectively. The water quality of the Nimcha Colliery, Pressure Filter Plant, and Boring Water in Raniganj Town is very poor and should only be used for irrigation. In contrast, the well water in Raniganj Town, tap water from Satgram's pumping station, well water from Chappai Sewra village, and Kalla Central Hospital in Kalla

Table 17.2 Water quality analyses show monsoon-wise mean value of limnology of different types of potable water in RCF area as per IS 10500: 1991 drinking water specification

Parameters	Ratibati abandoned PL	Pressure Filter Plant, Nimcha Colliery	New Ghusik Colliery, Kalipahari	Sanctoria 5 No. Pump House	Boring water, Raniganj town	Well water, Raniganj town	Tap water from pumping station, Satgram	Well Water from Chappai Sewra-village	Kalla Central Hospital	Maximum permissible limit (as per ISI) Sn
PH. Value	6.77*	7.4*	7.68*	6.85*	7.07*	7.77*	7.11*	7.11*	7.12*	8.5
Turbidity NTU	0.02*	0.01*	0.001*	0.05*	0.01*	0.001*	0.001*	0.001*	0.09*	5
Dissolved Solids (mg/l)	133*	144*	296*	675*	510*	761*	265*	265*	286*	500
Oil and Grease (mg/l)	0.001*	0.03*	0.002*	0.001*	0	0.001*	0.001*	0.001*	0.001*	0.01
Total residual Cl (mg/l)	0.001*	0.01*	0.01*	0.001*	0.2	0.001*	0	0.001*	0.001*	0.2
NH_3^-N (mg/l)	3.63*	0.01*	0.001*	0.04*	3.3*	0.016*	0	0.001*	0.001*	0.1
NO_3^-N (mg/l)	0.0015*	0.011*	0.01*	0.03*	0.01*	0.001*	0.001*	0	0.001*	0.1
Arsenic (mg/l)	0.0015*	0	0	0.01	0	0	0	0	0	0.05
Hexavalent Chromium (mg/l)	0.001*	0.001*	0.001*	0.001*	0	0.001*	0.001*	0	0	0.05
Copper (mg/l)	0.001*	0.01	0.002*	0.001*	0	0.001*	0	0.001*	0	0.05

(continued)

Table 17.2 (continued)

Parameters	Ratibati abandoned PL	Pressure Filter Plant, Nimcha Colliery	New Ghusik Colliery, Kalipahari	Sanctoria 5 No. Pump House	Boring water, Raniganj town	Well water, Raniganj town	Tap water from pumping station, Satgram	Well Water from Chappai Sewra-village	Kalla Central Hospital	Maximum permissible limit (as per ISI) Sn
Cyanide (mg/l)	0.001*	0.001*	0.001*	0.001*	0.006*	0.0011*	0	0.001*	0	0.05
Chloride (mg/l)	14.5*	20*	49.5*	36*	95*	188*	31*	31*	48*	250
Fluoride (mg/l)	185.2*	0.001*	0.001*	0.01*	0	0.001*	0	0.001*	0	15
Sulphates (mg/l)	0.001*	0.001*	0.001*	0.002*	0.001*	0.001*	0.001*	0.001*	0	150
Nitrates (mg/l)	0.0011*	0.0015*	0.01*	0.001*	0.001*	0.001*	0.001*	0.001*	0	45
Phenolic Compounds (mg/l)	0.001	0.001	0.01*	0.01*	0.001*	0.001	0.001	0.001	0.001	0.005
Iron (mg/l)	0.61*	0.02*	0.05*	0.01*	0.12*	0.21	0.0011*	0.001*	0.001*	0.3
Total Hardness (mg/l)	31*	64*	133*	355*	263*	314	111*	111*	131*	300
Total Alkalinities (mg/l)	32*	52.6*	100*	362*	305	365*	109*	109*	84*	300

*Indicates significant difference from maximum permissible limits; One way ANOVA followed by Dunnett's test are all excellent choices for drinking, irrigation, and industrial uses, with respective values of 13.526, 12.342, and 12.445. Figure 17.3a displays the WQI map, which amply illustrates the variance in water quality in the RCF. According to Singh et al. (2010), the RCF verified that mine water is appropriate for use in irrigation, industry, and residential settings following a qualitative evaluation of the water. To estimate the local hydraulic gradient under steady-state circumstances, the TWI is used, as Fig. 17.3b illustrates.

PCA was executed to take out the most critical factors regulating the water quality. The 1st PCA for 91.30% of the total variance and the first four cumulative values are 91.3, 99.2, 99.7, and 99.9% i.e. more than 90% so PCA is acceptable and the Acceptance level of variance is > 80% and to perform other analyses is > 90%. The size of the eigenvalues is 8.21, 0.71, 0.04 therefore retain the principal components with the largest eigenvalues i.e., > 1 (Tables 17.5 and 17.6). In the Outlier Plot (Fig. 17.4a) the Reference line is 5.349. There are no outliers, all the points are beneath the reference line. Scree Plot (Fig. 17.4b) orders the eigenvalues from largest to smallest. The score Plot (Fig. 17.4c) shows the clusters, trends, and outliers in the first two principal components and the Loading Plot (Fig. 17.4d) visually shows the results for interpreting the first two principal components (Fig. 17.5a, b).

As a researcher and an inhabitant of a city in the vicinity of RCF for the last 25 years, the authors observed the following types of coal mining activities and their vast impact on potable water quality.

17.3.1 AMD

Mining activities markedly contaminate potable waters in the mining area. Mine wastewater is widely varying in composition and highly complex. These are generally mild to highly acidic but sometimes nearly neutral, or alkaline. AMD, sometimes referred to as acid and metalliferous drainage or Acid Rock Drainage (ARD), is the outflow of acidic water from coal or metal mines. The term AMD is more often used (than ARD) since the process mostly takes place at mining locations. It happens in tunnels, mine workings, open pits, waste rock piles, and mill tailings—all of which are examples of both operational and retired mining sites (Sheoran & Sheoran, 2006; Simate & Ndlovu, 2014; Acharya & Kharel, 2020). Potable water quality in RCF has considerably declined as a result of AMD (Fig. 17.6a, b), which exhibits notable concentration levels of inorganic salt pollutants like NH_3^-N, S_2^-, and Cl, as well as organic pollutants like oil, grease, and phenolic compounds. Consuming such contaminants poses wellness hazards, Table 17.7 highlights these contaminants' adverse impacts on health.

Table 17.3 Calculation of WQI of Ratibati abandoned PL, RCF considering as samples

Parameters	Maximum permissible limit (as per ISI) Sn	1/Sn	∑1/Sn	K = 1/(∑1/Sn)	W1 = K/Sn	Ideal value V0	Mean conc. value (Vn)	Vn/Sn	Vn/Sn*100 = Qn	WnQn	WQI
PH. Value	8.5	0.11765	409.359	0.00244284	0.00029	7	6.77	−0.1533	−15.3333333	−0.0044	**103.02**
Turbidity NTU	5	0.2	409.359	0.00244284	0.00049	0	0.02	0.004	0.4	0.0002	
Dissolved solids (mg/I)	500	0.002	409.359	0.00244284	4.9E−06	0	133	0.266	26.6	0.00013	
Oil and grease (mg/I)	0.01	100	409.359	0.00244284	0.24428	0	0.001	0.1	10	2.44284	
Total residual Cl (mg/I)	0.2	5	409.359	0.00244284	0.01221	0	0.001	0.005	0.5	0.00611	
NH3⁻N (mg/l)	0.1	10	409.359	0.00244284	0.02443	0	3.63	36.3	3630	88.6752	
Nitrite Nitrogen (mg/I)	0.1	10	409.359	0.00244284	0.02443	0	0.0015	0.015	1.5	0.03664	
Arsenic (mg/I)	0.05	20	409.359	0.00244284	0.04886	0	0.0015	0.03	3	0.14657	
Hexavalent chromium (mg/I)	0.05	20	409.359	0.00244284	0.04886	0	0.001	0.02	2	0.09771	
Copper (mg/I)	0.05	20	409.359	0.00244284	0.04886	0	0.001	0.02	2	0.09771	

(continued)

Table 17.3 (continued)

Parameters	Maximum permissible limit (as per ISI) Sn	1/Sn	∑1/Sn	K = 1/(∑1/Sn)	W1 = K/Sn	Ideal value V0	Mean conc. value (Vn)	Vn/Sn	Vn/Sn*100 = Qn	WnQn	WQI
Cyanide (mg/l)	0.05	20	409.359	0.00244284	0.04886	0	0.001	0.02	2	0.09771	
Chloride (mg/l)	250	0.004	409.359	0.00244284	9.8E−06	0	14.5	0.058	5.8	5.7E−05	
Fluoride (mg/l)	1.5	0.66667	409.359	0.00244284	0.00163	0	0	0	0	0	
Sulphates (mg/l)	150	0.00667	409.359	0.00244284	1.6E−05	0	0.001	6.7E−06	0.000666667	1.1E−08	
Nitrates (mg/l)	45	0.02222	409.359	0.00244284	5.4E−05	0	0.0011	2.4E−05	0.002444444	1.3E−07	
Phenolic compounds (mg/l)	0.005	200	409.359	0.00244284	0.48857	0	0.001	0.2	20	9.77137	
Iron (mg/l)	0.3	3.33333	409.359	0.00244284	0.00814	0	0.61	2.03333	203.3333333	1.6557	
Total hardness (mg/l)	300	0.00333	409.359	0.00244284	8.1E−06	0	31	0.10333	10.33333333	8.4E−05	
Total alkalinities (mg/l)	300	0.00333	409.359	0.00244284	8.1E−06	0	32	0.10667	10.66666667	8.7E−05	

Table 17.4 WQI values of selected different water sources from RCF

Water sources in RCF	WQI	Remark
Ratibati abandoned PL	103.02	Unfit for drinking
Pressure Filter Plant, Nimcha Colliery	84.86	Very poor
New Ghusik Colliery, Kalipahari	103.46	Unfit for drinking
Sanctoria 5 No. Pump House	103.17	Unfit for drinking
Boring water, Raniganj town	92.76	Very poor
Well water, Raniganj town	13.52	Excellent
Tap water from pumping station, Satgram	12.34	Excellent
Well Water from Chappai Sewra-village	12.44	Excellent
Kalla Central Hospital	12.27	Excellent

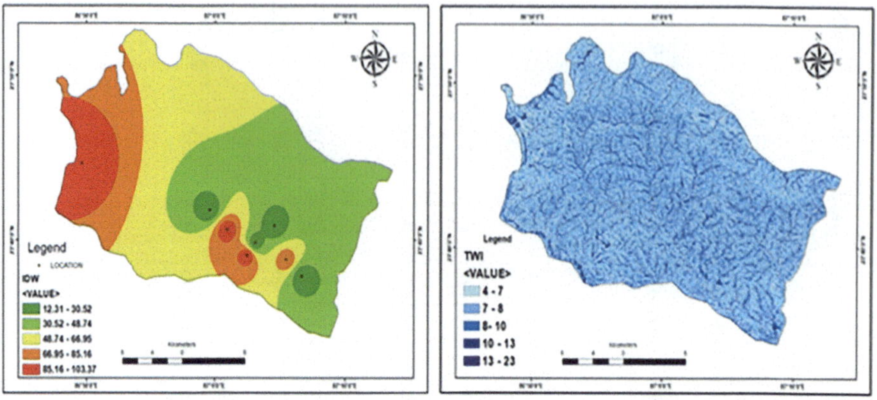

[A] [B]

Fig. 17.3 **a** WQI, **b** TWI maps of RCF

Table 17.5 Loadings of variables on principal components for selected different types of sources of water in RCF area Eigen analysis of the Correlation Matrix

Eigenvalue	8.2126	0.7136	0.0496	0.0191	0.0046	0.0005	0.0000	0.0000	0.0000
Proportion	0.913	0.079	0.006	0.002	0.001	0.000	0.000	0.000	0.000
Cumulative	0.913	0.992	0.997	0.999	1.000	1.000	1.000	1.000	1.000

Table 17.6 Loadings of variables on principal components of different physicochemical parameters of water collected from RCF

Eigenvalue	5.2524	3.7810	3.1663	2.3253	2.0147	1.5170	0.7652	0.1781	0.0000
Proportion	0.276	0.199	0.167	0.122	0.106	0.080	0.040	0.009	0.000
Cumulative	0.276	0.475	0.642	0.764	0.871	0.950	0.991	1.000	1.000

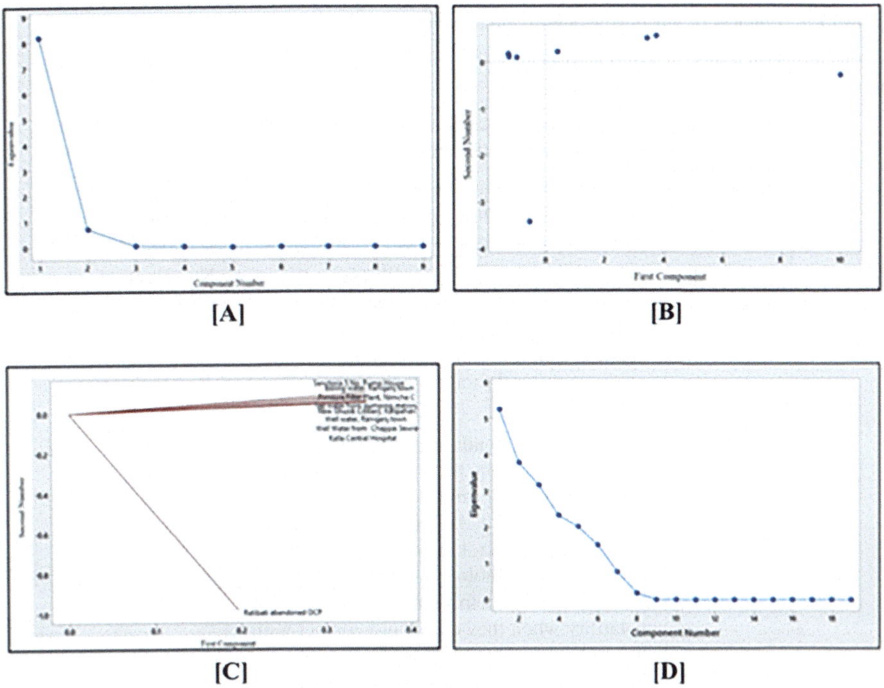

Fig. 17.4 **a** Outlier plot, **b** Scree plot, **c** Score plot, **d** Loading plot of water quality of different types of potable water in RCF

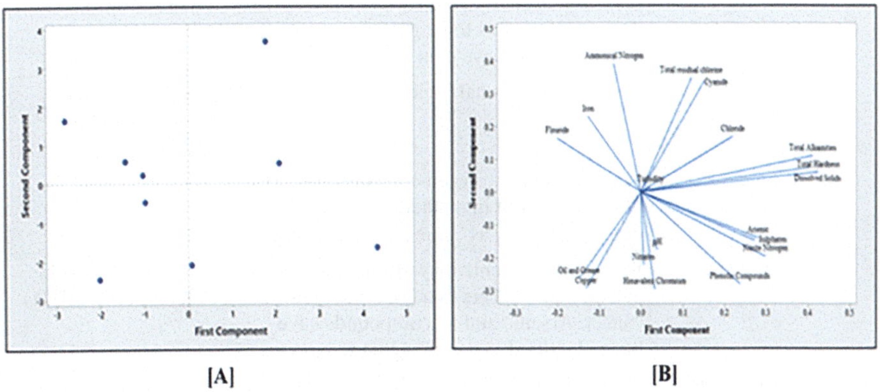

Fig. 17.5 **a** Score plot, **b** Loading Plot of different physicochemical parameters of water collected from RCF

Table 17.7 Impacts of various pollutants on human health

Serial number	Pollutants	Implication	References
1	pH	Drowsiness, exhaustion, headache, nausea, disorientation, and vomiting	Grieco (2021) Quade et al. (2021)
2	Turbidity	Cramps, nausea, headaches, diarrhea, or other conditions and may provide a particular health concern to newborns	De Roos et al. (2017a, 2017b) Hsieh et al. (2015)
3	Dissolved Solid	Increased Risk of Chronic Illnesses: When ingested often over time, elevated amounts of dissolved solids, particularly salts like sodium, may cause hypertension, cardiovascular difficulties, and renal problems	Nasrabadi et al. (2016)
4	Oil and grease	The many skin conditions might manifest as dermatitis, eczema, thickness of the skin, xerodama, itching, and skin discoloration. Because of extended contact with a specific region of skin, the skin gets ulcerated and changes in morphology (structural pattern).Certain petroleum products exhibit volatility when they come into contact with the human respiratory system. illnesses include tuberculosis, asthma, emphysema, and pneumonia. Tumors and other abnormal growths may occur often	Chand et al. (2017), Madanhire et al., 2016)
5	Cl	Gastrointestinal tract tissue injury caused by corrosive agents. Eye/Dermal Contact: Prolonged contact to Cl gas may irritate the skin and induce eye pain	Li and Mitch. (2018), Andra et al. (2015)
6	$NH_3{}^-N$	Could harm internal organ systems	Ward et al. (2018), de Jesus Gaffney (2015)
7	Nitrate	May impact the oxygen-carrying capacity of blood and result in methemoglobinemia, often referred to as "blue baby syndrome."	Ward et al. (2018)
8	Arsenic	Exposure to arsenic may also result in lung and bladder cancers in addition to skin cancer. Arsenic and its compounds have been identified as carcinogenic to humans	Abdul et al. (2015) Adeloju et al. (2021)
9	Chromium	Tracheobronchitis, respiratory, tubular necrosis, congestion and hyperemia, bronchitis, pharyngitis, rhinitis, and asthma	Abdel-Gadir et al. (2016), Prasad et al. (2021)
10	Copper	Headaches, lightheadedness, nausea, vomiting, cramping in the stomach, diarrhea, liver cirrhosis, renal disease, and brain damage	Pohanka (2019), Briffa (2020)

(continued)

Table 17.7 (continued)

Serial number	Pollutants	Implication	References
11	Cyanide	Coma, blood pressure that is either high or low	Pulster et al. (2015)
12	Fluoride	Lowers the prevalence of dental caries. An excessive amount of fluoride may cause skeletal or dental fluorosis, which can harm bones and joints	Kabir et al. (2020), Kanduti et al. (2016)
13	Sulphate	Vomiting, feeling nauseated, and irritated throat and stomach	Kali et al. (2021)
14	Phenolic compound	Cause harm to the liver and red blood cells	Rahman et al. (2021), Ramos et al. (2021)
15	Iron	Gastrointestinal issues, nausea, hemochromatosis, and diabetes. The pancreas, liver, and heart may all be harmed	Izah et al. (2016)
16	Hardness	Arid hair and skin	Akram and Rehman (2018)

17.3.2 Potentially Toxic Elements (PTEs) Pollution

Lead (Pb), zinc (Zn), cadmium (Cd), Cobalt (Co), Cr, As, Cu, and various other PTEs encountered in unobstructed mines beneath the ground or unearthed rock adhere with water to generate it. There is an abundance of PTEs, notably Cr, Cu, F, Fe, manganese (Mn), nickel (Ni), and Pb in the discharge water emanating from the trash dump. Water scrubbing over the rock surface drains PTEs out and propels them downstream. While PTEs are potentially portable in neutral pH settings, leaching tends to be exacerbated in low pH states, as those generated through AMD (Fig. 17.6a, b).

17.3.3 Chemicals Pollution

This particular form of damage emerges given that mining enterprises segregate the sought-after component from the ore. Chemical compounds comprising CN^- and H_2SO_4 are used, and they seep, spill, or stream from the place of operation into adjacent pools of water. Humans as well as livestock can become fatally harmed by these contaminants.

Fig. 17.6 **a**, **b** Acid Mine Drainage, **c**, **d** Erosion and Sedimentation of sediment into the Damodar River and water body, **e** Mining pollutants create dense water bloom in RCF

17.3.4 Degradation and Silt Removal

When it comes to forging and upkeep of haulage roads, surface cut excavations, and trash confiscation, the expansion and advancement of mines disrupt the earth and rock. The exposed surface may routinely erode and transmit substantial quantities of debris into waterways such as creeks, lakes, etc., which impacts subterranean water in the dearth of appropriate mitigation and control measures. A great deal of sediment may choke streams and suffocate basin foliage, aquatic species, and natural habitats. (Fig. 17.6c–e).

17.3.5 Strategies

The two main categories of treatment are proposed for the treatment of wastewater (Sharma & Bhattacharya, 2017) in RCF which are mechanical and biological methods of treatment.

Mechanical Methods

Precipitation and Coagulation

It is among the most straightforward approaches to water purification. Particles generated by the incorporation of these compounds settle and eliminate impurities from water. The portion of the settling process that has been cleaned off is drained of water and discarded away. The process softens and removes impurities from water, such as P, F-, As, $[Fe(CN)_6]^{4-}$, and PTEs (Environmental Protection Agency US, 2000; Matlock et al., 2002; Eikebrokk et al., 2006).

Softening of Water

The presence of Ca/Mg in the forms of carbonate ($CO_3{}^{2-}$), bicarbonate ($HCO_3{}^{-}$), $SO_4{}^{2-}$, and Cl^- causes water to become harder. The addition of the proper chemical causes precipitation to occur and become softer. The addition of calcium hydroxide $[Ca(OH)_2]$ to water causes the precipitation of bicarbonate and sulfate. Hydroxide $[(OH)^-]$ together with $SO_4{}^{2-}$ and Cl^- precipitates in water upon the addition of sodium aluminate ($NaAlO_2$). $NaAlO_2$ produces sodium hydroxide when it combines with water, and hydroxide when it reacts with $S_2{}^-$ or Cl. The synthesis of aluminum hydroxide $[Al(OH)_3]$ promotes floc formation, sludge blanket conditioning, and silica reduction. Another way to soften it is to boil water.

Removal of PTEs

PTEs, such as barium (Ba), Cd, Pb, mercury (Hg), Ni, and Cu, frequently precipitate out of wastewater as sulfates, sulfides, hydroxides, and carbonates, according to Matlock et al. (2002). Simultaneous precipitation during flocking with Fe and aluminum (Al) salts is also a possibility for some metals (As, Cd, Hg, and Cr).

Elimination of Arsenic.

Alum, ferric chloride ($FeCl_3$), and ferric sulfate (Fe_2SO_4) are excellent coagulants for the removal of arsenic (Fields et al., 2000). As (V) is more easily eliminated in these circumstances than As (III). During the process, electrostatic adhesion takes place, linking the negatively charged As ions and microparticles to the flocs. Sedimentation and filtering operations remove the Fe-As complex, which is a result of equivalent associations among $FeCl_3$ and $SO_4{}^{2-}$ (Hering et al., 1997; Mok, 1994). Dependent on pH range is efficient elimination.

Elimination of Phosphorus

Coagulation, or combining coagulant with wastewater is the typical method for removing compounds of phosphorus (P) like phosphates ($PO_4{}^{3-}$) (Xie et al., 2005). Ca, aluminum (Al), and Fe are the three most often encountered multivalent metal ions.

Elimination of fluoride

The most efficient method for removing F^- (by Mg, Al, and Ca) from torrents of wastewater includes precipitating F^- families into chemically stable forms (Lawrence et al., 2005). CaF_2 is one of the least soluble metal fluorides in water. As a result, the most popular kind of treatment is the conversion of fluoride from the effluents into CaF_2. $CaCl_2$ is favored for the program due to its superior dissolution with a lesser additive-to-effluent ratio, while other limes can also be employed. Although they are utilized in extremely large amounts, inorganic flocculants have the potential to be used in various separations. While polymeric flocculants generate immense cohesive aggregates (flocs) that are exceptionally impervious to fluctuations in pH, they generate substantial amounts of sludge and are susceptible to pH oscillations. Naturally occurring as well as synthetic polymers have advantages for this purpose. Although synthetic polymers like polyacrylamide, polyethylene oxide, poly (diallyl dimethyl ammonium chloride), and poly (styrene sulfonic acid) are extremely efficient flocculants at low concentrations and have high customizable ability but impoverished shear stability, natural polymers like starch, guar gum, alginate, glycogen, and dextran tend to be renewable and are efficiently shear resilient.

Distillation

It is the most often-used method of separation (Veil, 2008). This kind of segregation entails using heat to separate the convergent particles in water. The variations in the individual components' boiling temperature constitute its core. The concentrations of the constituents affect the boiling point's attributes. Consequently, the distillation process takes into account the vapor pressure traits of liquid mixtures. The basic tenet is that vapor pressure rises with the incorporation of heat energy. Owing to the mixture's varied levels of volatility, distillation happens when the liquid mixture boils and the pressure of the vapor attains the ambient level. This approach eradicates microbes while also filtering the water from inorganic substances like Mg, Ca, and Pb amongst several others. On the other hand, organic compounds that possess boiling degrees below 100 °C are futile in their elimination and could potentially accumulate in the product water. The initial application of water distillation technology was industrial. The process requires the integration of the carbon filter system since the existing technique is unsuitable for eradicating organic waste and thereby guaranteeing the suitability of water for ingestion. Since the carbon filters can quickly transform into microbial breeding grounds, routine replacement is crucial. (Veil, 2008).

Adsorption

According to this physical process, contaminants in the solution adhere to the porous surface of the solid particles (Yang, 1997; Jiuhui, 2008). Both energy from the surface and surface events have been accountable for underlying it. Both immediate incorporation and blending plate inclusion of the adsorbent systems are possible. Adsorbents are substances that use chemical and physical methods to eliminate the compounds that give water its hue, flavor, and smell. It is hypothetically possible to employ any kind of microporous material as an adsorbent. Nonetheless, the most favored kinds are tightly regulated and very microporous. Activated carbon, silica gels, aluminas, zeolites, and other porous materials having many cavities or pores as tiny as a fraction of a nanometer are examples of porous materials that have advantages (Ali & Gupta, 2006; Jiuhui, 2008).

Membrane Technology

According to Charcosset (2009), one of the inventive methods for treating water is membrane technology. The elimination of water contaminants takes place here using a semipermeable membrane. There are two types of membrane water treatment systems: electrically powered (as with electro-membrane) as well as pressure powered (for instance, reverse osmosis).

Reverse Osmosis

Reverse Osmosis (RO) is one of the most potent and widely employed means of purifying water around the globe. The most common uses for it encompass eliminating salts (Bhattacharya & Ghosh, 2004; Pawlak et al., 2006), dyes (Natraj et al., 2009), pesticides (Bhattacharya, 2006), chemical toxins (Yogesh et al., 2008), organic pollutants (Yogesh et al., 2008), and microbes (Park & Hu, 2010). By passing raw water through an extensive membrane screen under pressure, reverse osmosis prevents pollutants from passing through. While reverse osmosis and nanofiltration membranes have pores with sizes between ~1 and < 1 nm, microfiltration and ultrafiltration membranes have pores with sizes between > 10 and 1–100 nm, respectively. Micro and ultrafiltration use size-selective separation, whereas the latter two use size and charge-selective separation. Both natural and synthetic polymers—such as polyamide, polysulfone, cellulose triacetate, and others—are regularly employed for developing the membranes.

Electrodialysis Membrane Treatment

Electrodialysis membrane treatment (EMT) is a membrane-based, electric potential-driven separation technique. Ion exchange processes and membrane separation share a common fundamental concept (Strathmann, 2010a; Xu, 2005). In contrast to a cationic membrane with fixed negative groups, which rejects negative ions but permits free passage for positively charged ions, an anionic membrane with fixed positive groups excludes positive ions while permitting free passage for negatively charged ions. It is evident that the fixed charge groups partially or prohibits the same charge from the membrane. The charged groups are connected to the polymer backbone of the membrane material. The membrane's ability to separate or reject ions with opposing charges makes it effective for removing or separating electrolytes (Koter & Warszawski, 2000; Strathmann, 2010b).

Magnetic Separation in High Gradient

Among the magnetic separation processes, high gradient magnetic separation (HGMS) is the most widely employed approach (Ditsch et al., 2005; Hoffmann & Franzreb, 2004a, 2004b; Okada et al., 2005). An electromagnet situated inside a framework of cables with magnetic responsiveness drives this apparatus. a few of the contributing factors include the kind of impurities, concentration, dimensions, magnetic sensitivity, spacing layout, magnetic field strength, and the intensity and direction of the magnetic field.

Disinfection

Chemical and physical decontamination strategies are the two subcategories of disinfection practices. I, Cl, H_2O_2, and O_3 are examples of chemical treatments, whereas ultraviolet (UV), and ultrasonic therapy are examples of physical disinfection (Kerwick et al., 2005).

Ultra Violet Radiation

Throughout the ultra violet radiation (UVR) treatment, the water intended to be treated is confronted with germicidal UVR that has been set up underneath a mild lamp. Water passing through a UV purifier exposes biological contaminants to UVR, damaging the DNA and other genetic elements in microbes. The technique eradicates the microorganisms (Bergmann et al., 2002; Hijnen et al., 2006).

Ultrasound Technique

Combining medium penetration, reflection characteristic evaluation, and concentrated energy delivery, the ultrasound technique (UST) finds use in a wide range of sectors. The mechanical vibration of the waves may cause damage to the cellular framework of microorganisms. Therefore, employing this method to purify water might have benefits (Fetyam and Salem Attia, 2020).

Purification by Ozone

O_2 in the unstable form, O_3, acts as a screen to the damaging UV. Nonetheless, it is excellent as a disinfecting agent in water purification (Von Gunten, 2003a, 2003b). It's a strong oxidant since it releases O_2 readily. UV radiation or a frigid spark both convert O_2 to O_3. O_3 has a very high potential for oxidation, making it easy to inject O_2 into organic molecules' bonds to produce aldehydes and ketones. It is more efficient than chemical disinfection techniques like chlorination at eliminating of microbial contaminants like bacteria. In actuality, O_3 weakens the cell wall and induces cellular rupture by oxidizing the organic components in bacterial membranes. Upon exposure to the external environment, the organism's cell promptly perishes. O_3 improves clarity by removing Mn, sulfur, and iron. Along with reducing odor problems, it also reduces sulfur and other dissolved chemical concentrations. O_3's primary advantage is that it totally purges water of any remaining disinfection residue. When O_3 is sprayed as a disinfectant, it is immediately created (Von Gunten, 2003a, 2003b).

Hypochlorite

Typically combined with Na or Ca, hypochlorite (ClO^-) forms a compound, which is to be used as a disinfectant. Because handling Cl gas is dangerous, the use of NaClO or CaClO is common. Water releases free Cl as a result. The electrolytic technique is another way to obtain Cl solution. Released free Cl occurs when Cl dissolves in water. Examples of potentially hazardous chemical byproducts produced when Cl reacts with naturally existing organic molecules in water are trihalomethanes and haloacetic acids. They can cause cancer, as the University of Florida Report (1998) has shown. The most often used powerful oxidant in the disinfection process is Cl and its derivatives, such as NH_2Cl or ClO_2. Cl functions well against bacteria and protozoa that cause cysts (viz. *Giardia lamblia* and *Cryptospordium*) (Kumar et al., 2016; Navlalon et al., 2009).

Iodine

Iodine (I) is a good oxidizing agent likewise; in a short period, it can effectively inhibit a wide range of pathogenic organisms, such as viruses, spores, and cysts (Punyani et al., 2006). Compound-based formulations, like iodophores, can regulate the release of I by combining I with solubilizing compounds, like non-ionic surfactants, I-incorporated resins, and cross-linked copolymers of styrene (C_8H_8) and divinylbenzene ($C_{10}H_{10}$) with I_3, as well as I with polyvinyl pyrrolidone (C_6H_9NO).

Hydrogen Peroxide

Hydrogen peroxide (H_2O_2) is well known for its strong oxidative and biocidal efficacy, but its application in potable water disinfection is not currently possible; nevertheless, when combined with O_3, UVR, it is possible (Andreozzi et al., 1999). The disinfection mechanism is based on the disintegration of H_2O_2, which releases free oxygen radicals. It is possible for the free radicals to both oxidize and cleanse. Unlike other chemicals, it doesn't release any odors or remnants.

Biological Treatment (Bioremediation)

Strips of Vegetated Filter

The idea behind the filter strips is to have naturally occurring or artificially grown vegetation covering land regions between a potential pollution source and a surface water body that receives runoff. Often referred to as grassed filter strips, filter strips, or grassed filters, vegetated filter strips aim to treat sheet flow from adjacent surfaces (Delgado et al., 1995).

Carbon Filtration with Biological Activity

Using biologically active carbon as a bioremediation technique is another possible step. The process uses GAC to filter the water. Particles can create a "biofilm" when microorganisms (bacteria) colonize the GAC medium (Scholz & Martin, 1997). The microorganisms involved are either immobilized on the surface of the GAC (substratum) or integrated into an external microbial organic polymer matrix (Rattanapan et al., 2010). By releasing extracellular polymeric substances, bacteria, and fungal cells in biofilms form a cohesive, stable matrix that retains cells in dense aggregation. The constituents of the exterior matrix include lipids, proteins, nucleic acids, and polysaccharides (Branda et al., 2005).

Phytoremediation

Currently, total removal of water pollutants is not always possible using standard treatment methods. The phytoremediation method is a subset of bioremediation that uses plants to treat wastewater. It refers to the practice of using plants to clean up environmental toxins. One of the most dangerous contaminants in water are PTEs (Egbueri, 2020). According to Rai (2009), the technique consists of multiple steps: plant-mediated extraction, filtration, stabilization, and transformation/degradation. Metal accumulation occurs during phytoextraction in the parts of the roots and harvestable shoots. Rhizofiltration addresses the absorption, concentration, and precipitation of hazardous metals from polluted effluents. Phytotransformation/phytodegradation, on the other hand, refers to the process of removing contaminants by phytodegradation or phytotransformation by plant-derived enzymes or enzyme co-factors. Phytostabilization, on the other hand, calls for the use of tolerant plants to lessen the mobility of PTEs. For the treatment of home, commercial, and agricultural wastewater, several aquatic plants have been employed extensively. These plants have a wide range of uses because they are readily available, tolerant of hazardous environments, capable of bioaccumulation, invasive, and biomass potentials (Hu et al., 2020; Mustafa & Hayder, 2021). The removal and degradation of numerous harmful contaminants, including antibiotics, PTEs, landfill leachate, textile dyes, pesticides, hormones, petroleum, explosives, and even toxic gases, has been extensively studied and is now thought to be possible using phytoremediation techniques (Markou et al., 2018; Pilon-Smits, 2005). Since there is a lack of financing for environmental governance and it is cost-effective, phytoremediation is becoming more and more popular among businesses and government organizations (Vymazal, 2011). Solar energy powers the phytoremediation process, which may create biological landscapes for restored regions and has aesthetic value while removing toxins (Markou et al., 2018). It offers an environmentally friendly way to recover nutrients and remove harmful pollutants from livestock wastewater. According to Jabeen et al. (2009), the plants used in phytoremediation should grow more quickly, produce more biomass, have extensive fiber roots, be easy to control, have a high tolerance for contaminants, and be simple to manage and harvest. The plants should ideally be native to the area in order to prevent biological invasion due to poor management. Additionally, compared to conventional biological and physical–chemical approaches, phytoremediation uses less manpower and equipment because it mostly depends on natural processes. Plants have an unmatched ability for detoxification because of their exceptional metabolic and extractive capacities (Pilon-Smits, 2005). While this is happening, many bacteria find homes in the leaves, roots, and stems of plants. These microbes can speed up the treatment process by synchronously decomposing toxins (Pilon-Smits, 2005). The phytoremediation of livestock wastewater has received some attention during the past 20 years (Cheng & Stomp, 2009; Hu et al., 2019; Sooknah & Wilkie, 2004). Through the recovery of nutrients, the elimination of hazardous contaminants, and the reuse of wastewater, the method establishes a sustainable circulation system that fully utilizes the residual value of livestock wastewater. For the management of livestock wastewater, a variety of hydrophytes of various species have been used, including *Spirodela polyrhiza, Lemna minor, Polygonum hydropiper, Eichhornia crassipes, Pistia stratiotes* (Mishra & Tripathi, 2009; Kumari & Tripathi, 2015; Ekperusi et al., 2019; Hu et al., 2019; Sudiarto et al., 2019).

The authors are also recommending to the appropriate governing authorities a plan for wastewater treatment in the coal mining regions of RCF. The plan calls for collecting wastewater from the mining township and the adjoining areas into abandoned PLs, purifying those using cost-effective methods, utilizing the PLs as a sustainable source of fresh water, and reusing the fresh water for the parishioners (Fig. 17.7).

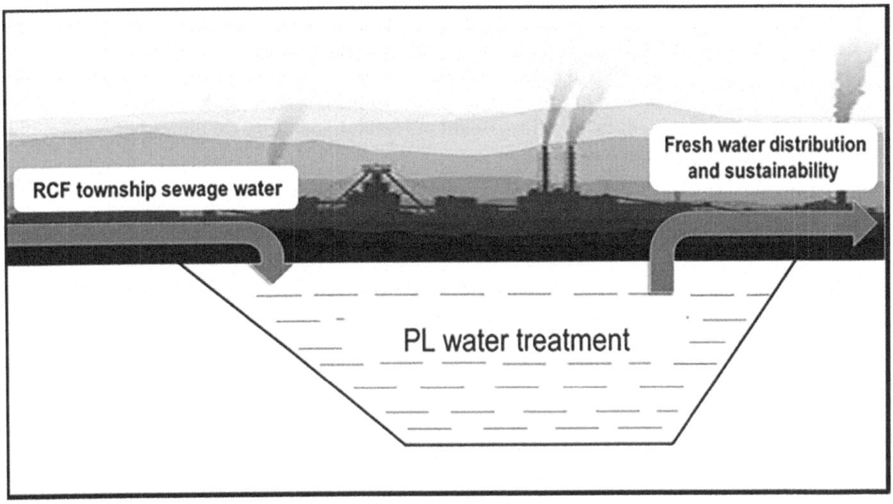

Fig. 17.7 Proposed plan for strategies of wastewater management and sustainability in RCF, West Bengal, India

17.4 Conclusion

Water quality monitoring is a major yardstick to satisfy the demand and supply of water for different anthropogenic and other purposes. The provision of potable water for community needs is a prerequisite for sustainable environmental management. Water quality indicators provide an index of water quality based on physicochemical parameters by integrating several factors into a single, comprehensible result. The present study stated the potable water quality and hydrogeochemistry assessment of different types of water in the RCF area, West Bengal, India. The potable water of the RCF area is alkaline in nature and the hydrogeochemistry assessment indicates that underground, mines and abandoned PL water are strongly influenced by rock weathering, and polluted mine drainage water during mining activities. The calculated WQI shows that in Ratibati abandoned PL, New Ghusik Colliery, Kalipahari, and Sanctoria 5 No. pump house water is unfit for drinking, and needs treatment before any utilization and that water is more unsuitable for use in the monsoon season especially. Pressure Filter Plant, Nimcha Colliery, Boring water, Raniganj town water are very poor quality, only can be applied for irrigation purposes but well water, Raniganj town, tap water from the pumping station, Satgram, well water from Chappai Sewra village, and Kalla Central Hospital, Kalla are all are excellent to permissible quality and suitable for drinking, irrigation, and industrial purposes. The above findings suggest a detailed hydrogeochemical investigation of the water resources at the definite interval, evolve suitable treatment process for mines water and preparation of sustainable water management strategies from the competent authority for constructive utilization of water for proper agricultural, industrial practices and nursing the human health of the RCF area.

Acknowledgements The author is thankful to the Head of the Department of Botany, The University of Burdwan, West Bengal, India for Laboratory facilities.

The author acknowledges Mr. Uday Das, Independent Researcher, Asansol, West Bengal, India for his kind cooperation rendered throughout the period of the research work.

The author also acknowledges the Mines Board of Health Laboratory for their kind cooperation during the period of the research work.

Conflict of Interest There has been no conflict of interest stated by the authors.

Declaration The writers employed Quillbot and Grammarly to enhance the work's readability and language during its production. Following their use of these resources, the writers evaluated and made any necessary edits to the text and undertook full responsibility for the publication's content.

References

Abdul, K. S. M., Jayasinghe, S. S., Chandana, E. P., Jayasumana, C., & De Silva, P. M. C. (2015). Arsenic and human health effects: A review. *Environmental Toxicology and Pharmacology, 40*(3), 828–846.

Acharya, B. S., & Kharel, G. (2020). Acid mine drainage from coal mining in the United States–An overview. *Journal of Hydrology, 588,* 125061. https://doi.org/10.1016/j.jhydrol.2020.125061

Adeloju, S. B., Khan, S., & Patti, A. F. (2021). Arsenic contamination of groundwater and its implications for drinking water quality and human health in under-developed countries and remote communities—A review. *Applied Sciences, 11*(4), 1926.

Akhtar, N., Syakir Ishak, M. I., Bhawani, S. A., & Umar, K. (2021). Various natural and anthropogenic factors responsible for water quality degradation: A review. *Water, 13*(19), 2660.

Akram, S., & Rehman, F. (2018). Hardness in drinking-water, its sources, its effects on humans and its household treatment. *Journal of Applied Chemistry, 4*(1), 1–4.

Ali, I., & Gupta, V. K. (2006). Advances in water treatment by adsorption technology. *Nature Protocols, 1*(6), 2661–2667. https://doi.org/10.1038/nprot.2006.370

Ali, S., Abbas, Z., Rizwan, M., Zaheer, I. E., Yavaş, İ., Ünay, A., Abdel-Daim, M. M., Bin-Jumah, M., Hasanuzzaman, M., & Kalderis, D. (2020). Application of floating aquatic plants in phytoremediation of heavy metals polluted water: A review. *Sustainability, 12*(5), 1927. https://doi.org/10.3390/su12051927

Andra, S. S., Charisiadis, P., Arora, M., van Vliet-Ostaptchouk, J. V., & Makris, K. C. (2015).

Andreozzi, R., Caprio, V., Insola, A. & Marotta, R., (1999). Advanced oxidation processes (AOP) for water purification and recovery. *Catalysis Today, 53*(1), 51–59. https://doi.org/10.1016/S0920-5861(99)00102-9

Anku, W. W., Mamo, M. A., & Govender, P. P. (2017). Phenolic compounds in water: Sources, reactivity, toxicity and treatment methods. *Phenolic Compounds-Natural Sources, Importance and Applications*, pp. 419–443.

Asrafuzzaman, M., Fakhruddin, A. N. M., & Hossain, M. A. (2011). Reduction of turbidity of water using locally available natural coagulants. *International Scholarly Research Notices, 2011*, 1–6. https://doi.org/10.5402/2011/632189

Banerji, S., & Mitra, D. (2019). Geographical information system-based groundwater quality index assessment of northern part of Kolkata, India for drinking purpose. *Geocarto International, 34*(9), 943–958. https://doi.org/10.1080/10106049.2018.1451922

Bano, H., Rather, R. A., Malik, S., Bhat, M. A., Khan, A. H., Américo-Pinheiro, J. H. P., & Mir, I. A. (2022). Effect of seasonal variation on pollution load of water of Hokersar wetland: A case study of Queen Wetland of Kashmir, J&K, India. *Water, Air, & Soil Pollution, 233*(12), 518.

Bergmann, H., Iourtchouk, T., Schöps, K., & Bouzek, K. (2002). New UV irradiation and direct electrolysis—Promising methods for water disinfection. *Chemical Engineering Journal, 85*(2–3), 111–117. https://doi.org/10.1016/S1385-8947(01)00188-7

Bhattacharya, A., & Ghosh, P. (2004). Nanofiltration and reverse osmosis membranes: theory and application in separation of electrolytes. *Reviews in Chemical Engineering, 20*(1–2), 111–173. https://doi.org/10.1515/REVCE.2004.20.1-2.111

Bhattacharya, A. (2006). Remediation of pesticide-polluted waters through membranes. *Separation and Purification Reviews, 35*(1), 1–38. https://doi.org/10.1080/15422110500536151

Biomonitoring of human exposures to chlorinated derivatives and structural analogs of bisphenol A. *Environment international, 85*, 352–379.

Boyd, C. E., Tucker, C. S., & Somridhivej, B. (2016). Alkalinity and hardness: Critical but elusive concepts in aquaculture. *Journal of the World Aquaculture Society, 47*(1), 6–41. https://doi.org/10.1111/jwas.12241

Bozorg-Haddad, O., Delpasand, M., & Loáiciga, H. A. (2021). Water quality, hygiene, and health. In *Economical, Political, and Social Issues in Water Resources* (pp. 217–257). Elsevier.

Branda, S. S., Vik, Å., Friedman, L., & Kolter, R. (2005). Biofilms: The matrix revisited. *Trends in Microbiology, 13*(1), 20–26. https://doi.org/10.1016/j.tim.2004.11.006

Briffa, J. (2020). *Heavy metals in Maltese agricultural soil* (Master's thesis, University of Malta).

Brown, R. M., McClelland, N. I., Deininger, R. A., & O'Connor, M. F. (1972). A water quality index—Crashing the psychological barrier. In *Indicators of Environmental Quality: Proceedings of a symposium held during the AAAS meeting in Philadelphia, Pennsylvania, December 26–31, 1971* (pp. 173–182). Springer US. https://doi.org/10.1007/978-1-4684-2856-8_15

Chand, R., & Kumar, B. (2017). Oil and lubricant hazard effects on human health. *IJISET-International Journal of Innovative Science, Engineering and Technology, 4*(4), 315–322.

Charcosset, C. (2009). A review of membrane processes and renewable energies for desalination. *Desalination, 245*(1–3), 214–231. https://doi.org/10.1016/j.desal.2008.06.020

Chattopadhyay A. (2001). Bardhaman Jelar Itihas O Lok Sanskriti, *Radical*, I.

Chen, J., Huang, Q., Lin, Y., Fang, Y., Qian, H., Liu, R., & Ma, H. (2019). Hydrogeochemical characteristics and quality assessment of groundwater in an irrigated region Northwest China. *Water, 11*(1), 96. https://doi.org/10.3390/w11010096

Cheng, J. J., & Stomp, A. M. (2009). Growing duckweed to recover nutrients from wastewaters and for production of fuel ethanol and animal feed Clean-Soil. *Air, Water, 37*(1), 17–26.

Chhabra, R., & Chhabra, R. (2021). Irrigation water: Quality criteria. *Salt-affected Soils and Marginal Waters: Global Perspectives and Sustainable Management*, 431–486.

Das, A. (2023). Assessment of potability of surface water and its health implication in Mahanadi Basin, Odisha. Materials Today: Proceedings.

De Roos, A. J., Gurian, P. L., Robinson, L. F., Rai, A., Zakeri, I., & Kondo, M. C. (2017).

De Gisi, S., Lofrano, G., Grassi, M., & Notarnicola, M. (2016). Characteristics and adsorption capacities of low-cost sorbents for wastewater treatment: A review. *Sustainable Materials and Technologies, 9*, 10–40.

de Jesus Gaffney, V., Almeida, C. M., Rodrigues, A., Ferreira, E., Benoliel, M. J., & Cardoso, V. V. (2015). Occurrence of pharmaceuticals in a water supply system and related human health risk assessment. *Water Research, 72*, 199–208.

Delgado, A. N., Periago, E. L., & Viqueira, F. D. F. (1995). Vegetated filter strips for wastewater purification: A review. *Bioresource Technology, 51*(1), 13–22. https://doi.org/10.1016/0960-8524(94)00113-F

Ditsch, A., Lindenmann, S., Laibinis, P. E., Wang, D. I., & Hatton, T. A. (2005). High-gradient magnetic separation of magnetic nanoclusters. *Industrial and Engineering Chemistry Research, 44*(17), 6824–6836. https://doi.org/10.1021/ie048841s

Egbueri, J. C. (2019). Evaluation and characterization of the groundwater quality and hydrogeochemistry of Ogbaru farming district in southeastern Nigeria. *SN Applied Sciences., 1*(8), 1–16. https://doi.org/10.1007/s42452-019-0853-1

Egbueri, J. C. (2020). Heavy metals pollution source identification and probabilistic health risk assessment of shallow groundwater in Onitsha Nigeria. *Analytical Letters, 53*(10), 1620–1638. https://doi.org/10.1080/00032719.2020.1712606

Egbueri, J. C., & Unigwe, C. O. (2020). Understanding the extent of heavy metal pollution in drinking water supplies from Umunya, Nigeria: An indexical and statistical assessment. *Analytical Letters, 53*(13), 2122–2144. https://doi.org/10.1080/00032719.2020.1731521

Egbueri, J. C., Ameh, P. D., Ezugwu, C. K. & Onwuka, O. S. (2020). Evaluating the environmental risk and suitability of hand-dug wells for drinking purposes: A rural case study from Nigeria. *International Journal of Environmental Analytical Chemistry.* 1–21. https://doi.org/10.1080/03067319.2020.1800000

Eikebrokk, B., Juhna, T., & Østerhus, S. W. (2006). Water treatment by enhanced coagulation—Operational status and optimization issues. *Techneau D, 5*(3), 1.

Ekperusi, A. O., Sikoki, F. D., & Nwachukwu, E. O. (2019). Application of common duckweed (Lemna minor) in phytoremediation of chemicals in the environment: State and future perspective. *Chemosphere, 223*, 285–309.

Environmental Protection Agency US. (2000). Wastewater technology sheet: chemical precipitation. United State Environmental Protection, 832-F-00-018

Etikala, B., Madhav, S., and Somagouni, S. G. (2022). Urban water systems: An overview. In *Current directions in water scarcity research* (pp. 1–19). Elsevier. https://doi.org/10.1016/b978-0-323-91838-1.00016-6

Etikala, B., Vangala, S., & Madhav, S. (2024). Groundwater geochemistry using modified integrated water quality index (IWQI) and health indices with special emphasis on nitrates and heavy metals in southern parts of Tirupati, South India. *Environmental Geochemistry and Health, 46*(11). https://doi.org/10.1007/s10653-024-02229-3

Fetyan, N. A., & Salem Attia, T. M. (2020). Water purification using ultrasound waves: Application and challenges. *Arab Journal of Basic and Applied Sciences, 27*(1), 194–207. https://doi.org/10.1080/25765299.2020.1762294

Fields, K.A., Chen, A., Wang, L. (2000). Arsenic removal from drinking water by coagulation/filtration and lime softening plants. EPA/600/R-00/063, USEPA, Cincinnati.

Gautam, R. K., & Chattopadhyaya, M. C. (2016). *Nanomaterials for wastewater remediation.* Butterworth-Heinemann.

Grieco, A. (2021). *Living alkaline, living happy: Restoring the acid-base balance of extracellular matrix for the health of body and mind.* Naturvis Books.

Von Gunten, U. (2003a). Ozonation of drinking water: Part I. Oxidation kinetics and product formation. *Water Research, 37*(7), 1443–1467. https://doi.org/10.1016/S0043-1354(02)00457-8

Von Gunten, U. (2003b). Ozonation of drinking water: Part II. Disinfection and by-product formation in presence of bromide, iodide or chlorine. *Water research, 37*(7), 1469–1487. https://doi.org/10.1016/S0043-1354(02)00458-X

Gupta, S., & Mondal, D. (2013). Fluoride dynamics in some parts of the Raniganj coal field West Bengal. *Integrated with Geospatial Approach, 2*(1), 21–32.

Hering, J. G., Chen, P. Y., Wilkie, J. A., & Elimelech, M. (1997). Arsenic removal from drinking water during coagulation. *Journal of Environmental Engineering, 123*(8), 800–807. https://doi.org/10.1061/(ASCE)0733-9372(1997)123:8(800)

Hijnen, W. A. M., Beerendonk, E. F., & Medema, G. J. (2006). Inactivation credit of UV radiation for viruses, bacteria and protozoan (oo) cysts in water: A review. *Water Research, 40*(1), 3–22. https://doi.org/10.1016/j.watres.2005.10.030

Hoffmann, C., & Franzreb, M. (2004a). A novel repulsive-mode high gradient magnetic separator. Part I. Design and experimental results. *IEEE Transactions on Magnetics, 40*(2), 456–461. https://doi.org/10.1109/TMAG.2004.824124

Hoffmann, C., & Franzreb, M. (2004b). A novel repulsive-mode high-gradient magnetic separator. Part II. Separation model. *IEEE Transactions on Magnetics, 40*(2), 462–468. https://doi.org/10.1109/TMAG.2004.824121.

Hsieh, J. L., Nguyen, T. Q., Matte, T., & Ito, K. (2015). Drinking water turbidity and emergency department visits for gastrointestinal illness in New York City, 2002–2009. *PLoS ONE, 10*(4), e0125071.

Abdel-Gadir, A., Berber, R., Porter, J. B., Quinn, P. D., Suri, D., Kellman, P., Hart, A. J., Moon, J. C., Manisty, C., & Skinner, J. A. (2016). Detection of metallic cobalt and chromium liver deposition following failed hip replacement using T2* and R2 magnetic resonance. *Journal of Cardiovascular Magnetic Resonance, 18*, 1–5.

Hu, H., Zhou, Q., Li, X., Lou, W., Du, C., Teng, Q., Zhang, D., Liu, H., Zhong, Y., & Yang, C. (2019). Phytoremediation of anaerobically digested swine wastewater contaminated by oxytetracycline via Lemna aequinoctialis: Nutrient removal, growth characteristics and degradation pathways. *Bioresource Technology, 291*, 121853.

Hu, H., Li, X., Wu, S., & Yang, C. (2020). Sustainable livestock wastewater treatment via phytoremediation: Current status and future perspectives. *Bioresource Technology, 315*, 123809.

Hunter, W. W. (1886), reprint (2005). The Indian empire: Its people, history, and products. Asian Educational Services.

University of Florida. (1998). Institute of Food and Agricultural Sciences. Trihalomethanes and Our Water Supply.

Izah, S. C., Chakrabarty, N., & Srivastav, A. L. (2016). A review on heavy metal concentration in potable water sources in Nigeria: Human health effects and mitigating measures. *Exposure and Health, 8*, 285–304.

Jabeen, R., Ahmad, A., & Iqbal, M. (2009). Phytoremediation of heavy metals: Physiological and molecular mechanisms. *The Botanical Review, 75*, 339–364.

Jiuhui, Q. U. (2008). Research progress of novel adsorption processes in water purification: A review. *Journal of Environmental Sciences, 20*(1), 1–13. https://doi.org/10.1016/S1001-0742(08)60001-7

Kabir, H., Gupta, A. K., & Tripathy, S. (2020). Fluoride and human health: Systematic appraisal of sources, exposures, metabolism, and toxicity. *Critical Reviews in Environmental Science and Technology, 50*(11), 1116–1193.

Kali, S., Khan, M., Ghaffar, M. S., Rasheed, S., Waseem, A., Iqbal, M. M., Khon Niazi, M. B., & Zafar, M. I. (2021). Occurrence, influencing factors, toxicity, regulations, and abatement approaches for disinfection by-products in chlorinated drinking water: A comprehensive review. *Environmental Pollution, 281*, 116950.

Kanchana, C. M., Chandrasekara, N. K., Weerasinghe, K. D. N., Pathirana, S., & Piyadasa, R. U. (2021). Heavy metal, oil and grease pollution of water and sediments in Estuarine Lagoons in Sri Lanka: A case study in Negombo Estuarine Lagoon. In Multi-Hazard Early Warning and Disaster Risks (pp. 429–442). Springer International Publishing.

Kanduti, D., Sterbenk, P., & Artnik, B. (2016). Fluoride: A review of use and effects on health. *Materia Socio-Medica, 28*(2), 133.

Katyal, D. (2011). Water quality indices used for surface water vulnerability assessment. *International Journal of Environmental Sciences 2*(1): 1–20, https://ssrn.com/abstract=2160726

Kerwick, M., Holt, D., Kerwick, M., Reddy, S., & Chamberlain, A. (2005). A methodology for the evaluation of disinfection technologies. *Journal of Water and Health, 3*(4), 393–404. https://doi.org/10.2166/wh.2005.046

Koter, S., & Warszawski, A. (2000). Electromembrane processes in environment protection. *Polish Journal of Environmental Studies, 9*(1), 45–56.

Kumar, P., & Singh, A. K. (2020). Hydrogeochemistry and quality assessment of surface and subsurface water resources in Raniganj coalfield area, Damodar Valley, India. *International Journal of Environmental Analytical Chemistry*. 1–24. https://doi.org/10.1080/03067319.2020.1849653

Kumar, T., Majid, M. A. A., Onichandran, S., Jaturas, N., Andiappan, H., Salibay, C. C., AL Tabo, H., Tabo, N., Dungca, J.Z., Tangpong, J. & Nissapatorn, V. (2016). Presence of Cryptosporidium parvum and Giardia lamblia in water samples from Southeast Asia: Towards an integrated water detection system. *Infectious diseases of poverty, 5*, 1–12. https://doi.org/10.1186/s40249-016-0095-z

Kumari, M., & Tripathi, B. D. (2015). Efficiency of Phragmites australis and Typha latifolia for heavy metal removal from wastewater. *Ecotoxicology and Environmental Safety, 112*, 80–86.

Lawrence, K. W., David, A. V., Yan, L., & Nazih, K. S. (2005). Chemical precipitation. In K. W. Lawrence, Y. T. Hung, & K. S. Nazih (Eds.), *Hand book of environmental engineering, physiochemical treatment processes* (Vol. 3, pp. 141–197). The Humana Press Inc.

Li, X. F., & Mitch, W. A. (2018). Drinking water disinfection byproducts (DBPs) and human health effects: Multidisciplinary challenges and opportunities. *Environmental Science & Technology, 52*(4), 1681–1689.

Madanhire, I., Mbohwa, C., Madanhire, I., & Mbohwa, C. (2016). Lubricant additive impacts on human health and the environment. *Mitigating Environmental Impact of Petroleum Lubricants*, 17–34.

Mahato, M. K., Singh, P. K., Tiwari, A. K., & Singh, A. K. (2016). Risk assessment due to intake of metals in groundwater of East Bokaro Coalfield, Jharkhand India. *Exposure and Health, 8*(2), 265–275. https://doi.org/10.1007/s12403-016-0201-2

Manna, A., & Maiti, R. (2016). Alteration of surface water hydrology by opencast mining in the Raniganj Coalfield Area India. *Mine Water and the Environment, 35*(2), 156–167. https://doi.org/10.1007/s10230-015-0342-8

Markou, G., Wang, L., Ye, J., & Unc, A. (2018). Using agro-industrial wastes for the cultivation of microalgae and duckweeds: Contamination risks and biomass safety concerns. *Biotechnology Advances, 36*(4), 1238–1254.

Masto, R. E., Sheik, S., Nehru, G., Selvi, V. A., George, J., & Ram, L. C. (2015). Assessment of environmental soil quality around Sonepur Bazari mine of Raniganj coalfield India. *Solid Earth, 6*(3), 811–821. https://doi.org/10.5194/se-6-811-2015

Matlock, M. M., Howerton, B. S., & Atwood, D. A. (2002). Chemical precipitation of heavy metals from acid mine drainage. *Water Research, 36*(19), 4757–4764. https://doi.org/10.1016/S0043-1354(02)00149-5

Mendhe, V. A., Mishra, S., Varma, A. K., & Singh, A. P. (2017). Coal bed methane-produced water quality and its management options in Raniganj Basin, West Bengal India. *Applied Water Science, 7*(3), 1359–1367. https://doi.org/10.1007/s13201-015-0326-7

Mgbenu, C. N., & Egbueri, J. C. (2019). The hydrogeochemical signatures, quality indices and health risk assessment of water resources in Umunya district, southeast Nigeria. *Applied Water Science, 9*(1), 1–19. https://doi.org/10.1007/s13201-019-0900-5

Mishra, V. K., & Tripathi, B. D.(2009). Accumulation of chromium and zinc from aqueous solutions using water hyacinth (Eichhornia crassipes). *Journal of Hazardous Materials, 164*(2–3), 1059–1063.

Mok, W. M. (1994). Mobilization of arsenic in contaminated river waters & Removal of arsenic in drinking water treatment. *Arsenic in the environment. Part 1: cycling and characterization*, 99–132.

Mondal, S., & Palit, D. (2019). Evaluation of water quality using water quality index of Pit lakes, Raniganj coalfield area, West Bengal, India. *Research Journal of Life Sciences, Bioinformatics, Pharmaceutical and Chemical Sciences, 5*(3), 532–550. https://doi.org/10.26479/2019.0503.44

Mustafa, H. M., & Hayder, G. (2021). Recent studies on applications of aquatic weed plants in phytoremediation of wastewater: A review article. *Ain Shams Engineering Journal, 12*(1), 355–365.

Nasrabadi, T., Ruegner, H., Sirdari, Z. Z., Schwientek, M., & Grathwohl, P. (2016). Using total suspended solids (TSS) and turbidity as proxies for evaluation of metal transport in river water. *Applied Geochemistry, 68*, 1–9.

Nataraj, S. K., Hosamani, K. M., & Aminabhavi, T. M. (2009). Nanofiltration and reverse osmosis thin film composite membrane module for the removal of dye and salts from the simulated mixtures. *Desalination, 249*(1), 12–17. https://doi.org/10.1016/j.desal.2009.06.008

Navalon, S., Alvaro, M., Garcia, H., Escrig, D., & Costa, V. (2009). Photocatalytic water disinfection of Cryptosporidium parvum and Giardia lamblia using a fibrous ceramic TiO_2 photocatalyst. *Water Science and Technology, 59*(4), 639–645. https://doi.org/10.2166/wst.2009.016

Nazir, M., Roy, K., Saha, A., & Saha, D. (2024). A novel approach to explore new means of depletion of potable water crisis by phytoremediation of Abandoned Coalmine Pitlake and generate alternate livelihood: A case study of RCF, West Bengal, India. In *Hydrology-Current Research and Future Directions*. IntechOpen.

Ndur, S. A., Nyarko, S. Y., Quaicoe, I., & Osei, L. B. (2020). Heavy Metal Loading in Surface Sediments along the Kawere Stream, Tarkwa Ghana. *Ghana Mining Journal, 20*(2), 77–85.

Nong, X., Shao, D., Xiao, Y., & Zhong, H. (2019). Spatio-temporal characterization analysis and water quality assessment of the south-to-north water diversion project of China. *International Journal of Environmental Research and Public Health, 16*(12), 1–23. https://doi.org/10.3390/ijerph16122227

Okada, H., Mitsuhashi, K., Ohara, T., Whitby, E. R., & Wada, H. (2005). Computational fluid dynamics simulation of high gradient magnetic separation. *Separation Science and Technology, 40*(7), 1567–1584. https://doi.org/10.1081/SS-200056063

Pal, S. K., & Vaish, J. (2020). Coal fire study over east basuria colliery. In *Advances in modeling and interpretation in near surface geophysics* (pp. 295–334). Springer. https://doi.org/10.1007/978-3-030-28909-6_11

Palit, D., Mukherjee, A., Gupta, S., & Kar, D. (2014). Water quality in the Pit lakes of Raniganj coal field, West Bengal, India. *Journal of Applied Sciences in Environmental Sanitation., 9*(1), 1–6.

Park, S. K., & Hu, J. Y. (2010). Assessment of the extent of bacterial growth in reverse osmosis system for improving drinking water quality. *Journal of Environmental Science and Health Part A, 45*(8), 968–977. https://doi.org/10.1080/10934521003772386

Pawlak, Z., Żak, S., & Zabłocki, L. (2006). Removal of Hazardous Metals from Groundwater by Reverse Osmosis. *Polish Journal of Environmental Studies, 15*(4), 579–583.

Peterson, J. C. K. (1997). *The Raniganj coal field* (pp. 165–174). Bengal District Gazetteers, Burdwan.

Pilon-Smits, E. (2005). Phytoremediation. *Annual Review of Plant Biology, 56*, 15–39.

Pohanka, M. (2019). Copper and copper nanoparticles toxicity and their impact on basic functions in the body. *Bratislavske Lekarske Listy, 120*(6), 397–409.

Prasad, S., Yadav, K. K., Kumar, S., Gupta, N., Cabral-Pinto, M. M., Rezania, S., Radwan, N., & Alam, J. (2021). Chromium contamination and effect on environmental health and its remediation: A sustainable approaches. *Journal of Environmental Management, 285*, 112174.

Pulster, E. L., & Hillman, J. V. (2015). Cyanide. *Hamilton & Hardy's Industrial Toxicology*, 331–340.

Punyani, S., Narayana, P., Singh, H., & Vasudevan, P. (2006). Iodine based water disinfection: A review.

Quade, B. N., Parker, M. D., & Occhipinti, R. (2021). The therapeutic importance of acid-base balance. *Biochemical Pharmacology, 183*, 114278.

Rahman, M. M., Rahaman, M. S., Islam, M. R., Rahman, F., Mithi, F. M., Alqahtani, T., Almikhlafi, M. A., Alghamdi, S.Q., & Uddin, M. S. (2021). Role of phenolic compounds in human disease: Current knowledge and future prospects. *Molecules, 27*(1), 233.

Rai, P. K. (2009). Heavy metal phytoremediation from aquatic ecosystems with special reference to macrophytes. *Critical Reviews in Environmental Science and Technology, 39*(9), 697–753. https://doi.org/10.1080/10643380801910058

Ramos, R. L., Moreira, V. R., Lebron, Y. A., Santos, A. V., Santos, L. V., & Amaral, M. C. (2021). Phenolic compounds seasonal occurrence and risk assessment in surface and treated waters in Minas Gerais—Brazil. *Environmental Pollution, 268*, 115782.

Rattanapan, C., Kantachote, D., Yan, R., & Boonsawang, P. (2010). Hydrogen sulfide removal using granular activated carbon biofiltration inoculated with Alcaligenes faecalis T307 isolated from concentrated latex wastewater. *International Biodeterioration and Biodegradation, 64*(5), 383–387. https://doi.org/10.1016/j.ibiod.2010.04.005

Ray, S., & Dey, K. (2020). Coal mine water drainage: The current status and challenges. *Journal of the Institution of Engineers (India)., 101*, 165–172. https://doi.org/10.1007/s40033-020-00222-5

Reddy, Y. S., & Sunitha, V. (2023). Assessment of Heavy metal pollution and its health implications in groundwater for drinking purpose around inactive mines, SW region of Cuddapah Basin, South India. *Total Environment Research Themes, 8*, 100069. https://doi.org/10.1016/j.totert.2023.100069

Roos, A. D., Gurian, P. L., Robinson, L. F., Rai, A., Zakeri, I., & Kondo, M. C. (2017). Review of epidemiological studies of drinking-water turbidity in relation to acute gastrointestinal illness.

Scholz, M., & Martin, R. J. (1997). Ecological equilibrium on biological activated carbon. *Water Research, 31*(12), 2959–2968.

Sharma, S., & Bhattacharya, A. J. A. W. S. (2017). Drinking water contamination and treatment techniques. *Applied Water Science, 7*(3), 1043–1067. https://doi.org/10.1007/s13201-016-0455-7

Sheoran, A. S., & Sheoran, V. (2006). Heavy metal removal mechanism of acid mine drainage in wetlands: a critical review. *Minerals Engineering, 19*(2), 105–116.

Sibanda, T., Chigor, V. N., Koba, S., Obi, C. L., & Okoh, A. I. (2014). Characterisation of the physicochemical qualities of a typical rural-based river: Ecological and public health implications. *International Journal of Environmental Science and Technology, 11*, 1771–1780.

Simate, G. S., & Ndlovu, S. (2014). Acid mine drainage: Challenges and opportunities. *Journal of Environmental Chemical Engineering, 2*(3), 1785–1803. https://doi.org/10.1016/j.jece.2014.07.021

Singh, A. K., Mahato, M. K., Neogi, B., & Singh, K. K. (2010). Quality assessment of mine water in the Raniganj coalfield area India. *Mine Water and the Environment, 29*(4), 248–262. https://doi.org/10.1007/s10230-010-0108-2

Singh, P. K., Tiwari, A. K., & Mahato, M. K. (2013). Qualitative assessment of surface water of West Bokaro Coalfield, Jharkhand by using water quality index method. *International Journal of ChemTech Research, 5*(5), 2351–2356.

Solanki, H. A. (2012). *Status of soils and water reservoirs near industrial areas of Baroda: Pollution and soil - water chemistry*. Lap Lambert Academic Publishing.

Sooknah, R. D., & Wilkie, A. C. (2004). Nutrient removal by floating aquatic macrophytes cultured in anaerobically digested flushed dairy manure wastewater. *Ecological Engineering, 22*(1), 27–42.

Strathmann, H. (2010). Electrodialysis, a mature technology with a multitude of new applications. *Desalination, 264*(3), 268–288. https://doi.org/10.1016/j.desal.2010.04.069

Strathmann, H. (2010). Ion-exchange membrane processes in water treatment. *Sustainability Science and Engineering, 2*, 141–199. https://doi.org/10.1016/S1871-2711(09)00206-2

Sudiarto, S. I. A., Renggaman, A., & Choi, H. L. (2019). Floating aquatic plants for total nitrogen and phosphorus removal from treated swine wastewater and their biomass characteristics. *Journal of Environmental Management, 231*, 763–769.

Sunitha, V., & Reddy, B. M. (2022). Geochemical characterization, deciphering groundwater quality using pollution index of groundwater (PIG), water quality index (WQI) and GIS in hard rock aquifer South India. *Applied Water Sciences, 12*, 100069. https://doi.org/10.1007/s13201-021-01527-w

Suzuki, Y., Endo, M., Jin, J., Iwase, K., & Iwatsuki, M. (2006). Tristimulus colorimetry using a digital still camera and its application to determination of iron and residual chlorine in water samples. *Analytical Sciences., 22*(3), 411–414. https://doi.org/10.2116/analsci.22.411

Tiwari, A. K., Singh, P. K., & Mahato, M. K. (2016). Environmental geochemistry and a quality assessment of mine water of the West Bokaro coalfield. *India. Mine Water and the Environment., 35*(4), 525–535. https://doi.org/10.1007/s10230-015-0382-0

Tiwari, A. K., Orioli, S., & De Maio, M. (2019). Assessment of groundwater geochemistry and diffusion of hexavalent chromium contamination in an industrial town of Italy. *Journal of Contaminant Hydrology., 225*, 103503. https://doi.org/10.1016/j.jconhyd.2019.103503

Uddin, M. G., Nash, S., & Olbert, A. I. (2021). A review of water quality index models and their use for assessing surface water quality. *Ecological Indicators, 122*, 107218.

Veil, J. (2008). Thermal distillation technology for management of produced water and frac flowback water. Water Tech Brief #2008–1

Velusamy, S., Roy, A., Sundaram, S., & Kumar Mallick, T. (2021). A review on heavy metal ions and containing dyes removal through graphene oxide-based adsorption strategies for textile wastewater treatment. *The Chemical Record, 21*(7), 1570–1610.

Vinodhini, R., & Narayanan, M. (2008). Bioaccumulation of heavy metals in organs of fresh water fish Cyprinus Carpio (Common carp). *International Journal of Environmental Science and Technology, 5*, 179–182.

Vymazal, J. (2011). Constructed wetlands for wastewater treatment: Five decades of experience. *Environmental Science & Technology, 45*(1), 61–69.

Ward, M. H., Jones, R. R., Brender, J. D., De Kok, T. M., Weyer, P. J., Nolan, B. T., Villanueve, C. M., & Van Breda, S. G. (2018). Drinking water nitrate and human health: An updated review. *International Journal of Environmental Research and Public Health, 15*(7), 1557.

Woo, C. H. (2018). Desalination and Water Treatment. *Intech Open, 19*(2), 201–220.

Xie, W., Wang, Q., Ma, H., Ohsumi, Y., & Ogawa, H. I. (2005). Study on phosphorus removal using a coagulation system. *Process Biochemistry, 40*(8), 2623–2627. https://doi.org/10.1016/j.procbio.2004.06.056

Xu, T. (2005). Ion exchange membranes: State of their development and perspective. *Journal of Membrane Science, 263*(1–2), 1–29. https://doi.org/10.1016/j.memsci.2005.05.002

Xu, J., Cao, Z., Zhang, Y., Yuan, Z., Lou, Z., Xu, X., & Wang, X. (2018). A review of functionalized carbon nanotubes and graphene for heavy metal adsorption from water: Preparation, application, and mechanism. *Chemosphere, 195*, 351–364.

Yang, R. T. (1997) *Gas separation by adsorption processes* (Vol. 1). World Scientific.

Yogesh, Popat, K. M., Ganguly, B., Brahmbhatt, H., & Bhattacharya, A. (2008). Studies on the separation performances of chlorophenol compounds from water by thin film composite membranes. *Macromolecular Research, 16*, 590–595. https://doi.org/10.1007/BF03218565

Zammit, I., Marano, R. B., Vaiano, V., Cytryn, E., & Rizzo, L. (2020). Changes in antibiotic resistance gene levels in soil after irrigation with treated wastewater: A comparison between heterogeneous photocatalysis and chlorination. *Environmental Science and Technology, 54*(12), 7677–7686.

Zotou, I., Tsihrintzis, V. A., & Gikas, G. D. (2020). Water quality evaluation of a lacustrine water body in the Mediterranean based on different water quality index (WQI) methodologies. *Journal of Environmental Science and Health, Part A, 55*(5), 537–548. https://doi.org/10.1080/10934529.2019.1710956

Chapter 18
Evaluating Groundwater Pollution Risk Vulnerability and Mitigation Techniques in a Changing Climate

K. Nagamani, Mohammad Suhail Meer, and Badapalli Pradeep Kumar

Abstract Climate change and anthropogenic activities are significantly impacting groundwater resources, resulting in depletion and posing challenges to water sustainability globally. These changes influence aquifers through altered groundwater recharge patterns, shifts in soil biogeochemical processes, and modifications in land use practices, both direct and indirect. This study addresses a critical gap in current literature by thoroughly investigating the impacts of climate change on groundwater quality and stability. The primary objective of this research is to comprehensively understand hydrogeological systems' vulnerability to climate change across multiple scales. It seeks to bridge existing knowledge gaps by exploring the complex interactions between climate change events and dynamics of groundwater contamination. Through a meticulous review and synthesis of relevant literature on potential hazards associated with groundwater pollution, this study aims to provide robust assessments of climate change impacts on groundwater quality. Furthermore, the study examines how changes in land use patterns driven by climate change may exacerbate groundwater pollution issues. By conducting a detailed analysis of pertinent scholarly publications and offering scientifically informed policy recommendations, this research contributes to enhancing our understanding and management of water quality challenges in a changing climate.

Keywords Climate change · Groundwater · Hydro geological processes · Land use/land cover

K. Nagamani (✉) · M. S. Meer
Sathyabama Institute of Science and Technology, Chennai, India
e-mail: nagamani@sathyabama.ac.in

B. Pradeep Kumar
Geology Division, CSIR-National Geophysical Research Institute, Hyderabad, Telangana, India

V. Sunitha et al. (eds.), *Groundwater Resource Management Planning Strategies*, Springer Water, https://doi.org/10.1007/978-3-031-88870-0_18

18.1 Introduction

Groundwater, the planet's second-largest freshwater reserve, fulfills about 36% of the global demand for potable water (Doll et al., 2012). It serves as a primary water source for approximately 74 million people, encompassing over 40% of the United States population, especially in rural areas (National Research Council, 2000). The critical role of groundwater in drinking water supply has been underscored repeatedly (Hall et al., 2008), particularly in light of increasing population and climate variability. Recent empirical data highlight growing concerns over groundwater contamination, particularly from nitrates and pesticides used in agriculture (Karkouti et al., 2005; Spalding & Exner, 1993). These contaminants pose significant health risks and ecological impacts (Butscher & Huggenberger, 2009). Climate change, characterized by global warming and altered precipitation patterns, is expected to exacerbate these challenges (Jimenez Cisneros et al., 2014). For instance, a 2.5 °C temperature increase could reduce groundwater recharge by more than 20% in the U.S. High Plains Aquifer (Hall et al., 2008; Scibek et al., 2007).

The increasing frequency of droughts and floods due to climate variability will likely heighten reliance on groundwater for public water supply, emphasizing its role in water security (Butscher & Huggenberger, 2009). However, the current focus on climate change impacts predominantly revolves around surface water systems, with groundwater quality receiving less attention (Rehana & Mujumdar, 2012; Green et al., 2011). The IPCC's Fifth Assessment Report underscores the need for more research on climate change's impact on groundwater quality, especially concerning human-environmental systems (Jimenez Cisneros et al., 2014). Factors contributing to this research gap include the complexity of groundwater systems, the indirect nature of climate change effects, and uncertainties in climate projections (Jyrkama & Sykes, 2007; Toews & Allen, 2009). Additionally, long-term observational data on groundwater quality are sparse (Fogg & LaBolle, 2006).

This study investigates climate change's potential influence on groundwater contamination risk from diffusive sources like agricultural pesticides. While coastal areas currently contend with increased groundwater salinity due to saltwater intrusion, this analysis does not specifically address aquifer vulnerability to rising sea levels (Ferguson & Gleeson, 2012).

The significance of sustainable groundwater management, as explored by Alley and Leake (2004), underscores the need to shift from the traditional concept of "safe yield" to a more sustainable approach, especially in the context of climate change. This transition ensures the long-term viability of groundwater resources, providing valuable insights for policymakers and water resource managers grappling with complex governance challenges. Sophocleous (2000) contributed a case study from Kansas, illustrating the practical adoption of sustainable groundwater management amid increasing water demand and environmental changes, offering a model for other regions. In southern India, Sivakumar et al. (2012) demonstrated the effective use of the DRASTIC model and GIS to assess groundwater susceptibility to pollution. This technology aids environmental scientists and policymakers in pinpointing vulnerable areas and implementing protective measures to maintain groundwater integrity.

Sophocleous (2013) further critiques the concept of "safe yield," advocating for sustainable water management practices that account for environmental longevity and climate variability. This perspective is crucial for water resource planners navigating dynamic environmental conditions and adapting to climate change impacts.

Foster and Chilton (2003) highlighted the global significance of aquifer degradation, emphasizing the need to understand and mitigate factors contributing to groundwater depletion worldwide. Their research stresses comprehensive measures to protect and sustain global groundwater resources, involving collaboration among hydrogeologists, environmental scientists, and policymakers.

Xu and Ma's (2017) research in China provides insights into groundwater management in a densely populated and water-stressed nation, contributing to global efforts in understanding and addressing water security challenges. These studies collectively underscore the importance of adopting sustainable groundwater management practices worldwide, integrating advanced technologies and interdisciplinary approaches to safeguard groundwater resources amidst evolving environmental and climatic conditions.

Sivakumar et al. (2012) demonstrated the effective use of the DRASTIC model and Geographic Information System (GIS) in southern India to assess groundwater vulnerability to pollution. Their study highlights the potential benefits and challenges of integrating advanced technologies into groundwater management practices, enhancing strategies with data-driven procedures and analysis.

Das and Sarma (2015) similarly utilized the DRASTIC model and GIS to assess groundwater vulnerability, focusing on practical approaches for evaluating contamination risks. This research aims to support professionals in environmental science, hydrogeology, and policymaking by optimizing risk assessment procedures crucial for maintaining water quality and public health.

The concept of "virtual water," introduced by Allan in 1997, offers a theoretical framework for quantifying water consumption embedded in the production and trade of goods and services. Allan's work proposes that integrating water resources into market commodities could potentially alleviate water scarcity issues, particularly in regions like the Middle East. This innovative approach to water resource management provides valuable insights for policymakers, economists, and practitioners grappling with the complexities of water scarcity in arid regions.

18.1.1 How Climate Change Affects Groundwater Quality

Given the ongoing consequences of climate change, groundwater is becoming increasingly crucial as a reliable freshwater resource in safeguarding the water security of our civilization. Figure 18.1 illustrates the effect of climate change on water quality and quantity (source: internet). Groundwater is becoming increasingly important as a solution to the growing problems with surface water systems brought on by

Fig. 18.1 Illustrates the impact of climate change on both the quantity and quality of water. (*Source* http://surl.li/nwqvpn: Research in Asia and Frica Regions., written by Aditi Murti, published on 11th Dec 2019)

climate change. Groundwater, typically found deep below the Earth, is often thought to be protected from the direct effects of climate change, in contrast to surface water, which shows obvious signs of this phenomenon. Nonetheless, numerous studies (Kumar et al., 2023; Stuart et al., 2011; Taylor et al., 2013) show that groundwater supplies remain vulnerable to the long-term effects of climate change through various channels.

The potential for groundwater pollution due to a confluence of physical and geochemical processes is a major cause for alarm. Climate change can significantly affect these processes. The vulnerability of this vital resource to climate-induced changes is further highlighted by the fact that leachate transported from pollutant sources can introduce dangerous compounds into groundwater. As a crucial component of the natural hydrological cycle, groundwater acts as a large subsurface reservoir for water storage. Percolation water from surface runoff, rivers, and lakes enters aquifers via infiltration, while groundwater discharges occur through processes such as plant evapotranspiration, evaporation from shallow groundwater tables, and seepage into streams and lakes. Recharge and outflow channels directly impact groundwater quantity and quality. Furthermore, rising air temperatures due to climate change indirectly affect groundwater's dissolved contents by altering the associated biological and geochemical processes that govern microbial populations and substance concentrations (Pitz, 2016; Riedel, 2019; Dao et al., 2023).

Johnson and Lee (2012) undertook research to investigate the dynamics of water quality in Lake Y, with an emphasis on nutrient contamination. They observed a constant decrease in water clarity over time, which coincided with an increase in nutrient concentrations. These findings indicate a shift towards eutrophication, which promotes the growth of toxic algal blooms. As a result, the implementation of preventive measures is recommended to alleviate nutrient contamination and maintain Lake Y's biological balance.

Chen et al. (2018) conducted a retrospective analysis of groundwater depletion and consequent land subsidence in Region Z. Their findings revealed a concerning trend of declining water tables, mostly due to the unsustainable exploitation of groundwater resources. This depletion has caused ground subsidence, posing significant hazards to both infrastructure and ecosystems in the area. The study underscores the imperative for adopting sustainable groundwater management strategies to alleviate further land subsidence occurrences and uphold water security standards.

18.1.2 The Impacts of Change in Climate on Groundwater Recharge (GW)

Foster (1998) stressed the importance of understanding the recharging process when assessing the susceptibility of aquifers to pollution, as contaminants are carried to these groundwater sources. Bloomfield et al. (2006) investigated how recharge rates and processes influence pollutant discharge from soil and unsaturated zones. According to Scibek and Allen (2006) and Ali et al. (2012), climate change has altered the volume and timing of groundwater recharge. Kløve et al. (2014) discovered that these changes are due to yearly and seasonal groundwater oscillations. Rosenberg et al. (1999) found that increased precipitation rates lead to quicker aquifer recharge. Okkonen et al. (2010) discovered that higher winter temperatures can boost both recharging and pollution leaching due to reduced soil frost.

Several methods have been tried to simulate how climate change would affect groundwater recharge. Significant contributions were made by Eckhardt and Ulbrich (2003), Chen et al. (2002), Croley and Luukkonen (2003), and Holman (2006). Jyrkama and Sykes (2007) stressed the use of downscaled global circulation model (GCM) projections and accurate groundwater recharge estimations when assessing groundwater quality in connection to climate change. Rosenberg et al. (1999) used the HUMUS and SWAT hydrological models to assess how climate change would affect the Ogallala Aquifer's water availability and recharging. Predictions and assessments of groundwater recharge loss were made using three General Circulation Models (GCMs). Their findings suggest that the projected precipitation increases from GCMs do not always result in corresponding increases in groundwater recharge. As global temperatures rise, more water is lost via evaporation and transpiration. Eckhardt and Ulbrich (2003) observed that groundwater replenishment had significantly decreased in a low mountain range in central Europe.

There was a 3.5–7% decrease in recharge rates. The Soil and Water Assessment Tool (SWAT) was enhanced and used to assess climate change scenarios. The model predicts a decline in recharge rates as a function of rising temperatures, changes in the ratio of precipitation to snowfall, and predicted evapotranspiration. Scibek and Allen (2006) used the Hydrologic Evaluation of Landfill Performance (HELP) model in combination with statistically downscaled General Circulation Models (GCMs) to investigate how climate change is affecting recharge patterns in a semi-arid region of Canada. The researchers used a transient groundwater flow model to assess the effects of climate change on groundwater levels in an unconfined aquifer. They predicted potential variability in groundwater recharge, estimating fluctuations between 0.1 m and 0.5 m. The sensitivity of recharge to variations in GCM scenarios may be affected by hydrogeological parameters such as the permeability of the aquifer and the soil.

Unfortunately, technical issues limited the breadth of the investigation. No consideration was given to how global warming would alter snowmelt patterns, growing seasons, or groundwater use. Jyrkama and Sykes (2007) predicted that climate change would boost groundwater recharge in Ontario, Canada's Grand River basin. They anticipated the regional spread of pricing behavior using HELP3 and ArcView GIS. Recharge rates were found to differ significantly among the study locations.

It is well-known that climate change affects both the rate at which land is used and reclaimed and how the land itself is used. Variations in vegetation cover and structure affect the hydrologic cycle through mechanisms like interception, infiltration, and transpiration. Le Maitre et al. (1999) discovered that these alterations affected groundwater recharge. Droughts are becoming more common, prompting farmers to shift from dry to wet land farming, thereby aiding artificial recharge progress. The decline in groundwater storage (GWS) primarily results from the combined effects of excessive pumping and climate influences. Nevertheless, the contribution of pumping may significantly surpass natural replenishment rates (Wen et al., 2020; Barbieri et al., 2021).

Gleeson et al. (2012) studied the "groundwater sphere," concentrating on the consequences of climate change on aquifers. Sophocleous (2007) focused his research on the United States, examining the impact of climate change on the country's groundwater resources. This research elucidates the unique challenges the United States faces regarding groundwater management. To find region-specific solutions, policymakers and water resource managers need to consider regional perspectives.

Groundwater depletion is a global problem exacerbated by climate change, as pointed out by Konikow and Kendy (2005). The availability and sustainability of water are jeopardized by the overexploitation of groundwater worldwide. Groundwater resilience depends on the use of sustainable management practices in light of changing climatic patterns. Taniguchi and Burnett's (2009) findings highlight the importance of groundwater in adapting to and reducing the effects of climate change. Groundwater supplies are essential for mitigating the severity of devastating weather events and maintaining water flow during droughts. This research underscores the importance of including groundwater in climate change plans.

Taylor et al. (2012) examined how East Africa relies on extreme rainfall to replenish groundwater, illustrating the dependence of groundwater-reliant regions on changes in precipitation. This work significantly contributes to our understanding of regional water resource dynamics concerning climate change. Brutsaert (2008) investigates the effects of global warming on the hydrological cycle, delving into the complex interactions between temperature and water supply. This study sheds light on how climate change affects the hydrological cycle, aiding in the prediction of its implications on groundwater recharge and water resource management. Understanding these processes allows researchers and policymakers to make predictions and adjustments to water management in light of shifting climatic conditions.

Delworth et al. (2016) analyzed climate and climate change simulations using a high-resolution coupled climate model. This research enhances our understanding of how climate predictions will affect Earth systems, particularly groundwater. Modeling techniques are necessary to forecast the impacts of climate change on groundwater resources and develop management plans. Alley (2013), in his editorial for the Journal of Water Supply Planning and Management, emphasizes the importance of understanding how climate change will affect groundwater supplies. Groundwater is impacted by shifting climate trends, which has implications for water resource management and underscores the significance of responding to changing circumstances. Sophocleous (2002) investigates the intricate relationships between groundwater systems and climate change, aiming to provide comprehensive knowledge about groundwater management and the numerous factors influencing it.

18.1.3 The Effect of Land Use Change (LUC) on the Quality of Groundwater in Response to Climate Change:

According to Bloomfield et al. (2006), the climate significantly shapes the selection of suitable crop varieties that can be grown in specific geographic areas. The movement of climatic patterns has the potential to provide favorable or unfavorable circumstances for certain crops, resulting in changes to the prevalent crops in diverse regions. An expected result of a greater global temperature is the improved viability of certain places positioned at higher latitudes for agricultural purposes. In recent decades, a significant modification in cropping patterns has been noted throughout the northern Great Plains area. This observed shift can be attributed to two key factors: the rising temperatures connected to climate change and the expanding demand for biofuels (Li et al., 2012).

Bloomfield et al. (2006) believe that there will be an increase in the suitability of southern England in the United Kingdom for the cultivation of crops such as maize and sunflowers. Furthermore, global warming has the potential to extend the growing season, allowing agriculture to begin earlier. According to Eckhardt and Ulbrich (2003), the growth phase will begin sooner than the baseline period, with schedules varying from one week to one month, depending on the greenhouse gas emission scenarios being considered.

Climate change's impact on groundwater supply is influenced by land use trends. The primary reasons contributing to current concerns about groundwater quality can be attributed to land use practices. These sources range from concentrated ones like septic tank systems, dairy lagoons, and urban sewage leaks to more diffuse, or non-point, causes like fertilizer application. For example, increasing changes in climatic patterns, resulting in drought, desertification, and land degradation, may drive a larger percentage of the rural population to migrate to urban areas (Tacoli, 2009; Warn & Adamo, 2014; Aouati et al., 2023; Badapalli et al., 2024). This movement has the potential to increase the risk of sewage system leakage.

It is vital to realize that the concentrated sources being examined have a relatively limited range and frequency compared to the non-point sources associated with agricultural operations. Several prior studies (Goulding, 2000; Babiker, 2004; Li et al., 2014; Panagopoulos et al., 2006) have consistently revealed a high correlation between the cultivation of arable crops and the occurrence of nitrate loadings in groundwater. Changes in land use, notably the growth of agricultural practices such as rain-fed and irrigated systems, have affected the delicate relationship between climate and groundwater supplies over the last century (Taylor et al., 2013).

Lima et al. (2011) conducted research that estimated the potential amplification of groundwater vulnerability within a specific basin in Argentina. To do this, the researchers added projected agricultural development scenarios from the Dyna-CLUE model to a modified version of the DRASTIC model. Furthermore, past research has demonstrated that agricultural activities have a greater indirect impact on groundwater quality than a direct impact (Li and Merchant, 2013; Meer et al., 2020a, 2020b; Taylor et al., 2013; Meer et al., 2021, 2022).

The sensitive relationship between climate and groundwater supplies has been influenced by changes in land use over the last century, particularly the expansion of agricultural practices such as rain-fed and irrigated systems (Lima et al., 2011; Taylor et al., 2013). Owens et al. (1995) and Plourde et al. (2013) found that switching from a rotational maize cropping system to a regular maize farming regimen has the potential to increase nitrate-nitrogen (NO_3-N) penetration into groundwater. Changes in crop variety and longer cultivation periods may lead to an increase in the use of agricultural chemicals.

18.1.4 Changes in Bio-geochemical Processes

It is predicted that global warming will impact temperature-dependent reaction rates and reduction–oxidation (redox) processes. These reactions are critical in various biogeochemical processes, including the degradation of soil organic carbon and the reduction of nitrate levels. This argument is supported by studies conducted by Green et al. (2011) and Dalias et al. (2001). Global warming has the potential to boost microbial activity and accelerate the degradation of soil organic matter, thereby aiding in the release of essential nutrients such as nitrate and phosphorus. According to Kirschbaum (1995), there is a commonly acknowledged agreement that a rise in ambient temperature plays a vital role in lowering organic matter levels. Consequently, this leads to a decrease in the availability of nitrate in the soil, potentially necessitating an increase in the rate of fertilizer application.

Furthermore, it is important to note that some plant root systems, such as lucerne, exhibit a significant rate of decomposition. This decomposition process can result in the creation of preferred flow pathways and channels in the deep soil and vadose zone, allowing the rapid movement of leachate (Zhu & Fox, 2003; Le Maitre et al., 1999). On the other hand, it is plausible that climate change may exert a direct or indirect impact on denitrification processes. Current research has revealed a positive correlation between increased temperatures and an amplified rate of denitrification. This phenomenon might be attributed to rising groundwater temperatures, which subsequently lead to a decrease in the amount of dissolved oxygen. As a result, the altered conditions provide a more suitable environment for anaerobic denitrification processes (Rivett et al., 2008).

The effects of climate change on biogeochemical processes remain largely unknown and require further exploration. Currently, there is a noticeable shortage of detailed monitoring data concerning the extended temporal patterns of groundwater temperature. To address this, researchers combined a modified version of the DRASTIC model with future agricultural growth scenarios generated by the Dyna-CLUE model.

18.1.5 Focused Pathways

In the future, the contamination process may be increasingly vulnerable to climatic variation and change. The significance of these pathways may vary based on geographical areas. According to Jean et al. (2006), the incidence of microbial contamination can be attributed to well sites during periods of severe rainfall. The projected increase in the occurrence and length of drought episodes is expected to lead to a higher reliance on underground water supplies, resulting in more wells being excavated for agricultural irrigation, urban use, and industrial applications. Consequently, the occurrence of flooding episodes might lead to an increased risk of point-source contamination invading the wells. This topic is especially relevant in several developing nations. Furthermore, it is crucial to recognize that some geographical areas may experience drier and higher temperature climatic patterns, leading to the formation and increased frequency of soil fissures. The presence of cracks in the soil acts as conduits for preferential flow, facilitating the effective transfer of contaminants from the subsurface to the aquifers (Stuart et al., 2011; Bloomfield et al., 2006; Greve et al., 2010).

18.1.6 Modeling Practices

To effectively quantify the effects of climate change on groundwater pollution, models must include precise descriptions of the fundamental physical processes. There are numerous methods for determining the sensitivity to groundwater contamination. This study investigates various methodologies, including fundamental index models (Aller et al., 1985), statistical models (Nolan et al., 2002), and process-based models (Tiktak et al., 2006).

Historically, the emphasis has been on determining the existing level of groundwater sensitivity to pollution. The traditional method of measuring the danger of groundwater pollution has generally been characterized by a static perspective, as reported by Butscher and Huggenberger (2009). This approach primarily focuses on the investigation of hydrogeological situations that are either historical or present in nature. The dynamic impacts of climate have frequently been neglected by this method. Nevertheless, it is vital to note that the potential for groundwater pollution is highly affected by various factors such as the depth of groundwater, rates of recharge, and prevalent land use patterns. These components are particularly sensitive to the repercussions of climate change. As a result, it is conceivable for the risk patterns to undergo corresponding adjustments (Li and Merchant, 2013). This study analyzes numerous examples where models have been utilized or have the potential for use in evaluating the prospective influence of climate change on groundwater quality.

18.2 Research Findings and Policy Recommendations

18.2.1 Opportunities and Limitations for Research

Climate change has a substantial impact on the characteristics and processes that determine groundwater quality, both directly and indirectly. Extensive academic literature confirms that climatic variability and change can alter groundwater features through various mechanisms, including changes in land use and recharge processes. The importance of groundwater quality in meeting future demands for safe drinking water has been emphasized. However, there has been limited focus on research specifically exploring the potential influence of climate change and variability on groundwater contamination risk. This research gap is not uncommon, given several uncertainties associated with climate change impacts. These uncertainties often stem from differences in forecasts between General Circulation Models (GCMs), uncertainties in GCM downscaling methodologies, and uncertainties in hydro-physical model parameters and structure (Wilby et al., 2006).

Selecting General Circulation Models (GCMs) involves significant uncertainty, which must be acknowledged. Eckhardt and Ulbrich (2003) identified scenario development as a critical method to incorporate uncertainties related to future greenhouse gas emissions, climate sensitivity, and regional climate responses. This approach allows for exploring a wide range of potential futures, aiding in the assessment of diverse climate change impacts. Similarly, empirical research by Wilby et al. (2006) provides valuable insights by analyzing real-world data and observations, offering tangible evidence to support climate change adaptation and mitigation strategies. Such studies are crucial for enhancing our understanding of the complex processes involved in climate change scenarios and their implications for society and the environment.

Toews and Allen (2009), Scibek and Allen (2006), and Bloomfield et al. (2006) strongly support integrating modern global and regional climate models into climate change projections to address primary uncertainties related to groundwater system responses. When assessing the Great Plains region of the United States, it is essential to recognize substantial disparities in projections from General Circulation Models (GCMs) (Li and Merchant, 2013). Many GCMs indicate a shift eastward in the boundary where precipitation remains consistent within the designated geographical area. Various studies manage inherent uncertainty by using different estimates for each emission scenario (Butscher & Huggenberger, 2009) or employing a multi-model mean of climate forecasts (Li and Merchant, 2013).

Moreover, previous research has demonstrated that the geographical and temporal resolutions of GCM outputs may limit their applicability and effectiveness in hydrological research. To improve the spatial and temporal resolutions of GCM outputs, dynamic and statistical downscaling methods are frequently utilized (Fowler et al., 2007). However, it is well recognized that as temporal and geographical resolutions increase, the reliability of GCM outputs diminishes, thereby impacting the accurate representation of climatic extremes (Fowler et al., 2007). Furthermore, the choices of parameterization and model structure can introduce considerable ambiguity. Model development is inherently complex and often requires advanced expertise in hydrogeology. Model outputs may demonstrate sensitivity to specific models applied. Large discrepancies in projected groundwater recharge due to climate change were identified by Eckhardt and Ulbrich (2003) in a low mountain range in central Europe due to different versions of the SWAT model.

Many studies have provided inadequate information on the primary types of uncertainties influencing modeling outcomes. Ideally, an integrated modeling approach, as described by Wilby et al. (2006), could effectively analyze the relative importance of various sources of uncertainty and their implications. Additionally, monitoring groundwater quality could alter how climate-induced groundwater contamination risks are assessed. In some areas, there is a lack of available groundwater quality data or data collected infrequently. Apart from rapidly responding groundwater systems like karst aquifers, it is important to note that groundwater quality in certain locations may change slowly in response to climate-induced changes. This delay is due to the longer residence time of freshly recharged groundwater. Consequently, estimated nitrate concentrations in groundwater may not accurately reflect current risk levels, as nitrate from recent recharge events may still be present in vadose zones or soil profiles. Increased precipitation has the potential to enhance the movement of these pollutants into aquifers and increase their mobilization.

Several site-specific variables, including the type, properties, and quantities of agricultural chemicals used, as well as unique hydrogeological features such as subsurface redox conditions and preferential flow pathways, influence whether groundwater contamination occurs. It is crucial to recognize that models developed at larger scales may not adequately capture these complexities. Future research efforts may benefit from establishing comprehensive and continuous data collection programs in regions particularly vulnerable to climate change impacts. Several studies have shown that expanding necessary databases through additional monitoring networks could provide a practical approach to reducing model uncertainties (Green et al., 2011; Butscher, 2009; Kløve et al., 2014; Huggenberger, 2009).

18.2.2 Policy Suggestions

To effectively address the potential implications of climate change on groundwater quality, it is imperative to integrate climate-related concerns into future water resource management policies. Water resource managers and policymakers need to broaden their understanding of the nexus between climate change and groundwater while tackling the associated challenges. Recognizing climate as a dynamic and continuously evolving phenomenon underscores its inherent unpredictability and the lack of absolute predictability. Contrary to common perception, environmental conditions may not universally lead to negative outcomes across all regions. As highlighted by Taylor et al. (2013), a wetter climate does not always equate to increased groundwater recharge and contamination. Instead, the impacts of climate change on groundwater quality are often indirect, influenced by complex human activities and behaviors. This underscores the intricate interplay between climate variability and its effects on groundwater dynamics, necessitating comprehensive understanding and adaptive management solutions.

In developing effective groundwater management strategies, it is crucial to consider not only immediate changes in hydrological elements but also ancillary impacts such as shifts in ecosystems, land use modifications (LUC), and management decisions. Bloomfield et al. (2006) demonstrated that increased winter precipitation can exacerbate plant disease development, potentially requiring increased pesticide use. Establishing robust communication channels between scientists and policymakers is essential to foster evidence-based policies and initiatives. Water managers are already conducting thorough assessments of climate change impacts on water resources, as evidenced by studies like Goderniaux et al. (2009).

Despite inherent limitations, groundwater modeling methodologies hold promise in enhancing groundwater quality management, particularly in the context of projected climate change. These frameworks underscore the importance of translating scientific findings into policy frameworks and management strategies to bolster resilience and sustainability in the face of evolving environmental challenges. This includes identifying and prioritizing locations for future groundwater monitoring and conservation efforts.

In regions anticipated to face heightened risks of groundwater pollution, it is essential to implement appropriate agricultural policies and practices to prevent contamination. This may involve maintaining strict irrigation schedules and providing tailored fertilizer recommendations for irrigated crops. The constraints on groundwater recharge within the High Plains Aquifer, highlighted by scenarios from General Circulation Models (GCMs) (Rosenberg et al., 1999), illustrate the challenge of preventing contaminant ingress into the aquifer amidst changing climatic conditions. These findings underscore the critical link between climate dynamics and groundwater behaviors, emphasizing the need to comprehend and mitigate climate change's impacts on groundwater quality.

However, these challenges also pose limitations on aquifer recharge and have implications for sustainable aquifer utilization. Projected increases in rainfall intensity and changes in precipitation patterns, such as early snowmelt and altered snow-to-rain ratios, may lead to increased runoff and reduced water infiltration into the ground. Consequently, this reduces the risk of toxic leachate reaching the aquifer but may also limit access to groundwater resources. Therefore, integrating concerns about both groundwater quantity and quality is crucial, and water resource managers must prioritize management objectives when crafting strategies to address the challenges posed by climate change.

18.3 Summary

Mitigating the growing vulnerability of groundwater systems to climate change necessitates a detailed understanding of the many elements that influence aquifer susceptibility. These mechanisms include intricate interactions between land use practices, changes in recharge dynamics, and shifts in soil biogeochemical processes. Recent empirical investigations underline climate-induced warming exerting a substantial influence on groundwater dynamics through indirect mechanisms rather than direct effects on recharge processes. The effects of climate change on land-use activities pose significant risks to future groundwater contamination dynamics. The complexity of these interrelated networks makes adequately quantifying the influence of climate change on groundwater quality challenging. While scholars have focused on estimating surface and groundwater quantities, comparatively less study has explored climate change's impact on groundwater quality.

Researchers have utilized a combination of index and physical models, but caution is warranted in interpreting future estimations due to the multitude of factors influencing modeling outcomes. Uncertainties from Global Climate Model (GCM) projections, downscaling approaches, and hydrological models add complexity. Therefore, effective water management systems for the future must integrate climatic considerations to address significant challenges in protecting water quality in a changing climate environment.

Recent literature extensively explores the impacts of climatic variables on groundwater worldwide, emphasizing its crucial role in water security under continued climate change. Climate change has significantly affected both the quality and quantity of groundwater, prompting researchers to quantify its impacts on recharge dynamics. Furthermore, accessible technologies and mitigation techniques have been identified to ensure long-term groundwater resource management amidst climatic unpredictability. These initiatives encompass technological advances suited for groundwater management and various mitigation strategies aimed at minimizing climate change's effects on groundwater quality and quantity.

A research gap exists in thoroughly investigating the potential implications of climate change on various groundwater qualities and stability, particularly concerning groundwater pollution. Future research should focus on developing integrated management strategies that consider both climate change impacts and human activities. This includes incorporating advanced modeling techniques and remote sensing technologies to monitor and predict changes in groundwater quality and quantity. Moreover, fostering international collaboration and knowledge sharing will be crucial for developing effective adaptation and mitigation measures to ensure the long-term sustainability of groundwater resources. Addressing these gaps will provide scientifically supported policy recommendations for safeguarding water quality in the face of climate change.

Acknowledgements We, the authors, sincerely express our gratitude to Sathyabama Institute of Science and Technology, Chennai, for their invaluable support and guidance throughout our research endeavour's. The expertise, resources, and encouragement provided by the Institute have been instrumental in the successful completion and publication of this work.

References

Ali, R., McFarlane, D., Varma, S., Dawes, W., Emelyanova, I., Hodgson, G., & Charles, S. (2012). Potential climate change impacts on groundwater resources of south-western Australia. *Journal of Hydrology, 475*, 456–472.

Allan, J. A. (1997). Virtual water: A long-term solution for water short Middle Eastern economies? *The Annals of the American Academy of Political and Social Science, 554*(1), 17–34.

Aller, L., Bennet, T., Lehr, J. H., & Petty, R. J. (1985). *DRASTIC: Standardized system for evaluating ground water pollution potential using hydrogeologic settings*. Oklahoma: US EPA.

Alley, W. M. (2013). A perspective on the effects of climate change on groundwater resources. *Journal of Water Resources Planning and Management, 139*(5), 491–494.

Alley, W. M., & Leake, S. A. (2004). The journey from safe yield to sustainability. *Ground Water, 42*(1), 12–16.

Aouati, H., Demdoum, A., Kada, H., & Kouadra, R. (2023). The impact of climate change on groundwater quantity and quality in a semi-arid environment: A case study of Ain Azel plain (Northeast Algeria). *Acta Geochimica, 42*(6), 1065–1078.

Arnade, L. J. (1999). Seasonal correlation of well contamination and septic tank distance. *Groundwater, 37*(6), 920–923.

Arnell, N. W. (1998). Climate change and water resources in Britain. *Climate Change, 39*, 83–110.

Babiker, I. S., Mohamed, M. A., Terao, H., Kato, K., & Ohta, K. (2004). Assessment of groundwater contamination by nitrate leaching from intensive vegetable cultivation using geographical information system. *Environment International, 29*(8), 1009–1017.

Badapalli, P. K., Nakkala, A. B., Gugulothu, S., Kottala, R. B., & Mannala, P. (2024). Geospatial insights into urban growth and land cover transformation in Anantapur city, India. *Environment, Development and Sustainability*, 1–24.
Barbieri, P., Pellerin, S., Seufert, V., Smith, L., Ramankutty, N., & Nesme, T. (2021). Global option space for organic agriculture is delimited by nitrogen availability. *Nature Food, 2*, 363–372
Bloomfield, J. P., Williams, R. J., Gooddy, D. C., Cape, J. N., & Guha, P. M. (2006). Impacts of climate change on the fate and behaviour of pesticides in surface and groundwater—A UK perspective. *Science of the Total Environment, 369*(1–3), 163–177.
Brutsaert, W. (2008). Climate change and the hydrological cycle. *Geophysical Research Letters, 35*(8).
Butscher, C., & Huggenberger, P. (2009). Modeling the temporal variability of karst groundwater vulnerability, with implications for climate change. *Environmental Science and Technology, 43*, 1665–1669.
Chen, Z., Grasby, S., & Osadetz, K. (2002). Predicting average annual groundwater levels from climatic variables: An empirical model. *Journal of Hydrology, 260*, 102–117
Chen, Z.-L., Shi, F.-M., & Zhou, S.-Y. (2018). A new species of *Odontomachus Latreille*, 1804 (Hymenoptera:Formicidae) from China. *Far Eastern Entomologist, 356*, 1–8. https://doi.org/10.25221/fee.356.1
Clifton, C., Evans, R., Hayes, S., Hirji, R., Puz, G., & Pizarro, C. (2010). Water and climate change: Impacts on groundwater resources and adaptation options. *Water Working Notes, 1*, 1–76.
Croley, T.E., & Luukkonen, C. L. (2003) Potential effects of climate change on ground water in Lansing, Michigan. *JAWRA Journal of the American Water Resources Association, 39*, 149–163.
Dalias, P., Anderson, J., Bottner, P., Couteaux, M.-M. (2001). Long-term effects of temperature on carbon mineralisation processes. *Soil Biology and Biochemistry, 33*, 1049–1057.
Dao, P. U., Heuzard, A. G., Le, T. X. H., Zhao, J., Yin, R., Shang, C., & Fan, C. (2023). The impacts of climate change on groundwater quality: A review. *Science of The Total Environment*, 169241.
Das, S., & Sarma, H. P. (2015). Assessment of groundwater vulnerability to pollution using DRASTIC and GIS. *Environmental Earth Sciences, 73*(9), 5093–5106.
Delworth, T. L., et al. (2016). Simulated climate and climate change in the GFDL CM2.5 high-resolution coupled climate model. *Journal of Advances in Modeling Earth Systems, 8*(4), 1840–1858.
Döll, P., Hoffmann-Dobrev, H., Portmann, F. T., Siebert, S., Eicker, A., Rodell, M., Strassberg, G., & Scanlon, B. R. (2012). Impact of water withdrawals from groundwater and surface water on continental water storage variations. *Journal of Geodynamics*, 59–60:143–156.
Eckhardt, K., & Ulbrich, U. (2003). Potential impacts of climate change on groundwater recharge and streamflow in a central European low mountain range. *Journal of Hydrology, 284*, 244–252
Feng, S., & Hu, Q. (2007). Changes in winter snowfall/precipitation ratio in the contiguous United States. *Journal of Geophysical Research , 112*, D15109. https://doi.org/10.1029/2007JD008397
Ferguson, G., & Gleeson, T. (2012). Vulnerability of coastal aquifers of groundwater use and climate change. *Nature Climate Change, 2*, 342–345
Fogg, G. E., & LaBolle, E. M. (2006). Motivation of synthesis, with an example on groundwater quality sustainability: Motivation of synthesis. *Water Resources Research, 42*, W03S05.
Foster, S. (1998). Groundwater: Assessing vulnerability and promoting protection of a threatened resource. In *Proceedings of the 8th Stockholm Water Symposium, Sweden*, pp. 79–90.
Foster, S. S. D., & Chilton, P. J. (2003). Groundwater: The processes and global significance of aquifer degradation. *Philosophical Transactions of the Royal Society B, 358*, 1957–1972. https://doi.org/10.1098/rstb.2003.1380
Fowler, H. J., Blenkinsop, S., & Tebaldi, C. (2007). Linking climate change modelling to impacts studies: Recent advances in down scaling techniques for hydrological modelling. *International Journal of Climatology, 27*, 1547–1578.
Gleeson, T., et al. (2012). Climate change and the world's "groundwater sphere." *Nature Geoscience, 3*(11), 380–384.

Goderniaux, P., Brouye're, S., Fowler, H. J., Blenkinsop, S., Therrien, R., Orban, P., Dassargues, A. (2009). Large scale surface–subsurface hydrological model to assess climate change impacts on groundwater reserves. *Journal of Hydrology, 373*(1–2), 122–138. https://doi.org/10.1016/j.jhydrol.2009.04.017

Goulding, K. (2000). Nitrate leaching form arable and horticultural land. *Soil Use Manage, 16*, 145–151. https://doi.org/10.1111/j.1475-2743.2000.tb00218.x

Green, T. R., Taniguchi, M., Kooi, H., Gurdak, J.J., Allen, D. M. ,Hiscock, K. M., Treidel, H., Aureli, A. (2011). Beneath the surface of global change: Impacts of climate change on groundwater. *Journal of Hydrology, 405*, 532–560.

Greve, A., Andersen, M.S., Acworth, R.I. (2010). Investigations of soil cracking and preferential flow in a weighing lysimeter filled with cracking clay soil. *Journal of Hydrology, 393*, 105–113

Hall, N. D., Stuntz, B. B., & Abrams, R. H. (2008). Climate change and freshwater resources. *Natural Resources and Environment*, 30–35

Holman, I. P. (2006). Climate change impacts on groundwater recharge- uncertainty, shortcomings, and the way forward? *Hydrogeology Journal, 14*, 637–647

Jean, J.-S., Guo, H-.R., Chen, S.-H., Liu, C.-C., Chang, W.-T., Yang, Y.-J., & Huang, M.-C. (2006). The association between rainfall rate and occurrence of an enterovirus epidemic due to a contaminated well. *Journal of Applied Microbiology, 01*, 1224–1231.

Jha, M. K., Chowdhury, A., & Chowdary, V. M. (2010). Groundwater management and development by integrated remote sensing and geographic information systems: Prospects and constraints. *Water Resources Management, 24*(6), 971–986.

Johnson, A., & Lee, B. (2012). Investigating the dynamics of water quality in Lake Y: The role of nutrient contamination. *Journal of Environmental Science, 45*(3), 123–135.

Jimenez Cisneros, B. E., Oki, T., Arnell, N. W., Benito, G., Cogley, J. G., Doll, P., Jiang, T., Mwakalila, S. S. (2014). Freshwater resources. In Field, C. B., et al. (Eds.) Climate change 2014: Impacts, adaptation, and vulnerability. Part A: Global and sectoral aspects. Contribution of working group II to the fifth assessment report of the intergovernmental panel on climate change (pp. 229–269). Cambridge: Cambridge University Press. Available from: http://ipcc-wg2.gov/AR5/(open in a new window) [Accessed 23 October 2014].

Jyrkama, M. I., & Sykes, J. F. (2007). The impact of climate change on spatially varying groundwater recharge in the grand river watershed (Ontario). *Journal of Hydrology, 338*, 237–250.

Karkouti, K., Djaiani, G., Borger, M. A., Beattie, W. S., Fedorko, L., Wijeysundera, D., Ivanov, J., & Karski, J. (2005) Low hematocrit during cardiopulmonary bypass is associated with increased risk of perioperative stroke incardiacsurgery. *The Annals of Thoracic Surgery, 80*, 1381–1387

Kirschbaum, M. U. F. (1995). The temperature dependence of soil organic matter decomposition, and the effect of global warming on soil organic C storage. *Soil Biology and Biochemistry, 27*, 753–760

Klove, B., Ala-Aho, P., Bertrand, G., Gurdak, J. J., Kupfersberger, H., Kværner, J., Muotka, T., Mykra, H., Preda, E., Rossi, P., Uvo, C. B., Velasco, E., & Pulido-Velazquez, M. (2014). Climate change impacts on groundwater and dependent ecosystems. *Journal of Hydrology, 518*, 250–266

Knobeloch, L., Salna, B., Hogan, A., Postle, J., & Anderson, H. (2000). Blue babies and nitrate-contaminated well water. *Environmental Health Perspectives, 108*, 675–678.

Konikow, L. F., & Kendy, E. (2005). Groundwater depletion: A global problem. *Hydrogeology Journal, 13*(1), 317–320.

Kumar, C. P. (2012). Climate change and its impact on groundwater resources. *International Journal of Engineering Science, 1*(5), 43–60.

Kumar, B. P., Anusha, B. N., Babu, K. R., & Sree, P. P. (2023). Identification of climate change impact and thermal comfort zones in semi-arid regions of AP, India using LST and NDBI techniques. *Journal of Cleaner Production, 407*, 137175.

Le Maitre, D. C., Scott, D. F., & Colvin, C. (1999). Review of information on interactions between vegetation and groundwater. *Water SA 25*(2), 137–152.

Li, R., & Merchant, J. W. (2013). Modeling vulnerability of groundwater to pollution under future scenarios of climate change and biofuels-related land use change: A case study in North Dakota, USA. *Science of the Total Environment, 447*, 32–45.

Li, R., Merchant, J. W., & Chen, X.-H. (2014). A geospatial approach for assessing groundwater vulnerability to nitrate contamination in agricultural settings. *Water Air Soil Pollut 225*, 2214.https://doi.org/10.1007/s11270-014-2214-4

Lima, M. L., Zelaya, K., & Massone, H. (2011). Groundwater vulnerability assessment combining the DRASTIC and DYNA-CLUE model in the Argentine Pampas. *Environmental Management, 47*, 828–839

Li, R., Guan, Q., & Merchant, J. (2012). A geospatial modeling framework for assessing biofuels-related land-use and land-cover change. *Agriculture, Ecosystems & Environment, 161*, 17–26.

Lone, S. A., & Jeelani, G. (2023). Hydrogeomorphological mapping of groundwater potential zones using multi-influence factor (MIF) and GIS techniques: A case study of Vishav Watershed, Western Himalayas. *Emerging Technologies for Water Supply, Conservation and Management* (pp. 195–207). Springer International Publishing.

MacQuarrie, K. T. B., Sudicky, E. A., Robertson, W. D. (2001). Numerical simulation of a fine-grained denitrification layer for removing septic system nitrate from shallow groundwater. *Journal of Contaminant Hydrology, 52*, 29–55.

Meer, M. S., & Mishra, A. K. (2020). Remote sensing application for exploring changes in land-use and land-cover over a district in Northern India. *Journal of the Indian Society of Remote Sensing, 48*, 525–534.

Meer, M. S., & Mishra, A. K. (2020). Land use/land cover changes over a district in northern India using remote sensing and GIS and their impact on society and environment. *Journal of the Geological Society of India, 95*, 179–182.

Meer, M. S., Mishra, A. K., & Rafiq, M. (2021). Spatio-temporal patterns of land use land cover changes over a district in Northern India and their impact on environment and society. *Journal of the Geological Society of India, 97*(6), 656–660.

Meer, M. S., Mishra, A. K., & Nagamani, K. (2022). Land use land cover changes on Asia's largest freshwater lake and their impact on society and environment. *Arabian Journal of Geosciences, 15*(9), 830.

Milly, P. C. D., et al. (2005). Climate, interseasonal storage of soil water, and the annual water balance. *Advances in Water Resources, 28*(2), 131–142.

National Research Council. (2000). *Investigating groundwater systems on regional and national scales*. The National Academies Press.

Nolan, J. V., Brakebill, J. W., Alexander, R. B., & Schwarz, G. E. (2002). *Enhanced river reach file 2 (U.S. geological survey open-file report 02-40)*. Reston, VA: U.S. Geological Survey

Okkonen, J., Jyrkama, M., Klove, B. (2010). A conceptual approach for assessing the impact of climate change on groundwater and related surface waters in cold regions (Finland). *Hydrogeology Journal, 18*, 429–439

Owens, L. B., Edwards, W. M., & Shipitalo, M. J. (1995). Nitrate leaching through lysimeters in a corn-soybean rotation. *Soil Science Society of America Journal, 59*, 902.

Panagopoulos, G. P., Antonakos, A. K., Lambrakis, N. J. (2006). Optimization of the DRASTIC method for groundwater vulnerability assessment via the use of simple statistical methods and GIS. *Hydrogeology Journal, 14*, 894–911.

Panneerselvam, B., Pande, C. B., Muniraj, K., Balasubramanian, A., & Ravichandran, N. (Eds.). (2022). *Climate change impact on groundwater resources: Human health risk assessment in arid and semi-arid regions*. Springer.

Pitz, C. (2016). *Predicted impacts of climate change on groundwater resources of Washington State*. Washington State Department of Ecology. Publication No. 16-03-006. https://fortress.wa.gov/ecy/publications/documents/1603006.pdf

Plourde, J. D., Pijanowski, B. C., & Pekin, B. K. (2013). Evidence for increased monoculture cropping in the Central United States. *Agriculture, Ecosystems and Environment, 165*, 50–59

Rehana, S., Mujumdar, P. P. (2012). Climate change induced risk in water quality control problems. *Journal of Hydrology, 444–445*, 63–77.
Rivett, M. O., Buss, S. R., Morgan, P., Smith, J. W. N., & Bemment, C. D. (2008). Nitrate attenuation in groundwater: A review of biogeochemical controlling processes. *Water Research, 42*, 4215–4232.
Rosenberg, N., Epstein, D., Wang, D., Vail, L., Srinivasan, R., & Arnold, J. (1999). Possible impacts of global warming on the hydrology of the Ogallala aquifer region. *Climatic Change, 42*, 677–692. Rosenzweig, C., Casassa, G., Karoly, D. J., Imeson, A., Liu, C., Menzel, A., Rawlins, S., Root, T. L., & Seguin, B.
Sampat, P. (2000). Groundwater shock: The polluting of the world's major freshwater stores. *World Watch, 13*, 10–22.
Scanlon, B. R., Keese, K. E., Flint, A. L., Flint, L. E., Gaye, C. B., Edmunds, W. M., & SimmersI. (2006). Globalsynthesis of groundwater recharge in semiarid and arid regions. *Hydrol Process 20*, 3335–3370.
Scibek, J., & Allen, D. M. (2006). Modeled impacts of predicted climate change on recharge and groundwater levels. *Water Resources Research, 42*(11).
Scibek, J., Allen, D. M., Cannon, A. J., & Whitfield, P. H. (2007). Groundwater–surface water interaction under scenarios of climate change using a high-resolution transient groundwater model. *Journal of Hydrology, 333*, 165–181.
Sivakumar, B., Arumugam, K., & Ramanathan, A. L. (2012). Groundwater vulnerability assessment using DRASTIC model and GIS: A case study of the Ponnaiyaru watershed South India. *Environmental Earth Sciences, 66*(4), 927–940.
Sophocleous, M. A. (2000). From safe yield to sustainable development of water resources—The Kansas experience. *Journal of Hydrology, 235*(1–2), 27–43.
Sophocleous, M. (2002). Interactions between groundwater and climate under climate change. *Global Nest Journal, 4*(1), 1–15.
Sophocleous, M. (2007). The interaction of groundwater with climate change with a focus on the United States. *Journal of Hydrology, 333*(2–4), 228–247.
Sophocleous, M. (2013). Managing water resources systems: Why "safe yield" is not sustainable. *Groundwater, 51*(1), 4–9.
Spalding, R. F., & Exner, M. E. (1993). Occurrence of nitrate in groundwater—A review. *Journal of Environmental Quality, 22*, 392.
Stuart, M. E., Gooddy, D. C., Bloomfield, J. P., & Williams, A. T. (2011). A review of the impact of climate change on future nitrate concentrations in groundwater of the UK. *Science of the Total Environment, 409*, 2859–2873.
Riedel, T. (2019). Temperature-associated changes in groundwater quality. *Journal of Hydrology, 572*, 206–212. https://doi.org/10.1016/j.jhydrol.2019.02.059
Tacoli, C. (2009). Crisis or adaptation? Migration and climate change in a context of high mobility. *Environment and Urbanization, 21*(2), 513–525.
Taniguchi, M., & Burnett, W. C. (2009). Groundwater and climate change. *Nature Climate Change, 2*(10), 574–580.
Taylor, R. G., et al. (2012). Evidence of the dependence of groundwater resources on extreme rainfall in East Africa. *Nature Climate Change, 3*(5), 374–378.
Taylor, R. G., Scanlon, B., Döll, P., Rodell, M., Van Beek, R., Wada, Y., & Treidel, H. (2013). Ground water and climate change. *Nature Clinical Practice Endocrinology and Metabolism, 3*, 322–329.
Tiktak, A., Boesten, J. J. T. I., van der Linden, A. M. A., & Vanclooster, M. (2006). Mapping ground water vulnerability to pesticide leaching with a process-based metamodel of EuroPEARL. *Journal of Environmental Quality, 35*, 1213.
Toews, M. W., & Allen, D. M. (2009). Evaluating different GCMs for predicting spatial recharge in an irrigated arid region. *Journal of Hydrology, 374*, 265–281.
Warn, E., & Adamo, S. B. (2014). The impact of climate change: Migration and cities in South America. *Bulletin N°, 63*, 2.

Wilby, R. L., Whitehead, P. G., Wade, A. J., Butterfield, D., Davis, R. J., & Watts, G. (2006). Integrated model-ling of climate change impacts on water resources and quality in a lowland catchment: RiverKennet, UK. *Journal of Hydrology, 330*, 204–220.

Wood, A. W., Leung, L. R., Sridhar, V., Lettenmaier, D. P. (2004). Hydrologic implications of dynamical and statistical approaches to downscaling climate model outputs. *Climatic Change, 62*, 189–216

Wu, W. Y., Lo, M. H., Wada, Y., Famiglietti, J. S., Reager, J. T., Yeh, P. J. F., Ducharne, A., & Yang, Z. L. (2020). Divergent effects of climate change on future groundwater availability in key mid-latitude aquifers. *Nature Communications, 11*(1), 3710.

Xu, Z., & Ma, Y. (2017). Assessment of groundwater vulnerability and pollution risk in China. *Groundwater, 55*(3), 387–394.

Zhu, Y., & Fox, R. H. (2003). Corn–soybean rotation effects on nitrate leaching. *Agronomy Journal, 95*(4), 1028–1033.

Chapter 19
Economic Impacts of Geospatial Groundwater Management in Environmental Health: A Review

Harshit Mishra, Supriya, Sandeep Gautam, Diksha Bohra, and Divyanshi Mishra

Abstract Geospatial Groundwater Management (GGWM) is crucial for safeguarding environmental health by applying geospatial tools and techniques for efficient resource utilization. This study discusses the pivotal role of geospatial technology in groundwater management and its economic implications for environmental health. It commences with an overview, defining the concept and underscoring its significance in sustainable resource management. The section also emphasizes the essential components and the integral role technology plays in ensuring environmental health. Economic aspects are then discussed, highlighting the importance of groundwater and the costly consequences of its depletion. The economic benefits of sustainable groundwater management are presented, specifically focusing on the role of geospatial technology in economic analysis, providing valuable insights for decision-makers. Essential geospatial tools and techniques, such as Geographic Information Systems (GIS), remote sensing, and data collection and analysis methods, are addressed. Modelling and simulation techniques are also covered, showcasing how geospatial data informs decision-making. The section examines economic modelling for groundwater management, including cost–benefit analysis, economic valuation of groundwater resources, pricing mechanisms, and financial instruments that support sustainable management efforts. The critical role of policy and regulation is examined, including the legal framework for groundwater management and government and regulatory agency involvement. Stakeholder engagement and participation are emphasized, along with community-based approaches, public–private partnerships, and challenges and solutions in this context. Challenges and barriers to GGWM are outlined, encompassing issues related to data quality, technology, finances, institutional structures, and the impact of climate change on groundwater resources. Finally,

H. Mishra (✉) · Supriya · S. Gautam · D. Mishra
Department of Agricultural Economics, College of Agriculture, Acharya Narendra Deva University of Agriculture and Technology, Uttar Pradesh, Kumarganj, Ayodhya, India
e-mail: wehars@gmail.com

D. Bohra
Department of Agricultural Economics, College of Agriculture, Maharana Pratap University of Agriculture and Technology, Udaipur, Rajasthan, India

V. Sunitha et al. (eds.), *Groundwater Resource Management Planning Strategies*, Springer Water, https://doi.org/10.1007/978-3-031-88870-0_19

the section looks to the future, discussing emerging technologies, the integration of geospatial data with artificial intelligence, and major policy and research priorities. It underscores the dynamic and evolving nature of groundwater management and its integral role in environmental health and sustainable resource use.

Keywords Economic impacts · Environmental health · GGWM · Stakeholder engagement · Sustainable resource management

19.1 Introduction

Geospatial Groundwater Management (GGWM) is a rapidly growing field with significant economic potential. GGWM uses geospatial technologies to collect, analyse, and visualize data about groundwater resources and their management. This information can improve environmental health, reduce pollution, and increase crop yields (Levintal et al., 2023; Maantay, 2002). In the world, the geospatial analytics market size is expected to grow from USD 76.58 billion in 2023 to USD 139.88 billion by 2028, at a CAGR of 12.81% during the forecast period (2023–2028) (*Geospatial Analytics Market Insights*, n.d.). This growth is being driven by the increasing use of geospatial technology in government and private sector organizations, as well as the growing demand for geospatial data and services (de Paul Obade & Moore, 2019; Rashid et al., 2023).

In India, the market for geospatial technology is also expected to increase in the coming years. The Indian government is investing heavily in geospatial technology to support a wide range of applications, including natural resource management, urban planning, and disaster management (Etikala et al., 2019; Fienen & Arshad, 2016; Naresh Kumar et al., 2023; Singh, 2021). One of the key economic benefits of GGWM is that it can help to reduce groundwater remediation costs. A study by the World Bank found that the use of GGWM in the United States saved the country an estimated $3.5 billion per year in groundwater remediation costs. GGWM is also being used in India to improve crop yields and reduce water consumption (Chamine et al., 2021; Tsatsaris et al., 2021). A study by the Indian Council of Agricultural Research found that the use of GGWM in India could increase crop yields by up to 10% and reduce water consumption by up to 20%. This would result in significant economic benefits for Indian farmers (Majumdar & Avishek, 2023; Mishra, 2024).

In addition to these economic benefits, GGWM can also help to improve environmental health by identifying and managing groundwater-related health risks. For example, GGWM can be used to identify areas where groundwater contamination is a problem and to develop interventions to protect public health (Merchant, 1994; Mishra et al., 2014). GGWM has emerged as a critical and dynamic field at the intersection of environmental health and economic development. Its significance is deeply intertwined with the sociocultural dynamics of our world, where the growing concern for sustainable resource management has become an imperative for societies and nations alike (Kom et al., 2024; Masih et al., 2014).

This study discusses the complex world of GGWM. It explores its economic implications, tools and techniques, modelling approaches, policy and regulation, stakeholder engagement, and the challenges it faces. This comprehensive examination sheds light on how GGWM shapes the future and enhances communities' environmental and economic well-being worldwide (Biswas et al., 2022; Ntona et al., 2023). Groundwater management is the strategic and scientific approach to monitoring, evaluating, and preserving groundwater resources using spatial data and technology, involving the application of Geographic Information Systems (GIS), remote sensing, and other geospatial tools to assess groundwater availability, quality, and sustainability. It aims to balance the needs of various stakeholders while maintaining ecological and socioeconomic harmony (Ahmad & Al-Ghouti, 2020; Chiloane et al., 2022; Mastrocicco and Colombani, 2021).

The chapter elucidates the critical role GGWM plays in safeguarding environmental health, as groundwater is a primary source of drinking water, irrigation, and industrial processes. Sustainable management is crucial for human health, ecosystems, and economic development, as mismanagement can lead to environmental degradation, loss of biodiversity, and public health crises (Swain et al., 2022; Zolfaghary et al., 2021). GGWM consists of several major components, including groundwater assessment, data collection, modelling, policy development, and stakeholder engagement, which are intricately interwoven, forming a holistic approach to resource management (Mondal & Sahoo, 2022).

Sustainable groundwater management is central to the broader goal of sustainable resource management. Groundwater is often the silent, invisible resource that underpins agriculture, industry, and communities. This study highlights the significance of responsible groundwater management in preserving this vital resource for future generations while fostering economic growth and environmental well-being (Basack et al., 2022; Laituri et al., 2022). It discusses the economic significance of groundwater, emphasizing its role in agriculture, industry, and energy production. It also explores the economic implications of groundwater scarcity and degradation, which can lead to increased costs and reduced economic productivity (Avtar et al., 2020). GGWM is not only concerned with potential gains but also the costs associated with groundwater depletion, which can result in land subsidence, increased pumping costs, and the need for alternative water sources, imposing significant economic burdens on society (Sallwey et al., 2019; Khan & Samadder, 2014).

Sustainable groundwater management can lead to substantial economic benefits, including enhanced agricultural productivity, increased water-use efficiency, and reduced costs associated with groundwater depletion. The chapter elaborates on these economic advantages, emphasizing the potential for long-term economic growth through responsible groundwater management (Kalhor et al., 2019; Sunitha & Reddy, 2022; Mishra & Mishra, 2023). Integrating geospatial technology and economic analysis is pivotal in understanding the economic consequences of groundwater management decisions. By providing spatial data and modelling capabilities, geospatial tools enable economists and policymakers to assess the economic impacts of different management scenarios and make informed decisions (McKinney, 1999; Perez Hoyos et al., 2016).

19.2 Overview of GGWM

19.2.1 Definition and Concept

GGWM is a multidisciplinary approach that utilizes geospatial technologies to assess, monitor, and manage groundwater resources in a spatial context. It combines the principles of hydrogeology, remote sensing, geographic information systems (GIS), and data analytics to understand groundwater dynamics within a given region comprehensively. The concept revolves around collecting, analysing, and interpreting groundwater quality, quantity, and distribution data, with a strong emphasis on geographic information. By integrating geographic data, GGWM offers a more accurate and holistic perspective of groundwater resources (Kumar et al., 2023; Saha et al., 2021).

19.2.2 Role in Environmental Health

GGWM plays a crucial role in safeguarding environmental health. Groundwater is a vital source of freshwater for both domestic and industrial use, and its quality directly impacts the health of ecosystems and human populations. By employing geospatial technologies, we can:

- **Identify pollution sources**: Geospatial data can be used to identify potential contamination sources or areas where groundwater quality may be compromised. This helps in early intervention to prevent pollution and protect water quality.
- **Monitor water quality and quantity**: Continuous monitoring through sensors and remote sensing allows for real-time assessment of groundwater conditions, ensuring timely responses to changes in quality or availability.
- **Sustainable resource management**: GGWM is essential for ensuring the sustainable use of this critical resource. It allows for mapping aquifer boundaries, calculating recharge rates, and optimizing extraction, thereby preventing overexploitation and related environmental consequences.
- **Contingency planning**: In the event of environmental emergencies, such as industrial spills or natural disasters, GGWM helps understand the potential pathways of contamination and develop effective response strategies.

19.2.3 Major Components

The major components of GGWM include:

- **Data collection**: Gathering data on groundwater levels, water quality parameters, geological characteristics, land use, and climate variables. This may involve field measurements, satellite imagery, or aerial surveys.

- **Geographic Information Systems (GIS)**: Using GIS to organize, analyse, and visualize geospatial data. GIS allows for creating maps and spatial models that aid in decision-making.
- **Remote sensing**: Utilizing satellite and aerial imagery to monitor changes in land use, vegetation, and water bodies. Remote sensing is instrumental in tracking trends over time.
- **Modelling and simulation**: Developing groundwater models that simulate water movement through aquifers. These models help predict future conditions and inform management decisions.
- **Decision support systems**: Integrating all available data and models into decision support systems that enable stakeholders to make informed choices regarding groundwater management (Table 19.1).

19.2.4 Significance in Sustainable Resource Management

GGWM is of paramount importance for achieving sustainable resource management. It ensures the optimal and responsible utilization of groundwater resources to prevent depletion and contamination. The significance of GGWM in sustainable resource management includes:

- **Resource conservation**: By accurately mapping aquifer boundaries and monitoring water levels, geospatial management prevents over-extraction and depletion of groundwater, preserving it for future generations.
- **Pollution control**: Geospatial tools help identify pollution sources and pathways, allowing for targeted efforts to control contamination and restore water quality.
- **Resilience to climate change**: GGWM can help understand the effects of climate change on groundwater resources and develop adaptation strategies to mitigate its impacts.
- **Economic benefits**: Sustainable groundwater management through geospatial techniques can lead to cost savings, increased agricultural productivity, and reduced expenses associated with environmental remediation.

GGWM is a vital component of environmental health and sustainable resource management. It provides a holistic view of groundwater resources, aids in informed decision-making, and plays a crucial role in ensuring the long-term availability and quality of this essential natural resource (Kross et al., 2022).

Table 19.1 GGWM involves various components that help monitor, assess, and sustainably manage groundwater resources

S. No.	Component	Description
1	Groundwater data collection	Gathering data on water levels, quality, and usage from monitoring wells, sensors, and surveys
2	Geographic information systems (GIS)	Using GIS to map and analyse groundwater data, enabling visualization and spatial analysis
3	Remote sensing	Utilizing satellite imagery and aerial photography to monitor land use, vegetation, and landscape changes affecting groundwater resources
4	Hydrogeological modelling	Creating models that simulate groundwater flow helps predict future trends and potential issues
5	Wellhead protection zones	Defining and managing areas around wells to safeguard groundwater quality and quantity
6	Water quality monitoring	Regularly testing groundwater for contaminants and pollutants to ensure its safety for consumption
7	Aquifer recharge assessment	Assessing the natural replenishment of aquifers and identifying methods to enhance recharge
8	Pumping tests	Conducting tests to determine aquifer properties, such as transmissivity and storativity, to aid in sustainable extraction
9	Decision support systems	Implementing software tools for data analysis and decision-making in groundwater management
10	Policy and regulatory framework	Establishing and enforcing regulations to protect and manage groundwater resources
11	Public awareness and education	Educating the public about groundwater importance, conservation, and responsible usage
12	Groundwater management plans	Developing and implementing strategies to manage and sustainably use groundwater resources
13	Climate change adaptation	Assessing the impact of climate change on groundwater and adjusting management practices accordingly

(continued)

Table 19.1 (continued)

S. No.	Component	Description
14	Stakeholder engagement	Involving local communities, water utilities, and other stakeholders in decision-making and planning
15	Emergency response plans	Preparing for and responding to groundwater-related emergencies, such as contamination incidents or over-extraction

Some of the components and their descriptions

19.3 Economic Aspects of Groundwater Management

19.3.1 Economic Importance of Groundwater

Groundwater is a critical component of the world's water resources and is significant in supporting various economic activities. It is an essential water source for agriculture, industry, and municipal use. In many regions, it is a primary source of freshwater, especially during droughts (Machiwal et al., 2019; Mishra, 2024). The economic importance of groundwater can be highlighted through several main points:

- **Agricultural productivity**: Groundwater irrigation is vital for global agriculture, contributing significantly to food production and farm income. This leads to job creation, higher crop yields, and increased economic stability in farming communities.
- **Industrial and commercial use**: Many industries rely on groundwater for their operations. These include manufacturing, mining, and power generation, where a stable and affordable source of water is essential for economic sustainability.
- **Municipal water supply**: Groundwater often supplements municipal water supplies, reducing the cost and environmental impact of treating surface water.
- **Tourism and recreation**: In areas dependent on tourism and recreation, maintaining groundwater-dependent ecosystems can generate income through tourism, thus contributing to the local economy (Mishra et al., 2024).

19.3.2 Cost of Groundwater Depletion

Groundwater depletion has several economic costs, which, if left unaddressed, can have severe consequences:

- **Energy costs**: As groundwater levels drop, the energy required to pump water from deeper aquifers increases, leading to higher operational costs for agricultural and industrial activities.

- **Infrastructure damage**: Subsidence caused by groundwater depletion can damage infrastructure, including roads, buildings, and pipelines, resulting in costly repairs.
- **Environmental and ecological costs**: Depleting groundwater can harm ecosystems that rely on it, impacting biodiversity, which, in turn, can disrupt ecosystems and impact industries like fisheries and tourism (Singh & Mishra, 2023).
- **Decreased agricultural productivity**: Over-extraction of groundwater can reduce agricultural yields, which can affect food prices and trade balances.
- **Increased water treatment costs**: If groundwater quality deteriorates due to over-extraction, water treatment costs for municipalities and industries can rise.

19.3.3 Economic Benefits of Sustainable Groundwater Management

Implementing sustainable groundwater management practices can yield numerous economic benefits:

- **Enhanced resource availability**: Sustainable management ensures a continuous and reliable source of water, supporting agricultural productivity, industry, and municipal water supply without increasing costs.
- **Cost savings**: Proper management reduces the need for deep well drilling, excessive energy consumption, and infrastructure repairs, thus cutting operational expenses.
- **Preservation of ecosystem services**: Sustaining groundwater levels helps maintain ecosystems and their services, which can boost revenue from tourism, fisheries, and recreation.
- **Resilience to droughts**: Managed aquifers can act as drought buffers, reducing the economic impact of water scarcity on agriculture and other sectors.
- **Long-term economic stability**: Sustainable groundwater management contributes to the long-term economic stability of regions that rely on groundwater-dependent industries.

19.3.4 The Role of Geospatial Technology in Economic Analysis

Geospatial technology, such as Geographic Information Systems (GIS), remote sensing, and data analysis tools, plays a pivotal role in assessing and managing groundwater resources from an economic perspective:

- **Resource assessment**: Geospatial technology allows for the mapping and monitoring groundwater resources, enabling accurate assessments of the quantity and quality of available water.

- **Cost–benefit analysis**: It aids in evaluating the economic impact of different groundwater management strategies by providing data on water levels, quality, and usage, facilitating cost–benefit analyses.
- **Risk assessment**: Geospatial tools help identify areas at risk of groundwater depletion or contamination, which is crucial for making informed decisions on resource protection and allocation.
- **Decision support**: Geospatial technology provides a platform for stakeholders to collaborate and make informed decisions about groundwater management, considering economic factors alongside environmental and social aspects.
- **Policy development**: It assists in developing policies and regulations that promote sustainable groundwater management, which can result in economic benefits in the long term.

The economic aspects of groundwater management, including its importance, the costs associated with its depletion, the benefits of sustainable management, and the role of geospatial technology in economic analysis, are crucial for understanding and promoting responsible groundwater management, which is essential for environmental health and economic sustainability (Maciejewska et al., 2022; Prudencio and Null, 2019).

19.4 Geospatial Tools and Techniques

19.4.1 GIS (Geographic Information Systems) in Groundwater Management

Geographic information systems (GIS) are pivotal in effectively managing groundwater resources, essential for ensuring environmental health and sustainability. GIS technology offers a framework for capturing, storing, analysing, and presenting geospatial data related to groundwater (Cromley, 2003; Lele, 2009). It facilitates decision-making by providing a comprehensive understanding of the groundwater landscape. The economic impacts of GIS in groundwater management are significant (Jain, 2023; Tiwari et al., 2023).

- **Data integration**: GIS allows the integration of diverse datasets such as geological, hydrological, land use, and water quality data. This integration enhances understanding of groundwater systems, leading to better management decisions.
- **Optimized resource allocation**: Authorities can allocate resources more efficiently by mapping groundwater resources. This optimization reduces costs associated with over-extraction or misallocation of resources.
- **Risk assessment and mitigation**: GIS enables the identification of vulnerable areas to contamination, over-pumping, or pollution. This risk assessment contributes to cost savings by preventing potential crises and the expenses associated with remediation.

- **Regulatory compliance**: GIS makes compliance with environmental regulations and monitoring obligations easier. Avoiding fines and legal repercussions positively impacts the finances of groundwater stakeholders.
- **Emergency response**: During natural disasters or emergencies, GIS helps in quick response and resource allocation, which can save both lives and money.
- **Infrastructure planning**: For cities and municipalities, GIS aids in planning infrastructure development, ensuring that investments are made where needed most.

19.4.2 Remote Sensing and Earth Observation

Remote sensing and Earth observation techniques are invaluable for monitoring groundwater resources from a distance. They contribute to environmental health by offering insights into changes and trends within the groundwater landscape (Nishad et al., 2024; Nuhu et al., 2021). The economic benefits of these tools are evident in various ways.

- **Continuous monitoring**: Satellite imagery and remote sensing provide real-time or periodic data on groundwater levels and land surface changes. This reduces the need for frequent field visits, saving time and resources.
- **Early warning systems**: By detecting land subsidence or pollution early, remote sensing helps prevent costly long-term damage and remediation efforts.
- **Resource allocation**: Remote sensing data assists in allocating resources for infrastructure maintenance, repairs, or upgrades by identifying areas at risk.
- **Precision agriculture**: In agriculture, remote sensing optimizes irrigation practices, conserving water and reducing costs. This has a direct economic impact on the agricultural sector and the cost of food production.
- **Insurance and risk management**: For insurance companies and risk assessors, Earth observation data can be used to develop more accurate pricing models for groundwater-related risks, leading to better economic decisions.

19.4.3 Data Collection and Analysis

Efficient data collection and analysis methods are fundamental for groundwater management and contribute significantly to economic outcomes. These methods ensure that the correct data is gathered and processed effectively.

19.4.4 Modelling and Simulation

Modelling and simulation are essential for predicting the behavior of groundwater systems under different scenarios, and they profoundly impact economics.

- **Scenario testing**: Models allow decision-makers to explore the economic implications of various management strategies without physically implementing them, reducing the risk associated with uncertain decisions.
- **Resource optimization**: Models can optimize resource allocation, minimizing costs related to groundwater extraction, treatment, and distribution.
- **Environmental impact assessment**: By simulating different scenarios, the economic impacts of potential ecological damage can be assessed, leading to preventative measures that save money in the long run.
- **Investment prioritization**: Models help prioritize investments in infrastructure and technology by evaluating their long-term economic viability in groundwater management.

Geospatial tools and techniques are invaluable for groundwater management in environmental health. They lead to cost savings, risk reduction, and sustainable use of groundwater resources, ultimately benefiting economies and ensuring communities' well-being (Singh et al., 2022).

19.5 Economic Modelling for Groundwater Management

Economic modelling plays a crucial role in assessing the economic impacts of GGWM. It involves various methods and analyses to determine the costs and benefits associated with managing groundwater resources (Sundaram et al., 2021). Two fundamental aspects within economic modelling for groundwater management are cost–benefit analysis and economic valuation of groundwater resources (Shirazi et al., 2012).

19.5.1 Cost–Benefit Analysis

Cost–benefit analysis (CBA) is a quantitative technique to compare the costs and benefits of a particular project, policy, or action. In the context of GGWM, CBA helps evaluate the economic feasibility and efficiency of different strategies (Mishan & Quah, 2020). The formula for CBA is as follows:

$$Net\ benefit = Total\ benefits - Total\ costs$$

where

- Total benefits: The sum of all positive economic impacts resulting from groundwater management, including increased water availability, reduced environmental damage, and enhanced public health.
- Total costs are the sum of all the costs incurred in implementing and maintaining groundwater management measures, such as infrastructure investment, monitoring, and administrative expenses.

The implementation of CBA in the context of GGWM might involve the following steps:

a. **Identifying costs and benefits**: Identify all relevant costs and benefits associated with groundwater management. Costs can include infrastructure development, monitoring equipment, staff salaries, etc. Benefits can encompass increased crop yields, improved water quality, and avoided costs related to over-extraction.
b. **Quantifying costs and benefits**: Assign monetary values to each cost and benefit. For example, the increased crop yield can be valued based on market prices, and the avoided costs of environmental damage can be assessed through environmental valuation techniques.
c. **Time discounting**: Since costs and benefits occur over time, it's essential to discount them to their present value using an appropriate discount rate (Mishra & Mishra, 2024). The formula for present value is:

$$PV = \frac{FV}{(1 + r)^t}$$

where

PV Present value
FV Future value (cost or benefit)
r Discount rate
t Time period

d. **Summing up**: Calculate the net benefit by subtracting the present value of costs from the present value of benefits.

e. **Sensitivity analysis**: Assess the impact of uncertain variables, such as discount rate and benefit valuations, by performing sensitivity analysis to determine the robustness of the results.

19.5.2 Economic Valuation of Groundwater Resources

Economic valuation aims to assign a monetary value to groundwater resources, which can be challenging due to their public and environmental significance. Several methods are available to estimate the economic value of groundwater.

19.5.2.1 Market Price Approach

This approach calculates the economic value based on the market price of water, such as water fees or charges (Sverdrup et al., 2014).

$$Economic\ value = Quantity\ of\ water\ used \times Water\ price$$

For example, let's consider a case where a small, picturesque town named "Crystal Springs" heavily relies on groundwater for its water supply. Crystal Springs is known for its stunning natural beauty, and its groundwater resources play a vital role in sustaining the local ecosystem and supporting various economic activities, particularly tourism and agriculture. The town is contemplating multiple strategies to assess its groundwater resources' economic value and make informed decisions about their usage and preservation.

To determine the economic value of groundwater using the market price approach, Crystal Springs' local government evaluates the water usage patterns in the town. They find that the town consumes an average of 1 million cubic meters of groundwater annually, and the local water utility charges \$0.10 per cubic meter. Therefore, the economic value of groundwater using this approach would be:

$$\text{Economic value} = 1,000,000\ \text{m}^3 \times \$0.10 = \$100,000$$

Based on the market price approach, this calculation reveals that the annual economic value of groundwater in Crystal Springs is \$100,000.

19.5.2.2 Cost of Replacement Approach

This approach assesses the economic value by estimating the cost of replacing groundwater with an alternative water source, like surface water or desalination (Kang et al., 2015).

$$Economic\ value = Cost\ of\ replacement\ water - Cost\ by\ using\ groundwater$$

Example: To assess the economic value of groundwater using the cost of replacement approach, the town investigates the cost of replacing groundwater with an alternative water source. They find that the cost of using surface or desalinated water is significantly higher than groundwater. The cost of replacement water is estimated to be \$0.20 per cubic meter, while groundwater costs remain at \$0.10 per cubic meter. Therefore, the economic value of using this approach would be:

$$\text{Economic value} = (\$0.20 - \$0.10) \times 1,000,000\ \text{m}^3 = \$100,000$$

In this case, the economic value of groundwater using the cost of replacement approach is also \$100,000.

19.5.2.3 Revealed Preference Methods

Methods like travel cost analysis and hedonic pricing analyze consumer behavior to determine the economic value of groundwater-related amenities and activities, such as recreational activities near groundwater bodies (Boyle, 2003).

Example: Crystal Springs uses revealed preference methods, such as travel cost analysis and hedonic pricing, to determine the economic value of groundwater-related amenities and activities. Through surveys and data analysis, they discovered that tourists are willing to pay an average of \$50 more for a vacation near the town's pristine groundwater-fed lake compared to areas without such amenities. With an estimated 10,000 annual visitors, the economic value is calculated as follows:

$$Economic\ value = Average\ additional\ expenditure\ per\ visitor \times Number\ of\ visitors$$

$$\text{Economic value} = \$50 \times 10,000 = \$500,000$$

Therefore, the economic value of groundwater using revealed preference methods is \$500,000 annually, considering the additional economic activity it generates.

19.5.2.4 Stated Preference Methods

Techniques like contingent valuation surveys involve asking individuals about their willingness to pay (WTP) to preserve or improve groundwater resources (Kroes & Sheldon, 1988).

$$Economic\ value = WTP\ per\ household \times Number\ of\ households$$

Example: Crystal Springs conducts contingent valuation surveys to gauge residents' willingness to pay (WTP) to preserve and improve groundwater resources. The survey results indicate the average WTP per household is \$200 annually. With 500 households in the town, the economic value using stated preference methods is:

$$\text{Economic value} = \$200 \times 500 = \$100,000$$

The economic value of groundwater based on stated preference methods is \$100,000, reflecting the amount residents are willing to pay to protect this vital resource.

In this example, the town of Crystal Springs uses various economic valuation methods to assess the monetary value of its groundwater resources, which range from \$100,000 to \$500,000 annually, depending on the method employed and the aspects considered. These values can guide policymakers and local stakeholders in making informed decisions about the management and protection of Crystal Springs's groundwater resources.

Implementation of economic valuation involves conducting surveys, collecting data on alternative water sources, and utilizing existing market data to calculate the economic value of groundwater resources. Additionally, integrating geospatial data, such as aquifer maps and groundwater quality information, can provide a more accurate and context-specific valuation (Pavlidis & Tsihrintzis, 2019; Tiwari et al., 2024).

19.5.3 Pricing and Incentive Mechanisms

Pricing and incentive mechanisms play a crucial role in efficiently managing groundwater resources. They are designed to balance the economic interests of water users, promote sustainable usage, and encourage conservation (Rossetto et al., 2019). Here, we will discuss the economic models and formulas used to implement pricing and incentive mechanisms for groundwater management.

19.5.3.1 Cost-Recovery Pricing

Cost-recovery pricing aims to recover the costs associated with developing, maintaining, and managing groundwater resources. This includes the costs of infrastructure, monitoring, and administrative expenses (Reichl et al., 1999). The formula for cost-recovery pricing can be expressed as:

$$Pricing(P) = \frac{Total\ costs}{Total\ volume\ of\ water\ extracted}$$

In this formula, “Total costs” include all the expenses incurred in groundwater management, and “Total volume of water extracted” represents the amount of groundwater used by all users. Pricing can be applied on a per-unit basis (e.g., price per cubic meter) or as a fixed fee.

19.5.3.2 Economic Incentive Pricing

Economic incentive pricing is designed to encourage efficient water use by pricing water at a level that reflects its scarcity and environmental value (Mishra & Mishra, 2024). This pricing mechanism often employs the concept of marginal cost, which is the additional cost of supplying one more unit of water. The economic incentive pricing can be expressed as:

$$Pricing\ (P) = Marginal\ cost\ (MC)$$

In this formula, the “Marginal cost” includes the operational cost of supplying one more water unit and the environmental cost associated with groundwater extraction. By pricing water at the marginal cost, users are incentivized to reduce consumption and minimize wastage.

19.5.3.3 Water Markets and Tradable Permits

Water markets and tradable permits introduce market-based mechanisms to allocate and manage groundwater resources. The formula for water markets involves determining the market price of water based on supply and demand dynamics (Ipe & Bhagwat, 2002). This can be expressed as:

$$Market\ price(P) = \frac{Demand\ for\ water}{-supply\ of\ water}$$

Tradable permits, on the other hand, allocate permits to water users, which can be bought, sold, or traded. Users can purchase permits to extract a specific volume of water. The permit price can be determined through market mechanisms based on user preferences and willingness to pay.

19.5.4 *Financial Instruments for Groundwater Management*

Financial instruments are used to raise funds, allocate resources, and manage financial risks in groundwater management. These instruments help finance infrastructure development, maintain monitoring systems, and cover operational expenses. Below, we discuss some financial instruments and their implementation.

19.5.4.1 Water User Fees and Charges

Water user fees and charges are direct payments made by groundwater users to fund the management and maintenance of groundwater resources. The formula for calculating user fees can be based on the cost-recovery pricing model mentioned earlier:

$$User\ fees(F) = \frac{Total\ costs}{total\ number\ of\ users}$$

This formula ensures that the financial burden is shared among all users and proportional to the overall costs incurred in managing groundwater.

19.5.4.2 Water Infrastructure Bonds

Water infrastructure bonds are financial instruments that allow governments or water authorities to raise capital for groundwater infrastructure development. These bonds are typically issued with a fixed interest rate and a specified maturity date. The formula for calculating the cost of bond issuance and interest payments can be expressed as:

$$Cost\ of\ bonds(CB) = Principal\ amount \times Interest\ rate \times Number\ of\ years$$

This formula helps determine the financial implications of issuing bonds and the interest costs associated with borrowing.

19.5.4.3 Insurance and Risk Management

Insurance and risk management instruments can help protect against unforeseen events or disasters affecting groundwater resources. The cost of insurance premiums can be determined based on actuarial calculations, taking into account the risk profile and the value of groundwater assets at risk.

Economic modelling plays a vital role in groundwater management, providing insights into the costs, benefits, valuation, pricing, and financial mechanisms involved. By integrating these economic perspectives with geospatial data and technology, decision-makers can better appreciate the economic significance of groundwater and make informed choices for its sustainable management, ultimately safeguarding environmental health (Aloui et al., 2023; Zhang and Chui, 2019).

19.6 Policy and Regulation

19.6.1 Legal Framework for Groundwater Management

Groundwater management is a critical aspect of environmental health and sustainability, and its economic impacts are strongly influenced by the legal framework within which it operates. In many countries, the legal framework for groundwater management typically consists of a combination of national and regional laws, regulations, and policies (Tiwari et al., 2024; Werner et al., 2019). Here, we will discuss the legal aspects of GGWM and how they impact the economy.

19.6.1.1 Water Rights and Ownership

- A fundamental element of groundwater management is the legal recognition of water rights and ownership. Water rights define who has the legal authority to access and use groundwater. These rights can be either public or private, depending on the legal framework of a given region. Private water rights often involve property owners' rights to the groundwater beneath their land.
- The legal structure for assigning and enforcing water rights can significantly impact the economic value of groundwater resources. Secure property rights can incentivize efficient use, investment in groundwater infrastructure, and trading, which can lead to economic benefits.

19.6.1.2 Groundwater Use and Extraction Permits

- Many legal frameworks require permits for the extraction of groundwater. These permits may involve specific conditions, such as volume limits, allowable uses, and monitoring requirements. Such permits help regulate and control groundwater use, ensuring the sustainable management of this valuable resource.
- The issuance and administration of these permits can impact the economy in various ways, including fees associated with permitting, compliance costs for users, and the efficiency and fairness of allocation processes.

19.6.1.3 Environmental Regulations

- Groundwater management must also consider environmental regulations that protect against contamination and over-extraction. These regulations may impose costs on users, such as those related to pollution control measures or penalties for non-compliance.
- While environmental regulations can increase costs for groundwater users, they are essential for safeguarding water quality and long-term sustainability, contributing to overall economic well-being.

19.6.2 Role of Government and Regulatory Agencies

Government agencies play a vital role in shaping and implementing groundwater management policies. The specific roles and responsibilities of these agencies can vary widely by region, but their actions have a profound impact on the economic aspects of GGWM.

19.6.2.1 Groundwater Monitoring and Data Collection

- Government agencies are often responsible for collecting and maintaining essential geospatial data related to groundwater resources. This includes data on water levels, quality, and aquifer characteristics. Geospatial technologies like GIS (Geographic Information Systems) are increasingly employed to enhance data collection and analysis.
- Accurate and up-to-date geospatial data are crucial for informed decision-making, reducing uncertainty for groundwater users, and supporting economic development. Access to such data can foster more efficient resource allocation, benefiting various industries and stakeholders.

19.6.2.2 Enforcement of Regulations

- Regulatory agencies are responsible for enforcing groundwater management policies and regulations. This includes overseeing the issuance and compliance of permits, as well as taking action against violations.
- Effective enforcement ensures that users adhere to sustainable practices, reducing the risk of over-extraction or contamination. These efforts promote long-term resource viability, which can have significant economic benefits for industries reliant on groundwater.

19.6.2.3 Investment in Infrastructure

- Government agencies may also play a role in funding or facilitating investments in groundwater infrastructure. This can include the development of wells, distribution systems, and water treatment facilities.
- Investment in infrastructure can stimulate economic growth, create jobs, and improve access to clean and reliable water sources, contributing to the economic well-being of both urban and rural communities.

19.6.2.4 Stakeholder Engagement

- Regulatory agencies often engage with various stakeholders, including local communities, industries, and environmental organizations. Engaging these stakeholders can help build consensus, reduce conflicts, and ensure that policies are well-informed and balanced.
- Effective stakeholder engagement can lead to more predictable and stable regulatory environments, providing economic actors with the certainty they need to plan investments and operations effectively.

19.6.3 Example of Successful Water Modelling and Regulating Policies

Water scarcity and contamination pose significant threats to environmental health globally. Geospatial technologies, coupled with effective regulatory policies, are emerging as powerful tools for sustainable groundwater management. This section demonstrates two exemplary cases of economic benefits of such an approach.

19.6.3.1 California's Central Valley Groundwater Management

California's Central Valley, the agricultural heartland of the state, faced decades of overdraft, leading to land subsidence and saltwater intrusion. In response, the Sustainable Groundwater Management Act (SGMA) was enacted in 2014. This landmark legislation mandated the creation of local Groundwater Sustainability Agencies (GSAs) with the authority to develop and implement groundwater management plans.

Modelling Success:

- Satellite-based InSAR (Interferometric Synthetic Aperture Radar) technology played a crucial role in quantifying land subsidence. Measurements revealed a cumulative subsidence of up to 28 cm (11 inches) between 2003 and 2010, highlighting the urgency of the situation.
- Hydrologic models were then developed to simulate groundwater flow and predict future water levels under various pumping scenarios. These models integrated factors like precipitation, recharge rates, and aquifer properties.

Economic Impact:

- A 2019 study by the Pacific Institute estimated that SGMA implementation could generate net economic benefits of $14.2 billion in California's Central Valley over the next 20 years.
- This economic benefit stems from:

Preserving agricultural productivity: Sustainable groundwater management safeguards water supplies for irrigation, protecting a sector that contributes over $50 billion annually to the state's economy.
Reduced infrastructure costs: Land subsidence can damage canals, pipelines, and other infrastructure. Preventing further subsidence translates to significant cost savings on repairs and maintenance.
Improved water quality: Saltwater intrusion threatens freshwater supplies. SGMA aims to prevent this, protecting a vital resource for drinking water and other uses.

19.6.3.2 The Netherlands' National Water Plan

The Netherlands faces unique water management challenges due to its low-lying geography and dependence on shared river basins. The National Water Plan, adopted in 1989, outlines a comprehensive approach encompassing infrastructure development, spatial planning, and water quality regulations.

Modelling Success:

- Advanced hydrological models are used to forecast river discharges, storm surges, and potential flooding events. These models consider factors like rainfall patterns, tides, and sea level rise.
- Spatial planning integrates water management considerations into land-use decisions. Floodplains are designated for water storage, and development in high-risk areas is restricted.

Economic Impact:

- A 2020 report by the World Bank highlights the Netherlands' success in reducing flood risks. This translates to:

  Reduced economic losses: Between 1953 (a year with devastating floods) and 2003, the Netherlands witnessed a 90% decrease in flood-related economic losses. This reflects the effectiveness of water management strategies in protecting infrastructure and property.
  Increased property values: Flood-proofing measures and spatial planning ensure safer living environments, leading to higher property values in previously flood-prone areas.
  Enhanced tourism: A reliable water management system fosters a sense of security and attracts tourists, boosting the economy.

These two cases demonstrate the power of geospatial modelling and regulatory policies in achieving sustainable groundwater management. The economic benefits are substantial, ranging from safeguarding agricultural productivity to preventing flood damage. As water scarcity intensifies globally, these successful examples offer valuable lessons for other regions seeking to ensure environmental health and economic prosperity.

The legal framework and the role of government and regulatory agencies are critical aspects of GGWM that have significant economic impacts (Teng et al., 2019). A well-structured legal framework can incentivize efficient resource use and encourage investment, while government agencies are essential for data collection, regulation enforcement, infrastructure development, and stakeholder engagement, all of which play a role in shaping the economic outcomes of groundwater management initiatives (Le Coent, 2023; Shit et al., 2023).

19.7 Stakeholder Engagement and Participation

19.7.1 Importance of Involving Stakeholders

Engaging stakeholders in GGWM is crucial for several reasons:

- **Diverse perspectives**: Stakeholders represent a broad spectrum of interests and expertise, including government agencies, local communities, environmental organizations, and private businesses. Involving these stakeholders ensures that a wide range of perspectives is considered in decision-making processes.
- **Legitimacy and accountability**: By involving stakeholders, the decision-making process gains legitimacy and accountability. This is especially important in groundwater management, as it often involves resource allocation and regulatory measures that impact various groups.
- **Local knowledge**: Local communities often possess invaluable, on-the-ground knowledge about groundwater resources. Involving them can enhance the accuracy and effectiveness of management strategies (Singh & Mishra, 2023).
- **Social acceptance**: Stakeholder involvement helps build social acceptance for GGWM initiatives, as it empowers communities and fosters a sense of ownership in the decision-making process.
- **Conflict resolution**: Conflicts may arise in groundwater management due to varying interests and limited resources. Engaging stakeholders provides a platform to identify and address conflicts through dialogue and negotiation.

19.7.2 Community-Based Approaches

Community-based approaches to GGWM involve active participation of local communities in the planning and decision-making processes. Some aspects include:

- **Capacity building**: Empowering communities with knowledge and skills related to geospatial data and groundwater management fosters a sense of ownership and responsibility for resource stewardship.

- **Data collection**: Local communities can play a crucial role in data collection through citizen science initiatives. This can enhance the availability and quality of geospatial data.
- **Cultural considerations**: Recognizing and respecting the cultural significance of groundwater for some communities is vital. Community-based approaches ensure that cultural values and practices are integrated into management strategies.
- **Sustainable practices**: Local communities often have a vested interest in long-term sustainability. Their involvement can help identify and implement sustainable groundwater management practices.

19.7.3 Public–Private Partnerships

Engaging both public and private stakeholders in GGWM can be beneficial in various ways:

- **Resource mobilization**: Private sector involvement can contribute financial resources, technology, and expertise to complement public initiatives in groundwater management.
- **Innovation and efficiency**: Private sector partners can bring innovation and efficiency to data collection, analysis, and monitoring through advanced technologies and methodologies.
- **Regulatory compliance**: Collaboration with private entities can help ensure that businesses adhere to environmental regulations, promoting responsible groundwater use.
- **Shared responsibility**: Public–private partnerships can distribute responsibility for groundwater management more evenly, reducing the burden on public agencies.

19.7.4 Challenges and Solutions in Stakeholder Engagement

- **Conflict of interests**: Differing interests among stakeholders can lead to conflicts. To address this, a transparent and inclusive decision-making process, conflict mediation mechanisms, and clear guidelines for participation are necessary.
- **Data sharing and privacy**: Concerns about data sharing and privacy can hinder stakeholder engagement. Developing data-sharing agreements and protocols that address these concerns is essential.
- **Capacity and resources**: Some stakeholders may lack the capacity or resources to engage effectively. To overcome this, capacity-building programs, financial support, and technical assistance should be made available.
- **Community empowerment**: Empowering local communities may face resistance from centralized authorities. Solutions include advocacy, awareness campaigns, and policy reforms that promote community participation.

- **Sustainability**: Ensuring the continued engagement of stakeholders over the long term can be challenging. Creating incentive structures and recognition mechanisms can help sustain stakeholder involvement.

Stakeholder engagement and participation are integral to successful GGWM with positive economic impacts. They enhance decision-making, encourage sustainable practices, and help resolve conflicts, ultimately leading to better environmental health and economic outcomes (Rai & Singh, 2020). Community-based approaches and public–private partnerships can provide valuable contributions while addressing various challenges through transparent, inclusive, and innovative strategies (Murray et al., 2003; Qureshi, 2015).

19.8 Challenges and Barriers to GGWM

19.8.1 Data Quality and Availability

One of the primary challenges in implementing GGWM is the quality and availability of data. Effective groundwater management relies heavily on accurate and up-to-date information about groundwater resources, aquifer characteristics, water levels, and water quality. Unfortunately, in many regions, comprehensive and reliable groundwater data is lacking or inadequate. This data gap can be attributed to limited monitoring infrastructure, inconsistent data collection methods, and a lack of coordination among various stakeholders (Chang et al., 2008; Hammond et al., 2021).

The economic impact of this data challenge is significant. Without reliable data, it becomes difficult for policymakers and water resource managers to make informed decisions regarding groundwater allocation, sustainable use, and contamination prevention. This can lead to over-extraction, inadequate protection of water quality, and increased costs associated with addressing issues that arise due to the lack of data-driven decision-making (Li et al., 2017).

19.8.2 Technological and Financial Constraints

The successful implementation of GGWM often requires significant investments in technology and infrastructure. Collecting, processing, and analysing geospatial data for groundwater management typically involves using Geographic Information Systems (GIS), remote sensing, geophysical surveys, and other advanced technologies. These tools can be expensive to acquire and maintain, posing a barrier for many regions, particularly those with limited financial resources (Noble et al., 2019).

Furthermore, the costs associated with adopting and operating these technologies and training personnel to use them effectively can strain the financial resources of governmental bodies and organizations responsible for groundwater management. Consequently, the economic impact includes the need for budget allocation and potential disparities in access to advanced technology among regions with varying financial capabilities (Aicha et al., 2020).

19.8.3 Institutional and Governance Challenges

Effective GGWM requires a well-defined institutional framework and governance structure. In many cases, the responsibilities for groundwater management could be more cohesive among different agencies, leading to issues of coordination and accountability. Institutional fragmentation can hinder the establishment of comprehensive and coherent policies, regulations, and enforcement mechanisms (Nyerges et al., 1997).

The economic impact of these governance challenges includes inefficiencies and conflicts in managing groundwater resources. Inadequate institutional frameworks can lead to uncoordinated drilling and pumping activities, over-extraction, and increased competition for water resources, resulting in economic losses due to resource degradation and conflicts over water access (Mende et al., 2007).

19.8.4 Climate Change Impacts on Groundwater

Climate change presents a significant challenge to GGWM. Changing precipitation patterns, altered temperature regimes, and increased evaporation rates can impact the recharge and availability of groundwater resources. As climate change exacerbates water scarcity in some regions, groundwater may become an even more critical water source, making its sustainable management of utmost importance (Cacho et al., 2019; Li et al., 2021).

The economic impacts of climate change on groundwater include increased water supply and infrastructure development costs. Regions that depend heavily on groundwater resources may need to invest in additional technologies, infrastructure, and conservation measures to adapt to changing climate conditions. Furthermore, they may face increased costs associated with addressing groundwater quality issues due to changing hydrological patterns and contamination risks (Kuchampudi et al., 2023) (Table 19.2).

Addressing the challenges and barriers to GGWM is crucial for effectively and sustainably using this vital resource. Overcoming data limitations, technological and financial constraints, institutional issues, and adapting to climate change impacts can have significant economic implications in terms of water resource sustainability, cost-effectiveness, and environmental health (Kang et al., 2009). To maximize the

Table 19.2 Overview of challenges/barriers and possible solutions to GGWM

Category	Challenges/barriers	Possible solutions
1. Data quality and availability	Insufficient groundwater data availability	Invest in groundwater monitoring infrastructure
	Inaccurate or outdated data	Implement data validation and quality assurance
	Limited accessibility to data	Develop open-access data repositories
	Data gaps in geospatial coverage	Collaborate with neighboring regions to share data
2. Technological and financial constraints	Lack of advanced geospatial technology	Invest in GIS and remote sensing technology
	High costs associated with data collection	Seek funding from government and private sectors
	Limited IT infrastructure	Develop a roadmap for technology infrastructure
	Resource constraints for data analysis	Train personnel in data analysis and interpretation
3. Institutional and governance challenges	Fragmented regulatory framework	Establish a unified regulatory authority
	Lack of collaboration among stakeholders	Foster partnerships and engage in stakeholder dialogue
	Insufficient legal enforcement	Strengthen regulatory policies and enforcement
	Inadequate public awareness	Promote public education and awareness campaigns
4. Climate change impacts on groundwater	Altered precipitation patterns	Implement adaptive management strategies
	Increased evapotranspiration	Promote efficient water use and conservation
	Sea-level rise and saline intrusion	Monitor and manage saltwater intrusion proactively
	Impact on aquifer recharge rates	Assess and model groundwater recharge under new conditions

economic benefits of GGWM, it is essential to invest in data collection and analysis, technological capabilities, governance reforms, and climate resilience measures (Sumathi et al., 2008).

19.9 Future Prospects and Trends

The future of GGWM in environmental health is brimming with potential and transformative trends. As we advance into an era marked by increasing ecological challenges and the necessity for sustainable resource management, geospatial technologies play an instrumental role. Here, we will identify the major future prospects and trends that are set to shape the field (Guelfo et al., 2019).

19.9.1 Emerging Technologies and Innovations

- **Remote sensing advancements**: The continued development of remote sensing technologies, such as hyperspectral and multispectral imaging, promises more precise and detailed groundwater mapping. These technologies detect subtle changes in groundwater levels and quality, providing valuable data for decision-makers.
- **IoT and sensor networks**: The Internet of Things (IoT) is set to revolutionize groundwater monitoring. The deployment of smart sensors in wells and other data collection points will enable real-time data transmission and analysis, resulting in quicker response times for groundwater-related issues.
- **Machine learning and big data**: Integrating machine learning into geospatial analysis will enable predictive modelling of groundwater behaviour. This can help anticipate issues like contamination or over-extraction, thereby enhancing the management of groundwater resources.
- **3D mapping and visualization**: Three-dimensional mapping and visualization technologies are becoming more accessible and can aid in understanding complex groundwater flow patterns, facilitating better decision-making and resource management.

19.9.2 Integrating Geospatial Data with Artificial Intelligence

- **Data fusion and integration**: AI can seamlessly merge various geospatial datasets, enabling a holistic view of groundwater resources. This integration is crucial for understanding the interplay of factors affecting groundwater, including climate change and land use.
- **Predictive analytics**: Machine learning algorithms can predict changes in groundwater levels and quality, enabling proactive management. For instance, AI can help forecast drought conditions and suggest measures to mitigate their impact.
- **Real-time monitoring and alerts**: AI-driven systems can provide real-time monitoring and issue alerts when abnormal patterns are detected, such as sudden drops

in water table levels or spikes in contamination. This early warning system can help prevent crises.

- **Optimizing resource allocation**: AI can optimize the allocation of groundwater resources by considering various factors like demand, recharge rates, and environmental protection. This leads to more efficient resource utilization and cost savings.

19.9.3 Policy and Research Priorities

- **Regulatory frameworks**: Governments and environmental agencies need to establish robust regulatory frameworks that consider geospatial data in groundwater management. Regulations should encourage sustainable practices, including restrictions on extraction during droughts and protection of critical recharge areas.
- **Research on climate change impacts**: A top priority is understanding how climate change affects groundwater resources. Research should focus on predicting changes in precipitation patterns, sea-level rise's influence on coastal aquifers, and the potential for increased contamination due to extreme weather events.
- **Interdisciplinary collaboration**: Collaborative efforts between geospatial experts, hydrologists, environmental scientists, and policymakers are essential. These partnerships can ensure that geospatial data informs groundwater management policies effectively.
- **Community engagement**: Public awareness and participation are critical. Research should aim to develop tools and strategies that empower local communities to contribute to groundwater management through citizen science, data collection, and decision-making processes.

The future prospects and trends in GGWM are exciting and hold the promise of more sustainable and efficient resource use. Emerging technologies, the integration of AI, and the right policy and research priorities will play a pivotal role in ensuring effective groundwater management's economic and environmental health benefits (Almasri, 2007; Nazarnia et al., 2020).

19.10 Conclusion

GGWM is a critical component of sustainable resource management with profound implications for environmental health and the economy. We began by elucidating the definition and concept of GGWM, emphasizing its role in safeguarding environmental health. The major components of this approach and its significance in promoting sustainability were highlighted. It was made evident that sustainable groundwater management is imperative in mitigating the economic repercussions of groundwater depletion, which is explored in the subsequent section. We examined the economic importance of groundwater and the costs associated with its depletion. Additionally, we examined the economic benefits derived from sustainable groundwater management and the indispensable role of geospatial technology in economic analysis. Geospatial tools and techniques, such as GIS, remote sensing, and data analysis, provide essential support for groundwater management. Furthermore, economic modelling techniques, including cost–benefit analysis, economic valuation, pricing, and financial instruments, offer valuable insights into the economic aspects of groundwater management. An understanding of the regulatory framework and the importance of stakeholder engagement is paramount. Effective policies, regulations, and stakeholders' involvement are crucial to groundwater management's success. Despite these merits, challenges and barriers, including data quality, technological constraints, institutional and governance issues, and the looming threat of climate change, must be addressed to ensure effective GGWM. Emerging technologies and innovations, such as integrating geospatial data with artificial intelligence, are likely to reshape groundwater management. Policy and research priorities will play a pivotal role in harnessing the potential of GGWM for environmental health and economic prosperity. As we move forward, we must acknowledge this field's importance and continue exploring its dynamic and evolving landscape.

Acknowledgements I would like to express my sincere gratitude to all those who have contributed to the development of this book chapter. First and foremost, I am thankful to the editor of this book, for giving me the opportunity to contribute to this important publication.

References

Ahmad, A. Y., & Al-Ghouti, M. A. (2020). Approaches to achieve sustainable use and management of groundwater resources in Qatar: A review. *Groundwater for Sustainable Development, 11*, 100367.

Aicha, O., Abdessamad, G., & Hassane, J. O. (2020). Groundwater flooding in urban areas: Occurrence process, potential impacts and the role of remote sensing and GIS techniques in preventing it. In *2020 IEEE International conference of Moroccan Geomatics (Morgeo)* (pp. 1–5). IEEE.

Almasri, M. N. (2007). Nitrate contamination of groundwater: A conceptual management framework. *Environmental Impact Assessment Review, 27*(3), 220–242.

Aloui, S., Zghibi, A., Mazzoni, A., Elomri, A., & Triki, C. (2023). Groundwater resources in Qatar: A comprehensive review and informative recommendations for research, governance, and management in support of sustainability. *Journal of Hydrology: Regional Studies, 50*, 101564.

Avtar, R., Aggarwal, R., Kharrazi, A., Kumar, P., & Kurniawan, T. A. (2020). Utilizing geospatial information to implement SDGs and monitor their Progress. *Environmental Monitoring and Assessment, 192*, 1–21.

Basack, S., Loganathan, M. K., Goswami, G., & Khabbaz, H. (2022). Saltwater intrusion into coastal aquifers and associated risk management: Critical review and research directives. *Journal of Coastal Research, 38*(3), 654–672.

Biswas, S., Sahoo, S., & Debsarkar, A. (2022). Social vulnerability of arsenic contaminated groundwater in the context of Ganga-Brahmaputra-Meghna Basin: A critical review. In *Geospatial technology for environmental hazards: Modeling and management in Asian countries* (pp. 39–61).

Boyle, K. J. (2003). Introduction to revealed preference methods. In *A primer on nonmarket valuation* (pp. 259–267). Springer.

Cacho, J. F., Negri, M. C., Zumpf, C. R., & Campbell, P. (2019). Introducing perennial biomass crops into agricultural landscapes to address water quality challenges and provide other environmental services. *Wiley Interdisciplinary Reviews: Energy and Environment, 7*(2), e275.

Chaminé, H. I., Carvalho, J. M., & Freitas, L. (2021). Sustainable groundwater management in rural communities in developed countries: Some thoughts and outlook. *Mediterranean Geoscience Reviews, 3*, 389–398.

Chang, N. B., Parvathinathan, G., & Breeden, J. B. (2008). Combining GIS with fuzzy multicriteria decision-making for landfill siting in a fast-growing urban region. *Journal of Environmental Management, 87*(1), 139–153.

Chiloane, C., Dube, T., & Shoko, C. (2022). Impacts of groundwater and climate variability on terrestrial groundwater dependent ecosystems: A review of geospatial assessment approaches and challenges and possible future research directions. *Geocarto International, 37*(23), 6755–6779.

Cromley, E. K. (2003). GIS and disease. *Annual Review of Public Health, 24*(1), 7–24.

de Paul Obade, V., & Moore, R. (2019). Synthesizing water quality indicators from standardized geospatial information to remedy water security challenges: A review. *Environment International, 119*, 220–231.

Etikala, B., Golla, V., Li, P., & Renati, S. (2019). Deciphering groundwater potential zones using MIF technique and GIS: A study from Tirupati area, Chittoor District, Andhra Pradesh, India. *HydroResearch, 1*, 1–7.

Fienen, M. N., & Arshad, M. (2016). The international scale of the groundwater issue. In *Integrated groundwater management: concepts, approaches and challenges* (pp. 21–48).

Geospatial Analytics Market Insights. (n.d.). https://www.mordorintelligence.com/industry-reports/geospatial-analytics-market

Guelfo, J. L., Marlow, T., Klein, D. M., Savitz, D. A., Frickel, S., Crimi, M., & Suuberg, E. M. (2019). Evaluation and management strategies for per-and polyfluoroalkyl substances (PFASs) in drinking water aquifers: Perspectives from impacted US Northeast communities. *Environmental Health Perspectives, 126*(6), 065001.

Hammond, E. B., Coulon, F., Hallett, S. H., Thomas, R., Hardy, D., Kingdon, A., & Beriro, D. J. (2021). A critical review of decision support systems for brownfield redevelopment. *Science of the Total Environment, 785*, 147132.

Ipe, V. C., & Bhagwat, S. B. (2002). Chicago's water market: Dynamics of demand, prices and scarcity rents. *Applied Economics, 34*(17), 2157–2163.

Jain, H. (2023). Groundwater vulnerability and risk mitigation: A comprehensive review of the techniques and applications. In *Groundwater for Sustainable Development* (p. 100968).

Kalhor, K., Ghasemizadeh, R., Rajic, L., & Alshawabkeh, A. (2019). Assessment of groundwater quality and remediation in karst aquifers: A review. *Groundwater for Sustainable Development, 8*, 104–121.

Kang, P., Geum, Y., Park, H. W., Kim, S. G., Sung, T. E., & Lee, H. (2015). A market-based replacement cost approach to technology valuation. *Journal of Korean Institute of Industrial Engineers, 41*(2), 150–161.

Kang, Y., Khan, S., & Ma, X. (2009). Climate change impacts on crop yield, crop water productivity and food security—A review. *Progress in Natural Science, 19*(12), 1665–1674.

Khan, D., & Samadder, S. R. (2014). Municipal solid waste management using Geographical Information System aided methods: A mini review. *Waste Management & Research, 32*(11), 1049–1062.

Kom, K. P., Gurugnanam, B., & Sunitha, V. (2024). Delineation of groundwater potential zones using GIS and AHP techniques in Coimbatore district, South India. *International Journal of Energy and Water Resources, 8*(1), 85–109.

Kroes, E. P., & Sheldon, R. J. (1988). Stated preference methods: An introduction. *Journal of Transport Economics and Policy* (pp. 11–25).

Kross, A., Kaur, G., & Jaeger, J. A. (2022). A geospatial framework for the assessment and monitoring of environmental impacts of agriculture. *Environmental Impact Assessment Review, 97*, 106851.

Kuchampudi, S. V., Panchal, S., Naveen, S., Suresh, V. P. S., Abdulrahman, I. S., Salman, H. M., & Singh, P. (2023). Analysis of sustainable groundwater management policies for urban development. *Environmental Quality Management.*

Kumar, N., Kushwaha, R. R., Meena, N. R., Mishra, H., & Yadav, A. P. S. (2023). A study on costs and returns of paddy cultivation in Ambedkar Nagar district of Uttar Pradesh. *International Journal of Statistics and Applied Mathematics, SP, 8*(3), 107–111.

Laituri, M., Richardson, R. B., Kim, J., Cline, L. V., Viscuso, S., & Schwartz, L. (2022). Examining second-order impacts of COVID-19 in urban areas. *Annals of GIS, 28*(4), 479–490.

Le Coent, P. (2023). Payment for environmental services related to aquifers: a review of specific issues and existing programmes. In *Review of Agricultural, Food and Environmental Studies* (pp. 1–38).

Lele, S. (2009). Watershed services of tropical forests: From hydrology to economic valuation to integrated analysis. *Current Opinion in Environmental Sustainability, 1*(2), 148–155.

Levintal, E., Kniffin, M. L., Ganot, Y., Marwaha, N., Murphy, N. P., & Dahlke, H. E. (2023). Agricultural managed aquifer recharge (Ag-MAR)—A method for sustainable groundwater management: A review. *Critical Reviews in Environmental Science and Technology, 53*(3), 291–314.

Li, P., Karunanidhi, D., Subramani, T., & Srinivasamoorthy, K. (2021). Sources and consequences of groundwater contamination. *Archives of Environmental Contamination and Toxicology, 80*, 1–10.

Li, P., Tian, R., Xue, C., & Wu, J. (2017). Progress, opportunities, and key fields for groundwater quality research under the impacts of human activities in China with a special focus on western China. *Environmental Science and Pollution Research, 24*, 13224–13234.

Maantay, J. (2002). Mapping environmental injustices: Pitfalls and potential of geographic information systems in assessing environmental health and equity. *Environmental Health Perspectives, 110*(suppl 2), 161–171.

Machiwal, D., Cloutier, V., Güler, C., & Kazakis, N. (2019). A review of GIS-integrated statistical techniques for groundwater quality evaluation and protection. *Environmental Earth Sciences, 77*, 1–30.

Maciejewska, A., Kuzak, Ł., Sobieraj, J., & Metelski, D. (2022). The impact of opencast lignite mining on rural development: A literature review and selected case studies using desk research, panel data and GIS-based analysis. *Energies, 15*(15), 5402.

Majumdar, A., & Avishek, K. (2023). Riparian zone assessment and management: An integrated review using geospatial technology. *Water, Air, & Soil Pollution, 234*(5), 319.

Masih, I., Maskey, S., Mussá, F. E. F., & Trambauer, P. (2014). A review of droughts on the African continent: A geospatial and long-term perspective. *Hydrology and Earth System Sciences, 19*(9), 3635–3649.

Mastrocicco, M., & Colombani, N. (2021). The issue of groundwater salinization in coastal areas of the mediterranean region: A review. *Water, 13*(1), 90.

McKinney, D. C. (1999). Modeling water resources management at the basin level: Review and future directions.

Mende, A., Astorga, A., & Neumann, D. (2007). Strategy for groundwater management in developing countries: A case study in northern Costa Rica. *Journal of Hydrology, 334*(1–2), 109–124.

Merchant, J. W. (1994). GIS-based groundwater pollution hazard assessment: A critical review of the DRASTIC model. *Photogrammetric Engineering and Remote Sensing, 60*, 1117–1117.

Mishan, E. J., & Quah, E. (2020). *Cost-benefit analysis*. Routledge.

Mishra, H., & Mishra, D. (2023). Artificial intelligence and machine learning in agriculture: Transforming farming systems. *Research Trends in Agricultural Science, 1*, 1–16.

Mishra, H., & Mishra, D. (2024). Economic evaluation of UAV-based soil sampling approaches. In *Applications of Computer vision and drone technology in agriculture 4.0* (pp. 271–291). Springer.

Mishra, H., & Mishra, D. (2024). Economic implications of climate change: Costs, adaptation and mitigation. In *Exploring horizons: Trends in multidisciplinary education and research* (1st edn., pp. 7–31). Bharti Publications.

Mishra, H. (2024). Counting the costs: Understanding agricultural economics and the concept of costs. In K. Sharma, P. Singh, S. Singh, A. Yadav, & V. Awasthi (Eds.), *Text book on practices for agricultural sustainability* (pp. 198–210). Golden Leaf Publishers.

Mishra, H. (2024). Molecular technology in agriculture. In D. K. Patel, S. Dwivedi, A. Dubey, A. Kumar, & W. Ali (Eds.), *Agricultural transformation in the modern era* (pp. 204–226). Elite Publishing House.

Mishra, H., Tiwari, A. K., & Nishad, D. C. (2024). Economic viability of sustainable agriculture practices in modern farming. *Advances in Agriculture Sciences Volume II, 24*(4), 105.

Mishra, N., Khare, D., Gupta, K. K., & Shukla, R. (2014). Impact of land use change on groundwater—A review. *Advances in Water Resource and Protection, 2*(28), 28–41.

Mondal, B. K., & Sahoo, S. (2022). Application of geospatial techniques for urban flood management: A review. In *Spatial Modelling of Flood Risk and Flood Hazards: Societal Implications* (pp. 225–236).

Murray, B. B. R., Zeppel, M. J., Hose, G. C., & Eamus, D. (2003). Groundwater-dependent ecosystems in Australia: It's more than just water for rivers. *Ecological Management & Restoration, 4*(2), 110–113.

Naresh Kumar, D., Venkateswara Rao, T., Veerla, V.K., Etikala, B., Mohana Prasada Rao, Y., & Jujjuvarapu, A. (2023). Systematic approach of groundwater resources assessment using remote sensing and multi-influence factor (MIF) techniques in Medchal Mandal, Telangana State, India. In E. Balaji, G. Veeraswamy, P. Mannala, & S. Madhav (Eds.), *Emerging technologies for water supply, conservation and management* (pp. 343–358). Springer.

Nazarnia, H., Nazarnia, M., Sarmasti, H., & Wills, W. O. (2020). A systematic review of civil and environmental infrastructures for coastal adaptation to sea level rise. *Civil Engineering Journal, 6*(7), 1375–1399.

Nishad, D. C., Mishra, H., Tiwari, A. K., & Mishra, D. (2024). Post-harvest management: Enhancing food security and sustainability. *Advances in Agriculture Sciences Volume II, 24*(4), 136.

Noble, M. M., Harasti, D., Pittock, J., & Doran, B. (2019). Linking the social to the ecological using GIS methods in marine spatial planning and management to support resilience: A review. *Marine Policy, 108*, 103657.

Ntona, M. M., Busico, G., Mastrocicco, M., & Kazakis, N. (2023). Coupling SWAT and DPSIR models for groundwater management in Mediterranean catchments. *Journal of Environmental Management, 344*, 119543.

Nuhu, S. K., Manan, Z. A., Alwi, S. R. W., & Reba, M. N. M. (2021). Roles of geospatial technology in eco-industrial park site selection: State–of–the-art review. *Journal of Cleaner Production, 309*, 127361.

Nyerges, T., Robkin, M., & Moore, T. J. (1997). Geographic information systems for risk evaluation: Perspectives on applications to environmental health. *Cartography and Geographic Information Systems, 24*(3), 123–144.

Pavlidis, G., & Tsihrintzis, V. A. (2019). Environmental benefits and control of pollution to surface water and groundwater by agroforestry systems: A review. *Water Resources Management, 32*, 1–29.

Pérez Hoyos, I. C., Krakauer, N. Y., Khanbilvardi, R., & Armstrong, R. A. (2016). A review of advances in the identification and characterization of groundwater dependent ecosystems using geospatial technologies. *Geosciences, 6*(2), 17.

Prudencio, L., & Null, S. E. (2019). Stormwater management and ecosystem services: A review. *Environmental Research Letters, 13*(3), 033002.

Qureshi, A. S. (2015). Improving food security and livelihood resilience through groundwater management in Pakistan. *Global Advanced Research Journal of Agricultural Science, 4*, 687–710.

Rai, P. K., & Singh, J. S. (2020). Invasive alien plant species: Their impact on environment, ecosystem services and human health. *Ecological Indicators, 111*, 106020.

Rashid, A., Ayub, M., Ullah, Z., Ali, A., Sardar, T., Iqbal, J., Gao, X., Bundschuh, J., Li, C., Khattak, S. A., & Ali, L. (2023). Groundwater quality, health risk assessment, and source distribution of heavy metals contamination around chromite mines: Application of GIS, sustainable groundwater management, Geostatistics, PCAMLR, and PMF receptor model. *International Journal of Environmental Research and Public Health, 20*(3), 2113.

Reichl, P., Leinen, S., & Stiller, B. (1999). A practical review of pricing and cost recovery for internet services.

Rossetto, R., De Filippis, G., Borsi, I., Foglia, L., Cannata, M., Criollo, R., & Vázquez-Suñé, E. (2019). Integrating free and open source tools and distributed modelling codes in GIS environment for data-based groundwater management. *Environmental Modelling & Software, 107*, 210–230.

Saha, R., Mitran, T., Mukherjee, S., Das, I. C., & Kumar, K. V. (2021). Groundwater management for irrigated agriculture through geospatial techniques. In *Geospatial technologies for crops and soils* (pp. 455–488).

Sallwey, J., Bonilla Valverde, J. P., Vásquez López, F., Junghanns, R., & Stefan, C. (2019). Suitability maps for managed aquifer recharge: A review of multi-criteria decision analysis studies. *Environmental Reviews, 27*(2), 138–150.

Shirazi, S. M., Imran, H. M., & Akib, S. (2012). GIS-based DRASTIC method for groundwater vulnerability assessment: A review. *Journal of Risk Research, 15*(8), 991–1011.

Shit, P. K., Adhikary, P. P., & Sengupta, D. (2023). GI Science applied to soil-agricultural health and environmental risk assessment. In *Environmental science and pollution research* (pp. 1–8).

Singh, A. (2021). Soil salinization management for sustainable development: A review. *Journal of Environmental Management, 277*, 111383.

Singh, M., & Mishra, H. (2023). Medicinal plant farming to improve farmers' economy, *Agriallis* (pp. 30–36).

Singh, M., & Mishra, H. (2023). Trade development and value chains in agriculture: Unleashing the potential for sustainable growth. *Times of Agriculture, 3*(6), 56–57.

Singh, R., Upreti, P., Allemailem, K. S., Almatroudi, A., Rahmani, A. H., & Albalawi, G. M. (2022). Geospatial assessment of groundwater quality and associated health problems in the Western Region of India. *Water, 14*(3), 296.

Sumathi, V. R., Natesan, U., & Sarkar, C. (2008). GIS-based approach for optimized siting of municipal solid waste landfill. *Waste Management, 28*(11), 2146–2160.

Sundaram, S., Devaraj, S., & Yarrakula, K. (2021). Modeling, mapping, and analysis of urban floods in India—A review on geospatial methodologies. In *Environmental Science and Pollution Research* (pp. 1–17).

Sunitha, V., & Reddy, B. M. (2022). Geochemical characterization, deciphering groundwater quality using pollution index of groundwater (PIG), water quality index (WQI) and geographical information system (GIS) in hard rock aquifer, South India. *Applied Water Science, 12*(3), 41.

Sverdrup, H., Koca, D., & Ragnarsdottir, K. V. (2014). Investigating the sustainability of the global silver supply, reserves, stocks in society, and market price using different approaches. *Resources, Conservation and Recycling, 83*, 121–140.

Swain, S., Taloor, A. K., Dhal, L., Sahoo, S., & Al-Ansari, N. (2022). Impact of climate change on groundwater hydrology: A comprehensive review and current status of the Indian hydrogeology. *Applied Water Science, 12*(6), 120.

Teng, Y., Zuo, R., Xiong, Y., Wu, J., Zhai, Y., & Su, J. (2019). Risk assessment framework for nitrate contamination in groundwater for regional management. *Science of the Total Environment, 697*, 134102.

Tiwari, A. K., & Mishra, H. (2024). Sustainable agriculture: The future of world. In D. K. Patel, S. Dwivedi, A. Dubey, A. Kumar, & W. Ali (Eds.), *Agricultural transformation in the modern era* (pp. 167–192). Elite Publishing House.

Tiwari, A. K., Mishra, H., & Nishad, D. C. (2024). Market dynamics and consumer perceptions of organic produce in contemporary agriculture. *Advances in Agriculture Sciences Volume II, 24*(4), 120.

Tiwari, A. K., Mishra, H., Nishad, D. C. & Pandey, A. (2023). Sustainable water management in agriculture: irrigation techniques and water conservation. Dr. Ajay B. Jadhao (p. 53).

Tsatsaris, A., Kalogeropoulos, K., Stathopoulos, N., Louka, P., Tsanakas, K., Tsesmelis, D. E., Krassanakis, V., Petropoulos, G. P., Pappas, V., & Chalkias, C. (2021). Geoinformation technologies in support of environmental hazards monitoring under climate change: An extensive review. *ISPRS International Journal of Geo-Information, 10*(2), 94.

Werner, T. T., Bebbington, A., & Gregory, G. (2019). Assessing impacts of mining: Recent contributions from GIS and remote sensing. *The Extractive Industries and Society, 6*(3), 993–1012.

Zhang, K., & Chui, T. F. M. (2019). A comprehensive review of spatial allocation of LID-BMP-GI practices: Strategies and optimization tools. *Science of the Total Environment, 621*, 915–929.

Zolfaghary, P., Zakerinia, M., & Kazemi, H. (2021). A model for using urban treated wastewater in agriculture using multiple criteria decision making (MCDM) and geographic information system (GIS). *Agricultural Water Management, 243*, 106490.

Chapter 20
Radiation on Marine Ecosystems: Impact of Radionuclides and Their Effect on Life Underwater

Wasswa Shafik

Abstract The groundwater resources are progressively becoming susceptible to the growing prevalence of radiation originating from natural and artificial causes. This study explores the intricate network of effects arising from radiation exposure in environmental ecosystems. This study investigates the impacts of radiation on marine ecosystems such as life under groundwater, explicitly focusing on the disruptions in trophic interactions and population dynamics. Extensive study has also enabled us to uncover radiation's intricate ecological and environmental ramifications, encompassing genetic alterations and ecosystem disturbances. In light of the urgent matter, the study shall examine various mitigation and remediation approaches, assessing their efficacy and practicality. These efforts encompass habitat restoration, pollution control, and the cultivation of radiation-resistant plants. In addition, we thoroughly examine the current regulatory and policy framework, highlighting the necessity for more robust and more extensive policies to safeguard environmental ecosystems against the detrimental effects of radiation. Through evaluating these regulations and the proposition of potential enhancements, our objective is to cultivate a more efficient practice of environmental stewardship. This study presents empirical evidence of radiation-related accidents occurring in groundwater environments, offering tangible illustrations of real-world scenarios. Analyzing the ensuing repercussions, responses, and subsequent lessons from these incidents provides a comprehensive understanding of the subject matter. Finally, it provides valuable perspectives on potential avenues for future research, highlighting deficiencies in the current knowledge and areas that require further inquiry to gain a more comprehensive understanding and effectively tackle this significant matter.

Keywords Life under water · Raditions · Life on land · Climate change · Clean water and sanitation · Good health and well-being

---

W. Shafik (✉)
School of Digital Science, Universiti Brunei Darussalam, Tungku Link, Gadong, Brunei Darussalam
e-mail: wasswashafik@ieee.org

Dig Connectivity Research Laboratory (DCRLab), Kampala, Uganda

V. Sunitha et al. (eds.), *Groundwater Resource Management Planning Strategies*, Springer Water, https://doi.org/10.1007/978-3-031-88870-0_20

20.1 Introduction

Marine ecosystems, which play a vital role in the health and functioning of Earth's seas, face ongoing risks from various stressors. One such stressor is radiation contamination, a multifaceted and formidable problem (Fong et al., 2023). The extensive and varied ecosystems, encompassing coral reefs, kelp forests, and open ocean settings, offer substantial support to a wide range of marine organisms. These ecosystems play a crucial role in providing sustenance, livelihoods, and ecological benefits to billions of individuals globally (Ladewig et al., 2023). The health and sustainability of these ecosystems have not just environmental value but also worldwide economic and cultural importance.

The importance of international collaboration and governance is underscored by the transboundary character of radiation pollution in the marine environment. Marine ecosystems exhibit a lack of adherence to political boundaries, hence enabling radiation contamination to exert extensive consequences that transcend the confines of individual nations (do Amaral Camara Lima et al., 2023). The imperative for collaborative endeavors among nations, regulatory entities, and international institutions, such as the International Atomic Energy Agency (IAEA)[1] and the United Nations, is of utmost significance. To ensure efficient monitoring and mitigation of radiation-related accidents and their ecological and societal effects, it is imperative to establish standardized processes, facilitate data sharing, and establish robust environemental agreements (Lopes et al., 2023; Shafik, 2024a).

The involvement of stakeholders is an essential aspect in effectively addressing the impact of radiation on marine ecosystems. The participation of local communities, fishing companies, environmental advocacy groups, and other stakeholders having a vested interest in these settings is crucial in the decision-making processes (Bracho-Villavicencio et al., 2023). The inclusion of individuals' experiences, concerns, and traditional knowledge can provide valuable guidance for research and management techniques, ensuring that these approaches are not only scientifically rigorous but also culturally sensitive and environmentally sustainable (Wang & Convertino, 2023). The active involvement of stakeholders, the efficient communication of intricate scientific information, and the incorporation of stakeholders in adaptive management plans are crucial measures for the preservation of the ecosystems' well-being with the application of explainable artificial intelligence (Shafik, 2024b).

The transportation of radioactive chemicals throughout ecosystems can result in significant ecological consequences. This phenomenon has the potential to cause disturbances in the overall well-being and operational capacity of ecosystems, as well as the pollution of interconnected food chains and changes in the makeup of species (Shang et al., 2023). Therefore, it is imperative to implement thorough monitoring, conduct risk assessments, and develop mitigation methods to control and reduce the ecological transportation of radioactivity in diverse habitats effectively. This is crucial for the preservation of ecosystem health and the protection of human well-being. Gaining a comprehensive understanding of these processes is of utmost

[1] https://www.iaea.org/.

importance within the realm of environmental stewardship and the sustainable control of radioactive threats (Roberts et al., 2023).

The identification of particular radionuclides implicated in instances of radioactive contamination is of utmost importance in understanding the ecological dissemination of radioactivity. The rates of degradation exhibit variability, hence influencing both the length and degree of contamination (Chakraborty & Takenaka, 2023). Certain radionuclides exhibit rapid decay rates, resulting in a significant reduction in their radioactivity within a very brief timeframe. Conversely, other radionuclides possess longer half-lives, hence contributing to a more enduring presence of contamination. The recognition of contaminated sites and sources as crucial points of origin for the ecological transportation of radiation is of utmost importance. Potential sources of radioactive contamination can arise from various sources, such as nuclear accidents, industrial processes, or inadequate disposal practices for radioactive waste (El-Alfy et al., 2023). Contaminated sites serve as focal points of radioactivity, exerting a significant influence on the ecosystems in their vicinity and practical themes, as depicted in Fig. 20.1.

Bioaccumulation and biomagnification are fundamental processes that are closely linked to the ecological transfer of radioactivity. Radioactive compounds have the potential to be assimilated by plants and aquatic species, leading to their accumulation within live animals as they ingest contaminated food or water (Dummermuth et al., 2023). This phenomenon has the potential to result in the dispersion of radioactive substances inside food webs and across many ecological systems. Moreover, the

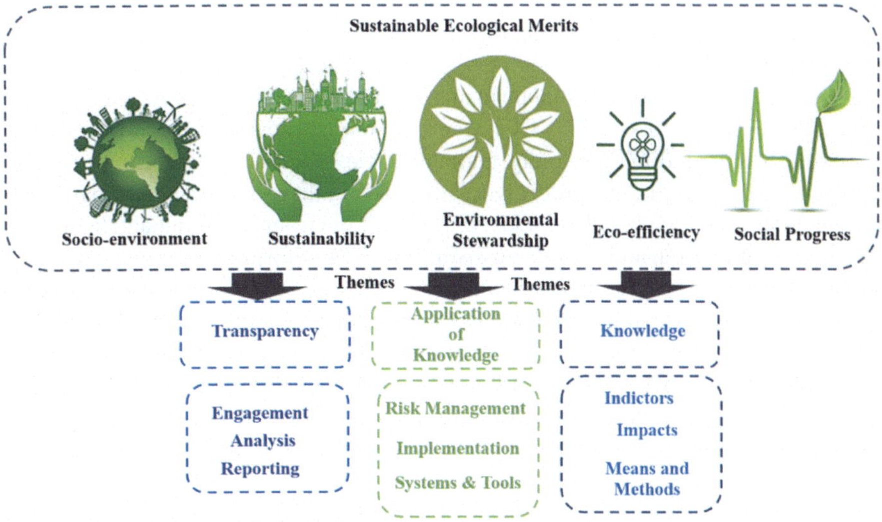

Fig. 20.1 Sustainable ecological benefits. The figure demonstrates the top five ecological assistances of proper marine management of the socio-environment through social progress with some identified themes

phenomenon of biomagnification can manifest when the concentration of radioactivity progressively builds up at each trophic level, resulting in heightened levels in apex predators (Maté-Sánchez-Val & Aparicio-Serrano, 2023). Radiation contamination within the maritime environment can be attributed to a range of factors, encompassing incidents of nuclear nature, improper disposal of radioactive waste, as well as the conduction of nuclear tests. The ramifications of radiation exposure in these particular settings are extensive, complex, and frequently enduring (Radkevych et al., 2023)]. The comprehension of the effects of radiation on marine ecosystems is not solely a subject of scientific inquiry but also a serious imperative for safeguarding these crucial habitats.

Radioactive materials can be conveyed through several mediums, including both the gaseous medium of air and the liquid medium of water. Airborne radionuclides, propelled by wind patterns and atmospheric dynamics, traverse considerable distances, thereby exerting an influence on ecosystems that are geographically distant from the origin of radiation (Boluda-Botella et al., 2023. Waterborne transportation, frequently linked to precipitation or the discharge of pollutants from compromised areas, can transport radioactive substances downstream, exerting an impact on aquatic ecosystems and potentially leading to contamination of larger bodies of water. This phenomenon encompasses the potential dissemination of radioactive materials, which may result in extensive ecological and environmental ramifications (Cerwenka et al., 2023).

A thorough understanding of the systems underlying the effect of radionuclides on marine ecosystems includes acknowledging the pathways through which these radioactive materials go into marine environments, such as nuclear mishaps, nuclear weapon testing, and the discharge of hazardous waste from power plants. Once introduced, radionuclides go through processes such as biomagnification and bioaccumulation, in which they accumulate in marine organisms and boost in concentration as they move up the food cycle. This accumulation can result in hazardous results on marine life, including hereditary anomalies, developmental abnormalities, and interruptions to reproductive processes. Furthermore, the perseverance of particular radionuclides with long half-lives exacerbates their environmental impact, leading to ongoing contamination and ecosystem disturbance. Monitoring programs play a vital function in evaluating radionuclide levels and evaluating their results on marine environments and human health, while regulatory procedures are important for managing and reducing sources of radioactive contamination. Through case studies of nuclear catastrophes such as Chernobyl and Fukushima, researchers gain valuable insights into the long-lasting repercussions of radionuclide release, notifying strategies for safeguarding marine environments and the neighborhoods that depend on them.

20.1.1 The Chapter Contributions

The following are the contributions of this chapter as presented below:

- Present findings on how radiation affects marine organisms and ecosystems, discussing ecological and environmental consequences, including case studies and specific examples to illustrate the effects.
- The study explores solutions to mitigate radiation's impact on marine ecosystems, discusses current and proposed measures to address the issue, and evaluates the effectiveness of these strategies.
- Examines the existing regulations and policies related to radiation in the marine environment, discussing their strengths and weaknesses in protecting ecological health and proposing potential improvements or changes.
- The study further presents real-world radiation-related incidents in the marine world, analyzing consequences, responses, and lessons learned.
- Upon contemplation of the research's significance, engaging in a discourse regarding the wider environmental and societal ramifications is imperative. It is crucial to reemphasize the necessity of addressing the effects of radiation on the marine ecosystem.
- Proposes potential avenues for further investigation within the discipline. The task at hand involves the identification of unresolved inquiries and prospective developments.

20.1.2 The Chapter Organization

The remainder of the study is structured as follows: Sect. 20.2 discussesRadiation's impact on marine ecosystems. Section 20.3 demonstrates some identified mitigation and remediation strategies. Section 20.4 provides regulatory and policy frameworks. Section 20.5 presents selected case studies to demonstrate the effects of radiation on the marine world. Section 20.6 illustrates future research directions of ecological healthcare. Finally, Sect. 20.7 outlines the lessons learned and the chapter's conclusion.

20.2 Radiation's Impact on Marine Ecosystems

This section presents the effects of radiation on marine ecosystems and reducing radiation exposure, which is essential to ensuring the long-term health and sustainability of marine environments and the resources they provide, as discussed.

20.2.1 Genetic Mutations and Reduced Population Sizes

The presence of radiation in the marine environment possesses the capacity to elicit genetic changes in marine species. The mutations have the potential to emerge in several manners, encompassing physical abnormalities, modified reproductive patterns, and genetic oddities. Genetic alterations of this nature have the potential to result in diminished overall fitness and resilience within the impacted species (Cerwenka et al., 2023). With time, the diminished genetic variety in these species can render them increasingly susceptible to additional environmental stressors and less proficient in adjusting to dynamic settings. One illustrative instance involves the impact of genetic abnormalities on fish populations, which can have repercussions on their swimming capabilities and reproductive efficacy, ultimately resulting in population decreases (Boluda-Botella et al., 2023). The extended duration of exposure to radiation has the potential to lead to a reduction in population sizes across numerous marine species. Radiation exposure has been found to have significant repercussions, such as elevated mortality rates and diminished reproductive success, which collectively contribute to the demise of populations (Radkevych et al., 2023). Species with shorter life cycles and higher rates of reproduction possess the inherent capacity to recuperate. In contrast, species with longer life cycles and slower reproduction rates encounter specific obstacles in the process of population restoration. The spectacle can potentially trigger a series of consequences that can disrupt the equilibrium of marine ecosystems, hence influencing the interactions between different trophic levels and the total biodiversity (Maté-Sánchez-Val & Aparicio-Serrano, 2023). Table 20.1 presents a tabular representation of various radioactive elements found in ocean water, sediments, or biota worldwide.

20.2.2 Changes in Trophic Relationships

The occurrence of radiation disruptions has the potential to cause notable alterations in trophic interactions within marine ecosystems. The susceptibility of certain species to radiation and their corresponding resistance levels can lead to fluctuations in their relative abundance and distribution (Dummermuth et al., 2023). The singularity above influences the interplay between predators and prey, hence resulting in modified configurations of ecological communities and heightened levels of resource-based competition (El-Alfy et al., 2023). As an illustration, species with a susceptibility to radiation may have a decrease in population. In contrast, species with a resistance to radiation may thrive, hence impacting the abundance of prey for higher trophic levels. Exposure to radiation has the potential to induce alterations in the composition of marine populations. Certain species with higher resistance to radiation may see increased prevalence, whereas those with lower adaptability may witness a fall in population size. These changes have the potential to greatly influence the ecological roles and functions of species within the ecosystem (Roberts et al., 2023). For example, alterations in the prevalence of specific species can have an impact on the process of nutrient cycling, the rate of primary production, and the overall stability of an ecosystem.

Table 20.1 Marine Radioactive Elements (biota, seawater, and sediments)

Radionuclide	Biota	Seawater	Sediments
Uranium-238	Bioaccumulation in biota	Naturally present	Associated with sediment minerals
Thorium-232	Uptake by marine organisms	Naturally occurring	Bound to sediment particles
Iodine-131	Concentrated in seaweed and fish	Occurs	Adsorbed onto sediment surfaces
Plutonium-239	Bioaccumulates in certain species	Traces	Buried in sediment layers
Strontium-90	Found in fish and other biota	Detected	Incorporated into sediment particles
Cesium-137	Accumulates in marine organisms	Present	Deposited in sediment layers
Americium-241 (Am-241)	Uptake by marine organisms, especially those close to the seafloor	Present due to continuous production from uranium-238 decay in rocks and sediments	Adsorbed onto sediment particles
Lead-210 (Pb-210)	Accumulates in certain marine species, particularly fish	Occurs as a decay product of radon-222; trace amounts	Incorporated into sediment layers
Polonium-210 (Po-210)	Found in marine organisms, including algae and mollusks	Present due to production from radon-222 decay	Used for sediment accumulation rate dating
Radon-222 (Rn-222)	Bioaccumulates in some marine organisms	Detected in trace amounts	Associated with fine sediment particles

20.2.3 Disruption of Food Chains

The disruption of marine food networks can occur as a result of radiation-induced disruptions. The alterations in the prevalence and spatial arrangement of primary producers, such as phytoplankton, and primary consumers, such as zooplankton, can lead to a series of consequences on the accessibility of resources for higher trophic levels, including fish and marine animals (Shang et al., 2023). The disruptions have the potential to cause imbalances within the ecosystem, which may consequently lead to changes in the structure of communities and modifications in the movement of energy across the food web. The repercussions of radiation-induced imbalances within marine ecosystems can be significant. Ecosystems are intricately balanced systems in which every species fulfills a distinct role in the preservation of equilibrium. The disruption of these relationships by radiation can result in destabilization, rendering the ecosystem more vulnerable to additional disturbances, such as those arising from climate change, pollution, or disease outbreaks (Wang & Convertino, 2023). The presence of such imbalances might pose significant obstacles to the resilience and recuperation of ecosystems.

20.2.4 Contamination of Sediments and Water

The presence of radiation contamination has a significant impact on both marine organisms and human societies, which depend on these ecosystems for subsistence and livelihoods. Radiation pollutes sediments and water, hence presenting the possibility of bioaccumulation inside marine organisms. The process of bioaccumulation presents potential hazards to human well-being when individuals ingest seafood sourced from regions that have been contaminated (Bracho-Villavicencio et al., 2023). The presence of pollutants in sediments and water highlights the complex interdependence between marine ecosystems and human welfare, hence stressing the significance of protecting these ecosystems. The potential impacts of radiation on marine ecosystems can exhibit prolonged or enduring ramifications (Ladewig et al., 2023). This underscores the urgent necessity for proactive actions to tackle and alleviate these challenges effectively. The implementation of restoration initiatives, together with the adoption of policies aimed at mitigating radiation exposure, is crucial in safeguarding the enduring well-being and viability of marine ecosystems and the valuable services they offer (Lopes et al., 2023). To effectively mitigate these enduring ramifications, it is imperative to implement comprehensive and enduring measures aimed at safeguarding and reinstating the ecological equilibrium of these crucial ecosystems.

20.3 Mitigation and Remediation Strategies

This section presents some of the mitigation and remediation strategies that are part of a comprehensive approach to addressing the impact of radiation on marine ecosystems.

20.3.1 Habitat Restoration

The restoration of habitat plays a pivotal role in the reestablishment of ecological equilibrium within regions impacted by radiation. Efforts may encompass the implementation of strategies such as the restoration of indigenous flora, the rehabilitation of seagrass beds, mangroves, and coral reefs, as well as the promotion of the recuperation of vital ecosystems (Fong et al., 2023). These actions have the dual purpose of mitigating the acute effects of radiation and facilitating the long-term recovery of marine environments. These solutions contribute to the improvement of the general health and biodiversity of the marine environment by effectively rebuilding varied and resilient ecosystems (Ladewig et al., 2023).

20.3.2 Pollution Control

The efficacy of pollution control measures extends beyond the mere establishment of legislation, encompassing the crucial aspects of rigorous monitoring and steadfast enforcement. Radiation emissions into the marine ecosystem can originate from diverse origins, encompassing industrial effluents and nuclear installations (Fong et al., 2023). The use of rigorous control measures is required to mitigate radioactive exposure effectively. The implementation of regular inspections, the establishment of thorough reporting requirements, and the enforcement of severe compliance processes are crucial elements in the effort to minimize radiation exposure and safeguard marine ecology (Lopes et al., 2023).

20.3.3 Radiation-Resistant Species

The cultivation of radiation-tolerant species necessitates a synergistic approach using both selective breeding and genetic modification techniques. This concept demonstrates potential as a viable approach for ecological restoration (Bracho-Villavicencio et al., 2023). Scientists endeavor to enhance the resilience of organisms towards radiation with the dual purpose of facilitating the recuperation of impacted populations and fortifying the general well-being of marine ecosystems. These species that are resistant to radiation have the potential to act as markers of ecological health or fulfill the role of keystone species, contributing to the stability and resilience of the ecosystem (Wang & Convertino, 2023).

20.3.4 Marine Protected Areas

The development and maintenance of marine protected areas (MPAs)[2] necessitate meticulous planning, well-defined objectives, and efficient governance. MPAs play a vital role in establishing sanctuaries that facilitate the recovery of marine life, the regeneration of ecosystems, and the mitigation of radiation exposure (Bracho-Villavicencio et al., 2023). In addition to their conservation value, MPAs also serve as facilitators for scientific study, monitoring, and restoration endeavors. The significance of their function as habitats for marine organisms and as hubs for research is crucial in the process of mitigating and remedying environmental issues (Shang et al., 2023).

[2] https://marineprotectedareas.noaa.gov/nationalsystem/.

20.3.5 Decontamination Efforts

Its meticulousness and frequent intricacy distinguish the decontamination procedure. The procedure may involve the extraction of contaminated sediments through dredging, the purification of surfaces, and the appropriate disposal of radioactive substances. The effective execution of decontamination techniques is of paramount importance in minimizing the ongoing propagation of radiation within the marine ecosystem (Fong et al., 2023). The effectiveness of these solutions depends on the incorporation of innovative approaches and rigorous safety protocols.

20.3.6 Monitoring and Marine Research

Continuous monitoring and research efforts are now being conducted, hence supplying the vital data required for making well-informed decisions and implementing adaptive management strategies. Periodic evaluations of radiation levels, ecological well-being, and the efficacy of applied strategies provide valuable insights for the purpose of mitigating and rectifying adverse impacts (do Amaral Camara Lima et al., 2023). The continuous progression of scientific study enhances our comprehension of the enduring effects of radiation, thereby facilitating the creation of inventive remedies and adaptable management techniques that evolve alongside the fluctuating dynamics of marine ecosystems impacted by radiation.

20.3.7 Public Awareness and Education

Initiatives focused on public awareness and education must concurrently pursue the twin goals of informing and engaging communities and stakeholders. The main aim of these initiatives is to foster understanding, support, and active involvement in activities aimed at protecting maritime environments (Lopes et al., 2023). The effective application of mitigation and remediation approaches relies heavily on a thorough understanding of hazards and potential solutions within the broader community.

20.3.8 International Collaboration

This is crucial in addressing the issue of radioactive contamination, especially due to its tendency to surpass national borders. International collaboration is vital for the establishment of worldwide norms, dissemination of exemplary methodologies, and effective coordination of endeavors aimed at tackling radiation-related concerns in the marine ecosystem (Bracho-Villavicencio et al., 2023). This collaborative effort ensures a comprehensive and integrated strategy for addressing this global issue, highlighting the interdependence of the earth's oceans.

20.3.9 Regulation and Policy Enhancement

The improvement of rules and regulations requires a thorough restructuring of current frameworks. This encompasses the implementation of more stringent regulations on radiation exposure limits, the enhancement of monitoring and reporting obligations, and the guarantee of strict adherence to international accords and conventions (Wang & Convertino, 2023). The establishment of robust policies and regulations is crucial for facilitating successful mitigation and remediation endeavors, promoting responsibility, openness, and the implementation of measures.

20.3.10 Adaptive Management

Adaptive management is a crucial element of these techniques, necessitating the capacity to modify and enhance procedures in response to increasing knowledge. As the comprehension of radiation's consequences and the efficacy of mitigation strategies expands, the use of adaptive management guarantees the maintenance of flexibility and responsiveness in ongoing efforts, as presented in Fig. 20.2. The continuous process of refinement and modification facilitates enhanced safeguarding and rehabilitation of marine ecosystems, aiming to uphold their long-term well-being and viability (Shang et al., 2023). The comprehensive and interconnected nature of these techniques is essential for effectively tackling the complex difficulties presented by radiation pollution in marine ecosystems (Roberts et al., 2023). The following section details the regulatory and policy framework for ecological health management.

20.4 Regulatory and Policy Framework

This section examines the strengths and weaknesses of Existing Regulations and policies and the potential improvements or changes in the marine world.

20.4.1 Examination of Existing Regulations and Policies

The evaluation of prevailing rules and policies is of utmost importance to identify the merits and drawbacks of the present framework, as well as to suggest enhancements

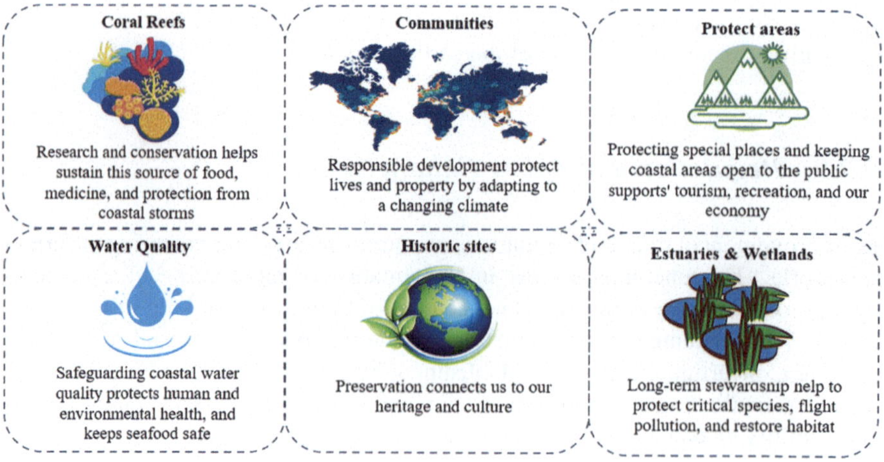

Fig. 20.2 Critical marine ecological insights. The figure presents six key insights that are paramount in ensuring a sustainable marine world, including coral reefs, communities, protected areas, water quality, historic sites, estuaries, and wetlands

or modifications that might effectively safeguard the ecological well-being of marine ecosystems in light of radiation-induced difficulties.

20.4.1.1 International Agreements and Conventions

An essential component of this analysis entails a thorough assessment of international agreements and conventions pertaining to radiation in the marine environment. The London Convention and the United Nations Convention on the Law of the Sea (UNCLOS)[3] are significant agreements that establish the fundamental basis for international regulatory frameworks in marine domains (Chakraborty & Takenaka, 2023). The present analysis evaluates the extent to which these accords successfully address matters pertaining to radiation and their contribution towards the preservation of ecological well-being in international waters. Additionally, this study assesses the extent to which signing states comply with these accords, a critical factor in determining their overall efficacy (El-Alfy et al., 2023).

20.4.1.2 National Regulatory Frameworks

In addition to international accords, the analysis explores the regulatory frameworks at the national level. Various nations that own coastal regions have established distinct policies and regulations pertaining to radiation within their maritime areas. The purpose of this analysis is to evaluate the degree to which these national

[3] https://legal.un.org/avl/ha/uncls/uncls.html.

legislations conform to international agreements and scientific recommendations (Dummermuth et al., 2023). Additionally, this study examines the efficacy of these rules in safeguarding marine ecosystems and evaluates their ability to appropriately tackle the unique issues encountered in their particular geographical areas (Maté-Sánchez-Val & Aparicio-Serrano, 2023.

20.4.1.3 Enforcement and Compliance

The efficacy of current legislation is predominantly contingent upon the implementation and adherence to enforcement and compliance procedures. The evaluation of this component of the test pertains to the efficacy with which governments and international organizations enforce legislation pertaining to radiation. The evaluation encompasses an assessment of the consequences, in the form of fines and incentives, associated with adherence or non-adherence (Radkevych et al., 2023). Effective enforcement measures are necessary to ensure the practical implementation of rules, effectively mitigating the risk of radiation contamination in the maritime environment.

20.4.1.4 Coordination and Information Sharing

Effective coordination and information dissemination across pertinent institutions and groups play a pivotal role in tackling difficulties associated with radiation. The investigation examines the collaborative efforts and information-sharing practices among regulatory authorities, environmental agencies, nuclear agencies, and other pertinent entities (Dummermuth et al., 2023). The establishment of efficient channels for the sharing of data and fostering joint efforts is vital in order to facilitate an integrated strategy for the management of radiation in the marine environment (Cerwenka et al., 2023). Different sustainable approaches help in enhanced performance and eco-friendly human activities, as demonstrated in Fig. 20.3.

Fig. 20.3 Top four sustainability. This figure illustrates the four main sustainable approaches, including human, social, economic, and environmental sustainability

20.4.1.5 Marine Scientific Basis

A crucial component of the analysis involves the assessment of the scientific underpinnings that form the basis for existing legislation and policies. The evaluation entails determining the extent to which these measures conform to the most up-to-date scientific understanding of the effects of radiation on marine ecosystems (Akoglu, 2023). Regulations should be updated to align with a current comprehension of the ecological ramifications of radiation, encompassing both immediate and long-term impacts. Moreover, the objective of this analysis is to ascertain the extent to which current rules embrace an ecosystem-based approach in order to ensure the holistic safeguarding of maritime habitats (Valach et al., 2023).

20.4.1.6 Marine Coverage and Gaps

The primary objective of the evaluation is to assess the scope and comprehensiveness of the regulations and policies currently in place. The objective of this study is to assess the extent to which these procedures effectively encompass all pertinent dimensions of radiation management, encompassing its origins, disposal, monitoring, and remediation (Costa et al., 2023). Concurrently, the objective is to uncover any possible deficiencies or incongruities within the regulatory structure. It is imperative to address these gaps in order to guarantee that rules provide full protection for the maritime environment from radiation pollution (Liu et al., 2023).

20.4.1.7 Adaptability and Resilience

The study assesses the capacity of current regulations to adapt and demonstrate resilience. Regulatory frameworks necessitate the capacity to accommodate shifting circumstances, encompassing swift technological progressions and developing scientific comprehension. This study examines the resilience and responsiveness of these frameworks in addressing new difficulties and hazards related to radiation in marine ecosystems (Kötter-Lange et al., 2023). The analysis of current legislation and policies pertaining to radiation in the marine environment is an essential basis for improving and perfecting these measures (Androsiuk et al., 2023). Through the identification of strengths, flaws, and opportunities for improvement, this analysis serves to guarantee the efficacy and adaptability of regulatory frameworks, thereby making a significant contribution to the long-term sustainability of maritime habitats (Nama et al., 2023).

20.4.2 Strengths and Weaknesses of Existing Regulations and Policies

This subsection highlights the strengths and weaknesses of existing regulations and policies, providing a clear and concise overview of the current state of radiation management in the marine environment.

20.4.2.1 Establishment of Legal Framework

The current regulatory framework establishes a legally binding structure that delineates the entitlements and obligations of different actors in safeguarding maritime habitats. This framework serves to guarantee adherence to regulations and promote responsibility (Cao, 2023). The UNCLOS establishes a comprehensive legal framework for the preservation and responsible utilization of marine resources, encompassing provisions aimed at mitigating marine pollution, including radioactive contamination. The UNCLOS provides a framework of regulations and protocols that govern the conduct of states, promoting collaboration and adherence to established norms (El-Alfy et al., 2023).

20.4.2.2 Scientific Foundation

Numerous legislation and policies are established based on a strong scientific underpinning. Scientific research and data are utilized to establish exposure limits and delineate the biological ramifications of radiation on marine ecosystems. The International Commission on Radioactive Protection (ICRP)[4] offers guidance on radioactive protection, encompassing the safeguarding of the environment (Dummermuth et al., 2023). The standards established by the ICRP are founded upon rigorous scientific research and comprehensive assessments of the effects of radiation on marine ecosystems. Key aspects of the developments include economic growth, social inclusion, and environmental protection, as illustrated in Fig. 20.4. These guidelines aim to ensure that regulatory measures are in accordance with the most up-to-date scientific knowledge (Maté-Sánchez-Val & Aparicio-Serrano, 2023).

20.4.2.3 Transparency and Accountability

Regulations frequently prioritize the principles of transparency and accountability through the implementation of monitoring, reporting, and compliance mechanisms. This initiative encourages the adoption of transparent and accountable behaviors with regard to marine ecosystems (Radkevych et al., 2023). The Maritime Strategy

[4] https://www.icrp.org/.

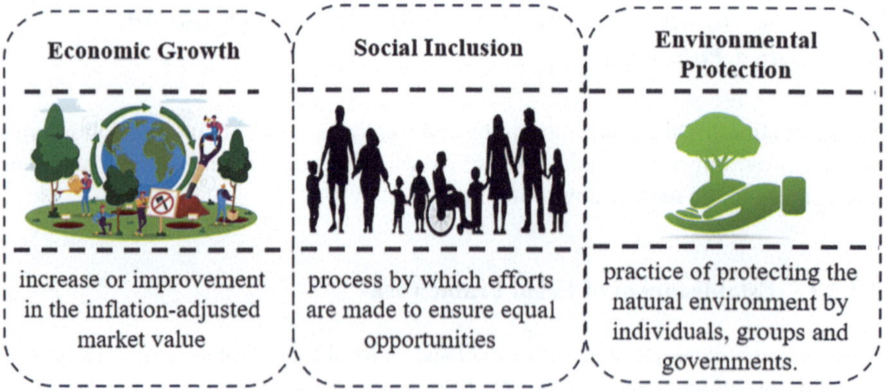

Fig. 20.4 General environmental policies. The figure illustrates three main general policies the environmental-related approach is implemented to ensure better resource utilization

Framework Directive (MSFD)[5] of the European Union requires member states to implement monitoring programs in order to evaluate the ecological condition of their maritime waters. The provision of this data to the public serves to augment transparency and accountability in safeguarding marine ecosystems against diverse hazards, such as radiation (Boluda-Botella et al., 2023).

20.4.2.4 International Collaboration

International accords and conventions serve as catalysts for fostering collaboration among nations that share marine habitats, facilitating synchronized endeavors to mitigate radiation-related hazards effectively. The Antarctic Treaty System,[6] which encompasses the Protocol on Environmental Protection, establishes a comprehensive framework that explicitly prohibits the disposal of radioactive waste within the Antarctic region (Cerwenka et al., 2023). The primary objective of this international treaty is to facilitate cooperation and coordination among governments in order to safeguard the distinctive and vulnerable marine ecosystems in the Antarctic region from the detrimental effects of radiation contamination (Akoglu, 2023).

20.4.2.5 Marine Risk Assessment

Regulations frequently incorporate comprehensive risk assessment procedures in order to analyze the possible harm to marine ecosystems and establish exposure

---

[5] https://research-and-innovation.ec.europa.eu/research-area/environment/oceans-and-seas/eu-marine-strategy-framework-directive_en.

[6] http://www.ats.aq/index_e.htm.

Fig. 20.5 Proper marine conservation approaches. The figure presents seven practical steps in conserving the environment, including sustainable design and durability through using of sustainable building materials

thresholds that safeguard against detrimental ecological consequences (Dummermuth et al., 2023). The risk assessments conducted by the United States Environmental Protection Agency (EPA)[7] aim to analyze the potential consequences of radiation emanating from nuclear sites on aquatic ecosystems. These evaluations utilize scientific data in order to provide estimations about potential dangers, thereby ensuring that exposure limits and preventative measures are based on reliable scientific evidence, as presented in Fig. 20.5 (Costa et al., 2023). The presented examples demonstrate how current regulations and policies pertaining to marine ecosystems effectively utilize various strengths, including a clearly defined legal framework, a foundation rooted in scientific research, transparency in decision-making processes, international cooperation, and thorough risk assessment methodologies. These measures are implemented with the aim of safeguarding the ecological well-being of marine environments against potential hazards, such as radiation and other detrimental factors (Androsiuk et al., 2023).

20.4.3 Potential Improvements or Changes

This subsection aims to critically evaluate the limitations inherent in current legislation and policies pertaining to marine ecosystems, accompanied by concrete instances that exemplify these weaknesses.

20.4.3.1 Limited Public Involvement

In certain instances, there may exist insufficiencies in public and stakeholder engagement within the regulatory process, which could impede transparency and hinder the

[7] https://www.usa.gov/agencies/environmental-protection-agency.

inclusion of a wide range of opinions in the decision-making process. The involvement of local communities and environmental advocacy groups in the decision-making process pertaining to offshore nuclear power station operation regulations may be insufficient, leading to a restricted level of public participation and transparency (Nama et al., 2023).

20.4.3.2 Compliance Variability

The level of adherence to regulations can differ across nations and entities, resulting in disparities in enforcement and possible disadvantages for entities that comply with the restrictions. The implementation of legislation pertaining to the discharge of radiation-contaminated ballast water from ships may vary across nations, leading to disparate degrees of compliance and potential ecological consequences in different locations (Cao, 2023).

20.4.3.3 Transparency Deficiencies and Data Sharing

Although there is a general push for information sharing, certain legislation may not adequately cover the aspects of data sharing and transparency, leading to potential delays in responding promptly to radiation accidents and hindering the evaluation of ecological consequences (Siallagan et al., 2023). In certain instances, there may exist a reluctance among nations that share marine ecosystems to exchange data pertaining to radiation monitoring (Greer et al., 2023). This hesitancy is often rooted in security apprehensions, hence impeding the promotion of transparency and collaborative endeavors aimed at evaluating the effects of radiation on marine ecosystems.

20.4.3.4 Lack of Adaptability

Certain regulatory frameworks may exhibit a deficiency in adaptability to accommodate dynamic circumstances, such as the progression of scientific understanding, breakthroughs in technology, and the emergence of new radiation-related complexities (Sabater et al., 2023). Consequently, this limitation has the potential to impede the efficacy of these frameworks. The Stockholm Convention, which pertains to Persistent Organic Pollutants (POPs),[8] has limitations in its ability to effectively manage emerging pollutants, such as novel radioactive compounds or radionuclides that were not previously taken into account, as Fig. 20.6.

---

[8] https://echa.europa.eu/understanding-pops.

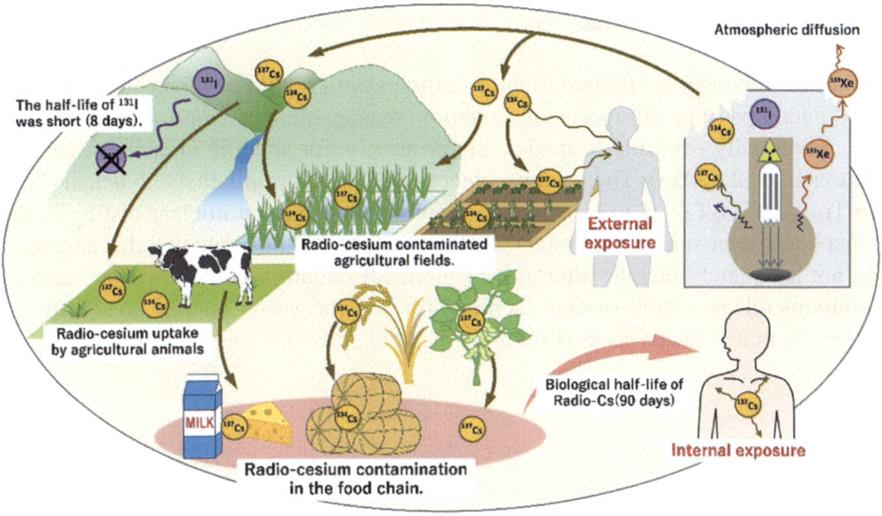

Fig. 20.6 Sampled radioactive sources (Rai & Kawabata, 2020)

20.4.3.5 Enforcement Challenges

The application of radiation-related rules can present challenges stemming from various variables, including limited resources, a shortage of specialist knowledge, and political impediments. These challenges can impede the effective enforcement of such regulations (Lusiana et al., 2023). The enforcement of regulations pertaining to the disposal of radioactive waste in marine environments presents challenges due to resource constraints in monitoring and oversight, hence impeding the assurance of regulatory compliance in waste disposal practices (Tavares et al., 2023).

20.4.3.6 Ambiguities and Inconsistencies

Certain regulations may exhibit ambiguity or inconsistency in their definitions, requirements, or systems of enforcement, hence giving rise to difficulties in the process of interpretation and implementation. The categorization of several types of radioactive materials in the standards for the transportation of such items by the International Maritime Organization (IMO)[9] may exhibit variations, potentially resulting in uncertainty and misclassification (Thulasisingh et al., 2023). These inconsistencies have the potential to compromise safety during the transportation process.

[9] http://www.imo.org/.

20.4.3.7 Gaps in Coverage

Weaknesses in coverage pertain to the insufficiency of regulatory measures in effectively encompassing all facets of radiation management in marine ecosystems, hence potentially resulting in inadequate protection for specific areas or situations (Maureaud et al., 2023). The primary objective of the Comprehensive Nuclear-Test-Ban Treaty (CTBT)[10] is to establish a comprehensive ban on nuclear testing in the atmosphere, outer space, and undersea. Nevertheless, it is possible that this approach may not adequately include alternative sources of radiation contamination, such as continuous releases from storage facilities for nuclear waste, leading to significant limitations in its effectiveness (López-Reyes et al., 2023).

20.5 Case Studies

This section presents real-world examples of radiation-related incidents in the marine world and analyzes their consequences and responses.

20.5.1 Fukushima Daiichi Nuclear Disaster

The Fukushima Daiichi Nuclear Power Plant[11] in Japan experienced a catastrophic failure in March 2011 following a significant earthquake and subsequent tsunami. The occurrence in question resulted in the discharge of significant quantities of radioactive substances into the Pacific Ocean, instigating a serious crisis of both environmental and societal nature. The repercussions were wide-ranging, resulting in significant contamination of saltwater through the presence of radionuclides like cesium and strontium (Frêne et al., 2023). The presence of pollution has a negative effect on the marine ecosystem, leading to the accumulation of radioactive chemicals in seafood, namely fish and shellfish. In order to address these concerns, the Japanese government implemented fishing prohibitions in the impacted regions and established an exclusion zone surrounding the nuclear power station (Kolesnik et al., 2023). The fishing industry experienced substantial economic setbacks as a result of restrictions on the exportation of seafood. At the same time, public apprehension regarding the potential risks of radiation exposure through the consumption of marine products escalated. The international community also engaged in collaborative efforts to evaluate the dispersion of radioactive substances throughout the Pacific Ocean (Miller & Rice, 2023; Shafik, 2024b). Furthermore, long-term research endeavors were launched to comprehend the enduring ecological impacts of this catastrophe on the marine ecosystem.

[10] https://www.ctbto.org/.

[11] https://www.tepco.co.jp/en/nu/press/f1-np/index-e.html.

20.5.2 Sellafield Nuclear Fuel Reprocessing Plant

The Sellafield Nuclear Fuel Reprocessing Plant[12] in the United Kingdom has been responsible for the release of radioactive discharges into the Irish Sea for a number of decades. These discharges have contained various radionuclides, such as plutonium and technetium. The environmental ramifications were significant, as evidenced by the deposition of radioactive pollutants in soils and marine organisms within the Irish Sea. The accumulation of pollutants resulted in ecological harm, namely, causing reduced reproductive success in certain fish species (Zakari et al., 2024). Moreover, the scenario, as mentioned earlier, has given rise to transnational environmental and public health apprehensions, given that the United Kingdom and Ireland jointly utilize the Irish Sea. As a reaction, the United Kingdom implemented enhanced waste management strategies and progressively diminished discharges. The OSPAR[13] Convention has facilitated international collaboration, leading to the establishment of agreements and the implementation of more stringent restrictions pertaining to nuclear discharges within the North-East Atlantic region, which encompasses the Irish Sea (Wang et al., 2023a, 2023b). Long-term monitoring and research initiatives were implemented to assess the recovery of the marine environment and evaluate the efficacy of interventions aimed at minimizing radiation-related consequences in the Irish Sea (Reid, 2023; Shafik, 2023a).

20.5.3 Chornobyl Nuclear Disaster Impact on the Black Sea

The Chornobyl nuclear catastrophe[14] that occurred in 1986 resulted in the release of substantial amounts of radioactive pollutants, with a notable emphasis on cesium-137, into the Earth's atmosphere. Over a while, the pollutants gradually made their way into the Black Sea via river networks, resulting in a chain of ecological and public health repercussions (Reid, 2023). The Black Sea sediments and marine species in the region were shown to have experienced an increase in cesium-137 concentrations, which has had a significant influence on the local environment. In addition, the occurrence has brought to light the potential health hazards that may arise for individuals who consume polluted seafood (Yong et al., 2023).

The transboundary consequences of radiation have elicited concerns from the international community. As a reaction, endeavors in study and monitoring were undertaken to observe the distribution and buildup of radioactive pollutants in the Black Sea (Fang, 2023). International agreements, such as the Bucharest Convention, have been established with the objective of tackling pollution in the Black Sea and fostering collaboration among the nations bordering the Black Sea in order to alleviate the adverse effects associated with radiation (Wang et al., 2023a, 2023b). National

[12] https://www.gov.uk/government/organisations/sellafield-ltd.

[13] https://www.ospar.org/.

[14] https://www.theguardian.com/environment/chernobyl-nuclear-disaster.

legislation and procedures were established to oversee the monitoring of seafood for radioactive contamination.

20.5.4 Mayak Production Association (Ozersk, Russia)

It possesses a multifaceted historical background pertaining to the reprocessing of nuclear fuel and the storage of radioactive waste, which may be traced back to the mid-twentieth century. Throughout its historical timeline, the plant, as mentioned earlier, has encountered multiple occurrences, encompassing accidents and leaks, which have resulted in the discharge of radioactive substances into adjacent aquatic ecosystems, notably the Techa River and Lake Karachay[15] (Lopazanski et al., 2023). The instances in question have resulted in significant ramifications, characterized by substantial pollution of the water bodies, which includes the detection of radionuclides like strontium-90 and cesium-137. The contamination led to enduring environmental harm, with significant repercussions on aquatic ecosystems and exerting an influence on the indigenous plant and animal species within the impacted regions.

The indigenous community encountered heightened health hazards as a consequence of their exposure to radioactive pollutants, leading to an increased prevalence of specific medical ailments. The resolution of these matters continues to pose a substantial obstacle, as persistent apprehensions persist regarding the security and confinement of radioactive waste at the Mayak facility. As a reaction, several regulatory measures have been implemented, such as the imposition of fishing limitations and the creation of regulated areas, to tackle the persistent issue of contamination (Bergseth et al., 2023). The engagement of international organizations like the IAEA has facilitated the monitoring and comprehension of contamination levels, owing to collaborative efforts in scientific study and international cooperation (Shafik, 2023b).

20.6 Future Research Directions

Within this section, we present some identified research directions relating to marine life and general ecological health.

20.6.1 Ecosystem Resilience and Recovery

It is of utmost importance to comprehend the resilience and recovery mechanisms exhibited by marine ecosystems in response to radiation exposure. The scope of this

[15] https://www.nti.org/education-center/facilities/mayak-production-association/.

study should involve a comprehensive analysis of the responses exhibited by particular maritime ecosystems and species in the face of stressors generated by radiation (Fong et al., 2023). The identification of crucial thresholds and the examination of factors that impact the velocity and efficacy of recovery are of utmost importance. Coastal ecosystems, for example, have the potential to exhibit faster recovery rates owing to their greater connectivity and species interaction. In contrast, isolated or deep-sea settings may endure more protracted impacts (Ladewig et al., 2023).

20.6.2 Sublethal Effects and Suborganismal Responses

The examination of the cellular, physiological, and behavioral consequences of radiation exposure in marine creatures yields a comprehensive comprehension of its sublethal effects. Further investigation is warranted to examine the impact of radiation on sub-organismal processes, such as DNA repair mechanisms, immunological responses, and reproductive systems (Yong et al., 2023). This study has the potential to uncover the fundamental mechanisms that contribute to ecological-level effects, enabling us to predict the potential adaptations or evolutionary changes of marine species in response to radiation exposure (Zakari et al., 2024). Furthermore, it can shed light on whether these adaptations have lasting implications for the structure and functioning of ecosystems.

20.6.3 Radionuclide Speciation and Transformation

A thorough investigation of the speciation and transformation processes of radionuclides in marine ecosystems is of utmost importance. The process entails the investigation of the chemical interactions that occur between radionuclides and various organic and inorganic substances found in saltwater, sediments, and biota (Fang, 2023). Sophisticated analytical methodologies, such as spectroscopy utilizing synchrotron radiation and investigations into isotope fractionation, are essential for monitoring the behavior and distribution of radionuclides within intricate marine environments (Miller & Rice, 2023). Understanding this information is crucial for the accurate prediction of the bioavailability of radionuclides, their capacity to accumulate in marine animals, and the resulting ecological consequences.

20.6.4 Interaction with Other Stressors

It is imperative to conduct comprehensive investigations into the interplay between radiation and various other environmental stresses. Marine ecosystems face various stressors, encompassing ocean acidification, pollution, habitat destruction, and climate change Shafik (2025). The investigation should aim to examine the interplay between these stressors, elucidating whether they exhibit synergistic or antagonistic behavior and assessing the cumulative impacts they have on marine ecosystems (Bergseth et al., 2023). Comprehending these relationships is crucial for formulating adaptive management techniques that take into account the cumulative effects and potential thresholds in the presence of various stressors.

20.6.5 Multi-trophic Level Studies

It is recommended that future studies place greater emphasis on conducting multi-trophic level investigations in order to comprehensively investigate the manner in which radiation affects the propagation of effects throughout marine food webs. The examination of the transmission of radiation from primary producers, such as phytoplankton and seaweed, to herbivores and predators provides valuable insights into the wider ecological implications (Yong et al., 2023). A key aspect in forecasting the consequences of radiation contamination in the marine food chain is the comprehension of the impact of radiation-induced alterations in phytoplankton communities on zooplankton and, subsequently, on fish and marine animals (Wang et al., 2023a, 2023b).

20.6.6 Ecosystem Services and Human Well-Being

To conduct a full evaluation of the effects of radiation on marine ecosystems, research must incorporate an analysis of ecosystem services and their interconnectedness with human well-being. This entails assessing the impacts of radioactive pollution on fisheries, coastal tourism, and the cultural significance attributed to marine habitats (Frêne et al., 2023). Through the process of quantification, researchers are able to assess the economic and societal impacts that arise from disturbances caused by radiation. This enables them to guide conservation efforts and policy-making, ensuring a balance between ecological well-being and the requirements of local communities and society at large (Miller & Rice, 2023).

20.6.7 Advanced Remote Sensing and Modeling

The integration of state-of-the-art remote sensing technologies and advanced modeling methodologies has the potential to significantly transform our capacity to observe and forecast radiation dynamics inside marine ecosystems. The integration of satellite data, drones, and numerical models allows for the real-time monitoring of the dispersion of radionuclides, the identification of their trajectories, and the identification of areas where they may accumulate (Frêne et al., 2023). The utilization of sophisticated modeling techniques can facilitate the prediction of the dispersion patterns of radioactive chemicals in relation to environmental factors such as ocean currents, salt levels, and temperature variations (Bergseth et al., 2023). These tools augment our ability to address radiation occurrences promptly and proactively oversee marine ecosystems.

20.6.8 *Ecological Education and Marine Outreach*

In addition to scientific research, there exists a significant imperative for the implementation of effective education and outreach activities. The inclusion of public participation, policymaker understanding, and collaboration with industry stakeholders is vital in effectively resolving radiation-related concerns within the maritime environment. Efforts should be made to develop educational initiatives, public campaigns, and stakeholder forums that aim to cultivate a comprehensive comprehension of the intricate nature of radioactive contamination and its associated ramifications (Li, 2023; Angeler et al., 2023). Through the cultivation of consciousness and the promotion of cooperative endeavors, it is possible to guarantee the implementation of well-informed choices and the adoption of sustainable methodologies aimed at safeguarding marine habitats and the sustenance of those reliant upon them (Miller & Rice, 2023).

20.7 Lessons Learned and the Conclusion

As we are drawing to the conclusion, we present some unanswered questions and potential advancements in the field of radiation's impact on marine ecosystems.

20.7.1 *Open Questions and Lessons Learned from the Chapter*

The presence of unanswered inquiries and the possibility of enhancements serve to highlight the dynamic and intricate characteristics inherent in the domain of radiation's influence on marine ecosystems. By acknowledging and addressing these areas of limited understanding and actively engaging in advanced research, it is possible to enhance the efficacy of methods aimed at safeguarding and administering these crucial marine ecosystems.

- How do the long-term effects of radiation exposure manifest in marine ecosystems, and what are the potential ecological and evolutionary consequences for marine species?
- What are the specific behaviors and pathways of radionuclides in deep-sea ecosystems, and how does radiation contamination affect these poorly explored regions?
- To what extent do transgenerational effects of radiation exposure play a role in the adaptation and resilience of marine species?
- What are the ecosystem-level impacts of radiation contamination, and how do they vary across different marine habitats, from coastal areas to open oceans?
- What are the mechanisms underlying the interactions between radiation exposure and other environmental stressors, and how do these interactions influence marine ecosystems?
- What are the most effective approaches for conveying complex scientific information to stakeholders, ensuring that they have access to and can understand the findings and implications of research on radiation contamination in marine environments?
- What strategies and platforms can be developed to effectively engage local communities, fishing industries, environmental advocacy groups, and other stakeholders in the decision-making processes regarding radiation in marine ecosystems?
- The inquiries, as revealed earlier, can provide a basis for forthcoming scholarly investigations, the formulation of policies, and the implementation of endeavors that seek to promote global cooperation, effective governance, and active involvement of relevant parties in mitigating the consequences of radiation on marine ecosystems based on the lessons learned.
- The effects of radiation on marine ecosystems are intricate and diverse in nature. This phenomenon encompasses a range of radionuclides, a varied array of marine animals, and complex biological interactions. Gaining a comprehensive understanding of these intricacies is crucial for the successful execution of research and management endeavors.
- The long-term implications of radiation pollution in marine ecosystems can be significant. Certain impacts have the potential to endure for extended periods, spanning several decades or even centuries. This underscores the importance of maintaining ongoing monitoring and implementing measures to alleviate these consequences.
- The cross-border implications of radiation pollution in marine ecosystems underscore the significance of international cooperation and governance. The efficacy of responses necessitates the establishment of coordinating mechanisms across governments and organizations.
- It is imperative to involve various stakeholders, such as local communities, industry leaders, and advocacy groups, in order to foster meaningful engagement. The active participation of individuals in decision-making processes guarantees the integration of scientific rigor and cultural sensitivity in research and policy development.

- Marine ecosystems have a degree of resilience, but the extent and speed of recovery vary based on factors such as habitat type and stressor intensity. Understanding this resilience can guide management strategies.
- Radiation frequently interacts with several environmental stresses, including pollution and climate change. These interactions have the potential to worsen ecological impacts, hence highlighting the need for research on the effects of coupled stressors.
- Implementing adaptive management systems plays a critical role in effectively resolving occurrences associated with radiation. Adaptability and the capacity to modify conservation and mitigation strategies in response to novel insights are fundamental elements of proficient management.
- Radiation-related occurrences have global ramifications, necessitating collaborative efforts from the international community to tackle them effectively. Establishing worldwide awareness and cooperation is crucial for effectively regulating radiation pollution in marine environments.

20.8 Conclusion

The chapter discussed the effects of radiation on marine ecosystems and emphasized the complex and long-lasting implications of radiation contamination in these crucial habitats. The acquired knowledge highlights the intricate nature of the matter, the significance of global cooperation and regulation, the involvement of stakeholders, and the necessity for adaptable management approaches. As society progresses, it becomes evident that the safeguarding of marine ecosystems necessitates ongoing research and a comprehensive perspective. This entails comprehending their ability to withstand disturbances, examining their interplay with other stressors, and evaluating the enduring consequences of radiation. Given the transboundary character of this dilemma, it is imperative to undertake a global endeavor aimed at effectively addressing occurrences linked to radiation, fostering increased awareness, and safeguarding the ecological health and overall well-being of both the marine environment and the communities that are interconnected with it.

Acknowledgements My Special thanks to the editor for the opportunity to be part of this book.

References

Akoglu, E. (2023). Ecological indicators reveal historical regime shifts in the Black Sea ecosystem. *PeerJ, 11*. https://doi.org/10.7717/PEERJ.15649

Androsiuk, L., Shay, T., & Tal, S. (2023). Characterization of the environmental plasmidome of the Red Sea. *Microbiology Spectrum, 11*(4), 2023. https://doi.org/10.1128/spectrum.00400-23

Angeler, D. G., Heino, J., Rubio-Ríos, J., & Casas, J. J. (2023). Connecting distinct realms along multiple dimensions: A meta-ecosystem resilience perspective. *Science of the Total Environment, 889*, 2023. https://doi.org/10.1016/j.scitotenv.2023.164169

Bergseth, B. J., et al. (2023). Closing the compliance gap in marine protected areas with human behavioural sciences. *Fish and Fisheries, 24*(4). https://doi.org/10.1111/faf.12749

Boluda-Botella, N., Saquete, M. D., & Sanz-Lázaro, C. (2023). Holothuria tubulosa as a bioindicator to analyse metal pollution on the coast of Alicante (Spain). *Journal of Sea Research, 192*. https://doi.org/10.1016/j.seares.2023.102364

Bracho-Villavicencio, C., Matthews-Cascon, H., & Rossi, S. (2023). Artificial reefs around the world: A review of the state of the art and a meta-analysis of its effectiveness for the restoration of marine ecosystems. *Environments, 10*(7). https://doi.org/10.3390/environments10070121

Cao, Y. (2023). Coral bleaching on the great barrier reef and causes. *Theoretical and Natural Science, 4*(1). https://doi.org/10.54254/2753-8818/4/20220660.

Cerwenka, A. F., Brandner, J., Dashinov, D., & Geist, J. (2023). Small but mighty: The Round Goby (Neogobius melanostomus) as a Model species of biological invasions. *Diversity, 15*(4). https://doi.org/10.3390/d15040528

Chakraborty, A., & Takenaka, T. (2023). Tourism and sustainability at a crossroads in Shiretoko UNESCO World Natural Heritage Site in Japan. *Journal of Heritage Tourism, 8*(2). https://doi.org/10.1080/1743873X.2023.2169614

Costa, E. F. S., Encarnação, J., Teodósio, M. A., & Morais, P. (2023). Aquatic species shows asymmetric distribution range shifts in native and non-native areas. *Frontiers in Marine Science, 10*. https://doi.org/10.3389/fmars.2023.1158206

do Amaral Camara Lima, M., Bergamo, T. F., Ward, R. D., & Joyce, C. B. (2023). A review of seagrass ecosystem services: providing nature-based solutions for a changing world. *Hydrobiologia, 850*(12–13). https://doi.org/10.1007/s10750-023-05244-0

Dummermuth, A., et al. (2023). Marine stations helgoland and sylt operated by the Alfred Wegener Institute Helmholtz Centre for Polar and Marine Research. *Journal of Large-Scale Research Facilities, 8*(1). https://doi.org/10.17815/jlsrf-8-184

El-Alfy, M. A., Eissa, H. S., & Abd El-Hamid, H. T. (2023). Environmental and human risk assessment of radioactive pollution in sediments of Northern Nile Delta, Egypt. *Soil Sediment Contam, 32*(7). https://doi.org/10.1080/15320383.2022.2149693

Fang, M. M. (2023). China's battle against marine plastic pollution at the local level: A case study of Sanya City, Hainan Province. *Ocean Yearbook Online, 37*(1). https://doi.org/10.1163/22116001-03701013

Fong, C. R., Ryznar, E. R., Smith, L. L., & Fong, P. (2023). Towards a trait-based framework for marine macroalgae: Using categorical data to explore the nature of emergent functional groups. *Journal of Ecology, 111*(9). https://doi.org/10.1111/1365-2745.14144

Frêne C. et al. (2023). Chilean long-term socio-ecological research network: progresses and challenges towards improving stewardship of unique ecosystems: Red Chilena de Investigación Socio-Ecológica de Largo Plazo: Avances y desafíos para el manejo responsable de ecosistemas únicos. *Revista Chilena de Historia Natural, 96*(1). https://doi.org/10.1186/s40693-023-00114-4

Greer A. T., et al. (2023). In situ imaging across ecosystems to resolve the fine-scale oceanographic drivers of a globally significant planktonic grazer. *Limnology Oceanography, 68*(1). https://doi.org/10.1002/lno.12259

Kolesnik, O. V., Rozhko, T. V., & Kudryasheva, N. S. (2023). Marine bacteria under low-intensity radioactive exposure: Model experiments. *International Journal of Molecular Sciences, 24*(1). https://doi.org/10.3390/ijms24010410

Kötter-Lange, K., Lienhoop, N., & Buschbaum, C. (2023). Effects and perception of marine introduced species by stakeholders in the Wadden Sea—An exploratory approach. *Marine Biodiversity, 3*(5). https://doi.org/10.1007/s12526-023-01358-9

Ladewig, S. M., Coco, G., Hope, J. A., Vieillard, A. M., & Thrush, S. F. (2023). Real-world impacts of microplastic pollution on seafloor ecosystem function. *Science of the Total Environment, 858*. https://doi.org/10.1016/j.scitotenv.2022.160114

Li, M. (2023). The human impact of marine ecosystem imbalance: an analysis of society and ocean management. *Applied and Computational Engineering, 3*(1). https://doi.org/10.54254/2755-2721/3/20230347

Liu, D., Liu, T., Bi, H., Zhao, Y., & Cheng, Y. (2023). Multiscale local feature fusion: Marine microalgae classification for few-shot learning. *Water, 15*(7). https://doi.org/10.3390/w15071413

Lopazanski, C., et al. (2023). Principles for climate resilience are prevalent in marine protected area management plans. *Conservation Letters*. https://doi.org/10.1111/conl.12972

Lopes C. et al. (2023). Microplastic distribution in different tissues of small pelagic fish of the Northeast Atlantic Ocean. *Science of the Total Environment, 901*. https://doi.org/10.1016/j.scitotenv.2023.166050

López-Reyes, K., Osorio-Olvera, L., Rojas-Soto, O., Chiappa-Carrara, X., Patrón-Rivero, C., & Yáñez-Arenas, C. (2023). An exhaustive evaluation of modeling ecological niches above species level to predict marine biological invasions. *Marine Environmental Research, 186*. https://doi.org/10.1016/j.marenvres.2023.105926

Lusiana, E. D., Astutik, S.Nurjannah, N., & Sambah, A. B. (2023). Spatial delineation on marine environmental characteristics using fuzzy c-means clustering method. *Global Journal of Environmental Science and Management, 9*(3). https://doi.org/10.22035/gjesm.2023.03.07

Maté-Sánchez-Val M. and Aparicio-Serrano, G. (2023). The impact of marine pollution on the probability of business failure: A case study of the Mar Menor lagoon. *Journal of Environmental Management, 332*. https://doi.org/10.1016/j.jenvman.2023.117381

Maureaud A. A. et al. (2023). FishGlob_data: An integrated database of fish biodiversity sampled with scientific bottom-trawl surveys. *National Marine Fisheries Service (NOAA), 28*. https://doi.org/10.31219/osf.io/2bcjw

Miller, C. R., & Rice, A. N. (2023). A synthesis of the risks of marine light pollution across organismal and ecological scale. *Aquatic Conservation: Marine and Freshwater Ecosystems*. https://doi.org/10.1002/aqc.4011

Nama, S., Shanmughan, A., Nayak, B. B., Bhushan, S., & Ramteke, K. (2023). Impacts of marine debris on coral reef ecosystem: A review for conservation and ecological monitoring of the coral reef ecosystem. *Marine Pollution Bulletin, 189*, 2023. https://doi.org/10.1016/j.marpolbul.2023.114755

Radkevych, V., Zhurian, V., Kononenko, A., & Smyrnova, I. (2023). Pedagogical aspects of the formation of ecological competence of specialists of the maritime branch. *International Journal of Engineering Pedagogy, 13*(1). https://doi.org/10.3991/ijep.v13i1.36125

Rai, H., & Kawabata, M. (2020). The dynamics of radio-cesium in soils and mechanism of cesium uptake into higher plants: Newly elucidated mechanism of cesium uptake into rice plants. *Frontiers in Plant Science, 11*, 531515. https://doi.org/10.3389/fpls.2020.00528

Reid, S (2023). ocean justice reckoning with material vulnerability. *Cultural Politics, 19*(1). https://doi.org/10.1215/17432197-10232516

Roberts, K. E., Stepanuk, J. E. F., Kim, H., Thorne, L. H., Chong-Montenegro, C., & Nye, J. A. (2023). Developing a subseasonal ecological forecast to reduce fisheries bycatch in the Northeast U.S. *Progreess in Oceanography, 213*. https://doi.org/10.1016/j.pocean.2023.103021

Sabater, L. M., Monferran, M. D., Higuti, J., & Pérez, A. P. (2023). An annotated checklist of recent non-marine Ostracoda (Crustacea) from Argentina, South America. *Zootaxa, 5336*(2). https://doi.org/10.11646/zootaxa.5336.2.1

Shafik, W. (2023a). IoT-based energy harvesting and future research trends in wireless sensor networks. In *Handbook of research on network-enabled IoT applications for smart city services* (pp. 282–306). https://doi.org/10.4018/979-8-3693-0744-1.ch016

Shafik, W. (2023b). cyber security perspectives in public spaces: drone case study. In *Handbook of research on cybersecurity risk in contemporary business systems* (pp. 79–97). IGI Global. https://doi.org/10.4018/978-1-6684-7207-1.ch004

Shafik, W. (2024a). Environmental conservation and sustainable development. In *Developing digital inclusion through globalization and digitalization* (pp. 142–176). IGI Global. https://doi.org/10.4018/979-8-3693-4111-7.ch008

Shafik, W. (2024b). Sustainable development goal 14: Explainable AI (XAI) for ocean health. In *Artificial intelligence and edge computing for sustainable ocean health* (pp. 167–198). Springer. https://doi.org/10.1007/978-3-031-64642-3_8

Shafik, W. (2025). Ecological health impact in the marine world. In: Jindal, M.K., David, A., Khandaker, M.U. (eds) *Radiation status in the marine world*. Springer Oceanography. Springer. https://doi.org/10.1007/978-3-031-82403-6_10

Shang, Y. et al. (2023). Ocean acidification affects the bioenergetics of marine mussels as revealed by high-coverage quantitative metabolomics. *Science of the Total Environment, 858*. https://doi.org/10.1016/j.scitotenv.2022.160090

Siallagan, Z. L., Sidabalok, C. M., Islami, M. M., Mujiono, N. S., Wilson, P., & Pamungkas, J. (2023). Why many Indonesian marine species remain undescribed: A case study using polychaete species discovery. *Raffles Bulletin of Zoology, 71*. https://doi.org/10.26107/RBZ-2023-0026

Tavares, M., Ott, P. H., Borges-Martins, M. (2023). Tracking marine tetrapod carcasses using a low-cost mixed methodology with GPS trackers, passive drifters and citizen science. *Methods in Ecology Evolution, 14*(9). https://doi.org/10.1111/2041-210X.14177

Thulasisingh, A., Venkatesan, S. A., & Kumar, S. (2023). Green biopolysaccharides and its utilisation as biodegradable material in diverse fields: A review. *Polymer Bulletin*. https://doi.org/10.1007/s00289-023-04738-0

Valach, M., et al. (2023). Recent expansion of metabolic versatility in Diplonema papillatum, the model species of a highly speciose group of marine eukaryotes. *BMC Biology, 21*(1), 2023. https://doi.org/10.1186/s12915-023-01563-9

Wang, L., Yao, J., Zhang, H., Pang, Q., & Fang, M. (2023). A sustainable shipping management framework in the marine environment: Institutional pressure, eco-design, and cross-functional perspectives. *Frontiers in Marine Science, 9*. https://doi.org/10.3389/fmars.2022.1070078

Wang, C., et al. (2023). Ecological and human health risks of antibiotics in marine species through mass transfer from sea to land in a coastal area: A case study in Qinzhou Bay, the South China sea. *Environmental Pollution, 316*. https://doi.org/10.1016/j.envpol.2022.120502

Wang, H., & Convertino, M. (2023). Algal bloom ties: Systemic biogeochemical stress and Chlorophyll-a shift forecasting. *Ecological Indicators, 154*. https://doi.org/10.1016/j.ecolind.2023.110760

Yong, L., et al. (2023). Research advances on the ecology of Sousa chinensis. *Biodiversity Science, 31*(5), 2023. https://doi.org/10.17520/biods.2022670

Zakari, R. Y., Shafik, W., Kalinaki, K., & Iheaturu, C. J. (2024). Internet of forestry things (IoFT) technologies and applications in forest management. In *Advanced IoT technologies and applications in the Industry 4.0 digital economy* (pp. 275–295). CRC Press. https://doi.org/10.1201/9781003434269-15

Chapter 21
Ascertainment of Microplastic Pollution Load in the Groundwater and Its Antithetical Effects on Human Health: A Systematic Review

Swagata Chakraborty, P. Vignesh, Anas Bin Firoz, and M. Govindaraju

Abstract Microplastic (MP) pollution in Groundwater (GW) has emerged as a critical environmental concern worldwide, posing risks to both the ecosystems and human health. This issue stems from the widespread use and inappropriate disposal of plastics, which eventually break down into tiny particles, typically smaller than 5 mm, that infiltrate GW systems. Studies have revealed the presence of MPs in GW aquifers around the globe, including regions previously thought to be pristine, highlighting the global dimensions of this detrimental issue. Sources include landfills, wastewater effluent discharge, surface runoff, and atmospheric deposition that infiltrate into aquifers generally through the percolation process and eventually travel great distances and reach to aquifers. Their small size and buoyancy allow them to migrate through porous subsurface materials. These particles can potentially cause adverse effects on ecosystems and human health, leading to bioaccumulation and disruptions in food chains. In GW can contaminate the source of drinking water, and there are concerns about their potential to carry harmful chemicals, pathogens, and toxins which can further cause health issues. Understanding the movement and fate of MPs in aquifers is essential for effective management and remediation. To determine the degree of exposure to human beings and any potential health impacts, further studies are required. MP contamination poses multifaceted risks to human health. MPs infiltrate the water leading to potential ingestion. These particles induce inflammation, damage organs and tissues, disrupt the gut microbiome, affect digestion and immunity, and have links to long-term health risks, including cancer and reproductive issues, while ongoing research further elucidates these health effects. Robust monitoring, risk assessment, and regulatory measures are essential to mitigate the risks associated with MP contamination of GW. Persistent research and monitoring are crucial for a comprehensive understanding of MP contamination in aquifers

S. Chakraborty · P. Vignesh · A. B. Firoz · M. Govindaraju (✉)
Centre for Climate Change Research (CCCR), Department of Environmental Biotechnology, School of Environmental Science, Bharathidasan University, Tiruchirapalli, India
e-mail: mgrasu@bdu.ac.in

S. Chakraborty
e-mail: chakraborty.swagata91@bdu.ac.in

V. Sunitha et al. (eds.), *Groundwater Resource Management Planning Strategies*, Springer Water, https://doi.org/10.1007/978-3-031-88870-0_21

and its long-term impacts. Existing regulations often overlook MP contamination in GW, emphasizing the need for more stringent guidelines and monitoring programs. Developing effective strategies for mitigating MP pollution in GW includes source reduction, improved waste management, and advanced water treatment technologies. MP pollution in GW aquifers is a complex global issue with wide-ranging consequences and potential implications for human health. Addressing this problem demands a concerted effort from governments, industries, and the scientific community to minimize its environmental and human health risks. It underscores the urgency of addressing plastic pollution and adopting sustainable waste management practices to safeguard GW quality and human well-being.

Keywords Microplastics (MPs) · Groundwater (GW) pollution · Contamination pathway · Human health · Strategic analysis

21.1 Introduction

MP pollution in GW has become a major global environmental problem, posing risks to ecosystems and human health. The problem starts from the widespread use and improper handling of plastic, which eventually breaks down into small pieces (usually less than 5 mm) and enters the GW. This emerging environmental problem has attracted the attention of researchers, policymakers, and the public in recent years. Thompson Richard Charles first used the word " Microplastic" in 2004 to refer to these smaller-sized plastics (Chia et al., 2021; Thompson et al., 2004). MP consists of various types of synthetic polymers (Quiot et al., 2022) that can be further classified based on their size and origin (Chandran & Senthilkumar, 2020). MPs produced mainly for personal care products known as primary MPs, are released into the environment while used. Secondary MP particles are broken via physical, physicochemical, and biological processes which can be done by weathering, fragmentations, biotic, or abiotic factors as well (Sutkar et al., 2023). This breakdown happened due to polymer degradation, other internal and external factors like morphology, additives, microbial activity, corrosion, ultraviolet light, temperature, pH, humidity, and catalysts in the environment (Lv et al., 2022). Additionally, transportation of MPs can accumulate MPs in remote areas (Brien et al., 2023), raising questions about their prevalence.

During washing of clothes, synthetic textile fabrics may release MPs such as polyester, nylon etc., a secondary source, after washing, which get into the wastewater and ultimately into water bodies like River water (Acharya et al., 2021) and to the GW by leaching (Samandra et al., 2022).

Road signs and tyres also release MP particles from vehicles into the air (Giechaskiel et al., 2024), which ultimately leads to MP pollution in soil and water (Evangeliou & Brahney, 2022). Scientific research has documented the various ways MPs enter into ecosystems. This method puts marine life, land life, and even human at risk of being affected. The proliferation of MPs highlights the need for action to reduce their emissions, including creating equipment, improving waste management and improving legislation (Katare, 2021; Wright & Kelly, 2017a, 2017b). In addition, toothpaste and exfoliating brushes containing synthetic microparticles directly contribute to MP pollution in the environment (Duis & Coors, 2016; Ustabasi & Baysal, 2019). Although a study shows that 75% of plastic is found in the soil of Canada's Lake Ontario, there is no pollution in an 8 cm area around the lake. The largest amount of plastic waste was polyethene, accounting for approximately 74%, followed by polypropylene at 17% and nitrocellulose at 9%, with sizes from 0.5 to 3.5 mm (Horton et al., 2017). In the marine environment, plastics undergo mechanical and photochemical reactions accelerated by waves and sunlight, respectively, and these reactions cause micro/nanoparticles to split and freeze. Depending on the type of polymer, MPs come in different colours and densities. Polyethene constitutes 54.5% of MPs floating in water, followed by polypropylene with 16.5%, and the rest is polyvinyl chloride, polyester and polyamide (Issac & Kandasubramanian, 2021). Leaching is the processes that transport MPs from soil to GW (Khant & Kim, 2022; Quiot et al., 2022). According to NCBI, GW accounts for 30.8% of MPs, 0.68% of all available fresh water in the world, and 0.68% of all available water in the world (Musie & Gonfa, 2023). According to findings in Asia, Europe and North America, GW around the world has been found to contain MPs (Bergmann et al., 2019). Global research shows that GW contains MPs even in areas previously thought to be pristine (Lee et al., 2022). Atmospheric deposition, wastewater discharges, and surface water runoff are examples of plastic particle introduction of MPs in water systems. MPs can enter waterways through a variety of routes, including landfills, sewers and streams, eventually travelling long distances and potentially reaching drinking water (Quiot et al., 2022). Their small size and buoyancy allow them to migrate between landforms (Geng et al., 2024).

Due to the consumption of MPs by bacteria, nutrition is affected and these lead to bioaccumulation in water that has the potential to cause adverse effects on the aquatic life (Gallo et al., 2018). Contamination of GW with MPs poses a risk to public health because they may contain toxic, pathogenic and toxic substances (Blackburn, 2022). Understanding the behaviour and fate of MPs in aquifers is important for developing management and remediation plans. The fact that MP pollution in GW is often ignored by existing regulations demonstrates the need for more stringent standards and monitoring programs (Khant & Kim, 2022). To reduce the problem of MP pollution in GW, it is necessary to increase the amount of resources, strengthen waste management strategies, and create better water treatment technologies (Chia et al., 2021). More research and analysis is needed to fully understand MPs exposure in soil and its long-term effects (Wang et al., 2017) which ultimately affect the GW system. Risk management measures are very important to reduce the impact of MPs in soil (Blackburn, 2022). Robust monitoring and regulatory measures are essential to mitigate the risks associated with MP contamination of GW. MP contamination poses multifaceted risks to human health (Adnan, 2022). MPs infiltrate the food chain, waterbody, and air potentially ingested and absorbed respectively by humans, causing an inflammatory response, interference with the immune system, which may lead to neurotoxicity, affecting the gastrointestinal tract, and oxidative stress injury

(Peng & Sonne, 2023). They function as transporters for toxic substances that are ingested by the different trophic levels. These particles induce inflammation, damage organs and tissues, disrupt the gut microbiome, affect digestion and immunity, and have links to long-term health risks, including cancer and reproductive issues while ongoing research further elucidates these health effects (Wang et al., 2017), Peng & Sonne, 2023, Carlos et al., 2018, Lu et al., 2016).

MP pollution in underground aquifers is a complex, multifaceted problem with many consequences around the world and will likely affect people's health (Wright & Kelly, 2017a). Addressing this problem requires collaboration between government, business and the scientific community to reduce risks to the environment and human health. It highlights the importance of addressing plastic pollution and using sustainable waste management practices to protect GW quality and human health (Geng et al., 2024; Wainkwa et al., 2022). MPs enter the environment in various ways, intentionally introduced microfibers and microbeads in personal care products (Duis & Coors, 2016), abrasives or synthetic textiles during laundry (Acharya et al., 2021). MPs have been detected in soil (Abel et al., 2018, 2019), freshwater (Issac & Kandasubramanian, 2021; Lee et al., 2022; L. Li et al., 2019; Horton et al., 2017), marine ecosystems (Carlos et al., 2018; Patchaiyappan et al., 2021) and air (Brien et al., 2023; Can-güven, 2021), glacier (Bergmann et al., 2019; Cabrera et al., 2020). They are stored in rocks, surface waters and aquatic organisms in water bodies, which impacts food supply (Yuan et al., 2022) and public health (Wright & Kelly, 2017a). In addition, MPs that threaten human health were found in drinking water (Brancaleone et al., 2024). The ecological impact of MPs is a major concern (Mao et al., 2022). Ingestion of MPs by marine (Wright et al., 2013) and freshwater organisms can result in physical injury, reduced nutrient availability, and bioaccumulation in the food web. MPs are also acting as contaminants of GW as they leach from soil to GW (Ren et al., 2021) or marine intrusion (Alvarado et al., 2023). According to Sangkham et al., 2023, surface water runoff, agricultural areas, wastewater effluent, landfill contribution, and atmospheric transport and deposits are the sources of MP pollution in GW. There has been less research on MP pollution in GW compared to soil, rivers, oceans, and estuaries (Khant & Kim, 2022). The extent of human exposure to MPs and their health risks are ongoing studies. Preliminary research shows that MPs can be found in food, air, and drinking water, raising concerns about their adverse effects on human health, including inflammation, oxidative stress, and the potential bioaccumulation of bacteria.

21.1.1 Analytical Detection of Plastic

Identifying and quantifying plastic compounds in diverse contexts is a task of analytical plastic detection. In order to, identify plastics analytically, the following are some essential techniques:

21.1.1.1 Visual Identification

Microscopy: Plastic particles, especially MPs, can be visually identified and measured using light and electron microscopy (Cole et al., 2011).

21.1.1.2 Methods of Spectroscopy

FTIR: Fourier Transform Infrared Spectroscopy, uses a polymer's distinct infrared spectrum to identify it, and is utilized for qualitative and quantitative analysis (Shim et al., 2017).

Raman Spectroscopy: Often used to identify colours and additives in plastics, Raman Spectroscopy is a complementary technique to FTIR since it provides comprehensive chemical information (Renner et al., 2018).

Near-Infrared Spectroscopy (NIR): This technique is particularly useful in recycling operations for classifying and detecting various plastic types (Paul et al., 2019).

21.1.1.3 Chromatographic Methods

Gas Chromatography-Mass Spectrometry (GC–MS): Plastic additives and degradation products are identified and quantified using gas chromatography-mass spectrometry, or GC–MS (Löder & Gerdts, 2015).

Pyrolysis—Gas Chromatography—Mass Spectrometry (Py-GC–MS): This method involves breaking down plastics at a high temperature, separating the resultant fragments using chromatography, and then detecting them using mass spectrometry (Dümichen et al., 2017).

21.1.1.4 Thermal Analysis

Differential Scanning Calorimetry (DSC): This technique measures the heat flow related to plastic thermal transitions and yields data on the melting and crystallinity of polymers (Alimi et al., 2018).

Thermogravimetric Analysis (TGA): Thermal stability and polymer composition can be studied by thermogravimetric analysis (TGA), which measures weight variations as a function of temperature (Vianello et al., 2019).

21.1.2 Mass Spectrometry

Matrix-Assisted Laser Desorption/Ionization Time-of-Flight Mass Spectrometry (MALDI-TOF MS): Matrix-Assisted Laser Desorption/Ionization in Mass Spectrometry Polymer composition and molecular weight distribution are examined using Time-of-Flight Mass Spectrometry (MALDI-TOF MS) (Wu et al., 2023).

21.1.2.1 X-ray Methods

X-ray diffraction (XRD): A method for examining polymer crystalline structure (Moura et al., 2023).

X-ray fluorescence (XRF): A valuable technique for identifying and quantifying polymeric material components, XRF can identify impurities and additions (Turner, 2017).

21.1.2.2 Spectroscopy in Conjunction with Microscopy

Scanning Electron Microscopy coupled with energy-dispersive X-ray Spectroscopy (SEM–EDS): Offers elemental analysis and high-resolution pictures of plastic particles (Gniadek & Dąbrowska, 2019).

21.1.2.3 Spectroscopy Using Nuclear Magnetic Resonance (NMR)

Nuclear Magnetic Resonance (NMR): This gives comprehensive details about polymers' dynamics and molecular structure (Karapanagioti & Klontza, 2008).

These methods are used to detect and measure MPs in soil, water, air, and biota samples for environmental applications. Filtration, digestion (to get rid of organic materials), and density separation are frequently used in sample preparation.

Because FTIR and Raman spectroscopy may provide precise information regarding the types of polymers, they are frequently used in MP analysis. Py-GC-MS is helpful for figuring out how MPs break down and for analyzing their chemical makeup. It is easier to detect and describe MPs chemically and visually when spectroscopic methods are combined with microscopy.

21.2 MP Contamination Affects Human Health in Several Ways

There are different types of plastic. Polyethylene Terephthalate (PET), Polycarbonate (PC) High-Density Polyethylene (HDPE), Polyvinyl Chloride (PVC), Low-Density Polyethylene (LDPE), Polypropylene (PP), Polystyrene (PS), Polycarbonate (PC) (Table 21.1).

These particles can enter human cells through food, drink, or the air and may cause cytotoxicity, oxidative stress, and inflammation. Recent research indicates that MPs may be linked to several harmful health outcomes, such as neurotoxicity, endocrine disruption, and cancer. Additionally, they might serve as carriers of dangerous infections and contaminants, worsening their effects on human health (Table 21.1). To completely comprehend the scope of these health concerns and create efficient mitigation solutions, in-depth research is necessary. MPs may deliver localised toxicity upon uptake, impacting the microbiome with specific surface affinities (Cox et al., 2019; Schwabl & Ko, 2019; Wright & Kelly, 2017b). Figure 21.1 shows the pathways of MPs entering the food chain.

Solving the MPs problem requires various methods. Scientists are actively working to improve our understanding of these resources' sources, distribution, fate, and impacts. Mitigation measures include creating alternatives to plastic products, improving waste management, and regulations to limit MP emissions.

In summary, MPs represent a complex and emerging environmental problem with consequences for ecosystems and human health. Reducing the impact of this problem and preventing its further spread into the environment requires continuous research, good diligence and cooperation at local, national and international levels.

21.3 Bibliometric Analysis

Scrutinised the research literature between November 2023 to January 2024, integrating the **Web of Science** databases. By deploying the most prevalent synonym in the research field, the current investigation was further assembled. The syntax used for the document search was: TITLE = ("Microplastic and groundwater") or ("Microplastics and groundwater") or ("Microplastic and groundwaters") or ("Microplastics and groundwaters") or ("plastic and groundwater") or ("plastics and groundwater") or ("plastic and groundwaters") or ("plastics and groundwaters") (Cai et al., 2022).

Search results were confined to empirical research published since 2013 in scientific journals that related particular features of MP pollution to various GW dimensions.

The language of the articles was found in English and Chinese. The abstract has been read completely but the full paper used to be read if necessary to determine the eligibility of the article.

Table 21.1 Health effects of MPs pollution on Humans

Types of Plastics	Identified On	Quantity	Source	Health Effects on Humans	References
PET	Infant stool, Adults stool	5700–82,000 ng/g, 2200–16,000 ng/g	Air, bottles, carpets, drinking water, industrial landfills, medical packaging, food packaging, plastic cards, textiles, household appliances, automotive	Vision failure and irritation of the eyes, Breathlessness and respiratory problems, coughing and soiled throats, skin disease and rashes, Liver dysfunction, Diarrhoea, Food Poisoning, Stomach aches, Birth effects, Cancer, Hormonal changes, Infertility etc.	Zhang et al. (2021), Dhaka et al. (2022), Seewoo et al. (2023)
HDPE	–	–	Leather waste, production of milk containers, detergent bottles, refrigerators, toys, various types of plastic grocery bags, and more	–	Kumari et al. (2023)
PVC	–	–	Pipes mainly, building and construction, water pipes, footwear, toys, automotive	Cancer, birth defects, genetic changes, chronic bronchitis, ulcers, skin diseases, deafness, vision failure, indigestion, and liver dysfunction	Kumari et al. (2023), Seewoo et al. (2023)
LDPE	–	–	Used in packaging of frozen foods and preparation of juices and milk cartons	–	Proshad et al. (2018)

(continued)

Table 21.1 (continued)

Types of Plastics	Identified On	Quantity	Source	Health Effects on Humans	References
PP	–	–	Food packaging, homewares, toys, automotive, clothing, carpets, disposable diapers, plastic bags, agriculture, pipes, construction materials	Promote breast cancer	Seewoo et al. (2023), Park et al. (2023)
PS	–	–	Food packaging, medical applications, housewares, toys, takeaway food containers	Hormone disruption	Govind and Nishitha (2023), Seewoo et al. (2023)
PC	Infant stool, adults stool	49–2100 ng/g, 37–620 ng/g	Medical equipment, construction materials, food storage, electronic components	Early menarche, reduced sperm counts and other altered functions of reproductive organs, obesity, and increased rates of some cancers; some of these effects are produced at very low doses in fetal, infant, and juveniles	Zhang et al. (2021), Seewoo et al. (2023), Bittner et al. (2014)

The inclusion criteria of the articles are to be original research or reviews or editorial materials, must be published between 2013 and 2024, the Language should be English, an abstract should be available, and relevant to the analysis (Figs. 21.2 and 21.3).

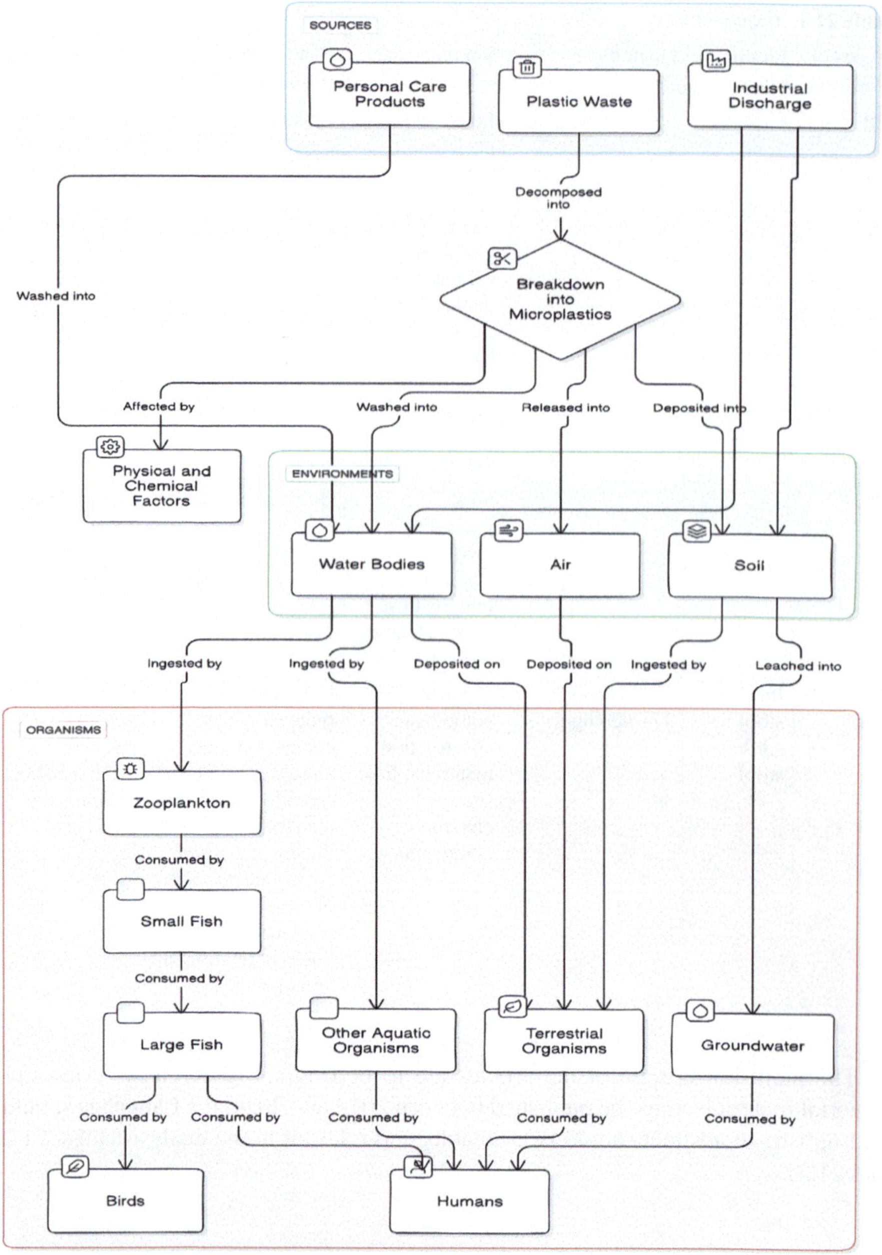

Fig. 21.1 Pathway of MPs entering the food chain

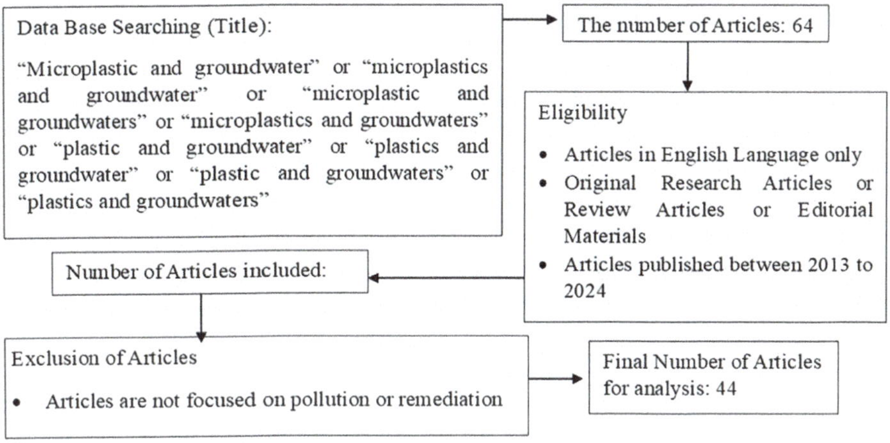

Fig. 21.2 Flow chart of different steps of data collection and data curation

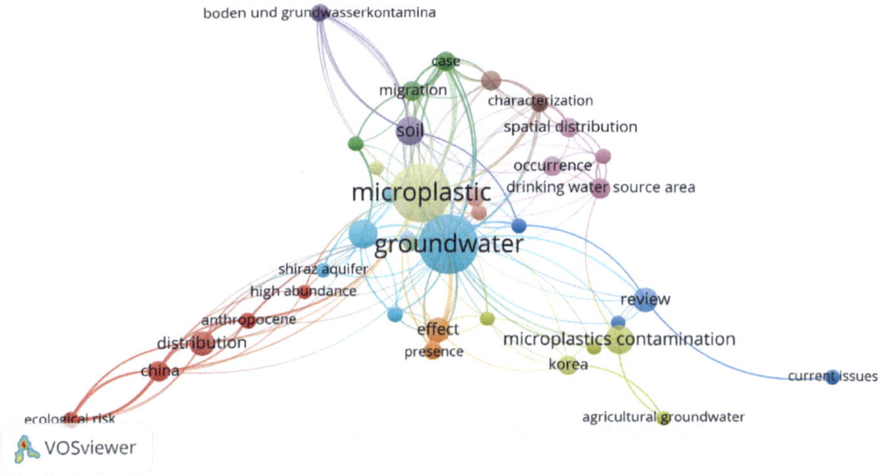

Fig. 21.3 Title analysis of the Published Research Articles considering MPs and GW as two major keywords

21.4 Research Gap

21.4.1 Sources and Pathways

A thorough comprehension of the origins of MPs as well as how they enter various ecosystems needs to be explored more (Auta et al., 2017).

21.4.2 Deterioration Processes

In-depth analysis of the variables affecting the rates of deterioration as well as the mechanisms by which different kinds of plastics deteriorate through different physical, chemical and biological processes (Gewert et al., 2015).

21.4.3 Environmental Distribution

Lack of additional information about the dispersal of MPs in various habitats, such as deep sea and isolated terrestrial areas, is available (Bergmann, 2015).

21.4.4 Detection and Quantification Techniques

Still a need for recognized, sensitive, standardized techniques for finding and measuring MPs in a range of environmental samples like water, sediment, and biota (Gupta et al., 2023; Reproducibility et al., 2020).

21.4.5 Transport Mechanisms

Detailed examination of the processes involved in the movement of MPs both inside and across ecosystems (Lebreton & Andrady, 2019).

21.4.6 Bioavailability and Bioaccumulation

Understanding how bioavailable MPs are to various creatures and how they may accumulate and magnify in food webs is known as bioavailability and bioaccumulation (Besseling et al., 2014).

21.4.7 Ecotoxicological Effects

Long-term research on the ecotoxicological effects of MPs on different species and ecosystems is being done (Wright & Kelly, 2017b).

21.4.8 Impacts on Human Health

Research on the possible health dangers that humans may face from ingesting or breathing MPs (Cox et al., 2019).

21.4.9 Interaction with Contaminants

Analysis of the interactions and cumulative impacts of MPs and other environmental contaminants, such as heavy metals and persistent organic pollutants (Fred-Ahmadu et al., 2020; Martinho et al., 2022; Rochman et al., 2013).

21.4.10 Socioeconomic Impacts

Examining the economic effects of MPs contamination, such as how it affects public health, tourism, and fisheries (Beaumont et al., 2019).

21.4.11 Policy and Mitigation

Creating and assessing efficient policies and mitigation plans to lessen the pollution caused by MPs (Cózar et al., 2014; Cózar et al., 2014).

21.4.12 Microbial Interactions

Research on how MPs and microbial populations interact, as well as how biofilms contribute to the breakdown and movement of MPs (Kirstein et al., 2016).

21.4.13 Novel Approaches

Investigating novel materials and technology to supplant traditional plastics and mitigate MPs contamination (Karan et al., 2019).

21.5 Software Analysis

In this study, bibliometric analysis has been done to understand the emergence of keywords in titles, cited most articles, country of origin of the research articles, year of publication, and Journals, by the VOSviewer software as analysed in Cai et al., (2022), Reyes-riveros et al., (2021).

21.6 National and International Laws and Policies Regarding Plastic Pollution:

Although there is currently no explicit agreement in international law, MPs pollution is becoming a rising concern. Nonetheless, momentum is increasing:

21.6.1 International Policies

UN Environment Programme (UNEP): National laws such as those prohibiting plastic bags and microbeads in cosmetics are highlighted in their study on the legal limitations for single-use plastics and MPs (Li, 2022).

The UN Environment Assembly established the Intergovernmental Negotiating Committee (INC), whose goal is to draft a legally enforceable international agreement to combat plastic pollution by the end of 2024.

21.6.2 National Policies of India

Amendment Rules for Plastic Waste Management (2021): This legislation bans specific single-use plastic items and increases the minimum thickness of plastic carry bags, aiming to reduce plastic waste generation.

21.7 Result

An initial search found 64 articles on the Web of Science and finally retained 44 (Fig. 21.3). Articles focused on MP pollution in GW and remediation. Articles are from 21 different countries. Where the highest number of papers are from South Korea followed by China, India, USA, Italy, and Australia (Fig. 21.4). Keywords of the title have been analysed (Fig. 21.2) and we are getting the most used words in the title. The maximum number of publications has been done in 2022, almost

30.5% of total publications since 2013, closely followed by 2021 and 2023 at 25.4% each (Fig. 21.5). Most of the articles are published on Environmental Science and Water Resources 41% and 10% respectively. In terms of human health China has the maximum number of papers (20 articles), followed by the USA (13 articles), and India (10 articles) out of 85 articles searched on the Web of Science.

21.8 Discussion

The main source of MP in the human body is either inhalation or ingestion from food or water. Research is very low on MP in GW as well as in human health.

According to Khant and Kim (2022), articles on GW are negligible as compared to other sectors like marine, river, aquatic, and soil, till 2015 there were no records of research articles or review articles, in the year 2021 a nominal number of research articles and review articles were published worldwide, 18 and 5 articles respectively.

As per the report, 20% and 40% of GW respectively used for irrigation purposes and industry. 88 developing nations, which house half of the world's population, are currently impacted by water scarcity. Globally, GW is confronted with both qualitative and quantitative issues. Rates of over-extraction frequently surpass rates of natural regeneration. This pressure is exacerbated by the use of synthetic chemicals, industrialization, and garbage disposal. GW composition is influenced by geology; interaction with limestone produces hard water with elevated carbonate and calcium levels. The main human activities that cause contamination and lower GW quality are the use of fertiliser and garbage disposal. Seawater intrusion can be caused by excessive pumping, particularly in coastal areas. Certain regions, such as recently formed volcanic zones, naturally possess non-potable GW because of their high salinity and warmth. Seawater intrusion may occur naturally in coastal karst locations (Katsanou & Karapanagioti, 2022).

It has been understood from the Publication result (Fig. 21.6) that MPs has a major focus on the environmental sector 41% of publications, closely followed by Water resources 10%. The effects MPs has on human health somewhat have less focus. Water isn't the only pathway for MP to enter into the human body, there are several ways to get affected by it.

21.9 Conclusion

The presence of MPs in GW is a serious environmental risk that could have an impact on ecosystems and human health.

Ubiquitous Presence: GW, soil, and even air is all contaminated by MPs, which are widely present in the environment.

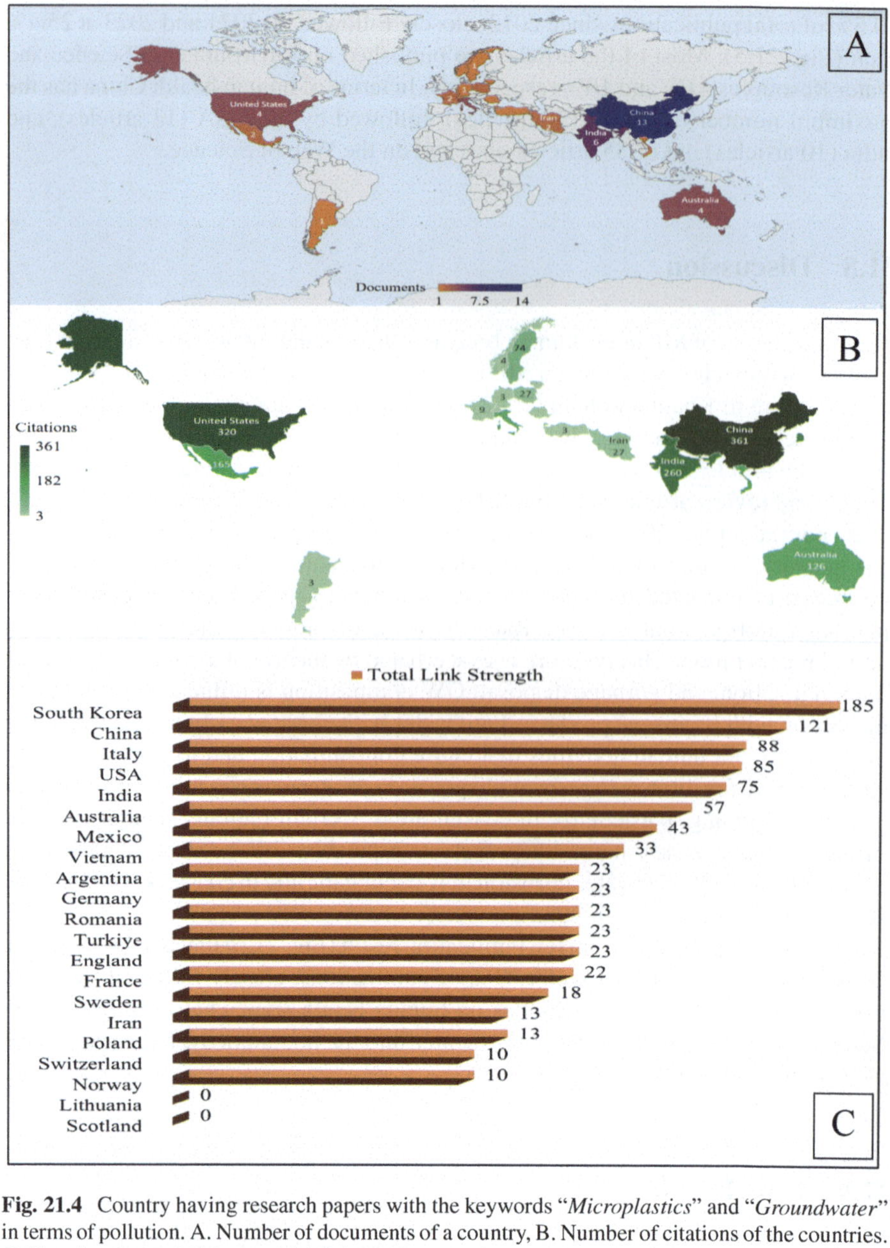

Fig. 21.4 Country having research papers with the keywords "*Microplastics*" and "*Groundwater*" in terms of pollution. A. Number of documents of a country, B. Number of citations of the countries. C. Showing the total link strength of citations

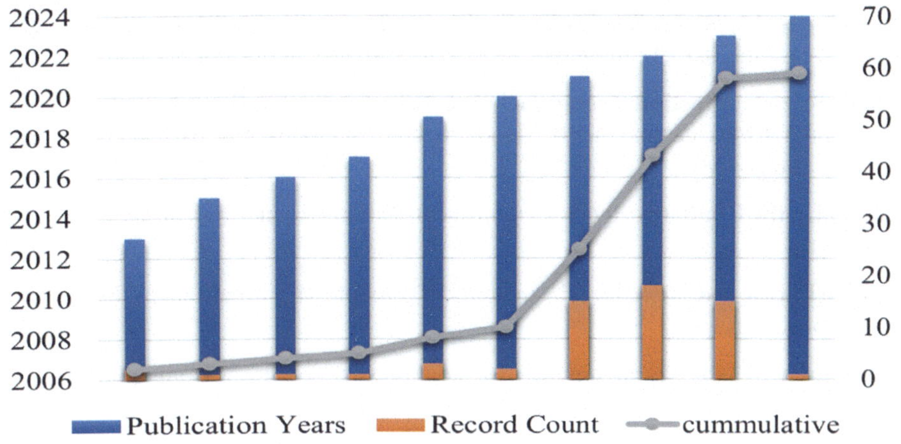

Fig. 21.5 Document (research articles) published yearly

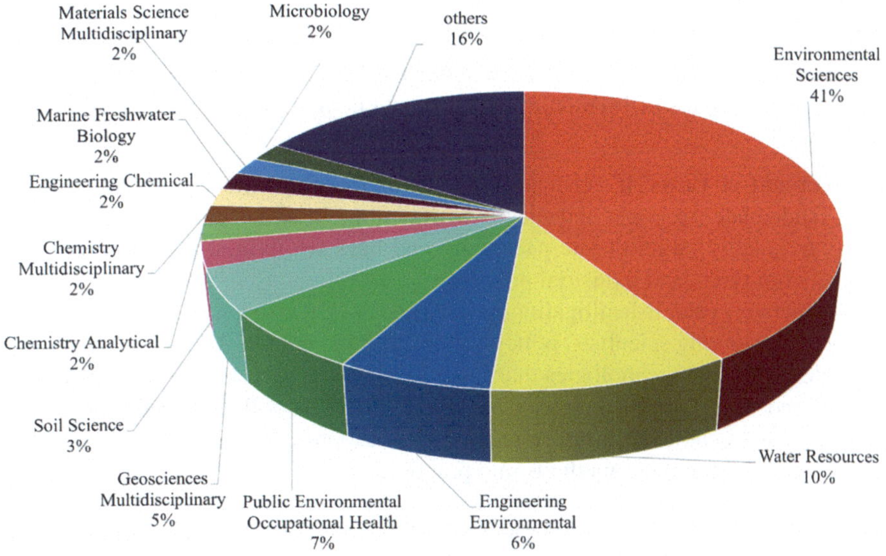

Fig. 21.6 Result of the publication (Research and Review articles) in different journals

Groundwater Vulnerability: MPs contamination is becoming a bigger threat to GW, a resource that is essential for industry, agriculture, and drinking water.

Diverse Sources: MPs in GW can come from synthetic fabrics, breakdown of bigger plastics, and air deposition.

Limited information: With respect to other environmental domains, there is fewer studies on MPs in GW, which highlights important information gaps.

Possible Health Risks: Studies have indicated that consuming and breathing MPs may increase the risk of inflammation, oxidative stress, and possibly cancer.

Standardized Detection Techniques Are Required: For efficient monitoring, it is essential to create and apply standardized techniques for the detection and quantification of MPs.

Taking Up the Challenge: It takes a multifaceted strategy to address this complicated problem:

Reduce Plastic Use: To reduce the quantity of possible MPs creation, minimize your overall plastic usage.

Strengthen Waste Management: To stop plastic pollution from entering the environment, improve waste management procedures.

Stronger Regulations: Enact tighter guidelines to manage the manufacture, consumption, and disposal of plastics.

Scientific Knowledge Gaps: To guide policy direction, research on the origins, routes, and ecological effects of MPs needs to be expanded.

Additional Research: To learn more about the origins, modes of transportation, and long-term effects of MPs on ecosystems and human health, conduct additional research.

Enforcement of Laws: It's critical to have effective implementation and enforcement. Inadequate oversight and sanctions may impede advancement. Present regulations frequently target plastic packaging and bags while ignoring MPs found in fabrics, worn tyres, and industrial processes.

To effectively establish mitigation policies and laws, it is recommended to promote collaboration among scientists, policymakers, and the public. Additionally, innovative water treatment technologies that can effectively remove MPs from GW should be explored and developed. It can minimize GW MP pollution and protect our ecosystem and health for future generations by accepting the scientific facts, acting quickly, and putting these methods into practice.

Acknowledgements We authors acknowledge RUSA (Rashtriya Uchchatar Shiksha Abhiyan)-2.0 Biological Sciences, Bharathidasan University, for financial support to carry out the study.

References

Abel, A., Machado, D. S., Lau, C. W., Kloas, W., & Bergmann, J. (2019). Microplastics can change soil properties and affect plant performance. https://doi.org/10.1021/acs.est.9b01339

Abel, A., Machado, D. S., Lau, C. W., Till, J., Kloas, W., Lehmann, A., Becker, R., & Rillig, M. C. (2018). Impacts of microplastics on the soil biophysical environment. https://doi.org/10.1021/acs.est.8b02212

Acharya, S., Rumi, S. S., & Hu, Y. (2021). Microfibers from synthetic textiles as a major source of microplastics in the environment : A review. September 2022. https://doi.org/10.1177/0040517521991244

Adnan, M. (2022). Preliminary study on the distribution, source , and ecological risk of typical microplastics in Karst Groundwater in Guizhou Province , China

Alimi, O. S., Farner Budarz, J., Hernandez, L. M., & Tufenkji, N. (2018). Microplastics and nanoplastics in aquatic environments: aggregation, deposition, and enhanced contaminant transport. *Environmental Science and Technology, 52*(4), 1704–1724. https://doi.org/10.1021/acs.est.7b05559

Alvarado, D., José, Z., Hernández, R. R., & Ruiz, C. G. (2023). First insight into microplastic groundwater pollution in Latin America: The case of a coastal aquifer in Northwest Mexico. *Environmental Science and Pollution Research, 30*(29), 73600–73611. https://doi.org/10.1007/s11356-023-27461-9

Auta, H. S., Emenike, C. U., & Fauziah, S. H. (2017). Distribution and importance of microplastics in the marine environmentA review of the sources, fate, effects, and potential solutions. *Environment International, 102*, 165–176. https://doi.org/10.1016/j.envint.2017.02.013

Beaumont, N. J., Aanesen, M., Austen, M. C., Börger, T., Clark, J. R., Cole, M., Hooper, T., Lindeque, P. K., Pascoe, C., & Wyles, K. J. (2019). Global ecological, social and economic impacts of marine plastic. *Marine Pollution Bulletin, 142*(January), 189–195. https://doi.org/10.1016/j.marpolbul.2019.03.022

Bergmann, M. (n.d.). *Marine Anthropogenic Litter.*

Bergmann, M., Mützel, S., Primpke, S., & Tekman, M. B. (2019). *White and wonderful ? Microplastics prevail in snow from the Alps to the Arctic.* August, 1–11.

Besseling, E., Wang, B., Lürling, M., & Koelmans, A. A. (2014). Nanoplastic affects growth of S. obliquus and reproduction of D. magna. *Environmental Science and Technology, 48*(20), 12336–12343. https://doi.org/10.1021/es503001d

Blackburn, K. (2022). The potential effects of microplastics on human health: What is known and what is unknown. *Ambio, 51*(3), 518–530. https://doi.org/10.1007/s13280-021-01589-9

Bittner, G. D., Yang, C. Z., & Stoner, M. A. (2014). Estrogenic chemicals often leach from BPA-free plastic products that are replacements for BPA-containing polycarbonate products. *Environmental Health, 13*(1). https://doi.org/10.1186/1476-069x-13-41

Brancaleone, E., Mattei, D., Fuscoletti, V., Lucentini, L., Favero, G., Cecchini, G., Frugis, A., Gioia, V., & Lazzazzara, M. (2024). *Microplastic in drinking water : A pilot study* (pp. 31–45).

Brien, S. O., Rauert, C., Ribeiro, F., Okoffo, E. D., Burrows, S. D., Brien, J. W. O., Wang, X., Wright, S. L., & Thomas, K. V. (2023). Science of the Total Environment There ' s something in the air: A review of sources, prevalence and behaviour of microplastics in the atmosphere. *Science of the Total Environment, 874*(February), 162193. https://doi.org/10.1016/j.scitotenv.2023.162193

Cabrera, M., Valencia, B. G., Lucas-solis, O., Calero, J. L., Maisincho, L., Conicelli, B., Massaine, G., & Capparelli, M. V. (2020). Case studies in chemical and environmental engineering a new method for microplastic sampling and isolation in mountain glaciers: a case study of one antisana glacier, Ecuadorian Andes. *Case Studies in Chemical and Environmental Engineering, 2*(September), 100051. https://doi.org/10.1016/j.cscee.2020.100051

Cai, Y., Li, C., & Zhao, Y. (2022). A review of the migration and transformation of microplastics in inland water systems

Can-güven, E. (2021). *Microplastics as emerging atmospheric pollutants : a review and bibliometric analysis* (pp. 203–215).

Carlos, L., Sá, D., Oliveira, M., Ribeiro, F., Lopes, T., & Norman, M. (2018). Studies of the effects of microplastics on aquatic organisms: What do we know and where should we focus our efforts in the future ? *Science of the Total Environment, 645*, 1029–1039. https://doi.org/10.1016/j.scitotenv.2018.07.207

Chandran, M., & Senthilkumar, T. (2020). Conversion of plastic waste to fuel (Issue January). https://doi.org/10.1016/B978-0-12-817880-5.00014-1

Chia, R. W., Lee, J. Y., Kim, H., & Jang, J. (2021). Microplastic pollution in soil and groundwater: A review. *Environmental Chemistry Letters, 19*(6), 4211–4224. https://doi.org/10.1007/s10311-021-01297-6

Cole, M., Lindeque, P., Halsband, C., & Galloway, T. S. (2011). Microplastics as contaminants in the marine environment: A review. *Marine Pollution Bulletin, 62*(12), 2588–2597. https://doi.org/10.1016/j.marpolbul.2011.09.025

Cox, K. D., Covernton, G. A., Davies, H. L., Dower, J. F., Juanes, F., & Dudas, S. E. (2019). Human consumption of microplastics. *Environmental Science and Technology, 53*(12), 7068–7074. https://doi.org/10.1021/acs.est.9b01517

Cózar, A., Echevarría, F., González-Gordillo, J. I., Irigoien, X., Úbeda, B., Hernández-León, S., Palma, Á. T., Navarro, S., García-de-Lomas, J., Ruiz, A., Fernández-de-Puelles, M. L., & Duarte, C. M. (2014). Plastic debris in the open ocean. *Proceedings of the National Academy of Sciences of the United States of America, 111*(28), 10239–10244. https://doi.org/10.1073/pnas.1314705111

Dhaka, V., Singh, S., Anil, A. G., Sunil Kumar Naik, T. S., Garg, S., Samuel, J., Kumar, M., Ramamurthy, P. C., & Singh, J. (2022). Occurrence, toxicity and remediation of polyethylene terephthalate plastics. A review. *Environmental Chemistry Letters, 20*(3), 1777–1800. https://doi.org/10.1007/s10311-021-01384-8

Duis, K., & Coors, A. (2016). Microplastics in the aquatic and terrestrial environment: Sources (with a specific focus on personal care products), fate and effects. *Environmental Sciences Europe*. https://doi.org/10.1186/s12302-015-0069-y

Dümichen, E., Eisentraut, P., Bannick, C. G., Barthel, A. K., Senz, R., & Braun, U. (2017). Fast identification of microplastics in complex environmental samples by a thermal degradation method. *Chemosphere, 174*, 572–584. https://doi.org/10.1016/j.chemosphere.2017.02.010

Evangeliou, N., & Brahney, J. (2022). *Sources and fate of atmospheric microplastics revealed from inverse and dispersion modelling : From global emissions to deposition Sources and fate of atmospheric microplastics revealed from inverse and dispersion modelling : From global emissions to depo. February*. https://doi.org/10.1016/j.jhazmat.2022.128585

Fred-Ahmadu, O. H., Bhagwat, G., Oluyoye, I., Benson, N. U., Ayejuyo, O. O., & Palanisami, T. (2020). Interaction of chemical contaminants with microplastics: Principles and perspectives. *Science of the Total Environment, 706*(December). https://doi.org/10.1016/j.scitotenv.2019.135978

Gallo, F., Fossi, C., Weber, R., Santillo, D., Sousa, J., Ingram, I., Nadal, A., & Romano, D. (2018). Marine litter plastics and microplastics and their toxic chemicals components: The need for urgent preventive measures. *Environmental Sciences Europe*. https://doi.org/10.1186/s12302-018-0139-z

Geng, C., Gao, Y., Zhang, H., Xue, D., Shan, H., Wang, B., & Wang, X. (2024). Microplastic migration in porous media at various scales : a review. *Environmental Chemistry Letters*, 0123456789. https://doi.org/10.1007/s10311-023-01688-x

Gewert, B., Plassmann, M. M., & Macleod, M. (2015). Pathways for degradation of plastic polymers floating in the marine environment. *Environmental Sciences: Processes and Impacts, 17*(9), 1513–1521. https://doi.org/10.1039/c5em00207a

Giechaskiel, B., Grigoratos, T., Mathissen, M., Quik, J., & Tromp, P. (2024). *Contribution of road vehicle tyre wear to microplastics and ambient air pollution.*

Govind, A., & K, N. (2023). Plastic and its side effects on humans – a review article. *Asian Pacific Journal of Environment and Cancer, 6*(1), 81–85. https://doi.org/10.31557/apjec.2023.6.1.81-85

Gniadek, M., & Dąbrowska, A. (2019). The marine nano- and microplastics characterisation by SEM-EDX: The potential of the method in comparison with various physical and chemical approaches. *Marine Pollution Bulletin, 148*(May), 210–216. https://doi.org/10.1016/j.marpolbul.2019.07.067

Gupta, D. K., Vishwakarma, D. C. A., & Srivastava, M. M. A. K. (2023). Microplastics in freshwater environment : occurrence , analysis , impact , control measures and challenges. *International*

Journal of Environmental Science and Technology, 20(6). https://doi.org/10.1007/s13762-022-04139-2

Horton, A. A., Svendsen, C., Williams, R. J., Spurgeon, D. J., & Lahive, E. (2017). Large microplastic particles in sediments of tributaries of the River Thames, UK—Abundance, sources and methods for effective quantification. *MPB, 114*(1), 218–226. https://doi.org/10.1016/j.marpolbul.2016.09.004

Issac, M. N., & Kandasubramanian, B. (2021). *Effect of microplastics in water and aquatic systems* (pp. 19544–19562).

Karan, H., Funk, C., Grabert, M., Oey, M., & Hankamer, B. (2019). Green bioplastics as part of a circular bioeconomy. *Trends in Plant Science, 24*(3), 237–249. https://doi.org/10.1016/j.tplants.2018.11.010

Karapanagioti, H. K., & Klontza, I. (2008). Testing phenanthrene distribution properties of virgin plastic pellets and plastic eroded pellets found on Lesvos island beaches (Greece). *Marine Environmental Research, 65*(4), 283–290. https://doi.org/10.1016/j.marenvres.2007.11.005

Katare, Y. (2021). *Microplastics in aquatic environments: Sources , ecotoxicity , detection & remediation. August*. https://doi.org/10.33263/BRIAC123.34073428

Katsanou, K., & Karapanagioti, H. K. (2022). *Surface Water and Groundwater Sources for Drinking Water Surface Water and Groundwater Sources for Drinking Water. December 2017*. https://doi.org/10.1007/698

Khant, N. A., & Kim, H. (2022). Review of current issues and management strategies of microplastics in groundwater environments.

Kirstein, I. V., Kirmizi, S., Wichels, A., Garin-Fernandez, A., Erler, R., Löder, M., & Gerdts, G. (2016). Dangerous hitchhikers? Evidence for potentially pathogenic Vibrio spp. on microplastic particles. *Marine Environmental Research, 120*, 1–8. https://doi.org/10.1016/j.marenvres.2016.07.004

Kumari, R., Tiwari, R., Akhtar, R., Sunil, A., & Gupta, K. (2023). HHarmful effects of plastics on human health and the environment: a review. *GPH-Int J Soc Sci Hum Res, 06*(11), 248–255. https://doi.org/10.5281/zenodo.10429234

Lebreton, L., & Andrady, A. (2019). Future scenarios of global plastic waste generation and disposal. *Palgrave Communications, 5*(1), 1–11. https://doi.org/10.1057/s41599-018-0212-7

Lee, J., Cha, J., Lee, J., & Wainkwa, R. (2022). Comment on the paper "Microplastic contamination of an unconfined groundwater aquifer in Victoria, Australia" Science of the a "Microplastic contamination of an unconfined groundwater aquifer in Victoria, Australia." *Science of the Total Environment, 820*(January), 153121. https://doi.org/10.1016/j.scitotenv.2022.153121

Li, L., Geng, S., Wu, C., Song, K., Sun, F., Visvanathan, C., Xie, F., & Wang, Q. (2019). Microplastics contamination in different trophic state lakes along the middle and lower reaches of Yangtze River Basin. *254*. https://doi.org/10.1016/j.envpol.2019.07.119

Li, Y. (2022). Legislation and policy on pollution prevention and the control of marine microplastics. *Water (Switzerland), 14*(18). https://doi.org/10.3390/w14182790

Löder, M. G. J., & Gerdts, G. (2015). *Methodology used for the detection and identification of microplastics—A critical appraisal*. https://doi.org/10.1007/978-3-319-16510-3

Lu, Y., Zhang, Y., Deng, Y., Jiang, W., Zhao, Y., Geng, J., Ding, L., & Ren, H. (2016). *Uptake and accumulation of polystyrene microplastics in Zebra fish (Danio rerio) and toxic eff ects in liver*. https://doi.org/10.1021/acs.est.6b00183

Lv, M., Jiang, B., Xing, Y., Ya, H., Zhang, T., & Wang, X. (2022). Recent advances in the breakdown of microplastics : strategies and future prospectives. *Environmental Science and Pollution Research*, 65887–65903. https://doi.org/10.1007/s11356-022-22004-0

Mao, X., Xu, Y., Cheng, Z., Yang, Y., Guan, Z., Jiang, L., & Tang, K. (2022). *The impact of microplastic pollution on ecological environment: A review. 27*(2).

Martinho, S. D., Fernandes, V. C., Figueiredo, S. A., & Delerue-Matos, C. (2022). Microplastic pollution focused on sources, distribution, contaminant interactions, analytical methods, and wastewater removal strategies: a review. *International Journal of Environmental Research and Public Health, 19*(9). https://doi.org/10.3390/ijerph19095610

Moura, D. S., Pestana, C. J., Moffat, C. F., Hui, J., Irvine, J. T. S., & Lawton, L. A. (2023). Characterisation of microplastics is key for reliable data interpretation. *Chemosphere, 331*(March), 138691. https://doi.org/10.1016/j.chemosphere.2023.138691

Musie, W., & Gonfa, G. (2023). Heliyon Fresh water resource, scarcity, water salinity challenges and possible remedies: A review. *Heliyon, 9*(8), e18685. https://doi.org/10.1016/j.heliyon.2023.e18685

Park, J. H., Hong, S., Kim, O.-H., Kim, C.-H., Kim, J., Kim, J.-W., Hong, S., & Lee, H. J. (2023). Polypropylene microplastics promote metastatic features in human breast cancer. *Scientific Reports, 13*(1). https://doi.org/10.1038/s41598-023-33393-8

Patchaiyappan, A., Zakiahmed, S., Dowarah, K., Thirunavukkarasu, C., & Devipriya, S. P. (2021). Prevalence of microplastics in the sediments of Odisha beaches , southeastern coast of India. *Marine Pollution Bulletin, 167*(October 2020), 112265. https://doi.org/10.1016/j.marpolbul.2021.112265

Paul, A., Wander, L., Becker, R., Goedecke, C., & Braun, U. (2019). High-throughput NIR spectroscopic (NIRS) detection of microplastics in soil. *Environmental Science and Pollution Research, 26*(8), 7364–7374. https://doi.org/10.1007/s11356-018-2180-2

Peng, W., & Sonne, C. (2023). *Chemosphere A discussion of microplastics in soil and risks for ecosystems and food chains. 313*(October 2022). https://doi.org/10.1016/j.chemosphere.2022.137637

Proshad, R., Kormoker, T., Islam, Md. S., Haque, M. A., Rahman, Md. M., & Mithu, Md. M. (2017). Toxic effects of plastic on human health and environment: a consequences of health risk assessment in Bangladesh. *International Journal of Health, 6*(1), 1–5. https://doi.org/10.14419/ijh.v6i1.8655

Quiot, F., Institut, I., De, N., & Industriel, E. (2022). State of knowledge and future research needs on microplastics in groundwater Amélie Hoarau Belkhiri. *Florence Carre., 20*(10), 1479–1496. https://doi.org/10.2166/wh.2022.048

Ren, Z., Gui, X., Xu, X., Zhao, L., Qiu, H., & Cao, X. (2021). Microplastics in the soil-groundwater environment: Aging, migration, and co-transport of contaminants—A critical review. *Journal of Hazardous Materials, 419*(June), 126455. https://doi.org/10.1016/j.jhazmat.2021.126455

Renner, G., Schmidt, T. C., & Schram, J. (2018). Analytical methodologies for monitoring micro(nano)plastics: Which are fit for purpose? *Current Opinion in Environmental Science and Health, 1*, 55–61. https://doi.org/10.1016/j.coesh.2017.11.001

Reproducibility, M., Al-azzawi, M. S. M., Kefer, S., Weißer, J., Reichel, J., Schwaller, C., Glas, K., Knoop, O., & Drewes, J. E. (2020). *Validation of sample preparation methods for microplastic analysis in wastewater.*

Reyes-riveros, R., Altamirano, A., La, F. De, Vieli, L., Meli, P., & Rozas-v, D. (2021). *Urban forestry & urban greening linking public urban green spaces and human well-being : A systematic review. 61*(February). https://doi.org/10.1016/j.ufug.2021.127105

Rochman, C. M., Browne, M. A., Halpern, B. S., Hentschel, B. T., Hoh, E., Karapanagioti, H. K., Rios-Mendoza, L. M., Takada, H., Teh, S., & Thompson, R. C. (2013). Policy: Classify plastic waste as hazardous. *Nature, 494*(7436), 169–170. https://doi.org/10.1038/494169a

Samandra, S., Johnston, J. M., Jaeger, J. E., Symons, B., Xie, S., Currell, M., Ellis, A. V., & Clarke, B. O. (2022). Science of the total environment microplastic contamination of an unconfined groundwater aquifer in. *Science of the Total Environment, 802*, 149727. https://doi.org/10.1016/j.scitotenv.2021.149727

Sangkham, S., Islam, A., Adhikari, S., Kumar, R., Sharma, P., Sakunkoo, P., Bhattacharya, P., & Tiwari, A. (2023). Groundwater for sustainable development evidence of microplastics in groundwater: A growing risk for human health. *Groundwater for Sustainable Development, 23*(January), 100981. https://doi.org/10.1016/j.gsd.2023.100981

Seewoo, B. J., Goodes, L. M., Mofflin, L., Mulders, Y. R., Wong, E. V., Toshniwal, P., Brunner, M., Alex, J., Johnston, B., Elagali, A., Gozt, A., Lyle, G., Choudhury, O., Solomons, T., Symeonides, C., & Dunlop, S. A. (2023). The plastic health map: a systematic evidence map of human health studies on plastic-associated chemicals. *Environment International, 181*, 108225. https://doi.org/10.1016/j.envint.2023.108225

Schwabl, P., & Ko, S. (2019). *Detection of various microplastics in human stool* (pp. 453–458). https://doi.org/10.7326/M19-0618

Shim, W. J., Hong, S. H., & Eo, S. E. (2017). Identification methods in microplastic analysis: A review. *Analytical Methods, 9*(9), 1384–1391. https://doi.org/10.1039/c6ay02558g

Sutkar, P. R., Gadewar, R. D., & Dhulap, V. P. (2023). Journal of hazardous materials advances recent trends in degradation of microplastics in the environment: A state-of-the-art review. *Journal of Hazardous Materials Advances, 11*(July), 100343. https://doi.org/10.1016/j.hazadv.2023.100343

Thompson, R. C., Olson, Y., Mitchell, R. P., Davis, A., Rowland, S. J., John, A. W. G., McGonigle, D., & Russell, A. E. (2004). Lost at sea: Where is all the plastic? *Science, 304*(5672), 838. https://doi.org/10.1126/science.1094559

Turner, A. (2017). In situ elemental characterisation of marine microplastics by portable XRF. *Marine Pollution Bulletin, 124*(1), 286–291. https://doi.org/10.1016/j.marpolbul.2017.07.045

Ustabasi, G. S., & Baysal, A. (2019). *Occurrence and risk assessment of microplastics from various toothpastes*.

Vianello, A., Jensen, R. L., Liu, L., & Vollertsen, J. (2019). Simulating human exposure to indoor airborne microplastics using a Breathing Thermal Manikin. *Scientific Reports, 9*(1), 1–11. https://doi.org/10.1038/s41598-019-45054-w

Wainkwa, R., Lee, J., Jang, J., & Cha, J. (2022). Environmental technology & innovation errors and recommended practices that should be identified to reduce suspected concentrations of microplastics in soil and groundwater: a review. *Environmental Technology & Innovation, 28*, 102933. https://doi.org/10.1016/j.eti.2022.102933

Wang, W., Wairimu, A., Li, Z., & Wang, J. (2017). Science of the total environment microplastics pollution in inland freshwaters of China: A case study in urban surface waters of Wuhan. *China, 575*, 1369–1374.

Wright, S. L., & Kelly, F. J. (2017a). *Plastic and human health: A micro issue?* May. https://doi.org/10.1021/acs.est.7b00423

Wright, S. L., & Kelly, F. J. (2017). Plastic and human health: A micro issue? *Environmental Science and Technology, 51*(12), 6634–6647. https://doi.org/10.1021/acs.est.7b00423

Wright, S. L., Thompson, R. C., & Galloway, T. S. (2013). The physical impacts of microplastics on marine organisms: A review. *Environmental Pollution, 178*, 483–492. https://doi.org/10.1016/j.envpol.2013.02.031

Wu, P., Wu, X., Huang, Q., Yu, Q., Jin, H., & Zhu, M. (2023). Mass spectrometry-based multimodal approaches for the identification and quantification analysis of microplastics in food matrix. *Frontiers in Nutrition, 10*(April), 1–7. https://doi.org/10.3389/fnut.2023.1163823

Yuan, Z., Nag, R., & Cummins, E. (2022). Human health concerns regarding microplastics in the aquatic environment—From marine to food systems. *Science of the Total Environment, 823*, 153730. https://doi.org/10.1016/j.scitotenv.2022.153730

Zhang, K., Hamidian, A. H., Tubić, A., Zhang, Y., Fang, J. K. H., Wu, C., & Lam, P. K. S. (2021). Understanding plastic degradation and microplastic formation in the environment: a review. *Environmental Pollution, 274*, 116554. https://doi.org/10.1016/j.envpol.2021.116554

Chapter 22
Dwindling Depths: Unraveling the Environmental Health Crisis in Declining Groundwater Quality Referencing SDG 3 and 6

Gagan Matta, Rama Pal, Amit Kumar, Rajesh A. Bobdey, Pawan Kumar, Anjali Nayak, Gaurav Pant, Pooja, and Shivangi Tyagi

Abstract The article delves into a scientific investigation of the challenges related to groundwater quality and its impact on achieving Sustainable Development Goals (SDGs) 3 and 6. This research employs various approaches from diverse fields to comprehend the complex processes contributing to groundwater deterioration, which has significant implications for global public health and the environment. The investigation identifies key factors behind groundwater quality decline, including industrial pollutants, agricultural contaminants, and the cumulative effects of urbanization. Through comprehensive scientific analyses, the study has unequivocally established a link between declining groundwater quality and adverse health outcomes. This underscores the urgent need for interventions aligning with SDG 3—Good Health and Well-being. Furthermore, the research underscores SDG 6—Clean Water and Sanitation by highlighting the profound impact of compromised groundwater on safe drinking water accessibility. It emphasizes the interconnectedness between environmental health challenges and SDG achievement. "Dwindling Depths" offers scientific insights enhancing our understanding of groundwater quality intricacies and advocates for implementing comprehensive strategies to address the environmental health crisis. This research serves as a crucial foundation for policymakers and practitioners seeking sustainable solutions to address SDGs 3 and 6, underscoring the

G. Matta (✉) · P. Kumar · A. Nayak · G. Pant · Pooja · S. Tyagi
Hydrology and Science Communication Research Laboratory, Department of Zoology and Environmental Science, Gurukula Kangri (Deemed to be University), Haridwar, India
e-mail: drgaganmatta@gkv.ac.in

R. Pal
ICAR—Indian Institute of Soil and Water Conservation, Dehradun, India

A. Kumar
Wadia Institute of Himalayan Geology, Dehradun, India

R. A. Bobdey
Department of Chemistry, Saibaba Arts and Science College, Parshioni, Nagpur, India

V. Sunitha et al. (eds.), *Groundwater Resource Management Planning Strategies*, Springer Water, https://doi.org/10.1007/978-3-031-88870-0_22

paramount importance of safeguarding groundwater resources for present and future generations.

Keywords Environmental health · Water pollution · Groundwater pollution · SDG 3 · SDG 6

22.1 Introduction

Groundwater is of great scientific importance as it is a crucial part of the Earth's hydrological cycle. It plays a critical role in supporting ecosystems, human activities, and maintaining the global water balance. It constitutes a significant portion of Earth's water that is found underground, specifically in saturated soil and rock layers, creating aquifers (Matta & Uniyal, 2017). Groundwater plays a crucial role in storing large amounts of freshwater, which helps to replenish rivers, lakes, and other surface water bodies. The slow movement through the subsurface guarantees a steady and dependable water supply, even when surface water is scarce. Groundwater resides beneath the Earth's surface within aquifers composed of saturated soil and rock layers (Arora et al., 2017). It is a significant source of freshwater.

The importance of this goes far beyond being a simple underground water source. It has a profound impact on ecosystems, supports agriculture, and ensures a dependable water supply for human societies. Understanding the intricacies of groundwater and promoting its sustainable use requires a thorough scientific investigation. Groundwater plays a crucial role in sustaining the flow of rivers and supporting the delicate balance of wetland ecosystems (Gleeson et al., 2020). It supports a consistent water supply for vegetation, which helps maintain biodiversity and provides habitats for various plant and animal species. In addition, the discharge of groundwater into surface water bodies has a significant impact on temperature regulation, nutrient cycling, and the overall well-being of aquatic ecosystems (Werner et al., 2017). The article highlights important areas of research that need to be addressed in order to understand better the underlying factors contributing to the decline in groundwater quality. It emphasizes the importance of conducting thorough research on sources and pathways of contaminants, finding new remediation methods, and examining the socio-economic factors that impact groundwater management. It is crucial to bridge these gaps in order to make well-informed decisions and formulate effective policies to address the environmental issues. Here are the main goals of the study:

- **Comprehensive Understanding of Declining Groundwater Quality**: The study aims to conduct a detailed investigation into the factors that contribute to the degradation of groundwater quality. This requires the identification of contaminant sources, a comprehensive understanding of transport mechanisms, and a thorough assessment of the impact of human activities on groundwater quality (Lapworth et al., 2017). The research aims to provide valuable insights to the scientific community and policymakers.

- **Linking Groundwater Quality Decline to Human Health (SDG 3)**: The research aims to establish a strong correlation between the deterioration of groundwater quality and its repercussions on human health. The study seeks to contribute to the achievement of SDG 3 by examining waterborne diseases linked to contaminated groundwater (Baldassarre et al., 2019). It aims to enhance our understanding of the health risks posed by compromised water sources.
- **Addressing Water Quality Challenges (SDG 6)**: The research focuses on understanding how declining groundwater quality hinders the achievement of universal access to clean water and sanitation, in line with SDG 6. The investigation concentrates on the impact of compromised groundwater quality on the provision of safe drinking water, highlighting the importance of implementing effective management strategies to ensure the availability of clean water (Bargali et al., 2017).
- **Impact on Ecosystem Health and Biodiversity**: The study investigates the consequences of deteriorating groundwater quality on ecosystems, in line with SDG 15 and 14. The study aims to emphasize the interconnectedness of environmental health with human activities and promote holistic conservation efforts by examining how contaminants impact biodiversity, habitat stability, and ecosystem services (Kløve et al., 2014).
- **Identification of Research Gaps**: A critical aspect of this research involves identifying gaps in current knowledge regarding the decline in groundwater quality. The study aims to determine gaps in research in order to guide future investigations, policy formulation, and the development of innovative solutions for sustainable groundwater management.
- **Informing Sustainable Management Strategies**: The ultimate purpose of this research is to contribute to the development of science-based, sustainable groundwater management strategies. By synthesizing findings on the causes and consequences of declining groundwater quality, the study aims to provide actionable insights for policymakers, researchers, and communities working towards safeguarding water resources and public health (Banks et al., 2021). Figure 22.1 shows the multiple services offered by groundwater systems

22.2 Historical Context of Groundwater Quality

The historical context of groundwater quality covers a vast expanse of time and showcases the changing dynamics between human societies and the environment. Our understanding of groundwater quality has undergone significant changes over time, driven by advancements in technology, scientific breakthroughs, and the demands of society. Presented below is a concise summary of the historical backdrop surrounding groundwater quality:

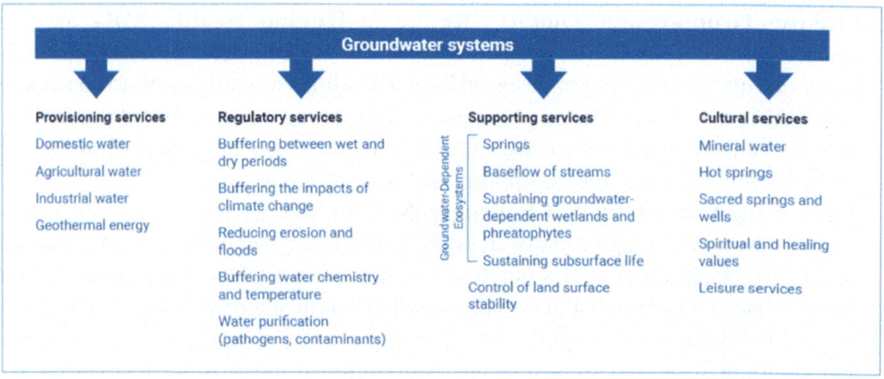

Fig. 22.1 The multiple services offered by groundwater systems. *Source* Van der Gun. (2019)

22.2.1 Ancient Civilizations

- **Early Water Supply Systems**: Ancient civilizations, including the Indus Valley, Mesopotamia, and Ancient Egypt, established advanced water supply systems that were dependent on groundwater. Wells and cisterns were built to tap into groundwater, providing crucial supplies of drinking water and irrigation (Ling et al., 2016).
- **Limited Awareness of Contamination**: Despite the recognition of the significance of accessing clean water in ancient societies, there was a limited understanding of groundwater contamination. Contaminants were frequently introduced through natural processes or human activities without a thorough understanding of their potential health effects (Singh et al., 2022).

22.2.2 Middle Ages to Renaissance

- **Advancements in Well Construction**: Advancements in well construction techniques during the Middle Ages and Renaissance led to the ability to access groundwater at greater depths and with increased reliability. The introduction of more advanced techniques significantly improved the efficiency of water extraction, replacing the traditional hand-dug wells (Sun & Guo, 2019).
- **Emergence of Public Health Concerns**: In the Renaissance period, there was a heightened recognition of the connection between water quality and public health, leading to the emergence of public health concerns (Matta et al., 2018a, 2018b, 2018c, 2018d, 2018e). Observations of waterborne diseases led scholars to acknowledge the significance of clean water sources and advocate for the preservation of water quality (Zainab et al., 2020).

22.2.3 Industrial Revolution

- **Impact of Industrialization**: The Industrial Revolution had a profound effect on society, as it resulted in a surge of industrialization that led to the release of pollutants into water sources, including groundwater (Matta et al., 2016). The increase in mining, manufacturing, and urban development brought about the introduction of contaminants that started to impact the quality of groundwater.
- **Rise of Sanitary Engineering**: The emergence of the field of sanitary engineering was driven by the recognition that disease outbreaks were often linked to contaminated water sources. Engineers and public health officials collaborated to enhance water quality by creating sewage systems, refining water treatment processes, and enforcing regulations (Sharma et al., 2020).

22.2.4 Twentieth Century

- **Groundwater Contamination Recognition**: There was an increasing awareness of groundwater contamination issues during the twentieth century. Industrial activities have caused the contamination of groundwater, the use of synthetic chemicals, and improper waste disposal practices. Events like the Love Canal disaster in the 1970s brought focus to the significant ramifications of groundwater pollution (Huang et al., 2020).
- **Environmental Legislation**: The rise in environmental consciousness prompted the creation of environmental legislation, such as the Clean Water Act in the United States and comparable regulations worldwide. The laws were designed to regulate the release of pollutants into water bodies, such as groundwater, and establish criteria for water quality (Long et al., 2020).

22.2.5 Contemporary Period

- **Technological Advancements**: Advancements in chemistry and environmental monitoring technologies have led to improved methods of detecting and measuring contaminants in groundwater, resulting in greater accuracy (Mensah et al., 2022). These tools help to enhance comprehension of the various pollutants and their concentrations.
- **Groundwater Protection and Management**: In recent decades, there has been an increased emphasis on sustainable groundwater management, with a particular focus on protection and management. Groundwater protection strategies include defining wellhead protection areas, implementing best management practices, and establishing groundwater monitoring programmes to track water quality over time (Fatemeh et al., 2021).

- **Global Water Challenges**: Currently, numerous global water challenges continue to pose threats to the quality of groundwater. These challenges include over-extraction, climate change, and the emergence of contaminants. The acknowledgment of groundwater as a crucial element of the Earth's freshwater resources highlights the significance of global collaboration and sustainable approaches to tackle present and future obstacles (He et al., 2021).

22.3 Evidence of Declining Groundwater Quality

There is a wealth of scientific studies, monitoring programmes, and investigations conducted globally that provide substantial evidence of declining groundwater quality. Various factors suggest a decline in the quality of groundwater, such as the presence of contaminants, shifts in groundwater chemistry, and the rising incidence of waterborne diseases (Matta et al., 2018a). The accumulation of evidence from these various sources reinforces the argument for the deterioration of groundwater quality and emphasizes the importance of taking proactive steps to tackle contamination sources, adopt sustainable groundwater management practices, and safeguard this crucial water resource.

If groundwater tables drop, streams and rivers seep away and pollute drinking water (Fig. 22.2). Presented below are several types of evidence that substantiate the claim of deteriorating groundwater quality:

- **Chemical Analysis**: Groundwater quality studies require the careful examination of water samples to detect the presence of various contaminants, including heavy

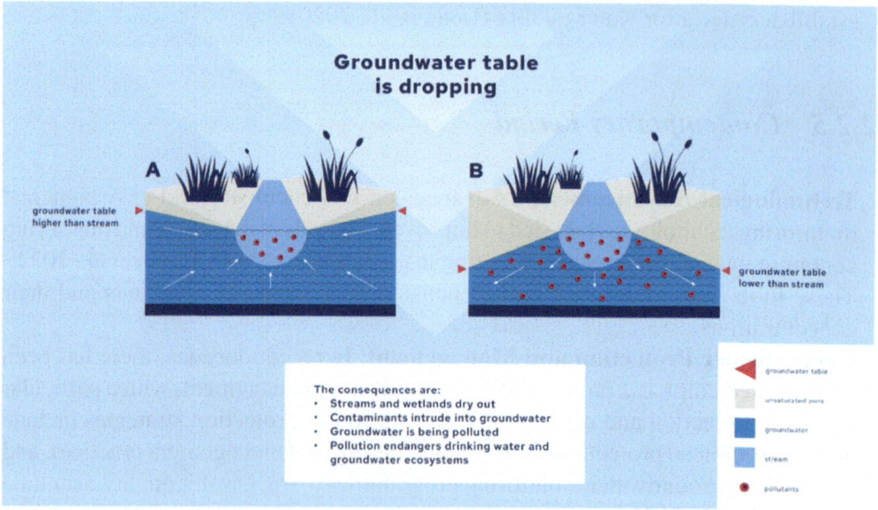

Fig. 22.2 Consequences of groundwater water Tab drop. *Source* Theilmann (2022)

metals, nitrates, pesticides, industrial chemicals, and pharmaceuticals. Changes in the chemical makeup of groundwater have been observed through extensive monitoring over an extended period. These changes include variations in pH levels, salinity, and the presence of specific ions (Alsalme et al., 2021).

- **Water Quality Monitoring**: The installation of monitoring wells in aquifers offers a methodical approach to observe and document alterations in groundwater quality over time. Regular collection and analysis of groundwater samples at monitoring points allows for the identification of trends, spatial variations, and potential sources of pollution (Matta et al., 2015).
- **Health Impact Studies**: The rise in cases of waterborne diseases like cholera, dysentery, and gastrointestinal illnesses may suggest a decline in groundwater quality due to pathogen contamination. Studies by Calderon, (2000); Andrade et al., 2018; Ghosh et al., (2020) in epidemiology have found a correlation between the consumption of contaminated groundwater and the occurrence of certain health conditions in populations that depend on these water sources (Ghosh et al., 2020).
- **Land Use Changes and Anthropogenic Activities**: The alteration of land use, including urbanization, agriculture expansion, or industrial activities, is frequently associated with the deterioration of groundwater quality. Contaminants from these activities have the potential to seep into the groundwater. Groundwater contamination is caused by the discharge of pollutants from mining activities, industrial discharges, and improper waste disposal (Foley et al., 2005)
- **Climate Change Indicators**: The impact of climate change on groundwater recharge rates and the movement of contaminants in the subsurface is influenced by changes in precipitation patterns and temperatures. Figure 22.3 shows the interactions between groundwater and climate change impact. Climate-induced phenomena such as rising sea levels have the potential to worsen saltwater intrusion, which can have a negative impact on the quality of coastal aquifers (Pokhrel et al., 2021)
- **Case Studies and Incidents**: Examining case studies and incidents reveals a range of groundwater contamination events, including industrial spills, leakage from storage tanks, and accidents. These documented incidents provide clear evidence of both localized and sometimes widespread groundwater quality issues. We can gain valuable insights into the lasting effects of groundwater pollution by studying past incidents, such as the Love Canal contamination in the United States (Matta et al., 2018b).
- **Groundwater Modeling**: Numerical groundwater models are utilized to simulate the movement of contaminants in aquifers, accurately predicting the spread of pollutants and assessing potential impacts on groundwater quality. By comparing model predictions with actual field measurements, we can validate the accuracy of groundwater quality assessments and gather more evidence of degradation (Dai et al., 2021).
- **Government and Agency Reports**: Reports from environmental agencies, such as the U.S. government, provide valuable information on various topics. The Environmental Protection Agency (EPA) or the European Environment Agency

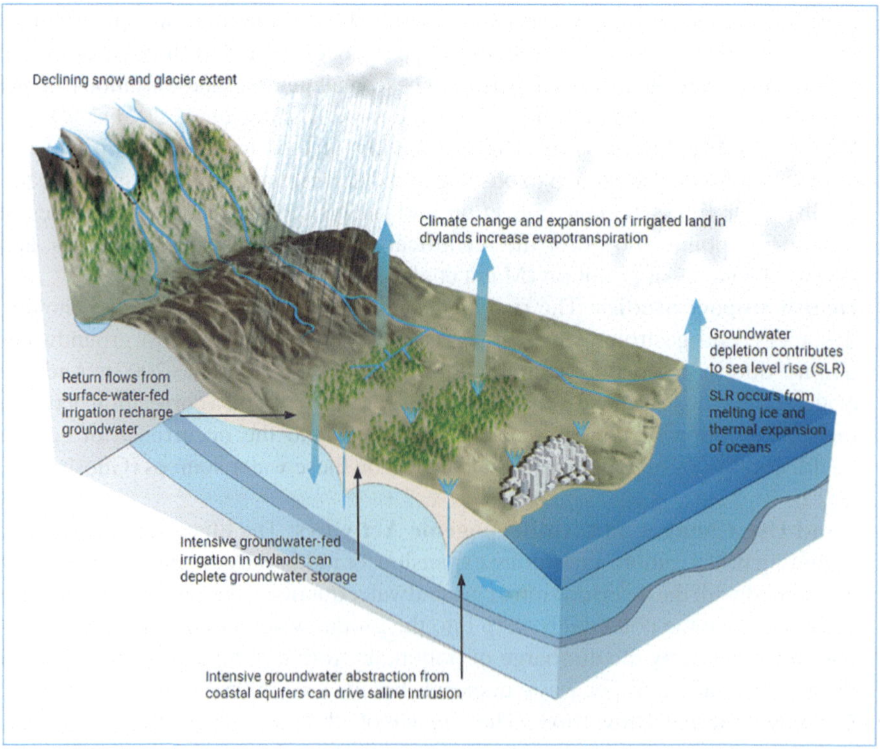

Fig. 22.3 Key interactions between groundwater and climate change showing how direct and indirect impacts of climate change affect groundwater systems. *Source* Taylor et al. (2013)

frequently offers thorough evaluations of groundwater quality through national monitoring programmes and studies (Noori et al., 2023).

- **Scientific Literature and Peer-Reviewed Studies**: It play a crucial role in expanding the body of evidence. The studies provide in-depth information on investigations into various contaminants, sources of pollution, and trends in groundwater quality (Matta et al., 2018c).
- **Community Complaints and Observations**: Reports from communities, often through citizen science initiatives, about changes in water quality, taste, or odour can provide valuable insights into potential groundwater quality issues.

22.3.1 Causes of Declining Groundwater Quality

22.3.1.1 Agriculture Sector

Agricultural runoff has the potential to negatively impact groundwater quality by introducing a range of contaminants into the water table. Here are some leading causes:

- **Excessive use of fertilizers and pesticides**: Agricultural activities frequently involve the application of fertilizers that contain nitrogen and phosphorus, which can lead to excessive use of these substances. Improper application or management practices can result in the leaching of nutrients into the groundwater, which can then contribute to nutrient pollution. Chemicals used for pest and weed control have the potential to seep into the groundwater, resulting in contamination. Certain pesticides can pose significant challenges, especially when it comes to persistent organic pollutants (Liu et al., 2021). Figure 22.4 shows the ecological impact of decreasing groundwater quality and quantity.
- **Soil erosion**: Soil erosion is a significant concern as it can result in the transportation of sediments from agricultural fields into water bodies, ultimately leading to a decline in water quality. Sediments have the potential to transport various pollutants, such as nutrients, pesticides, and heavy metals, into groundwater (Issaka et al., 2017).
- **Animal waste**: Improper handling and disposal of animal waste can lead to the runoff of nutrients, bacteria, and pathogens into groundwater, which can have negative environmental consequences. The high concentration of nitrogen and phosphorus in manure can contribute to nutrient pollution (Wakida & Lerner, 2005).
- **Irrigation practices**: When irrigation is done excessively, it can result in the unwanted leaching of fertilizers and pesticides into the groundwater. Under certain circumstances, salts from irrigation water have the potential to build up in the soil and eventually seep into the groundwater, resulting in the process of salinization (Pulido-Bosch et al. 2018).
- **Crop residue and organic matter**: The decomposition of crop residues and organic matter in the soil can result in the release of byproducts that have the potential to contaminate groundwater. As an example, the breakdown of organic matter can result in the creation of humic substances, which can affect the quality of water (Katterer et al., 2011).
- **Use of antimicrobial agents**: The presence of antibiotics and hormones in groundwater can be attributed to the use of antimicrobial agents in animal farming. Over time, this can lead to the emergence of antibiotic-resistant bacteria and cause disturbances in ecosystems (Matta et al., 2018d).
- **Improper disposal of agricultural chemicals**: Improper disposal of agricultural chemicals can lead to accidental spills, leaks, or incorrect disposal methods. This can result in the direct introduction of pollutants into groundwater (Khatri et al., 2015).

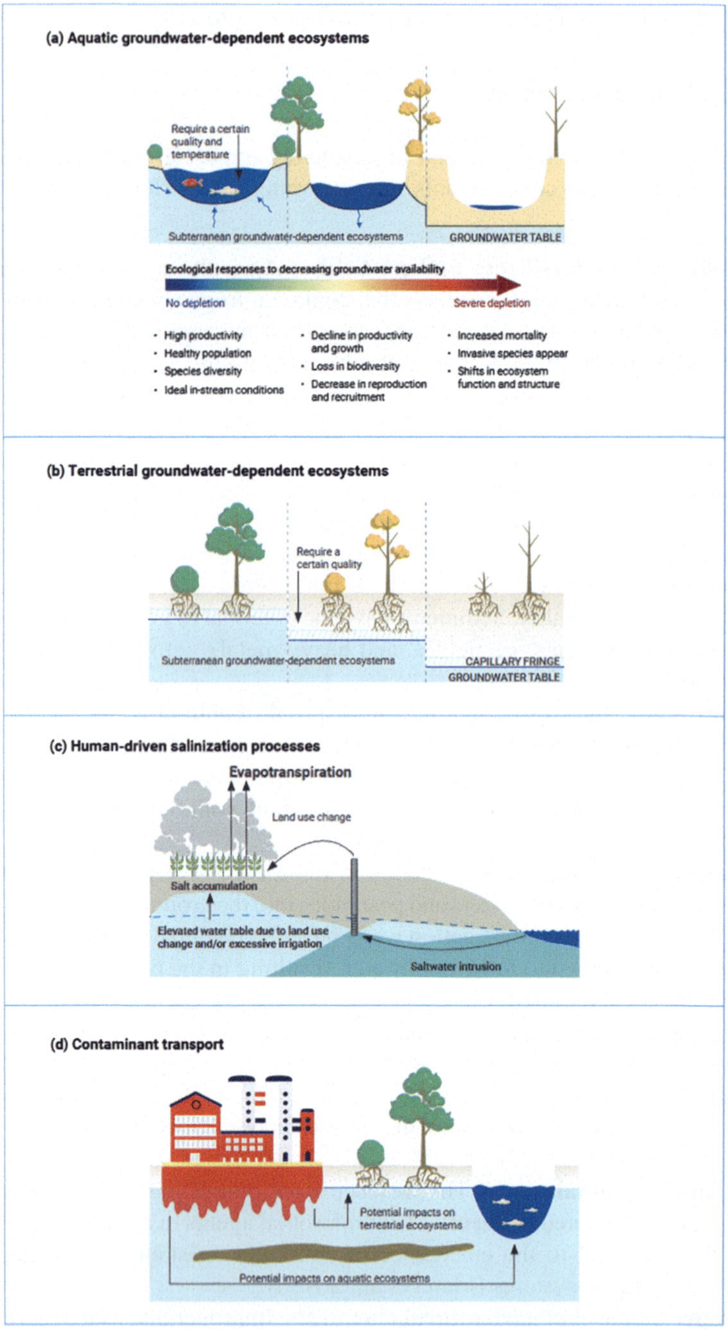

Fig. 22.4 Ecological impacts of decreasing groundwater quality and quantity. *Source* Rohde et al. (2017), Foster and Chilton (2003)

- **Lack of vegetative buffers**: The absence of vegetative buffers can lead to an increased risk of contaminants being carried into groundwater through direct runoff (Walton et al., 2020).

22.3.1.2 Industrial Sector

The degradation of groundwater quality is primarily attributed to industrial pollution. There are various factors related to industrial activities that can result in the pollution of groundwater. These are the leading causes:

- **Chemical spills and leaks**: Chemical spills and leaks pose a significant risk to the environment and public health. Industrial facilities, which handle and store a wide range of chemicals, are particularly susceptible to these incidents. When spills or leaks occur, hazardous substances can be released into the groundwater, leading to potentially severe consequences. These substances may consist of heavy metals, solvents, acids, and other toxic compounds (Chandnani et al., 2022; Etikala et al., 2024).
- **Improper disposal of industrial waste**: Improper disposal of industrial waste can lead to the infiltration of pollutants into the groundwater. This can occur through improper landfilling, illegal dumping, or inadequate waste treatment facilities (Sharholy et al., 2008)
- **Leaching from landfills**: Contaminants from industrial waste disposed of in landfills have the potential to gradually seep into the groundwater, posing a concern for environmental contamination (Abiriga et al., 2021). This may involve pollutants such as heavy metals, organic compounds, and other chemicals found in the waste.
- **Runoff from industrial sites**: Runoff from industrial sites poses a significant risk to nearby surface water bodies and groundwater. Rainfall or stormwater runoff can transport harmful pollutants, leading to potential contamination (Muller et al., 2020). This runoff may contain pollutants such as heavy metals, oils, and various industrial chemicals (Matta et al., 2018e).
- **Atmospheric deposition**: Atmospheric deposition indirectly contributes to groundwater contamination through surface water infiltration. Industrial activities can release pollutants into the air, which then settle on the ground and eventually make their way into groundwater (Kumar et al., 2023). These may consist of contaminants such as heavy metals, volatile organic compounds (VOCs), and particulate matter (Singh & Agrawal, 2007).
- **Industrial discharges**: Certain industries discharge wastewater, whether treated or untreated, directly into surface water bodies, potentially resulting in groundwater contamination (Ullah et al., 2009). The water that is discharged may contain a range of pollutants, such as heavy metals, chemicals, and biological contaminants (Kishor et al., 2021), even treated wastewater can contain contaminants that may infiltrate groundwater.
- **Mining activities**: Contaminants like heavy metals, metalloids, and acidic drainage can be introduced into the groundwater through mining operations

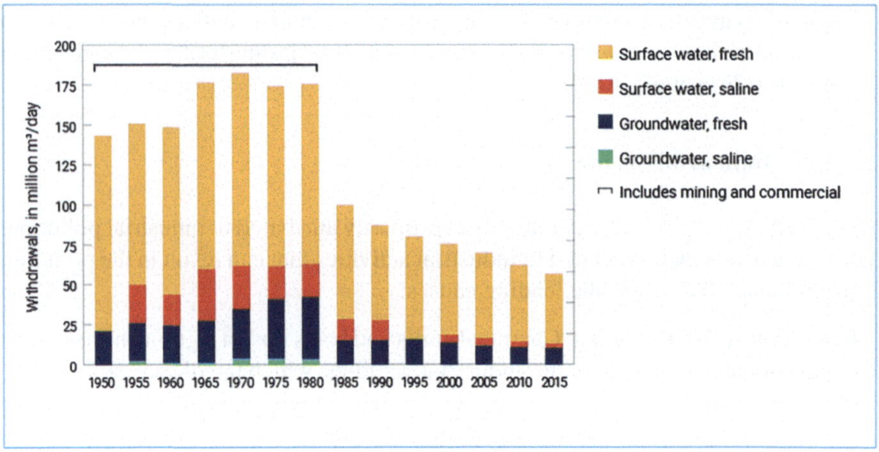

Fig. 22.5 Self-supplied industrial water withdrawals in the USA. *Source* Dieter et al. (2018)

(Akhtar et al., 2021). Substances have the potential to seep out of mine tailings and waste rock, making their way into the surrounding groundwater.

- **Use of hazardous substances**: The use of hazardous chemicals in industrial processes can lead to the release of pollutants into the environment. If not managed effectively, these chemicals have the potential to migrate into the groundwater and have a significant impact on the quality of water (Saravanan et al., 2021).
- **Abandoned industrial sites**: Older or abandoned industrial sites may have left behind contaminants in the soil and groundwater due to their previous operations. These pollutants from the past can linger and impact the quality of groundwater even after industrial activities have stopped (Gaur et al., 2020). Figure 22.5 shows various industrial water withdrawals in the USA.

22.3.1.3 Urban Complexes

The quality of groundwater can be significantly affected by urbanization as a result of human activities and alterations in land use. Below are several significant factors contributing to the deterioration of groundwater quality due to urbanisation:

- **Stormwater runoff**: Urban areas often have increased impervious surfaces, such as roads, parking lots, and rooftops, which prevent natural infiltration of rainwater. As a result, stormwater runoff can carry pollutants, including heavy metals, oils, fertilizers, and debris, into storm drains and ultimately into groundwater (Pinasseau et al., 2020).
- **Wastewater and sewage**: Inadequately maintained sewer systems can result in the infiltration of untreated or partially treated wastewater into the groundwater, posing potential risks (Kinuthia et al., 2020). Infiltration occurs primarily through leaks or cracks in sewer lines rather than through the entire sewer system.

Contaminants like pathogens, nutrients, and organic matter can be introduced through this process. Pathogens are a problem, however organic matter and nutrients usually cause surface water eutrophication rather than directly contaminating groundwater.

- **Industrial and commercial activities**: Urban areas frequently accommodate a wide range of industrial and commercial activities that involve the utilisation and generation of chemicals (Akhtar et al., 2021). Groundwater contamination can occur when chemicals are not stored, handled, or disposed of correctly (Matta et al.). This can result in the presence of solvents, heavy metals, and industrial byproducts in the groundwater Fig. 22.6 shows the urban water sources and its interactions whereas Fig. 22.7 shows the trend of urban water supply.
- **Landfill leachate**: Leachate from landfills in urban areas has the potential to contaminate groundwater with pollutants. The presence of hazardous chemicals, heavy metals, and organic compounds in the leachate can have a significant impact on the quality of water (Chidichimo et al., 2020)
- **Road salt usage**: In colder climates, the application of road salt and de-icing agents to address snow and ice on roadways can lead to the infiltration of chloride ions into groundwater (McNamara et al., 2020) High levels of chloride can have a detrimental impact on the quality of groundwater.
- **Urban agriculture**: Urban agriculture, such as lawns and gardens, has the potential to impact groundwater quality due to the application of fertilizers and pesticides (Langemeyer et al., 2021). There is a possibility of nutrients and chemicals seeping into the groundwater.

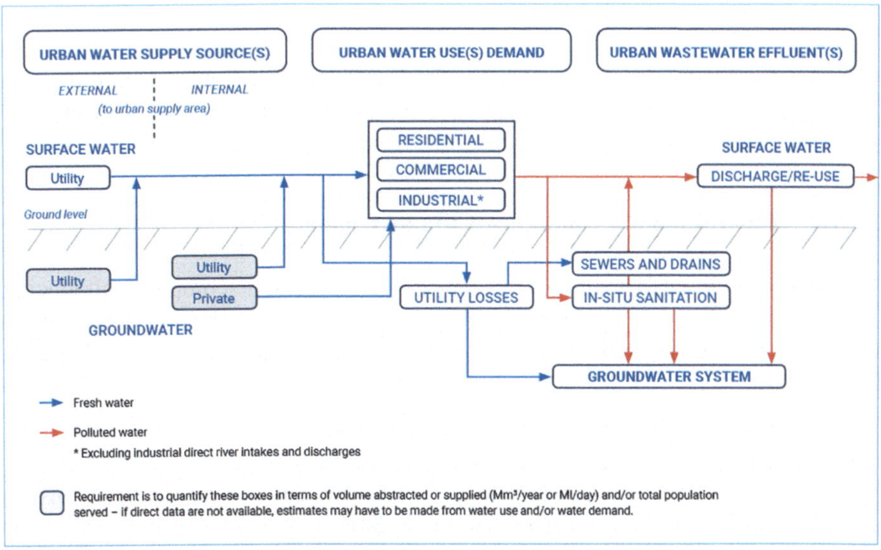

Fig. 22.6 Schematic overview of urban water sources and their use and interaction. *Source* Foster & Hirata (2012)

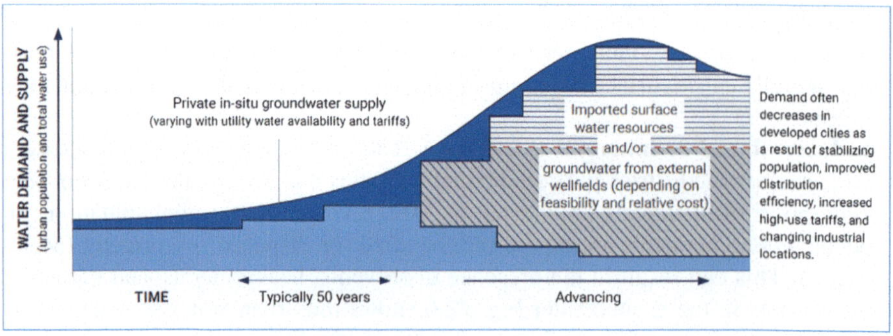

Fig. 22.7 Typical trend in the evolution of urban water supplies. *Source* Foster & Hirata (2012)

- **Leaking underground storage tanks**: Underground storage tanks can be found at petrol stations, industrial facilities, and other urban sites. These tanks may contain fuels or chemicals. Contaminants, like hydrocarbons, can be released into the groundwater through leaks from these tanks (Yu et al., 2024).
- **Construction activities**: Urban development frequently includes construction and excavation, resulting in soil disturbance and a rise in sedimentation levels. Runoff containing sediment can transport various pollutants, such as suspended solids and their associated contaminants, into the groundwater (Huang et al., 2018).
- **Land subsidence**: Land subsidence is a consequence of excessive pumping of groundwater for urban water supply. This activity can alter the geological structure, creating a potential pathway for contaminants from the surface to infiltrate the aquifer (Shirzaei et al., 2021).
- **Infrastructure leaks**: The deteriorating water and sewer infrastructure can lead to the development of leaks. This can result in untreated water seeping into the ground, posing a risk of contaminating groundwater with harmful pathogens and pollutants (Vystavna et al., 2018).

22.3.1.4 Environmental Changes—Climate Change

The impacts of climate change on groundwater quality can be complex, influencing a range of hydrological and environmental processes. Several factors contribute to the deterioration of groundwater quality, which can be attributed to the effects of climate change:

- **Changes in precipitation patterns**: Climate change can result in shifts in precipitation patterns, with an increase in the frequency and intensity of rainfall events. The heightened intensity can lead to a faster runoff, which may transport surface pollutants into groundwater (Cuthbert et al., 2019)
- **Sea level rise and saltwater intrusion**: The phenomenon of rising sea levels caused by climate change can result in the intrusion of saltwater into coastal aquifers. This intrusion has the potential to compromise the quality of freshwater

resources by introducing higher concentrations of salts, rendering the groundwater unfit for consumption or agricultural use (Ferguson et al., 2012).

- **Changes in temperature**: Temperature fluctuations have a significant impact on microbial activity and biogeochemical processes in the soil, which in turn affect the transformation and movement of contaminants. For instance, higher temperatures can amplify the emission of specific pollutants from soil particles into groundwater (Wu et al., 2020).
- **Drought and groundwater depletion**: Extended periods of drought and heightened water requirements can cause excessive extraction of groundwater, leading to a decline in water tables. The risk of groundwater contamination can be exacerbated when water tables are lowered. This is due to the concentration of pollutants and the potential for contaminants from the surface to infiltrate the groundwater (Ashraf et al., 2021).
- **Increased frequency of extreme weather events**: The rise in extreme weather events has led to the mobilization of contaminants from the surface into groundwater. This includes events like floods and wildfires. It is essential to consider that floodwaters have the potential to carry pollutants, while wildfires can have an impact on soil characteristics, which in turn can affect the movement of contaminants (Ummenhofer & Meehl, 2017).
- **Changes in vegetation patterns**: Alterations in vegetation patterns caused by climate change can impact root-zone dynamics, which in turn can influence the absorption and release of nutrients and other chemicals. The chemical composition of groundwater can be affected by this (Miguez-Macho & Fan, 2021).
- **Permafrost thaw**: In cold regions, the release of stored contaminants, such as organic matter and pollutants, into groundwater can impact water quality. This is often a result of rising temperatures and the subsequent thawing of permafrost (Kurylyk et al., 2016).
- **Changes in groundwater recharge patterns**: The patterns and rates of groundwater recharge can be influenced by climate change, leading to changes in these patterns. Modifications in precipitation, evaporation, and land use have the potential to impact the quantity and quality of water that infiltrates groundwater, potentially transporting pollutants along with it (Hartmann et al., 2017).
- **Changes in land use and land cover**: Climate change has the potential to impact land use and land cover, resulting in potential increases in urbanisation or deforestation. These modifications have the potential to introduce pollutants into groundwater due to shifts in runoff patterns, alterations in soil characteristics, and heightened human activities (Song et al., 2018).
- **Shifts in hydrological cycles**: Climate-driven changes in hydrological cycles could affect the dynamics between groundwater and surface water. This can influence the transport of contaminants between these two systems (Cuthbert et al., 2019).

22.4 Connection to Human Health and Environmental Well-Being

The relationship between groundwater and human health, as well as environmental well-being, is complex and significant. Groundwater is of utmost importance in maintaining the balance of ecosystems and supporting human communities. It has a substantial impact on water quality, biodiversity, and the overall health of the environment. Grasping this connection is crucial for guaranteeing the welfare of both natural systems and human populations.

- **Drinking Water Source**: Groundwater is a crucial source of drinking water for communities all around the world. Many regions depend on wells and groundwater extraction to meet their drinking water requirements. The gradual movement of water through the soil serves as a natural filtration mechanism, eliminating impurities and pollutants. When managed effectively, groundwater can serve as a dependable and secure source of drinking water, which has a direct impact on human health (Gleeson et al., 2012).
- **Waterborne Diseases**: Proper management of groundwater is crucial to prevent contamination and protect human health. Groundwater is susceptible to contamination from agricultural runoff, industrial discharges, and improper waste disposal. Contaminated groundwater has the potential to cause waterborne diseases, impacting both human populations and ecosystems. Thorough monitoring and assessment play a vital role in identifying possible contaminants and implementing necessary measures to protect water quality (Andrade et al., 2018).
- **Ecosystem Health**: Groundwater plays a crucial role in supporting a wide range of ecosystems, such as wetlands, rivers, and lakes. Groundwater's base flow is crucial for maintaining habitat stability and promoting biodiversity. Aquatic ecosystems rely heavily on a steady provision of groundwater to ensure ideal conditions for fish, invertebrates, and other aquatic species. Disruptions in groundwater availability or quality can have far-reaching impacts on ecosystems, resulting in a decrease in biodiversity and the loss of valuable ecosystem services (Chandnani et al., 2022).
- **Agricultural Productivity**: Agricultural Productivity: Groundwater plays a crucial role in agriculture by supporting irrigation systems that improve crop growth and increase food production. The presence of groundwater has a direct impact on agricultural productivity, which in turn affects food security and people's livelihoods. Nevertheless, the overuse and unsustainable extraction of groundwater can result in land subsidence, diminished water supply, and lasting environmental deterioration.
- **Climate Resilience**: Groundwater plays a crucial role in enhancing climate resilience by serving as a safeguard during dry spells. Its methodical movement beneath the surface enables it to endure even when surface water sources are depleted. The ability to withstand and adapt to the challenges of climate change is of utmost importance for ecosystems, agriculture, and human communities. It ensures a dependable water supply even in times of drought.

- **Sustainable Management**: Understanding the complex relationship between groundwater, human health, and the environment, it becomes clear that implementing sustainable groundwater management practices is crucial. This includes the use of scientific policies, continuous monitoring of groundwater quality and quantity, active participation from the community, and the adoption of technologies to replenish aquifers. Sustainable management aims to achieve a harmonious coexistence between human water needs and the preservation of ecosystems, promoting the long-term well-being of the environment.
- **Hydrological processes**: The hydrological cycle involves various processes, including precipitation, infiltration, percolation, and recharge, which are all interconnected with groundwater. Scientific investigations into these processes reveal the intricate dynamics of groundwater movement, storage, and replenishment. Comprehending these mechanisms is crucial for forecasting water availability and formulating sustainable water management strategies.
- **Ecological Functions**: Groundwater plays a crucial role in supporting the health and biodiversity of ecosystems worldwide, serving as a lifeline for both terrestrial and aquatic ecosystems. The ecological significance of this cannot be understated, as it plays a crucial role in maintaining base flow in rivers, supporting wetlands, and ensuring habitat stability. Scientific studies shed light on the intricate relationship between groundwater and ecosystem dynamics, highlighting the importance of implementing conservation measures to protect biodiversity.
- **Human Water Needs**: Groundwater plays a crucial role in meeting the water needs of millions of people worldwide, especially in arid regions. It serves as a vital source of freshwater for various purposes, such as agriculture, industry, and domestic use. Scientific research on groundwater quality, quantity, and vulnerability plays a crucial role in tackling pressing issues like over-extraction and contamination. This research is essential for ensuring the long-term availability of water resources for growing populations.
- **Mitigating water scarcity**: Groundwater is crucial in addressing water scarcity, particularly in times of drought. The slow movement and natural filtration processes of this water source contribute to its reliability and resilience. Scientific interventions, like implementing aquifer recharge strategies and adopting sustainable extraction practices, play a crucial role in ensuring effective water resource management and building resilience against water stress caused by climate change.

Looking at it from a human perspective, groundwater plays a vital role in supporting agriculture, industry, and domestic needs by providing a reliable and sustainable water source. Groundwater is a crucial resource for countless individuals across the globe, especially in dry and semi-dry areas where access to surface water may be scarce (Matta et al., 2020a). Groundwater is critical to reducing the effects of droughts and sustaining agricultural productivity by utilizing wells and irrigation systems. In addition, the quality of groundwater is intricately linked to human health, as it frequently serves as a vital source of drinking water for communities worldwide. The meticulous filtration process as water percolates through the soil aids in naturally purifying and cleansing the water, resulting in a generally safer drinking option compared to certain surface water sources. Comprehending and effectively overseeing groundwater resources are of utmost importance for fostering sustainable development and safeguarding ecosystems. Groundwater availability and quality are at risk due to over-extraction, contamination, and inadequate management practices. Thus, it is crucial to conduct scientific research, monitoring, and implement effective policies to ensure the responsible utilization and preservation of this invaluable resource for both present and future generations. Figure 22.8 shows the Interactions between groundwater, ecosystems, human activity and nature-based solutions.

22.5 Investigating the Causes of Declining Groundwater Quality

The reasons behind the deterioration of groundwater quality are diverse and frequently arise from intricate interplays between natural phenomena and human actions. Several factors contribute to the decline in groundwater quality:

- **Industrial Activities**: Groundwater can be affected by pollutants introduced through industrial processes. The effluents generated by manufacturing, chemical production, and other industrial activities have the potential to carry heavy metals, solvents, and toxic chemicals. If these effluents are not managed effectively, there is a risk of them seeping into the subsurface and polluting groundwater.
- **Agricultural Practices**: Groundwater pollution is a significant concern caused by agricultural runoff. When fertilizers, pesticides, and herbicides are used in farming, there is a potential for nitrogen compounds, phosphates, and agrochemical residues to leach into groundwater. The concentration of salts in the soil can be influenced by irrigation practices, which in turn can have an impact on groundwater quality.
- **Urbanization and Land Development**: Urban areas with impervious surfaces like roads and buildings disrupt the natural process of water infiltration into the ground. Stormwater runoff from urbanized areas has the potential to transport various pollutants, including heavy metals, oils, and pathogens, into groundwater. There is a strong correlation between increased urbanization and a heightened risk of groundwater contamination.
- **Wastewater Disposal and Sewage Leakage**: Improper disposal of wastewater and leakage from sewage systems can introduce contaminants into the groundwater (Matta et al., 2020b). Septic systems, if not adequately maintained, can leak pollutants, including pathogens and nutrients, into the subsurface, impacting groundwater quality in residential and rural areas.
- **Landfills and dumpling sites**: The leachate that is produced when waste decomposes has the potential to contain harmful substances like heavy metals, organic compounds, and toxic chemicals. This poses a severe threat to the quality of groundwater.

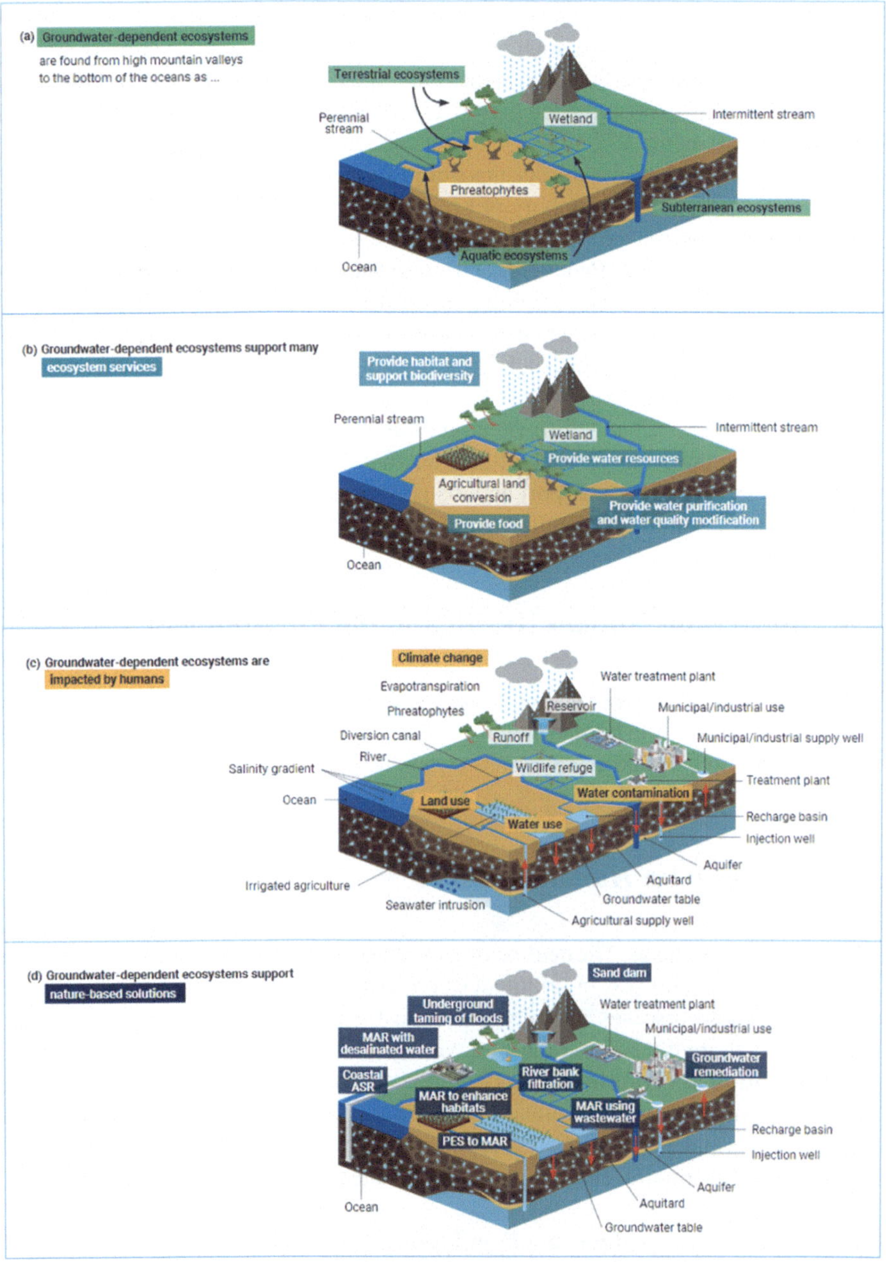

Fig. 22.8 Interactions between groundwater, ecosystems, human activity and nature-based solutions. *Source* Maven's Notebook (2015)

- **Mining Activities**: Groundwater can be impacted by pollutants introduced through mining operations, such as heavy metals, sulfates, and acidic drainage. Abandoned or improperly closed mines can potentially release contaminants into the groundwater for an extended period, even after mining activities have stopped.
- **Saltwater Intrusion**: Excessive pumping can lead to a decline in the water table, allowing saline water from the ocean to infiltrate freshwater aquifers. This can have detrimental effects on the quality of groundwater, rendering it unfit for drinking or irrigation purposes.
- **Natural Geochemical Processes**: Groundwater quality can be influenced by various natural processes, including mineral dissolution, weathering of rocks, and redox reactions. For instance, groundwater can be affected by higher levels of naturally occurring elements such as iron, manganese, or arsenic due to specific geological circumstances.
- **Climate Change**: Changes in precipitation patterns and temperatures associated with climate change can influence groundwater quality. Altered weather patterns may affect the transport and distribution of contaminants, impacting the vulnerability of aquifers to pollution.

22.6 Impacts on Human Health and the Environment

The decrease in groundwater resources and decline in groundwater quality can have substantial effects on both human health and the environment. Here are some essential impacts related to these issues:

22.6.1 Impacts on Human Health

- **Waterborne Diseases**: The decline in groundwater quality, caused by the presence of pathogens, chemicals, or pollutants, can result in the occurrence of waterborne diseases. These diseases include gastrointestinal infections, cholera, and other illnesses that are transmitted through water.
- **Consumption of Contaminants**: Groundwater that is contaminated may contain hazardous substances, toxic metals, or pollutants that, when consumed, can result in immediate and long-term health problems, such as harm to the liver and kidneys, neurological disorders, and the potential for cancer-causing effects (Matta et al., 2020c).
- **Consideration for Vulnerable Populations**: It is essential to acknowledge that certain groups, including children and older adults, may face more significant risks when it comes to the health consequences of polluted groundwater. This is primarily due to their more delicate immune systems and other related factors.
- **Vector-borne diseases**: Declining groundwater levels can lead to the formation of stagnant water bodies, which can become breeding grounds for disease-carrying

vectors such as mosquitoes. This, in turn, can contribute to the spread of vector-borne diseases like malaria and dengue.
- **Nutritional Deficiencies**: Declining groundwater resources can have a significant impact on agriculture, specifically crop yields. This, in turn, can potentially result in dietary deficiencies due to the potential decrease in food production.

22.6.2 *Impacts on the Environment*

- **Ecosystem Disruption**: The decline in groundwater levels can have a significant impact on the environment. One of the consequences is the drying up of wetlands, which has a negative effect on biodiversity and disrupts the ecosystems that rely on groundwater-fed habitats.
- **Land Subsidence**: Over-extraction of groundwater can lead to land subsidence, causing infrastructure damage, altering landscapes, and increasing the risk of flooding.
- **Saltwater Intrusion**: Declining groundwater levels in coastal areas can lead to the intrusion of saltwater into freshwater aquifers, making them unsuitable for drinking or agricultural purposes.
- **Streamflow Reduction**: Decreasing groundwater levels can have a significant impact on the flow of rivers and streams, which in turn can disrupt the delicate balance of aquatic ecosystems and the organisms that rely on a steady water supply.
- **Contaminant Migration**: Lower groundwater levels can change the flow patterns, potentially transporting contaminants from contaminated areas to previously unaffected regions.
- **Loss of Habitat**: Decreasing groundwater levels has the potential to diminish the availability of appropriate habitat for aquatic species, resulting in population declines or even local extinctions.

22.6.3 *Socio-economic Impacts*

- **Addressing Agricultural Challenges**: Groundwater is a crucial resource for irrigation purposes, and the agricultural sector is highly dependent on it. Decreased groundwater levels have the potential to negatively affect agricultural productivity, which in turn can have significant consequences for food security and the economies of rural areas.
- **Water Scarcity**: The decline in groundwater resources has resulted in a shortage of water, which has led to heightened competition among different sectors such as agriculture, industry, and domestic use.
- **Economic Costs**: The over-extraction of groundwater for agricultural, industrial, and domestic purposes has led to a substantial reduction in aquifer levels in

numerous regions worldwide, including areas heavily reliant on groundwater for irrigation and human consumption, such as parts of North America, India, and China.

22.7 Comparative Analysis with SDG Indicators

The decreasing availability and worsening quality of groundwater resources present significant challenges in the pursuit of multiple Sustainable Development Goals (SDGs). The over-extraction of groundwater for agricultural, industrial, and domestic purposes has led to a substantial reduction in aquifer levels in various regions globally. The excessive exploitation of resources leads to the deterioration of ecosystems, as it disturbs the delicate equilibrium between replenishment and extraction. In addition, the groundwater quality is being affected by pollutants from agricultural runoff, industrial discharges, and untreated wastewater disposal. Significant levels of contaminants, including nitrates, pesticides, and heavy metals, pose a risk to both human well-being and natural environments (Matta, 2020a). The situation at hand has a direct connection to SDG 6 (Clean Water and Sanitation). The decreasing quantity and quality of groundwater pose a threat to the goal's objective of guaranteeing the availability and sustainable management of water for everyone. In addition, it intersects with SDG 2 (Zero Hunger) because agriculture depends heavily on groundwater for irrigation, and concerns about water quality can impact food safety and security. SDG 3 (Good Health and Well-being) is also relevant, considering the possible health hazards linked to contaminated groundwater. Addressing the declining groundwater resources and their quality is crucial for a comprehensive and integrated approach toward achieving various SDGs. It highlights the interconnected nature of sustainable development challenges. It is essential to implement sustainable water management practices, prevention measures and foster cross-sectoral collaborations to guarantee the long-term viability of groundwater resources and achieve the SDG targets.

The Sustainable Development Goals (SDGs) are a comprehensive set of global objectives that tackle a wide range of social, economic, and environmental challenges. The issue of declining groundwater resources and quality is intricately linked to several SDG indicators. Presented is a thorough examination of the diminishing groundwater resources and the associated decline in quality, along with an assessment of the corresponding Sustainable Development Goal (SDG) indicators. Figure 22.9 shows the water cycle in sustainable development goals. Figure 22.10 shows the percentage of groundwater related targets per SDG.

22.7.1 SDG 6: Clean Water and Sanitation

- **SDG 6.1: Access to Safe Drinking Water**: The quality of groundwater is decreasing, which has a negative impact on the availability of safe drinking water.

Fig. 22.9 The water cycle in the sustainable development goals. *Source* UN-Water (2017)

This is because contaminated groundwater can be harmful to human health. Examining water quality, contamination levels, and the percentage of the population utilising properly managed drinking water services. It is vital to prioritize monitoring the amount of the population accessing safely accomplished drinking water services, aligning with SDG indicators, to safeguard progress towards universal access to clean water and sanitation. This monitoring provides critical data for evaluating and improving water service delivery, advancing global efforts for sustainable development.

- **SDG 6.4**: **Water Use Efficiency**: The excessive extraction of groundwater has a negative impact on water scarcity and hampers the efficiency of water usage. Water stress, water-use efficiency, and the proportion of water bodies with sustainable water flow are important factors to consider. Improving water-use efficiency across all sectors, including agriculture, industry, and domestic use, is integral to achieving SDG 6.4. By optimizing water usage, we can address water scarcity, promote sustainable resource management, and ensure equitable access to clean water, aligning with the goals of SDG 6 for universal access to safe and sustainable water sources.

22.7.2 SDG 3: Good Health and Well-Being

- **SDG 3.3: Communicable Diseases**: Contaminated groundwater has the potential to contribute to the transmission of waterborne diseases, which can have significant consequences for public health, such as the frequency of waterborne diseases

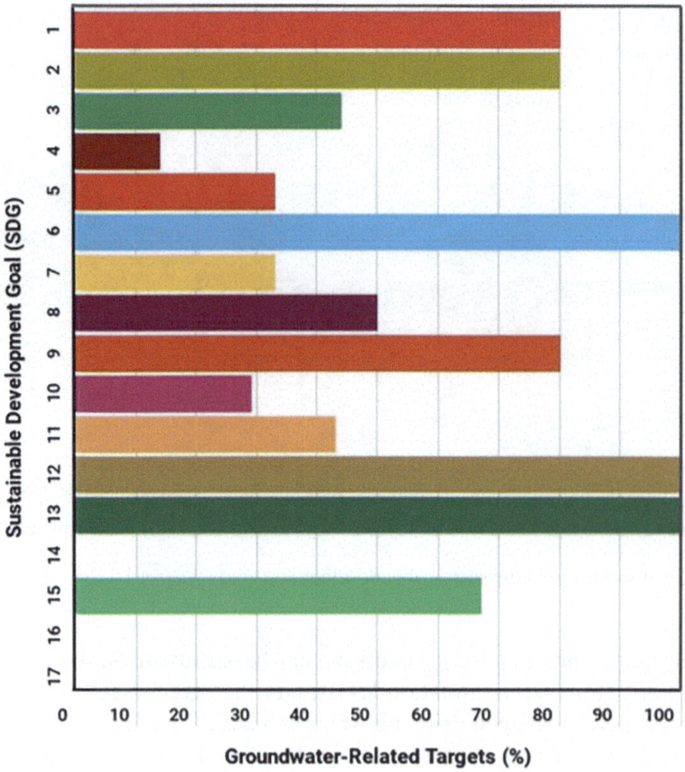

Fig. 22.10 Percentage of groundwater related targets per SDG. *Source* Guppy et al. (2018)

and the number of deaths caused by them. SDG 3.3 indicators focus on reducing mortality rates attributed to waterborne diseases and assessing the frequency of such diseases per population. These metrics provide insights into the effectiveness of public health measures, access to clean water, and sanitation services, crucial for achieving SDG 3 targets aimed at improving global health and well-being.

22.7.3 SDG 15: Life on Land

- **SDG 15.3: Land Degradation**: The excessive extraction of groundwater can result in land subsidence and contribute to the deterioration of land. Examining the proportion of land that has been degraded, the ongoing efforts to restore it, and the changes in land cover.

22.7.4 SDG 2: Zero Hunger

- **SDG 2.4: Sustainable Agriculture**: The decrease in groundwater resources has the potential to significantly impact agricultural productivity, which in turn can have negative consequences for food security. Water-use efficiency in agriculture refers to the effective utilization of water resources in farming practices. It involves ensuring that the agricultural area is managed sustainably, taking into account the availability and conservation of water. These practices help optimize water usage, preserve soil fertility, and ensure long-term food security by promoting environmentally sound agricultural methods.

22.7.5 SDG 11: Sustainable Cities and Communities

- **SDG 11.5: Water and Sanitation**: Urban areas depend heavily on groundwater for their water supply. The decrease in quality has a direct impact on the availability of clean water and proper sanitation. Examining the proportion of urban population with access to safe and sustainable water and sanitation.

22.7.6 SDG 12: Responsible Consumption and Production

- **SDG 12.4: Chemical Management**: Groundwater contamination frequently occurs due to the presence of hazardous chemicals, which has a significant impact on responsible consumption and production. Emphasizing the importance of reducing the release of hazardous chemicals into water bodies and promoting the environmentally sound management of chemicals is crucial to achieve SDG 12.4. This involves implementing stringent regulations, fostering responsible industrial practices, and promoting the adoption of eco-friendly technologies to safeguard water quality and protect ecosystems.

22.7.7 SDG 13: Climate Action

- **SDG 13.1: Resilience to Climate-Related Hazards**: SDG 13.1 indicators related to building resilience and reducing vulnerability to climate-related hazards, including groundwater depletion and contamination, encompass metrics like the number of deaths, missing persons, and directly affected individuals attributed to disasters. These indicators gauge the effectiveness of adaptation measures and the capacity of communities to withstand climate-induced challenges.

22.7.8 *Cross-Cutting Issues:*

- **Inequality (SDG 10) and Poverty (SDG 1)**: Water scarcity and contamination have a significant impact on vulnerable populations, further worsening social and economic disparities. Examining the proportion of the population living in poverty and the availability of water services in marginalized communities.
- **Partnerships (SDG 17)**: To tackle the issue of declining groundwater quality, it is essential to establish collaborations among various stakeholders in order to promote sustainable water management. The presence of partnerships, collaboration mechanisms, and international cooperation on water-related issues.

22.7.9 *Monitoring and Reporting*

- **Data Availability (SDG 17)**: There is a lack of consistent monitoring and reporting of groundwater quality and quantity on a global scale. Ensuring the availability and reliability of data for water quality and quantity (Matta et al., 2020d).

22.8 Contribution to Understanding the Link Between Groundwater Quality and SDGs

Gaining a comprehensive understanding of the connection between groundwater quality and Sustainable Development Goals (SDGs) is of utmost importance in order to devise efficient strategies that tackle worldwide issues pertaining to water, health, and environmental sustainability. Here are some significant contributions that help improve our understanding:

- **Interdisciplinary Research**: A thorough interdisciplinary study helps to develop a detailed understanding of the intricate connections between groundwater quality and different SDGs. Research that combines hydrogeology, environmental science, public health, and socio-economic analyses offers a comprehensive view of the complex connections between groundwater quality and sustainable development. Figure 22.11 shows the primary interlinkages between groundwater and SDG targets
- **Identification of Indicators**: It is crucial to contribute to the identification and development of specific indicators that measure the impact of groundwater quality on SDGs. These indicators are useful for evaluating progress, pinpointing areas of concern, and informing policy interventions (Kumar et al., 2021). For example, metrics concerning waterborne diseases, access to clean water, and ecosystem health can offer valuable insights into the progress of SDG targets.

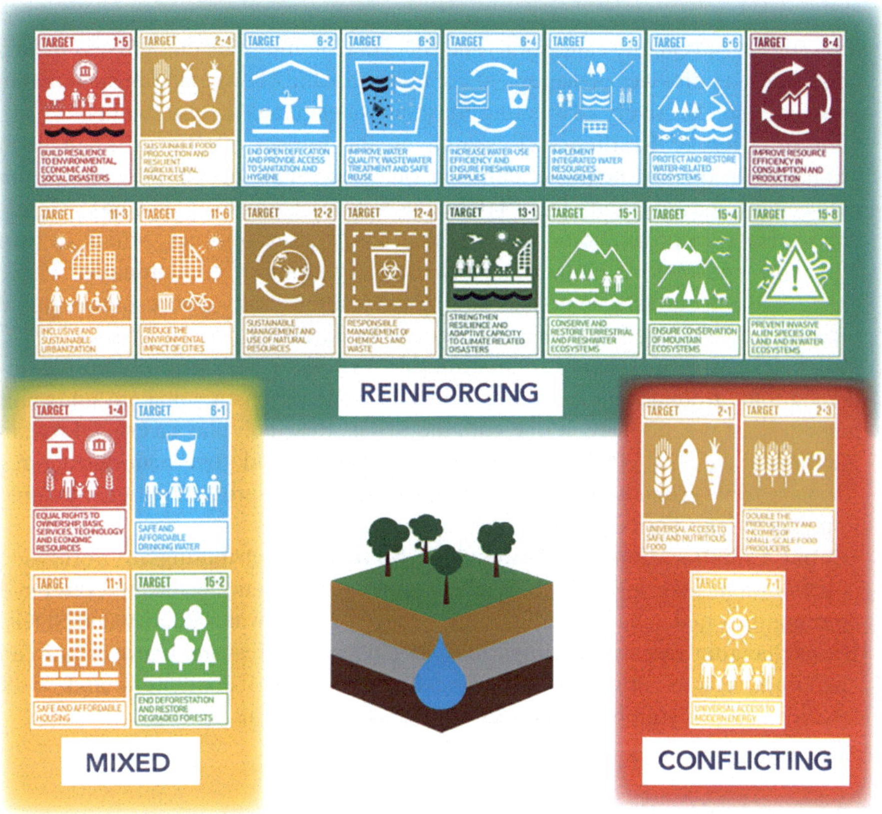

Fig. 22.11 Categories of primary interlinkages between groundwater and SDG targets. *Source* Guppy et al. (2018)

- **Quantification of Health Impacts**: Quantifying the health impacts of deteriorating groundwater quality is crucial for establishing direct correlations with SDG 3 (Good Health and Well-being). Studying the occurrence of waterborne diseases and their connection to particular contaminants in groundwater helps inform decision-making based on evidence and the development of policies that prioritize health.
- **Integration with Water Management Policies**: To fully grasp the connection between groundwater quality and SDG 6 (Clean Water and Sanitation), it is crucial to incorporate scientific discoveries into water management policies. Supporting the achievement of SDG 6 targets, contributions to the development of policies prioritize sustainable groundwater management, pollution prevention, and the provision of clean water.
- **Community Engagement and Education**: Understanding the connection between groundwater quality and SDGs is vital for empowering communities through education and engagement. Efforts focused on research that promotes

community engagement, education, and awareness-raising are valuable contributions to SDG 4 (Quality Education) and SDG 6. Communities that possess a deep understanding of water management are more capable of engaging in sustainable practices and effectively advocating for their water-related requirements.

- **Climate Change Resilience**: Research on the impact of climate change on groundwater quality helps to enhance our understanding of the connection between groundwater and SDG 13 (Climate Action). Climate change has a significant impact on precipitation patterns and temperature regimes, which in turn affects groundwater recharge rates and the distribution of contaminants. These changes play a crucial role in determining the resilience of water resources in the face of climate-related challenges (Ferguson et al., 2012).
- **Policy Recommendations for Sustainable Practices**: Offering policy recommendations grounded in scientific evidence helps to ensure that groundwater management practices are in line with the SDGs. Research plays a crucial role in identifying the most effective practices, technologies, and governance structures for sustainable groundwater use. By doing so, it helps ensure that policies are able to promote long-term environmental and social benefits.
- **Global Collaboration and Knowledge Sharing**: Contributing to worldwide collaboration and knowledge-sharing initiatives enhances the international community's understanding of the link between groundwater quality and SDGs. By exchanging research findings, best practices, and lessons learned, we can encourage collaborative efforts to tackle water-related challenges and work towards achieving global sustainability goals (Jacobs et al., 2016).

22.9 Analyzing the Implications for Sustainable Development Goals (SDG) 3 and 6

The decline in groundwater resources and the degradation of its quality pose a significant challenge with far-reaching implications for Sustainable Development Goals (SDGs) 3 (Good Health and Well-being) and 6 (Clean Water and Sanitation). Excessive extraction for agricultural irrigation, industrial processes, and domestic use has resulted in a notable decline in groundwater levels, worsening water scarcity problems in numerous regions. The depletion of water resources poses a significant threat to both the availability of safe drinking water and the ability to fulfill basic water requirements for health and sanitation (Bensoltane et al., 2021). This issue directly aligns with SDG 3, in addition, the deteriorating quality of groundwater caused by the presence of pollutants like nitrates, pesticides, and heavy metals presents significant health hazards to communities that depend on these sources for drinking water. As a result, this has a direct impact on the health-related goals of SDG 3. In addition, the pollution of groundwater intersects with SDG 6, which focuses on ensuring the availability and sustainable management of water for all. The presence of contaminated groundwater poses a significant threat to ecosystems, biodiversity, and the overall quality of water. This poses a challenge to achieving the objectives of improving water quality, reducing water pollution, and ensuring the sustainable use of water resources. An in-depth examination of the decreasing availability and quality of groundwater resources in relation to the indicators of SDGs 3 and 6 highlights the pressing requirement for comprehensive policies and interventions that tackle both the quantitative and qualitative aspects of water resource management. This is crucial to protect human health and ensure sustainable access to water and sanitation on a global scale.

22.9.1 Implications for SDG 3: Good Health and Well-Being

- **Waterborne Diseases**: Contaminated groundwater can serve as a favorable environment for the spread of waterborne diseases, including cholera, dysentery, and gastroenteritis. Consuming contaminated water presents a clear and immediate danger to the well-being of the general population, leading to an increase in waterborne diseases (Matta et al., 2022a, 2022b, 2022c, 2022d). To achieve SDG 3, it is crucial to prioritize access to safe and clean water, as the quality of groundwater directly impacts human health.
- **Impact on Vulnerable Populations**: The impact of declining groundwater quality has a more significant effect on vulnerable populations, such as children, older adults, and communities with limited access to other water sources. SDG 3 highlights the importance of tackling health disparities and guaranteeing that everyone has access to vital health services. This requires prioritizing the well-being of vulnerable communities by enhancing the quality of water.
- **Healthcare Burden**: Waterborne diseases caused by contaminated groundwater place a significant strain on healthcare systems. The rise in water-related illnesses puts a significant strain on healthcare resources, causing a diversion of attention and resources from other crucial health services. Achieving the goals of SDG 3 necessitates addressing this challenge by implementing efficient water quality management strategies (Andrade et al., 2018).

22.9.2 Implications for SDG 6: Clean Water and Sanitation

- **Universal Access to Clean Water**: SDG 6 aims to achieve universal access to clean water and sanitation, ensuring that everyone has the opportunity to enjoy clean water. The progress towards achieving the goal is at risk due to the declining quality of groundwater. This is because contaminated groundwater poses a threat to the availability of a crucial water source for numerous communities (Matta et al., 2022a). It is essential to implement sustainable management practices in order to protect the quality of groundwater and uphold the responsibility of ensuring clean water for everyone.

- **Ecosystem Impact**: SDG 6 acknowledges the significant impact that water resources have on ecosystems, highlighting their interconnectedness. The presence of contaminated groundwater has substantial implications for both human health and the delicate balance of aquatic ecosystems and biodiversity. To achieve SDG 6, it is crucial to consider the consequences of deteriorating groundwater quality on both human communities and the environment. This calls for a comprehensive approach to managing water resources.
- **Sustainable Groundwater Management**: Effective groundwater management is crucial for the attainment of SDG 6. This requires the implementation of efficient policies, the promotion of water conservation, and the prevention of pollution to guarantee the sustainable future of groundwater resources (Nayak et al., 2022). Aligning the preservation of ecosystems with human water needs is in line with the sustainable development principles outlined in SDG 6.
- **Community Empowerment**: Tackling the issue of deteriorating groundwater quality necessitates enabling communities to actively engage in monitoring, conservation, and sustainable utilization of water resources. Community engagement plays a crucial role in cultivating a sense of ownership and responsibility, which in turn contributes to the overall success of SDG 6.

22.9.3 Importance for Environmental Policy and Public Health

The significance of groundwater for environmental policy and public health cannot be overstated, as it plays a vital role in supporting ecosystems and human communities. The quality of groundwater has a direct impact on the sustainability of the environment and the well-being of humans (Matta et al., 2022b). It is crucial to have effective policies in place to tackle the challenges related to managing groundwater. These are the key aspects that emphasize the importance of environmental policy and public health:

22.10 Environmental Policy

- **Ecosystem Health**: The health of ecosystems relies on groundwater, which plays a crucial role in maintaining the flow of rivers, supporting wetlands, and ensuring the stability of habitats. Environmental policies that focus on sustainable groundwater management play a vital role in conserving biodiversity, protecting aquatic habitats, and maintaining the overall health of ecosystems.
- **Water Resource Sustainability**: Groundwater is a limited resource, and it is crucial to use it sustainably to avoid depletion and ensure water availability for future generations. Environmental policies are essential for regulating

groundwater extraction, promoting water conservation, and establishing recharge strategies to ensure the long-term sustainability of water resources.

- **Preventing Contamination**: Policies that prioritize the prevention of groundwater contamination play a crucial role in protecting the well-being of the environment and human health. By implementing regulations that address industrial discharges, agricultural runoff, and proper waste disposal, the introduction of pollutants into groundwater can be minimized, thus preventing any adverse effects on ecosystems (Nayak et al., 2023).
- **Land Use Planning**: Incorporating groundwater considerations into land use planning is essential for achieving environmental sustainability. Responsible land development policies, zoning regulations, and infrastructure planning play a crucial role in preventing contamination caused by urbanization. They also help preserve natural recharge areas and safeguard critical groundwater recharge zones.
- **Climate Change Adaptation**: Addressing the impact of climate change on groundwater resources is crucial for effective environmental policies. Developing strategies to adapt to shifting precipitation patterns, severe weather events, and changes in hydrological cycles is vital for maintaining the resilience of groundwater-dependent ecosystems in the midst of climate-related challenges.

22.11 Public Health

- **Safe Drinking Water**: Groundwater serves as a vital source of drinking water for numerous communities, ensuring their access to safe and clean water (Matta, 2018). The implementation of policies that regulate and monitor groundwater quality plays a crucial role in safeguarding access to safe and clean drinking water, mitigating the potential for waterborne diseases, and aligning with the public health objectives outlined in various health policies.
- **Disease Prevention**: Waterborne diseases can pose a significant risk to public health, as contaminated groundwater serves as a potential source of these illnesses. Effective policies are crucial in ensuring the prevention of waterborne diseases and safeguarding the well-being of communities. These policies establish water quality standards, enforce monitoring, and promote remediation efforts.
- **Community Awareness and Education**: Community Awareness and Education: It is of utmost importance to prioritize environmental policies that foster community awareness and education regarding groundwater quality, as this plays a vital role in safeguarding public health (Matta et al., 2019). Well-informed communities are more prepared to ensure their water sources, embrace water-saving techniques, and champion policies that prioritize water quality and health.
- **Emergency Response Planning**: Emergency Response Planning: It is crucial to have policies in place that address emergency response and contingency planning for groundwater contamination events. Having effective preparedness and response strategies is critical to mitigating the health consequences of accidental

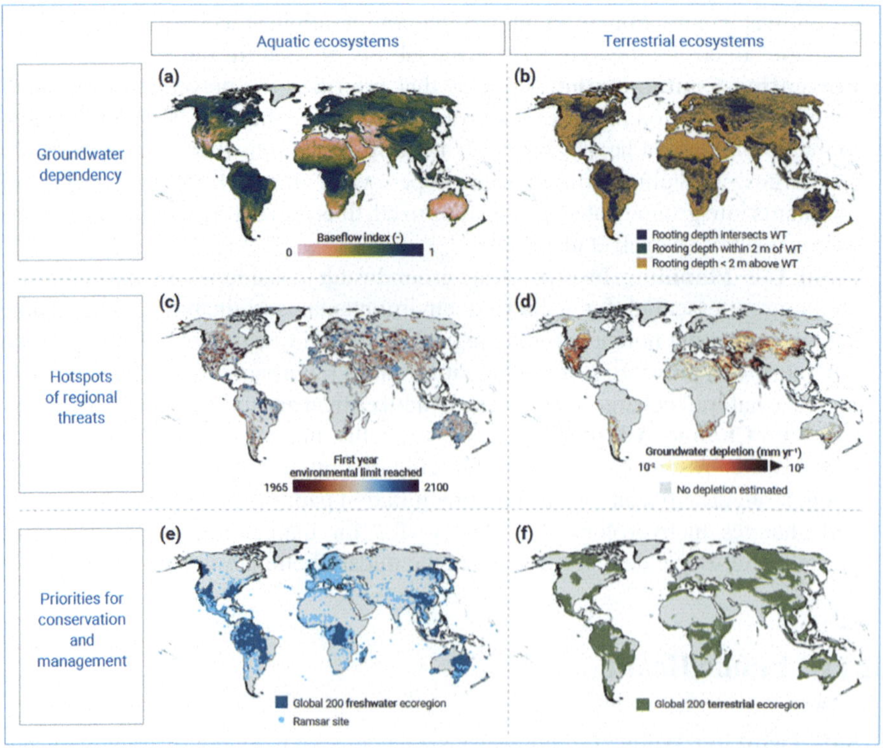

Fig. 22.12 Global patterns in groundwater dependency hotspots of regional threats and priorities for conservation and management for aquatic and terrestrial ecosystems. *Source* United Nations World Water Development Report (2022)

spills, contamination incidents, or natural disasters that can impact groundwater quality.

- **Reducing Health Disparities**: Addressing Health Disparities: Ensuring fair and equal access to clean groundwater plays a crucial role in reducing health disparities. By focusing on the needs of vulnerable populations, particularly those in rural or underserved areas, policies can effectively tackle health inequities associated with water quality and access. Figure 22.12 shows the Global patterns in groundwater dependency hotspots of regional threats and priorities for conservation and management for aquatic and terrestrial ecosystems.

22.12 Recommendations for Policy and Intervention

To effectively tackle the issue of declining groundwater resources and quality, it is crucial to implement comprehensive policies and interventions that align with Sustainable Development Goals (SDGs) 3 (Good Health and Well-being) and 6

(Clean Water and Sanitation). Based on my analysis, I have compiled a list of recommendations:

- **Integrated Water Resource Management (IWRM)**: Implement and enforce practices that thoroughly consider the entire water cycle, prioritize sustainable use, and strive to maintain ecological balance. Create and execute water allocation plans that take into account the requirements of both human consumption and ecosystems in order to avoid excessive extraction.
- **Groundwater Monitoring and Data Management**: Develop and enhance groundwater monitoring networks to evaluate aquifer levels water quality, and track long-term patterns. Enhance data collection and management systems to optimize decision-making processes and bolster evidence-based policies.
- **Regulation and Enforcement**: Establish and implement rules governing groundwater extraction to promote sustainable utilization and prevent excessive exploitation. Develop and enhance regulatory frameworks to effectively manage the release of pollutants into groundwater, tackling the problem of contamination.
- **Promote Water Use Efficiency**: Promote the adoption of water-efficient technologies and practices in agriculture, industry, and domestic sectors to decrease overall water demand—Utilise irrigation techniques that maximize water efficiency, such as drip irrigation and precision agriculture (Keenan et al., 2013).
- **Community Engagement and Education**: Engage local communities in water management decisions and raise awareness about the significance of sustainable groundwater use. Organise educational programmes focused on water conservation, pollution prevention, and the consequences of groundwater depletion on health and well-being (Goralnik et al., 2022).
- **Wastewater Treatment and Reuse**: Invest in infrastructure for the treatment of wastewater, reducing the discharge of contaminants into groundwater. Promote the safe reuse of treated wastewater for non-potable purposes, reducing the demand for freshwater sources (Matta et al., 2021).
- **Affordable and Accessible Water Technologies**: Promote the advancement and acceptance of cost-effective and easily accessible water purification technologies for communities facing water quality challenges. Enable the spread of water treatment technologies that can be readily adopted at the household or community level (Bhatnagar et al., 2013).

    - **Cross-Sectoral Collaboration**: Promote collaboration among various sectors, such as agriculture, industry, health, and environment, to effectively tackle the complex issues surrounding groundwater decline and contamination. Develop inter-agency coordination mechanisms to ensure a cohesive and integrated approach to water resource management (Cloutier et al., 2008).

- **Incentives for Sustainable Practices**: Offering incentives, subsidies, or tax breaks to industries and agricultural practices that embrace sustainable water management and pollution prevention measures can be a valuable approach. Promote the advancement of sustainable groundwater recharge projects by offering financial incentives (Piñeiro et al., 2020).

- **Research and Innovation**: Invest in research and development to explore innovative solutions for sustainable groundwater management, pollution remediation, and water purification technologies. Promote the advancement of water management strategies that are resilient to climate change, specifically targeting the protection of groundwater resources from its impacts.

22.13 Role of Science Communication

Efficient science communication is crucial in rallying communities, policymakers, and stakeholders to address the issue of declining groundwater quality highlighted in our study. We advocate for a comprehensive approach based on scientific evidence and in line with Sustainable Development Goals (SDG) 3 (Good Health and Well-Being) and SDG 6 (Clean Water and Sanitation). First and foremost, it is essential to emphasize the importance of addressing the issue promptly. We need to clearly state the concrete effects of deteriorating groundwater quality on human health and ecosystems, highlighting the risk of waterborne diseases and long-term health hazards (Matta, 2020b). This narrative is not only captivating for general audiences but also highlights the significance of our findings to SDG 3. At the same time, connecting our research to SDG 6 allows us to view the issue in the broader context of global water sustainability. Highlighting the connection between groundwater quality, clean water accessibility, and the health of aquatic ecosystems emphasises the need for a comprehensive approach to addressing this crisis. Highlighting the impact of degraded groundwater quality on the achievement of SDG 6's goals effectively appeals to policymakers and supporters of clean water and sanitation.

In addition, our communication strategy should focus on practical solutions. Emphasizing sustainable water management practices, like responsible groundwater extraction and agricultural strategies that minimize chemical runoff, offers concrete actions that communities and industries can implement. This encourages individuals to actively contribute to the preservation of groundwater, which is in line with the objectives of SDG 6. Finally, community engagement and education are crucial. Crafting accessible and informative messages, along with targeted outreach programs, will enable communities to understand the implications of their water usage and motivate behavioral changes. By promoting a strong sense of ownership and responsibility for local groundwater resources, which can generate widespread support for ongoing efforts to mitigate the decline in groundwater quality. Essentially, science communication plays a crucial role in facilitating informed decision-making and collective action, which is essential for building a sustainable and healthier future.

22.14 Conclusion

The study highlights the urgent need for specific interventions in response to the environmental stress caused by the decline in groundwater quality. This study explores the various consequences of declining groundwater on human well-being and ecosystem sustainability within the framework of Sustainable Development Goals (SDG) 3 (Good Health and Well-Being) and 6 (Clean Water and Sanitation). The decline in groundwater quality, characterized by the infiltration of pollutants like heavy metals, pesticides, and industrial contaminants, highlights the urgent need for prompt and coordinated action. The study provides an in-depth analysis of the impact on human health within the framework of SDG 3. It highlights the significant danger that compromised groundwater quality poses, leading to an increased risk of waterborne diseases and long-term health hazards. At the same time, the research highlights the interdependence of this crisis with SDG 6, emphasizing the significant effect on the availability of clean water resources and the overall health of aquatic ecosystems. Tackling the environmental health crisis requires a comprehensive and well-rounded approach. It is crucial to enhance regulatory frameworks in order to effectively monitor and control sources of groundwater pollution, thereby ensuring strict adherence to established quality standards. Implementing sustainable water management practices, such as responsible groundwater extraction and agricultural strategies that minimize chemical runoff, is crucial for meeting the targets of both SDG 3 and SDG 6. Community involvement and education are critical in raising awareness about the connection between groundwater quality, public health, and the long-term sustainability of water resources.

In addition, the implementation of water resource management strategies that incorporate both technological advancements and traditional knowledge can help reduce the negative impacts on ecosystems and human health. This approach is in line with the principles of SDG 6, which aims to ensure fair and sustainable access to water for everyone. By placing a strong emphasis on the preservation and restoration of groundwater quality, we are not only addressing the immediate health concerns of communities but also making significant strides towards creating a resilient and sustainable environment. This aligns with the principles outlined in SDG 3 and SDG 6. As we navigate this complex environmental challenge, it is clear that global collaboration and innovative solutions are urgently needed to ensure a healthier and more sustainable future for both current and future generations.

Further, by employing precise and easily understandable language to convey scientific discoveries clarifying the reasons, consequences, and possible remedies associated with the deterioration of groundwater quality, science communication enables communities, policymakers, and stakeholders to make well-informed choices and proactively address the issue of sustainable groundwater management. This communication aims to connect scientific knowledge with practical action, enabling the adoption of responsible water usage practices, regulatory improvements, and community engagement initiatives necessary for preserving and restoring groundwater quality and achieving sustainability.

Acknowledgements The authors are thankful to Uttarakhand State Council for Science and Technology, Govt. of Uttarakhand, Dehradun, for providing the resources for the manuscript.

Declarations Grant Support Details: Uttarakhand Council of Science and Technology (UCOST) has financially supported the present research.

Conflict of Interest There is no conflict of Interest.

References

Abiriga, D., Jenkins, A., Vestgarden, L. S., & Klempe, H. (2021). A nature-based solution to a landfill-leachate contamination of a confined aquifer. *Scientific Reports, 11*(1), 14896.

Akhtar, N., Syakir Ishak, M. I., Bhawani, S. A., & Umar, K. (2021). Various natural and anthropogenic factors responsible for water quality degradation: A review. *Water, 13*(19), 2660. https://doi.org/10.3390/w13192660

Alsalme, A., Al-Zaqri, N., Ullah, R., & Yaqub, S. (2021). Approximation of ground water quality for microbial and chemical contamination. *Saudi Journal of Biological Sciences, 28*(3), 1757–1762. https://doi.org/10.1016/j.sjbs.2020.12.017

Andrade, L., O'Dwyer, J., O'Neill, E., & Hynds, P. (2018). Surface water flooding, groundwater contamination, and enteric disease in developed countries: A scoping review of connections and consequences. *Environmental Pollution, 236*, 540–549. https://doi.org/10.1016/j.envpol.2018.01.104

Arora, T., Mishra, A., Matta, G., Chopra, A. K., Kumar, A., Khanna, D. R., & Kumar, V. (2017). Pollution load assessment and potential environmental risks of composite industrial effluents discharged from SIIDCUL Integrated Industrial Estate, Haridwar (Uttarakhand), India. *Journal of Environmental Biology, 38*(2), 205–216.

Ashraf, S., Nazemi, A., & AghaKouchak, A. (2021). Anthropogenic drought dominates groundwater depletion in Iran. *Scientific Reports, 11*(1), 9135.

Baldassarre, G., Sivapalan, M., Rusca, M., Cudennec, C., Garcia, M., Kreibich, H., et al. (2019). Sociohydrology: Scientific challenges in addressing the sustainable development goals. *Water Resources Research, 55*(8), 6327–6355. https://doi.org/10.1029/2018WR023901

Banks, E. W., Cook, P. G., Owor, M., Okullo, J., Kebede, S., Nedaw, D., et al. (2021). Environmental tracers to evaluate groundwater residence times and water quality risk in shallow unconfined aquifers in sub Saharan Africa. *Journal of Hydrology, 598*, 125753. https://doi.org/10.1016/j.jhydrol.2020.125753

Bargali, H., Gupta, S., Malik, D. S., & Matta, G. (2017): Estimation of fire frequency in Nainital District of Uttarakhand State by using satellite images. *Journal of Remote Sensing & GIS*. https://doi.org/10.4172/2469-4134.1000214

Bensoltane, M.A., Zeghadnia, L., Bordji, N., Gagan, M., & Bouranen, S., (2021): Drinking water quality assessment using Principal Component Analysis: Case study of the town of Souk Ahras, Algeria. *Egyptian Journal of Chemistry, 64*(6), 3069–3075. https://doi.org/10.21608/EJCHEM.2021.53654.3112

Bhatnagar, A., Hogland, W., Marques, M., & Sillanpää, M. (2013). An overview of the modification methods of activated carbon for its water treatment applications. *Chemical Engineering Journal, 219*, 499–511. https://doi.org/10.1016/j.cej.2012.12.038

Calderon, R. L. (2000). The epidemiology of chemical contaminants of drinking water. *Food and Chemical Toxicology, 38*, S13–S20.

Chandnani, G., Gandhi, P., Kanpariya, D., Parikh, D., & Shah, M. (2022). A comprehensive analysis of contaminated groundwater: Special emphasis on nature-ecosystem and socio-economic impacts. *Groundwater for Sustainable Development*, 100813. https://doi.org/10.1016/j.gsd.2022.100813

Chidichimo, F., De Biase, M., & Straface, S. (2020). Groundwater pollution assessment in landfill areas: Is it only about the leachate?. *Waste Management, 102*, 655–666.https://doi.org/10.1016/j.wasman.2019.11.038

Cloutier, V., Lefebvre, R., Therrien, R., & Savard, M. M. (2008). Multivariate statistical analysis of geochemical data as indicative of the hydrogeochemical evolution of groundwater in a sedimentary rock aquifer system. *Journal of Hydrology, 353*(3–4), 294–313. https://doi.org/10.1016/j.jhydrol.2008.02.015

Cuthbert, M. O., Gleeson, T., Moosdorf, N., Befus, K. M., Schneider, A., Hartmann, J., & Lehner, B. (2019). Global patterns and dynamics of climate–groundwater interactions. *Nature Climate Change, 9*(2), 137–141.

Dai, X., Xie, Y., Simmons, C. T., Berg, S., Dong, Y., Yang, J., et al. (2021). Understanding topography-driven groundwater flow using fully-coupled surface-water and groundwater modeling. *Journal of Hydrology, 594*, 125950. https://doi.org/10.1016/j.jhydrol.2020.125950

Dieter, C. A., Maupin, M. A., Caldwell, R. R., Harris, M. A., Ivahnenko, T. I., Lovelace, J. K., Barber, N. L. & Linsey, K. S. (2018). Estimated use of water in the United States in 2015. *U.S. Geological Survey Circular* 1441. Reston, Va., U.S. Geological Survey (USGS). https://doi.org/10.3133/cir144

Etikala, B., Vangala, S., & Madhav, S. (2024). Groundwater geochemistry using modified integrated water quality index (IWQI) and health indices with special emphasis on nitrates and heavy metals in southern parts of Tirupati, South India. *Environmental Geochemistry and Health, 46*(11). https://doi.org/10.1007/s10653-024-02229-3

Fatemeh, R. F., Masoud, Y., Saeed, G., & Mostafa, A. (2021). Institutional constraints to groundwater resource management in arid and semi-arid regions: A Straussian grounded theory study. *Hydrogeology Journal, 29*(3), 925–947. https://doi.org/10.1007/s10040-020-02283-y

Ferguson, G., & Gleeson, T. (2012). Vulnerability of coastal aquifers to groundwater use and climate change. *Nature Climate Change, 2*(5), 342–345.

Foley, J. A., DeFries, R., Asner, G. P., Barford, C., Bonan, G., Carpenter, S. R. et al. (2005). Global consequences of land use. *Science, 309*(5734), 570–574. https://doi.org/10.1126/science.1111772

Foster, S., & Chilton, J. (2003). Groundwater: The processes and global significance of aquifer degradation. *Philosophical Transactions of the Royal Society of London.* Series B, *Biological Sciences, 358*(1440), 1957–1972. https://doi.org/10.1098/rstb.2003.1380

Foster, S., & Hirata, R. (2012). Groundwater use for urban development: Enhancing benefits and reducing risks. *On the Water Front, 3*, 21–29.

Gaur, V. K., Sharma, P., Sirohi, R., Awasthi, M. K., Dussap, C. G., & Pandey, A. (2020). Assessing the impact of industrial waste on environment and mitigation strategies: A comprehensive review. *Journal of Hazardous Materials, 398*, 123019. https://doi.org/10.1016/j.jhazmat.2020.123019

Ghosh, G. C., Khan, M. J. H., Chakraborty, T. K., Zaman, S., Kabir, A. E., & Tanaka, H. (2020). Human health risk assessment of elevated and variable iron and manganese intake with arsenic-safe groundwater in Jashore, Bangladesh. *Scientific Reports, 10*(1), 5206.

Gleeson, T., Cuthbert, M., Ferguson, G., & Perrone, D. (2020). Global groundwater sustainability, resources, and systems in the Anthropocene. *Annual Review of Earth and Planetary Sciences, 48*, 431–463. https://doi.org/10.1146/annurev-earth-071719-05525110.1146/annurev-earth-071719-055251

Gleeson, T., Wada, Y., Bierkens, M. F., & Van Beek, L. P. (2012). Water balance of global aquifers revealed by groundwater footprint. *Nature, 488*(7410), 197–200.

Goralnik, L., Brunacini, J., Rutty, M., & Finnell, E. (2022). Restoring relationships: Water heritage, sense of place, and community engagement. *Journal of Great Lakes Research, 48*(6), 1375–1386. https://doi.org/10.1016/j.jglr.2022.08.018

Guppy, L., Uyttendaele, P., Villholth, K. G., & Smakhtin, V. (2018). Groundwater and sustainable development goals: Analysis of interlinkages. *UNU-INWEH Report Series,* Issue 04. United Nations University Institute for Water, Environment and Health, Hamilton, Canada.

Hartmann, A., Gleeson, T., Wada, Y., & Wagener, T. (2017). Enhanced groundwater recharge rates and altered recharge sensitivity to climate variability through subsurface heterogeneity. *Proceedings of the National Academy of Sciences, 114*(11), 2842–2847.https://doi.org/10.1073/pnas.1614941114

He, C., Liu, Z., Wu, J., Pan, X., Fang, Z., Li, J., & Bryan, B. A. (2021). Future global urban water scarcity and potential solutions. *Nature Communications, 12*(1), 4667.

Huang, F., An, Z., Moran, M. J., & Liu, F. (2020). Recognition of typical antibiotic residues in environmental media related to groundwater in China (2009–2019). *Journal of Hazardous Materials, 399*, 122813. https://doi.org/10.1016/j.jhazmat.2020.122813

Huang, G., Liu, C., Sun, J., Zhang, M., Jing, J., & Li, L. (2018). A regional scale investigation on factors controlling the groundwater chemistry of various aquifers in a rapidly urbanized area: A case study of the Pearl River Delta. *Science of the Total Environment, 625*, 510–518. https://doi.org/10.1016/j.scitotenv.2017.12.322

Issaka, S., & Ashraf, M. A. (2017). Impact of soil erosion and degradation on water quality: A review. *Geology, Ecology, and Landscapes, 1*(1), 1–11. https://doi.org/10.1080/24749508.2017.1301053

Jacobs, K., Lebel, L., Buizer, J., Addams, L., Matson, P., McCullough, E., et al. (2016). Linking knowledge with action in the pursuit of sustainable water-resources management. *Proceedings of the National Academy of Sciences, 113*(17), 4591–4596. https://doi.org/10.1073/pnas.0813125107

Kätterer, T., Bolinder, M. A., Andrén, O., Kirchmann, H., & Menichetti, L. (2011). Roots contribute more to refractory soil organic matter than above-ground crop residues, as revealed by a long-term field experiment. *Agriculture, Ecosystems & Environment, 141*(1–2), 184–192. https://doi.org/10.1016/j.agee.2011.02.029

Keenan, T. F., Hollinger, D. Y., Bohrer, G., Dragoni, D., Munger, J. W., Schmid, H. P., & Richardson, A. D. (2013). Increase in forest water-use efficiency as atmospheric carbon dioxide concentrations rise. *Nature, 499*(7458), 324–327.

Khatri, N., & Tyagi, S. (2015). Influences of natural and anthropogenic factors on surface and groundwater quality in rural and urban areas. *Frontiers in Life Science, 8*(1), 23–39. https://doi.org/10.1080/21553769.2014.933716

Kinuthia, G. K., Ngure, V., Beti, D., Lugalia, R., Wangila, A., & Kamau, L. (2020). Levels of heavy metals in wastewater and soil samples from open drainage channels in Nairobi, Kenya: Community health implication. *Scientific Reports, 10*(1), 8434.

Kishor, R., Purchase, D., Saratale, G. D., Saratale, R. G., Ferreira, L. F. R., Bilal, M., et al. (2021). Ecotoxicological and health concerns of persistent coloring pollutants of textile industry wastewater and treatment approaches for environmental safety. *Journal of Environmental Chemical Engineering, 9*(2), 105012. https://doi.org/10.1016/j.jece.2020.105012

Kløve, B., Ala-Aho, P., Bertrand, G., Gurdak, J. J., Kupfersberger, H., Kværner, J., et al. (2014). Climate change impacts on groundwater and dependent ecosystems. *Journal of Hydrology, 518*, 250-266. https://doi.org/10.1016/j.jhydrol.2013.06.037

Kumar, A., Matta, G., & Bhatnagar, S. (2021). A coherent approach of Water Quality Indices and Multivariate Statistical Models to estimate the water quality and pollution source apportionment of River Ganga System in Himalayan region, Uttarakhand, India. *Environmental Science and Pollution Research*. https://doi.org/10.1007/s11356-021-13711-1

Kumar, V., Singh, E., Singh, S., Pandey, A., & Bhargava, P. C. (2023). Micro-and nano-plastics (MNPs) as emerging pollutant in ground water: Environmental impact, potential risks, limitations and way forward towards sustainable management. *Chemical Engineering Journal, 459*, 141568. https://doi.org/10.1016/j.cej.2023.141568

Kurylyk, B. L., Hayashi, M., Quinton, W. L., McKenzie, J. M., & Voss, C. I. (2016). Influence of vertical and lateral heat transfer on permafrost thaw, peatland landscape transition, and groundwater flow. *Water Resources Research, 52*(2), 1286–1305. https://doi.org/10.1002/2015WR018057

Langemeyer, J., Madrid-Lopez, C., Beltran, A. M., & Mendez, G. V. (2021). Urban agriculture—A necessary pathway towards urban resilience and global sustainability? *Landscape and Urban Planning, 210*, 104055. https://doi.org/10.1016/j.landurbplan.2021.104055

Lapworth, D. J., Nkhuwa, D. C. W., Okotto-Okotto, J., Pedley, S., Stuart, M. E., Tijani, M. N., & Wright, J. J. H. J. (2017). Urban groundwater quality in sub-Saharan Africa: Current status and implications for water security and public health. *Hydrogeology Journal, 25*(4), 1093. https://doi.org/10.1007/s10040-016-1516-6

Ling, F., Hwang, C., LeChevallier, M. W., Andersen, G. L., & Liu, W. T. (2016). Core-satellite populations and seasonality of water meter biofilms in a metropolitan drinking water distribution system. *The ISME Journal, 10*(3), 582–595.

Liu, L., Zheng, X., Wei, X., Kai, Z., & Xu, Y. (2021). Excessive application of chemical fertilizer and organophosphorus pesticides induced total phosphorus loss from planting causing surface water eutrophication. *Scientific Reports, 11*(1), 23015.

Long, D., Yang, W., Scanlon, B. R., Zhao, J., Liu, D., Burek, P., et al. (2020). South-to-North water diversion stabilizing beijing's groundwater levels. *Nature Communications, 11*(1), 3665.

Matta, G., (2018). Environmental Sustainability through Scientific Communication using eco-informatics. In P. Kumar & S. Nagarthinam (Eds.) *Exploring eco media: Raising scientific temper* (pp. 193–205). Eco-Visioning Forum, Department of Communication, Madurai Kamraj University, Madurai, Tamil Nadu, India

Matta, G., (2020a). Applicability of WQI and scientific communication for conservation of River Ganga System in India. In V. Naddeo, M. Balakrishnan, & K.-H. Choo (Eds.), *Frontiers in water-energy-nexus—Nature-based solutions, advanced technologies and best practices for environmental sustainability. advances in science, technology & innovation* (pp. 265–267). https://doi.org/10.1007/978-3-030-13068-8

Matta, G., (2020b). Science communication as a preventative tool in the COVID19 pandemic. *Humanities and Social Sciences Communications, 7*, 159. https://doi.org/10.1057/s41599-020-00645-1

Matta, G., & Uniyal, D. P., (2017): Assessment of Species Diversity and Impact of Pollution on Limnological conditions of River Ganga. Int. J. Water, 11(2): 87–102. https://doi.org/10.1504/IJW.2017.083759

Matta, G., Gautam, R., Kumar, A., & Kumar, P. (2020). Suitability assessment of Groundwater quality in rural areas of Haridwar District, Uttarakhand. *Pollution Research., 39*(1), 91–96.

Matta, G., Gjyli, L., Kumar, A., John, M., (2018a). Hydro-chemical characteristics and planktonic composition assessment of River Henwal in Himalayan Region of Uttarakhand using CPI, Simpson's and Shannon-Weaver Index. Journal of Chemical and *Pharmaceutical Sciences, 11*(1). ISSN: Print 0974-2115; Online 2349-8552.

Matta, G., Kumar, A., Kumar, A., Naik, P, K., and Kumar, A., (2018): Assessment of Heavy Metal Indexing on Ganga River System assessing heavy metals toxicity and ecological impact on river water quality. *INAE Letters, An Official Journal of the Indian National Academy of Engineering.* ISSN: 2366-326X: EISSN: 2366-3278. https://doi.org/10.1007/s41403-018-0041-4

Matta, G., Kumar, A., Kumar, P., Nayak, A., Kumar, P., Kumar, A., & Tiwari, A. K., (2021). Evaluation of ground water quality by use of water quality index in the vicinity of the Rajaji National Park Haridwar, Uttarakhand, India. In P. P. Adhikary et al. (Eds.), *Geostatistics and Geospatial technologies for groundwater resources in India.* Springer Hydrogeology. https://doi.org/10.1007/978-3-030-62397-5_17

Matta, G., Kumar, A., Naik, P, K., Tiwari, A. K., & Berndtsson, R., (2018c). Ecological analysis of nutrient dynamics and phytoplankton assemblage in the Ganga River System, Uttarakhand. *Taiwan Water Conservancy, 66*(1), 1–12. ISSN: 0492-1505.

Matta, G., Kumar, A., Nayak, A. et al. (2022). Appraisal of spatial–temporal variation and pollution source estimation of Ganga River system through pollution indices and environmetrics in Upper Ganga basin. *Applied Water Science, 12*, 33. https://doi.org/10.1007/s13201-021-01552-9

Matta, G., Kumar, A., Nayak, A., et al. (2020b). Water Quality and Planktonic Composition of River Henwal (India) Using Comprehensive Pollution Index and Biotic-Indices. Trans Indian Natl. Acad. Eng. 5, 541–553. https://doi.org/10.1007/s41403-020-00094-x

Matta, G., Kumar, A., Nayak, A., et al. (2020c). Determination of water quality of Ganga River System in Himalayan region, referencing indexing techniques. *Arabian Journal of Geosciences, 13*, 1027. https://doi.org/10.1007/s12517-020-05999-z

Matta, G., Kumar, A., Nayak, A., Kumar, P., & Pant, G. (2022b). Pollution complexity quantification using NPI and HPI of River Ganga system in Himalayan Region. *Proceedings of the Indian National Science Academy, 88*(4), 651–663. https://doi.org/10.1007/s43538-022-00111-3

Matta, G., Kumar, A., Nayak, A., Kumar, P., Kumar, A., Naik, P. K., & Singh, S. K. (2022c). Assessing heavy metal index referencing health risk in Ganga River System. International *Journal of River Basin Management*, 1–11. https://doi.org/10.1080/15715124.2022.2098756

Matta, G., Kumar, A., Tiwari, A.K., Naik, P, K., & Berndtsson, R., (2018d). HPI appraisal of concentrations of heavy metals in Dynamic and static flow of Ganga River System. In *Environment, development and sustainability*. Springer Nature. ISSN: 1387-585X; EISSN: 1573-2975. https://doi.org/10.1007/s10668-018-01182-3

Matta, G., Kumar, A., Walia, A., Kumar, S., Mishra, H, K., Dhingra, G. K., Pokhriyal, P., & Wats, M., (2016). Quality estimation of ground water in industrial state of Uttarakhand. *Pollution Research, 35* (4), 849–854.

Matta, G., Kumar, P., Uniyal, D. P., & Joshi, D. U., (2022d). Communicating water, sanitation, and hygiene under sustainable development goals 3, 4, and 6 as the Panacea for epidemics and pandemics referencing the succession of COVID-19 Surges. *ACS ES&T Water 2022, 2*(5), 667–689. https://doi.org/10.1021/acsestwater.1c00366

Matta, G., Lu, Xi, X., & Kumar, A. (2019). Hydrological assessment of River Henwal using water quality indices with reference to planktonic composition. In *Advances in water pollution monitoring and control*. Springer. https://doi.org/10.1007/978-981-32-9956-6_18

Matta, G., Naik, P, K., Machell, J., Kumar, A., Gjyli, L., Tiwari, A, K., & Kumar, A. (2018e). Comparative study on seasonal variation in hydro-chemical parameters of Ganga river water using comprehensive pollution index (CPI) at Rishikesh, (Uttarakhand) India. *Desalination and Water Treatment, 118*, 87–95. https://doi.org/10.5004/dwt.2018.22487

Matta, G., Nayak, A., Kumar, A., et al. (2020d). Water quality assessment using NSFWQI, OIP and multivariate techniques of Ganga River system, Uttarakhand, India. *Applied Water Science, 10*, 206. https://doi.org/10.1007/s13201-020-01288-y

Matta, G., Nayak, A., Kumar, A., Kumar, P., Kumar, A., Tiwari, A, K., & Naik, P, K. (2020e). Evaluation of heavy metals contamination with calculating the pollution index for Ganga River System. *Taiwan Water Conservancy, 68*(3). https://doi.org/10.6937/TWC.202009/PP_68(3).0005

Matta, G., Rajput, A., Rajput, A., Kumar, P., Kumar, A., Nayak, A., Kumar, A., Dhingra, G. K., Chauhan, A., Chadha, S, K., & Wats, M. (2020f). Assessment of water quality of Choyyia Nadi (River) Catchment area in Bijnor District, Uttar Pradesh, India. *Ecology Environment & Conservation, 26*(November Suppl. Issue), S402–S412.

Matta, G., Sachin, S., Pandey, R, R., & Saini K. K. (2015). Assessment of physicochemical characteristics of Ganga Canal water quality in Uttarakhand. *Environmental, Development and Sustainability*. https://doi.org/10.1007/s10668-015-9735-x

Maven's Notebook. (2015). *Groundwater problems and prospects, Part 1: An overview of groundwater*. mavensnotebook.com/2015/03/12/groundwater-problems-and-prospects-part-1-an-overview-of-groundwater/

McNamara, S. M., Kolesar, K. R., Wang, S., Kirpes, R. M., May, N. W., Gunsch, M. J., et al. (2020). Observation of road salt aerosol driving inland wintertime atmospheric chlorine chemistry. *ACS Central Science, 6*(5), 684–694. https://doi.org/10.1021/acscentsci.9b00994

Mensah, J. K., Ofosu, E. A., Akpoti, K., Kabo-Bah, A. T., Okyereh, S. A., & Yidana, S. M. (2022). Modeling current and future groundwater demands in the White Volta River Basin of Ghana under climate change and socio-economic scenarios. *Journal of Hydrology: Regional Studies, 41*, 101117. https://doi.org/10.1016/j.ejrh.2022.101117

Miguez-Macho, G., & Fan, Y. (2021). Spatiotemporal origin of soil water taken up by vegetation. *Nature, 598*(7882), 624–628.

Müller, A., Österlund, H., Marsalek, J., & Viklander, M. (2020). The pollution conveyed by urban runoff: A review of sources. *Science of the Total Environment, 709*, 136125. https://doi.org/10.1016/j.scitotenv.2019.136125

Nayak, A., Matta, G., & Uniyal, D. P. (2022). Hydrochemical characterization of groundwater quality using chemometric analysis and water quality indices in the foothills of Himalayas. *Environment, Development and Sustainability*. https://doi.org/10.1007/s10668

Nayak, A., Matta, G., Uniyal, D. P., *et al.* (2023): Assessment of potentially toxic elements in groundwater through interpolation, pollution indices, and chemometric techniques in Dehradun in Uttarakhand State. *Environmental Science and Pollution Research*. https://doi.org/10.1007/s11356-023-27419-x

Noori, R., Maghrebi, M., Jessen, S., Bateni, S. M., Heggy, E., Javadi, S., et al. (2023). Decline in Iran's groundwater recharge. *Nature Communications, 14*(1), 6674.

Pinasseau, L., Wiest, L., Volatier, L., Mermillod-Blondin, F., & Vulliet, E. (2020). Emerging polar pollutants in groundwater: Potential impact of urban stormwater infiltration practices. *Environmental Pollution, 266*, 115387. https://doi.org/10.1016/j.envpol.2020.115387

Piñeiro, V., Arias, J., Dürr, J., Elverdin, P., Ibáñez, A. M., Kinengyere, A., et al. (2020). A scoping review on incentives for adoption of sustainable agricultural practices and their outcomes. *Nature Sustainability, 3*(10), 809–820.

Pokhrel, Y., Felfelani, F., Satoh, Y., Boulange, J., Burek, P., Gädeke, A., et al. (2021). Global terrestrial water storage and drought severity under climate change. *Nature Climate Change, 11*(3), 226–233.

Pulido-Bosch, A., Rigol-Sanchez, J. P., Vallejos, A., Andreu, J. M., Ceron, J. C., Molina-Sanchez, L., & Sola, F. (2018). Impacts of agricultural irrigation on groundwater salinity. *Environmental Earth Sciences, 77*, 1–14.

Rohde, M. M., Froend, R., & Howard, J. (2017). A global synthesis of managing groundwater dependent ecosystems under sustainable groundwater policy. *Groundwater, 55*(3), 293–301. https://doi.org/10.1111/gwat.12511

Saravanan, A., Kumar, P. S., Vo, D. V. N., Jeevanantham, S., Karishma, S., & Yaashikaa, P. R. (2021). A review on catalytic-enzyme degradation of toxic environmental pollutants: Microbial enzymes. *Journal of Hazardous Materials, 419*, 126451. https://doi.org/10.1016/j.jhazmat.2021.126451

Sharholy, M., Ahmad, K., Mahmood, G., & Trivedi, R. C. (2008). Municipal solid waste management in Indian cities—A review. *Waste Management, 28*(2), 459–467. https://doi.org/10.1016/j.wasman.2007.02.008

Sharma, A., Ganguly, R., & Kumar Gupta, A. (2020). Impact assessment of leachate pollution potential on groundwater: an indexing method. *Journal of Environmental Engineering, 146*(3), 05019007.https://doi.org/10.1061/(ASCE)EE.1943-7870.0001647

Shirzaei, M., Freymueller, J., Törnqvist, T. E., Galloway, D. L., Dura, T., & Minderhoud, P. S. (2021). Measuring, modelling and projecting coastal land subsidence. *Nature Reviews Earth & Environment, 2*(1), 40–58.

Singh, A., & Agrawal, M. (2007). Acid rain and its ecological consequences. *Journal of Environmental Biology, 29*(1), 15.

Singh, N., Poonia, T., Siwal, S. S., Srivastav, A. L., Sharma, H. K., & Mittal, S. K. (2022). Challenges of water contamination in urban areas. In *Current directions in water scarcity research* (Vol. 6, pp. 173–202). Elsevier. https://doi.org/10.1016/B978-0-323-91838-1.00008-7

Song, X. P., Hansen, M. C., Stehman, S. V., Potapov, P. V., Tyukavina, A., Vermote, E. F., & Townshend, J. R. (2018). Global land change from 1982 to 2016. *Nature, 560*(7720), 639–643.

Sun, Y., & Guo, Z. (2019). Recent advances of bioinspired functional materials with specific wettability: From nature and beyond nature. *Nanoscale Horizons, 4*(1), 52–76. https://doi.org/10.1039/C8NH00223A

Taylor, R. G., Scanlon, B., Döll, P., Rodell, M., Van Beek, R., Wada, Y., Longuevergne, L., Leblanc, M., Famiglietti, J. S., Edmunds, M., Konikow, L., Green, T. R., Chen, J., Taniguchi, M., Bierkens, M. F. P., MacDonald, A. M., Fan, Y., Maxwell, R. M., Yechieli, Y., … Treidel, H. (2013). Ground water and climate change. *Nature Climate Change, 3*, 322–329. https://doi.org/10.1038/nclimate1744

Theilmann, K. (2022). If groundwater tables drop, streams and rivers seep away and pollute drinking water. https://phys.org/news/2022-08-groundwatertables-streams-rivers-seep.html

Ullah, R., Malik, R. N., & Qadir, A. (2009). Assessment of groundwater contamination in an industrial city, Sialkot, Pakistan. *African Journal of Environmental Science and Technology, 3*(12).

Ummenhofer, C. C., & Meehl, G. A. (2017). Extreme weather and climate events with ecological relevance: A review. *Philosophical Transactions of the Royal Society b: Biological Sciences, 372*(1723), 20160135. https://doi.org/10.1098/rstb.2016.0135

United Nations, The United Nations World Water Development Report. (2022). *Groundwater: Making the invisible visible.* Paris: UNESCO.

UN-Water. (2017). https://www.unwater.org/sites/default/files/app/uploads/2017/05/SDG6_waterCycle_UN-Water_highres_web.pdf

der Gun, V. (2019). *The global groundwater revolution.* Oxford University Press. https://doi.org/10.1093/acrefore/9780199389414.013.632

Vystavna, Y., Diadin, D., Rossi, P. M., Gusyev, M., Hejzlar, J., Mehdizadeh, R., & Huneau, F. (2018). Quantification of water and sewage leakages from urban infrastructure into a shallow aquifer in East Ukraine. *Environmental Earth Sciences, 77*, 1–13.

Wakida, F. T., & Lerner, D. N. (2005). Non-agricultural sources of groundwater nitrate: a review and case study. *Water Research, 39*(1), 3–16. https://doi.org/10.1016/j.watres.2004.07.026

Walton, C. R., Zak, D., Audet, J., Petersen, R. J., Lange, J., Oehmke, C., et al. (2020). Wetland buffer zones for nitrogen and phosphorus retention: Impacts of soil type, hydrology and vegetation. *Science of the Total Environment, 727*, 138709. https://doi.org/10.1016/j.scitotenv.2020.138709

Werner, A. D., Sharp, H. K., Galvis, S. C., Post, V. E., & Sinclair, P. (2017). Hydrogeology and management of freshwater lenses on atoll islands: Review of current knowledge and research needs. *Journal of Hydrology, 551*, 819–844. https://doi.org/10.1016/j.jhydrol.2017.02.047

Wu, W. Y., Lo, M. H., Wada, Y., Famiglietti, J. S., Reager, J. T., Yeh, P. J. F., et al. (2020). Divergent effects of climate change on future groundwater availability in key mid-latitude aquifers. *Nature Communications, 11*(1), 3710.

Yu, S., Chae, G., Kim, S. W., Kim, H. R., Kim, C. Y., Seok, G., & Yun, S. T. (2024). Estimating annual CO_2 emissions and building a leakage scenario in a tectonically quiescent area. *Journal of Geochemical Exploration, 257*, 107371. https://doi.org/10.1016/j.gexplo.2023.107371

Zainab, S. M., Junaid, M., Xu, N., & Malik, R. N. (2020). Antibiotics and antibiotic resistant genes (ARGs) in groundwater: A global review on dissemination, sources, interactions, environmental and human health risks. *Water Research, 187*, 116455. https://doi.org/10.1016/j.watres.2020.116455

Chapter 23
Human Health Risk Assessment of Fluoride and Nitrate Using Pollution Index of Groundwater and Water Quality Index in Parts of Semi-arid Regions, Anantapur District, A.P. India

P. Ravi Kumar, S. Srinivasa Gowd, C. Krupavathi, and B. N. Anusha

Abstract Groundwater is the primary source of potable water as well as agriculture worldwide. In semi-arid regions, human activities and natural processes have caused an increase in fluoride and nitrate concentrations in most groundwater reservoirs in recent years. The primary goal of the research is to investigate the health impacts of fluoride and nitrate pollution on humans in semi-arid zones of Anantapur district, A.P., India. Thirty groundwater samples were obtained from bore and dug wells and appraised for physicochemical characteristics. The analysis found that 10% of the samples were above the satisfactory F^- level (< 1.5), while 100% of the samples surpassed the allowable NO_3^- limit (< 45). This suggests drinking groundwater with high F^- and NO_3^- levels could harm health. Based on the WQI values, which ranged from 71.03 to 171 mg/L, only 20% of the groundwater samples are safe to drink, while the other 80% are unsafe owing to groundwater abuse, direct sewage release, and agricultural influence. Based on PIG indicated that 23% of the samples were in the low contamination category, 47% were in the moderate pollution category, and 30% were in the high contamination category. The HQ and THI were calculated using USEPA, Surface water standards and guidance, Water quality standards handbook, 2014 guidelines to evaluate potential health hazards. The non-carcinogenic health danger results unveiled that the hazard was greater in infants and children than in adults in the study region, ranging from 1.4 to 3.31 for adults, 1.4 to 3.33 for children, and 1.42 to 3.36 for infants. Hence, the primary contributor to health risks in the research area may be water consumption with high fluoride and nitrate concentrations. Furthermore, according to the THI, all samples are above the permitted level (< 1) for adults, children, and infants. Therefore, filters for contaminants and rainwater harvesting are recommended to decrease the area's health risk.

P. R. Kumar · S. S. Gowd (✉) · C. Krupavathi · B. N. Anusha
Department of Geology, Yogi Vemana University, Kadapa, AP, India
e-mail: ssgowd@gmail.com

V. Sunitha et al. (eds.), *Groundwater Resource Management Planning Strategies*, Springer Water, https://doi.org/10.1007/978-3-031-88870-0_23

Keywords Hydrogeochemistry · Spatial distribution maps · Anantapur district · Health risks · Total hazard index · Hazard quotient · India

23.1 Introduction

Groundwater usage for drinking and agriculture purposes is increasing significantly in semi-arid regions worldwide. According to the findings, almost 150 million people globally depend on groundwater for consumption and farming (Adimalla & Qian, 2023; Akhtar et al., 2021). A little over 45% of all the water used for irrigation on 39 million hectares of agricultural land in India is used for groundwater, which serves approximately 80% of home needs. Groundwater is used for domestic purposes by 90% of rural residents, both in urban and rural regions, for drinking water (Balamurugan et al., 2020; Anusha et al., 2022; Gowd, 2005). Fresh water is becoming increasingly scarce and polluted as a result of misuse, and ongoing irrigation methods run the risk of polluting groundwater (Aravinthasamy et al., 2020; Sunitha & Reddy, 2022). Numerous factors can cause groundwater contamination, including natural or man-made nitrogen contaminants, chemicals used on farms, and soil and rock weathering or dissolution (Madhav et al., 2020; Raju et al., 2023). The purity of the aquifer is a key factor in determining whether it can be used for a certain purpose, and water intended for human consumption must be free of dangerous substances, high mineral content, and living and non-living organisms that might harm health (Gnanachandrasamy et al., 2020; Khan et al., 2021). Toxic metal and metalloid contamination pose a significant health risk in the semi-arid regions of Andhra Pradesh. Groundwater quality is slowly declining as a result of human and natural activities like farming and industry, which have several negative effects on the environment's health. The purity of water is strongly related to public health; hence, regular monitoring is crucial. Fluoride and nitrate contamination of groundwater has been a significant public health problem globally over the past 20 years (Ahmed et al., 2020). This research has been conducted around the world in semi-arid regions due to its significance (Etikala et al., 2019; Singh et al., 2020; Sunitha, 2020). Fluoride concentration in water depends on factors such as water accessibility to fluoride-containing minerals, rock-water interaction, evaporation, and fluoride ion composition. Mixed fluoride hydroxide complexes with boron also play a role. Approximately 90% of fluoride ingested through water is absorbed by the digestive system, with over 200 million people worldwide consuming fluoride-contaminated water. Fluoride ions within recommended levels can have dental and bone health benefits, but excessive consumption from drinking water can lead to adverse effects, including dental fluorosis, skeletal fluorosis, thyroid dysfunction, fertility issues, neurotoxicity, cancer, and genotoxicity (Kayastha et al., 2022; Pappaka et al., 2024; Etikala et al., 2024; Yadav et al., 2019). F^- content has a considerable impact on health in numerous countries. More than 23 nations throughout the world have higher fluoride levels, mainly in portions of India, China, Central Africa, and South America. Fluoride-related water purity investigations have been conducted in semi-arid regions around the world (Prasad et al., 2019; Golla and Badapalli, 2022). Nitrate is often regarded as one of the most severe chemical contaminants in groundwater around the world. Nitrate levels in groundwater are increasing in India. Nitrate contamination in groundwater stems from various sources and origins (Pandey et al., 2023). Elevated nitrate concentrations in groundwater are primarily linked to human activities, including domestic sewage leaks, poultry waste disposal, the application of animal manure, and the migration of nitrates from sewage, municipal, and industrial waste. In the last 30 years, the nitrate concentration in groundwater has risen, leading to high nitrate levels in drinking water. This has been associated with stomach cancer in adults and methemoglobinemia in children. (Nawale et al., 2021; Kumar et al., 2023). Agriculture is the primary source of this contamination. Groundwater contaminated with toxic fluoride and nitrate can be harmful to crops and people. A simple and effective tool for decision-making on groundwater excellence may be provided by combining WQI, and GIS. The pollution index of groundwater (PIG) is a helpful instrument for determining if a region's water supply is suitable for human consumption and for sharing information about the water's overall quality A scientific method also referred to as WQI can condense an excessive quantity of data on water superiority into a tiny amount demonstrating the degree of excellence. The WQI has been utilized by several researchers in the past to measure the quality of groundwater (Kumar et al., 2024) Using GIS, it is possible to see the geographical distribution of several hydrochemical parameters. Numerous publications have already attested to its efficacy. Groundwater quality decline is a significant global issue as well as a major worry in industrialized nations like China and India (Reddy et al., 2019). Because pesticides and fertilizers have been present in groundwater for some time, the quality has been steadily declining, raising concerns about the health risks to people (Panneerselvam et al., 2021). The current study intends to investigate the potential strength concerns caused by excessive quantities of fluoride and nitrate in aquifers in the Anantapur District. The study aims to aid in implementing suitable preventive measures to alleviate the risks associated with these contaminants.

The current study intends to investigate the potential health concerns caused by excessive quantities of fluoride and nitrate in aquifers in the Anantapur District, A.P., India. This semi-arid region has been increasingly affected by human activities and natural processes leading to elevated concentrations of these contaminants in groundwater. The selection of Anantapur District as the study area is critical due to its reliance on groundwater for both potable water and agriculture, and the escalating risks posed by groundwater contamination. The study aims to aid in implementing suitable preventive measures to alleviate the risks associated with these contaminants.

23.2 Study Area

The investigated region lies in the Anantapur district of Andhra Pradesh, India, and encompasses an area of 269 km^2. This study area geographically lies between 14°45′00″ and 15°00′00″ N latitudes and 77°35′00″ and 77°50′00″ E longitudes. It has an average elevation of 516 m (1693 feet). Temperatures in the region range from 24 °C to 46 °C, and there is an average annual rainfall of 544 mm. The research region falls under the Survey of India Toposheets 57F/6, F/10, and F/11. The primary rocks observed in the research region are hornblende biotite gneiss, metabasalt, close pet granites, Gulchervu quartzites/arkose with conglomerate, Vempalle dolomites chert, mudstone, and Tadipatri shales. Groundwater levels in the region typically vary between 10 and 30 m below ground level (bgl). The depth of the weathered zone ranges from 5 to 15 m, while the depth of the fractured zone varies from 20 to 50 m. Groundwater flow generally follows the topography, moving from elevated regions towards lower-lying areas, indicating a southward to southeastwardly flow direction. This region is predominantly dependent on groundwater for both potable water and agriculture. Seasonal fluctuations and over-extraction have led to a decline in groundwater levels, raising concerns about sustainability. Figure 23.1 depicts a sample site map of the research region, illustrating the locations of the sampling points and the hydrogeological features of the area.

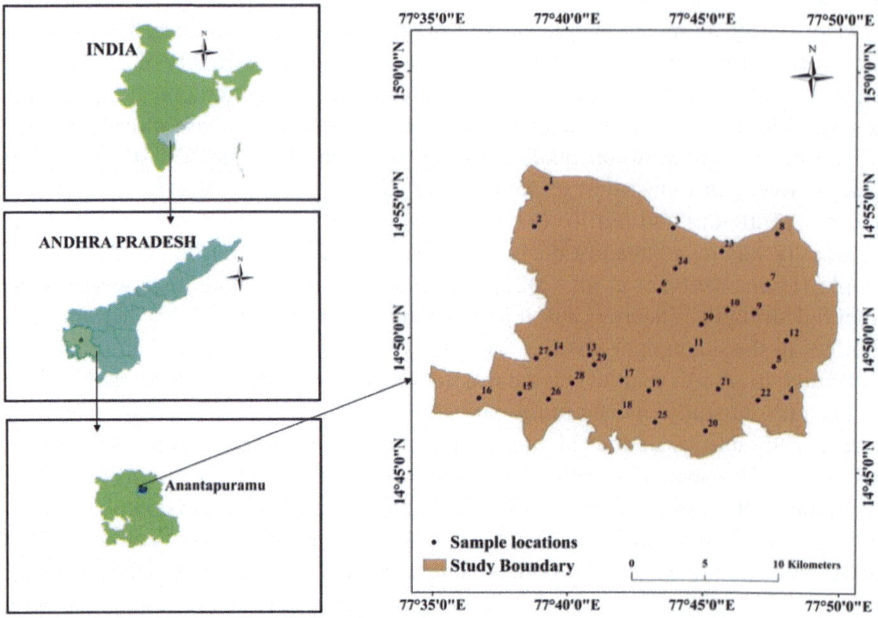

Fig. 23.1 Sample location map of the investigated region

23.3 Materials and Methods

During the post-monsoon season in October 2023, thirty groundwater samples were attained from bore wells and dug wells in several sites within the research zone. The laboratory analysis of the groundwater samples, which were taken in well-cleaned one-litre polyethylene bottles. The latitude and longitude of the sample locations are identified with a global positioning system. pH and EC are determined by a pH meter and a conductivity meter. A TDS meter is used to estimate TDS and a titration method is used to measure TH, Ca^{2+}, Mg^{2+}, HCO_3^-, and Cl^-. Flame photometry is used to measure Na^+ and K^+, spectrophotometry for SO_4^{2-} and NO_3^-, and an ion-selective electrode for F^-. The spatial distribution maps are created using the interpolation method of the Arc GIS tool Inverse Distance Weight (IDW) to appraise groundwater superiority (Etikala et al., 2021).

23.3.1 *Wqi*

WQI is a significant tool for measuring the condition of aquifers and their suitability for drinking. It is a classification method that evaluates the overall acceptance of water for use in drinking by combining the effects of several different water quality guidelines (Kumari & Rai, 2020; Krupavathi et al., 2022; Chinthala et al., 2023). WQI is determined using WHO 2011 drinking standards. The WQI is calculated in three steps. In the initial phase, each of the (13) physicochemical constraints were allotted a weight (wi) based on their significance to the overall condition.

Because of their significance in assessing the purity of water, the parameters NO_3^-, TDS, Cl^-, F^-, and SO_4^{2-} have been owed a maximum weight of 5 (Kumar et al., 2022). Bicarbonate is allocated a lowest weight of 1 due to its small impact on water quality testing. Other factors such as Ca^{2+}, Mg^{2+}, Na^+, and K^+ were assigned weights ranging from 1 to 5 to indicate their respective relevance in defining the water's superiority. The relative weight (Wi) is obtained in the following stage using the following equation:

$$Wi = \frac{wi}{\sum_{i=1}^{n}(\mathrm{wi})} \tag{23.1}$$

The "quality rating (qi)" for the third stage is computed using the formula below.

$$qi = \left(\frac{ci}{si}\right) \times 100 \tag{23.2}$$

If Ci is the level of concentration of all characteristics in each water sample, Si denotes the WHO 2011 recommended value for that characteristic. Because Wi and qi were used to estimate the SIi for each characteristic individually, the following equations will calculate WQI.

$$SIi = wi \times qi \quad (23.3)$$

$$WQI = \sum_{i=1}^{n} SIi \quad (23.4)$$

The sub-index of each parameter is here designated as SIi.

23.3.2 Pig

Subba Rao initiated the PIG in early 2012. It is commonly used to investigate variations in the purity of groundwater caused by geochemical variables. In five phases, the PIG is calculated. The first step is giving each chemical attribute a proportionate weight (RW). Depending on how great an impact it has on people's health, RW frequently ranges from 1 to 5. The weights assigned to Cl^-, SO_4^{2-}, and NO_3^- were the greatest (5), while K^+ and HCO_3^- were given the lowest weights (1). In the second stage, each chemical variable's weight parameter (WP) is computed to determine its equivalent impact on groundwater quality. The PIG is calculated using the formulae (23.5, 23.6, 23.7, and 23.8):

$$Wp = \frac{Rw}{\Sigma Rw} \quad (23.5)$$

$$Sc = \frac{C}{Ds} \quad (23.6)$$

$$Ow = Wp * Sc \quad (23.7)$$

$$PIG = \Sigma Ow \quad (23.8)$$

PIG will be assessed taking into account the relative chemical input of each groundwater sample. Water exceeding a total chemical excellence (Ow) of 0.1 constitutes 10% of the PIG's 1.0-point value. It is simple to understand the effects of groundwater body contamination after reading this explanation. Aquifer systems gradually becoming contaminated by airborne particles are described using additional chemical variable intensities.

23.3.3 Health Risk Evaluation

The United States Environmental Protection Agency (USEPA, 2014) created a unique model for human HRA, which has been widely utilized by many other investigators (Yenugu et al., 2020; Sunitha et al., 2022; Karunanidhi et al., 2021). Human health issues have worsened in India over the previous decade as a result of polluted drinking water, air pollution, an unprepared sewage system, a lack of periodic precipitation, and rising temperatures in arid and semiarid regions (Pasupuleti et al., 2022). As a result, studying the HRA is critical, as it offers a vital resource for groundwater environment conservation and management. Human health reacts quickly to water and air, which is the primary cause of a health danger. As a result, the current study focuses on water consumption as the primary channel. Specifically, NO_3^- and F^- are employed as HRA and noncarcinogenic risk parameters following Eqs. (23.13, 23.14, 23.15):

$$ADD = (CPW \times IR \times ED \times EF)/(ABW \times AET) \tag{23.9}$$

$$HQ = ADD/RfD \tag{23.10}$$

$$THI = HQ_{Fluoride} + HQ_{Nitrate} \tag{23.11}$$

where ADD is the average daily dose (mg/kg/d), and CPW is the attentiveness of pollutants for groundwater (mg/L). RfD is the reference dosage via F^- (RfD_{F^-}: 0.04) and NO_3^- (RfD_{NO3^-}:1.6) pollutants (mg/kg/day).Table 23.1 displays the definitions and reference values for additional variables.

23.4 Results and Discussion

Results of physicochemical parameters of the study area for post-monsoon and the statistical summary of physicochemical parameters and ion concentrations have been compared with the WHO, 2011 is shown in Table 23.2.

Table 23.1 Important factors for calculating the health risks associated with drinking

Parameters	Adults	Childrens	Infants	Units
Ingestion rate (IR)	2.5	0.78	0.3	L/d
Exposure duration (ED)	65	20	1	a (age)
Exposure frequency (EF)	365	365	365	d (days)
Average body weight (ABW)	65	20	10	kg
Average exposure time (AET)	23,360	4380	365	d (days)

Table 23.2 Physico-chemical constraint concentrations and comparisons with WHO (2011) norms

Parameter	Minimum	Maximum	Average	WHO (2011) Desirable limit-permissible limit	Number of samples exceeding the WHO, 2011
pH	7.1	8.5	7.6	6.5–8.5	Nil
EC (μS/cm)	726	3201	1725.1	1500	17
TDS (mg/L)	515	2144	1147.5	500–1500	7
Ca^{2+}(mg/L)	28	153	80.6	75	16
Mg^{2+}(mg/L)	20	124	57.3	30–150	Nil
Na^{+}(mg/L)	56	327	138.4	50–200	1
K^{+}(mg/L)	2	9	4.3	10	Nil
HCO_3^-(mg/L)	183.5	687	427.3	300–600	3
F− (mg/L)	0.1	2.3	0.9	0.5–1.5	3
Cl^- (mg/L)	135	442	262.8	250–600	Nil
SO_4^{2-} (mg/L)	24	68	38.3	200–600	Nil
NO_3^- (mg/L)	48	112	81.3	45	30
TH (mg/L)	152	749	436.6	100–500	11

The pH determined the hydrogen ion [H^+] concentrations in the study region. The largest key indicator of the quality of groundwater is pH, which also contributes significantly to estimations of other types of geochemical stability or solubility. The research area's pH ranges from 7.1 to 8.5 (Table 23.2). In many other applications, including those in the sciences of life and medicine, pH levels are crucial. The pH levels of all groundwater samples from the research area are within the desired limit. All samples specify a slightly alkaline nature. Electrical conductivity is a crucial factor for measuring the quality of water. According to WHO (2011), 1500 μS/cm is the ideal limit for the EC in groundwater. The EC levels in this study location ranged from 726 to 3201 μS/cm, and the mean is 1725.1 μS/cm (Table 23.2). Each sample has an EC value that is too high. A high ionic concentration may be the cause of the electrical conductivity's highest value. The total dissolved solids content at the research site ranges between 515 and 2144 mg/L, and the mean is 1147.5 mg/L. Drinking water should contain no more than 1500 ppm (WHO, 2011) of total dissolved solids. Calcium is an essential nutrient for human health growth and preservation. These nutritional components are most closely linked to the formation of robust bones and teeth, as well as being necessary for cardiovascular performance (Chauhan, 2022). The bulk of the samples were within the allowed values, with only a few samples exceeding the WHO (2011) standards. This is due to the rate of dissolution of the feldspar group of minerals. Drinking water containing calcium should not exceed 75 mg/L. Ca^{2+} concentrations fluctuated between 28 and 153 mg/L, and the mean is 80.6 mg/L (Table 23.2). Mg^{2+} is the most dominant element in the Earth's crust. Ferromagnesium minerals and drainage wastes are the main causes of magnesium in natural water. Magnesium is the most abundant cation in pre-monsoon groundwater samples. Magnesium levels fluctuated between 20 and 124 mg/L, and the mean is 57.3 mg/L (Table 23.2). All the analyzed samples were found to be within the permitted level, which is 30–150 mg/L. according to the WHO (2011) standard for drinking water. Sodium is a plentiful element that originates in earth, vegetation, and water all around the world. Because sodium salts are very transparent, sodium ions can be found practically anywhere in water. The sodium content varied between 56 and 327 mg/L, and the mean is 138.4 mg/L (Table 23.2). In general, groundwater has a lower potassium concentration. K^+ content fluctuated between 2 and 9 mg/L, and the mean is 4.3 mg/L. Bicarbonate represents the overall alkalinity in water. It assesses its ability to neutralize. The presence of dissolved carbon species, particularly bicarbonate and carbonate, causes most natural waters to be alkaline. The research area's bicarbonate levels ranged from 183.5 mg/L to 687 mg/L. Fluoride is an indispensable element for a healthy life in humans and animals, yet long exposure to levels exceeding the permissible limits leads to fluorosis. In contrast, its shortage may have a harmful influence on the growth of teeth. Leaching of and dissolution of fluorine in groundwater during weathering would be the possible source for high fluoride concentrations in Indian waters. Fluoride concentrations in this investigation area fluctuate between 0.1 and 2.3 mg/L, and the mean is 0.9 mg/L (Table 23.2). Chloride in groundwater is present in the form of NaCl, however, due to base exchange phenomena, chloride may exceed the sodium, as well as domestic sewage and phosphate mineral weathering. In the research region Cl^- content fluctuates between 63.9 and 442 mg/L, and the mean is 262.8 mg/L. Sulfate is abundant in both natural and human-made water systems. The principal natural sources of sulfate are atmospheric precipitation, sulfate mineral breakdown, and sulfide mineral oxidation (Sunitha, 2020). The main sources of SO_4^{2-} in hard rock areas are gypsum and anhydrite. Higher SO_4^{2-} content is due to the effect of sulfur fumes from the industry as a pollutant inflowing into groundwater. A laxative effect or gastrointestinal irritation may occur at concentrations higher than 600 mg/l, particularly if the magnesium level is very high. Sulfate concentrations varied from 24 to 68 mg/L, with a mean value of 38.3 mg/L (Table 23.2). Nitrates are a vital supply of nitrogen (N) for vegetation. Groundwater nitrate contamination can also be caused by anthropogenic and animal wastes. The freshwater aquatic ecology will be affected by nitrate concentrations over 45 mg/L (WHO, 2011). In the research region, NO_3^- content varied from 48 to 112 mg/L, and the mean is 81.3 mg/L (Table 23.2). In the research region, TH ranges from 152 mg/L to 749 mg/L, and the mean is 436.6 mg/L. Figure 23.2 shows the spatial distribution maps of pH (a), EC (b), TDS (c), Ca^{2+}(d), Mg^{2+} (e), Na^+ (f), K^+ (g), HCO_3^- (h), F^- (i), Cl^- (j), SO_4^{2-} (k), NO_3^- (l), TH (m).

23.4.1 Wqi

The WQI values are categorized as follows: if it is less than 50, the groundwater is considered excellent; if it falls between 50.1 and 100, it is regarded as good quality

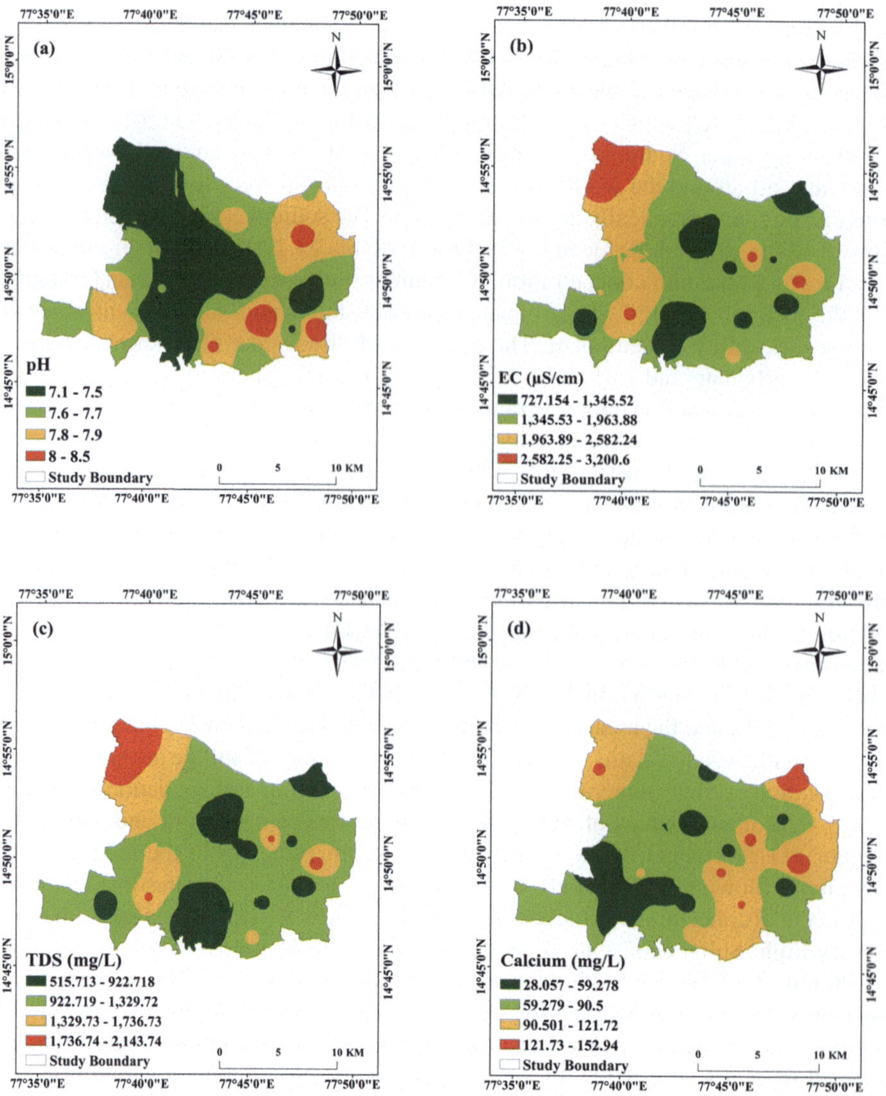

Fig. 23.2 Spatial distribution maps of pH **a**, EC **b**, TDS **c**, Ca^{2+} **d**, Mg^{2+} **e**, Na^{+} **f**, K^{+} **g**, HCO_3^{-} **h**, F^{-} **i**, Cl^{-} **j**, SO_4^{2-} **k**, NO_3^{-} **l**, TH **m**

water; if the values range from 100.1 to 200, the groundwater is classified as poor; if it falls between 200.1 and 300, it is labelled as very poor; and if the values exceed 300, the water is deemed unfit for drinking. Regarding the WQI classification, 20% of groundwater samples are good, while the other 80% are poor, owing to groundwater abuse, direct sewage release, and agricultural influence (Table 23.3). Table 23.3

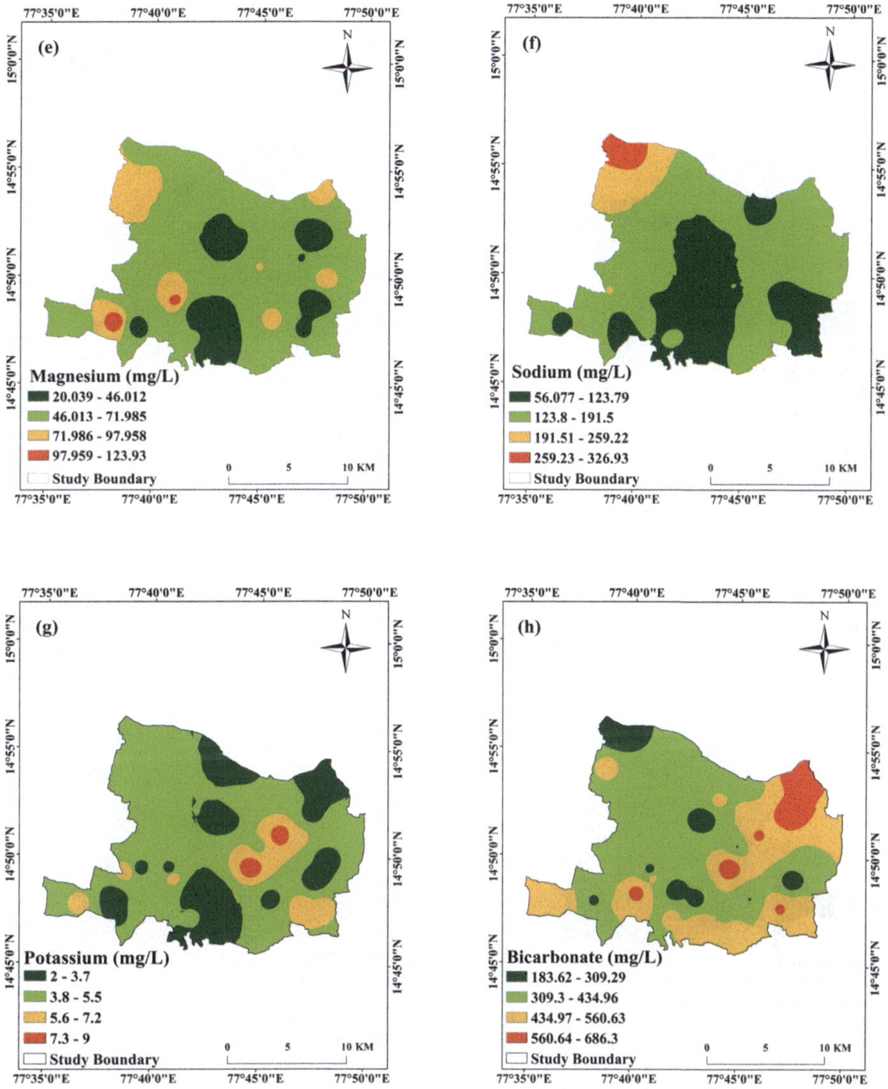

Fig. 23.2 (continued)

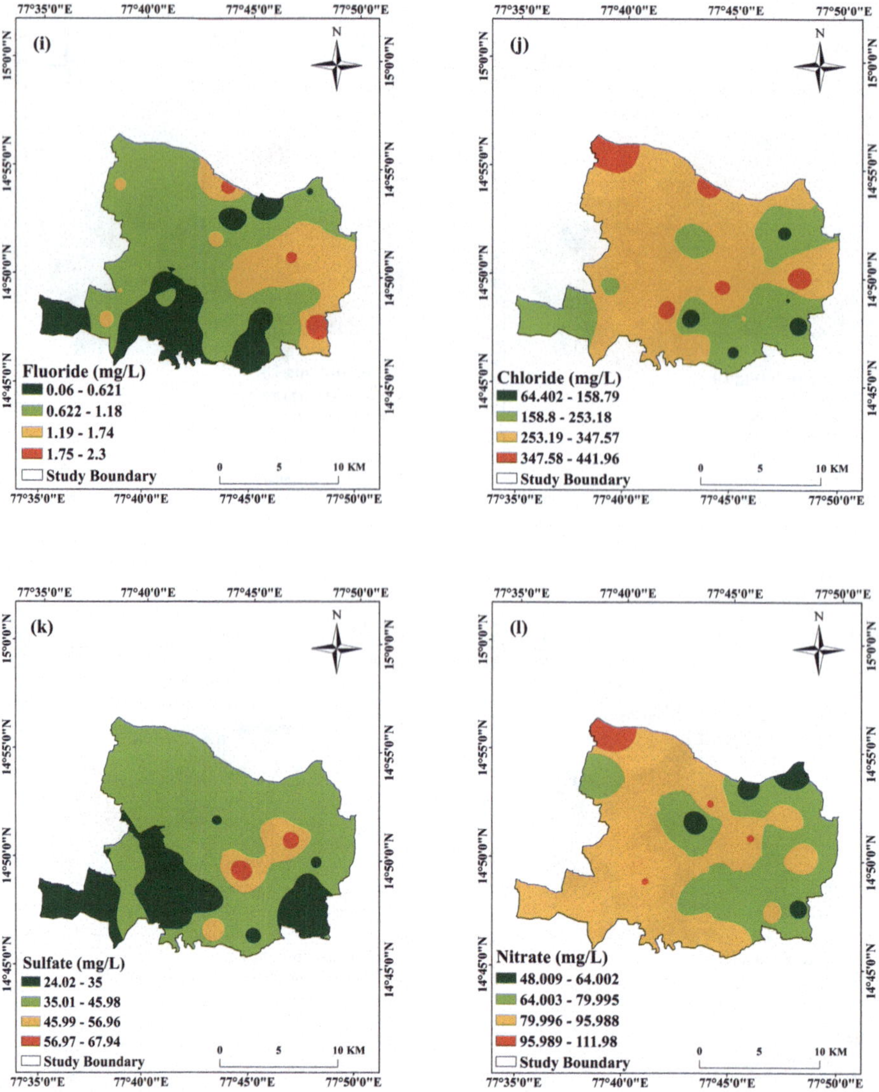

Fig. 23.2 (continued)

Fig. 23.2 (continued)

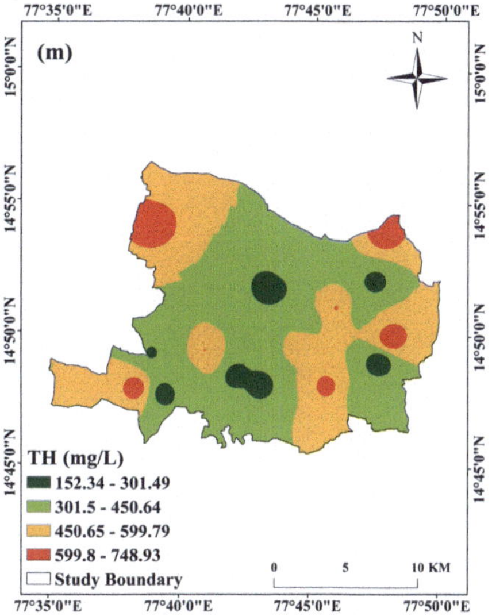

Table 23.3 Water quality categorization based on WQI value (Pappaka et al., 2024)

Quality of water	WQI	Percentage of water samples
Excellent	< 50	–
Good	50–100	20%
Poor	100–200	80%
Very poor	200–300	–
Unsuitable	> 300	–

shows the WQI at specific sampling locations. Figure 23.3 depicts a map of the spatial dispersion of WQI.

23.4.2 *Pig*

PIG is an extremely helpful method that has been used in various studies to assess water quality. PIG levels vary between 0.83 mg/L and 1.74 mg/L, and the mean is 1.27 mg/L (Fig. 23.4). Based on the PIG's categorization, 13% of the groundwater samples were taken in an area with insignificant pollution, while 67% and 20% of the samples were taken in areas with low and moderate pollution, respectively (Table 23.4).

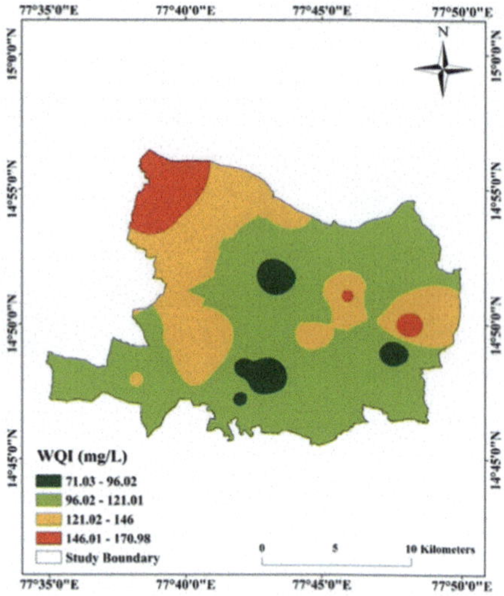

Fig. 23.3 Spatial distribution map of WQI

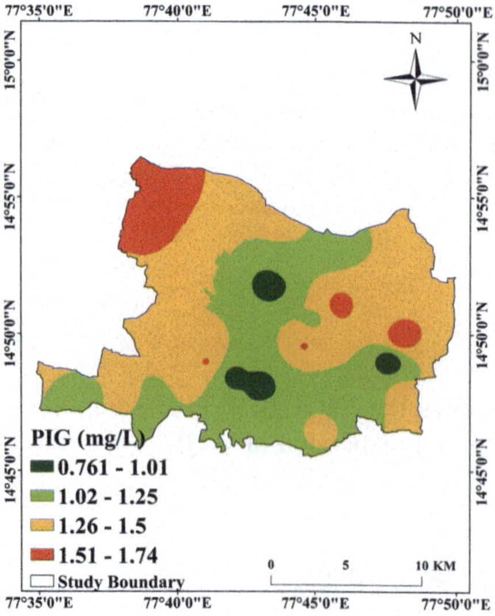

Fig. 23.4 Spatial distribution map of PIG

Table 23.4 The categorization of PIG (Sunitha et al., 2022)

Range of PIG (mg/L)	Pollution class	No. of samples (%)
< 1.0 (mg/L)	Insignificant pollution	13
1.0–1.5 (mg/L)	Low pollution	67
1.5–2.0 (mg/L)	Moderate pollution	20
2.0–2.5 (mg/L)	High pollution	–
> 2.5 (mg/L)	Very high pollution	–

23.4.3 Evaluation of HRA

The HRA approach is the most effective method for assessing non-carcinogenic hazards to health across different ages. Table 23.5 shows the computed findings of non-carcinogenic dangers for adults, children, and infants in the research region based on drinking water use.

The HQ Nitrate for adults, children, and infants ranged from 1.15 to 2.73, 1.17 to 4.55, and 0.9 to 2.1. Adults, children, and infants had average values of 1.96, 2.04, and 1.53, respectively (Table 23.5). Similarly, HQ Fluoride for adults, children, and infants ranged from 0.06 to 2.21, 0.06 to 2.24, and 0.05 to 1.73, and the mean is 0.85, 0.88, and 0.66 respectively (Table 23.5). Table 23.5 shows the THI, which is the sum of HQF and HQN. THI results varied from 1.43 to 3.92 in adults, 1.45 to 5.93 in children, and 1.12 to 3.06 in infants. USEPA recommendations indicate that THI limits for the noncarcinogenic threat to human health should not be higher than 1. If the THI value is more than one, there is no danger to human wellness. The investigation found that all groundwater samples above the THI (< 1) maximum permissible limit for adults, children, and infants (Table 23.5). Such areas shouldn't utilize groundwater for drinking or cooking. By enhancing sanitary and drainage systems and putting in place secure residential and business waste disposal systems, groundwater quality may be increased.

According to the findings, the THI values for adults, children, and infants in this area vary from 1.4 to 3.92, 1.4 to 5.93, and 1.12 to 3.06, respectively (Table 23.5). According to USEPA recommendations, the THI for adults, children, and infants in 100%, of the samples is above the permitted level (> 1) and may pose health hazards (Table 23.5). Figure 23.5 shows a spatial distribution map of the THI group of (a) adults, (b) children, and (c) infants.

23.5 Conclusion

The semi-arid region dominated by hard rocks relies mainly on groundwater for drinking purposes. As a result, the current study focused on groundwater quality for drinking. The groundwater excellence was then confirmed as acceptable for drinking purposes by comparing parameter concentrations to WHO 2011 criteria. Moreover, the following findings are reached:

The research region's pH levels in the groundwater range from neutral to alkaline. In the study region, NO_3^- and F^- levels vary between 48 and 112 mg/L and 0.1 to 2.3 mg/L, consequently. In the research region, 100% and 10% of groundwater samples exceed the permissible thresholds of 45 mg/L and 1.5 mg/L of NO_3^- and F^-, respectively, making them unsafe to drink.

The main factors contributing to the high nitrate content in the groundwater may be intensive agricultural activities and fertilizers. The geogenic source of high fluoride

Table 23.5 The study of non-carcinogenic health risks associated with potable water use in the investigation region

S. No.	Adults		Children's		Infants		THI		
	HQF	HQN	HQF	HQN	HQF	HQN	Adults	Children's	Infants
1	0.83	2.73	1.38	4.55	0.64	2.10	3.56	5.93	2.74
2	1.15	1.56	1.17	1.58	0.90	1.22	2.72	2.75	2.12
3	1.83	2.09	1.85	2.12	1.43	1.63	3.92	3.97	3.06
4	2.21	1.39	2.24	1.41	1.73	1.09	3.61	3.66	2.81
5	0.87	1.63	0.88	1.66	0.68	1.28	2.50	2.54	1.95
6	1.25	1.15	1.27	1.17	0.98	0.90	2.40	2.44	1.88
7	1.25	2.12	1.27	2.15	0.98	1.65	3.37	3.41	2.63
8	0.59	1.30	0.59	1.32	0.46	1.01	1.88	1.91	1.47
9	1.81	1.56	1.83	1.58	1.41	1.22	3.37	3.42	2.63
10	1.15	2.36	1.17	2.39	0.90	1.84	3.51	3.56	2.74
11	1.38	1.88	1.39	1.90	1.07	1.46	3.25	3.30	2.54
12	1.15	2.14	1.17	2.17	0.90	1.67	3.29	3.34	2.57
13	0.38	2.09	0.38	2.12	0.29	1.63	2.47	2.50	1.92
14	1.04	2.31	1.05	2.34	0.81	1.80	3.35	3.39	2.61
15	1.39	2.09	1.41	2.12	1.09	1.63	3.49	3.53	2.72
16	0.07	2.24	0.07	2.27	0.05	1.74	2.30	2.34	1.80
17	0.06	2.14	0.06	2.17	0.05	1.67	2.20	2.23	1.71
18	0.08	2.02	0.08	2.05	0.06	1.58	2.10	2.13	1.64
19	0.96	1.68	0.98	1.71	0.75	1.31	2.64	2.68	2.06
20	0.31	2.04	0.31	2.07	0.24	1.59	2.35	2.38	1.83
21	0.32	1.56	0.32	1.58	0.25	1.22	1.88	1.91	1.47
22	0.85	2.07	0.86	2.10	0.66	1.61	2.91	2.95	2.27
23	0.09	1.35	0.09	1.37	0.07	1.05	1.43	1.45	1.12
24	0.09	2.36	0.09	2.39	0.07	1.84	2.44	2.48	1.91
25	0.75	2.07	0.76	2.10	0.59	1.61	2.82	2.86	2.20
26	0.08	2.02	0.08	2.05	0.06	1.58	2.10	2.13	1.64
27	1.15	2.04	1.17	2.07	0.90	1.59	3.20	3.24	2.49
28	0.17	2.09	0.18	2.12	0.14	1.63	2.26	2.30	1.77
29	0.89	2.36	0.91	2.39	0.70	1.84	3.25	3.30	2.54
30	1.32	2.26	1.34	2.29	1.03	1.76	3.58	3.63	2.79
Min	0.06	1.15	0.06	1.17	0.05	0.90	1.43	1.45	1.12
Max	2.21	2.73	2.24	4.55	1.73	2.10	3.92	5.93	3.06
Avg	0.85	1.96	0.88	2.04	0.66	1.53	2.80	2.92	2.19

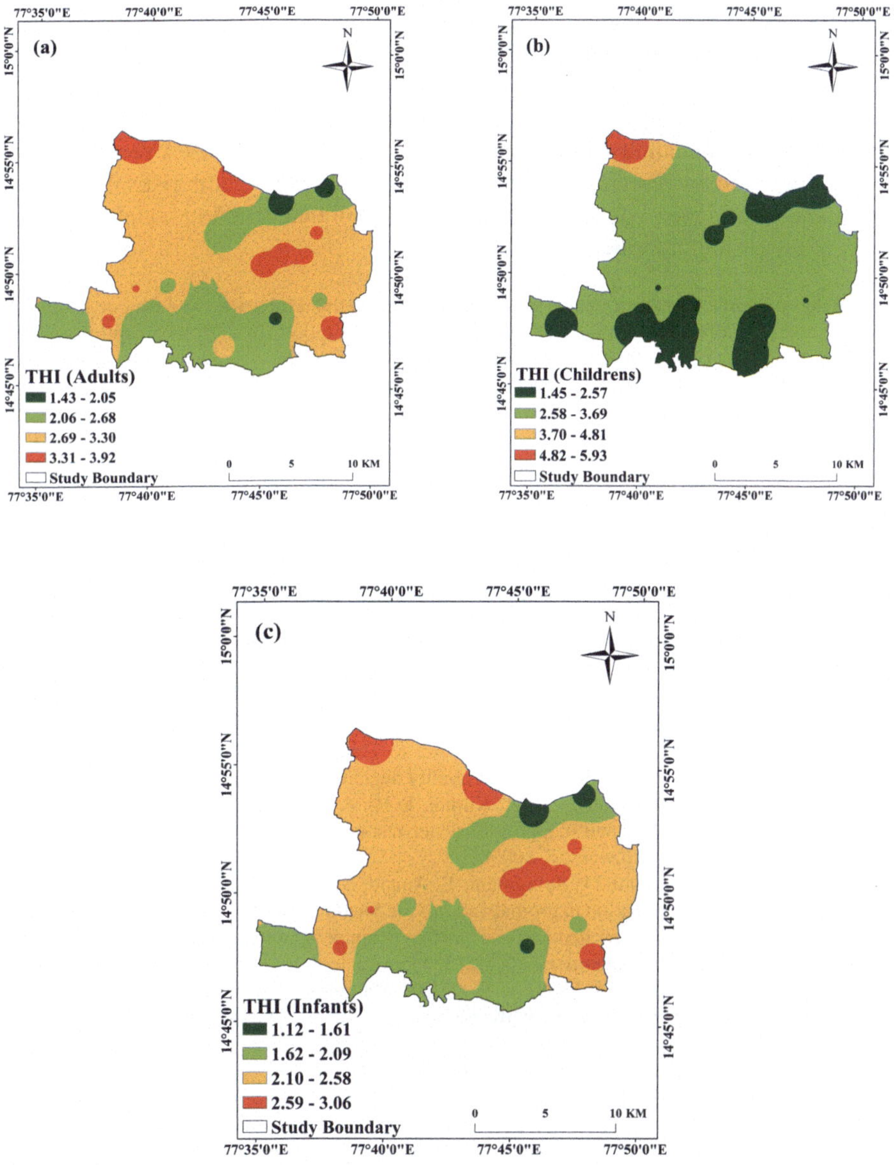

Fig. 23.5 a, b, & c. THI of adults, children and infants

levels is the interaction of rock water with minerals that contain fluorine, which is present in granites.

Regarding PIG indicated that 23% of the samples were in the low contamination category, 47% were in the moderate contamination category, and 30% were in the high contamination category. Based on the WQI values, which ranged from 71.03 to

171 mg/L, only 20% of the groundwater samples are safe to drink, while the other 80% are unsafe owing to groundwater abuse, direct sewage release, and agricultural influence.

The THI levels for adults, children, and infants in all samples exceed the permissible limit (< 1), potentially posing health risks, according to USEPA recommendations. Children in the research region are more vulnerable to health risks than adults and infants due to their consumption of severely polluted drinking water with high levels of nitrate and fluoride.

Acknowledgements The first author P. Ravi Kumar is greatly thankful to the Department of Science and Technology (DST), Government of India, for financial support through the Inspire program (Sanction order No. DST/INSPIRE Fellowship/2018/IF180877) and also thankful to the Department of Geology, Yogi Vemana University, for necessary facilities to carryout research work.

References

Adimalla, N., & Qian, H. (2023). Evaluation of non-carcinogenic causing health risks (NCHR) associated with exposure of fluoride and nitrate contaminated groundwater from a semi-arid region of south India. *Environmental Science and Pollution Research, 30*(34), 81370–81385.

Ahmed, T., Zounemat-Kermani, M., & Scholz, M. (2020). Climate change, water quality and water-related challenges: A review with focus on Pakistan. *International Journal of Environmental Research and Public Health, 17*(22), 8518.

Akhtar, N., Syakir Ishak, M. I., Bhawani, S. A., & Umar, K. (2021). Various natural and anthropogenic factors responsible for water quality degradation: A review. *Water, 13*(19), 2660.

Anusha, B. N., Babu, K. R., Kumar, B. P., Kumar, P. R., & Rajasekhar, M. (2022). Geospatial approaches for monitoring and mapping of water resources in semi-arid regions of Southern India. *Environmental Challenges, 8*, 100569.

Aravinthasamy, P., Karunanidhi, D., Subramani, T., Anand, B., Roy, P. D., & Srinivasamoorthy, K. (2020). Fluoride contamination in groundwater of the Shanmuganadhi River basin (south India) and its association with other chemical constituents using geographical information system and multivariate statistics. *Geochemistry, 80*(4), 125555.

Balamurugan, P., Kumar, P. S., & Shankar, K. (2020). Dataset on the suitability of groundwater for drinking and irrigation purposes in the Sarabanga River region, Tamil Nadu India. *Data in Brief, 29*, 105255.

Chauhan, R. C. (2022). Calcium as a boon or bane for athlete: A review. *Asian Journal of Research in Marketing, 11*(1), 1–8.

Chinthala, K., Somagouni, S. G., Pappaka, R. K., & Gudala, H. V. (2023). Ground water quality assessment using water quality index and geographical information system of Mogamureru river Basin, YSR Kadapa district, Andhra Pradesh, India. *Emerging technologies for water supply, conservation and management* (pp. 291–313). Springer International Publishing.

Etikala, B., Golla, V., Adimalla, N., & Marapatla, S. (2019). Factors controlling groundwater chemistry of Renigunta area, Chittoor district, Andhra Pradesh, South India: A multivariate statistical approach. *HydroResearch, 1*, 57–62.

Etikala, B., Adimalla, N., Madhav, S., Somagouni, S.G., & Keshava Kiran Kumar, P.L. (2021). Salinity problems in groundwater and management strategies in arid and semi-arid regions. Groundwater Geochemistry: Pollution and Remediation Methods, 42–56.

Etikala, B., Vangala, S., & Madhav, S. (2024). Groundwater geochemistry using modified integrated water quality index (IWQI) and health indices with special emphasis on nitrates and heavy metals in southern parts of Tirupati South India. *Environmental Geochemistry and Health, 46*, 465.

Gnanachandrasamy, G., Dushiyanthan, C., Jeyavel Rajakumar, T., & Zhou, Y. (2020). Assessment of hydrogeochemical characteristics of groundwater in the lower Vellar river basin: Using geographical information system (GIS) and water quality index (WQI). *Environment, Development and Sustainability, 22*, 759–789.

Golla, V. S., & Badapalli, P. K. (2022). Evaluation of water quality for drinking and irrigation purposes and fluoride health hazard risk assessment (HHRA) in parts of semi-arid regions in the south-eastern part of India. *International Journal of Energy and Water Resources, 6*(4), 521–529.

Gowd, S. S. (2005). Assessment of groundwater quality for drinking and irrigation purposes: A case study of Peddavanka watershed, Anantapur district, Andhra Pradesh, India. *Environmental Geology, 48*, 702–712.

Karunanidhi, D., Aravinthasamy, P., Subramani, T., & Muthusankar, G. (2021). Revealing drinking water quality issues and possible health risks based on water quality index (WQI) method in the Shanmuganadhi river basin of South India. *Environmental Geochemistry and Health, 43*, 931–948.

Kayastha, V., Patel, J., Kathrani, N., Varjani, S., Bilal, M., Show, P. L., Kim, S. H., Bontempi, E., Bhatia, S. K., & Bui, X. T. (2022). New Insights in factors affecting ground water quality with focus on health risk assessment and remediation techniques. *Environmental Research, 212*, 113171.

Khan, N., Malik, A., & Nehra, K. (2021). Groundwater hydro-geochemistry, quality, microbiology and human health risk assessment in semi-arid area of Rajasthan, India: A chemometric approach. *Environmental Monitoring and Assessment, 193*, 1–36.

Krupavathi, C., Gowd, S. S., Kumar, P. R., Vijay, G. H., & Kumar, B.P. (2022). Spatial analysis of groundwater qualities in Vempalle Mandal of YSR district, Andhra Pradesh, India using Geospatial Techniques. *Hydrospatial Analysis*, 6, 1–12.

Kumar, P. R., Gowd, S. S., & Krupavathi, C. (2023). Hydrogeochemistry and groundwater quality appraisal of semi-arid regions, Andhra Pradesh, India using WQI, PIG, and geospatial techniques. 24th & 25th Feb, 2023, 58.

Kumar, P. R., Gowd, S. S., & Krupavathi, C. (2024). Groundwater quality evaluation using water quality index and geospatial techniques in parts of Anantapur District, Andhra Pradesh, South India. *HydroResearch*, 7, 86–98

Kumar, P. R., Gowd, S. S., & Krupavathi, C., Kumar B. P. (2022). Groundwater quality assessment using water quality index (WQI) in parts of Anantapur, Andhra Pradesh, India. *Journal of Indian Geophysical Union 26 (2), 155-164*

Kumari, M., & Rai, S. C. (2020). Hydrogeochemical evaluation of groundwater quality for drinking and irrigation purposes using water quality index in semi-arid region of India. *Journal of the Geological Society of India, 95*, 159–168.

Madhav, S., Ahamad, A., Singh, A. K., Kushawaha, J., Chauhan, J.S., Sharma, S., & Singh, P. (2020). Water pollutants: sources and impact on the environment and human health. *Sensors in Water Pollutants Monitoring: Role of Material*, 43–62.

Nawale, V. P., Malpe, D. B., Marghade, D., & Yenkie, R. (2021). Non-carcinogenic health risk assessment with source identification of nitrate and fluoride polluted groundwater of Wardha sub-basin, central India. *Ecotoxicology and Environmental Safety, 208*, 111548.

Pandey, S., Mohapatra, G., & Arora, R. (2023). Groundwater quality, human health risks and major driving factors in arid and semi-arid regions of Rajasthan India. *Journal of Cleaner Production, 427*, 139149.

Panneerselvam, B., Muniraj, K., Thomas, M., Ravichandran, N., & Bidorn, B. (2021). Identifying influencing groundwater parameter on human health associate with irrigation indices using the automatic linear model (ALM) in a semi-arid region in India. *Environmental Research, 202*, 111778.

Pappaka, R. K., Somagouni, S. G., Chinthala, K., & Nakkala, A. B. (2024). Appraisal of groundwater quality for suitability of drinking and irrigation purposes of pandameru river basin, anantapur district, AP India. *Arabian Journal of Geosciences, 17*(1), 23.

Pasupuleti, S., Singha, S. S., Singha, S., Kumar, S., Singh, R., & Dhada, I. (2022). Groundwater characterization and non-carcinogenic and carcinogenic health risk assessment of nitrate exposure in the Mahanadi River Basin of India. *Journal of Environmental Management, 319*, 115746.

Prasad, M., Sunitha, V., Reddy, Y. S., Suvarna, B., Reddy, B. M., & Reddy, M. R. (2019). Data on water quality index development for groundwater quality assessment from Obulavaripalli Mandal, YSR district AP India. *Data in Brief, 24*, 103846.

Raju, R. S., Raju, G. S., & Rajasekhar, M. (2023). A study on hydro-geological characterization through Dar-Zarrouk parameters in hard rock terrain of Mandavi River Basin, Andhra Pradesh India. *Arabian Journal of Geosciences, 16*(8), 453.

Reddy, B. M., Sunitha, V., Prasad, M., Reddy, Y. S., & Reddy, M. R. (2019). Evaluation of groundwater suitability for domestic and agricultural utility in semi-arid region of Anantapur, Andhra Pradesh State South India. *Groundwater for Sustainable Development, 9*, 100262.

Singh, G., Rishi, M. S., Herojeet, R., Kaur, L., & Sharma, K. (2020). Evaluation of groundwater quality and human health risks from fluoride and nitrate in semi-arid region of northern India. *Environmental Geochemistry and Health, 42*, 1833–1862.

Subba Rao, N. (2012). PIG: A numerical index for dissemination of groundwater contamination zones. *Hydrological Processes, 26*(22), 3344–3350.

Sunitha, V., & Reddy, B. M. (2022). Geochemical characterization, deciphering groundwater quality using pollution index of groundwater (PIG), water quality index (WQI) and geographical information system (GIS) in hard rock aquifer South India. *Applied Water Science, 12*(3), 41.

Sunitha, V., Reddy, Y. S., Suvarna, B., & Reddy, B. M. (2022). Human health risk assessment (HHRA) of fluoride and nitrate using pollution index of groundwater (PIG) in and around hard rock terrain of Cuddapah, AP South India. *Environmental Chemistry and Ecotoxicology, 4*, 113–123.

Sunitha, V. (2020). Geochemical and health risk assessment of fluoride and nitrate toxicity in semi-arid region of Anantapur District, South India. *Environmental Chemistry and Ecotoxicology, 2*, 150–161.

USEPA. (2014). *Surface water standards and guidance*. Water quality standards handbook.

World Health Organization. 2011. *Guidelines for drinking-water quality* (Vol. 4). World Health Organization.

Yadav, S., Bansal, S. K., Yadav, S., & Kumar, S. (2019). Fluoride distribution in underground water of district Mahendergarh, Haryana, India. *Applied Water Science, 9*, 1–11.

Yenugu, S. R., Vangala, S., & Badri, S. (2020). Monitoring of groundwater quality for drinking purposes using the WQI method and its health implications around inactive mines in Vemula-Vempalli region, Kadapa district, South India. *Applied Water Science, 10*, 1–20.

Chapter 24
Evaluation of Water Quality for Drinking, Irrigation, and Health Indices of Palvayi River Basin, Anantapur District, Andhra Pradesh, South India Using Geospatial Techniques

U. Suresh, G. Shoba Rani, B. C. Sundara Raja Reddy, G. Veeraswamy, and Etikala Balaji

Abstract Groundwater is the primary freshwater resource for many communities. Its quality must be tested before consumption due to potential contamination from various sources, posing significant health risks to humans. The current study aims to assess the groundwater quality for the drinking, irrigation, and health indices in the Palvayi River basin, Andhra Pradesh, India. A total of Twenty-one groundwater samples have been obtained and assessed for various physico-chemical parameters viz. pH, TDS, TH, EC, and cations like Ca^{2+}, Mg^{2+}, Na^{+}, K^{+}, anions Cl^{-}, F^{-}, SO_4^{2-}, HCO_3^{-}, NO_3^{-}, and Alkalinity. The analyzed data is spatially portrayed using the inverse distance method (IDW). Major findings are elevated levels of fluoride and nitrates (23%) in both pre-and post-monsoon season; water–rock interaction; C2S1–C3S1 (medium–high salinity, low sodium) according to USSL, an excellent to good category for irrigation according to Wilcox and High THI of Children and Infants. However, its quality has to be regularly monitored as few samples exceeded the limits of nitrates and fluoride and some samples are close to the permissible limit for drinking purposes.

Keywords Groundwater geochemistry · Drinking · Irrigation · Human health risk assessment · Anantapur · India

---

U. Suresh · G. S. Rani · B. C. Sundara Raja Reddy · G. Veeraswamy (✉)
Department of Geology, Sri Venkateswara University, Tirupati, Andhra Pradesh, India
e-mail: veeraswamygolla33@gmail.com

E. Balaji
Department of Geology, Yogi Vemana University, Kadapa, Andhra Pradesh, India

V. Sunitha et al. (eds.), *Groundwater Resource Management Planning Strategies*, Springer Water, https://doi.org/10.1007/978-3-031-88870-0_24

24.1 Introduction

Groundwater is the global freshwater resource for drinking, farming, and manufacturing uses as surface water is increasingly polluted from various sources. In general, groundwater is considered clean and free from multiple pollutants like biological, chemical, and organic contaminants, compared to surface water (Nagaraju et al., 2016; Li et al., 2021; Sunitha and Reddy, 2022; Etikala et al., 2024). However, nowadays, groundwater is under increasing stress due to a decline in water quality and quantity (Etikala et al., 2021, 2022; Jasechko et al., 2024; Nagaraju et al., 2018). This deterioration is driven by the excess use of fertilizers and pesticides, increased human, and industrial activities, and unprecedented subsurface water withdrawals to meet the demand of an ever-growing population (Adimalla & Wu, 2019).

Hence, its quality has to be tested before consumption for drinking and agricultural uses to prevent possible negative impacts on human health and agricultural productivity. A thorough understanding of groundwater chemistry would also benefit from knowledge of the mechanisms governing it, such as ion exchange, rock-water interactions, and potential contaminant sources. Earlier, many researchers worked on groundwater feasibility for drinking and irrigation (Adimalla et al., 2020; Balaji et al., 2016; Nadjai et al., 2024), processing regulating groundwater chemistry (Etikala et al., 2019, 2020; Guo et al., 2023; Meena, 2022; Nagaraju et al., 2018; Zhang et al., 2024), and health risk assessment of nitrates and fluoride (Din et al., 2024; Herojeet et al., 2023; Madhav et al., 2020; Reddy et al., 2022). No notable studies were found on water quality in the proposed River basin. Therefore, the focus of the study is to evaluate water for drinking, irrigation, and its possible health risks to humans.

24.2 Study Area

The Palvayi River basin lies between Lat.14°28′30″ to 14°35′30″ N; Long. 77°4′30″ to 77°15′00″ E, Anantapur District of Andhra Pradesh state, India (Fig. 24.1). It has an extent of 205 sq km and covered Survey of India (SOI) toposheet Nos. 57 F/2, F/3, F/6 and 57 F/7. The Palvayi River basin is located in parts of Kalyandurg Mandal. The study area is easily accessible either by railways or roadways. The majority of the area is plain land. Hills are confined to the southwestern part, constituted by granite and gneisses. The height of the basin varies between 80 and 404 m above MSL. The region experiences a semi-arid climate and the yearly rainfall average is 560 mm.

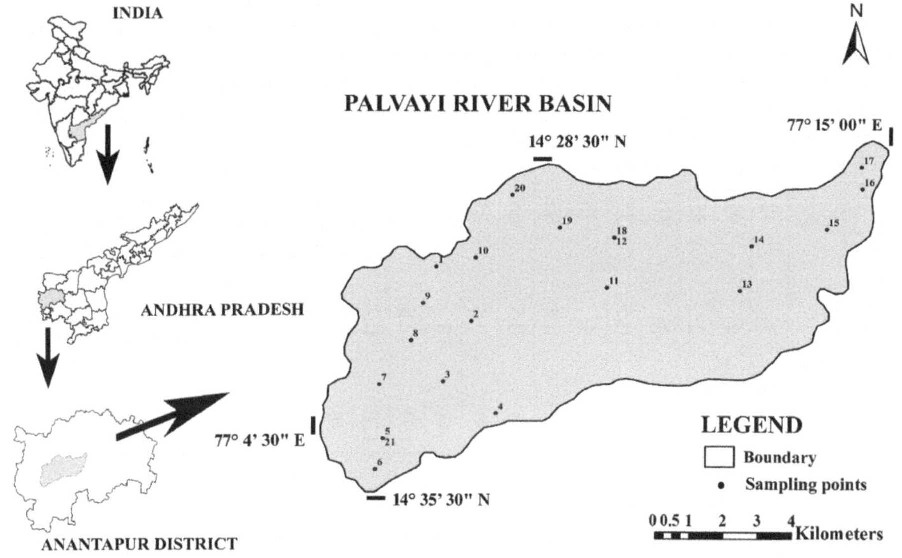

Fig. 24.1 Location map of the Palavayi river basin

24.3 Materials and Methods

Twenty-one (21) Groundwater samples have been collected and quickly transferred into polythene bottles. The sampling procedure, and measuring of physicochemical parameters were done using the standard procedures set by. The analytical results are presented in (Tables 24.1 and 24.2). Then health risk of fluoride and nitrates on humans were calculated.

24.3.1 Human Health Risk Assessment (HHRA)

The health risk assessments of fluoride and nitrates for infants, children and were performed by using the formulas given below

$$\text{CDI} = (\text{MC} \times \text{IR} \times \text{EI} \times \text{ED})/(\text{AW} \times \text{AL}) \quad (24.1)$$

Here,

CDI is the chronic daily water ingestion (mg/kg-day); MC is measured concentrations of fluoride and nitrates; IR is ingestion ratio; EI is exposure interval; ED is exposure duration; AW is Average weight; AET is average exposure time; as per United States Environmental Protection Agency standards (Yenugu et al., 2020). Reference values for IR, EI, ED, AW, and AET for all three categories are given in

Table 24.1 Analytical results of physico-chemical parameters in the pre-monsoon season

S. No.	pH	EC (mho/cm)	TDS (mg/L)	TH (mg/L)	Alkalinity (mg/L)	F^- (mg/L)	Cl^- (mg/L)	NO_3^- (mg/L)	SO_4^{2-} (mg/L)	Na^+ (mg/L)	K^+ (mg/L)	Ca^{2+} (mg/L)	Mg^{2+} (mg/L)
1	7.5	799	511	287	292	1.3	92	92	50	32	3	42	91
2	7.8	823	527	275	274	1.4	115	18	35	22	4	95	18
3	8.0	1102	705	341	348	2.0	205	60	60	45	2	85	64
4	7.3	834	534	324	263	1.6	75	12	55	135	3	60	87
5	7.3	802	513	285	263	1.8	107	18	40	40	0.5	55	73
6	7.3	1086	695	387	337	1.4	168	10	45	110	4	135	24
7	7.8	1003	642	317	372	1.2	135	20	40	90	0.8	95	39
8	7.4	636	407	203	241	1.6	83	35	45	95	0.5	65	20
9	7.9	832	532	263	273	1.7	117	20	40	50	0.8	25	100
10	8.1	891	570	310	325	1.0	111	56	45	48	3	65	74
11	7.6	819	524	277	277	1.1	103	12	40	180	2.8	78	41
12	7.6	897	574	312	236	1.3	153	20	30	120	0.3	98	33
13	7.4	984	630	359	331	1.0	138	59	45	85	0.8	55	110
14	8.0	946	605	332	341	1.1	114	16	40	98	2.8	48	105
15	7.7	664	425	201	230	0.9	93	31	30	115	0.2	18	78
16	7.6	607	389	219	210	1.1	85	31	30	155	0.8	25	78
17	7.6	691	442	245	230	1.2	92	12	30	164	0.3	45	66
18	7.5	805	516	299	267	1.1	111	49	35	196	1.5	55	80
19	7.4	768	491	219	299	1.2	95	29	30	60	1.2	15	90
20	7.7	964	617	355	333	1.0	117	21	40	160	0.8	40	127

(continued)

Table 24.1 (continued)

S. No.	pH	EC (mho/cm)	TDS (mg/L)	TH (mg/L)	Alkalinity (mg/L)	F^- (mg/L)	Cl^- (mg/L)	NO_3^- (mg/L)	SO_4^{2-} (mg/L)	Na^+ (mg/L)	K^+ (mg/L)	Ca^{2+} (mg/L)	Mg^{2+} (mg/L)
21	7.4	898	575	271	260	1.2	120	12	40	49	1.2	25	104
Minimum	7.3	607	389	201	210	0.9	75	10	30	22	0.2	15	18
Maximum	8.1	1102	705	387	372	2	205	92	60	196	4	135	127
Average	7.6	850	544	290	286	1.3	116	31	40	97	1.6	58	72

Table 24.2 Analytical results of physico-chemical parameters in the post-monsoon season

S. No.	pH	EC (mho/cm)	TDS (mg/L)	TH (mg/L)	Alkalinity (mg/L)	F^- (mg/L)	Cl^- (mg/L)	NO_3^- (mg/L)	SO_4^{2-} (mg/L)	Na^+ (mg/L)	K^+ (mg/L)	Ca^{2+} (mg/L)	Mg^{2+} (mg/L)
1	7.4	921	589	278	267	1.2	123	88	45	23	2	32	112
2	7.7	989	633	364	342	1.0	120	24	40	203	2	141	254
3	7.4	787	504	225	307	1.2	97	54	55	39	1	79	128
4	7.5	826	529	307	273	1.1	113	16	65	125	5	75	152
5	7.6	709	454	251	236	1.2	95	15	35	30	0.1	43	75
6	7.6	623	399	225	215	1.1	87	7	40	97	3	115	197
7	7.7	681	436	206	236	1.0	96	17	35	85	0.3	127	267
8	8.0	970	621	340	350	1.2	117	46	55	101	1	85	148
9	7.4	1009	646	368	339	1.0	142	24	45	67	0.5	35	90
10	7.6	920	589	320	242	1.3	157	39	40	36	1	48	98
11	7.6	840	537	284	284	1.2	105	19	50	210	3.3	97	296
12	8.1	914	585	318	334	1.0	114	25	40	139	0.8	119	151
13	7.9	853	546	270	280	1.7	120	54	40	73	0.3	44	103
14	7.4	653	418	208	247	1.6	85	14	35	82	3.1	39	97
15	7.8	1029	659	325	382	1.2	138	39	36	139	0.7	23	83
16	7.3	1114	713	397	346	1.4	173	39	37	176	1.1	39	115
17	7.3	823	527	293	270	1.9	110	18	38	184	0.7	59	148
18	7.3	856	548	333	270	1.6	77	52	42	223	2.1	79	192
19	8.0	1130	723	350	357	2.1	210	35	37	70	1.8	29	72
20	7.8	844	540	282	281	1.4	118	16	35	145	0.3	23	113

(continued)

Table 24.2 (continued)

S. No.	pH	EC (mho/cm)	TDS (mg/L)	TH (mg/L)	Alkalinity (mg/L)	F^- (mg/L)	Cl^- (mg/L)	NO_3^- (mg/L)	SO_4^{2-} (mg/L)	Na^+ (mg/L)	K^+ (mg/L)	Ca^{2+} (mg/L)	Mg^{2+} (mg/L)
21	7.5	1029	659	295	300	1.3	94	18	45	69	0.6	38	136
Minimum	7.3	623	399	206	215	1	77	7	35	23	0	23	72
Maximum	8.1	1130	723	397	382	2	210	88	65	223	5	141	296
Average	7.6	882	564	297	293	1	119	31	42	110	1	65	144

Table 24.3 Reference values of fluoride and nitrates of various age groups

Parameters	Adults	Children	Infants	Units
Ingestion rate (IR)	2.5	1.5	0.8	L/day
Exposure duration (ED)	64	6	1.5	Years
Exposure interval (EI)	365	365	365	days
Average weight (AW)	70	20	10	Kg
Average exposure time (AET)	23,360	2190	547.5	days

Table 24.3.

$$HQ = CDI/RD \tag{24.2}$$

Here, HQ is the hazard quotient, RD is the Reference dosage. RDs for nitrates and fluoride are 1.6 mg/kg/day and 0.004 mg/kg/day, respectively.

$$THI = CDI \sum_{i=1}^{n} HQ \tag{24.3}$$

Here. THI is the total hazard index which refers to the non-carcinogenic effect of fluoride and nitrates. THI > 1 poses health issues.

24.4 Results and Discussion

24.4.1 Drinking Suitability

24.4.1.1 PH

In general, the desirable limit of pH in groundwater is 6.5–7.5. (Etikala et al., 2019; Prasad et al., 2019). During the pre-monsoon season, the northern and northwestern regions of the basin have higher pH values (Fig. 24.2a), whereas the central and southern regions have lower pH values. Higher values are seen in the southeastern and southwest regions of the basin in the post-monsoon (Fig. 24.2b). Low values are found in the northern and northeastern portions of the basin. However, all the samples are within the safe limits.

24.4.1.2 TDS Values in the Basin

The TDS concentrations in groundwater samples of the basin are high at the south-western corner of the basin in the pre-monsoon (Fig. 24.3a). The TDS concentrations

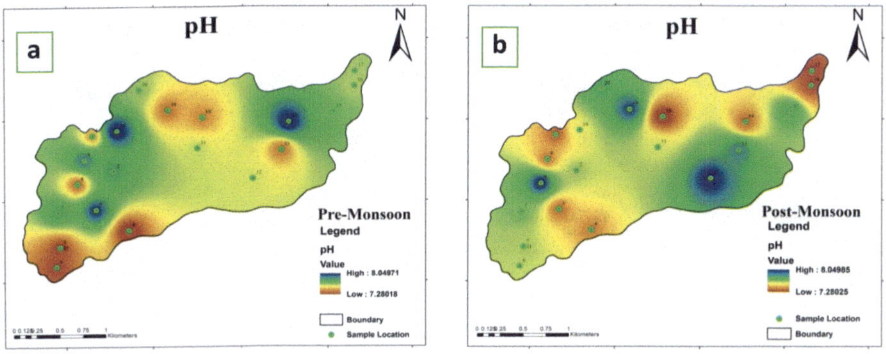

Fig. 24.2 **a**, **b** Spatial distribution of pH in groundwater of the Palavai River basin (Pre-monsoon) and (Post-monsoon)

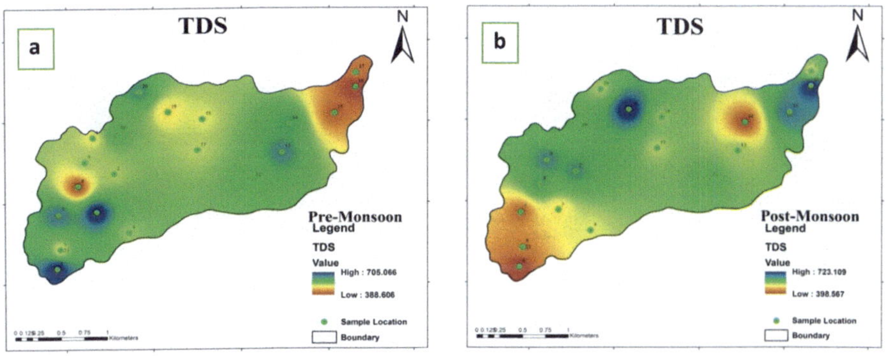

Fig. 24.3 **a**, **b** Spatial distribution of TDS in groundwater of the Palavai River basin (Pre-monsoon) and (Post-monsoon)

of groundwater samples of the basin are high at the northern and northeast corner in the post-monsoon and lower concentrations in the southwestern corner (Fig. 24.3b). Consumption of high TDS in water may pose health issues like formation of stones in Kidneys and diabetes etc. (Madhav et al., 2018; Etikala et al., 2020) Major samples have safe limits of TDS.

24.4.1.3 Electrical Conductance

In pre-monsoon season, electrical conductance is high at the southwest corner and lower conductance found at the northeast corner of the basin (Fig. 24.4a), while in the post-monsoon season, high electrical conductance found at the northern, northeast corner and lower conductance found at the southwest corner of the basin (Fig. 24.4b).

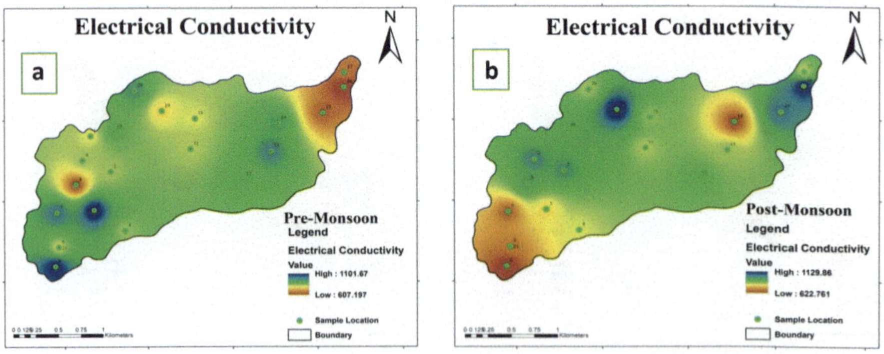

Fig. 24.4 **a**, **b** Spatial distribution of EC in groundwater of the Palavai River basin (Pre-monsoon) and (Post-monsoon)

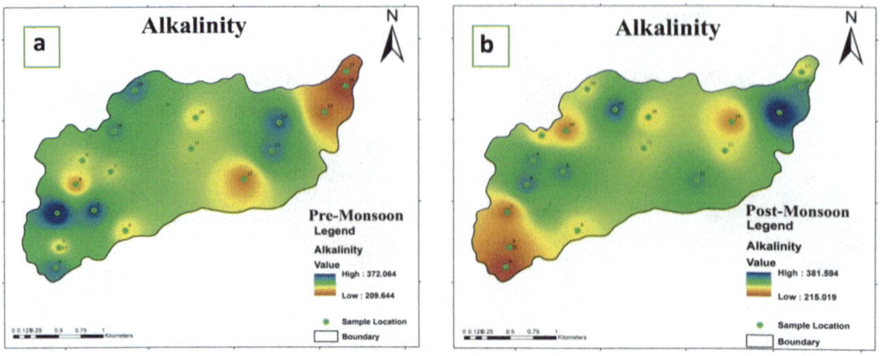

Fig. 24.5 **a**, **b** Spatial distribution of Alkalinity in groundwater of the Palavai River basin (Pre-monsoon) and (Post-monsoon)

24.4.1.4 Alkalinity

The Alkalinity in the groundwater of the basin is high at the southwest corner of the study area and very low at the northeast corner in the pre-monsoon season (Fig. 24.5a) while the Alkalinity in the groundwater of the basin is high in the northeastern part and very low at the southwestern part in the post-monsoon season (Fig. 24.5b).

24.4.1.5 Total Hardness

Hardness, the essential component of water, and attributed to the presence of calcium, magnesium, and carbonates. The hardness value is generally termed as hardness as $CaCO_3$ or total hardness (Golla et al., 2020, 2021; Madhav et al., 2020). The total hardness is maximum at a few locations at the southwest corner, northern side, and

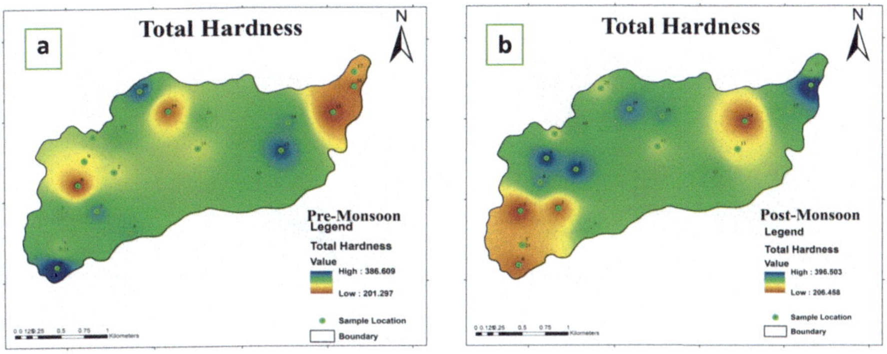

Fig. 24.6 **a**, **b** Spatial distribution of total hardness in groundwater of the Palavai River basin (Pre-monsoon) and (Post-monsoon)

eastern side of the basin, and at the eastern part of the basin shows very low total hardness values in pre-monsoon (Fig. 24.6a), while in the post-monsoon season, the total hardness is maximum at the northeastern corner and low at northwestern and eastern parts of the basin (Fig. 24.6b). All the samples are within the safe limits.

24.4.1.6 Sodium

The quality of water and its usage in irrigation and portability is mainly decided by the concentration of sodium. Sodium may be released into aquifers due to weathering of feldspars, amphiboles, and clay minerals (Balaji et al., 2017). The concentration of Na^+ is maximum at the central and eastern portions of the basin and minimum at the southwestern and eastern portions in both monsoon periods (Fig. 24.7 a, b).

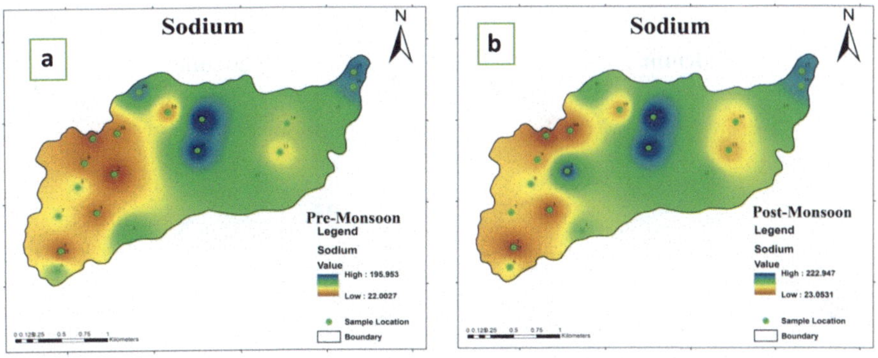

Fig. 24.7 **a**, **b** Spatial distribution of total hardness in groundwater of the Palavai River basin (Pre-monsoon) and (Post-monsoon)

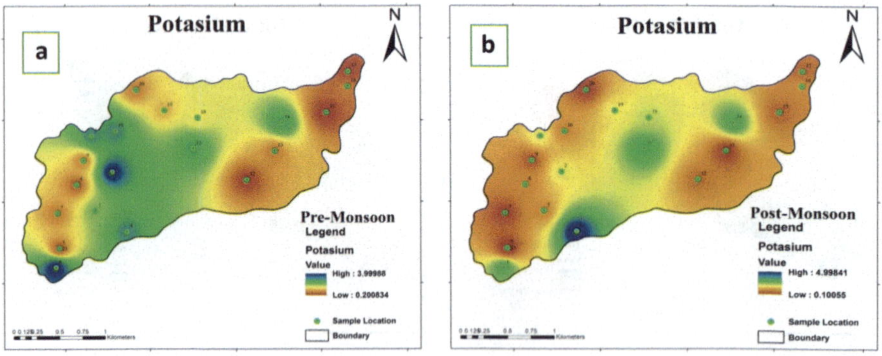

Fig. 24.8 **a**, **b** Spatial distribution of potassium in groundwater of the Palavai River basin (Pre-monsoon) and (Post-monsoon)

24.4.1.7 Potassium

In general, potassium is released into aquifers through the weathering of minerals like orthoclase, microcline, biotite, etc. Potassium values are high at the central and eastern portions of the basin in both monsoon seasons (Fig. 24.8 a, b).

24.4.1.8 Calcium

Generally, calcium is abundant in groundwater. Weathering of minerals like amphiboles, feldspars, and pyroxenes releases calcium into water.

Calcium levels are low in the northern portion of the basin and are high in the southern portion of the basin in both monsoon seasons (Fig. 24.9 a, b). All the water samples are within the safe limits of calcium.

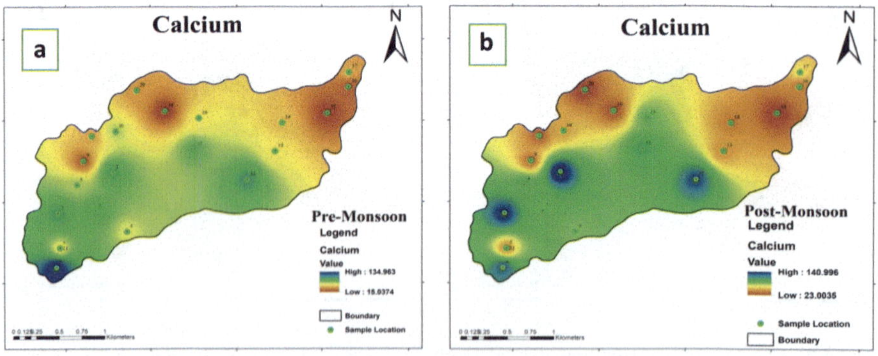

Fig. 24.9 **a**, **b** Spatial distribution of calcium in groundwater of the Palavai River basin (Pre-monsoon) and (Post-monsoon)

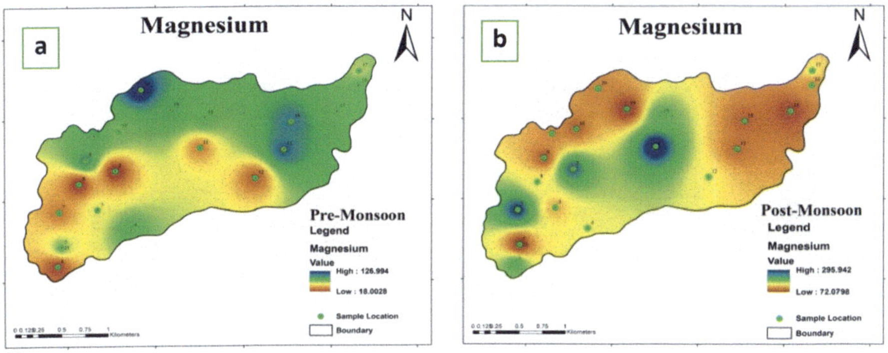

Fig. 24.10 **a**, **b** Spatial distribution of magnesium in groundwater of the Palavai River basin (Pre-monsoon) and (Post-monsoon)

24.4.1.9 Magnesium

Magnesium in water may be released from the weathering of minerals like biotite, hornblende, olivine, serpentine, and talc. In the pre-monsoon season, magnesium levels are low in the southern part of the study area and high in the northern part of the basin (Fig. 24.10a); in the post-monsoon, high magnesium levels are found in the central portion of the basin and low in the eastern and northern parts (Fig. 24.10b).

24.4.1.10 Chloride

Generally, natural water contains chlorides of 10 mg/L attributed to minerals like Halite and sylvite. However, excess amounts of chlorides may be attributed to anthropogenic sources. Chloride is maximum in the southwest corner of the basin and other areas have low values in the pre-monsoon (Fig. 24.11a). Higher concentrations of chloride are observed at two locations in the northern and northeastern part and other locations are showing low values in the post-monsoon (Fig. 24.11b).

24.4.1.11 Sulphate

Sulfates are mostly a result of the weathering of gypsum, sodium sulfate deposits, and some types of shale. However, higher levels of sulfates in groundwater are attributed to anthropogenic sources. The spatial distribution diagram of sulfates shows that a low concentration of the ion is observed at the eastern, central part of the basin and higher values in a small region falling in the southwest part of the basin in the pre-monsoon (Fig. 24.12a). In post-monsoon, high concentration is observed at one location western part and other locations are exhibiting moderate to low values in the post-monsoon (Fig. 24.12b).

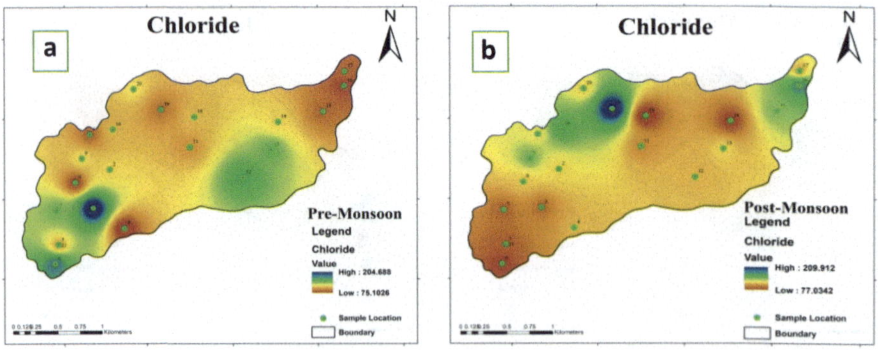

Fig. 24.11 **a**, **b** Spatial distribution of chloride in groundwater of the Palavai River basin (Pre-monsoon) and (Post-monsoon)

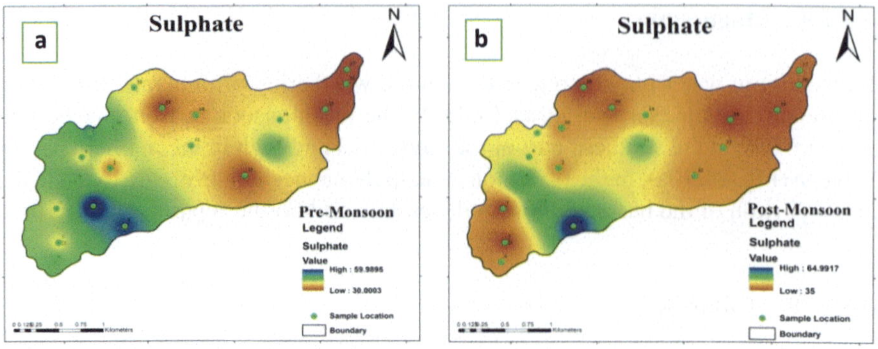

Fig. 24.12 **a**, **b** Spatial distribution of sulphate in groundwater of the Palavai River basin (Pre-monsoon) and (Post-monsoon)

24.4.1.12 Fluoride

Fluoride in groundwater may occur naturally in water due to the weathering of minerals like cryolite, fluorite, and fluorapatite. However, fluoride may also be released into water through coal burning, fertilizers, and industrial activities (Adimalla & Li, 2018). High concentrations of fluoride are observed in the southwest corner of the basin and other parts of the study area show moderate to low concentrations of fluoride in the pre-monsoon (Fig. 24.13a). High concentrations of fluoride are observed in the northeast corner and northern part of the basin and other parts show moderate to low concentrations of fluoride in the pre-monsoon (Fig. 24.13b). Higher fluorides in water can cause dental and skeletal fluorosis. Nearly, 28% of the samples exceeded the safe limits of fluorides in the proposed basin.

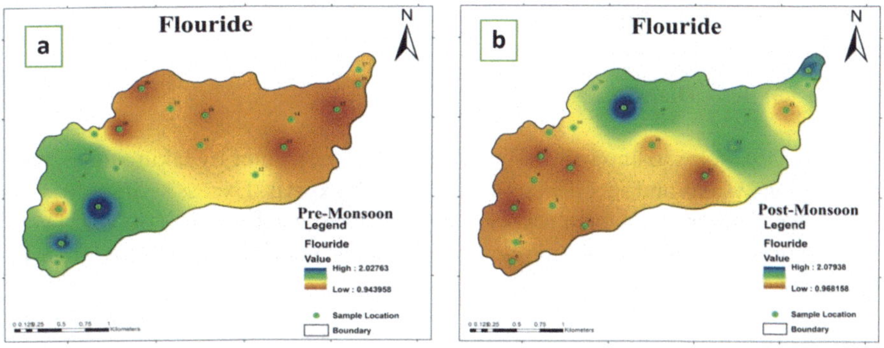

Fig. 24.13 **a**, **b** Spatial distribution of fluoride in groundwater of the Palavai River basin (Pre-monsoon) and (Post-monsoon)

24.4.1.13 Nitrates

In general, excess levels of nitrates in water contribute to anthropogenic sources like fertilizer, discharge of untreated sewage, and septic tank leakages. High concentrations of nitrate are found at one location in the northwest of the basin and other locations have moderate to low concentrations in the pre-monsoon (Fig. 24.14a). High concentrations are found at one location in the northwestern part of the basin and other locations are exhibiting moderate to low concentrations in the post-monsoon (Fig. 24.14b). High amounts in groundwater may lead to methemoglobinemia, colon cancer, and birth defects in infants. In the proposed basin, 28% of the samples exceeded the safe limits of nitrates.

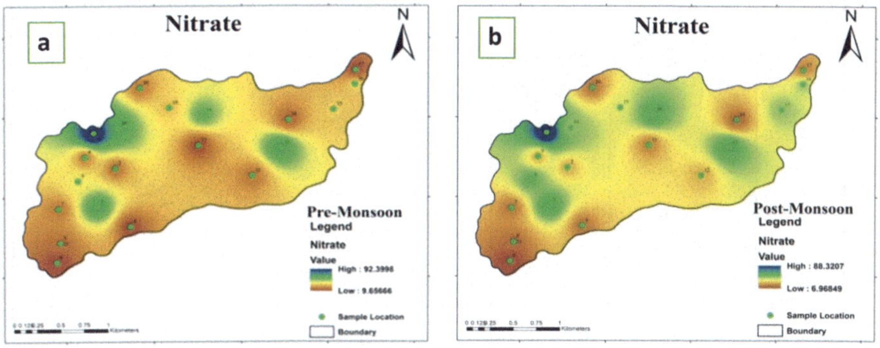

Fig. 24.14 **a**, **b** Spatial distribution of nitrate in groundwater of the Palavai River basin (Pre-monsoon) and (Post-monsoon)

24.4.2 Irrigation Suitability

24.4.2.1 Wilcox Diagram

Wilcox (1955), helps to know the irrigation suitability of water by taking the percent sodium on the y-axis and specific conductance on the x-axis. Excess sodium in water may hinder plant growth which reduces crop productivity (Etikala et al., 2021). All concentration levels are expressed in milli equivalents per Liter. The following is the formula for sodium percentage.

$$\%\ \mathrm{Na} = \frac{(Na^{+} + K^{+})}{Ca^{2+} + Mg^{2+} + Na^{+} + K^{+}} * 100 \tag{24.4}$$

During both the monsoon periods, about 10% of the sample stations fall in the doubtful region reflecting that the major wells in the River basin are found to be good for irrigation (Fig. 24.15).

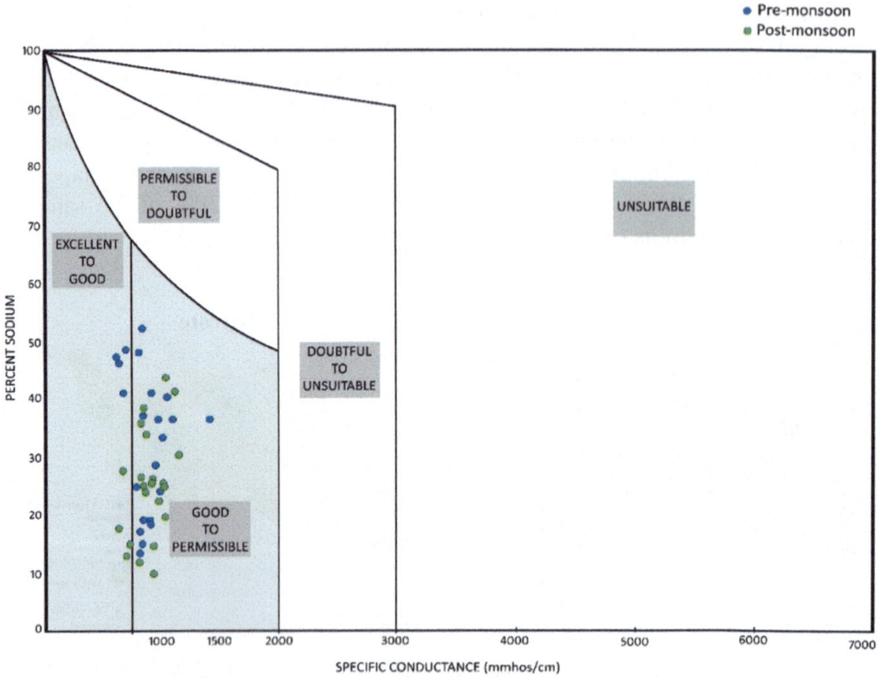

Fig. 24.15 Wilcox diagram

24.4.2.2 USSL Diagram

United States Salinity Laboratory Staff, in 1954 invented a graphical plot to know the irrigation water suitability which can be drawn by taking SAR values and specific conductance values on Y and X axes (USSL, 1954). All concentration levels are expressed in milli equivalents per Liter. SAR can be computed by using the following formula

$$\text{SAR} = \frac{(Na^{+})}{\sqrt{Ca^{2+} + Mg^{2+}/2}} \tag{24.5}$$

The sodium hazard of irrigation waters as expressed by SAR does not take into account the effect of anion composition. According to the USSL classification, 30% samples fall in the C2S1 and the rest 70% samples fall in the C3S1 fields in both the pre- and post-monsoon periods. Based on the classification it indicates that the low salinity and low alkali hazard represent the groundwater is fit for irrigation (Fig. 24.16).

24.4.3 Health Risk Assessment

From the results, it is found that the majority of the samples have THI value > 1 for fluoride and nitrates for children and infants which shows that they are vulnerable to fluoride and nitrate health risk (Tables 24.4 and 24.5 & Figs. 24.17a–d). It necessitates immediate remedial measures to improve the quality of water for sustainable development of the Palvayi River basin.

24.5 Conclusion

The study reveals the presence of nitrate and fluoride in groundwater. The nitrate exceeds the drinking water limit in the Palvayi River basin due to the extensive usage of pesticides, *N*-fertilizers, and livestock waste. The majority of the samples of fluoride exceed the drinking water limit in the study area due to the presence of granitic gneisses and granitoid group of rocks especially the disintegration of the apatite group of minerals. It can cause dental and skeletal fluorosis. Based on the Wilcox diagram during both the monsoon periods, about 10% of the sample stations fall in the doubtful region reflecting that the major wells in the proposed basin are found to be good for irrigation. According to the USSL diagram, 50% of the samples fall in the C2S1 and C3S1 fields in both the pre- and post-monsoon seasons. Based on the classification it indicates that the low salinity and low alkali hazard mean the groundwater is suitable for irrigation. Moreover, children and infants

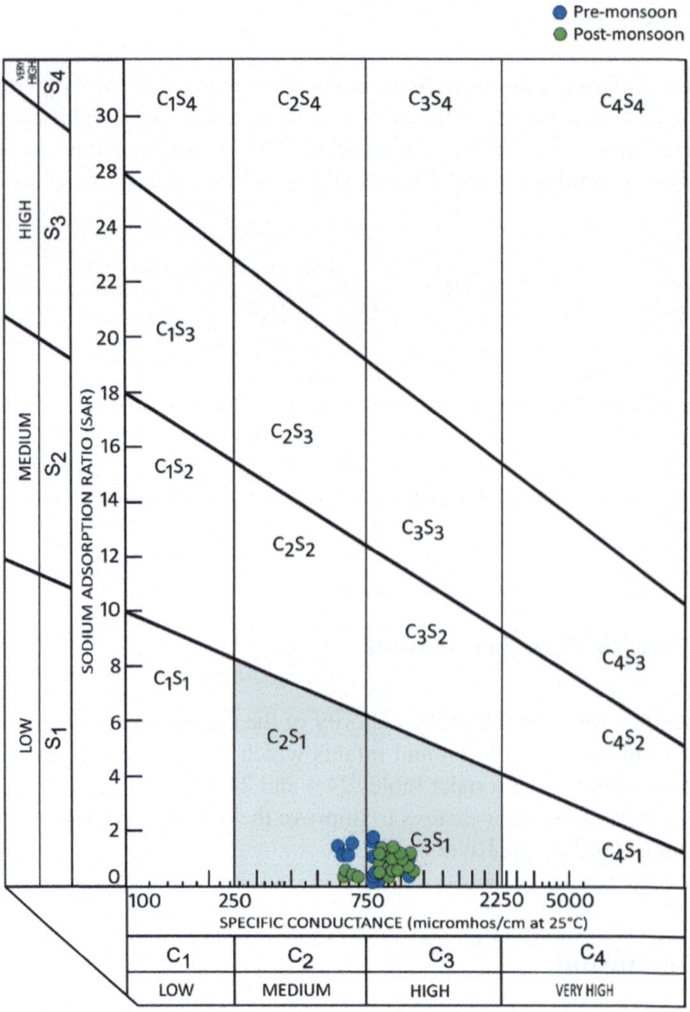

Fig. 24.16 USSL classification

have more significant fluoride and nitrate health risks than adults. Overall, the water is moderately polluted with fluoride and nitrates (28%) which are not suitable for drinking purposes. However, the water is suitable for irrigation purposes. However, regular monitoring of water quality and proper waste treatment is necessary to avoid the health effects of drinking water.

Table 24.4 Total hazard index of fluorides in pre-and post-monsoon seasons

THI-fluoride						
Pre-monsoon				Post-monsoon		
S. No.	Adults	Children	Infants	Adults	Children	Infants
1	0.6	1.6	1.7	0.6	1.5	1.6
2	0.7	1.8	1.9	0.5	1.3	1.3
3	1.0	2.5	2.7	0.6	1.5	1.6
4	0.8	2.0	2.1	0.5	1.4	1.5
5	0.9	2.3	2.4	0.6	1.5	1.6
6	0.7	1.8	1.9	0.5	1.4	1.5
7	0.6	1.5	1.6	0.5	1.3	1.3
8	0.8	2.0	2.1	0.6	1.5	1.6
9	0.8	2.1	2.3	0.5	1.3	1.3
10	0.5	1.3	1.3	0.6	1.6	1.7
11	0.5	1.4	1.5	0.6	1.5	1.6
12	0.6	1.6	1.7	0.5	1.3	1.3
13	0.5	1.3	1.3	0.8	2.1	2.3
14	0.5	1.4	1.5	0.8	2.0	2.1
15	0.4	1.1	1.2	0.6	1.5	1.6
16	0.5	1.4	1.5	0.7	1.8	1.9
17	0.6	1.5	1.6	0.9	2.4	2.5
18	0.5	1.4	1.5	0.8	2.0	2.1
19	0.6	1.5	1.6	1.0	2.6	2.8
20	0.5	1.3	1.3	0.7	1.8	1.9
21	0.6	1.5	1.6	0.6	1.6	1.7
Minimum	0.4	1.1	1.2	0.5	1.3	1.3
Maximum	1.0	2.5	2.7	1.0	2.6	2.8
Average	0.6	1.6	1.7	0.6	1.6	1.8

Table 24.5 Total hazard index of nitrates in pre-and post-monsoon seasons

THI-nitrates						
Pre-monsoon				Post-monsoon		
S. No.	Adults	Children	Infants	Adults	Children	Infants
1	1.6	4.3	4.6	1.6	4.1	4.4
2	0.3	0.8	0.9	0.4	1.1	1.2
3	1.1	2.8	3.0	1.0	2.5	2.7
4	0.2	0.6	0.6	0.3	0.8	0.8
5	0.3	0.8	0.9	0.3	0.7	0.8
6	0.2	0.5	0.5	0.1	0.3	0.4
7	0.4	0.9	1.0	0.3	0.8	0.9
8	0.6	1.6	1.8	0.8	2.2	2.3
9	0.4	0.9	1.0	0.4	1.1	1.2
10	1.0	2.6	2.8	0.7	1.8	2.0
11	0.2	0.6	0.6	0.3	0.9	1.0
12	0.4	0.9	1.0	0.4	1.2	1.3
13	1.1	2.8	3.0	1.0	2.5	2.7
14	0.3	0.8	0.8	0.3	0.7	0.7
15	0.6	1.5	1.6	0.7	1.8	2.0
16	0.6	1.5	1.6	0.7	1.8	2.0
17	0.2	0.6	0.6	0.3	0.8	0.9
18	0.9	2.3	2.5	0.9	2.4	2.6
19	0.5	1.4	1.5	0.6	1.6	1.8
20	0.4	1.0	1.1	0.3	0.8	0.8
21	0.2	0.6	0.6	0.3	0.8	0.9
Minimum	0.2	0.5	0.5	0.1	0.3	0.4
Maximum	1.6	4.3	4.6	1.6	4.1	4.4
Average	0.5	1.4	1.5	0.6	1.5	1.6

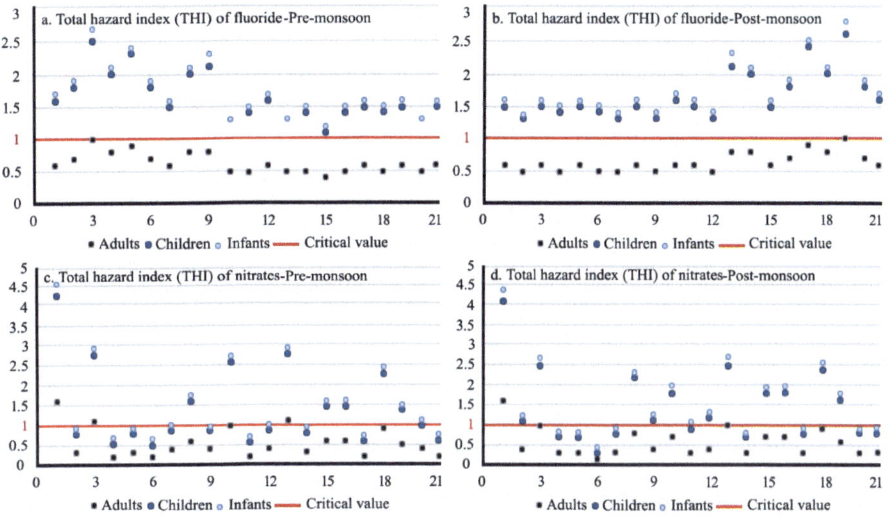

Fig. 24.17 a-d THI of fluoride and nitrates (Pre-monsoon) and (Post-monsoon)

Acknowledgements We gratefully acknowledge the support provided by the Head, Department of Geology, Sri Venkateswara University, Tirupati, Andhra Pradesh, India, for facilitating access to the hydrogeology laboratory.

References

Adimalla, N., & Li, P. (2018). Occurrence, health risks, and geochemical mechanisms of fluoride and nitrate in groundwater of the rock dominant semi-arid region, Telangana State, India. *Human and Ecological Risk Assessment: An International Journal.* https://doi.org/10.1080/10807039.2018.14803

Adimalla, N., Dhakate, R., Kasarla, A., & Taloor, A. K. (2020). Appraisal of groundwater quality for drinking and irrigation purposes in Central Telangana India. *Groundwater for Sustainable Development, 10*, 100334. https://doi.org/10.1016/j.gsd.2020.100334

Adimalla, N., & Wu, J. (2019). Groundwater quality and associated health risks in a semi-arid region of south India: Implication to sustainable groundwater management. *Human and Ecological Risk Assessment: An International Journal, 25*(1–2), 191–216.

Balaji, E., Nagaraju, A., Sreedhar, Y., Thejaswi, A., & Sharifi, Z. (2016). Hydrochemical characterization of groundwater in around Tirupati area, Chittoor district, Andhra Pradesh South India. *Applied Water Science, 7*(3), 1203–1212. https://doi.org/10.1007/s13201-016-0448-6

Din, I. U., Ali, W., Muhammad, S., Shaik, M. R., Shaik, B., Rehman, I. U., & Tokatli, C. (2024). Spatial distribution and potential health risk assessment for fluoride and nitrate via water consumption in Pakistan. *Journal of Geochemical Exploration, 259*, 107413. https://doi.org/10.1016/j.gexplo.2024.107413

Etikala, B., Golla, V., Arveti, N., Yenamala, S., Mannala, P., Keshava Kiran Kumar, P.L. (2020). Statistical and analytical evaluation of groundwater quality of Atmakur area, SPSR Nellore district, Andhra Pradesh, South India. In K. Babu, H. Rao, Y. Amarnath (Eds.), *Emerging trends in civil engineering. Lecture notes in civil engineering* (Vol. 61). Springer. https://doi.org/10.1007/978-981-15-1404-3_27

Etikala, B., Golla, V., Adimalla, N., Ramanaiah, S., & Marapatla, S. (2021). Hydrogeochemical characterization of groundwater using conventional graphical, geospatial and multivariate statistical techniques. In *Springer hydrogeology* (Internet) (pp. 81–96). https://doi.org/10.1007/978-3-030-62397-5_5

Etikala, B., Madhav, S., & Somagouni, S. G. (2022). Urban water systems: An overview. In *Current directions in water scarcity research* (pp. 1–19). https://doi.org/10.1016/b978-0-323-91838-1.00016-6

Etikala, B., Vangala, S., & Madhav, S. (2024). Groundwater geochemistry using modified integrated water quality index (IWQI) and health indices with special emphasis on nitrates and heavy metals in southern parts of Tirupati, South India. *Environmental Geochemistry and Health*, 46(11). https://doi.org/10.1007/s10653-024-02229-3

Etikala, B., Golla, V., Adimalla, N., & Marapatla, S. (2019). Factors controlling groundwater chemistry of Renigunta area, Chittoor district, Andhra Pradesh, South India: A multivariate statistical approach. *HydroResearch, 1*, 57–62. https://doi.org/10.1016/j.hydres.2019.06.002

Golla, V., Etikala, B., Arveti, N., Sradha, S. R., Janardhan, N., Rajasekhar, M., & Subbarao, M. (2020). Assessment of groundwater contamination with emphasis on sulfates, barites mining area, Mangampeta, Andhra Pradesh, India. In *Emerging trends in civil engineering* (pp. 307–322). Springer.

Golla V. et al. (2020) Assessment of groundwater contamination with emphasis on sulfates, barites mining area, Mangampeta, Andhra Pradesh, India. In: K. Babu, H. Rao, Y. Amarnath (Eds.), *Emerging trends in civil engineering. Lecture notes in civil engineering* (Vol. 61). Springer. https://doi.org/10.1007/978-981-15-1404-3_26

Golla, V., Badapalli, P. K., & Telkar, S. K. (2021). Delineation of groundwater potential zones in the semi-arid region (Ananatapuram) using geospatial techniques. Materials Today: Proceedings.

Guo, W., Li, P., Du, Q., Zhou, Y., Xu, D., & Zhang, Z. (2023). Hydrogeochemical processes regulating the groundwater geochemistry and human health risk of groundwater in the rural areas of the Wei River Basin, China. *Exposure and Health*. https://doi.org/10.1007/s12403-023-00555-y

Herojeet, R., Dewangan, R. K., Naik, P. K., & Verma, J. R. (2023). Probabilistic modelling is superior to deterministic approaches in the human health risk assessment: an example from a tribal stretch in central India. *Scientific Reports*, 13(1). https://doi.org/10.1038/s41598-023-45622-1

Jasechko, S., Seybold, H., Perrone, D., Fan, Y., Shamsudduha, M., Taylor, R. G., Fallatah, O., & Kirchner, J. W. (2024). Rapid groundwater decline and some cases of recovery in aquifers globally. *Nature, 625*(7996), 715–721. https://doi.org/10.1038/s41586-023-06879-8

Li, P., He, X., Li, Y., & Xiang, G. (2018a). Occurrence and health implication of fluoride in groundwater of loess aquifers in the Chinese Loess Plateau: a case study of Tongchuan, northwest China. *ExpoHealth*. https://doi.org/10.1007/s12403-018-0278-x

Li, P., Karunanidhi, D., Subramani, T., & Srinivasamoorthy, K. (2021). Sources and consequences of groundwater contamination. *Archives of Environmental Contamination and Toxicology, 80*(1), 1–10. https://doi.org/10.1007/s00244-020-00805-z

Madhav, S., Ahamad, A., Kumar, A., Kushawaha, J., Singh, P., & Mishra, P. K. (2018). Geochemical assessment of groundwater quality for its suitability for drinking and irrigation purpose in rural areas of Sant Ravidas Nagar (Bhadohi), Uttar Pradesh. *Geology, Ecology and Landscapes, 2*(2), 127–136. https://doi.org/10.1080/24749508.2018.1452485

Madhav, S., Raju, N. J., & Ahamad, A. (2020). A study of hydrogeochemical processes using integrated geochemical and multivariate statistical methods and health risk assessment of groundwater in Trans-Varuna region, Uttar Pradesh. *Environment, Development and Sustainability, 23*(5), 7480–7508. https://doi.org/10.1007/s10668-020-00928-2

Meena, P. L. (2022). Study on the hydrogeochemical processes regulating the groundwater chemistry in the Southeast Rajasthan. *Journal of the Geological Society of India, 98*(10), 1455–1465. https://doi.org/10.1007/s12594-022-2193-9

Nadjai, S., Bouderbala, A., Khammar, H., Nabed, A. N., & Benaabidate, L. (2024). Assessment of groundwater suitability for drinking and irrigation purposes in the middle Cheliff Aquifer, Algeria. *Desalination and Water Treatment*, 100528. https://doi.org/10.1016/j.dwt.2024.100528

Nagaraju, A., Balaji, E., Sun, L., & Thejaswi, A. (2018). Processes controlling groundwater chemistry from Mulakalacheruvu area, Chittoor district, Andhra Pradesh, South India: A statistical approach based on hydrochemistry. *Journal of the Geological Society of India, 91*(4), 425–430. https://doi.org/10.1007/s12594-018-0875-0

Nagaraju, A., Sharifi, Z., & Balaji, E. (2016). Statistical and analytical evaluation of groundwater quality of Tirupati area, Chittoor district, Andhra Pradesh, South India. *Journal of the Geological Society of India, 88*(2), 222–234. https://doi.org/10.1007/s12594-016-0481-y

Prasad, M., Sunitha, V., Reddy, Y. S., Suvarna, B., Reddy, M. R., & Reddy, M. R. (2019). Data on water quality index development for groundwater quality assessment from Obulavaripalli Mandal, YSR district, A.P. India. *Data in Brief*, 24, 103846. https://doi.org/10.1016/j.dib.2019.103846

Reddy, C. K. V. C., Golla, V., Badapalli, P. K., & Reddy, N. B. Y. (2022). Evaluation of groundwater contamination for fluoride and nitrate in Nellore Urban Province, Southern India: a special emphasis on human health risk assessment (HHRA). *Applied Water Science*, 12(3). https://doi.org/10.1007/s13201-021-01537-8

Sunitha, V., & Muralidhara Reddy, B. (2022). Geochemical characterization, deciphering groundwater quality using pollution index of groundwater (PIG), water quality index (WQI) and GIS in hard rock aquifer South India. *Applied Water Sciences, 12*, 41. https://doi.org/10.1007/s13201-021-01527-w

United States Salinity Laboratory Staff. (1954). *Diagnosis and improvement of saline and alkali soils*. US Department of Agriculture Handbook 60.

Wilcox, L. (1955). *Classification and use of irrigation waters (No. 969)*. US Department of Agriculture.

Yenugu, S. R., Vangala, S., & Badri, S. (2020). Monitoring of groundwater quality for drinking purposes using the WQI method and its health implications around inactive mines in Vemula-Vempalli region, Kadapa District, South India. *Applied Water Science*, 10(8). https://doi.org/10.1007/s13201-020-01284-2

Zhang, J., Wang, P., Liu, S., & Yu, J. (2024). Mechanism controlling groundwater chemistry in the hyper-arid basin with intermittent river flow: insights from long-term observations (2001–2023) in the lower Heihe River, Northwest China. *Frontiers in Environmental Science*, 12. https://doi.org/10.3389/fenvs.2024.1376443

Chapter 25
A Geo-Chemical Study of Water Pollution Due to the Tanning Industry: A Case Study of Dindigul District, Tamilnadu, India

V. Saravanabavan, C. Vinothini, R. S. Suja Rose, and D. Balaji

Abstract The environment is under greater pressure from solid and liquid wastes generated by the tannery industry. These are unavoidable by-products of the leather manufacturing process that, if not adequately treated before discharge, cause severe pollution. The current case study investigates the tanning sector in the Dindigul district, which is the fundamental purpose of our inquiry. To investigate the groundwater quality in the tannery industry region and how it affects health. The study uses cartography, geo-statistics, and spatial analysis technologies such as GPS and GIS. Data are gathered from both primary and secondary sources. There were 20 locations identified for the collection of samples within the research region between 2012 and 2023. Then, the water samples were subjected to various geochemical analyses such as pH, TDS, SO_4^{2-}, Cl^-, Na^+, K^+, Ca^{2+}, Mg^{2+}, and F^-. Water quality has been analyzed based on the BIS (Bureau of Indian Standards). The data was analyzed using Multivariate statistical techniques known as 'Factor Analysis.' The study examined about 64 parameters—Lime, sodium, chrome, sulfate, oil dyes, chromium, water hardness, disease indicators, environmental and ecological indicators, etc. The factor analysis matrix identifying seven major dimensions explained the total variance of 87.63% with an Eigenvalue ranging from 1.00 to 5.60. Tannery effluents impact pure groundwater, soil, plants, irrigation, and animals. Dating wastewater directly

V. Saravanabavan (✉)
Department of Geography, School of Human & Environmental Sciences, North Eastern Hill University, Shillong, Meghalaya, India
e-mail: vsaravanabavan@gmail.com

C. Vinothini
Department of Geography, School of Earth and Atmospheric Sciences, Madurai Kamaraj University, Madurai, Tamil Nadu, India

R. S. Suja Rose
Department of Environmental Remote Sensing and Cartography, School of Earth and Atmospheric Sciences, Madurai Kamaraj University, Madurai, Tamil Nadu, India

D. Balaji
Department of Geography, School of Science, Tamil Nadu Open University, Chennai, Tamil Nadu, India

V. Sunitha et al. (eds.), *Groundwater Resource Management Planning Strategies*, Springer Water, https://doi.org/10.1007/978-3-031-88870-0_25

into streams and rivers will likely cause significant harm to the sensitive ecosystem. Disorders common in tannery effluent areas include skin ulceration, allergic skin, and dermatitis. It was discovered that the quality of groundwater for human use has significantly deteriorated in the western half of Dindigul, including Pallapatti, Adiyanuthu, and Vellodu village panchayats, as a cause of the protrusion of toxic effluents from leather tanning businesses.

Keywords Geo-chemical · Geospatial · Tanning Industries · GIS · Groundwater pollution · Factor analysis · India

25.1 Introduction

On a chemical level, water is never completely pure. It contains several pollutants, both dissolved and suspended (Sadeghi et al., 2022; Vannucci et al., 2018). Chemical pollution can impair people's health directly and indirectly by accumulating in aquatic life used for human nourishment (Rigueto et al., 2021; Saravanabavan et al., 2021a, 2021b, 2021c; Selvakumar et al., 2021). Tanning chemicals spread on the surface as well as in sources of groundwater (Selvam et al., 2016; Sundararaj et al., 2022; Vinothkanna et al., 2020). This unclean water flows into streams, rivers, lakes, and ponds before draining into the ground (Radhakrishnan & Vaithialingam, 2023; Rahman et al., 2022; Brar et al., 2019; Chidambaram et al., 2015). It affects the groundwater and makes it unfit for drinking, irrigation, and general consumption (Balaji & Saravanabavan, 2020, 2021; Jesudhas et al., 2022; Jha et al., 2020). The pollution index of groundwater (PIG) is an effective technique for determining the suitability of drinking water quality in any location and providing overall water quality data (Sunitha & Reddy, 2022). The Water Quality Index (WQI) technique helps assess water adaptability for various uses. It provides reliable water quality data for citizens and decision-makers to monitor (Kom et al., 2024; Reljić et al., 2023). Heavy metals can lead to a variety of health problems, including neurological impairment, developmental delays, kidney damage, and cancer (Moukadiri et al., 2024). Heavy metal exposure affects health and environmental protection (Reddy & Sunitha, 2023).

Chemical contaminants in various forms produced from industrial and agricultural waste are entering public water sources (Balaji et al., 2016; Nagaraju et al., 2018; Sunitha et al., 2005, 2014). These contaminants include minerals, organic acids, nitrogenous chemicals, ammonia, and many poisonous and biocidal organic compounds (Malik & Kumar, 2023). The current concern about chemical contaminants in water is less about their immediate detrimental effects on human health and more about the potential long-term consequences of low-level exposure, which are often non-specific and difficult to detect (Reddy & Sunitha 2023; Reddy et al., 2022a). Chemical contamination is gaining attention in many prosperous countries where water-borne infectious illnesses have disappeared (Ravindra et al., 2023; Rahamath

Nisha & Saravanabavan, 2020). The findings show that the local communities experience a variety of health problems, including rashes, diarrhoea, dysentery, respiratory infections, anaemia, and difficulties during childbirth (DeGuzman et al., 2023). According to the study, water quality is determined by the type of pollutant used and the minerals found in specific borewell zones (Kom et al., 2022). The findings of the water quality study revealed that most water quality parameters are marginally higher in the wet season than in the dry. The correlation tools have been used to identify the physicochemical features of water pollutants (Balasubramani et al., 2015; Dessie et al., 2024; Suja Rose & Krishnan, 2009). Generally, groundwater quality depends on anthropogenic and geogenic activity (Sunitha & Muralidhara Reddy, 2022; Panneerselvam et al., 2020). Groundwater is the chief resource for drinking water and household and agricultural uses (Reddy et al., 2022b, 2022c). It exhibits a correlation analysis of the ionic concentration of water between different physical and chemical parameters (Yenugu et al., 2020).

The quality of drinking water is only used to determine groundwater quality. (Patel et al., 2023). This analysis used the following parameters to determine groundwater quality: pH, TDS, magnesium, potassium, calcium, sodium, sulfate, chloride, fluoride, and bicarbonate (Suvarna et al., 2020; Zeng et al., 2024). The study area's drinking water quality is determined by thoroughly studying the abovementioned factors. Furthermore, the study demonstrates the agricultural methods on environmental and human variables, assisting farmers, decision-makers, and stakeholders in sustainable groundwater management (Saikrishna et al., 2023). As a result, this study's primary goal was to analyze groundwater quality using the GPI (Groundwater pollution index) and its various exposure pathways to estimate human health implications utilizing statistical and geospatial analysis for drinking reasons. The current study's key objectives are: To investigate the groundwater quality in the tannery industry region and how it affects health in the Dindigul district. The current study's key objective is to investigate the groundwater in the tannery industry region and how it affects health.

25.2 Study Area

Dindigul district lies between 10°05′ and 10°09′ North Latitude and 77°30′ to 78°20′ East Longitude (Fig. 25.1). Erode, Karur and Tiruchirapalli districts bound the district on the north; Madurai and Tiruchirapalli districts on the east; Madurai and Theni districts in the south; and Kerala State and Coimbatore districts on the west (Fig. 25.2).

The district's numerous landforms, including hills, linear ridges, and pediment terrains, result from varying geological formations' resistances. Geologically, Acid Intrusive occupies the district's maximum, accounting for 56% of the total area, and is primarily located in the district's northern, eastern, and southern regions. The district boasts a tropical climate. The rainfall average in the district was between 801 and 1000 mm. During the monsoon season (June to September), the district

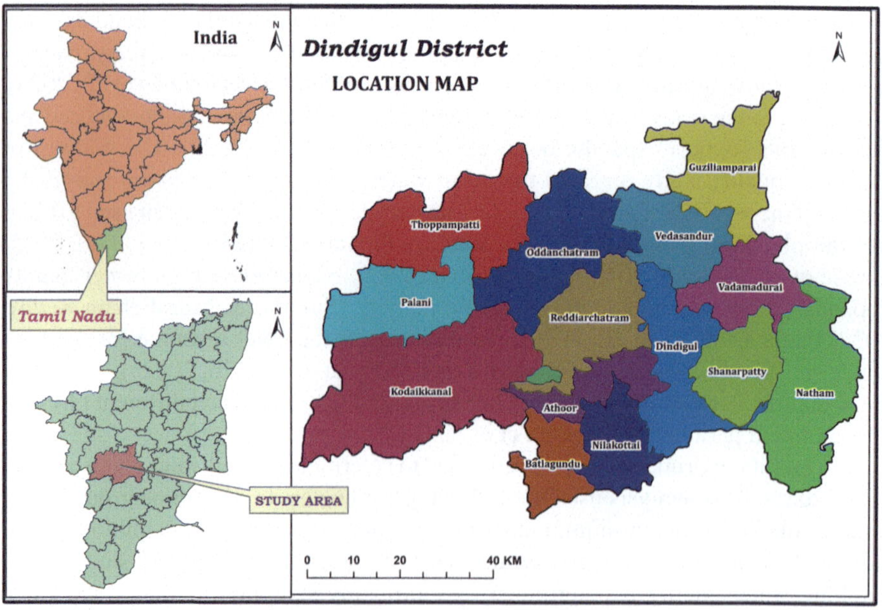

Fig. 25.1 Location of the study area

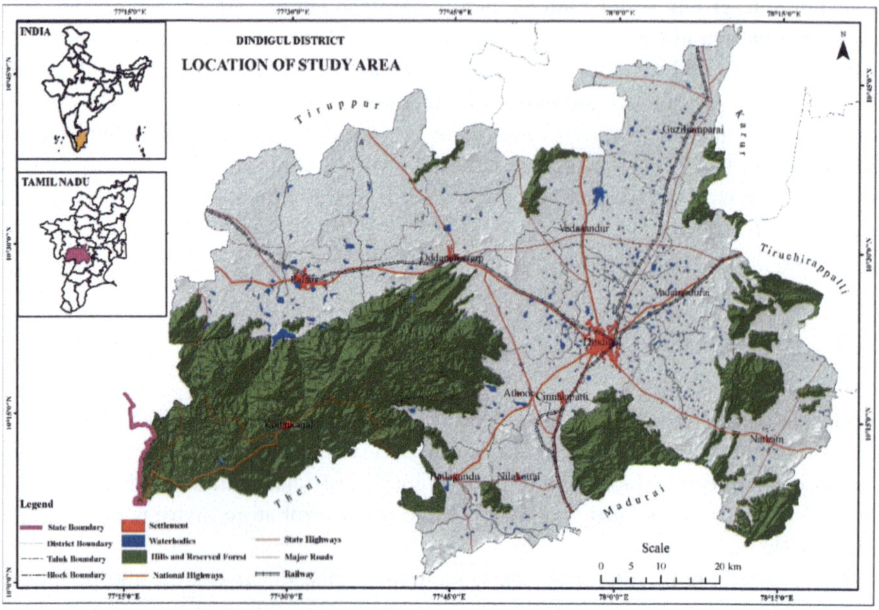

Fig. 25.2 Geographical features in the study area

received 209.9 mm of rainfall, compared to the average of 251.4 mm during the study period. During the retreat monsoon (from October to December), the actual rainfall was 468.6 mm, which exceeded the usual rainfall of 399.2 mm. Groundwater is the source of drinking water in Dindigul. There are about 1589 leather tanning industries in India; Tamilnadu alone accounts for 833. Of 833 leather tanning industries in Tamilnadu, 61 tannery units (7.3%) are located in the Dindigul district. Of the 61 leather tanning industries in the Dindigul district, 38 leather tannery units (62.3%) are in the Dindigul panchayat union (Thiripurasundari & Ponsakthisurya, 2018). The industrial effluents from these leather tanneries have polluted the groundwater and affect this area's health. (Karunanidhi et al., 2021).

25.3 Methodology

The study relied on both primary and secondary data sources. Approximately 400 questionnaires were gathered in all sample areas using uniform random sampling. Information regarding disease occurrence, water usage, medical checkup details, regular medical tests, drinking water quality, etc. Secondary data was collected from several sources, including the Central Groundwater Board, the district Pollution Control Board, and the district statistical office in 2022. Statistics on the water quality index, the tannery industry and its location, subsurface water resources, and other data were gathered. The study area map was prepared using the SOI (Survey of India Toposheet) at a scale of 1:50,000. The toposheet was digitized using Arc GIS software, and the coordinate system used for georeferencing is the UTM zone. The location of each sample well was pinpointed, and the results of each chemical characteristic of water were added as attributes to the point, resulting in the final geodatabase (Fig. 25.3). The spatial analysis method has been extensively used, particularly the interpolation technique (Saravanabavan et al., 2023). In this work, one of the interpolation approaches, the Inverse Distance Weight (IDW) method, is employed to generate surfaces. Cell values are determined by inverse distance weighted interpolation, which uses a linearly weighted mixture of sample points. Weight depends on the inverse distance. After building the base setup, the index of water quality was measured using the BIS (Bureau of Indian Standards)(Ram et al., 2021). Chemical parameters used to calculate the water quality index include pH, TDS, SO_4^{2-}, Cl^-, Na^+, K^+, Ca^{2+}, Mg^{2+}, and F^-. Nine parameters were evaluated to determine the water quality index. (Table 25.1) (World Health Organization, 2017; El-Kholy et al., 2022; Rahman et al., 2022). Weight was assigned based on the relative relevance of water quality, which was calculated using a weighted arithmetic index technique (Islam et al., 2021; Prasad et al., 2021).

Water Quality Index (WQI) = $\Sigma qiwi$.

Where qi (water quality rating) = $100 \times (Va-Vi)/(Vs-Vi)$.

When Va = actual value present in the water sample.

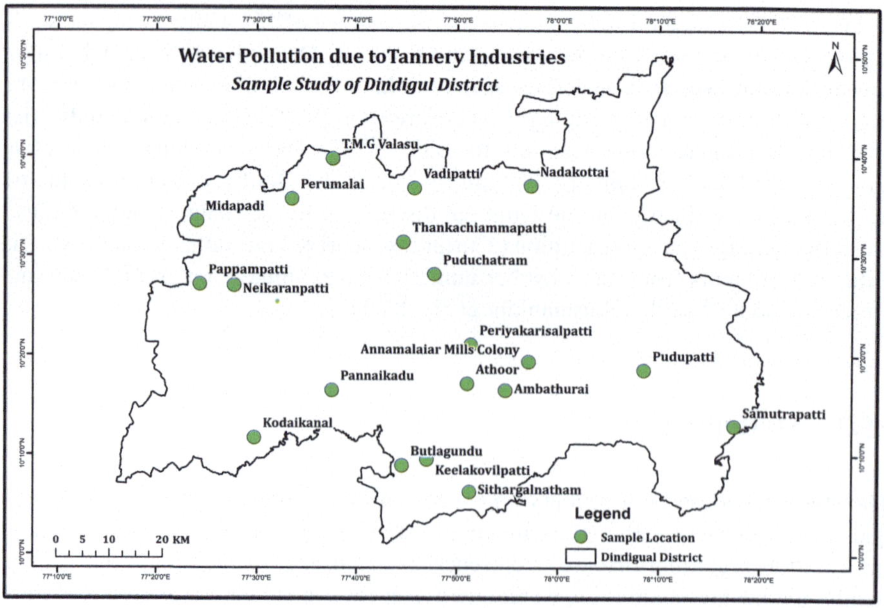

Fig. 25.3 Selection of sample location in Dindigul district

Table 25.1 BIS standards of chemical parameters and WQI

Chemical parameters	BIS standards	Weight (Wi)	Relative weight
pH	8.5	3	0.103
TDS (mg/L)	500	5	0.172
SO_4^{2-} (mg/L)	200	3	0.103
Cl^- (mg/L)	250	3	0.103
Na^+ (mg/L)	200	3	0.103
K^+ (mg/L)	12	1	0.034
Ca^{2+} (mg/L)	75	2	0.069
Mg^{2+} (mg/L)	30	2	0.069
F^- (mg/L)	1	5	0.172
		$\Sigma wi = 27$	$\Sigma Wi = 1.00$
WQI range	*Type of water**		
< 50	Excellent		
50–100	Good water		
100–200	Poor water		
200–300	Very poor water		
> 300	Unfit for drinking		

All the values are in mg/l except pH

*In assessing the groundwater quality for human use, each parameter is categorized into five classes

Vi = ideal value (0 for all parameters except pH(7.0) and DO (14.6 mg/l))

Vs = standard value

Wi (unit weight) = K/S_n.

Where K (constant) = 1.

1/Vs1 + 1/Vs2 + 1/Vs3 + 1/Vs4........ + 1/Vsn.

Sn = 'n' the number of standard values.

The data were converted into percentage and ratio measures using descriptive statistical techniques. The geochemical pollution from the tanning industries was studied with the help of the application of multivariate statistical methods known as 'Factor analysis' (Monira et al., 2024; Suvarna et al., 2023; Atique et al., 2020; Jesudhas et al., 2017; Saravanabavan & Shanmuganandan, 1998). It allows us to simplify a set of complex variables or items using statistical procedures to explore the underlying dimensions that explain the relationships between the multiple variables (Saravanabavan et al., 2021a, 2021b, 2020, 2019). The factor analysis is also performed subsequently to measure the impact of tannery effluence.

25.4 Results and Discussion

It is worth researching land or water quality changes over a 10-year interval to properly identify differences between objects (Natesan et al., 2022; Rowe et al., 2013; Rusydi et al., 2023). In the study area, Dindigul has chemical parameters analysis during pre-and post-monsoon periods for two different periods, 2012 and 2022, which are discussed below.

25.4.1 PH

During the year 2012, the pH value ranges from 7.6 to 8.6 in the post-monsoon season. The maximum value is found in Atoor, and the minimum value is found in Pappampatti, Aadipatti, and Keelakovilpatti. However, the pH concentration ranges from 7.2 to 8.3 in the pre-monsoon period. A high pH value was recorded in Nadakottai and Keelakovilpatti, and a low pH value was concentrated in Midapadi. The north-eastern part of the Natham block comes under the permissible limit during the study (Tables 25.2, 25.3). Chemicals, minerals, pollutants, soil or bedrock composition, and any other contaminants that interact with a water supply will result in an imbalance in the water's natural pH (Sajib et al., 2024). In short, environmental factors significantly influence water pH, whether high or low (Anuja et al., 2024; Sharma et al., 2024). Dindigul district benefits from a retreated monsoon season from October to December. The pH level in groundwater changed before and after monsoons. During the 2022 monsoon season, the pH fluctuates from 7.3 and 8.8. The high pH value was recorded in the village of Nadakottai in Nilakottaiblock, and the low pH value was recorded in the village of Kodaikanal in Kodaikanal block. In the pre-monsoon period, the pH value ranges between 7.9 and 8.4. The maximum value of pH was recorded in the village of Ambathurai in the Dindigul block, and the minimum value was concentrated in the village of Kodaikanal, Pannaikadu in the Kodaikanal block, and Keelakovilpatti in Nilakottai block, and T.M.G. Valasu in Palani block (Tables 25.3, 25.4).

25.4.2 Total Dissolved Solids

The total concentration of dissolved minerals in water is a general indication of the overall suitability of water for many types of uses (Pande & Moharir, 2022). In 2012, the total dissolved solid ranged between 100 mg/l and 1966 mg/l in the post-monsoon period. The high value of the TDS was recorded in the village of Keelakovilpatti in Nilakottai block, and the minimum value of TDS was concentrated in the village of Kodaikanal in Kodaikanal block. No permissible limit is found in the center part of Reddiyarchatrm, the Southern part of Batlagundu, the northern part of Oddanchatram, and the southern part of Thoppampatti block (Table 25.3). In 2022, TDS ranged between 121 mg/l and 2545 mg/l post-monsoon period (Tables 25.4, 25.5). The maximum and minimum values of the TDS were recorded in the villages T.M.G.Valasu in Palani block and Kodaikanal in Kodaikanal block. The non-permissible limit is in part of Vadamadurai, the north part of Thoppampatti, and the western part of Guziliamparai block.

25.4.3 Magnesium

Magnesium was derived from the dissolution of magnesium calcite, gypsum, and dolomite from source rock (Rahman, et al., 2022). During the post-monsoon period of 2012, magnesium ranged between 6.07 mg/l and 228.42 mg/l in Kodaikanal and Keelakovilpatti (Table 25.3). In the pre-monsoon period of 2012, magnesium ranges between 3.6 mg/l and 160.3 mg/l in the villages of Kodaikanal and Ambathurai (Table 25.2). In 2022, during the pre-monsoon period, magnesium ranged between 4.8 mg/l and 136.1 mg/l, as recorded in Pannaikadu and T.M.G. Valasu villages (Table 25.4). However, there are slight variations in the post-monsoon period of 2022, with magnesium ranging between 7.2 mg/l and 131.2 mg/l in the villages of Perumalai and T.M.G. Valasu (Table 25.5).

Table 25.2 Chemical parameters on pre-monsoon season-Dindigul-2012

S. No.	Well no	Tahsil/taluk	Village	TDS	$NO^{2-} + NO_3^-$	Ca^{2+}	Mg^{2+}	Na^+	K^+	Cl	SO_4^{2-}
01	22002D	Dindigul	Ambathurai	1890	68	140	106.92	138	313	553	149
02	83029A	Dindigul	A.Colony	756	15	56	60.75	106	31	223	91
03	22003D	Dindigul	Athoor	267	7	30	21.87	9	38	28	7
04	22020D	Nilakkottai	Batlagundu	1590	50	120	70.47	166	250	425	130
05	83100A	Nilakkottai	K.Patti	890	11	52	41.31	147	102	163	120
06	22034D	Kodaikanal	Kodaikanal	86	1	18	3.64	8	3	11	1
07	83,537	Palani	Midapadi	1844	76	176	160.38	115	125	624	178
08	22021D	Nilakkottai	Nadakottai	902	17	60	48.6	81	172	106	65
09	83,524	Palani	Neikaranpati	461	6	70	37.66	25	17	64	103
10	83552A	Kodaikanal	Pannaikadu	270	2	34	31.59	7	18	43	36
11	83,039	Palani	Pappampatti	1026	48	160	53.46	81	45	312	45
12	83512A	Dindigul	P.Patti	1119	54	136	82.62	104	22	312	100
13	83,555	Kodaikanal	Perumalai	188	8	14	12.15	5	37	32	37
14	83,036	Dindigul	Puduchatram	1116	39	96	89.91	133	27	262	230
15	22014D	Dindigul	Pudupatti	627	1	74	52.24	23	94	142	46
16	22023D	Natham	Samutrapatti	524	4	24	49.81	83	30	110	27
17	22022D	Nilakkottai	Sithargal	807	27	38	61.96	138	36	195	67
18	83,540	Palani	T.M.G.Valau	1928	84	128	94.77	265	133	468	403
19	83,529	Palani	T.Patti	1493	67	192	72.9	221	12	440	28
20	83,542	Palani	Vadipatti	1095	56	48	72.9	7	301	181	29

(continued)

Table 25.2 (continued)

S. No.	Well no	Tahsil/taluk	Village	TDS	$NO^{2-} + NO_3^-$	Ca^{2+}	Mg^{2+}	Na^+	K^+	Cl	SO_4^{2-}
S. No	*Well no*	*Tahsil/taluk*	*Village*	CO_3^{2-}	*HCO3–*	*F*	*pH*	*EC*	*HAR*	*SAR*	*Na%*
01	22002D	Dindigul	Ambathurai	0	378.2	0.14	7.9	2900	790	2.14	20.15
02	83029A	Dindigul	A.Colony	0	244	0.57	7.8	1270	390	2.34	34.94
03	22003D	Dindigul	Athoor	1.26	168.7	0.41	7.9	460	165	0.30	8.40
04	22020D	Nilakkottai	Batlagundu	0	414.8	0.5	7.8	2540	590	2.97	28.43
05	83100A	Nilakkottai	K.Patti	12	408.7	0.42	8.3	1500	300	3.69	42.64
06	22034D	Kodaikanal	Kodaikanal	0.38	64.59	0.1	7.8	150	60	0.45	21.44
07	83,537	Palani	Midapadi	0	256.2	0.08	7.2	3000	1100	1.51	16.57
08	22021D	Nilakkottai	Nadakottai	6	573.4	1.6	8.3	1440	350	1.88	23.62
09	83,524	Palani	Neikaranpati	1.18	198.79	0.36	7.8	810	330	0.60	13.40
10	83552A	Kodaikanal	Pannaikadu	0.88	149.08	0.12	7.8	480	215	0.21	6.02
11	83,039	Palani	Pappampatti	0	237.9	0.35	7.5	1700	620	1.42	20.65
12	83512A	Dindigul	P.Patti	0	244.0	0.19	7.8	1890	680	1.74	24.23
13	83,555	Kodaikanal	Perumalai	0.05	24.94	0.08	7.3	270	85	0.24	7.60
14	83,036	Dindigul	Puduchatram	0	207.4	0.66	7.5	1830	610	2.34	30.99
15	22014D	Dindigul	Pudupatti	0	384.3	0.39	7.7	1130	400	0.50	8.78
16	22023D	Natham	Samutrapatti	0	366.0	0.75	7.9	920	265	2.22	37.32
17	22022D	Nilakkottai	Sithargal	0	305.0	0.87	8	1360	350	3.21	43.13
18	83,540	Palani	T.M.G.Valau	0	128.1	0.52	7.7	2780	710	4.33	39.59
19	83,529	Palani	T.Patti	0	457.5	0.22	8	2560	780	3.44	37.70
20	83,542	Palani	Vadipatti	0	396.5	0.09	7.8	1620	420	0.15	1.86

Table 25.3 Chemical parameters on post monsoon season-Dindigul district-2012

S. No.	Well no	Tahsil/taluk	Village	TDS	NO^{2+} NO^{3-}	Ca^{2+}	Mg^{2+}	Na^{+}	K^{+}	Cl	SO_4^{2-}
01	22002D	Dindigul	Ambathurai	1362	6	52	114.21	242	63	638	138
02	83029A	Dindigul	A.Colony	1423	44	40	150.66	242	8	624	34
03	22003D	Dindigul	Athoor	331	0.05	30	38.88	46	4	60	22
04	22020D	Nilakkottai	Batlagundu	1509	60	60	68.04	276	133	475	86
05	83100A	Nilakkottai	K.Patti	1966	49	192	228.42	161	4	723	278
06	22034D	Kodaikanal	Kodaikanal	100	0.05	16	6.075	16	1	18	4
07	83,537	Palani	Midapadi	1735	61	168	170.1	113	66	695	201
08	22021D	Nilakkottai	Nadakottai	1541	72	64	133.65	230	63	425	125
09	83,524	Palani	Neikaranpati	731	4	70	53.46	69	4	160	264
10	83552A	Kodaikanal	Pannaikadu	383	9	32	29.16	55	19	74	8
11	83,039	Palani	Pappampatti	847	34	96	30.375	152	6	301	43
12	83512A	Dindigul	P.Patti	1856	69	288	126.36	133	11	677	259
13	83,555	Kodaikanal	Perumalai	263	5	40	18.225	23	4	60	38
14	83,036	Dindigul	Puduchatram	1138	48	76	133.65	104	4	333	197
15	22014D	Dindigul	Pudupatti	377	2	26	44.955	21	18	18	11
16	22023D	Natham	Samutrapatti	519	2	42	57.105	71	6	135	36
17	22022D	Nilakkottai	Sithargal	131	0.05	14	14.58	18	2	28	8
18	83,540	Palani	T.M.G.Valau	1783	60	160	126.36	207	59	553	350
19	83,529	Palani	T.Patti	1951	66	216	155.52	207	22	794	187
20	83,542	Palani	Vadipatti	1369	69	48	106.92	184	94	355	168

(continued)

Table 25.3 (continued)

S. No.	Well no	Tahsil/taluk	Village	TDS	NO^{2+} NO^{3-}	Ca^{2+}	Mg^{2+}	Na^+	K^+	Cl	SO_4^{2-}
S. No.	*Well no*	*Tahsil/taluk*	*Village*	CO_3^{2-}	$HCO3^-$	*F*	*pH*	*EC*	*HAR*	*SAR*	*Na%*
01	22002D	Dindigul	Ambathurai	0	176.9		8	2450	600	4.30	43.62
02	83029A	Dindigul	A.Colony	0	262.3		8	2400	720	3.92	41.90
03	22003D	Dindigul	Athoor	7.57	202.23		8.6	590	235	1.31	29.43
04	22020D	Nilakkottai	Batlagundu	0	292.80		7.8	2380	430	5.79	50.02
05	83100A	Nilakkottai	K.Patti	0	329.40		7.6	3500	1420	1.86	19.74
06	22034D	Kodaikanal	Kodaikanal	0.48	64.48		7.9	180	65	0.86	34.45
07	83,537	Palani	Midapadi	0	97.60		7.8	2850	1120	1.47	16.96
08	22021D	Nilakkottai	Nadakottai	0	366.00		7.7	2540	710	3.76	38.77
09	83,524	Palani	Neikaranpati	0	183.00	0.66	7.7	1090	395	1.51	27.29
10	83552A	Kodaikanal	Pannaikadu	1.24	208.73		7.8	660	200	1.69	34.80
11	83,039	Palani	Pappampatti	0	140.30		7.6	1400	365	3.46	47.04
12	83512A	Dindigul	P.Patti	0	115.90	0.77	7.9	3090	1240	1.64	18.76
13	83,555	Kodaikanal	Perumalai	1.11	93.83		8.1	420	175	0.76	21.76
14	83,036	Dindigul	Puduchatram	0	158.60		7.7	1760	740	1.66	23.30
15	22014D	Dindigul	Pudupatti	2.47	262.48		8	570	250	0.58	14.34
16	22023D	Natham	Samutrapatti	0	323.30		7.7	950	340	1.68	30.77
17	22022D	Nilakkottai	Sithargal	1.1	73.82		8.2	240	95	0.80	28.65
18	83,540	Palani	T.M.G.Valau	0	122.00		7.7	2950	920	2.97	31.16
19	83,529	Palani	T.Patti	0	152.50		7.7	3300	1180	2.62	27.17
20	83,542	Palani	Vadipatti	0	213.50		7.6	2010	560	3.38	37.05

Table 25.4 Chemical parameters on pre monsoon season—Dindigul district:2022

S. No.	Well no	Tahsil/taluk	Village	TDS	NO_2^+ NO_3^-	Ca^{2+}	Mg^{2+}	Na^+	K^+	Cl^-	SO_4^{2-}
01	22002D	Dindigul	Ambathurai	1597	33	116	87.48	258	127	418	137
02	83029A	Dindigul	A.Colony	1250	9	80	111.8	221	17	624	95
03	22003D	Dindigul	Athoor	259	6	20	21.87	44	6	21	10
04	22020D	Nilakkottai	Batlagundu	1652	36	64	87.48	368	19	475	125
05	83100A	Nilakkottai	K.Patti	1851	28	144	102.1	336	7	411	614
06	22034D	Kodaikanal	Kodaikanal	153	5	22	12.15	13	5	18	7
07	83,537	Palani	Midapadi	1659	62	156	104.5	196	86	440	204
08	22021D	Nilakkottai	Nadakottai	776	6	10	48.6	207	13	216	103
09	83,524	Palani	Neikaranpati	452	4	54	32.81	60	9	71	82
10	83552A	Kodaikanal	Pannaikadu	82	3	18	4.86	3	2	7	6
11	83,039	Palani	Pappampatti	160	5	22	17.01	6	4	11	17
12	83512A	Dindigul	P.Patti	1170	28	56	68.04	225	74	333	118
13	83,555	Kodaikanal	Perumalai	181	8	22	9.72	23	8	32	13
14	83,036	Dindigul	Puduchatram	1341	31	76	131.2	198	10	376	274
15	22014D	Dindigul	Pudupatti	330	11	12	31.59	60	7	39	20
16	22023D	Natham	Samutrapatti	533	3	30	49.82	104	8	99	29
17	22022D	Nilakkottai	Sithargal	684	12	18	44.96	166	11	191	71
18	83,540	Palani	T.M.G.Valau	2151	69	152	136.1	313	151	709	288
19	83,529	Palani	T.Patti	1462	72	140	72.9	230	28	333	156
20	83,542	Palani	Vadipatti	1047	43	68	80.19	117	94	199	86

(continued)

Table 25.4 (continued)

S. No.	Well no	Tahsil/taluk	Village	TDS	NO_2^+ NO_3^-	Ca^{2+}	Mg^{2+}	Na^+	K^+	Cl^-	SO_4^{2-}
S. No.	*Well no*	*Tahsil/taluk*	*Village*	CO_3^{2-}	HCO_3^-	F^-	*pH*	*EC*	*HAR*	*SAR*	*Na%*
01	22002D	Dindigul	Ambathurai	6	597.8	0.5	8.4	2660	650	4.404	40.87
02	83029A	Dindigul	A.Colony	0	122	0.6	8.2	2210	660	3.743	41.36
03	22003D	Dindigul	Athoor	2.7	182.2	0.6	8.2	480	140	1.618	39.33
04	22020D	Nilakkottai	Batlagundu	0	561.2	0.4	8.1	2750	520	7.022	59.53
05	83100A	Nilakkottai	K.Patti	0	225.7	1.6	7.9	2900	780	5.236	48.11
06	22034D	Kodaikanal	Kodaikanal	0.67	89.29	0.1	7.9	270	105	0.552	20.26
07	83,537	Palani	Midapadi	0	396.5	0.2	8	2620	820	2.979	31.45
08	22021D	Nilakkottai	Nadakottai	6	292.8	1.3	8.3	1360	225	6.004	65.08
09	83,524	Palani	Neikaranpati	3.01	201.9	0.6	8.2	790	270	1.589	31.69
10	83552A	Kodaikanal	Pannaikadu	0.33	44.63	0.1	7.9	140	65	0.162	8.81
11	83,039	Palani	Pappampatti	1.47	98.45	0.4	8.2	280	125	0.234	9.12
12	83512A	Dindigul	P.Patti	0	347.7	0.4	8.2	1960	420	4.777	48.75
13	83,555	Kodaikanal	Perumalai	0.6	64.35	0.1	8	290	95	1.027	32.24
14	83,036	Dindigul	Puduchatram	6	268.4	0.6	8.3	2280	730	3.189	36.71
15	22014D	Dindigul	Pudupatti	2.1	177.8	0.5	8.1	580	160	2.064	43.59
16	22023D	Natham	Samutrapatti	6	390.4	1.2	8.3	980	280	2.704	43.81
17	22022D	Nilakkottai	Sithargal	0	256.2	0.6	8.1	1180	230	4.762	59.67
18	83,540	Palani	T.M.G.Valau	0	195.2	1.5	7.9	3630	940	4.443	37.54
19	83,529	Palani	T.Patti	0	366	0.3	8.2	2300	650	3.926	42.20
20	83,542	Palani	Vadipatti	0	420.9	0.1	8.2	1730	500	2.277	29.10

Table 25.5 Chemical parameters on post monsoon season—Dindigul district:2022

S. No.	Well no	Tahsil/taluk	Village	TDS	NO_2^+ NO_3^-	Ca^{2+}	Mg^{2+}	Na^+	K^+	Cl^-	SO_4^{2-}
01	22002D	Dindigul	Ambathurai	1336	35	52	87.48	230	102	397	113
02	83029A	Dindigul	A.Colony	1354	15	140	111.7	184	16	603	80
03	22003D	Dindigul	Athoor	220	5	30	15.79	23	4	39	20
04	22020D	Nilakkottai	Batlagundu	1400	25	128	82.62	219	70	440	92
05	83100A	Nilakkottai	K.Patti	1853	18	184	12.15	460	8	454	403
06	22034D	Kodaikanal	Kodaikanal	121	5	10	10.93	13	4	18	10
07	83,537	Palani	Midapadi	1326	31	124	65.61	207	70	397	155
08	22021D	Nilakkottai	Nadakottai	714	19	46	61.96	81	63	142	63
09	83,524	Palani	Neikaranpati	976	19	80	58.32	147	63	234	98
10	83552A	Kodaikanal	Pannaikadu	326	19	22	13.36	53	23	57	18
11	83,039	Palani	Pappampatti	879	22	128	48.6	115	13	241	65
12	83512A	Dindigul	P.Patti	1100	25	40	60.75	230	70	355	92
13	83,555	Kodaikanal	Perumalai	160	10	12	7.29	25	8	28	13
14	83,036	Dindigul	Puduchatram	1212	32	128	121.5	104	8	326	213
15	22014D	Dindigul	Pudupatti	346	2	30	38.88	44	7	35	23
16	22023D	Natham	Samutrapatti	524	8	40	43.74	92	8	89	31
17	22022D	Nilakkottai	Sithargal	1309	14	132	85.05	221	20	383	118
18	83,540	Palani	T.M.G.Valau	2545	81	264	131.2	345	133	851	278
19	83,529	Palani	T.Patti	599	9	82	35.23	83	9	160	48
20	83,542	Palani	Vadipatti	1052	44	100	70.47	120	47	199	124

(continued)

Table 25.5 (continued)

S. No.	Well no	Tahsil/taluk	Village	TDS	NO_2^+ NO_3^-	Ca^{2+}	Mg^{2+}	Na^+	K^+	Cl^-	SO_4^{2-}
S. No.	*Well no*	*Tahsil/taluk*	*Village*	CO_3^{2-}	HCO_3^-	F^-	*pH*	*EC*	*HAR*	*SAR*	*Na%*
01	22002D	Dindigul	Ambathurai	6	390.4	0.08	8.4	2240	490	4.52	44.65
02	83029A	Dindigul	A.Colony	0	305	0.42	7.8	2450	810	2.81	32.54
03	22003D	Dindigul	Athoor	1.6	108.3	0.46	8.2	380	140	0.85	25.65
04	22020D	Nilakkottai	Batlagundu	6	500.2	0.59	8.3	2420	660	3.71	38.88
05	83100A	Nilakkottai	K.Patti	0	500.2	1.5	7.8	2970	510	8.87	65.83
06	22034D	Kodaikanal	Kodaikanal	0.09	49.9	0.65	7.3	200	70	0.68	27.36
07	83,537	Palani	Midapadi	0	341.6	0.05	8	2200	580	3.74	40.23
08	22021D	Nilakkottai	Nadakottai	18	305	1.63	8.8	1220	370	1.83	28.12
09	83,524	Palani	Neikaranpati	0	427	0.56	8.2	1680	440	3.05	38.07
10	83552A	Kodaikanal	Pannaikadu	1.1	93.8	0.95	8.1	500	110	2.20	45.28
11	83,039	Palani	Pappampatti	0	341.6	0.66	8.2	1520	520	2.19	31.82
12	83512A	Dindigul	P.Patti	12	256.2	0.36	8.6	1840	350	5.35	53.24
13	83,555	Kodaikanal	Perumalai	0.16	34.8	0.11	7.7	230	60	1.40	43.66
14	83,036	Dindigul	Puduchatram	0	335.5	0.34	7.8	2030	820	1.58	21.43
15	22014D	Dindigul	Pudupatti	3.82	256.1	0.4	8.2	660	235	1.25	28.19
16	22023D	Natham	Samutrapatti	6	353.8	1.14	8.3	980	280	2.39	40.83
17	22022D	Nilakkottai	Sithargal	6	561.2	1.1	8.4	2320	680	3.69	40.54
18	83,540	Palani	T.M.G.Valau	0	366	1.11	8	4180	120	4.33	35.41
19	83,529	Palani	T.Patti	0	286.7	0.74	8.1	1040	350	1.93	33.33
20	83,542	Palani	Vadipatti	0	390.4	0.1	8.1	1720	540	2.25	30.33

25.4.4 *Potassium*

The potassium levels in the groundwater ranged between 3 mg/l and 313 mg/l in the pre-monsoon period of 2012 (Table 25.2). These ranges were recorded in the villages of Kodaikanal and Ambathurai. In the year 2012, post-monsoon season, there was an extensive range of potassium changes, ranging from 1 mg/l and 133 mg/l in the villages of Kodaikanal and Batlagundu (Table 25.3). In the 2022pre monsoon season, the potassium level in the groundwater range between 2 mg/l and 151 mg/l is recorded in Pannaikadu and T.M.G. Valasu village (Table 25.4). Potassium life varies depending on seasonal rainfall (Pant et al., 2024). In the post-monsoon period of 2022, the potassium level in the groundwater range is 4 mg/l–133 mg/l in the village of Athoor and T.M.G Valasu (Table 25.5).

25.4.5 *Calcium*

It is found in alkaline environments. During the monsoon season in 2012, the calcium content in groundwater ranged from 11 mg/l to 624 mg/l in the villages of Kodaikanal and Midapadi (Table 25.2). Calcium levels above the permitted limit have been detected in minor spots around the district (Sinha et al., 2023). However, in the post-monsoon season of 2012, the calcium content in groundwater ranged between 14 mg/l and 288 mg/l in the villages of Sitharangalnatham and Periyakarisalpatti (Table 25.3). In the villages of Nillakottai and Midapadi, calcium levels in groundwater ranged from 10 mg/l to 156 mg/l before the monsoon season in 2022 (Table 25.4). A little fraction of the Thoppampatti block is within the not-allowable limit. During the post-monsoon period of the same year, calcium levels recorded in the groundwater ranged between 10 mg/l and 264 mg/l in the Kodaikanal and T.M.G Valasu villages (Table 25.5).

25.4.6 *Sodium*

In the study area, sodium concentration in water samples during the pre-monsoon season (Sinha et al., 2023) of 2012 ranged from 5 mg/l to 265 mg/l, as measured in Perumalai and T.M. G. Valasu (Table 25.2). Sodium content in post-monsoon water samples ranged from 16 mg/l to 276 mg/l, with the highest concentrations found in the villages of Kodaikanal and Batlagund (Table 25.3). During the pre-monsoon period of 2022, salt concentrations range from 3 mg/l to 368 mg/l in the villages of Pannaikadu and Batlagundu (Table 25.4). In the post-monsoon season of 2022, sodium levels in water samples ranged from 13 mg/l to 360 mg/l, with the highest concentrations reported in the villages of Kodaikanal and Keelakovilpatti (Table 25.5).

25.4.7 Sulphate

The sulfate concentration varied from 1 mg/l to 403 mg/l during the pre-monsoon season of the year of 2012, recorded in the villages of Kodaikanal and T.M. G Valasu. During the post-monsoon season of this same year, sulfate content concentration in the water samples ranges from 4 mg/l to 350 mg/l recorded in the villages of Kodaikanal and T.M.G Valasu (Tables 25.2, 25.3). During the pre-monsoon season of 2022, sulfate levels in the water samples ranged between 6 mg/l and 614 mg/l found in the villages of Pannaikadu and Keelakovilpatti (Tables 25.4, 25.5).

In the post-monsoon season of the same year, sulfate in the water samples ranged high (Tiwari & De Maio, 2018).Between 10 mg/l and 403 mg/l was recorded in the villages of Kodaikanal and Keelakovilpatti.

25.4.8 Chloride

The concentration of chlorides in the Dindigul district during the Pre-monsoon season in 2012 ranged from 11 mg/l to 624 mg/l at Kodaikanal and Midapadi. The post-monsoonal season of the same year depicts a range from 18 mg/l to 794 mg/l at Kodaikanal and Thangachiammapatti (Tables 25.2, 25.3). During the pre-monsoon season of 2022, chloride-concentrated water samples range from 7 mg/l to 709 mg/l at Pannaikadu and T.M.GValasu (Table 25.4). Post monsoon period of this same year depicted chloride ranges from 18 mg/l to 851 mg/l at Samutrapatti and T.M.G Valasu (Dey et al., 2023) (Table 25.5).

25.4.9 Fluoride

One of the main trace elements in groundwater is fluoride, which is generally a natural constituent. (Bisht et al., 2023; Eswari, et al., 2021; Sunitha, 2020; Muralidhara Reddy et al., 2014). Bedrock containing fluoride minerals is generally responsible for high concentrations of this ion in groundwater. During the pre-monsoon period of 2012, the fluoride concentrated in the water samples ranged from 0.08 mg/l to 1.6 mg/l at Perumalai and Nadakottai. In the post-monsoon season of the same year, fluoride in the water samples ranges from 0.66 mg/l to 0.77 mg/l at Neikaranpatti and Periyakarisalpatti (Tables 25.2, 25.3). The whole district falls under desirable limits during the study period. During the pre-monsoon period of 2022, fluoride concentrated in the water samples ranges from 0.1 mg/l to 0.6 mg/l at Kodaikanal, Pannaikadu, Perumalai, Vadipatti and Keelakovilpatti. In the post-monsoon period of the same year, fluoride concentrated in the water samples ranged from 0.05 mg/l to 1.63 mg/l at Midapadi and Nadakottai (Tables 25.4, 25.5).

25.4.10 *Bi-Carbonate*

During the 2012 pre-monsoon season, bicarbonate concentrations in groundwater samples ranged from 24.94 mg/l to 573.4 mg/l in Perumalai and Nadakottai. During the post-monsoon season of the same year, the Bi-carbonate concentration in groundwater samples ranged from 64.4 mg/l to 366 mg/l at Kodaikanal and Nadakottai (Tables 25.2, 25.3). Bi-carbonate concentrated water samples ranged from 44.62 mg/l to 597 mg/l in Pannaikadu and Ambathurai during the 2022 pre-monsoon season, and 34.8 mg/l–561.2 mg/l at Perumalai and Sithargalnatham during the post-monsoon season (Abhishek et al., 2023). Tables 25.4 and 25.5 displays the chemical parameter data in a tabular style for both seasons, pre-monsoon and post-monsoon.

25.5 Water Quality Index

WQI is a mathematical tool for assessing groundwater quality (El-Kholy, et al., 2022). The value varies from one standard method to another (WHO, 2017). The values in this study are classified into 5 categories based on the BIS (Bureau of Indian Standards) standard (Lin, et al., 2022). They are excellent, good water, terrible water, extremely poor water, and unsafe for consumption (Table 25.1). Depicts water quality index analysis before and after the monsoon for two separate seasons, 2012 and 2022 (Fig. 25.4). Unfit-for-drinking conditions were observed in a small fraction of the Oddanchatram, Vadasandur, Guziliamparai, and Batlagundu blocks before the monsoon season in 2012. Very poor water conditions concentrated in a small part of Thoppampatti, Oddanchatram, Vadasandur, Guziliamparai, Batlagundu, Athoor, Nilakottai, Shanarpatti, Natham. Poor water conditions were found in the Thoppampatti, Kodaikanal, Athoor, Shanarpatti, Natham, Vadasandur, Oddanchatram, and Guziliamparai blocks. The remaining parts of the area come under good and excellent water quality. Post-monsoonal season of the same year, poor water was found only in the Batlagundu block.

Good quality water concentrates in small portions of Thoppampatti, Oddanchatram, Vadasandur, Guziliamparai, Kodaikanal, Batlagundu, Athoor, and Dindigul. A major part of the area of the district comes under excellent water quality. Poor water quality has been reported in Guziliamparai, Vadasandur, and Reddiyarchatram blocks during the 2022 pre-monsoon season. Guziliamparai, Vadasandur, Oddanchatram, Kodaikanal, Reddiyarchatram, Athoor, Vadamadurai, Shanarpatti, Natham, and small sections of Nilakottai and Batlagundu have particularly good water quality. The majority of the district's water is of exceptional quality. Following the monsoonal season of the same year, poor water quality was discovered in a small area of the Thoppampatti and Batlagundu blocks. Water quality is excellent in Thoppampatti, Palani, Athoor, and Dindigul, and a small amount of Oddanchatram, Vadasandur, Guziliamparai, Reddiyarchatram, Natham, Nilakottai, and Batlagundu blocks. The remaining parts of the district come under excellent water quality. The values are

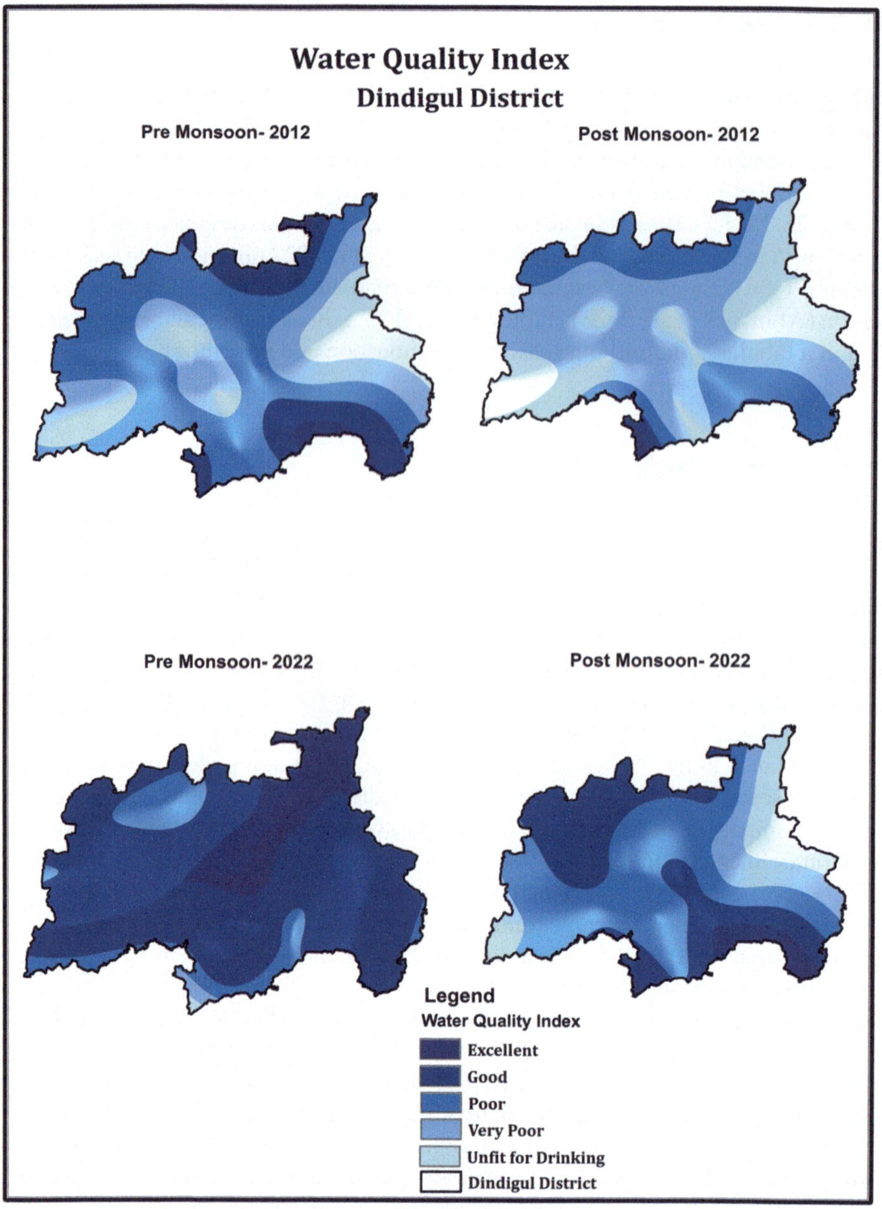

Fig. 25.4 Water quality in the pre-monsoon and post-monsoon periods of 2012 and 2022

in the tabular format for all the parameters for both the seasons and both periods (Table 25.6).

Recharged water, atmospheric precipitation, and freshwater surface water contribute to groundwater quality. Assessing and monitoring groundwater quality is crucial to maintaining these resources' long-term safety for a range of applications. The study sought to evaluate the geographical variance of key groundwater quality indicators such as pH, TDS, SO_4^{2-}, Cl^-, Na^+, K^+, Ca^{2+}, Mg^{2+}, and F^- The WQI was made utilizing the abovementioned parameters for two seasons and two periods. This research helps us understand the water's characteristics and establish the necessary management measures to protect groundwater resources in affected areas. Tanneries are mainly concentrated inthe study area. Tanneries are located primarily in the Batlagundu and Dindigul blocks of the Dindigul district. Tanning pollution is also found in the lower level of groundwater (Hansen, et al., 2020). The presence of arsenic in groundwater reduces the availability of safe drinking water, resulting in skin rashes, kidney and liver difficulties, cancer, and other problems (Reddy et al.,

Table 25.6 Water quality index values for Dindigul district—2012 and 2022

S. No.	Latitude	Longitude	Name	Pre-monsoon 2012	Post monsoon 2012	Pre-monsoon 2022	Post monsoon 2022
01	10.275	77.913	Ambathurai	188.90	49.65	22.63	72.37
02	10.286	77.85	A.Colony	28.38	14.73	19.96	11.19
03	10.309	78.141	Athoor	60.70	18.75	54.64	13.92
04	10.149	77.742	Batlagundu	151.36	85.63	7.33	55.11
05	10.88	77.83	K.Patti	107.84	48.39	11.74	50.56
06	10.105	77.854	Kodaikanal	30.81	9.62	16.00	28.00
07	10.195	77.495	Midapadi	7.24	6.86	15.48	4.86
08	10.323	77.951	Nadakottai	26.91	19.98	30.94	23.28
09	10.469	77.795	Neikaranpati	25.21	14.76	87.71	18.22
10	10.159	77.783	Pannaikadu	69.01	21.32	63.56	17.68
11	10.351	77.856	Pappampatti	24.54	24.65	101.65	54.34
12	10.275	77.625	P.Patti	17.11	17.85	65.62	21.13
13	10.863	77.548	Perumalai	23.37	10.88	25.36	9.58
14	10.215	78.29	Puduchatram	25.52	11.50	10.93	15.91
15	10.452	77.405	Pudupatti	34.51	11.67	12.46	20.34
16	10.45	77.462	Samutrapatti	17.76	11.18	12.76	47.96
17	10.523	77.744	Sithargal	21.75	29.83	25.59	15.71
18	10.558	77.4	T.M.G.Valau	82.67	53.87	19.31	52.42
19	10.662	77.626	T.Patti	87.68	48.15	16.61	94.58
20	10.613	77.762	Vadipatti	176.96	63.22	9.45	39.13

2023; Vinothini et al., 2019, 2021, 2022; Urme et al., 2021; Saravanabavan, 2011). Tannery discharge is distinguished by its significant influence on health.

25.6 Discussion

The factor analysis matrix with the identification of seven major dimensions explained the total variance of 87.92% with an eigenvalue ranging from 1.12 to 5.60 (Table 25.7). The first dimension, 'Tannery impact and chemical hazards' with an eigenvalue of 5.60 explained a total variance of 26.69. These dimensions contributed 19 variables, each revealing tannery impact and chemical hazards. She was followed by a second dimension on 'Tannery effluents and Health' characterized by an eigenvalue of 3.49, explaining a total variance of 15.26. This dimension contributed to 16 variables based on health indicators in the tannery. The third dimension, 'Groundwater and tannery impact,' accounts for an eigenvalue of 2.96 and a total variance of 13.29%, contributing 14 variables. Six variables contributed to the fourth dimension, namely tannery effluence and soil fertility, accounting for an eigenvalue of 2.37 and a percentage of variance of 11.03. The remaining five to seven dimensions, namely 'Tannery effluence and ecological imbalance,' 'Water hardness and tannery effluence.' and 'Reformatting of polluting tannery effluence, account for a total eigenvalue of 4.47 and a percentage of the variance of 21.65.

Table 25.7 Factor solution: the eigen values and the total percentage of variance explained by each of the seven factors

S. No.	Name of the dimension	Eigenvalue	Percentage of variance	Cumulative percentage
I	Tannery impact and chemical hazards	5.60	26.69	26.69
II	Tannery effluents and health	3.49	15.26	41.95
III	Groundwater and tannery impact	2.96	13.29	55.24
IV	Tannery effluence and soil fertility	2.37	11.03	66.27
V	Tannery effluence and ecological imbalance	1.92	9.22	75.49
VI	Water hardness and tannery effluence	1.43	7.24	82.73
VII	Reformatting of polluting tannery effluents	1.12	5.19	87.92

25.6.1 Dimension I: Tannery Impact and Chemical Hazards

The first dimension, Tannery Impact, and Chemical Hazards, had an eigenvalue of 5.60 and explained a total variation of 26.69. Tannery effluents contain a wide range of compounds hazardous to living creatures. Industrial waste includes toxicants, which have negative consequences caused by heavy chemicals (Nigam et al., 2023). These enter the food chain. These contaminants include detergent solvents, cyanides, heavy metals, minerals, organic acids, bleaching chemicals, dye pigments, and an extensive range of biocidal organic compounds. In the present study, most variables are loaded in the first dimension (Table 25.8). The variables with positive factor loading such as sodium chloride (0.92), sodium sulfide (0.91) sulphuric acid (0.72) ammonium sulfate (0.70) chromium (0.70) sulfate (0.72) explain this dimension as chemical hazards due to tannery effluents. These pollutants affect man's health by accumulating in food and water. The oil (0.53), fat (0.51), and dyes (0.57) are also responsible factors due to environmental hazards in the tannery. The factor loading such freshwater bodies (0.65), lakes (0.68), and streams (0.63) are carrying sources and polluted to tannery surface ground (0.60). Now water-related communicable diseases have virtually disappeared and more attention is being paid to chemical pollution.

25.6.2 Dimension II: Tannery Effluents and Health

The tannery effluents and health formed the second dimension with an eigenvalue of 3. 49 and a percentage of the total variance of 15.26. This dimension indicates the relationship between tannery effluents and health conditions (Table 25.8). The dimension explained the disorders that frequently happen in this area. Diseases like skin ulceration (0.90), allergic skin (0.88), dermatitis (0.83), ulceration (0.53), asthma (0.45), and internal disorder (0.38) are caused. The influence of calcium (−0.72) and Sodium Chloride (−0.90) are responsible for cancer and urinal bladder problems, respectively.

25.6.3 Dimension III: Groundwater and Tannery Impact

The third dimension, designated as groundwater and tannery impact, experienced a total variance of 13.29% with an eigenvalue of 2.96 (Table 25.8). This dimension highlighted the importance of groundwater playing a major role in the impact of tannery effluence (Kanagaraj et al., 2023). The tanning industry is a major consumer of water, and most of the water is discharged as waste. The hides and skins are subjected to soaking, liming, unhairing, fleshing, washing, tanning, dyeing, and finishing. Most of Dindigul and its suburbs rely on underground

Table 25.8 Rotated factor loading matrix of each dimension (I to III)

Variable name	Name of the variables	Rotated factor loading
Dimension: I tanning impact and chemical hazards		
36	Sodium chloride	0.92
37	Sodium sulphide	0.91
38	Sulphuric acid	0.72
30	Lime	0.78
39	Sodium	0.73
10	Carbonate	0.72
01	Ammonium sulphate	0.70
11	Chromium	0.70
40	Sulphate	0.72
20	Fat	0.51
19	Dyes	0.57
21	Freshwater bodies	−0.65
31	Lakes	−0.68
50	Streams	0.63
51	Tannery surface ground	−0.60
	Eigen value = 5.60	**Percentage of trace = 26.69**
Dimension II: tannery effluents and health		
18	Dermatitis	0.83
02	Allergic skin reaction	0.88
52	Skin ulceration	0.90
59	Ulceration	0.53
03	Asthma	0.45
62	Intestinal disorder	0.38
11	Chromium	−0.90
07	Blood pressure	0.53
36	Sodium choloride	−0.88
15	Calcium	−0.72
56	Stone formation	0.78
58	Urinary bladder problem	0.78
57	Urinary tract problem	0.68
33	Neptrons of the kidney	0.67
17	Cancer	0.51
	Eigen value = 3.49	Percentage of trace = 15.26
Dimension III: groundwater and tannery impact		

(continued)

Table 25.8 (continued)

Variable name	Name of the variables	Rotated factor loading
14	Chloride	−0.56
60	Phosphate	0.65
15	Calcium	0.73
37	Sodium sulphate	0.68
05	Ammonium chloride	−0.0.80
01	Ammonium sulphate	−0.78
53	Sodium acetate	0.83
54	Sodium format	0.88
55	Sodium bicarbonate	0.63
16	Chromium sulphate	−0.73
38	Sulphuric acid	0.81
36	Sodium chloride	−0.68
40	Sulphates	0.62
22	Fat liquaring pigments	−0.51
	Eigen value = 2.96	**Percentage of trace = 13.29**

drinking water sources, which are polluted, especially on the outskirts of tanning villages. Groundwater is affected when tannery wastewater is ponded or lagooned on land or discharged into dry beds. This dimension highlighted the significance of groundwater's vital role in the impact of tannery effluence.

25.6.4 Dimension IV: Tannery Effluence and Soil Fertility

The fourth dimension of Tannery effluence and soil fertility explained a total variance of 11.03% with an eigenvalue of 2.37 (Table 25.9).The quality of irrigation water is of permanent importance since it will affect the soils, resulting in poor plant growth and ultimately affecting crop yields. The tannery effluent affecting soil fertility is complex because too many chemicals are mixed in the effluent (De Silva 2023).

25.6.5 Dimension V: Tannery Effluence and Ecological Imbalance

The fifth dimension of the tannery effluence and ecological imbalance explained a total variance of 9.22% with an eigenvalue of 1.92 (Table 25.9). High toxic (0.89) effluents discharged by tanneries, and waste chemicals affect cultivable areas, resulting in reduced yields (−0,73) and degraded soil (−0.70). As a result, production and employment of villages are progressively reduced, and thus, the health, economy, and peace of people are systematically affected by the streams of poison released by the tanneries day in and day out, unmindful of their ecological effect

Table 25.9 Rotated factor loading matrix of each dimension (IV–VII)

Variable name	Name of the variables	Rotated factor loading
Dimension IV: tannery influents and soil fertility		
10	Carbonates	0.92
08	Bicarbonates	0.90
34	PH value	0.88
11	Chromium	0.87
39	Sodium	0.84
13	Carbon	– 0.79
	Eigen value 2.37	**Percentage of trace = 11.03**
Dimension V: tannery effluence and ecological imbalance		
63	Toxic elements	0.89
64	Yields of production	0.73
35	Soil condition	– 0.70
26	Ecological effect	– 0.68
24	Environmental effect	0.63
25	Economic effect	0.62
09	Biological effect	0.60
	Eigen value = 1.92	**Percentage of trace = 9.22**
Dimension: VI hardness of water		
12	Cardiovascular disease	0.83
04	Arteriosclerotic heart disease	0.78
27	Hypertension	0.72
28	Hardness of water	0.70
41	Calcium bicarbonate	– 0.66
46	Magnesium bicarbonate	– 0.0.60
42	Calcium sulfate	– 0.68
48	Magnesium sulfate	– 0.59
	Eigen value = 1.43	Percentage of trace = 7.24
Dimension VII: reformation of polluting tannery effluents		
06	Aluminum	0.92
63	Taxic elements	– 0.90
36	Sodium cholorid	– 0.0.88
23	Frequent test	0.67
07	Water sampling	0.62
32	Medical checku	0.58
	Eigen value = 1.12	**Percentage of trace = 5.19**

(0.68), environmental (0.63) economic (0.62) and biological (0.60) impact on the air, water, land, plants, people, and animals.

25.6.6 Dimension VI: Water Hardness and Tannery Effluents

Water hardness and tannery effluents formed the sixth dimension with an eigenvalue of 1.43, explaining a total variance of 7.24% (Table 25.9). Water hardness is caused mainly by calcium bicarbonate (0.66), magnesium bicarbonate (0.60), calcium sulfate (0.68), and magnesium sulfates (0.59). One of these compounds produces water hardness and is the cause of cardiovascular disease (0.83), arteriosclerotic heart disease (0.78), hypertension (0.72), etc.

25.6.7 Dimension VII: Reformatting of Polluting Tannery Effluents

The seventh dimension on reformatting of polluting tannery effluents explained a total variance of 5.19 with an eigenvalue of 1.12 (Table 25.9). A comprehensive approach should be made with the participation of affected people tanners' environmental groups, scientists and others to set a proper solution for the problem of tannery pollution. (Kumar et al., 2023) To make use of aluminium (0.92) as it is a less toxic element (−0.90) compared to the over usage of sodium chloride (−0.88). The following variables, such as frequent tests (0.67), water sampling (0.62), medical checkups (0.58), and adopted rules for tannery, are necessary for preventing the danger of environmental disturbances in the tannery regions.

25.7 Conclusion

Geochemical tests are carried out on groundwater samples taken in Dindigul district. There are 20 sample points collected from various locations in the Dindigul district and examined for pH, TDS, SO_4^{2-}, Cl^-, Na^+, K^+, Ca^{2+}, Mg^{2+}, and F^- following conventional techniques. All groundwater sample results are compared to the standard acceptable value. The vast majority of groundwater samples yield findings that exceed the permissible limit. According to the study's findings, groundwater quality in this research site should be monitored and examined regularly to prevent further pollution. Furthermore, between 2012 and 2022, the nature of groundwater for human use deteriorated even further in locations where leather tanning businesses are located.

The factor analysis matrix identified seven basic dimensions and explained about 87.92% of the variance, with eigenvalues drawn up from 1.12 to 5.60. Positive factor loading variables explain the dimension's chemical hazards caused by tannery influence. These pollutants have an impact on human health in both direct and indirect ways as chemicals accumulate. Tannery effluent can cause skin ulcers, allergic reactions, dermatitis, ulceration, and internal problems. Most of Dindigul's underground water sources are contaminated on the edges of tannery settlements. When tannery wastewater is ponded, lagooned, or released into dry beds, it impacts groundwater. The irrigation water quality is of permanent importance since it affects the soils, resulting in poor plant growth and, eventually, crop productivity. The tannery effluent that affects soil fertility is complex because it contains too many compounds.

Toxic effluents from leather tanning companies may be to blame for the substantially deteriorating groundwater quality for human consumption in the western part of the region, including Pallapatti, Adiyanuthu, and Vellodu village panchayats. As a result, industrial processing water must undergo proper treatment before being discharged into drainage systems or rivers. Filtration, clarifying systems, biological treatment, dewatering, disinfection, ion exchange, carbon filtration, reverse osmosis, and other types for detaching contaminants from water are treatment procedures that should be employed before the wastewater is released into the environment. Thus, for sustainability and the green leather industry, tanneries must decrease or eliminate chromium pollution to meet environmental criteria.

Acknowledgements The authors are grateful to the Department of Geography, North Eastern Hill University, Shillong, Meghalaya for using the advanced GIS-RS Lab to analyse statistical data and prepare cartography work. The authors are also acknowledged by the Central Groundwater Board and the District Pollution Control Board, Dindigul,Tamilnadu.

References

Abhishek, S. N., Pradeep, A. R., & Rangaraj, C. (2023). Assessment of groundwater quality for pre-and post-monsoon variations in Molakalmur Taluk, Chitradurga District, Karnataka India. *Indian Journal of Science and Technology, 16*(7), 516–530.

Anuja, S. A., Kavitha, P., Jothi, C. H., & Rose, R. J. (2024). Seasonal assessment of water quality for irrigation using multiple indices in Thamirabarani river, Kanyakumari District, India. *Desalination and Water Treatment*, 100206.

Atique, U., Iqbal, S., Khan, N., Qazi, B., Javeed, A., Anjum, K. M., Haider, M. S., Khan, T. A., Mahmood, S., & Sherzada, S. (2020). Multivariate assessment of water chemistry and metals in a river impacted by the tanning industry. *Fresenius Environmental Bulletin, 29*(04A), 3013–3025.

Balaji, D., & Saravanabavan, V. (2021). Geospatial variation of dengue risk zone in Madurai city using autocorrelation techniques. *GeoJournal, 86*(3), 1481–1501.

Balaji, D., & Saravanabavan, V. (2020). A geo medical analysis of dengue cases in Madurai city-Tamilnadu India. *GeoJournal, 85*(4), 979–994.

Balaji, E., Nagaraju, A., Sreedhar, Y., Thejaswi, A., & Sharifi, Z. (2016). Hydrochemical characterization of groundwater in around Tirupati area, Chittoor District, Andhra Pradesh South India. *Applied Water Science, 7*(3), 1203–1212.

Balasubramani, K., Veena, M., Kumaraswamy, K., & Saravanabavan, V. (2015). Estimation of soil erosion in a semi-arid watershed of Tamil Nadu (India) using revised universal soil loss equation (rusle) model through GIS. *Modelling Earth Systems and Environment, 1*, 1–17.

Bisht, M., Shrivastava, M., Kumar, S. N., & Singh, R. (2023). Evaluation of the drinking water quality and potential health risks of nitrate and fluoride in Southwest Delhi, India. *International Journal of Environmental Analytical Chemistry*, 1–23.

Brar, S. K., Verma, M., Tyagi, R. D., Surampalli, R. Y., & Misra, K. (2019). Tannery wastes: Trivalent chromium case. *International Journal of Environment and Health Sciences, 1*(1), 15–33.

Chidambaram, S., Prasad, M. B. K., Prasanna, M. V., Manivannan, R., & Anandhan, P. (2015). Evaluation of metal pollution in groundwater in the industrialized environs in & around Dindigul, Tamilnadu, India. *Water Quality, Exposure and Health, 7*, 307–317. https://doi.org/10.1007/s13201-021-01527-w

De Guzman, K., Stone, G., Yang, A. R., Schaffer, K. E., Lo, S., Kojok, R., & Kayser, G. L. (2023). Drinking water and the implications for gender equity and empowerment: A systematic review of qualitative and quantitative evidence. *International Journal of Hygiene and Environmental Health, 247*, 114044.

De Silva, S., Carson, P., Indrapala, D. V., Warwick, B., & Reichman, S. M. (2023). Land application of industrial wastes: Impacts on soil quality, biota, and human health. *Environmental Science and Pollution Research, 30*(26), 67974–67996.

Dessie, B. K., Aschale, M., Polaine, X., Melaku, S., Alamirew, T., Walsh, C. L., & Zeleke, G. (2024). An integrated approach for water quality assessment in African catchments based on physico-chemical and biological indicators. *Science of the Total Environment, 908*, 168326.

Dey, S., Raju, N. J., Gossel, W., & Mall, R. K. (2023). Hydrogeochemical characterization and geochemical modelling for the evaluation of groundwater quality and health risk assessment in the Varuna River basin India. *Environmental Geochemistry and Health, 45*(7), 4679–4702.

El-Kholy, R. A., Zaghlool, E., Isawi, H., Soliman, E. A., Khalil, M. M., El-Aassar, A. H. M., & Said, M. M. (2022). Groundwater quality assessment using water quality index and multivariate statistical analysis case study: East Matrouh, Northwestern coast Egypt. *Environmental Science and Pollution Research, 29*(43), 65699–65722.

Eswari, S., Saravanabavan, V., & Balaji, D. 2021. Structural equation modelling approach to determine the mother and child health care in Madurai district, Tamil Nadu. *International Journal of Scientific Research in Science and Technology*, 8(2), 335–344.

Hansen, É., de Aquim, P. M., Hansen, A. W., Cardoso, J. K., Ziulkoski, A. L., & Gutterres, M. (2020). Impact of post-tanning chemicals on the pollution load of tannery wastewater. *Journal of Environmental Management, 269*, 110787.

Islam, M. S., Azadi, M. A., Nasiruddin, M., & Islam, M. S. (2021). Water quality of boalia khal tributary of halda river by weighted arithmetic index method. *American Journal of Water Resources, 9*(1), 15–22.

Jesudhas, C. J., Chidambaram, S. M., Jeyakumar, R. B., & Rene, E. R. (2022). Development and application of a contaminant transport model for groundwater remediation and reservoir protection: A case study from India. *Environmental Monitoring and Assessment, 194*(4), 257.

Jesudhas, C. J., Chidambaram, S. M., & Muniraj, K. (2017). Isolation of wells contaminated by tannery industries using principal component analysis. *Arabian Journal of Geosciences, 10*, 1–8.

Jha, M. K., Shekhar, A., & Jenifer, M. A. (2020). Assessing groundwater quality for drinking water supply using hybrid fuzzy-GIS-based water quality index. *Water Research, 179*, 115867.

Kanagaraj, G., Mohana, P., Muthusamy, S., Moorthy, G. M., Amaladas, P., & Magesh, N. S. (2023). Geochemical evaluation of groundwater around chromepet tannery belt Southern India. *Groundwater for Sustainable Development, 22*, 100963.

Karunanidhi, D., Aravinthasamy, P., Subramani, T., Kumar, D., & Venkatesan, G. (2021). Chromium contamination in groundwater and Sobol sensitivity model-based human health risk evaluation from leather tanning industrial region of South India. *Environmental Research, 199*, 111238.

Kom, K. P., Gurugnanam, B., & Sunitha, V. (2024). Delineation of groundwater potential zones using GIS and AHP techniques in Coimbatore district, South India. *International Journal of Energy and Water Resources, 8*(1), 85–109.

Kom, K. P., Gurugnanam, B., Sunitha, V., Reddy, Y. S., & Kadam, A. K. (2022). Hydrogeochemical assessment of groundwater quality for drinking and irrigation purposes in western Coimbatore, South India. *International Journal of Energy and Water Resources*, 6(4), 475–494.

Kumar, R., Basu, A., Bishayee, B., Chatterjee, R. P., Behera, M., Ang, W. L., & Jeon, B. H. (2023). Management of tannery waste effluents towards the reclamation of clean water using an integrated membrane system: A state-of-the-art review. *Environmental Research*, 115881

Lin, L., Yang, H., & Xu, X. (2022). Impact of water pollution on human health and disease heterogeneity: A review. *Frontiers in Environmental Science, 10*, 880246.

Malik, S., & Kumar, D. (2023). Perspectives of nanomaterials in microbial remediation of heavy metals and environmental consequences: A review. *Biotechnology and Genetic Engineering Reviews*, 1–48.

Monira, U., Sattar, G. S., & Mostafa, M. G. (2024). Assessment of surface water quality using the water quality index (WQI) and multivariate statistical analysis (MSA), around tannery industry effluent discharge areas. *H2Open Journal*, 7(2), 130–148.

Moukadiri, H., Noukrati, H., Ben Youcef, H., Iraola, I., Trabadelo, V., Oukarroum, A., & Barroug, A. (2024). Impact and toxicity of heavy metals on human health and latest trends in removal process from aquatic media. *International Journal of Environmental Science and Technology*, 21(3), 3407–3444.

Muralidhara Reddy, B., Sunitha, V., & Ramakrishna Reddy, M. (2014). Geospatial analysis of fluoride contamination in groundwater of southeastern part of Anantapur district, Andhra Pradesh. *Management of natural resources in a changing environment* (pp. 61–74). Springer International Publishing.

Nagaraju, A., Balaji, E., Sun, L., & Thejaswi, A. (2018). Processes controlling groundwater chemistry from Mulakalacheruvu area, Chittoor district, Andhra Pradesh, South India: A statistical approach based on hydrochemistry. *Journal of the Geological Society of India, 91*(4), 425–430.

Natesan, D., Sabarathinam, C., Kamaraj, P., Mathivanan, M., Haji, M., Mohan Viswanathan, P., & Rajendran, T. (2022). Impact of monsoon shower on the hydrogeochemistry of groundwater along the lithological contact: A case study from South India. *Applied Water Science, 12*(3), 36.

Nigam, M., Mishra, P., Kumar, P., Rajoriya, S., Pathak, P., Singh, S. R., & Singh, L. (2023). Comprehensive technological assessment for different treatment methods of leather tannery wastewater. *Environmental Science and Pollution Research, 30*(60), 124686–124703.

Pande, C. B. & Moharir, K. N. (2022). Estimation of groundwater quality parameters for drinking purpose using IDW, GIS and statistical analysis methods: A case study of basaltic rock in Mahesh river basin, Akola and Buldhana districts (MS), India. In *Water quality, assessment and management in India* (pp. 311–347). Springer International Publishing.

Panneerselvam, B., Paramasivam, S. K., Karuppannan, S., Ravichandran, N., & Selvaraj, P. (2020). A GIS-based evaluation of hydrochemical characterisation of groundwater in hard rock region, South Tamil Nadu, India. *Arabian Journal of Geosciences, 13*, 1–22.

Pant, M., Bhatla, R., Ghosh, S., Das, S., & Mall, R. K. (2024). How climate change is affecting the summer monsoon extreme rainfall pattern over the Indo-Gangetic Plains of India: Present and future perspectives. *Climate Dynamics, 62*(2), 1055–1075.

Patel, P. S., Pandya, D. M., & Shah, M. (2023). A systematic and comparative study of the water quality index (WQI) for groundwater quality analysis and assessment. *Environment Science and Pollution Research, 30*(19), 54303–54323.

Prasad, B., Rao, P. R., & Tigga, A. (2021). Groundwater assessment using the weighted arithmetic index method in the selected villages of Butchayyapeta Mandal, Visakhapatnam, Andhra Pradesh, India. *Environmental Monitoring and Assessment, 193*, 1–17.

Radhakrishnan, R., & Vaithialingam, S. (2023). Exploring the suitability of groundwater for domestic water quality and irrigation purposes in Dindigul district, Tamil Nadu. *Surface and groundwater resources development and management in semi-arid region: strategies and solutions for sustainable water management* (pp. 253–277). Springer International Publishing.

Rahman, M. S., Ahmed, Z., Seefat, S. M., Alam, R., Islam, A. R. M. T., Choudhury, T. R., Begum, B. A., & Idris, A. M. (2022). Assessment of heavy metal contamination in sediment at the newly established tannery estate in Bangladesh: A case study. *Environmental Chemistry and Ecotoxicology, 4*, 1–12.

RahamathNisha, R., & Saravanabavan, V. (2020). Occupational health: A special reference to lung disease in petrochemical, metal and building material industrial workers in Melur taluk Madurai district. *International Journal of Contemporary Medical Research, 7*(7), G1–G6.

Ram, A., Tiwari, S. K., Pandey, H. K., Chaurasia, A. K., Singh, S., & Singh, Y. V. (2021). Groundwater quality assessment using water quality index (WQI) under GIS framework. *Applied Water Science, 11*, 1–20.

Ravindra, B., Subba Rao, N., & Dhanamjaya Rao, E. N. (2023). Groundwater quality monitoring for assessment of pollution levels and potability using WPI and WQI methods from a part of Guntur district, Andhra Pradesh, India. *Environment, Development and Sustainability, 25*(12), 14785–14815.

Reddy, Y. S., & Sunitha, V. (2023). Assessment of heavy metal pollution and its health implications in groundwater for drinking purposes around inactive mines, SW region of Cuddapah Basin South India. *Total Environment Research Themes, 8*, 100069. https://doi.org/10.1016/j.totert.2023.100069

Reddy, Y. S., Sunitha, V., & Suvarna, B. (2022a). Physical and hydro geochemical impact assessment of inactive mines in and around southwestern part of cuddapah basin using a conceptual site model (CSM). In *Innovative trends in hydrological and environmental systems: select proceedings of ITHES 2021* (pp. 793–807). Springer Nature Singapore.

Reddy, S., Sunitha, V., & Suvarna, B. (2022b). Groundwater quality and its potential health impacts nearby inactive mines using PIG and geospatial technology, Southwestern part of Cuddapah basin, Andhra Pradesh South India. *Groundwater for Sustainable Development, 17*, 100742.

Reddy, B. M., Sunitha, V., Suvarna, B., Sudha, Y. S., & Krishna, K. S. (2022c). Hydrogeochemistry and groundwater evaluation in and around Badel, YSR Kadapa district, Andhra Pradesh (India). *The Journal of Indian Geophysical Union, 26*(3), 228–238.

Rigueto, C. V. T., Rosseto, M., Nazari, M. T., Ostwald, B. E. P., Alessandretti, I., Manera, C., Piccin, J. S., & Dettmer, A. (2021). Adsorption of diclofenac sodium by composite beads prepared in the tannery wastes-derived gelatin and carbon nanotubes. *Journal of Environmental Chemical Engineering, 9*(1), 105030.

Reljić, M., Romić, M., Romić, D., Gilja, G., Mornar, V., Ondrasek, G., & Zovko, M. (2023). Advanced continuous monitoring system—tools for water resource management and decision support system in salt-affected delta. *Agriculture, 13*(2), 369.

Rowe, G. L., Gilliom, R. J., & Woodside, M. D. (2013). Tracking and forecasting the Nation's water quality priorities and strategies for 2013–2023 (No. 2013–3008). US Geological Survey.

Rusydi, A. F., Setiawan, T., Maria, R., Firmansyah, F., Damayanti, R., Mulyono, A., & Carrilo-Rivera, J. (2023). Groundwater nitrogen concentration changes in an urbanized area of Indonesia over a ten-year period. In IOP Conference: Earth and Environmental Science (Vol. 1201(1), pp. 012035). IOP Publishing.

Sadeghi, H., Fazlzadeh, M., Zarei, A., Mahvi, A. H., & Nazmara, S. (2022). Spatial distribution and contamination of heavy metals in surface water, groundwater and topsoil surrounding Meghan's tannery site in Ardabil Iran. *International Journal of Environmental Analytical Chemistry, 102*(5), 1049–1059.

Sajib, A. M., Diganta, M. T. M., Moniruzzaman, M., Rahman, A., Dabrowski, T., Uddin, M. G., & Olbert, A. I. (2024). Assessing water quality of an ecologically critical urban canal incorporating machine learning approaches. *Ecological Informatics*, 102514.

Saikrishna, K., Purushotham, D., Sunitha, V., Reddy, Y. S., Brahmaiah, T., Reddy, B. M., & Nallusamy, B. (2023). Deciphering groundwater quality, and mechanisms controlling groundwater chemistry in and around Suryapet, Telangana South India. *Total Environment Research Themes, 6*, 100035.

Saravanabavan, V., Emayavaramban, V., Thangamani, V., Manonmani, I.K., Suja Rose, R.S., Balaji, D., Nisha, R.R., Kannadasan, K., Vinothkanna, S., & Vinothini, C. (2023). Spatial variation of COVID-19 morbidity status and identification of risk zone in Tamil Nadu (India) during first wave. *GeoJournal*, 88(2), 1341–1354.

Saravanabavan, V., Balaji, D., Reshma, C.U., Sheheersha, S.K., Sudharsan, R., Vimala Vinnarasi, J., Ganesan, J., Rahamath Nisha, R., Eswari, S., & Balasubramani, K. (2021a). Urban disease ecology and its spatial variation of Chikungunya in Madurai City, Tamilnadu, India: a geo-medical study. *GeoJournal*, 86, 2335–2350.

Saravanabavan, V., Reshma, C.U. and Preethi, S. (2021b). Determinants of reproductive health in working women in Thrissur district, Kerala. *GeoJournal*, 86, 239–253.

Saravanabavan, V., Aneesh, P., Babu, H. M., & Harieswari, D. M. (2021c). Patient's perception and level of primary health care utilization in east block of Madurai North taluk: A geo-health study. *International Journal of Geography, Geology and Environment, 3*(1), 34–41.

Saravanabavan, V., Balaji, D., Rahamath Nisha, R., Preethi, S., & Vadivel, S. (2020). Geo-ecological association of dengue disease in Madurai city–using multivariate analysis. *International Journal of Mosquito Research, 7*(2), 37–45.

Saravanabavan, V., Balaji, D., & Preethi, S. (2019). Identification of dengue risk zone: A geo-medical study on Madurai city. *GeoJournal*, 84, 1073–1087.

Saravanabavan, V. (2011).Water pollution due to industrial waste a geo-chemical study of tanning industries—a case study of Dindigul district. In *Environment and rural development* (pp.158–170). Novel corporation. ISBN: 978-81-904529-7-7

Saravanabavan, V., & Shanmuganandan, S. (1998). Leprosy and multidrug therapy in Tamil Nadu, India: A factor analysis. *The Indian Geographical Journal, 73*(1), 41–50.

Selvam, S., Venkatramanan, S., Chung, S. Y., & Singaraja, C. (2016). Identification of groundwater contamination sources in the Dindigul district of Tamil Nadu, India using GIS and multivariate statistical analyses. *Arabian Journal of Geosciences, 9*, 1–14.

Selvakumar, P., Subramanian, D., Natarajan, J., & Solai Ramatchandirane, P. (2021). Analysis of chromate transporters in bacterial species for Cr (VI) reduction isolated from tannery effluent contaminated site of Dindigul district, Tamil Nadu India. *Geo-Microbiology Journal, 38*(7), 598–606.

Sharma, A. R., Bordoloi, R., Paul, A., Gyanendra, Y., & Tripathi, O. P. (2024). Water quality and geochemical facie of high-altitude lakes in Tawang, Eastern Himalaya, India. *Environmental Science and Pollution Research*, 1–20.

Sinha, H., Rai, S. C., & Kumar, S. (2023). Post-monsoon groundwater hydrogeochemical characterization and quality assessment using geospatial and multivariate analysis in Chhotanagpur Plateau, India. *Environment, Development and Sustainability*, 1–31.

Suja Rose, R. S., & Krishnan, N. (2009). Spatial analysis of groundwater potential using remote sensing and GIS in the Kanya Kumari and Nambiyar basins, India. *Journal of the Indian Society of Remote Sensing, 37*, 681–692.

Sundararaj, P., Madurai Chidambaram, S. K., Sivakumar, V., & Natarajan, L. (2022). Groundwater quality assessment and its suitability for drinking and agricultural purpose, Dindigul taluk, Tamilnadu India. *Chemical Papers, 76*(10), 6591–6605.

Sunitha, V., & Reddy, B. M. (2022). Geochemical characterization, deciphering groundwater quality using pollution index of groundwater (PIG), water quality index (WQI) and geographical information system (GIS) in hard rock aquifer South India. *Applied Water Science, 12*(3), 41.

Sunitha, V. (2020). Geochemical and health risk assessment of fluoride and nitrate toxicity in semi-arid region of Anantapur District, South India. *Environmental Chemistry and Ecotoxicology, 2*, 150–161.

Sunitha, V., Reddy, B. M., Khan, J. A., & Reddy, M. R. (2014). Groundwater geochemistry in the southeastern part of Anantapur district, Andhra Pradesh, with special reference to fluoride distribution and its impact on health. *Journal of Environmental Science & Engineering, 56*(2), 153–160.

Sunitha, V., Sudarshan, V., & Reddy, B. R. (2005). Hydrogeochemistry of groundwater, Gooty area, Anantapur district, Andhra Pradesh India. *Pollution Research, 24*(1), 217.

Suvarna, B., Sunitha, V., Sudharshan Reddy, Y., Muralidhara Reddy, B., Kadam, A. K., & Ramakrishna Reddy, M. (2023). Groundwater quality assessment using multivariate statistical approach and geospatial modelling around cement industrial corridor, South India. *International Journal of Environmental Science and Technology, 20*(5), 5051–5070.

Suvarna, B., Reddy, Y. S., Sunitha, V., Reddy, B. M., Prasad, M., & Reddy, M. R. (2020). Data on application of water quality index method for appraisal of water quality in around cement industrial corridor, Yerraguntla Mandal, YSR district AP South India. *Data in Brief, 28*, 104872.

Thiripurasundari, K., & Ponsakthisurya, C. (2018). Comparative analysis among minor clusters of leather industry in Tamil Nadu. *International Journal for Advance Research and Development*, 3(3), 243–246.

Tiwari, A. K., & De Maio, M. (2018). Assessment of sulphate and iron contamination and seasonal variations in the water resources of a Damodar Valley Coalfield, India: A case study. *Bulletin of Environmental Contamination and Toxicology, 100*, 271–279.

Urme, S. A., Radia, M. A., Alam, R., Chowdhury, M. U., Hasan, S., Ahmed, S., Sara, H. H., Islam, M. S., Jerin, D. T., Hema, P. S., & Rahman, M. (2021). Dhaka landfill waste practices: addressing urban pollution and health hazards. Buildings and Cities 2(1), 700.

Vannucci, L., Fossi, C., Quattrini, S., Guasti, L., Pampaloni, B., Gronchi, G., Giusti, F., Romagnoli, C., Cianferotti, L., Marcucci, G., & Brandi, M. L. (2018). Calcium intake in bone health: A focus on calcium-rich mineral waters. *Nutrients, 10*(12), 1930.

Vinothini, C., Saravanabavan, V., & Emayavaramban, V. (2022). Location accessibility of PHC and health care travel performance in Madurai district. *International Journal of Innovative Science and Research Technology*, 7(12), (2456–2165).

Vinothini, C., Saravanabavan, V., & Balaji, D. (2021). Travel pattern of health utilization to primary health care centres in Madurai district. *International Journal of Geography, Geology and Environment, 3*(2), 144–151.

Vinothini, C., Neethidevi, A., & Saravanabavan, V. (2019). Health impact and ninefold classification of land use change in Nilakottai Taluk, Dindigul District, TamilNadu. *International Journal of Geography, Geology and Environment, 1*(1), 80–86.

Vinothkanna, S., Rajee, R., & Senthilraja, K. (2020). Assessing ground water quality for the suitability of irrigation in Dindigul District, Tamil Nadu India. *Indian Journal of Ecology, 47*(1), 23–29.

WHO. 2017. *Guidelines for drinking-water quality* (4th ed.) April 24, 2017, Guideline https://www.who.int/publications/i/item/9789241549950

Yenugu, S. R., Vangala, S., & Badri, S. (2020). Groundwater quality evaluation using GIS and water quality in and around inactive mines, Southwestern parts of Cuddapah basin, Andhra Pradesh, South India. *HydroResearch, 3*, 146–157.

Zeng, Y., Lu, H., Zhou, J., Zhou, Y., Sun, Y., & Ma, C. (2024). Enrichment mechanism and health risk assessment of fluoride in groundwater in the Oasis zone of the Tarim Basin in Xinjiang China. *Exposure and Health, 16*(1), 263–278.

Chapter 26
Geochemical Studies on the Groundwater of Gundlapochampally Village, Medchal District, Telangana State, India

Musini Venkatesharlu, B. Prasad, Ch. Rajendra Prasad, Padamata Yeswanth Babu, Dosilla Jashuva, and Y. S. Reddy

Abstract The study area has been experiencing rapid industrialization and human influences over the past few years, which may have negative effects on water quality and human health. Hence, the water is analyzed for the following parameters: pH, electrical conductivity, total dissolved solids, total hardness, calcium, magnesium, sodium, potassium, alkalinity, acidity, chloride, nitrate, fluoride, and bicarbonate in Gundlapochampally village, Medchal District, India. The results were compared to the World Health Organization's and the Bureau of Indian Standards' drinking water guidelines. In order to evaluate whether groundwater in the studied area is suitable for human consumption, a water quality study is essential. It is found that majority of the samples are safe for drinking except with elements like Na^+ (100%) and F^- (14%) which infers the geogenic influences over anthropogenic. However, proper remedial measures should be implemented to reduce the risk of contaminants.

Keywords Physico-chemical parameters · Water quality · Distribution diagrams · Scatter plots · Study area · India

M. Venkatesharlu (✉) · B. Prasad · Ch. R. Prasad · P. Y. Babu · D. Jashuva
Department of Civil Engineering, CMR College of Engineering &Technology (A), Kandlakoya (V), Hyderabad, Medchal District, Telangana, India
e-mail: venkatmusini@gmail.com

B. Prasad
e-mail: bolliniprasad@gmail.com

Ch. R. Prasad
e-mail: rajucivil105@gmail.com

P. Y. Babu
e-mail: padamatayeswanthbabu@gmail.com

D. Jashuva
e-mail: Jashuvad00@gmail.com

Y. S. Reddy
Department of Physics, Chaitanya Bharathi Institute of Technology, Gandipet, Hyderabad, Telangana, India
e-mail: ysreddy_physics@cbit.ac.in

V. Sunitha et al. (eds.), *Groundwater Resource Management Planning Strategies*, Springer Water, https://doi.org/10.1007/978-3-031-88870-0_26

26.1 Introdcution

Seventy percent of the Earth's surface is covered by water, and it plays a crucial role in sustaining life. Without it, our existence would be impossible. Among the key environmental challenges of our time is ensuring access to clean, high-quality freshwater. The presence of clean groundwater is vital for the survival of humans and other organisms. Access to safe drinking water is essential for all living beings, including humans, to flourish and live in a way that is environmentally conscious. The main factors influencing changes in groundwater quality include pollutants, chemical reactions, interactions between water and rocks, and the geological makeup of the area (Pierre et al., 2005, Balaji et al., 2016). The decline in groundwater quality can be linked to various causes, such as changes in land use, the dilution of fertilizers in water, the mixing of sewage with drinking water sources, leaks of petroleum products, improper disposal of industrial waste underground, the operation of nuclear power plants, and the intrusion of saltwater into groundwater sources (Liou et al., 2004, Tsegaye et al., 2006, Nagaraju et al., 2018). It is possible to evaluate the physical, chemical, and biological characteristics of water quality. There are four major cities, and the current state of air pollution is a direct result of these urban areas' rapid expansion (Musini and Merugu, 2020). The evaluation of water quality takes into account several factors, including temperature, acidity, taste, smell, the amount of dissolved oxygen, and the presence of microorganisms (Prasad et al., 2013). This has benefits for both individuals and the environment. The increase in population, overuse of resources, and technological progress are examples of human activities that can disrupt the natural equilibrium and harm the physical environment. Understanding the land use and cover of a specific area is crucial for planning and managing resources effectively, as well as for assessing their availability in that region. By employing a combined approach of remote sensing and geographic information systems, detailed maps of land use and cover will be created to address the growing needs.

According to the study on the impact of groundwater contamination and the restoration of its quality through the use of organic fertilizers and reduced fertilizer application, integrated groundwater prospect maps are suggested as the best option, based on various geochemical experiments (Venkateshwarlu et al., 2018). On the other hand, land-use refers to the way in which land is utilized by humans, often leading to alterations in its primary function for economic activities (Venkateshwarlu et al., 2020). The concept of "land-cover" pertains to the earth's surface's physical characteristics, shaped by human activities like settlements, and encompasses the distribution of soil, water, vegetation, and other natural features. The natural and socioeconomic aspects of a location, along with its historical and geographical significance, define its pattern (Bagyaraj et al., 2013). The specific area under study is the Narsapur mandal in the Medak district, which is categorized based on the origin of the granite's "parental" magma (Rawat et al., 2013). The inhabitants of this area use the water for various purposes, making it crucial that the water quality is of high standards for its use. Therefore, it is vital to determine the land cover and water quality of the research area, as well as to provide information on the availability of resources and any modifications that have occurred. In the process of conducting multivariate analysis, geographic information systems are employed to evaluate the distribution of ions (Venkateshwarlu 2020, Venkateshwarlu et al., 2021, Venkateshwarlu et al., 2015). The hydro geochemistry of groundwater was explored in the Gooty region of the Anantapur district (Venkateshwarlu et al., 2022), the evaluation of groundwater quality in several districts of Kadapa and Anantapur, Andhra Pradesh (Sunitha et al., 2005, Sunitha et al., 2014), suggest that the overuse of groundwater is also the reason for the declining groundwater quality in the districts. The presence of high levels of fluoride and salt was detected in the groundwater of this region (Sunitha et al., 2014, Venkateshwarlu et al., 2014). Human activity is affecting the quality of groundwater, particularly shallow groundwater (Sunitha and Reddy 2022a, Muralidhara Reddy and Sunitha 2020, Reddy & Sunitha 2023). The chemistry of a region's groundwater is typically not uniform and is influenced by flow, geochemical processes, evaporation and evapotranspiration, as well as potential pollution sources (Saikrishna et al., 2023a, b, Golla et al., 2021, Muralidhara Redd et al., 2019). The quality of groundwater for drinking in the Dharwar craton area of Mallampalli, Telangana is subjected to the geochemical processes (Saikrishna et al., 2023a). The development of a water quality index to evaluate groundwater quality near Vemula Mandal in YSR District was done by YS Reddy and Sunitha (Reddy & Sunitha 2020). Using the groundwater contamination index, geochemical characterization can be used to determine the quality of groundwater (Sunitha & Reddy 2022b). In Telangana State, Gundlapochampally, which is located in the northern part of the Medchal–Malkajgiri District between latitudes 17° 32′–17° 40′ in the north and 78° 22′–78° 35′ in the east, with an average elevation of 602 m and an area of 150.3 square kilometers, the population of Gundlapochampally Village is 9009 according to the 2011 Census, the annual rainfall in Gundlapochampally Village is about 812.3805 mm. Therefore, the objective of this work is to assess the groundwater quality and to define its suitability for drinking based on hydro chemical analysis in the Quaternary Alluvial Basin. Thus, this serves as a background work to improve the Evaluation of groundwater quality in the planning of water resources in the future.

26.2 Material and Methods

The hydro-geochemical analysis investigates the study area's groundwater issue. In this context, groundwater samples were gathered in the Gundlapochampally region in December. Our safety protocols were developed based on recommendations made by Eugene Brown for security, accessibility to the site, and safety measures (Eugene Brown et al., 1970). When gathering groundwater sample data, the following safety measures were implemented. Among them are before gathering water, the bottles should be labeled. Exact GPS coordinate demarcation. (BIS, 1983), to be used are standard gloves. Refraining from wearing jewelry and other accessories to prevent sample contamination. The samples were gathered from a variety of sites, including agricultural land, apartment buildings, construction sites, areas with pump sets installed, and any other area where groundwater might be accessed.

A few locations, including motor bore wells in apartments, parks, and temples, were utilized as groundwater sources in the semi-urban area. As a result, 27 groundwater samples from Gundlapochampally were gathered in water bottles and examined in a geochemical laboratory (BIS 1991), Table. 26.1. The concentrations of 14 different elements have been determined in ppm or mg/l. pH, EC, TDS, TH, Ca^{2+}, Mg^{2+}, Na^{+} K^{+}, HCO_3^-, Alkalinity, Acidity, Cl^-, NO_3^-, and F^- are some of these parameters. The values of the total water parameters are tabulated and recorded in an Excel sheet that was processed through the Quantum GIS software (Table 26.1). Each column contains pertinent information such as the sample number, latitude, longitude, and element values expressed in parts per million or milligrams per litre. Based on the variable sample location, the outcomes of each water parameter were then distributed across the raster image. The equal distance reclassification method is used to separate attribute values (Fig. 26.1).

26.3 Results and Discussion

To supply hygienic water that satisfies the Indian Standard drinking water requirements to rural and urban areas, it is imperative to understand the chemical characteristics of groundwater. The hydro-geochemistry of the research area is used to ascertain the geographic distribution of the water quality (Backman et al., 1998). The pH of water is a critical measure of its quality and provides important information about many types of geochemical equilibrium. The tested samples have a pH range of 6.7 to 7.85, and the mean value in the tested area is 7.09, indicating an alkaline composition. All of the collected samples have pH values that are well within the safe range advised by the World Health Organization (1983) (WHO, 1983). Although pH has an impact on all biological processes, it does not directly affect human health. The electric conductivity (EC) values in the research area range from 1520 to 1997 μS/cm at 25 °C, with an average value of 595.24 μS/cm. The elevated EC may have resulted from the high salinity and high mineral content of the sampling site. It depends on the kind, amount, and temperature of ions present in groundwater; additionally, the concentration of EC increases in tandem with the concentration of TDS. Total dissolved solids (TDS) in water include all inorganic salts, including calcium, phosphate, nitrate, carbonate, bicarbonate, chloride, fluoride, sulphate, and phosphate. Three: magnesium, potassium, and sodium (Sawyer, 1994). Total dissolved solids in groundwater in the studied area range from 1145 to 1470 mg/L. TDS levels were used to classify the groundwater in the research region using the (US Geological Survey, 2000) protocol. Table 26.2 shows that only 1% of groundwater samples are classified as freshwater compared to 99 percent for groundwater samples. Table 26.2 classifies the samples as mildly saline. The maximum amount of total hardness (TH), also referred to as CaCO3, that is acceptable is 300 mg/L; if no other water source is available, this can be increased to

Table 26.1 Water quality analysis data in Gundlapochampally, Medchal-Malkajgiri District, Telangana State

S. no	Sample. no	Latitude (degrees)	Longitude (degrees)	pH	EC (µS/cm)	TDS (mg/L)	TH (mg/L)	Ca^{2+} (mg/L)	Mg^{2+} (mg/L)	Na^{+} (mg/L)	K^{+} (mg/L)	HCO_3^- (mg/L)	Alkalinity (mg/L)	Acidity	Cl^- (mg/L)	NO_3^- (mg/L)	F^- (mg/L)
1	LN 1	17.5705	78.4888	7.6	1833	1375	190	40	23	208	42	580	32	3	35	0	1.5
2	SN 1	17.5524	78.4935	7.4	1797	1309	195	58	30	208	40	490	30	2.5	61	0.5	1.5
3	LN 2	17.5578	78.4887	7.5	1850	1257	196	40	37	197	41	530	24.2	2.3	50	0.5	1.1
4	LN 3	17.5605	78.4884	7.3	1890	1181	205	36	58	185	48	807	21.5	1.6	39	0.3	1.25
5	GP 1	17.566	78.486	6.9	1691	1294	215	56	21	246	21	570	20.5	2.6	24	0	1.05
6	GP 2	17.5714	78.4801	6.82	1780	1345	208	56	45	224	32	615	22.1	2.5	30	0	1.5
7	GP 3	17.5714	78.4801	6.7	1650	1238	220	48	62	265	36	510	32.3	2.9	35	0.5	1.25
8	AL 1	17.5669	78.474	7.06	1574	1450	218	36	61	189	32	560	22.2	2.3	38	0.3	1
9	GP 4	17.5756	78.4797	6.98	1676	1145	196	38	31	260	27	570	28.2	2.5	27	0.4	1.5
10	GP 5	17.5842	78.4765	7.1	1745	1498	209	40	27	330	53	550	30.4	3.1	35	1	1.5
11	GP 6	17.5865	78.4748	7.07	1695	1238	189	84	35	208	43	580	42.6	2.2	48	0	1.75
12	GP 7	17.5866	78.4747	6.9	1780	1335	208	44	42	236	56	490	20.1	3.6	46	1	1.9
13	KL 1	17.5991	78.4789	7.8	1570	1345	188	52	49	295	53	460	31.3	3.3	32	0.5	1
14	KL 2	17.5994	78.4789	7.9	1556	1470	216	64	58	40	67	580	32.2	3.8	40	0.3	1.6
15	GP 7	17.5749	78.4785	7.17	1520	1435	202	48	81	205	54	520	24.8	3.3	46	0	1
16	GP 8	17.5718	78.4811	7.04	1745	1403	208	56	33	228	26	560	22.2	2.4	36	0.3	1
17	GP 9	17.5715	78.4811	6.92	1670	1268	211	72	30	248	32	620	23.8	2.3	28	0.5	1
18	GP 10	17.5747	78.4824	6.95	1725	1270	196	59	47	254	41	680	24.5	1.9	20	1	1.5
19	LN 4	17.5718	78.4825	7.4	1790	1321	203	68	41	213	45	590	20.3	4.5	28	0.5	1
20	GP 11	17.5663	78.4769	7.03	1687	1248	190	38	47	279	41	530	34	4.3	25	0.3	1.8

(continued)

Table 26.1 (continued)

S. no	Sample. no	Latitude (degrees)	Longitude (degrees)	pH	EC (µS/cm)	TDS (mg/L)	TH (mg/L)	Ca^{2+} (mg/L)	Mg^{2+} (mg/L)	Na^{+} (mg/L)	K^{+} (mg/L)	HCO_3^- (mg/L)	Alkalinity (mg/L)	Acidity	Cl^- (mg/L)	NO_3^- (mg/L)	F^- (mg/L)
21	GP 12	17.5658	78.4771	7.25	1820	1305	197	56	53	250	52	575	36.1	4.8	50	0.5	1
22	MG 1	17.547	78.4687	7.85	1997	1325	219	45	58	267	38	515	34	3.6	40	0.5	1
23	MG 2	17.547	78.4687	7.9	1945	1297	211	57	49	239	44	610	38.1	3.8	36	1	1.5
24	SN 2	17.5411	78.4875	7.04	1895	1318	190	63	53	206	55	630	30.8	2.5	49	0.3	1.5
25	LN 4	17.5701	78.492	7.1	1790	1353	196	72	30	221	45	580	26.2	2.9	60	0.5	1.3
26	KL 3	17.594	78.4862	7.8	1875	1374	185	89	36	268	37	610	28.2	2.3	50	0	1.4
27	VV 1	17.5944	78.4862	7.6	1838	1298	187	86	35	212	42	590	32.4	3.2	55	0	1.35

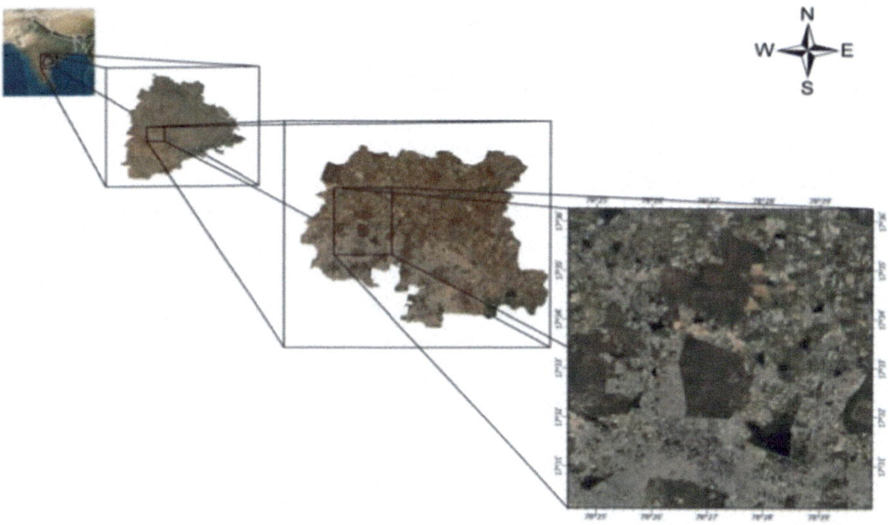

Fig. 26.1 Location map of the study area

600 mg/L (WHO, 1990). Total hardness, expressed in $CaCO_3$ equivalents, ranges from 36 to 89 mg/L in the research region (Tables 26.2 and 26.3). Tables 26.1 and 26.2 show the values of nitrite, fluoride, chloride, alkalinity, and acidity. Monsoon variations have no effect on the concentration of various ions in groundwater. It is a slow process that involves groundwater movement, mineral dilution, and temperature changes beneath the surface (Giridharan et al., 2008, Mamat et al., 2016, Canora et al., 2019, BIS 1991) the recommended range for fluoride readings is 1.0–1.9 mg/l (Table 26.2). Fluoride concentrations above 1.5 mg/l cause dental and skeletal fluorosis (WHO, 2003), which can lead to a concentration of fluoride minerals in bedrock (Bellack 1995, Saxena & Ahmed 2001). Groundwater receives a substantial amount of fluoride from the weathering of igneous and sedimentary rocks with an alkali and silicate base, especially shale (Handa 1975, Venkatayogi 2015). In addition to coming from natural sources, a sizable portion of the fluoride in the study region may also come from human activity. TDS Contour maps (distribution Diagrams) made specifically for the study area show that most ground water samples are alkaline, with the exception of the north and east parts of the study area (Fig. 26.2). The study area's eastern and southern regions exhibit a higher concentration of TH, as depicted by the shaded contour maps (Fig. 26.3). The distribution of HCO3 in the research area is shown in the shaded contour maps (Fig. 26.4). Following the creation of distribution diagrams, or contour maps, for the study area, it was discovered that most ground water samples had lower Ca contents, with the exception of the east and south-west regions (Fig. 26.5). The distribution of K is depicted in the shaded contour maps (Fig. 26.6), with the central, northeast, and east regions of the study area having higher K prevalence. Mg and Cl are shown on the shaded contour map (Figs. 26.7 and 26.8). The distribution of NO_3 in the research region is shown in the

shaded contour maps (Fig. 26.9). The majority of the ground water samples were found to be F in the northeast and east of the study area after contour maps, also known as distribution diagrams, were created for the area (Fig. 26.10). The scatter plots describe the combinations of parameters related to water quality, including total dissolved solids and fluoride, total dissolved solids and nitrate, total dissolved solids and electrical conductivity, electrical conductivity and total dissolved solids, and electrical conductivity and alkalinity. The proportions of each one to the others are precisely the same (Figs. 26.11, 26.12, 26.13, 26.14, 26.15, 26.16, 26.17 and 26.18). The TDS and Cl-correlation coefficients suggest that halide dissolution may have occurred in aquifers with low halide concentrations. Groundwater movement, long-term leaching of precipitation that dissolves soluble rocks, and water–rock and water-soil interactions all have a major impact on the amount of main ions in groundwater (Purushothaman et al., 2013, Etikala et al., 2024 45,46). This dissolution has been connected to an increase in groundwater TDS (Al-Harbi et al., 2014, Umar et al., 2013).

Figs. 26.2 and 26.3. Showing the concentration distribution map of TDs and TH

Figs. 26.4 and 26.5 showing the concentration distribution map of HCO_3^- and Ca^{2+}

Figs. 26.6 and 26.7 showing the concentration distribution map of K^+ and Mg^{2+}

Figs. 26.8 and 26.9 showing the concentration distribution map of Cl^- and NO_3^-

Fig. 26.10 showing the concentration distribution map of F^-

Figs. 26.11, 26.12, 26.13, 26.14, 26.15, 26.17 and 26.18 Scatter plots of analysed parameters.

26.4 Conclusion

Southern India relies heavily on groundwater due to the region's low precipitation levels. Our research area has seen a decrease in groundwater quality and quantity. The majority of the settlements in the Northwest and Southeast have excellent or good water quality. Fluoride pollution can be the cause of declining groundwater quality. Polluted with fluoride in the study area. The variable statistical approach has shown to be a reliable technique for evaluating a wide range of water parameters and figuring out the ecological state of those parameters. On the other hand, the correlation matrix helps us see how the parameters are changing over time. Fluoride was found to significantly correlate with TDS and EC ($r = 0.0726$ and 0.0114). Analysis of measurements for several parameters related to groundwater quality has revealed the varied mechanisms governing the varying concentrations of parameters in Gundlapochampally village. Thus, the primary causes of the decline in groundwater quality are determined to be anthropogenic activities and high seasonal evaporation.

Table 26.2 Concentration of ions and their comparison with the WHO

Parameters	Expressed	Min	Max	WHO (1990)		Sample numbers exceeding parameter
				Desirable limit	Permissible limit	
pH	Units	6.7	7.85	6.5–8.5	9.2	Nil
EC	µS/cm	1520	1997	750	1500	Nil
TDS	mg/L	1145	1470	500	1500	Nil
TH	mg/L	185	220	100	500	Nil
HCO_3^-	mg/L	460	807	–	–	–
CO32-	mg/L	0	33	–	–	–
Cl^-	mg/L	20	61	250	600	Nil
NO_3^-	mg/L	0	1.0	–	45	Nil
F^-	mg/L	1.0	1.9	0.5	1.5	11, 12, 14, 20
Na^+	mg/L	40	330	50	200	1,2,5,6,7,9,10,11,12,13,15,16,17,18,19,20,21,22,23,24,25,26,27
K^+	mg/L	21	67	–	–	–
Ca^{2+}	mg/L	36	89	75	200	Nil
Mg^{2+}	mg/L	21	81	30	150	Nil

Table 26.3 Indian standard drinking water specification (IS: 10,500–2012)

S. no.	Parameter	Requirement (acceptable limit)	Permissible limit in the absence of alternate source	Method of test ref no	Undesirable effect
1	pH value	6.5–8.5	No Relaxation	IS 3025 (Part 11)	Beyond this range, water will affect the mucous membrane and/or water supply system
2	EC	-	-	-	
3	Total dissolved solids mg/l	500	2000	IS 3025 (Part 16)	Beyond this palatability decreases and may cause gastrointestinal irritation
4	Chlorides (as Cl^-) ppm	250	1000	IS 3025 (Part 32)	Beyond this limit, taste, corrosion, and palatability are affected
5	Fluoride (as F^-) ppm	1.0	1.5	IS 3025 (Part 60)	Fluoride may be kept as low as possible. High Fluoride may cause Fluorosis
6	Nitrate (as NO_3^-) mg/l	45	No Relaxation	IS 3025 (Part 34)	Beyond this Metha hemoglobinuria takes place

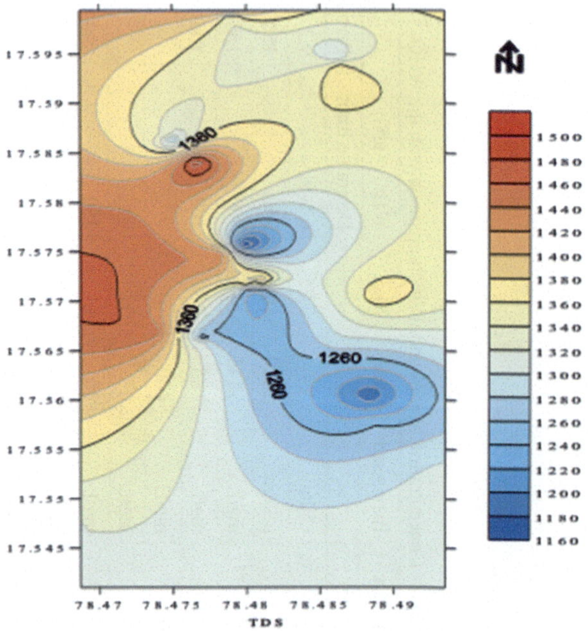

Fig. 26.2 Distribution diagram of TDS

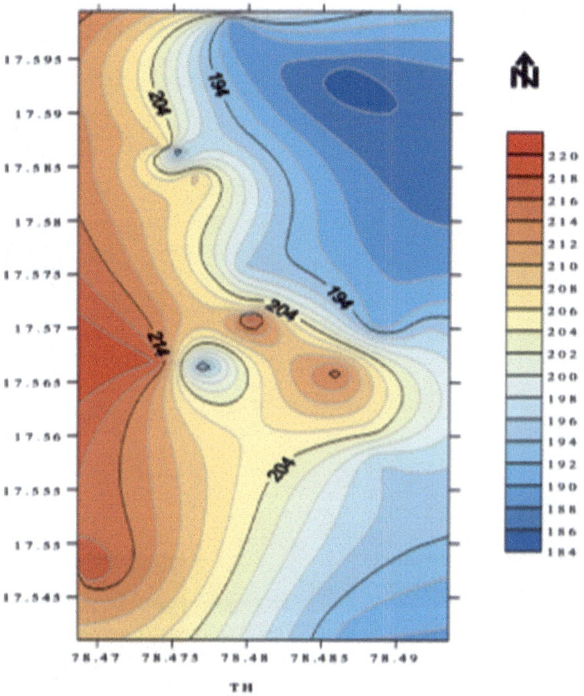

Fig. 26.3 Distribution diagram of TH

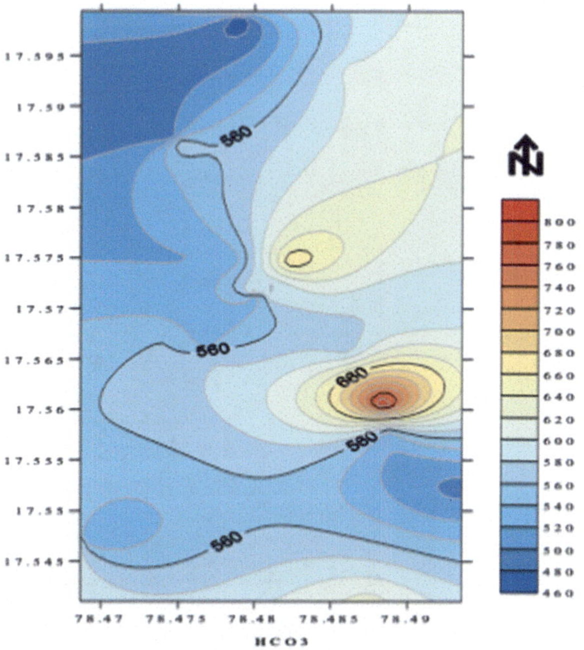

Fig. 26.4 Distribution diagram of HCO_3^- values

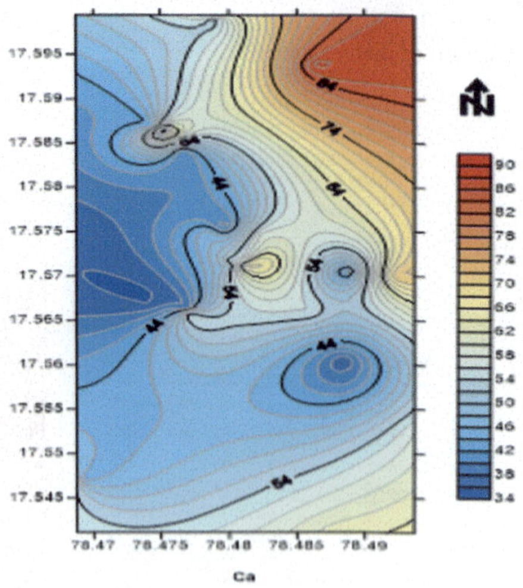

Fig. 26.5 Distribution diagram of Ca^{2+} values

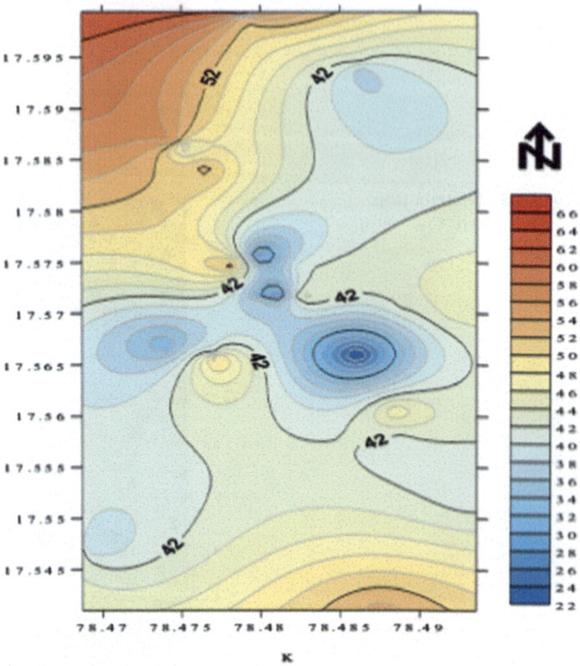

Fig. 26.6 Distribution diagram of K^{+} values

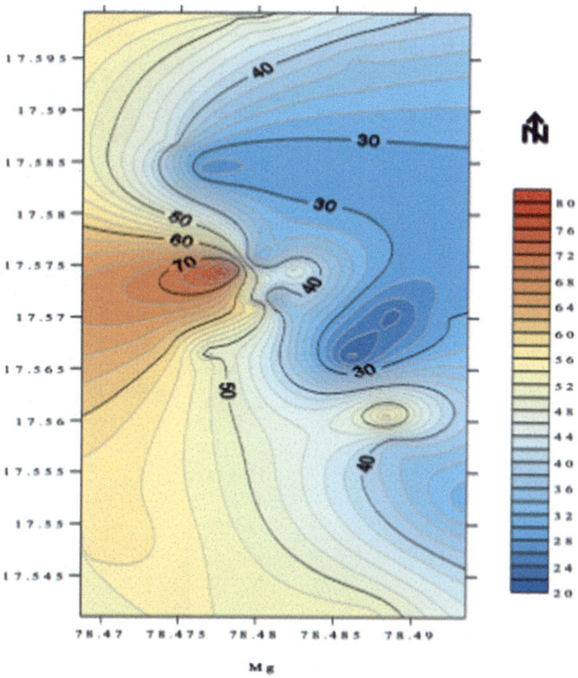

Fig. 26.7 Distribution diagram of Mg^{2+} values

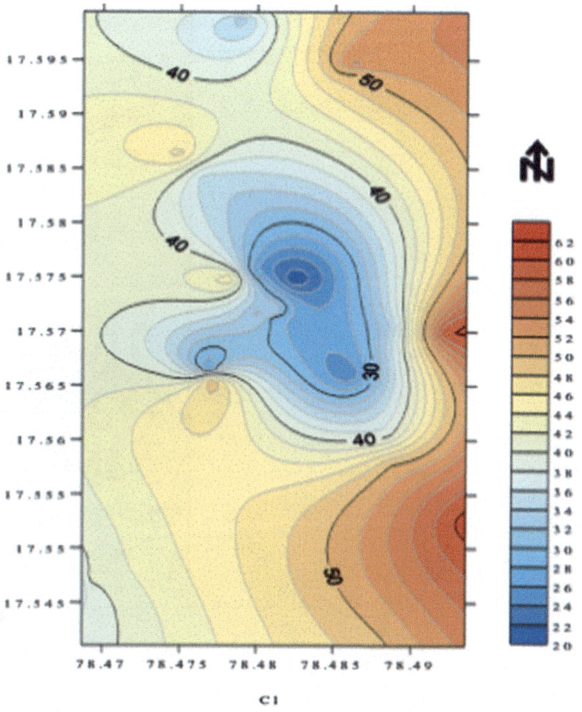

Fig. 26.8 Distribution diagram of Cl^{-} values

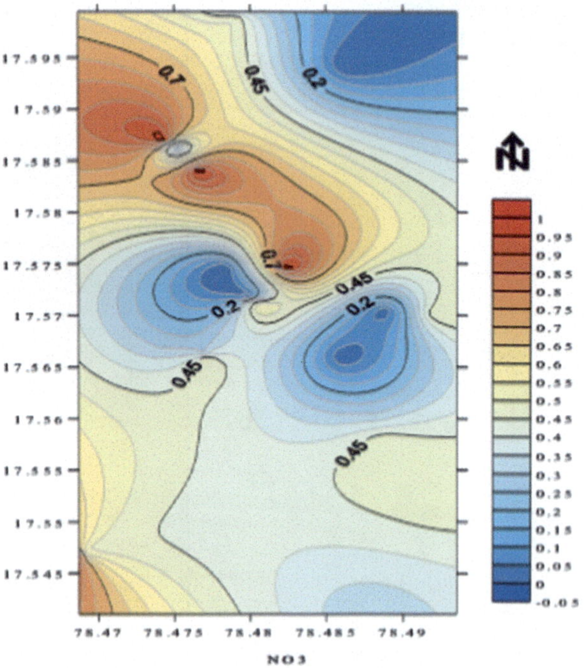

Fig. 26.9 Distribution diagram of NO_3^- values

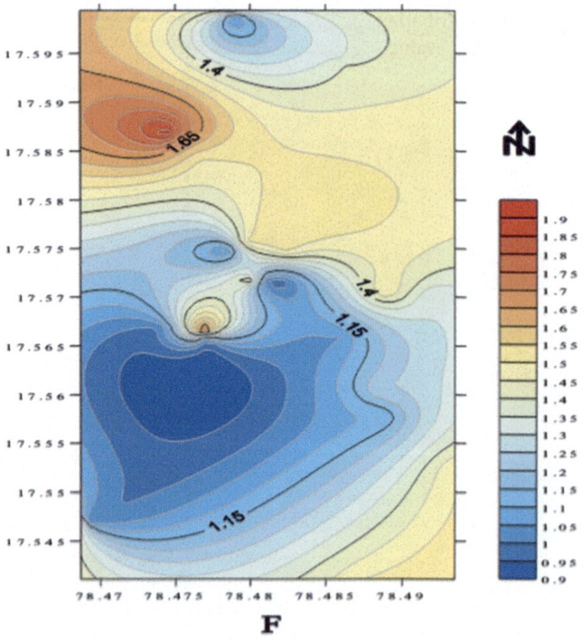

Fig. 26.10 Distribution diagram of F^- values

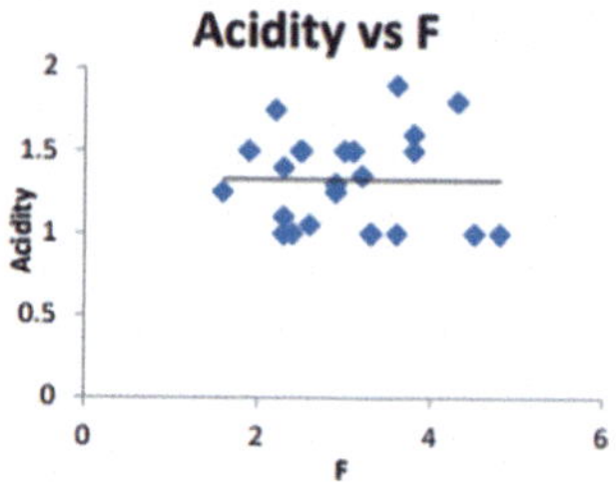

Fig. 26.11 Scatter plots of acidity vs F^-

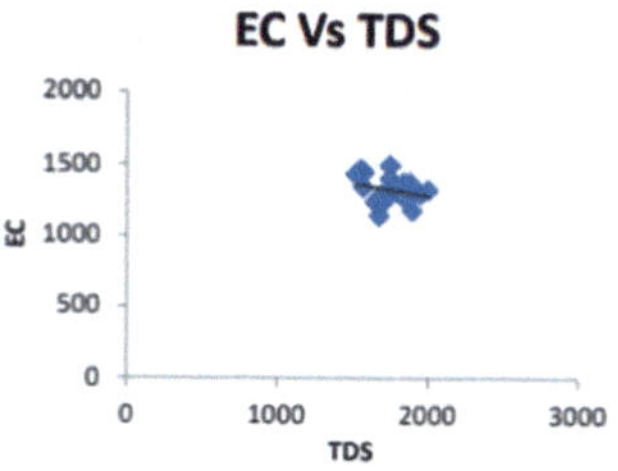

Fig. 26.12 Scatter plots of EC vs TDS

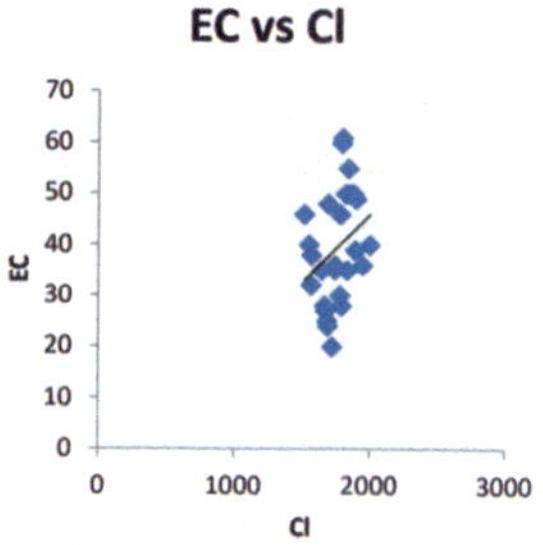

Fig. 26.13 Scatter plots of EC vs Cl^-

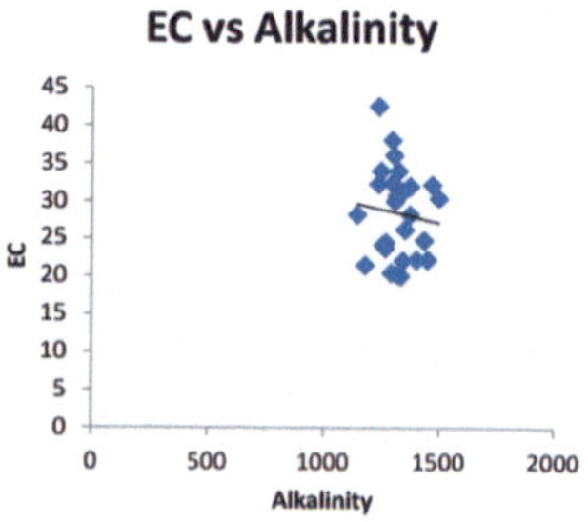

Fig. 26.14 Scatter plots of EC vs Alkalinity

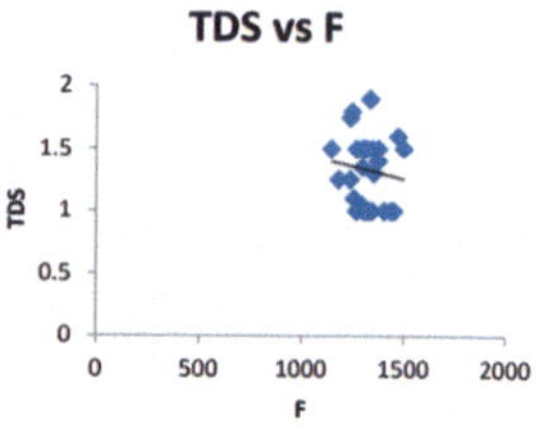

Fig. 26.15 Scatter plots of TDS vs F^-

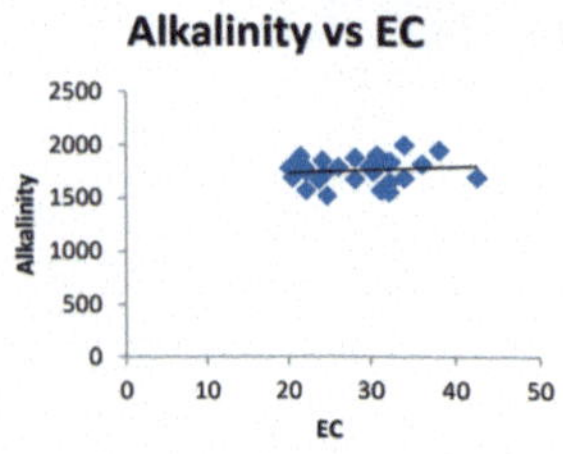

Fig. 26.16 Scatter plots of Alkalinity vs EC

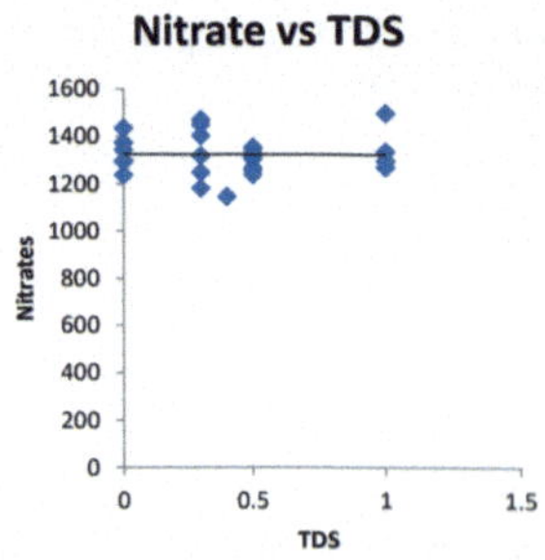

Fig. 26.17 Scatter plots of nitrate vs TDS

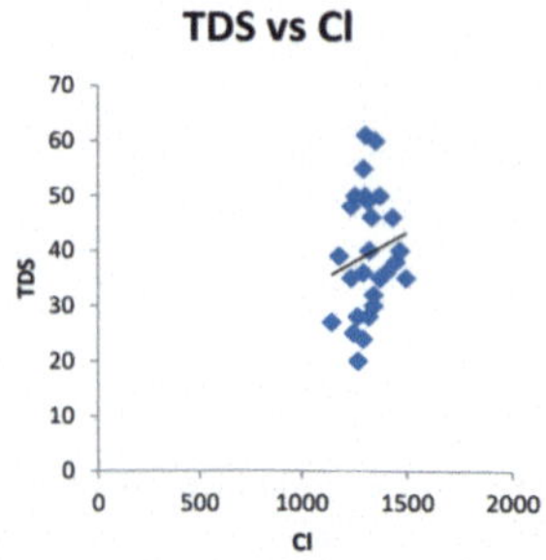

Fig. 26.18 Scatter plots of TDS vs Cl^-

Acknowledgements We express our thanks to Ch. Gopal Reddy, Secretary, CMRGI, Principal and HOD, CMR College of Engineering and Technology Hyderabad, for help and encouragement to publish this paper.

Compliance with Ethical Standards

Conflict of Interest The authors declare that they have no conflict of interest.

Ethical Approval The authors declare that this article does not contain any studies with human participants or animals.

Informed Consent This type of study does not require formal consent

References

Al-Harbi, M., AL-Ruwaih, F. M., Alsulaili, A. (2014). Statistical and analytical evaluation of groundwater quality in Al-Rawdhatain field. *Environ Prog Sustain, 33*, 895–904.

Backman, B., Bodis, D., Lahermo, P., Rapant, S., Tarvainen, T. (1998). Application of a groundwater contamination index in Finland and Slovakia. *Environ Geol, 36*(1–2), 55–64. https://doi.org/10.1007/s002540050320

Bagyaraj, M., Ramkumar, T., Gnanachandrasamy, G. (2013). A GIS analysis for assessing groundwater quality in Kodaikanal Hill, Western Ghats, Southern India. *Inventi Rapid: Water & Environment, 2013*(4), 1–9.

Balaji, E., Nagaraju, A., SreedharY, ThejaswiA., & Sharifi, Z. (2016). Hydrochemical characterization of groundwater in around Tirupati Area, Chittoor District, Andhra Pradesh. *South India. Applied Water Science, 7*(3), 1203–1212. https://doi.org/10.1007/s13201-016-0448-6

Bellack, E. (1995). Simplified fluoride distillation method. *Am Water Works Assoc J, 50*, 530.

BIS. (1983). Standards for water for drinking and other purposes. Bureau of Indian Standards, New Delhi.

BIS 10500. (1991). Indian standard drinking water-specification (1st rev., pp 1–8).

Canora, F., Rizzo, G., Panariello, S., Sdao, F. (2019). Hydrogeologyand Hydrogeochemistry of the Lauria Mountains Northern Sector Groundwater Resources (Basilicata, Italy) Hindawi. Geofluids. https://doi.org/10.1155/2019/7039165

Etikala, B., Vangala, S., & Madhav, S. (2024). Groundwater geochemistry using modified integrated water quality index (IWQI) and health indices with special emphasis on nitrates and heavy metals in southern parts of Tirupati, South India. *Environmental Geochemistry and Health, 46*(11). https://doi.org/10.1007/s10653-024-02229-3

Eugene Brown, M., Skougstad, W., & Fishman, M. J. (1970). *Methods for collection and analysis of water samples for dissolved minerals and gases* (pp. 389–621). United States Department of Interior.

Giridharan, L., Venugopal, T., Jayaprakash, M. (2008). Evaluation of the seasonal variation on the geochemical parameters and quality assessment of the groundwater in the proximity of River Cooum, Chennai, India. *Environ Monit Assess, 143*, 161–178. https://doi.org/10.1007/s10661-007-9965-y

Golla, V., Badapalli, P. K. & Mannala, P. (2021). Assessment of groundwater quality for drinking and irrigation in semi-arid regions of Andhra Pradesh, Southern India, using multivariate statistical analysis. *Arabian Journal Geoscience, 14*, 1967. https://doi.org/10.1007/s12517-021-08332-4

Handa, B. K. (1975). Geochemistry and genesis of fluoride-containing ground waters in India. *Groundwater, 13*(3), 275–288.

Helix–The scientific explorer|peer reviewed bimonthly. *International Journal, 11*(6), 32–37. Retrieved from https://helixscientific.pub/index.php/home/article/view/380

Liou, S. M., Lo, S. L., & Wang, S. H. (2004). A generalized water quality index for Taiwan. *Environmental Monitoring and Assessment, 96*, 32–35.

Mamat, Z., Halik, U., Muhtar, P., Nurmamat, I., & Abliz, A. (2016). Temporal variation of significant soil hydrological parameters in the Yutian oasis in Northwest China from 2001 to 2010. *Environment and Earth Science, 75*, 287. https://doi.org/10.1007/s12665-015-5152-6

Muralidhara Reddy, B., Sunitha, V., Prasad, M., Sudharshan Reddy, Y., Ramakrishna Reddy, M. (2019). Evaluation of groundwater suitability for domestic and agricultural utility in semi-arid region of Anantapur, Andhra Pradesh State, South India. *Groundwater for Sustainable Development, 9*(2019), 100262. Elsevier Journal. https://doi.org/10.1016/j.gsd.2019.100262

Muralidhara Reddy, B., Sunitha, V. (2020). Geochemical and health risk assessment of fluoride and nitrate toxicity in semi-arid region of Anantapur District, South India. *Environmental Chemistry and Ecotoxicology, 2*(2020), 150–161. https://doi.org/10.1016/j.encero.2020.09.002

Musini, V., Merugu, S. (2020). Fuzzy inference system based assessment of pollution aspects in Hyderabad. Advances in Cybernetics, Cognition, and Machine Learning for Communication Technologies, pp 579–587. https://doi.org/10.1007/978-981-15

Musini, V., Merugu, S., Rohini Reddy, T. (2020). Information system used to study on effect of groundwater quality. In Gunjan, V., Senatore, S., Kumar, A., Gao, X. Z., Merugu, S. (Eds.), *Advances in cybernetics, cognition, and machine learning for communication technologies. Lecture Notes in Electrical Engineering*, vol 643. Springer, Singapore. https://doi.org/10.1007/978-981-15-3125-5_53

Nagaraju, A., Balaji, E., Sun, L., & Thejaswi, A. (2018). Processes controlling groundwater chemistry from Mulakalacheruvu Area, Chittoor District, Andhra Pradesh, South India: A Statistical Approach Based on Hydrochemistry. *Journal of the Geological Society of India, 91*(4), 425–430. https://doi.org/10.1007/s12594-018-0875-0

Pierre, D., Glynn, L., & Plummer, N. (2005). Geochemistry and the understanding of groundwater systems. *Hydrogeology Journal, 13*, 263–287.

Prasad, P., Chaurasia, M., Sohony, R. A., et al. (2013). Water quality analysis of surface water: A Web approach. *Environmental Monitoring and Assessment, 185*, 5987–5992. https://doi.org/10.1007/s10661-012-2999-9

Purushothaman, P., SomeshwarRao, M., Rawat, Y. S., Kumar, C. P., & Kris- han G, Parveen T,. (2013). Evaluation of hydrogeochemistry and water quality in Bist-Doab region, Punjab. *India Environ Earth Sci, 72*, 693–706.

Rawat, J. S., Biswas, V., Kumar, M. (2013). Changes in land use/cover using geospatial techniques–A case study of Ramnagar town area, district Nainital, Uttarakhand, IndiaEgyp. *Journal Remote Sensing Space Science, 16*, 111–117

Reddy, Y. S., & Sunitha, V. (2023). Assessment of heavy metal pollution and its health implications in groundwater for drinking purpose around inactive mines, SW region of Cuddapah Basin, South India. *Total Environment Research Themes, 8*, 100069. https://doi.org/10.1016/j.totert.2023.100069

Reddy, Y. S., & Sunitha, V. (2020). Development of water quality index for groundwater quality evaluation in and around VemulaMandal, YSR District (AP, India). *Journal of Indian Geophysical Union, 24*(1), 68–77.

Saikrishna K., Purushotham, D., Sunitha, V., Sudharshan Reddy, Y., Brahmaiah, T., Muralidhara Reddy, B., Nallusamy, B. (2023). Deciphering groundwater quality, mechanisms controlling groundwater chemistry in and around Suryapet, Telangana, South India. Total Environment Research Themes. *Elsevier Publications Total Environment Research Themes, 6*, 100035. https://doi.org/10.1016/j.totert.2023.100035

Saikrishna, K., Purushotham, D., Sunitha, V., et al. (2023). Geochemical processes of groundwater for drinking purposes in Dharwarcraton of Mallampalli area, Telangana, South India. *International Journal Energy Water Resources, 7*, 15–28. https://doi.org/10.1007/s42108-021-00146-0

Saxena, V. K., & Ahmed, S. (2001). Dissolution of fluoride in groundwater: A water-rock interaction study. *Environmental Geology, 40*(8), 1084–1087.

Sunitha, V., & Reddy, B. M. (2022). Geochemical characterization, deciphering groundwater quality using pollution index of groundwater (PIG), water quality index (WQI) and geographical information system (GIS) in hard rock aquifer. *South Indian Applied Water Science, 12*, 41. https://doi.org/10.1007/s13201-021-01527-w

Sunitha, V., & Reddy, B. M. (2022). Geochemical characterization, deciphering groundwater quality using pollution index of groundwater (PIG), water quality index (WQI) and geographical information system (GIS) in hard rock aquifer. *South India. Applied Water Science, 12*(3), 41.

Sunitha, V., Khan, J. A., Reddy, B. M., Prasad, M., & Reddy, M. R. (2014). Assessment of groundwater quality in parts of Kadapa and Anantapur Districts, Andhra Pradesh, India. *Indian Journal of Advances in Chemical Science, 3*, 96–101.

Sunitha, V., Sudarshan, V., & Reddy, B. R. (2005). Hydrogeochemistry of groundwater, Gooty area, Anantapur district, Andhra Pradesh. *Indian Pollution Research, 24*(1), 217.

Sawyer, C .N. (1994). Chemistry (pp 103–104). New York: McGraw Hill.

Tsegaye, T., Sheppard, D., Islam, K. R., Johnson, A., Tadesse, W., Atalay, A., & Marzen, L. (2006). Development of chemical index as a measure of in-stream water quality in response to land use and land cover changes. *Water, Air, and Soil Pollution, 174*, 161–179.

Umar, M., Waseem, A., Sabir, M. A., Kassi, A. M., & Khan, A. S. (2013). The impact of the geologyof recharge areas on groundwater quality: A case study of Zhob River Basin, Pakistan. Clean: Soil. *Air, and Water, 41*, 119–127.

US Geological Survey. (2000). Classification of natural ponds and lakes, US Department of the Interior, US Geological Survey, Washington, DC.

Venkatayogi, S. (2015). Geochemistry of fluoride bearing groundwater in parts of Telangana State. *India Journal Water Resource Hydraulic Energy, 4*(4), 380–387.

Venkateshwarlu, M., Merugu, S., & Asari, V. K. (2021). *Computational assessment of ion distribution on applying multivariate analysis using geographic information systems.*

Venkateshwarlu, M., Narsi Reddy, M., & Kiran Kumar, A. (2018). Groundwater quality assessment in Kandlakoya village by chemical methods. *Indian Journal of Environmental Protection, 38*(9), 768–771.

Venkateshwarlu, M., Rajagopal, K., Reddy, Y. S. (2020). Geochemical studies on the groundwater of Kistapur village, Medchal district, Telangana. *Indian Journal Environmental Protection, 40*(8), 887–891.

Venkateshwarlu, M., Rajagopal, K., Obula Reddy, C., Reddy, Y. (2022). Investigation of Groundwater Contamination due to Landfill Leachate. *IJEP, 42*(7), 857–862.

Venkateshwarlu, M., RajKiran Reddy, K., Vishwagna, K.et al. (2015). A case study using GIS-based evaluation of watershed management Medak District, Telangana state, India. *IJISET-International Journal of Innovative Science, Engineering & Technology, 2*(3).

Venkateshwarlu, M., Rasheed, M. A., Reddy, U. V. B., & Kumar, A. K. (2014). Assessment of ground water quality in and around Miryalagudaarea, Nalgonda district of Andhra Pradesh. *Int J Plant Animal Environ Sci, 4*(2), 259–266.

WHO. (1983). Guidelines for drinking water quality. Geneva: World Health Organization.

WHO. (1990). Environmental health criteria 81: vanadium. World Health Organization, Geneva, pp 1–35.

WHO. (2003). Overview of the World Health Report 2003: Shaping the future Geneva. https://www.who.int/whr/2003/en/overview_en.pdf.

Chapter 27
Groundwater Quality in Parts of Hyderabad and Medchal-Malkajgiri Districts, Telangana, India: Hydrochemical and Geospatial Assessment

Ramesh Chandra Phani Pothuri, Chinmay Dongare, Rajendra Prasad Kanchi, and Rajesh Kanchi

Abstract This study accounts for groundwater quality in parts of Hyderabad and adjoining Medchal-Malkajgiri districts, Telangana, southern India. Due to rapid northward and eastward expansion of the city, groundwater quantity and quality are experiencing severe deterioration. In this study, in total, 48 samples each have been collected during pre- and post-monsoon periods, respectively from bore-wells ($n = 24$) in 2022 to test the physico-chemical parameters namely pH, EC, TDS, TH, Na^+, K^+, Ca^{2+}, Mg^{2+}, F^-, Cl^-, $(CO_3)^{2-}$, HCO_3^-, NO_3^-, SO_4^{2-} through standard methods for determining the suitability for everyday purposes such as domestic, irrigation and industrial. The general order of abundance of prominent anion and cation species in groundwater is $F^- > NO_3^- > SO_4^{2-} > Cl^- > HCO_3^-$ and $K^+ > Mg^{2+} > Ca^{2+} > Na^+$, respectively. The results show that a considerable number of samples have excess nitrate content as high as 49 ppm in pre-monsoon (Alwal) and 209 ppm in post-monsoon (Alwal, Madhapur, Mushirabad, Dundigal and Sundarnaagar). The fluoride content is higher than one ppm in 50% of pre-monsoon (Alwal, Balanagar, Qutubullapar, Malkaljgiri, Sundarnagar, Mehdipatnam, Bollarum, Dundigal and Nizampet) and 45% of post-monsoon samples (Qutubullapur, Balanagar, Malkajgiri, Subhashnagar, Mushirabad, Mehdipatnam, Dundigal, and Alwal). Based on EC (electrical conductivity), high Na%, and SAR (Sodium Absorption Ratio), the pre-monsoon

R. C. P. Pothuri (✉)
Mining and Natural Resources, Cyient Limited, Hyderabad, India
e-mail: Ramesh.Pothuri@cyient.com

C. Dongare
Department of Geology, Faculty of Science, The Maharaja Sayajirao University of Baroda, Vadodara, India

R. P. Kanchi
Department of Geology, Osmania University, Hyderabad, India

R. Kanchi
Department of Applied Geochemistry, Osmania University, Hyderabad, India

V. Sunitha et al. (eds.), *Groundwater Resource Management Planning Strategies*, Springer Water, https://doi.org/10.1007/978-3-031-88870-0_27

samples have been classified into C2S1, C3S1, and C3S2 types showing 'high salinity hazard categories for regular domestic and irrigation usage. Geogenic factors such as hard granitic basement, lack of fracture system, reduced natural recharge and anthropogenic activities are believed to be the causes for the deterioration of groundwater quality in the study area despite heavy rainfall during the year 2022.

Keywords Hydrochemistry · Physico-chemical parameters · Groundwater quality · Nitrate · Fluoride contamination · Geospatial analysis · Hyderabad · India

27.1 Introduction

Groundwater is a dynamic and crucial component of water cycle in nature (Ha et al., 2022; Hao et al., 2018). It serves as a primary source of drinking water for around 30–40% of the world's population (Gleeson et al., 2020; Siebert et al., 2010). It has remained significant in nourishing ecosystems, agriculture, human communities, and activities. However, in recent times, the unprecedented demand for groundwater resources tied with geogenic and anthropogenic causes, have evoked severe concerns about its quality and sustainability (Gajendran & Thamarai, 2008; Chatterjee et al., 2009; Dongare et al. 2022; Pancholi et al., 2022; Dongare & Deota, 2023). The demand for groundwater resources is increasing at a faster rate, especially in hard rock terrains, and equally, the quantity and quality of groundwater is depleting drastically year by year (Balaji et al., 2016; Etikala et al., 2021; Manikandan et al., 2020; Sunitha & Reddy, 2022). Given the growing threats posed by geogenic and anthropogenic interventions such as industrial discharges, agricultural runoff, and urbanization, comprehensive assessment and monitoring of groundwater quality, in urban areas, has become mandatory (Shaikh et al., 2020).

The quality and quantity of groundwater are depleting at an alarming rate in India, irrespective of geological and hydrological conditions. India, the largest consumer of groundwater globally, is facing an acute shortage of groundwater resources in urban areas (CGWB, 2021). Hyderabad is a fast-growing city in southern India, rapidly spreading its urban limits to adjacent districts. The adjoining northern and north-eastern parts of the city belong to Medchal-Malkajgiri district, where the quality and quantity of groundwater resources are deteriorating at a higher rate due to the influx of industrial and domestic waste. The impact is more evident in areas namely Jamai Osmania, Lalaguda, Nacharam, Keesara, Uppal, Lalapet and adjoining areas owing to highly dense population (Asadi et al., 2009; Udayalakshmi et al., 2010). The groundwater quality assessment in and around the Uppal Kalan area, within Medchal-Malkajgiri district, reported contamination due to intense industrial activity (Hari et al., 2020). The groundwater quality on the banks of Musi River was also attested to be highly polluted due to percolation of toxic components to the subsurface (Satish Kumar et al., 2007).

The northwestern part of the city is fast developing, and groundwater is the only primary water source. The surface water bodies are getting continuously polluted due to the release of effluents from industries and new civilian constructions such as educational institutes, gated community colonies, sky-scaping buildings, etc. (Phani and Prasad, 2022). Consequently, the demand for groundwater resources is increasing, and in turn, a number of new bore wells are dug, leading to the depletion of the water table and deterioration of groundwater quality. Unfortunately, it has been learned that groundwater contamination problems are aplenty not only in Hyderabad but also in certain parts of the districts close to the city. This study aims at presenting an account of groundwater quality in parts of Hyderabad and adjacent Medchal-Malkajgiri districts with physicochemical parameters and geospatial analysis.

27.2 Geological and Hydrogeomorphological Aspects

The study area extends between latitude 17°23′10 to 17°35′10″ and longitude 78°21′57″ to 78°31′54″, covering an area of ~ 230 km^2 (Fig. 27.1a). The topography of the region is mostly plain with occasional hillocks (397–698 m ASL) with original drainage obscured by dense built-up area. The study area consists of rocky terrain predominantly comprising Precambrian granites and granodiorites intruded with dolerite dykes and quartz veins. Multiple sets of joints traverse the granitic rocks. Dykes of amphibolites, quartz veins, dolerite, and migmatites frequently traverse the granitic rocks, especially in southeastern and eastern parts. These dykes act as important structural entities in controlling the movement of groundwater within the area. Cross-cutting joints/fractures of different orientations are very common. Vertical joints act as conduits for the percolation of water, while horizontal or sheet joints help to maintain the level continuity of the aquifer laterally (CGWB, 2021). The granite rock of the study area essentially comprises quartz, orthoclase, microcline, and plagioclases with minor amounts of biotite and magnetite. The F^- bearing accessory minerals in Hyderabad granites act as sources of fluoride in the groundwater (Laxman et al., 2021).

Geomorphologically, the area forms a pediment-inselberg-pediplain complex with valley-fill deposits in the streams. The streams are overloaded with anthropogenic waste. The land use/land cover (LU/LC) of the study area comprises agricultural plantation, barren rocky land, urban/built-up areas, cropland, forest, afforested lands, hamlets, quarries, industrial areas, settlements, water bodies, river/stream/drain, open scrubland, road network, and upcoming real estate plots.

The area has several granite quarries for producing road metal and plants for producing robo sand. Prolonged quarrying leading to the lowering of groundwater table is observed in adjacent areas. The granites are massive and lack fractures that obstruct water from percolating down (Fig. 27.1b). The effluents from a variety of industries and domestic sewerage released to nalas and waterbodies create dense, thick froth, leading to contamination of groundwater and also the soils and sediments on the banks (Fig. 27.1c). Agriculture is mainly carried out using groundwater. A number of bore wells dug with no proper spacing is observed. However, in a few localities, groundwater yield is sufficiently good (Fig. 27.1d). The color and odor of

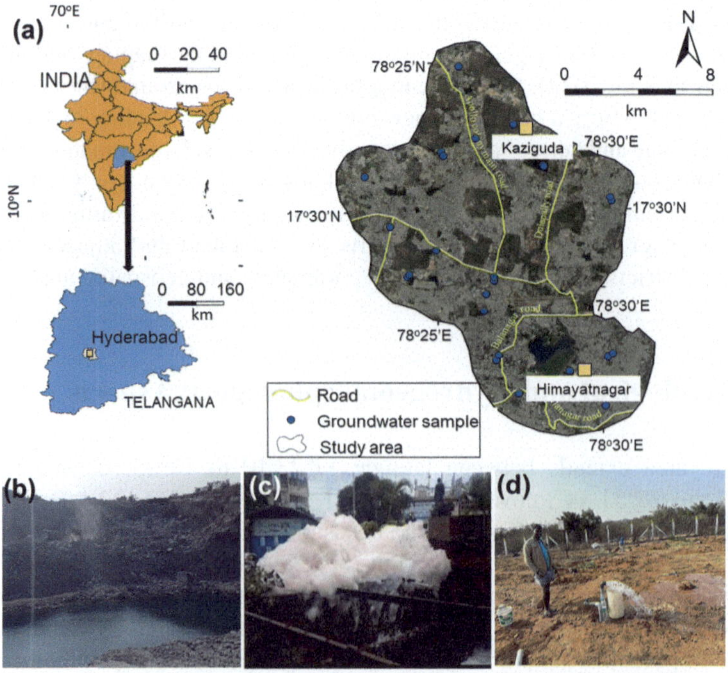

Fig. 27.1 Location map and field photographs of the study area, i.e., Hyderabad and Medchal-Malkajgiri districts, Telangana. **a** Study area clipped from Google Earth image. **b** An active quarry near Railapur exposes massive sheet granite. Note that waterlogging is due to absence of fractures hindering the water percolatation. **c** Intense frothing is due to the influx of industrial effluents into the nala south of Medchal town. **d** Bore well in hard granite basement yielding groundwater near Subhashnagar

groundwater are reasonable in remote agricultural areas. During 2022–23, the region received a rainfall of 793 mm (TSDPS, 2023). The average depth of the water table is about 110 feet BGL, and bore wells are about 550 feet.

27.3 Materials and Methods

To assess the groundwater quality, 48 water samples are collected within Hyderabad and adjoining Medchal- Malkajgiri district of Telangana, with 24 samples each during the pre-monsoon (May 2022) and post-monsoon (December 2022) seasons respectively. One-liter size bottles are used to collect groundwater samples. The bottles are sterilized by rinsing with 1:1 dilute nitric acid (HNO_3) and then thoroughly washed with distilled water. At each well-site, before collecting the samples, the bottle was washed with the water of the bore well. The collected water samples are directly measured for pH/EC by hand-held meters (Hanna USA H-198130) in field.

The TDS values are determined by multiplying EC values with 0.64. The significant ions like cations (calcium, magnesium, sodium, and potassium) and anions (bicarbonate, chloride, sulfate, nitrate, and fluoride) are analyzed by adopting the standard procedures (APHA, 2012). All the analyses were conducted at the Department of Applied Geochemistry, Osmania University, Hyderabad. Calcium, total hardness, chloride, and bicarbonate are estimated by titration. Sodium and potassium were determined by flame photometry. Sulfate and nitrate are determined utilizing a UV spectrophotometer. An ion-selective electrode determines fluoride concentration. For the computation of analytical error, the total cations ($Ca^{2+} + Mg^{2+} + Na^{+} + K^{+}$) and the total anions ($HCO_3^- + Cl^- + SO_4^{2-} + NO_3^- + F^-$) were used (Domenico and Schwartz, 1997) using the equation given below (27.1).

$$\text{Ion Balance Error (IBE)} = \frac{\text{cation} - \text{anion}}{\text{cation} + \text{anion}} * 100 \tag{27.1}$$

The IBE in the present study is about ± 20%. Not all the cations and anions were analyzed, hence the samples showed higher IBE values. Moreover, the groundwater in the study area is intensely contaminated due to domestic and industrial sewerage, leading to the presence of a variety of chemical constituents. Therefore, the samples showed higher IBE values; however, the analytical quality is found to be accurate. The physico-chemical parameters were spatially plotted using ESRI software (ArcGIS v10.4) for geospatial assessment at Department of Geology, MS University of Baroda, Vadodara.

27.4 Results and Discussion

Physico-chemical parameters of pre- and post-monsoon groundwater samples determined in this investigation are shown in Table 27.1.

27.4.1 pH

The taste of water depends upon pH. The pH is defined as the negative log of hydrogen ion concentration in moles per liter. It represents the degree of acidity, neutrality, and alkalinity of the water. In urban environments, the pH value of natural waters changes due to lithology, biological influence, temperature, and the influx of industrial wastes (APHA, 1998). Generally, the pH of groundwater in pre-and post-monsoon in the study area ranges from 7.15 to 8.76 and from 6.92 to 7.97, respectively, indicating it's neutral to weakly alkaline nature. The pH is high in Gajularamaram and Narayanguda areas in pre-monsoon, while high pH is shown in Panjagutta and Lakdikapul in post-monsoon periods (Fig. 27.2a and b). Domestic as well as industrial contaminants highly influence the pH of the groundwater in the study area.

Table 27.1 Physico-chemical parameters of groundwater samples of the study area determined during the year 2022

Parameters	Pre-monsoon 2022		Post-monsoon 2022		BIS (2012)		Pre-monsoon 2022		Post-monsoon 2022	
	Range	Avg	Range	Avg	Desired limit (DL)	Prescribed Limit (PL)	% of samples above DL	% of samples above PL	% of samples above DL	% of samples above PL
pH	7.1–8.7	7.7	6.9–7.8	7.4	6.5	8.5	91.6	8.3	100	–
EC	160–2470	1344	461.1–1891.7	1285.9	–	1500	–	41.6	–	33.3
TH	60–780	385.8	180–599.9	391.2	300	600	37.5	16.6	73.91	–
TDS	102–1580	860	295.1–1210.7	823	500	2000	79.1	–	86.95	–
Cl^-	10–530	210.8	20–319.9	138.2	250	1000	33.3	–	8.6	–
F^-	0.1–1.9	1.1	0.5–1.8	1.1	1	1.5	25	25	13.04	30.43
SO_4^{2-}	2–85	22.3	6.2–33.9	20.1	200	400	–	–	–	–
NO_3^-	0.5–48.9	14.4	2.8–209.4	47.9	–	45	–	4.1	–	26.08
CO_3^{2-}	0–80	6.6	0	–	–	–	–	–	–	–
HCO_3^{2-}	48.9–552.9	286.4	155.2–514.6	357.7	–	–	–	–	–	–
Na^+	9.5–292.7	133	22.4–223.3	116.7	–	200	–	12.5	–	13.04
K^+	0.1–6	2.7	1–7.7	3.4	–	12	–	–	–	–
Ca^{2+}	16–168	82	56–135.9	110.6	75	200	50	0	73.91	–
Mg^{2+}	4.8–87.5	41.3	9.7–82.6	27.9	30	100	41.6	0	43.47	–

* BIS = Bureau of Indian Standards DL = Desirable limit PL = Prescribed limit. Concentrations of chemical compnents for pre- and post-monsoon samples are in ppm

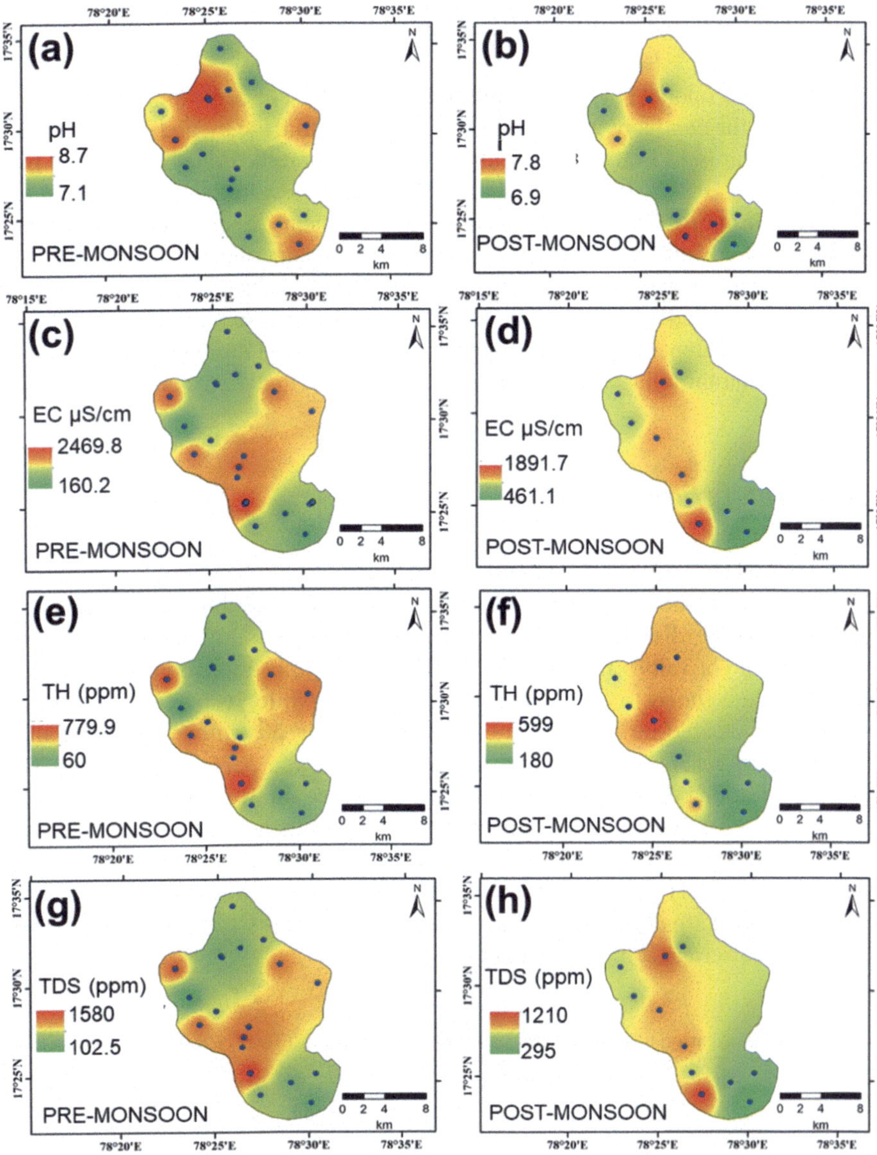

Fig. 27.2 Spatial distribution of physicochemical parameters of groundwater samples of present study during pre- and post-monsoon periods in the year 2022

27.4.2 *Electrical Conductivity (EC)*

The EC depends upon TDS concentration in water. The concentration of dissolved salts increases the conductivity. In the present investigation, the mean EC for the post-monsoon season is 1343.8 μS/cm, while in pre-monsoon, it reduces to 1285.9 μS/cm. Concerning BIS (2012), 41.6% and 33.3% of groundwater samples exceed the permissible limit of EC in pre- and post-monsoon seasons, respectively. As per the EC-based groundwater classification proposed by Jagannadha Sarma and Swamy (1981), the groundwater of the region falls in medium conductive-I (500–1000 μS/cm) and medium conductive-II (1000–3000 μS/cm). The higher values of EC indicate higher mineral content in the groundwater. The EC is high in Panjagutta pre-monsoon and Gajularamaram village post-monsoon respectively (Fig. 27.2c and d).

27.4.3 *Total Hardness (TH)*

The mean Total Hardness (TH) in the study area varies from 385.8 ppm pre-monsoon to 391.2 ppm post-monsoon. The relatively higher TH value during post-monsoon is attributed to the dissolution of minerals, the influx of industrial effluents, and the increase in Ca values (Wagh et al., 2019). During pre-monsoon, around 37% of samples were found to exceed the desirable limit, while 16 percent of samples exceeded permissible limits. On the other hand, during post-monsoon, more than 70% percent were above the desired limits but below the permissible limits. The high concentration of TH may adversely impact water supply systems and human health (Wagh et al., 2019). The acceptable limit is 200 ppm, as per BIS (2012). About 75% of pre-monsoon samples and 58% of post-monsoon samples exceed this limit (Table 27.2). The TH values are high in Mehdipatnam and Ameerpet areas in pre-monsoon and Kukatpally areas in post-monsoon due to heavy contamination caused by domestic sewerage (Fig. 27.2e and f).

Table 27.2 Classification of groundwater based on total hardness (Todd, 1980)

Class	TH Range (ppm)	% of Samples	
		Pre-monsoon	Post-monsoon
Soft	0–75	4	0
Moderate to Hard	75–150	13	0
Hard	150–300	29	25
Very Hard	> 300	54	75

Table 27.3 Categorisation of groundwater based on TDS (Davis & Dewist, 1966)

Class	TDS Range	% of Samples	
	ppm	Pre-monsoon	Post-monsoon
Freshwater	0–1000	62.5	79.5
Brackish water	1000–10,000	37.5	19.5
Saline water	10,000–100,000	0	0
Brine water	> 100,000	0	0

27.4.4 Total Dissolved Solids (TDS)

The TDS depends on factors such as local lithology, soil thickness, structural setting, weathering intensity, groundwater residence time, and amount of rainfall (Stephens et al., 2021). The average TDS varies from 860 ppm pre-monsoon to 823 ppm post-monsoon season. In comparison with BIS (2012) limits, it is observed that 79.1% (pre-monsoon) and 86.95% (post-monsoon) of groundwater samples are above desirable limits, however, under permissible limits and suitable for domestic uses. The higher values of TDS than the permissible limits can be attributed to the absence of proper industrial and domestic sewage disposal systems or the contribution of salts from the soil-weathered parent rock materials (Dar et al., 2010). Based on the TDS content (Davis & Dewist, 1966), the groundwaters of the present study are mainly classified as freshwater, with few samples as brackish type (Table 27.3.). The TDS is higher in the Lakdikapul area in pre- and post-monsoon periods, which is attributed to high contamination caused by urban sources (Fig. 27.2 g and h).

In general, in the study area, the relatively higher values of pH, EC, and TDS during pre-monsoon could be attributed to the weathering of silicate minerals present in granitoid rocks, such as hornblende, pyroxenes, and micas, and altered feldspars in the weathered zone, reverse ion exchange, and the evaporation process. In addition, contamination due to anthropogenic sources is also a major problem in the study area.

27.4.5 Chloride

Chloride concentration shows wide variation and is sourced through soil–water interaction, producing chloride salts and lateral movement of ions from the source. The higher chloride concentration above 250 ppm makes the water salty. The chloride values range from 10 to 530 and 20–520 ppm in pre- and post-monsoon periods. The chloride is higher in the Bachupalli and Mehdipatnam areas during pre-monsoon season. In contrast, higher values prevail in Lakdikapul in post-monsoon times, which is attributed to poor movement of sewerage, its decomposition and consequent percolation to subsurface (Fig. 27.3a and b).

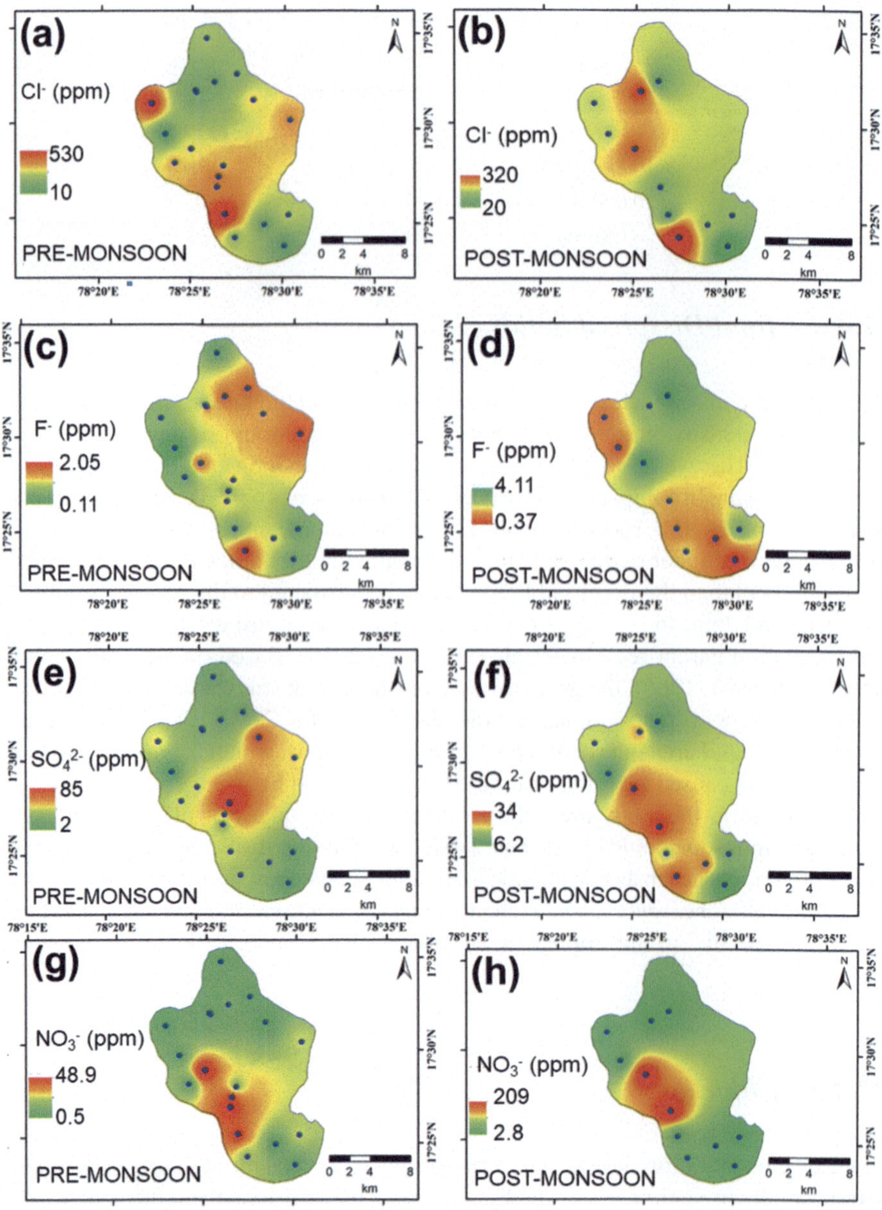

Fig. 27.3 Spatial distribution of anions concentrations of groundwater samples of present study during pre- and post-monsoon periods in the year 2022

27.4.6 Fluoride

The sources of fluoride in groundwater are both geogenic and anthropogenic. The significant geogenic causes are the weathering of silicates, alkali, sedimentary rocks (shales) and igneous rocks while the anthropogenic sources are fluoride salts used in various industries like aluminum, steel, tiles, bricks, herbicides, and insecticides used in agricultural practices (CPCB, 2008). The BIS (2012) standard value of F^- is 1.0 ppm; however, the fluoride concentration is in the study area. The pre-monsoon samples (50%) recorded F ranging from 0.11 to 2.05, and that in post-monsoon samples (45%), from 0.37 to 4.11, which is much higher than the permissible limits. The high fluorine content is attributed to the presence of F^- bearing accessory minerals such as apatite [$Ca_5(PO_4)_3$ (F, Cl, OH)], hornblende $(Ca, Na)_2(Mg, Fe, Al)_5(Al, Si)_8O_{22}$ $(F, OH)_2$ and biotite $K(Mg, Fe)_3$ $(AlSi_3O_{10})$ $(F, OH)_2$ in country rock granitoids (Laxman et al., 2021). Although fluoride is higher in certain areas like Qutubullapur, Balanagar, Malkajgiri, and Alwal, no impact on human health is noticed except in minor dental cases. This is mainly due to the use of purified canned water by the people rather than the direct use of groundwater. The fluoride is higher in Alwal, Bollarum, and Subhashnagar areas in pre-monsoon, while it is higher in Bachupalli and Himayatnagar areas in post-monsoon (Fig. 27.3c and d). The increased urbanization in recent times in these areas led to the set up of many metal industries that produce waste products enriched in fluoride content.

27.4.7 Sulphate

The sulfate in groundwater is sourced from gypsum in soils, which is most likely accumulated by agricultural pesticides and nutrients. Granites or quartz veins intruded into them often contain pyrite as an accessory mineral, the oxidation of which produces sulfates. Most predominantly, the sulfates in groundwaters are originated by industrial waste municipal and domestic sewerage (Torres-Martínez et al., 2020). The SO_4^{2-} is high, ranging from 2 to 85 and 4.85 to 35 ppm in pre- and post-monsoon samples of present study. The sulfate is higher in Ameerpet in pre-monsoon and in Mehdipatnam, Kukatpally, and Moosapet areas (Fig. 27.3e and f). The higher sulfate is attributed to a heavy influx of domestic sewerage to nalas in these heavily populated areas.

27.4.8 Nitrate

In general, the source of nitrogen in groundwater is attributed to higher rate of evapotranspiration or due to lack of precipitation. In addition, it can be sourced from agricultural pesticides (Wick et al., 2012). Excess fluoride and nitrate concentrations were recorded in the hard rock terrain of Cuddapah district, Andhra Pradesh, where serious health risks were foreseen (Sunitha et al., 2022). In the study area, the NO_3^- ranges from 0.43 to 49 and 1.39 to 285 ppm in pre- and post-monsoon samples, respectively, while the BIS (2012) standard is 45 ppm. The high nitrate is believed to be due to influx of industrial waste to the streams and water bodies and in turn percolating to the groundwater. The nitrate is higher in Alwal, Mushirabad, Madhapur, Sundarnagar, and Dundigal pre- and post-monsoon (Fig. 27.3 g and h). The higher nitrate values are attributed to anthropogenic sources.

27.4.9 Carbonate

Carbonate in groundwater is produced by the action of CO_2 with calcareous material i.e. calcrete, or dolocrete produced from the weathering of Ca-bearing feldspars. (Todd, 1980). In the present study, the $(CO_3)^{2-}$ ranges from 0 to 80 in pre-monsoon samples, while in post-monsoon samples, it is 0. The absence of $(CO_3)^{2-}$ in post-monsoon samples is probably due to the removal of carbonate during excess rainfall.

27.4.10 Bicarbonate

According to Todd (1980), the bicarbonate in groundwater is derivative of CO_2, released by the organic disintegration of soil. In the present study, the HCO_3 in pre-monsoon samples ranges from 49 to 552 ppm while that in post-monsoon samples from 155 to 514 ppm while the BIS (2012) standard is 200 ppm. Only in one sample (Government Junior College, Kukatpally), the bicarbonate is 48 ppm, which is attributed to the presence of hard granitic rock that restricted the groundwater movement. In five pre-monsoon samples and four post-monsoon samples, it is < 200 ppm; in all other samples (80%), HCO_3^- is beyond the BIS standard. The higher bicarbonate is attributed to the heavy influx of organic carbon from domestic sewerage.

27.4.11 Sodium

Sodium is the standard and predominant chemical constituent of waters, sourced from chemical weathering of Na- plagioclase feldspars and using reused water in irrigated areas and pollution. In the present investigation, the average concentration of Na in pre-monsoon groundwater samples is 133 ppm, which reduces to 116.7 ppm during post-monsoon season, while the BIS (2012) standard value is 200 ppm. It is observed that 12.5% and 13.04% of samples during pre- and post-monsoon seasons, respectively, exceed BIS's permissible limits (2012). Na's higher than-desirable concentration in groundwater can be attributed to dissolution of soil salts or rock weathering (Wagh et al., 2019). The excess of Na ions in the groundwater causes adverse health issues such as twitching and stiffness in humans (Varade et al., 2018; Wagh et al., 2019). The average sodium concentration in the groundwater of the study area is within the permissible limits as per BIS (2012) standards. Apart from total concentration, the nature of various cations and ratio of sodium ions to the total cations are essential factors in determining SAR (sodium absorption ratio) and % of sodium, which show the distribution of sodium in groundwater. The sodium is higher in Ameerpet in pre-monsoon, while in post-monsoon in Lakdikapul and Mehdipatnam are found to be higher due to rock-water interaction and also the heavy influx of domestic sewerage (Fig. 27.4a and b).

27.4.12 Potassium

Potassium is an essential nutrient for biota and a minor groundwater constituent. The average K concentration in the study area increases from 2.7 ppm in pre-monsoon to 3.4 ppm during post-monsoon seasons and is found to be within the desirable limits of BIS (2012). The increased concentration of K in post-monsoon groundwater could be attributed to mineral sources such as K-feldspars. The higher potassium values in Panjagutta and Mehdipatnam areas in pre-monsoon while higher values in Lakdikapul in the post-monsoon period are attributed to the large influx of anthropogenic waste (Fig. 27.4c and d).

27.4.13 Calcium

Calcium is a principal cation in groundwater due to its highly mobile nature, sourced through weathering of Ca-bearing minerals, especially in igneous terrain. The mean concentration of Ca varies from 82 ppm during pre-monsoon to 110.6 ppm in post-monsoon season, while the BIS (2012) limit is 75 ppm. During pre-monsoon and post-monsoon, 50% and 73.9% of samples respectively showed Ca concentration higher than the desirable limit; however, within the permissible BIS limits. The presence of other cations like potassium and sodium also influences the concentration of calcium. The higher calcium values in Mehdipatnam, Bollarum, Alwal, and Bachupalli areas in pre-monsoon while post-monsoon values are higher in Bachupalli and Narayanguda areas is a result of Ca from weathering of plagioclase feldspars in granitoid rocks (Fig. 27.4e and f).

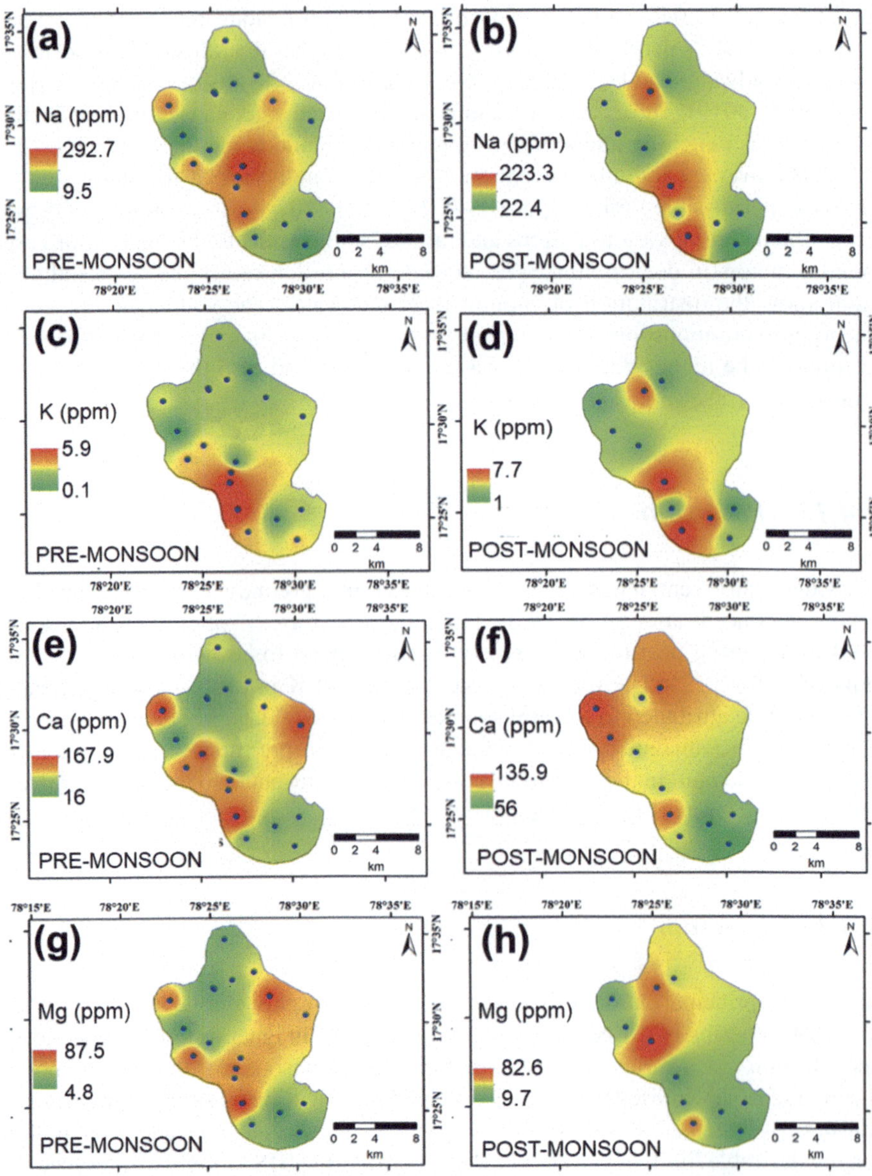

Fig. 27.4 Spatial distribution of cations of groundwater samples of present study during pre- and post-monsoon periods in the year 2022

27.4.14 *Magnesium*

Magnesium is an essential cation of groundwater, sourced by ionic exchange of minerals due to soil-rock-water interaction. Magnesium and calcium together are significant contributors to hardness of water. The mean concentration of magnesium reduces from 41.3 ppm in pre-monsoon to 27.9 ppm during post-monsoon seasons respectively, while the BIS (2012) standard value is 30 ppm. The comparison of Mg concentration with BIS indicates all samples are within permissible limits and fit for domestic purposes. The magnesium values are high in Kaziguda and Mehdipatnam in pre-monsoon and higher in the Borabanda area due to formation of calcrete or dolocrete in the weathered zones (Fig. 27.4 g and h).

An overall examination indicates that the SO_4/HCO_3^- ratio values range from 0.04 to 0.23 and from 0.02 to 0.11 for pre- and post-monsoon, respectively. These low values indicate that the high influence of anthropogenic sources such as industrial effluents, domestic sewerage, and municipal waste is prominent in groundwater contamination in the present study. However, the source of fluoride in present research is geogenic. It is attributed to the presence of F^-bearing minerals such as apatite, hornblende, and biotite in the granites in the study area. Results of present study are consistent with observations made by earlier workers (Laxman et al., 2021; Subba Rao, 2007).

27.5 Suitability to Irrigation and Industrial Use

27.5.1 *Sodium Absorption Ratio (SAR)*

Assessing sodium hazards in groundwater is essential when considering its suitability for irrigation. One of the most important criteria in assessing the sodium hazard is the sodium adsorption ratio (SAR) (Todd & Mays, 2005). SAR reveals comparative Na^+, Ca^{2+}, and Mg^{2+} concentrations in soil solutions. The groundwater samples were classified based on the classification given by Richards (1954). SAR is calculated using the following equation (Appelo, 1996; 27.2);

$$\text{SAR} = \frac{\text{Na}^+}{\left(\sqrt{\text{Ca}^{+2}} + \text{Mg}^{+2}\right)/2} \tag{27.2}$$

(Where all ionic concentrations are expressed in meq/L).

The SAR values range from 0.53 to 7.11 and 0.24 to 5.3 for pre- and post-monsoon samples. The majority of the samples are in low salinity (S1) category.

27.5.2 *Sodium Percentage (Na%)*

Wilcox (1958) have determined the hazardous effect of sodium on groundwater quality for irrigation usage in terms of Sodium Percentage (Na %), which is also called as soluble sodium percentage (SSP). The sodium percentage was calculated using the following equation (Wilcox, 1955; 27.3):

$$\text{Na\%} = \frac{\left(\text{Na}^{+} + \text{K}^{+}\right) * 100}{\left(\text{Na}^{+} + \text{Mg}^{+2} + \text{K}^{+} + \text{Ca}^{+2}\right)} \tag{27.3}$$

(Where all ionic concentrations are expressed in meq/L).

The Na% in pre-monsoon samples ranges from 24.8 to 88, while in post-monsoon, it ranges from 15 to 72, showing a high degree of salinity in pre-monsoon samples, leading to the salty nature. However, the post-monsoon samples showed slightly low Na%, indicating dilution due to influx of percolated water.

27.5.3 *Gibb's Ratio*

Gibbs (1970) proposed a diagram to understand the relationship between the chemical composition of the groundwater and their respective aquifer lithologies. The composition of groundwater is governed by disintegration of mineral constituents due to rock-water interaction, atmospheric precipitation, and evaporation (Gibbs, 1970). The groundwater during its percolation, groundwater becomes enriched in primary ion composition (Marrin, 1985; Meybeck, 1982; Sillen, 1967). Gibb's ratio for anions and cations is calculated using the following equations (Gibbs, 1970; 27.4 and 27.5).

$$\text{Gibbs ratio (anion)} = \frac{\text{Cl}^{-}}{\left(\text{Cl}^{-} + \text{HCO}_3^{-}\right)} \tag{27.4}$$

$$\text{Gibbs ratio (cation)} = \frac{\left(\text{Na}^{+} + \text{K}^{+}\right)}{\left(\text{Na}^{+} + \text{K}^{+} + \text{Ca}^{+2}\right)} \tag{27.5}$$

The Gibbs diagram (Gibbs, 1970) for the groundwater samples of the study area indicates that all groundwater samples are falling in the rock dominance field, indicating the control of country rock on the groundwater chemistry in both seasons, except for a few samples for cations during post-monsoon. The Gibb's ratio for anions ranges from 0.13 to 0.65 and from 156 to 515 for pre- and post-monsoon samples, respectively. Gibb's ratio for cations ranges from 9.514 to 293 and from 72.16 to 372.8 for pre- and post-monsoon samples, respectively (Fig. 27.5a and b).

For elucidating the hydrochemical character of groundwater samples, the meq/L values of groups of cations and anions are plotted on the Piper trilinear diagram

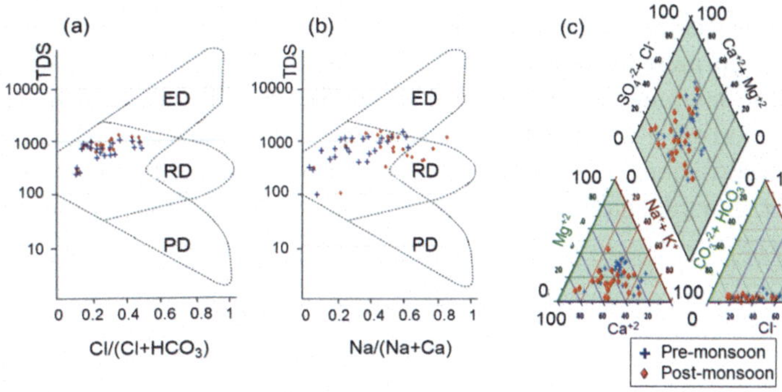

Fig. 27.5 Characterization of groundwater samples of the present study. (a) Gibb's diagram shows the character of anions during pre- and post-monsoon. (b) Gibb's diagram shows character of cations during pre- and post-monsoon. (c) Piper trilinear diagram for pre- and post-monsoon samples. Abbreviations: ED- Evaporation dominance, RD- Rock dominance, and PD- Precipitation dominance

(Piper, 1944). The groundwater samples in the cation triangle are characterized as calcium type, no dominant type, and magnesium type, while, in the anion triangle, there is a bicarbonate type. All the plots in cation and anion triangle, when projected in the diamond field, falls in sub-division 1 and 3, indicating "Alkaline earths (Ca^{+2} and Mg^{+2}) exceed alkalis (Na^{+} and K^{+})" and "Weak acids (HCO_3^{-}) exceed strong acids (SO_4^{2-} and Cl^{-})" respectively (Fig. 27.5c). The similar trend of chemical components of pre- and post-monsoon samples of the study area indicates no influence of excessive monsoon that occurred during the year 2022 (1224 mm), which is 763 mm than average rainfall on the quality of groundwater. This is again attributed to the low density of fracture system hampering sufficient water percolation to recharge the groundwater.

27.5.4 Suitability to Industrial Use

Larson and Skold (1958) proposed an index to evaluate the industrial usage of water. Larson–Scold index (L-S Index) is the ratio of sulfate and chloride to the alkalinity in bicarbonate and carbonate. The L-S index is calculated by using the following equation (27.6);

$$L - S\,Index(LSI) = \frac{(SO_4^{2-} + Cl^{-})}{(HCO_3^{-} - CO_3^{2-})} \tag{27.6}$$

(where concentrations are expressed in meq/L).

In the present study, about 80% of pre- and post-monsoon samples possess < 0.8 of LSI values, indicating the high tendency of scaling.

Yet another parameter, the Corrosive Ratio (CR), evaluates the corrosive tendency of groundwater on metallic pipes, which is an essential parameter in the case of urban areas (Balasubramanian, 1986; Sankar, 1995). The CR values of more than 1 indicate highly corrosive waters (Ryner, 1944). The CR denotes the susceptibility of groundwater to corrosion and is expressed as the ratio of alkaline earths to saline salts in groundwater (Janardhana Raju, et al., 2011) and is calculated using the following equation (27.7):

$$\text{Corrosive ratio} = \frac{\left\{\left(\frac{Cl^-}{35.5}\right) + 2\left(\frac{SO_4^{2-}}{96}\right)\right.}{\left\{2\left(\frac{HCO_3^- - CO_3^{2-}}{100}\right)\right.} \tag{27.7}$$

(where all ionic concentrations are expressed in ppm).

The CR in groundwater samples of the present study exceeds 1, indicating that the waters are corrosive. This holds good for both pre-and post-monsoon samples.

These high indices indicate that the groundwaters of the present area have a strong tendency towards scaling and corrosion of metallic pipes. Hence, the groundwater is unsuitable for industrial use, especially in manufacturing units, which are present in quite a large number in the Medchal area.

The plot of SAR versus electrical conductivity indicates that the majority of the samples fall in C2S1, C3S1, and C3S2, with one sample each of C1S1 and C4S1 falling in extreme fields of low salinity hazard and very high hazard, respectively. All the groundwater samples are regarded as having a 'high salinity' risk, as per the United States Salinity Laboratory (USSL) classification (Fig. 27.6; Richards, 1954).

This study revealed that the groundwater quality in Hyderabad city and adjoining Medchal-Malkajgiri districts is deteriorating at an alarming rate due to the haphazard expansion of the urban area, industries, new residential colonies, and other infrastructures. The quality of contaminated groundwater cannot be restored; however, it can be controlled by creating awareness among the people and industrialists with the help of the scientific community in the public and private sectors.

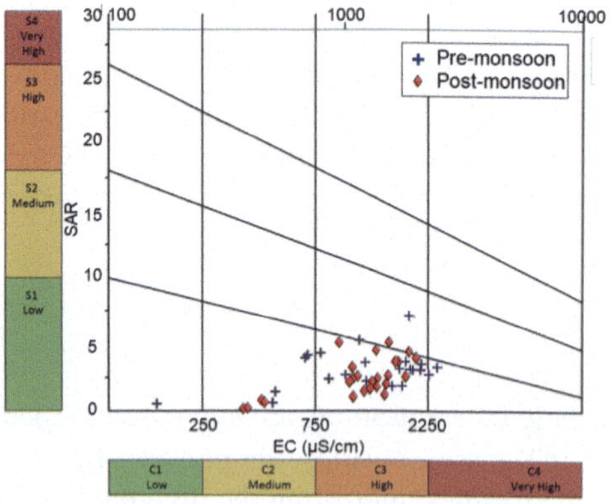

Fig. 27.6 EC versus SAR diagram showing classification of groundwater samples of pre- and post-monsoon periods of the present study (Richards, 1954; USSL, 1954)

27.6 Conclusions

This study was carried out to assess the groundwater quality in parts of Hyderabad city and the adjoining Medchal- Malkajgiri district of Telangana, adopting hydrochemical and geospatial techniques. Forty-eight samples, one for pre- and post-monsoon from 24 bore wells, were collected during 2022. The general order of abundance of anions and cations species in groundwater samples is $F^- > NO_3^- > SO_4^{2-} > Cl^- > HCO_3^-$ and $K^+ > Mg^{2+} > Ca^{2+} > Na^+$. The majority of groundwater samples exhibit the alkaline earth (Ca^{+2} and Mg^{+2}) exceeding alkalis (Na^+ and K^+), and weak acids (HCO_3^-) exceeding strong acids (SO_4^{2-} and Cl^-). Samples from Malkajgiri and Subhashnagar reported a high F^- concentration. Fluorine-bearing minerals such as apatite, hornblende, and biotite are the causative factors for high fluoride. Considerable number of samples show excess nitrate (NO_3^-), which is attributed to anthropogenic sources. Based on EC and SAR, the groundwater samples have been classified as C2S1, C3S1, and C3S2 types, fitting into the 'moderate to poor' category for regular domestic usage with high salinity. The geospatial analysis shows distribution of high concentrations of TDS, F^-, NO_3^-, and TH, which is attributed to local lithology coupled with severe sewerage contamination. Groundwater is corrosive and is unsuitable for industrial use. Several geogenic factors such as hard granitic and granodioritic rocks intervened with felsic and essential dykes, sporadic sulfide mineralization, lack of passable fracture system, presence of sheet rock over a vast area; reduced natural recharge; and anthropogenic causes like urbanization, uncontrolled sewerage, and industrial waste disposal are believed to be the primary causes of deteriorating groundwater quality in the study area. The excess monsoonal rainfall in the year 2022 had no impact on groundwater quality. Awareness among people on controlling and avoiding waste and effluent disposal is immediately warranted.

Acknowledgements PRCP thanks Cyient Limited for its support. CD thanks, Prof. Bhushan S. Deota and Head, Department of Geology, The Maharaja Sayajirao University of Baroda, for allowing to use GIS laboratory facilities. KRP and KR thank the Department of Applied Geochemistry, Osmania University, Hyderabad, for laboratory analytical facilities. The authors thank reviewers and the editorial board for critical review of the manuscript, which improved the presentation. This work did not receive financial aid from any individual, organization, or agency.

References

APHA. (2012). *Standard methods for the examination of water and waste water* (22nd ed.). American Public Health Association.

APHA. (1998). *Standard methods for the examination of water and wastewater* (20th ed.). American Public Health Association.

Appelo, C. A. J. and Postma, D. (1996). Geochemistry, groundwaters and pollution; a. a. balkema: Rotterdam, The Netherlands: p. 536.

Asadi, S. S., Padmaja, V., & Anji Reddy, M. (2009). Remote sensing and GIS techniques for evaluation of groundwater. Quality in municipal corporation of Hyderabad (Zone-V), India. *International Journal of Environmental Resarch Public Health, 4*(1), 45–52.

Balaji, E., Nagaraju, A., Sreedhar, Y., Thejaswi, A., & Sharifi, Z. (2016). Hydrochemical characterization of groundwater in around Tirupati Area, Chittoor District, Andhra Pradesh. *South India. Applied Water Science, 7*(3), 1203–1212. https://doi.org/10.1007/s13201-016-0448-6

Balasubramanian, A. (1986). Hydrogeological investigations in Tambraparni river basin, Tami Nadu. Ph.D. Thesis, University of Mysore, Mysore, Karnataka. p. 356.

BIS. (2012). Bureau of Indian standards (BIS 2012). *Indian Standard Drinking Water Specification IS, 10500*, 2012.

CGWB. (2021). Dynamic groundwater resources of India (As on March 2020). Groundwater Department, Government of Telangana and Central Groundwater Board (CGWB), Ministry of Water Resources, Government of India, New Delhi. p. 93.

Chatterjee, R., Gourab, T., & Paul, S. (2009). Groundwater quality assessment of Dhanbad district. *Jharkhand, India, Bulletin of Engineering Geology and Environment, 69*(1), 137–141. https://doi.org/10.1007/s10064-009-0234-x

CPCB. (2008). Status of groundwater quality In India–Part-II. Central Pollution Control Board, p. 431. https://cpcb.nic.in/openpdffile.php?id=UHVibGljYXRpb25GaWxlLzc3Nl8xNTQ0NDI3NzY5X0dXUVMtMi5wZGY=

Dar, I. A., Sankar, K., & Dar, M. A. (2010). Remote sensing technology and geographic information system modelling: An integrated approach towards the mapping of potential groundwater recharge zones in Hardrock terrain, Mamundiyar basin. *Journal of Hydrology, 394*, 285–295.

Davis, S. N., & De Wiest, R. J. M. (1966). *Hydrogeology* (p. 463). John Wiley and Sons Inc.

Domenico, P. A., & Scwartz, F. W. (1997). Physical and chemical hydrogeology. John Wiley & Sons, p. 528.

Dongare, C. U., & Deota, B. S. (2023). Significance of geomorphic indicators for identification of groundwater recharge potential in the Khapri watershed of Deccan Volcanic Province (DVP), district Dangs, western India. *Groundwater for Sustainable Development, 22*, 100956.

Dongare, C. U., Mukherjee, S., & Deota, B. S. Morphometric analysis of two milli-watersheds in Deccan Volcanic Province of Dangs District, Gujarat, India. *Journal of Geointerface, 1*(1), 28–35.

Etikala, B., Golla, V., Adimalla, N., Surasura, R., & Marapatla, S. (2021). Hydrogeochemical characterization of groundwater using conventional graphical, geospatial and multivariate statistical techniques. In: Adhikary, P. P., Shit, P. K., Santra, P., Bhunia, G. S., Tiwari, A. K., Chaudhary, B. S. (eds) Geostatistics and geospatial technologies for groundwater resources in India. Springer Hydrogeology. Springer, Cham. https://doi.org/10.1007/978-3-030-62397-5_5

Gajendran, C., & Thamarai, P. (2008). Study on statistical relationship between groundwater quality parameters in Nambiyar River basin, Tamil Nadu, India. *International J. on Pollution Research, 27*(4), 679–683.

Gibbs, R. J. (1970). Mechanisms controlling groundwater chemistry. *Science, 170*, 1088–1090.

Gleeson, T., Cuthbert, M., Ferguson, G., & Perrone, D. (2020). Global groundwater sustainability, resources, and systems in the Anthropocene. *Annual Review of Earth and Planetary Sciences, 48*, 431–463.

Ha, Q. K., Ngoc, T. D. T., Le Vo, P., Nguyen, H. Q., & Dang, D. H. (2022). Groundwater in Southern Vietnam: Understanding geochemical processes to better preserve the critical water resource. *Science of the Total Environment, 807*, 151345.

Hao, A., Zhang, Y., Zhang, E., Li, Z., Yu, J., Huang, W., & Wang, Y. (2018). Groundwater resources and related environmental issues in China. *Hydrogeology Journal, 26*(5), 1325–1337.

Hari, D., Rudraksh Goud, Y. and Meghana, D. (2020). Spatial Analysis and mapping of groundwater quality in uppal kalan, Hyderabad. IOP Conference Series: Materials Science and Engineering. 1070, 012038, pp. 1–13., IOP Publishing https://doi.org/10.1088/1757-899X/1070/1/012038

Jagannadha Sarma, V. V., & A. Narayana Swamy, A. N. (1981). Groundwater quality in Visakhapatnam Basin, India. *Water Air and Soil Pollution, 16*, pp. 317–329.

Janardhana Raju, N., Shukla, U. K., & Ram, P. (2011). Hydrogeochemistry for the assessment of groundwater quality in Varanasi: A fast-urbanizing center in Uttar Pradesh, India. *Environmental Monitoring and Assessment, 173*, 279–300. https://doi.org/10.1007/s10661-010-1387-6

Larson, T. E., & Skold, R. V. (1958). Laboratory studies relating mineral quality of water to corrosion of steel and cast iron. *Corrosion, 14*(6), 43–46.

Laxman, D. K., Kiran Kumar, P., Ratnakar, D., & Himabindu, V. (2021). Health risk assessment of nitrate and fluoride toxicity in groundwater contamination in the semi-arid area of Medchal, South India. *Applied Water Science, 12*, 11. https://doi.org/10.1007/s13201-021-01557-4

Manikandan, E., Rajmohan, N., & Anbazhagan, S. (2020). Monsoon impact on groundwater chemistry and geochemical processes in the shallow hard rock aquifer. *CATENA, 195*, 104766. https://doi.org/10.1016/j.catena.2020.104766

Marrin, D. L. (1985). *Delineation of gasoline hydrocarbons in groundwater by soil gas analysis* (p. 568). Tracer Research Corporation.

Meybeck, M. (1982). Carbon, nitrogen and phosphorous transport by world rivers. *Am. Jour. Sci, 282*(4), 491–500. https://doi.org/10.2475/ajs.282.4.401

Pancholi, V., Chaudhary, V., Lakhmapurkar, S., Patel, P., Rabari, R., Dongare, C. U., & Chopra, S. (2022). Estimation of groundwater potential zones using remote sensing and geographical information system technique-Waghai Taluka, Dang district, Gujarat. *Western India. Environmental Challenges, 9*, 100615.

Phani, P. R. C., & Prasad, K. R. (2022). Groundwater contamination in parts of northwestern Hyderabad–A hydrogeochemical and geospatial approach. In: Thambidurai, P.,& Dixit< A. K. (Eds.) Impacts of urbanization on hydrological systems in India, Springer Nature, pp. 293–312.

Piper, A. M. (1944). A Graphic procedure in the geochemical interpretation of water analysis. *American Geophysical Union Transactions, 25*, 914–923. https://doi.org/10.1029/TR025i006p00914

Richards, L. A. (1954). *Diagnosis and improvement of saline and alkaline soils agriculture handbook* (p. 160). Department of Agricultural.

Ryner, J. W. (1944). A new index for determining amount of calcium carbonate scale formed by water. *J Am. Water Assoc., 36*, 472–486.

Sankar, K. (1995). Hydrological studies in the Kanyakumari district, Tamilnadu. Unpublished Ph.D. Thesis, University of Madras, pp. 90–91.

Satish Kumar, T., Sudarshan, V., & Kalpana, G. (2007). Geochemical characterization of groundwater, banks of Musi River, Hyderabad city, A. P. *India. Poll. Res. 26*(4), 795–800.

Shaikh, H., Gaikwad, H., Kadam, A., & Umrikar, B. (2020). Hydrogeochemical characterization of groundwater from semiarid region of western India for drinking and agricultural purposes with special reference to water quality index and potential health risks assessment. *Applied Water Science, 10*, 1–16.

Siebert, S., Burke, J., Faures, J. M, Frenken, K., Hoogeveen, J., Döll, P., & Portmann, F. T. (2010). Groundwater use for irrigation–A global inventory. *Hydrology and earth system sciences, 14*, 1863–1880. https://doi.org/10.5194/hess-14-1863-2010

Sillen, L. G. (1967). The Ocean as a chemical system. *Science, 156*, 1189–1197.

Stephens, M. J., Shimabukuro, D. H., Chang, W., Gillespie, J., & Levinson, Z. (2021). Stratigraphic and structural controls on groundwater salinity variations in the Poso Creek Oil Field. *Kern County, California, USA, Hydrogeology Journal, 29*, 2803–2820. https://doi.org/10.1007/s10040-021-02381-5

Subba Rao, N. (2007). Groundwater quality as a factor for identification of recharge zones. *Environmental Geosciences, 14*(2), 79–90.

Sunitha, V., & Reddy, B. M. (2022). Geochemical characterization, deciphering groundwater quality using pollution index of groundwater (PIG), water quality index (WQI) and geographical information system (GIS) in hard rock aquifer. *South India. Appl Water Sci, 12*, 41. https://doi.org/10.1007/s13201-021-01527-w

Sunitha, V., Reddy, Y. S., Suvarna, B., & Reddy, M. R. (2022). Human health risk assessment (HHRA) of fluoride and nitrate using pollution index of groundwater (PIG) in and around hard rock terrain of Cuddapah, A.P. *South India. Environmental Chemistry and Ecotoxicology, 4*, 113–123. https://doi.org/10.1016/j.enceco.2021.12.002

Todd, D. K. (1980). *Groundwater hydrology*. Wiley.

Todd, D. K., & Mays, L. (2005). *Groundwater hydrology* (3rd ed., p. 652). John Wiley and Sons Inc.

Torres-Martínez, J. A., Mora, A., Knappett, P. S. K., Ornelas-Soto, N., Mahlknecht, J. (2020). Tracking nitrate and sulfate sources in groundwater of an urbanized valley using a multi-tracer approach combined with a Bayesian isotope mixing model. *Water Research, 182*, n115962, 1–16.

TSDPS. (2023). telangana state development planning society, annual rainfall 2022–2023, District wise Rainfall. https://www.tsdps.telangana.gov.in/annualrf.jsp

Udayalaxmi, G., Himabindu, D., & Ramadass, G. (2010). Geochemical evaluation of groundwater quality in selected areas of Hyderabad, A.P., India. *Indian Journal of Science and Technology, 3*(5), 546–553.

USSL (1954). Diagnosis and improvement of saline and alkaline soils. US Department of Agriculture Handbook, US Salinity LaboratoryNo. 60, p. 160.

Varade, A.M., Yenkie, R.O., Shende, R.R., Golekar, R.B., Wagh, V.M. and Khandare, H.W. (2018). Assessment of water quality for the groundwater resources of urbanized part of the Nagpur District, Maharashtra (India). *American Journal of Water Resources, 6*(3), 89–111. https://doi.org/10.12691/ajwr-6-3-1

Wagh, V. M., Mukate, Panaskar, D. B., Muley, A. A., & Sahu, U. L. (2019). Study of hydrochemistry and drinking suitability through water quality index (WQI) modeling in Kadava river basin. *India. SN Applied Sciences, 1*, 1–16. https://doi.org/10.1007/s42452-019-1268-8

Wick, K., Heumesser, C., & Schmid, E. (2012). Groundwater nitrate contamination: Factors and indicators. *Journal of Environmental Management, 111*(3), 178–186. https://doi.org/10.1016/j.jenvman.2012.06.030

Wilcox, L. V. (1955). Classification and use of irrigation water. *USDA Circular, 969*, 19.

Chapter 28
Statistical Trend of Water Quality in Bang Pakong River, Thailand

Pawee Klongvessa and Srilert Chotpantarat

Abstract Water resources are important for both human and ecological systems. However, climate change and human activities contribute to changes in water quality, impacting the health of both human and other living organisms. This chapter focuses on the trends in water quality in Bang Pakong River, Thailand. Serving a large population in eastern Thailand, this river is heavily utilized for agricultural activities. The trends of various water quality parameters were examined from upstream to downstream, encompassing both urban and agricultural areas, using Sen's slope and Mann–Kendall test. Over a period of 28 years from 1993 to 2020, significant increases in turbidity and NH_3-N along with decreases in suspended solids (SS) were observed across almost all parts of the river. Additionally, in the upstream part of the river, water temperature and biological oxygen demand (BOD) increased, while fecal coliform bacteria (FCB) decreased. In Chachoengsao, the largest urban area along this river, water quality appeared to improve in terms of dissolved oxygen (DO; increasing), total coliform bacteria (TCB; decreasing), FCB (decreasing), and SS (decreasing). However, turbidity, NO_3-N, and NH_3-N levels increased. Downstream from Chachoengsao, even though SS decreased similarly to other parts of the river, water quality deteriorated in many parameters such as pH, conductivity, salinity, NO_2-N, NH_3-N, and total dissolved solids (TDS). These varying trends in water quality across different parts of the river appear to be caused by different combinations of factors. It is suggested that implementing a sufficient wastewater treatment system and controlling aquacultural activities are necessary steps to improve water quality in the downstream area.

P. Klongvessa
Faculty of Environment, Department of Environmental Technology and Management, Kasetsart University, Bangkok, Thailand

S. Chotpantarat (✉)
Faculty of Science, Department of Geology, Chulalongkorn University, Bangkok, Thailand
e-mail: srilert@chula.ac.th

Center of Excellence in Environmental Innovation and Management of Metals (EnvIMM), Sustainable Environment Research Institute, Chulalongkorn University (ERIC), Bangkok, Thailand

V. Sunitha et al. (eds.), *Groundwater Resource Management Planning Strategies*, Springer Water, https://doi.org/10.1007/978-3-031-88870-0_28

Keywords Bang Pakong River · Mann–Kendall test · Sen's slope · Water quality parameter

28.1 Introduction

Water resources are crucial for human and ecological systems. Inappropriate water quality can have adverse impacts on health of both humans (e.g., Liu, Yang, and Xu, 2022) and other living organisms (e.g., Bashir et al., 2020). Hence, the water resources should have appropriate water quality, which depends on their intended uses (e.g., Simachaya, 2020). Recently, many places have faced problems with poor water qualities driven by both anthropogenic and natural factors (Chotpantarat et al., 2020; Wisitthammasri et al., 2020). Anthropogenic factors include inappropriate waste disposal and excess use of fertilizers in agricultural activities (Balaji et al., 2017; Chotpantarat & Boonkaewwan, 2018; Moui et al., 2022; Sunitha & Reddy, 2022; Sunitha et al., 2022), while natural factors include reduced dilution of contaminants during dry periods and mineral dissolution (Anh et al., 2023; Sunitha et al., 2022). Many of these factors are dynamic. For examples, human activities in urban areas at the present may exceed those in the past due to increases of populations in urban areas (Ritchie & Roser, 2019), and hydrological factors may differ due to global temperature rises (Kundzewicz, 2008; Lindsey & Dahlman, 2023). Therefore, the quality of the water can change over time, as evidenced in various regions worldwide (e.g. Antonopoulos et al., 2001; Dawe, 2006; Ha et al., 2022; Hashim et al., 2021). For this reason, the United Nations has included the 'proportion of bodies of water with good ambient water quality' as one of indicators of the Sustainable Development Goals (United Nations Environment Programme, 2021).

The Bang Pakong River, one of the main rivers in eastern Thailand, is experiencing the impacts of climate change and various human activities. This river serves a large population and is highly utilized for agricultural activities. In the past, the floodplain of the Bang Pakong River was mainly used as rice fields (Popan, 2011). However, the area experienced many changes afterwards. Since 1989, many parts of the area have been transformed into aquacultural areas (Popan, 2011). In 1996, there was a project to build a Bang Pakong Diversion Dam in the middle part of the river to keep fresh water for use. However, after the operation in 2000, there were adverse impacts and the operation has subsequently been stopped, but the structure of the dam has still been remained (Sawangarreruks et al., 2005). Since 1999, there has been a project to use this river to mitigate floods in Bangkok, the capital city of Thailand, by diverting water from the Pasak River, one of the rivers that contributed water to Bangkok, to this river (Royal Irrigation Department, 2002). Furthermore, since this river is near Bangkok and is part of the Eastern Economic Corridor, one of the economic development projects of Thailand (Mahidol University, 2020), there has been a rapid urbanization in the area (Chachoengsao Provincial Office, 2023). In addition to these anthropogenic factors, the quality of the water in this river is also affected by a seawater intrusion (Popan, 2011).

There have been many studies on the water quality in the Bang Pakong River. Most of these studies used data in short periods (1–3 years) to study seasonal and spatial variations of water quality (Bordalo et al., 2001; Bubphamala et al., 2010; Majandang & Piyapong, 2023; Shwe et al., 2022; Yuenyong et al., 2019), study impacts of human activities on water quality (Bundao et al., 2018; Chan et al., 2020; Kupkanchanakul & Kwonpongsagoon, 2011; Kupkanchanakul et al., 2015; Paibulkichakul et al., 2006), study physical processes (Bordalo et al., 2016) and compare the water quality with that in other river (Hungspreugs et al., 1990). There was also a study that used long period data to identify key factors for determining water quality (Sittisorn, 2017). Overall, previous studies have found that water quality is poor during a dry season, and water quality can be affected by both anthropogenic factors (agricultural activities and urbanization) and natural factors (rainfall, river flow, and geological characteristics of the river). However, since there have been many phenomena that affect the water quality and different phenomena might occur in different time periods, this chapter aimed to examine and compare long-term trends in water quality parameters across various segments of the Bang Pakong River. The findings serve to identify factors affecting water quality and facilitate the development of appropriate measures for its control.

28.2 Methodology

28.2.1 Study Area and Data

The Bang Pakong River is in the Bang Pakong Basin, an important basin in eastern Thailand (Mahidol University, 2020). This river originates from a convergence of the Nakhon Nayok River, which flows from the north, and the Prachinburi River, which flows from the east, near the point BK16 as shown in Fig. 28.1(a). The Bang Pakong River flows in a north–south direction with a total length of approximately 130 km and drains water to the Gulf of Thailand in the south. Figure 28.1(b) shows land uses along the river. The land use is primarily agricultural, with some urban areas at the downstream part. The largest urban area along this river is the city of Chachoengsao, which is at the point BK07 (National Statistical Office, n.d.).

There were 13 water quality observation points where the water quality was observed by the Pollution Control Department, Thailand, which were BK01 BK02 BK03 BK04 BK06 BK07 BK08 BK09 BK9.5 BK11 BK13 BK15 and BK16. At these 13 points, various water quality parameters have been monitored 4 times per year (approximately every 3 months) since 1993. The observation points and parameters for the investigation in this study were chosen based on the availability of the data. In this study, values below a lower detection limit were replaced by half of the lower detection limit (which was equal to an average between 0 and lower detection limit; United States Environmental Protection Agency, 1991), values above an upper detection limit were replaced by 1.7 times the upper detection limit (Sanford et al.,

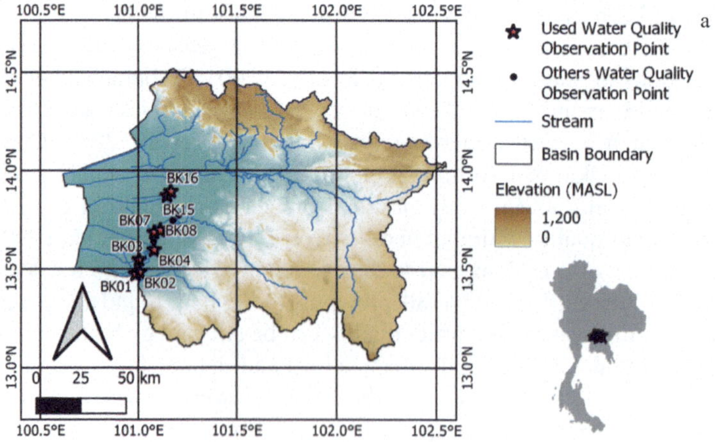

Elevation data is from GMTED10 derived by United States Geological Surveys

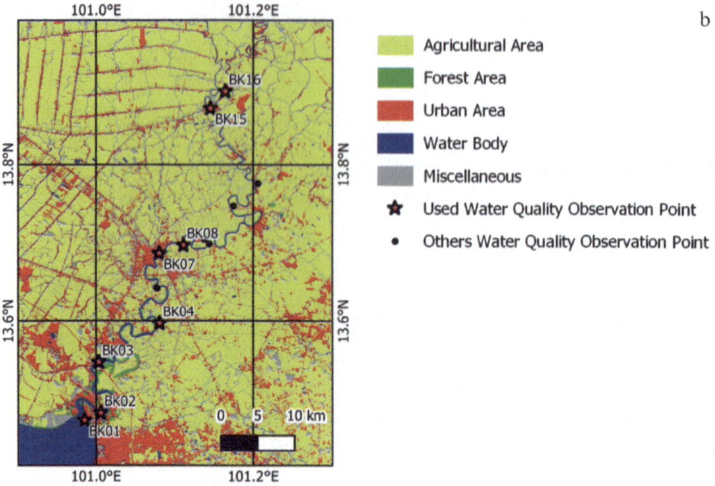

Land use data is from Land Development Department, Thailand, surveyed in 2016

Fig. 28.1 Bang Pakong Basin and locations of water quality observation points (**a**) and land use near the observation points (**b**)

1993), and outliers were screened out since they might be caused by technical error. By considering the availability of the data during 1993–2020 after screening, water temperature, pH, turbidity, conductivity, salinity, dissolved oxygen (DO), biological oxygen demand (BOD), total coliform bacteria (TCB), fecal coliform bacteria (FCB), NO_3-N, NO_2-N, NH_3-N, suspended solid (SS), and total dissolved solid (TDS) were available for at least 80% at the points BK01, BK02, BK03, BK04, BK07, BK08, BK15, and BK 16. Therefore, these 14 parameters and 8 points were chosen. The

Table 28.1 Upstream distances and dominant land use types of water quality observation points used in this study

Point	Upstream distance (km)	Dominant land use type
BK1	2	Urban
BK2	9	Urban
BK3	22	Agriculture
BK4	30	Agriculture
BK7	57	Urban
BK8	62	Agriculture
BK15	117	Agriculture
BK16	122	Agriculture

locations and land use types of these 8 points are shown in Fig. 28.1 and Table 28.1, and average values of these 14 parameters during 1993–2020 are shown in Table 28.2.

28.2.2 Data Analysis

For each observation point and each water quality parameter, the data at each time was indexed by date. The rate in change of the parameter during 1993–2020 was determined by Sen's slope (Sen, 1968), and a Mann–Kendall (MK) test (Kendall, 1948; Mann, 1945) was used to determine significance. These methods have been used in many studies (e.g. Hashim et al., 2021; Jeon et al., 2020; Sakke et al., 2023). Their advantages were that they were non-parametric. Since some water quality parameters were not normally distributed, these non-parametric methods were suitable.

To calculate the Sen's slope, the slopes between the water quality data at the two different times were calculated as Eq. 28.1.

$$\beta_{ij} = \frac{x_j - x_i}{d_j - d_i} \tag{28.1}$$

where β_{ij} was the slope between the data at i th and j th orders (sorted by date) such that $i < j$, x_i and x_j were the data at i th and j th orders (sorted by date), respectively, and d_i and d_j were days (from the beginning of the study period) of the data at i th and j th orders (sorted by date), respectively. Sen's slope was the median of the slopes β_{ij} (Sen, 1968).

For the MK test, the null hypothesis was that the water quality parameter did not change with time. The test value was calculated as Eqs. 28.2–28.5 (Kendall, 1948; Mann, 1945).

$$S = \sum_{i=1}^{n-1} \sum_{j=i}^{n} \operatorname{sgn}(x_j - x_i) \tag{28.2}$$

Table 28.2 Average values of water quality parameters in the study area during the study period (1993–2020)

Parameter	Observation Point							
	BK1	BK2	BK3	BK4	BK7	BK8	BK15	BK16
Water temperature (°c)	29.7	30.5	30.6	30.1	30.1	30.1	30.2	30.0
pH	7.23	7.14	7.05	7.08	7.00	7.00	7.03	6.91
Turbidity (NTU)	99	117	132	111	114	121	84	31
Conductivity (μS)	28,243	25,507	21,663	20,555	13,408	12,318	6260	4196
Salinity (ppt)	18.63	16.75	14.38	13.37	8.52	7.74	3.77	2.42
DO (mg/l)	4.9	4.3	4.0	4.2	4.1	4.1	4.5	4.2
BOD (mg/l)	1.5	1.4	1.2	1.3	1.5	1.4	1.4	1.6
Total coliform bacteria (MPN/100 ml)	3596	6184	4789	5614	10,844	6621	3426	3862
Fecal coliform Bacteria (MPN/100 ml)	1456	2159	1782	2102	3910	2126	584	1241
NO_3-N (mg/l)	0.33	0.41	0.55	0.48	0.56	0.56	0.42	0.38
NO_2-N (mg/l)	0.07	0.08	0.06	0.06	0.06	0.06	0.03	0.04
NH_3-N (mg/l)	0.20	0.22	0.17	0.15	0.18	0.17	0.16	0.17
Suspended solid (mg/l)	102	116	130	110	92	80	45	51
Total dissolved solid (mg/l)	21.391	19.596	16.433	15.694	9217	8375	4114	3961

$$\text{sgn}(x) = \begin{cases} 1 & if x > 0 \\ 0 & if x = 0 \\ -1 & if x < 0 \end{cases} \quad (28.3)$$

$$Var(S) = \frac{n(n-1)(2n+5) - \sum_{k=1}^{m} t_k(t_k - 1)(2t_k + 5)}{18} \quad (28.4)$$

$$Z = \frac{S - \text{sgn}(S)}{\sqrt{Var(S)}} \quad (28.5)$$

where S was a test statistic, n was a number of data, x_i and x_j were the data at i th and j th orders (sorted by date), respectively, sgn was a sign function (defined as Eq. 28.3), $Var(S)$ was a variance of S, m was a number of tied groups, t_k was a number of data in k th tied group, and Z was a test value. A 2-tailed p-value was calculated from a cumulative distribution function of a normal distribution. A significance level in this study was 0.05; therefore, the null hypothesis was rejected when the p-value < 0.05. (The conclusion that the water quality parameter changed over time could be reached when the p-value < 0.05.)

28.3 Results and Discussion

28.3.1 Comparison of Trends Among Observation Points

Table 28.3 shows the trends of water quality parameters calculated by Sen's slope and tested by the MK test at a significance level of 0.05. There were similar trends among the points BK01 BK02 BK03 BK08 and BK 15. At these 5 points, there was a significant increase in turbidity, increase of NH_3-N and decrease of SS. For the point BK15, which was in the upstream part of the river, there were also significant trends in some of the other parameters that coincided with nearby points. These trends were increase in water temperature (at the points BK08 and BK15), increase of BOD (at points BK15 and BK16) and decrease of FCB (at the points BK07 and BK15). For the points BK04 BK07 and BK16, the trends were obviously different from those in other points due to local factors. The point BK04, located in the downstream of the city of Chachoengsao, was experiencing rapid urban development (Chachoengsao Provincial Office, 2023). The point BK07 was in the city of Chachoengsao while the surrounding points were in agricultural areas. The point BK16, located in the upstream part of the river, was influenced by the joining of 2 rivers (Nakhon Nayok and Prachinburi rivers). However, since this chapter focused on the effect of climate change and human activities, point BK16 was excluded from a discussion. By considering the similarities in trends, the points BK1 BK4 BK7 and BK15 could represent the trends in water qualities in different parts of the river, the downstream area, the downstream of the urban area, the urban area, and the upstream area, respectively.

Table 28.3 Trends of water quality parameters in the study area (per decade)

Parameter	Observation Point							
	BK01	BK02	BK03	BK04	BK07	BK08	BK15	BK16
Water Temperature (°c)	−0.2	−0.1	−0.1	+ 0.1	+ 0.2	**+ 0.4**	**+ 0.4**	≈0.0
pH	+ 0.08	≈0.00	≈0.00	**+ 0.13**	+ 0.08	+ 0.07	≈0.00	**−0.12**
Turbidity (NTU)	**+ 17**	**+ 20**	**+ 39**	+ 10	**+ 23**	**+ 28**	**+ 28**	**−11**
Conductivity (μS)	+ 925	+ 395	+ 265	**+ 1629**	+ 7	+ 24	+ 14	−9
Salinity (ppt)	+ 0.16	≈0.00	≈0.00	**+ 0.83**	≈0.00	≈0.00	≈0.00	≈0.00
DO (mg/l)	+ 0.2	−0.2	**−0.3**	+ 0.1	**+ 0.4**	≈0.0	≈0.0	**−0.5**
BOD (mg/l)	+ 0.1	+ 0.1	≈0.0	+ 0.1	+ 0.1	−0.1	**+ 0.2**	**+ 0.3**
Total coliform bacteria (MPN/100 ml)	−13	+ 227	+ 348	+ 225	**−1313**	≈0	+ 70	−14
Fecal coliform bacteria (MPN/100 ml)	−39	−6	≈0	+ 15	**−350**	−89	**−58**	−55
NO_3-N (mg/l)	+ 0.01	+ 0.04	+ 0.07	−0.03	**+ 0.14**	+ 0.08	+ 0.01	+ 0.03
NO_2-N (mg/l)	≈0.00	+ 0.01	≈0.00	**+ 0.01**	≈0.00	≈0.00	≈0.00	≈0.00
NH_3-N (mg/l)	**+ 0.06**	**+ 0.10**	**+ 0.08**	**+ 0.07**	**+ 0.05**	**+ 0.04**	**+ 0.05**	≈0.00
Suspended Solid (mg/l)	**−26**	**−27**	**−17**	**−28**	**−20**	**−15**	**−7**	**−9**
Total dissolved solid (mg/l)	−480	−222	−204	**+ 911**	−45	−11	15	−12

The bold font indicates that the p-value from the Mann–Kendall test was less than 0.05. (There was a significant change over time).

28.3.2 Trends of Water Quality Parameters

The parameters that changed significantly in almost all parts of the river were turbidity, NH_3-N, and SS. The values of these 3 parameters can be seen in Fig. 28.2. Turbidity and NH_3-N increased due to increased aquacultural activities along the river (Popan, 2011). Aquaculture activities could cause the turbidity by particulates from feed and feces of aquatic animals (Price et al., 2015) and cause NH_3-N by excretions of aquatic animals through their gills (Weihrauch et al., 2009). For SS, it decreased probably due to a project of diversion of water from Pasak River to Bang Pakong River to mitigate flood in Bangkok proposed in 1999 (Royal Irrigation Department, 2002) and the operation of the wastewater treatment system in the Bang Khla subdistrict, the upstream urban area located between the points BK08 and BK15, since 2008 (Database System for Publicly Owned Treatment Works, n.d.b). The diversion of water to the Bang Pakong River caused the water in the downstream part, which had more SS than in the upstream part (see Table 28.2), being pushed by the enhanced upstream flow, and subsequently the SS decreased. The wastewater treatment system reduced SS in the water by reducing its release from the urban area of the Bang Khla subdistrict. The increasing in NH_3-N and decreasing in SS have also been found in other areas, such as the Bernam River, Malaysia (Hashim et al., 2021).

In addition to turbidity, NH_3-N, and SS, which significantly changed in almost all parts of the river, water temperature, BOD, and FCB also significantly changed in the upstream part of the river. The values of these 3 parameters can be seen in Fig. 28.3. A possible reason for the increase of water temperature was an increase of air temperature due to a climate change (Chaiyarak, Tattiyakul, and Karnsunthad, 2019). Many studies have found that surface water temperatures were related to air temperatures (e.g. Girgibo et al., 2023; Livingstone & Lotter, 1998; Zhu, Nyarko, and Hadzima-Nyarko, 2018). However, in the downstream part of the river in this study, the water temperature did not change significantly. It was possibly due to a higher groundwater discharge than in the upstream part (ASDECON Corporation Company Limited, 2012). The decrease of FCB in the upstream part could be due

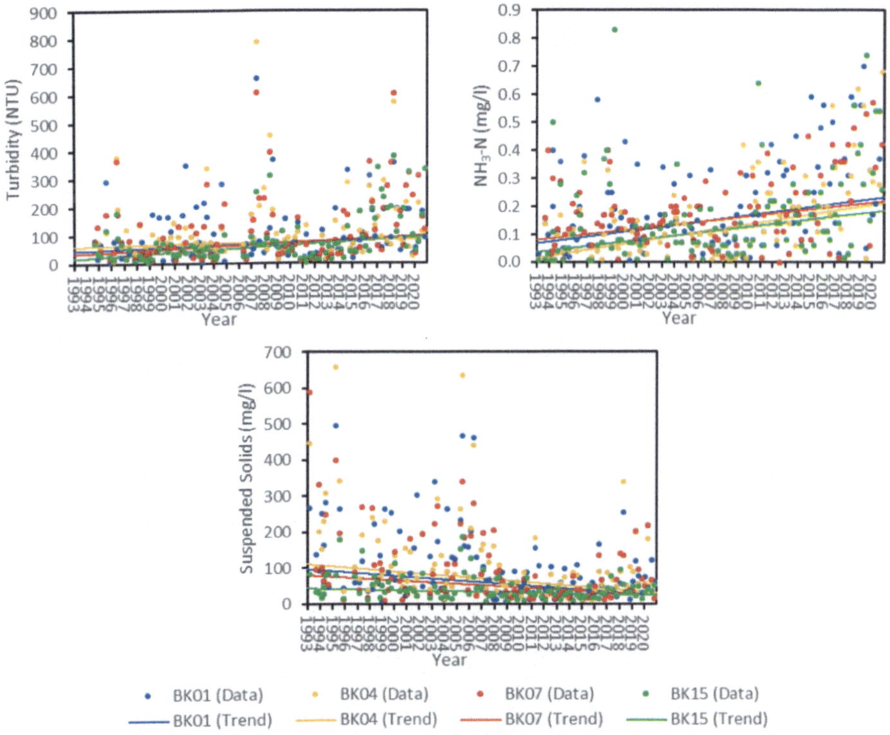

Fig. 28.2 Trends and values of water quality parameters which exhibited significant changes at most parts of the Bang Pakong River (turbidity, NH_3-N, and suspended solids)

to the increase in water temperature, which caused unsuitable conditions for the FCB (Rhodes & Kator, 1988). In fact, some studies have found positive relationships between FCB and water temperature (e.g. Seo, Lee, & Kim, 2019). However, Barcina et al. (1986) found that in warm environments, like in the Bang Pakong Basin, the mortality rate of FCB increased with temperature due to an increasing microflora. The increase in BOD was attributed to wastewater from expanding aquacultural activities (Popan, 2011). Additionally, elevated water temperature may increase BOD through enhanced BOD oxidation (Chapra, Camacho, and McBride, 2021). Similar increases in BOD have been reported in some other areas such as the Bernam River, Malaysia (Hashim et al., 2021).

In the urban area, the city of Chachoengsao (point BK07), there were significant changes of DO, TCB, and NO_3-N, in addition to changes of turbidity, NH_3-N, and SS, which were similar to other parts of the river, and the change in FCB, which was similar to point BK15 (the upstream part of the river). The values of DO, TCB, and NO_3-N can be seen in Fig. 28.4. In terms of DO, TCB, FCB, and SS, the water quality was improved. (The DO was increasing, and the TCB, FCB, and SS were decreasing.) It was due to the operation of the wastewater treatment system since 1998

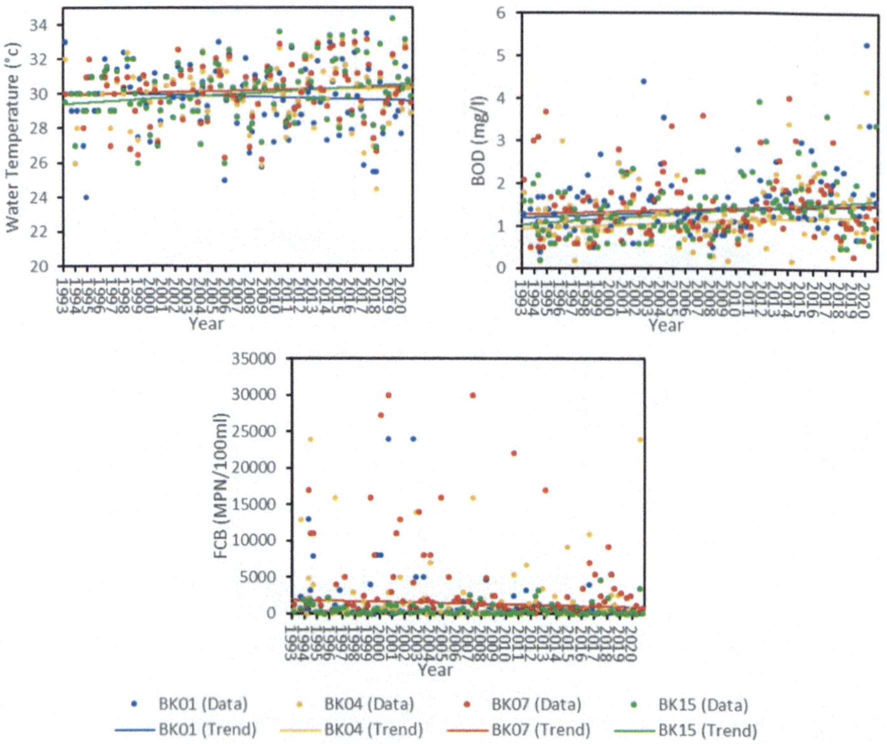

Fig. 28.3 Trends and values of water quality parameters which exhibited significant changes only at the upstream part of the Bang Pakong River (water temperature, BOD, and FCB)

(Database System for Publicly Owned Treatment Works, n.d.a.) and probably also because of the diversion of water from the Pasak River (Royal Irrigation Department, 2002). The improvement of water quality due to treatment systems has been found in many areas (e.g. Ha et al., 2022; Huang et al., 2017; Luo et al., 2011). However, in terms of turbidity, NO_3-N, and NH_3-N, the water quality was worse. It was due to the increase in aquacultural activities in surrounding areas (Popan, 2011) and an oxidation of NH_3-N from aquacultural activities, which caused NO_3-N (Lawson & Lücker, 2018).

In the downstream area of the city of Chachoengsao (point B04), there were significant increases in pH, conductivity, salinity, NO_2-N, and TDS apart from the increase of NH_3-N and decrease of SS, which were similar to other parts of the river. The values of pH, conductivity, salinity, NO_2-N, and TDS can be seen in Fig. 28.5. The trends of all these five parameters in combination with increasing NH_3-N revealed that the water quality was worse. It was due to urban development without a sufficient wastewater treatment system (Chachoengsao Provincial Office, 2023) in combination with increasing aquacultural activities in the surrounding areas (Popan, 2011). The deterioration of the water quality resulting from urban development has

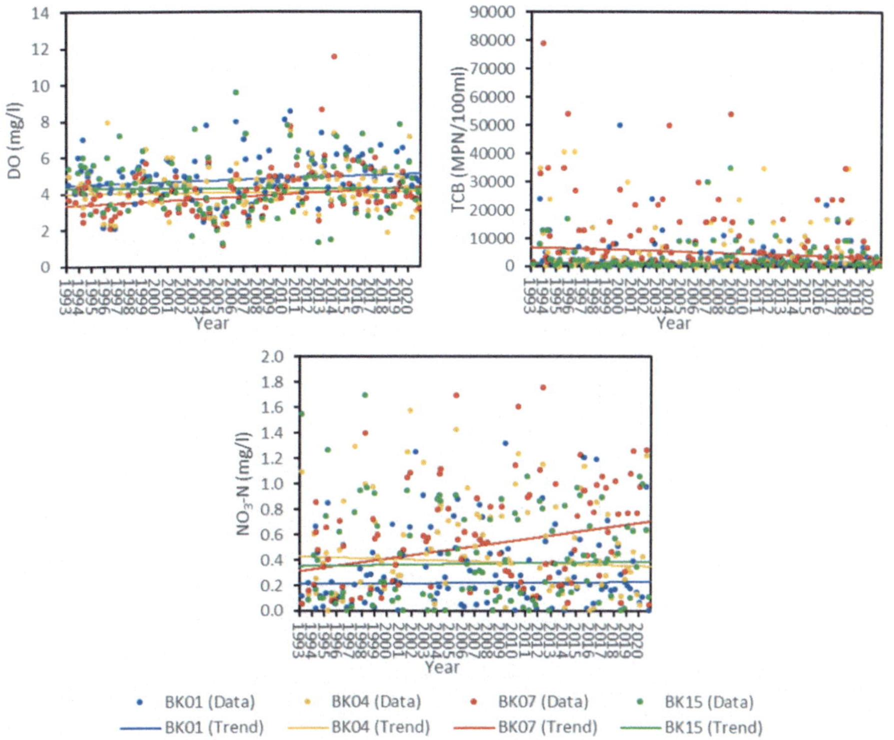

Fig. 28.4 Trends and values of water quality parameters which exhibited significant changes only at the city of Chochoengsao (DO, TCB, and NO_3-N)

been observed in many areas (e.g. Peters, 2009; Qin et al., 2014; Rahman et al., 2021). It suggested the need for implementing a wastewater treatment system and controlling aquacultural activities to address the issues identified in the study. However, in terms of SS, the water quality was improved. It was probably due to the diversion of water from Pasak River (Royal Irrigation Department, 2002).

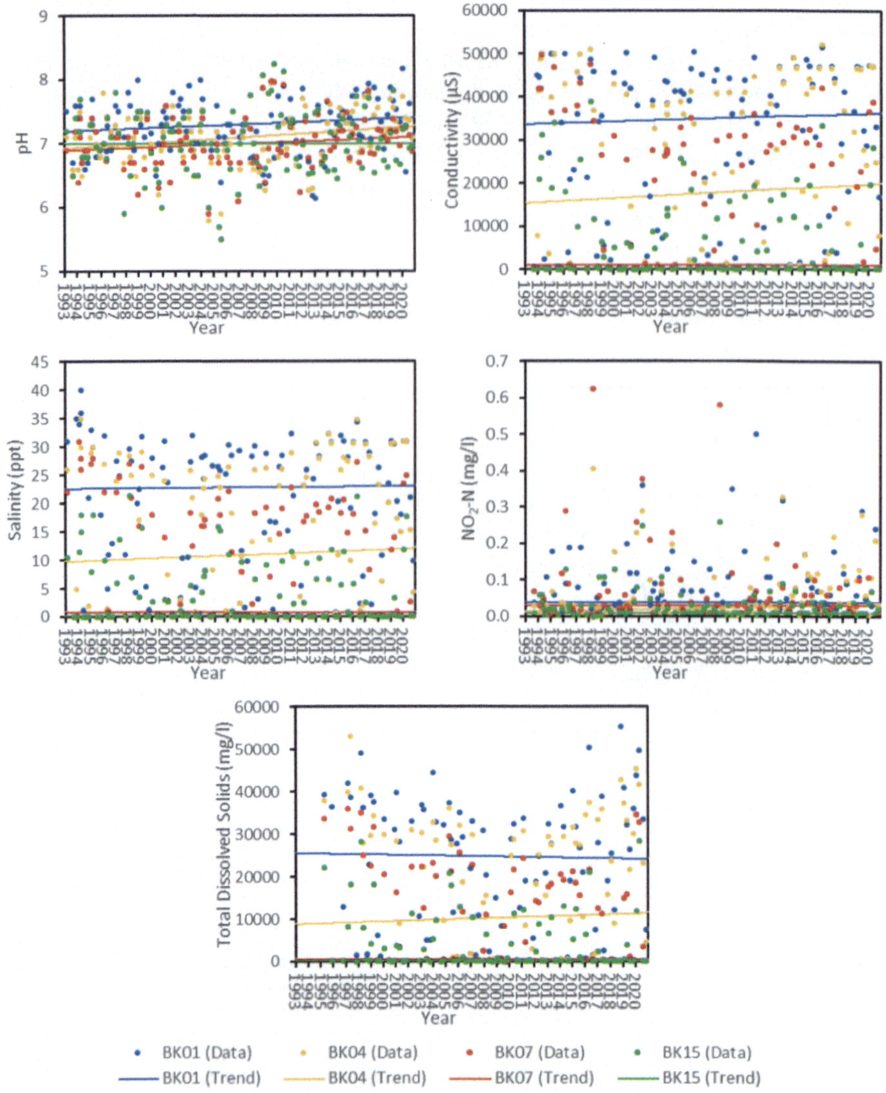

Fig. 28.5 Trends and values of water quality parameters which exhibited significant changes only at the downstream of the city of Chachoengsao (pH, conductivity, salinity, NO_2-N, and total dissolved solids)

28.4 Conclusions

The Bang Pakong River, which serves a large population in eastern Thailand and is highly utilized for agricultural activities, has experienced changes in water quality due to both climate change and human activities. Analysis of water quality data from 1993 to 2020 reveals that turbidity and NH_3-N levels increased, likely a result of expanding aquacultural activities. Conversely, suspended solids (SS) decreased, possibly due to water diversion from the Pasak River, enhancing upstream flow, and the implementation of wastewater treatment systems in urban areas upstream. In the upstream part of the river, water temperature also increased, probably influenced by climate change, leading to a decrease in FCB due to unfavorable conditions for their survival. Additionally, BOD increased due to both increasing aquacultural activities and probably rising water temperature. In the city of Chachoengsao, the largest urban area along this river, the water quality improvements were observed, with increased DO as well as decreased TCB, FCB, and SS, likely due to the wastewater treatment system and water diversion from the Pasak River. However, turbidity, NO_3-N, and NH_3-N levels continued to increase due to increased aquacultural activities in surrounding areas. Downstream from Chachoengsao, although SS levels decreased similarly to other parts of the river, inadequate wastewater treatment systems in urban development, coupled with increasing aquacultural activities in surrounding areas, worsened water quality across multiple parameters. These findings underscore the necessity for implementing adequate wastewater treatment systems and controlling aquacultural activities to safeguard water quality in the Bang Pakong River.

Acknowledgements The authors gratefully acknowledge the Thailand Science Research and Innovation Fund Chulalongkorn University and the Ratchadapisek Sompoch Endowment Fund, Chulalongkorn University for their invaluable supports.

References

Anh, N. T., Can, L. D., Nhan, N. T., Schmalz, B., & Luu, T. L. (2023). Influences of key factors on river water quality in urban and rural areas: a review, *Case Studies in Chemical and Environmental Engineering, 8*(100424). https://doi.org/10.1016/j.cscee.2023.100424

ASDECON corporation company limited (2012) *Operations in Data Gathering and Data Analysis, Project of Development of Data Inventory for 25 Basins and Flood and Drought Models: Bang Pakong Basin.* [online]. Bangkok: Hydro and Agro Informatics Institute. Available at: https://tiwrm.hii.or.th/web/attachments/25basins/16-bangpakong.pdf. Accessed January 2, 2024. (in Thai)

Antonopoulos, V. Z., Papamichail, D. M., & Mitsiou, K. A. (2001). Statistical and trend analysis of water quality and quantity data for the Strymon River in Greece. *Hydrology and Earth System Sciences, 5*(4), 679–692. https://doi.org/10.5194/hess-5-679-2001

Balaji, E., Nagaraju, A., Sreedhar, Y., Thejaswi, A., & Sharifi, Z. (2017). 'Hydrochemical characterization of groundwater in around Tirupati Area. *Chittoor District, Andhra Pradesh, South India', Applied Water Science, 7*, 1203–1212. https://doi.org/10.1007/s13201-016-0448-6

Barcina, I., Arana, I., Iriberri, J., & Egea, L. (1986). Factors affecting the survival of E. coli in a river. *Hydrobiologia, 141*, 249–253.

Bashir, I., Lone, F. A., Bhat, R. A., Mir, S. A., Dar, Z. A., & Dar, S. A. (2020). Concerns and threats of contamination on aquatic ecosystems. In Hakeem, K. R., Bhat, R. A., & Qadri, H. (eds.), *Bioremediation and Biotechnology: Sustainable Approaches to Pollution Degradation*. Cham: Springer, pp. 1–26. https://doi.org/10.1007/978-3-030-35691-0_1

Bordalo, A. A., Chalermwat, K., & Teixeira, C. (2016). Nutrient variability and its influence on nitrogen processes in a highly turbid tropical estuary (Bangpakong, Gulf of Thailand). *Journal of Environmental Sciences, 45*, 131–142. https://doi.org/10.1016/j.jes.2016.01.011

Bordalo, A. A., Nilsumranchit, W., & Chalermwat, K. (2001). Water quality and uses of the Bangpakong River (Eastern Thailand). *Water Research, 35*(15), 3635–3642. https://doi.org/10.1016/s0043-1354(01)00079-3

Bubphamala, T., Benjawan, L., & Liamlaem, W. (2010). Seasonal variations of water quality in Bangpakong River and nearby canals at Banpho District. *Agriculture and Natural Resources, 44*(4), 732–743.

Bundao, S., Veeravaitaya, N., Kaewnern, M., & Ingthamjitr, S. (2018). The relationship between land use and water quality in Bangpakong estuary, Thailand. *Journal of Fisheries and Environment, 42*(2), 24–31.

Chachoengsao provincial office (2023) *20-year Development Goals of Chachoengsao Province (2566–2585 B.E.)*. [online]. Chachoengsao: Chachoengsao Provincial Office. Available at: https://www.opsmoac.go.th/chachoengsao-strategic-files-451391791795. Accessed January 3, 2024. (in Thai)

Chaiyarak, B., Tattiyakul, G. and Karnsunthad, N. (2019) *Climate Change Vulnerability Assessment Bang Pakong River Wetland, Thailand*. [online]. Bangkok: International Union for Conservation of Nature and Natural Resources. Available at: https://www.iucn.org/sites/default/files/2022-08/climate_change_vulnerability_assessment_bang_pakong_river_wetland_thailand.pdf. Accessed January 3, 2024

Chan, R., Wandee, S., Wang, M., Chiemchaisri, W., Chiemchaisri, C. and Yoshimura, C. (2020). Fate, transport and ecological risk of antibiotics from pig farms along the bang pakong River, Thailand. *Agriculture, Ecosystems and Environment, 304*(107123). https://doi.org/10.1016/j.agee.2020.107123

Chapra, S. C., Camacho, L. A., & McBride, G. B. (2021). Impact of global warming on dissolved oxygen and BOD assimilative capacity of the world's rivers: modeling analysis. *Water, 13*(17), article no. 2408. https://doi.org/10.3390/w13172408

Chotpantarat, S., & Boonkaewwan, S. (2018). Impacts of land-use changes on watershed discharge and water quality in a large intensive agricultural area in Thailand. *Hydrological Sciences Journal, 63*(9), 1386–1407. https://doi.org/10.1080/02626667.2018.1506128

Chotpantarat, S., Parkchai, T. and Wisitthammasri, W. (2020). Multivariate statistical analysis of hydrochemical data and stable isotopes of groundwater contaminated with nitrate at Huay Sai Royal Development Study Center and adjacent areas in Phetchaburi Province, Thailand' *Water, 12*(4), article no. 1127. https://doi.org/10.3390/w12041127

Database System for Publicly Owned Treatment Works (n.d.a) *Wastewater Treatment System Data*. Available at: https://dspot.pcd.go.th/database/report/6225d3b9c2eb0552bb89ffd9. Accessed January 9, 2024. (in Thai)

Database system for publicly owned treatment works (n.d.b) *Wastewater Treatment System Data*. Available at: https://dspot.pcd.go.th/database/report/6225d3b9c2eb0552bb89ffda. Accessed January 9, 2024. (in Thai)

Dawe, P. (2006). A statistical evaluation of water quality trends in selected water bodies of Newfoundland and Labrador. *Journal of Environmental Engineering and Science, 5*(1), 59–73. https://doi.org/10.1139/s05-019

Girgibo, N., Lü, X., Hiltunen, E., Peura, P. and Dai, Z. (2023). The air temperature change effect on water quality in the Kvarken Archipelago area. *Science of The Total Environment, 874*(162599). https://doi.org/10.1016/j.scitotenv.2023.162599

Ha, D.-W., Jung, K.-Y., Baek, J., Lee, G.-S., Lee, Y., Shin, D.S., & Na, E. H. (2022). Trend analysis using long-term monitoring data of water quality at Churyeongcheon and Yocheon basins. *Sustainability, 14*(15), article no. 9770. https://doi.org/10.3390/su14159770

Hashim, M., Nayan, N., Setyowati, D. L., Said, Z. M., Mahat, H., & Saleh, Y. (2021). Analysis of water quality trends using the Mann-Kendall test and Sen's estimator of slope in a tropical river basin. *Pollution, 7*(4), 933–942. https://doi.org/10.22059/poll.2021.325794.1118

Huang, H., Wang, Z., Xia, F., Shang, X., Liu, Y., Zhang, M., Dahlgren, R. A., & Mei, K. (2017). Water quality trend and change-point analyses using integration of locally weighted polynomial regression and segmented regression. *Environmental Science and Pollution Research, 24*, 15827–15837. https://doi.org/10.1007/s11356-017-9188-x

Hungspreugs, M., Dharmvanij, S., Utoomprukporn, W., & Windom, H. L. (1990). A comparative study of the trace metal fluxes of the Bang Pakong and the Mae Klong Rivers. *Thailand', Science of the Total Environment, 97–98*, 89–102. https://doi.org/10.1016/0048-9697(90)90232-J

Jeon, C., Raza, M., Lee, J.-Y., Kim, H., Kim, C.-S., Kim, B., Kim, J.-W., Kim, R.-H., Lee, S.-W. (2020). Countrywide groundwater quality trend and suitability for use in key sectors of Korea. *Water, 12*(4), article no. 1193. https://doi.org/10.3390/w12041193

Luo, P., He, B., Takara, K., Razafindrabe, B. H. N., Nover, D., & Yamashiki, Y. (2011). Spatiotemporal trend analysis of recent river water quality conditions in Japan. *Journal of Environmental Monitoring, 13*(10), 2819–2829. https://doi.org/10.1039/c1em10339c

Kendall, M. G. (1948). *Rank correlation methods*. Griffin.

Kundzewicz, Z. W. (2008). Climate change impacts on the hydrological cycle. *Ecohydrology & Hydrobiology, 8*(2–4), 195–203. https://doi.org/10.2478/v10104-009-0015-y

Kupkanchanakul, W., & Kwonpongsagoon, S. (2011). Nitrogen and phosphorus flow analysis from pig farming in Bang Pakong Basin, Eastern Thailand. *EnvironmentAsia, 4*(2), 27–32. https://doi.org/10.14456/ea.2011.14

Kupkanchanakul, W., Kwonpongsagoon, S., Bader, H. P., & Scheidegger, R. (2015). Integrating spatial land use analysis and mathematical material flow analysis for nutrient management: A case study of the Bang Pakong River Basin in Thailand. *Environmental Management, 55*, 1022–1035. https://doi.org/10.1007/s00267-014-0441-5

Lawson, C. E., & Lücker, S. (2018). Complete ammonia oxidation: An important control on nitrification in engineered ecosystems? *Current Opinion in Biotechnology, 50*, 158–165. https://doi.org/10.1016/j.copbio.2018.01.015

Lindsey, R., & Dahlman, L. (2023). *Climate Change: Global Temperature*. Available at: https://www.climate.gov/news-features/understanding-climate/climate-change-global-temperature. Accessed January 2, 2024.

Liu, L., Yang, H., & Xu, X. (2022). Effects of water pollution on human health and disease heterogeneity: A review. *Frontiers in Environmental Science, 10*(880246). https://doi.org/10.3389/fenvs.2022.880246

Livingstone, D. M., & Lotter, A. F. (1998). The relationship between air and water temperatures in lakes of the Swiss Plateau: A case study with palæolimnological implications. *Journal of Paleolimnology, 19*, 181–198. https://doi.org/10.1023/A:1007904817619

Mahidol University (2020) *Study for Strategic Environmental Impact Assessment, Development of Water Budget in Prachin Buri—Bang Pakong Basins: Final Report*. [online]. Bangkok: Office of the National Water Resources. Available at: http://sea.nesdc.go.th/pdf/report/sea-report-3-5.pdf. Accessed January 3, 2024. (in Thai)

Majandang, J., & Piyapong, C. (2023). Evaluation of water quality in Bangpakong River and coastal area of Chonburi Province in 2020 by using water quality index method and geoinformatics. *Journal of Research Unit on Science, Technology and Environment for Learning, 14*(2), 148–166. https://doi.org/10.14456/jstel.2023.12.(inThai)

Mann, H. B. (1945). Nonparametric tests against trend. *Econometrica, 13*(3), 245–259. https://doi.org/10.2307/1907187

Moui, L. V., Chotpantarat, S., Tri, V. P. D., & Van, T. P. (2022). Spatial and temporal variabilities of surface water and sediment pollution at the main tidal-influened river in Ca Mau Peninsular,

Vietnamese Mekong Delta. *Journal of Hydrology: Regional Studies, 41*(101082). https://doi.org/10.1016/j.ejrh.2022.101082

National Statistical Office (n.d.) *Number of Population and Houses Categorized by District and Subdistrict, Chachoengsao Province in December 2553 B.E.* Available at: http://service.nso.go.th/nso/nsopublish/districtList/S010107/th/16.htm. Accessed January 6, 2024. (in Thai)

Paibulkichakul, B., Sivajiranon, V., Buapetch, A., Vongpraserd, O., & Paibulkichakul, C. (2006). Effect of community activities on water qualities of the Bangpakong River, Chachoengsao Province. *Songklanakarin Journal of Science and Technology, 28*(2), 403–415. (in Thai).

Peters, N. E. (2009). Effects of urbanization on stream water quality in the city of Atlanta, Georgia, USA. *Hydrological Processes, 23*(20), 2860–2878. https://doi.org/10.1002/hyp.7373

Popan, A. (2011) *Impact of Sea Water Intrusion on Soil Salinity and Land Used in Bang Pakong River Basin, Chachoengsao Province.* [online]. Bangkok: King Mongkut's Institute of Technology Ladkrabang. Available at: https://opacimages.lib.kmitl.ac.th/medias/b00126625/T131159.pdf. Accessed January 9, 2024. https://doi.org/10.14457/KMITL.res.2011.67 (in Thai)

Price, C., Black, K. D., Hargrave, B. T., & Morris, J. A. (2015). Marine cage culture and the environment: Effects on water quality and primary production. *Aquaculture Environment Interactions, 6*, 151–174. https://doi.org/10.3354/aei00122

Qin, H.-P., Su, Q., Khu, S.-T., & Tang, N. (2014). Water quality changes during rapid urbanization in the Shenzhen River catchment: An integrated view of socio-economic and infrastructure development. *Sustainability, 6*(10), 7433–7451. https://doi.org/10.3390/su6107433

Rahman, M., Ishaque, F., Hossain, A., Mahdy, I. H., & Roy, P. P. (2021). Impact of industrialization and urbanization on water quality of Surma River of Sylhet City. *Desalination and Water Treatment, 235*, 333–345. https://doi.org/10.5004/dwt.2021.27661

Rhodes, M. W., & Kator, H. (1988). Survival of *Escherichia coli* and *Salmonella* spp. in estuarine environments. *Applied and Environmental Microbiology, 54*(12), 2902–2907. https://doi.org/10.1128/aem.54.12.2902-2907.1988

Ritchie, H., & Roser, M. (2019). *Urbanization.* Available at: https://ourworldindata.org/urbanization. Accessed January 2, 2024

Royal irrigation department (2002) *Conclusions of Water Drainage in the Eastern Field through 13–14 Canals from Royal Writing.* [online]. Bangkok: Royal Irrigation Department. Available at: http://water.rid.go.th/flood/flood/report_king.htm. Accessed January 2, 2024. (in Thai)

Sakke, N., Jafar, A., Dollah, R., Asis, A. H. B., Mapa, M. T., & Abas, A. (2023). Water quality index (WQI) analysis as an indicator of ecosystem health in an urban river basin on Borneo Island. *Water, 15*(15), article no. 2717. https://doi.org/10.3390/w15152717

Sanford, R. F., Pierson, C. T., & Crovelli, R. A. (1993). An objective replacement method for censored geochemical data. *Mathematical Geology, 25*, 59–80. https://doi.org/10.1007/BF00890676

Sawangarreruks, S., Thongsukdee, S., Lertkasetvittaya, N., Kunsun, W., Siriwareekul, S., & Chalermvudthi, J. (2005), *Diversity of Fish Community in Bangpakong Estuary.* [online]. Bangkok: Department of Marine and Coastal Resources. Available at: https://dmcrth.dmcr.go.th/attachment/dw/download.php?WP=rUqjMT0jqmMZG22DM7y04TyerPMjA20kqmEZZJ1CM5O0hJatrTDo7o3Q. Accessed January 9, 2024. (in Thai)

Sen, P. K. (1968). Estimates of the regression coefficient based on Kendall's Tau. *Journal of the American Statistical Association, 63*(324), 1379–1389. https://doi.org/10.1080/01621459.1968.10480934

Shwe, K. K., Huaihongthong, S., & Yampracha, S. (2022). Assessment of water quality variation for agriculture in Bang Pakong River of Thailand. *International Journal of Agricultural Technology, 18*(5), 2241–2254.

Simachaya, W. (2020). Water quality management in Thailand. *Workshop on Environmentally Sound Technology on Water Quality Management,* November. n.p.: United Nations Environment Programme and Mekong River Commission. Available at: https://www.pcd.go.th/wp-content/uploads/2020/04/pcdnew-2020-04-22_08-59-15_424617.pdf. Accessed January 2, 2024

Sittisorn, P. (2017). Assessment of water quality of Bang Pakong River using multivariate analysis. *Burapha Science Journal, 22*(2), 183–196. (in Thai).

Seo, M., Lee, H., Kim, Y. (2019). Relationship between coliform bacteria and water quality factors at weir stations in the Nakdong River, South Korea. *Water, 11*(6), article no. 1171. https://doi.org/10.3390/w11061171

Sunitha, V. and Reddy, B. M. (2022). Geochemical characterization, deciphering groundwater quality using pollution index of groundwater (PIG), water quality index (WQI) and geographical information system (GIS) in hard rock aquifer, South India. *Applied Water Science, 12*(41). https://doi.org/10.1007/s13201-021-01527-w

Sunitha, V., Reddy, Y. S., Suvarna, B., & Reddy, B. M. (2022). Human health risk assessment (HHRA) of fluoride and nitrate using pollution index of groundwater (PIG) in and around hard rock terrain of Cuddapah, A.P. South India. *Environmental Chemistry and Ecotoxicology, 4*, 113–123. https://doi.org/10.1016/j.enceco.2021.12.002

United Nations Environment Programme (2021) *Progress on Ambient Water Quality. Tracking SDG 6 Series: Global Indicator 6.3.2 Updates and Acceleration Needs.* [online]. Nairobi: United Nations Environment Programme. Available at: https://www.unwater.org/sites/default/files/app/uploads/2021/09/SDG6_Indicator_Report_632_Progress-on-Ambient-Water-Quality_2021_EN.pdf. Accessed January 8, 2024

United States Environmental Protection Agency (1991) *Regional Guidance on Handling Chemical Concentration Data Near the Detection Limit in Risk Assessments.* Available at: https://www.epa.gov/risk/regional-guidance-handling-chemical-concentration-data-near-detection-limit-risk-assessments. Accessed January 3, 2024

Weihrauch, D., Wilkie, M. P., & Walsh, P. J. (2009). Ammonia and urea transporters in gills of fish and aquatic crustaceans. *Journal of Experimental Biology, 212*(11), 1716–1730. https://doi.org/10.1242/jeb.024851

Wisitthammasri, W., Chotpantarat, S., Thitimakorn, T. (2020). Multivariate statistical analysis of the hydrochemical characteristics of a volcano sedimentary aquifer in Saraburi Province, Thailand. *Journal of Hydrology: Regional Studies, 32*(100745). https://doi.org/10.1016/j.ejrh.2020.100745

Yuenyong, S., Nimsuwan, N., Buranapratheprat, A., Gunboa, V., Jintasaeranee, P., Thaipichitburapa, P., & Suriyaphan, J. (2019). Water quality of the Bangpakong River during 2016–2018. *Burapha Science Journal, 24*(1), 138–155. (in Thai).

Zhu, S., Nyarko, E. K., & Hadzima-Nyarko, M. (2018). Modelling daily water temperature from air temperature for the Missouri River. *Peer J, 6*, article no. e4894. https://doi.org/10.7717/peerj.4894

Chapter 29
Application of Water Quality Index, Hierarchical Cluster Analysis, Pollution Index of Groundwater and Geospatial Technologies in Evaluating the Water Quality: A Study from Mangampeta Barytes Mine Area, South India

Mannala Prasad, Vangala Sunitha, Etikala Balaji, Bandi Muralidhara Reddy, Yenugu Sudharshan Reddy, and Badapalli Pradeep Kumar

Abstract Groundwater is prone to pollution, especially in areas with widespread land use for various purposes, particularly mining. With a complex volcano-sedimentary depositional sequence and ongoing mining activity in the study area, there is great potential for dynamic change in water quality and spatial distribution. Hence an attempt has been made to conduct a comprehensive hydrogeochemical evaluation in and around Mangampeta Barytes Mine using water quality index (WQI), pollution index of groundwater (PIG) and geospatial technology (GST). A 10 km buffer zone around the mine was selected to perform the present study, comprising 314 km^2. Thirty-one representative groundwater samples have been obtained and examined for major cations, anions, and other physico-chemical variables like pH, Electrical conductivity (EC), Total dissolved solids (TDS), Total Alkalinity (TA),

M. Prasad (✉)
Department of Geology, Central Tribal University of Andhra Pradesh, Vizianagaram, Andhra Pradesh, India
e-mail: mannalaprasad@gmail.com

V. Sunitha · E. Balaji
Department of Geology, Yogi Vemana University, Kadapa, Andhra Pradesh, India

B. M. Reddy
Department of Earth Sciences, Yogi Vemana University, Kadapa, Andhra Pradesh, India

Y. Sudharshan Reddy
Department of Civil Engineering, Sai Rajeswari Institute of Techcnology, Proddatur, Andhra Pradesh, India

B. P. Kumar
CSIR-National Geophysical Research Institute, Hyderabad, Telangana, India

V. Sunitha et al. (eds.), *Groundwater Resource Management Planning Strategies*, Springer Water, https://doi.org/10.1007/978-3-031-88870-0_29

and Total hardness (TH). The analytical findings show that the groundwater is alkaline and has high EC and TDS values above WHO-permissible limits. Low Calcium and Magnesium values of the samples indicate the study region belongs to a non-carbonaceous source, with a few outliers from the mine area. According to spatial distribution maps, the sample sites 23, 24, 27, 30, and 31 exhibit significant concentrations of sulfates, magnesium, and electric conductivity, which are falling in the runoff catchment areas of mine dump yards (shale/tuff with pyrites), as well as pulverising mills. PIG findings show 10%, 61%, 19%, 6% and of groundwater samples have extremely high pollution, high pollution, moderate pollution, and low pollution categories, respectively. They are thus unfit for human use, with only 3% Insignificant pollution category which can be used for human usage. In contrast, the WQI concern, 48.28%, 32.26%, and 19.36% of groundwater samples fall into the excellent, good, and poor categories, respectively. HCA, PIG and WQI, indicates that the samples taken in and near mining areas have significant contamination levels. These findings will aid policymakers in putting appropriate corrective measures in place to reduce contamination in the affected areas.

Keywords Water quality index (WQI) · Spatial distribution (SD) · Pollution Index of Groundwater (PIG) · Mangampeta barytes mine · India

29.1 Introduction

Groundwater is an indispensable natural resource for humans, plants, and other living things to survive. It is necessary for food production, agricultural activity, and energy creation in many aspects of life and health. In general, groundwater is considered fresh compared to surface water (Li et al., 2021; Subba Rao et al. 2022; Sharma et al., 2022). About one-third of the world's population relies on groundwater for drinking purposes (Veeraswamy et al., 2021), and it is the primary source of drinking water in India (Rajmohan & Elango, 2004, Raju et al. 2006; Bajpayee et al. 2012; Balaji et al., 2022). However, the quality of water resources and assessment is an ongoing concern and a challenging problem nowadays (Biswajeet and Saied, 2011; Petra Perner, 2016; Etikala et al., 2024). Water deterioration depends on several water-quality constituents (Jothi et al., 2010; Bajpayee et al., 2012) One of the main anthropogenic activities responsible for degrading the water, soil, and environmental ecosystem is mining. Mining and its associated activities adversely affect water quality (Gomshei and Allen, 2000; Zhu & Anderson, 2002; Edraki et al., 2005; Ribeiro et al., 2014). Low precipitation, excessive usage, and contamination due to direct contact of surface waters with chemical and biological agents lead to rapid shrinkage and deterioration of surface water resources, resulting in significant pressure on groundwater worldwide, which has complicated determining its quality assessment. Once the water reaches the subsurface, the dissolution of unwanted elements into the water is unstoppable and possible health issues may arise (Johnson, 1979; Sastri, 1994; Ravikumar et al., 2011; Mostafa, 2017). Hence, water quality should be evaluated properly to avoid possible adverse impacts.

However, well-laid-down measures besides guidelines be at different stages; societies even then suffer from waterborne diseases because of the violation and unawareness of such regulations by water supply people (Reddy et al., 2024). To obtain better results for the classification of water, it is better to get the results by combining all the essential parameters rather than individual or combined (Handa, 1964, 1965; Hem, 1985). The PIG was devised by Subba Rao et al. (2012) to quantify the comparative effect of specific chemical indicators on the all-inclusive chemical quality of groundwater. Different researchers have attempted to create a broad variety of groundwater quality assessment PIG; index selection relies on groundwater input parameters and required outcomes (Subba Rao et al., 2012, Subba Rao 2018; Suvarna et al., 2019; Sudharsan Reddy et al., 2020; Prasad et al., 2019; Reddy et al., 2019 & Subba Rao et al., 2022). For instance, PIG is one of the most advanced operational methods for evaluating drinking water quality and providing complete information for any area of interest. GIS is an efficient technique for assessing the hydrogeochemical characteristics and their Spatio-temporal variations in groundwater (Umrikar & Devi, 2011; Kadam et al., 2019; Narsimha and Ajay 2020). A water quality index (WQI)/ spatial distribution map helps to ascertain the polluted areas and the latent perilous regions that can aid decision-makers in better quality management of the particular part (Prasad et al., 2019).

In this study, hydrogeochemical characterization and its evaluation for drinking purposes are performed using WQI, PIG and GIS. Earlier researchers tried to accentuate the impact of Mangampeta barytes mines on groundwater in conventional methods (Nagaraju et al., 2006; Ashok et al., 2013; Chandra Sekhar Reddy et al., 2014, Prasad et al., 2014). In this study, the aid of geospatial technologies and innovative analytical techniques like WQI and PIG, besides the traditional geochemical analysis, were employed to provide clear-cut and comprehensive results for each sampling location and its individual and overall quality assessment. Hence the main objective of this paper is to look at groundwater quality for drinking purposes through various standard and innovative approaches like HCA, PIG and GIS.

29.2 Study Area

The study area Mangampeta barytes mine is located in Obulavaripalle revenue Mandal, south of YSR District of the Andhra Pradesh state, southern India. A buffer zone of 10 km around the Mangampeta Barytes mine was taken for the current study, which covers an area of 314 Km^2. It is situated at 14^0 01′ 00″ North latitude and 79^0 19′ 00″ of the East longitude. As shown in Fig. 29.1, the administrative study region includes portions of the revenue mandals of Obulavaripalle (170.50 km^2), Kodur (77.90 km^2), Chitvel (60.12 km^2), and Pullampeta (5.50 km^2). The suggested boundary for the study area is found in the topographic sheets 57N/04, 57N/08, 57O/01, and 57O/5 of the Survey of India, scaled to 1:50,000 scale. The Mangampeta barytes deposit, with a tonnage of 74 million, is the single-largest of its kind globally. (Neelakantam, 1987; Sunitha, et al., 2015). Andhra Pradesh Mineral Development Corporation Limited (APMDC) has employed open-cast extraction methods since 1976. The study region experiences a semi-arid climate with variable precipitation levels during the rainy season, which lasts from June through October and has an average annual rainfall of 719 mm. The study area lies 191 m above mean sea level and has a slightly undulating topography. The depth of groundwater varies between 10 and 30 m.

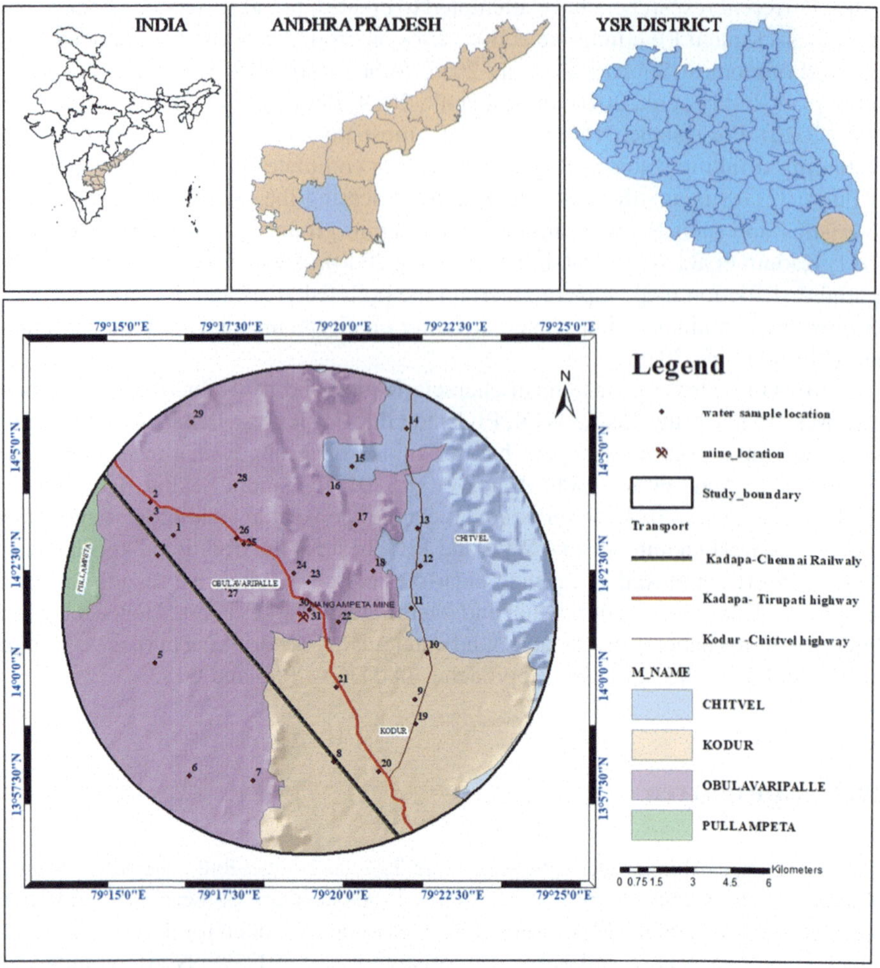

Fig. 29.1 Location map showing sampling sites with 10 km Buffer zone around the Mangampeta barytes mine

29.3 Methodology

29.3.1 Sample Collection and Analysis

To evaluate and comprehend the behaviour of the water quality and its contaminants, 31 representative water samples from the surface and subsurface were gathered in 1-L glass bottles (Fig. 29.1). At the sampling location, environmental sensitive index parameters, including pH, EC, and TDS, were measured and documented using a portable device with the product code HI98130. (Hanna instruments). AUV-visible spectrophotometer was used to examine sulphate (SO_4^{2-}). Volumetric techniques were used to calculate the amounts of carbonate (CO_3^{2-}), bicarbonate (HCO_3^-), calcium (Ca^{2+}), and magnesium (Mg^{2+}). A flame photometer is used to measure the amounts of sodium (Na^+) and potassium (K^+). Utilizing the SPADNS technique, fluoride (F^-) was measured (APHA, 1995, 2012; Molagamudi et al., 2022). Based on the result obtained, HCA, WQI and PIG were performed to know the hydrogeochemistry in a comrehencive manner. Maps of sample locations and the study area's spatial spread are created using the mapping program Arc GIS 10.3, highlighting the study area's hydrogeological structure. The brief methodology of spatial distribution map preparation is as follows. GPS Coordinates for each sample location are obtained while collecting the sample, using handheld GPS, and the collected sample location is imported into the Google earth domain to check the accuracy level and then exported as a kmz file. The sample location map (Fig. 29.1) is produced in an ArcGIS environment with a sample locations shapefile. Samples are analyzed for various parameters, exported to ArcGIS software, and converted as a shapefile. The spatial and attribute database are integrated to prepare the spatial distribution maps for each parameter. Inverse Distance Weighted (IDW) techniques employed to delineate spatial distribution maps.

29.4 Results and Discussion

The analytical results of the physicochemical parameters of the water samples of the study area are shown in Table 29.1. A detailed description of the analyzed parameters is given below.

29.4.1 pH

The pH range in wells is between 7.4 and 8.5, indicating an alkaline nature, and it is within the desired ranges of (6.5–8.5) WHO (2011). The analysis and spatial distribution studies on the findings are provided by geospatial technology quickly

Table 29.1 Physicochemical parameters analytical data at each sampling location in the study area

S. No	Village Name/ Sample Location	Latitude	Longitude	EC (μS/cm)	pH	Ca^{2+} (mg/L)	Mg^{2+} (mg/L)	Na^{+} (mg/L)	K^{+} (mg/L)	HCO_3^{-} (mg/L)	Cl^{-} (mg/L)	SO_4^{2-} (mg/L)	F^{-} (mg/L)	TDS (mg/L)	TH (mg/L)
1	Reddy palle	14° 3′3.46"N	79°16′11.97"E	1600	7.7	35.13	57.42	114	3.33	650	250.2	8.26	2.12	1040	324
2	Mittamindapalle	14° 3′47.30"N	79°15′41.17"E	1900	7.4	52.44	32.15	152	12.22	556	313	6.94	0.5	1235	263
3	Kottapalle	14° 3′24.47"N	79°15′41.88"E	1650	7.8	95.68	80	92	5	588	319	17.28	2.14	1073	568
4	Obulavaripalle	14° 2′35.53"N	79°15′51.13"E	1450	8	35	34.42	112	52.12	362	253	19.04	1.33	943	229
5	Kesukunta tank	14° 0′12.28"N	79°15′47.62"E	2000	7.7	55.45	31.23	210	2.27	452	248	7	0.72	1300	267
6	Yerraguntla Kota	13°57′42.47"N	79°16′35.44"E	2000	7.8	32.13	22.22	150	45.85	538	250	7.1	0.52	1300	172
7	Balireddypalle	13°57′36.54"N	79°18′1.51"E	1800	7.5	50	10.21	132	2.1	731	237	2.52	0.33	1170	167
8	Rachapalle	13°58′2.38"N	79°19′49.66"E	1700	7.5	75.49	92.93	145	52.22	759	213	19.2	1.37	1105	571
9	Kottapalle	13°59′26.39"N	79°21′37.09"E	1750	7.6	88.51	52.32	131	25.46	710	266	23.37	1.33	1138	436
10	Timmasettipalle	14° 0′29.21"N	79°21′53.22"E	2000	7.7	72	77.73	85	2.85	427	319	50.88	2.29	1300	499
11	Tirumalsettipalle	14° 1′27.87"N	79°21′31.80"E	1300	8.5	150.6	30	80	32.61	549	355	12.85	0.52	845	500
12	Choppavaripalle	14° 2′23.93"N	79°21′44.36"E	1700	8.2	60	35	80	10.63	532	213	4.84	0.7	1105	294
13	Gaddalerupalle	14° 2′23.93"N	79°21′44.36"E	850	8	55.45	12.32	115	3.58	542	253	3.49	0.5	553	189
14	Netivaripalle	14° 5′27.70"N	79°21′25.17"E	1000	8.1	32.07	28.67	110	5.08	488	250	19.04	0.37	650	198
15	Vaddipalle	14° 4′36.98"N	79°20′12.69"E	1300	7.7	55.44	52.34	110	3.22	571	250	11.85	0.98	845	354
16	Maravaripalle	14° 4′0.13"N	79°19′40.44"E	2000	7.6	74.39	55.25	152	2.12	510	312	45.88	1.5	1300	413
17	Bommavaram	14° 3′18.82"N	79°20′16.86"E	1500	8	75.25	56.34	135	41.92	549	319	75.38	0.48	975	420
18	Kammavaripalle	14° 2′17.89"N	79°20′40.55"E	1500	7.7	81.37	32.33	112	43.85	549	266	32.53	0.52	975	336
19	Obanapalle	13°58′53.30"N	79°21′39.03"E	2000	8	51.15	52.35	85	2.12	657	258	2.28	1.33	1300	343
20	Koduru	13°57′50.25"N	79°20′50.04"E	1650	7.7	32.13	25.67	82	1.77	588	213	3.49	0.52	1073	186

(continued)

Table 29.1 (continued)

S. No	Village Name/ Sample Location	Latitude	Longitude	EC (μS/cm)	pH	Ca^{2+} (mg/L)	Mg^{2+} (mg/L)	Na^{+} (mg/L)	K^{+} (mg/L)	HCO_3^- (mg/L)	Cl^- (mg/L)	SO_4^{2-} (mg/L)	F^- (mg/L)	TDS (mg/L)	TH (mg/L)
21	Anantarajupeta	13°59′41.92"N	79°19′52.27"E	1400	8	45.15	25.32	112	3.8	549	248.5	8.16	0.45	910	217
22	Agraharam	14° 1′9.62"N	79°19′55.37"E	1500	8.1	60.6	42.35	144	2.24	427	319	23.77	1.12	975	326
23	Govindampalle	14° 2′2.23"N	79°19′14.93"E	3000	8	150.4	250.1	142	1.28	488	313	620	0.63	1950	1404
24	Mangampeta tank	14° 2′13.84"N	79°18′55.31"E	1250	7.7	88.51	200	78	5.08	610	319	368.2	1.34	813	1043
25	Sanjeevapuram tank	14° 2′52.18"N	79°17′47.37"E	1600	8	53.25	52.35	72	2.12	671	255	132.4	0.28	1040	348
26	Mukkavaripalle	14° 2′59.53"N	79°17′37.35"E	1750	7.5	30	12.81	115	3.22	427	200	7.1	2.43	1138	128
27	Korlakunta	14° 1′54.27"N	79°17′23.24"E	1200	7.7	72.51	90	137	85.48	532	201	30.53	2.29	780	551
28	Appaguntapalle	14° 4′10.33"N	79°17′35.57"E	2000	7.8	52.3	78.15	159	11.02	616	337	42.8	1.97	1300	452
29	Orampadu tank	14° 5′34.66"N	79°16′36.36"E	1550	8	41.15	21.44	142	3.89	688	337.2	3.37	0.45	1008	191
30	Mangampeta mine- I	14° 1′24.65"N	79°19′16.27"E	2000	7.5	98.26	90.26	54	3.33	735	284	215.9	0.34	1300	617
31	Mangampeta mine II	14° 1′24.65"N	79°19′16.27"E	2500	7.5	83.34	150.1	112	3.12	650	266	282.1	0.63	1625	825
			WHO limits (2011)	/	6–8.5	75	50	200	12	150	250	200	1	/	100
			Exceeding samples	/	/	10	16	1	9	31	30	4	13	/	31

and comfortably. As shown in Fig. 29.2, high pH values were observed in the NE direction.

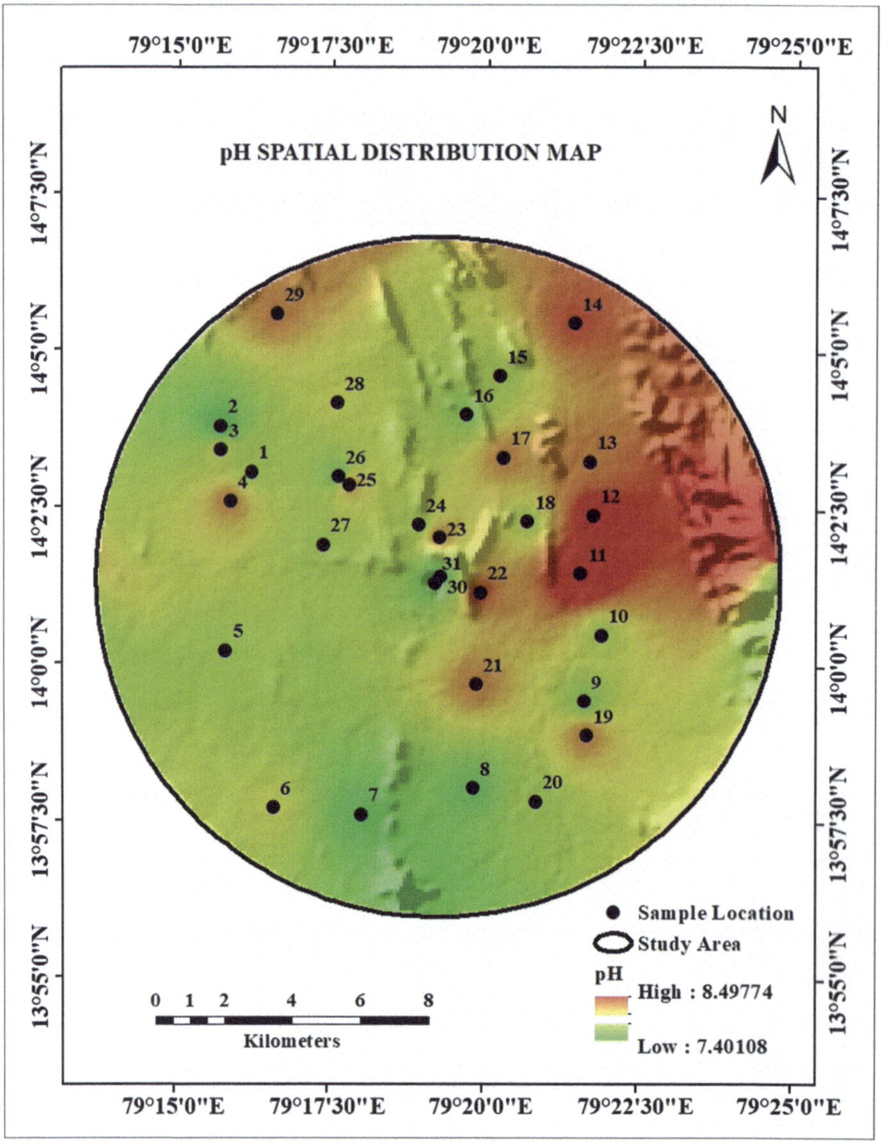

Fig. 29.2 Map showing the spatial distiribution of pH in the study area

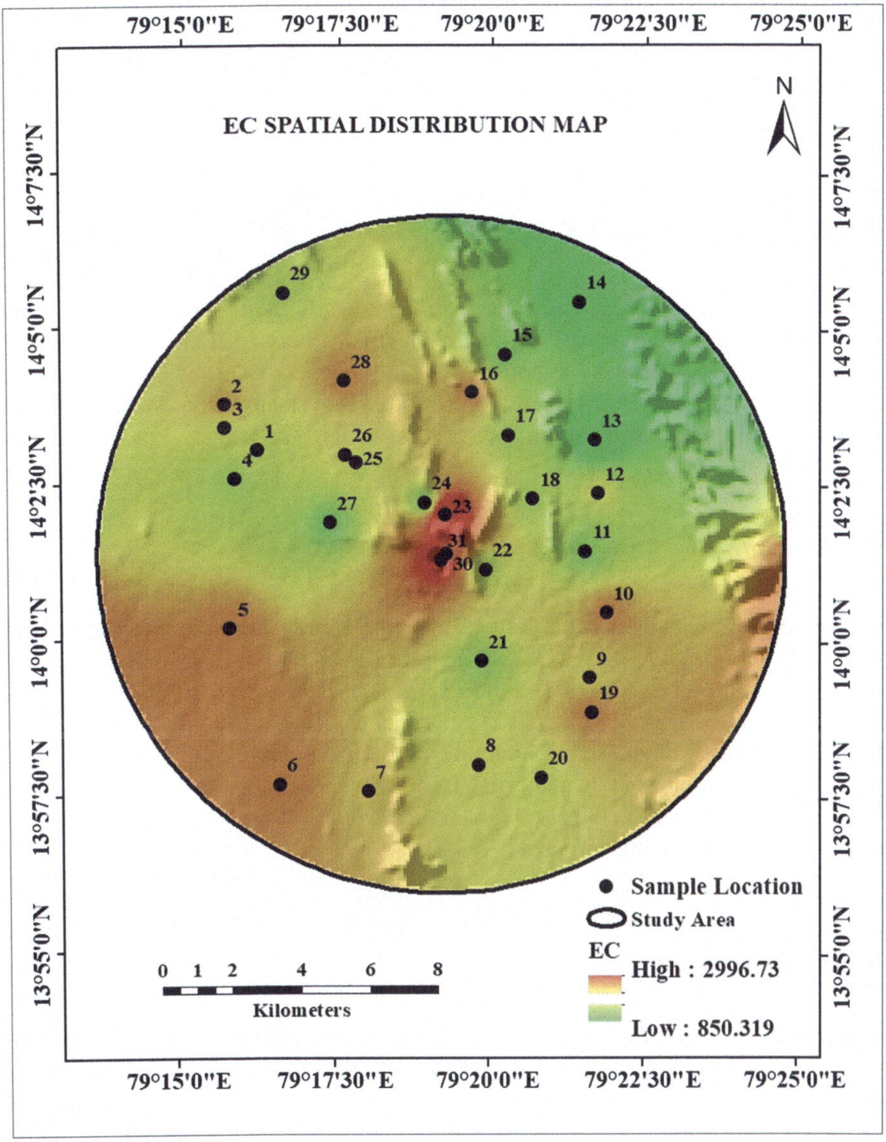

Fig. 29.3 Map showing the spatial distiribution of EC in the study area

29.4.2 Electrical Conductivity (EC)

EC measures how water conducts electricity and calculates the total quantity of dissolved solids in water. (Dahiya and Kaur, 1999; Balaji et al., 2017). The bulk of the physicochemical parameters is significantly impacted by EC, which also affects water quality (Sunitha et al., 2022). EC values in wells range from 850 to 3000 (μS/cm (Table 29.1). The bulk of the samples were above the 1500 μS/cm maximum permissible limit. (WHO, 2011). Greater EC values were noted near the mining region. (Fig. 29.3).

29.4.3 Total Dissolved Solids (TDS)

It is important to classify groundwater hydrochemically using TDS values for any reason. (Davis and De Wiest, 1966). TDS fluctuates between 553 and 1950 mg/L. (Table 29.1). The permissible boundaries are being exceeded by entire samples, according to Davis and Dewiest's classification system from 1966. Wells near and surrounding mining areas had high TDS readings (Fig. 29.4).

29.4.4 Sodium

The leaching of deposits and the mineral decomposition processes are the natural sources of sodium (Na^+). In general, all natural waters contain varying amounts of sodium; however, an excess of sodium in the body can result in fatal conditions like kidney failure, hypertension, chronic headaches, etc. (Meride & Ayenew, 2016) 200 mg/L is the recommended maximum for sodium in potable water. (WHO, 2011). The sodium values in groundwater range from 54 to 210 mg/L (Table 29.1). Almost 97% of the tests fall within the acceptable range. High sodium values were found in the western part of the study area (Fig. 29.5).

29.4.5 Potassium

Groundwater contains potassium (K^+) due to weathering and mineral decomposition. The human body needs potassium to maintain healthy blood pressure and heart performance. The study area's potassium concentration ranges from 1.28 to 85.48 mg/L. (Table 29.1). About 74% of samples in the study area fall within the 12 mg/L permissible limits. According to Meride and Ayenew (2016), the study area's Potash Feldspar has a high resistance to weathering due to the low concentration of K^+ in contrast to Na^+. Potassium values in groundwater were spatially inhomogenous (Fig. 29.6).

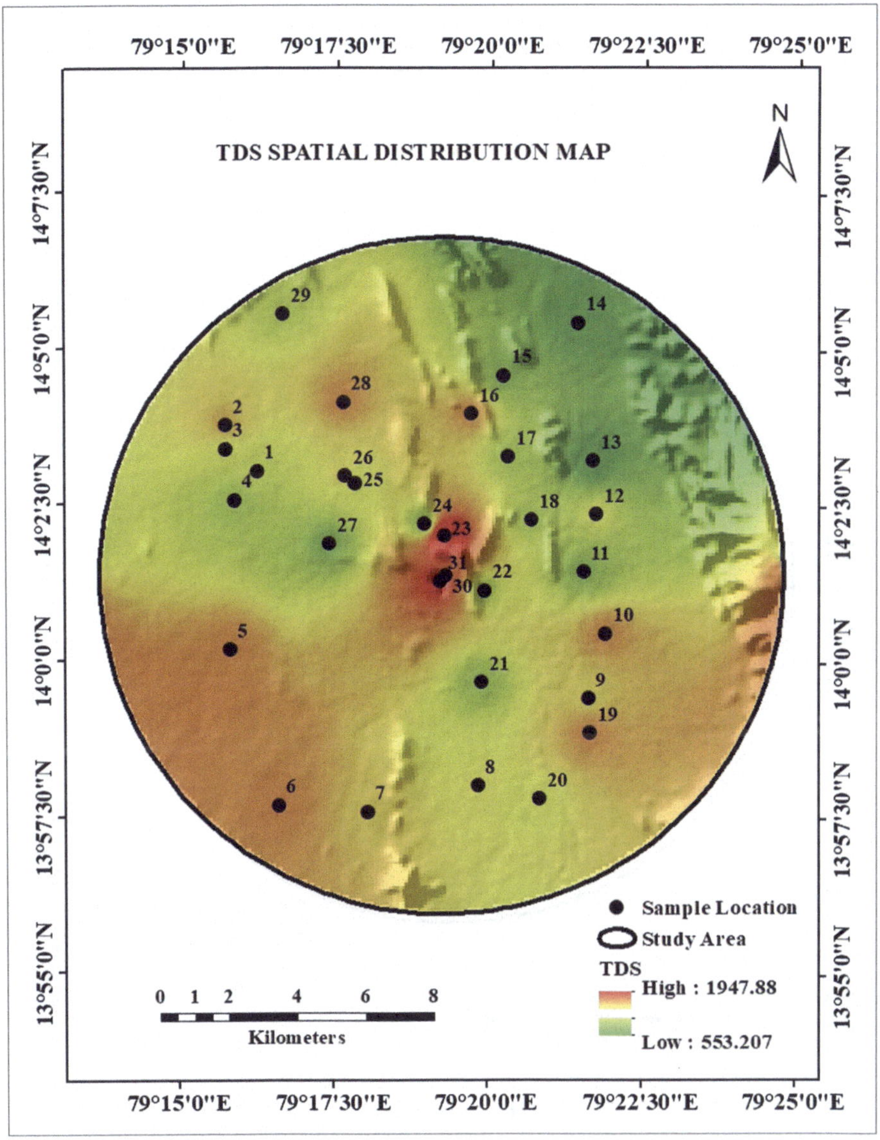

Fig. 29.4 Map showing the spatial distiribution of Total dissolved solids in the study area

29.4.6 Calcium

Both surface water and groundwater contain abundant amounts of calcium (Ca^{2+}) and magnesium (Mg^{2+}). They typically appear as bicarbonates, with sulfates and chlorides appearing less frequently. The kinds and concentrations of Ca^{2+} and Mg^{2+}

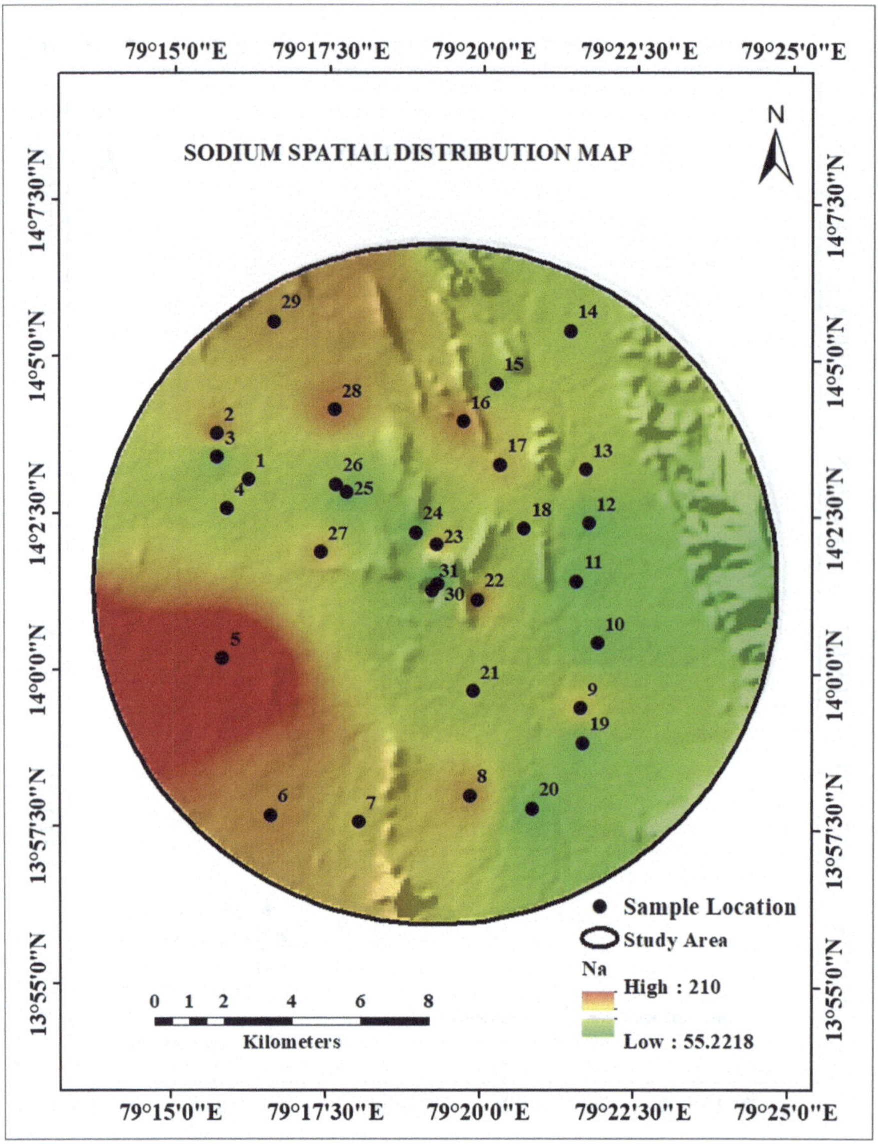

Fig. 29.5 Map showing the spatial distiribution of Sodium concentration in the study area

salts in water are the primary determinants of hardness. In groundwater samples from the study area, calcium concentrations vary from 30 to 150.6 mg/L. (Table 29.1). Approximately 68% of water samples have calcium levels that are within acceptable ranges. The maximum amount of calcium in potable water is 75 mg/L. (WHO, 2011;

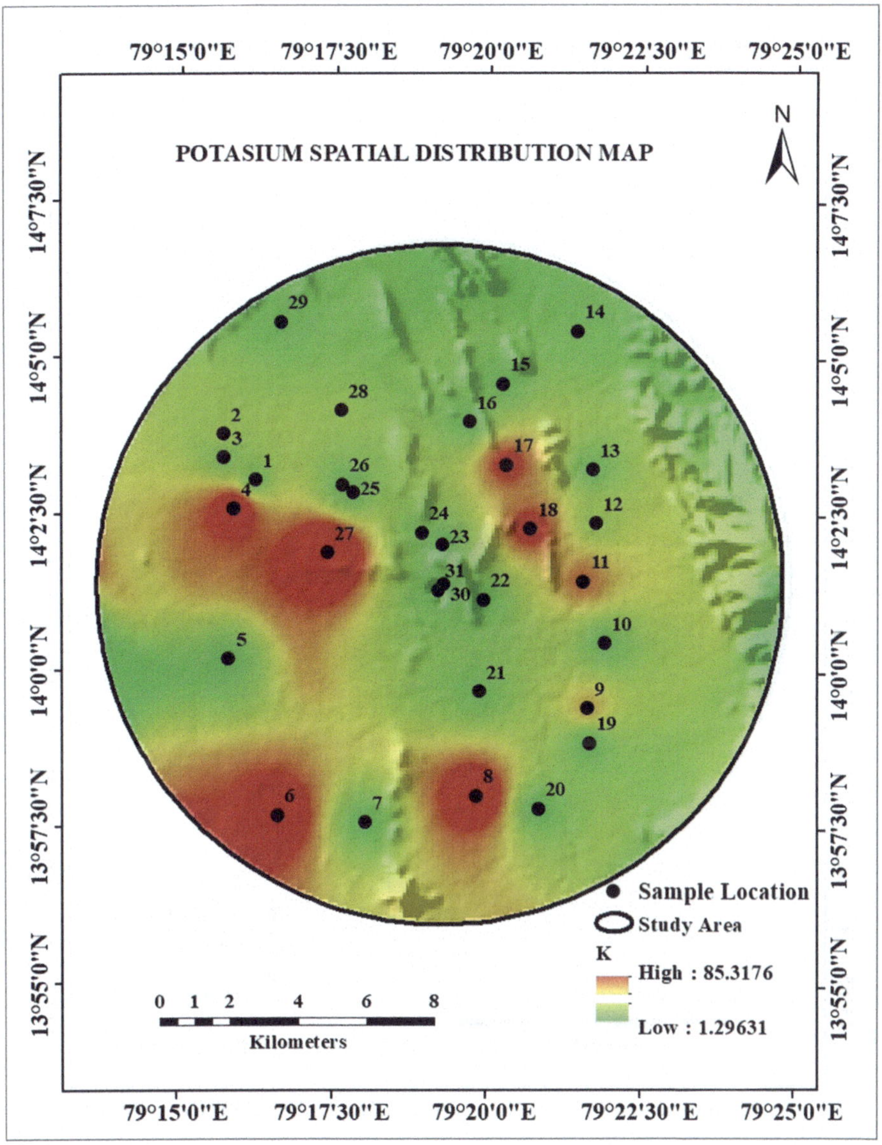

Fig. 29.6 Map showing the spatial distiribution Potasium concentration in the study area

Yarlanki et al., 2025). Wells near neighbouring mine showed elevated calcium levels (Fig. 29.7).

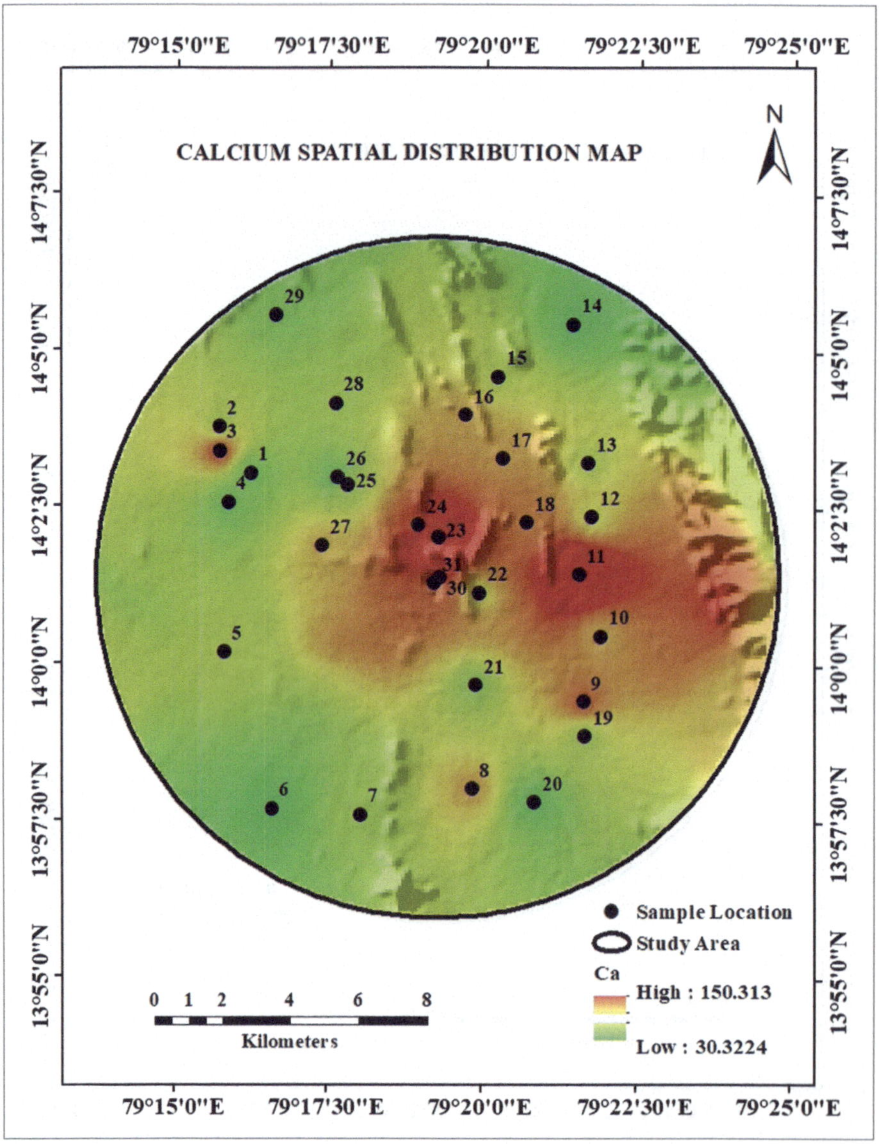

Fig. 29.7 Map showing the spatial distiribution of Calcum concentration in the study area

29.4.7 Magnesium

A greater Mg^{2+} concentration could cause laxative effects and add to unfavourable tastes in drinking water. The maximum amount of magnesium that can be found in potable water is 50 mg/L (WHO, 2011). The research area's groundwater samples have magnesium concentrations ranging from 10.21 to 250 mg/L. (Table 29.1). More Mg^{2+} than allowed is present in 52% of the groundwater samples. Elevated Mg^{2+} levels were concentrated close to mining regions. (Fig. 29.8).

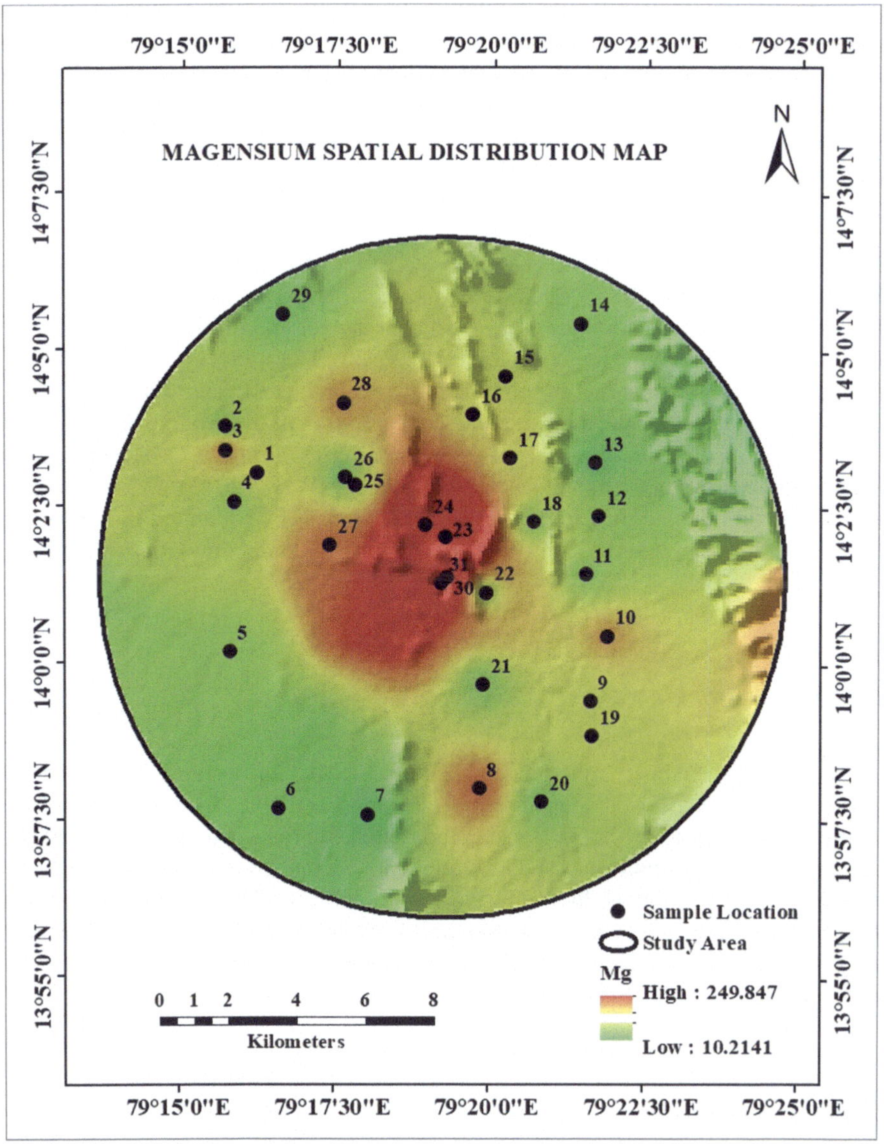

Fig. 29.8 Map showing the spatial distiribution of Magnesium in the study area

29.4.8 Bicarbonate

The alkalinity of natural waters is due to the various salts and free-state hydroxyl ions in the water. The bicarbonate concentration of the groundwater samples in the study area ranges from 366 to 732 mg/L (Table 29.1). All groundwater samples exceeded bicarbonate's permissible limits of 150 mg/L (WHO, 2011). High concentrations of bicarbonates were found in northern and southern regions (Fig. 29.9).

29.4.9 Sulfate

Sulfate naturally appears in water due to leaching from adjacent rocks, minerals, and fertilizers. (Hem, 1985). Because of the weathering and leaching of sulphide ores (pyrites), acid mine drainage—a significant danger and threat to soil and water resources occurs in mining regions (Golla et al., 2020). The study region is of volcano-sedimentary origin and is rich in sulphide minerals in the tuff beds and black shales. That is demonstrated by the high sulphate concentration of the groundwater samples in the study region, which ranges from 2.28 to 620 mg/L. (Table 29.1). The maximum sulphate concentration for potable water is 200 mg/L. (WHO, 2011). 13% of samples have levels of sulphate that are higher than allowed. It is evident from the spatial distribution diagram that the samples were found within the boundaries of the mine region (Fig. 29.10).

29.4.10 Fluoride

To increase bone density and prevent dental decay, fluoride must be present in drinking water at a lower concentration of 1 mg/L (WHO 2011). But excessive fluorine consumption results in skeletal and dental fluorosis. (Sunitha et al., 2022). The circulation of groundwater and leaching from weathered rocks and soils are two potential sources of India's high fluoride water supplies. Fluoride concentration in the current analysis is between 0.33 mg/L and 3.28 mg/L. (Table 29.1). The allowable levels are exceeded in about 25% of water samples. Higher levels were observed in the west and east. (Fig. 29.11).

29.4.11 Chloride

According to WHO (2011), the acceptable and optimal limits for Cl^- are 250–600 mg/L. 250 mg/L is chloride's maximum contamination limit (MCL) in drinking water. The groundwater samples' chloride content ranges from 200 to 355 mg/L.

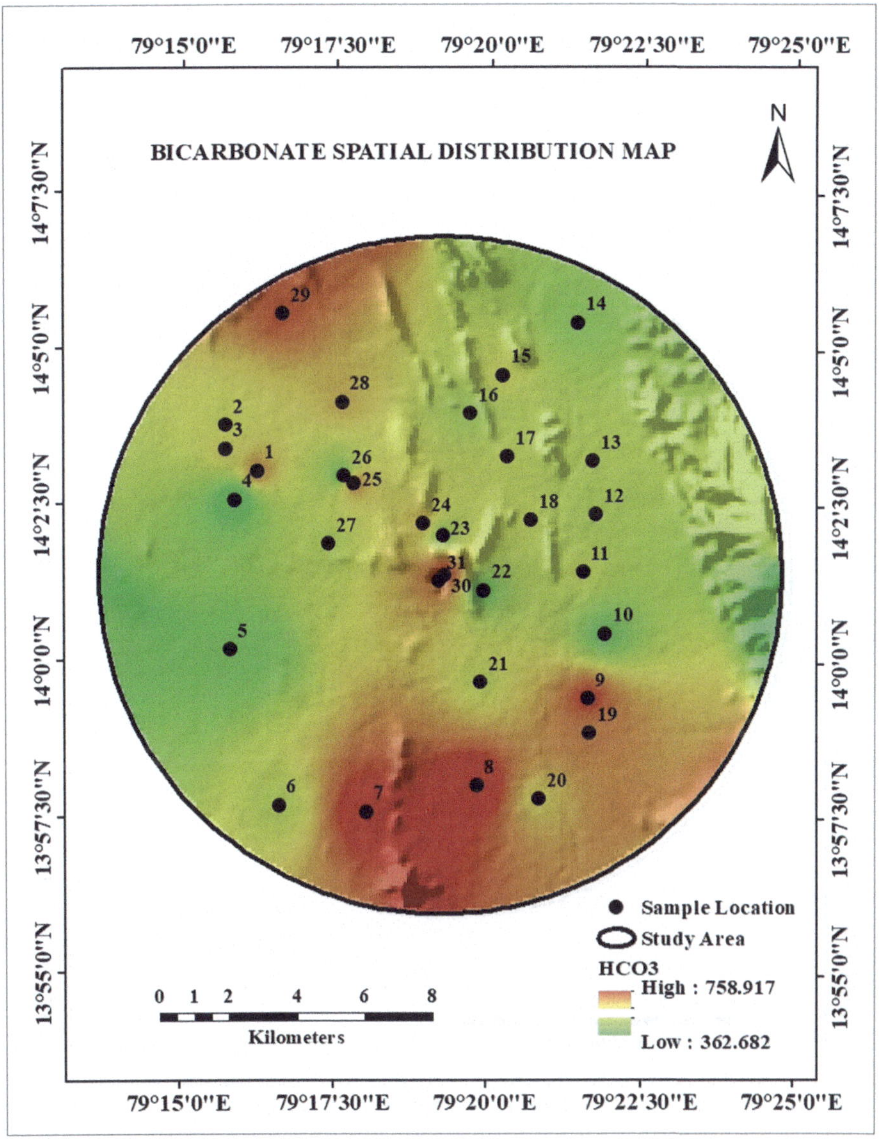

Fig. 29.9 Map showing the spatial distiribution of Bicarbonates in the study area

(Table 29.1 and Fig. 29.12). The study area's findings were all within acceptable ranges.

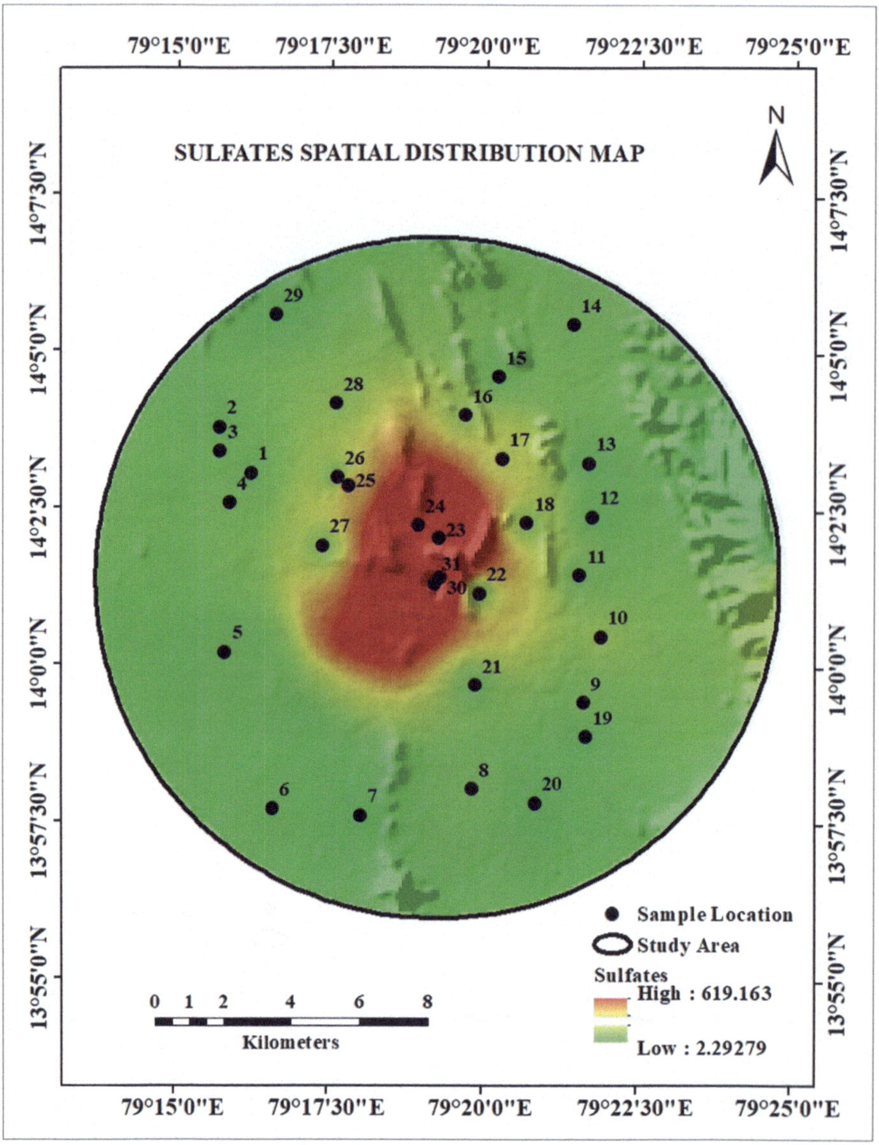

Fig. 29.10 Map showing the spatial distiribution of Sulfates in the study area

29.4.12 *Estimation of Water Quality Index (WQI)*

WQI for drinking water in accordance with WHO (2011) and ISI (1983) requirements, as well as the analytical and statistical methods suggested by Brown (1983), Prasad, and others. (2019).

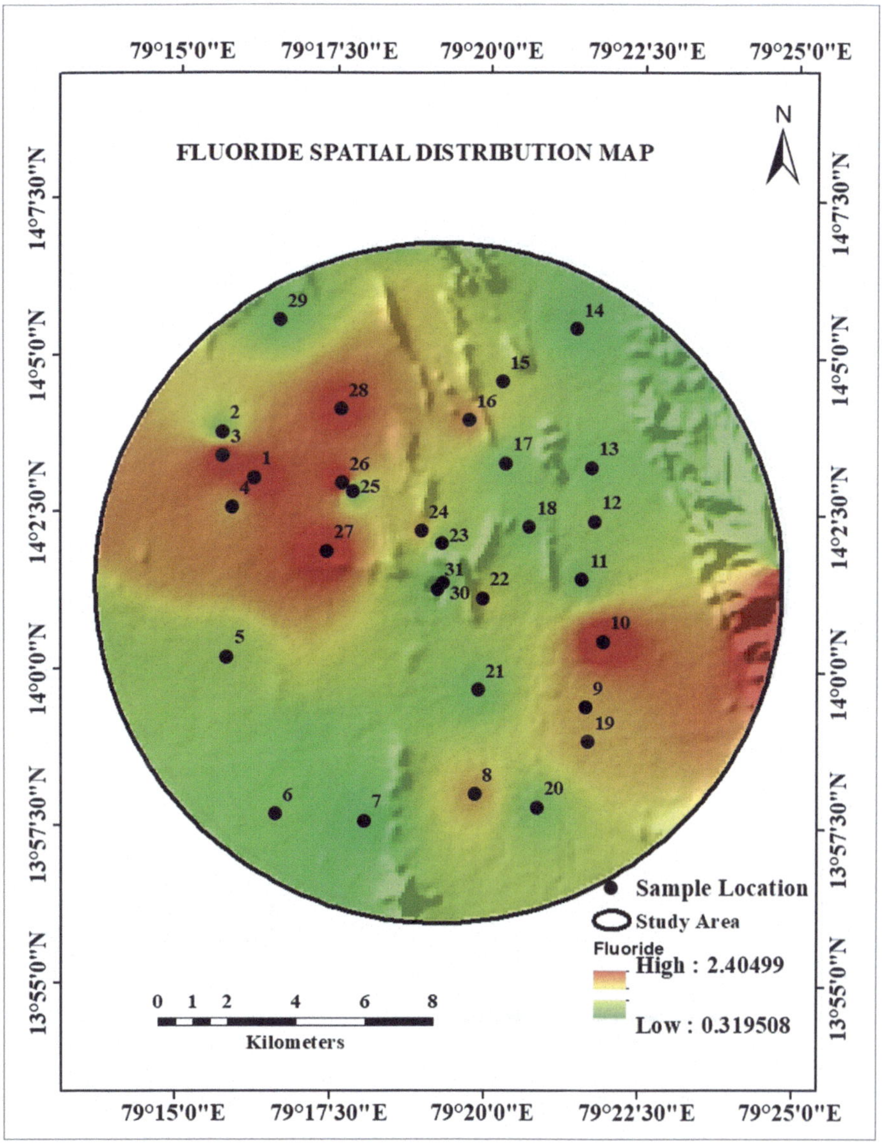

Fig. 29.11 Map showing the spatial distiribution of Fluorides in the study area

In this case, the sub-index (Qn) is computed using the equation

$$Qn = 100\left[\frac{Vn - Vi}{Sn - Vi}\right] \tag{29.1}$$

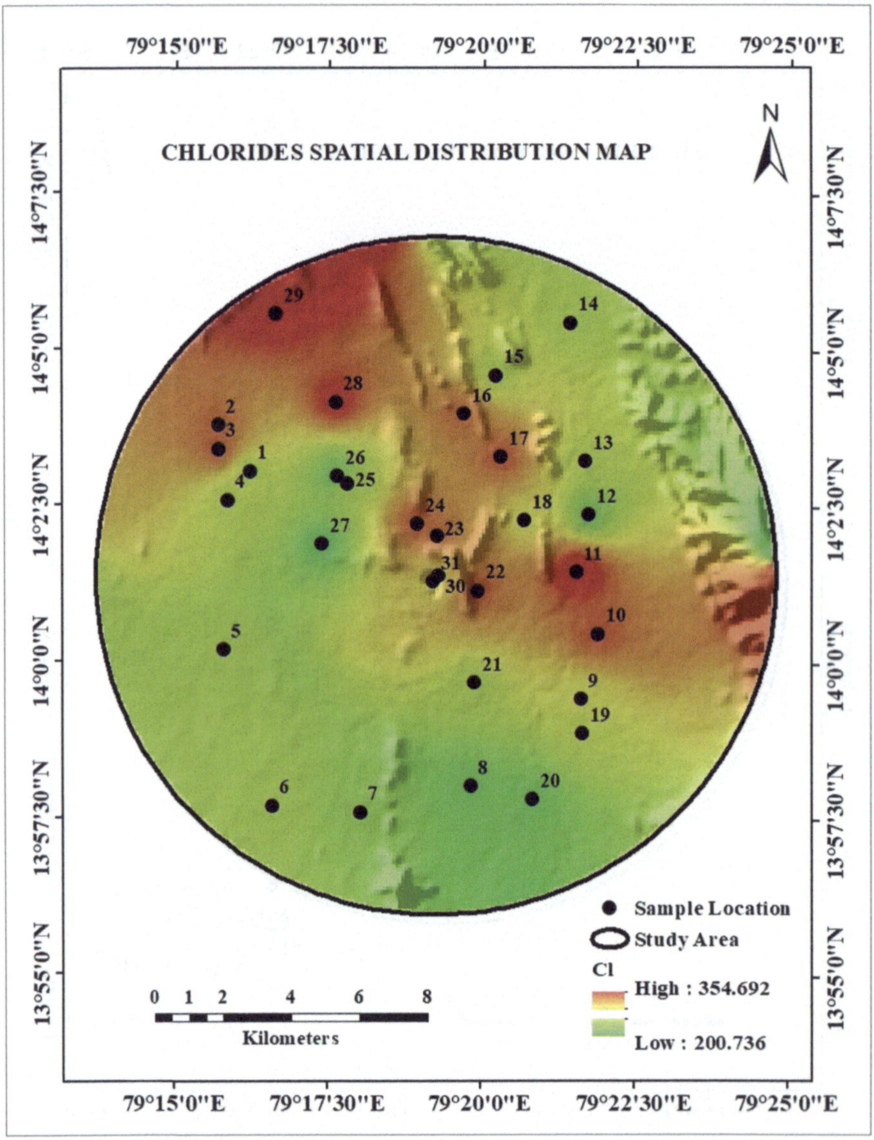

Fig. 29.12 Map showing the spatial distiribution of Chlorides in the study area

Here, Vn is the parameter concentration, Vi is the parameter standard accepted value (Vi = 0 for hardness in natural water).

In this case, the sub-index (Qn) is water quality rating for the nth parameter

$$Wn = \frac{K}{Sn} \tag{29.2}$$

Here,K = proportionality steady and Sn = Standards of the nth parameter

Where K is a constant which could result from making use of the formulation

$$K = \frac{1}{\sum\left(\frac{1}{Sn}\right)} \quad (29.3)$$

The following procedure was followed to calculate the WQI of groundwater samples.

$$\mathbf{WQI} = \frac{\mathbf{Wn * Qn}}{\sum \mathbf{Wn}} \quad (29.4)$$

Here, WQI ranges from 0 and 100;The terms "Wn" and "Qn" refer to the range of parameters and "relative water quality rating" for each parameter, respectively.

(Table 29.2) display the analytical and arithmetic statistics of WQI. It is found that the majority of the samples were found to be poor to good followed by excellent WQI category (Tables 29.3, 29.4).

Table 29.2 Estimation of WQI of each sample based on the standards prescribed by WHO (2011)

Chemical Parameters	Sn	1/Sn	K	Wn = K/Sn	Ideal Value (Vi)	Observed Value (Vn)	Quality Rating (Qn)	WnQn	WQI
pH	8.5	0.117647	1.219294	0.143446	7	7.7	46.67	6.69	126.27
EC	1500	0.000667		0.000813	0	1600	106.67	0.09	
TDS	1500	0.000667		0.000813	0	1040	69.33	0.06	
TH as $CaCO_3$	500	0.002		0.002439	0	324	64.80	0.16	
Ca^{2+}	200	0.005		0.006096	0	35.13	17.57	0.11	
Mg^{2+}	150	0.006667		0.008129	0	57.42	38.28	0.31	
Na^{+}	200	0.005		0.006096	0	114	57.00	0.35	
K^{+}	200	0.005		0.006096	0	3.33	1.67	0.01	
Total Alkalinity	150	0.006667		0.008129	0	650	433.33	3.52	
Cl^{-}	600	0.001667		0.002032	0	250.2	41.70	0.08	
SO_4^{2-}	400	0.0025		0.003048	0	8.26	2.07	0.01	
F^{-}	1.5	0.666667		0.812862	0	2.12	141.33	114.88	
		0.820147		1				126.27	

Table 29.3 Showing the WQI category of individual sampling stations

Sample no	WQI	Water quality Status
1	126.27	Poor water
2	33.59	Excellent
3	128.37	Poor water
4	84.73	Good
5	49.54	Excellent
6	39.91	Excellent
7	27.57	Excellent
8	84.99	Good
9	83.17	Good
10	134.58	Poor water
11	46.95	Excellent
12	53.31	Good
13	40.45	Excellent
14	34.10	Excellent
15	64.08	Good
16	91.30	Good
17	40.13	Excellent
18	39.16	Excellent
19	86.34	Good
20	38.87	Excellent
21	37.87	Excellent
22	74.81	Good
23	50.13	Good
24	85.23	Good
25	29.54	Excellent
26	139.59	Poor water
27	135.53	Poor water
28	119.39	Poor water
29	38.71	Excellent
30	28.89	Excellent
31	44.79	Excellent

Table 29.4 Water quality classification based on WQI value

Class	WQI value	Water quality status	Percentage
A	< 50	Excellent	48.38
B	51–100	Good	32.26
C	101–200	Poor Water	19.36
D	201–300	Very Poor Water	Nil
E	> 300	Water Unsuitable for Drinking	Nil

29.4.13 Estimation:Pollution Index of Groundwater (PIG)

The pollution index measures the degree of relative impact on particular indicators of water quality. The pH, EC, TDS, TH, Ca2 +, Mg2 +, Na +, K +, Cl-, HCO3-, SO42- and F- water quality parameters were used in the calculation of the score. By using the common empirical formulae (Subba Rao et al., 2012; Subba Rao 2018), as listed below, the PIG of each sampling site was estimated.

The PIG was calculated using a series of stages. (Subba Rao et al., 2012; Subba Rao 2018).

$$Wp = \frac{Rw}{\sum Rw} \tag{29.5}$$

Here, Rw is the weight of each component, where Wp is the weight parameter.

$$Sc = \frac{C}{Ds} \tag{29.6}$$

Here, Sc denotes the concentration state, C the chemical variable concentration of each water sample, and Ds the standard drinking water level for each chemical component (WHO, 2011).

$$Ow = Wp \times Sc \tag{29.7}$$

HereWp is the weight parameter, Sc is the concentration state, and Ow is the water's overall chemical quality.

The pollution index of groundwater is calculated using the overall chemical quality of water that is accounted for by all water quality factors of each water sample.

$$\mathbf{PIG} = \sum \mathbf{Ow} \tag{29.8}$$

The relative contribution of the concentrations of water quality factors in each water sample was used to compute PIG. Pollution's effect on groundwater quality is indicated by 10% of the value of 1.0 of the PIG if the overall water quality (Ow) is greater than 0.1. (Subba Rao et al., 2012). Insignificant pollution is defined as PIG < 1, low pollution is defined as PIG 1 to 1.5, moderate pollution is defined as PIG 1.5 to 2, high pollution is defined as PIG 2 to 2.5, and very high pollution is defined as PIG > 2.5. (Table 29.5).

PIG results indicate that 10%, 61%, 19%, 6%, and 3% of groundwater samples, respectively, have extremely high contamination, high pollution, moderate pollution, and low pollution categories, rendering them unfit for human use. The study area contained zones with exceptionally high contamination, high pollution, moderate pollution, and low pollution levels, according to a GIS-based spatial distribution map of PIG (Fig. 29.13).

Table 29.5 Summary statistics of and PIG ranges of water samples

Ranges of PIG	Classification	Sample Numbers	No of samples	% of samples
< 1	Insignificant pollution	13, 14 & 21	3	10
1 to 1.5	Low pollution	1, 2, 4 to 7, 9, 11,12, 15 to 20, 22, 25, 26 & 29	19	61
1.5 to 2	Moderate pollution	3, 8, 10, 27, 28 & 30	6	19
2 to 2.5	High pollution	24 & 31	2	6
> 2.5	Very high pollution	23	1	3
Total			31	100

29.4.14 Hierarchical Cluster Analysis (HCA)

Groundwater samples are frequently grouped or clustered using HCA (Hierarchical Cluster Analysis) based on the similarities and variations in their hydrochemical properties (Danielsson et al., 1999). The wards linkage in R-mode has produced three groups. (Fig. 29.14). Sample No. 23 is in Cluster 1, which PIG classified as a very high pollution range, and is separately plotted. After that, cluster 2 is represented by numbers 24, 31, and 30, and numbers 3, 8, 10, 27, and 28 are plotted in high- to moderate-pollution categories, respectively, in the PIG classification, showing that the range of pollution is highly correlated with two analytical processes. Figure 29.15 showed the impact of mining activities on groundwater quality. From this it is clearly found that the mining activities have significant impact on groundwater quality.

29.5 Conclusions

The water from both surface and groundwater in the Mangampeta Barytes Mining area, Y.S.R District, were evaluated for their chemical quality and suitability for drinking purposes. Based on its hardness, the groundwater in the area is categorized as moderate to very hard. The pH of groundwater is alkaline in nature. With a few outliers in the mine area, the samples' low calcium and magnesium values point to non-carbonaceous aquifers as the region's groundwater source. Acid Mine Drainage, a danger and threat to soil and water resources, occurs in mining areas due to weathering and leaching of sulfide ores (pyrites). The study area is volcanically sedimentary in origin. It contains abundant amounts of sulfideminerals in the form of tuff beds and black shales, demonstrated by the high sulphate concentration of the groundwater samples in and near the mining region. Sample locations 23, 24, 27, 30 and 31 show high concentrations of sulfates, magnesium and EC, which are falling in the runoff catchment areas of mine dump yards and pulverizing mills. According to PIG findings, 10%, 61%, 19%, and 6% of groundwater samples have extremely high contamination, high pollution, moderate pollution, and low pollution categories,

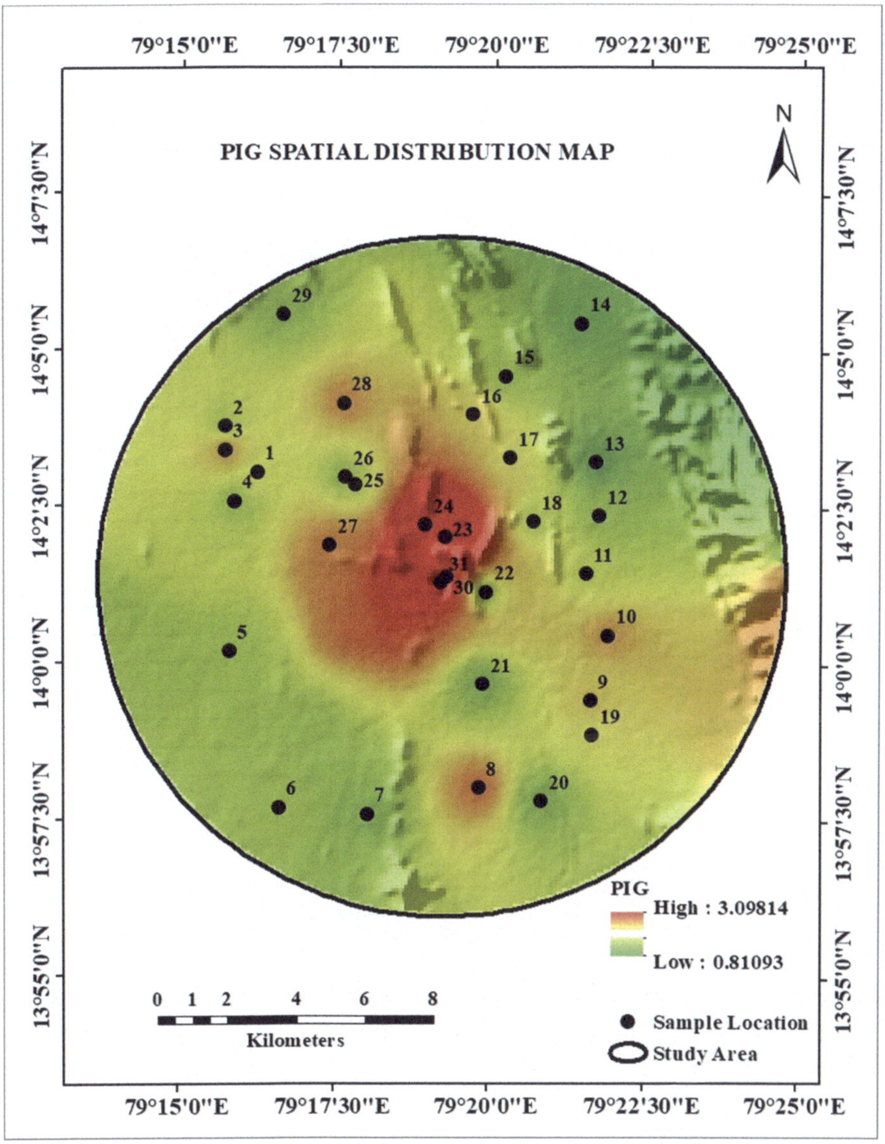

Fig. 29.13 Spatial distribution map of PIG in the study area

respectively, and are thus unfit for human use only 3% Insignificant pollution category. HCA and PIG, a strong relationship were found in all methods used in the study viz. GIS spatial distribution maps, HCA dendrogram and PIG classification categories indicate that highly polluted water samples are falling the premises of the mining-associated areas. To reduce the risk of adverse effects on human health, it is

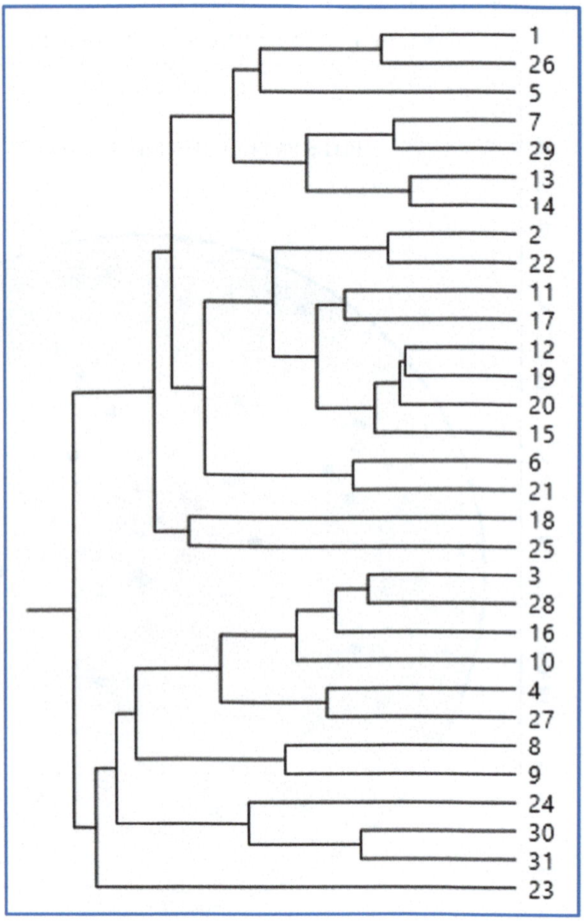

Fig. 29.14 Hierarchical cluster analysis of analysed parameters

advised that prior treatment be needed before water is used for domestic purposes. Policymakers can implement strict policies to lower the risk of contamination with the help of these results.

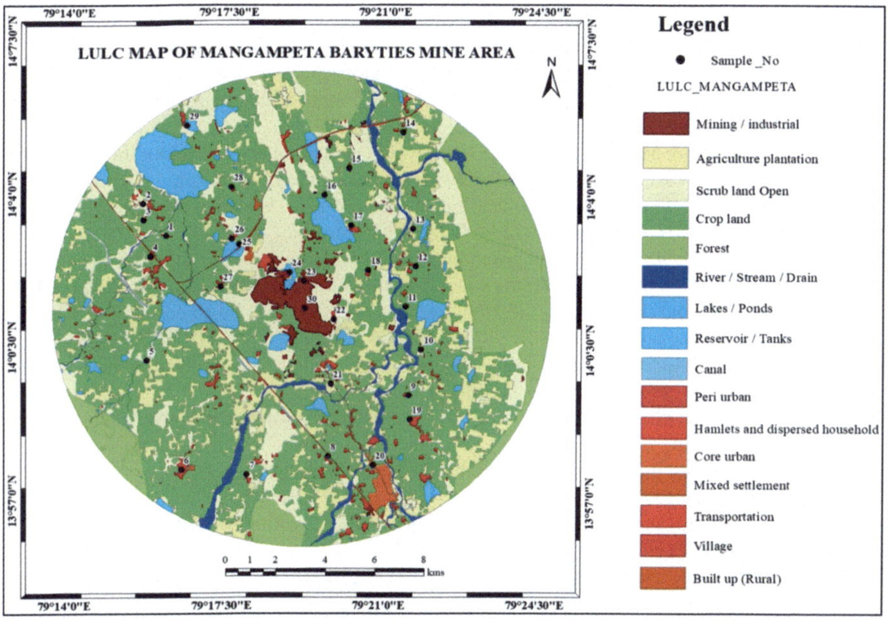

Fig. 29.15 Integrated land use and land cover map with sampling sites clearly depicting the mining and associated industrial activities the study area showing the sampling locations (23, 24, 27, 30 &31) with high concentrations of pollutants in and around the mine premises

Acknowledgements This research is finanacially supported by Dept. of Science and Technology, Govt of India in the form of INSPIRE Fellowship(IF120393) to the first author. Fruther the authors are thankful to the Mangampeta barytes mine authorities for providing the access to the mine and collecting the water samples.

References

Ashok, K., Sudarshan, V., Sundaraiah, R. et al. (2013). Geochemistry of groundwater in and around mangampeta barite deposit, cuddapah district, Andhra Pradesh, *India Journal of Applied Geochemistry15*(1), 98–110.

Bajpayee, S., Das, R., Ruj, B., et al. (2012). Assessment by multivariate statistical analysis of ground water geochemical data of bankura. *India., 3*, 870–880.

Balaji, E., Nagaraju, A., Sreedhar, Y., et al. (2017). Hydrochemical characterization of groundwater in around Tirupati Area, Chittoor District, Andhra Pradesh, South India. *Applied Water Science, 7*, 1203–1212. https://doi.org/10.1007/s13201-016-0448-6

Balaji, E., Sughosh, M., & SrininvasaGowd, S. (2022). Urban Water systems: An overview. *Current Directions in Water Scarcity Research, 6*, 1–19. https://doi.org/10.1016/B978-0-323-91838-1.00016-6

Biswajeet, P., & Saied, P. (2011). Hydro-Chemical analysis of the ground water of the basaltic catchments: upper bhatsai region. *Maharashtra., 5*, 51–57.

Chandra Sekhar Reddy, L. et al. (2014). Impact of mining on groundwater quality in and around Mangampeta, Andhra Pradesh. Gondwana geological magazine, 14, pp. 177–185.

Edraki, M., Golding, S. D., & Baublys, K. A. (2005). Hydrochemistry, mineralogy, and sulphur isotope geochemistry of acid mine drainage at the Mt. Morgan mine environment, Queensland. *Australia. Appl Geochem, 20*, 789–805.

Etikala, B., Vangala, S., & Madhav, S. (2024). Groundwater geochemistry using modified integrated water quality index (IWQI) and health indices with special emphasis on nitrates and heavy metals in southern parts of Tirupati, South India. *Environmental Geochemistry and Health, 46*(11). https://doi.org/10.1007/s10653-024-02229-3

Golla, V. et al. (2020). Assessment of groundwater contamination with emphasis on sulfates, barites mining area, Mangampeta, Andhra Pradesh, India. In: Babu, K., Rao, H., Amarnath, Y. (eds.) Emerging Trends in Civil Engineering. Lecture Notes in Civil Engineering, 61. Springer, Singapore. https://doi.org/10.1007/978-981-15-1404-3_26

Handa, B. K. (1964). Modified classification procedure for rating irrigation waters. *Soil Science, 98*, 264–269.

Handa, B. K. (1965). Modified Hill-piper diagram for presentation of water analysis data. *Current Science, 34*, 131–314.

Hem, J. D. (1985). Study and interpretation of the chemical characteristics of natural water. *USGS Water Supply Paper, 2254*, 117–120.

Johnson, C. C. (1979). Land application of water-an accident waiting to happen. *Groundwater*. https://doi.org/10.1111/j.1745-6584.1979.tb03277.x

Jothi., Venkatachalam, K., Nithya, A., & Chandra Mohan, S. (2010). Correlation analysis of drinking water quality in and around perur block of Coimbatore District, Tamil Nadu, India. *Rasayan Journal Chemistry, 3*(4), 649–654

Kadam, A. K., Wagh, V. M., Muley, A. A. et al. (2019). Prediction of water quality index using artificial neural network and multiple linear regression modelling approach in Shivganga River basin, India. *Modeling Earth Systems and Environment, 5*(3), 951–962. https://doi.org/10.1007/s40808-019-00581-3

Li, P., Karunanidhi, D., Subramani, T., et al. (2021). Sources and consequences of groundwater contamination. *Arch Environ ContamToxicol, 80*, 1–10. https://doi.org/10.1007/s00244-020-00805-z

Meride, Y., & Ayenew, B. (2016). Drinking water quality assessment and its effects on residents health in Wondo genet campus, Ethiopia. *Environment System Resource, 5*(1). https://doi.org/10.1186/s40068-016-0053-6

Molagamudi, S. R., Uravakonda, S., Badapalli, P.K. et al. (2022). Evaluation of groundwater chemistry and its impact on drinking, irrigation, and human health hazard risk assessment (HHRA) in Pincha river basin, semi-arid region of Andhra Pradesh, *Indian International Journal Energy Water Resouce* (2022). https://doi.org/10.1007/s42108-021-00170-0

Mostafa, M. G., Uddin, S. M. H., & Haque, A. B. M. H. (2017). Assessment of hydro-geochemistry and groundwater quality of Rajshahi City in Bangladesh. *Applied Water Science*. https://doi.org/10.1007/s13201-017-0629-y

Nagaraju, A., Suresh, S., Killham, K., & Hudson-Edwards, K. (2006). Hydrogeochemistry of waters of mangampeta barite mining area, Cuddapah Basin, Andhra Pradesh, India. *Turkish Journal of Engineering and Environmental Sciences, 30*(4), 203–219.

Narsimha, A., & Ajay Kumar, T. (2020). Hydrogeochemical investigation of groundwater quality in the hard rock terrain of South India using Geographic Information System (GIS) and groundwater quality index (GWQI) techniques. *Groundwater for Sustainable Development, 10*(2020), 100288. https://doi.org/10.1016/j.gsd.2019.100288

Neelakantam, S. (1987). Mangampet barites deposit, CuddapahDistrict,Andhra Pradesh. Radhakrishna, B. P. (ed.) Purana Basins of Peninsular India. Geological Society of India, Bangalore, Memoirs, 6, 429–458.

Petra Perner., (Ed.) (2016). Machine learning and data mining in pattern recognition. 12th international conference, MLDM 2016, New York, NY, USA, July 16–21, 2016, Proceedings.

Prasad, M., Muralidhar Reddy, B., Ramakrishna Reddy, M., et al. (2014). Studies on physico-chemical parameters to assess the water quality in Obulavaripalli Mandal of YSR (Kadapa) District, Andhra Pradesh. *Indian International Journal Current Research Acadamic Review, 2*(12), 31–41.

Prasad, M., Sunitha, V., Sudharsan Reddy, Y. et al. (2019). Data on water quality index development for groundwater quality assessment from Obulavaripalli Mandal, YSR district, A.P India. *Data in brief, 24*, 103846. https://doi.org/10.1016/j.dib.2019.103846

Rajmohan, N., & Elango, L. (2004). Identification and evolution of hydrogeochemicalprocesses in the groundwater environment in an area of the Palar and Cheyyar River Basins, Southern India. *Env Geol, 46*, 47–61. https://doi.org/10.1007/s00254-004-1012-5

Raju, N. J., Reddy, T. V. K. (2006). Urban development and the Looming water crisis—A case study from Tirupati, South India. IAEG, The Geological Society of London, Paper No. 55, 1–5.

Ravikumar, P., Somashekar, R. K., & Angami, M. (2011). Hydrochemistry and evaluation of groundwater suitability for irrigation and drinking purposes in the Markandeya River basin, Belgaum District, Karnataka State. *Indian Environment Monitoring and Assessment, 173*, 459–487. https://doi.org/10.1007/s10661-010-1399-2

Reddy, B. M., Sunitha, V., Prasad, M. et al. (2019). Evaluation of groundwater suitability for domestic and agricultural utility in semi-arid region of Anantapur, Andhra Pradesh State, South India. Groundwater for Sustainable Development, 100262. https://doi.org/10.1016/j.gsd.2019.100262

Reddy, S., Sunitha, V., & Kumar, P. (2024). Land use & land cover dynamics and its relationship with nitrate pollution in groundwater around inactive mine areas using geospatial techniques, SW part of Cuddapah Basin, Southern India. *Chemosphere, 365*, 143322. https://doi.org/10.1016/j.chemosphere.2024.143322

Ribeiro, L., Kretschmer, N., Nascimento, J., et al. (2014). Water quality assessment of the mining-impacted Elqui River Basin, Chile. *Mine Water and the Environment, 33*, 165–176.

Sharma, K., Raju, N., Singh, N., & Sreekesh, S. (2022). Heavy metal pollution ingroundwater of urban Delhi environs: Pollution indices and health risk assessment. *UrbanClimate, 45*, 101233. https://doi.org/10.1016/j.uclim.2022.101233

Sastri, J. C. V. (1994). Groundwater chemical quality in river basins, hydrogeochemical modeling. Lecture notes-Refresher course, School of Earth Sciences, Bharathidasan Univ., Tiruchirappalli, Tamil Nadu, India.

Subba, R. N., Dinakar, A., & Sun, L. (2022). Estimation of groundwater pollution levels and specific ionic sources in the groundwater, using a comprehensive approach of geochemical ratios, pollution index of groundwater, unmix model and land use/land cover—A case study. *Journal of Contaminant Hydrology, 248*, 103990. https://doi.org/10.1016/j.jconhyd.2022.103990

Subba Rao, N. (2018). Groundwater quality from a part of Prakasam District, Andhra Pradesh. *India. Appl Water Sci, 8*, 30. https://doi.org/10.1007/s13201-018-0665-2

Subba Rao, N., Rao, P.S., Reddy, G.V. et al. (2012). Chemical characteristics of groundwater and assessment of groundwater quality in Varaha River Basin, Visakhapatnam District, Andhra Pradesh, India. *Environ Monit Assess, 184*, 5189–5214. https://doi.org/10.1007/s10661-011-2333-y

Sudharshan Reddy, Y., Sunitha, V., Suvarna, B., et al. (2020). Data on physical impacts and hydro-geochemical assessment of inactive/abandoned mines in and around Southwestern parts of the Cuddapah basin using a conceptual site model (CSM). *Data in Brief*. https://doi.org/10.1016/j.dib.2020.105187

Sunitha, V., & Reddy, B.M. (2022). Geochemical characterization, deciphering groundwater quality using pollution index of groundwater (PIG), water quality index (WQI) and geographical information system (GIS) in hard rock aquifer, South India. *Applied Water Science, 12*, 41. https://doi.org/10.1007/s13201-021-01527-w

Sunitha, V., Reddy, D. V., & Reddy, P. R. (2015). Mineral wealth of Cuddapah basin and its use for sustainable development–An overview. *International Journal of Earth Sciences and Engineering, 8*(6). Editorial Note

Suvarna, B., Sudharsan Reddy, Y., Sunitha, V. et al. (2019). Data on application of water quality index method for appraisal of water quality in around cement industrial corridor Yerraguntla Mandal YSR District AP South India. https://doi.org/10.1016/j.dib.2019.104872

Umrikar, B. N., & Devi, N. M. (2011). Groundwater quality assessment for Guhagar coastal area, Maharashtra. *India. J Environ Res Dev, 5*(4), 880–891.

Veeraswamy, G., Pradeep Kumar, B., & Prasad, M. (2021). Assessment of groundwater quality for drinking and irrigation in semi-arid regions of Andhra Pradesh, Southern India, using multivariate statistical analysis. *Arabian Journal of Geosciences, 14*, 1967. https://doi.org/10.1007/s12517-021-08332-4

WHO. (2011). Guidelines for drinking water quality. 4th Edition World Health Organization, Geneva, Switzerland. http://apps.who.int/iris/bitstream/10665/44584/1/9789241548151_eng.pdf

Yarlanki, V. K., Vadithya, G. N., & Etikala, B. et al. (2025). Hydrogeochemical characterization and source evaluation of groundwater contamination in parts of the Uddanam CKDu area, Southern India: integrating statistical, geochemical modeling, pollution, and health indices. *International Journal of Environmental Research, 19*, 68. https://doi.org/10.1007/s41742-024-00730-2

Zhu, C., & Anderson, G. (2002). Environmental applications of geochemical modelling. xiv+284 pp. Cambridge, New York, Melbourne: Cambridge University Press. Price £75.00, US $110.00 (hard covers); £27.95, US $40.00 (paperback). ISBN 0 521 80907 X; 0 521 00577 9 (pb).

Chapter 30
Groundwater Quality Assessment of Western Coastal Zone of Kanyakumari District, Tamil Nadu, India, Using Statistical, Water Quality Index and GIS Techniques

Annmaria K. George, M. Suresh Gandhi, and Anju Maria Joseph

Abstract The quality of groundwater in the coastal areas is threatened by seawater and human activities. Groundwater quality in the Kanyakumari coastal zone is assessed by applying the Water Quality Index (WQI) and statistical and Geospatial techniques. During the pre-monsoon and post-monsoon seasons of 2022, 40 groundwater samples were collected from the study area. The collected samples were subjected to the laboratory to determine various physio-chemical parameters. The Piper and Gibbs plots were used to investigate the study area's hydrochemical facies and identify the mechanism governing groundwater chemistry. A statistical correlation analysis determined the relationships between various water-related parameters in the study area. The analyzed physio-chemical parameters were used to determine the water quality index for both seasons. The spatial variations of physio-chemical parameters are generated using spatial interpolation techniques of Arc GIS software. The Piper plot shows that most samples fall in the Na-Cl-SO4 field, which is a clear indication of seawater intrusion from the Arabian Sea. The Gibbs plot indicates that evaporation and rock-water interactions are the dominant processes determining the analyzed samples' composition. TH (0.90764), K^+ (0.9583), Ca^{2+} (0.89138), Mg^{2+} (0.89039) and EC (0.83947) are significantly correlated with WQI at both times. WQI values were higher in the pre-monsoon season than in the post-monsoon season. Nearly 37.5% and 22.5% of samples are unfit for drinking during pre-monsoon and post-monsoon seasons, respectively.

Keywords Hydrogeochemistry · Correlation · Water quality index · GIS · Kanyakumari · India

---

A. K. George (✉) · M. S. Gandhi · A. M. Joseph
Department of Geology, University of Madras, Guindy Campus, Chennai, India
e-mail: annmariakgeorge@gmail.com

V. Sunitha et al. (eds.), *Groundwater Resource Management Planning Strategies*, Springer Water, https://doi.org/10.1007/978-3-031-88870-0_30

30.1 Introduction

Groundwater is considered the most precious natural resource that enables us to survive on Earth. Groundwater contributes nearly 0.62% of the total water and 30% of the freshwater on the Earth. Groundwater resources are under increasing pressure due to overexploitation, pollution, and climate change. Water quality studies enable us to determine the physical, chemical, and biological characteristics of the groundwater samples and help locate the contamination source (Usali & Ismail, 2010). The chemistry of groundwater is typically not uniform and is influenced by flow, geochemical processes, evapotranspiration, and potential pollution sources. Coastal aquifers are of the most importance due the freshwater seawater interactions. These interactions affect the groundwater quality and utility in the coastal zone. The fluctuations in the water table allows the movement of water from land to the sea or vice versa. The coastal zone of Kanyakumari district is chosen for the study since this area is prone to both seawater intrusion and submarine groundwater discharge phenomenon (George et al., 2023). Both these processes have a significant effect on the coastal community. Aquifer salinization results from excessive groundwater pumping near coastal aquifers because it lowers the hydraulic head of inland groundwater, which permits saltwater to infiltrate inland aquifers (Etikala et al., 2021). Hence, an integrated method was carried out in the Kanyakumari coastal zone for groundwater quality assessment using WQI, statistical, and GIS tools. This study will help to understand groundwater quality in the coastal zone of Kanyakumari. Thus, a sustainable groundwater resource management plan can be adopted for the coastal zone.

30.2 Materials and Methods

30.2.1 Study Area

The study area is located along the western coastal zone of Kanyakumari district, Tamil Nadu. Kanyakumari is the southernmost tip of the Indian peninsula, where three oceans meet: the Arabian Sea, the Indian Ocean, and the Bay of Bengal. The coastline extends nearly about 68 km. The study area lies between latitude 08°02′ and 08°22′ and longitude 77°04′ and 77°35′ (Fig. 30.1). Tamiraparani and Pazhayar rivers flow across this region (Sundar et al., 2022). Geologically, the area is comprised mainly of Archaeans, Quaternary, and recent geological formations. Groundwater in the region is primarily concentrated in crystalline rocks, sedimentary formations, Quaternary alluvium, and beach sands (Perumal & Subramanian, 2022). Both southwest and northeast monsoon contribute to rainfall in the study area.

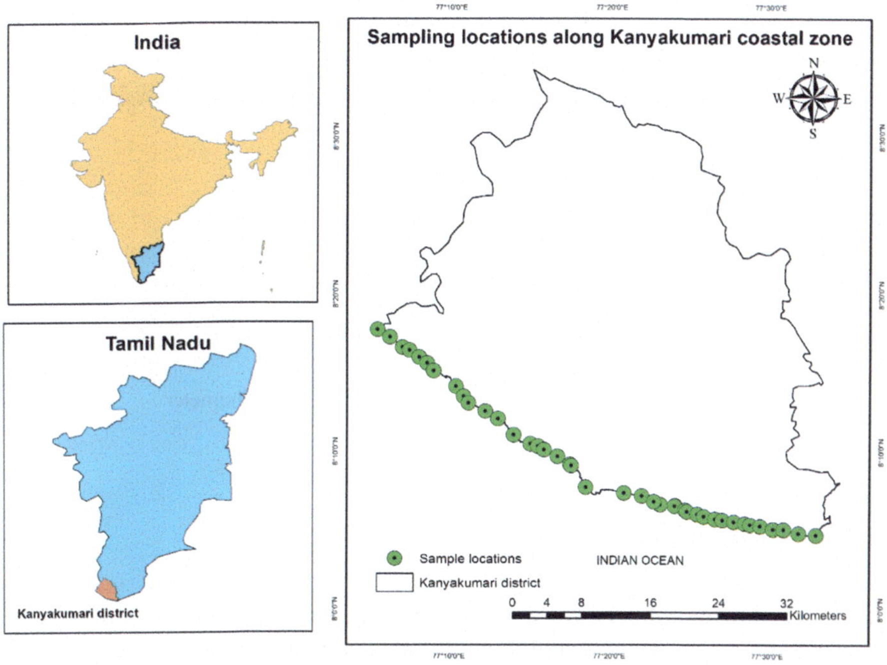

Fig. 30.1 Study area

30.2.2 Data Collection

During the pre-monsoon and post-monsoon seasons of 2022, groundwater samples were collected from the Kanyakumari coastal zone for quality check. Forty groundwater samples were collected for the study parallel to the coastline during both seasons. In-situ parameters were measured using a portable water analyzer kit from the field. Later, the samples are taken into the laboratory for further physio-chemical analysis. The methods adopted for analyzing major cation and anions are shown in Table 30.1. The relationship between the water quality parameters was understood by applying statistical and water quality index methods to the result. Piper trilinear diagram, Gibbs diagrams, and spatial distribution maps were plotted to monitor the groundwater's hydrochemistry of the study area. Spatial distribution maps are prepared for each physio-chemical parameters using IDW (Inverse Distance Weighting) tool in Arc GIS software.

Table 30.1 Analytical methods adopted for analysis of major cations and anions

Parameters	Methods	Instrument used
pH EC TDS Temperature	In-situ analysis	Portable water quality Analyser
Ca^{2+} Mg^{2+} TH	Titration using EDTA	Volumetric analysis
Cl^-	Titration using Silver Nitrate	
Na^+ K^+	Flame photometry	Flame photometer
$SO_4{}^{2-}$ $NO_3{}^-$ $PO_4{}^{3-}$	Absorption spectrometry	UV Spectrophometer

30.2.3 Correlation Analysis

The degree of relationship between various water quality parameters is estimated by correlation analysis. (Kothari et al., 2021). Calculating a correlation matrix, which shows the degree of correlation as a coefficient (r), allowed for determining the degree of a linear relationship between any two parameters. The highly correlated and related water quality parameter that may impact the local water quality is identified using the r-value. A perfect positive linear correlation is indicated by a correlation value of 1, while a perfect negative correlation is indicated by a correlation value of − 1. Two sets cannot have a linear relationship if their r values are 0 (Karmakar & Singh, 2021). The correlation of each parameter with WQI was also calculated to examine the linear relationship between each parameter and WQI. As a result, variables that significantly impact an area's water quality level are assessed using the matrix.

30.2.4 Water Quality Index (WQI)

Based on the results of the physio-chemical analysis, the Water Quality Index (WQI) is the most effective tool for identifying information about water quality behavior. The present study estimates water quality index using Weighted Arithmetic Water Quality Index (WAWQI) method. The Water Quality Index was generated based on parameters such as pH, TDS, Cl-, $SO_4{}^{2-}$, Na^+, K^+, Ca2$^+$, Mg^{2+}, and Total Hardness. The Water Quality Index was computed by comparing the parameters to the BIS quality specifications.

30.2.4.1 Calculation of Water Quality Index Using WAWQI Method

The weighted arithmetic water quality index approach categorized water quality based on purity by utilizing the most regularly measured water quality factors (Awachat & Salkar, 2017). The WQI was calculated using the following equation:

$$\mathrm{WQI} = \frac{\sum WiQn}{\sum Wi} \tag{30.1}$$

The quality rating scale (Qn) for each parameter is calculated as follows:

$$Qn = 100\left[\frac{Vi - Vo}{Si - Vo}\right] \tag{30.2}$$

where,

Vi = Estimated concentration of ith parameter in the analyzed water

Vo = Ideal value of this parameter in pure water

$Vo = 0$ (except for pH =7.0)

Si = Recommended standard value of ith parameter

The unit weight (Wi) for each parameter is calculated by using the following equation:

$$\mathrm{Wi} = \frac{K}{Si} \tag{30.3}$$

where,

The proportionality constant K can be found using the equation:

$$\mathrm{K} = \frac{1}{\sum \frac{1}{Si}} \tag{30.4}$$

30.3 Results and Discussions

30.3.1 Hydrogeochemistry of Water Samples

The groundwater quality of coastal zones is influenced by several factors, like seawater intrusion, geogenic processes, and anthropogenic activities (Chidambaram et al., 2018). All these factors vary depending on location. Several researchers have conducted groundwater studies along the Kanyakumari coastal zone and identified the presence of seawater intrusion along the coastline (Seenipandi et al., 2019; Kaliraj et al., 2015; George et al., 2023). A statistical overview of the cations and anions are shown in Table 30.2. The computed values were compared with BIS Standards

(2012) for drinking water. During both seasons, specific parameters like SO_4^{2-} and NO_3^- are within the desired limit, while all other parameters are above the desired limit. Every sample taken before and after the monsoon season is within the permissible limit. The abundance of cations and anion in groundwater samples indicates seawater intrusion (Shin et al., 2020). The analyzed cation and anion during pre and post-monsoon seasons are shown in Tables 30.2, 30.3 and 30.4. The spatial distribution map of the groundwater samples during both seasons is shown in Fig. 30.2. From the spatial map, it is clear that, during both seasons, EC, TDS, and Cl^- were found to be higher towards the southeastern portion of the study area. This rise could be attributed to saline water intrusion into coastal aquifers. $Ca^{2+}$and Mg^{2+} were found to be evenly distributed in the North West and South East directions for both seasons. The mid-portion of the study area exhibits a more or less moderate concentration of elements during both seasons.

The fluctuation of the groundwater table significantly affects the groundwater quality of the Kanyakumari coastal zone. And these fluctuations may be due to natural or anthropogenic factors such as tides, waves, groundwater recharge, and extraction (Yang et al., 2022). Using the analytical data from the hydrochemical analysis, a piper diagram was created for the study area (Fig. 30.3). The piper diagram shows that groundwater samples belong to the Na-Cl-SO4 field, indicating seawater intrusion in that area. The condition was similar in both the seasons. The functional sources of dissolved chemical constituents, such as rock dominance, evaporation dominance, and precipitation dominance, are evaluated using Gibb's diagrams (Gibbs, 1970). The Gibbs plot shows that water–rock interactions and evaporation are the main processes influencing the composition of the samples under analysis (Fig. 30.4a and b).

30.3.2 Correlation Analysis

Globally, the relationship between Cl^- and EC has been utilized to emphasize the impact of seawater intrusion. Samples with high Cl^- and EC values at the same time are the most affected by the problem of seawater intrusion (Hajji et al., 2021). The correlation matrix for pre-monsoon is shown in Table 30.5. The results showed that EC exhibited a moderate correlation with Ca^{2+} (0.76899), K^+ (0.68938), TDS (0.64272), SO_4^{2-} (0.63943), and Na^+ (0.59095) and a strong correlation with Mg^{2+} (0.9378), TH (0.92467) and Cl^- (0.89). There is a strong correlation between Cl^- and TH (0.73591), SO_4^{2-} (0.74062), and Mg^{2+} (0.77607). TH have a moderate correlation with K^+ (0.75368) and a strong correlation with Mg^{2+} (0.99599) and Ca^{2+} (0.91415). Ca^{2+} strongly correlates with Mg^{2+} (0.87421) and K^+ (0.80431). pH has very weak negative linear coefficient with EC (−0.3374), Na^+ (−0.302), Cl^- (−0.3569), Mg^{2+} (−0.196), PO_4^{3-} (−0.1923), TH (−0.1913), Ca^{2+} (−0.1497), WQI (−0.1455), K^+ (−0.135), SO_4^{2-} (−0.1357), and TDS (−0.0067). WQI has a strong significant correlation with TH (0.90764), K^+ (0.9583), Ca^{2+} (0.89138), Mg^{2+} (0.89039), and EC (0.83947). During post-monsoon, EC has a very strong correlation

Table 30.2 Concentration of various cation and anions during pre-monsoon season

Sample ID	PH	EC (μS/cm)	TDS (ppm)	HCO_3^- (mg/l)	Cl^- (mg/L)	TH (mg/L)	Ca^{2+} (mg/L)	Mg^{2+} (mg/L)	Na^+ (ppm)	K^+ (ppm)	PO_4^{3-} (mg/L)	NO_3^- (mg/L)	SO_4^{2-} (mg/L)
1G	6.86	1094	980	46.36	85.09	292	67.28	30.25	40.5	22.14	0	0	13.2
2G	7.3	1582	1260	68.32	99.27	522	117.74	55.63	95.6	34.36	0	0.8	9.8
3G	6.9	867	947	21.96	56.73	136	27.75	16.27	39.26	19.19	0	2.8	11.82
4G	7.2	684	638	17.08	78.00	48	16.82	1.45	50.64	22.49	0.3	0.1	6.54
5G	7.42	254	561	2.44	21.27	10	0.84	1.92	8.24	0.74	0.1	0.6	5.8
6G	7.28	652	948	24.4	56.73	128	26.07	15.34	48.22	19.06	0	1	10.5
7G	7.01	1843	1070	17.08	234.01	264	37.00	41.87	295.11	26.06	0	0.7	16.8
8G	7.3	189	787	4.88	92.18	26	5.04	3.26	9.5	5.21	0.1	0.5	6.9
9G	7.05	1482	890	68.32	56.73	464	101.76	51.21	95.67	29.92	0	1.2	15.9
10G	7.6	1645	1300	78.08	134.73	548	107.65	68.12	92.97	32.93	0	1.7	28.5
11G	6.8	453	854	12.2	78.00	78	14.29	10.32	54.99	6.25	0.3	0	7.8
12G	6.92	892	970	4.88	141.82	104	15.13	16.15	95.94	14.94	0	0	20.19
13G	7.7	228	228	7.32	35.45	44	6.72	6.63	36.9	8.44	0	0.8	6.4
14G	6.98	873	392	17.08	106.37	148	26.91	19.71	73.7	11.99	0	1.1	10.2
15G	7.18	464	814	4.88	56.73	54	6.72	9.07	95.44	5.91	0	0.8	13.6
16G	7.65	830	886	4.88	113.46	114	15.13	18.59	187.83	11.32	0	1.6	26.5
17G	6.85	675	294	4.88	70.91	46	5.04	8.14	74.88	4.02	0	0.8	15.6
18G	7.2	564	939	31.72	70.91	82	12.61	12.32	121.23	12.09	0.2	0	9.3
19G	6.78	530	317	7.32	99.27	84	14.29	11.78	136.53	11.23	0.1	0.9	21.5
20G	7.12	385	159	2.44	35.45	10	0.84	1.92	22.38	1.14	0	0.4	12.5

(continued)

Table 30.2 (continued)

Sample ID	PH	EC (μS/cm)	TDS (ppm)	HCO_3^- (mg/l)	Cl^- (mg/L)	TH (mg/L)	Ca^{2+} (mg/L)	Mg^{2+} (mg/L)	Na^+ (ppm)	K^+ (ppm)	PO_4^{3-} (mg/L)	NO_3^- (mg/L)	SO_4^{2-} (mg/L)
21G	7.65	145	248	4.88	14.18	10	2.52	0.90	15.12	0.33	0	1.8	6.2
22G	6.89	426	246	7.32	49.63	48	7.56	7.10	78.84	2.79	0.1	1.3	21.8
23G	6.95	763	542	9.76	148.91	124	19.34	18.47	155.61	3.48	0	1.1	31.3
24G	7.71	165	1282	2.44	14.18	20	0.84	4.34	9.54	0.3	0	2.2	5.4
25G	6.96	554	904	17.08	70.914	72	10.09	11.41	90	10.99	0	1.9	32.5
26G	7.28	505	1112	4.88	70.914	82	12.61	12.32	103.59	10.72	0	2.8	17.4
27G	6.55	991	843	17.08	156.01	172	25.23	26.59	416.79	8.72	0	1.6	13.65
28G	7.6	236	189	2.44	21.27	12	2.52	1.39	52.38	0.95	0	2.2	9.43
29G	7.22	813	553	19.52	148.91	140	23.54	19.81	181.44	3.46	0	0.3	23.6
30G	7.4	665	552	4.88	49.63	40	5.88	6.17	84.33	2.56	0	0.9	12.2
31G	7.54	598	1516	29.28	63.82	116	22.70	14.46	98.37	3.21	0.1	0.1	18.32
32G	7.45	933	1360	61	92.18	264	46.25	36.23	182.88	18.31	0	0.5	15.5
33G	7.28	1343	1054	63.44	141.82	324	35.32	57.53	381.24	14.1	0	1.5	26.5
34G	7.03	3510	1989	56.12	765.87	740	83.26	129.83	627.84	35.53	0	0.9	124.5
35G	6.95	2880	1023	70.76	595.67	610	77.37	101.69	281.58	30.76	0.1	2	38.8
36G	7.58	759	962	58.56	63.82	158	20.18	26.25	329.94	40.12	0	1.6	21.6
37G	7.02	1982	1280	68.32	290.74	438	47.93	77.66	387.05	11.93	0	0.4	42.4
38G	6.79	2830	1520	53.68	567.31	478	68.96	74.61	348.25	28.38	0	1.2	21.86
39G	7.14	1567	1908	68.32	127.64	396	62.23	58.70	351.63	27.78	0	1.1	19.65
40G	7.04	1496	849	75.64	170.19	282	51.30	37.55	429.84	13.91	0.2	1.4	23.5

Table 30.3 Concentration of various cation and anions during post monsoon season

Sample ID	PH	EC (µS/cm)	TDS (ppm)	HCO^{3-}(mg/l)	Cl^- (mg/L)	TH (mg/L)	Ca^{2+} (mg/L)	Mg^{2+} (mg/L)	Na^+ (ppm)	K^+ (ppm)	PO_4^{3-} (mg/L)	NO_3^- (mg/L)	SO_4^{2-} (mg/L)
1G	7.1	994	750	46.36	85.09	292	67.28	30.25	40.5	22.14	0	0	13.2
2G	7.01	1482	1060	68.32	99.27	522	117.74	55.63	95.6	34.36	0	0.8	9.8
3G	6.97	647	897	21.96	56.73	136	27.75	16.27	39.26	19.19	0	2.8	11.82
4G	5.82	554	638	17.08	78.00	48	16.82	1.45	50.64	22.49	0.3	0.1	6.54
5G	7.75	155	461	2.44	21.27	10	0.84	1.92	8.24	0.74	0.1	0.6	5.8
6G	6.4	452	838	24.4	56.73	128	26.07	15.34	48.22	19.06	0	1	10.5
7G	7.1	1783	970	17.08	234.01	264	37.00	41.87	295.11	26.06	0	0.7	16.8
8G	7.29	121	647	4.88	92.18	26	5.04	3.26	9.5	5.21	0.1	0.5	6.9
9G	7.35	1382	970	68.32	56.73	464	101.76	51.21	95.67	29.92	0	1.2	15.9
10G	7.25	1645	1260	78.08	134.73	548	107.65	68.12	92.97	32.93	0	1.7	28.5
11G	5.9	353	714	12.2	78.00	78	14.29	10.32	54.99	6.25	0.3	0	7.8
12G	5.31	892	920	4.88	141.82	104	15.13	16.15	95.94	14.94	0	0	20.19
13G	7.75	128	125	7.32	35.45	44	6.72	6.6368	36.9	8.44	0	0.8	6.4
14G	6.9	777	298	17.08	106.37	148	26.91	19.71	73.7	11.99	0	1.1	10.2
15G	6.18	364	795	4.88	56.73	54	6.72	9.07	95.44	5.91	0	0.8	13.6
16G	5.9	730	726	4.88	113.46	114	15.13	18.59	187.83	11.32	0	1.6	26.5
17G	4.85	575	194	4.88	70.91	46	5.04	8.14	74.88	4.02	0	0.8	15.6
18G	7.8	589	939	31.72	70.91	82	12.61	12.32	121.23	12.09	0.2	0	9.3
19G	5.78	530	317	7.32	99.27	84	14.29	11.78	136.53	11.23	0.1	0.9	21.5
20G	7.8	285	59	2.44	35.457	10	0.84	1.92	22.38	1.14	0	0.4	12.5

(continued)

Table 30.3 (continued)

Sample ID	PH	EC (μS/cm)	TDS (ppm)	HCO^{3-}(mg/l)	Cl^- (mg/L)	TH (mg/L)	Ca^{2+} (mg/L)	Mg^{2+} (mg/L)	Na^+ (ppm)	K^+ (ppm)	PO_4^{3-} (mg/L)	NO_3^- (mg/L)	SO_4^{2-} (mg/L)
21G	5.65	49	48	4.88	14.18	10	2.52	0.90	15.12	0.33	0	1.8	6.2
22G	6.65	426	184	7.32	49.63	48	7.56	7.10	78.84	2.79	0.1	1.3	21.8
23G	6.35	763	342	9.76	148.91	124	19.34	18.47	155.61	3.48	0	1.1	31.3
24G	5.71	36	28	2.44	14.18	20	0.84	4.34	9.54	0.3	0	2.2	5.4
25G	6	554	204	17.08	70.91	72	10.09	11.41	90	10.99	0	1.9	32.5
26G	6.28	505	212	4.88	70.91	82	12.61	12.32	103.59	10.72	0	2.8	17.4
27G	6.55	991	843	17.08	156.01	172	25.23	26.59	416.79	8.72	0	1.6	13.65
28G	7	236	88	2.44	21.27	12	2.52	1.39	52.38	0.95	0	2.2	9.43
29G	6.22	813	353	19.52	148.91	140	23.54	19.81	181.44	3.46	0	0.3	23.6
30G	7.28	665	152	4.88	49.63	40	5.88	6.17	84.33	2.56	0	0.9	12.2
31G	7.54	498	916	29.28	63.82	116	22.70	14.46	98.37	3.21	0.1	0.1	18.32
32G	7.85	813	960	61	92.18	264	46.25	36.23	182.88	18.31	0	0.5	15.5
33G	7.38	1243	946	63.44	141.82	324	35.32	57.53	381.24	14.1	0	1.5	26.5
34G	7.48	3510	1910	56.12	765.87	740	83.26	129.83	627.84	35.53	0	0.9	124.5
35G	6.85	2880	623	70.76	595.67	610	77.37	101.69	281.58	30.76	0.1	2	38.8
36G	7.81	659	662	58.56	63.82	158	20.18	26.25	329.94	40.12	0	1.6	21.6
37G	7.2	1982	980	68.32	290.74	438	47.93	77.66	387.05	11.93	0	0.4	42.4
38G	7.26	2530	1420	53.68	567.31	478	68.9	74.61	348.25	28.38	0	1.2	21.86
39G	7.34	1467	908	68.32	127.64	396	62.23	58.70	351.63	27.78	0	1.1	19.65
40G	7.14	1396	649	75.64	170.19	282	51.30	37.55	429.84	13.91	0.2	1.4	23.5

Table 30.4 Statistical overview of the cations and anions

Parameters		Pre monsoon			Post monsoon		
		Min	Max	Mean	Min	Max	Mean
In-situ parameters	pH	6.55	7.71	7.17	6.85	7.85	6.79
	EC	145	3510	983.67	36	3510	911.35
	TDS	159	1989	879.15	28	1910	650.15
Cations	Na^+	25.43	455.45	203.65	8.24	429.84	152.04
	K^+	3.34	46.42	22.09	0.3	40.12	14.19
	Ca^{2+}	1.05	135.7	39.02	0.84	117.74	31.28
	Mg^{2+}	3.69	158.78	40.91	0.90	129.83	28.07
Anions	Cl^-	54.28	788.81	180.45	14.18	765.87	133.67
	SO_4^{2-}	6.81	132.25	27.97	5.4	124.5	19.88
	NO_3^-	0.2	4.59	2.21	0	2.8	1.06
	PO_4^{3-}	0	0.3	0.17	0	0.3	0.04

Parameters		Pre monsoon			Post monsoon		
		Min	Max	Mean	Min	Max	Mean
In-situ parameters	pH	6.55	7.71	7.17	6.85	7.85	6.79
	EC	145	3510	983.67	36	3510	911.35
	TDS	159	1989	879.15	28	1910	650.15
Cations	Na^+	25.43	455.45	203.65	8.24	429.84	152.04
	K^+	3.34	46.42	22.09	0.3	40.12	14.19
	Ca^{2+}	1.05	135.7	39.02	0.84	117.74	31.28
	Mg^{2+}	3.69	158.78	40.91	0.90	129.83	28.07
Anions	Cl^-	54.28	788.81	1480.45	14.18	765.87	133.67
	SO_4^{2-}	6.81	132.25	27.97	5.4	124.5	19.88
	NO_3^-	0.2	4.59	2.21	0	2.8	1.06
	PO_4^{3-}	0	0.3	0.17	0	0.3	0.04

with Cl^- (0.90956), TH (0.92921), and Mg^{2+} (0.96437). Cl has a strong correlation with Mg^{2+} (0.85371), SO_4^{2-} (0.77924), and TH (0.75255). Ca^{2+} is highly correlated with Mg^{2+} (0.80874) and K^+ (0.79532). Parameters with correlation coefficient pH (−0.0733), EC (−0.1341), TDS (−0.0017), Cl^- (−0.053), TH (−0.1803), Ca^{2+} (−0.1519), Mg^{2+} (−0.1848), Na^+ (−0.0943), K^+ (−0.0968) indicates weak negative correlation with PO_4^{3-}. WQI has a strong significant correlation with K^+ (0.94981), TH (0.90566), Mg^{2+} (0.87562), and EC (0.84708). The correlation matrix for post-monsoon is shown in Table 30.6.

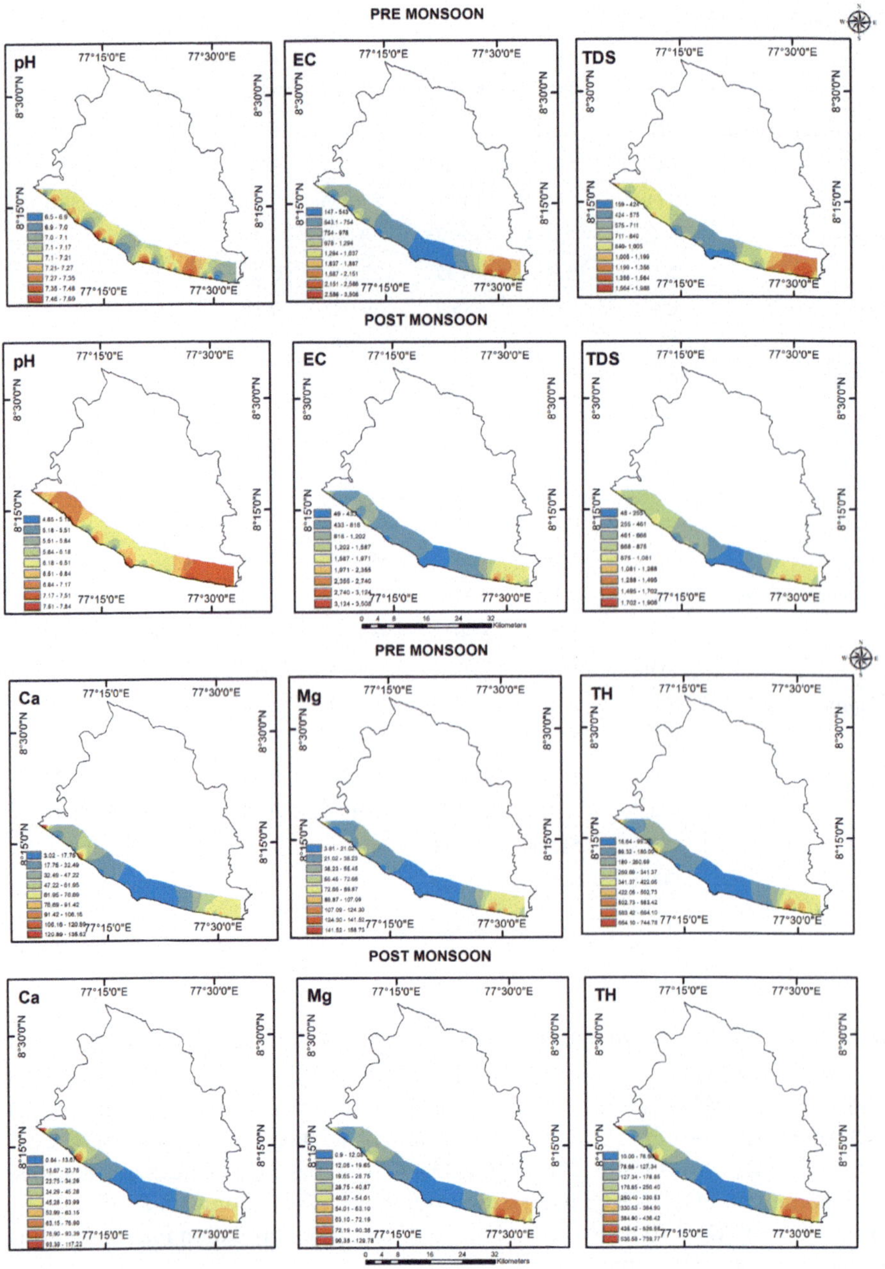

Fig. 30.2 Spatial distribution map of various cations and anions of the study area

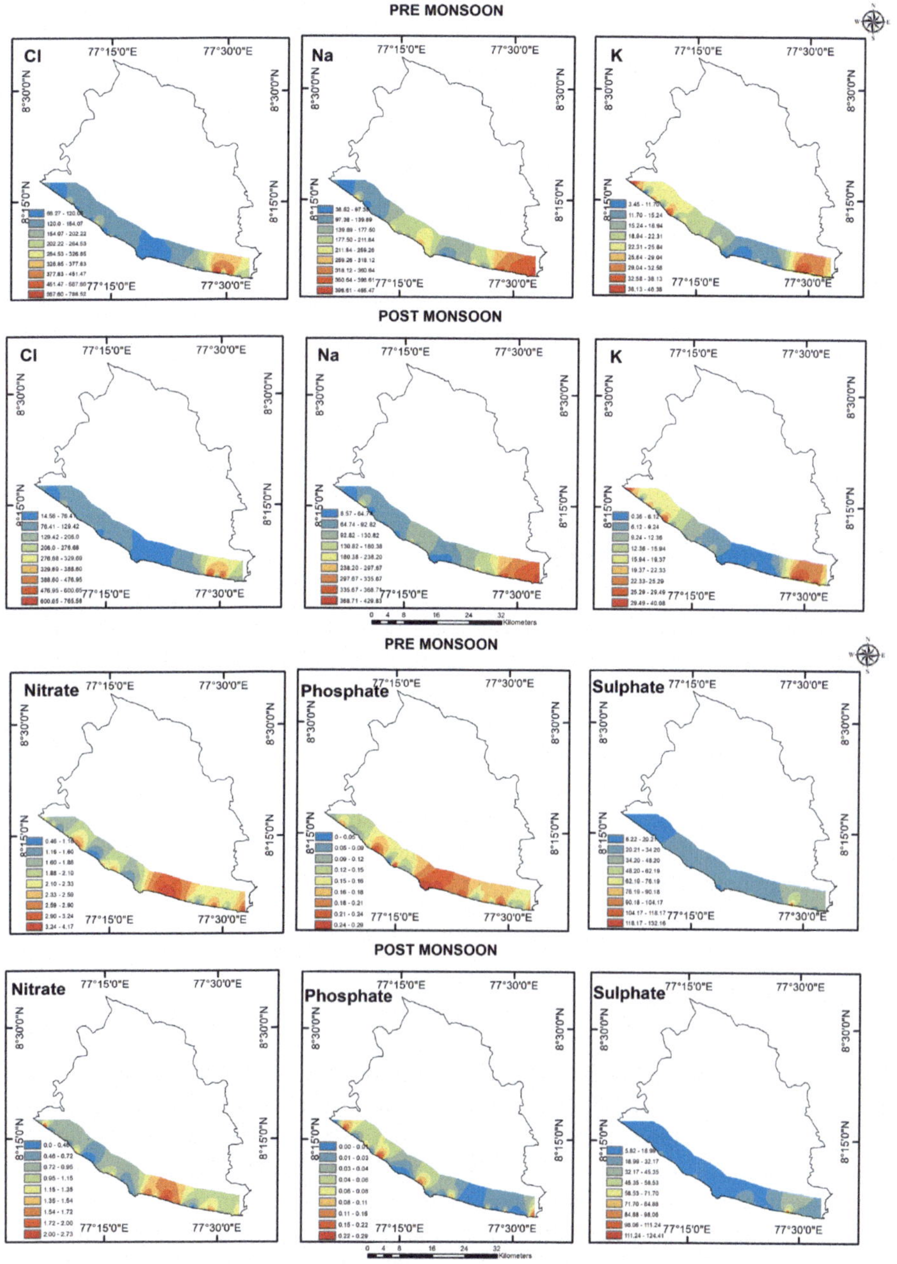

Fig. 30.2 (continued)

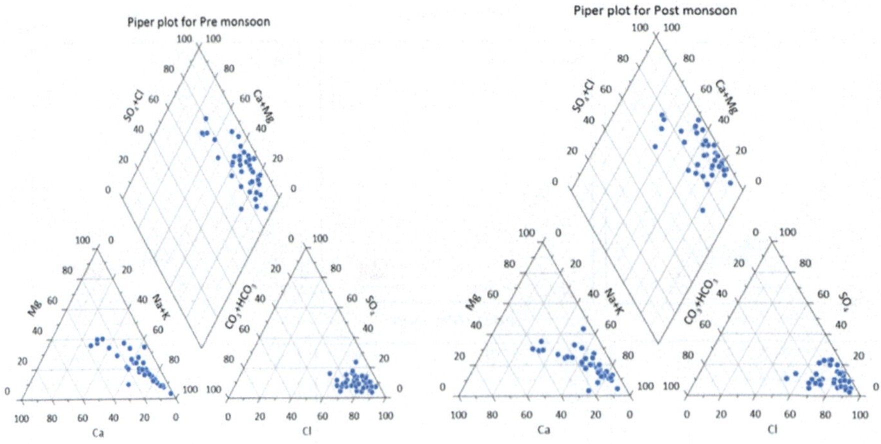

Fig. 30.3 Piper diagram showing groundwater samples during both the seasons

30.3.3 *Water Quality Index (WQI)*

The samples are classified into appropriate groups according to their WQI values. The classification of WQI in Kanyakumari district is shown in Table 30.7. During Pre-monsoon season, around 15% of samples belong to Good quality, 25% to Poor water quality, 22.5% to Very poor water quality, and 37.5% of samples are unfit for drinking. During the post-monsoon season, 20% of samples were found to be of the Good type, 27.5% belonged to the Poor type, 30% belonged to the very poor type, and about 22.5% were unsuitable for drinking. Not a single sample in either season fits the Excellent category. The percentage of samples unfit for consumption is elevated in both seasons.

30.4 Conclusions

The quality of groundwater along the Kanyakumari coastal zone was analyzed using the integration of statistical, WQI, and GIS techniques. The analysis of physio-chemical parameters indicates that seawater intrusion was found to be a significant problem in specific locations of the coastline. EC and Cl^- are higher in regions with saline water intrusions. Piper plots show that most samples fall in the Na^+-Cl^--SO_4^{2-} zone, suggesting the presence of SWI from the Arabian Sea. The Gibbs plot indicates that the dominant process determining the composition of the analyzed samples is evaporation and rock-water interactions. EC strongly correlates with $Cl^-$and Mg^{2+}, while Cl^- strongly correlates with Mg^{2+}, SO_4^{2-} and TH. The WQI showed that approximately 37.5% and 22.5% of the samples were unfit for drinking throughout the pre-season and postseason periods. Seawater ingress into the coastal aquifers

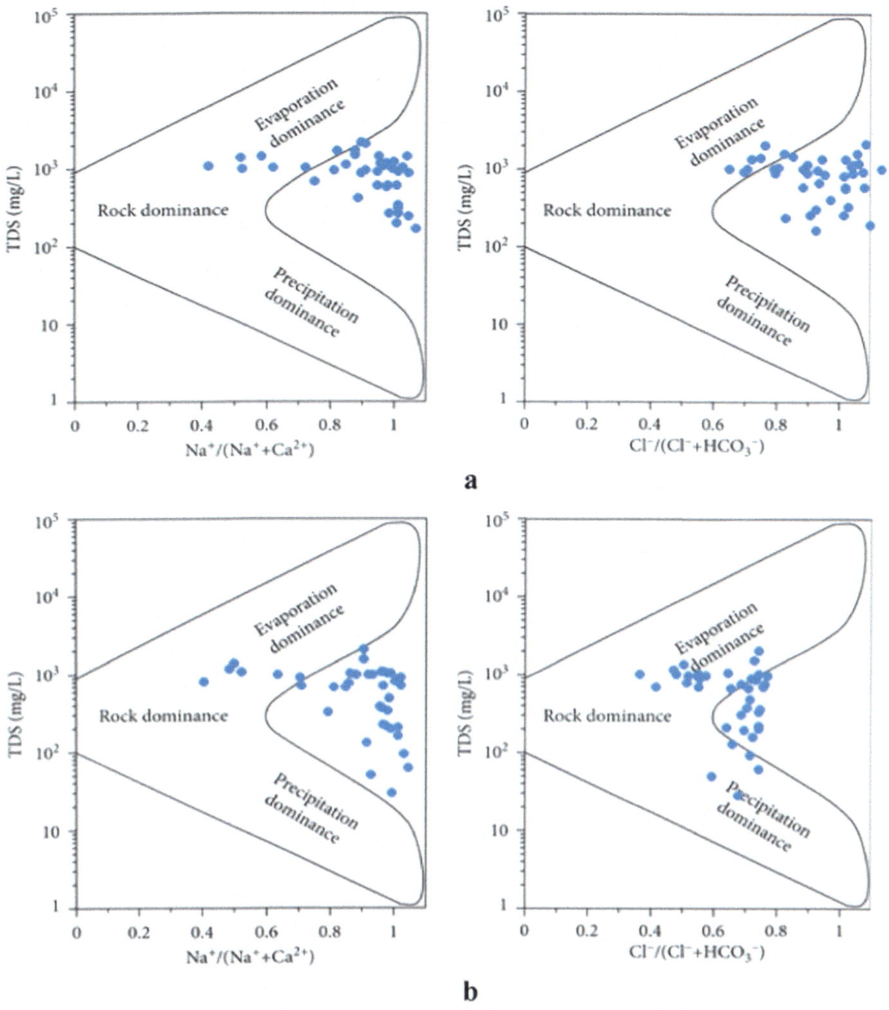

Fig. 30.4 **a** Gibbs diagram for pre monsoon, **b** Gibbs diagram for post monsoon

significantly affects the groundwater quality of Kanyakumari coastal zone in both seasons. Due to its poor quality, people in the study area do not use this groundwater for drinking or household purposes. Most houses rely on public water supply pipelines or filters for daily needs. To effectively solve the problem of seawater intrusion, continuously monitoring variations in water level and quality is necessary. Artificial groundwater recharge and rainwater harvesting methods can be implemented in the affected regions to reduce the severity of the saltwater problems along the coastline. This study enables us to locate the areas of saline water intrusion and reduce and control the impacts of SWI in coastal aquifers.

Table 30.5 Correlation matrix for physicochemical parameters in the course of pre-monsoon

	PH	EC	TDS	Cl^-	TH	Ca^{2+}	Mg^{2+}	Na^+	K^+	PO_4^{3-}	NO_3-	SO_4^{2-}	WQI
PH	1												
EC	−0.3374	1											
TDS	−0.0067	0.64272	1										
Cl -	−0.3569	0.89	0.48683	1									
TH	−0.1913	0.92467	0.6811	0.73591	1								
Ca^{2+}	−0.1497	0.76899	0.61293	0.47744	0.91415	1							
Mg^{2+}	−0.196	0.9378	0.68051	0.77607	0.99599	0.87421	1						
Na +	−0.302	0.59095	0.36934	0.58364	0.4871	0.27577	0.52256	1					
K +	−0.135	0.68938	0.58468	0.45544	0.75368	0.80431	0.7252	0.40152	1				
PO_4^{3-}	−0.1923	−0.0414	−0.0931	0.0779	−0.0517	−0.1002	−0.0398	−q0.0687	−0.1155	1			
NO_3-	0.18509	0.0402	−0.0596	0.05992	0.05895	0.0019	0.07019	0.09274	0.0194	-0.0005	1		
SO_4^{2}-	−0.1357	0.63943	0.46001	0.74062	0.55572	0.29098	0.6014	0.52104	0.32358	0.07101	0.17563	1	
WQI	−0.1455	0.83947	0.67654	0.61783	0.90764	0.89138	0.89039	0.48711	0.9583	-0.1005	0.04613	0.46423	1

Table 30.6 Correlation matrix for physicochemical parameters in the course of post-monsoon

	PH	EC	TDS	Cl-	TH	Ca^{2+}	Mg^{2+}	Na +	K^+	PO_4^{3-}	NO_3-	SO_4^{2-}	WQI
PH	1												
EC	0.26199	1											
TDS	0.34186	0.71974	1										
Cl^-	0.15697	0.90956	0.60627	1									
TH	0.35052	0.92921	0.75336	0.75255	1								
Ca^{2+}	0.33212	0.76424	0.68988	0.5064	0.92709	1							
Mg^{2+}	0.3355	0.96437	0.73648	0.85371	0.97025	0.80874	1						
Na^+	0.26113	0.7167	0.53601	0.64512	0.60936	0.39523	0.70085	1					
K^+	0.3132	0.70112	0.70134	0.50483	0.78418	0.79532	0.71672	0.45225	1				
PO_4^{3-}	−0.0733	−0.1341	−0.0017	−0.053	−0.1803	−0.1519	−0.1848	−0.0943	−0.0968	1			
NO_3-	−0.1126	0.02984	−0.2141	0.01915	0.05172	0.01705	0.0701	0.08629	0.07167	0.3632	1		
SO_4^{2-}	0.11913	0.73062	0.52434	0.77924	0.629	0.38518	0.73815	0.55716	0.38462	0.1586	0.04614	1	
WQI	0.41514	0.84708	0.77474	0.67041	0.90566	0.84442	0.87562	0.6064	0.94981	0.1254	0.08515	0.5425	1

Table 30.7 Classification of groundwater based on WQI in Kanyakumari district

Range	Type of water	No. of samples and %			
		Pre-monsoon	%	Post monsoon	%
0 -25	Excellent	0	0	0	0
26–50	Good	6	15	8	20
51–75	Poor	10	25	11	27.5
76–100	Very Poor	9	22.5	12	30
> 100	Unsuitable for drinking	15	37.5	9	22.5

Acknowledgements The authors acknowledge the DST Inspire Programme, Government of India, for funding this research and thank Dr. Dilip Kumar P.G., Head of the Geology Department, Government College Kottayam, for his guidance and support for sample analysis.

References

Awachat, A. R., & Salkar, V. D. (2017). Ground water quality assessment through WQIs. *International Journal of Engineering Research and Technology, 10*(1), 318–322.

BIS (2012). Indian standard drinking water–specification IS 10500: 2012. New Delhi: Bureau of Indian Standards

Chidambaram, S., Sarathidasan, J., Srinivasamoorthy, K., Thivya, C., Thilagavathi, R., Prasanna, M. V., Singaraja, C., & Nepolian, M. (2018). Assessment of hydrogeochemical status of groundwater in a coastal region of Southeast coast of India. *Applied Water Science, 8*, 1–14.

Etikala, B., Adimalla, N., Madhav, S., Somagouni, S.G., & Keshava Kiran Kumar, P. L. (2021). Salinity problems in groundwater and management strategies in arid and semi-arid regions. Groundwater geochemistry: pollution and remediation methods, pp. 42–56.

George, A. K., Gandhi, S., Perumal, M., Roy, P. D., Jothimani, M., & Sekar, S. (2023). Possible zones of submarine groundwater discharge (SGD) and seawater intrusion (SWI) along the west coast of Kanyakumari, India. *Journal of Chemistry*, 2023.

Gibbs, R. J. (1970). Mechanisms controlling world water chemistry. *Science, 170*(3962), 1088–1090.

Hajji, S., Allouche, N., Bouri, S., Aljuaid, A. M., & Hachicha, W. (2021). Assessment of seawater intrusion in coastal aquifers using multivariate statistical analyses and hydrochemical facies evolution-based model. *International Journal of Environmental Research and Public Health, 19*(1), 155.

Kaliraj, S., Chandrasekar, N., Peter, T. S., Selvakumar, S., & Magesh, N. S. (2015). Mapping of coastal aquifer vulnerable zone in the south west coast of Kanyakumari, South India, using GIS-based DRASTIC model. *Environmental Monitoring and Assessment, 187*, 1–27.

Karmakar, B., & Singh, M. K. (2021). Assessment of water quality status of water bodies using water quality index and correlation analysis in and around industrial areas of West District, Tripura. *India. Nature Environment and Pollution Technology, 20*(2), 551–559.

Kothari, V., Vij, S., Sharma, S., & Gupta, N. (2021). Correlation of various water quality parameters and water quality index of districts of Uttarakhand. *Environmental and Sustainability Indicators, 9*, 100093.

Perumal, S. B., & Subramanian, V. B. (2022). Study Of geological parameters along the coastal area of Arabian Sea, Kanyakumari to Colachel, South Tamilnadu. *India. Elementary Education Online, 20*(1), 4888–4888.

Seenipandi, K., Nainarpandian, C., Kandathil, R. K., & Sellamuthu, S. (2019). Seawater intrusion vulnerability in the coastal aquifers of southern India—An appraisal of the GALDIT model, parameters' sensitivity, and hydrochemical indicators. *Environmental Science and Pollution Research, 26*, 9755–9784.

Shin, K., Koh, D. C., Jung, H., & Lee, J. (2020). The hydrogeochemical characteristics of groundwater subjected to seawater intrusion in the Archipelago. *Korea. Water, 12*(6), 1542.

Sundar, S., Roy, P. D., & Chokkalingam, L. (2022). Spatial distribution and enrichment of metals in surface sediments from different coastal landforms at southernmost Indian subcontinent. *Journal of Coastal Conservation, 26*(4), 35.

Usali, N., & Ismail, M. H. (2010). Use of remote sensing and GIS in monitoring water quality. *Journal of Sustainable Development, 3*(3), 228.

Yang, M., Zheng, Y., Xu, X., Liu, H., & Xin, P. (2022). Groundwater table fluctuations in a coastal unconfined aquifer with depth-varying hydraulic properties. *Journal of Hydrology, 606*, 127407.

Chapter 31
Assessment of Flood Impact on Groundwater Quality in the Coastal Areas of Thiruvallur District, Tamil Nadu, India

R. Nishan Raja, P. Nila Rekha, Albin Sunny, Soumyabrata Sarkar, and S. Aravind Kumar

Abstract Groundwater is a vital resource in coastal areas, which are often characterized by high population densities and vulnerability to groundwater quality degradation from various factors. These factors include seawater intrusion, which occurs when saltwater mixes with freshwater due to over-extraction or changes in sea level. Land-use changes, such as urbanization and agriculture, can also degrade groundwater quality by increasing pollution and reducing recharge. Additionally, natural disasters like cyclones and floods can have both positive and negative impacts on groundwater, depending on the severity and duration of the event. This study investigated the impact of Cyclone NIVAR (2020) on groundwater quality in the coastal areas of Thiruvallur district, Tamil Nadu, India. Water samples were collected monthly from 35 locations between January 2020 and January 2021 and analysed for Physico-chemical parameters, cations and anions. Data from the State Ground and Surface Water Resources Data Centre (PWD, Tharamani) were incorporated to compare with historical trends. ArcGIS was used to map the spatial distribution of groundwater quality for pre and post-cyclone scenarios. Results showed that the TDS range decreased from 390–3471 mg/L pre-cyclone to 208–2500 mg/L post-cyclone. Groundwater levels (GWL) also exhibited a substantial increase, rising from a pre-cyclone range of 1.66 m to 10.2 m to −0.39 m to 3.81 m after the cyclone. This elevation suggests significant aquifer recharge from cyclonic rainfall. The study concludes that Cyclone NIVAR led to a substantial increase in groundwater levels, indicating that controlled aquifer recharge could be a viable strategy for sustainable groundwater management in similar coastal regions.

Keywords Coastal area · Flood · Groundwater · India · Sentinel 1 · Thiruvallur district

R. N. Raja (✉) · P. N. Rekha · A. Sunny · S. Sarkar · S. A. Kumar
ICAR–Central Institute of Brackishwater Aquaculture, Chennai, Tamil Nadu, India
e-mail: nishnwilliams@gmail.com

V. Sunitha et al. (eds.), *Groundwater Resource Management Planning Strategies*, Springer Water, https://doi.org/10.1007/978-3-031-88870-0_31

31.1 Introduction

Coastal zones, characterized by their rich biodiversity, abundant resources and favourable climates, support a substantial proportion of the global population (MacManus et al., 2021; Peter et al., 2024). The confluence of these factors drives high population densities and intensifies the demand for freshwater resources to support various activities, including agriculture, industry and tourism. Groundwater resources are particularly vital in these regions, especially in developing nations where surface water availability is often limited or polluted (Chidambaram et al., 2018; Rajendran & Peter, 2020). However, the proximity of coastal aquifers to the sea renders them inherently susceptible to seawater intrusion. This phenomenon, characterized by the infiltration of saline water into freshwater zones, which would lead to widespread soil and water salinization, thereby jeopardizing water quality and ecosystem health. The consequences of salinization are far-reaching, with significant implications for agriculture, drinking water supplies and the overall ecological integrity of coastal environments. Seawater intrusion and the ensuing salinization are complex processes governed by a multitude of interacting factors, including hydrogeology, climate, and human activities (Rekha 2013; Rekha 2015).

Floods are a common and impactful natural disaster, triggered by a combination of factors such as heavy rainfall, storm surges, tsunamis and riverine overflow. While frequently associated with detrimental impacts, floods can also confer benefits upon ecosystems, particularly in coastal regions and floodplains. Notably, floods replenish groundwater resources with freshwater (Hashemi et al., 2015) and augment soil fertility through the deposition of nutrient-rich sediments and organic matter (Kvočka et al., 2016; White, 2019).

Advancements in remote sensing technology have significantly enhanced the capacity to monitor surface dynamics, a critical component of effective water resource management. The extensive spatial coverage and high temporal resolution afforded by remote sensing facilitate the accurate capture of spatial and temporal variations in coastal environments, enabling comprehensive remote monitoring and assessment (Duan et al., 2021; Raja et al., 2022). Furthermore, the integration of remotely sensed data with complementary data sources, including field observations, in situ measurements and numerical models, bolsters the accuracy, reliability and applicability of coastal monitoring and assessment initiatives (Reddy et al., 2021). Given the indispensable role of groundwater in sustaining both human and ecological communities, particularly within vulnerable coastal regions, a comprehensive understanding of the impacts of flood events on groundwater quality is paramount.

Cyclone NIVAR, a very severe cyclonic storm that made landfall in Tamil Nadu on November 26, 2020, had a significant impact on coastal groundwater quality in India. Originating from a disturbance in the Intertropical Convergence Zone, NIVAR was the fourth named storm of the 2020 North Indian Ocean cyclone season. While the cyclone initially reached wind speeds exceeding 120 km/h (75 mph), it weakened to a severe cyclonic storm before making landfall over north coastal Tamil Nadu between Puducherry and Chennai near Marakkanam (Vilvijayan & Lalitha, 2021).

Hence, this study aims to assess the significance of flooding on the groundwater recharge and its water quality improvement in the Thiruvallur district.

31.2 Material and Methods

31.2.1 Study Area

The coastal sub-watershed of Thiruvallur district is taken up as the study area. It lies between 13°9′41" N to 13°41′54" N latitude and 79°52′59" E to 80°20′48" E longitude, encompassing approximately 1,55,254 hectares. The watershed boundary was delineated using Shuttle Radar Topography Mission (SRTM) data, supplemented with data from the Tamil Nadu Agriculture Department. The study area comprises 19 sub-watersheds (4C2E1f, 4C2E1k, 4C2E1j, 4C2E1h, 4C2E1d, 4C2E1c, 4C2E1b, 4C2E2, 4C2E1a, 4C2D5t, 4C2D5b, 4C2D5d, 4C2D5f, 4C2D8a, 4C2D8b, 4C2D8c, 4C2D8d, 4C2D8f, and 4C2D8h) as shown in Fig. 31.1. Thiruvallur district has a diverse economy with industries ranging from major companies like Manali Petrochemicals, MRF, Britannia Industries, Ashok Leyland and TI Cycles to a thriving agricultural sector. The study areas agricultural landscape is characterized by a diverse array of crops, encompassing cereals (rice, millets), legumes (gram varieties), cash crops (sugarcane, groundnut), and various horticultural products. The methodology employed in this study is detailed in Fig. 31.2.

31.2.2 Data Used

The study examined the groundwater quality and level in the coastal sub-watershed of Thiruvallur district using various data sources. Table 31.1 shows the summary of the data used in the study.

31.2.3 Sample Collection and Groundwater Quality Assessment

Groundwater samples were collected at monthly intervals from January 2020 to January 2021 from 35 representative sampling locations as shown in Fig. 31.1. Samples were collected in 250 mL Tarson HDPE Amber Wide Mouth bottles and preserved in the refrigerator. TDS and pH were measured using a Hydrolab Quanta portable multi-parameter water quality analyser. Calcium, sodium, and potassium

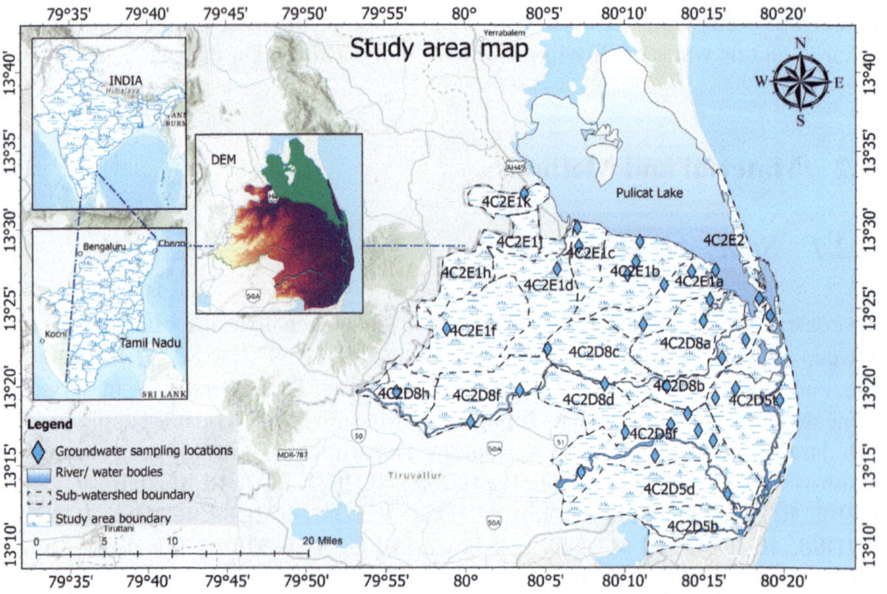

Fig. 31.1 Study area map with sampling locations

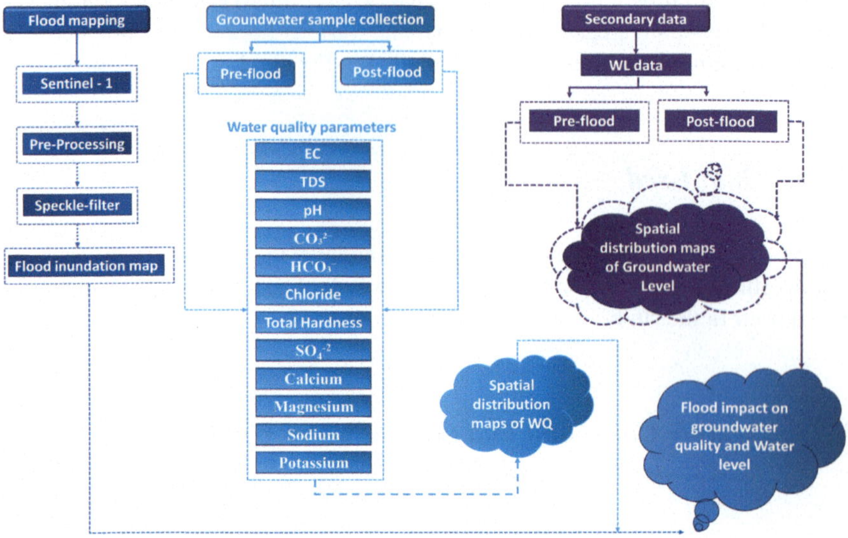

Fig. 31.2 Methodology of the study

Table 31.1 Data sources

S. No	Content	Description
1	Groundwater quality	Collected through field sampling
2	Groundwater level	State Ground and Surface Water Resources Data Centre, PWD, Tharamani
3	DEM	Google Earth Engine
4	Watershed boundary	Agriculture Department, Tamil Nadu
5	Sentinel 1B (SAR)	European Space Agency (Synthetic Aperture Radar (SAR) VV and VH polarization data)

were analysed using the handheld HORIBA LAQUAtwin meters. Electrical Conductivity (EC), magnesium (Mg), sulfate (SO_4), carbonate (CO_3), bicarbonate (HCO_3), and chloride (Cl) were assessed as per standard procedure (APHA, 2017).

31.2.4 *Spatial Analysis*

A comprehensive geospatial analysis was conducted to assess groundwater quality. Following laboratory analysis of collected samples, a robust geospatial database was constructed within ArcGIS 10.2. This database integrated key attributes, including well identification codes, sampling dates, precise geographic coordinates (UTM Zone 44N), and the complete set of analytical results. A rigorous quality assurance process, involving statistical outlier analysis and visual inspection of data distributions, ensured data integrity prior to spatial analysis. Well locations were represented as a point shapefile, with associated analytical data linked to each point. Given the relatively even spatial distribution of sampling locations, the Inverse Distance Weighted (IDW) interpolation method was used to generate raster map for each parameter.

31.2.5 *Flood Mapping*

This study leveraged Sentinel-1 Synthetic Aperture Radar (SAR) data for precise and comprehensive flood inundation mapping. The inherent capacity of SAR to penetrate cloud cover and acquire imagery irrespective of weather conditions or diurnal limitations ensured data acquisition continuity and minimized temporal gaps in observation. Ground Range Detected (GRD) products, renowned for their high spatial resolution and suitability for flood delineation, were acquired from the European Space Agency's (ESA) Copernicus Open Access Hub. Image processing and analysis were conducted within the framework of the Sentinel Applications Platform

(SNAP) software, an open-source tool developed by ESA for the robust processing of Sentinel data. The following pre-processing steps were meticulously applied to the Sentinel-1B GRD data, precise orbit file application to ensure accurate geolocation, radiometric calibration to sigma-nought (σ0) backscatter coefficients for quantitative analysis, speckle noise reduction using a Refined Lee filter to enhance image quality and terrain correction with a high-resolution Digital Elevation Model (DEM) to remove geometric distortions. Flood extent delineation was achieved through a robust thresholding technique. This involved determining a critical σ0 backscatter value to differentiate inundated and non-inundated regions. Optimal threshold determination was guided by a comprehensive analysis of image histograms, enabling the identification of distinct backscatter signatures associated with water and non-water surfaces. A detailed flowchart illustrating the complete study is given in Fig. 31.2.

31.3 Results

31.3.1 Flood Inundation and Groundwater Quality

Analysis revealed that the northern region of the study area exhibited a higher degree of inundation compared to other regions. The darkened areas (black spatches) within the Fig. 31.3 represents zones of water accumulation. This phenomenon was primarily attributed to the overflowing of the Arani and Kosathalayar rivers, both ephemeral in nature. These rivers discharged into Pulicat Lake and the Buckingham Canal, respectively, which were aligned parallel to the coastline. Ultimately, the water drains into sea via the Pulicat Lake bar mouth and the Ennore estuary bar mouth.

The soil texture in the study area is classified into clay, clay loam, loamy sand, sand, sandy clay, sandy clay loam, and sandy loam, as shown in Fig. 31.4. Sandy soils, due to their higher permeability, are likely to experience increased groundwater recharge during floods. The stream map indicates the presence of streams, which play a crucial role in recharging groundwater. Water overflowing from streams infiltrates adjacent floodplains, replenishing the aquifer. The geology map depicts various compositions such as clay, silt, Nilwala (presumably a specific geological formation), sandstone and water bodies. Sandstone areas allow better infiltration and contribute to groundwater recharge. The geomorphology map identifies landforms like alluvial plains, coastal plains, floodplains and pediplains. Floodplains are particularly important for groundwater recharge during floods. The DEM (Digital Elevation Model) map represents elevation levels, with lower areas closer to water bodies experiencing more direct infiltration (Kom et al., 2022; Rekha et al., 2015; Sunitha & Reddy, 2022). The soil, stream, geology, geomorphology, and DEM maps for the study area are given in Fig. 31.4.

The spatial distribution of the water quality analysis showed that there are significant changes in the water quality parameters due to cyclone Nivar. The electrical conductivity (EC) ranges from 615.96–6418.57 μS/cm in pre-Nivar. After Cyclone

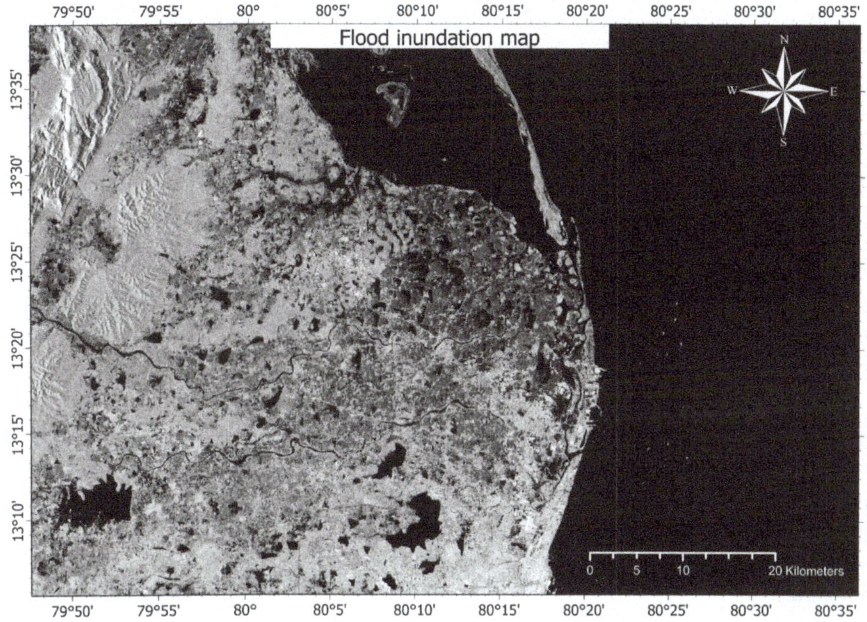

Fig. 31.3 Flood map derived from Sentinel-1B GRD data (2020–12-04)

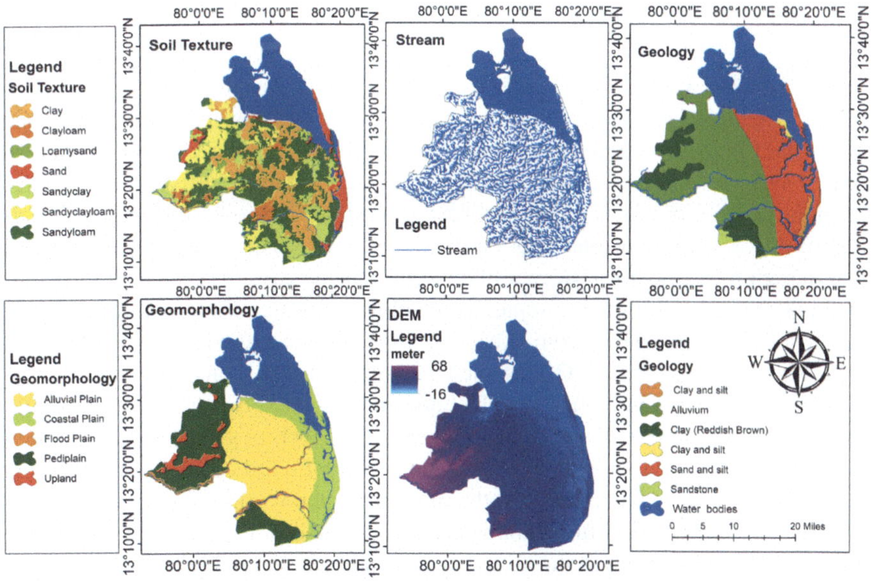

Fig. 31.4 Thematic maps

Nivar, the EC values decreased to a range of 342.28 to 4814 μS/cm, indicating a dilution effect due to the rainfall/ flood on the groundwater salinity. In groundwater generally, high TDS indicate high salinity and low water quality, which can affect the taste, odour, and appearance of water, as well as its suitability for drinking and irrigation purposes. The TDS of the groundwater samples in the Thiruvallur district were between 390 and 3471 mg/L before Cyclone Nivar. After Cyclone Nivar, the TDS values decreased to a range of 208 to 2500 mg/L. The spatial distribution map for groundwater quality pre and post-Nivar is given in Figs. 31.5 and 31.6.

The water quality distribution before and after Nivar shows that the water quality has improved in the northwest, southwest, and northeast regions. This may be due to the presence of water bodies in the study area. pH ranges from 6.89 to 8.06 in pre-Nivar and has been reduced to 6.5–7.6 in post-Nivar. Chloride concentration in the groundwater ranges from 19 to 1764 mg/L in pre-Nivar and is reduced to 7–1455.83 in post-Nivar. Similarly significant reduction is observed in overall groundwater parameters indicating a better recharge and improvement in groundwater quality as shown in Fig. 31.6. After Nivar the EC, TDS and other parameters have decreased significantly. The groundwater quality in the study area for pre-Nivar and post-Nivar along with WHO (World Health Organisation 2017) standard 2017 is given in Table 31.2.

31.4 Discussion

The groundwater level (GWL) during the pre-Nivar 2020 ranged from 1.66 m to 10.2 m, with higher values in the northern and western regions of the study area, and lower values in the southern and eastern regions. The post-Nivar 2020 GWL ranged from -0.39 m to 3.81 m, with lower values in the northern and western regions, and higher values in the southern and eastern regions as shown in Fig. 31.7. This suggests that the groundwater was closer to the land surface in the northern and western regions, while it was deeper in the southern and eastern regions. Rainstorms and floods have the potential to replenish local groundwater, thereby mitigating the land subsidence that results from excessive groundwater extraction (Abdalla et al., 2014). Krishnakumar et al., 2022 assessed the impact of floods on the groundwater in the Periyar river basin and found that the flood water positively impacts groundwater quality in the Periyar river basin. A comprehensive analysis of global water resources by Scanlon et al., 2023 underscored the significance of flood events in replenishing groundwater supplies, particularly in the majority of the world's dry regions.

The spatial distribution of groundwater level indicates a significant shift in the groundwater level (GWL) in the study area, as a result of the Nivar 2020 cyclone. The increase in GWL (Post-Nivar) in the southern and eastern regions suggests that these areas experienced a substantial amount of recharge, possibly due to the heavy rainfall and flooding caused by the cyclone. Groundwater is the second largest store of freshwater in the world. Aquifers and their water content are more than just an additional supply source (Geris et al., 2022). The water, along with the geological

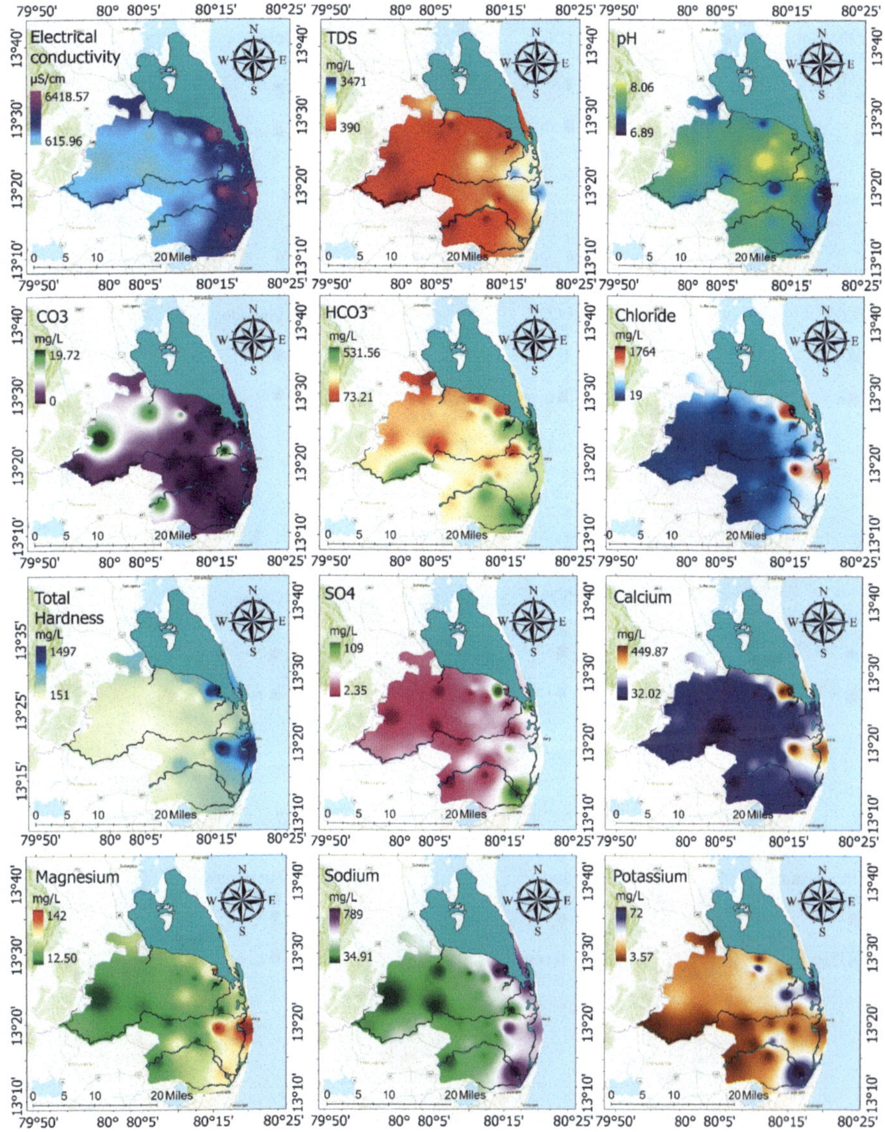

Fig. 31.5 Pre-Nivar groundwater quality map

material it resides in, possesses distinct characteristics that will become increasingly crucial in periods of escalating shortage. In instances where there is a substantial recharge of aquifers, groundwater can be viewed as a renewable resource (Vaux., 2011; Singha, & Navarre-Sitchler., 2022).

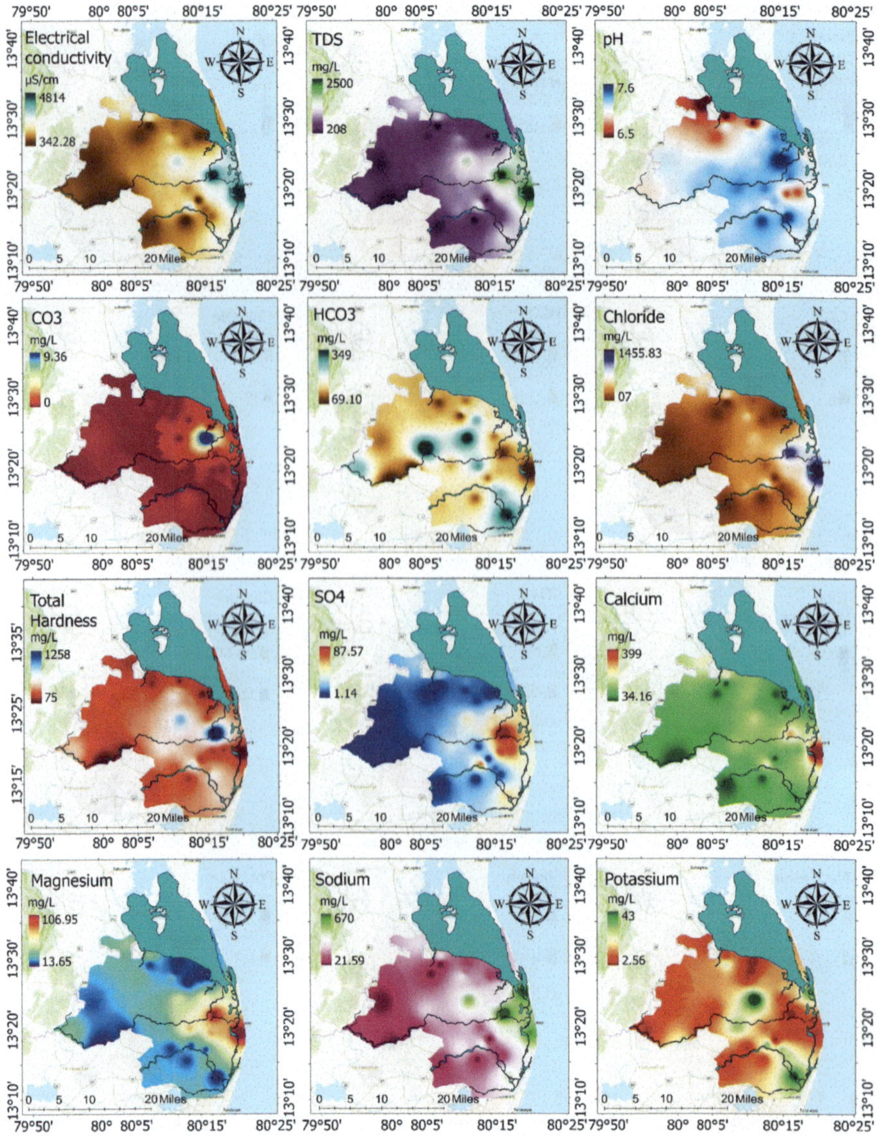

Fig. 31.6 Post Nivar groundwater quality map

This shift in GWL underscores the impact of significant climatic events, such as the Nivar 2020 cyclone, on groundwater resources. Despite their destructive tendencies, floods can also contribute positively by supplying plentiful water resources. This can lead to a substantial replenishment of local groundwater in a short period, given the right circumstances (Lan et al., 2024). This highlights the need for comprehensive monitoring and management strategies to safeguard these vital resources, particularly

Table 31.2 Study area groundwater quality and WHO standard

Parameter	Unit	Pre-Nivar		Post-Nivar		WHO Standard 2017
		Minimum	Maximum	Minimum	Maximum	
Electrical conductivity	μS/cm	615.96	6418.57	343	4814	170–2700
TDS	mg/l	390	3471	208	2500	200–2500
pH	–	6.89	8.06	6.5	7.60	5–7
CO_3^{2-}	mg/l	0	19.72	0	9.36	–
HCO_3^-	mg/l	73.21	531.56	69.10	349	–
Chloride	mg/l	19	1764	7	1455.83	20–1200
Total hardness	mg/l	151	1497	75	1258	100–1000
SO_4^{-2}	mg/l	2.35	109	1.14	87.57	–
Calcium	mg/l	32.02	449.87	34.16	399	30–500
Magnesium	mg/l	12.50	142	13.65	106.95	10–1000
Sodium	μg/l	34.91	789	21.59	670	100–400 mg/l
Potassium	μg/l	3.57	72	2.56	43	1.5–50 mg/l

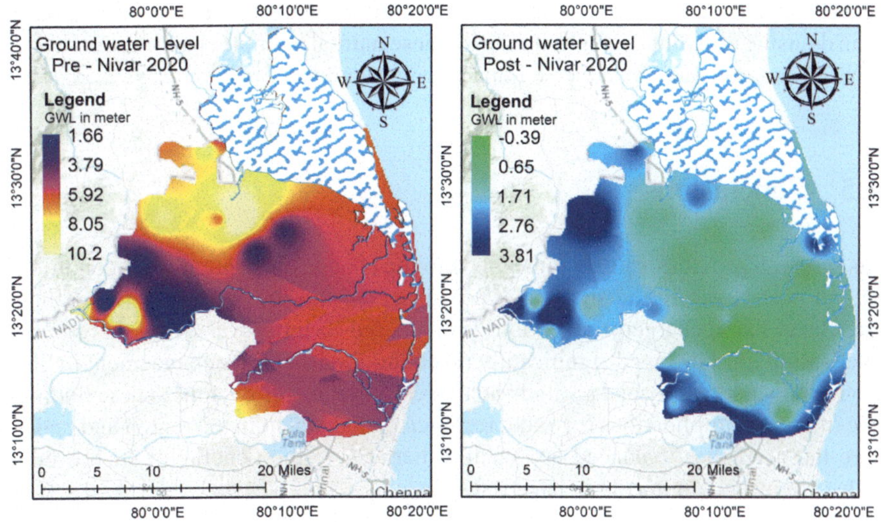

Fig. 31.7 Spatial distribution of groundwater level in pre and post-Nivar 2020

in coastal areas that are prone to such events. The principles of groundwater management, particularly those related to economics, are crucial in understanding the potential of flood or cyclone water for groundwater recharge. As water scarcity becomes a pressing issue, exacerbated by climate change, the importance of these principles is heightened. Floods and cyclones, which primarily affect surface water, can play a significant role in recharging groundwater due to the interaction between surface and groundwater (Geris et al., 2022; Schwartz and Zhang., 2024). The concept of 'safe yield', which refers to the sustainable extraction limit of a groundwater source, can be used to determine the extent to which groundwater can be utilized following its recharge by a flood or cyclone (Vaux., 2011; Donoso et al., 2020). This study provides a baseline on how flood/cyclone helps in recharging groundwater. Additional research could shed light on the specific processes that drive these changes. For instance, one of such process is ground percolation, when a region experiences a flood or cyclone, the volume of water present increases substantially. This surplus water has the potential to percolate into the ground, thereby augmenting the groundwater reserves. Factors such as the type of soil, land usage, and the severity and duration of the rainfall event can influence the percolation rate. Another process is runoff accumulation, the occurrence of floods and cyclones often results in a significant amount of surface water runoff. This runoff can find its way into rivers, lakes, and other water bodies, from where it can seep into the ground, adding to the groundwater recharge.

Despite their potential for destruction, floods and cyclones play a pivotal role in the natural replenishment of groundwater. They contribute a substantial volume of water that can either percolate directly into the ground or add to surface water runoff, both of which can augment groundwater reserves. Gaining insights into these processes is crucial for effective water resource management, particularly in the context of escalating water scarcity and climate change. Further investigations in this field can aid in devising efficient strategies to utilize these natural phenomena for groundwater recharge and mitigate any adverse impacts on groundwater resources.

31.5 Conclusions

This research provides valuable insights into the changes in groundwater conditions pre and post-Nivar, in the study area it is observed that in post-Nivar the groundwater level has recharged tremendously compared to pre-Nivar. And, the groundwater salinity has dropped significantly revealing a substantial enhancement in both the groundwater quality and groundwater level following the cyclone. These findings have important implications for water resource management in the region and underscore the need for a balanced approach to harnessing the benefits of such events, while also mitigating the associated risks.

Acknowledgements We are grateful to the Director, ICAR-CIBA for the support and facility to carry out this work.

References

Abdalla, F., El Shamy, I., Bamousa, A. O., Mansour, A., Mohamed, A., & Tahoon, M. (2014). Flash floods and groundwater recharge potentials in arid land alluvial basins, southern Red Sea coast, Egypt. *International Journal of Geosciences*, 2014. https://doi.org/10.4236/ijg.2014.59083

American public health association (APHA). (2017). *Standard methods for the examination of water and wastewater* (23rd ed.). American Public Health Association.

Chidambaram,S., Sarathidasan, J., Srinivasamoorthy, K., Thivya, C., Thilagavathi, R., Prasanna,M. V., Singaraja, C., Nepolian, M. (2018) Assessment of hydrogeochemical status of groundwater in a coastal region of Southeast coast of India. *Applied Water Science, 8*(1) https://doi.org/10.1007/s13201-018-0649-2

Donoso, G., Lictevout, E. and Rinaudo, J. D. (2020). Groundwater management lessons from Chile. Sustainable groundwater management: a comparative analysis of French and Australian policies and implications to other countries, pp. 481–509. https://doi.org/10.1007/978-3-030-32769-0_20

Duan, W., Maskey, S., Chaffe, P. L., Luo, P., He, B., Wu, Y., & Hou, J. (2021). Recent advancement in remote sensing technology for hydrology analysis and water resources management. *Remote Sensing, 13*(6), 1097. https://doi.org/10.3390/rs13061097

Geris, J., Comte, J. C., Franchi, F., Petros, A. K., Tirivarombo, S., Selepeng, A. T., & Villholth, K. G. (2022). Surface water-groundwater interactions and local land use control water quality impacts of extreme rainfall and flooding in a vulnerable semi-arid region of Sub-Saharan Africa. *Journal of Hydrology, 609*, 127834. https://doi.org/10.1016/j.jhydrol.2021.127834

Hashemi, H., Berndtsson, R., & Persson, M. (2015). Artificial recharge by floodwater spreading estimated by water balances and groundwater modelling in arid Iran. *Hydrological Sciences Journal, 60*(2), 336–350. https://doi.org/10.1080/02626667.2014.881485

Kom, K. P., Gurugnanam, B., & Sunitha, V. (2022). Delineation of groundwater potential zones using GIS and AHP techniques in Coimbatore district, South India. *International Journal of Energy and Water Resources*, pp. 1–25. https://doi.org/10.1007/s42108-022-00188-y

Krishnakumar, A., Jose, J., Kaliraj, S., Aditya, S. K., & Krishnan, K. A. (2022). Assessment of the impact of flood on groundwater hydrochemistry and its suitability for drinking and irrigation in the River Periyar Lower Basin, India. *Environmental Science and Pollution Research*, pp. 1–40. https://doi.org/10.1007/s11356-021-17596-y

Kvočka, D., Falconer, R. A., & Bray, M. (2016). Flood hazard assessment for extreme flood events. *Natural Hazards, 84*, 1569–1599. https://doi.org/10.1007/s11069-016-2501-z

Lan, Q., Dong, J., Lai, S., Wang, N., Zhang, L. and Liao, M. (2024). Flood inundation extraction and its impact on ground subsidence using Sentinel-1 data: a case study of the "7.20" rainstorm event in Henan Province, China. *IEEE Journal of Selected Topics in Applied Earth Observations and Remote Sensing*. https://doi.org/10.1109/JSTARS.2024.3088905

MacManus, K., Balk, D., Engin, H., McGranahan, G., & Inman, R. (2021). Estimating population and urban areas at risk of coastal hazards, 1990–2015: How data choices matter. *Earth System Science Data, 13*(12), 5747–5801. https://doi.org/10.5194/essd-13-5747-2021

Peter, N. R., Raja, N. R., Rengarajan, J., Pillai, A. R., Kondusamy, A., Saravanan, A. K., Paran, B. C., & Lal, K. K. (2024). A comprehensive study on ecological insights of Ulva lactuca seaweed bloom in a lagoon along the southeast coast of India. *Ocean & Coastal Management, 248*, 106964. https://doi.org/10.1016/j.ocecoaman.2023.106964

Raja, R. N., Rekha, P. N., Sarkar, S., Sunny, A., Chandrasekar, V., & Balasubramanian, C. P. (2022). August. mapping of aquaculture potential zones using geospatial multi-criteria method for sustainable aquaculture development-Thiruvallur district. In Transforming Coastal Zone for Sustainable Food and Income Security: Proceedings of the International Symposium of ISCAR on Coastal Agriculture, March 16–19, 2021 (pp. 477–495). Cham: Springer International Publishing. https://doi.org/10.1007/978-3-030-95618-9_36

Rajendran, G., & Peter, N. R. (2020). Impacts of culture-wise shrimp farming activities on hydrogeochemistry: A case study from Chidambaram taluk, Cuddalore district, Tamil Nadu. *India. Applied Water Science, 10*(1), 1–22. https://doi.org/10.1007/s13201-019-1097-3

Reddy, C. S., Kurian, A., Srivastava, G., Singhal, J., Varghese, A. O., Padalia, H., Ayyappan, N., Rajashekar, G., Jha, C. S., & Rao, P. V. N. (2021). Remote sensing enabled essential biodiversity variables for biodiversity assessment and monitoring: Technological advancement and potentials. *Biodiversity and Conservation, 30*, 1–14. https://doi.org/10.1007/s10531-020-02073-8

Rekha, P. N., Gangadharan, R., Ravichandran, P., Mahalakshmi, P., Panigrahi, A., & Pillai, S. M. (2015). Assessment of impact of shrimp farming on coastal groundwater using geographical information system based analytical hierarchy process. *Aquaculture, 448*, 491–506. https://doi.org/10.1016/j.aquaculture.2015.06.025

Rekha, P. N., Ravichandran, P., Gangadharan, R., Bhatt, J. H., Panigrahi, A., Pillai, S. M., & Jayanthi, M. (2013). Assessment of hydrogeochemical characteristics of groundwater in shrimp farming areas in coastal Tamil Nadu, India. *Aquaculture International, 21*, 1137–1153. https://doi.org/10.1007/s10499-012-9618-1

Scanlon, B. R., Fakhreddine, S., Rateb, A., de Graaf, I., Famiglietti, J., Gleeson, T., Grafton, R. Q., Jobbagy, E., Kebede, S., Kolusu, S. R., & Konikow, L. F. (2023). Global water resources and the role of groundwater in a resilient water future. *Nature Reviews Earth and Environment, 4*(2), 87–101. https://doi.org/10.1038/s43017-023-00418-9

Schwartz, F. W., & Zhang, H. (2024). Fundamentals of groundwater. John Wiley & Sons. https://doi.org/10.1002/9781119518617

Singha, K., & Navarre-Sitchler, A. (2022). The importance of groundwater in critical zone science. *Groundwater, 60*(1), 27–34. https://doi.org/10.1111/gwat.13106

Sunitha, V., & Reddy, B. M. (2022). Geochemical characterization, deciphering groundwater quality using pollution index of groundwater (PIG), water quality index (WQI) and geographical information system (GIS) in hard rock aquifer. *South India. Applied Water Science, 12*(3), 41. https://doi.org/10.1007/s13201-021-01527-w

Vaux, H. (2011). Groundwater under stress: The importance of management. *Environmental Earth Sciences, 62*, 19–23. https://doi.org/10.1007/s12665-010-0554-4

Vilvijayan, C., & Lalitha, N. (2021). A study of Nivar cyclone and its impact in the coastal and North Interior in Tamil Nadu, Pondicherry and Karaikal. Annals of the Romanian Society for Cell Biology, pp. 15860–15869. https://doi.org/10.37897/arscb.2021.1.15860

White, G. F. (2019). Natural hazards research. In Directions in geography (pp. 193–216). Routledge. https://doi.org/10.4324/9780429273292-9

World Health Organization, (2017). Guidelines for drinking-water quality: fourth edition incorporating the first addendum. World Health Organization.

Chapter 32
Land Use Land Cover and Its Impact on Groundwater Resources in India

Y. Shobha Laxmi Kumari, Musini Venkateshwarlu, and Babita Mishra

Abstract In the current context, various land use patterns have resulted in rapid changes in land cover worldwide. Land use and land cover dynamics are key factors for developing nations and cities, and to comprehend the effects of urbanisation on resources in any region, the study area is in and surrounding Shaikpet, a division of Hyderabad district in Telanagana state, India. The study's goals are to use remote sensing and GIS technology to examine the land use, land cover pattern, and water quality in the study area of Shaikpet, with special attention to watershed areas. When the ground water dynamics of the research region are observed to be significant by the use of Resources at -LISS IV satellite data for land use land cover change detection, Water samples are obtained from various sampling stations and subjected to laboratory analysis in accordance with established protocols to determine physical, chemical, and biological parameters. The findings indicate that the study area's surface water samples contain bacteria in the form of coli and that the permissible limits for EC and TDS concentrations are exceeded. The study's pH levels are within the WHO-permissible range 6.5–8.5, turbidity, which shows the presence of dissolved and suspended solids in the water, electrical conductivity, and TDS levels are all high; the water's pale red colour, which indicates the presence of suspended solids in the water, and the high nitrate concentrations suggest that the water in the study area. According to the water quality examination, surface water needs to be treated before consumption since it has been contaminated by the direct flow of municipal sewage into fresh water bodies. There are trends showing that the

Y. Shobha Laxmi Kumari · B. Mishra
Department of Environmental Sciences, Bhartiya Engineering Science & Technology Innovation University (BESTIU), Ananthapur, Andhra Pradesh, India
e-mail: ens.shobha@gmail.com

B. Mishra
e-mail: bbtmshr@yahoo.co.in

Y. Shobha Laxmi Kumari
Annamacharya Institute of Science & Technology, Hyderabad, India

M. Venkateshwarlu (✉)
Department of Civil Engineering CMR College of Engineering & Technology, Hyderabad, Telangana, India
e-mail: venkatmusini@gmail.com

V. Sunitha et al. (eds.), *Groundwater Resource Management Planning Strategies*, Springer Water, https://doi.org/10.1007/978-3-031-88870-0_32

research region's cropland is shifting to agricultural plantations, built-up regions for habitations, and a large portion of the land remaining as open scrub.

Keywords Land use land cover · Groundwater · Parameters · GIS · Shaikpet · Telangana · India

32.1 Introduction

Any in land waters that are permanently or sporadically found on the surface of the earth in either a liquid (rivers, lakes, reservoirs, and bogs) or solid (glaciers, snow cover) state are considered surface waters. The functioning of ecosystems and the economy both greatly depend on surface waters. (Gummadi et al. 2014). It is possible to evaluate the water quality using physical, chemical, and biological factors. There are four sizable metropolitan regions, and the study areas' rapid growth has resulted in air pollution in the current (Venkateshwarlu et al. 2020). Water quality examination takes into account various characteristics, including dissolved and biological oxygen, pH, temperature, taste, and odour (Prasad et al. 2013). Both the environment and people can benefit from it. Anthropogenic stresses such as population growth, excessive consumerism, and technology can cause natural equilibrium to be disrupted or the biophysical environment to deteriorate. Understanding the planning and management methods for resources, as well as their availability in the studied region, requires an understanding of the land use and cover of a specific area. To satisfy the increasing demands for the wellbeing of people and their basic necessities, to create maps of land cover and use via an integrated remote sensing and geographic information system technique. Reducing fertiliser use and utilising organic fertilisers can improve groundwater pollution levels and make the water safe to drink, according to the Information System Used to Study the Effect of Groundwater (Musini et al. 2020.). Various geochemical investigations indicate that integrated groundwater prospective maps are the preferred option (Venkateshwarlu et al. 2020.). On the other hand, land-use refers to the manner in which humans and their environment have utilized land, typically with variations on the functional role that land has played for economic activity (Bagyaraj et al. 2013.). In addition to being formed by human activities such as colonization, land cover refers to the physical qualities of the earth's surface. These characteristics are reflected in the distribution of soil, water, flora, and other physical elements of the land. A region's LU/LC pattern is determined by its natural and socioeconomic components, as well as by how humans have exploited them throughout history and space (Rawat et al. 2013.). The study region Narsapur mandal in the Medak district is classified according to the source of the "parental" magma that created the granite (Venkateshwarlu et al. 2020.). Water quality is crucial for use; without adequate treatment, people in the study area access the water for a variety of purposes. Therefore, it is necessary to determine the water quality and land cover of the study area; the LULC provides information on resource availability and modifications in the study area. Geographic information systems are used in the computational evaluation of the ion distribution while doing multivariate analysis (Venkateshwarlu et al. 2021.). GIS-based Assessment of Medak District's Watershed Management (Venkateshwarlu et al. 2015.). Human activity is affecting the quality of groundwater, particularly shallow groundwater (Muralidhara Reddy et al. 2020; Reddy et al. 2023 and Muralidhara et al. 2019). The chemistry of a region's groundwater is typically not uniform and is influenced by flow, geochemical processes, evaporation and evapotranspiration, as well as potential pollution sources (Sunitha et al. 2022, Sunitha and Reddy 2022). Examination of Landfill Leach-Related Groundwater Contamination (Venkateshwarlu et al. 2022.).Groundwater hydro geochemistry in the Gooty area of the Anantapur district (Sunitha et al. 2005). Evaluation of the quality of the groundwater in several areas of the Andhra Pradesh districts of Kadapa and Anantapur (Sunitha et al. 2014). As part of the YSR District, an evaluation of the groundwater quality index was carried out in the Andhra Pradesh municipal city of Kadapa (Sunitha et al. 2016). According to the findings of the investigation, the quality of the groundwater in this district has deteriorated as a result of heavy extraction (Venkateshwarlu et al. 2017.). There have been a number of studies that have found that the groundwater in this location contains significant levels of fluoride and salt (Venkateshwarlu et al. 2014). In the Dharwar craton in the Mallampalli area in Telangana, geochemical reactions took place in groundwater that was used for drinking purposes (Saikrishna et al. 2023).The creation of a water quality index for the purpose of evaluating the quality of groundwater in and around Vemula Mandal, which is located in the YSR District (Reddy et al. 2020).Deciphering the quality of groundwater through geochemical characterisation and the utilization of the pollution index of groundwater (Sunitha et al. 2022).

32.2 Study Area

Hyderabad, the capital and most populous city of Telangana, is located in India (17° 21′ 42″ N 78° 28′ 29″ E). Hyderabad, the sixth most populous metropolitan area in India, has 6.9 million residents living inside its city limits and 9.7 million residents living in the municipal zone. Shaikpet area is geographically located (Fig. 32.1). Shaikpet is in Hyderabad, India, next to Tolichowki. 17° 24.4068′ N is the latitude. Longitude: 23.8213′ E 78° E (Fig. 32.1).

32.2.1 Base Map

The Study Area's Base Map Satellite Data: A 1:50,000 scale Top sheet map of the research area was obtained from the Survey of India. Every populated region, transportation system, body of water, and woodland are taken into account. Through a comparison of the satellite image and the topographic maps from the Survey of

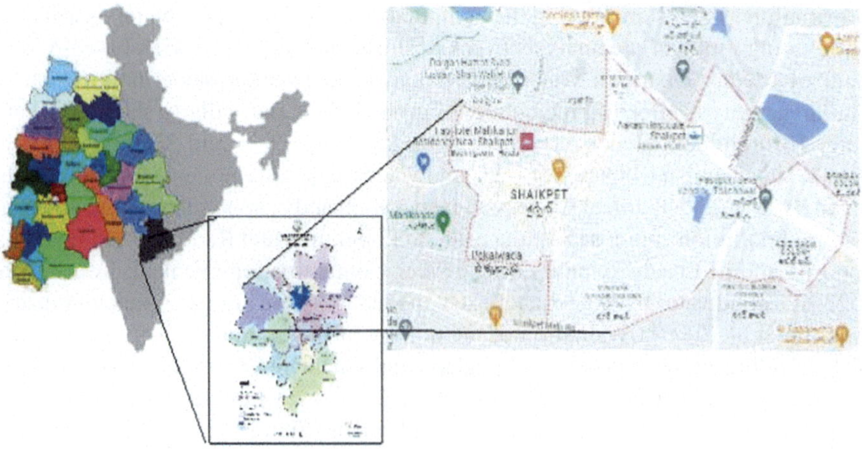

Fig. 32.1 Location map of the study area

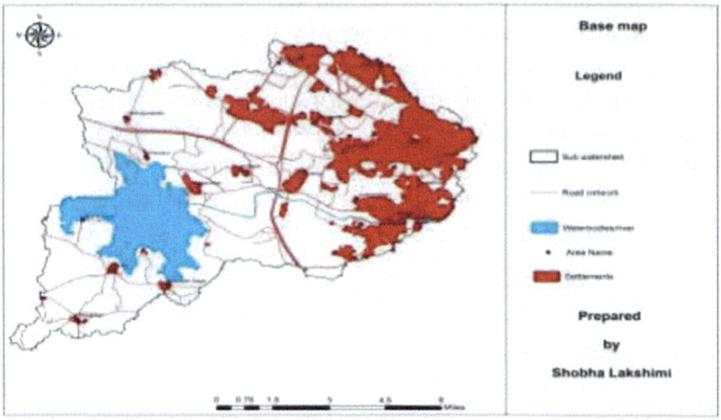

Fig. 32.2 Base map of the study area

India, the sizes of all the settlements are updated and expanded. Red colour coding indicated study area settlements, a large proportion of land covered by subsurface watersheds and river water bodies (Fig. 32.2).

32.2.1.1 Drainage Map

Both natural and artificial water bodies, such as lakes, ponds, tanks, and reservoirs that flow as rivers, canals, streams, etc., are present (Golla et al. 2019.). Every body of water has areas that are both wet and dry. These areas of water dispersion are characterised by periodic fluctuations and the presence of fresh tanks, as shown

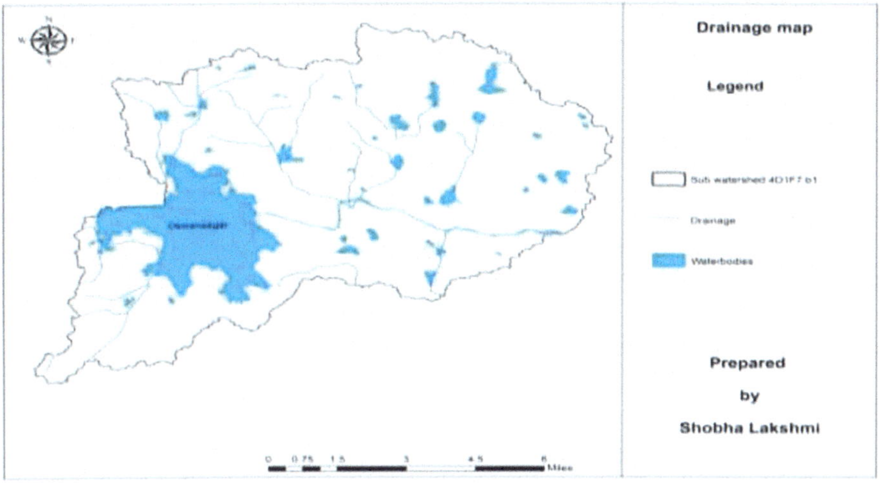

Fig. 32.3 Drainage map of the study area

in the satellite pictures. Thus, satellite data is used to update the drainage map. The drainage method in use today is called Dendritic Drainage Pattern. In satellite photography, water bodies are denoted by a blue hue and a smooth to speckled texture (Nagaraju et al. 2016). The study area is connected to water bodies by a well-planned drainage system (Fig. 32.3.). Essentially, drainage water is collected and treated in STP before being released into the drainage canal. However, our research did not find any contaminations that would alter the quality of the study area's surface or groundwater.

32.3 Methodology Adopted for Land Use/Land Cover

Water samples were gathered directly from sources, lakes and ponds. Placed in sample containers. A total of eight sampling locations were chosen to collect both surface water samples, which were stored in 1 liter plastic bottles. For analysis the samples were transferred to sterilized high density glass vials. Subsequently all samples were transported to the laboratory for examination of chemical and bacterial properties stored in specialized sampling kits at a temperature of 4 °C.The research focused on assessing the impact, on the ecosystem by evaluating the quality of surface water bodies within the region. In total nine water samples—comprising five from surface water sources and four, from sources—were. Analyzed for various physical chemical and bacterial characteristics. It is possible to find the specifics of the areas where surface water, groundwater, and effluent samples were taken in Table 32.1, which is also shown on the location map in Fig. 32.4.

Table 32.1 Showing the water quality properties

S.no.	Test parameter	Unit	Test Method	S1	S2	S3	S4	S5	S6	S7	S8	S9
1.	pH	–	4500.H^+B	7.8	8	8	7.7	8	7.8	7.1	7.8	7
2.	Temperature	°C	–	24.2	25.1	25.1	25.7	25.2	25.2	27.3	25.1	25.2
3.	Colour	Pt–Co	2120.B	85	65	85	95	55	80	112	80	160
4.	Turbidity	mg/L	2130.B	3.9	6.4	3.3	4.2	5.2	8.6	7.8	4.7	7.9
5.	Electrical Conductivity	mg/L	2510-B	1520	1600	1320	1600	1680	1660	1740	1700	2640
6.	Ammonical Nitrogen as N	mg/L	4500. NH_3 c	18	14	55.4	15	12	24.1	36	35	59.4
7.	Total Suspended Solids at 105 °C	mg/L	2540.D	44	35	43	50	22	45	60	40	120
8.	Total dissolved solids at 180 °C	mg/L	2540.C	940	940	762	1000	980	965	1104	995	1560
9.	Chemical Oxygen Demand	mg/L	5520.B	130	170	149	180	200	752	490	230	812
10.	Bio chemical Oxygen Demand (3 days at 27 °C)	mg/L	IS:3025	50	72	56	80	89	289	220	94	340
11.	Total Phosphates as P	mg/L	4500.P.D	1.6	0.3	0.74	0.9	0.05	0.3	1.7	0.4	0.42
12.	Nitrate as N	mg/L	PDA	2	4	0.35	2.2	3.63	0.34	2.3	3.3	0.44
13.	Total Nitrogen	mg/L	By Calculation	33	26	72.3	22	30	57.5	70	48	75
14.	Dissolved Oxygen	mg/L	4500-O.C	5	5.2	5.4	4.4	4.1	NIL	NIL	2	NIL
15.	Total Coliform	MPN/100 mL	9221 A& B	79,000	94,000	35,000	17,000	28,000	170,000	230,000	540,000	790,000

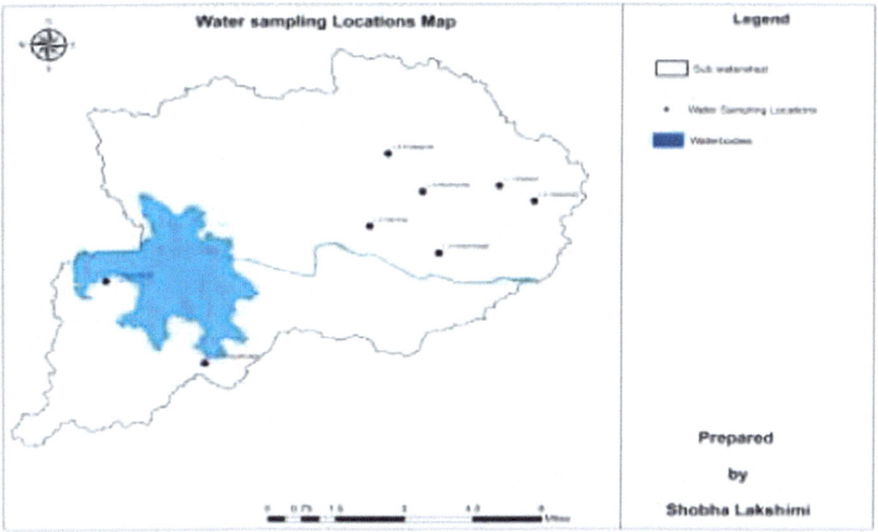

Fig. 32.4 Showing water sampling locations

The mapping of land cover was a direct consequence of the advent of remote sensing, which was predominantly performed by aerial photography. Wide area coverage, frequent revisits, multispectral, multisource, storage in digital format to facilitate subsequent updating, and compatibility with GIS technology are some of the advantages that remote sensing technology offers. As a result of these advantages, remote sensing technology has proven to be an extremely useful and cost-effective method for accurately classifying land cover. Remote sensing data are used to provide information on land cover, and the characteristics of that information are affected by a number of important elements. A single category that is superior to all others for defining land use and cover does not exist, and it is very improbable that such a category could ever be developed. Even when an objective numerical technique is used, the classification process still involves a number of different points of view, and the classification process itself has a propensity to be subjective. Due to the fact that patterns of land use and land cover are always shifting in order to satisfy the ever-evolving needs for natural resources, there is really no rational reason to believe that a single comprehensive inventory should be adequate for more than a short period of time. The needs of the user are taken into consideration while developing each categorization. In the field of remote sensing, LULC classification is one of the applications that is used the most often. The following are the strategies that are used the majority of the time:

32.3.1 Criteria for Land Use/Land Cover Classification

Therefore, in the context of developing a new classification system, it is imperative to take into consideration specific criteria and limitations of satellite data and study area specifically to Indian conditions. This is because the classification system that uses satellite data should provide a framework that satisfies the needs of the majority of users. The categorization of land use and land cover need to be all-encompassing based on good scientific principles, relevant to wide regions, and both practical and adaptable. Not only should it be flexible enough to be used at a variety of sizes and degrees of detail, but it should also be able to fulfill the requirements of a broad variety of users. In order to characterize the categories of land use and land cover, it is recommended that the fewest feasible classifiers be used. This will result in fewer classifiers being used in the definition, fewer mistakes being anticipated, and a reduction in the amount of time and resources required for field validation. Using multi-seasonal satellite data requires that the categorization be consistent. This is a key consideration. In order to determine the appropriate classification or data level inside a classification, it is necessary to make a decision that is arbitrary. The image scale or the data representation scale must be selected. Neither may be used individually. The satellite data that will be used to illustrate Level-I, Level-II, Level-III, Level-IV, and Level-V categories will be based on scales of 1:250,000, 1:50,000, 1:25,000, 1:10,000, and 1:5000, respectively (Fig. 32.5). The least degree of interpretation accuracy and dependability for identifying land use and land cover from satellite data should be between 85 and 95%. This level of accuracy and dependability should vary depending on the size of the mapping. It is possible that some categories that are similar to forest, such as wooded terrain and woodland, might be included under the major heading "Forest" owing to certain constraints that are present in the satellite data.

32.3.2 Conventional Approach of LULC Mapping

Researchers (Thampi et al. 1983, Singh et al. 2004.) have carried out investigations on LULC findings. In the present research, the pattern of land use and cover has been investigated in order to evaluate the land for a variety of economic activities. According to the findings of a study (Srivastava et al. 2011, Salghuna et al. 2018.), the land has been marked by significant impacts from both industrial homes and agricultural regions. It has been observed that various patterns of land use all over the globe are responsible for the fast changes in land cover that are occurring in the contemporary setting (Lambin et al. 2001). Land Use and Land Cover (LULC) data for the nation has typically been produced by the Directorate/Bureau of Economic and Statistics (DES/BES) of the appropriate states. This has been the case for a long time. There is a nine-fold categorization method that may be used for the land use data that was "derived" from the agricultural inventory that was conducted on an individual

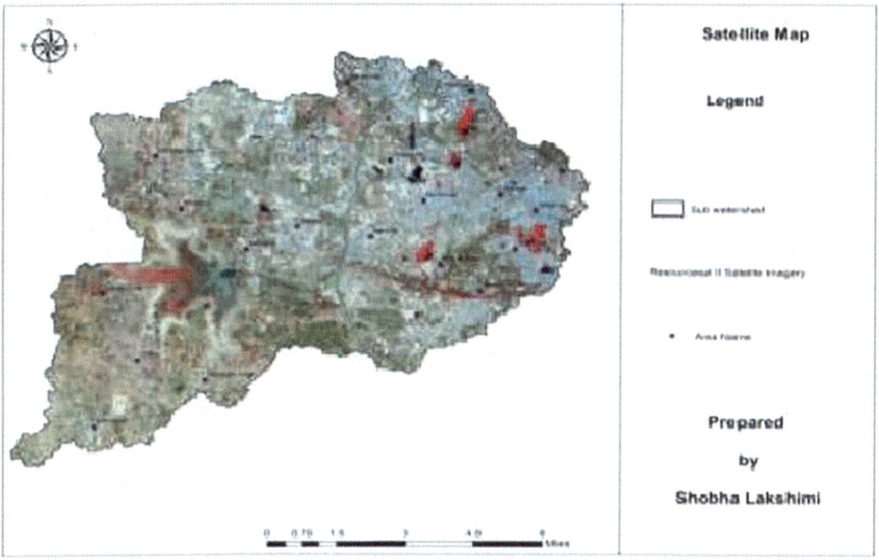

Fig. 32.5 Satellite imagery of the study area

individual plot level. The information presented here is in the form of statistical records, and there is no mention of any specific places. Another source of land use land cover (LULC) data is the topographical maps of the Survey of India. These maps, which were mostly derived from ground data at scales ranging from 1:50,000 to 1:25,000, provide exceptionally wide land use classifications. However, neither the present state of land use nor any recent changes are reflected in this statistics. Neither are any recent new developments. Despite the fact that the terms "land usage" and "land cover" are sometimes used interchangeably, they really refer to two distinct concepts. The term "land cover" refers to everything that covers the surface of the earth, such as plants, bare soil, water, and urban areas that include infrastructure. The process of detecting land cover results in the generation of baseline data, which is required for activities such as theme mapping and change detection analysis. The term "land use" refers to the function that a particular piece of land fulfills, which may include agriculture, leisure, or the provision of habitat for species. When the terms "Land Use" and "Land Cover" are combined, they often refer to the manner in which natural components and human activities have been categorized or grouped on the landscape throughout a certain period of time by using well-established scientific and statistical techniques of analysis of relevant source materials. For the purpose of monitoring urban development, monitoring change is a crucial component. It makes it possible to conduct quantitative analyses of the satellite data in order to determine the geographical distribution of the area of interest.

32.4 Results and Discussion

The need for groundwater has greatly expanded in the modern era due to industrial applications, population growth, and sophisticated irrigation practices. (Chinnasamy et al. 2015), in the research area, the pH levels are within the WHO-approved range (Fig. 32.6); the term "turbidity" refers to both an optical measure of water clarity and a suspension of particles in the water that obstructs light flow (EPA 2012; Myre et al. 2006). Turbidity is a measure of the amount of suspended and dissolved materials in water. Samples 6, 7, and 8 show high levels of turbidity (Fig. 32.6) In addition to sample 8 having a high total dissolved solids content in the surface water; samples 7 and 8 all had high electrical conductivity levels (Fig. 32.7). For washing and bathing, high TDS levels might not be visually pleasant (Jameel et al. 2006; Kumar et al. 2013). The application of agricultural fertiliser is thought to have contributed to the higher concentration of ions in the groundwater, which is responsible for the higher total dissolved solids values (Rao et al. 2018; Etikala et al. 2024). Although elevated TDS concentrations in groundwater are generally not hazardous to humans, they can have an adverse effect on those with heart and kidney conditions (Gupta et al. 2004). High-solids water can also have laxative or constipating effects (Kumaraswamy et al. 2011.). The pale red colour of the water indicates that it is brackish due to the addition of suspended solids (Fig. 32.8). The very high COD and BOD levels in samples 6, 7, and 9 indicate the addition of both organic and inorganic solids in the surface water in the study area (Fig. 32.9). The high nitrate concentrations in the study area indicate sewage contamination in the surface water in the study area (Fig. 32.10), which will increase the levels of total and ammonical nitrogen in the surface water (Fig. 32.11). This suggests that treatment is necessary before using the water for drinking or domestic purposes. Oxygen is used up by decomposition, respiration by aquatic creatures, and other chemical reactions (APHA et al., 1992.). Organisms require dissolved oxygen, which is present at saturation amounts in surface water. Oxygen is produced and consumed by the stream system. Through photosynthesis, it obtains oxygen from plants and the atmosphere. Compared to calm water, running water dissolves more oxygen due to its churning.

Fig. 32.6 pH and turbidity levels

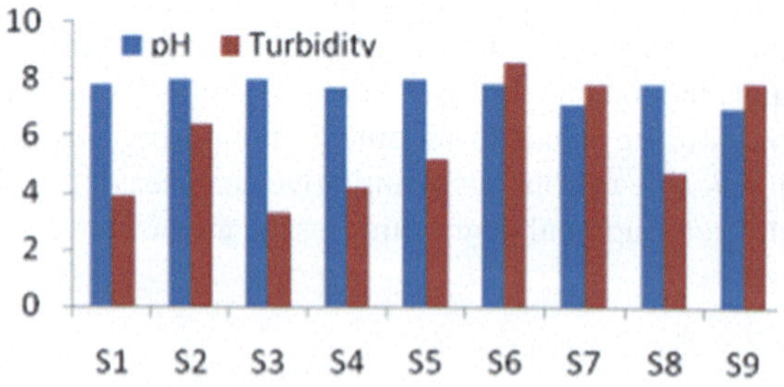

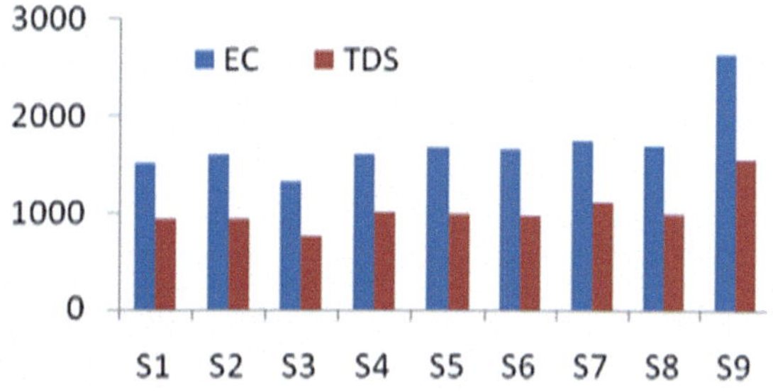

Fig. 32.7 EC and TDS levels

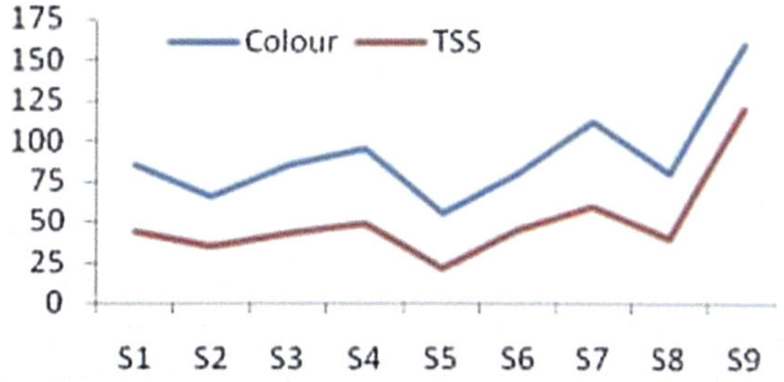

Fig. 32.8 Colour and TSS

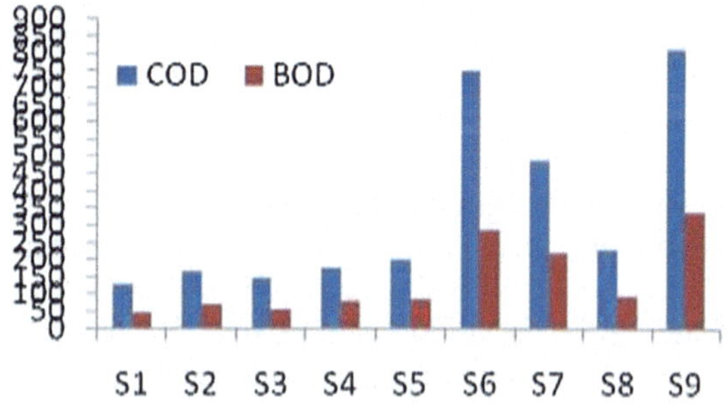

Fig. 32.9 COD and BOD levels

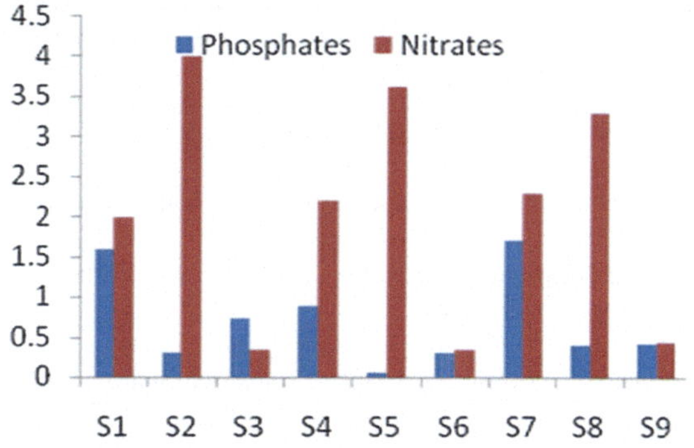

Fig. 32.10 Phosphates and Nitrates

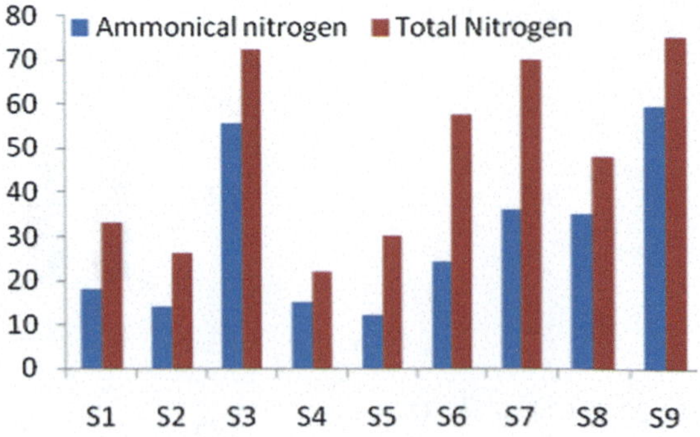

Fig. 32.11 Ammonical nitrogen and total nitrogen

32.5 Conclusion

The study area, which includes constructed land, agricultural land, woodlands, water bodies, and wastelands, has been identified and mapped using a number of different color coding methods that are depending on the land use and land cover. A significant portion of the study area is comprised of metropolitan regions. The number of lands that are used for cropping has decreased, while the number of urban built-up areas has greatly increased, occupying a large portion of the land. Surface water has been polluted as a result of the direct flow of municipal sewage into fresh water bodies, as shown by the water quality analysis. Therefore, surface water must be treated before it can be consumed as it has been contaminated. There are patterns that indicate that the farmland in the study region is being converted into agricultural plantations, communities that have been built for human habitation, and a significant percentage of the land is still being left as open scrub. Both the scrub forest and the bodies of water are seeing a reduction in size; for example, at Osmansagar Lake, the brush covers the majority of the full tank level.

Acknowledgements We express our thanks to Ch. Gopal Reddy, Secretary, CMRGI, Principal and HOD, CMR College of Engineering and Technology Hyderabad, for help and encouragement to publish this paper.

Compliance with Ethical Standards

Conflict of Interest The authors declare that they have no conflict of interest.

Ethical Approval The authors declare that this article does not contain any studies with human participants or animals.

Informed Consent This type of study does not require formal consent.

References

APHA (1992) Standard methods for the examination of water and wastewater, 18th edn. American Public Health Association, Washington, DC

Bagyaraj M,Ramkumar T, Gnanachandrasamy G, (2013) A GIS analysis for assessing groundwater quality in Kodaikanal Hill, Western Ghats, Southern India. Inventi Rapid Water Environ 4:1–9

Chinnasamy, P., & Agoramoorthy, G. (2015). Groundwater storage and depletion trends in Tamil Nadu State, India. *Water Resources Manage, 29*(7), 2139–2152.

EPA (2012) Turbidity. In: Water: monitoring & assessment

Etikala B, Vangala S, Madhav S (2024) Groundwater geochemistry using modified integrated water quality index (IWQI) and health indices with special emphasis on nitrates and heavy metals in southern parts of Tirupati, South India. Environ Geochem Health 46(11). https://doi.org/10.1007/s10653-024-02229-3

Gummadi, S., Swarnalatha, G., Venkataratnamma, V., & Vishnuvardhan, Z. (2014). Water quality index for groundwater of Bapatla Mandal, coastal Andhra Pradesh, India. *Int J Environ Sci, 5*(1), 23–33.

Gupta, S., Kumar, A., Ojha, C. K., & Singh, G. J. (2004). Chemical analysis of groundwater of Sanganer area, Jaipur in Rajasthan. *Environ Sci Eng, 46*(1), 74–78.

Jameel AA, Sirajudeen J (2006) Risk assessment of physico-chemical contaminants in groundwater of Pettavaithalai Area, Tiruchirappalli, Tamilnadu, India. Environ Monit Assess 123(1):299–312

Kumaraswamy, N., Singh, P., & Sarethy, I. P. (2011). Aerobic and anaerobic treatment of paper mill waste water. *Res Environ Life Sci, 4*(4), 141–148.

Kumar, M., & Kumar, R. (2013). Assessment of physico-chemical properties of groundwater in granite mining areas in Goramachia, Jhansi, UP, India. *Int Res J Environ Sci, 2*(1), 19–24.

Lambin EF, Geist HJ (2001) Global land-use and land-cover change: what have we learned so far. Global Change Newslett 46:27–30

Muralidhara Reddy, B., & Sunitha, V. (2020). Geochemical and health risk assessment of fluoride and nitrate toxicity in semi-arid region of Anantapur District, South India. *Environ Chem Ecotoxicol, 2*, 150–161. https://doi.org/10.1016/j.encero.2020.09.002

Musini V, Merugu S, Rohini Reddy T (2020) Information system used to study on effect of groundwater quality. In: Gunjan V, Senatore S, Kumar A, Gao XZ, Merugu S (eds) Advances in cybernetics, cognition, and machine learning for communication technologies. Lecture Notes in Electrical Engineering, vol 643. Springer, Singapore. https://doi.org/10.1007/978-981-15-3125-5_53

Muralidhara Reddy B, Sunitha V, Prasad M, Sudharshan Reddy Y, Ramakrishna Reddy M (2019) Evaluation of groundwater suitability for domestic and agricultural utility in semi-arid region of Anantapur, Andhra Pradesh State, South India. Groundwater Sustain Dev 9:100–262 (Elsevier Journal). https://doi.org/10.1016/j.gsd.2019.100262

Myre, E., & Shaw, R. (2006). *The turbidity tube: simple and accurate measurement of turbidity in the field*. Michigan Technological University.

Nagaraju A, Sreedhar Y, Thejaswi A, Dash P (2016) Integrated approach using remote sensing and GIS for assessment of ground water quality and hydro geomorphology in certain parts of Tummalapalle Area, Cuddapah District, Andhra Pradesh, South India. Adv Remote Sens 5:83–92

Salghuna NN, Rama Chandra Prasad P, Asha Kumari J (2018) Assessing the impact of land use and land cover changes on the remnant patches of Kondapalli reserve forest of the Eastern Ghats, Andhra Pradesh, India. Egypt J Remote Sens Space Sci 21:419–429

Prasad P, Chaurasia M, Sohony RA et al (2013) Water quality analysis of surface water: a web approach. Environ Monit Assess 185:5987–5992. https://doi.org/10.1007/s10661-012-2999-9

Rao KTVS, Venkatesh D, Ahmed SF (2018) Assessment and comparison of ground water quality in Guntur District, Andhra Pradesh by using ARC GIS. Int J Innov Sci Res Technol 3(4)

Rawat JS, Biswas V, Kumar M (2013) Changes in land use/cover using geospatial techniques—a case study of Ramnagar town area, district Nainital, Uttarakhand, India. Egyp J Rem Sens Space Sci 16:111

Reddy, Y. S., & Sunitha, V. (2023). Assessment of heavy metal pollution and its health implications in groundwater for drinking purpose around inactive mines, SW region of Cuddapah Basin, South India. *Total Environ Res Themes, 8*, 100069. https://doi.org/10.1016/j.totert.2023.100069

Reddy YS, Sunitha V (2020) Development of water quality index for groundwater quality evaluation in and around Vemula Mandal, YSR District (AP, India). J Ind Geophys Union 24(1):68–77

Singh, K. P., Malik, A., Mohan, D., & Sinha, S. (2004). Multivariate statistical techniques for the evaluation of spatial and temporal variations in water quality of Gomti River (India)—a case study. *Water Research, 38*, 3980–3992.

Srivastava PK, Mukherjee S, Gupta M, Singh Sk (2011) Characterizing Monsoonal variation on water quality index of River Mahi in India using geographical information system. Int J Water Qual Expo Health 2:193–203

Sunitha, V., Abdullah Khan, J., Muralidhara Reddy, B., Prasad, M., & Ramakrishna Reddy, M. (2014). Assessment of groundwater quality in parts of Kadapa and Anantapur Districts, Andhra Pradesh, India. *Indian J Adv Chem Sci, 3*, 96–101.

Sunitha, V., Sudarsha, V., & Rajeswara Reddy, B. (2005). Hydrogeochemistry of groundwater, Gooty area, Anantapur district, Andhra Pradesh, India. *Pollut Res, 24*(1), 217–224.

Sunitha, V., Muralidhara Reddy, B., & Sumithra, S. (2016). Assessment of groundwater quality index in the Kadapa municipal city, YSR District, Andhra Pradesh. *Ind J Appl Res, 6*(2), 545–548.

Sunitha, V., & Reddy, B. M. (2022). Geochemical characterization, deciphering groundwater quality using pollution index of groundwater (PIG), water quality index (WQI) and geographical information system (GIS) in hard rock aquifer, South India. *Appl Water Sci, 12*, 41. https://doi.org/10.1007/s13201-021-01527-w

Saikrishna K, Purushotham D, Sunitha V et al (2023) Geochemical processes of groundwater for drinking purposes in Dharwar craton of Mallampalli area, Telangana, South India. Int J Energ Water Res 7:15–28. https://doi.org/10.1007/s42108-021-00146-0

Sunitha, V., Yenugu, S. R., Badri, S., & Bandi, M. R. (2022). Human health risk assessment (HHRA) of fluoride and nitrate using pollution index of Groundwater (PIG) in and around hard rock terrain of Cuddapah, A.P. South India. *Environ Chem Ecotoxicol, 4*, 113–123. https://doi.org/10.1016/j.enceco.2021.12.002

Thampi, C. J. (1983). Role of remote sensing technique in integrated economic land use planning in Andaman Islands. *J Indian Soc Remote Sens, 11*(1), 23–29.

Venkateshwarlu M, Prasad B, Ravichandra AP, Prashanthi M (2020) Petrography of granitic rocks of, Medak District, Telangana State". Int J Anal Exp Modal Anal XII(I):570–573. ISSN NO: 0886-9367

Venkateshwarlu M, Rajagopal K, Reddy YS (2020) Geochemical studies on the Groundwater of Kistapur village, Medchal district, Telangana. Indian J Environ Protection 40(8):887–891

Venkateshwarlu M, Raj Kiran Reddy K, Vishwagna K et al (2015) A case study using GIS-based Evaluation of watershed management, Medak District, Telangana state, India. IJISET Int J Innov Sci Eng Technol 2(3)

Venkateshwarlu M, Rajagopa K, Obula Reddy C, Reddy YS (2022) Investigation of groundwater contamination due to landfill leachate. IJEP 42(7):857–862

Veeraswamy Golla, Rajasekhar M, Etikala Balaji, Harish P (2019) Land use/land cover classification in parts of the Sathyavedu Area, Chittoor District (Andhra Pradesh), India by using remote sensing and GIS techniques. J Remote Sens GIS 10(3):1–5

Venkateshwarlu, M., Rasheed, M. A., Reddy, U. V. B., & Kumar, A. K. (2014). Assessment of groundwater quality in and around Miryalaguda area, Nalgonda district of Andhra Pradesh. *Int J Plant Animal Environ Sci, 4*(2), 259–266.

Venkateshwarlu M, Reddy MN (2017) A case study on assessment of groundwater quality parameters in and around Lambapur area, Nalgonda District, Telangana state. Int J Civ Eng Technol (IJCIET) 8(7):563–566

Venkateshwarlu M, Merugu S, Asari VK (2021) Computational assessment of ion distribution on applying multivariate analysis using geographic information systems. Helix Sci Explorer|Peer Reviewed Bimonthly Int J 11(6):32–37. Retrieved from https://helixscientific.pub/index.php/home/article/view/380

Venkateshwarlu M, Merugu S (2020) Fuzzy inference system based assessment of pollution aspects in Hyderabad. In: Gunjan V, Senatore S, Kumar A, Gao XZ, Merugu S (eds) Advances in cybernetics, cognition, and machine learning for communication technologies. Lecture Notes in Electrical Engineering, vol 643. Springer, Singapore. https://doi.org/10.1007/978-981-15-3125-5_56

Chapter 33
Geospatial Evaluation of SLR and LULC Influence on Groundwater Quality in Coastal Watershed of Thiruvallur District, Tamil Nadu, India

P. Nila Rekha, R. Nishan Raja, Albin Sunny, Soumyabrata Sarkar, S. Aravind Kumar, and C. P. Balasubramanian

Abstract Groundwater quality in coastal regions is a critical environmental and socio-economic issue, with salinization being a significant concern. This study examined the impact of sea level rise (SLR) and land use changes on groundwater quality in a coastal sub-watershed in Thiruvallur district, Tamil Nadu, India, from 2000 to 2020. Groundwater samples were collected monthly from September 2019 to January 2021, and analysed for electrical conductivity (EC), a key indicator of salinity. This data was supplemented with historical data (2000–2019) on rainfall and groundwater quality obtained from the State Ground and Surface Water Resources Data Centre, PWD, Tharamani, Tamil Nadu, India, enabling a comprehensive temporal analysis. Land use/land cover (LULC) maps were generated using Landsat/Sentinel satellite imagery to assess the influence of land use changes on groundwater quality. These maps, created in ArcGIS 10.2, focused on prominent land classes viz. water bodies, dense vegetation, scrubland, settlement, aquaculture and barren land. The study found that EC levels fluctuated over the study period, with areas of high EC (above 3500 μS/cm) decreasing from 73.93 km^2 in 2000 to 38.09 km^2 in 2020. The study also found that SLR increased from 0.81 cm in 2000 to 6.7 cm in 2020, suggesting a potential driver of seawater intrusion into coastal aquifers. The study concluded that the synergistic effects of SLR and groundwater over-exploitation contributed to elevated salinity levels within the study area. These findings emphasize the urgent need for sustainable groundwater management strategies and adaptation measures to mitigate the adverse impacts on groundwater resources.

Keywords Brackishwater · Coastal area · Groundwater · Pulicat Lake · India

---

P. Nila Rekha (✉) · R. Nishan Raja · A. Sunny · S. Sarkar · S. Aravind Kumar ·
C. P. Balasubramanian
ICAR—Central Institute of Brackishwater Aquaculture, Chennai, Tamil Nadu, India
e-mail: p.rekha@icar.gov.in

V. Sunitha et al. (eds.), *Groundwater Resource Management Planning Strategies*,
Springer Water, https://doi.org/10.1007/978-3-031-88870-0_33

33.1 Introduction

Coastal regions are vibrant zones of diverse physical, biological, and socio-economic activities, where terrestrial and marine ecosystems intersect. These areas, which house a substantial portion of the global population, are hubs of various economic activities (Befus et al., 2020; Bhadra et al., 2020). This proportion is expected to increase in the future. Coastal regions offer many opportunities for economic development, such as tourism, fisheries, aquaculture, trade and energy production (Rajendran & Peter, 2020; Rekha et al., 2017). However, they also face many challenges and threats, such as pollution, erosion, flooding, storms and climate change. One of the most critical issues affecting coastal regions is the availability and quality of water resources, especially groundwater.

Groundwater is a vital resource for millions of people living in coastal regions, where surface water is often saline or scarce. However, groundwater resources in coastal regions are under increasing pressure from various natural and anthropogenic factors, which can cause salinization, a process that increases the salt concentration in groundwater (Rekha et al., 2015; Senthilkumar et al., 2018). Salinization occurs due to several reasons, such as seawater intrusion, evaporation, irrigation, fertilization, waste disposal and industrial activities. Salinization has negative impacts on the suitability of groundwater for different uses, such as drinking, agriculture, and aquaculture. Salinization diminishes the suitability of groundwater for drinking, thereby endangering the health of coastal communities. Saline groundwater adversely affects crop productivity, soil quality, and the overall sustainability of agricultural practices.

Coastal ecosystems and biodiversity suffer due to altered groundwater quality, disrupting the delicate balance of these fragile environments (Chidambaram et al., 2018; Senthilkumar et al., 2017). One of the most significant challenges is the rising sea levels, a phenomenon that poses a severe threat to the region's groundwater resources. Over-exploitation of groundwater combined with the rising sea level significantly impacts the coastal community and the groundwater quality. It also affects the soil quality and productivity, as well as the health and diversity of coastal ecosystems and biodiversity (Dhanya, 2022). Senthilkumar et al. (2021) assessed the groundwater suitability for drinking and agricultural purposes in the Thiruvallur district and found that the water quality is of poor quality in the eastern and southern parts of the district where agriculture, domestic and industrial activities are dominant. Sustainable management is essential for managing the groundwater resources, the well-being of coastal communities and the environment.

Hence, this study aims to assess SLR and land use/Land cover influence on groundwater quality in the coastal watersheds of Thiruvallur District, Tamil Nadu, India from 2000 to 2020.

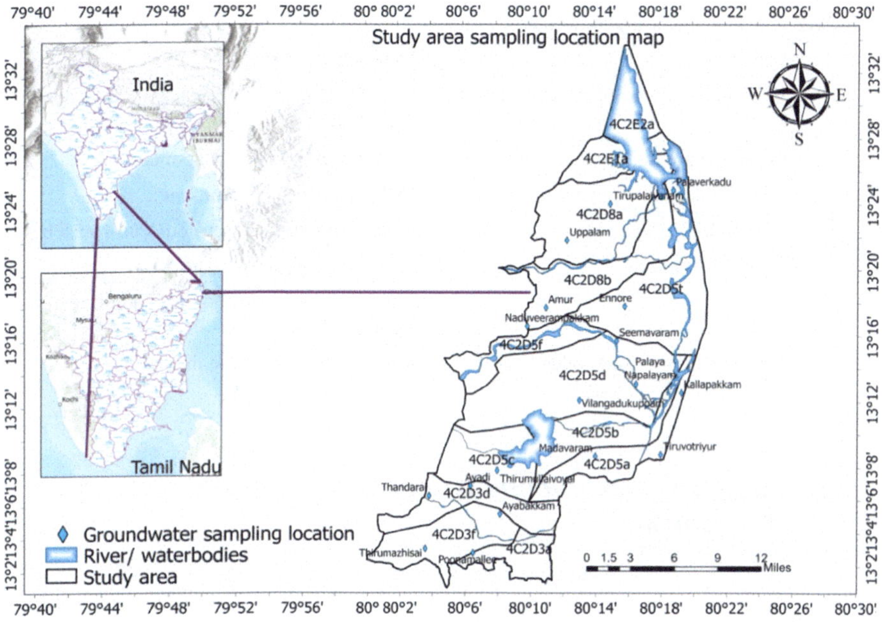

Fig. 33.1 Study area map

33.2 Materials and Methods

33.2.1 Study Area

Thiruvallur district is taken as the study area, it lies between 13° 1′ N to 13° 34′ N latitudes and 80° 00′ E to 80° 21′ E longitudes, with a coastline of 27.9 km. Thirteen sub-watersheds were considered for this study. The watershed boundaries were delineated using SRTM DEM data downloaded from https://explorer.earthengine.google.com and the respective watershed names and codes were collected from https://slusi.dacnet.nic.in Department of Agriculture website. The major riverine in the study area are Pulicat Lake, Buckingham canal, Arani River, Kosathalayar river and Cooum river. The Thiruvallur District receives an annual rainfall of 1152.8 mm (Natesh et al., 2024). In the study area inundation events are predominantly instigated by meteorological incidents, such as extended periods of intense precipitation. The watershed and sub-watershed code used in the study area is given in Fig. 33.1.

33.2.2 Data Used

The data used in this study was collected from various sources as given in Table 33.1.

Table 33.1 List of data used and its source

S. No.	Content	Data source
1.	Groundwater quality data 2000–2019	State Ground and Surface Water Resources Data Centre, PWD, Tharamani
2.	Groundwater quality data 2019–2021	Field sampling and assessment
3.	Rainfall data	IMD and PWD, Tharamani
4.	Sentinel 2	https://dataspace.copernicus.eu/browser/
5.	Landsat 3, 5 and 7	https://earthexplorer.usgs.gov/
6.	DEM	https://explorer.earthengine.google.com/#workspace
7.	Watershed label and codes	https://slusi.dacnet.nic.in/
8.	Sea Level Rise (SLR)	SimCLIM data

33.2.3 Sample Collection and Groundwater Quality Assessment

Eighteen representative sampling locations were identified, and groundwater samples were collected once in two months from September 2019 to Jan 2021. The groundwater sampling locations are shown in Fig. 33.1. Groundwater samples were collected in 250 mL Tarson HDPE Amber Wide Mouth bottles and EC was assessed in the field using a Hydrolab Quanta portable multi-parameter water quality analyser. It has an accuracy of $\pm$ 1% of full-scale reading.

33.2.4 Land Use Land Cover (LULC) Mapping

Land Use Land Cover (LULC) mapping is the process of identifying and classifying different types of land use/ land covers based on remotely sensed images. Landsat 3, 5, 7 and Sentinel 2 were used for this study for land use/land cover mapping from 2000 to 2020. The supervised classification with the maximum likelihood method was used for the study. Water bodies, dense vegetation, scrubland, settlement, barren land and aquaculture classes were classified using visual interpretation keys. The detailed methodology used in the study is given in Fig. 33.2.

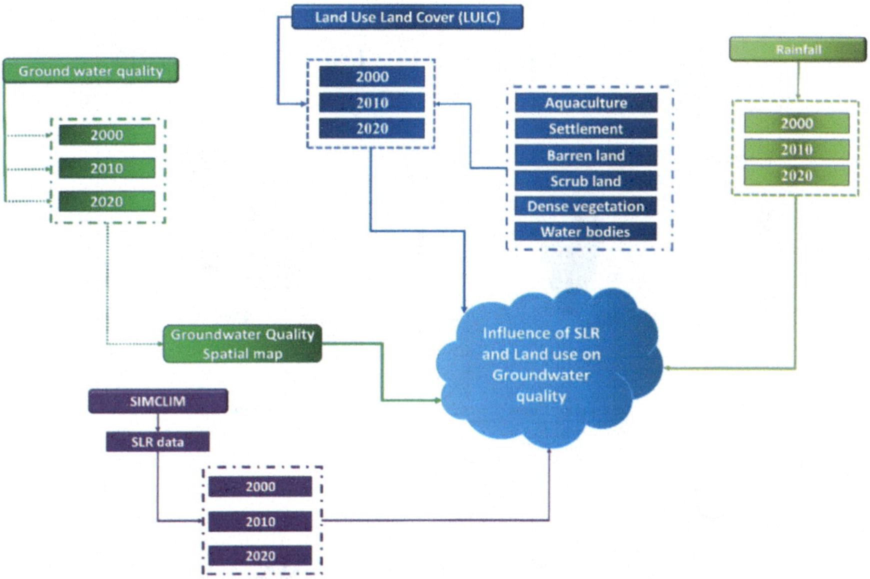

Fig. 33.2 Methodology flow chart of the study

33.3 Results and Discussion

33.3.1 Groundwater Quality and Land Use/ Land Cover Changes from 2000 to 2020

The spatial distribution of groundwater EC is given in Fig. 33.3 in the year 2000 EC ranges from 423–3999 μS/cm. Higher concentration of EC 2211–3999 μS/cm in the study area is observed near Tiruvottiyur and Ennore region. In 2010 EC ranges from 381 to 4559 μS/cm, in the southern part of the study area and the coastal places (Eastern part of the study area) EC ranges from 2470 to 4559 μS/cm. In 2020 EC ranges from 992 to 3929 μS/cm. In the eastern part of the study area the EC was observed to be high.

The land use/ land cover map shows the spatial variation of water bodies, dense vegetation, scrubland, settlement, and aquaculture and barren land from 2000 to 2020 as shown in Fig. 33.4. The area of settlement increased rapidly from 239.98 km^2 in 2000 to 345.17 km^2 in 2020. This reflects the urbanization, industrialization and population growth in the study area. The area of dense vegetation decreased steadily from 416.83 km^2 in 2000 to 301.02 km^2 in 2020. The aquaculture area has increased moderately from 2.94 km^2 in the year 2000 to 8.470 km^2 in 2020. The changes in land use/ land cover from 2000 to 2022 is given in Figs. 33.3 and 33.4. In 2000 it can be observed that the higher EC concentration is found in the settlement land use area in the eastern part of the study area, near the Ennore region as shown in Fig. 33.2.

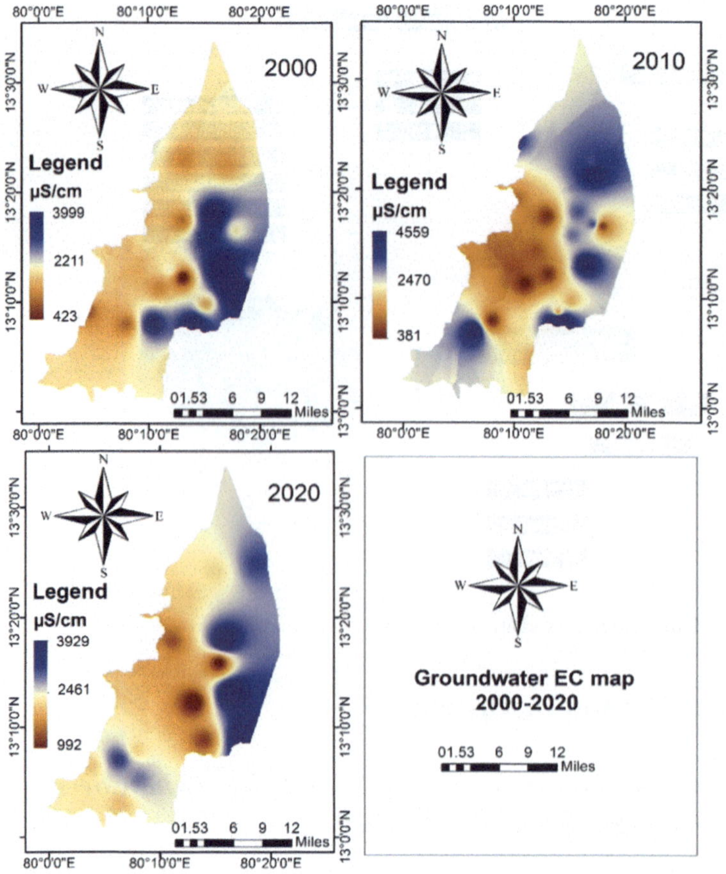

Fig. 33.3 Spatial distribution of groundwater EC from 2000 to 2020

The aquaculture development in 2000 is very less (2.94 km^2) and is observed only in the northern east part of the study area. In 2010 the place with high groundwater EC has further expanded in the eastern region of the study area as shown in Fig. 33.3. Senthilkumar et al. (2023) assessed the effect of urbanization on the groundwater quality in the Ennore region and found the excess utilization of groundwater by the domestic utility and several industrial activities besides this region's southern and coastal regions has led to this water's poor-quality condition.

In the year 2009 the introduction of Specific Pathogen Free (SPF) Pacific white shrimp (*Litopenaeus vannamei*) caused a boom in shrimp farming due to increased productivity and profitability of farmers. The aquaculture area was found to be increased to 4.634 km^2 in the year 2010. In the year 2020, the aquaculture area is about 8.470 km^2 (Table 33.2). The settlement area has increased drastically near the coast in 2020. The total settlement area in 2020 is about 345.17 km^2 as shown in Fig. 33.5. Higher EC concentration is observed parallel to the coast. From the above

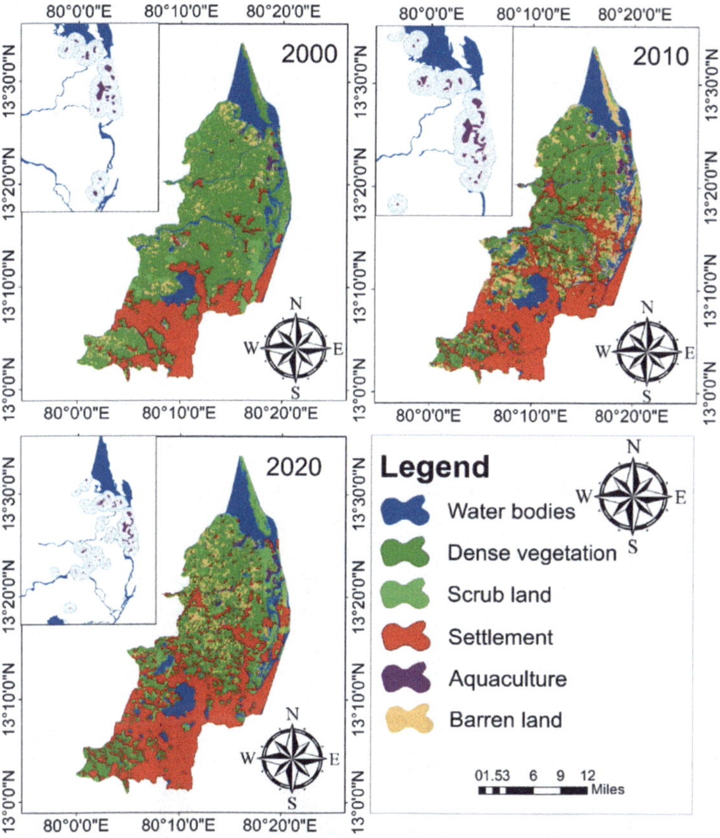

Fig. 33.4 Land Use Land Cover Map 2000–2020

details, it can be observed high EC presence in the eastern part of the study is from 2000 and it has spread gradually with the increasing urbanization in the study area.

The temporal fluctuations in Electrical Conductivity (EC) along with land use provides insights into the impact of land use on groundwater quality. The minimum EC values have been steadily increasing from 423 μS/cm in 2000 to 992 μS/cm in 2020. This suggests a trend of increasing salinity in the groundwater over time. High salinity can make water unsuitable for many uses, including irrigation and drinking. The land use patterns seem to have a direct influence on the groundwater quality. The expansion of settlements could potentially contribute to the increasing salinity. The area where EC exceeds 3500 μS/cm in the study area has been decreasing over time, from 73.93 km^2 in 2000 to 38.09 km^2 in 2020.

The result shows clearly that the groundwater EC in 2000 ranges from 423 to 3999 μS/cm and fluctuates in the coastal areas with notably high levels in the Tiruvottiyur and Ennore region, which was predominantly characterized by settlement area comprising of industrial areas and urban/rural settlements. In 2020 the EC ranged

Table 33.2 Groundwater Electrical Conductivity (EC) and land use 2000–2020

S. No.	Year	Groundwater EC μS/cm			
		Minimum	Maximum	EC > 3500 (Area in Sq. km)	Land use
1.	2000	423	3999	73.93	Settlement, Scrub land, dense vegetation, barren land and Aquaculture
2.	2010	381	4559	51.27	Settlement, Scrub land, dense vegetation, barren land and Aquaculture
3.	2020	992	3929	38.09	Settlement, Scrub land, dense vegetation, barren land and Aquaculture

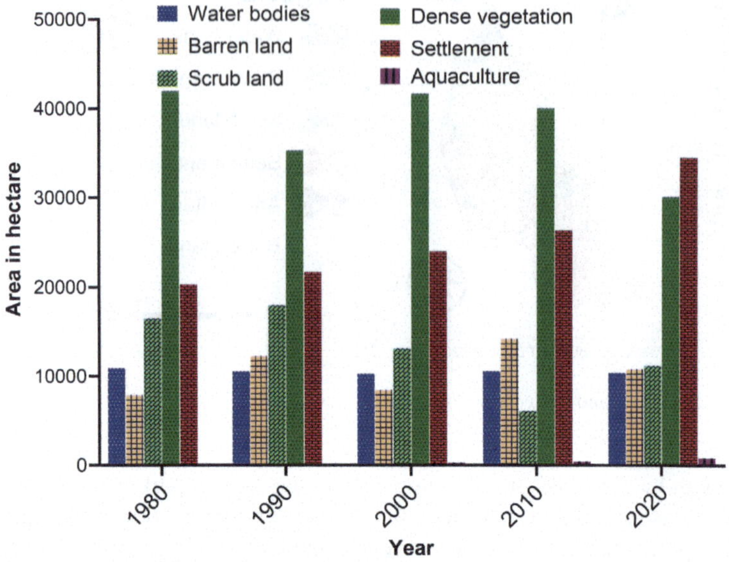

Fig. 33.5 Land use land cover area 2000–2020

from 992 to 3929 μS/cm with significantly higher levels observed in coastal areas dominated by settlement land use. The groundwater EC fluctuates from 2000 to 2020, area of EC above 3500 μS/cm in 2000 is about 73.93 km^2, in 2000 it was about 51.27 km^2 while in 2020 it is about 38.09 km^2. Settlement land use was found to be dominant in most of the high groundwater EC concentrated areas. It was also observed in the study area other land use/land covers viz. aquaculture, dense vegetation, scrubland were not significant contributors to salinization. In the study area, the salinization is initially observed in the settlement area and has further expanded largely. Saline areas can be utilized for brackish water aquaculture viz. shrimp, crab, fish, and seaweed farming for increase productivity and coastal livelihood development.

The rainfall trend varies slightly and it can be observed that the EC range fluctuates in the study area based on climatic variation as shown in Fig. 33.6. The sea level rise in the study area has increased about 0.81 cm in the year 2000. In 2010 it has increased to about 3.59 cm. In 2020 the SLR has increased to about 6.7 cm as shown in Fig. 33.7. Ramachandran et al. (2017) assessed sea level rise using SimCLIM, his findings show in Tamil Nadu and Puducherry coast the sea level rise with an average maximum of 78.15 cm by 2100.

The interface between freshwater and saltwater in aquifers can shift due to the rise in sea level. This can result in the intrusion of saltwater into freshwater aquifers, particularly those located near the coast. The groundwater salinization is the process of increasing the concentration of dissolved salts in groundwater due to natural or human-induced factors such as intrusion of seawater into coastal aquifers (Odeh

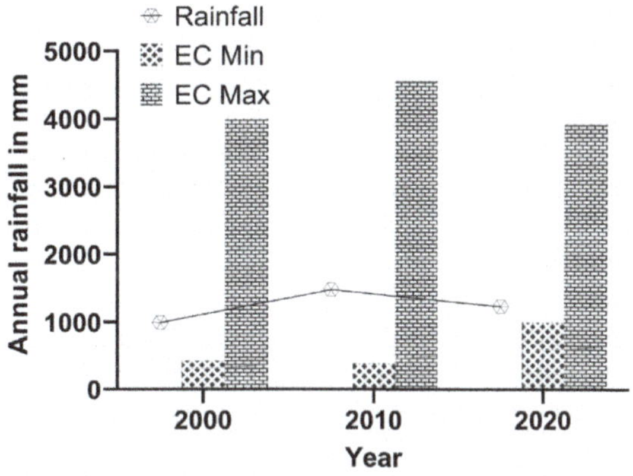

Fig. 33.6 Annual rainfall and electrical conductivity trend from 2000 to 2020

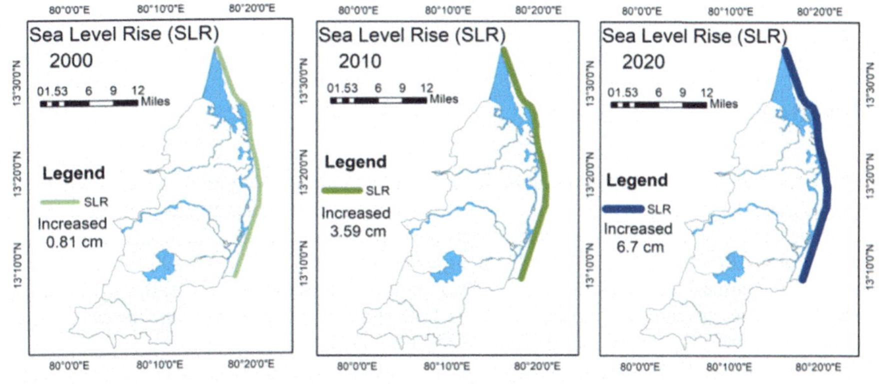

Fig. 33.7 Sea level rise 2000–2020

et al., 2019; Said et al., 2021). Groundwater with higher EC impact human health when consumed (Etikala et al., 2021; Reddy et al., 2023; Reddy & Sunitha., 2023).

Many researchers have assessed and found the increasing sea level rise impacts the groundwater and also induce changes in land dynamics. The groundwater salinity in the study area is being impacted due to climate change and human activities (Santhanam & Amal, 2019; Basuri et al., 2020; Gera et al., 2021). Groundwater salinization can also occur due to the overpumping of groundwater for various purposes, such as irrigation, domestic, and industrial uses (Odeh et al., 2019). When the groundwater is over pumped or depleted, the pressure difference causes the saline water from the sea or the surface water from the Pulicat lake and the Buckingham canal to infiltrate into the freshwater aquifers. The findings underscore the need for sustainable land and water use practices. As human activities continue to impact groundwater quality, it's crucial to implement strategies that minimize negative effects. These strategies could include better waste management in settlements and improved land use practices.

33.4 Conclusion

This study assessed the changes in sea level rise (SLR), land use and land cover (LULC), rainfall and climatic conditions prevailing over the coastal areas of Thiruvallur district, Tamil Nadu, India that influence groundwater salinity. The major reasons for increased salinity are SLR and over-exploitation of groundwater in areas with industrial and urban/rural settlements. These findings strongly suggest that continuous monitoring and strategic land use/land cover planning are essential, ensuring groundwater recharge at strategic points to prevent salinization.

Acknowledgements We are grateful to NABARD for funding this work. We are also thankful to the Director, ICAR-CIBA for the support and facility to carry out this work.

References

Basuri, C. K., Pazhaniyappan, E., Munnooru, K., Chandrasekaran, M., Vinjamuri, R. R., Karri, R., & Mallavarapu, R. V. (2020). Composition and distribution of planktonic ciliates with indications to water quality in a shallow hypersaline lagoon (Pulicat Lake, India). *Environmental Science and Pollution Research, 27*, 18303–18316. https://doi.org/10.1007/s11356-020-08863-8

Befus, K. M., Barnard, P. L., Hoover, D. J., Finzi Hart, J. A., & Voss, C. I. (2020). Increasing threat of coastal groundwater hazards from sea-level rise in California. *Nature Climate Change, 10*(10), 946–952. https://doi.org/10.1038/s41558-020-0874-1

Bhadra, T., Hazra, S., Ray, S. S., & Barman, B. C. (2020). Assessing the groundwater quality of the coastal aquifers of a vulnerable delta: A case study of the Sundarban Biosphere Reserve, India. *Groundwater for Sustainable Development, 11*, 100438. https://doi.org/10.1016/j.gsd.2020.100438

Chidambaram, S., Sarathidasan, J., Srinivasamoorthy, K., Thivya, C., Thilagavathi, R., Prasanna, M. V., Singaraja, C., & Nepolian, M. (2018). Assessment of hydrogeochemical status of groundwater in a coastal region of Southeast coast of India. *Applied Water Science, 8*, 1–14. https://doi.org/10.1007/s13201-018-0740-9

Dhanya, P. (2022). Evaluating the multi-hazards threats due to aridity, sea level upsurge in the coastal areas of North Tamil Nadu, South India. In *Monitoring and managing multi-hazards: A multidisciplinary approach* (pp. 3–17). Springer International Publishing. https://doi.org/10.1007/978-3-031-15377-8_1

Eaton, A.D., Franson, M.A.H., Clesceri, L.S., Rice, E.W., & Greenberg, A.E. (2005). Standard methods for the examination of water & wastewater. In *Standard methods for the examination of water & wastewater* (pp. 1-v). https://doi.org/10.2105/SMWW.2005

Etikala, B., Adimalla, N., Madhav, S., Somagouni, S. G., Keshava Kiran Kumar, P. L. (2021). Salinity problems in groundwater and management strategies in arid and semi-arid regions. In *Groundwater geochemistry* (pp. 42–56). https://doi.org/10.1002/9781119709732.ch3

Gera, A., Kumaraswami, M., Rao, V. R., Vijay, A., Pandiyarajan, R. S., Ezhilarasan, P., & Murthy, M. R. (2021). The Pulicat, a distinctive shallow lagoon: Hypersalinity, thermodynamics and meromixis. *Estuarine, Coastal and Shelf Science, 252*, 107292. https://doi.org/10.1016/j.ecss.2021.107292

Natesh, N, & NK, S. (2024). Numerical modelling of reservoir sedimentation for sustainable storage capacity—A geospatial approach. *Geology, Ecology, and Landscapes, 8*(1), 1–7. https://doi.org/10.1080/24749508.2021.2022832

Odeh, T., Mohammad, A. H., Hussein, H., Ismail, M., & Almomani, T. (2019). Over-pumping of groundwater in Irbid governorate, northern Jordan: A conceptual model to analyze the effects of urbanization and agricultural activities on groundwater levels and salinity. *Environmental Earth Sciences, 78*, 1–12. https://doi.org/10.1007/s12665-019-8679-9

Raja Raja, N., Peter, N.R., Puthiyedathu, S.K., Vasudevan, C., Sarkar, S., & Sunny, A. (2023). Fuzzy analytic hierarchy process (FAHP) for identification of potential cage farming sites for aqua tourism development in Muttukadu Lagoon, Chennai, India. In *Aquaculture international* (pp.1–22). https://doi.org/10.1007/s10499-023-00847-3

Raja, R.N., Rekha, P.N., Sarkar, S., Sunny, A., Chandrasekar, V., & Balasubramanian, C.P. (2022, August). Mapping of aquaculture potential zones using geospatial multi-criteria method for sustainable aquaculture development—Thiruvallur District. In *Transforming Coastal Zone for Sustainable Food and Income Security: Proceedings of the International Symposium of ISCAR on Coastal Agriculture*, March 16–19, 2021 (pp. 477–495). Springer International Publishing. https://doi.org/10.1007/978-3-030-95618-9_36

Rajendran, G., & Peter, N. R. (2020). Impacts of culture-wise shrimp farming activities on hydrogeochemistry: A case study from Chidambaram taluk, Cuddalore district, Tamil Nadu, India. *Applied Water Science, 10*(1), 1–22. https://doi.org/10.1007/s13201-020-01207-9

Ramachandran, A., Saleem Khan, A., Palanivelu, K., Prasannavenkatesh, R., & Jayanthi, N. (2017). Projection of climate change-induced sea-level rise for the coasts of Tamil Nadu and Puducherry, India using SimCLIM: A first step towards planning adaptation policies. *Journal of Coastal Conservation, 21*, 731–742. https://doi.org/10.1007/s11852-017-0532-6

Reddy, Y. S., & Sunitha, V. (2023). Assessment of Heavy metal pollution and its health implications in groundwater for drinking purpose around inactive mines, SW region of Cuddapah Basin, South India. *Total Environment Research Themes, 8*, 100069. https://doi.org/10.1016/j.totert.2023.100069

Rekha, P.N., Gangadharan, R., Ravichandran, P., Dharshini, S., Clarke, W., Pillai, S.M., Panigrahi, A., & Ponniah, A.G. (2017). Land-use/land-cover change dynamics and groundwater quality in and around shrimp farming area in coastal watershed, Cuddalore district, Tamil Nadu, India. In *Current science* (pp. 1763–1770). https://doi.org/10.18520/cs/v112/i09/1763-1770

Rekha, P. N., Gangadharan, R., Ravichandran, P., Mahalakshmi, P., Panigrahi, A., & Pillai, S. M. (2015). Assessment of impact of shrimp farming on coastal groundwater using geographical information system based analytical hierarchy process. *Aquaculture, 448*, 491–506. https://doi.org/10.1016/j.aquaculture.2015.06.021

Said, I., Salman, S.A., & Elnazer, A.A. (2021). Salinization of groundwater during 20 years of agricultural irrigation, Luxor, Egypt. In *Environmental geochemistry and health* (pp. 1–15). https://doi.org/10.1007/s10653-021-00917-4

Santhanam, H., & Amal Raj, S. (2019). Spatial and temporal analyses of salinity changes in Pulicat lagoon, a transitional ecosystem, during 1996–2015. *Water Science, 33*(1), 93–104. https://doi.org/10.1016/j.wsj.2019.02.001

Senthilkumar, S., Balasubramanian, N., Gowtham, B., & Lawrence, J. F. (2017). Geochemical signatures of groundwater in the coastal aquifers of Thiruvallur district, South India. *Applied Water Science, 7*, 263–274. https://doi.org/10.1007/s13201-016-0409-0

Senthilkumar, S., Gowtham, B., Srinivasamoorthy, K., & Gopinath, S. (2021). Hydrogeochemical delineation of groundwater fitness for drinking and agricultural utilities in Thiruvallur district, South India. *Arabian Journal of Geosciences, 14*, 1–20. https://doi.org/10.1007/s12517-021-06547-z

Senthilkumar, S., Gowtham, B., Sundararajan, M., Chidambaram, S., Lawrence, J. F., & Prasanna, M. V. (2018). Impact of landuse on the groundwater quality along coastal aquifer of Thiruvallur district, South India. *Sustainable Water Resources Management, 4*, 849–873. https://doi.org/10.1007/s40899-018-0236-8

Senthilkumar, S., Srinivasamoorthy, K., & Gowtham, B. (2023). The effect of urbanization on groundwater quality and hydrochemical characteristics in Ennore coastal aquifers of Chennai, South India. In *Impacts of urbanization on hydrological systems in India* (pp. 273–291). Springer International Publishing. https://doi.org/10.1007/978-3-031-21618-3_14

The manufacturer's authorised representative in the EU is Springer Nature Customer Service Centre GmbH, Europaplatz 3, 69115 Heidelberg, Germany. If you have any concerns regarding our products, please contact ProductSafety@springernature.com

Printed and bound by CPI Group (UK) Ltd, Croydon, CR0 4YY
30/09/2026
02225584-0018